World Spice Plants

Johannes Seidemann

World Spice Plants

With 93 Figures

Dr. Johannes Seidemann
Neuendorferstrasse 25/26
D - 14480 Potsdam
E-Mail: j.seidemann@arcor.de
Germany

ISBN 3-540-22279-0   Springer-Verlag Berlin Heidelberg New York

Library of Congress Control Number: 2004112586

This work is subject to copyright. All rights are reserved, whether the whole or part of the material is concerned, specifically the rights of translation, reprinting, reuse of illustrations, recitation, broadcasting, reproduction on microfilm or in any other way, and storage in data banks. Duplication of this publication or parts thereof is permitted only under the provisions of the German Copyright Law of September 9, 1965, in its current version, and permission for use must always be obtained from Springer. Violations are liable to prosecution under the German Copyright Law.

Springer is a part of Springer Science+Business Media

springeronline.com

© Springer-Verlag Berlin Heidelberg 2005
Printed in the European Union

The use of registered names, trademarks, etc. in this publication does not imply, even in the absence of a specific statement, that such names are exempt from the relevant protective laws and regulations and therefore free for general use.

Product liability: The publishers cannot guarantee the accuracy of any information about the application of operative techniques and medications contained in this book. In every individual case the user must check such information by consulting the relevant literature.

Editor: Thomas Mager, Heidelberg
Development Editor: Andrew Spencer, Heidelberg
Production Editor and Typesetting: Frank Krabbes, Heidelberg
Cover design: Erich Kirchner, Heidelberg

SPIN: 10986578   14/2109 fk – 5 4 3 2 1 0 – Printed on acid-free paper

Foreword

It is a pleasant fact that general consciousness of health and diet is growing, thus enabling the public, by means of a balanced diet, to participate in preserving global health. Spices and aromatic plants have a role to play in this which should not be underestimated. It involves not only the use of spices in preparing good tasting meals, but equally the increasing use of spices in all areas of the food industry, and also in pharmacy and medicine. The observations of our ancestors that certain herbs and their parts (leaves, fruits and seeds) not only improve the flavor of foods but have a positive effect on our health are are being turned to good use.

Initially, our forefathers collected only local plants. But via foreign trade, voyages of discovery, migration and even wars they came into contact with new plants, including many spices from other geographical areas. The oldest spice discoveries were made in Mexico. The native Mexicans were spicing their meals with chilli as early as 7000 B.C. For the western world the origins of spices was India. Five thousand years ago there was already an extensive network of trade routes ranging from China to India, Persia, Mesopotania and Egypt. The remains of certain spices (aniseed, fenugreek, fennel, cardamom, caraway, saffron and cinnamon) were found in the pyramids, indicating their use as burial gifts. In the world of the pharaohs, spices were not reserved for the ruling classes, the meals of the slaves were also spiced. This was not an act of charity but was intended to preserve the health of the workforce. Egyptian construction workers rebelled around 1600 B.C. because garlic, which was important for counteracting the negative health effects of building the pyramids, had been removed from their diet plan (the first strike in history!). The significance of spices at that time can be gleaned from the collection of recipes entitled „Papyrus Ebers", which was written in 1500 B.C. at the River Nile.

For centuries spices were transported along the caravan trails from China and India to Europe. The Silk Road was one of the more famous of these trade routes.

Despite their long tradition spices are subject to international modern scientific research. This has made it necessary to disseminate knowledge to a large audience of interest. This book addresses this need by informing the reader about the complex worldwide use of spice plants.

The many spice and aromatic plants are arranged in alphabetical order of their botanical relevance. It includes all species which have been cultivated for the above purposes. It also covers species whose usage has long ceased or which are used only rarely or have become wild. In this respect we have aimed at comprehensiveness. Furthermore, we have included plants used as pot herbs or whose ashes are used as a substitute for cooking salt, or those used as a basis of various condiments such as mustard, soy sauce, ketchup, etc., or are used to add aroma to foods, like spirits. In total over 1400 plants have been collated. Some of them are relevant only to a certain region. Plants and their extracts or ethereal oils used exclusively in perfumes and cosmetics have not been included.

The register of literature has been designed to facilitate intensive study of a specific plant or spice. The analysis of scientific literature – particularly over the last 40 years – was achieved by means of original works, specialist journals, textbooks, the secondary literature mentioned in all of these as well as our own database. Works both on botany and agriculture, and on chemistry, pharmacodynamics and usage have been considered. It can be seen clearly that individual spices have been treated very differently. Whilst the literature on many spices is almost never ending, e.g. *Allium*, *Capsicum*, *Glycine*, *Mentha*, *Ocimum*, *Origanum*, *Vanilla* and *Zingiber*, a few spices are completely lacking any significant information to fall back on. In a few individual cases our own personal experiences are included.

The book is aimed principally at spice and aroma experts, pharmacists, botanists and interested lay persons. But we also had in mind food chemists, dieticians and agricultural scientists, for whom botany, chemistry and pharmacological aspects may be of interest. It is hoped that those occupied in the spice and aroma industry involved in creating spice blends and the like will also take inspiration from the book.

The many color illustrations will hopefully acquaint the reader with both well and lesser known spice plants and with the spices from various geographical areas.

It is with great pleasure that I thank all those colleagues who offered advice and were happy to answer my questions. Without the generous assistance of the Botanical Garden and the Botanical Museum in Berlin-Dahlem this book would not have been possible. In particular I would like to thank the Directors Drs. Hakki and Th. Raus for their interest and discussions. They granted me permission to use their very extensive library and archive as well as the herbarium and the large collections which acquainted me with many unknown spices and aromatic plants and their various parts. My particular thanks go to Dr. Peter Scharf, formerly of the Botanical Institute of the University of Potsdam, for reading through the manuscript. I am also grateful to Mr. H.-D. Neuwinger St Leon for invaluable information particularly on African plants used as spices. And last but not least I thank Dr. Th. Mager and Mr. A. Spencer of Springer Verlag who accommodated many of my wishes concerning the layout and content of the book.

Author and publisher are fully aware that a work of this kind can never be exhaustive and that possible inaccuracies and gaps are unavoidable. Hints for additions and improvements by the readership will be very welcome.

Potsdam April 2004
Johannes Seidemann

Structure of Entries

The spice plants, plant spices and aromatic plants – hereafter referred to collectively as spices – are arranged in alphabetical order of their scientific genus. In the case of multiple species of a genus the arrangement of these is also alphabetical. Synonyms have been included with the species names. Of the myriad local names for spices only the more common have been included. They are followed by the usage of plants and which parts are used as spices, together with information on their geographical distribution and areas of cultivation, and finally a list of the relevant literature.

Many lesser known plants have been included whose usage is based on personal experiences on travels and those of our colleagues. In these cases there is often no literature available.

Scientific Names
The scientific plant names with their author names are in accordance with the International Code of the Botanical Nomenclature (ICBN) of 1978 (cf. Zander (2002) Handwörterbuch der Pflanzennamen, 17. edn. Ulmer, Stuttgart, or Hanelt P (2001) Mansfeld's Encyclopedia of Agricultural and Horticultural Crops. Springer, Heidelberg Berlin New York). It is emphasized that the oldest publication determines the nomenclature of a taxonomical group. Varieties, types and sorts have been included only in rare cases.

Synonyms
The most important plant synonyms are included for each spice. These are listed alphabetically with a cross reference to the official plant name.

Common Names
Listed after the English or anglicized name are other common names listed alphabetically by language or region of usage.

Inclusion of the various vernacular plant names in different languages was achieved by reference to the literature, both books and journals on the subject and dedicated dictionaries of plant names (Hanelt 2001, Quattrocchi 1999/2000, Kays/Dias 1995 and Wiersma/Leon 1999, etc). This was not always easy, particularly with Chinese, Indian and African names due to their sheer number, the size of the countries and the many tribes and peoples. Only the more common names have been included.

Usage
Here there are short details of usage as spice, pot herb, ingredient for condiments and the flavoring of food and dishes. Plant ashes used as a substitute for salt are listed under condiments.

Plant Parts

In the course of studying the literature it became evident that details of the used plant parts is often missing. In the case of better known plants these are already known, but it is a mistake to assume that the parts are always the same for different species of a genus. For instance, pepper does not always provide its fruit for flavouring but in some cases only the leaves. Conversely, multiple parts of some plants can be used, as with for example coriander and parsley, where the fruit and the seeds, leaves and sometimes even the roots are used.

Distribution

Details of distribution are given, wherever possible, uniformly – in some cases in the general area, like North America with Canada, South or Latin America from Mexico to Terra del Fuego; Europe; East Asia; the Near East; Africa; Australia and New Zealand, but at other times confined to a certain zone, like North or West Africa, or Central America. Where the distribution of a species is limited to a very small area, details of the countries, states or provinces have also been considered.

Notes

Here we have given special notes on individual plants, such as on taste, on types, e.g. green and white pepper with *Piper nigrum*, or on discovery, e.g. with *Mentha piperita*.

Literature

A selection of publications on which the information presented in the book is based is given with each entry. Here we have reduced the reference to name and year of publication for single works, and in the case of larger multi-volume works we have added the volume number. A complete literature list, sorted alphabetically by author, is included at the end of the book. Where more than one work has appeared by the same author in a year these have been marked a and b, e.g. Seidemann 1995a. With the abbreviations of journals we have adhered to the international standard, whereby clarity has been given priority over conciseness.

Abelmoschus Medik. - Malvaceae

 Abelmoschus ficulneus *Wight et Arn.*

▶ *Abelmoschus moschatus Medik.*

 Abelmoschus moschatus *Medik.* **ssp. moschatus** *Medik.*

Synonyms ▶ *Abelmoschus ficulneus* Wight et Arn., *Hibiscus abelmoschus* L., *Hibiscus moschatus* Salisb.

Common Names ▶ abelmosk seed, ambrette seed, amber seed, musk mallow, musk okra; *Arabic:* abu-el-misk (father of moschus), abu-el-mosk, bamia; *Brazil (Portuguese):* abelmosco, ambarino; *French:* ambrette, ketmie musquée; *German:* Abelmoschussamen, Ambramalve, Ambrette, Bisamkörner, Muskateller-Eibisch; *Hindi:* algalia, mushk dana; *Italian:* abelmosco, ambretta; *Japanese:* ryû kyûtororo-aoi; *Mexican:* Doña Elvira; *Pilipino:* kastuli, dalupang, daopang, marapoto; *Russian:* bamija, hibiskus; *Sanskrit:* latakasturi; *Spanish:* abelmosco, algalia, ambarcillo, ambarina, café extranjero, *Thai:* chamot-ton, somchaba, mahakadaeng

Usage ▶ spice, flavoring of coffee in Arabian regions; **product:** essential oil (liqueurs, tobacco industry)

Parts Used ▶ seed

Distribution ▶ E Africa, Indian to N Asia, Malaysia, N Australia, widely cultivated in tropical and warmer countries

▫ **Abelmoschus moschatus, flowering**

Bois 1934; Burkill 4, 1997; Cheers 1998; Cravo et al. 1992; Dastur 1954; Erhardt et al. 2002; Hager 4, 1992; Hanelt 2001; Hiller/Melzig 1999; Hoppe 1949; Nee et al. 1986; Oyen/Dung 1999; Schultze-Motel 1986; Seidemann 1993c; Sharma 2003; Shiva et al. 2002; Small 1997; Täufel et al. 1993; Turova et al. 1987; Uphof 1968; Villamar et al. 1994; Wiersema/León 1999; Wüstenfeld/Haensel 1964; Zeven/de Wet 1982

Abrus Adans. - Indian Liquorice - Fabaceae (Leguminosae)

Abrus precatorius L.

Synonyms ▶ *Glycine abrus* L.
Common Names ▶ Crab's eye vine, Indian liquorice, jequirity, rosary pea; *Chinese:* xiang si zi, hsiang ssu, hung tou, tzu, *French:* liane reglisse, pois à Chapelet, reglisse sauvage; *German:* Paternostererbse; *India:* glumachi, gunchi, gunja, kunni, kunri, rati; *Malaysian:* akar, belimbing, saga betina; *Pilipino:* laga, kaloo, saga; *Vietnamese:* cam thao day, day chi chi
Usage ▶ substitute for liquorice
Parts Used ▶ root
Distribution ▶ tropical aera, native in Florida
Note ▶ The seeds contain abrin, a toxalbumin, an exceedingly poisonous substance.

Ayensu 1978; Erhardt et al. 2002; Uphof 1968

Acanthopanax (Decne. et Planch.) Miq. - Araliaceae

Acanthopanax aculeatum Seem.

▶ *Acanthopanax trifoliatum (L.) Voss.*

Acanthopanax trifoliatum (L.) Voss.

Synonyms ▶ *Acanthopanax aculeatum* Seem.
Common Names ▶ Dreiblättrige Fingeraralie
Usage ▶ pot-herb
Parts Used ▶ leaf
Distribution ▶ India: NE hills, Assam

Arora/Pandey 1996; WHO Manila 1990

Acer L. - Maple - Aceraceae

Acer macrophyllum Pursh

Common Names ▶ broad-leaved maple, big leaf maple, large leaf maple, Oregon maple, Pacific maple; *German:* Großblättriger Ahorn, Oregon Ahorn
Usage ▶ spic (rarely)
Parts Used ▶ cambium
Distribution ▶ SE Alaska, USA: Washington, Oregon, California, also cultivated
Note ▶ Native Americans used the maple cambium to savor meat.

Cheers 1998; Erhardt et al. 2002; Hanelt 2001; Pirc 1994; Seidemann 1993c; Small 1997; Uphof 1968; Wiersema/León 1999

Acetosa pratensis Mill.

▶ *Rumex acetosa L.*

Acetosa scutata (L.) Mill.

▶ *Rumex scutatus L.*

Achasma walang (Blume) Val.

▶ *Etlingera walang (Blume) R.M. Smith*

Achillea L. - Yarrow - Asteraceae (Compositae)

Achillae ageratum L.

Synonyms ▶ *Achillae decolorans* Schrad.
Common Names ▶ garden mace, sweet nancy, sweet yarrow; *German:* Garten-Schafgarbe, Süße Schafgarbe
Usage ▶ spice
Parts Used ▶ leaf, herb
Distribution ▶ W Europe: Iberia, France, England, Croatia?

Davidson 1999; Erhardt et al. 2002; Schultze-Motel 1986; Uphof 1968

Achillea atrata L.

Synonyms ▶ *Ptarmica atrata* (L.) DC.
Common Names ▶ black-edged yarrow; *French:* achillée noire; *German:* Schwarze Schafgarbe, Schwarzrandige Schafgarbe
Usage ▶ spice; **product** of Iva liqueurs
Parts Used ▶ herb
Distribution ▶ Europe: Alpine regions of Germany, Austria, France, Italy, Switzerland, Austria, Slovenia

Aichele/Schwegler 4, 1995; Berger 4, 1954; Fleischhauer 2003; Hager 4, 1992; Schönfelder 2001; Uphof 1968

Achillea clavenae L.

Synonyms ▶ *Ptarmica clavenae* DC.
Common Names ▶ bitter yarrow; *French:* achillée amère; *German:* Bittere Schafgarbe, Steinraute, Weißer Speik; *Italian:* essenzio ombrellifera
Usage ▶ spice
Parts Used ▶ leaf
Distribution ▶ Alpine and subalpine regions of S, SE Europe: lower Austria, Croatia, Dalmatia, Bulgaria, Italian Peninsula.

Aichele/Schwegler 4, 1995; Erhardt et al. 2002; Hager 4, 1992; Uphof 1968; Wiersema/León 1999

Achillae decolorans Schrad.

▶ *Achillae ageratum* L.

Achillea erba-rotta All. ssp. moschata (Wulf.) I. Richardson

Synonyms ▶ *Achillae moschata* Wulf.
Common Names ▶ alpine yarrow, musk milfoil, musk yarrow; *French:* achillée, genepi blanc; *German:* Bisamkraut, Feldgarbe, Genipikraut, Ivakraut, Moschusschafgarbe; *Italian:* genepi
Usage ▶ spice, flavoring

Parts Used ▶ leaf, herb
Distribution ▶ S, C Europe

Abraham/Seeber 1995; Aichele/Schwegler 4, 1995; Berger 4, 1954; Hager 4, 1992; Hanelt 2001; Hiller/Melzig 1999; Hoppe 1949; Schönfelder 2001; Uphof 1968; Wiersema/León 1999

Achillea millefolium L.

Common Names ▶ common yarrow, fragrant yarrow, milfoil, nose bleed, Western yarrow, wounderwort, yarrow (herb); *Brazil (Portuguese):* milfolhas; *Chinese:* yang shi cao; *French:* achilée millefeuille, herbe aux charpentiers, millefeuille; *German:* Achilleskraut, Gemeine Scharfgarbe, Gewöhnliche Schafgarbe, Wiesen-Schafgarbe, Tausendblatt; *India:* biranjasif, gandama, rooamari, rojmari; *Italian:* achillea millefoglio; *Japanese:* nokogirosô-zoku; *Korean:* chonipthopphul; *Portuguese:* milefólia; *Russian:* tysjatschelistnik obyknownenyj; *Spanish:* cientoenrama, milenrama, milhojas, flor de pluma
Usage ▶ spice; flavoring
Parts Used ▶ fresh leaf
Distribution ▶ Europe, temperate Asia, India, Canada, USA, Mexico, Meso-America, Australia, New Zealand; widely native elsewhere

Agarwal 1990; Aichele/Schwegler 4, 1995; Anon. 1998; Berger 4, 1952; Bilgri/Adam 2000; Bremness 2001; Cheers 1998; Coiciu/Racz (no year); Czygan 1994; Dudtschenko et al. 1989; Erichsen-Brown 1989; Hager 4, 1992; Haggag et al. 1975; Hanelt 2001; Heeger 1956; Hiller/Melzig 1999; Hoffmann et al. 1992; Hofmann/Fritz 1995; Hohmann et al. 2001; Hoppe 1949; Miller/Chow 1954; Newall et al. 1996; Orth et al. 1994, 1999; Pschyrembel 1998; Roth/Kormann 1997; Schaller 1995; Schönfelder 2001; Schultze-Motel 1986; Sharma 2003; Shawl et al. 2002; Suleiman et al. 1996; Täufel et al. 1993; Turova et al. 1987; Ubillos 1989; Uphof 1968; Villamar et al. 1994; Wiersema/León 1999; Wyk et al. 2004; Zirvi/Ikram 1975

Achillea moschata Wulf.

▶ *Achillae erba-rotta All. ssp. moschata (Wulf.) I. Richardson*

Achillea nobilis L.

Common Names ▶ noble milfoil, noble yarrow; *German:* Edel-Schafgarbe
Usage ▶ spice (sporadically), flavoring
Parts Used ▶ leaf
Distribution ▶ Europe, Turkey, Caucasus, frequently cultivated

Aichele/Schwegler 4, 1995; Erhardt 2000; Hager 4, 1992; Hoppe 1949; Schultze-Motel 1986

Achyranthes L. - Amaranthaceae

Achyranthes aspera L.

Common Names ▶ pricklychaff flower, rough chaff tree, Washerman's plant; *Chinese:* dao kou cao; *German:* Rauhklette; *India:* agava, apamarg, apamargamu, apang; *Pilipino:* hangod, hangor, higadhigad, nimikitan; *Vietnamese:* co nha ngu, nguru tat nam
Usage ▶ pot-herb
Parts Used ▶ leaf
Distribution ▶ tropical area, India, Pakistan, native in Spain
Note ▶ Used by the Shari Chad races (Africa) for making salt over the plant.

Dastur 1954; v. Koenen 1996; Uphof 1968

Achyrocline (Less.) DC. - Asteraceae (Compositae)

Achyrocline satureioides DC.

Common Names ▶ *Brazil:* alecrim da parede, macela, marcala do campo
Usage ▶ flavoring for bitter spirits
Parts Used ▶ flower
Distribution ▶ S America: Brazil, Uruguay

Hanelt 2001; Schultze-Motel 1986

Acinos Mill. - Lamiaceae (Labiatae)

Acinos alpinus (L.) Moench.

Common Names ▶ alpine calamint; *French:* calament des Alpes; *German:* Alpen-Steinquendel
Usage ▶ spice (rarely)
Parts Used ▶ herb
Distribution ▶ European Alps, Balkans, Turkey, NW Africa

Erhardt et al. 2002; Fleischhauer 2003; Schnelle 1999; Small 1997

Acinos arvensis (Lam.) Dandy

Synonyms ▶ *Acinos thymoides* Moench; *Calaminthe acinos* (Clairv.) Man., *Calamintha arvensis* Lam., *Satureja acinos* (L.) Scheele
Common Names ▶ basil thyme, mother of thyme, spring savory; *French:* basilic sauvage; *German:* Feldquendel, Steinquendel; *Russian:* duschewka, stschebruschka
Usage ▶ spice (sporadically), flavoring
Parts Used ▶ herb
Distribution ▶ Europe, Turkey, Caucasus, N Iran, Asia introduced in N America

Aichele/Schwegler 4, 1995; Cheers 1998; Erhardt et al. 2002; Fleischhauer 2003; Small 1997; Tucker 1986; Uphof 1968

Acinos thymoides Moench

 *Acinos arvensis (Lam.) Dandy*

Acmella oleracea (L.) R.K. Jansen

▶ *Spilanthes oleracea L.*

Acorus L. - Sweet flag - Acoraceae

 Acorus asiaticus *Nakai*

▶ *Acorus calamus L.*

 Acorus calamus *L.*

Synonyms ▶ *Acorus asiaticus* Nakai; *Acorus terrestris* Spreng.

Common Names ▶ calamus, gladdon, lag root, myrtle flag, myrtle grass, sweet flag, sweet myrtle; *Arabic:* vash, vaj; *Chinese:* shi chang pu; *Dutch:* kalmoeswortel; *French:* acore calame, acore odorant, acore vrai, jonc odorant, lis de marais; *German:* Kalmus, Deutscher Ingwer; *Hindi:* bacc, baccha, gorbach; *India:* buch; *Italian:* calamo aromatico; *Japanese:* shôbu; *Malaysian:* deringu, jeringan, jeringau; *Russian:* air, air trostnikowyj, irnyj koren', gair, jawer, tatarskoe sel'e, sabel'nik, kalmus; *Sanskrit:* vaca ugragandha; *Spanish:* acoro, acoro verdadero, calamís, cálamo arómatico

Usage ▶ spice; **product:** essential oil (calamus oil)

Parts Used ▶ rhizome

Distribution ▶ SE Asia, India, Himalaya, Sri Lanka to N China, Siberia, Japan, Malaysia; introduced to E, C Europe, Canada, USA; native elsewhere

Note ▶ The essential oil of the American tribe *Acorus calamus* L. var. *americanus* (Raf.) Wulff. are free of ß-asarone. The asiatic tribe has a high ß-asarone content.

Agarwal 1990; Aichele/Schwegler 5, 1996; Alberts/Muller 2000; Bilgri/Adam 2000; Bois 1934; Bournot 1968; Cheers 1998; Coiciu/Racz (no year); Dastur 1954; Dudtschenko et al. 1989; Erhardt et al. 2002; Erichsen-Brown 1989; Fleischhauer 2003; Hager 3, 1992; Hanelt 2001; Heeger 1956; Hepper 1992; Hiller/Melzig 1999; Hohmann et al. 2001; Hoppe 1949; Keller/Stahl 1982, 1993; Koschtschejew 1990; Mazza 1985; Melchior/Kastner 1974; Motley 1994; Newall et al. 1996; Opdyke 1977; Özcan et al. 2002; Pochljobkin 1974, 1977; Pschyrembel 1998; Raina et al. 2003; Schneider/Jurenitsch 1992; Schönfelder 2001; Seidemann 1993c; Seidemann/Siebert 1987; Sharma 2003; Shiva et al. 2002; Siewek 1990; Small 1997; Staesche 1972; Stahl/Keller 1981; Täufel et al. 1993; Turova et al. 1987; Uphof 1968; Wiersema/León 1999; Wüstenfeld/Haensel 1964; Wyk et al. 2004; Zeven/de Wet 1982

◘ **Acorus calamus, flowering**

 Acorus gramineus *Sol.*

Synonyms ▶ *Acorus humilis* Salisb.

Common Names ▶ Chinese sweet grass, Japanese sweet flag, grass leaf sweet flag; *Chinese:* shi chang pu; *French:* acore à feuilles de graminée; *German:* Aniskalmus, Japanischer Kalmus, Lakritze-Kalmus; *India:* pahari buch; *Japanese:* sekisho; *Korean:* sokchangpho; *Vietnamese:* lay yam, xinh pau chu

Usage ▶ spice

Parts Used ▶ rhizome

Distribution ▶ E Asia: India, China, Thailand, Japan, Korea, Malaysia

Cheers 1998; Erhardt et al. 2002; Hanelt 2001; Leung 1991; Roth/Kormann 1997; Uphof 1968; WHO 1990; Wiersema/León 1999

 Acorus humilis *Salisb.*

▶ *Acorus gramineus Sol.*

A

  Acorus terrestris *Spreng*

▶ *Acorus calamus* L.

Acrodiclidium Nees et Martius - Lauraceae

 Acrodiclidium camara *Schomb.*

Common Names ▶ Akawai nutmeg, Guayana nutmeg; *French:* bois amer; *German:* Guayana Muskatnuss; *Spanish:* cajueiro, camarão, cambará de cheiro
Usage ▶ spice (locally)
Parts Used ▶ seed
Distribution ▶ Guyana

Hoppe 3, 1987

 Acrodiclidium puchury-major *Mez*

▶ *Ocotea puchury-major* Mart.

Acronychia Forster et G. Forster - Rutaceae

 Acronychia laurifolia *Bl.*

▶ *Acronychia pedunculata* (L.) Miq.

 Acronychia odorata *(Lour.) Baill.*

Synonyms ▶ *Cyminosma odorata* DC., *Jambolifera odorata* Lour.
Usage ▶ condiment (in Malaysia)
Parts Used ▶ leaf
Distribution ▶ tropical Asia, especially Malaysia

Uphof 1968

 Acronychia pedunculata *(L.) Miq.*

Synonyms ▶ *Acronychia laurifolia* Bl., *Cyminosma resinosa* DC., *Jambolifera resinosa* Lour.
Usage ▶ condiment; **product:** essential oil
Parts Used ▶ tender leaves
Distribution ▶ India: W, E Ghats, NE Himalayan hills

Arora/Pandey 1996; Hager 4, 1992; Oyen/Dung 1999

Adasonia L. - Monkeybread, Baobab - Bombacaceae

 Adansonia baobab *L.*

▶ *Adansonia digitata* L.

 Adansonia digitata *L.*

Synonyms ▶ *Adansonia baobab* L.
Common Names ▶ monkey bread tree, cork tree, (common) baobab, African calabash, Judas's bag; *Arabic:* hijid; *French:* baobab, calebassier du Sénégal, pain de singe *German:* Afrikanischer Affenbrotbaum, Affenbrotbaum, Baobab; *India:* gorakh amli, gorakh chinch, magimavu, kalp; *Italian:* baobab, calabaceira; *Portuguese:* baobab; *Russian:* baobab; *Spanish:* baobab; *Swahili:* mbuyu
Usage ▶ pot-herb
Parts Used ▶ leaf
Distribution ▶ tropical Africa, Madagascar, Arabian Peninsula; widely native in the Tropics

Aké Assi/Guinko 1991; Bendel 2002; Cheers 1998; Chevalier 1906; Dastur 1954; Hanelt 2001; Ijomah et al. 2000; v. Koenen 1996; Kremer/Jaeggi 1995; Lewington 1990; Schenck/Naundorf 1966; Schultze-Motel 1986; Seidemann 1993c; Täufel et al. 1993; Toury et al. 1957; Uphof 1968; Wickens 1979, 1982; Wiersoma/León 1999

Adenandra Willd. - Rutaceae

 Adenandra fragrans *(Sims) Roem. et Schult.*

Synonyms ▶ *Diosma fragrans* Sims

Common Names ▶ breath of heaven; *German:* Himmelsduft; *S Africa:* anysboegoe, anysbuchu, klipsissie, sissie
Usage ▶ flavoring
Parts Used ▶ leaf
Distribution ▶ S Africa, especially Cape region

Erhardt et al. 2002; Hanelt 2001; Uphof 1968; Usher 1967

 Adenanthera tetraptera Schum. et Thonn.

▶ *Tetrapleura tetraptera (Schum. et Thonn.) Taub.*

Adenocalymna Mart. ex Meissner - Bignonaceae

 Adenocalymna alliaceum Miers

Common Names ▶ *Spanish:* cipó d'alho
Usage ▶ spice
Parts Used ▶ herb
Distribution ▶ S America

Hanelt 2001; Terra 1966

Adiantum L. - Maidenhair Fern - Adiantaceae (Pteridaceae)

 Adiantum nigrum L.

Common Names ▶ black spleenwort; *Bolivia:* quilquina; *French:* capillaire boire, doradille noire; *German:* Schwarzer Frauenhaarfarn, Schwarzstieliger Krullfarn; *Russian:* adiant(um) tschjornaja
Usage ▶ spice for sauces (Llajhua, Sarsa)
Parts Used ▶ leaf
Distribution ▶ S America: Andes; native in SE Asia

Aegle Corrêa - Rutaceae

 Aegle marmelos (L.) *Corrêa ex Roxb.*

Synonyms ▶ *Crataeva marmelos* L., *Crataeva religiosa* Ainslie
Common Names ▶ bael, bel, Bengal quince, golden apple, Indian bael; *Arabic:* shul; *Chinese:* yiu-tu-chih; *Dutch:* slijim appelboom; *French:* bel Indien, oranger du Malabar; *German:* Baelbaum, Belbaum, Bengalische Quitte; *India:* beli, belethi, bilva, marmel, siriphal; *Javanese:* maja, modjo; *Portuguese:* marmelos; *Spanish:* bela, milv; *Thai:* ma tum
Usage ▶ flavoring
Parts Used ▶ young leaf and shoot
Distribution ▶ India, Indochina, Myanmar, cultivated in Malaysia

Arora/Pandey 1996; Berger 3, 1952; Blancke 2000; Bose 1985; Daniel/Rajendran 1996; Dastul 1954; Davidson 1999; Hanelt 2001; Hiller/Melzig 1999; Hoppe 1949; Kumar 2003; Schultze-Motel 1986; Sharma 2003; Täufel et al. 1993; Uphof 1968; Verheij/Coronel 1991; Wiersoma/León 1999; Zeven/de Wet 1982

Aegopodium L. - Apiaceae (Umbelliferae)

 Aegopodium podagraria L.

Common Names ▶ ashweed, bishop's weed, dwarf elder, goutweed, ground elder, herb Gerard; *French:* egopode, podagraire; *German:* Ackerholler; Geißfuß, Gewöhnlicher Giersch, Giersch, Podagra; *Italian:* castalda, girardina silvestre; *Russian:* snyt'obyknovennija
Usage ▶ pot-herb
Parts Used ▶ herb
Distribution ▶ temperate Asia, Europe, Turkey, Caucasus to C Asia, W Siberia, native elsewhere
Note ▶ The fruits used as falsification of *Carum carvi*-fruits.

Aichele/Schwegler 3, 1995; Bilgri/Adam 2000; Cheers 1998; Davidson 1999; Dudtschenko et al. 1989; Erhardt et al. 2002; Fleichhauer 2003; Hager 4, 1992; Hanelt 200; Hiller/Melzig 1999; Hoppe 1949;

Melchior/Kastner 1974; Schönfelder 2001; Schultze-Motel 1986; Staesche 1972; Täufel et al. 1993; Uphof 1968; Wiersoma/León 1999

Aeollanthus Mart. ex Spreng. - Lamiaceae (Labiatae)

Aeollanthus buettneri *Gürke*

▶ *Aeollanthus pubescens Benth.*

Aeollanthus heliotropoides *Oliv.*

Synonyms ▶ *Aeollanthus suaveolens* Mart. ex Spreng.
Common Names ▶ *Portuguese:* chegadinka, macassa
Usage ▶ pot-herb (locally), for flavoring soups
Parts Used ▶ leaf
Distribution ▶ tropical Africa: Nigeria, Cameroon

Burkill 3, 1995; Hanelt 2001; Irvine 1948; Schultze-Motel 1986; Uphof 1968

Aeollanthus pubescens *Benth.*

Synonyms ▶ *Aeollanthus buettneri* Gürke
Usage ▶ spice, analog basil
Parts Used ▶ leaf, flower (local)
Distribution ▶ tropical Africa: Nigeria, Sierra Leone, from Mali to Cameroon, Central African Republic

Burkill 3, 1995; Dalziel 1957; Hanelt 2001; Schultze-Motel 1986; Sohounhloue et al. 2002; Uphof 1968

Aeollanthus suaveolens *Mart. ex Spreng.*

▶ *Aeollanthus heliotropoides Oliv.*

Aethusa mutellina *St. Lag.*

▶ *Ligusticum mutellina (L.) Crantz*

Aframomum K. Schum. - Zingiberaceae

Aframomum alboviolaceum *(Ridley) K. Schum.*

Synonyms ▶ *Aframomum glatifolium* K. Schum., *Amomum alboviolaceum* Ridley
Common Names ▶ Cameroon cardamom; *Cameroon:* odjom; *German:* Kamerun-Kardamom, Sierra Leone-Kardamom
Usage ▶ spice
Parts Used ▶ leaf
Distribution ▶ tropical Africa: Sierra Leone, Cameroon also cultivated

Berger 3, 1952; Berger 1964/65; Duke et al. 2003; Hanelt 2001

Aframomum angustifolium *(Sonn.) K. Schum.*

Synonyms ▶ *Aframomum sanguineum* (K. Schum.) K. Schum., *Amomum angustifolium* Sonn., *Amomum sanguineum* K. Schum.
Common Names ▶ Madagascar cardamom, dark Crimson cardamom, wild cardamom; *French:* cardamome de Madagaskar; *German:* Blutroter Kardamom, Madagaskar-Kardamom
Usage ▶ spice, the seeds are used like black pepper
Parts Used ▶ seed
Distribution ▶ E Africa, Madagascar, Zanzisbar, Mauritius, Pemba and the Seychelles; cultivated on the Pemba Islands
Note ▶ In Near East and Egypt as addition to coffee.

Berger 3, 1952; 1964/65; Bois 1934; Coomes et al. 1955; Hager 5, 1993; Hanelt 2001; Hari et al. 1994; Melchior/Kastner 1974; Overdieck 1992; Peter 2001; Pruthi 1976; Schultze-Motel 1986; Seidemann 1993c; Siewek 1990; Täufel et al. 1993; Teuscher 2003; Uphof 1968

Aframomum cororima *(Braun) P.C.M. Jansen*

▶ *Aframomum korarima (Peireira) Engl.*

Aframomum daniellii *(Hook.f.) K. Schum.*

Synonyms ▸ *Amomum danielli* J.D. Hook.
Common Names ▸ bastard melegueta, Cameroon cardamom; *French:* cardamome de Cameroun; *German:* Bastard-Meleguetapfeffer; Kamerun-Kardamom
Usage ▸ spice
Parts Used ▸ seed
Distribution ▸ tropical W Africa: Cameroon, also cultivated

Adegoke/Shura 1994; Berger 1964/65; Bois 1934; Hager 5, 1993; Hanelt 2001; Schultze-Motel 1986; Seidemann 1993c; Siewek 1990; Täufel et al. 1993; Teuscher 2003

 Aframomum exscapum *(Sims) Hepper*

Common Names ▸ alligator pepper, grains of paradise; *German:* Alligatorpfeffer, Stengelloser Alligartorpfeffer, Paradieskörner
Usage ▸ spice
Parts Used ▸ fruit
Distribution ▸ tropical W Africa

Duke et al. 2003; Seidemann 1993c; Täufel et al. 1993; Teuscher 2003

 Aframomum hanburyi *K. Schum.*

Synonyms ▸ *Amomum clusii* Hanb.
Common Names ▸ Cameroon cardamom; *French:* cardamome de Cameroun; *German:* Kamerun-Kardamon, Bastard Malagetta
Usage ▸ spice, condiment
Parts Used ▸ seed
Distribution ▸ tropical W Africa: Cameroon

Berger 1964/65; Bois 1934; Melchior/Kastner 1974; Overdieck 1992; Peter 2001; Pruthi 1976; Seidemann 1993c; Teuscher 2003; Uphof 1968; Zollo et al. 2002

 Aframomum korarima *Pereira ex Engl.*

Synonyms ▸ *Aframomum cororima* (Braun) P.C.M. Jansen, *Amomum korarima* Pereira;
Common Names ▸ Ethiopan cardamom, korarima cardamom, false cardamom, nutmeg cardamom, Guragi spice; *Arabic:* habhal hobashi, heil; *French:* cardamome d'Ethiopie, poivre d'Ethiopie; *German:* Abessinischer Kardamom, Ethiopischer Kardamom, Guragi-Gewürz, Korarima-Kardamom
Usage ▸ spice, e.g. flavoring for coffee (or tea), bread, butter (Kefa Province)
Parts Used ▸ seed
Distribution ▸ Ethiopia, Somalia; also cultivated
Note ▸ In Arabia they serve the flavoring of coffee. The aroma is nutmeg-like.

Berger 3, 1952; Berger 1964/65; Biftu 1981; Hager 5, 1993; Hanelt 2001; Jansen 1981; Melchior/Kastner 1974; Norman 1990, 1991; Overdieck 1992; Peter 2001; Pruthi 1976; Schultze-Motel 1986; Seidemann 1993c; Siewek 1990; Täufel et al. 1993; Teuscher 2003; Zeven/de Wet 1982

 Aframomum latifolium *K. Schum.*

▸ *Aframomum alboviolaceum (Ridley) K. Schum.*

 Aframomum letestuanum *Gagnep.*

Usage ▸ spice
Parts Used ▸ seed
Distribution ▸ tropical Africa: Cameroon, Congo; also cultivated

Hanelt 2001; Poulsen/Lock 1999; Schultze-Motel 1986; Seidemann 1993c; Täufel et al. 1993; Teuscher 2003; Zollo et al. 2002

 Aframomum macrospermum *(Smith) Burkill*

Common Names ▸ Guinea cardamom; *French:* cardamome de Guinea; *German:* Großsamiger Kardamom, Guinea-Kardamom
Usage ▸ spice (locally)
Parts Used ▸ fruit
Distribution ▸ Guinea

Berger 1964/65

Aframomum K. Schum. - Zingiberaceae

Aframomum malum Schum.

Common Names ▶ East African cardamom; *German:* Ostafrikanischer Kardamom
Usage ▶ condiment
Parts Used ▶ seed
Distribution ▶ E and tropical Africa

Uphof 1968

Aframomum melegueta K. Schum.

Synonyms ▶ *Aframomum meleguetella* K. Schum.; *Amomum melegueta* Rosc.
Common Names ▶ Guinea pepper, Guinea grains, alligator pepper, grains of paradise, melegueta pepper; *Chinese:* hsi sha tou; *French:* graine de paradis, graine de maniguette, melegueta, poivre d'Afrique, poivre de Guinée, semance de paradis; *German:* Melegueta-Pfeffer, Paradieskörner, Grana paradisi, Alligatorpfeffer, Guineapfeffer, Piper Melegueta; *India:* malgoat; *Italian:* grani di meleguetta, grani paradisi, mani guetta; *Russian:* amomum, ili malagetta, rauskoe serno, mallawetskij, gwinejskij perez; *Slovakian:* guinejské korenie; *Spanish:* granos de paraíso, malagueta, maniguette
Usage ▶ spice, e.g. admixture for "Ras el Hanout"
Parts Used ▶ seed
Distribution ▶ W tropical Africa; also cultivated from Guinea, Sierra Leone through Ivory Coast, Ghana, Benin, Nigeria, Cameroon, Gabon to Angola
Note ▶ In 1876 the plant was tried in the Singapore Botanic Gardens, but seems not to have succeeded.

Adegoke et al. 2003; Aedo et al. 2001; Berger 3, 1952; Berger 1964/65; Blancke 2000; Bois 1934; Boisvert/Huber 2000; Burkill 5, 2000; Craze 2002; Davidson 1999; Duke et al. 2003; Erhardt et al. 2002; Govindarajan et al. 1982; Griebel 1943; Hanelt 2001; v.Harten 1970; Hoppe 1949; Lock et al. 1977; Melchior/Kastner 1974; Norman 1990, 1991; Overdieck 1992; Peter 2001; Pochljobkin 1974, 1977; Pruthi 1976; Schenck/Nauendorf 1966; Schultze-Motel 1986; Seidemann 1993c; Small 1997; Staesche 1972; Täufel et al. 1993; Teuscher 2003; Uphof 1968; Wiersoma/León 1999; Wüstenfeld/Haensel 1964; Zeven/de Wet 1982

Aframomum melegueta: a **flowering** b **fruit and seeds**

Aframomum pruinosum Gagn.

Common Names ▶ Cameroon cardamom; *German:* Bereifter Kardamom, Kamerun-Kardamom
Usage ▶ spice (rarely)
Parts Used ▶ seed
Distribution ▶ W Africa; Cameroon

Zollo et al. 2002

Aframomum meleguetella K. Schum.

▶ *Aframomum melegueta K. Schum.*

Aframomum sanguineum *(K. Schum.) K. Schum.*

▶ *Aframomum angustifolium (Sonn.) K. Schum.*

 Aframomum sceptrum *(Oliv. et Hanb.) K. Schum.*

Common Names ▶ black amomum, Guinea grains; *German:* Schwarzer Kardamom, Guinea-Körner
Usage ▶ spice (essential oil), flavoring for food; **product:** essential oil
Parts Used ▶ leaf
Distribution ▶ W Africa
Note ▶ The seeds are camphoraceous in taste.

Burkill 5, 2000; Dalziel 1937; Duke et al. 2003

 Aframomum subsericum *(Oliv. et Hanb.) K. Schum.*

Common Names ▶ alligator pepper; *German:* Alligatorpfeffer
Usage ▶ condiment
Parts Used ▶ seed
Distribution ▶ W Africa
Note ▶ A smell of lemon.

Burkill 5, 2000

Afrostyrax Perkins et Gilg - Styracaceae (Huaceae)

 Afrostyrax kamerunensis *Perkins et Gilg*

Common Names ▶ Cameroon garlic tree; *German:* Kamerun-Knoblauchbaum
Usage ▶ local as garlic-like condiment, special of sauces
Parts Used ▶ seed, bark
Distribution ▶ tropical Africa: Cameroon, Gabon
Note ▶ Bark and seeds have a garlic taste.

Burkill 5, 2000; Neuwinger 1999; Walker 1952

 Afrostyrax lepidophyllus *Mildbraet*

Common Names ▶ garlic bark (tree); *German:* Schuppenblättiger Knoblauchbaum
Usage ▶ spice, condiment
Parts Used ▶ bark, seed
Distribution ▶ tropical Africa: Cameroon, Congo, the Central Africa Republic, Zaire
Note ▶ Bark and seeds have a garlic taste.

Burkill 5, 2000; Milbread 1913; Neuwinger 1998; Sandberg/Cronlund 1982

 Agaricus edodes *Berk.*

▶ *Lentinus edodes (Berk.) Sing.*

Agastache Gronov. - Mexican Hyssop - Lamiaceae (Labiatae)

 Agastache anisata *hort.*

▶ *Agastache foeniculum (Pursh) Kuntze*

 Agastache foeniculum *(Pursh) Kuntze*

Synonyms ▶ *Agastache anisata* hort., *Hyssopus anisatus* Nutt., *Stachys foeniculum* Pursh
Common Names ▶ anise hyssop, anise mint, blue giant hyssop, fragrant giant hyssop, licorice mint; *French:* hysope anisée; *German:* Duftnessel, Anisysop
Usage ▶ spice, e.g. for cakes and sweets; **product:** essential oil
Parts Used ▶ leaf
Distribution ▶ N America: S Canada to the NC states of USA, Rocky Mountains, former Soviet Union (Crimea, Moldavia), S Finland

Cheers 1998; Erhardt et al. 2002; Hanelt 2001; Mazza/Kiehn 1992; Mikus et al. 1997; Nykänen et al. 1989; Schultze-Motel 1986; Small 1997; Svoboda 1995; Tucker 1986; Wiersoma/León 1999; Wilson et al. 1992

Agastache mexicana *(Kunth) Lint. et Epling*

Synonyms ▶ *Dracocephalum mexicanum* H.B.K., *Cedronella mexicana* (Kunth) Benth.
Common Names ▶ Mexican giant hyssop, lemon hyssop; *Chinese:* huo xiang; *French:* agastache; *German:* Mexikanische Duftnessel, Mexikanischer or Lemon Ysop; *Mexico:* toronjil; *Spanish:* totoji
Usage ▶ spice, flavoring
Parts Used ▶ leaf
Distribution ▶ Mexico; cultivated in N America

Cheers 1998; Erhardt et al. 2002; Hanelt 2001; Mikus/Schaser 1995; Ochoa/Alonso 1996; Omidbaigi/Sefidkon 2003; Schultze-Motel 1986; Seidemann 1993c; Small 1997; Svoboda et al. 1995; Teuscher 2003; Tucker 1986

 Agastache rugosa *(Fischer et C.A.Meyer) Kuntze*

Synonyms ▶ *Cedronella japonica* Hassk., *Elsholtzia monostachys* H. Lévl. et Van., *Lophanthus rugosus* Fischer et C.A. Meyer
Common Names ▶ Chinese giant hyssop, Korean mint, wrinkled giant hyssop; *Chinese:* huo xiang; *German:* Koreanische Minze, Minz-Agastache, Runzliger Ysop; *Japanese:* kawa-midori; *Korean:* bangah, paechohyang , pangaphul
Usage ▶ spice; **product**: essential oil (perfumery)
Parts Used ▶ leaf
Distribution ▶ E Siberia, Russian Far E, Korea, China, Japan, N Vietnam; introduced to N America after World War II, also cultivated in China
Note ▶ The essential oil contains more than 90% methyl chavicol.

Ahn/Yang 1991; Boo-Yong Lee et al. 2001; Charles et al. 1991; Cheers 1998; Dung et al. 1996; Hanelt 2001; Hee-Juhn Park et al. 2000; Hiller/Melzig 1999; Kim et al. 2001; Moon Jung Lee et al. 2001; NICPBP 1987; Schultze-Motel 1986; Seidemann 1993c; Small 1997; Svoboda et al. 1995; Tae Hwan Kim et al. 2001; Teuscher 2003; Tucker 1986; Weyerstahl et al. 1992; Wiersoma/León 1999; Wilson et al. 1992

 Agathophyllum aromaticum *(Sonn.) Willd.*

▶ *Ravensaria aromatica Sonn.*

 Agathosma betulina *Pillans*

▶ *Barosma betulina (Bergius) Bartl. et Wendl.*

 Agathosma crenulata *(L.) Pillans*

▶ *Barosma crenulata (L.)*

 Agathosma serratifolia *(Curtis) A.D. Spreeth*

▶ *Barosma serratifolia (Curtis) Will.*

Aglaia Lour. - Meliaceae

 Aglaia affinis *Merr.*

▶ *Aglaia odoratissima Blume*

 Aglaia heterophylla *Merr.*

▶ *Aglaia odoratissima Blume*

 Aglaia odorata *Lour.*

▶ *Aglaia odoratissima Blume*

 Aglaia odoratissima *Blume*

Synonyms ▶ *Aglaia affinis* Merr., *Aglaia heterophylla* Merr., *Aglaie odorata* Lour.
Common Names ▶ mock lime, orchid tree; *Chinese:* mi zan lau, mi sui lan, san yeh lan; *German:* Duft-Glanzbaum; *Indonesian:* pancal kidang, tanglu; *Japanese:* ju-ran, muran; *Malaysian:* chulan, kasai, telur belangkas, merlimau; *Thai:* prayong paa, sangkhriat
Usage ▶ flavoring of tea; **product**: essential oil
Parts Used ▶ flower
Distribution ▶ India, Indonesia: Sumatra, Java, Philippines, China

Note ▶ On Java a dwarf form (var. *microphylla* DC.) occasionally cultivated in gardens.

Artander 1960; Berger 2, 1950; Burkill 1966; Hanelt 2001; Heyne 1953; Corner 1988; Oyen/Dung 1999; Storrs 1997; Uphof 1968

Agrimonia L. - Agrimony - Rosaceae

Agrimonia eupatoria L. ssp.odorata *(Gouan)* Hook.

Synonyms ▶ *Agrimonia procera* Wallr.
Common Names ▶ odour agrimony, cock ley bur, liverwort; *Arabic:* ghafith, terfaq; *French:* aigremoine gariot; *German:* Duft-Odermennig, Großer Odermennig; *Italian:* agrimonia; *Spanish:* agrimonia
Usage ▶ flavoring (rarely)
Parts Used ▶ leaf, herb
Distribution ▶ Germany to W Russia, rarely cultivated

Bremness 2001; Hanelt 2002; Heeger 1956; Hoppe 1949; Schönfelder 2001; Schultze-Motel 1986; Small 1997; Uphof 1968; Wyk et al. 2004

Agrimonia procera *Wallr.*

▶ *Agrimonia eupatoria L. ssp. odorata (Gouan) Hook.*

Ajuga L. - Bugle - Lamiaceae (Labiatae)

Ajuga macrosperma *Wall.*

Common Names ▶ bugle; *French:* petite consoule; *German:* Großsamiger Günzel; *Russian:* shibutschka dubhiza
Usage ▶ pot-herb
Parts Used ▶ herb
Distribution ▶ India: temperate and tropical Himalaya

Arora/Pandey 1996; Hager 4, 1992; Shen, X.Y. et al. 1993

Ajuga reptans L.

Common Names ▶ creeped bugle; *French:* bugle rampante; *German:* Kriechender Günzel; *Russian:* shibutschka polsutschij
Usage ▶ pot-herb
Parts Used ▶ herb, leaf
Distribution ▶ Europe, Turkey, Caucasus, N Iran, Algeria, Tunesia; nat in N America

Erhardt et al. 2002; Fleischhauer 2003; Loch 1993

Alaraia pinnatifida *Harvey*

▶ *Undaria pinnatifida (W.H. Harvey) W.F.R. Suringar*

Alchemilla L. - Lady's Mantle - Rosaceae

Alchemilla conjucta *Bab.*

Common Names ▶ silver lady's mantle; *German:* Silber-Frauenmantel, Verwachsener Frauenmantel
Usage ▶ pot-herb
Parts Used ▶ leaf
Distribution ▶ Europe, France, Switzerland, SW Alps; native in England

Erhardt et al. 2002; Fleischhauer 2003

Alchemilla fissa *Günther et Schummel*

Common Names ▶ *German:* Spaltblättriger Frauenmantel, Zerschlitzter Frauenmantel
Usage ▶ pot-herb
Parts Used ▶ leaf
Distribution ▶ Europe: Spain, France, C Europe, Poland, Alps, Sudetenland

Erhardt et al. 2002; Fleischhauer 2003

Alchemilla splendens Christ ex Favrat

Common Names ▶ glittering lady's mantle; *German:* Schimmernder Frauenmantel
Usage ▶ pot-herb
Parts Used ▶ leaf
Distribution ▶ Europe: France, Switzerland, Alps, Jura

Erhardt et al. 2002; Fleischhauer 2003

 Alchemilla vulgaris L.

Common Names ▶ lady's mantle; *French:* alichimille, manteau de Notre-Dame; *German:* Gemeiner Frauenmantel, Sinau; *Italian:* alchemilla; *Russian:* manshetka; *Spanish:* pie de leon
Usage ▶ pot-herb
Parts Used ▶ leaf
Distribution ▶ C Europe, W Asia, E America

Erhardt et al. 2002; Fleischhauer 2003; Hanelt 2001; Hoppe 1, 1948; Koschtschejew 1990; Schönfelder 2001; Wyk et al. 2004

Aleurites J.R. Forst. et G. Forst. - Euphorbiaceae

 Aleurites javanica Gand.

▶ *Aleurites moluccana* (L.) Willd.

 Aleurites moluccana (L.) Willd.

Synonyms ▶ *Aleurites javanica* Gand., *Aleurites triloba* J.R. Forst. et G. Forst., *Jatropha moluccana* L., *Juglans camirium* Lour.
Common Names ▶ candle berry, candle nut, Indian walnut; *Chinese:* shi li zi; *French:* noix des Indes, noix de Bancoul, noix des Moluques; *German:* Kandeln, Bankuln, Kekunanuss, Kemiri, Kemiri-, Kerzen- or Lichternuss; *Hindi:* jangli akhrot; *Indonesian:* kemiri, miri, muncang; *Javanese:* kemiri; *Malaysian:* kemiri, kembiri, buah keras; *Laos:* kôk namz man; *Pilipino:* kami, lumbang, biao; *Portuguese:* noz da Índia; *Spanish:* calumbán, camirio, lumban; *Thai:* phothisat, kue-ra, purat, mayao; *Vietnamese:* cây lai
Usage ▶ spice; before used the seeds are roasted, specially for soups and sauces
Parts Used ▶ seed
Distribution ▶ India, China, SE Asia, Polynesia, New Zealand, origin. Malaysia; widely cultivated in the Tropics
Note ▶ Befor beeing used the seeds must be roasted. Raw seeds are poisonous. It is an indispensable spice in Indonesian cuisine.

Blank R.J. et al. 1997; Engelbeen 1946; Erhardt et al. 2002; Foster 1962; Guzman/Siemonsma 1999; Hanelt 2001; Seidemann 1993c; Seidemann/Siebert 1987; Staesche 1972; Strauß 1969c; Täufel et al. 1993; Wealth of India 1, 1948; Wiersoma/León 1999; Zeven/de Wet 1982

 Aleurites triloba J.R. Forst. et G. Forst.

▶ *Aleurites mollucana* (L.) Willd.

 Alga marina Lam.

▶ *Zostera marina* L.

Alliaria Heist. ex Fabr. - Brassicaceae (Cruciferae)

 Alliaria officinalis Andr. ex M. Bieb.

▶ *Alliaria petiolata* (M. Bieb.) Cavara ex M. Bieb.

 Alliaria petiolata (M. Bieb.) Cavara et Grande

Synonyms ▶ *Alliaria officinalis* Andr. ex M. Bieb., *Sisymbrium alliaria* Scop.
Common Names ▶ garlic mustard, hedge garlic, Jack-by-the-hedge, onion nettle; *Arabic:* hashisha thawmiyah; *French*: alliairé; *German:* Lauchkraut, Knoblauchsrauke, Lauchhederich; *Italian:* agliaria, alliaria; *Russian:* tschesnotschnik, tschesnotschniza, tschesnotschnaja trawa, lesnoj tschesnok; *Spanish:* erísimo

Alliaria petiolata, flowering

Usage ▶ pot-herb, condiment
Parts Used ▶ fresh and dried leaves
Distribution ▶ N Africa, temperate Asia, India, E Europe, SE and SW Europe, native elsewhere

Aichele/Schwegler 3, 1995; Arora/Pandey 1996; Bois 1934; Bremness 2001; Davidson 1999; Dudtscheko et al. 1989; Erhardt et al. 2002; Fleischhauer 2003; Hager 4, 1992; Hiller/Melzig 1999; Pochljobkin 1974, 1977; Schönfelder 2001; Seidemann 1993c, 1995a; Seidemann/Siebert 1987; Täufel et al. 1993; Tucker 1986; Uphof 1968; Wiersema/León 1999

Alliaria wasabi *Prantl*

▶ *Wasabi japonica (Miq.) Matsum.*

Allium L. – Garlic, Leek, Onion - Alliaceae (Liliaceae)

Abraham et al. 1976; Bernhardt 1970; Block 1985, 1992; Block et al. 1992 a, b; Brewster 1994; Burba/Galmarini 1997; Carson 1987; Davis 1992; Fenwick/Hanley 1985/1986; Hager 4, 1992; Hanelt 1994, 2001; Helm 1956; Herrmann 1995; Jones/Mann 1963; Maggioni et al. 2001; Mathew 1996; Rabinowitch/Brewster 1990; Rabinowitch/Currah 2002; Schwartz/Mohan 1994; Seidemann 1993c; Täufel et al 1993; Vvedensky 1944; Whitaker 1976; Widder/Sabater 2002; Woodward 1996; Xiao-Jia et al. 1994

Allium altaicum *Pall.*

Synonyms ▶ *Allium ceratophyllum* Bess. & Ledeb., *Allium microbulbum* Prokh., *Allium sapidissimum* Hedw., *Allium saxatile* Pall.
Common Names ▶ Altai onion; *German:* Altai-Zwiebel, Sibirische Zwiebel; *Russian:* altajskij luk, sibirskij dinkij luk, kamennij luk, gorowoj luk, sontschina, mongol'skij luk, luk altajskij
Usage ▶ spice
Parts Used ▶ bulb
Distribution ▶ Altai, mountains of S Siberia and Mongolia to the Transbaical region, SE Kasakhstan, NW China

Dudtschenko et al. 1989; Hanelt 2001; Pochljobkin 1974, 1977; Schultze-Motel 1986; Täufel et al. 1993; Zeven/de Wet 1982

Allium ampeloprasum L.

▶ *Allium ampeloprasum* L. var. *ampeloprasum*

Allium ampeloprasum L. var. ampeloprasum

[or Great-Headed Garlic Group]

Synonyms ▶ *Allium ampeloprasum* L., *Allium holmense* Mill., *Allium lineare* Mill., *Porrum ampeloprasum* (L.) Mill.
Common Names ▶ great round-headed garlic, levant garlic, wild leek, leek; *Arabic:* tum-zu el-raas; *Chinese:* da tou Suan; *Dutch:* wilde prei, wild look; *French:* ail d'Orient, ail àcheval, faux poireau, poireau d'été, petit oignon; *German:* Acker-Knoblauch, Sommer-Knoblauch, Pferde-Knoblauch, Weinlauch; *Italian:* porrandello, porraccio, porro selvatico; *Portuguese:* alho porro bravo; *Russian:* luk shemtschushnuj, luk vinogradnji; *Spanish:* cebollino, puerro agreste, puerro silvestre
Usage ▶ spice
Parts Used ▶ bulb, clove, leaf
Distribution ▶ Greece, Russia, Near East, Middle East:

Iran, NW India, Transcaucasus, Turkey; SE Asia, Mexico, Chile, USA, N Africa, also cultivated
Note ▶ The plant has a strong garlic-like smell.

Aichele/Schwegler 5, 1996; Blancke 2000; Cheers 1998; Davidson 1999; Erhardt et al. 2002; Fattorusso et al. 2002; Hager 4, 1992; Hanelt 2001; Hepper 1992; Täufel et al. 1993; Uphof 1968; Wiersema/León 1999; Zeven/de Wet 1982

 Allium ampeloprasum L. **var. porrum** (L.) J. Gray

▶ *Allium porrum L.*

 Allium angolense Baker

Common Names ▶ African onion, African shallot; *French:* echalotte d'Angole, echalotte d'Afrique tropicale; *German:* Afrikanische Zwiebel, Angolanische Zwiebel; *Russian:* luk angol'skij
Usage ▶ spice
Parts Used ▶ bulb
Distribution ▶ Africa: Angola, Congo, Gabun, Zaire, also cultivated

Schultze-Motel 1986; Teuscher 2003

 Allium anguinum Bubani

▶ *Allium victoralis L.*

 Allium angulosum Lour.

▶ *Allium ramosum L.*

 Allium angulosum L.

Common Names ▶ edged garlic; *German:* Kantiger Lauch
Usage ▶ spice
Parts Used ▶ bulb, leaf
Distribution ▶ Europe, Siberia

Erhardt et al. 2002; Fleischhauer 2003; Loch 1993

 Allium ascalonicum L.

▶ *Allium cepa L. var. ascalonicum Baker*

 Allium ascalonicum auct. non L

▶ *Allium cepa L. var. ascalonicum Baker*

 Allium bakeri Hoop. non Regel

▶ *Allium fistulosum L.*

 Allium bakeri Regel

▶ *Allium chinense G. Don*

 Allium bouddhae O. Debeaux

▶ *Allium fistulosum L.*

 Allium canadense L.

Synonyms ▶ *Allium continuum* Small; *Allium mutabile* Mich.
Common Names ▶ Canada onion, American wild onion; *Cuban:* ajo montaña, ajo porro; *German:* Kanadische Zwiebel
Usage ▶ spice (rarely), especially by the Cheyenne Indians
Parts Used ▶ bulb
Distribution ▶ Canada; cultivated in Cuba
Note ▶ Perhaps also taxonomic derivatives of the species cultivated in Cuba.

Hanelt 2001; Small 1997

 Allium carinatum L.

Common Names ▶ keeled garlic; *German:* Gekielter Lauch
Usage ▶ spice
Parts Used ▶ leaf, bulb
Distribution ▶ C Europe

Note ▶ The ssp. *carinatum* and ssp. *pulchellium* (G. Don) Bonn. et Layens (nice garlic; *French:* ail à carène élegant; *German:* Schöner Lauch are only rarely spreaded and used.

Erhardt et al. 2002; Fleischhauer 2003; Loch 1993

Allium cepa L. var. ascalonicum Baker

Synonyms ▶ *Allium ascalonicum* auct. non L.
Common Names ▶ eschalot, shalott, Spanish garlic; *Dutch:* esjalot; *French:* ail stérile, échalotte, ciboule oignon patate, oignon sous terre; *German:* Askalonzwiebel, Aschlauch, Batatenzwiebel, Eschlauch, Frühlingszwiebel, Frühlingslauch, Kartoffelzwiebel, Klöben, Levantelauch, Schalotte, Schlotte, Syrische Zwiebel; *Hindi:* kanda lasum; *India:* gundhun; *Indonesian:* bawang erah; *Italian:* scalogno; *Malaysian:* bawang merah; *Pilipino:* sibuyas tagalog; *Portuguese:* cebolha roxa; *Russian:* luk nemezkii, luk schalot; *Spanish:* chalote, ascalonia, ascaloña; *Thai:* horm daeng, horm lek; *Vietnamese:* kanh kho, kanh ieu
Usage ▶ spice; seeds for flat bread and curry powder
Parts Used ▶ bulb, seed
Distribution ▶ only cultivated

Bärtels 1997; Bendel 2002; Blancke 2000; Bois 1934; Cheers 1998; Davidson 1999; Dudtschenko et al. 1989; Duke et al. 2003; Fattorusso et al. 2002; Hager 4, 1992; Hanelt 2001; Hiller/Melzig 1999; Hoppe 1949; Hutton 1989; Körber-Grohne 1989; Melchior/Kastner 1974; Pochljobkin 1974, 1977; Seidemann 1993c; Seidemann/Siebert 1987; Täufel et al. 1993; Teuscher 2003; Tindall 1983; Vogel 1995a

Allium cepa L. – Cepa-Group *(Common Onion Group)*

Allium cepa L. var. cepa

Common Names ▶ onion, bulb onion, garden onion, potato onion, Spanish onion; *Arabic:* bas(s)al, besla; *Chinese:* hsieh pai, hu cong, xie ba, yang cong, yang ts'ung; *Dutch:* ui; *French:* ciboule, oignon; *German:* Küchenzwiebel, Sommerzwiebel, Speisezwiebel, Rams, Zip(p)eln, Zipolle; *Hindi:* piyaz,

 Allium cepa var. ascalonicum, aerial onions

pyaj; *Italian:* cipolla; *Japanese:* atasugi, tamanegi; *Korean:* okpha, tunggulpha, yangpha; *Malaysian:* bawang; *Pilipino:* lasona, sibuyas; *Portuguese:* cebola; *Russian:* luk reptschatyj; *Sanskrit:* palandu, durandha; *Spanish:* cebolla; *Thai:* yai; *Vietnamese:* khan ko, hanh la
Usage ▶ spice and food (vegetable)
Parts Used ▶ bulb
Distribution ▶ only cultivated

Adam et al. 2000; Aichele/Schwegler 5, 1996; Aksoy 1983; Bärtels 1997; Bendel 2002; Block 1985, 1992; Boelens et al. 1971; Bois 1934; Bremness 2001; Brodnitz et al. 1969; Carson 1987; Davidson 1999; Dudtschenko et al. 1989; Duke et al. 2003; Faheid 1998; Farkas et al. 1992; Farrell 1985; Fossen et al. 1998; Hager 4, 1992; Hanelt 2001; Heeger 1956; Hepper 1992; Hiller/Melzig 1999; Hoffmann et al. 1992; Hoppe 1949; Järvenpää et al. 1998; Kaak et al. 2004; Koch 1994; Kodera et al. 2002; Körber-Grohne 1989; Laul et al. 1984; Lewington 1990; Marotti/Piccaglia 2002; Martine 1980; Martin-Lagos et al. 1992; Matheis/Lösung 1999; Mazza et al. 1980; Melchior/Kastner 1974; Nefisa et al. 1994; Patil et al. 1995; Peter 2001; Pochljobkin 1974, 1977; Pruthi 1976; Pschyrembel 1998; Raine 1978; Rosengarten 1969; Roth/Kormann 1997; Saini/Davis 1969; Schönfelder 2001; Seidemann 1993c; Seidemann/Siebert 1987; Sfikas 1994; Sharma 2003; Täufel et al. 1993; Teuscher 2003; Tindall 1983;

Uhl 2000; Uphof 1968; Villamara et al. 1994; Wiersema/León 1999; Wyk et al. 2004; Zeven/de Wet 1982

 Allium cepa L. **ssp. vivipara** Metzg. Alef.

▸ *Allium x proliferum (Moench) Schrad. ex Willd.*

 Allium cepa L. **var. cepa**

▸ *Allium cepa L. – Cepa-Group.*

 Allium cepa L. **var. proliferum** (Moench) Targ.-Tozz

▸ *Allium x proliferum (Moench) Schrad. ex Willd.*

 Allium cepa L. **var. sylvestre** Regel

▸ *Allium oschaninii B. Fedtsch.*

 Allium cepa L. **var. viviparum** (Metzg.) Alef.

▸ *Allium x proliferum (Moench) Schrad. ex Willd.*

Allium cepa L. – Every-ready onion Group

 Allium cepa L. **var. perutile** Stearn

Common Names ▸ Ever-ready onion
Usage ▸ spice
Parts Used ▸ bulb, leaf
Distribution ▸ cultivated in France

Chopra et al. 1956; Hanelt 1985, 2001

 Allium ceratophyllum Bess. & Ledeb.

▸ *Allium altaicum Pall.*

 Allium cernuum Roth.

Common Names ▸ lady's leek, nodding onion, wild onion; *French:* ail penché; *German:* Sommerzwiebel
Usage ▸ spice, flavoring
Parts Used ▸ bulb, leek
Distribution ▸ Canada, USA: Florida, California, Mexico

Cheers 1998; Sánchez-Monge/Parellada 1981; Täufel et al. 1993; Teuscher 2003; Wiersema/León 1999

 Allium chinense Maxim.

▸ *Allium ramosum L.*

 Allium chinense G. Don

Synonyms ▸ *Allium bakeri* Regel, *Allium splendens* Miq., *Allium triquetrum* Lour.
Common Names ▸ Baker' garlic, Chinese chives, Chinese onion, Chinese scallion, Japanese scallion, Oriental onion; *Arabic:* tum rlkhabazeen; *Chinese:* jiao tou, qiao tou, xie; *French:* echalotte chinoise; *German:* Chinesischer Schnittlauch, Chinesische Zwiebel, Schnittknoblauch; *Japanese:* rakkyõ; *Korean:* junggukpuchu, yompuchu; *Portuguese:* chalota chinesa; *Spanish:* chalote chinesa; *Thai:* mee yoi
Usage ▸ spice (pickles); **product:** essential oil
Parts Used ▸ bulb
Distribution ▸ China, cultivated elsewhere, specially Asia, also cultivated in SE Asia, Hawaii, Australia, California, Cuba

Arora/Pamdey 1996; Davidson 1999; Hager 4, 1992; Hanelt 2001; Hiller/Melzig 1999; Kameoka et al. 1984; Kumar 2003; Mann/Stearn 1960; Peng et al. 1996; Pino et al. 2001; Schultze-Motel 1986; Seidemann 1993c; Siemonsma/Piluek 1993; Small 1997; Täufel et al. 1983; Teuscher 2003; Uphof 1968; Wiersema/León 1999; Zeven/de Wet 1982

 Allium consanguineum Kunth

Synonyms ▸ *Allium stracheyi* Baker
Common Names ▸ *German:* Blutsverwandter Lauch; *Hi-*

malayas: dunna, pharan; *India:* chollang (NE India)
Usage ▶ spice
Parts Used ▶ leaf
Distribution ▶ W, C Himalayas, N India

Arora/Pandey 1996; Hanelt 2001; Negi/Plant 1992; Sharma et al. 1997

 Allium continuum *Small*

▶ *Allium canadense L.*

 Allium ellipticum *Wall.*

▶ *Allium victorais L.*

 Allium fistulosum *L.*

Synonyms ▶ *Allium bakeri* Hoop. non Regel; *Allium bouddhae* O. Debeaux
Common Names ▶ cibol, Japanese bunching onion, Japanese leek, spring onion, silverskin onion, Welsh onion; *Arabic:* bassal el-mustatere, bassal el-ankudy; *Chinese:* bai, cong, mu ts'ung, ts'ung; *Dutch:* grof bieslook; *French:* ciboule, ail fistuleux, oignon d'hiver; *German:* Schnittzwiebel, Winterzwiebel, Heckenzwiebel, Hohllauch, Jakobslauch, Klöben, Silberzwiebel; *Indonesian:* aun bawang; *Italian:* cipoletta, cipolla d'inverno; *Japanese:* negi; *Korean:* pha; *Malaysian:* daun bawang; *Pilipino:* sibuyasna mura; *Russian:* kitajskij luk, luk batun, luk tatarka; *Spanish:* cebolleta, cebolleta francesa, cebollino inglés; *Thai:* ton horm; *Vietnamese:* hanh, hombua, thong bach
Usage ▶ spice; **product:** essential oil
Parts Used ▶ leek
Distribution ▶ only cultivated, probable origin in Asia

Aichele/Schwegler 5, 1996; Blancke 2000; Bois 1934; Bremness 2201; Cheers 1998; Davidson 1999; Dudtschenko et al. 1989; Hager 4, 1992; Hanelt 2001; Helm 1956; Hutton 1998; Kameoka et al. 1984; Körber-Grohne 1898; Kuo et al. 1990; Leung 1991; Lück 2004; Melchior/Kastner 1974; Pochljobkin 1974, 1977; Pruthi 1976; Schultze-Motel 1986; Seidemann 1993c; Seidemann/Siebert 1987; Siemonesma/Piluek 1993; Täufel et al. 1983; Teuscher 2003; Tindall

Allium fistolosum, flowering

1983; Tucker 1986; Uphof 1968; Vogel 1995b; WHO Manila 1990; Wiersema/León 1999; Zeven/de Wet 1982

 Allium giganteum *Regel*

Common Names ▶ giant onion; *French:* ail géant de l'Himalaya; *German:* Riesen-Zwiebel
Usage ▶ spice
Parts Used ▶ bulb
Distribution ▶ in the former Soviet Union, Himalayas, C Asia

Cheers 1997; Dudtschenko et al. 1989; Erhardt et al. 2002; Hanelt 2001; Täufel et al. 1993; Teuscher 2003; Wiersema/León 1997

 Allium grayi *Regel*

Common Names ▶ Chinese garlic, Japanese garlic, water garlic; *Chinese:* siao suan, xiao, xie bai, yeh-suan, yeh-tsin-tsai; *French:* ail du Japon; *German:* Chinesischer Knoblauch, Japanischer Knoblauch; Wasserknoblauch; *Italian:* aglio giapponense;

Japanese: no-biru; *Korean:* tallae; *Mongolian:* zerleg sarmis; *Portuguese:* alho do Japão; *Russian:* luk krupnotytschinkovyj; *Spanish:* ajo del Japon

Usage ▸ spice, flavoring (sporadically)

Parts Used ▸ bulb, leek

Distribution ▸ E Asia, China, Mongolia; Russia, Far and Middle East

Helm 1956; Ho JongYoul et al. 2001; Peng et al. 1995; Schultze-Motel 1986; Täufel et al. 1983; Teuscher 2003; Wiersema/León 1997

Allium holmense Mill.

▸ *Allium ampeloprasum* L. ssp. *ampeloprasum*

Allium hookeri Thwaites

Synonyms ▸ *Allium wallichii* Regel, *Allium tsoongii* Wang et Tang

Common Names ▸ Hooker garlic; *German:* Hooker-Zwiebel; *Chinese:* kuan ye jiu; *Thai:* su

Usage ▸ flavoring for soups

Parts Used ▸ leaf

Distribution ▸ E Himalayas, Tibet, SW China, Assam, N Thailand, Sri Lanka

Note ▸ The fleshy root is used as a vegetable; the cultivated forms are seed-sterile.

Hanelt 1994, 2001; Kumar 2003; Sharma et al. 1997

Allium karataviense Regel

Common Names ▸ *French:* ail du Turkestan; *German:* Blauzungenlauch, Turkestanische Zwiebel

Usage ▸ spice

Parts Used ▸ bulb, leaf

Distribution ▸ C Asia, especially Turkestan

Erhardt et al. 2002; Hanelt 2001

Allium kunthii G. Don

Synonyms ▸ *Allium longifolium* (Kunth) Humb., *Schoenoprasum longifolium* Kunth

Common Names ▸ Mexican onion; *German:* Mexikanische Zwiebel

Usage ▸ spice

Parts Used ▸ bulb

Distribution ▸ N Mexico, New Mexico; Texas

Hanelt 2001; Schultze-Motel 1986; Seidemann 1993c; Täufel et al. 1993

Allium kurrat Schweinf. ex K. Krause

[or *Allium kurrat* Group]

Common Names ▸ Egypt leek, salad leek; *Arabic:* kurrat, kurrat baladi, kurrat nabati; *French:* kurrat; *German:* Ägyptischer Lauch, Kurratlauch, Arabischer Schnittlauch, Salatlauch; *Japanese:* kurrat nabati; *Portuguese:* kurrat; *Russian:* luk kurrat, luk salatnji; *Spanish:* kurrat

Usage ▸ spice, condiment, food (vegetable)

Parts Used ▸ leaf

Distribution ▸ cultivated in N Africa: Egypt, also in Arabia and Near and Middle East: Palestine

Davidson 1999; Hanelt 2001; Hepper 1992; Schultze-Motel 1986; Seidemann 1993c; Täufel et al. 1993; Teuscher 2003; Tucker 1986; Uphof 1968; Wiersema/León 1999; Zeven/de Wet 1982

Allium lacteum Sm.

▸ *Allium neopolitanicum* Cyr.

Allium latissimum Prokh.

▸ *Allium victoralis* L.

Allium ledebourianum Schult.

Common Names ▸ ledebour onion; *German:* Ledebour-Zwiebel

Synonyms ▸ *Allium uliginosum* Ldb.

Usage ▸ spice, condiment

Parts Used ▸ bulb, leaf

Distribution ▸ Far East, Russia: Siberia, Japan

Hanelt 2001; Uphof 1968

Allium lineare Mill.

▷ *Allium ampeloprasum* L. ssp. *ampeloprasum*

Allium longifolium (Kunth) Humb.

▷ *Allium kunthii* G. Don

Allium longicuspis Regel

[Longicuspis Group]

Common Names ▸ *German:* Spitzer Lauch; *Russian:* luk dlinno-ostrokonetschji
Usage ▸ spice
Parts Used ▸ bulbil
Distribution ▸ C, W Asia, in the former Soviet Union; since 1952 cultivated in Kazakhastan

Aichele/Schwegler 5, 1996; Blancke 2000; Dudtschenko et al. 1989; Hanelt 2001; Hepper 1992; Schultze-Motel 1986; Täufel et al. 1993; Teuscher 2003; Wiersema/León 1999; Zeven/de Wet 1982

Allium macrostemon Bunge

▷ *Allium grayi* Regel

Allium margaritaceum Moench

▷ *Allium scorodoprasum* L.

Allium microbulbum Prokh.

▷ *Allium altaicum* Pall.

Allium microdictyum Prokh.

▷ *Allium victoralis* L.

Allium moly L.

Common Names ▸ lily leek, moly, yellow onion; *French:* ail doré; *German:* Goldlauch, Molyzwiebel, Spanischer Lauch
Usage ▸ spice
Parts Used ▸ leek, bulb
Distribution ▸ SW Europe: Spain, SW France, N Africa; also frequently cultivated in gardens as an ornamental plant
Note ▸ This plant is not identical with the moly-plant of antiquity.

Bärtels 1997; Cheers 1998; Erhardt et al. 2002; Koch 1995; Seidemann 1992b

Allium mutabile Mich.

▷ *Allium canadense* L.

Allium neglectum Wender

▷ *Allium scorodoprasum* L.

Allium neopolitanum Cyr.

Synonyms ▸ *Allium cowanii* Lindl., *Allium lacteum* Sm., *Allium sulcatum* DC.
Common Names ▸ Daffodil garlic, naples garlic, false garlic, flowering onion, Neapolitan garlic; *French:* ail de Naples; *German:* Neapel-Zwiebel, Neopolitanische Zwiebel; *Mexico (Nahua):* xonacat
Usage ▸ spice, condiment
Parts Used ▸ bulb, leaf
Distribution ▸ Europe: Iberia, Turkey; C Mexico, grown as an ornamental plant in several warm temperate countries; N Africa, W Asia

Bärtels 1997; Cheers 1998; Erhardt et al. 2002; Hanelt 2001; Wiersema/León 1997

Allium nigrum L.

Common Names ▸ black garlic, broad leaf garlic; *German:* Schwarzer Lauch, Zwiebelreicher Lauch

Usage ▶ spice
Parts Used ▶ bulbil, leaf
Distribution ▶ Europe, Turkey, Syria, N Africa, W Asia SE Europe, SW Europe

Erhardt et al. 2002; Fleischhauer 2003; Loch 1993; Wiersema/León 1997

Allium nutans L.

Common Names ▶ *German:* Nickender Lauch; *Russian:* slizun
Usage ▶ spice
Parts Used ▶ young leaves
Distribution ▶ Siberia, from S Ural to the Yenissei and Kazakhastan

Erhardt et al. 2002; Hanelt 1994, 2001

Allium obliquum L.

Common Names ▶ oblique onion, twisled leaf; *German:* Schiefe Zwiebel, Ziegenlauch; *Mongolian:* sarmisan songino; *Russian:* dikij tschesnok, luk kosoj
Usage ▶ spice
Parts Used ▶ bulb
Distribution ▶ Romania, S Europe, Russia, SW and C Siberia, Kazakhastan, Uzbekistan; NW Mongolia, NW China; cultivated in W Siberia
Note ▶ The plant is a wild substitute for true garlic.

Hanelt 2001; Mansfeld 1986; Schultze-Motel 1986; Seidemann 1993c; Täufel et al. 1993; Teuscher 2003; Uphof 1968

Allium ochotense Prokh.

▶ *Allium victoralis L. ssp. plytyphyllum*

Allium odorum L.

▶ *Allium ramosum L.*

Allium odorum auct. non L.

▶ *Allium tuberosum Rottl. ex Spreng.*

Allium oleraceum L.

Common Names ▶ field garlic; *German:* Feldzwiebel, Kohllauch, Rosslauch
Usage ▶ spice, flavoring
Parts Used ▶ bulb, leaf
Distribution ▶ Europe, especially Iberia; Caucasus, also naturalized elsewhere in temperate regions

Aichele/Schwegler 5, 1996; Dudtschenko et al 1989; Erhardt et al. 2002; Fleischhauer 2003; Hanelt 2001; Schnelle 1999; Täufel et al 1993; Uphof 1968

Allium ophioscorodon Link

▶ *Allium sativum var. ophioscorodon (Link) Döll*

Allium oschaninii B. Fedtsch.

Synonyms ▶ *Allium cepa* L. var. *sylvestre* Regel
Common Names ▶ *German:* Oschanin-Zwiebel; *Russian:* luk oschanina
Usage ▶ spice
Parts Used ▶ bulb
Distribution ▶ Tajikistan, N, W and C Afghanistan, NE Iran, Uzbekistan
Note ▶ According to molecular and isoenzyme data the cultivar group of French shallots belong to the species.

Hanelt 2001; Schultze-Motel 1986; Wiersema/León 1999; Zeven/de Wet 1982

Allium paradoxum (M. Bieb.) G. Don

Common Names ▶ few-flowered leek; *German:* Seltsamer Lauch
Usage ▶ spice
Parts Used ▶ bulb, leaf
Distribution ▶ EC Europe, Caucasus, N Iran, C Asia, W Asia, native elsewhere

Erhardt et al. 2002; Fleischhauer 2003; Hanelt 2001; Loch 1993; Wiersema/León 1997

Hanelt 1994, 2001; Schultze-Motel 1986; Seidemann 1993c; Teuscher 2003

 Allium porrum L.

[or Leek Group respectively Porrum Group]

Synonyms ▸ *Allium ampeloprasum* L. var. *porrum* (L.) J. Gay, *Porrum commune* Rchb., *Porrum sativum* Mill.

Common Names ▸ leek, common leek, purret; *Arabic:* kourrath; *French:* poireau, porreau; *German:* Porree, Breitlauch, Suppenlauch, Welschlauch, Winterlauch; *Hindi:* vilayaiti lasson; *India:* paru; *Indonesian:* bawang pere; *Italian:* porro; *Japanese:* nira-negi; *Portuguese:* alho francês, alho porro; *Russian:* luk porej; *Spanish:* ajo porro, ajo puerro, puerro;

Usage ▸ spice (greens), and food (vegetable)

Parts Used ▸ shaft of the plant, leaf

Distribution ▸ Egypt, Mediterranean region, cultivated in many European countries, especially in W Europe, N America

Aichele/Schwegler 5, 1996; Bendel 2002; Bonnet 1976; Bremness 2001; Cheers 1998; Davidson 1999; Dudtschenko et al. 1989; Fattorusso et al. 2000, 2001; Hager 4, 1992; Hanelt 1994, 2001; Heeger 1956; Hepper 1992; Hiller/Melzig 1999; Körber-Grohne 1989; Pochljobkin 1994, 1977; Schulz et al 1999; Schultze-Motel 1986; Seidemann 1993c; Siemonesma Piluek 1993; Starke/Herrmann 1976; Stephanie/Baltes 1991, 1992; Täufel et al. 1993; Teuscher 2003; Uphof 1968; Zeven/de Wet 1982

 Allium porrum L. **var. sectivum** F.H. Lueder

[or Pearl–Onion Group]

Common Names ▸ pearl onion, sand leek, Spanish garlic; *German:* Natternknoblauch, Perllauch, Perlzwiebel, Rockenbole, Rokambole, Schlangenlauch, Silberzwiebel

Usage ▸ spice and for pickling

Parts Used ▸ small bulb and bulblet

Distribution ▸ cultivated in Germany, the Netherlands, Italy

Note ▸ The Indonesian prei anak cultivars also belong to this group.

 Allium x proliferum *(Moench) Schrad. ex Willd.*

Synonyms ▸ *Allium cepa* L. var. *proliferum* (Moench) Targ.-Tozz., *Allium cepa* L. ssp. *vivipara* Metzg., *Allium cepa* L. ssp. *viviparum* (Metztg.) Alef., *Allium wakegi* Araki; *Cepa prolifera* Moench

Common Names ▸ Beltsville bunching onion, Catawissa onion, Egyptian onion, top onion, tree onion, Wakegi onion; *French:* oignon d'Egypte, oignon catawissa; *German:* Ägyptische Zwiebel, Catawissazwiebel, Etagenzwiebel, Luftzwiebel; *Japanese:* wakegi; *Russian:* luk mnogojarusnyj

Usage ▸ spice

Parts Used ▸ inflorescence bulbils, bulb, leaf

Distribution ▸ Altai, N Caucasus, W Siberia; only cultivated

Note ▸ Only cultivated; a hybrid between *A fistulosum* and *A. cepa*.

Bärtels 1997; Davidson 1999; Faray et al 1981; Hager 4, 1992; Hanelt 2001; Lück 2004; Pochljobkin 1974, 1977; Schultze-Motel 1986; Seidemann 1993c; Seidemann/Siebert 1987; Täufel et al. 1993; Teuscher 2003; Zeven/de Wet 1982

 Allium pskemense B. Fedtsch

Common Names ▸ *German:* Pskemenser Zwiebel; *Russian:* pskemskij luk, pies-ansyr; gornyj luk

Usage ▸ spice

Parts Used ▸ bulb

Distribution ▸ former Soviet Union: Tien-Schan, Taskentian, Alatau; C Asia

Dudtschenko et al. 1989; Hanelt 2001; Pochljobkin 1974, 1977; Wiersema/León 1999; Zeven/de Wet 1982

 Allium ramosum L.

Synonyms ▸ *Allium angulosum* L., *Allium chinense* Maxim, *Allium odorum* L., *Allium tartaricum* L.f., *Allium uliginosum* G. Don

Common Names ▸ Chinese chive, Chinese leek, fragrant

garlic, fragment flowered garlic; *Chinese:* chiu kieou, feng pen; *French:* ail civette de Chine, ail chinoise; *German:* Ästiger Lauch, Chinalauch, Chinesischer Lauch, Duftlauch; *Japanese:* nira; *Malaysian:* kuchai; *Korean:* puchu; *Russian:* luk dutschistyj; *Vietnamese:* rau he;

Usage ▶ spice, flavoring
Parts Used ▶ leaf (leek)
Distribution ▶ S Siberia, Mongolia, former Soviet Far East, Korea, N, NE and NW China, native Japan and other countries in E and S Asia, also cultivated

Erhardt et al. 2002; Hanelt 2001; Schultze-Motel 1986; Täufel et al. 1993; Teubner 2001; Uphof 1968; WHO Manila 1990; Zeven/de Wet 1982

Allium rotundum L.

▶ *Allium scorodoprasum L. ssp. rotundum (L.) W.T. Stearn*

Allium sapidissemum Hedw

▶ *Allium altaicum Pall.*

Allium sativum L.

[Sativum Group]

Synonyms ▶ *Porrum sativum* (L.) Rchb.
Common Names ▶ common garlic, garlic; *Arabic:* thom, thum, tôm; *Chinese:* da suan; *Dutch:* knoflook; *French:* aïl, aïl blanc, ail ordinaire; *German:* Knoblauch, Knobloch, Knofel; *Hindi:* lahsan, lahsun, lasan; *Indonesian:* bawang putih; *Italian:* aglio, aglio domestico; *Japanese:* ninniku; *Korean:* manul; *Malaysian:* bawang puteh; *Mexican:* xiito ajo; *Pilipino:* bawang; *Portuguese:* alho; *Russian:* tschesnoik; *Sanskrit:* lasuna, rasona; *Spanish:* ajo, ajo común; *Thai:* krathiem; *Vietnamese:* toí, dai toan, sluon
Usage ▶ spice; **product:** essential oil
Parts Used ▶ bulb
Distribution ▶ widely grown in Eurasia and America, perhaps originating from the Mediterranean region, cultivated world-wide
Note ▶ The People's Republic of China (N China) is the greatest exporter of garlic in the world.

Aichele/Schwegler 5, 1996; Balcke 2000; Bärtels 1997; Block 1985, 1992; Bocchini et al. 2001; Bois 1934; Boss-Teichmann/Richter 2002; Bremness 2001; Brodnitz et al. 1971; Carson 1987; Cheers 1998; Davidson 1999; Duke et al. 2003; Edris et al. 2002; Farnsworth/Bunyaprophatsara 1992; Farrell 1985; Gaßmann 1992; Güntzel-Lingner 1941; Gyung Hdon et al. 1999a, b; Hager 4, 1992; Hamon 1987; Hanelt 2001; Heeger 1956; Hepper 1992; Hiller/Melzig 1999; Hoffmann et al. 1992; Hoppe 1949; Hutton 1998; Keusgen 2001; Koch 1995, 1996; Koch/Hahn 1988; Konvička 1983; Konvička/Würfl 2001; Larkcom 1991; Lewington 1990; MacCarthy 2002; Madamba et al. 1995; Mazza et al. 1992; Mazza/Oomah 2000; McElnay et al. 1991; Melchior/Kastner 1974; Morris/Mackley 1999; Newall et al. 1996; Peter 2001; Pino et al. 1991; Pochljobkin 1974, 1977; Pschyrembel 1998; Reuter 1986; Rosengarten 1969; Roth/Kormann 1997; Seidemann 1993c; Seidemann/Siebert 1987; Sharma 2003; Siemonsma/Piluek 1993; Siewek 1990; Singh/Tiwari 1995; Sing et al. 1998; Small 1997; Starke/Herrmann 1976; Täufel et al. 1993; Teleky-Vámossy/Petró-Turza 1986; Teuscher 2003; Tokitomo/Kobayashi 1992; Tucker 1986; Uphof 1968; Villamara et al. 1994; WHO 1990; Weinberg et al. 1993a; Wyk et al. 2004; Yu et al. 1989; Zeven/de Wet 1982

Allium sativum L. var. ophioscorodon (Link) Döll

Synonyms ▶ *Allium ophioscorodon* Link
Common Names ▶ garden rocambole, giant garlic, serpent garlic; *French:* ail rocambole; *German:* Perlzwiebel, Rockenbolle, Schlangen(knob-)lauch, Natternknoblauch; *Italian:* aglio d'India
Usage ▶ spice
Parts Used ▶ bulb, leaf
Distribution ▶ only cultivated

Balcke 2000; Davidson 1999; Ehrhardt et al. 2002; Hager 4, 1992; Hanelt 2001; Hanelt/Ohle 1978; Mansfeld 1986; Melchior/Kastner 1974; Seidemann 1993c; Täufel et al. 1993; Teuscher 2003; Tucker 1986; Wiersema/León 1997

Allium sativum L. var. pekinense (Prokh.) Maekawa apud Makino

Common Names ▶ Peking garlic; *German:* Peking-Knoblauch; *Korean:* manul
Usage ▶ spice
Parts Used ▶ bulbil

Distribution ▶ China, Japan, cultivated in gardens

Mansfeld 1986; Teuscher 2003

Allium saxatile Pall.

▶ *Allium altaicum* Pall.

Allium schoenoprasum L.

Synonyms ▶ *Allium montanum* Schrank, *Allium raddeanum*, *Allium sibiricum* L., *Cepa schoenoprasa* (L.) Moench, *Schoenoprasum vulgare* Fourr.
Common Names ▶ chive, chive garlic, cive, civet; *Arabic:* bassal el-shifee; *Dutch:* bieslook; *French:* civette, ciboulette, petit porreau, cipoletta, cipollina, fausse échalote; *German:* Binsenlauch, Graslauch, Schittling, Schnittlauch, Suppenlauch; *Italian:* cipoletta, cipollina; *Japanese:* asatsuki, ezonegi; *Portuguese:* cebolinha francesa, cebolleta, cebollina común; *Russian:* reseanez, skoroda, schnittluk; *Spanish:* cebollino, cebollino francés, cebolleta
Usage ▶ spice
Parts Used ▶ fresh leaf, dry or deep-freezing
Distribution ▶ temperate Asia, India, Europe, N America, also cultivated
Note ▶ This type is a very variable species.

Aichele/Schwegler 5, 1996; Blancke 2000; Bois 1934; Bremness 2001; Cheers 1998; Davidson 1999; Fleischhauer 2003; Hager 4, 1992; Hanelt 2001; Hashimoto et al 1983; Heeger 1956; Hiller/Melzig 1999; Hoppe 1949; Kameoka/Hasimoto 1983; Leino 1992; Pochljobkin 1974, 1977; Pruthi 1976; Rosengarten 1969; Schönfelder 2001; Schultze-Motel 1986; Seidemann 1993c; Seidemann/Siebert 1987; Small 1997; Starke/Herrmann 1976; Täufel et al. 1993; Teuscher 2003; Tucker 1986; Tyndall 1983; Uphof 1968; Villamar et al. 1994; Wahlroos 1965; Wiersema/León 1999; Zeven/de Wet 1982

Allium schoenoprasum L. var. sibiricum (L.) Garcke

Common Names ▶ alpine chives, large chives; *German:* Alpen-Schnittlauch, Sibirischer Schnittlauch
Usage ▶ spice
Parts Used ▶ leaf

Distribution ▶ Europe, Siberia

Erhardt et al. 2002; Hanelt 2001; Täufel et al. 1993;

Allium scorodoprasum L.

Synonyms ▶ *Allium margaritaceum* Moench, *Allium neglectum* Wender., *Porrum scorodoprasum* (L.) Rchb.
Common Names ▶ sand leek, giant garlic, Spanish garlic rocambole; *Arabic:* tum el-emlak; *Chinese:* hu suan, ta suan, xiao suan; *Dutch:* slangenlook; *French:* ail rocambole, oignon d'Espagne, rocambole; *German:* Falscher Schlangen-(Knob-)lauch, Graslauch, Rockenbolle, Sandlauch, Wilder Porree; *Italian:* aglio d'India, aglio romana, rocambola; *Portuguese:* alho grosso de Espanha, alho rocambole; *Russian:* pritschecnotschnyj, luk rokambol'; *Spanish:* ajo pardo, rocambola
Usage ▶ spice
Parts Used ▶ bulbul, leaf
Distribution ▶ C, S Europe, Caucasus, Turkey, Asia minor

Aichele/Schwegler 5, 1996; Bremness 2001; Davidson 1999; Fleischhauer 2003; Hager 4, 1992; Hanelt 2001; Schultze-Motel 1986; Täufel et al. 1993; Teuscher 2003; Tucker 1986; Uphof 1968; Zeven/de Wet 1982

Allium scorodoprasum L. ssp. rotundum (L.) W.T. Stearn

Synonyms ▶ *Allium rotundum* L.
Common Names ▶ sand and leek, giant garlic, puroleflower garlic, Spanish garlic; *German:* Graslauch, Rundköpfiger Lauch
Usage ▶ spice (sporadically)
Parts Used ▶ leaf
Distribution ▶ N, C, E and SE Europe, Turkey, Levante, Caucasus, N Iran
Note ▶ In E Anatolia with other plants for the preparation of special herbal cheese.

Aichele/Schwegler 5, 1996; Dudtschenko et al. 1989; Hanelt 2001; Schultze-Motel 1986; Wiersema/León 1999; Zeven/de Wet 1982

Allium senescens L.

Common Names ▶ mountain leek; *German:* Berglauch, Trügerischer Lauch; *Russian:* mansur, starejuschij luk
Usage ▶ spice
Parts Used ▶ leaf
Distribution ▶ E Siberia (former Burjat ASSR), N Mongolia, Japan

Fleischhauer 2003; Hager 4, 1992; Loch 1993; Pochljobkin 1974, 1977; Täufel et al. 1993; Uphof 1968; Wiersema/León 1997

 Allium sphaerocephalon L.

Common Names ▶ ball leek; *German:* Kugellauch, Kugelköpfiger Lauch
Usage ▶ spice, bulb
Parts Used ▶ leaf
Distribution ▶ S Europe: especially the Balkan States, Turkey, N Africa

Dudtschenko et al. 1989; Fleischhauer 2003; Loch 1993; Štajner et al. 2003

 Allium splendens Miq.

▶ *Allium chinense* G. Don

 Allium stipitatum Regel

Synonyms ▶ *Allium hirtifolium* Boiss.
Common Names ▶ drumstick allium; *German:* Anzurzwiebel, Gestielte Zwiebel, Stiellauch; *Russian:* luk stebel'tschatyj
Usage ▶ spice (singel)
Parts Used ▶ bulb
Distribution ▶ C Asia, in the former Soviet Union, mountains from Altai, N India, Pakistan, Afghanistan

Erhardt et al. 2002; Hanelt 2001; Täufel et al. 1993; Teuscher 2003

 Allium stracheyi Baker

▶ *Allium consanguineum* Kunth

 Allium strictum Schrad.

Common Names ▶ *German:* Steifer Lauch
Usage ▶ spice
Parts Used ▶ leaf, bulb
Distribution ▶ Europe, Russia, W, E Siberia, Amur, Sakhalin, Kamchatka, Mongolia, C Asia

Erhardt et al. 2002; Fleischhauer 2003; Loch 1993

 Allium sulcatum DC.

▶ *Allium neopolitanicum* Cyr.

 Allium suvorovii Regel

Common Names ▶ *German:* Anzurzwiebel
Usage ▶ spice
Parts Used ▶ bulb
Distribution ▶ in the former Soviet Union

Hanelt 2001; Täufel et al. 1993; Teuscher 2003

 Allium tartaricum L.f.

▶ *Allium ramosum* L.

 Allium tricoccum Blanco

Common Names ▶ ramp, wild leek, wood leek; *French:* ail des bois, ail sauvage; *German:* Wilder Lauch, Holzlauch
Usage ▶ spice, flavoring
Parts Used ▶ leek
Distribution ▶ N America

Erichsen-Brown 1989; Hanelt 1994; Small 1997; Uphof 1968; Wiersema/León 1997

 Allium triquetrum Lour.

▶ *Allium chinense* G. Don

Allium tsoongii Wang et Tang

▶ *Allium hookeri* Thaites

Allium tuberosum Rottl. ex Spreng.

Synonyms ▶ *Allium odorum* auct. non L.
Common Names ▶ Chinese chives, Chinese leek, oriental garlic, garlic chives; *Arabic:* kurrat seeny; *Chinese:* cuchay, juzii, gow choy, kiu ts'ai; *French:* ail civette de Chine, ciboulette Chinoise, civette; *German:* Chinesischer Lauch, Schnittknoblauch; *India:* bunganana; *Indonesian:* kucai; *Japanese:* nira; *Korean:* buchu, tongibuchu; *Malaysian:* bawang kucai, kucai; *Pilipino:* kutsay; *Portuguese:* alho chinês; *Russian:* luk medweshij, luk tscheremscha; *Spanish:* cive chino; *Thai.:* bai kuichai, dok kuichai, kui chaai
Usage ▶ spice, flavoring (food); **product:** essential oil
Parts Used ▶ leek
Distribution ▶ China, India, Mongolia, Japan, native or perhaps native elsewhere in E Asia
Note ▶ The flowers also used as spice with garlic and honey taste.

Arora/Pandey 1996; Blancke 2000; Bremness 2001; Cheers 1998; Chung Hee Don/Youn Sun Joo 1996; Erhardt et al. 2002; Hager 4, 1992; Hanelt 1999; Hiller/Melzig 1999; Hutton 1998; Kumar 2003; Larkcom 1991; Lück 2000; Ooi et al. 2002; Oyen/Dung 1999; Pino et al. 2001; Sang et al. 1999; Schultze-Motel 1986; Seidemann 1993c; Siemonsma/Piluek 1993; Small 1997; Täufel et al. 1993; Teuscher 2003; Tucker 1986; Wiersema/león 1999; Zeven/de Wet 1982

Allium uliginosum G. Don

▶ *Allium ramosum* L.

Allium uliginosum Ldb.

▶ *Allium ledebourianum* Roem. et Schult.

Allium ursinum L.

Common Names ▶ bear's garlic, buckrams, gipsy onion, hog's garlic, ramsons, broad-leaved garlic, wild garlic, wood garlic, German Zigeunerknoblauch; *French:* ail des bois, ail des ours; *German:* Bär(en)-lauch, Hexenzwiebel, Judenzwiebel, Rams, Ramsons, Wald-(knob-) lauch, Zigeunerlauch, Wilder Knoblauch, Wilder Knofel, Teufelschnoblech; *Italian:* aglio orsino; *Russian:* tschermscha, medveshij luk, dikij luk, genseli
Usage ▶ spice
Parts Used ▶ leaf
Distribution ▶ C, N Europe, Caucasus, N, W Asia, temperate Europe
Note ▶ In Germany and W Caucasus (Svanetia) ssp. *ucrainicum* Kleopow et Oxner represent the garden forms.

Aichele/Schwegler 5, 1996; Boss-Teichmann/Richter 2002; Carotenuto et al. 1996; Cheers 1998; Davidson 1999; Dudtschenko 1989; Erhardt et al. 2002; Fleischhauer 2003; Hager 4, 1992; Hanelt 2001; Heeger 1956; Hiller/Melzig 1999; Hoppe 1949; Pochljobkin 1974, 1977; Pschyrembel 1998; Richter 1999; Schönfelder 2001; Seidemann 1993c; Täufel et al. 2000; Teuscher 2003; Uphof 1968; Wagner/Sendl 1990; Wiersema/León 1997

Allium vavilovii M. Popov

Common Names ▶ Vavilov leek; *German:* Vavilov-Lauch; *Russian:* luk vavilova
Usage ▶ spice
Parts Used ▶ bulb, leaf
Distribution ▶ Russia, C Asia: Kopet-Dag mountains at the borderline between Turkmenia and Iran, N Iran
Note ▶ Perhaps ancestral species of *Allium cepa* L.

Dudtschenko et al. 1989; Hanelt 2001; Wiersema/León 1999; Zeven/de Wet 1982

Allium victoralis L.

Synonyms ▶ *Allium anguinum* Bubani, *Allium ellipticum* Wall., *Allium latissimum* Prokh., *Allium microdictyum* Prokh., *Cepa victoralis* Moench
Common Names ▶ alpine leek, longroot onion, long-rooted garlic, wild garlic; *Chinese:* ge con, shan cong, tse suan; *French:* ail de ceuf, herbe à neuf chemises; *German:* Allermannsharnisch, Lange Siegwurzel; *Japanese:* gyôja-ninniku; *Russian:* kolba, luk pobednij, luk sibirskij, sibirskaja tschermscha

◘ Allium victoralis, flowering

Usage ► spice
Parts Used ► leek, bulb
Distribution ► mountains of Europe from N Portugal to S Ural, Caucasus, W Himalayas; S, C Siberia (Taiga), N Mongolai
Note ► Russian autors split *A. victoralis* L. into three species.

Aichele/Schwegler 5, 1996; Dudtschenko et al. 1989; Hager 4, 1992; Hanelt 2001; Heeger 1956; Hoppe 1949; Loch 1993; Pochljobkin 1974, 1977; Schultze-Motel 1986; Schnelle 1999; Seidemann 1993c; Täufel et al. 1993; Uphof 1968; Wijaya et al. 1991

 Allium victorialis L. ssp. platyphyllum *Hulten*

Synonyms ► *Allium ochotense* Prokh.
Common Names ► *German:* Breitblättriger Allermannsharnisch
Usage ► spice
Parts Used ► leaf
Distribution ► Russian, Far East, Korea, China, Japan

Hanelt 1994, 2001

 Allium vineale *L.*

Common Names ► crow garlic, false garlic, field garlic, slag garlic, wild onion; *German:* Koch's Lauch, Perlknoblauch, Weinberg-Lauch
Usage ► spice, flavoring
Parts Used ► leaf (leek)
Distribution ► Eurasia, N Africa, Turkey, Caucasus, native E USA, widely native elsewhere

Aichele/Schwegler 5, 1996; Davidson 1999; Fleischhauer 2003; Small 1997; Teubner 2001; Wiersema/León 1997

 Allium wakegi *Araki*

❯ *Allium x proleferum (Moench) Shrad. ex Willd.*

 Allium wallichii *Regel*

❯ *Allium hookeri Thaites*

 Aloe guineensis *Jacq.*

❯ *Sansevieria trifasciata Prain*

ALOYSIA Palau - Verbenaceae

 Aloysia gratissima *(Gillies et Hook.) Tronc.*

Synonyms ► *Lippea lycioides* (Cham.) Steud.
Common Names ► Texas white bush, common bec brush; *Mexico:* niñarupá, arrayan del campo; *Spanish:* cedrón de monte, hierba de la princesa, palo amerillo, reseda de campo, romerillo
Usage ► flavoring
Parts Used ► leaf
Distribution ► New Mexico, Arizona, Texas, Mexico, Brazil

Wiersema/León 1997

Aloysia citriodora *(Lam.) Humb.*

▶ *Lippea triphylla (L'Hérit.) Kuntze*

 Aloysia citriodora *Ortega ex Pers.*

▶ *Lippea triphylla (L'Hér.) Kuntze*

 Aloysia triphylla *(L.'Herit.) Britton*

▶ *Lippea triphylla (L'Herit.) Kuntze*

ALPINIA Roxb. - Galanga(l), Ginger Lily - Zingiberaceae

 Alpinia aromatica *AubL.*

▶ *Renealmia aromatica (Aubl.) Griseb.*

◧ Alpinia calcarata, flowering

 Alpinia calcarata *Rosc.* ◧

Common Names ▶ Indian ginger, snap ginger; *German:* Gespornter Galgant, Indischer Galgant, Indische Ingwerlilie; *India:* toroni
Usage ▶ spice
Parts Used ▶ rhizome
Distribution ▶ India, China; Papua New Guinea; in Indian cultivated
Note ▶ The rhizome used as a substitute of *Alpinia galanga* (L.) Willd.

Chopra 1956; Erhardt et al. 2002; Hanelt 2001; Schultze-Motel 1986

 Alpinia cardamomum *Roxb.*

▶ *Elettaria cardamomum (L.) Maton*

 Alpinia caerulea *Kuntz*

Common Names ▶ Australian blue ginger; common ginger of Australia

Usage ▶ seasoning
Parts Used ▶ herb
Distribution ▶ Australia

 Alpinia conchigera *Griff.*

Synonyms ▶ *Languas conchigera* Burkill
Common Names ▶ mussel galanga; *German:* Muschelgalgant; *SE Asia*: lĕngkuas, ranting, lĕngkuas kĕchil, lĕngkuas padang, jĕrnuang, rumput, kelemoyang; *Vietnamese:* rieng rung, rieng nuoc
Usage ▶ seasoning for dishes and alcoholic drinks
Parts Used ▶ rhizome
Distribution ▶ Bangladesh, Thailand, Vietnam, Malaysia, Sumatra
Note ▶ The fruits are used as vegetables.

Arora/Pandey 1996; Bois 1934; Burkill 1966; Hanelt 2001; Ogle et al. 2003; Oyen/Dung 1999; Schultze-Motel 1986; Täufel et al. 1993; Uphof 1968; Wong 1999; Yusuf et al. 2002; Zeven/de Wet 1982

Alpinia eliator *Jack*

> *Etlingera elatior (Jack) R.M. Sm.*

Alpinia galanga (L.) *Willd.*

Synonyms ▶ *Amomum galanga* L.; *Galanga major* Rumph; *Languas galanga* (L.) Stunz., *Languas vulgare* Koenig, *Maranta galanga* L.

Common Names ▶ galanga, galangal, the greater galangal, Java galangal, lesser galangal, Siamese ginger; *Arabic:* el galangal, el adkham; *Chinese:* gao liang jiang; *Dutch:* geelwortel; *French:* galanga, galanga de l'Inde, galanga majeur, grand galanga; *German:* Echter Galgant, Großer Galgant, Siam-Galgant, Siam-Ingwer, Thai-Ingwer; *Hindi:* kulanjan, barakulanjar; *Indonesian:* langkuas, laos, laja; *Italian:* galanga; *Japanese:* koryokyo, ukon; *Javanese:* laos; *Laos:* kha: x ta: dè: ng; *Malaysian:* lengkuas, meranang, puar; *Pilipino:* langkawas, palla langkuas; *Portuguese:* galanga maior; *Russian:* galangowyj koren', bol'schoj koren'; *Sanskrit:* malaavaca, sugandha; *Slovakian:* galgan jávsky; *Spanish:* galanga, galagana, galanga mayor, garengal; *Thai:* kha; *Vietnamese:* rieng, rieng am, rieng nep, hau kha

Usage ▶ spice, condiment; **product:** essential oil

Parts Used ▶ rhizome, seed

Distribution ▶ India, China, SE Asia; Sri Lanka, Malaysia; cultivated in the Paleotropics.

Arora/Pandey 1996; Bois 1934; Boisvert/Hubert 2000; Charles et al. 1992; Cheah/Abu Hasim 2000; Cheers 1998; Dalby 2000; Davidson 1999; Duke et al. 2003; Guzman/Siemonsma 1999; Hanelt 2001; Herklots 1972; Hiller/Melzig 1999; Holttum 1950; Kondo et al. 1993; Kumar 2003; Larsen et al. 1999; Lawrence et al. 1969; Melchior/Kastner 1974; Mitsui et al. 1976; Mori et al. 1995; Murakami et al. 2000; Noro et al. 1988; Norman 1991; Ogle et al. 2003; Oyen/Dung 1999; Pochljobkin 1974, 1977; Pooter et al. 1985; Pschyrembel 1998; Raina et al. 2002; Roth/Kormann 1997; Scheffer et al. 1981; Schultze-Motel 1986; Siewek 1990; Seidemann 1993c; Seidemann/Siebert 1987; Tao Guoda 1998; Täufel et al. 1993; Teuscher 2003; Tram Ngoc Ly et al. 2001; Uphof 1968; WHO 1990; Wiersema/León 1997; Wong 1999; Wüstenfeld/Haensel 1964; Wyk et al. 2004

Alpinia galanga var. pyramidata *(Blume) K. Schum.*

Synonyms ▶ *Alpinia pyramidata* Blume

Common Names ▶ langkauas, pal-ha; *German:* Pyramiden-Galgant

Usage ▶ spice

Parts Used ▶ rhizome

Distribution ▶ SE Asia: Java, Borneo, Philippines; in Java and the Philippines also cultivated

Burkill 1966; Chopra et al. 1956; Hanelt 2001; Schultze-Motel 1986; Täufel et al. 1993; Uphof 1968; Wealth of India 1, 1948

Alpinia globosa *(Lour.) Horan*

> *Amomum globosum Lour.*

Alpinia malaccensis *(Burm.f.) Rosc.*

Synonyms ▶ *Maranta malaccensis* Burm.

Common Names ▶ Malacca galangal; *German:* Malakka-Galgant ; *Indonesian:* laja gowah; *Javanese:* puar, laja goa, polang; *Malaysian:* puar, bangle; *Moluccas:* langkuas malaka; *Pilipino:* taglak babae; *Thai:* kha paa

Usage ▶ spice (rarely); **product:** essential oil (essence d'Amali, essence of Amali)

Parts Used ▶ rhizome

Distribution ▶ India, Bangladesh, Myanmar, SE Asia: Malaysia, Java, also cultivated

Note ▶ The fruits are fragrant and are used for washing clothes and hair.

Burkill 1966; Chopra et al. 1956; Erhardt et al. 2002; Hanelt 2001; Holttum 1951, 1971; Kumar 2001; Oyen/Dung 1999; Schultze-Motel 1986; Seidemann 1993c; Tao Guoda 1998; Täufel et al 1993; Wealth of India 1, 1948; Wiersema/León 1999; Wong 1999; Yusuf et al. 2002; Zeven/de Wet 1982

Alpinia nigra *(Gaertn.) Burtt.*

Synonyms ▶ *Zingiber nigrum* Gaertn.

Common Names ▶ black galangal; *French:* noir galanga; *German:* Schwarzer Galgant

Usage ▶ flavoring, e.g. curry in Bangladesh

Parts Used ▶ central part of stem

Distribution ▶ Bangladesh; Pacific Islands: Hawaii, Jamaica

Kumar 2001; Newman et al. 2004; Tao Guoda 1998; Yusuf et al. 2002

 Alpinia officinarum *Hance*

Synonyms ▶ *Languas officinarum* (Hance) Farw.

Common Names ▶ Chinese ginger, lesser galangal, lesser galingale, small galangal, East Indian root; *Arabic:* hodengal, khulingan; *Chinese:* hua ha, gao liang jiang; *French:* galangal officinal, galangal de la China, petit galangal; *German:* Echter Galgant, Kleiner Galgant, Siam-Ingwer, Thai-Ingwer; *India:* kúlinján; *Italian:* galanga minore; *Russian:* al'piniaaptetschnaja; malyj koren'; *Slovakian:* galgan liečivý; *Spanish:* galanga; *Vietnamese:* rieng thuoc

Usage ▶ spice, condiment

Parts Used ▶ rhizome

Distribution ▶ S China, Hainan, Japan

Bois 1934; Chopra et al. 1956; Craze 2002; Davidson 1999; Dudtschenko et al. 1989; Duke et al. 2003; Hanelt 20001; Hohmann et al. 2001; Hoppe 1949; Ly et al. 2002, 2003; Melchior/Kastner 1974; Morris/Mackley 1999; Norman 1991; Pochljobkin 1974, 1977; Schenck/Naundorf 1966; Schröder 1991; Schultze-Motel 1986; Seidemann 1993c; Täufel et al. 1993; Teuscher 2003; Uphof 1968; Wiersema/León 1999; Zeven/de Wet 1982

 Alpinia purpurata *(Vieill.) K. Schum.*

Common Names ▶ red ginger; *German:* Roter Ingwer; *Indonesian:* alpinia merah; *Malaysian:* lengkuas; *Thai:* khing daeng; *Vietnamese:* rièng tiá

Usage ▶ flavoring, e.g. coconut oil

Parts Used ▶ leaf

Distribution ▶ Moluccas, Caroline Islands, New Caledonia, Papua New Guinea, in SE Asia native

Cheers 1998; Engler/Phummai 2000; Erhardt et al. 2002; Hanelt 2001; Kottegota 1994; Larsen et al. 1999; Small 1997; Wiersema/León 1997; Wong 1999

 Alpinia pyramidata *Blume*

▶ *Alpinia galanga* var. *pyramidata* (Blume) K. Schum.

 Alpinia speciosa *D. Dietr.*

▶ *Etlingera elatior* (Jack) R.M. Sm.

 Alpinia speciosa *(J.W. Wendl.) K. Schum.*

▶ *Alpinia zerumbet* (Pers.) B.L. Burtt et R.M. Sm.

 Alpinia zerumbet *(Pers.) B.L. Burtt et R.M. Sm.*

Synonyms ▶ *Alpinia speciosa* (J.W. Wendl.) K. Schum., *Costus zerumbet* Pers., *Zerumbet speciosum* J.W. Wendl.

Common Names ▶ bright ginger, pink porcelain lily, light galangal, shell ginger; *Bengali:* punag champa; *German:* Martineque-Ingwer, Porzellan-Ingwerlilie; *Indonesian:* galoba merah, goloba koi; *Pilipino:* langkuas na pula;

Usage ▶ spice, utilized the rhizome like of *Alpinia galanga* (L.) Willd.

Parts Used ▶ leaf, rhizome (substitute for ginger)

Distribution ▶ NE India, also cultivated in tropical areas of Asia: India, Sri Lanka, Malaysia, Myanmar and Tonga Islands and Cuba

Note ▶ The aromatic leaves are used to wrap rice or fish for cooking.

Arora/Pandey 1996; Burkill 1966; Cheers 1998; Chopra et al. 1956; Dung et al. 1994a, b; Erhardt et al. 2002; Fujita et al. 1994; Hanelt 2001; Holttum 1950, 1971; Itokawa et al. 1981; Kottegota 1994; Larsen et al. 1999; Oyen/Dung 1999; Prudent et al. 1993; Schultze-Motel 1986; Seidemann 1993a; Small 1997; Täufel et al. 1993; Wealth of India 1, 1948; Wong 1999

ALSTONIA R.Br. - Apocynaceae

 Alstonia acuminata *Miq.*

Common Names ▶ ajooras, poole batoo

Usage ▶ flavoring

Parts Used ▶ roasted root

Distribution ▶ Moluccas, Bali, Java

Note ▶ The root is used to impart a bitter flavor to palm vine.

◘ Alpinia zerumbet, flowering

Uphof 1968

 Alstonia scholaris (L.) *R.Br.*

Synonyms ▸ *Echites malabarica* Lam., *Echites scholaris* L.
Common Names ▸ devils' tree, dita bark, palmira alstonia, shiatan wood, Australian fever, white cheese wood, milk wood; *Chinese:* tang jiao shu, xiang pi mu; *French:* echite; *German:* Ditabaum, Teufelsbaum, Zitronen-Mahagoni; *Hindi:* chaitan, chatwan, chitvan, chattiyan, eta-kula, lettok, pali-mari, shaitan; *Indonesian:* pulai; *Javanese:* pulé; *Malaysian:* basong, jelutong, pulai, rejang; *Nepal:* chatiwa; *Pilipino:* dita tangitang; *Sanskrit:* saptaparna, visalatvak; *Thai:* phayaa sattaban, tin ped; *Vietnamese:* may man, mu cua
Usage ▸ spice, flavoring liqueurs
Parts Used ▸ bark
Distribution ▸ E India, China, Himalayas, Sri Lanka, Malaysia, Australia; in Egypt, India, Java, Vietnam also cultivated

Note ▸ The plant serves as support for *Piper nigrum* L.

Arora/Pandey 1996; Berger 1, 1949; Cheers 1998; Dastur 1954; Ehrhardt et al. 2002; Engel/Phummai 2000; Gandhi/Vinayak 1990; Hanelt 2001; Hiller/Melzig 1999; Hoppe 1949; Kottegoda 1994; Rätsch 1998; Schultze-Motel 1986; Seidemann 1993c; 1998/2000; Uphof 1968; Wen-Ian Hu et al. 1989; WHO 1990; Wiersema/León 1997

ALYXIA Banks ex R.Br. - Apocynaceae

 Alyxia gynopogon *Roem. et Schult.*

▸ *Alyxia lucida* Wall.

 Alyxia lucida *Wall.*

Synonyms ▸ *Alyxia gynopogon* Roem. et Schult., *Gynopogon alyxia* J.R. Forst
Common Names ▸ alyxia cinnamon; *German:* Alyxia-Zimtrinde; *Malaysian:* akar bagan, mĕmpĕlasari, pulasari; *Thai:* cha-loot
Usage ▸ spice (locally)
Parts Used ▸ bark
Distribution ▸ Malaysia, Indonesia, Madagascar, also cultivated
Note ▸ Substitute for *Cinnamomum zeylanicum* Bl. – In Madagascar bark and leaves are employed in the manufacture of rum.

Hanelt 2001; Schultze-Motel 1986; Seidemann 1993c; Seidemann/Siebert 1987; Täufel et al. 1993; Uphof 1968

 Amaracus syriacus (L.) *Stokes*

▸ *Origanum syriacum* L.

AMBLYGONOCARPUS Harms - Fabaceae (Leguminosae)

 Amblygonocarpus andogensis *(Oliv.) Exell et Torre*

> *Amblygonocarpus schweinfurthii Harms*

 Amblygonocarpus schweinfurthii *Harms*

Synonyms ▶ *Amblygonocarpus andogensis* (Oliv.) Exell et Torre
Common Names ▶ *Ghana:* dagbani nanzidow; *Nigerian:* fula-fulfulde
Usage ▶ condiment
Parts Used ▶ seed
Distribution ▶ W Africa: Ghana, Nigeria, Ubangi
Note ▶ Pulverized seeds are boiled and fermented.

Burkill 3, 1995; Neuwinger 1999

AMBROSIA L. – Ragweed - Asteraceae (Compositae)

 Ambrosia maritima *L.*

Common Names ▶ sea ambrosia; *Arabic:* damsisa, ghobbeira, tannoun; *German:* Strand-Ambrosie
Usage ▶ flavoring of soups and liqueurs
Parts Used ▶ plant
Distribution ▶ W Africa: N Nigeria; Mediterranean Region: Iberia, France, Turkey

Burkill 1, 1985; Erhardt et al. 2002; Uphof 1968

 Amburana cearrensis *(Fr. Allem.) A.C. Smith*

> *Torresea cearensis Fr. Allem.*

AMELANCHIER Medik. - Rosaceae

 Amelanchier alnifolia *(Nutt.) Nutt.*

Synonyms ▶ *Amelanchier macrocarpa* Lunell, *Aronia alnifolia* Nutt., *Pyrus alnifolia* Lindl.
Common Names ▶ alderleaf berry, Pacific berry, saskatoon, western serviceberry; *German:* Erlenblättrige Felsenbirne; *Russian:* irga ol'cholistnaja
Usage ▶ spice
Parts Used ▶ fruit
Distribution ▶ N America: W coast (Colorado, Idaho, Nebrasca, New Mexico, California), Canada

Hanelt 2001; Schultze-Motel 1986; Uphof 1968

 Amelanchier macrocarpa *Lunell*

> *Amchelanchier alnifolia (Nutt.) Nutt.*

AMMI L. - Bullwort - Apiaceae (Umbelliferae)

 Ammi copticum *L.*

> *Trachyspermum ammi (L.) Sprague*

 Ammi majus *L.*

Synonyms ▶ *Apium ammi* Crantz
Common Names ▶ false bishop's weed, greater ammi, bullwort, lady's lace; *Arabic:* khilla shitani, killa, kille sheitani; *French:* ammi élevé; *German:* Bischofskraut, Großer Ammi, Große Knorpelmöhre; *Italian:* rizzomolo, visnaga maggiore; *Portuguese:* ertrudes (Brazil); *Russian:* ammi bol'schaja; *Spanish:* ameo bastardo, ameo mayor
Usage ▶ spice (of meat)
Parts Used ▶ fruit
Distribution ▶ N and NE Africa, Caucasus, W Asia, India, SE and SW Europe, Australia, New Zealand, N America, native elsewhere
Note ▶ Tastes like thyme.

◘ Ammi visnaga, fruiting

Aichele/Schwegler 3, 1995; Bärtels 1997; Berger 5, 1952; Cheers 1998; Erhardt et al. 2002; Fahmy et al. 1947; Hanelt 2001; Hiller/Melzig 1999; Lück 2000; Melchior/Kastner 1974; Schönfelder 2001; Schultze-Motel 1986; Sharma 2003; Täufel et al. 1993; Turova et al. 1987; Uphof 1968; Wiersema/León 1999; Zeven/de Wet 1982

 Ammi visnaga (*L.*) *Lam.* ◘

Synonyms ▶ *Apium visnaga* (L.) Crantz, *Daucus visnaga* L.
Common Names ▶ khella, lesser bishop's weed, pick tooth, visnaga; *Arabic:* khelal, khella, khilla baladi, noukha; *French:* fruit aux cure-dents, fruit de khella; *German:* Ammi, Zahnstocher-Kraut; *Italian:* kella, visnaga; *Korean:* ammi; *Russian:* ammi zubnaja; *Spanish:* escuradentis, viznaga
Usage ▶ spice, source of khellin and xanthotoxin
Parts Used ▶ fruit
Distribution ▶ SW Europe: Iberia, France, Turkey, Cyprus, N Iraq, Iran, Canary Islands, N Africa, W Asia, native elsewhere

Akačić/Kuštrak 1960; Bärtels 1997; Berger 3, 1952; Czupor 1970; Erhardt et al. 2002; Hager 8, 1996; Hanelt 2001; Hiller/Melzig 1999; Hohmann et al. 2001; Pschyrembel 1998; Schindler 1953; Schönfelder 2001; Schultze-Motel 1986; Sharma 2003; Turova et al. 1987; Uphof 1968; Wiersema/León 1997; Wyk et al. 2004

AMMODAUCUS Coss. et Durieu - Apiaceae (Umbelliferae)

 Ammodaucus leucotrichus *Coss. et Durrau*

Common Names ▶ *Arabic:* cafoun
Usage ▶ spice (similar caraway), cultivated as a condiment
Parts Used ▶ fruit, seed
Distribution ▶ Canary Islands, N and W Africa: Algeria, Tunesia, Nigeria, Mauritania, Upper Nile Valley
Note ▶ Cultivated in the oases of the Sahara, in Maritania and the upper Niger valley.

Adegoke et al. 1968; Bois 1934; Burkill 5, 2000; Dalziel 1937; Erhardt et al. 2002; Hanelt 2001; Melchior/Kastner 1974; Schnell 1957; Schultze-Motel 1986; Seidemann 1993c; Uphof 1968; Zeven/de Wet 1982

AMOMUM Roxb. - Cardamom - Zingiberaceae

 Amomum acre *VaL.*

Common Names ▶ *German:* Beißender Kardamom; *Indonesian:* panasa, panasan, pane
Usage ▶ pungent condiment
Parts Used ▶ fruit, inner part of petioles
Distribution ▶ Indonesia (S Sulawesi)
Note ▶ The seed-coat and very young stem are sometimes directly used as a pungent condiment

Guzman/Siemonsma 1999; Oyen/Dung 1999

 Amomum alboviolaceum *Ridley*

▶ *Aframomum alboviolaceum (Ridley) K. Schum.*

AMOMUM Roxb. - Cardamom - Zingiberaceae

 Amomum alpinia *Rottboel*

> *Renealmia alpinina (Roetbboel) Maas*

 Amomum angustifolium *Salisb.*

> *Zingiber officinale Rosc.*

 Amomum angustifolium *Sonn.*

> *Aframomum angustifolium (Sonn.) K. Schum.*

 Amomum aromaticum *Roxb.*

Common Names ▶ Bengal cardamom, Nepal cardamom, large cardamom; *French:* cardamome du Bengale; *German:* Bengalischer Kardamom, Nepal-Kardamom, Langer Kardamom; *India:* morang elaichi; *Vietnamese:* mac hau, thao qua
Usage ▶ spice; **product:** essential oil
Parts Used ▶ capsule, seed
Distribution ▶ N India (Assam), Bangladesh, Nepal; cultivated also in Bengal, India (Assam) to N Vietnam

Arora/Padney 1996; Berger 3, 1952; Berger 1964/65; Blancke 2000; Chopra et al. 1956; Erhardt et al. 2002; Guzman/Siemonsma 1999; Hager 4, 1992; Hanelt 2001; Hiller/Melzig 1999; Holttum 1951; Kumar 2001; Melchior/Kastner 1974; Ogle et al. 2003; Overdieck 1992; Pruthi 1976; Schultze-Motel 1986; Siewek 1990; Täufel et al. 1993; Teuscher 2003; Turova et al. 1987; Uphof 1968; Wealth of India 1, 1948; WHO Manila 1990; Zeven/de Wet 1982

 Amomum cardamomum *L.*

> *Elettaria cardamomum (L.) Maton*

 Amomum cardamom *auct., non L.*

> *Amomum compactum Sol. ex Maton*

 Amomum cardamomum *Roxb.*

> *Amomum compactum Sol. ex Maton*

 Amomum cevuga *Seemann*

Usage ▶ scenting coconut oil
Parts Used ▶ seed
Distribution ▶ Pacific Islands: Tahiti, Marquesas, Fiji

Uphof 1968

 Amomum clusii *Hanb.*

> *Aframomum hanburyi K. Schum.*

 Amomum compactum *SoL. ex Maton*

Synonyms ▶ *Amomum cardamomum* auct., non L., *Amomum kepulaga* Sprague et Burkill, *Amomum cardamomum* Roxb.
Common Names ▶ round cardamom, cluster cardamomum; Java cardamom; false cardamom, Indonesian cardamom, Siam cardamom, chester cardamom; *Cambodian:* cravanh; *Dutch:* ronde kardemom; *French:* amome à grappe; cardamome rond; *German:* Runder Kardamom, Trauben-Kardamom, Java-Kardamom, Siam-Kardamom; *Indonesian:* kapulaga, kapol, puwar pelaga; *Malaysian:* batang pelaga, pelaga, kardamoenggve, kĕpulaga, puar; *Vietnamese:* bach dâu khâu
Usage ▶ spice; **product:** essential oil
Parts Used ▶ capsule, seed
Distribution ▶ endemic in W Java; cultivated: Java, Sumatra, Moluccas, Malaysia, S China
Note ▶ The fruits have a camphor taste and the seeds serve as a condiment in cakes.

Berger 3, 1952; Berger 1964/65; Burkill 1966; Duke et al. 2003; Erhardt et al. 2002; Guzman/Siemonsma 1999; Hager 4, 1992; Hanelt 2001; Hiller/Melzig 1999; Larsen et al. 1999; Melchior/Kastner 1974; Oberdieck 1992; Oyen/Dung 1999; Peter 2001; Pruthi 1976; Schultze-Motel 1986; Seidemann 1993c; Täufel et al. 1993; Teuscher 2003; Uphof 1968; Wiersema/León 1999; Zeven/de Wet 1982

 Amomum curcuma *Jacq.*

> *Cucuma longa L.*

Amomum daniellii J.D. Hook.

> *Aframomum daniellii (J.D. Hook.) K. Schum.*

Amomum dealbatum Roxb.

> *Amomum maximum Roxb.*

Amomum echinosphaera K. Schum. ex Gagnep.

> *Amomum villosum Lour.*

Amomum gagnepainii T.L. Wu

Synonyms ▸ *Amomum thyrsoideum* Gagnep
Common Names ▸ *Vietnamese:* riêng âm
Usage ▸ spice
Parts Used ▸ fruit (capsule)
Distribution ▸ tropical Asia, especially Indochina

Hanelt 2001

Amomum galanga L.

> *Alpinia galanga (L.) Will.*

Amomum globosum Lour.

Synonyms ▸ *Alpinia globosa* (Lour.) Horan
Common Names ▸ round Chinese cardamom, cardamom nigra; *Chinese:* cao kou, *German:* Bitterer Kardamom, Chinesischer Kardamom; Schwarzer Kardamom, Bastard-Kardamom; *Korean:* choduguphul; *Vietnamese:* mè tré
Usage ▸ spice, condiment
Parts Used ▸ fruit, seed
Distribution ▸ S China, E Himalayas, Vietnam; Cambodia, Laos, N Thailand, in China and Korea also cultivated

Guzman/Siemonsma 1999; Hager 4, 1992; Hager 5, 1993; Hanelt 2001; Norman 1991; Overdieck 1992; Oyen/Dung 1999; Schultze-Motel 1986; Täufel et al. 1993; Teuscher 2003; Turova et al. 1987; Uphof 1968; Zeven/de Wet 1982

Amomum gracile Blume

Common Names ▸ slender cardamom; *German:* Schlanker Kardamom, Graziler Kardamom; *Malaysian:* serkom; *Sanskrit:* ela-ela, serkkom, parahuln
Usage ▸ spice, flavoring.
Parts Used ▸ leaf (leek)
Distribution ▸ Java, Malaysia

Burkill 1966; Hanelt 2001; Uphof 1968; Schultze-Motel 1986; Wiersema/León 1999

Amomum kepulaga Sprague et Burkhill

> *Amomum compactum Sol. et Maton*

Amomum korarima Pereira

> *Aframomum korarima Pereira ex Engl.*

Amomum krervanh Pierre ex Gagnep.

Synonyms ▸ *Amomum verum* Blackw.
Common Names ▸ Cambodian cardamom, chester cardamom, Bengal cardamom, Siam cardamom; *Cambodian:* krevanh, krervanh, kreko shmol; *Chinese:* pai tou kou; *French:* cardamome krervanh; *German:* Indochina-Kardamom, Kambodscha-Kardamom; *Thai:* krawan; *Vietnamese:* sa nhon, kreko krervanh
Usage ▸ spice (locally) e.g. for curry and cakes; in Europe different for sauces and liqueurs
Parts Used ▸ capsule, seed
Distribution ▸ SE Asia, Cambodia, China and Thailand also cultivated

Bois 1934; Burkill 1966; Dalby 2000; Erhardt et al. 2002; Hager 4, 1992; Hanelt 2001; Hiller/Melzig 1999; Holttum 1951; Melchior/Kastner 1974; Oyen/Dung 1999; Peter 2001; Schultze-Motel 1986; Seidemann 1993c; Siewek 1990; Teuscher 2003; Uphof 1968; Zeven/de Wet 1982

Amomum maximum *Roxb.*

Synonyms ▶ *Amomum dealbatum* Roxb., *Cardamomum dealbatum* Kuntze
Common Names ▶ Java cardamom; *Chinese:* guo gu; *German:* Java-Kardamom; *Javanese:* rĕsah, wrĕsah; hanggasa
Usage ▶ spice
Parts Used ▶ capsule, seed
Distribution ▶ India (Sikkim), Bangladesh, Malaysia; in India, China and Java also cultivated

Berger 3, 1952; Berger 1964/65; Burkill 1966; Hager 4, 1992; Hanelt 2001; Hiller/Melzig 1999; Kumar 2001; Morton 1976; Ochse/van den Brink 1931; Schultze-Motel 1986; Tao Guoda 1998; Täufel et al. 1993; Teuscher 2003; Uphof 1968; Yusuf et al. 2002; Zeven/de Wet 1982

 Amomum melegueta *Rosc.*

▶ *Aframomum melegueta* K. Schum.

 Amomum mioga *Thunb.*

▶ *Zingiber mioga (Thunb.) Rosc.*

 Amomum montanum *Koenig*

▶ *Zingiber cassumunar* Roxb.

 Amomum ochreum *Ridley*

Common Names ▶ *Malaysian:* tepus batu
Usage ▶ spice, substitute for true cardamom
Parts Used ▶ capsule (fruit, seed)
Distribution ▶ Malaysia

Guzman/Siemonsma 1999; Holttum 1950; Larsen et al. 1999; Oyen/Dung 1999; Wong 1999

 Amomum racemosum *Lam.*

▶ *Elettaria cardamomum (L.) Maton*

 Amomum roseum *K. Schum.*

▶ *Etlingera rosea* Burtt et Smith

 Amomum sanguineum *K. Schum.*

▶ *Aframomum angustifolium (Sonn.) K. Schum.*

 Amomum stenoglossum *Baker*

▶ *Amomum xanthophlebium* Baker

 Amomum subulatum *Roxb.*

Synonyms ▶ *Cardamomum subulatum* Kuntze
Common Names ▶ Bengal cardamom, Nepal cardamom, brown cardamom, greater cardamom, large cardamom, wringed cardamom, Indian cardamom; *French:* cardamome du Népal; *German:* Langer Kardamom, Großer Kardamom, Nepal-Kardamom, Sikkim-Kardamom; *Indian*: bara ilachai, elcho, motéveldodé, kátte-yelak-káy; bara alachi (Bengal)
Usage ▶ spice (locally), specially for curries, soups, sausages, sweets; **product:** essential oil
Parts Used ▶ seed
Distribution ▶ tropical Asia: Himalayas (Nepal, Bhutan), India (Sikkim), also cultivated
Note ▶ Also a substitute for true cardamom.

Annamalai et al. 1988; Berger 3, 1952; Berger 1964/65; Bois 1934; Chopra et al. 1956; Govindarajan et al. 1982; Gupta et al. 1984; Gurudutt et al. 1996; Guzman/Siemonsma 1999; Hager 4, 1992; Hanelt 2001; Hiller /Melzig 1999; Holttum 1951; Kumar 2001; Kumar/Raju 1989; Melchior/Kastner 1974; Mortin 1976; Mukherji 1973; Oyen/Dung 1999; Para Naik et al. 2000; Peter 2001; Pruthi 1976; Rao et al. 1993; Rätsch 1998; Schultze-Motel 1986; Singh 1978; Täufel et al 1993; Teuscher 2003; Wealth of India 1, 1948

 Amomum sylvestre *Lam.*

▶ *Zingiber zerumbet (L.) Rosc. ex Sm.*

AMOMUM Roxb. - Cardamom - Zingiberaceae

 Amomum testaceum *Ridley*

Common Names ▶ *Malaysian:* ka tepus, *Thai:* krawan, pla ko
Usage ▶ like true cardamom
Parts Used ▶ seed
Distribution ▶ Malaysia, S Thailand

Guzman/Siemonsma 1999; Holttum 1951; Larsen et al. 1999; Oyen/Dung 1999; Wong 1999

 Amomum thyrsoideum *Ruiz et Pav.*

▶ *Renealmia thyrsoidea (Ruiz et Pav.) Poepp. et Endl.*

 Amomum tsao-ko *Crevost et Lemaire*

Common Names ▶ tsao-ko cardamom, large cardamom; *Chinese:* cao guo, tsao ko; *French:* cardamome tsao-ko; *German:* Tsao-ko Kardamom, Langer Kardamom, Nepal-Kardamom; *Vietnamese:* dòho, sanhân âm
Usage ▶ spice
Parts Used ▶ fruit, seed
Distribution ▶ Vietnam, S China, also cultivated

Hager 4, 1992; Hanelt 2001; Wiersema/León 1999; Wu et al. 2000

 Amomum verum *Blackw.*

▶ *Amomum krervanh Pierre ex Gagnep.*

 Amomum villosum *Lour.*

Synonyms ▶ *Amomum echinosphaera* K. Schum. ex Gagnep.
Common Names ▶ bastard cardamom, false cardamom, Malabar cardamom, Tavoy cardamom, wild Siamese cardamom, *Chinese:* yang ch'un sha, sha ren; *French:* cardamome poilu de la Chine; *German:* Siam-Kardamom, Bastard-Kardamom; *Thai:* reo; *Vietnamese:* me te ba, pa dooc, sa nhan, xuân xa;
Usage ▶ spice
Parts Used ▶ seed

Distribution ▶ SE Asia, India, China, Korea, Thailand; Vietnam cultivated
Note ▶ Also a substitute for true cardamom.

Berger 3, 1952; Berger 1964/65; Burkill 1966; Chopra et al. 1956; Hager 4, 1992; Hanelt 2001; Hiller/Melzig 1999; Oyen/Dung 1999; Rehm/Espig 1984; Turova et al. 1987; Wealth of India 1, 1948; Wu et al. 2000

 Amomum xanthioides *Wall. ex Baker*

Common Names ▶ bastard Siamese cardamom, false cardamom, wild Siamese cardamom, Tavoy cardamom, Malabar cardamom; *Chinese:* chunsha, sha jin ko; *German:* Bastard-Kardamom, Malabar-Kardamom; *Korean:* chuksa; *Malaysian:* tepus, bubga tantan, bunga tanjong
Usage ▶ spice
Parts Used ▶ seed
Distribution ▶ SW India (Malabar coast), Myanmar, Thailand; cultivated in India and China
Note ▶ Probably the most predominantly cultivated var. is *xanthioides* (Wall. ex Baker) T.L. Wu et S.J Chen.

Berger 3, 1952; Berger 1964/65; Bois 1934; Erhardt et al. 2002; Guzman/Siemonsma 1999; Hager 4, 1992; Holttum 1951; Melchior/Kastner 1974; Schultze-Motel 1986; Siewek 1990; Täufel et al. 1993; Teuscher 2003; WHO 1990; Zeven/de Wet 1982

 Amomum xanthioides var. x xanthioides *(Wall. ex Baker) T.L. Wu et S.J. Chen*

▶ *Amomum xanthioides Wall. ex Baker*

 Amomum xanthophlebium *Baker*

Synonyms ▶ *Amomum stenoglossum* Baker
Common Names ▶ *India:* elach; *Malaysian:* bubga tatan, bunga tanjong, tepus
Usage ▶ flavoring in curries
Parts Used ▶ flower
Distribution ▶ Malaysian Peninsula, Borneo

Guzman/Siemonsma 1999; Oyen/Dung 1999

Amomum walang *(Blume) Val.*

▶ *Etlingera walang (Blume) R.M. Smith*

 Amomum zedoaria *Christm.*

▶ *Curcuma zedoaria (Christm.) Rosc.*

 Amomum zerumbet *L.*

▶ *Zingiber zerumbet (L.) Rosc. ex Sm.*

 Amomum zingiber *L.*

▶ *Zingiber officinale Rosc.*

AMPELOCISSUS Planch. - Vitaceae

 Ampelocissus africanus *(Lour.) Merr.*

Synonyms ▶ *Ampelocissus grantii* Planch.
Usage ▶ condiment (in Ubangi)
Parts Used ▶ root
Distribution ▶ W Africa

Burkill 5, 2000

 Ampelocissus grantii *Planch.*

▶ *Ampelocissus africanus (Lour.) Merr.*

ANACYCLUS L. - Asteraceae (Compositae)

 Anacyclus pyrethrum *(L.) Lag.*

Common Names ▶ Roman pellitory, Spanish pellitory, Spanish chaemomila, pyrethrum; *Arabic:* agargarha; *Brazil (Portuguese):* piretro da África; *French:* anacycle, pyréthre d'Afrique; *German:* Deutscher Bertram, Römischer Bertram; *Hindi:* akarkara; *India:* akahara; *Italian:* piretro romano; *Russian:* romaschka nemezkaja; *Sanskrit:* akarakarabha, akallaka; *Spanish:* pelitre
Usage ▶ spice; **product:** essential oil (for liqueurs; rarely)
Parts Used ▶ root
Distribution ▶ W Mediterranean region: SE Spain, Morocco, Algeria; cultivated in Pakistan, India; also in Austria, Germany, the Netherlands and Hungary

Aichele/Schwegler 4, 1995; André 1998; Chopra et al. 1956; Erhardt et al. 2002; Hanelt 2001; Hiller/Melzig 1999; Seidemann 1993c; Täufel et al. 1993; Wüstenfeld/Haensel 1964; Zeven/de Wet 1982

ANCISTROCLADUS Wall. - Ancistrocladaceae

 Ancistrocladus extensus *Wall. ex Planch.*

▶ *Ancistrocladus tectorius Merr.*

 Ancistrocladus tectorus *(Lour.) Merr.*

Synonyms ▶ *Ancistrocladus extensus* Wall. ex Planch.
Usage ▶ flavoring
Parts Used ▶ tender leaf
Distribution ▶ India, sea coast in the Andamans Islands

Arora/Pandey 1996; Oyen/Dung 1999

 Andropogon citratus *DC. ex Nees*

▶ *Cymbopogon citratus (DC.) Stapf*

 Andropogon citriodorum *Desf.*

▶ *Cymbopogon citratus (DC.) Stapf*

Andropogon confertiflorus Steud.
- *Cymbopogon nardus* (L.) Rendle var. *confertiflorus* Stapf ex Bor.

 Andropogon festucoides J.S. Presl.
- *Vetiveria zizaniodes* (L.) Nash

 Andropogon flexuosus Nees ex Steud.
- *Cymbopogon flexuosus* (Nees ex Steud.) Stapf

 Andropogon giganteus Hochst.
- *Cymbopogon giganteus* (Hochst.) Chiov.

 Andropogon iwarancusa W. Jones
- *Cymbopogon iwarancusa* (Jones) Schult. ex Roem. et Schult.

 Andropogon martinii Roxb.
- *Cymbopogon martinii* (Roxb.) J.F. Wats. ex Atkinson

 Andropogon muricatus Retz.
- *Vetiveria zizaniodes* (L.) Nash

 Andropogon nardus L.
- *Cymbopogon nardus* (L.) Rendle

 Andropogon nardus (L.) Hook f. **var. mahapangiri** auct.
- *Cymbopogon winterianus* Jowitt ex Bor

 Andropogon schoenanthus L.
- *Cymbopogon schoenanthus* (L.) Spreng.

 Andropogon squarrosus Hackel
- *Vetiveria zizanioides* (L.) Nash

 Andropogon zizanioides Urban
- *Vetiveria zizanioides* (L.) Nash

ANETHUM L. - Apiaceae (Umbelliferae)

 Anethum foeniculum L.
- *Foeniculum vulgare* Mill.

 Anethum graveolens L.

Synonyms ▸ *Anethum sowa* Roxb. ex Fleming, *Peucedanum graveolens* (L.) Hiern., *Peucedanum sowa* (Roxb. ex Fleming) Kurz

Common Names ▸ dill, Indian dill; *Arabic:* ainjarada, chibt, shebet, shibit; *Chinese:* shih luo zi; *French:* aneth, aneth odorant, fenouil bâtard, fenouil puant; *Dutch:* dille; *German:* Dill, Dille, Gurkenkraut, Bergkümmel, Tille; *Hindi:* sowa, soi, soy, surva; *India:* suva ni bhaji; *Indonesian:* adas manis, adas sowa ender; *Italian:* aneto; *Japanese:* inondo; *Korean:* sira, sohoehyang; *Malaysian:* adas china, adas pudus, ender; *Portuguese:* endro; *Russian:* ukron, koper, zan, schiwit; *Sanskrit:* satapuspi, chatra; *Slovakian:* kôpor; *Spanish:* aneldo, aneto, eneldo; *Thai:* thian khaopluak, thian tatakkataen, phakchi lao; *Vietnamese:* rau thia la

Use spice; **product:** ess. herb oil and essential seed oil

Parts Used ▸ leaf, seed

Distribution ▸ origin is not known; native in the Mediterranean and to S and SW Asia; cultivated world wide

Note ▸ Cultivated: *Anethum graveolens* L. var. *hortorum* Alef. and 'Sowa' (= *sowa* hort.).

Aggarwal et al. 2002; Aichele/Schwegler 3, 1995; Badoc/Lamarti 1991; Bärtels 1997; Baslas/Baslas 1971; Berger 3, 1952; Blank/Grosch 1991; Blancke 2000; Bois 1934; Bonnländer/Winterhalter 2000; Bremness 2001; Brunke et al. 1991b; Burkill 5, 2000; Craze 2002; Davidson 1999; Dudtschenko 1989; Erhardt et al. 2002; Faber et al. 1997; Farrell 1985; Fincke 1963; Garrabrants/Craker 1987; Gebczynski et al 2001; Guzman/Siemonsma 1999; Hälva et al. 1988, 1992, 1993; Hammer/Krüger 1995; Hanelt 2001; Hay/Waterman 1993; Heeger 1956; Hendry 1982; Hepper 1992; Hiller/Melzig 1999; Hornok 1980; Hondelmann 2002; Hoppe 1949; Huopalahti 1985; Huopalahti/Lathinen 1988; Jansen 1981; Jirovetz et al. 1994, 2002; Kmiecik et al. 2001; Krüger/Hammer 1996; McNavy Wood 2003; Melchior/Kastner 1974; Mhen van der/Bosch 1994; Morris/Mackley 1999; Norman 1991; Peter 2001; Pino et al. 1995; Pochljobkin 1974, 1977; Poggendorf et al. 1977; Pschyrembel 1998; Rätz 1974; Raghavan et al. 1994; Rosental 1969; Roth/Kormann 1997; Schönfelder 2001; Schröder 1991; Schultze-Motel 1986; Seidemann 1993c; Seidemann/Siebert 1987; Sharma 2003; Shiva et al. 2002; Siewek 1990; Singh 1987; Small 1997; Staesche 1972; Strunz et al. 1992; Su/Hovart 1988; Tainter/Grenis 1993; Täufel et al. 1993; Teuber/Herrmann 1978; Teuscher 2003; Tucker 1986; Tugrul et al. 2001; Ubillos 1989; Villamar et al. 1994; Wiersema/León 1999; Wyk et al. 2004; Zawirska-Wojesiak/Wąsowicz 2002; Zeven/de Wet 1982

■ *Angelica archangelica*, flowering

 Anethum pastinaca *Wibel*

▶ *Pastinaca sativa* L.

 Anethum piperitum *Ucria*

▶ *Foeniculum vulgare* Mill. ssp. *piperitum* (Ucria) Cout.

 Anethum sowa *Roxb. ex Fleming*

▶ *Anethum graveolens* L.

ANGELICA L. - Angelica, Archangel - Apiaceae (Umbelliferae)

 Angelica acutilobe (Sieb. et Zucc.) Kitagawa

Synonyms ▶ *Ligusticum acutilobum* Sieb. et Zucc.) Kitag.
Common Names ▶ *Chinese:* dong dang gui; *German:* Spitzlappige Angelika; *Japanese:* tang-kuei
Usage ▶ flavoring
Parts Used ▶ root, leaf

Distribution ▶ E Asia: China, Japan

Hanelt 2001; Schultze-Motel 1986; Wiersema/León 1999

 Angelica archangelica L. ■

Synonyms ▶ *Angelica major* Gilib., *Angelica officinalis* (Moench) Hoffm., cultivated chiefly *Angelica archangelica* L. ssp. *archangelica* var. *Sativa*, *Archangelica officinalis* Hoffm.
Common Names ▶ archangel, garden angelica, wild parsnip; *Arabic:* hashesha almalak; *Chinese:* bai zhi; *Dutch:* engelwortel; *French:* angélique, angélique vraie, archangélique; *German:* Angelika, Brustwurz, Echte Engelwurz(el), Erzengelwurzel, Theriakwurzel; *India:* chora; *Italian:* angelica, archangelica; *Korean:* padaganghwal; *Portuguese:* angélica, erva do Espírito Santo; *Russian:* djagilL, djagil'nik, angelika, anschelika, korownik sladkijctwol; ptetschnyj, *Spanish:* angélica, raíz del Epíritu Santo
Usage ▶ spice, flavoring; **product:** essential oil

ANGELICA L. - Angelica, Archangel - Apiaceae (Umbelliferae)

Parts Used ▶ root, fruit, stem
Distribution ▶ Caucasus, Siberia, E, C and N Europe (Scandinavia), widely native elsewhere
Note ▶ The stem are often candied and used in confectionery.

Aichele/Schwegler 3, 1995; Berger 3, 1952; Bois 1934; Bournot 1968; Bremness 2001; Bruyan et al. 1954; Cheers 1998; Coiciu/Racz (no year); Doneanu/Anitescu 1998; Dudtschenko et al. 1989; Erhardt et al. 2002; Erichsen-Brown 1989; Fleischhauer 2003; Genius 1981; Hager 4, 1992; Hanelt 2002; Heeger 1956; Hiller/Melzig 1999; Hohmann et al. 2001; Holm et al. 1997; Hoppe 1949; Kerrola et al. 1994; Newall et al. 1996; Nykänen et al. 1991; Opdyke 1975a; Paroul et al. 2002; Pochljobkin 1974, 1977; Pruthi 1976; Pschyrembel 1998; Roth/Korzmann 1997; Schönfelder 2001; Schultze-Motel 1986; Seidemann 1993c; Seidemann/Siebert 1987; Sharma 2003; Small 1997; Staesche 1972; Taskinen/Nykänen 1975; Täufel et al. 1993; Tucker 1986; Ubillos 1989; Wiersema/León 1999; Wolski et al. 2003; Wüstenfeld/Haensel 1964; Wyk et al. 2004; Zeven/de Wet 1982; Zobel/Brown 1991

Angelica glauca Edgew.

Common Names ▶ *German:* Blaugrüne Engelwurz; *India:* gandrayan, chohore, chora
Usage ▶ flavoring
Parts Used ▶ root
Distribution ▶ W Himalayas
Note ▶ The roots have a celery-like flavor.

Arora/Pandey 1996; Dhar/Dhar 2000; Sharma 2003; Wealth of India 1, 1948

Angelica japonica A. Gray

Synonyms ▶ *Angelica kiusiana* Maxim, *Angelica seiboldii* Miq.
Common Names ▶ Japanese archangel; *Chinese:* ma ch'ing; *German:* Japanische Engelwurz(el); *Japanese:* hama-udo, oni-udo
Usage ▶ spice, flavoring
Parts Used ▶ root, fruit
Distribution ▶ China, Taiwan, Japan, Korea

Hanelt 2001; Schultze-Motel 1986; Seidemann 1993c; Täufel et al. 1993

Angelica kiusiana Maxim

▶ *Angelica japonica A. Gray*

Angelica keiskei (Miq.) Koidz.

Common Names ▶ *Japanese:* ashitaba
Usage ▶ spice, flavoring applied *Angelica archangelica*
Parts Used ▶ root
Distribution ▶ Japan, especially Izu Island

Hanelt 2001; Schultze-Motel 1986; Teuscher 2003

Angelica lucida L.

Common Names ▶ seaside angelica, sea watsch
Usage ▶ spice (rarely)
Parts Used ▶ root
Distribution ▶ Western N America from Oregon to Alaska, N Canada to Siberia, Newfoundland, Labrador

Hoppe1, 1975; Small 1997

Angelica major Gilib.

▶ *Angelica archangelica L.*

Angelica officinalis (Moench) Hoffm.

▶ *Angelica archangelica L.*

Angelica polymorpha Maxim var. sinensis Oliv.

▶ *Angelica sinensis (Oliv.) Diels*

Angelica seiboldii Miq.

▶ *Angelica japonica A. Gray*

Angelica sinensis *(Oliv.) Diels*

Synonyms ▶ *Angelica polymorpha* Maxim var. *sinensis* Oliv.

Common Names ▶ Chinese angelica; *Chinese:* dang gui, dong gui, dang guai, dong guai; *German:* Chinesische Angelikawurz(el); *Japanese:* shirane-senkyū, suzuka-zeri

Usage ▶ spice for soups and poultry meat

Parts Used ▶ root

Distribution ▶ China, India, Indochina, Malaysia, Australia, Africa, native elsewhere in tropical regions

Hanelt 2001; Lück 2000; Schultze-Motel 1986; Wiersema/León 1999

 Angelica sylvestris *L.*

Common Names ▶ wild angelica, ginseng; *German:* Wald-Engelwurzel, Brustwurzel

Usage ▶ spice, flavoring

Parts Used ▶ root

Distribution ▶ C and N Europe, Turkey, Caucasus, Siberia

Aichele/Schwegler 3, 1995; Dudtschenko et al. 1989; Erichsen-Brown 1989; Fleischhauer 2003; Hager 4, 1992; 5, 1993; Hanelt 2001; Heeger 1956; Hoffmann et al. 1992; Krzaczek 1998; Krzaczek/Nowak 2002; Schönfelder 2001; Small 1997; Täufel et al. 1993; Wiersema/León 1999

 Angelica tenuissima *Nakai*

Synonyms ▶ *Ligusticum tenuissimum* (Nakai) Kilag

Common Names ▶ *German:* Zarte Angelikawurzel, Zarte Mutterwurz; *Korean:* kobon

Usage ▶ flavoring (for traditional Asian dishes)

Parts Used ▶ herb

Distribution ▶ Asia: Korea, NE China, also cultivated

Baik et al. 1986; Hanelt 2001; Hyang-Sook Choi et al. 2001; Pu 1991

ANIBA Aubl. - Lauraceae

 Aniba canellila *(H.B.K.) Mez*

Synonyms ▶ *Cryptocaria canellila* H.B.K.

Common Names ▶ Oriniko cinnamom; *Brazil (Portuguese):* canellila, casca preciosa, lauro precios, pereirorá; *German:* Orinoko-Zimt

Usage ▶ spice

Parts Used ▶ bark, leaf

Distribution ▶ Brazil (Amazonas region), Columbia, Venezuela, Guayana

Note ▶ The leaves and the bark have a cinnamon-like flavor.

Gottlieb/Magalhães 1959, 1960; Hanelt 2001; Kumar 2001; Mors et al. 2000; Mors/Rizzini 1961; Schultze-Motel 1986; Täufel et al. 1993; Uphof 1968

 Aniba duckei *Kosterm.*

 *Aniba rosaeodora* Ducke

 Aniba fragrans *Ducke*

Common Names ▶ *Brazil (Portuguese):* macacaporanga

Usage ▶ flavoring

Parts Used ▶ bark, wood

Distribution ▶ S America: Brazil

Kumar 2001

 Aniba rosaedora *Ducke*

Synonyms ▶ *Aniba duckei* Kosterm.

Common Names ▶ rose wood; *Brazil (Portuguese):* bois de rose; *Dutch:* echt rozenhout; *French:* bois de rose fewmelle, bois de rose de Cayenne; *German:* Brasilianisches Rosenholz; *Portuguese:* páo rosa

Usage ▶ flavoring ; **product:** essential oil

Parts Used ▶ wood

Distribution ▶ tropical S America: Brazil, Mexico; cultivated in French Guyana, Surinam

Hanelt 2001; Kumar 2001; Roth/Kormann 1997; Schultze-Motel 1986; Wiersema/León 1999

ANISOCHILUS Wall. ex Benth. - Lamiaceae (Labiatae)

 Anisochilus siamensis *Ridl.*

Usage ▸ spice
Parts Used ▸ leaf
Distribution ▸ NW Thailand

Hanelt 2001

 Anisum officinarum *Moench*

▸ *Pimpinella anisum* L.

 Anisum vulgare *Gaertn.*

▸ *Pimpinella anisum* L.

 Annona myristica *Gaertn.*

▸ *Monodora myristica (Gaertn.) Dunal*

ANREDERA Juss. - Basellaceae

 Anredera cordifolia *(Ten.) van Steenis*

Synonyms ▸ *Boussingaultia cordifolia* Ten.
Common Names ▸ bridal wreath, cascade creeper, Madeira vine; *Argentina:* papilla, zarza; *Brazil (Portuguese):* bertalha; *French:* patae d'Amériqua; *German:* Madeira Wein, Peru-Portulak, Resedenwein; *Spanish:* enredadera del mosquito, parra de Madeira, Madeira vine
Usage ▸ spice (sporadically)
Parts Used ▸ leaf
Distribution ▸ subtropical S America: Paraguay, S Brazil, N Argentina, native in Iberia, France, Malta

◘ Anthoxanthum odoratum, flowering

Erhardt et al. 2002; Hanelt 2001; Haumann 1925

 Anthadenia sesamoides *Lem.*

▸ *Sesamum indicum* L.

 Anthemis nobilis *L.*

▸ *Chamaemelum nobile (L.) All.*

ANTHOXANTHUM L. - Poaceae (Gramineae)

Anthoxanthum odoratum *L.* ◘

Common Names ▸ scented vernal grass, sweet vernal grass; *French:* flouve odorante, chiendent odorante; *German:* Gewöhnliches Ruchgras; *Italian:* paleino odoroso, paléo odorosa; *Russian:* kolosok

duschistyj; *S Africa:* heuníng gras; *Spanish:* alesta olorosa, grama de olor, grama olorosa
- Usage ▶ spice, flavoring
- Parts Used ▶ leaf
- Distribution ▶ Europe, Turkey, Caucasus, W, E Siberia, C Asia, NW Africa; native in N America, Australia, Tasmania

Aichele/Schwegler 5, 1996; Ashton/Davies 1962; Chamisso 1987; Dudtschenko et al. 1989; Erhardt et al. 2002; Fleischhauer 2003; Hanelt 2001; Heeger 1956; Heeger/Poethke 1954; Hiller/Melzig 1999; Rätsch 1998; Schnelle 1999; Schönfelder 2001; Seidemann 1993c; Täufel et al. 1993; Wiersema/León 1999; Zeven/de Wet 1982

ANTHRISCUS Pers. - Chervil - Apiaceae (Umbelliferae)

 Anthriscus cerefolium (L.) *Hoffm.*

- Synonyms ▶ *Anthriscus cerefolium* (L.) Hoffm. var. *sativus* (Lam.) Endl., *Scandix cerefolium* L.
- Common Names ▶ chervil, charviel, garden chervil; *Arabic:* maqdunis afranji; *Chinese:* san lo-po; *Dutch:* kervel; *French:* cerfeuil, cerfeuil cultivé; *German:* Deutscher Kerbel, Gartenkerbel, Kerbel, Körbel, Kufel; *Italian:* cerfoglio; *Portuguese:* cerefólio; *Russian:* kerbel, kun'r', snedok, *Spanish:* cerefolio, parifollo; shurnizo
- Usage ▶ spice; **product:** essential oil
- Parts Used ▶ fresh herb
- Distribution ▶ Europe, Turkey, N Iraq, Caucasus, Iran, C Asia, N Africa, Libya, cultivated in many countries (garden)

Berger 4, 1954; Bilgri/Adam 2000; Bois 1934; Bremness 2001; Cheers 1998; Davidson 1999; Dudtschenko et al. 1989; Farrell 1985; Fleischhauer 2003; Hanelt 2001; Heeger 1956; Hiller/Melzig 1999; Hoppe 1949; Lemberkovics et al. 1994; Melchio/Kastner 1974; Petri et al. 1994; Philisoph-Hadas et al. 1993; Pochljobkin 1974, 1977; Pruthi 1976; Robbins/Greenhalgh 1979; Rosengarten 1969; Schönfelder 2001; Schultze-Motel 1986; Šedo/Krejča 1993; Seidemann 1993c; Seidemann/Siebert 1987; Siewek 1990; Simandi et al. 1996; Small 1997; Täufel et al. 1993; Teuscher 2003; Tucker 1986; Wiersema/León 1999; Zeven/de Wet 1982

 Anthriscus cerefolium (L.) *Hoffm.* **var. sativus** (Lam.) Endl.

▶ *Anthriscus cerefolium* (L.) *Hoffm.*

 Anthriscus sylvestris (L.) *Hoffm.*

- Common Names ▶ cow parsley, Queen Anne's lace, woodland chervil; *German:* Gewöhnlicher Wiesenkerbel
- Usage ▶ pot-herb
- Parts Used ▶ young leaf
- Distribution ▶ Europe, N Africa, Caucasus, Siberia; native in N America

Erhardt et al. 2002; Facciola 1990; Hjanelt 2001; Small 1997

 Anthrophyllum pinnatum (Lam.) *Clarke*

▶ *Polyscias cumingiana* (Presl.) F.-Villar

ANTIDESMA L. - Euphorbiaceae

 Antidesma acidum *Ruiz*

▶ *Antidesma ghaesembilla* Gaertn.

 Antidesma bunius (L.) *Spreng.*

- Synonyms ▶ *Antidesma dallachyanum* Baill., *Antidesma rumphii* Tulasne, *Stilago bunius* L.
- Common Names ▶ Chinese laurel, Guinese laurel, currant tree, salamander tree; *French:* antidesme; *German:* Lorbeerblättriger Flachsbaum, Salamanderbaum; *India:* amati, anepu, nolaiali; *Javanese:* boeni, buni, buneh; *Indonesian:* daoen hoeni; *Malaysian:* berunai, buni, lundoh, lundu; *Pilipino:* bignai, búndei, dokdoko, isip; *Spanish:* bignai; *Vietnamese:* chói mói, liên tu
- Usage ▶ spice (rarely)
- Parts Used ▶ leaf

Distribution ▸ Himalayas, India, Sri Lanka, SW China, Malaysia, Philippines; Australia: Queensland
Note ▸ The plant in Malaysia elsewhere cultivated for its fruits.

Arora/Pandey 1996; Davidson 1999; Erhardt et al. 2002; Hanelt 2001; Hiller/Melzig 1999; Ochse et al. 1961; Schultze-Motel 1986; Täufel et al. 1993; Wiersema/León 1999; Zeven/de Wet 1982

Antidesma dallachyanum Baill.

▸ *Antidesma bunius (L.) Spreng.*

Antidesma ghaesembilla Gaert.

Synonyms ▸ *Antidesma acidum* Ruiz, *Antidesma pubescens* Roxb.
Common Names ▸ black currant, chinese laurel; *German:* Antidesmablätter; *Hindi:* januprulisaru, umtao; *India:* januprulisaru, pullamurasi gida, umtao; *Java:* děmpul, sepat; *Malaysian:* gunchak, guchek, gunchin; *Pilipino:* baniyuyo; *Sumatra:* kunchir; *Thai:* mao soi
Usage ▸ spice, pot-herb
Parts Used ▸ leaf
Distribution ▸ Asia: India, Nepal, China, Bhutan, Myanmar, Sri Lanka, Philippines, Java; tropical Africa, cultivated in the Tropics

Arora/Pandey 1996; Hanelt 2001; Schultze-Motel 1986; Seidemann 1993c; Täufel et al. 1993; Usher 1968; Wiersema/León 1999

Antidesma pubescens Roxb.

▸ *Antidesma ghaesembilla Gaertn.*

Antidesma rumphii Tulasne

▸ *Antidesma bunius (L.) Spreng.*

APIUM L. - Celery - Apiaceae (Umbelliferae)

Apium ammi Crantz

▸ *Ammi majus L.*

Apium anisum (L.) Crantz

▸ *Pimpinella anisum L.*

Apium celleri Gaertn.

▸ *Apium graveolens L.*

Apium crispum Mill.

▸ *Petroselinum crispum (Mill.) Nym.*

Apium dulce Mill.

▸ *Apium graveolens L.*

Apium graveolens L.

Synonyms ▸ *Apium cellerie* Gaert., *Apium dulce* Miller, *Apium maritimum* Salisb.
Common Names ▸ celery; *Arabic:* krafis; *Chinese:* qin cai; *Dutch:* selderij; *French:* céleri; *German:* (Echter) Sellerie, Eppich, Stangensellerie; *Hindi:* shalari; *India:* ajmud; *Italian:* apio, sedano, sedano da coste; *Japanese:* oranda mutsuba; *Korean:* patminari; *Portuguese:* aipo, aipo hortense; *Russian:* sel'derej, sellerej, duschistaja petruschka; *Slovakian:* zeler; *Spanish:* apio, apio blanco, apio de tallo
Usage ▸ spice, pot-herb, flavoring for salt and liqueurs; **product:** essential oil (fruit)
Parts Used ▸ leaf seed: for spice salt (celery salt)
Distribution ▸ W, C Europe, Caucasus to W Himalayas and C Asia, Turkey, Levante, Iraq, Iran, Madeira, N Africa; native, cultivated worldwide
Note ▸ *Apium graveolens* L. var. *dulce* (Mill.) Pers. = German: Bleichsellerie, Stangen- or Stielsellerie.

APIUM L. - Celery - Apiaceae (Umbelliferae) 47

Aichele/Schwegler 3, 1995; Bärtels 1997; Bartschat/Mosandl 1997; Beier/Oertli 1983; Beier et al. 1983; Bendel 2002; Berger 4, 1954; Bjeldanes/Kim 1978; Bois 1934; Bosabalidis 1996; Bremness 2001; Cheers 1998; Davidson 1999; Dovicovico et al 2004; Dudtschenko et al. 1989; Duke et al. 2003; Farrell 1985; Fehr 1979; Habegger/Schnitzler 2000; Hager 4, 1992; Hanelt 2001; Heeger 1956; Helm 1972; Hiller/Melzig 1999; Hoppe 1949; Ilyas 1980; v. Koenen 1996; Körber-Grohne 1989; MacLeod/James 1989; MacLeod et al 1988; Melchior/Kastner 1974; Morris/Mackley 1999; Newall et al. 1996; Nigg et al. 1997; Opdyke 1974a; Philippe et al. 2002; Pochljobkin 1974, 1977; Pruthi 1976; Pschyrembel 1998; Rao et al. 2000; Rosengarten 1969; Roth/Kormann 1997; Schenck/Naundorf 1966; Schönfelder 2001; Schultze-Motel 1986; Seidemann 1993c; Seidemann/Siebert 1987; Sfikas 1994; Sharma 2003; Short 1979; Siemonsma/Piluek 1993; Small 1997; Staesche 1972; Täufel et al. 1993; Teuscher 2003; Uhlig et al. 1987; Villamar et al. 1994; Wassenhove et al. 1990; Wiersema/León 1999; Zeven/de Wet 1982

Apium graveolens L. var. rapaceum *(Mill.) Gaudin*

Synonyms ▸ *Apium rapaceum* Mill.
Common Names ▸ celeriac, turnip rooted celery, root celery; *Arabic:* cherafes; *Chinese:* gen qin cai; *Dutch:* knolselderij; *French:* céleri rave, céleri tubéreux, ache-douce; *German:* Knollensellerie, Wurzelsellerie; *Italian:* sedano rapa; *Japanese:* mitsuba, seruiatku; *Malaysian:* selderi; *Portuguese:* aipo de raíz, aipo de cabeça; *Russian:* sel'derej korneplodnyj; *Spanish:* apio de bulbo, apio nabo, apio rábano
Usage ▸ spice; flavoring in salads, soups or vegetable dishes, food (vegetable)
Parts Used ▸ tuber; leaf: used dried for spice salt
Distribution ▸ mainly cultivated in C, E Europe, the Netherlands for its tuber

Aichele/Schwegler 3, 1995; Bärtels 1997; Bendel 2002; Farrell 1985; Hager 4, 1992; Hanelt 2001; Morris/Mackley 1999; Norman 1991; Schultze-Motel 1986; Seidemann 1993c; Seidemann/Siebert 1987; Shiva et al. 2002; Short 1979; Täufel et al. 1993; Teuscher 2003; Wiersema/León 1999; Wüstenfeld/Haensel 1964

Apium graveolens L. var. secalinum *Alef.*

Common Names ▸ Chinese celery, leaf celery, smallage, soup celery; *French:* céleri couper; *German:* Sellerie, Blattsellerie, Schnittsellerie; *Indonesian:* selederi; *Italian:* sedanino, sedano da erbucce, sedano da taglio; *Malaysian:* daun sop, selederi; *Pilipino:* kintsay; *Russian:* sel'derej listovoj; *Spanish:* apio de cortar, apio pequeño; *Thai:* khen chaai; phak chee lom
Usage ▸ spice
Parts Used ▸ leaf, tuber
Distribution ▸ only cultivated

Aichele/Schwegler 3, 1995; Bärtels 1997; Bosabaldis 1996; Erhardt et al. 2002; Fehr 1974; Habegger/Schnitzler 2000; Hager 4, 1992; Hanelt 2001; Hutton 1998; Pino et al. 1997; Seidemann 1993c; Seidemann/Siebert 1987; Short 1979; Täufel et al. 1993; Teuscher 2003; Wiersema/León 1999;

Apium involuncratum *Roxb. ex Flem.*

▸ *Trachyspermum roxburghianum (DC.) Craib*

Apium latifolium *Mill.*

▸ *Petroselinum crispum (Mill.) Nym. convar. radicosum (Alef.) Danert*

Apium maritinum *Salisb.*

▸ *Apium graveolens L.*

Apium petroselinum *L.*

▸ *Petroselinum crispum (Mill.) Nym. convar. radicosum (Alef.) Danert*

Apium rapaceum *Mill.*

▸ *Apium graveolens L.var. rapaceum (Mill.) Gaudin*

Apium tuberosum *Bernh. ex Rchb.*

▸ *Petroselinum crispum (Mill.) Hoffm. convar. radicosum (Alef.) Danert*

Apium visnaga *(L.) Crantz*

▸ *Ammi visnaga (L.) Lam.*

AQUILARIA Lam. - Thymelaeaceae

 Aquilaria agallocha Roxb.

Common Names ▶ agar wood, eagle wood; *Arabic:* oud kameira; *German:* Adlerholz
Usage ▶ condiment (ingredients of "Rasel-Hanout")
Parts Used ▶ fragrant wood
Distribution ▶ Orient, Morocco

Davidson 1999; Ding Hou 1964; Hepper 1992; Rätsch1998

 Aquilaria khasiana H. Hallier

Common Names ▶ Khasi-eagle wood; *German:* Khasi-Adlerholz
Usage ▶ condiment (rarely)
Parts Used ▶ fragrant wood
Distribution ▶ India: Assam: Khasi mountains

Hepper 1992

 Aquilaria malaccensis Lam.

Common Names ▶ Indian aloewood, Malacca eagle wood; *German:* Malacca-Adlerholz; *SE Asia:* alim, balok
Usage ▶ condiment (rarely)
Parts Used ▶ fragant wood
Distribution ▶ Myanmar to Malacca, Malaysia Arch.
Note ▶ The wood has an inferior quality. It is the aloe of the bible.

Burkill 1966; Ding Hou 1964; Hanelt 2001; Hepper 1992; Rätsch 1998; Wiersema/León 1999

ARACHIS L. - Peanut - Fabaceae (Leguminosae)

 Arachis hypogaea L.

Common Names ▶ goober, ground nut, monkey nut, peanut; *Brazil (Portuguese):* amendoim mandobi, manobi; *Chinese:* chang slang kuo, fan tou, luo hua shoag, tu tou; *French:* arachide, cacahouète; *German* Erdnuss; *Hindi:* mung phali; *India:* mongphali; *Italian:* arachide; *Japanese:* yatcha-sei, jimami; *Korean:* tangkhong; *Mexico:* cacahuate; *Malaysian:* kachang china, kachang gori; *Pilipino:* batung-china, mani, *Portuguese:* amendoine; *Russian:* arachis, zeml anojorech; *Spanish:* avellana americana, cacahuete, maní
Usage ▶ spice (meat)
Parts Used ▶ nut (fatty oil)
Distribution ▶ origin tropical S America; only cultivated in temperate regions S America and Africa

Blancke 2000; Cheers 1998; Hager 4, 1992; Hanelt 2001; Hiller/Melzig 1999; Menninger 1977; Neuwinger 1999; Schenck/Naundorf 1966; Schönfelder 2001; Schultze-Motel 1986; Täufel et al. 1993; Zeven/de Wet 1982

 Aralia quilfoylei Cogn. et Marché

Polyscias guilfoylei (Cogn. et Marché) Bailey

 Archangelica officinalis Hoffm

▶ *Angelica archangelica* L.

ARCHIDENDRON F. Muell. - Fabaceae (Leguminosae)

 Archidendron fagifolium (Blume ex Miq.) I. Nielson

Synonyms ▶ *Pithecellobium angulatum* auct. non Benth., *Pithecellobium fagifolium* Blume ex Miq., *Pithecellobium mindanaense* Merr.
Common Names ▶ *Indonesian:* jengkolan, *Javanese:* jering goleng; *Pilipino:* kulikul, lalatan, tomanag
Usage ▶ spice (locally)
Parts Used ▶ seed
Distribution ▶ Thailand, Myanmar, Borneo, Java *(var. borneense Nielsen)*, Philippines *(var. Mindanaense [Merr.] Nielsen)*, Sumatra *(var. fagifolium)*

Guzman/Siemonsma 1999; Oyen/Dung 1999

Archidendron jiringa (Jack) I. Nielsen

Synonyms ▶ *Mimosa jiringa* Jack, *Pithecellobium jiringa* (Jack) Prain ex King, *Pithecellobium lobatum* Benth.
Common Names ▶ jering tamarind; *German:* Malayischer Affenohrring; *Indonesian:* jengkol, jering; *Malaysian:* jiring, djering; *Thai:* jawang nieng
Usage ▶ condiment, flavoring
Parts Used ▶ seed (cooked green and ripe)
Distribution ▶ India, Myanmar, Malaysia

Bois 1934; Erhardt et al. 2000, 2002; Täufel et al. 1993; Uphof 1968; Wiersema/León 1999; Zeven/de Wet 1982

ARDISIA Sw. - Myrsinaceae

 Ardisia boissierri A. DC.

▶ *Ardisia squamulosa* Presl

 Ardisia drupacea (Blanco) Merr.

▶ *Ardisia squamulosa* Presl.

 Ardisia humilis auct., *non Vahl*

▶ *Ardisia squamulosa* Presl

 Ardisia squamulosa Presl

Synonyms ▶ *Ardisia boissieri* A. DC., *Ardisia drupacea* (Blanco) Merr., *Ardisia humilis* auct., non Vahl.
Common Names ▶ spicy berry; *German:* Schuppige Spitzenblume; *Pilipino:* babagion, butau, tagpo
Usage ▶ flavoring (fish)
Parts Used ▶ flower, fruit (cooked)
Distribution ▶ endemic(?) Philippines

Guzman/Siemonsma 1999; Hoppe 3, 1987; Oyen/Dung 1999

 Arduina edulis Spreng

▶ *Carissa edulis* Vahl

ARISTOLOCHIA L. - Aristolochiaceae

 Aristolochia hastata Nutt.

▶ *Aristolochia serpentaria* L.

 Aristolochia officinalis Nees

▶ *Aristolochia serpentaria* L.

 Aristolochia serpentaria L.

Synonyms ▶ *Aristolochia hastata* Nutt., *Aristolochia officinalis* Nees
Common Names ▶ Virginia serpentary, Virginia snakeroot; *German:* Virginische Schlangenwurzel; *India:* sarpmoola; *Russian:* zmejuyj koren
Usage ▶ spice, flavoring of liqueurs and bitter, e.g. angostura, boonekamp
Parts Used ▶ rhizom
Distribution ▶ N America: Florida, Virginia, cultivated in India

Charalambous 1994; Erhardt et al. 2002; Hanelt 2001; Hiller/Melzig 1999; Seidemann 1993c; Wüstenfeld/Haensel 1964

ARMENIACA Scopoli - Rosaceae

 Armeniaca dasycarpa (Ehrh.) Borkh.

Synonyms ▶ *Armeniaca fusca* Turp. et Poit. ex Duhamel, *Prunus dasycarpa* Ehrh., *Prunus nigra* Desf.
Common Names ▶ black apricot, purple apricot; *German:* Rauhfrüchtige Marille, Schwarze Marille; *Russian:* abrikos tschernij, abrikos volosistoplodnyj
Usage ▶ spice, condiment

- **Parts Used** ▶ fruit
- **Distribution** ▶ Asia: Iran Afghanistan, Kashmir, Europe: France to Ukraine, Transcaucasia, southern USA
- **Note** ▶ The plant is a spontaneous hybrid of *Armeniaca vulgais* Lam. x *Prunus cerasifera* Ehrh.

Hanelt 2001; Schultze-Motel 1986

Armeniaca fusca *Turp. et Poit. ex Duhamel*

❯ *Armeniaca dasycarpa (Ehrh.) Borkh.*

Armeniaca vulgaris *Lam. ex* Prunus cerasifera *Ehrh.*

❯ *Armeniaca dasycarpa (Ehrh.) Borkh.*

ARMORACIA Gaertn., B. Mey. et Schreb. - Horseradish - Brassicaceae (Cruciferae)

Armoracia lapathifolia *Gilib.*

❯ *Armaoracia rusticana P. Gaertn.*

Armoracia rusticana *Gaertn. et B. Mey. et Scherb.*

- **Synonyms** ▶ *Armoracia lapathifolia* Gilib., *Cochlearia armoracia* Lam., *Nasturtium armoracia* (L.) Fries
- **Common Names** ▶ horseradish; *Arabic:* feegel, fidgel; *Chinese:* la gen; *Dutch:* mierikswortel; *French:* cran, cranson, mérédic, raifort sauvage; *German:* Bauernsenf, Meerrettich, Mährrettich, Kren (in Austria, S Germany and Slavonic countries), Pfefferwurzel; *Italian:* barbaforte, cren, rafano rusticano; *Portuguese:* armorácia, rábano picanto, rábano rusticano, raiz forte; *Russian:* chren; *Slovakian:* chren; *Spanish:* rábano picante, rábano rustico, rábano silvestre
- **Usage** ▶ spice
- **Parts Used** ▶ (rasped) fresh root
- **Distribution** ▶ possible origin in E Europe, Russia, native in Europe, Caucasus, N America; widely cultivated in many countries

Aichele/Schwegler 3, 1995; Bendel 2002; Bois 1934; Bremness 2001; Cheers 1998; Courter/Rhodes 1969; Davidson 1999; Dudtschenko et al. 1989; Due et al. 2003; Farrell 1985; Galacci 1989; Gilbert/Nursten 1972; Grob/Matile 1980; Guzman/Siemonsma 1999; Hager 4, 1992; Hanelt 2001; Hansen 1974; Herrmann 1997b; Hiller/Melzig 1999; Hoppe 1949; Kotow 1978; Kraxner et al. 1987; Lin et al. 2001; Melchior/Kastner 1974; Morris/Mackley 1999; Newall et al. 1996; Pochljobkin 1974, 1977; Pschyrembel 1998; Pursglove 1968; Rhodes et al. 1969; Rosengarten 1969; Sahasrabudhe/Mullin 1980; Schönfelder 2001; Schultze-Motel 1986; Seidemann 1993c; Small 1997; Täufel et al. 1993; Teuscher 2003; Tucker 1986; Usher 1968; Weber 2002; Wiersema/León 1999; Zeven/de Wet 1982; Winter/Hornborstel 1993; Wyk et al. 2004

Armoracia sisymbrioides *(DC.) N. Busch ex Ganesch.*

- **Common Names** ▶ Russian wild horseradish; *German:* Wilder russischer Meerrettich; *Russian:* chren guljavnikovyi, chren lugovoi
- **Usage** ▶ spice
- **Parts Used** ▶ root
- **Distribution** ▶ SW and E Siberia to Sakhalin, also cultivated in gardens

Hager 4, 1992; Hanelt 2001; Mansfeld 1986; Schultze-Motel 1986; Teuscher 2003

ARNICA L. - Asteraceae (Compositae)

Arnica chamissonis *Less.*

- **Synonyms** ▶ *Arnica foliosa* Nutt.
- **Common Names** ▶ leafy leopardsbane; *German:* Amerikanische Arnika
- **Usage** ▶ spice (flavoring) for liqueurs and bitters (Abtei, Aromatique, Kartaeser)
- **Parts Used** ▶ flower
- **Distribution** ▶ Alaska, Canada, USA: NW, SW, California, Rocky mountains, also cultivated in many countries

Hager 4, 1992; Hanelt 2001; Hiller/Melzig 1999; Schönfelder 2001; Schultze-Motel 1986; Turova et al. 1987; Willuhn 1985a; Willuhn/Leven 1991

Arnica foliosa *Nutt.*

 Arnica chamissonis Less.

 Arnica montana *L.*

Common Names ▶ European arnica, leopards bane; mountain arnica, mountain tobacco; *French:* arnica, fleur d'arnica, panacée des Montagnes; *German:* Arnika; Bergwohlverleih, Echte Arnika; *India:* arnica; *Italian:* arnica, fiore de arnica; *Russian:* arnica gornaja; *Spanish:* árnica de montaña, flor de árnica

Usage ▶ spice, flavoring for liqueurs and bitters; **product:** essential oil: arnica (flower) oil, celtic nard oil

Parts Used ▶ flower

Distribution ▶ Europe (Alps): Austria, Germany, Switzerland, France

Note ▶ In Austria, Germany and Switzerland the plant has been placed under nature protection.

Aichele/Schwegler 4, 1995; Berger 1, 1949; Crocilius 1956; Coiciu/Rasz (no year); Dudtschenko et al. 1989; Erhardt et al. 2002; Galambosi et al. 1998; Hager 4, 1992; Hahn/Mayer 1985; Hanelt 2001; Heeger 1956; Hiller/Melzig 1999; Hohmann et al. 2001; Hoppe 1949; Merfort/Wendisch 1988; Meyer/Berge 1991; Meyer-Chlond 1999; Newall et al. 1996; Pschyrembel 1998; Roth/Kormann 1997; Saukel 1984; Schönfelder 2001; Schultze-Motel 1986; Seidemann 1993c; Täufel et al. 1993; Turova et al. 1987; Willuhn 1981, 1983, 1991; Wüstenfeld/Haensel 1964; Wyk et al. 2004

ARONIA Medik. - chokeberry - Rosaceae

 Aronia alnifolia *Nutt.*

 Amelanchier alnifolia (Nutt.) Nutt.

Aronia melanocarpa *(Michx.) Elliot*

Common Names ▶ black chokeberry; *French:* aronica du noir; *German:* Apfelbeere; *Russian:* rjabina cernoplodnaja

Usage ▶ condiment, especially for sauces of venison,

Parts Used ▶ fruit

Distribution ▶ E Canada, NE, NCE, SE USA; cultivated in E Europe, Germany

Erhardt et al. 2002; Hanelt 2001; Hirvi/Honkanan 1985; Lehmann 1982, 1990; Oszmianski/Sapis 1988; Plocharski/Smolarz 1987; Rosa/Krugty 1987; Schultze-Motel 1986; Seidemann 1993c, e; Wilska-Jeszka et al. 1988

ARTABOTRYS R.Br. - Tail grape - Annonaceae

 Artabatrys hexapetalus *(L.f.) Bhandri*

Synonyms ▶ *Artabotrys odoratissimus* R.Br. ex Ker-Gawl., *Artabotrys unicatus* (Lam.) Merr.

Common Names ▶ tail grape; *German:* Falscher Ylang-Ylang; *India:* madan mast, manoranjitan; *Sanskrit:* harachampala

Usage ▶ flavoring (tea); **product:** essential oil

Parts Used ▶ flower

Distribution ▶ S India, Sri Lanka

Burkill 1966; Erhardt et al. 2002; Hanelt 2001; Uphof 1968; Usher 1974; Wealth of India 1, 1984

 Artabotrys odoratissimus *R.Br. ex Ker-Gawl*

▶ *Artabotrys hexapetalus (L.f.) Bhandri*

 Artabotrys uncinatus *(Lam.) Merr.*

▶ *Artabotrys hexapetalus (L.f.) Bhandri*

 Artabotrys scytophyllus *(Diels) Cavaco et Kerandren*

Common Names ▶ tail grape; *German:* Lederblättrige Klimmtraube

Usage ▶ spice for sauces, flavoring (flower)

Parts Used ▶ herb

Distribution ▶ SE Asia; Madagascar also cultivated

Artanthe adunca (L.) Miq.

▶ *Piper aduncum* L.

ARTEMISIA L. - Mugwort - Asteraceae (Compositae)

 Artemisia abrotanum L.

Common Names ▶ lady's love, old man storenwood, southern (worm) wood; *Arabic:* chissum; *Dutch:* citroenkruid; *French:* abrotone, armoise, aurone, aurone des jardins aurone mâle, citronelle aurone, ivrogne, garde robes; *German:* Eberraute, Eberreis, Gartenheil, Zitronenkraut; *Italian:* abrotano; *Portuguese:* abrotano; *Russian:* bosh'e; *Spanish:* abrótano macho, ajenjo común, boja

Usage ▶ spice, flavoring and in liqueur manufacture
Parts Used ▶ herb
Distribution ▶ S, SE Europe, W Asia, Siberia, also cultivated

Aichele/Schwegler, 4, 1995; Bärtels 1997; Bilgri/Adam 2000; Bois 1934; Bremness 2001; Cheers 1998; Clair 1961; Davidson 1999; Erhardt et al. 2002; Hager 4, 1992; Hanelt 2001; Heeger 1956; Hiller/Melzig 1999; Hoffmann et al. 1992; Hoppe 1949; Neumerkel 2001; Roth/Korzmann 1997; Schönfelder 2001; Schultze-Motel 1986; Seidemann 1993c; Small 1997; Täufel et al. 1993; Teuscher 2003; Tucker 1986; Vostrowsky et al. 1984; Wiersema/León 1999; Wright 2000; Wüstenfeld/Haensel 1964; Zeven/de Wet 1982

 Artemisia absinthium L.

Synonyms ▶ *Artemisia officinale* Brot., *Artemisia vulgare* Lam.
Common Names ▶ absinthe, common wormwood, mugwort, warmnot; *Arabic:* afsantin, eshbet mariam, mamitsa, shadjrat mariam, shi rumi; *Brazil (Portuguese):* absinthe losna; *Dutch:* absint-alsem; *French:* absinthe, grande absinthe, armoise absinthe; *German:* Absinth, Bitterer Beifuß, Wermut; *India:* vilayati afsantin; *Italian:* assenzio, assenzio Romano, assenzio maggiore; *Korean:* pukyakssuk; *Portuguese:* absinto, losna; *Russian:* polyn' gor'kaja, wermut; *Spanish:* absintio, ajenjo, ajenja mayor

Usage ▶ spice, flavoring; **product:** vermuth bitters and absinthe liqueur and aperitifs
Parts Used ▶ herb
Distribution ▶ temperate Eurasia; N, NW Africa, adventive in N and S America, cultivated in the Mediterranean area, in the USA (Indiana, Michigan), Kashmir and S Siberia, also in C, S Mexico, Brazil, Europe

Aichele/Schwegler 4, 1995; Alberts/Muller 2000; Ariño et al. 1999; Arnold 1989; Bärtels 1997; Berger 4, 1954; Bielenberg 2002; Bois 1934; Böttchers/Günther 2002; Bornot 1968; Bremness 2001; Cheers 1998; Chialva et al. 1983; Clair 1961; Coiciu (no year); Dastur 1954; Davidson 1999; Dudtschenko et al. 1989; Erichsen-Brown 1989; Gebhardt 1977b; Giebelmann 2001; Gurim-Fakim/Brendler 2004; Hager 4, 1992; Hanelt 2001; Heeger 1956; Hese 2002; Hiller/Melzig 1999; Hoppe 1949; Hose 2002; Hutton 2002; Kennedy et al. 1993; Lachenmeier et al. 2004; Larkcom 1991; Neumerkel 2001; Nin et al. 1995; Pochljobkin 1974, 1977; Pschyrembel 1998; Rätsch 1995; Roth/Korzmann 1997; Schneider/Mielke 1978, 1979; Schönfelder 2001; Schultze-Motel 1986; Seidemann 1993c; Seidemann/Siebert 1987; Sharma 2003; Staesche 1972; Stahl/Gerard 1982; Tateo/Riva 1991; Täufel et al. 1993; Teuscher 2003; Tucker 1986; Turova et al. 1987; Ubillos 1989; Villamar et al. 1994; Vostrowsky et al. 1981; Wiersema/León 1999; Wright 2002; Wüstenfeld/Haensel 1964; Wyk et al. 2004; Zeven/de Wet 1982

 Artemisia afra Jacq. ex Willd.

Common Names ▶ African wormwood, *Afrikaans:* wildeals; *French:* armoise d'Afrique; *German:* Afrikanischer Wermut
Usage ▶ flavoring; **product:** vermuth bitters
Parts Used ▶ herb
Distribution ▶ E Africa, from Ethiopia to S Africa

Wyk et al. 2004

 Artemisia alba Turra

Synonyms ▶ *Artemisia camphorata* Vill.
Common Names ▶ camphor absinthe; *French:* armaise camphrée; *German:* Kampfer-Wermut; *Italian:* abrotano camforata
Usage ▶ spice, flavoring
Parts Used ▶ herb
Distribution ▶ S ans S Central Europe, NW Africa
Note ▶ Cultivated in Germany in the 16th Century.

ARTEMISIA L. - Mugwort - Asteraceae (Compositae)

Berger 4, 1954; Hanelt 2001; Schultze-Motel 1986; Wiersema/León 1999; Wright 2002

Artemisia arborescens L.

Common Names ▶ silver wormwood, tree wormwood; *Arabic:* shagaret mariam, shiba; *French:* absinthe arborscente; *German:* Baum-Wermut; *Cyprus:* genia ou gerru
Usage ▶ spice
Parts Used ▶ herb
Distribution ▶ N Africa, E Mediterranean Region cultivated in gardens

Bremness 2001; Hanelt 2001; Schultze-Motel 1986; Wright 2002

Artemisia balchanorum H. Krasch

Synonyms ▶ *Serephidium balchanorum* (Krasch) Polj.
Common Names ▶ *German:* Zitronen-Wermut; *Russian:* polyn' balchanov, polyn' limmonaja
Usage ▶ spice, pot-herb
Parts Used ▶ herb
Distribution ▶ Turkmenia; cultivated: Ukraine, Georgia, Armenia, Moldavia
Note ▶ Various cultivars are existing herbs.

Hager 3, 1992; Hanelt 2001; Schultze-Motel 1986; Seidemann 1993c; Täufel et al. 1993; Teuscher 2003

Artemisia balsamita Willd.

▶ *Artemisia pontica L.*

Artemisia camphorata Vill.

▶ *Artemisia alba Turra*

Artemisia dracunculus L.

Synonyms ▶ *Artemisia glauca* Pall.
Common Names ▶ tarragon, dragon sagewort, French tarragon; *Arabic:* tarkhun; *Chinese:* xai ye qing hao; *Dutch:* dragon, estragon; *French:* dragon, estragon; *German:* Bertram, Bertramskraut, Dragun-Wermut, Eierkraut, Estragon, Kaisersalat; *Italian:* dragone, dragoncello, estragone, herbe dragonne, targone; *Japanese:* esutoragon; *Portuguese:* estragão; *Russian:* draguntrawa, estragon, polyn' stragon, stragon, tarchun; *Spanish:* draconcillo, estragón
Usage ▶ spice, flavoring for salads, cucumber, tomato conserves, soups etc., manufacture of tarragon vinegar
Parts Used ▶ herb
Distribution ▶ Europe, W, E Siberia, C Asia, W Himalayas, Mongolia, China, Alaska, Canada, USA; native in France, S Europe, Balcan; cultivated in many countries (garden)
Note ▶ f. *dracunculus* = German or French Estragon; f. *redowskii* hort = Russian dragon.

Aichele/Schwegler 4, 1995; Bärtels 1997; Berger 4, 1954; Bilgri/Adam 2000; Bois 1934; Bremness 2001; Burkill 1966; Cheers 1998; Clair 1961; Coiciu/Rascz (no year); Cotton et al. 1991; Davidson 1999; Dudtschenko et al. 1989; Erichsen-Brown 1989; Farrell 1985; Gebhardt 1977a; Guzman/Siemons 1999; Hager 4, 1992; Hagmann/Knauss 1988; Hanelt 2001; Heeger 1956; Hiller/Melzig 1999; Hoppe 1949; Mackay/Kitto 1987; Melchior/Kastner 1974; Neumerkel 2001; Pochljobkin 1974, 1977; Prakash 1990; Rosengarten 1969; Rosenthal 1954; Roth/Korzmann 1997; Satar 1986; Schönfelder 2001; Schultze-Motel 1986; Seidemann 1993 c; Seidemann/Siebert 1987; Small 1997; Siewek 1990; Staesche 1972; Täufel et al. 1993; Teuscher 2003; Thieme/Nguyen Thi Tam 1968 a, b; Tucker 1986; Tunman/Mann 1968; Ubillos 1989; Uhl 2000; Vostrowsky et al. 1981a, b; Wealth of India 1, 1948; Werker et al. 1994; Wiersema/León 1999; Wright 2002; Zeven/de Wet 1982

Artemisia genipi Weber

Common Names ▶ *French:* genépi noir; *German:* Schwarze Edelraute; *Italian:* genepi
Usage ▶ spice, flavoring for herb liqueurs
Parts Used ▶ leaf
Distribution ▶ C Europe, S Europe especially Italy, SE Europe, SW Europe

Erhardt et al. 2002; Wiersema/León 1997; Wright 2002

54 ARTEMISIA L. - Mugwort - Asteraceae (Compositae)

Artemisia glacialis L.

Common Names ▸ glacier wormwood; *German:* Gletscher-Edelraute, Gletscher-Beifuß; *Italian:* genipi
Usage ▸ spice, flavoring for herb liqueurs
Parts Used ▸ leaf
Distribution ▸ Europe: Alps of France, N Italy, Switzerland

Erhardt et al. 2002; Hoppe 1949; Wright 2002

Artemisia glauca Pall.

 *Artemisia dracunculus* L.

Artemisaia herba-alba Asso

Common Names ▸ white mugwort; *German:* Weißer Wermut
Usage ▸ spice (rarely)
Parts Used ▸ leaf
Distribution ▸ Palestine, Israel, Arabia: desert regions
Note ▸ The plant is the wormwood of the bible.

Hepper 1992; Mabberly 1997; Rätsch 1998; Wright 2002

Artemisia indica Willd.

Synonyms ▸ *Artemisia vulgaris* L. var. *indica* (Willd.) Maxim.
Common Names ▸ Indian mugwort; *German:* Indischer Beifuß, Indisches Moxakraut
Usage ▸ spice, flavoring
Parts Used ▸ leaf
Distribution ▸ S, E Asia: Himalayas, Maabmar, Sri Lanka, China, Korea, Taiwan, Indochina, Malaysia, Indonesia cultivated mainly in gardens

Hanelt 2001; Neumerkel 2001; Teuscher 2003; Wright 2002

Artemisia judaica L.

Common Names ▸ *French:* graines à vers, zédoire; *German:* Palästinenser Beifuß

Usage ▸ spice, flavoring for liqueurs
Parts Used ▸ herb
Distribution ▸ N Africa, Egypt, Arabia, Israel, cultivated E Mediterranean region

Hanelt 2001; Hepper 1992; Ravid et al. 1994; Schultze-Motel 1986; Täufel et al. 1993; Tucker 1986; Uphof 1968; Usher 1974; Zeven/de Wet 1982

Artemisia lactiflora Wall. ex DC.

Common Names ▸ white mugwort; *French:* armoise à fleurs laiteuses; *German:* Weißer Chinabeifuß, Weißer Beifuß
Usage ▸ spice, flavoring
Parts Used ▸ herb, leaf
Distribution ▸ W China

Bremness 2001; Erhardt et al. 2002

Artemisia laxa (Lam.) Fritzsch

 Artemisia mutellina Vill.

Artemisia ludoviciana Nutt.

Synonyms ▸ *Artemisia purshiana* Besser
Common Names ▸ cudweed, western mugwort, white sage; *German:* Mexikanischer Wermut, Silberraute, Weiße Raute; *Mexico:* estafiate
Usage ▸ spice, used like absinthe
Parts Used ▸ herb
Distribution ▸ Alaska, W Canada; USA, Mexico, also cultivated

Berger 4, 1954; Bremness 2001; Erichsen-Brown 1989; Hanelt 2001; Neumerkel 2001; Rätsch 1998; Tull 1999; Villamar et al. 1994; Wiersema/León 1999

Artemisia maritima L.

Common Names ▸ sea wormwood; *French:* absinthe de mer, armoise maritime; *German:* Strandbeifuß; Meerbeifuß, Meerwermut; *Hindi:* kirmala, kirmani; *India:* gandha; *Korean:* santhoninssuk;

Sanskrit: cauhara, kitamari; *Spanish:* ajenjo marino
Usage ▶ spice, flavoring; **product:** santonine
Parts Used ▶ herb
Distribution ▶ Europe: coasts from the Faroes to Scandinavia and the Baltics

Berger 4, 1954; Coiciu/Racz (no year); Fleischhauer 2003; Hager 4, 1992; Hanelt 2001; Heeger 1956; Hoppe 1949; Jeffrey 1965; Schultze-Motel 1986; Sharma 2003; Täufel et al. 1993; Tucker 1986; Wiersema/León 1999; Zeven/de Wet 1982

Artemisia mexicana *Willd.*

Common Names ▶ Mexican mugwort; *German:* Mexikanischer Beifuß; *Guatamala:* tsin tsin; *Mexico:* estafiata, artamisa Méjico
Usage ▶ spice
Parts Used ▶ herb
Distribution ▶ SW USA to Mexico, also cultivated in Mexico

Ochoa/Alonso 1996; Rätsch 1998; Schultze-Motel 1986

Artemisia mutellina *Vill.*

Synonyms ▶ *Artemisia laxa* (Lam.) Fritzsch, *Artemisia splendens* Willd. var. *splendens*, *Artemisia umbelliformis* Lam.
Common Names ▶ alpine wormwood, white genipi; *French:* genépi blanc; *German:* Echte Edelraute, Echter Alpenbeifuß; *Russian:* polyn' al'nijskaja
Usage ▶ spice, flavoring of liqueurs and other drinks
Parts Used ▶ herb
Distribution ▶ Alps and mountains of C Europe, cultivated in C and S Europe

Aichele/Schwegler 4, 1995; Berger 4, 1954; Hanelt 2001; Hoppe 1949; Pochljobkin 1974, 1977; Rey/Slacanin 1999; Schultze-Motel 1986; Seidemann 1993c; Täufel et al. 1993; Teuscher 2003; Wiersema/León 1999; Wüstenfeld/Haensel 1964; Zeven/Zhukovsky 1975

Artemisia officinale *Brot.*

 *Artemisia absinthium* L.

Artemisia ordosica *Krasch.*

Usage ▶ flavoring; **product:** essential oil
Parts Used ▶ herb
Distribution ▶ N and NW China

Hanelt 2001; Rätsch 1998; Wright 2004

Artemisia pallens *Wall. ex DC.*

Common Names ▶ *India:* davana; *German:* Blasser Beifuß; *Russian:* polyn' blednejusschtschaja
Usage ▶ spice, flavoring for cake, pastries; **product:** essential oil (davana oil)
Parts Used ▶ herb
Distribution ▶ NE India, N Thailand

Charalambous 1994; Hanelt 2001; Rätsch 1998; Schultze-Motel 1986; Sharma 2003; Shiva et al. 2002; Wealth of India 1, 1948; Wiersema/León 1999; Wright 2002

Artemisia pontica *L.*

Synonyms ▶ *Artemisia balsamita* Willd.
Common Names ▶ Roman wormwood, small absinth; *French:* absinthe romaine, petite absinthe; *German:* Pontischer Beifuß, Pontischer Wermut, Römischer Beifuß; *Russian:* polyn' pontijskaja, polyn'rumskaja, alekcandrijskaja polyn', tschernomorskaja polyn'
Usage ▶ spice, flavoring
Parts Used ▶ leaf, herb
Distribution ▶ S, E and Europe to W Siberia, C Asia, Kazakhstan and NW China; cultivated in France, Italy, Austria, Germany, Switzerland, S Canada and USA
Note ▶ The plant is used like *Artemisia absinthium* L.

Aichele/Schwegler 4, 1995; Berger 4, 1954; Bremness 2001; Erhardt et al. 2002; Hanelt 2001; Hiller/Melzig 1999; Hoppe 1949; Pochljobkin 1974, 1977; Rätsch 1998; Schönfelder 2001; Schultze-Motel 1986; Seidemann 1993c; Small 1997; Täufel et al. 1993; Teuscher 2003; Tucker 1986; Usher 1974; Wiersema/León 1999

ARTEMISIA L. - Mugwort - Asteraceae (Compositae)

 Artemisia princeps *Pamp.*

Common Names ▸ Japanese mugwort, ticed bub; *German:* Japanischer Beifuß
Usage ▸ flavoring, and coloring of rice cakes
Parts Used ▸ leaf
Distribution ▸ Japan

Davidson 1999; Larkcom 1991; Small 1997; Umano et al. 2000

 Artemisia procera *Willd.*

Common Names ▸ *German:* Rispen-Beifuß, *Russian:* polyn' metel'tschataja
Usage ▸ spice
Parts Used ▸ herb
Distribution ▸ former Tatarian SSR

Berger 4, 1954; Hager 4, 1992; Hanelt 2001; Heeger 1956; Hiller/Melzig 1999; Pochljobkin 1974, 1977

 Artemisia purshiana *Besser*

▸ *Artemisia ludoviciana Nutt.*

 Artemisia splendens *Willd.* **var. splendens**

▸ *Artemisia mutellina Vill.*

 Artemisia umbelliformis *Lam.*

▸ *Artemisia mutellina Vill.*

 Artemisia vallesiaca *All.*

Common Names ▸ alpine wormwood, mountain wormwood, Valais wormwood; *French:* armoise du Valais, pityouda absinthe; *German:* Beng-Genipopp, Walliser Beifuß, Walliser Genipp, Walliser Wermut
Usage ▸ flavoring for liqueurs; **product** of santonine
Parts Used ▸ herb
Distribution ▸ Europe: N Italy, SE France, Switzerland

▣ Artemisia vulgaris, flowering

Hanelt 2001; Hoppe 1949; Schultze-Motel 1986

 Artemisia verlotiorum *Lamotte*

Common Names ▸ Kamshatka wormwood; *German:* Kamtschatka Beifuß
Usage ▸ flavoring
Parts Used ▸ herb
Distribution ▸ Japan, Kamtschatka, native in EC Europe

Carnat et al. 2001; Erhardt et al. 2002

 Artemisia vulgare *Lam.*

▸ *Artemisia absinthium L.*

 Artemisia vulgaris *L.* ▣

Common Names ▸ mugwort, motherwort, sagebrush, Sailor's tobacco, wormwood; *Chinese:* ai hao, bei ai, bai hao, xi ye ai; *French:* armoise commune, couronne de Saint Jean; *German:* Gemeiner Beifuß, Gewürz-Beifuß, Gänsekraut, Wilder Wermut; *Hindi:* dauna, nagdauna; *India:* nilum, nagadamani, titipati; *Italian* amarella; *Japanese:* yomogi, nishi-yomogo, moxa; *Pilipino:* damong Maria, erbaka, tinisas; *Portuguese:* artemísia; *Russian:* polyn'obyknovennaja, prostaja polyn', tschernobyl'nik; *Sanskrit:* damanaka, tapodhana; *Spanish:* artemisia, hierba de San Juan; *Vietnamese:* ngai diep, nha ngai, qua su

Usage ▶ spice, flavoring

Parts Used ▶ herb

Distribution ▶ Europe, N, C Asia, adventive in America and Australia; cultivated in W and C Europe, Asia and elsewhere, e.g., in the Philippines, Thailand, Malaysia, Vietnam, China and Brazil

Aichele/Schwegler 4, 1955; Berger 2, 1950; Berger 4, 1954; Burkill 1966; Cheers 1998; Clair 1961; Davidson 1999; Dudtschenko et al. 1989; Dürbeck 1996; Erichsen-Brown 1989; Gebhardt 1977c; Hager 4, 1992; Hanelt 1956, 2001; Heeger 1956; Hiller/Melzig 1999; Hoppe 1949; Jerkovic et al. 2003; Larkcom 1991; Melchior/Kastner 1974; Michaelis et al. 1982; Nano et al. 1976; Neumerkel 2001; Pochljobkin 1974, 1977; Pschyrembel 1998; Roth/Korzmann 1997; Schönfelder 2001; Schultze-Motel 1986; Seidemann 1993c; Seidemann/Siebert 1987; Siewek 1990; Small 1997; Staesche 1972; Täufel et al. 1993; Teuscher 2003; Tucker 1986; Turova et al. 1987; Villamar et al. 1994; WHO 1990; Wiersema/León 1999; Wright 2002; Wörner et al. 1991c; Wyk et al. 2004: Zeven/de Wet 1982

ARTOCARPUS J.R. Forst. et Forst. - Breadfruit - Moraceae

 Artocarpus lakoochus Roxb.

Synonyms ▶ *Artocarpus yunnanensis* Hu

Common Names ▶ monkey, monkey jack; *German:* Lakoocha; *India:* barhal, lakooch, lakeichi; *Malaysian:* tampang, *Thai:* lokhat, mahád

Usage ▶ condiment, compound for some curry mixtures

Parts Used ▶ male inflorescence

Distribution ▶ tropical S Asia: India, Sri Lanka, Myanmar, S China, Thailand, Malaysia, also cultivated

Arora/Pandey 1996; Hanelt 2001; Schultze-Motel 1986; Seidemann 1993c; Thomas 1960; Zeven/de Wet 1982

 Artocarpus yunnanensis Hu.

▶ *Articarpus lakoochus* Roxb.

ASARUM L. - Aristolochiaceae

 Asarum canadense L.

Common Names ▶ wild ginger; *German:* Kanadische Haselwurz

Usage ▶ spice (of the native Americans)

Parts Used ▶ rhizom

Distribution ▶ Canada, USA

Note ▶ The rhizome has a flavor and taste of ginger.

Erhardt et al. 2002; Rätsch 1998; Small 1997

 Asarum europaeum L.

Common Names ▶ asarabacca, European wild ginger; *French:* asaret d'Europe, cabaret, oreille d'home; *German:* Braune Haselwurz, Gewöhnliche Haselwurz, Hasenpfeffer; *Italian:* asaro baccaro; *Russian:* kopyten'europeyckij; *Spanish:* ásarocomún, ásaro de Europa

Usage ▶ spice (very rarely)

Parts Used ▶ rhizom

Distribution ▶ S Europe: Iberia, W Sibera, native in C Europe

Aichele/Schwegler 2, 1994; Cheers 1998; Erhardt et al. 2002; Fleischhauer 2003; Hager 4, 1992; Heeger 1956; Hiller/Melzig 1999; Hoppe 1949; Schnelle 1999; Schönfelder 2001

 Asperula odorata L.

▶ *Galium odoratum (L.) Scop.*

ASPLENIUM L. – Nest Fern - Aspleniaceae

 Asplenium acrobryum *Christ*

Common Names ▶ New Guinea salt fern; *German:* New Guinea Salzfarn
Usage ▶ condiment, spice
Parts Used ▶ plant ash
Distribution ▶ In inland New Guinea
Note ▶ A source of salt.

Croft/Leach 1985

 Aster helenium *Scop.*

▶ *Inula helenium L.*

 Aster officinalis *All.*

▶ *Inula helenium L.*

ASYSTASIA Bl. - Acanthaceae

 Asystasia coromandeliana *Wight ex Nees*

▶ *Astysia gangetica (L.) T. Anders*

 Asystasia gangetica *(L.) T. Anderson*

Synonyms ▶ *Asystasia coromandeliana* Wight ex Nees, *Justicia gangetica* L.
Common Names ▶ ganges river asystasia, asystasia, Chinese violet; *Congo:* ondoko, ondo; *Pilipino:* asistania zambo angenita
Usage ▶ pot-herb (in India)
Parts Used ▶ leaf
Distribution ▶ India, Orissa, W Bengal; also cultivated, tropical Africa, tropical America cultivated as a medicinal plant
Note ▶ In Kenya, Tanzania and W Africa grown as a traditional leaf vegetable.

Arora/Pandey 1996; Hanelt 2001; Wiersema/León 1999

 Asystasia neesiana *Nees*

Synonyms ▶ *Asystasiella neesiana* Lindau
Usage ▶ pot-herb
Parts Used ▶ leaf
Distribution ▶ India: NE region

Arora/Pandey 1996

 Asystasiella neesiana *Lindau*

▶ *Asystasia neesiana Nees*

 Atalandia hindsii *Oliv. (Champ.)*

▶ *Fortunella hindsii (Champ. ex Benth.) Swingle*

 Atalantia polyandra *Ridl.*

▶ *Fortunella polyandra (Ridl.) Tanaka*

 Athamantha chinensis *Lour.*

▶ *Cnidium monnierieri (L.) Cuss. ex Juss.*

ATHEROSPERMA Lab. - Monimiaceae

 Atherosperma moschatum *Lab.*

Common Names ▶ plum nutmeg; *German:* Pflaumenmuskat
Usage ▶ spice, flavoring
Parts Used ▶ seed (bark)
Distribution ▶ Australia, Tasmania

Hoppe 1949; Mabberly 1997; Teuscher 2003

Athroandra atrovirens (Pax) Pax et Hoffm.

> *Erythrococca atrovierens (Pax) Prain*

ATRIPLEX L. - Orach, Saltbush - Chenopodiaceae

 Atriplex canescens (Pursh) Nutt.

Synonyms ▸ *Atriplex nuttallii* S. Wats., *A. occidentalis* Dietr.; *Calligonum canescens* Pursh, *Obione canescens* (Pursh) Moq.
Common Names ▸ hoary salt bush, fourwing salt bush, shad scale, wing scale; *German:* Salzmelde; *Mexico:* chamere, chamizo, laza, nocuana
Usage ▸ condiment; **product:** plant ash
Parts Used ▸ herb
Distribution ▸ W America to Mexico
Note ▸ The plant salt is a substitute for normal salt.

Davidson 1999; Hanelt 2001; Uphof 1968

 Atriplex loccidentalis Dietr.

> *Atriplex canescens (Pursh) Nutt.*

Atriplex nuttallii S. Wats.

> *Atriplex canescens (Pursh) Nutt.*

AVERRHOA L. - Oxalidaceae

 Averrhoa bilimbi L.

Synonyms ▸ *Averrhoa obtusangula* Stokes, *Blimbigum teres* Rumph.
Common Names ▸ bilimbi, cucumber tree, tree sorrel; *Brazil (Portuguese):* lunao de caiena; *Chinese:* san nien; *French:* zibeline, bilimbi; *German:* Bilimbi, Gurkenbaum; *India:* bilim, bilimbi, bimblee, tamarang; *Indonesian:* belimbing wuluh; *Malaysian:* belimbing asam; *Pilipino:* balambing, camias kalanias, kilingiba; *Portuguese:* limão de caiena; *Spanish:* bilimbí, grosella China, mimbro, vinagrillo; *Thai:* taling pling;
Usage ▸ spice (cooked) for pickles and sambals or fish curries, relish
Parts Used ▸ fruit
Distribution ▸ origin unknown; widely cultivated in the tropics, specially in India and Malaysia

Blancke 2000; Burkill 4, 1997; Cheers 1998; Davidson 1999; Hanelt 2001; Hiller/Melzig 1999; Lück 2000; Schultze-Motel 1986; Täufel et al. 1993; Wiersema/León 1999; Zeven/de Wet 1982

 Averrhoa obtusangula Stokes

> *Averrhoa bilimbi L.*

B

 Baccharoides anthelmintica *(L.) Moench*

◉ *Vernonia anthelmintica (L.) Willd.*

BACKHOUSIA Hook et Harvey - Myrtaceae

 Backhousia citriodora *F. v. Muell.*

Common Names ► Australian lemon myrtle, citron backhousia, citron myrtle, lemon scented myrtle, scrub myrtle, native myrtle, sweet verbene myrtle; *French:* myrte citronée; *German:* Australische Zitronenmyrte; *Italian:* mirto dal ptofumi di limone; *Portuguese:* limón myrto; *Spanish:* limon myrto

Usage ► flavoring; **product:** essential oil and citral

Parts Used ► leaf, fresh and dried as pot-herb and as adjunct in food

Distribution ► Australia: Queensland, cultivated since 1990.

Note ► The oil resembles lemon grass oil.

Arctander 1960; Groom 1997; Hayes/Markovic 2002; Oyen/Dung 1999; Uphof 1968; Wilkinson et al. 2003

 Badianifera officinarum *Kuntze*

◉ *Illicium verum Hook. f.*

BALANITES Delile - Balanitaceae (Zygophyllaceae)

 Balanites aegyptiaca *(L.) Del.*

Synonyms ► *Ximenia aegyptica* L.

Common Names ► desert date, Egyptian balsam, Egyptan myrobalane, soap berry; *Arabic:* balah harara, betu, heglig, teborak zachun; *French:* balanites, dattier du désert, myrobalan d'Egypt; *German:* Ägyptischer Zahnbaum, Zachunbaum; *Hindi:* hingan; *Sanskrit:* ingudi

Usage ► spice for souce, seasoning

Parts Used ► leaf, shoot

Distribution ► Palestine, Arabia, tropical Africa: Ethiopia and Angola; cultivated in Egypt

Ayensu 1978; Burkill 1, 1985; Cufodontis 1957; Erhardt et al. 2002; Hanelt 2001; Hepper 1992; Schultze-Motel 1986; Täufel et al. 1993; Terra 1967; Uphof 1968; Wealth of India 1, 1948; Zeven/de Wet 1982; Zeven/Zhukovsky 1975

 Ballota suaveolens *L.*

◉ *Hyptis suaveolens L.*

 Balsamita major *Desf.*

◉ *Tanacetum balsamite L.*

Balsamodendron gileadense *Kunth*

▸ *Commiphora opobalsam* (L.) Engl.

BARBAREA R.Br. - Barbara herb - Brassicacea (Cruciferae)

 Barbarea praecox *(R.Br.) Dul.*

▸ *Barbarea verna* (Mill.) Aschers.

 Barbarea verna *(Mill.) Aschers.*

Synonyms ▸ *Barbarea praecox* (R.Br.) Dul., *Erysimum vernum* Mill.
Common Names ▸ American cress, early yellow rocket, Bello-Isle cress, land cress, upland Cress; *French:* cresson de terre, roquette des jardins; *German:* Barbarakraut, Frühe Winterkresse; *Italian:* barbarea; *Spanish:* oruga del jardin
Usage ▸ pot-herb
Parts Used ▸ fresh herb
Distribution ▸ W, SW Europe: France, England, also cultivated

Fleischhauer 2003; Hanelt 2001; Koschtschejew 1990; Loch 1993; Mansfeld 1962; Schnelle 1999; Schultze-Motel 1986; Small 1997; Zeven/de Wet 1982

 Barbarea vulgaris *R.Br.*

Synonyms ▸ *Erysimum barbarea* L., *Sisymbrium barbarea* (L.) Cr.
Common Names ▸ bitter cress, common winter cress, rocket or upland cress, winter cress, yellow rocket; *French:* barbarée herbe, cresson de jardin, herbe de Sainte Barbe; herbe aux charpentiers; *German:* Barbenkraut, Gewöhnliches Barbarakraut, Winterkresse; *Italian:* barbarea, erba Santa Barbara; *Portuguese:* agrião de horta; *Russian:* surepiza, surepka obyknowennaja; *Spanish:* berrillo, berro, mastuerzo, hierba de Santa Bárbara
Usage ▸ pot-herb
Parts Used ▸ fresh herb
Distribution ▸ Europe, Turkey, Levante, Caucasus, Iran, W and E Siberia, Himalayas, Tibet, Mongolia, China: Sinkiang; Tunesia, Algeria; naturalized in S Africa, N America, Australia, New Zealand

Aichele/Schwegler 3, 1995; Bilgri/Adam 2000; Cheers 1998; Erhardt et al. 2002; Fleischhauer 2003; Hanelt 2001; Hiller/Melzig 1999; Koschtschenjew 1990; Körber-Grohne 1989; Loch 1993; Lück 2004; Rust 2003; Schnelle 1999; Schultze-Motel 1986; Senatore et al. 2000; Small 1997; Täufel et al. 1993; Teuscher 2003; Uphof 1968; Wiersema/León 1999; Zeven/de Wet 1982

BAROSMA Willd. - Rutaceae

 Barosma betulina *(Bergius) Bartl. et Wendl.f.*

Synonyms ▸ *Agathosma betulina* Pillans
Common Names ▸ buchu, round-leaf buchu (S Africa); *French:* buc hu; *German:* Birkenblättriger Bukkostrauch, Bucco; *Italian:* buchu
Usage ▸ spice, flavoring; **product:** essential oil
Parts Used ▸ leaf
Distribution ▸ S Africa; also cultivated
Note ▸ In Tunesia the leaves are used for the prepartion of brandy and the leaves oil in America for flavoring baked goods, sweets and spices. In 1821 the drug was introduced into England.

Berger 2, 1950; Hager 4, 1992; Hiller/Melzig; Hoppe 1949; Morton 1976; Pschyrembel 1998; Schultze-Motel 1986; Seidemann 1993c; Täufel et al. 1993; Teuscher 2003; Uphof 1968; Wyk et al. 2004; Zeven/de Wet 1982

 Barosma crenulata *(L.) Hook.*

Synonyms ▸ *Agathosma crenulata* Pillans; *Diosma odorata* (Wendl.) DC.
Common Names ▸ oval buchu; short buchu, buch (S Africa); *German:* Gekerbtblättriger Bukkostrauch
Usage ▸ spice; **product:** essential oil
Parts Used ▸ leaf
Distribution ▸ S Africa; also cultivated
Note ▸ The leaves and the essential oil are utilized like those of *Barosma betulina* (Bergius) Bartl. et Wendl.

Berger 2, 1950; Fuchs et al. 2001; Hager 4, 1992; Hanelt 2001; Hiller/Melzig; Morton 1976; Rust 2003; chultze-Motel 1986; Seidemann 1993c; Täufel et al. 1993; Uphof 1968

Barosma serratifolia *(Curtis) Willd.*

Synonyms ▸ *Agathosma serratifolia* (Curtis) Spreeth, *Barosma serratifolia* Roem. et Schult.
Common Names ▸ long buchu, buchu; *German:* Gesägtblättriger Bukkostrauch
Usage ▸ spice; **product:** essential oil
Parts Used ▸ leaf
Distribution ▸ S Africa, also cultivated
Note ▸ The leaves and the oil are utilized like those of *Barosma betulina* (Bergius) Bartl. et Wendl.

Berger 2, 1950; Hager 4, 1992; Hanelt 2001; Hiller/Melzig 1999; Rust 2003; Sánchez-Monge/Parellada 1981; Schultze-Motel 1986; Seidemann 1993c; Uphof 1968

Barosma serratifolia *Roem. et Schult.*

▸ *Barosma serratifolia (Curtis) Willd.*

BARRINGTONIA J.R. Forst. et G. Forst. - Lecythidaceae

Barringtonia scortechninii *King*

Common Names ▸ *Malaysian:* putat gajah, putat hutan, putat tuba; *Borneo:* tempalang, langsat burung
Usage ▸ flavoring
Parts Used ▸ fruit
Distribution ▸ Thailand, Malaysia, Borneo, Sumatra
Note ▸ Bark and pounded seeds are used as a fish poison.

Guzman/Siemonsma 1999; Uphof 1968

Barosma serratifolia *Roem. et Schult.*

▸ *Barosma serratifolia (Curtis) Willd.*

Baryosma tonga *Gaertn.*

▸ *Dipteryx odorata (Aubl. Willd.*

BASELLA L. - Basellaceae

Basella alba *L.*

Synonyms ▸ *Basella rubra* L.
Common Names ▸ Indian or Ceylon spinach, Malabar nightshade; *Chinese:* chuan ts'ai, lo-kuei; *French:* épinard de Malabar, frède de Angola, baselle; *German:* Indischer Spinat, Malabar Spinat; *Hindi:* poi; lalbachlu; *Indonesian:* gendola; *India:* poi; *Korean:* rakkyu; *Portuguese:* bacela espinaca blanca, espinaca de Malabar; *Russian:* malabarskij schpinat; *Spanish:* espinaca de Ceilán
Usage ▸ pot-herb (in India)
Parts Used ▸ tender stem, leaf
Distribution ▸ E S India, Sri Lanka; Africa, native in the Neotropics

Arora/Pandey 1996; Bendel 2002; Erhardt et al. 2000, 2002; Hanelt 2001; Schultze-Motel 1986; Wiersema/León 1999

Basella rubra *L.*

▸ *Basella alba L.*

BASILICUM Moench - Lamiaceae (Labiatae)

Basilicum citratum *Rumph*

▸ *Ocimum basilicum L.*

Basilicum polystachon *(L.) Moench*

Synonyms ▸ *Moschosma polystachon* (L.) Benth.
Common Names ▸ African curry powder, polystachyous basilicum; *Chinese:* xiao guan xun; *German:* Vielähriger Basilikum

Usage ▶ spice
Parts Used ▶ herb
Distribution ▶ W Africa: Ghana, Nigeria

Burkill 3, 1995; Dalziel 1957; Irvine 1948

BASSIA All. - Chenopodiaceae

 Bassia scoparia (L.) A.J. Scott

Synonyms ▶ *Kochia scoparia* (L.) Schrader
Common Names ▶ summer cypres, burning bush, belvedere; *Chinese:* sao chou tsao, ti fu; *German:* Sommer-Zypresse; *Japanese:* hôki-gi
Usage ▶ pot-herb
Parts Used ▶ leaf
Distribution ▶ W, E Siberia, Mongolia, China: Sinkiang, Japan; N Africa, cultivated in China and Japan

Erhardt et al 2002; Rätsch 1998; Zeven/de Wet 1982

BAUHINIA L. - Caesalpiniaceae (Leguminosae)

 Bauhinia malabarica Roxb.

Common Names ▶ Malabar-Bauhinia, Malabar orchid; *German:* Malabar-Bauhinie; *India:* amli, kattra, sehara, shadloo; *Pilipino:* alibangbang, balibamban
Usage ▶ flavoring meat and fish *(Philippines)*
Parts Used ▶ leaf
Distribution ▶ tropical Asia, especially the Philippines

Uphof 1968

 Bauhinia purpurea L.

Common Names ▶ orchid tree, butterfly tree, camel's foot tree purple, Bauhinia; *German:* Purpurne Bauhinie, Schmetterlings-Bauhinie; *Hindi:* kachnar, keolar, khairwal; *Indonesian:* bunga kupu-kupu; *Japanese:* murasaki soshin ka; *Malaysian:* tapak kud; *Pilipino:* alibangbang; *Sri Lanka:* kolar; *Thai:* chong kho; *Vietnamese:* móng bò hoa tíma
Usage ▶ pot-herb, condiment, admixture curries or pickles
Parts Used ▶ flowers, floral bud
Distribution ▶ India, China, Myanmar, rarely cultivated

Arora/Pandey 1996; Burkill 3, 1994; Cheers 1998; Chopra et al. 1956; Dastur 1954; Engel/Phummai 2000; Hanelt 2001; Kumar 2003; Schultze-Motel 1986; Storrs 1997; Uphof 1968; Wealth of India 1, 1948; Zeven/de Wet 1982

BEGONIA L. - Begonia - Begoniaceae

 Begonia tuberosa Lam.

Common Names ▶ *German:* Knollenbegonie
Usage ▶ as sorrel, also cooked with fish and in sauce (of the Moluccas)
Parts Used ▶ leaf
Distribution ▶ Molluccas; worldwide cultivated in numerous hybrids

Cheers 1998; Uphof 1968

BEILSCHMIEDIA Nees - Lauraceae

 Beilschmiedia madagascariensis (Baill.) Kosterm.

Common Names ▶ *German:* Madagaskar-Beilschmiedie
Usage ▶ spice (limited)
Parts Used ▶ leaf
Distribution ▶ Madagascar, S Africa

Burkill 3, 1995; Neuwinger 1999

 Beilschmiedia mannii (Meisn.) Benth. et Hook f.

Synonyms ▶ *Tylostemon mannii* Staph.

Common Names ► cedar, spicy cedar; *Nigerian:* gbako nisa
Usage ► flavoring (for rice)
Parts Used ► flower, leaf
Distribution ► W Africa: Guinea, Ivory Coast, Liberia, Nigeria, Sierra Leone

Adegoke et al. 1968; Burkill 3, 1995; Dalziel 1957

Benzoin aestivale *(L.) Nees*

▶ *Lindera benzoin (L.) Bl.*

Bergora koenigii *L.*

▶ *Murraya koenigii (L.) Spreng.*

BETONICA L. - Betony - Lamiaceae (Labiatae)

Betonica officinalis *L.*

Synonyms ► *Stachys officinalis* (L.) Trevis.
Common Names ► betony, Bishop's word, woody betony; *Chinese:* yao shui su; *French:* bétoine, bétoine officinale; *German:* Echter Ziest, Flohblume, Gemeine Betonie, Heilziest; *Italian:* betonica; *Russian:* bukwiza lekarstwennaja; *Spanish:* betónica oficiale
Usage ► spice (very rarely)
Parts Used ► herb
Distribution ► Europe, Turkey, Caucasus, N Iran, C Asia, NW Africa

Aichele/Schwegler 4, 1995; Berger 2, 1950; Berger 4, 1954; Fleischhauer 2003; Hiller/Melzig 2000; Hoppe 1949; Jeker et al. 1989; Schönfelder 2001; Schultze-Motel 1986; Seidemann 1993c; Small 1997; Wyk et al. 2004

BIDENS L. - Bur-Marigold - Asteraceae (Compositae)

Bidens bipinnata *L.*

Common Names ► black jack, marygold, Spanish needles, sweet hearts; *Chinese:* gui zhan cao; *German:* Fiederblättriger Zweizahn
Usage ► pot-herb (in India)
Parts Used ► fleshy shoot
Distribution ► India, USA: California; S, C Europe; worldwide weed, propable origin the Neotropics and Subtropics

Arora/Pandey 1996; Erhardt et al 2000; Wiersema/León 1999

BIFORA Hoffm. - Apiaceae (Umbelliferae)

Bifora radians *M. Bieb.*

Common Names ► *German:* Strahlen-Hohlsame; *Russian:* dwojtschatka
Usage ► spice (rarely)
Parts Used ► root
Distribution ► S, C Europe, Turkey, Caucasus, Iran; Germany: Thuringia

Dudtschenk et al. 1989; Erhardt et al. 2002; Fleischhauer 2003; Loch 1993; Rothmaler 1987

BIXA L. - Bixaceae

Bixa orellana *L.*

Synonyms ► *Bixa urucurana* Willd.; *Orellana americana* L.
Common Names ► an(n)atto, atsuete, chiote tree, lipstick tree; *Brazil (Portuguese):* açafroa, açafroeira-da-terra, urucu, urucum; *French:* rocouyer; *German:* Achote, Achiote, An(n)otta, Baja, Orleansamen, Roucon-Samen, Rukusamen; *Hindi:* latkan, rangamali, sedri, thidin; *Indonesian:* kesumba, kes-

◧ Bixa orellana, flowering and fruiting

umba keling, pacar keling; *Malaysian:* jarak belanda, kesumba; *Pilipino:* atsuete, chuete, annato; *Russian:* biksa; *Spanish:* achiote, bija, bixa; *Thai:* kham-ngo, kham saet

Usage ▶ spice, dye stuff
Parts Used ▶ seed
Distribution ▶ Amazonia, Mexico, C America, West Indies, trop S America; native in tropical Africa, tropical Asia, Australia: Queensland

Aedo et al. 2001; Anderson Simon et al. 1997; Baer 1977; Bahl et al. 1971; Barnicoat 1950; Bhalkar/Dubash 1983; Cheers 1998; Davidson 1999; Farrell 1985; Galindo-Cuspinera et al. 2002; Dastur 1954; Degnan et al. 1991; Duke et al. 2003; Engel/Phummai 2000; Erhardt et al. 2002; Ferrão 1992; Hanelt 2001; Hiller/Melzig 1999; Hoppe 1949; Ingram/Francis 1969; Jondiko/Pattenden 1989; Lewington 1990; Magiatis et al. 1999; Morris/Mackley 1999; Mors et al. 2002; Ohler 1968; Preston/Rickard 1980; Rath et al. 1990; Satyanarayana et al. 2003; Schultze-Motel 1986; Scotter et al. 1998; Seidemann 1993c; Shuhama et al. 2003; Srinivasulu/Mahapatra 1982, 1989; Srivastava et al. 1999; Strorrs 1997; Täufel et al. 1993; Teuscher 2003; Uhl 2000; Uphof 1968; Villamara et al. 1994; Warren 1998; Wealth of India 2B, 1988; Zeven/de Wet 1982

Bixa urucurana *Willd.*

▶ *Bixa orellana* L.

BLIGHIA Koenig - Sapindaceae

Blighia welwitscha *(Hiern.) Radlk.*

Common Names ▶ *Cameroon:* onkom, toko; *Nigerian:* nofua, ukpe
Usage ▶ flavoring (for soups in Liberia)
Parts Used ▶ young leaves
Distribution ▶ W Africa: Ghana, Ivory Coast, Nigeria, Sierra Leone

Burkill 5, 2000; Irvine 1961; Uphof 1968

Blimbigum teres *Rumph.*

▶ *Averrhoa bilimbi* L.

BLUMEA DC. - Asteraceae (Compositae)

Blumea chinensis *DC.*

Synonyms ▶ *Blumea riparia* DC., *Conyza riparia* Bl.
Common Names ▶ tombak-tombak, djonge areuj
Usage ▶ flavoring
Parts Used ▶ leaf
Distribution ▶ SE Asia, Malaysia, Indonesia
Note ▶ The succulent leaves have a mustard taste.

Uphof 1968

Blumea lanceolaria *(Roxb.) Druce*

Synonyms ▶ *Blumea myriocephala* DC.
Common Names ▶ lanceolated buffalo ear; *German:* Lanzettliche Blumée, Lanzettliches Büffelohr; *Indonesian:* xu ong song, xang song
Usage ▶ spice for fish, condiment

Parts Used ▸ leaf

Distribution ▸ India, Himalayas, S China, Myanmar, Taiwan, Malaysia, Philippines, cultivated in Vietnam and Malaccas

Bois 1934; Schultze-Motel 1986; Seidemann 1993c; Tucker 1986; WHO 1990

 Blumea myriocephala *DC.*

▸ *Blumea lanceolaria (Roxb.) Druce*

 Blumea riparia *DC.*

▸ *Blumea chinensis DC.*

BOERHAVIA L. - Nyctaginaceae

 Boerhavia erecta *L.*

Common Names ▸ erect boerhavia, erect spiderling, hogweed; *German:* Aufrechte Boerhavie

Usage ▸ pot-herb (sporadically)

Parts Used ▸ fresh leaf

Distribution ▸ C and N Mexico, Caribbean, Meso-America

Burkill 4, 1997; Diarra 1977; Wiersema/León 1999

BOESENBERGIA KUNTZE - Zingiberaceae

 Boesenbergia pandurata *(Roxb.) Schlechter*

▸ *Boesenbergia rotunda (L.) Mansf.*

 Boesenbergia rotunda *(L.) Mansf.*

Synonyms ▸ *Boesenbergia pandurata* (Roxb.) Schlechter, *Curcuma rotunda* L., *Kaempferia ovata* Rosc., *Kaempferia pandurata* Roxb., *Zingiber xanthorhizum* Moon, *Gastrochilus pandurata* (Roxb.) Ridley

Common Names ▸ galingale, Chinese key, lesser ginger, tropical crocus; *Chinese:* soh shi; *French:* petits doigts; *German:* Runde Gewürzlilie, Fingerwurzel; *India:* chekkur, *Indonesian/Javanese:* temu kunchi; *Malaysian:* tamoe koentji, toemae kontji, temu kunchi; *Thai:* kae-aen, kra chai, wan-phraa-thit, krachai; *Vietnamese:* ngai num kho, bong nga truat, cu ngai

Usage ▸ spice, condiment; often also leaves and sprouts

Parts Used ▸ rhizome and root

Distribution ▸ Indian, Sri Lanka, SE Asia: Thailand, Malaysia, Indochina, China, probably native to Java and Sumatra

Note ▸ The taste is mild like galanga.

Arora/Pandey 1996; Davidson 1999; Duke et al. 2003; Erhardt et al. 2002; Guzman/Siemonsma 1999; Hager 4, 1992; Herklots 1972; Hiller/Melzig 1999; Janran et al. 2001; Larsen et al. 1999; Lemmens/Bunyapraphatsara 2003; Schultze-Motel 1986; Seidemann 1993c; Sirirugsa 1992; Täufel et al. 1993; Wong 1999; Zeven/de Wet 1982

▪ Boesenbergia rotunda, flowering

Boldus boldus (Mol.) *Lyons*

▷ *Peumus boldus Molina*

BORAGO L. - Borage - Boraginaceae

 Borago laxiflora *Poir.*

▷ *Borago pygmaea (DC.) Chater ex Greuter*

 Borago officinalis *L.*

Common Names ▶ borage, talewort; *Arabic:* bou shenaf, bousassal, harsha; *Chinese:* bo li qu; *Dutch:* bernagie, borago; *French:* bourrache (officinale), bourroche; *German:* Bor(r)etsch, Blauhimmelstern, Borgelkraut, Gurkenkraut, Wohlgemut; *Italian:* bor(r)agine, borrana, buglossa vera; *Russian:* buratschnik, buratschnik lekarstwennyj; *Spanish:* borracha, borraja

Usage ▶ spice

Parts Used ▶ leaf, flores

Distribution ▶ Europe: Spain, France, Germany, Cyprus; Syria, Iran, Libya; native in C and E Europe, Turkey

Bärtels 1997; Beaubaire/Simon 1987; Berger 1, 1949; Bilgri/Adam 2000; Bremness 2001; Cheers 1998; Coiciu/Racz (no year); Davidson 1999; Dudtschenko et al. 1989; Erhardt et al. 2002; Fleischhauer 2003; Guzman/Siemonsma 1999; Hager 4, 1992; Hager 2001; Heeger 1956; Heron et al. 1995; Hiller/Melzig 1999; Hoffmann et al. 1992; Hoppe 1949; Marini 1989; Newall et al. 1996; Ochoa/Alonso 1996; Pschyrembel 1998; Schönfelder 2001; Schultze-Motel 1986; Seidemann 1993c; Seidemann/Siebert 1987; Sharma 2003; Small 1997; Täufel et al. 1993; Teuscher 2003; Uphof 1968; Villamar et al. 1994; Wyk et al. 2004; Zeven/de Wet 1982

 Borago pymaea *(DC.) Chater et Greuter*

Synonyms ▶ *Borago laxiflora* Poir.

Common Names ▶ dwarf borage; *German:* Ausdauernder Boretsch, Zwerg-Boretsch

Usage ▶ spice

Parts Used ▶ herb

Distribution ▶ Corsica, Sardinia

Borago officinalis: a **flowering,** b **separate flower**

Cheers 1997; Erhardt et al. 2002; Hager 4, 1992

BORONIA Sm. - Rutaceae

 Boronia megastigma *Nees ex Bart.*

Common Names ▶ boronia, Melbourne boronia, scented boronia, brown boronia; *German:* Großnarbige Boronie, Duftende Korallenraute

Usage ▶ spice, flavoring (rarely); **product:** essential oil (oil of Boronia)

Parts Used ▶ flower

Distribution ▶ W Australia; near Melbourne also cultivated

Bourton 1968; Bussell et al. 1995; Charalambous 1994; Cheers 1998; Davies/Menary 1984; Erhardt et al. 2000, 2002; Hanelt 2001; Morton 1976; Porsch 1906; Schultze-Motel 1986; Täufel et al. 1993; Uphof 1968; Usher 1974

BOSWELLIA Roxb. ex Colebr. - Burseraceae

 Boswellia sacra *Flueck*

Common Names ▶ frankincense, bible frankincense, olibanum tree; *French:* arbre à encens; *German:* Weihrauch-Baum; *Italian:* incensi; *Spanish:* incienso
Usage ▶ flavoring, used in baking goods, candies, gelatines, ice creams, puddings and soft drinks
Parts Used ▶ resin
Distribution ▶ Egypt, Somalia, S Arabia

Duke et al. 2003; Erhardt et al. 2002; Rätsch 1998h; Wyk et al. 2004

 Boussingaultia cordeifolia *Ten.*

▶ *Anredera cordifolia (Ten.) van Steenis*

BRASSICA L. - Brassicaceae (Cruciferae)

 Brassica cernua *(Thunb.) Forb. et Hemsl.*

Synonyms ▶ *Sinapis cernua* Thunb.
Common Names ▶ *German:* Chinesischer Senf, Japanischer Senf
Usage ▶ spice
Parts Used ▶ seed
Distribution ▶ China, also cultivated

Hager 4, 1992; Heeger 1956; Melchior/Kastner 1974; Seidemann 1993c; Seidemann/Siebert 1987; Staesche 1972

 Brassica carinata *A. Br.*

Common Names ▶ Abessinian mustard; *German:* Abessinischer Senf
Usage ▶ pot-herb
Parts Used ▶ herb
Distribution ▶ Ethiopia

Small 1997; Uphof 1968

 Brassica eruca *L*

▶ *Eruca sativa Mill.*

 Brassica hirta *Moench*

▶ *Sinapis alba L.*

 Brassica integrifolia *(West) Rupr.*

▶ *Brassica juncea (L.) Czern.*

 Brassica juncea *(L.) Czern.*

Synonyms ▶ *Brassica integrifolia* (West) Rupr., *Sinapis juncea* L.
Common Names ▶ brown mustard, Chinese mustard, Indian mustard, leaf mustard, mustard green Sarepta mustard; *Chinese:* jie cai; *Dutch:* mosterd; *French:* moutarde à feuilles, moutarde de Chine, moutarde de Sarepte; *German:* Chinesischer Senf, Indischer Senf, Russischer Senf, Rutenkohl, Sareptasenf; *Hindi:* mohari, rai, rajika, rayan, sarson, sarsva; *Indonesian:* sesawi; *Italian:* senape indiana; *Korean:* kat; *Russian:* russkaja gortschiza, gortschiza sareptskaja, sisaja gortschiza; *Sanskrit:* asuri, rajika; *Spanish:* mostaza de Indias, repollo chino; *Vietnamese:* rau cai
Usage ▶ spice, condiment; **product:** mustard
Parts Used ▶ seed
Distribution ▶ E, W Siberia, N China, Mongolia, C Asia, native in Spain, Bulgaria, Romania, Russia, Turkey, Caucasus, Iran, Afghanistan, India
Note ▶ In Guinea the leaves are used as pot-herb and in Zaire for pepper condiment.

Aichele/Schwegler 3, 1995; Bois 1934; Bournot 1968; Bremness 2001; Burkill 1, 1985; Cheers 1998; Dalziel 1957; Davidson 1999; Dudtschenko et al. 1989; Farrell 1985; Fournier/Vangheesdaele 1980; Hager 4, 1992; Hanelt 2001; Heeger 1956; Hermey/Ludi 1994; Hiller/Melzig 1999; Leung 1991; Mazza 1998; Melchior/Kastner 1974; Morris/Mackley 1999; Neuwinger 1999; Norman 1991; Pochljobkin 1974, 1977; Pruthi 1976; Pursglove 1968; Rong Tsao et al. 2002; Schönfelder 2001; Schröder 1991; Schultze-Motel 1986; Seidemann 1993c; Seidemann/Siebert 1987; Shankaranarayana et al. 1972; Siemonesma/Piluek 1993; Small 1997; Staesche 1972; Tainter/Grenis 1993; Täufel et al. 1993; Teuscher 2003; Tucker 1986;

Uhl 2000; Uphof 1968; Vaugham/Gordon 1973; Vaugham/Hemingway 1959; Zeven/de Wet 1982

Brassica juncea *(L.)* ssp. integrifolia *O. Schulz*

Common Names ▸ *Chin:* gaai choy; *German:* Indischer Senf
Usage ▸ pot-herb
Parts Used ▸ leaf
Distribution ▸ N America; Asia: China

Davidson 1999

Brassica nigra *(L.) Koch*

Synonyms ▸ *Sinapis nigra* L.
Common Names ▸ black mustard, true mustard; *Arabic:* lisban, khardal; *Brazil (Portuguese):* mostarda preta; *Dutch:* swarte mosterd; *French:* moutarde noire, sénevé noire; *German:* Brauner, Senf, Französischer Senf, Roter Senf, Schwarzer Senf, Senfkohl; *Hindi:* benarasi rai; *Italian:* senape vera, senavra; *Pilipino:* mustasa; *Russian:* gortschiza tschernaja, nastojastschaja gortschiza, ranzusskaja gortschiza; *Slovenian:* horčica; *Spanish:* mostaza negra; *Turkish:* chordal
Usage ▸ spice, condiment; **product:** essential oil (mustard oil)
Parts Used ▸ seed
Distribution ▸ Asia minor: Iran; wild in the Mediterrean region, throughout C Europe, Middle East; cultivated in many countries, mainly in the garden
Note ▸ Soiling or falsification with 1. charlock, field mustard; *French:* moutarde sauvage, sanve, sénevé; *German:* Ackersenf, *Russian:* gortschiza polewaja, (*Sinapis arvensis* L.) 2. cabbage, rape; *French:* chou navet *German:* Raps; *Russian:* repa (*Brassica napus* L.) and 3. field rape, winter turnip rape; *German:* Rüben; *Russian:* turneps (*Brassica rapa* L.)

Aichele/Schwegler 3, 1995; Bois 1934; Bournot 1968; Cheers 1998; Clair 1961; Craze 2002; Coiciu (no year); Davidson 1999; Dudtschenko et al. 1989; Erhardt et al. 2002; Farrell 1985; Fleischhauer 2003; Guzman/Siemonsma 1999; Hager 4, 1992; Hanelt 2001; Heeger 1956; Hepper 1992; Hiller/Melzig 1999; Hohmann et al. 2001; Hondelmann 2002; Hoppe 1949; Leung 1991; Mazza 1998; Melchior/Kastner 1974; Morris/Mackley 1999; Norman 1991; Pruthi 1976; Pschyrembel 1998; Pursglove 1968; Roth/Korzmann 1997; Schönfelder 2001; Schröder 1991; Schultze-Motel 1986; Seidemann 1993c; Seidemann/Siebert 1987; Shankaranarayana et al. 1971; Siewek 1990; Small 1997; Staesche 1972; Täufel et al. 1993; Teuscher 2003; Tucker 1986; Uhl 2000; Uphof 1968; Zeven/et Wet 1982

Brindonia Indica *Thouars*

▸ *Garcinia indica (Thouars) Choisy*

BROSIMUM O. Sw. Cow tree, milk tree – Moraceae

Brosimum gaudichaudii *Trec.*

Common Names ▸ mama cadela
Usage ▸ flavoring of tobacco
Parts Used ▸ root
Distribution ▸ S America: Brazil

Kumar 2001

BRUCEA Mill. - Simaroubaceae

Brucea antidysenterica *Lam.*

Usage ▸ spice, flavoring
Parts Used ▸ bark
Distribution ▸ W Africa: W Cameroon, in Ethiopia also cultivated
Note ▸ The bark is a substitute for angustura bark and one of the constituents of "Angostura Bitter".

Burkill 5, 2000

BUNIUM L. - Apiaceae (Umbelliferae)

Bunium aromaticum *L.*

▸ *Trachyspermum ammi (L.) Sprague*

Bunium bulbocastanum L.

Synonyms ▶ *Apium bulbocastanum* Caruel, *Bunium bulbosum* Dulac, *Bunium majus* Vill., *Bunium majus* S.F. Gray, *Carum bulbocastanum* Koch, *Carvi bulbocastanum* Bub., *Pimpinella bulbocastanum* Jessen

Common Names ▶ earth chestnut, great earthnut or pignut; *French:* chateigne de terre, cumin tubéreux, dardilon, gernotte, moinson, noix de terre, terrenoix; *German:* Knollenkümmel, Erdeichel, Erdkastanie, *Italian:* bulbocastano, castagne di terra, terra noce; *Russian:* schischnik, schischetschnik

Usage ▶ spice (rarely)
Parts Used ▶ fruit, seed
Distribution ▶ SW Europe: Iberia, France, C Europe, Slovenia, native in Denmark

Aichele/Schwegler 3, 1995; Erhardt et al. 2002; Fleischhauer 2003; Hager 4, 1992; Hanelt 2001; Melchior/Kastner 1974; Schultze-Motel 1986; Seidemann 1993c; Sharma 2003; Small 1997; Täufel et al. 1993; Uphof 1968; Zeven/de Wet 1982

 Bunium bulbosum *Dulac*

▶ *Bunium bulbocastanum* L.

 Bunum carvi *Bieb.*

▶ *Carum carvi* L.

 Bunium majus *S.F. Gray*

▶ *Bunium bulbocastanum* L

 Bunium majus *Vill.*

▶ *Bunium bulbacastanum* L.

 Bunium copticum *Spreng.*

▶ *Trachyspermum ammi (L.) Sprague*

 Bunium persicum *(Boiss.) Fedch.*

Common Names ▶ black caraway, black cumin, black zira; *German:* Persischer Kreuzkümmel, Schwarzer Kreuzkümmel; *India:* siah zeera; *Persian:* zireh siyah
Usage ▶ spice
Parts Used ▶ fruit
Distribution ▶ Iran, Afghanistan, Pakistan, India

Dhar/Dhar 2000; Foroumadi et al. 2002; Hanelt 2000; Mansfeld 1986; Schultze-Motel 1986; Sharma 2003; Small 1997; Teuscher 2003

 Buphthalmum oleraceum *Lour.*

▶ *Eupatorium chinense* L.

BURASAIA Thouars - Menispermaceae

 Burasaia madagascariensis *DC.*

Common names ▶ *Madagascar:* ambora, amborasaha, odiandro, borasahr
Usage ▶ a source of an aromatic bitter used principally for the production of beer
Parts Used ▶ leaf
Distribution ▶ Madagascar

Uphof 1968

 Bursa pastoris *Weber*

▶ *Capsella bursa-pastoris* (L.) Medik.

C

 Cacalia porophyllum L.

▸ *Porophyllum ruderale (Jacq.) Cass.*

CAKILE Mill. – Sea Rocket – Brassicaceae (Crucuferae)

 Cakile edentula (Bigel.) Hook.

Common Names ▸ American sea rocket; *German:* Amerikanischer Meersenf
Usage ▸ pot-herb, for salads
Parts Used ▸ fresh leaf
Distribution ▸ along beaches of N America

Uphof 1968

 Cakile maritima Scop.

Common Names ▸ sea rocket; *French:* caquillier, roquette de mer; *German:* Europäischer Meersenf, Meersenf; *Russian:* gortschiza morskaja
Usage ▸ pot-herb, for salads
Parts Used ▸ fresh herb
Distribution ▸ coasts of Europe, N Africa, SW Asia, native in N America, Australia

Aichele/Schwegler 1995; Erhardt et al. 2002; Fleischhauer 2003; Schnelle 1999; Teubner 2001

CALAMINTHA Mill. - Calamint - Lamiaceae (Labiatae)

 Calamintha acinos (Clairv.) Man.

▸ *Acinos arvensis (Lam.) Dandy*

 Calamintha albiflora Van.

▸ *Nepeta cataria L.*

 Calamintha arvensis Lam.

▸ *Acinos arvensis (Lam.) Dandy*

 Calamintha clinopodium Benth.

▸ *Clinopodium vulgare L.*

 Calamintha cretica Benth.

Synonyms ▸ safüreja cretica *(L.) Brig.*
Common Names ▸ dwarf calamint; *Dutch:* bergsteentijm; *German:* Kreta-Bergminze
Usage ▸ spice, flavoring
Parts Used ▸ herb
Distribution ▸ Crete

Jahn/Schönfelder 1995; Türland et al. 1993

© Springer-Verlag Berlin Heidelberg 2005

Calamintha grandiflora *(L.) Moench*

Synonyms ▶ *Melissa grandiflora* L., *Satureja grandiflora* (L.) Scheele, *Thymus grandiflorus* (L.) Scop.
Common Names ▶ showy calamint, showy savory, large flowered calamint; *French:* calament à grande fleur, sarriette à grandes fleurs; *German:* Großblütige Bergminze, Großblütiger Steinquendel; *Italian:* mentuccio montana; *Russian:* duschevik krupnocvetkovyj
Usage ▶ spice (flavoring); **product:** essential oil
Parts Used ▶ herb
Distribution ▶ SW, W Europe: Iberia, France, Romania, Crimea, Caucasus, Anatolia, NW Iran

Aichele/Schwegler 4, 1995; Bärtels 1997; Başer/Özek 1993c; Carnat et al. 1991b; Cheers 1998; Erhardt et al. 2002; Hanelt 2001; Small 1997; Soulélès/Argyriadou 1990; Soulélès et al. 1991; Wiersema/León 1999

 ### Calamintha hederacea *(L.) Scop.*

▶ *Glechoma hederaceae L.*

 ### Calamintha menthifolia *Host*

Synonyms ▶ *Calamintha sylvatica* Bromf., *Satureja calamintha* (L.) Scheele, *Thymus calamintha* Scop.
Common Names ▶ calamint; *French:* sariette népéta; *German:* Waldbergminze, Aufsteigende Bergminze, Wald-Steinquendel; *Italian:* mentuccio maggiore, calamenta, nepitella; *Russian:* duschevik mjatolistnyj
Usage ▶ spice (flavoring); **product:** essential oil
Parts Used ▶ herb
Distribution ▶ W, S and S C Europe, Turkey, Crimea, Caucasus, Syria, N Iran, NW, Africa, Algeria

Başer/Özek 1993; Davidson 1999; Erhardt et al. 2002; Giannetto et al. 1979; Hanelt 2001; Hanlidou et al. 1991; Hidalgo et al. 2002; Romeo et al. 1980; Schultze-Motel 1986; Seidemann 1993c; Small 1997; Täufel et al. 1993; Tucker 1986

 ### Calamintha nepeta *(L.) Savi*

Synonyms ▶ *Calamintha nepetoides* Jord., *Calamintha officinalis* Moench, *Calamintha parviflora* Lam., *Melissa calamintha* L., *Melissa nepeta* L.
Common Names ▶ common calamint, lesser calamint; *French:* calament, menthe de montagne, sarriette népéta, *German:* Echte Bergminze, Drüsige Bergminze, Echter Steinquendel, Katzen-Bergthymian, Waldbergminze; *Italian:* calaminta; mentuccio comune; *Russian:* duschevik kotovnikovyj; *Spanish:* calamento
Usage ▶ spice (flavoring), mint-like, especially in Tuscany and Turkey
Parts Used ▶ herb
Distribution ▶ S, W and C Europe: SE England, Crimea, Caucasus, W Anatolia, native N America, cultivated in Europe and N America

Ahmed et al. 1975; Aichele/Schwegler 4, 1995; Berger 4, 1954; Cheers 1998; Dagab et al. 1983a, b; Erhardt et al. 2002; Fleischhauer 2003; Fraternale et al. 2001; Hager 4, 1992; Hanelt 2001; Hiller/Melzig 1999; Hoppe 1949; Kyong Hi Tcha et al. 1976; Mikus et al. 1997; Pagni et al. 1990; Pooter et al. 1987; Seidemann 1993c; Small 1997; Soulélès/Shammas 1985; Soulélès et al. 1987; Trucker 1986; Wiersema/León 1999

 ### Calamintha nepeta *(L.) Savi* **ssp. nepeta**

Common Names ▶ lesser calamint; *French:* menthe de montagne; *German:* Kleinblütige Katzenminze
Usage ▶ flavoring
Parts Used ▶ herb
Distribution ▶ E Europe, Turkey, Caucasus

Erhardt et al. 2002

 ### Calamintha nepetoides *Jord.*

▶ *Calamintha nepeta (L.) Savi*

 ### Calamintha officinalis *Moench*

▶ *Calamintha nepeta (L.) Savi*

Calamintha parviflora Lam.

▶ *Calamintha nepeta (L.) Savi*

 Calamintha sylvatica Bromf.

▶ *Calamintha menthifolia Host*

CALENDULA L. - Marigold - Asteraceae (Compositae)

 Calendula officinalis L.

Common Names ▶ pot marigold, marigold flower, calendula, ruddles; *Arabic:* ajamir, djoúmaira; *Chinese:* chin chan hua, jin zhan ju; *French:* fleur de souci (des jardins), souci officinal, fleur de tous les mois; *German:* Goldblume, Ringelblume, Ringelrose, Sonnenwende; *Italian:* calendola, calta, fior d'ogni, fiorrancio dei gardine; *Japanese:* tô kin sen ka; *Korean:* kŭmjanhwa; *Russian:* nogotki; *Spanish:* caléndula, flor de merte, maravilla

Usage ▶ spice (rarely); **product:** essential oil

Parts Used ▶ flower

Distribution ▶ Europe: Iberia, France, C Europe, Hungary, Romania, W Russia, Crime, Caucasus, Turkey, Iran, Afghanistan, NW Africa

Note ▶ Floral leaves used as falsification of saffron.

Bärtels 1997; Berger 1, 1949; Bois 1934; Bomme 1989; Bremness 2001; Cheers 1998; Clair 1961; Coiciu/Racz (no year); Davidson 1999; Dudtschenko et al 1989; Gracza 1987; Hager 4, 1992; Hanelt 2001; Heeger 1956; Heisig /Wichtl 1990; Hiller/Melzig 1999; Hoffmann et al. 1992; Hohmann et al. 2001; Hoppe 1949; Isaac 1992, 1994; Janssens/Vernooij 2001; Masterova et al. 1991; Meyer 2000; Ochoa/Aloso 1996; Pschyrembel 1998; Schmidt 1993; Schönfelder 2001; Schultze-Motel 1986; Seidemann 1993c; Small 1997; Staesche 1972; Täufel et al. 1993; Tucker 1986; Turova et al. 1987; Uphof 1968; Villamar et al. 1994; Willuhn 1987; Wyk et al. 2004; Zeven/de Wet 1982

 Calligonum canescens Pursh

▶ *Atriplex canescens (Pursh) Nutt.*

 Callopsima amplexifolium Mart.

▶ *Deianira nervosa Cham. et Schlecht.*

CALTHA L. - Rananculaceae

 Caltha cornuta Schott, Nyman et Kotschy

▶ *Caltha palustris L.*

 Caltha palustris L.

Synonyms ▶ *Caltha cornuta* Schott, Nyman et Kotschy

Common Names ▶ kingcup, marsh marigold, mayblob, meadow bright; *Brazil (Portuguese):* cabudula; *Chinese:* ma ti ye; *French:* caltha des marais, populage, souci d'eau; *German:* Butterblume, Dotterblume, Sumpfdotterblume; *Italian:* calta palustre, farferugine; *Russian:* kalushniza bolotnaja; *Spanish:* calta, hierba centella;

Usage ▶ condiment, substitute for capers

Parts Used ▶ flower-puds

Distribution ▶ Europe, Caucasian region, W, E Siberia, E Amur, Sachalin, Kamschatka, C Asia, Mongolia, China, Japan, Alaska, Canada, NC, NE, NCW, NW and SE USA

Aichele/Schwegler 2, 1994; Berger 1, 1949; Cheers 1998; Czygan 1999; Dudtschenko et al. 1989; Erhardt et al. 2002; Hager 4, 1992; Hanelt 2001; Hiller/Melzig 1999; Schönfelder 2001; Seidemann 1993c; Seidemann/Siebert 1987; Sharma 2003; Staesche 1972; Täufel et al. 1993; Teuscher 2003; Tucker 1986; Uphof 1968; Wiersema/León 1999

CALYCANTHUS L. - Spicebush - Calycanthaceae

 Calycanthus floridus L.

Synonyms ▶ *Calycanthus glaucus* Willd., *Calycanthus sterilis* Walter

Common Names ▶ Carolina allspice; *French:* arbre aux anémones; *German:* Echter Gewürzstrauch, Nelkenpfeffer

◘ **Calycanthus floridus, flowering**

Usage ▶ spice by the Indians of N America
Parts Used ▶ bark
Distribution ▶ USA: Florida, Virginia
Note ▶ Formerly the Indians used the flowers as a substitute for cinnamom and cloves.

Cheers 1998; Collin/Halim 1971; Erhardt et al. 2002; Hager 4, 1992; Hiller/Melzig 1999; Uphof 1968

 Calycanthus glaucus *Willd.*

❯ *Calycanthus floridus L.*

 Calycanthus sterilis *Walter*

❯ *Calycanthus floridus L.*

CALYPTRANTHES Sw. - Myrtaceae

 Calyptranthes aromatica *St. Hill.*

Common Names ▶ Brazil pimento; *Brazil (Portuguese)*: craveiro-da-terra; *German*: Brasilpiment
Usage ▶ spice, similar allspice or cloves; condiment (bark)
Parts Used ▶ fruit, bark
Distribution ▶ S America: Brazil

Mors et al. 2000; Overdieck 1989; Teuscher 2003; Uphof 1968

 Calyptranthes schiedeana *Berg.*

Usage ▶ spice (locally)
Parts Used ▶ leaf
Distribution ▶ Mexico

Uphof 1968; Wiersema/León 1999

 Calyptranthes variabilis *Berg*

Common Names ▶ *Brazil (Portuguese)*: craveiro-do-campo, cravo-do-campo
Usage ▶ flavoring (locally)
Parts Used ▶ plant
Distribution ▶ Brazil: Minas Gerais to Rio Grande

Mors et al. 2000

 Camara vulgaris *Benth.*

❯ *Lantana camara L.*

 Canagium odoratum *(Lam.) King*

❯ *Cananga odorata (Lam.) Hook.f. et Thomps.*

CANANGA (DC.) Hook.f. et Thomson - Annonaceae

 Cananga fruticosum *Craib.*

❯ *Cananga odorota (DC.) Hook.f. et Thomas*

 Cananga odorata *(DC.) Hook.f. et Thomps.*

Synonyms ▶ *Cananga fruticosum* Craib., *Canagium odoratum* (Lam.) Baill., *Uvaria odorata* Lam.
Common Names ▶ cananga, ilang-ilang; *Brazil (Portuguese)*: cananga, ilanga; *French*: canang odorant, ylang-ylang; *German*: Ilang-Ilang, Ylang-Ylang;

Indonesian: kananga; *Malaysian:* chenanga, kenaga, nyaix; *Myanmar:* sagasein

Usage ▸ flavoring (of confectionary goods and the like, and also coconut oil); **product:** essential oil (*Annonae aetheroleum*, oil of ylang-ylang, German: Cangaöl)

Parts Used ▸ flower

Distribution ▸ SE Asia: Malaysia, Philippines, N Australia, Pacific Islands, in tropical regions cultivated, e.g. Seychelles, Kenya, Angola, Tanzania

Anon. 2001; Bournot 1968; Burkill 1966; Charalambous 1944; Dastur 1951; Davidson 1999; Erhardt et al. 2002; Gurib-Fakim/Brendler 2004; Hunnius 1993; Lawrence 1999; Randriamiharisoa 1983; Roth/Kormann 1997; Schultze-Motel 1986; Sharma 2003; Shiva et al. 2002; Teisseire/Galfre 1974; Uphof 1968; Wealthof India 2, 1950; Zeven/de Wet 1982

CANARIUM L. - Chinese olive - Burseraceae

Canarium schweinfurthii *Engl.*

Common Names ▸ African elemi, incense tree, Papo Canary tree; *Cameroon:* abé, abel, bele, mbili, toumba; *French:* elemier d'Afrique; *Ghana:* bedi wunue

Usage ▸ condiment

Parts Used ▸ fruit

Distribution ▸ tropical Africa: Senegal to E Africa, Ethiopa, Angola, Zimbabwe

Burkill 1, 1965; Facciola 1990; Irvin 1961; Neuwinger 1998; Sawadogo et al. 1985; Uphof 1968; Wiersema/León 1999

CANAVALIA DC. - Jack bean - Fabaceae (Leguminosae)

Canavalia maritima *(Aubl.) Thou.*

Synonyms ▸ *Canavalia rosea* (Sw.) DC.

Common Names ▸ seaside sword bean, seaside jack bean; *French:* pois bord de mer, pois maritime; *German:* Meeres-Jackbohne, Schwertbohne; *Portuguese:* feijão, feijão bravo

Usage ▸ flavoring

Parts Used ▸ flower

Distribution ▸ India, Malaysia, Mexico Gulf, West Indies, W Africa

Burkill 3, 1995; Hanelt 2001; Zeven/de Wet 1982

Canavalia rosea *(Sw.) DC.*

▸ *Canavalia maritima (Aubl.) Thou*

CANELLA P. Browne - Canellaceae (Winteranaceae)

Canella alba *Murr.*

▸ *Canella winterana (L.) Gaertn.*

Canella winterana *(L.) Gaertn.*

Synonyms ▸ *Canella alba* Murr., *Winterana canella* L.

Common Names ▸ white caneel, white cinnamon, wild caneel, wild cinnamon; *French:* canella blanche; *German:* Kaneelinde, Weißer Kaneel, Weißer Zimt

Usage ▸ spice, flavoring liqueur and tobacco

Parts Used ▸ bark

Distribution ▸ West Indies, USA: S Florida

Note ▸ The bark is used as cinnamon.

Bois 1934; Brockmann 1979; Brown 1994; Hager 4, 1992; Hanelt 20001; Hiller/Melzig 1999; Hoppe 1949; McNavy Wood 2003; Rätsch 1992; Seidemann 1993c; Staesche 1972; Täufel et al. 1993; Uphof 1968; Wolters 1994; Wüstenfeld/Haensel 1964

CANNA L. - Canna Lily - Cannaceae

Canna coccinea *Mill.*

▸ *Canna indica L*

Canella winterana, flowering

índia, birú manso, merú; *Russian:* kanna indijkaja; *Spanish:* achera, achira, caña comestible, caña de las Indias

Usage ▶ spice

Parts Used ▶ rhizome

Distribution ▶ S America: Venezuela, Chile, Columbia, Brazil, West Indies, C America, native in USA, SE Asia, also cultivated in India, Philippines, tropical Africa: Sierra Leone, Gabon

Note ▶ In the literature *Canna edulis* Ker.-Gawl. treated as separate species the starch of the tubers is the Queensland arrowroot.

Erhardt et al. 2002; Hanelt 2001; Mansfeld 1962; Mors et al. 2000; Schultze-Motel 1986; Uphof 1968

 Canna edulis *Ker-Gawl.*

❯ *Canna indica L.*

 Canna indica *L.*

Synonyms ▶ *Canna coccinea* Mill., *Canna edulis* Ker-Gawl., Canna lutea Mill., *Canna ovatus* Moench, *Canna patens* (Ait.) Rosc.

Common Names ▶ edible canna, Indian canna, Indean bread shot, Indian shot, African turmeric, Queensland Arrowroot, wild tapioca; *Brazil (Portuguese):* achira; *Chinese:* mei ren jiao ger, yang di li; *Dutch:* eetbar bloemriet, Indisch bloemriet; *French:* balisier comestible, balisierou faux-sucrier, canna d'Inde; *German:* Indisches or Westindisches Arrowroot (starch); Indisches Blumenohr; *Italian:* canna commestibile canna dolce, canna d'India; *Japanese:* kanna dandoku, shokuyô kanna; *Malaysian:* kenyong, pisang sebiak, sebeh, ubi gereda, zemba; *Pilipino:* bendera Española, kolintaso, tikas tikas; *Portuguese:* canna comestível, canada-

 Canna lutea *Mill*

❯ *Canna indica L.*

 Canna ovatus *Moench*

❯ *Canna indica L.*

 Canna patens *Rosc.*

❯ *Canna indica L.*

CAPPARIS L. - Caper - Capparaceae

 Capparis aegyptica *Lam.*

❯ *Capparis spinosa L.*

 Capparis aphylla *Roth.*

❯ *Capparis decidua Edgew.*

Capparis atlantica *Inocencio, D. Rivera, Obón and Alcaraz*

Common Names ▶ *Arabic:* kabar, quabar, lassaf, shafellah
Usage ▶ condiment
Parts Used ▶ floral bud
Distribution ▶ Morocco

Rivera et al. et al. 2003

Capparis cartilaginea *Decne.*

▶ *Capparis inermis Forssk.*

Capparis cordifolia *Lam.* ssp. cordifolia

Synonyms ▶ *Capparis mariana* DC.
Usage ▶ condiment
Parts Used ▶ floral bud
Distribution ▶ Solomon Islands, Archipelago of Tuamotu, Cook Island, Fiji Island, Hawaii, Islands of Nauru, Rurutu; Philippines, Marianas

Rivera et al. 2003

Capparis cynophallophora *L.*

▶ *Capparis flexuosa L.*

Capparis decidua *(Forssk.) Edgew.*

Synonyms ▶ *Capparis aphylla* Roth.
Common Names ▶ siwak; *French:* câprier sans feuilles; *Hindi:* kair, ker, delha, kurrel
Usage ▶ condiment
Parts Used ▶ floral bud, fruit
Distribution ▶ Morocco, W Africa: Mauritiana, Guinea, Mali, Niger, Nigeria, Senegal, Sudan
Note ▶ Cultivated for the edible fruits.

Arora/Pandey 1996; Burkill 1, 1985; Fageria et al. 2003; Schultze-Motel 1986; Sushila 1987; Täufel et al. 1993; Teuscher 2003; Wealth of India 2, 1950

Capparis flexuosa *L.*

Synonyms ▶ *Capparis cynophallophora* L., *Morisonia flexuosa* L.
Common Names ▶ bay-leaved caper, caper tree; *German:* Jamaika Kaper; *Mexico:* xpayumak, pan y agua, burro, palo de burro
Usage ▶ condiment
Parts Used ▶ floral bud
Distribution ▶ West India, Mexico to southern USA
Note ▶ Cultivated for the edible fruits. The flavor of the fruits is horse radish.

Hanelt 2001; Sánchez-Monge/Parellada 1981; Uphof 1968

Capparis florantesii *DC.*

Synonyms ▶ *Capparis ovata* Desf.
Common Names ▶ *Arabic:* khabbar, soukoum
Usage ▶ condiment
Parts Used ▶ floral bud
Distribution ▶ N Africa: Algeria, Morocco, Chad

Rivera et al. 2003

Capparis inermis *Forssk.*

Synonyms ▶ *Capparis cartilaginea* Decne.
Common Names ▶ *Arabic:* lassaf, lattssaf; *India:* karat
Usage ▶ condiment
Parts Used ▶ floral bud
Distribution ▶ Egypt, Saudi Arabia, Kenya, Yemen, India

Rivera et al. 2003

Capparis herbacea *Willd.*

Synonyms ▶ *Capparis ovata* Desf. var. *herbacea* (Willd.) Zohary
Common Names ▶ *Armenian:* aggenko (fruit); *Russian:* kapersi; *Turkish:* keber
Usage ▶ condiment
Parts Used ▶ floral bud, fruit

Distribution ▶ Turkey, Armenia, Georgia, Azerbaidschan, Turkmenistan

Rivera et al. 2003

Capparis mariana DC.

▶ *Capparis cordifolia* Lam. ssp. *cordifolia*

Capparis napaulensis DC.

Synonyms ▶ *Capparis himalayensis* Jafri
Common Names ▶ *German:* Nepal-Kaper, Himalaya-Kaper; *India:* kabra, karil; *Pakistan:* kakri, kander, kabra
Usage ▶ condiment
Parts Used ▶ floral bud, fruit
Distribution ▶ India, Pakistan

Rivera et al. 2003

Capparis himalayensis Jafri

▶ *Capparis napaulensis* DC.

Capparis ovata M. Bieb.

▶ *Capparis spinosa* L.

Capparis obovata Royle

Common Names ▶ *India:* kabar, kabara; *Pakistan:* khwarg, pahinro kirap, panetro khafkhader
Usage ▶ condiment
Parts Used ▶ floral bud, unripe and ripe fruit
Distribution ▶ India, Pakistan

Rivera et al. 2003;

Capparis orientalis Duhamel

Common Names ▶ Eastern caper, oriental caper, *German:* Orientalische Kaper

Usage ▶ condiment
Parts Used ▶ floral bud
Distribution ▶ Mediterranean region: Spain, Italy, Greece, Portugal, Turkey and Morocco

Inocencio et al. 2000, 2002; Özcan 1999; Özcan/Akgül 1998

Capparis ovata Desf.

▶ *Capparis fontanesii* DC.

Capparis rupestris Sibth et Sm.

▶ *Capparis spinosa* L.

Capparis sicula Duhamel ssp. sicula

Common Names ▶ Sicili caper; *German:* Sizilianische Kaper; *Turkish:* kapari, kebere çiçegi (flower)
Usage ▶ condiment
Parts Used ▶ floral-bud, unripe fruit
Distribution ▶ Atlantic coasts of Europe and N Africa to the Mediterranean coasts, especially Sicily, Morocco; also cultivated

Inocencio et al. 2000, 2002; Rivera et al. 2003

Capparis spinosa L.

Synonyms ▶ *Capparis aegyptica* Lam., *Capparis ovata* M. Bieb., *Capparis rupestris* Sibth et Sm.
Common Names ▶ caper, common caper, spineless caper; *Arabic:* habbar, kabar, kalvari, lasafa; *Chinese:* ci shan gan; *Dutch:* kapper boom; *French:* câprier (commun); *German:* Kaper, Echte Kaper; *Hindi/India:* basari, bauri, kabbar, kebir, kakadani, kabra; *Italian:* càppero, càparo; *Portuguese:* alcaparra, alcaparreira; *Russian:* kaperzy; *Slovenian:* kapary; *Spanish:* alcaparna, acaparrón, tápana *Turkish:* kebéré
Usage ▶ condiment
Parts Used ▶ floral bud unripened fruit, caper berries; *French:* cornichons de câprier, *German:* Kapernfrucht, Kaperngurke
Distribution ▶ Mediterranean region to Caucasus, Turkestan, W Himalayas, also cultivated
Note ▶ Trade sorts: I. quality: 'Nonpareilles', 'Surfines',

◘ Capparis spinosa, flowering

'Capucines'; II quality: 'Capotes', 'Fines', 'Hors calibres'. The name of the great Italian trade sort: 'Capperoni'. The fruit (French: cornichons de câprier, cucunci) are also used. The plant of *Capparis flexuosa* L. is considered as the wild progenitor of the domesticated species. The name of unripened fruits is 'Cornichons de câprier', used in pickles.

Akgül 1996; Alvarruiz et al. 1990; Al-Said et al. 1988; Arora/Padney 1996; Barbera 1991; Bärtelös 1997; Berger 1, 1949; Bois 1934; Calis et al. 2002; Castro/Nosti 1987; Cheers 1998; Craze 2002; Davidson 1999; Duke et al. 2003; Farrell 1985; Gerhardt 1979; Giuffrida et al. 2002; Hanelt 2001; Heeger 1956; Hepper 1992; Hiller/Melzig 1999; Hondelmann 2002; Inocencio et al. 2000, 2002; Kjaer/Thomsen 1963; Melchior/Kastner 1974; Morris/Mackley 1999; Neuwinger 1999; Özcan/Akgül 1995, 1998, 2000, 2001; Pascual et al. 2003; Rodrigo et al. 1992; Sánchez et al. 1992; Schönfelder 2001; Schröder 1991; Schultze-Motel 1986; Seidemann 1993c; Seidemann/Siebert 1987; Sharma 2003; Siewek 1990; Small 1997; Staesche 1972; Täufel et al. 1993; Teuscher 2003; Tucker 1986; Uphof 1968; Vega/Ramos 1987; Zeven/de Wet 1982; Zohary 1960

 Capparis spinosa L. **var. inermis** *Turra*

Common Names ▶ caper spurge, spineless caper
Usage ▶ condiment
Parts Used ▶ floral bud
Distribution ▶ S Europe: Italy (Tuscany), N Africa

 Capparis spinosa L. **var. mariana**

Common Names ▶ Mariana caper; *German:* Mariannen-Kaper
Usage ▶ condiment
Parts Used ▶ floral bud
Distribution ▶ Mediterranean regions

Guzman/Siemonsma 1999

 Capparis spinosa L. **ssp. rupestris** *(Sm.) Nyman*

Common Names ▶ *French:* câprier; *German:* Dornenloser Kapernstrauch; *Italian:* capparo, cappari, chiappara; *Portuguese:* alcoparras, alcaparreira; *Spanish:* tàpara, alcaparra (flower buds), gorrinets (fruits)
Usage ▶ condiment
Parts Used ▶ floral bud
Distribution ▶ Mediterranean regions, N Africa; N India

Erhardt et al. 2002; Rivera et al. 2003

CAPSELLA Medik. – Shepherd's Purses - Brassicaceae (Cruciferae)

 Capsella bursa-pastoris *(L.) Medik.*

Synonyms ▶ *Bursa pastors* Weber, *Thlaspi bursa-pastoris* L.
Common Names ▶ capsell, farmer mustard, case weed, mother heart, shepherd's purse, paniquesillo, toy wort; *Chinese:* chi tsai, ji cai; *French:* bourse à pasteur, capselle, capselle à pasteur; *German:* Bauernsenf, Gemeines Hirtentäschel, Herzkraut, Täschel, Taschenkraut; *Italian* borsacchina, borsa del pastore; *Korean:* naeng-i; *Portuguese:* bolsadepastor, erra do bompastor; *Russian:* pastuschija sumka; *Spanish:* bolsa de pastor, zurrón de pastor; *Vietnamese:* co tam giac, dinh lich, te thai
Usage ▶ spice for soups and stews (rarely)
Parts Used ▶ pods, young leaves
Distribution ▶ Europe, origin Mediterranean area(?), today spread worldwide, except tropical areas; cultivated in China
Note ▶ The green pods have a peppery taste; different cultivars exist.

Dhar/Dhar 2000; Dudtschenko et al. 1989; Erhardt et al. 2002; Fleischhauer 2003; Hanelt 2001; Heeger 1956; Hoppe 1949; Koschtschejew 1990; Loch 1993; Schönfelder 2001; Schultze-Motel 1986; Tull 1999; Turova et al. 1987; Uphof 1968; Wyk et al. 2004; Zeven/de Wet 1982

CAPSICUM L. - Chillis, Paprika, Bird's Eye Chilli, Pepper - Solanaceae

Alkämper 1972; Andrews 1985, 1998; D'Arcy/Eshbaugh 1974; Bois 1934; Davidson 1999; Ernö 1985; Ferrão 1992; Ferrari/Aillaud 1971; Forster/Cordell 1992; Gordon-Smith 1996; Govindarajan 1985, 1986; Govindaraja et al. 1988; Govindarajan/Sathyanarayan 1990; Hager 4, 1992; Heiser/Pickersgill 1969; Heiser/Smith 1953; Herrmann 1999; Jurenitsch et al. 1979; Long-Solis 1986; Maga 1975; Newall et al. 1996; Oberdieck 1988; Palevitch/Craker 1995; Peter 2001; Rabinowitsch/Brewster 1999; Raunert 1939; Schönfelder 2001; Schratz/Rangoonwala 1966; Seidemann 1997a; Shin et al. 2001; Somos 1981; Somos/Kundt 1984; Terpó 1966; Teuner et al. 1993; Vaupel 2002 a, b; Zoschke 1997;

Capsicum annuum L. var. annuum

Common Names ▶ bell pepper, capsicum pepper, Cayenne pepper, cherry pepper, Chile pepper, chile (pepper), cone pepper, green capsicum, green pepper, paprika, red pepper, sweet pepper, Pandrón-pepper; *Arabic:* filfil ahmar; *Chinese:* la chiao; *Dutch* spaanse peper; *French:* piment doux, poivre de Cayenne, poivre d'Espagne, poivre rouge; *German:* Beißbeere, Einjährige Paprika, Cayenne Pfeffer, Spanischer or Ungarischer Pfeffer, Türkischer Pfeffer, Schotenpfeffer; *Italian:* peperone; *Hindi:* lalmica; *Indonesian:* cabe besar, lombok besar; *Japanese:* hsiug ya-li chiao; *Korean:* kochu; *Malaysian:* lada besar; *Portuguese:* pimento; *Russian:* perez strukovyj; *Sanskrit:* kutavira, raktamarica, suédois; *Slovenian:* ročná paprika; *Spanish:* ají, chile, guindilla, pimiento; *Thai:* pulang sili

Usage ▶ spice
Parts Used ▶ fruit
Distribution ▶ only cultivated form; Mexico; tropical America
Note ▶ This is the most widely spread and highly cultivated of all *Capsicum* species; probably first domesticated in Mexico. The very hot, peppery varieties, with high capsaicin content, are used as spices and seasoning; also in medicine. Fruits with small capsaicin con- tent are used as vegetables, cooked or raw. They are subdivided into the following sorts:

cerasiforme-group: cherry pepper; *German:* Zier-Paprika
conioides-group: cone pepper; *German:* cone pepper
fasciculatum-group: red cone pepper; *German:* Büschel-Paprika
grossum-group: bell pepper, pimento, sweet pepper; *German:* Gemüse- or Süßer Pfeffer
longum group: Cayenne pepper, Chili (pepper); *German:* Cayennepfeffer, Peperoni.

Aichele/Schwegler 4, 1995; Alberts/Muller 2000; Ananda Nayaki/Natarajan 2000; Bärtels 1997; Bendel 2002; Berger 3, 1952; Berta 2000; Bois 1934; Burkill 5, 2000; Buttery et al. 1969; Cheers 1998; Coiciu/Racz (no year); Craze 2002; Csiktusnádi et al. 2000; Dalby 2000; Dudtschenko et al. 1989; Erhardt et al. 2000, 2002; Estrada et al. 2000; Farrell 1985; Forgacs et al. 1996; Gnayfeed et al. 2001; Hager 4, 1992; Hanelt 2001; Heeger 1956; Hiller/Melzig 1999; Jansen 1981; Jurenitsch et al. 1979; Kirschbaum-Titze et al. 2002; Kobata et al. 1998; Krstic et al. 2001; Leung 1991; Lunig et al. 1995; Maoka et al. 2001; Melchior/Kastner 1974; Molnár et al. 2000; Morais et al. 2001; Morris/Mackley 1999; Norman 1991; Oruña-Conda et al. 1996; Perucka/Oleszek 2000; Peter 2001; Pochljobkin 1974, 1977; Pruthi 1976; Pursglove 1968; Rosengarten 1969; Rubio et al. 2002; Schröder 1991; Schultze-Motel 1986; Seidemann 1993c; Seidemann/Siebert 1987; Sekiwalijima et al. 2001; Sharma 2003; Siewek 1990; Simioan et al. 2004; Small 1997; Soo Hyun et al. 1997; Staesche 1972; Surk et al. 1996; Täufel et al. 1993; Terpó 1966; Teuscher 2003; Tindall 1983; Tucker 1986; Tull 1999; Villalón et al. 1994; Villamar et al. 1994; Wolters 1996; Yayeh Zewdie/Bosland 2000; Zeven/de Wet 1982

Capsicum annuum L. var. glabriusculum (Desf.) C.B. Heiser et Pickersgill

Synonyms ▶ *Capsicum minimum* Mill., *Capsicum annuum* L. var. *minimum* (Mill.) Heiser, *Capsicum hispidum* Dunal var. *glabriusculum* Dunal
Common Names ▶ American bird pepper, bird pepper; *German:* Vogelpfeffer; *India:* gach mirichi, lalmirichi, marcha, marichiphalam, perangimulik; *Japanese:* kôreigus, tô gara shi; *Pilipino:* sili; *Russian:* ptitschii perez, melkij perez, stolowij perez; *Spanish:* ají, chilipiquin, chiltepe, chiltepin
Usage ▶ spice
Parts Used ▶ fruit
Distribution ▶ wild form; cultivated: Mexico, USA, Caribbean, Meso-America, W, S America

Hager 4, 1992; Hanelt 2001; Pochljobkin 1974, 1977; Schultze-Motel 1986; Terpó 1966; Tucker 1986

Capsicum annuum L. **var. minimum** *(Mill.)* Heiser

▶ *Capsicum annuum L. var. glabriusculum (Dunal.) C.B. Heiser et Pickersgill*

 Capsicum baccatum L. **var. baccatum**

Synonyms ▶ *Capsicum microcarpum* Cav.
Common Names ▶ Peruvian pepper; *Spanish:* locoto
Usage ▶ spice
Parts Used ▶ fruit
Distribution ▶ wild form; S America: Peru, Bolivia, S Brazil, Paraguay, N Argentina

Eshbaugh W.H. 1970; Hager 4, 1992; Hanelt 2001; Melchior/Kastner 1974; Schultze-Motel 1986; Seidemann 1993c; Seidemann/Siebert 1987; Terpó 1966; Tucker 1986; Zeven/de Wet 1982

 Capsicum baccatum L. **var. pendulum** *(Willd.) Eshbaugh*

Synonyms ▶ *Capsicum pendulum* Willd.
Common Names ▶ Brown's pepper, Peruvian pepper; *French:* piment chien; *German:* Peruanischer Pfeffer; *Spanish:* ají, escabeche, piris, uchu (*Peruvian*)
Usage ▶ spice
Parts Used ▶ fruit
Distribution ▶ only cultivated, mostly in S America, also in C America, USA, Hawai, India, Japan, S Europe
Note ▶ Archaeological findings in the late precaramic (ca. 2500 BC) from Peru.

Dalby 2000; Eshbough 1970; Hanelt 2001; Jurenitsch et al. 1979; Melchior/Kastner 1974; Pruthi 1976; Schultze-Motel 1986; Small 1997; Terpó 1966; Tucker 1986

 Capsicum cardenasii *C.B. Heiser et P.G. Sm*

Usage ▶ spice, flavoring
Parts Used ▶ fruit
Distribution ▶ W, S America

Hager 4, 1992; Terpó 1966; Zeven/de Wet 1982

 Capsicum chinense *Jacq.*

Synonyms ▶ *Capsicum luteum* Lam.
Common Names ▶ bonnet pepper, Chinese pepper, datil pepper, Habañero (pepper), piri-piri pepper, scotch bonnet, squash pepper, tabasco pepper, yellow squash pepper; *Brazil (Portuguese):* piri-piri; *German:* Chinesische Paprika, Habanero, Tabasco-Pfeffer; *Spanish:* rocotillo
Usage ▶ spice
Parts Used ▶ fruit
Distribution ▶ W, S Amazonas: Brazil; cultivated; wild forms unknown; widely cultivated in the Neotropics
Note ▶ Archaeological remains from Peru suggest domestication 10,000 years ago.

Burkill 5, 2000; Hager 4, 1992; Hanelt 2001; Jurenitsch et al. 1979; Melchior/Kastner 1974; Schultze-Motel 1986; Seidemann 1993c; Small 1997; Smith/Heiser 1957; Terpó 1966; Tucker 1986; Zeven/de Wet 1982

 Capsicum fastigiatum *Bl.*

▶ *Capsicum frutescens L.*

 Capsicum frutescens *L.*

Synonyms ▶ *Capsicum fastigiatum* Bl., *Capsicum minimum* Roxb.,
Common Names ▶ Cayenne pepper, hot pepper, bird pepper, red chilli, spur pepper, tabasco (pepper); *Arabic:* schatta; *Dutch:* chili peper, spaanse peper; *French:* poivre de Cayenne, piment de cayenne, piment oiseau, poivre rouge; *German:* Cayennepfeffer, Chayenne, Chili(s), Tabasco (Pfeffer), Vogelpfeffer; *Indonesian:* cabe, cili, cili rawit, lombok; *Italian:* capsico, diavoletto; *Malaysian:* lada, cili, cili padi; *Portuguese:* pimenta malagueta; *Russian:* perez struzkovij; *Slovenian:* paprika čili; *Spanish:* ají, chile, fruto de cápsico, guindilla, pimienta de Cayena, pimienta picante; *Thai:* phrik kheefa
Usage ▶ spice
Parts Used ▶ fruit
Distribution ▶ tropical America: S Mexico to Costa Rica; India to Polynesia, worldwide in tropical and subtropical regions

CARDAMINE L. - Bitter Cress - Brassicaceae (Cruciferae)

Berger 3, 1952; Bois 1934; Burkill 5, 2000; Cheers 1998; Craze 2002; Cserháti et al. 2000; Davidson 1999; Dudtschenko et al. 1989; Duke et al. 2003; Farrell 1985; Hager 4, 1992; Hanelt 2001; Heeger 1956; Hiller/Melzig 1999; Hohmann et al. 2001; Howard et al. 1994; Leung 1991; Melchior/Kastner 1974; Morris/Mackley 1999; Norman 1991; Peter 2001; Pruthi 1976; Purseglove 1968; Rosengarten 1969; Schönfelder 2001; Schröder 1991; Schultze-Motel 1986; Seidemann 1993c; Seidemann/Siebert 1987; Small 1997; Staesche 1972; Täufel et al. 1993; Terpó 1966; Tindall 1983; Tucker 1986; Usher 1968; Villalón et al. 1994; Villamar et al. 1994; Wyk et al. 2004; Zeven/de Wet 1982

 Capsicum guatemalense *Bitter*

▶ *Capsicum pubescens* Ruiz et Pav.

 Capsicum hispidum *Dunal* **var. glabriusculum** *Dunal*

▶ *Capsicum annuum* L. var. *glabriusculum* (Dunal) C.B. Heiser et Pickersgill

 Capsicum luteum *Lam.*

▶ *Capsicum chinense* Jacq.

 Capsicum microcarpum *Cav.*

▶ *Capsicum baccatum* L. var. *baccatum*

 Capsicum minimum *Mill.*

▶ *Capsicum annuum* L. var. *glabriusculum* (Dunal) C.B. Heiser et Pickersgill

 Capsicum minimum *Roxb.*

▶ *Capsicum frutescens* L.

 Capsicum pendulum *Willd.*

▶ *Capsicum baccatum* L. var. *pendulum* (Willd.) Eshbaugh

 Capsicum pubescens *Ruiz et Pav.*

Synonyms ▶ *Capsicum guatemalense* Bitter
Common Names ▶ apple chilli, chilli manzana; *Dutch:* paprika; *German:* Filziger Paprika, Apfel-Chili, Baum-Chili; *Indonesian:* cabe bendot, cabe gondol; *Javanese:* cabe dieng; *Spanish:* chamburoto, chile japonés, chile manzana, escabeche, lacoto, siete caldos, rocoto (Peru)
Usage ▶ spice (used like *Capsicum annuum* L. var. *annuum*)
Parts Used ▶ fruit
Distribution ▶ not wild; cultivated from Mexico to S Bolivia, especially in the Andes (Bolivia, Peru, Columbia)
Note ▶ Probable origin in Bolivia; the fruits are said to be of mild to strong pungency.

Almela et al 1991; Erhardt et al. 2002; Guzman/Siemonsma 1999; Hager 4, 1992; Hanelt 2001; Heiser/Smith 1953; Pruthi 1976; Seidemann 1993c; Small 1997; Täufel et al. 1993; Terpó 1966; Tucker 1986; Wilkins 1992; Zeven/de Wet 1982; Zimmermann/Schieberle 2000

CARDAMINE L. - Bitter Cress - Brassicaceae (Cruciferae)

 Cardamine parviflora *L.*

Common Names ▶ *German:* Kleinblütiges Wiesenschaumkraut
Usage ▶ pot-herb (rarely)
Parts Used ▶ young herb
Distribution ▶ Europe, native in Algeria

Erhardt et al. 2002; Fleischhauer 2003

 Cardamine pratensis *L.*

Common Names ▶ cuckoo flower, lady's smock, bitter cress, meadow cress, spinks; *Dutch:* pinksterbloem; *French:* cardamine des prés, cresson élégant, cressonnette, cresson des prés; *German:* Wiesenkresse, Wiesenschaumkraut; *Italian:* cardamino dei prati, billeri, crescione dei prati, violo da pesci; *Portuguese:* agrião dos prados, cardamina dos prados; *Russian:* lugowoj kress, polewaja

gortschiza, serdetschnik, smojanka; *Spanish:* cardamina, berros de prado; mastuerzo de prado
Usage ▶ pot-herb, substitute of water cress
Parts Used ▶ young fresh herb
Distribution ▶ Europe, British Isles, W, E Siberia, Kamchatka, W Tibet, Korea, Alaska, Canada, NE USA

Aichele/Schwegler 3, 1995; Bremness 2001; Cheers 1998; Dudtschenko et al. 1989; Erhardt et al. 2002; Fleischhauer 2003; Heger 5, 1993; Hiller/Melzig 1999; Koschtschejew 1990; Pochljobkin 1974, 1977; Schnelle 1999; Schönfelder 2001; Seidemann 1993c; Stobert 1978; Täufel et al. 1993; Tucker 1986; Uphof 1968

Cardamine resedifolia *L.*

Common Names ▶ *German:* Resedablättriges Schaumkraut
Usage ▶ pot-herb (rarely)
Parts Used ▶ young herb
Distribution ▶ Europe, Caucasus, Iran, W, E Siberia, Amur, Mongolia, Tibet, E Asia

Erhardt et al. 2002; Fleischhauer 2003

Cardamine trifolia *L.*

Common Names ▶ trifoliate bitter cress; *French:* cardamine à trois feuilles; *German:* Dreiblättrige Zahnwurz, Kleeblättriges Schaumkraut
Usage ▶ pot-herb (rarely)
Parts Used ▶ young herb
Distribution ▶ EC Europe, Slovenia, Croatia, Bosnia

Erhardt et al. 2002; Fleischhauer 2003

Cardaminum nasturtium *Moench*

▶ *Nasturtium officinale R. Br.*

Cardamomum dealbatum *Kuntze*

▶ *Amomum maximum Roxb.*

Cardamomum officinale *Salisb.*

▶ *Elettaria cardamomum (L.) Maton*

Cardamomum subulatum *Kuntze*

▶ *Amomum subulatum Roxb.*

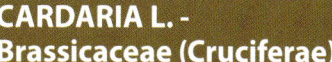
CARDARIA L. - Brassicaceae (Cruciferae)

Cardaria draba *(L.) Desv.*

Synonyms ▶ *Lepidium draba* L.
Common Names ▶ hoary cress, whitetop; *French:* cardaire, passerage drave; *German:* Pfeilkresse, Türkenkresse, *Russian:* chrinnizja krupkowidna
Usage ▶ spice, condiment
Parts Used ▶ leaf, pungent seed
Distribution ▶ N Africa: Egypt, temperate Asia, India, E, SE and SW Europe, widely native elsewhere

Aichele/Schwegler 3, 1995; Dudtschenko et al. 1989; Erhardt et al. 2002; Fleischhauer 2003; Schultze-Motel 1986; Seidemann 1993c; Teuscher 2003; Upof 1968

CARDIOSPERMUM L. - Sapindaceae

Cardiospermum halicacabum *L.*

Common Names ▶ ballon vine, heart pea, heart-leaved pea, heart seed, winter cherry; *Chinese:* jia hu gua; *French:* pois de coeur; *German:* Ballonpflanze, Ballonrebe, Herzerbse, Herzsame; *India:* bodha, indravalli, laftaf, malmai, paravati padi, uzina; *Japanese:* papaia, manjui; *Malaysian:* peria bulan, uban kayu; *Pilipino:* patul parolan, bangkolon
Usage ▶ pot-herb (in India)
Parts Used ▶ young leaf
Distribution ▶ USA: Florida, SE tropical America, tropical Africa, India, also cultivated

Arora/Pandey 1996; Erhardt et al. 2000, 2002; Hager 4, 1992; v. Koenen 1996; Pschyrembel 1998; Wiersoma/León 1999

Sharma 2003; Storrs 1997; Täufel et al. 1993; Tschirch 1892; Uphof 1968; Verheij/Coronel 1991; Villamar et al. 1994; Wolters 1994; Wyk et al. 2004; Zeen/de Wet 1982

 Carduus marianus *L.*

 Silybum marianum (L.) Gaertn.

 Carduus tinctorius *(L.) Falk.*

 Carthamus tinctorius L.

CARICA L. - Papaya, Pawpaw - Caricaceae

Carica papaya *L.*

Synonyms ▸ *Papaya communis* Noronha, *Papaya vulgaris* DC.
Common Names ▸ papaya, paw-paw, melone tree; *Arabic:* fafy, pawpaw; *Chinese:* fan kua, jia hu gua, shu kua, wan shou guo; *French:* arbre de melon, melon de tropiques, papayer; *German:* Melonenbaum, Papaya; *Hindi:* papita; *India:* papeeta; *Indonesian:* pepaya; *Italian:* carica, papaia, papaja; *Japanese:* papaia, manjui; *Malaysian:* betek, betik, papaya; *Mexico:* melón zapote; *Pilipino:* papaya; *Portuguese:* mamoeiro (plant), mamão (fruit); *Russian:* papjja, dunnoe; *Sanskrit:* erandakarkati; *Spanish:* higo de mastuero, lechosa, mamón, papayo, papayero, fruta bomba; *Swahili:* mpapai; *Thai:* malakaw; *Vietnamese:* du du xan
Usage ▸ spice, source and **product** of chymopapin and papain
Parts Used ▸ seed
Distribution ▸ native likely Mexico, C America, West Indies, E Andes
Note ▸ The seeds are used instead of black pepper, sporadically as falsification.

Aedo et al. 2001; Aké Assi/Guinko 1991; Bendel 2002; Bhattacharjee et al. 2003; Blancke 2000; Cheers 1998; Dastur 1954; Davidson 1999; Engel/Phummai 2000; Erhardt et al. 2002; Hanelt 2001; Hiller/Melzig 1999; Hoppe 1949; Leung 1991; Lindner 1971; Meisenbacher 1998; Morris/Mackley 1999; Mors et al. 2000; Paradkar et al. 2001; Pschrymbel 1998; Pursglove 1968; Schenck/Naundorf 1966; Schultze-Motel 1986; Seidemann 1993c, 1994b, 1998/2000;

CARISSA L. - Apocynaceae

 Carissa edulis *Vahl*

Synonyms ▸ *Arduina edulis* (Vahl) Spreng., *Carissa pubescens* A. DC.,
Common Names ▸ carandas plum, Egyptian carissa; *Chinese:* tian jia hu ci; *French:* arbre à cire, carisse; *German:* Essbare Wachsbaumwurzel, Karandapflaume; *Spanish:* ciruela de Natal; *Uganda:* muyonza
Usage ▸ spice, condiment, flavoring for bitters
Parts Used ▸ root
Distribution ▸ S Arabia, tropical Africa, India, Indochina
Note ▸ The fruit pulp is used to produce juice and jam.

Ayensu 1978; Bose 1985; Burkill 1, 1985; Erhardt et al. 2002; Hanelt 2001; v. Koenen 1996; Schultze-Motel 1986; Seidemann 1993c; Täufel et al. 1993; Thompson 1976; Wiersema/León 1999

 Carissa pubescens *DC.*

Carissa edulis Vahl

CARTHAMUS L. - Safflower - Asteraceae (Compositae)

 Carthamus glaber *Burm.*

Carthamus tinctorius L.

 Carthamus tinctorius *L.*

Synonyms ▸ *Carduus tinctorius* (L.) Falk., *Carthamnus glaber* Burm.
Common Names ▸ dyer's saffron, false saffron, safflower, saffron thistle; *Arabic:* asfar, shawrina, gurdum; *Chinese:* hong hua, hong lan hua, hong hua cai;

■ Carissa edulis, flowering

French: carthame, carthame de teinturiers, safre, safranon, safran bâtard; *German:* Färberdistel, Saflor, Bastardsafran, Falscher Safran, Touristen-Safran; *India:* kusum, kusumba, kushumba; *Italian:* cartamo, zaffrone; *Japanese:* benibana, kûkwa; *Korean:* itkkot; *Portuguese:* açafrão bastardo; *Russian:* saflor krasil'nyj; *Slovenian:* požlt farbiarsky; *Spanish:* alazor, azafrán bastardo, azafrán canario, azafran del pais, cártamo, Spanish saffron

Usage ▶ spice, exchange for saffron; dry fruits to put into chutneys, dye stuff; **product:** greasy oil of saffron (fruit)

Parts Used ▶ flower, fruit

Distribution ▶ probable origin W Asia; widely cultivated in N Africa, China, India, and the USA

Aichele/Schwegler 4, 1995; Bärtels 1997; Berger 1, 1949; Bois 1934; Boisvert/Hubert 2000; Burkill 1, 1985; Cheers 1998; Coiciu/Racz (no year); Craze 2002; Dastur 1954; Davidson 1999; Erhardt et al. 2002; Hanelt 1961, 1963; 2001; Heeger 1956; Hepper 1992; Hiller/Melzig 1999; Hoppe 1949; Hyun-Jung Kim et al. 2000; Kazuma et al. 2000; Knowles 1955, 1958; Körber-Grohne 1989; Lewington 1990; Melchior/Kastner 1974; NICPBP 1987; Pruthi 1976; Pursglove 1968; Schenck/Naundorf 1966; Schönfelder 2001; Schultze-Motel 1986; Seidemann 1993c, 2000; Sharma 2003; Small 1997; Smith 1996; Staesche 1972; Täufel et al. 1993; Teuscher 2003; Uphof 1968; Wealth of India 2. 1950; Weiss 1971; Wyk et al. 2004; Zeven/de Wet 1982

 Carthartocarpus fistula Pers.

▶ *Cassia fistula* L.

CARUM L. - Caraway - Apiaceae (Umbelliferae)

 Carum anisum (L.) Baill.

▶ *Pimpinella anisum* L.

 Carum aromaticum Druce

▶ *Trachyspermum ammi* (L.) sprague

 Carum bulbocastanum Koch

▶ *Bunium bulbocastanum* L.

 Carum carvi L.

Synonyms ▶ *Carum velenovskyi* Rohlena

Common Names ▶ (common) caraway; *Arabic:* karauya; *Chinese:* fang feng, se lu zi, shan-chu-tsai, yuan sui, zang hui xiang; *French:* anis des vosges, carvi, cumin de prés; *German:* (Echter) Kümmel, Brotkümmel, Feldkümmel, Wiesenkümmel, Carbe, Karbensamen; *Hindi:* syahjira; *India:* siyah jira, siah seerah; *Italian:* cumino dei prati; cumino tedesco, karawiya; *Korean:* kharum; *Portuguese:* alcarávia; *Russian:* tmin, timon, tmin obyknowennyj; *Sanskrit:* krsnajiraka; *Slovenian:* rasca; *Spanish:* alcaravea, carvi, comino de prado, hinojo de prade; *Turkish:* frenk kimionou

Usage ▶ spice; **product:** essential oil (caraway oil) and for production of alcoholic beverages, e.g. "Allasch"

Parts Used ▶ fruits

Distribution ▶ N Africa, temperate Asia, India, Europe, widely cultivated in temperate regions

Note ▶ In Germany used with anise, coriander and fennel fruits are sold as "bread spice".

Ahro et al. 2001; Aichele/Schwegler 3, 1995; Bärtels 1997; Baysal/Starmans 1999; Berger 3, 1952; Bois 1934; Boisvert/Hubert 2000; Bournot 1968; Bouwmeester et al. 1995; Bremness 2001; Cabizza et al. 2001; Cheers 1998; Coiciu/Racz (no year); Craze 2002; Dalby 2000; Davidson 1999; Dudtschenko et al. 1989; Erhardt et al. 2002; Farrell 1985; Fleischhauer 2003; Guzman/Siemonsma 1999; Hager 4, 1992; Hanelt 2001; Heeger 1956; Hiller/Melzig 1999; Hohmann et al. 2001; Hoppe 1949; Krüger/Zeiger 1993; Matsumura et al. 2002; Morris/Mackley 1999; Norman 1991; Partanen et al. 2002; Pochljobkin 1974, 1977; Pruthi 1976; Pschyrembel 1998; Rosengarten 1969; Roth/Korzmann 1997; Salveson/Svendsen 1976; Schönfelder 2001; Schröder 1991; Schultze-Motel 1986; Sedlakova et al. 2001; Seidemann 1993c; Seidemann/Siebert 1987; Sharma 2003; Shiva et al. 2002; Siewek 1990; Small 1997; Staesche 1972; Täufel et al. 1993; Teuscher 2003; Ubillos 1989; Uphof 1968; Wüstenfeld/Haensel 1964; Wyk et al. 2004; Zeven/de Wet 1982

Carum copticum (L.) Benth.

▶ *Trachyspermum ammi (L.) Sprague*

Carum involuncratum (Roxb.) Baill.

▶ *Trachyspermum roxborghianum (DC.) Craib.*

Carum petroselinum Benth. et Hook.

▶ *Petroselinum crispum (Mill.) Nym.*

Carum roxburghianum Benth.

▶ *Trachyspermum roxburghianum (DC.) Craib.*

Carum velenovskyi Rohlena

▶ *Carum carvi L.*

Carvi bulbocastanum Bub.

▶ *Bunium bulbocastanum L.*

Caryophyllata officinalis Moench

▶ *Geum urbanum L.*

Caryophyllata urbana Scop.

▶ *Geum urbanum L.*

Caryophyllus aromaticus L.

▶ *Syzygium aromaticum (L.) Merr. et L.M. Perry*

CASSIA L - Shower Tree - Fabaceae (Leguminosae)

Cassia fistula L.

Synonyms ▶ *Cathartocarpus fistula* Pers.

Common Names ▶ golden shower, Indian laburnum, pudding pipe tree, purging cassia, purging fistula; *Brazil (Portuguese):* canafistule, tapira coinana; *Chinese:* guo mai long lang, huai hua ching; *French:* bâton casse, canéficier, casse fistuleuse; *German:* Röhrenkassia, Indischer Goldregen; *Hindi:* amaltas, jagaruwa, sinara, sundali; *Indonesian:* bereksa, kasia sena, tengguli, trangguli; *Japanese:* nanban saikachi; *Javanese:* keyok, klohor; *Malaysian:* bereksa, rajah kayu, těngguli, kasia sena; *Mexico:* canafistula, canapistola; *Pilipino:* caña pistula; *Sanskrit:* arabadha, suvarnaka; *Spanish:* canafístula; *Thai:* khuun, kun, raja pruk

Usage ▶ spice

Parts Used ▶ fruit

Distribution ▶ possible origin tropical Asia (India, Sri Lanka, Indonesia), widely cultivated and native in the Tropics.

Note ▶ Dadawa kalva, a local spice preparation in northern Nigeria, obtained from carob bean.

Berger 3, 1952; Cheers 1998; Dashak et al. 2001; Dastur 1954; Engel/Phummai 2000; Erhardt et al. 2002; Hager 4, 1992; Hanelt 2001; Hiller/Melzig 1999; Hoppe 1949; Kottegoda 1994; Schenck/Naundorf 1966; Schultze-Motel 1986; Seidemann 1993c; Sharma 2003; Storrs 1997; Täufel et al. 1993; Uphof 1968; Villamar et al. 1994; Warren 1998; Wiersema/León 1999; Wüstenfeld/Haensel 1964; Yueh-Hsing Kuo et al. 2002; Zewen/de Wet 1982

Cassia obtusifolia L.

▸ *Senna obtusifolia* (L.) Irwin et Barneby

CASSINE L. - Celastraceae

 Cassine crocea *(Thunn.) Kuntze*

Synonyms ▸ *Crocoxylon croceum* (Thunb.) N. Robson
Common Names ▸ small-leaved saffron, saffron cassine; *German:* Schmalblättiger Safran; *S Africa:* fynblaar-saffraan
Usage ▸ spice (rarely)
Parts Used ▸ flower
Distribution ▸ Southern Africa

Quattrocchi 1999

 Cassine metabelica *(Loes) Steedman*

Synonyms ▸ *Elaeodendron metabelica* Loes
Common Names ▸ *German:* Cassineholz
Usage ▸ spice for meat (at barbecue)
Parts Used ▸ wood
Distribution ▸ Southern Africa: Kalahari
Note ▸ Barbecue (meat) with the wood give a spicy aroma.

Neuwinger 1999; Quattrocchi 1999

 Cassumunar roxburghii *Colla*

▸ *Zingiber cassumunar* Roxb.

 Cathartocarpus fistula *Pers.*

▸ *Cassia fistula* L.

CECROPIA Loeffl. - Snake Wood - Cecropiaceae (Moraceae)

 Cecropia asperrima *Pitt.*

▸ *Cecropia peltata* L.

 Cecropia peltata *L.*

Synonyms ▸ *Cecropia asperrima* Pitt.
Common Names ▸ Congo pump, trumpet tree, pop-a-gun, snake wood; *Brazil (Portuguese):* ambaúba, embaúba, abaiba, imbaúba; *French:* bois cannon, tranpetier; *German:* Karibischer Ameisenbaum, Kanonenbaum, Trompetenbaum; *Portuguese:* ambaitinga, imbaúba; *Spanish:* changrarro, guarumo, yagrumo
Usage ▸ pot-herb
Parts Used ▸ young pods
Distribution ▸ C America, West Indies, tropical S America

Erhardt et al 2002; Rätsch 1998; Uphof 1968; Wiersema/León 1999

CEDRONELLA Moench - Lamiaceae (Labiatae)

 Cedronella canariensis *(L.) Webb et Berth.*

Synonyms ▸ *Cedronella triphylla* Moench,
Common Names ▸ balm of Giliad, canary balm; *French:* baume de galaad; *German:* Balsamstrauch; *Portuguese:* hortelã de burro, meutastro; *Spanish:* thé des Canaries
Usage ▸ spice
Parts Used ▸ leaf
Distribution ▸ Canary Islands, Azores, Madeira.
Note ▸ The plant is very frost-sensitive.

Bremness 2001; Carreiras et al. 1987; Cheers 1998; Engel et al. 1991; Erhardt et al 2002; López-García et al. 1991, 1992

Cedronella japonoca *Hassk.*

▸ *Agastache rugosa (Fisch. et C.A. Mey.) Kuntze*

 Cedronella mexicana *(Kunth) Benth.*

▸ *Agastache mexicana (Kunth) Lint et Epling*

 Cedronella triphylla *Moench*

▸ *Cedronella canariensis (L.) Webb et Berth.*

 Celastrus senegalensis *Lam.*

▸ *Maytenus senegalensis (Lam.) Exell*

 Cellus edodes *(Berk.) Ito et Imai*

▸ *Lentinus edodes (Berk.) Sing.*

CELOSIA L. - Cockscomb, Woolflower - Amaranthaceae

 Celosia argentea L. **var. cristata** *(L.) Kuntze*

Synonyms ▸ *Celosia cristata* L., *Celosia splendens* Schum. et Thonn.
Common Names ▸ cockscomb, red fox, red spinach; *Arabic:* katifa orf el-deek; *Chinese:* ji guan hua, qing xiang zi, quimg hsiang, ye chi kuan; *French:* crête de coq, célosie; *German:* Silber-Brandschopf, Silber-Hahnenkamm; *Hindi:* lal murghka, kokan, pile murghka, sufaid murga, sarwari, silara; *Indonesian:* borotyo. jenggar ayam; *Italian:* cresta di gallo; *Japanese:* no-geito, keitó; *Malaysian:* balung ayam; *Portuguese:* crista de galo; *Spanish:* cresta de gallo
Usage ▸ pot-herb
Parts Used ▸ leaf
Distribution ▸ Asia: India, Sri Lanka

Burkill 1985; Erhardt et al. 2001; Kays/Dias 1995; Uphof 1968; Wiersema/León 1999; Zeven/de Wet 1982

 Celosia cristata *L.*

▸ *Celosia argentea L. var. cristata (L.) Kuntze*

 Celosia isertii *C.C. Townsand*

Synonyms ▸ *Celosia laxa* Dalziel
Common Names ▸ cockscomb; *French:* célosie, crête de coq; *German:* Brandschopf, Hahnenkamm; *Russian:* petuschij predeschok
Usage ▸ flavoring for sauces and soups, pot-herb
Parts Used ▸ herb
Distribution ▸ W Africa: from Senegal, the Gambia to Nigeria, Angola, Cameroon, Gabon, Zambia

Burkill 1, 1985; Irvine 1956

 Celosia laxa *Dalziel*

▸ *Celosia isertii C.C. Townsand*

 Celosia splendens *Schum. et Thonn.*

▸ *Celosia argentea L. var. cristata (L.) Kuntze*

 Celosia trigyna *L.*

Synonyms ▸ *Celosia triloba* E. Meyer ex Meisn.
Common Names ▸ *German:* Dreigriffeliger Hahnenkamm; *S Africa:* hanekam, isihlaza
Usage ▸ flavoring (sauces, soups), seasoning
Parts Used ▸ leaf
Distribution ▸ W Africa: Senegal, the Gambia, Nigeria, Upper Volta, Sierra Leone, Madagascar, Arabia; also cultivated in Africa

Burkill 1, 1985; Uphof 1968; Zeven/de Wet 1982

 Celosia triloba *E. Meyer ex Meisn.*

▸ *Celosia trigyna L.*

CELTIS L. - Nettle tree - Ulmaceae

 Celtis intigrifolia *Lam.*

Common Names ▶ African false elm, African nettle tree; *French:* micocoulier, micocoulier africain, micocoulier d'Afrique; *German:* Afrikanischer or Ganzblättriger Zürgelbaum
Usage ▶ pot-herb in soups and sauces
Parts Used ▶ leaf
Distribution ▶ W Africa: the Gambia, Guinea, Mali, Senegal, Arabia

Dalziel 1937; Erhardt et al. 2002; Irvin 1948; Mabberly 1997; Usher 1974

 Centaurea benedicta *(L.) L.*

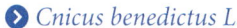 *Cnicus benedictus L.*

CENTAURIUM Hill. - Centaury - Gentianaceae

 Centaurium erythraea *Rafn.*

Synonyms ▶ *Centaurium minus* (L.) Pers., *Centaurium umbellatum* Gilib., *Erythraea centaurum* (L.) Pers.
Common Names ▶ common centaury, centaury, feverwort, pink centaury; *French:* erythrée, petite centaurée, herbe à mille florins; *German:* Centorelle, Magenkraut, Echtes Tausendgüldenkraut, Kopfiges Tausendgüldenkraut, Sanktoriakraut; *Italian:* biondella, caccia febbre, centaurea minore; *Russian:* solototycjatschnik malyj; *Spanish* centaurea menor
Usage ▶ spice, flavoring
Parts Used ▶ herb
Distribution ▶ N Africa, temperate Europe to W Asia
Note ▶ The herb is used for seasoning brandy in Hungary; often wild plants are used.

Aichele/Schwegler 3, 1995; Aquino et al. 1985; Bellarita et al. 1974; Berger 4, 1954; Bilgri/Adam 2000; Charalambous 1994; Clair 1961;

◘ **Centaurum erythrea, flowering**

Erhardt et al. 2002; Fleischhauer 2003; Hager 1992; Hanelt 2001; Heeger 1956; Hiller/Melzig 1999; Hohmann et al. 2001; Hoppe 1949; Kaouadji/Mariotte 1986; Länger 1990; Newall et al. 1996; Pschyrembel 1998; Schimmer/Mauthner 1994; Schönfelder 2001; Schultze-Motel 1986; Seidemann 1993c; van der Sluis/Labadie 1978; Turova et al. 1987; Valentão et al. 2001; Wüstenfeld/Haensel 1964; Wyk et al. 2004

 Centaurium minus *(L.) Pers.*

▶ *Centaurium erythreae Rafn.*

 Centaurium umbellatum *Gilib.*

▶ *Centeurium erythreaea Rafn.*

 Cepa prolifera *Moench*

▶ *Allium x proliferum (Moench) Schrad. ex Willd.*

Cepa victoralis *Moench*

▸ *Allium victoralis* L.

CERASTIUM L. - Mouse Ear - Caryophyllaceae

 Cerastium semidecandrum *L.*

Common Names ▸ mouse-ear chickweed; *German:* Sand-Hornkraut
Usage ▸ pot-herb (rarely)
Parts Used ▸ herb
Distribution ▸ Europe, Cyprus, Turkey, Caucasus, N Iran

Erhardt et al. 2002; Fleischhauer 2003; Uphof 1968

 Cerasus laurocerasus *(L.) Loisel*

▸ *Prunus laurocerasus* L.

 Cerasus mahaleb *(L.) Mill.*

▸ *Prunus mahaleb* L.

CERATONIA L. – Carob - Caesalpiniaceae (Fabaceae)

 Ceratonia coriacea *Salisb.*

▸ *Ceratonia siliqua* L.

 Ceratonia inermis *Stokes*

▸ *Ceratonia siliqua* L.

 Ceratonia siliqua *L.*

Synonyms ▸ *Ceratonia coriaceae* Salisb., *Ceratonia inermis* Stokes

Ceratonia siliqua: a flowering, b fruits

Common Names ▸ carob, locust bean, St. John's bread; *Arabic:* charrûb, chiruub, klarroub, nubti; *Chi-*

nese: jiao dou shu, huai shu; *Dutch:* karobbe, karube, St. Jansbrood; *French:* carobier, caroube, carouge; *German:* Johannisbrot, Judasbaum, Karobe; *Italian:* carrubio, pane di santo Giovanni; *Malaysian:* gĕlenggang; kĕchil, *Portuguese:* alfarrobeira; *Russian:* roshkoboe, judino derevo; *Spanish:* algarrobo, caroba, garrofa; *Thai:* chum het tai

Usage ▶ spice
Parts Used ▶ fruit
Distribution ▶ SE and SW Europe: Iberia, France, N Africa, W Asia, widely cultivated and native. The area of origin is presumably the E Mediterranean region or the Arab Peninsula

Anon. 1993; Avallone et al. 1997; Bärtels 1997; Berger 3, 1952; Cheers 1998; Coit 1951; Davidson 1999; Davis 1970; Duke et al. 2003; Erhardt et al. 2002; Grainger /Winer 1980; Hager 4, 1992; Hanelt 2001; Hepper 1992; Hiller/Melzig 1999; Hills 1980; Hoppe 1949; Koebnick/Zunft 2004; Lewington 1990; Lück 2000; Schultze-Motel 1986; Seidemann 1993c; Uphof 1968; Winer 1980; Wüstenfeld/Haensel 1964; Zeven/de Wet 1982

CERCIS L. - Red Bud - Caesalpiniaceae (Fabaceae)

Cercis florida *Salisb.*

▶ *Cercis siliquastrum L.*

Cercis siliquastrum *L.*

Synonyms ▶ *Cercis florida* Salisb.
Common Names ▶ Judas tree, love tree, red bud, St. John's bread; *Arabic:* argûân, khazrîq, zamzariq; *French:* arbre de Judée, gainier; *German:* (Gewöhnlicher) Judasbaum, Wildes Johannisbrot; *Italian:* albero di Giuda, chiantanella, tregano; *Japanese:* seiyo hanazuo; *Portuguese:* olaia; *Russian:* bagrjannik, judino derevo; *Turkish:* arcuan
Usage ▶ spice: substitute for capers.
Parts Used ▶ flower bud
Distribution ▶ Mediterranean region, S Europe to Iran, N Africa, often cultivated

Bärtels 1997; Cheers 1998; Coit 1951; Erhardt et al. 2002; Hanelt 2001; Hepper 1992; Hiller/Melzig 1999; Schultze-Motel 1986; Seidemann 1993c; Täufel et al. 1993; Zeven/de Wet 1982

CEROPEGIA L. - Asclepiadaceae

Ceropegia acuminata *Roxb.*

▶ *Ceropegia bulbosa Roxb.*

Ceropegia bulbosa *Roxb.*

Synonyms ▶ *Ceropegia acuminata* Roxb., *Ceropegia lushii* Graham
Common Names ▶ *German:* Knollen-Leuchterblume; *Hindi:* khapparkadu; *India* patal tumbari
Usage ▶ pot-herb
Parts Used ▶ herb, leaf
Distribution ▶ C, W India

Arora/Padney 1996; Erhardt et al. 2002; Hanelt 2001; Wealth of India 2, 1950, 3 1992

Ceropegia hirsuta *Wight et Arn.*

Synonyms ▶ *Ceropegia hispida* Blatter et McCann.
Common Names ▶ *German:* Borstige Leuchterblume, Rauhhaarige Leuchterblume
Usage ▶ pot-herb (in Indian deserts)
Parts Used ▶ herb
Distribution ▶ India: Deccan Peninsula

Arora/Pandey 1996

Ceropegia hispida *Blatter et MacCann.*

▶ *Ceropegia hirsuta Wight et Arn.*

Ceropegia *lushii Graham*

▶ *Ceropegia bulbosa Roxb.*

CHAEROPHYLLUM L. - Apiaceae (Umbelliferae)

 Chaerophyllum aromaticum L.

Common Names ▶ chervil, spice cow; *French:* cerfeuil des fous; *German:* Aromatischer Kälberkropf, Gewürz-Kälberkropf
Usage ▶ flavoring (rarely)
Parts Used ▶ leaf, seed
Distribution ▶ C Europe to E Europe
Note ▶ The leaves and seeds have a carrot-parsley taste.

Aichele/Schwegler 3, 1995; Erhardt et al. 2002; Hager 4, 1992

 Chaerophyllum aureum L.

Common Names ▶ *German:* Gold-Kälberkropf
Usage ▶ flavoring (rarely)
Parts Used ▶ leaf
Distribution ▶ C Europe to E Europe, Turkey, Caucasus, N Iran
Note ▶ The leaves have a characteristic parsley aroma.

Aichele/Schwängler 3, 1995; Erhardt et al. 2002

 Chaerophyllum bulbosum L.

Synonyms ▶ *Chaerophyllum caucasicum* (Hoffm.) Schischk.
Common Names ▶ bulbous chervil, garden chervil, parsnip chervil, turnip rooted chervil; *German:* Kerbelrübe, Knollenkerbel, Knolliger Kälberkropf
Usage ▶ flavoring (rarely)
Usage ▶ root
Distribution ▶ C, E Europe, Turkey, Iran, Caucasus Altai, Cult in C and SE Europe

Erhardt et al. 2002; Fleischhauer 2003; Hanelt 2001; Loch 1993; Manfeld 1986; Schnelle 1999; Schönfelder 2001; Small 1997; Teuscher 2003; Uphof 1968

 Chaerophyllum byzantinum Boiss.

Common Names ▶ Byzantinischer Kälberkropf

Usage ▶ spice
Parts Used ▶ leaf
Distribution ▶ SE parts of Balkan Peninsula to NW Anatolia
Note ▶ Occasionally cultivated as a spice plant by Turks in Germany (Hessen).

Hanelt 2001

 Chaerophyllum caucasicum (Hoffm.) Schischk.

 Chaerophyllium bulbosum L.

 Chalcas koenigii Kurz

 Murraya koenigii (L.) Spr.

 Chalcas paniculata L.

 Murraya paniculata (L.) Jack

 Chamaedrys marum Moench

▶ *Teucrium marum* L.

 Chamaedrys officinalis Moench

▶ *Teucrium chamaedrys* L.

Chavica peepuloides Miq.

▶ *Piper peepuloides* (A. Dietr.) Roxb.

Chavica sarmentosa (Roxb. ex Hunter) Miq.

▶ *Piper sarmentosum* Roxb. ex Hunter

CHAMAEMELUM Mill. - Chamomile - Asteraceae (Compositae)

 Chamaemelum nobile (L.) All.

Synonyms ▶ *Anthemis nobilis* L.
Common Names ▶ flowerhead, garden chamomile, noble chamomile, Roman chamomile; *Brazil (Portuguese):* camomila nobre, camomila verdadeira, camomila odoranta, macela dourada; *French:* camomille romain; *German:* Römische Kamille; *Italian:* camomilla romana; *Russian:* rimskaja romaschka; *Spanish:* camomila de jardin, manzanilla fina, manzanilla romana
Usage ▶ spice; **product:** essential oil (Roman chamomile oil) for liqueurs
Parts Used ▶ flower
Distribution ▶ Canary Islands, N Africa, N, SW Europe, formerly Soviet Union, widely native and cultivated

Bärtels 1997; Bicchi/Frattini 1987; Bremness 2001; Coiciu/Racz (no year); Erhardt et al. 2002; Hager 4, 1992; Hanelt 2001; Hiller/Melzig 1999; Hohmann et al. 2001; Hoppe 1949; Isaac 1993; Lewington 1990; Newall et al. 1996; Opdyke 1974b; Pschyrembel 1998; Roth/Korzmann1997; Schönfelder 2001; Schultze-Motel 1096; Seidemann 1993c; Shiva et al. 2002; Small 1997; Ubillos 1989; Wyk et al. 2004; Zeven/de Wet 1982

 Chavica retrofracta (Vahl) Miq.

▶ *Piper retrofractum* Vahl

 Chavica sarmentosa (Roxb. ex Hunter) Miq.

▶ *Piper sarmentosa* Roxb. ex Hunter

CHENOPODIUM L. - Goosefoot - Chenopodiaceae

 Chenopodium album L.

Synonyms ▶ *Chenopodium candidans* Lam., *Chenopodiumn viride* L.,
Common Names ▶ blue weed, goose foot, lamb's quarters, pigweed; *Chinese:* hsien, hui t'iao, *French:* ansérine blanche; *German:* Weißer Gänsefuß, Weiße Melde; *Hindi:* bethu sag; *Japanese:* shiro akoza; *Portuguese:* anserina branca; *Russian:* mar belaja, mar obyknowennaja; *Spanish:* cenizo blanco
Usage ▶ pot-herb (rarely)
Parts Used ▶ leaf, young tops
Distribution ▶ Europe, cosmopolital

Erhardt et al. 2002; Fleischhauer 2003; Hanelt 2001; Schnelle 1999; Schultze-Motel 1986; Small 1997; Uphof 1968

 Chenopodium ambrosioides L.

Common Names ▶ American epazote, goosefoot, Jerusalem or Mexican tea, wormseed; *Arabic:* lajouma, natna, sianama; *Dutch:* wormdrijvende ganzevoet; *French:* ambroisie du Mexique, ansérine, ambroisine, herbe à vers, thé du Mexique, *German:* Ambrose, Epazote, Kartäusertee, Wohlriechender Gänsefuß, Mexikanischer Tee, Mexikanisches Traubenkraut, Lima-Tee; *Italian:* ambrosia, chenopodio ambrosioide, tè del Méssico; *Korean:* yaknŭngjaengi; *Portuguese:* ambrósia do México; anserine vermifuge, ervate Santa Maria; mar'ambrosiewidnaja; *Spanish:* apazote, aposote, epazote, paico, pazote, hierba hormi uera, té de Méjico; *Vietnamese:* co hoi, dau giun, kinh gioi dat, rau nuoi dai
Usage ▶ spice, flavoring, condiment; **product:** essential oil
Parts Used ▶ fresh leaf and herb
Distribution ▶ cosmopolitica, probable origin Neotropics, widespread in tropical and subtropical America, ruderal in Brazil, W Africa: Sierra Leone, Ghana, Niger; native S Europe
Note ▶ This polymorphus species has the two cultivated racea: var. *ambrosioides* and var. *anthelminticum*.

Aichele/Schwegler 2, 1994; Bauer/Silva 1973; Berger 4, 1954; Bornot 1968; Burkill 1, 1985; Coiciu/Racz (no year); Davidson 1999; Erhardt et al. 2002; Erichsen-Brown 1989; Hanelt 2001; Heeger 1956; Hiller/Melzig 1999; Hoffmann et al. 1992; Hoppe 1949; v. Koenen 1996; Mors et al. 2000; Ochoa/Alonso 1996; Rätsch 1992; Roth/Korzmann 1997; Schönfelder 2001; Schultze-Motel 1986; Seidemann 1993c; Sharma 2003; Shiva et al. 2002; Small 1997; Täufel et al. 1993; Tucker 1986; Tull 1999; Turova et al. 1987; Uphof 1968; WHO Manila 1990; Wolters 1996; Wyk et al. 2004; Zeven/de Wet 1982

CHENOPODIUM L. - Goosefoot - Chenopodiaceae

 Chenopodium bonus-henricus *L.*

Synonyms ▶ *Chenopodium esculentus* Salisb.

Common Names ▶ allgood, good king Henry; *French:* bon Henri, épinard sauvage, toute-bonne; *German:* Guter Heinrich; *Italian:* buona Enrico, colubrina, tuttabuona

Usage ▶ pot-herb

Parts Used ▶ leaf

Distribution ▶ native to the temperate Old World; spread throughout Europe and Siberia

Bremness 2001; Erhardt et al. 2001; Fleischhauer 2003; Hanelt 2001; Schultze-Motel 1986; Uphof 1968; Zeven /de Wet 1982

 Chenopodium botrys *L.*

Common Names ▶ Jerusalem oak, feather geranium, slimy anserine herb; *French:* herbe à printemps, piment botris; *German:* Bertholdskraut, Klebriger Gänsefuß, Traubenkraut; *Italian:* botri; *Portuguese:* ambródia dos boticas; *Russian:* mar duschistaja; *Spanish:* biengranada

Usage ▶ spice

Parts Used ▶ fresh herb

Distribution ▶ temperate Asia, India, Himalayas, C Asia introduced to C Europe, S Europe native elsewhere in temperate regions, Turkey, Cyprus, S Africa, Australia, S, N America and locally native

Note ▶ The herb is an important spice for the S America Indians for meat, fish and vegetables.

Aichele/Schwegler 2, 1994; Berger 4, 1954; Erhardt et al. 2002; Erichsen-Brown 1989; Fleischhauer 2003; Hanelt 2001; Hiller/Melzig 1999; Schultze-Motel 1986; Seidemann 1993c; Small 1997; Wiersema/León 1999; Zeven/de Wet 1982

 Chenopodium candidans *Lam.*

▶ *Chenopodium album L.*

 Chenopodium capitatum *(L.) Asch.*

Common Names ▶ strawberry blight; *German:* Kopfiger Gänsefuß, Kopfiger Erdbeerspinat

Usage ▶ pot-herb

▶ **Chenepodium bonus-henricus, flowering**

Parts Used ▶ leaf

Distribution ▶ Eurasia, variously cultivated in Europe

Erhardt et al. 2002; Hanelt 2001; Mansfeld 1962; Schultze-Motel 1986; Small 1997

 Chenopodium esculentus *Salisb.*

▶ *Chenopodium bonus-henricus L.*

 Chenopodium viride *L.*

▶ *Chenopodium album L.*

CHIMAPHILA Pursh - Prince's Pine - Pyrolaceae

 Chimaphila umbellata (L.) Barton

Synonyms ▶ pirola umbellata L.
Common Names ▶ prince's pine; *German:* Doldiges Winterlieb, Nabelkraut, Walddolde
Usage ▶ spice for beer: substitute for hops (locally)
Parts Used ▶ leaf, herb
Distribution ▶ Europe, W Siberia, Sakhalin, Amur, Japan, N. America: Canada, C America: Mexico

Erhardt et al. 2002; Fleischhauer 2003; Hiller/Melziger 1997; Rätsch 1998; Wiersema/León 1997

CHLORANTHUS Sw. - Chloranthaceae

 Chloranthus inconspicuus Sw.

▶ *Chloranthus spicatus (Thunb.) Makino*

 Chloranthus indicus Wight

▶ *Chloranthus spicathus (Thunb.) Makino*

 Chloranthus obtusifolius Miq.

▶ *Chloranthus spicathus (Thunb.) Makino*

 Chloranthus spicatus (Thunb.) Makino

Synonyms ▶ *Chloranthus inconspicuus* Sw., *Chloranthus indicus* Wight, *Chloranthus obtusifolius* Miq., *Nigrina spicifera* Lam.
Common Names ▶ pearl orchid flower, chulan, cha ran; *Chinese:* zhu lem, zu lan hua; *German:* Chloranthusblüten, Perlen-Orchidee; *Indonesian:* barlen; *Japanese:* charan; *Thai:* niam om, foi faa, raam; *Vietnamese:* hoa soi
Usage ▶ spice (sporadically), flavoring for tea
Parts Used ▶ flower, leaf

Distribution ▶ wild in China; cultivated in E Asia: S China, Vietnam, Japan, Malaysia, Java and Sumatra.

Arctander 1960; Burkill 1966; Hanelt 2001; Heyne 1953; Oyen/Dung 1999; Schultze-Motel 1986; Seidemann 1993c; Zeven/de Wet 1982

 Chlorocodon whitei Hook.f.

▶ *Mondia whitei (Hook.f.) Skeels*

CHRYSANTHEMUM L. - Chrysanthemum - Asteraceae (Compositae)

 Chrysanthemum balsamite L.

▶ *Tanacetum balsamite L.*

 Chrysanthemum coronarium L.

Synonyms ▶ *Xanthophthalmum coronarium* (L.) Trehane
Common Names ▶ crown daisy, cooking chrysanthmum, garland chrysanthemum, garland, garland daisy; *Chinese:* penghao, t'ung hao; *French:* chrysanthème a couronne; *German:* Goldblume, Kronen-Wucherblume; *Italian:* crisantemo, fior d'oro; *Japanese:* kikunori kiku-na, shungiku; *Russian:* chrizantema uventschannaja; *Sanskrit:* chandramallika
Usage ▶ condiment, pot-herb
Parts Used ▶ leaf, germinated seedlings
Distribution ▶ S Europe, Italy, Turkey, Israel, N Africa, cultivated in numerous countries, especially in China and Japan
Note ▶ In Japan leaves of various *Chrysanthemum* species are used to produced the seasoning "Shungiku".

Ahmed et al. 1999; Alvarez-Castellanos et al. 2001; Bois 1934; Chuda et al. 1996, 1998; Davidson 1999; Dudtschenko et al. 1989 El-Masry et al. 1984; Erhardt et al. 2000, 2002; Flamini et al. 2003; Hepper 1992; Sanz et al. 1990; Schultze-Motel 1986; Sulas/Caredda 1997; Takenaka et al. 2000; Tucker 1986; Uphof 1968; Zeven/de Wet 1982

Chrysanthemum leucanthemum L.

 Leucanthemum vulgare Lam.

 Chrysanthemum majus *(Desf.) Aschers.*

 Tanacetum balsamite L.

 Chrysanthemum vulgare *(L.) Bernh.*

 Tanacetum vulgare L.

CHYTRANTHUS Hook.f. - Sapindaceae

 Chytranthus cauliflorus *(Hutch et Dalg.) Wickens*

Usage ▶ flavoring
Parts Used ▶ leaf
Distribution ▶ W Africa: Liberia

Burkill 5, 2000

 Chytranthus talbotii *(Bak.f.) Keay*

Usage ▶ condiment
Parts Used ▶ fruit pulp
Distribution ▶ W Africa: Gabon

Burkill 5, 2000

 Cicuta amomum *Crantz*

 Sison amomum L.

 Cienskowskia aethiopica *Schweinf.*

 Siphonochilus aethiopicus (Schweinf.) B.L. Burtt.

CINCHONA L. - Quinine - Rubiaceae

 Cinchona calisaya *Wedd.*

 Cinchona officinalis L.

 Cinchona cordifolia *Mutis*

 Cinchona pubescens Vahl.

 Cinchona glabra *Ruiz*

 Cinchona officinalis L.

 Cinchona grandifolia *Mut.*

 Cinchona pubescens Vahl 'succirubra'

 Cinchona hirsuta *Ruiz et Pav.*

 Cinchona pubescens Vahl

 Cinchona lancifolia *Mut.*

 Cinchona officinalis L.

 Cinchona ledgrerana *Moens et Trim.*

 Cinchona officinalis L.

 Cinchona nitida *Ruiz et Pav.*

 Cinchona officinalis L.

 Cinchona officinalis *L.*

Synonyms ▶ *Cinchona calisaya* Wedd., *Cinchona glabra* Ruiz, *Cinchona lancifolia* Mut., *Cinchona ledgrerana* Moens et Trim.; *Cinchona nitida* Ruiz et Pav.

Common Names ▸ chinabark, Peruvian bark, yellow cinchona, ledger bark, yellow bark, quinine tree; *Bolivia:* calisaya, calisaya blanca, calisaya morala; quina amarilla; *Ecuador:* cascarilla del rey, quina negra; *French:* quinquina (jaune); *German:* Calisya Chinarinde, Gelbe Chinarinde, Echte Königschinarinde; *Hindi:* kunain; *Italian:* corteccia di China; *Japanese:* aka kina no ki; *Peru:* cascarilla crespilla, cascarilla de la lomas, cascarilla verde; *Portuguese:* quina amarela; *Russian:* chinnaja kora, hinnos; *Sanskrit:* kunayana; *Spanish:* árbol dela quina, quina, quinquina

Usage ▸ spice, flavoring (for soft drinks as a bitter, principally for tonic water)

Parts Used ▸ bark

Distribution ▸ W S America, also cultivated

Berger 1, 1949; Duke et al. 2003; Erhardt et al. 2002; Ferrão 1992; Flückiger 1883; Gorkom 1869; Hager 4, 1992; Hanelt 2001; Hermann 2001a, b; Hiller/Melzig 1999; Hobhouse 2000; Lewington 1990; Ochse et al. 1961; Plotkin 1994; Pursglove 1968; Schultze-Motel 1986; Seidemann 1993c; Sharma 2003; Täufel et al. 1993; Tschirch 1892; Turora et al. 1987; Uphof 1968; Villamar et al. 1994; Westermann 1909; Wolters 1994; Wüstenfeld/Haensel 1964; Zeven/de Wet 1982

Cinchona pubescens *Vahl*

Synonyms ▸ *Cinchona cordifolia* Mutis; *Cinchona hirsuta* Ruiz et Pav.,

Common Names ▸ quinine, red cinchona, red Peruvian, red bark, Jesuit bark; *Chinese:* chin chi lo, jin ji le; *French:* quinquina rouge, écorce de quinquina; *German:* Fieberrinde, Rote Chinarinde; *Italian:* china rossa, corteccia di china; *Portuguese:* quina do Amazonas; *Spanish:* cascarilla, cinchona, corteza de quina

Usage ▸ spice, flavoring

Parts Used ▸ bark

Distribution ▸ C America, tropical S America, Costa Rica to Bolivia, India, Java, also cultivated

Duke et al. 2003; Erhardt et al. 2002; Ferrão 1992; Flückiger 1883; Hager 4, 1992; Hanelt 2001; Hermann 2001a, b; Hiller/Melzig 1999; Hohmann et al. 2001; Pschyrembel 1998; Pursglove 1968; Rätsch 1992; Schultze-Motel 1986; Seidemann 1993c; Täufel et al. 1993; Uphof 1968; Wolters 1994; Wüstenfeld/Haensel 1964; Wyk et al. 2004; Zeven/de Wet 1982

Cinchona pubescens *Vahl* 'succiruba'

Synonyms ▸ *Cinchona grandifolia* Mut.; *Cinchona succirubra* Pav. ex Klotzsch

Common Names ▸ red cinchona; *German:* Rote Chinarinde; *Spanish:* cascarilla amarga, cascarilla gallingo

Usage ▸ spice (flavoring)

Parts Used ▸ bark

Distribution ▸ Meso-America, C, NS and WS America, tropical S America, cultivated

Berger 1, 1949; Hager 4, 1992; Hanelt 2001; Hiller/Melzig 1999; Hoppe 1949; Lewington 1990

Cinchona succirubra *Pav. ex Klotzsch*

▸ *Cinchona pubescens Vahl 'succiruba'*

CINERARIA L. - Cineraria - Asteraceae (Compositae)

Cineraria lyrata *L.*

Common Names ▸ African marigold, wild parsley; *French:* cinéraire d'Afrique; *German:* Afrikanische Ringelblume, Wilde Petersilie; *S Africa:* geelblom, boerelusern

Usage ▸ pot-herb (locally, very sporadically)

Parts Used ▸ leaf

Distribution ▸ S Africa

Wiersema/León 1999

CINNAMODENDRON Endl. - Canellaceae

Cinnamodendron corticosum *Miers*

Common Names ▸ false Winter's bark, kinnamon; *German:* Falsche Winterrinde

Usage ▸ spice (rarely)

Parts Used ▸ bark

Distribution ▸ West India

Erhardt et al. 2002; Uphof 1968

CINNAMOMUM Schaeff. - Cinnamon - Lauraceae

 Cinnamomum aromaticum Nees

Synonyms ▶ *Cinnamomum cassia* Nees ex Bl., *Laurus cassia* L., *Laurus cinnamomum* Andr.,
Common Names ▶ cassia, Chinese cinnamon, Chinese cassia, cassia vera (strictly speaking); *Chinese:* gui zhi, mo kuei, rou gui; *Dutch:* kassie; *French:* casse, canelle casse, canelle de Chine (plante), canelle fausse, canéfice, canellier de Chine; casse ligneuse; *German:* Kaneel, Kassia, Chinesische Zimtrinde, Holzkassia, Holzzimt, Kanton-Zimt, Mutterzimt, Zimtkassia; *Indonesian:* kayu manis cina; *Italian:* canella cinese; *Japanese:* kashia-keihi; *Malaysian:* kayu manis; *Portuguese:* cássia; *Russian:* kitajskaja oritscha, koritschnik kitajskij; *Slovenian:* kasia; *Spanish:* canelo de la China, canelero chino, cassia lignea; *Thai:* ob cheuy; *Vietnamese:* que, may que
Usage ▶ spice; **product:** essential oil (leaves: *Folia malabathri*); chewing pan
Parts Used ▶ bark
Distribution ▶ China: Kwangsi

Berger 1, 1949; Boivert/Hubert 2000; Bournot 1968; Bozan et al. 2003; Brown 1956; Clair 1961; Craze 2002; Dalby 2000; Davidson 1999; Dudtschenko et al. 1989; Duke et al. 2003; Erhardt et al. 2002; Farrell 1985; Guzman/Siemonsma 1999; Hager 4, 1992; Hanelt 2001; Hepper 1992; Hiller/Melzig 1999; Hoppe 1949; Jiroretz et al. 2000; Karig 1975; Kojoma et al. 2002; Lallemand et al. 2000; Landry 1985; Leung 1991; Liu/Ou 1969; Lockwood 1979; Maistre 1964; Melchior/Kastner 1974; Morris/Mackley 1999; Mors/Rizzini 1961; Norman 1991; Peter 2001; Pruthi 1976; Rosengarten 1969; Rothe/Korzmann 1997; Schröder 1991; Schultze-Motel 1986; Seaforth 1962; Seidemann 1961, 1993c; Seidemann/Siebert 1987; Shiva et al. 2002; Siewek 1990; Staesche 1972; Täufel et al. 1993; Teuscher 2003; Turova et al. 1987; Uphof 1968; Way 1985; WHO 1990; Wijesekera 1978; Wüstenfeld/Haensel 1964; Wyk et al. 2004; Zeven/de Wet 1982

 **Cinnamomum assamicum** S.C. Nath et A. Baruah

Common Names ▶ Assam cinnamom; *German:* Assam-Zimt

Usage ▶ fragrant, preservative (rarely)
Parts Used ▶ bark, leaf
Distribution ▶ NE India

Baruah/Nath 2001

 Cinnamomum bejolghata (Buch.-Ham.) Sweet

Common Names ▶ cinnamon tree; *Thai:* op choei; *Vietnamese:* qué lá trà
Usage ▶ spice, fragrant
Parts Used ▶ bark, leaf
Distribution ▶ Thailand, Vietnam

Engel/Phummai 2000; Oyen/Dung 1999

 Cinnamomum burmanii (Nees) Bl.

Synonyms ▶ *Cinnamomum burmanii* (Nees et T. Nees) Nees ex Bl., *Lauris burmanii* Nees, *Laurus dulcis* Roxb.
Common Names ▶ Batavia cinnamon, cassia, Chinese cinnamon, false cinnamon, Indonesian cassia, Padang cassia, Batavia cassia, Korintje or Korinjii cassia; *French:* canelle, canelle d'Indonésie, canelle de Batavia, canelle de Padang, canelle de Timor; *German:* Chinesischer Zimt, Padang-Zimt, Fagot-Zimt, Batavia-Kassie, Holzzimt, Indonesischer Zimt, Korintji Kassie; *Indonesian:* kaya manis; *Pilipino:* kalingag, kami; *Russian:* kitajckaja koriza; *Thai:* suramarit
Usage ▶ spice, flavoring; **product** essential oil
Parts Used ▶ bark, leaf
Distribution ▶ Malaysia, Thailand

Bois 1934; Erhardt et al. 2002; Guzman/Siemonsma 1999; Hager 4, 1992; Hanelt 2001; Hiller/Melzig 1999; Hoppe 1949; Jiroretz et al. 2000; Kojoma et al. 2002; Lallemand et al. 2000; Lindner 1951; Lück 2000; Maistre 1964; Melchior/Kastner 1974; Pruthi 1976; Sangat-Roemantyo 1990; Schröder 1991; Schultze-Motel 1986; Seidemann 1993c; Seidemann/Siebert 1987; Siewek 1990; Staesche 1972; Tainter/Grenis 1993; Täufel et al. 1993; Teuscher 2003; Uphof 1968; Zeven/de Wet 1982

Cinnamomum cambodianum Hance

Common Names ▶ tep pirou; Cambodia cinnamon; *German:* Kambodscha-Zimt
Usage ▶ spice
Parts Used ▶ bark
Distribution ▶ Cambodia

Uphof 1968

 Cinnamomum cassia *Bl.*

▶ *Cinnamomum aromaticum Nees*

 Cinnamomum culilaban *(L.) J.S. Presl*

▶ *Cinnamomum culilawan (L.) Klosterm.*

 Cinnamomum culilawan *(Roxb.) J.S. Presl*

▶ *Cinnamomum culilawan (L.) Klosterm.*

 Cinnamomum culilawana *Bl.*

Common Names ▶ Culilawan cassia; *German:* Kuliwan-Zimt, Lavang-Zimt
Usage ▶ flavoring (locally), bud, substitute for clove bud; **product:** essential oil (Culilawan oil) in cake and perfumery
Parts Used ▶ bark, bud
Distribution ▶ China, Moluccas, Amboyana
Note ▶ Substitute vor clove bud (*Syzygium aromaticum* [L.] Merr et L.M. Perry).

Bois 1934; Hager 4, 1992; Hanelt 2001; Melchior/Kastner 1974; Peter 2001; Pochljobkin 1974, 1977; Schultze-Motel 1986; Seidemann 1993c; Täufel et al. 1993; Teuscher 2003

 Cinnamomum culilawana *(L.) Klosterm.*

Synonyms ▶ *Cinnamomum culilaban* (L.) J.S. Presl, *Cinnamomum culilawan* (Roxb.). J.S. Presl, *Laurus culitlawan* L.
Common Names ▶ *German:* Culilawan-Zimt, Lavang-

rinde, Cortex Caryophylloides ruber; *Indonesian:* kaya teja, kulitlawang, salakat
Usage ▶ spice
Parts Used ▶ bark
Distribution ▶ Indonesia, cultivated also in Malaysia and India
Note ▶ The bark smells of cloves, the wood bark has a fennel (*Foeniculum vulgare* Mill.) flavor and has been used as a substitute for sassafras bark (*Sassafras albidium* [Nutt.] Nees.

Berger 1, 1949; Guzman/Siemonsma 1999; Hiller/Melzig 1999; Lallemand et al. 2000; Peter 2001; Staesche 1972; Uphof 1968

 Cinnamomum deschampsii *Gamble*

Common Names ▶ *Malaysian:* kayu manis
Usage ▶ spice (flavoring)
Parts Used ▶ bark
Distribution ▶ Malakka, Malaysia (Penang), cultivated in Singapore
Note ▶ Substitute for Ceylon cinnamon (*Cinnamomum zeylanicum* Bl.).

Guzman/Siemonsma 1999; Oyen/Dung 1999; Täufel et al. 1993; Teuscher 2003

 Cinnamomum eucalytoides *T. Nees*

▶ *Cinnamomum iners Reinw. ex Bl.*

 Cinnamomum impressinervium *Meissner*

Usage ▶ spice
Parts Used ▶ bark
Distribution ▶ Sikkim Himalayas, China
Note ▶ Substitute for, or adultant of *Cinnamom zeylanicum* Bl.

Burkill 1966; Oyen/Dung 1999

Cinnamomum iners Reinw. ex Bl.

Synonyms ▸ *Cinnamomum eucalyptoides* T. Nees, *Cinnamomum nitidum* Bl., *Cinnamomum paraneuron* Miq.
Common Names ▸ wild cinnamon (of Japan); *German:* Wilder (japanischer) Zimt; *Malaysian:* kayu manis hutan, lelang, mědang kěmangi, teja
Usage ▸ flavoring (locally)
Parts Used ▸ bark
Distribution ▸ SE Asia, Indonesia, Philippines, also cultivated

Burkill 1985; Hanelt 2001; Lallemand et al. 2000; Morton 1976; Peter 2001; Pruthi 1976; Schultze-Motel 1986; Teuscher 2003; Uphof 1968;

 Cinnamomum japonicum Siebold

Common Names ▸ Japanese cinnamon; *German:* Japanischer Zimt
Usage ▸ spice (locally)
Parts Used ▸ bark
Distribution ▸ China, Japan

Dudtschenko et al. 1989; Liu/Ou 1969

 Cinnamomum loureirii Nees

Synonyms ▸ *Cinnamomum obtussifolium* (Roxb.) Nees var. *loureiri* Nees ex Watt
Common Names ▸ Saïgon cinnamon, Saïgoncassia, Vietnamese cassia *French:* canelle de Saïgon, canelle de Cochinchine, canelle de Viêtnam; *German:* Annamzimt, Saïgon-Zimt, Tonkin-Zimt, Vietnam-Zimt; *Spanish:* canela de Saïgón; *Thai:* op choei
Usage ▸ spice (flavoring)
Parts Used ▸ bark, flower
Distribution ▸ Indochina, Japan, China, S Vietnam, Java

Dudtschenko et al. 1989; Erhardt et al. 2002; Guzman/Siemonsma 1999; Hager 4, 1992; Hanelt 2001; Hoppe 1949; Hiller/Melzig 1999; Lallemand et al. 2000; Liu/Ou 1969; Maistre 1964; Melchior/Kastner 1974; Peter 2001; Pruthi 1976; Schröder 1991; Schultze-Motel 1986; Seidemann 1993c; Seidemann/Siebert 1987; Staesche 1972; Tainter/Grenis 1993; Täufel et al. 1993; Teuscher 2003

 Cinnamomum massoia Schew.

▸ *Cryptocaria aromaticum* (Becc.) Kosterm.

 Cinnamomum micranthum (Hayata) Hayata

Common Names ▸ Taiwan cinnamon; *German:* Taiwanischer Zimt
Usage ▸ spice (locally)
Parts Used ▸ bark
Distribution ▸ Taiwan

Erhardt et al. 2002; Liu/Ou 1969

 Cinnamomum mindanaense Elmer

Common Names ▸ mindanao cinnamon; *German:* Mindanao-Zimt
Usage ▸ spice (locally)
Parts Used ▸ bark
Distribution ▸ Philippines
Note ▸ Substitute for Ceylon cinnamon (*Cinnamomum zeylanicum* Bl.).

Lawrence/Hogg 1974; Seidemann 1997b; Uphof 1968; Usher 1974

 Cinnamomum nitidum Bl.

▸ *Cinnamomum iners* Reinw. ex Bl.

 Cinnamomum obtusifolium Nees

Common Names ▸ Bengali cinnamon; *French:* canelle du Bengale; *German:* Annam-Zimt, Saïgon-Zimt
Usage ▸ spice (locally)
Parts Used ▸ bark, flower
Distribution ▸ E Himalayas, Vietnam, Myanmar, Assam
Note ▸ Substitute for Ceylon cinnamon (*Cinnamomum zeylanicum* Bl.)

Hager 4, 1992; Melchior/Kastner 1974; Peter 2001; Pruthi 1976; Schultze-Motel 1986; Seidemann 1993c; Staesche 1972; Täufel et al. 1993; Teuscher 2003

Cinnamomum obtusifolium *(Roxb.) Nees var. loureiri* Nees ex Watt

▷ *Cinnamomum loureirii Nees*

 Cinnamomum olivera *Bailey*

Common Names ▸ Australian cinnamon, black sassafras, Oliver's bark; *French:* canelle l'Australie; *German:* Australischer Zimt; *Japanese:* nikkai
Usage ▸ spice
Parts Used ▸ bark
Distribution ▸ Australia: Queensland
Note ▸ Substitute for Ceylon cinnamon (*Cinnamomum zeylanicum* Bl.)

Karig 1975; Peter 2001; Rätsch 1998; Uphof 1968

 Cinnamomum paraneuron *Miq.*

▷ *Cinnamomum iners Reinw. ex Bl.*

 Cinnamomum parthenoxylon *Meisn.*

Usage ▸ spice (locally); **product:** essential oil (bark and leaf)
Parts Used ▸ bark, leaf
Distribution ▸ N India, Bangladesh, Myanmar, Malaysia

Baruah/Nath 2000; Hanelt 2001; Uphof 1968; Usher 1974

 Cinnamomum philippinense *(Merr.) C.E. Chang*

Common Names ▸ Philippine cinnamon; *French:* canelle de les Philippines; *German:* Philippinischer Zimt
Usage ▸ spice (locally)
Parts Used ▸ bark
Distribution ▸ Philippines

Lallemand et al. 2000; Melchior/Kastner 1974; Seidemann 1993c; Teuscher 2003

 Cinnamomum porrectum *(Roxb.) Kosterm.*

Synonyms ▸ *Cinnamomum parthenoxylon* (Jack) Nees, *Laurus porrecta* Roxb.
Common Names ▸ *Chinese:* zong guan, zong hai, mai zong
Usage ▸ spice
Parts Used ▸ bark
Distribution ▸ China, Myanmar, Malaysia, in S China cultivated in the garden

Hanelt 2001; Yu et al. 1985; Zeven/Zhukovsky 1975

 Cinnamomum puberulum *Ridley*

Common Names ▸ *Malaysian:* teja, medang kemangi
Usage ▸ spice, e.g. to flavor curries (locally)
Parts Used ▸ bark
Distribution ▸ Malaysian Peninsula

Guzman/Siemonsma 1999; Oyen/Dung 1999

 Cinnamomum rhynchophyllum *Miq.*

Common Names ▸ *Indonesian:* kayu lawang, kayu salangan, modang sanggar; *Malaysian:* teja
Usage ▸ spice (locally)
Parts Used ▸ bark
Distribution ▸ Malaysia Peninsula, Sumatra, Borneo
Note ▸ The bark smells like cloves and nutmeg.

Guzman/Siemonsma 1999; Oyen/Dung 1999

 Cinnamom sintok *Bl.*

Synonyms ▸ *Cinnamomum camphoratum* Bl.
Common Names ▸ Java cassia; *German:* Java-Holzzimt; *Javanese:* sintok
Usage ▸ flavoring (locally)
Parts Used ▸ bark
Distribution ▸ Indonesia: Java, Sumatra

Backer/Bakhuizen van den Brink 1963; Peter 2001; Sangat-Roemantyo 1990; Uphof 1968

CINNAMOMUM Schaeff. - Cinnamon - Lauraceae

Cinnamomum tamala (Buch.-Ham.) Nees et Eberm.

Common Names ▶ Indian cassia lignea, Indian bark, Malabathri bark; *French:* canelle d'Inde; *German:* Malabarzimt, Indischer Holzzimt; *Hindi:* tejpat; *Russian:* Malabarskaja koriiza
Usage ▶ spice (locally), condiment
Parts Used ▶ leaf, bark
Distribution ▶ India, Malaysia
Note ▶ The leaves are substitute for bay leaves (*Laurus nobilis* L.).

Arora/Pandey 1996; Bois 1934; Dalby 2000; Dastur 1954; Erhardt et al. 2002; Hager 4, 1992; Ilyas 1976; Kumar 2003; Lallemand et al. 2000; Pochljobkin 1974, 1977; Pruthi 1976; Schultze-Motel 1986; Seidemann 1993c; Sharma 2003; Staesche 1972; Täufel et al. 1993; Teuscher 2003; Uphof 1968; Zeven/de Wet 1982

Cinnamomum verum J. Presl

▶ *Cinnamomum zeylanicum* Bl.

Cinnamomum zeylanicum Bl.

Synonyms ▶ *Cinnamomum verum* J. Presl, *Laurus cinnamomum* L.
Common Names ▶ Cinnamon, Ceylon cinnamon, Indian cassia, Sri Lanka cinnamon, true cinnamon; *Arabic:* irfa, qurfa; *Chinese:* jou-kwei; *Dutch:* kaneel; *French:* canelle, canellier, canelle de Ceylan; *German:* Ceylon-Zimt, Echter Zimt, Kaneel, Kanehl, Seychellen-Zimt; *Hindi:* dal-chini, darcini; *Italian:* canella, cannella di Ceylon; *Malaysian:* kayu manis; *Mexico:* guiña castilla, cicanaga latyaga, guiña xtilla ticanaca; *Pilipino:* cinnamon canela; *Portuguese:* canela, caneleira em ceilão; *Russian:* zejsonskaja koriza; *Sanskrit:* durasita, tvak; *Slovenian:* škorica; *Spanish:* canela, canela de Ceilán
Usage ▶ spice (flavoring, (bark); **product:** essential oils (leaves)
Parts Used ▶ bark, leaf
Distribution ▶ Sri Lanka, S West Indies; widely cultivated in the Tropics: Indonesia Java, Brazil, Martinique, Jamaica, Madagascar

Cinnamomum zeylanicum: a flowering, b dry bark

Berger 1, 1949; Bicking 1986; Bois 1934; Boisvert/Hubert 2000; Bournot 1968; Brown 1955; Cheers 1998; Craze 2002; Dalby 2000; Dastur 1954; Davidson 1999; Domrös 1973; Dudtschenko et al. 1989; Duke et al. 2003; Erhardt et al. 2002; Farrell 1985; Ferrão 1992; Guzman/Siemonsma 1999; Hager 1992; Hanelt 2001; Hepper 1992; Hiller/Melzig 1999; Hohmann et al. 2001; Hoppe 1949; Jayaprakasha et al. 2003; Jiroretz et al. 2000, 2001; Karig 1975; Kojoma et al. 2002; Krützfeld 2002; Lallemand et al. 2000; Lewington 1990; Liu/Ou 1969; Mallavarapu et al. 1995; Melchior/Kastner 1974; Morris Mackley 1999; Mors/Rizzini 1961; Nath et al. 1996a; Newall et al. 1996; Norman 1991; Parain 1986; Peter 2001; Pochljobkin 1974, 1977; Poole/Poole 1994; Pruthi 1976; Pschyrembel 1998; Pursglove 1968; Raina et al. 2001; Rau 1994; Rosengarten 1969; Schenck/Naundorf 1966; Schneider 1988; Schröder 1991; Schultze-Motel 1986; Seidemann 1993c, 1998/2000; Seidemann/Siebert 1987; Sharma 2003; Shiva et al. 2002; Siewek 1990; Staesche 1972; Täufel et al. 1993; Teuscher 2003; Vaupel 2002b; Villamar et al. 1994; Wijesekera 1978; Wüstenfeld/Haensel 1964; Wyk et al. 2004; Zeven/de Wet 1982

CISSUS L. - Grape Ivy - Vitaceae

 Cissus arborea *Forrsk.*

▶ *Salvadora persica* L.

 Cissus quadrangularis *L.*

Synonyms ▶ *Vitis quadrangularis* Wall.
Common Names ▶ edible stemmed vine; *French:* vigne de Bakel, cissus de Galam, *German:* Vierkantige Klimme; *India:* vajra valli, cannalam paranta, harjora, hadsandhi
Usage ▶ condiment
Parts Used ▶ young shoot
Distribution ▶ Arabia, Ethiopia, tropical Africa, S Africa: Natal; Madagascar, Iraq, India, Sri Lanka, Philippines, Malaysian Archipelago
Note ▶ The young shoots are a curry ingredient in India and Malaysia.

Ambasta 1986; Burkill 5, 2000; Erhardt et al. 2002; v. Koenen 1996; Sharma 2003; Uphof 1968; Wealth of India 2, 1988

CITRUS L. - Lemon, Lime, Orange - Rutaceae

Davis/Albrigo 1994; Dugo/di Giacomo 2002; Klock 1998, 2001; Klock/Klock 2002; Mazza 1998; Ochse et al. 1961; Ramón-Laca 2003; Saunt 2000; Schirarend/Heilmeyer 1996; Shi et al. 2002; Steiner/Hochhausen 1952

 Citrus amblycarpa *(Hassk.) Ochse*

Synonyms ▶ *Citrus limonellus* Hassk. var. *amblycarpa* Hassk., *Citrus nobilis* Lour. var. *amblycarpa* (Hassk.) Ochse et de Vries
Common Names ▶ *Indonesian:* jeruk limau, jeruk limu, jeruk sambal
Usage ▶ juice: condiment (e.g. sambal, soto and bahmie)
leaf: substitute for the leaves of *Citrus hystrix* DC., also to perfume washing water
Parts Used ▶ juice of immature fruits; leaf

Distribution ▶ cultivated in Indonesia (Java)

Guzman/Siemonsma 1999

 Citrus aurantiifolia *(Christm. et Panz.) Swingle*

Synonyms ▶ *Citrus javanica* Bl., *Citrus lima* Lunan, *Citrus notissima* Blanco, *Limonia acidissima* Houtt., *Limonia aurantiifolia* Christm. et Panz.
Common Names ▶ Adams apple, key lime, lime, limon, Mexican lime, sour lime, Persian lime, West Indian lime; *Chinese:* suan ning mêng, limau nipis; *Dutch:* limmetje; *French:* limettier, limette acide, citron vert; *German:* Saure Limette, Limone, Limonelle; *Hindi:* kaghzi mimboo; *India:* erumichinarakam, nimma, limbe; *Indonesian:* jeruk niois; *Italian:* limette; *Javanese:* djěrook pětjěl; *Malaysian:* limau amkian, limau kapas; *Mexico:* guela castilla, guela lima, guela xtilla; *Pilipino:* dayap; *Portuguese:* lima, limão, lima ácida; *Spanish*: lima, limero, limón ceuti; *Thai:* magrood, manao; *Vietnamese:* cam
Usage ▶ spice (peel), flavoring (juice); **product**: essential oil (peel)
Parts Used ▶ peel, juice
Distribution ▶ widely cultivated in tropical and subtropical, probable origin tropical Asia
Note ▶ The unripe peel is termed Curacao peel.

Ayensu 1978; Bärtels 1997; Bovill/Reeve 2003; Bournot 1968; Charalambous 1994; Cheers 1998; Chisholm et al. 2003; Davidson 1999; Dugo/di Giacomo 2002; Engel/Phummai 2000; Hanelt 2001; Hiller/Melzig 1999; Kumar 2001; Minh Tu et al. 2002; Pursglove 1968; Robards et al. 1997; Roth/Korzmann 1997; Schultze-Motel 1086; Seidemann 1993c; Sharma 2003; Täufel et al. 1993; Uphof 1968; Venkateshwarlu/Selvaraj 2000; Verheij/Coronel 1991; Villamar et al. 1994; Zeven/de Wet 1982

 Citrus aurantium *L.*

Synonyms ▶ *Citrus bigaradia* Loisel; *Citrus vulgaris* Risso
Common Names ▶ bigarade, bitter orange, Seville orange, orange (American) (ssp. *amara* (L.), English: sour orange); *Arabic:* arenddj, narenddj; *Brazil (Portuguese):* laranja da terra; *Chinese:* suan-chêng, zhishi; *Dutch:* oranjeappel; *French:* bigardier, d'orange amère; *German:* Bigarde, Bitter-orange,

Pomeranze; *India:* khatta, nebu, limu, nibu, dantaharshana; *Italian:* arancio amaro, arancia fiori, arancia forte, meolangolo; *Japanese:* dai-dai, ka busu, kaisei to; *Korean:* kwanggyulnamu; *Portuguese:* laranjeira azêda; *Russian:* apelcin kuslyj, pomeranez; *Spanish:* naranjo agrio, naranjo amargo, toronja, flor de azahar

Usage ▸ spice, flavoring (juice); **product:** essential oil (peel): Bigarad(i) oil

Parts Used ▸ flower, peel, juice

Distribution ▸ : widely cultivated in tropical and subtropical regions; Indochina

Bärtels 1997; Berger 1, 1949; 3, 1952; Boelens 1991; Boelens et al. 1989; Bournot 1968; Cheers 1998; Dudtschenko et al. 1989; Dugo/di Giacomo 2002; Hager 4, 1992; Hanelt 2001; Hiller/Melzig 1999; Hoffmann et al. 1992; Hohmann et al. 2001; Hoppe 1949; Kirbaslar et al. 2001; Kirbaslar/Kirbaslar 2003; Kumar 2001; Lewington 1990; Lin Zheng-kui et al. 1986; Melchior/Kastner 1974; Minh Tu et al. 2002; Mondello et al. 2003; Oyen/Dung 1999; Pschyrembel 1998; Pursglove 1968: Schönfelder 2001; Schultze-Motel 1986; Seidemann 1993c; Seidemann/Siebert 1987; Sharma 2003; Shiva et al. 2002; Staesche 1972; Täufel et al. 1993; Teuscher 2003; Ubillos 1989; Uphof 1968; Villamar et al. 1994; Wagner et al. 1975; Wüstenfeld/Haensel 1964; Wyk et al. 2004; Zeven/de Wet 1982

Citrus aurantium L. ssp. bergamia *(Risso et Poit) Wight et Arn. ex Engl.*

Common Names ▸ curassao peel; *German:* Jacmalschalen, Curassaoschalen

Usage ▸ spice, flavoring

Parts Used ▸ peel of unripened fruits

Distribution ▸ cultivated in tropical regions

Hoppe 1949; Kumar 2001; Seidemann 1993c; Täufel et al. 1993; Wüstenfeld/Haeseler 1964

Citrus aurantium L. var. curassaviensis

▸ *Citrus aurantium L. (note)*

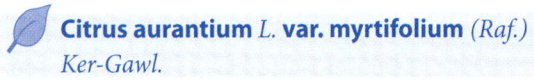

▸ *Citrus myrtifolia Raf.*

Citrus aurantium L. var. sinensis L.

▸ *Citrus sinensis (L.) Osbeck*

Citrus bergamia Risso et Poit.

Synonyms ▸ *Citrus aurantium* L. ssp. *bergamia* (Risso et Poit.) Wight et Arn. ex Engl.

Common Names ▸ bergamot, bergamot orange; *Dutch:* bergamot; *French:* bergamot(t)e, bergamotier; *German:* Bergamotte, Bergamot-Orange; *Indonesian:* bergamet; *Italian:* bergamotto (albero); *Portuguese:* (baranjeira) bergamota; *Russian:* bergamot (plod); *Spanish:* bergamota

Usage ▸ **product:** essential oil (bergamot oil)

Parts Used ▸ peel of ripe fruits

Distribution ▸ S Italy, cultivated in Turkey

Baser et al. 1995; Bendel 2002; Bournot 1968; Casabianca 1994; Charalambous 1994; Davidson 1999; Dudtschenko et al. 1989; Dugo/di Giacomo 2002; Erhardt et al. 2002; Hager 3, 1992; Hanelt 2001; Hiller/Melzig 1999; Huang et al. 1990; Kirbaslar et al. 2000; Kumar 2001; Lewington 1990; Mazza 1986; Mondello et al. 2003; Mosandl/Juchelka 1997; Oyen/Dung 1999; Poiana et al. 1994; Roth/Korzmann 1997; Schönfelder 2001; Schultze-Motel 1986; Seidemann 1993c; Small 1997; Spinelli 1951; Uphof 1968; Verzera et al. 2000, 2003

Citrus bigaradia Loisel

▸ *Citrus aurantium L.*

Citrus cedra Link

▸ *Citrus medica L.*

Citrus crassa Hassk.

▸ *Citrus medica L.*

Citrus deliciosa Tenore

▸ *Citrus reticulata Blanco*

Citrus fragrans Salisb.

▸ *Citrus medica* L.

**Citrus hystrix** DC.

Common Names ▸ Indian lemon, kaffir lime, leech lime, makrut lime, Mauritius papeda; *Chinese:* eabuyao; *French:* citron combara, combavas; *German:* Indisches Zitronenblatt, Kaffir-Limette, Langdorniger Orangenbaum; *Indonesian:* daoen djeroek poeroet; daun jeruk purut; *Malaysian:* limau purut; *Pilipino:* amontau, cabuyau; *Russian:* lajm, limetta; *Thai:* bai ma-krut; *Vietnamese:* la chanch

Usage ▸ spice

Parts Used ▸ leaf, fruit

Distribution ▸ India, S China, Indonesia, Sumatra, Philippines, sporadically cultivated in Mediterranean region

Bendel 2002; Davidson 1999; Dugo/de Giacomo 2002; Kumar 2003; Melchior/Kastner 1974; Morris/Mackley 1999; Murakami et al. 1995; Norman 1991; Schultze-Motel 1986; Seidemann 1993c, 1998/2000; Seidemann/Siebert 1987; Siewek 1990; Strauß 1969; Staesche 1972; Täufel et al. 1993; Uphof 1968; Zeven/de Wet 1982

Citrus japonica Thunb.

▸ *Fortunelle japonica (Thunb.) Swingle*

Citrus javanica Bl.

▸ *Citrus aurantiifolia (Christm. ex Panz.) Swingle*

Citrus junos Sieb. ex Tanaka

Common Names ▸ *Japanese:* yuzu
Usage ▸ flavoring
Parts Used ▸ fruit
Distribution ▸ cultivated in China and Japan

Wiersema/León 1999

Citrus latifolia Tanaka

Common Names ▸ limon, Persian lime, Tahiti lime, lime, bear's lime, sour lime; *French:* limettier, limette acide, Perse limette, Mexican limette, citron vert; *German:* Breitblättige Limette, Persische Limette

Usage ▸ spice (peel), flavoring (juice) **product**: essential oil (peel)

Parts Used ▸ fruit (juice), peel

Distribution ▸ Tahiti; widely cultivated in tropical and subtropical regions: S America: Brazil, Mexiko, The Argentine; USA: Florida; Europe: Spain

Note ▸ The fruits are larger than the fruits of the Mexican or key lime (→*Citrus aurantiifolia* (Christm. et Panz.) Swingle

Bovill/Reeve 2003; Dugo/di Giacomo 2002; Erhardt et al. 2000, 2002; Hanelt 2001; Schultze-Motel 1986; Wiersema/León 1999

Citrus lima Lunan

▸ *Citrus aurantiifolia (Christm. et Panz.) Swingle*

Citrus limon (L.) Burm.f.

Synonyms ▸ *Citrus limonum* Risso

Common Names ▸ lemon, limum; *Arabic:* laimun, loom; *Brazil (Portuguese):* limão galego; *Chinese:* li mung, ning mêng; *Dutch:* citroen; *French:* citron, citronnier, limonier; *German:* Zitrone, Limone; *Hindi:* jambiri, nimbu; *India:* bijapura, idalimbu; *Italian:* citreno, limone; *Japanese:* remon; *Portuguese:* limão, limoeiro azedo; *Russian:* limona; *Sanskrit:* jambira; *Spanish:* limón, limonero

Usage ▸ spice, flavoring; **product**: essential oil (peel)
Parts Used ▸ peel, fruit
Distribution ▸ widely cult in tropical and subtropical regions

Bärtels 1977; Bendel 2002; Boise 1934; Bose 1985; Bournot 1968; Braunsdorf et al. 1993; Chamblee et al 1991; Cheers 1998; Davidson 1999; Dudtschenko et al. 1989; Dugo 1994; Dugo/di Giacomo 2002; Dugo et al. 1995; Hager 3, 1992; Hanelt 2001; Hiller/Melzig 1999; Hoppe 1949; Ikeda et al. 1962; Juchelka 1997; Kumar 2001; Lawrence 1982; Lewington 1990; Lund/Bryan 1976; MacHale/Sheridan 1988; Melchior/Kastner 1974; Miyake et al. 1997, 1999; Mondello et al. 2003; Philipp/Isengard 1995; Pino et al. 1993; Pochljobkin 1974, 1977; Risch/Herrmann 1988; Robards et al. 1997;

Roth/Korzmann 1997; Schönfelder 2001; Schultze-Motel 1986; Seidemann 1993c; Seidemann/Siebert 1987; Sfikas 1994; Sharma 2003; Sheppard/Boyd 1970; Shiva et al. 2002; Staesche 1972; Staley/Vannier 1957; Täufel et al. 1993; Teuscher 2003; Uphof 1968; Verheij/Coronel 1991; Verzera et al. 2001; Villamar et al. 1994; WHO 1990; Wiersema/León 1999; Wyk 2004; Zeven/de Wet 1982

Citrus limonellus Hassk. var. amyblycarpa Hassk.

> *Citrus amblycarpa* (Hassk.) Ochse

Citrus limonum Risso

> *Citrus limon* (L.) Burm.f.

Citrus madurensis Lour.

> *Fortunella japonica* (Thunb.) Swingle

Citrus margarita Lour.

> *Fortunelle margarita* (Lour.) Swingle

Citrus medica L.

Synonyms ▶ *Citrus cedra* Link. *Citrus crassa* Hassk, *Citrus fragrans* Salisb., *Citrus odorota* Roussel

Common Names ▶ cedrat, citron, citron peel, succade; *Arabic:* gares; *Chinese:* kou-yüan, hsiang yuan, xiang; *Dutch:* cedraat, sukade; *French:* cédrat, cédratier; citron de Medie; *German:* Zedratzitrone, Zitronat, Zitronatzitrone, Zedernfrucht; *Hindi:* bara nimbu, bijaura, turanj; *India:* gilam, mahalunga, mahaphala, kadaranarathai; *Italian:* cedro (frutto), cedrato; *Japanese:* musan ô maru Bushukan; *Javanese:* djerool bodong; *Malaysian:* limau susu; *Mexico:* bihuii nayi xtilla, pehuij nayi castillo; *Portuguese:* cidra, cidreira; *Russian:* zitron; *Spanish:* cidra, cidro, poncil; *Thai:* som mu'

Usage ▶ spice, flavoring

Parts Used ▶ fruit peel

Distribution ▶ only cultivated in Italy, Greece, Corsica, USA (California), Brazil; possible origin in India

◘ Citrus medica, half candied fruit

Note ▶ The etrog citron (*Citrus medica* L. var. *ethrog* Engl.) is used by Jews at the Feast of Tabernacles.

Bärtels 1997; Bendel 2002; Berger 1, 1949; Bois 1934; Bournot 1968; Camperi et al. 2002; Cheers 1998; Davidson 1999; Dudtschenko et al. 1989; Dugo/di Giacomo 2002; Hanelt 2001; Hepper 1992; Hiller/Melzig 1999; Hoppe 1949; Kumar 2001, 2003; Melchior/Kastner 1974; Pursglove 1968; Schönfelder 2001; Schultze-Motel 1986; Seidemann 1993c; Seidemann/Siebert 1987; Shiva et al. 2002; Staesche 1972; Täufel et al. 1993; euscher 2003; Uphof 1968; Verheij/Coronel 1991; Villamar et al. 1994; Wiersema/León 1999; Zeven/de Wet 1982

Citrus medica L. var. ethrog Engl.

> *Citrus medica* L.

Citrus medica L. var. sarcodactylis (Noot.) Swingle

Synonyms ▶ *Citrus sarcodactylis* Noot.

Common Names ▶ 'Buddah's hand, finger citron; *Chinese:* fo shou kan, shu kann; *French:* main de Bouddha, sacrodactyle, cédrat digité, *German:* Gefingerte

Zitrone, Finger-Zitrone; *Indonesian:* phât tu; *Malaysian:* jěruk tangan, limau jari

Usage ▶ spice
Parts Used ▶ fruit peel
Distribution ▶ cultivated E, S Asia
Note ▶ The citron did not reach China until the 4th Century A.D. When it did, a freak form developed in which the fruit was separated into five (or more) lobes like the fingers of the hand. The fruit was not suitable unless fresh.

Davidson 1999; Dugo/di Giacomo 2002; Erhardt et al. 2002; Hanelt 2001; Mazza 1998; Schirarend/Heilmeyer 1996; Schultze-Motel 1986

Citrus medica ssp. limonum Hook.

▶ *Citrus limon Burm.*

Citrus myrtifolia Raf.

Synonyms ▶ *Citrus aurantium* L. var. *myrtifolium* (Raf.) Ker-Gawl.
Common Names ▶ myrtle leaf orange; *French:* chinois; *German:* Duftorange, Myrtenblättrige Pomeranze; *Italian:* chinotto; *Spanish:* naranja mirtifolia
Usage ▶ flavoring
Parts Used ▶ flower, fruit peel
Distribution ▶ ?China

Facciola 1990; Klock 2001;

Citrus nobilis Lour. var. amblycarpa (Hassk.) Ochse et De Vries.

▶ *Citrus amblycarpa (Hassk.) Ochse*

Citrus notissima Blanco

▶ *Citrus aurantiifolia (Christm. ex Panz.) Swingle*

Citrus odorota Roussel

▶ *Citrus medica L.*

Citrus reticulata Blanco

Synonyms ▶ *Citrus deliciosa* Tenore, *Citrus tangarine* hort. ex Tan.
Common Names ▶ clementine, santara orange, Swato orange, tangerine; *Chinese:* chen pi, chü, jui-sheng-nu, kan chü; *French:* tangarine; *German:* Mandarine, Tangarine; *Hindi:* santara; *Japanese:* ponkan; *Javanese:* djeroek garoet; *Pilipino:* narangita; *Portuguese:* tangarine; *Vietnamese:* may cam chia, quit, quat thuc
Usage ▶ flavoring, in candy and soft drinks; **product**: essential oil (from peel)
Parts Used ▶ peeled fruit
Distribution ▶ SE Asia, cultivated worldwide in tropical countries.
Note ▶ The tangerine fruit has a sweet, fresh quality.

Chisholm 2003; Dugo/di Giacomo 2002; Freeburg et al 1994; Hanelt 2001; Uphof 1968; Zeven de Wet 1982

Citrus sarcodactylis Noot.

▶ *Citrus medica L. var. sarcodactylis (Noot.) Swingle*

Citrus sinensis (L.) Osbeck

Synonyms ▶ *Citrus aurantium* L. var. *sinensis* L.
Common Names ▶ orange, blood orange, navel orange, sweet orange, Valencia orange; *Arabic:* bordguene; *Chinese:* cheng, tian-cheng; *Dutch:* zoete djeroek; *French:* orange (douce), oranger doux, pomme d'orange; *German:* Apfelsine, Blutorange, Nabelorange, (süße) Orange; *Hindi:* musambi, narangi; *India:* battavinarinja, kittile, sathagudi; *Italian:* arancio dolce, orangio dolce; *Korean:* kamkyulnamu, langgyulnamu; *Mexico:* yaga naraxo; *Portuguese:* laranja doce, laranjeira; *Russian:* apel'sina (plod); *Spanish:* naranja (dulce); *Vietnamese:* cáy cam
Usage ▶ spice (peel), flavoring; **product**: essential oil (peel, flowers; neroli oil, sweet orange oil)
Parts Used ▶ fruit, peel, flower
Distribution ▶ widely cultivated in tropical and subtropical regions, probable origin SE Asia

Bärtels 1997; Bendel 2002; Bournot 1968; Camperi et al. 2002;

Cheers 1998; Dudtschenko et al. 1989; Dugo 1995; Dugo/di Giacomo 2002; Hanelt 2001; Hiller/Melzig 1999; Hoffmann et al. 1992; Hoppe 1949; Kumar 2001; Lewington 1990; Lund et al. 1972; Matthews 1987; Minh Tu et al. 2002; Mondello et al. 2003; Moyler/Stephens 1992; Pino et al. 1992b; Pochljobkin 1974, 1977; Pschyrembel 1998; Pursglove 1968; Roth/Korzmann 1997; Schönfelder 2001; chultze-Motel 1986; Seidemann 1993c; Seidemann/Siebert 1987; Sharma 2003; Täufel et al. 1993; Uphof 1968; Verheij/Coronel 1991; Villamar et al. 1994; Zeven/de Wet 1982

Citrus trifolia Thunb.

▶ *Poncirus trifiliata (L.) Raf.*

Citrus trifoliata L.

▶ *Poncirus trifoliata (L.) Raf.*

Citrus vulgaris Risso

▶ *Citrus aurantium L.*

CLAOXYLON A. Juss. - Euphorbiaceae

Claoxylon indicum (Reinw. ex Bl.) Hassk.

Synonyms ▶ *Claoxylon polot* Merr., *Erythrochilus indicus* Reinw. ex Bl.
Common Names ▶ *Cambodian:* chhe:tô:ch; *Indonesian:* bleketupuk, talingkup, katerbik; *Malaysian:* lampin budak, laping pudak, sitampu; *Thai:* kha ka ai, khang namphung, ngun phung khao
Usage ▶ spice (for sauces)
Parts Used ▶ young leaf
Distribution ▶ India, throughout continental SE Asia and S China, Malaysia, Indonesia

Guzman/Siemonsma 1999; Oyen/Dung 1999; Uphof 1968

Claoxylon polot Merr.

▶ *Claoxylon indicum (Reinw. ex Bl.) Hassk.*

CLAUSENA Burm. - Rutaceae

Clausena anisata Hook. ex Benth.

Synonyms ▶ *Clausena inaequalis* (DC.) Benth.
Common Names ▶ horsewood, clausena; *German:* Anis-Clausenie, Anisblätter; *Kenya:* mutasia, siunya
Usage ▶ spice
Parts Used ▶ leaf
Distribution ▶ native in tropical W Africa, naturalized in Java, Philippines, Indian Ocean Islands; cultivated in Indonesia

Ayensu 1978; Gundidza et al. 1994; Gurib-Fakim/Brendler 2004; Molino 1993, 2000; Oyen/Dung 1999; Schultze-Motel 1986; Seidemann 1993c; Täufel et al. 1993; Teuscher 2003; Uphof 1968; Westphal/Jansen 1989; Wiersema León 1999

Clausena anisum-olens (Blanco) Merr.

Synonyms ▶ *Clausena laxiflora* Quis et Mer.; *Clausena sanki* (Perr.) Molino; *Cookia anisumolens* Blanco
Common Names ▶ *Pilipino:* anis, kayumanis, danglais
Usage ▶ condiment, flavoring cigarettes; **product:** essential oil, substitute for anise oil
Parts Used ▶ leaf
Distribution ▶ Endemic to the Philippines and Borneo; all over the archipelago also cultivated, in China, Taiwan, Vietnam and Indonesia sporadically cultivated in gardens.
Note ▶ The essential oils are used for the preparation the Philippines drink 'Anisado'.

Hanelt 2001; Molino 1993, 1995; Oyen/Dung 1999; Schultze-Motel 1986; Uphof 1968

Clausena excavata Burm.f.

Synonyms ▶ *Clausena punctata* Wight et Arb.
Common Names ▶ pink lime-berry, Hollywood clausena; *Chinese:* shan huang pi; *Hindi:* agnijal, *Javanese:* bagal tikus, *Malaysian:* sěmeru, pokok cherek, pokok kěmantu; *Thai:* fia fan, *Vietnamese:* giôi, ti-kusan

Usage ▶ spice
Parts Used ▶ leaf
Distribution ▶ India, China, lower Himalayan ranges and in NE hills, E Asia, Indochina, Malaysia
Note ▶ The leaves used like ❯ *Murraya koenigii (L.) Spreng.* leaves.

Arora/Pandey 1996; Burkill 1960; Hanelt 2001; Wealth of India 2, 1950; Wiersema/León 1999

Clausena inaequalis *(DC.) Benth.*

❯ *Clausena anisata Hook. ex Benth.*

Clausena indica *Oliv.*

Common Names ▶ Indian clausena; *German:* Indische Clausenie
Usage ▶ flavoring curries
Parts Used ▶ leaf
Distribution ▶ India: W Ghats

Arora/Pandey 1996; Wiersema/León 1999

Clausena lansium *(Lour.) Skeels*

Synonyms ▶ *Clausena wampi* (Blanco) Oliv.
Common Names ▶ Chinese wampee, wampi; *Chinese:* huang pi gua, wong pei, womg pa, wong poi; *French:* vampi; *German:* Clausenieblätter; *Korean:* hwangphinamu; *Malaysian:* wampee, wampi, wampoi; *Pilipino:* huampit, uampi galumpi; *Thai:* som ma fai; *Vietnamese:* hong bi
Usage ▶ spice; **product:** essential oil
Parts Used ▶ leaf
Distribution ▶ S China, Indochina, widely cultivated in tropical and subtropical Asia: Philippines, Vietnam; Australia, Hawaii, Cuba
Note ▶ The essential oil is a substitute for anise oil. The name of the fruit is wampi or wampee.

Cheers 1998; Davidson 1999; Erhardt et al. 2002; Hanelt 2001; Schultze-Motel 1986; Uphof 1068; Verheij/Coronel 1991; Westphal/Jansen 1989; Wiersema/León 1999; Zeven/de Wet 1982

Clausena laxiflora *Quis et Merr.*

❯ *Clausena anisum-olens (Blanco) Merr.*

Clausena punctata *Wight et Arn.*

❯ *Clausena excavata Burm.f.*

Clausena sanki *(Perr.) Molino*

❯ *Clausena anisum-olens (Blanco) Merr.*

Clausena wampi *(Blanco) Oliv.*

❯ *Clausena lansium (Lour.) Skeels*

Claytonia perfoliata *Donn*

❯ *Montia perfoliata (Donn) Howell*

CLEOME L. - Spider Flower - Capparaceae

Cleome gynandra *L.*

Synonyms ▶ *Cleome pentaphylla* (L.) DC., *Gynandropsis gynandra* (L.) Briq., *Gynandropis pentaphylla* (L.) DC.
Common Names ▶ African mustard, African spider flower, bastard mustard, cat's whiskers, kaffir caggage, spider herb, spider wisp; *Arabic:* abu qaru; *Chinese:* bai hua cai; *French:* feuilles caya, mozambé, mouzambe; *German:* Afrikanische Spinnenpflanze, Bastardsenf, Katzenschnurbart, Massarubee, Senfkaper; *India:* hulul, hurhur, karaila; *Japanese:* fû-chô-sô; *Malaysian:* manam; *Pilipino:* apuy-apuyan, balabalanoyan, tantandok, kulaya; *Portuguese:* mussambé cor de rosa, mussambé de cinco folhas; *Spanish:* caya mouzambi, volatín, volantincillo; *Thai:* phak sian; *West Indies:* massarubee
Usage ▶ spice, condiment (in Gabun), pot-herb (in India)
Parts Used ▶ leaf, seed, seed capsule

Distribution ▶ tropical Africa, tropical Asia: India, China, Indochina; native in tropical America

Note ▶ Ground as mustard.

Arora/Pandey 1996; Burkill 1, 1985; Davidson 1999; Erhardt et al. 2000, 2002; v. Koenen 1996; Oome/Grubben 1978; Schultze-Motel 1986; Seidemann 1993c; Siemonsma/Piluek 1993; Täufel et al. 1993; Wiersema/León 1999; Zeven/de Wet 1982

 Cleome monophylla L.

Common Names ▶ single-leave cleome, spindlepod; *German:* Einblättrige Spinnenpflanze

Usage ▶ pot-herb (by the Santals)

Parts Used ▶ leaf

Distribution ▶ W Africa: Senegal, Nigeria; tropical Asia, especially India

Burkill 1, 1985; Irvine 1952; Uphof 1968

 Cleome pentaphylla (L.) DC.

 *Cleome gynandra* L.

CLINOPODIUM L. - Calamint - Lamiaceae (Labiatae)

 Clinopodium vulgare L. ssp. vulgare

Synonyms ▶ *Calamintha clinopodium* Benth., *Melissa clinopodium* Benth., *Melissa vulgare* Trev.

Common Names ▶ cushion calamint, dog mint, basil weed, wild basil; *French:* calamint commun, clinopode; *German:* Wilder Basililkum, Wilde Melisse, Wirbeldost, Gewöhnliche Clinopode

Usage ▶ flavoring

Parts Used ▶ leaf

Distribution ▶ Europe, Caucasus, W, E Siberia

Erhardt et al. 2002; Small 1997; Uphof 1968

CNICUS L. - Blessed Thistle - Asteraceae (Compositae)

 Cnicus benedictus L.

Synonyms ▶ Centaurea benedicta (L.) L.

Common Names ▶ blessed thistle, holy thistle; *French:* chardon bénit; *German:* Benediktenkraut, Bitterdistel, Kardobenediktenkraut; *Italian:* cardo santo, cnicus; *Portuguese:* cardo bento; *Russian:* benedikt aptetschnyj, kardobenedikt; boltschez blagocloblennyj; *Spanish:* cardo bendito, cardo santo

Usage ▶ spice, flavoring for liqueur and spirits

Parts Used ▶ herb

Distribution ▶ S and SW Europe, Caucasus, Iran, C, W Asia, Afghanistan, S Russia, Romania, N Africa

Aichele/Schwegler 4, 1995; Bärtels 1997; Berger 3, 1952; Berger 4, 1954; Coiciu/Racz (no year); Dudtschenko et al. 1989; Erhardt et al. 2002; Hanelt 2001; Heeger 1956; Hiller/Melzig 1999; Hohmann et al. 2001; Hoppe 1949; Newall et al. 1996; Pschyrembel 1998; Schneider/Lachner 1987; Schönfelder 2001; Schultze-Motel 1986; Seidemann 1993c; Täufel et al. 1993; Ulubelen/Berkan 1977; Uphof 1968; Vanhaelen-Fastré 1973, 1974; Vanhaelen-Fastré/Vanhaelen 1974; Wiersema/León 1999; Wyk et al. 2004; Zeven/de Wet 1982

CNIDIUM Cuss. ex Juss. - Apiaceae (Umbelliferae)

 Cnidium confertum Moench.

▶ *Cnidium monnieri* (L.) Cuss. ex Juss.

 Cnidium monnieri (L.) Cuss. ex Juss.

Synonyms ▶ *Athamantha chinensis* Lour., *Cnidium confertum* Moench., *Selinum monnieri* L.

Common Names ▶ *Chinese:* she ch'uang tzu, giêng sàng, ta hui hsiang ts'ao, xà sàng; *French:* cnide, ivêche de Chine; *German:* Monnier's Brenndolde; *Japanese:* oka-zeri

Usage ▶ spice, condiment (in Vietnam)

Parts Used ▶ fruit

Distribution ▶ E Asia: E Siberia, Mongolia, Korea; in S

Europe the plant grows wild; cultivated in the former Soviet Union, China, N Vietnam and Laos

Hanelt 2001; Schultze-Motel 1986; Uphof 1968

COCCINIA Wight et Arn. - Cucurbitaceae

 Coccinia cordifolia Cogn.

Synonyms ▶ *Coccinia grandis* (L.) Voigt, *Coccinia indica* Wight et Arn.
Common Names ▶ ivy gourd; *Chinese:* hong gua; *German:* Indische Scharlachranke; *Hindi:* kundri, telacucha; *Malaysian:* kabare kindi, pepasan; *Thai:* tam lung;
Usage ▶ pot-herb (pensinsular India)
Parts Used ▶ tender shoots
Distribution ▶ tropical Asia, tropical Africa, Australia, native in tropical America

Arora/Pandey 1996; Hanelt 2001; Täufel et al. 1993; Wiersema/León 1999; Zeven/de Wet 1982

 Coccinia grandis (L.) Voigt

▶ *Coccinia cordifolia* Cogn.

 Coccinia indica Wight et Arn.

▶ *Coccinia cordifolia* Cogn.

COCHLEARIA L. - Scurvygrass - Brassicacea (Cruciferae)

 Cochlearia armoracia Lam.

▶ *Armoracia rustican* Gaertn. et B. Mey. et Scherb.

 Cochlearia danica L.

Common Names ▶ Danish scurvy grass; *German:* Dänisches Löffelkraut
Usage ▶ pot-herb
Parts Used ▶ fresh leaf
Distribution ▶ Europe, especially Iberian coasts

Erhardt et al. 2001; Small 1997

 Cochlearia officinalis L.

Common Names ▶ (common) scurvy grass, scorbute grass, spoonwort; *Brazil (Portuguese):* cochleária; *French:* cochléaire, cranson, herbe à la cuiller, herbe aux cuillers, herbe aux scorbut, raifort officinal; *German:* Bitterkresse, Echtes Löffelkraut, Löffelkresse, Scharbockskraut, Skorbutkraut; *Italian:* coclearia, erba cocchiara; *Russian:* gor'kij kress, loschetschnik, warucha, morskoj calat; *Spanish:* coclearia, hierba de las cucharas
Usage ▶ pot-herb
Parts Used ▶ fresh leaf
Distribution ▶ Europe, Siberia, Kamchatka, N Japan, sporadically cultivated

Aichele/Schwegler 3, 1995; Berger 4, 1954; Bilgri/Adam 2000; Cheers 1998; Erhardt et al. 2002; Fleischhauer 2003; Hager 4, 1992; Hanelt 2001; Heeger 1956; Hiller/Melzig 1999; Hoppe 1949; Pochljobkin 1974, 1977; Pruthi 1976; Schönfelder 2001; Schultze-Motel 1986; Seidemann 1993c; Small 1997; Täufel et al. 1993; Teuscher 2003; Uphof 1968; Vanhaelen-Fastre 1973; Zeven/de Wet 1982

 Cochlearia wasabi Sieb.

▶ *Wasabia japonica (Miq.) Matsum.*

CODIAEUM A. Juss. - Croton - Euphorbiaceae

 Codiaeum variegatum (L.) Bl.

Synonyms ▶ *Croton variegatum* L.
Common Names ▶ Bombay laurel; *German:* Indischer Lorbeer, Indischer Wunderstrauch; *Japanese:* henyô-boku; *Malaysian:* puding mas; *Thai:* kohson

Usage ▶ spice
Parts Used ▶ young leaf
Distribution ▶ Moluccas, India, also cultivated in the tropics
Note ▶ The plant contains toxic and allergy-producing compounds (Phorbolesters?).

Erhardt et al. 2002; Hanelt 2001; Schultze-Motel 1986; Seidemann 1993c; Storrs 1997; Teuscher 2003; Wealth of India 2, 1950

COELOCARYON Warb. - Myristicaceae

Coelocaryon sphaerocarpum *Fouilloy*

Common Names ▶ Wild nutmeg; *German:* Wilder Muskat
Usage ▶ substitute for nutmeg (locally)
Parts Used ▶ seed, aril
Distribution ▶ Africa: Cameroon
Note ▶ The arils and seeds have a slightly spicy fragrance.

Burkill 4, 1997; Neuwinger 1999

Coleus amboinicus *Lour.*

▶ *Plectranthus amboinicus (Lour.) Spreng.*

Coleus aromaticus *Benth.*

▶ *Plectranthus amboinicus (Lour.) Spreng.*

Coleus barbartus *(Andr.) Benth.*

▶ *Plectranthus barbartus Andr.*

COLOCASIA Schott - Taro - Araceae

Colocasia gigantea *(Blume) Hook.*

Synonyms ▶ *Colocasia indica* auct. non (Lour.) Kunth; *Leucocasia gigantea* (Blume ex Hassk.) Schott
Common Names ▶ great dasheen, great taro; *German:* Großer Taro; *Hindi:* talas padang; *Thai:* khuun
Usage ▶ condiment
Parts Used ▶ fruit
Distribution ▶ India, Indochina, Malaysian Archipelago

Burkill 1966; Erhardt et al. 2000, 2002; Hanelt 2001; Oyen/Dung 1999; Wiersema/León 1999

Colocasia indica auct. non *(Lour.) Kunth*

▶ *Colocasia gigantea (Blume) Hook*

COLURIA R. Br. - Rosaceae

Coluria geoides *(Pall.) Ledeb.*

Synonyms ▶ *Coluria potentilloides* R. Br.; *Geum laxmanni* Gaertn, *Geum potentilloides* Ait.
Common Names ▶ Siberian avens, clove root, clove oil plant; *German:* Sibirische Nelkenwurzel; *Russian:* koljurija gravilatovidnaja
Usage ▶ spice, condiment for alcoholic beverages; **product:** essential oil
Parts Used ▶ root
Distribution ▶ Altai mountains: W, E Siberia, Mongolia, cultivated in Russia and Ukraine
Note ▶ The content of the essential oil is up to 90% Eugenol.

Dudtschenko et al. 1989; Erhardt et al. 2002; Hanelt 2001; Schultze-Motel 1986; Seidemann 1993c

Coluria potentilloides *R. Br.*

▶ *Coluria geoides (Pall.) Ledeb.*

COMBRETUM Loefl. - Combretaceae

Combretum racemosum *F. Beauv.*

Common Names ▶ *Congo:* usonsumbi; *German:* Traubiger Langfaden
Usage ▶ spice
Parts Used ▶ herb
Distribution ▶ Africa: Congo

Ayensu 1978; Terra 1966

COMMIPHORA Jacq. - Myrrh - Burseraceae

Commiphora abyssinica *Engl.*

Common Names ▶ Arabian myrrh; *German:* Arabische Myrrhe, Fadhi-Myrrhe, Weihrauch
Usage ▶ spice, flavoring, joss drug
Parts Used ▶ gummi resin
Distribution ▶ N Ethiopia, S Arabia

Erhardt 2002; Hepper 1992; Hoppe 1949; Lohs/Martinez 1987; Seidemann 1993c; Uphof 1968; Usher 1974; Wüstenfeld/Haensel 1964

Commiphora molmol *(Engl.) Engl.*

▶ *Commiphora myrrha (Nees) Engl. var. molmol Engl.*

Commiphora mukul *(Hook ex Stokes) Engl.*

▶ *Commiphora myrrha (Nees) Engl. var. molmol Engl.*

Commiphora myrrha *(Nees) Engl.* **var. molmol** *Engl.*

Synonyms ▶ *Commiphora molmol* (Engl.) Engl., *Commiphora mukul* (Hook ex Stokes) Engl.
Common Names ▶ African myrrh, Somali myrrh, common myrrh, herabol myrrh, myrrh, *Arabic:* myrrh; *Chinese:* mo yao, mu yao; *French:* myrrhe; *German:* Echte Myrrhe, Somalia-Myrrhe, Herabol-Myrrhe, Weihrauch; *Italian:* mirra; *Spaish:* mirra
Usage ▶ spice, flavoring, joss drug
Parts Used ▶ gummi resin
Distribution ▶ Somalia, Arabia,

Duke et al. 2003; Erhardt et al. 2002; Hepper 1992; Hoppe 1949; Lohs/Martinez 1987; Morteza-Semnani/Saeedi 2003; Peters 1927; Pschyrembel 1998; Qédan 1974; Seidemann 1993c; Täufel et al. 1993; Wüstenfeld/Haensel 1964; Wyk et al. 2004

Commiphora opobalsamum *(L.) Engl.*

Synonyms ▶ *Balsamodendron gileadense* Kunth.
Common Names ▶ balm of Gilead, balsam tree, Mecca myrrh; *French:* balsamier de la Mecque; *German:* Balsammyrrhe, Mekka-Myrrhe, Süße Myrrhe, Opoponax-Harz; *Italian:* balsamo della Mecca; *Turkish:* balsam Makkah
Usage ▶ flavoring, joss drug
Parts Used ▶ gummi resin
Distribution ▶ Sudan, Ethiopia, Somalia, Arabia
Note ▶ Formerly (11th–17th Century) cultivated in Egypt and Palestine

Erhardt et al. 2002; Hanelt 2001; Hoppe 1949; Peters 1927; Qédan 1974; Schultze-Motel 1986; Uphof 1968; Usher 1974; Zeven/de Wet 1982

CONVOLVULUS L. - Bindeweed - Convolvulaceae

Convolvulus dissectus *Jacq.*

▶ *Merremia dissecta (Jacq.) Hall.*

Convolvulus scoparius *L.*

Common Names ▶ Canary convolvulus; *German:* Kanarische (Besen-)Winde
Usage ▶ spice (flavoring); **product:** essential oil
Parts Used ▶ whole plant
Distribution ▶ Canary Islands (Teneriffe)

Note ▶ Often used as seasoning for olive oil.

Erhardt et al. 2002; Hoppe 1949; Oyen/Dung 1999

Conyza riparia DC.

▶ *Blumea chinensis* DC.

Cookia anisum-olens Blanco

▶ *Clausena anisum-olens* (Blanco) Merr.

CORDYLINE Comm. ex R. Br. - Cabbage Tree - Dracaenaceae (Agavaceae)

Cordyline fruticosa (L.) A. Chev.

▶ *Cordyline terminalis* (L.) Kunth

Cordyline terminalis (L.) Kunth

Synonyms ▶ *Cordyline fruticosa* (L.) A.Chev.
Common Names ▶ good luck plant, Chinese fire leaf, palm lily, Hawaiian good luck plant, tree of kings; *Brazil (Portuguese):* croton; *Chinese:* ya zhu na; *Dutch:* limiestruik; *German:* Endständige Keulenlilie, Endständige Kolbenlilie; *Hawaiian:* ki; *Malaysian:* andong, daun juang, juang, jenjuang, senjuang; *Pilipino:* tungkadi pari; *Tahiti:* auti; *Thai:* maak phuu, maak mia
Usage ▶ pot-herb, flavoring (of rice)
Parts Used ▶ young shoot
Distribution ▶ Malaysia. Archipelago, N Guinea, India, Indonesia, Oceania, NE Australia, New Zealand, Polynesia, Hawaii
Note ▶ During the 17th and early 18th Century, roots were fermented and distilled to produce spirits.

Barrau 1961; Burkill 3, 1995; Davidson 1999; Engler/Phummai 2000; Erhardt et al. 2002; Hanelt 2001; Uphof 1968; Zeven/de Wet 1982

CORIANDRUM L. – Coriander - Apiaceae (Umbelliferae)

Coriandrum globosum Salisb.

▶ *Coriandrum sativum* L.

Coriandrum majus Gouan

▶ *Coriandrum sativum* L.

Coriandrum sativum L.

Synonyms ▶ *Coriandrum majus* Gouan; *Coriandrum globosum* Salisb.
Common Names ▶ *Fruit*: coriander, collender; *Arabic:* kuzbara, tabel; *Chinese:* yuen sai, hsing sui, hu sui; *Dutch:* koriander; *French:* coriandre, punaise mâle; *German:* Koriander, Schwindelkorn, Stinkdill, Wanzendill, Wanzenkümmel; *Hindi:* dhania, dhanya; *India:* leela dhana; *Indonesian:* ketumbar *Italian:* coriandro, coriandolo; *Japanese:* koendroro, kûshibâ; *Javanese:* tumbar; *Korean:* kosu; *Malaysian:* kjintan, ketumbar, penjilang, wansui; *Portuguese:* coentro; *Russian:* coriandr, kischnez; *Sanskrit:* dhanyaka; *Slovenian:* koriander; *Spanish:* coriandro, cilandrio, cilantro, culantro; *Thai:* phakchi, phakhom, phakhom-noi; *Turkish:* kişniş; *Vietnamese:* ngo;
Leaf: Chinese parsley, Mexican parsley; *French:* persil arabe; *German:* Korianderblätter, Cilantro; *India:* leela dhana; *Indonesian:* ketumbar, daun ketumbar; *Malaysian:* ketumbar, daun ketumbar; *Pilipino:* unsuy; *Russish:* kinsa, koriandr, kischnez, koljandra, kindsi klopownik; *Sanskrit:* chatra, dhanyaka; *Spanish:* cilantro, culantro; *Thai:* bai pak, phak chee
Usage ▶ spice, pot-herb (leaves); **product**: essential oil (Coriander oil)
Parts Used ▶ fruit, leaf
Distribution ▶ S Europe, native; C, S America (Mexico), cultivated in Egypt, Morocco, France, Turkey, Hungary, Germany, Italy, Ukraine, China, India, Iran, USA
Note ▶ Wild unknown, but it has widely escaped cultivation and naturalization in S Europe and else-

⬛ Coriandrum sativum, flowering

Sharma 2003; Siewek 1990; Small 1997; Staesche 1972; Stoyanova et al. 2002; Taniguchi et al. 1996; Täufel et al. 1993; Teuscher 2003; Tindall 1983; Tainter/Grenis 1993; Tucker 1986; Ubillos 1989; Uphof 1968; Villamar et al. 1994; Wyk et al. 2004; Zeven/de Wet 1982

Coriandrum sativum L. var. microcephalum

Common Names ▶ Russian coriander, wild coriander; *German:* Russischer Koriander, Kleinköpfiger Koriander; *Russian:* jantar, coriandr
Usage ▶ spice
Parts Used ▶ fruit
Distribution ▶ S Russia, the Ukraine

Heeger 1956

Coridothymus capitatus *(L.) Rchb.*

▶ *Thymus capitatus (L.) Hoffm. et Link*

Cortinellus shiitake *P. Henn.*

▶ *Lentinus edodes (Berk.) Sing.*

COSMOS Cav. - Cosmos, Mexican Aster - Asteraceae (Compositae)

Cosmos atrosanguineus *(Hook.) Voss.*

Common Names ▶ black cosmos, chocolate cosmos; *German:* Schokoladenblume; Schwarze Kosmee, Schwarzes Schmuckkörbchen
Usage ▶ flavoring (rarely)
Parts Used ▶ flower
Distribution ▶ Mexico, also cultivated
Note ▶ On warm days, the plant has a chocolate fragrance. In Europe frequently cultivated as an ornamental plant.

Cheers 1998; Erhardt et al. 2002; Wiersema/León 1999

where. In Germany the fruits used with anise, fennel and caraway fruits are sold as "bread spice". In Thailand the roots (raak pak chee) are used similarly to *Petroselinum* roots.

Ahmed et al. 2001; Aichele/Schwegler 3, 1995; Arganosa et al. 1998; Bärtels 1997; Berger 3, 1952; Bois 1934; Boisvert/Hubert 2000; Bournot 1968; Bremness 2001; Burkill 5, 2000; Carruba et al. 2002; Chunhui Deng et al. 2003; Clair 1961; Coiciu/racz (no year); Coleman/Lawrence 1992; Cortes-Eslava et al. 2001; Craze 2002; Dalby 2000; Davidson 1999; Deng et al. 2003; Diederichsen 1996; Diederichsen/Hammer 1994; Dudtschenko et al. 1989; Fan et al. 2003; Farell 1985; Frank et al. 1995; Gil et al. 1999, 2002; Gupta et al. 1991; Guzman/Siemonsma 1999; Hager 4, 1992; Hanelt 2001; Heeger 1956; Hepper 1992; Herklots 1972; Hiller/Melzig 1999; Hohmann et al. 2001; Hondelmann 2002; Hoppe 1949; Illes et al. 2000; Jansen 1981; Kallio/Kerrola 1992; Kerrola/Kallio 1992; Ksandopulo et al 1995; Lenardis et al. 2000; Leung 1991; Lewington 1990; Lück 2000; Mac-Leod/Islam 1976; Mazza 2002; Melchior/Kastner 1974; Minija/Thoppil 2001; Misharina 2001; Morris/Mackley 1999; Norman 1991; Oliver 2003; Pande et al 2000; Perineau et al. 1991; Pino et al. 1993a; Pochljobkin 1974, 1977; Potter 1996; Potter/Fagerson 1990; Pruthi 1976; Pschyrembel 1998; Ramadan/Mörsel 2002c; Rosengarten 1969; Roth/Korzmann 1997; Schönfelder 2001; Schratz/Quadry 1996; Schröder 1991; Schultze-Motel 1986; Seidemann 1993c; Seidemann/Siebert 1987; Shankarachary/Natarajan 1971;

Cosmos sulphureus *Cav.*

Common Names ▶ orange cosmos, yellow cosmos; *German:* Schwefelgelbes Schmuckkörbchen, Gelbe Kosmee; *Indonesian:* kenikir; *Japanese:* kibana kosumosu; *Mexico:* xochipelli; *Russian:* kosmos tscheltys

Usage ▶ spice; **product:** essential oil

Parts Used ▶ plant

Distribution ▶ C and N Mexico, Meso-America, cultivated and native elsewhere in the Tropics

Note ▶ In Indonesia a strong aromatic plant.

Cheers 1998; Schultze-Motel 1986; Wiersema/León 1999

COSTUS L. - Costaceae (Zingiberaceae)

 Costus afer *Ker-Gawl.*

◘ Costus speciosus, flowering

Synonyms ▶ *Costus pterometra* K. Schum.

Common Names ▶ ginger lily, spiral ginger; *Congo:* moussanga-voulou, mussanga-vulu; *German:* Ingwerlilie, Spiralingwer

Usage ▶ spice (rarely), flavoring

Parts Used ▶ rhizom

Distribution ▶ E, W and W, C tropical Africa

Cheers 1998; Erhardt et al. 2002; Hager 4, 1992; Hanelt 2001; Schultze-Motel 1986; Uphof 1968

 Costus pterometra *K. Schum.*

❯ *Costus afer Ker-Gawl.*

 Costus speciosus *(J.G. König) Sm.*

Common Names ▶ cane reed, crape ginger, Malaysian ginger, white costas; *Chinese:* zhang liu tou; *German:* Prächtige Kostwurz, Malayischer Ingwer; *Hindi:* keu; *Indonesian:* pacing; *Javanese:* statjing; *Malaysian:* setawar, tawar tawar, tebu; *Sanskrit:* kemika; *Thai:* kushta, ueang phet maa; *Vietnamese:* cat loi, cây cu chóc, cú chóc, mia do, se vona

Usage ▶ spice, like ginger

Parts Used ▶ rhizom

Distribution ▶ Himalayas, India, Sri Lanka, Vietnam, Malaysia, New Guinea; in SE Asia and India also cultivated

Chauhan 1999; Cheers 1998; Duke et al. 2003; Engel/Phummai 2000; Ehrhardt et al. 2002; Hager 4, 1992; Hanelt 2001; Kottegota 1994; Kumar 2001, 2003; Küster 1987; Larsen et al. 1999; Ogle et al. 2003; Rätsch/Müller-Ebeling 2003; Santapau 1951; Schultze-Motel 1986; Seidemann 1998/2000; Sharma 2003; WHO 1990; Wong 1999

Costus zerumbet *Pers.*

❯ *Alpinia zerumbet (Pers.) B.L. Burtt et R.M. Sm.*

COTULA L. - Buttonweed - Asteraceae (Compositae)

 Cotula cinerea *Del.*

Common Names ▶ ash buttons, ashy cotula; *French:* cotule à gris cendre; *German:* Aschgraue Laugenblume
Usage ▶ condiment (from the Tuareg)
Parts Used ▶ herb
Distribution ▶ W Africa: Mali

Burkill 1, 1985

COULA Baill. - Oleaceae

 Coula edulis *Baill.*

Common Names ▶ Gaboon nut; *Congo:* kumumu; *German:* Gabunnuss
Usage ▶ condiment
Parts Used ▶ fermented seeds
Distribution ▶ tropical Africa

Mabberly 1997; Uphof 1968

 Coumarouna odorota *Aubl.*

▶ *Dipteryx odorota (Aubl.) Willd.*

COURBONIA Brongn. - Capparidaceae

 Courbonia virgata *Brongn.*

Usage ▶ spice
Parts Used ▶ plant ash from stems and leaves
Distribution ▶ tropical Africa

Uphof 1968

CRAMBE L. - Kale - Brassicaceae (Cruciferae)

 Crambe arborea *Webb. ex Christ*

Common Names ▶ *German:* Baumförmiger Meerkohl
Usage ▶ spice, rarely in the Canaries
Parts Used ▶ leaf
Distribution ▶ Canary Islands

Bramwell 1997

 Crambe cordata *Willd.*

▶ *Crambe cordifolia Steven*

 Crambe cordifolia *Steven*

Synonyms ▶ *Crambe cordata* Willd.
Common Names ▶ *German:* Herzblättriger Meerkohl; *Russian:* katran serdcelistnyj
Usage ▶ pot-herb (in India)
Parts Used ▶ young leaf
Distribution ▶ NW Himalayas (India), endemic of the N Margarin of the Caucasus

Arora/Pandey 1996; Erhardt et al. 2000, 2002; Hanelt 2001; Schultze-Motel 1986; Zeven/de Wet 1982

CRASSOCEPHALUM Moench - Asteraceae (Compositae)

 Crassocephalum rubens *(Juss.) S. Moore*

Common Names ▶ *Sierra Leone:* kikpoi
Usage ▶ pot-herb (for sauces and soups)
Parts Used ▶ young plants
Distribution ▶ W Africa: Ghana, Liberia, Nigera, Sierra Leone

Burkill 1, 1985

CRATAEVA L. - Capparaceae

Crataeva adansonii DC.

Synonyms ▶ *Crataeva religiosa* auct. non G. Forst.
Common Names ▶ temple plant; *German:* Tempelpflanze
Usage ▶ spice, pot-herb
Parts Used ▶ leaf
Distribution ▶ tropical Africa: Ghana, Mali, Niger, Upper Volta; China, E Asia, India, Indochina, Malaysia, cultivated elsewhere

Burkill 1, 1985; Chopra et al. 1956; Erhardt et al. 2002; Hanelt 2001

Crataeva marmalos L.

▶ *Aegle marmalos (L.) Corrêa ex Roxb.*

Crataeva religiosa G. Forst.

Synonyms ▶ *Crataeva tapia* Burm.
Common Names ▶ sacred garlic pear, sacred barma, temple tree; *German:* Spinnenbaum, Tempelbaum; *Hindi:* barna, baruna; *India:* barna bidasi, tikasag, titka sak, varuna; *Laos:* kunz; *Sanskrit:* varuna; *Thai:* kum, kum bok
Usage ▶ spice
Parts Used ▶ fruit
Distribution ▶ E Asia, India, Indochina, Malaysia, Australia, cultivated in India, S China, S Africa, SE Asia
Note ▶ The fruit has a garlic taste.

Cheers 1998; Chopra et al 1956; Dhar/Dhar 2000; Erhardt et al. 2002; Hanelt 2001; Lorenzi 1998; Mansfeld 1962; Schultze-Motel 1986; Seidemann 1993c

Crataeva religiosa Ainslie

▶ *Aegle marmelos (L.) Corrêa et Roxb.*

Crataeva religiosa auct. non G. Forst.

▶ *Crataeva adansonii DC.*

Crataeva tapia Burm.

▶ *Crataeva religiosa G. Forst.*

Crataeva tapia L.

Common Names ▶ tapia fruit, payagua; *Brazil (Portuguese):* cabaceira, pau-d'alho, tapiá, trapiá; *German:* Knoblauchbirne, Naranjille, Payagua, Tapiafrucht
Usage ▶ spice
Parts Used ▶ fruit
Distribution ▶ tropical America
Note ▶ The fruit has a garlic taste.

Chopra et al. 1956; Hanelt 2001; Lorenzi 1992; Lück 2004; Mansfeld 1962; Mors et al. 2000; Seidemann 1993c; Siemonsma/Piluek 1993; Villamara et al. 1994

CRITHMUM L. - Shampire - Apiaceae (Umbelliferae)

Crithmum maritimum L.

Common Names ▶ sea fennel, Peter's cress, rock samphire, (sea) samphire; *Arabic:* shanar bahariya; *Dutch:* zeevenkel; *French:* bacile, fenouil marin, fenouil de mer, herbe de Saint-Pierre, perre pierres; *German:* Seefenchel, Meerfenchel, Wasserfenchel; *Italian:* bacicci, critmo, erba San Pietro, finocchio marino; *Portuguese:* funcho do mar, funcho marino; *Russian:* serpnik, kritmum morskoj; *Spanish:* hinojo marino
Usage ▶ pot-herb (e.g. USA)
Parts Used ▶ fresh leaves, very occasionally the fruit
Distribution ▶ Europe: Iberia, Madaira, Greece, France, Krim, Turkey, Caucasus, W Asia, N Africa, Canaria Islands, native USA
Note ▶ The leaves have a bitter-salty taste.

dorf 1966; Seidemann 1993c, 2003; Staesche 1972; Teuscher 2003; Uphof 1968; de Vos 1984

◘ Crataeva tapia, fruits

Bärtels 1997; Baser et al. 2000; Bois 1934; Davidson 1999; Dudtschenko et al. 1989; Ehrhardt et al. 2002; Famini et al. 1999; Franke 1982; Lück 2004; Katsouri et al. 2001; Özcan et al. 2001; Schultze-Motel 1986; Schwenck/Naundorf 1966; Seidemann 1993c; Senatore/de Feo; 1994; Täufel et al. 1993; Tucker 1986; Uphof 1968; Usher 1974; Zeven/de Wet 1982

CROCOSMIA Planch. - Montbretia, Tritonia - Iridaceae

Crocosmia aurea *(Pappe ex Hook.) Planch.*

Synonyms ▸ *Tritonia aurea* (Hook.) Planch.
Common Names ▸ golden monbretia, tritonia; *German:* Crocus-Tritonie, Goldmontbretie, Kapsafran
Usage ▸ spice, sporadic substitute for saffron.
Parts Used ▸ flower
Distribution ▸ E Africa, S Africa: Cap region, Natal
Note ▸ This plant is frequently collected with

Batten/Bokelmann 1966; Berger 1, 1949; Burkill 2, 1994; Cheers 1998; Erhardt et al. 2002; Fenwick 2003; Joffe 1993; Schenck/Naun-

Crocosmia x crocosmiiflora *(Burbridge et Dean) N.E. Br.* ◘

Synonyms ▸ *Crocosmia aurea* x *Crocosmia pottsii*), *Tritonia x crocosmiiflora* (Lemoine ex E. Morr.) N.E Br.
Common Names ▸ cape saffron, crocus tritonia, garden montbretia; *German:* Garten-Montbretie, Fackellilie, Kapsafran
Usage ▸ spice, sporadic alternative for saffron
Parts Used ▸ flower
Distribution ▸ only cultivated
Note ▸ The hybrid originated 1880 in France.

Cheers 1998; Batten/Bokelmann 1966; Erhardt et al. 2002; Fenwick 2003; Seidemann 1993c, 2003; de Vos 1984

Crocoxylon croceum *(Thun.) N. Robinson*

▶ *Cassine crocea (Thun.) Kuntze*

CROCUS L. - Crocus - Iridaceae

Crocus biflorus *Mill.*

Common Names ▸ Scotch crocus; *German:* Frühlings-Krokus
Usage ▸ falsification of saffron (sporadically)
Parts Used ▸ stigma and rest of flowers
Distribution ▸ Europe, Russia, Turkey, Caucasus, N Iran, N Iraq

Erhardt et al. 2002; Hepper 1992; Melchior/Kastner 1974; Seidemann/Siebert 1987

Crocus cartwrightianus *Herb.*

Common Names ▸ Cartwright saffron, wild saffron; *German:* Cartwright-Safran
Usage ▸ spice (rarely)

CROTOLARIA L. - Rattlebox - Fabaceae (Leguminosae)

 Crocosmia x crocosmiflora, flowering

Parts Used ▶ stigma
Distribution ▶ Mediterranean Region: Greece

Hepper 1992; Negbi 1999; Small 1997

Crocus haussknechtii *Boiss. et Reut. ex Boiss.*

Common Names ▶ Hausknecht saffron, wild saffron; *German:* Hausknecht-Safran
Usage ▶ spice, instead of saffron
Parts Used ▶ stigma
Distribution ▶ Iran (Hamedan, Kermanshah, Fars provinces): dry fields and rocky hillsides of W and S areas

Radjabian et al. 2001

 Crocus officinalis *Martyn.*

▶ *Crocus sativus* L.

 Crocus sativus L.

Synonyms ▶ *Crocus officinalis* Martyn.
Common Names ▶ saffron, Spanish saffron; *Arabic:* asafar, kruku, sa'afaran; zaafraran, za'ferān; *Chinese:* fan hung hua, zan hunh hua; *Dutch:* saffraan; *French:* saf(f)ran, crocus; *German:* Safran, Echter Safran; *Hindi:* kesar; *Indonesian:* kunyit kering; *Italian:* zafferano, zafferano domestica, zaffrone, gaiallone; *Japanese:* safuran; *Korean:* saphuran; *Malaysian:* sāfarān; *Portuguese:* açafrão; *Russian:* shafran, shafran posewnoj; *Sanskrit:* kunkuma; *Slovenian:* šafran; *Spanish:* azafrán; *Thai:* ya faran; *Turkish:* safrān, za'ferān; *Vietnamese:* nghe
Usage ▶ spice, flavoring and coloring (rice, sauces); dye stuff
Parts Used ▶ stigma
Distribution ▶ Mediterranean region (Spain, France, Switzerland), Israel cultivated, N India, Brazil, and New Zealand only cultivated
Note ▶ Frequently adulterated. Trade sorts: 'Coupé' (1. quality with 5% flower rests), 'Mancha' (rest of 5% flowers + 10–15% styles), 'Rio' (rest of 10% flowers + 20–25% styles, 'Sierra' (rest of 15% flowers + 25–35 styles), 'Molido': pulverized product of all sorts.

Alonso et al. 1996, 1998, 2001, 2003; Anon. 1999; Bärtels 1997; Basker/Negbi 1983, 1985; Berger 1, 1949; Bhat/Broker 1953; Bois 1934; Boisvert/Aucante 1993; Boisvert/Hubert 2000; Bowles 1952; Burkill 3, 1995; Cheers 1998; Corti et al. 1996; Crecchio 1960; Dalby 2000; Dastur 1954; Davidson 1999; Dhar et al. 1988; Dudtschenko et al. 1989; Duke et al. 2003; Farrell 1985; Freiburglaus et al. 1998; Gonso et al. 2003; Greven 1992; Hanelt 2001; Hepper 1992; Hiller/Melzig 1999; Himeno/Sano 1987; Hohmann et al. 2001; Hoppe 1949; Ingram 1969; ISO-Standard 1980; Leung 1991; Lewington 1990; Madan et al. 1966; Madan et al 1966; Mathew 1977; McGimpsey et al. 1996; Melchior/Kastner 1974; Morris/Mackley 1999; Munshi et al 1989; Negbi 1999; Negbi et al. 1989; NICPBP 1987; Norman 1991; Oberdieck 1991; Omidbaigi et al. 2002; Palmer 1983, 1990; Pardo et al. 2002; Peter 2001; Peters 1927; Pfänder/Wittmer 1975; Pfister et al. 1996; Pochljobkin 1974, 1977; Pschyrembel 1998; Radjabian et al. 2001; Rosengarten 1969; Sampathu et al. 1984; Schönfelder 2001; Schröder 1991; Schultze-Motel 1986; Seidemann 1993c; 2001a; Seidemann/Siebert 1987; Sharma 2003; Shiva et al. 2002; Small 1997; Souret/Weathers 2000; Staesche 1972; Straubinger et al 1998; Tammaro/Marra 1990; Tarantalis/Polissiou 1997; Täufel et al. 1993; Teuscher 2003; Tucker 1986; Uphof 1968; Vaupel 2002b; Vogt 1987; Wiersema/León 1999; Winterhalter/Straubinger 2000; Wüstenfeld/Haensel 1964; Wyk et al. 2004; Zamboni et al. 1995; Zareena et al. 2001; Zeven/de Wet 1982

CROTOLARIA L. - Rattlebox - Fabaceae (Leguminosae)

Crotolaria guatemalensis *Benth.*

Synonyms ▸ *Crotolaria carmioli* Polak.
Usage ▸ pot-herb
Parts Used ▸ leaf, young branches
Distribution ▸ Panama to Mexico, C America

Uphof 1968

Crotolaria longirostrata *Hook et Arn.*

Common Names ▸ castanet plant, long beaked rattlepod; *Guatamala:* cap-in chop; *Spanish:* chipil(e), chipilin de comerc, panajachel
Usage ▸ pot-herb
Parts Used ▸ leaf
Distribution ▸ S America: Guatamala, Mexico, native Hawaii

Morton 1994; Uphof 1968

CROTON L. - Croton - Euphorbiaceae

Croton cascarilla *Benn.*

Synonyms ▸ *Croton niveus* Jacq.
Common Names ▸ wild rosemary; *German:* Wilder Rosmarin, Kaskarille; *Spanish:* cascarilla de Cuba, cascarilla de Trinidad
Usage ▸ spice, like rosemary(?)
Parts Used ▸ bark
Distribution ▸ Bahamas

Berger 1, 1949; Erhardt et al. 2002; Hanelt 2001; Hiller/Melzig 1999; Hoppe 1949; Ochoa/Alonso 1996; Seidemann 1993c; Villamar et al. 1994

Croton eluteria *(L.) Sw.*

Common Names ▸ cascarilla, cascarille; *German:* Saftlose Kaskarille; *Spanish:* cascarilla
Usage ▸ spice; **product:** essential oil. In the liqueur and tobacco industries
Parts Used ▸ bark
Distribution ▸ West Indies, Mexico, Caribbean, Meso-America, W, S America

Berger 1, 1949; Charalambous 1994; Erhardt et al. 2002; Hanelt 2001; Hiller/Melzig 1999; Hoppe 1949; Seidemann 1993c; Spapiro/Frances 2001; Täufel et al. 1993; Wolters 1994; Wüstenfeld/Haensel 1964

Croton mubange *Muell.*

Usage ▸ spice
Parts Used ▸ bark
Distribution ▸ Congo

Terra 1966

Croton niveus *Jacq*

▸ *Croton cascarilla Benn.*

Croton variegatum *L.*

▸ *Codiaeum variegatum (L.) Bl.*

Croton zambesicus *Müll.-Arg.*

Common Names ▸ Egyptian wood tree;
Usage ▸ spice (in the Adamawa region of Nigeria)
Parts Used ▸ fruit
Distribution ▸ tropical Africa, from the Gambia to S Nigeria

Burkhill 2, 1994

Crucifera fontana *E.H.L. Krause*

▸ *Nasturtium officinale R.Br.*

Crucifera latifolia *E.H.L. Krause*

▶ *Lepidium latifolium* L.

 Crucifera nasturtium *E.H.L. Krause*

▶ *Ledpidum sativum* L.

 Cryphiospermum repens *Pal.*

▶ *Enydra fluctuans* Lour.

CRYPTOCARIA R. Br. - Lauraceae

 Cryptocaria canellila *H.B.K.*

▶ *Aniba canellila* (H.B.K.) Mez

 Cryptocaria aromatica *(Becc.) Kosterm.*

Synonyms ▶ *Cinnamomum massoia* Schew.; *Massoia aromatica* Becc.
Common Names ▶ massoy; *German:* Massoy, Massoi
Usage ▶ spice; earlier liqueur
Parts Used ▶ bark
Distribution ▶ New Guyana
Note ▶ Alternative for clove and a cinnamon substitute.

Griebel/Freymuth 1916; Hanelt 2001; Schultze-Motel 2003; Seidemann 1993c; Staesche 1972; Uphof 1968; Wüstenfeld/Haensel 1964

 Cryptocaria guianensis *Meissn.*

Common Names ▶ *Brazil (Portuguese):* casca-acucena; coaxicó
Usage ▶ flavoring
Parts Used ▶ wood
Distribution ▶ S America: Brazil

Kumar 2001; Mors et al. 2000

 Cryptocaria massoy *(Becc.) Kosterm.*

Common Names ▶ massoy; *German:* Massoy
Usage ▶ spice; **product:** essential oil
Parts Used ▶ bark
Distribution ▶ Malaysia

Guzman/Siemonsma 1999; Wiersema/León 1997

 Cryptocaria moschata *Nees et Mart.*

Synonyms ▶ *Cryptocaria aschersonia* Mez.
Common Names ▶ Brazilian nutmeg, (South) American nutmeg; *Brazil (Portuguese):* canela de porco, canela fogo, canela pimenta, canela pururuca, noz-moscada-do-brasil; *German:* Amerikanische Muskatnuss, Brasilianische Muskatnuss; *Spanisch:* moscado do Brasil
Usage ▶ spice
Parts Used ▶ seed
Distribution ▶ S America: Brazil
Note ▶ Substitute for nutmeg.

Erhardt et al. 2002; Hager 5, 1993; Lorenzi1992; Mors et al. 2000; Rätsch 1998; Seidemann 1993c; Teuscher 2003; Warburg 1897

 Cryptocaria pretiosa *Mart.*

Synonyms ▶ *Ocotea pretiosa* Benth. et Hook.
Usage ▶ spice, flavoring
Parts Used ▶ bark
Distribution ▶ Guyana to Brazil

Boiss 1934

CRYPTOTAENIA DC. - Apiaceae (Umbelliferae)

Cryptotaenia canadense *(L.) DC.*

Common Names ▶ honewort, wild chervil, wild stone parsley; *German:* Kanadische Steinpetersilie
Usage ▶ pot-herb
Parts Used ▶ fresh leaf
Distribution ▶ N America: Georgia, Alabama, Arkansas

Erhardt et al. 2002; Mansfeld 1968; Small 1997

Cryptotaenia canadense (L.) DC. **var. japonica** (Hassk.) Makino

▸ *Cryptotaenia japonica* Hassk.

Cryptotaenia japonica Hassk.

Synonyms ▸ *Cryptotaenia canadense* (L.) DC. var. *japonica* (Hassk.) Makino
Common Names ▸ Japanese parsley, (Japanese) honewort, Japanese wild chervil, mitsuba; *Chinese:* ya êrh chin; *German:* Dreiblätterkraut, Japanische Petersilie, Steinpetersilie; *Japanese:* mitsuba, mitzuba; *Korean:* padudugnamul; *Russian:* skryptnica japonskaja
Usage ▸ flavoring, pot-herb; **product:** essential oil seasoning (seeds)
Parts Used ▸ fresh leaf (herb); seed
Distribution ▸ Japan, China, E Asia, Russia, also cultivated in Korea, China, N America; sporadically in Europe
Note ▸ The plant comes in two varieties, *kansai* (green) and *kanto* (white).

Davidson 1999; Hanelt 2001; Kling 1999; Lee et al. 1993; Meijer 1940; Schultze-Motel 1986; Small 1997; Staesche 1972; Täufel et al. 1993; Teuscher 2003; Tucker 1986; Zeven/de Wet 1982

Cubeba officinalis Raff.

▸ *Piper cubeba* L.

CUMINUM L. - Caraway - Apiaceae (Umbelliferae)

Cuminum cyminum L.

Synonyms ▸ *Cuminum odorum* Salisb., *Ligusticum cuminum* (L.) Crantz
Common Names ▸ cumin, Egyptian caraway, Roman caraway; *Arabic:* kam(m)oun, kamun; *French:* cumin, cumin des prés; faux aneth, faux anis; *German:* Kreuzkümmel, Kumin, Mutterkümmel, Römischer or Türkischer Kümmel, Wanzenkümmel, Kala; *Hindi:* jira; *India:* jeera, jirakasafed zeera; *Indonesian:* djintan, jinten putih; *Italian:* cumino vero, cumino proprio, cumino romano, cumino di Malta; *Javanese:* jinten bodas, jinten poteh; *Malaysian:* jintan puteh; *Mexico:* bere lele; *Russian:* kumin, timon, prjanij tmin, tmin tminowyj, rimskij tmin; *Sanskrit:* ajaji, jiraka; *Slovenian:* rascovec; *Spanish:* comino; *Thai:* yeera; *Turkish:* kimion, siyah zira
Usage ▸ spice, condiment; **product:** essential oil
Parts Used ▸ seed
Distribution ▸ Mediterranean region (Turkey), Egyptian, S Russia, Asia: India, Sri Lanka, Pakistan, China, Japan, Indonesia, Iran

Baser et al. 1992; Berger 3, 1952; Borges/Pino 1993a; Burkill 5, 2000; Clair 1961; Craze 2002; Dalby 2000; Davidson 1999; Dudtschenko et al. 1989; El-Hamadi/Richter 1965; Erhardt et al. 2002; Farrell 1985; Guzman/Siemonsma 1999; Hager 4, 1992; Hanelt 2001; Heeger 1956; Heikes et al. 2001; Hepper 1992; Hiller/Melzig 1999; Hondelmann 2002; Hoppe 1949; Jansen 1981; Melchior/Kastner 1974; Morris/MacKley 1999; Norman 1991; Peter 2001; Pochljobkin 1974, 1977; Pruthi 1976; Quadry/Atal 1963; Rosengarten 1969; Roth/Kormann 1997; Schröder 1991; Schultze-Motel 1986; Seidemann 1993c; Seidemann/Siebert 1987; Shah/Ray 2003; Shetty et al. 1994; Shiva et al. 2002; Siewek 1990; Small 1997; Staesche 1972; Tassan/Russel 1975; Täufel et al. 1993; Tainter/Grenis 1993; Teuscher 2003; Tucker 1986; Ubillos 1989; Uphof 1968; Verghese 1991; Weber 1951; Wüstenfeld/Haensel 1964; Zeven/de Wet 1982

Cuminum odorum Salisb.

▸ *Cuminum cyminum* L.

CUNILA Royen ex L. - Lamiaceae (Labiatae)

Cunila origanoides (L.) Britton

Synonyms ▸ *Satureja origanoides* L.
Common Names ▸ American dittany, dittany, frost flower, frost mint, Maryland dittany, mountain dittany, stone mint; *German:* Amerikanischer Diptam, Maryland Diptam; Steinminze
Usage ▸ spice, as pizza herb
Parts Used ▸ herb

Distribution ▶ N America

Duke 2002; Duke et al. 2003; Small 1997; Uphof 1968

 Cunila pulegioides L.

▶ *Hedeoma pulegioides* (L.) Pers.

 Cunila spicata Benth.

Common Names ▶ (American) stone mint; *German:* (Amerikanische) Steinminze; *Portuguese:* poejo
Usage ▶ flavoring, use similar to the oregano herb
Parts Used ▶ herb
Distribution ▶ S America: Brazil: São Paulo to Rio Grande

Baren et al. 2001; Mabberly 1997; Mors et al. 2000; Teubner 2001; Villamar et al. 1994

CURCUMA L. - Saffron root - Zingiberaceae

 Curcuma amada Roxb.

Common Names ▶ mango ginger; *German:* Mango-Ingwer, Blockzitwer, Gelber Zitwer; *India:* aam haldi; *Sanskrit:* karpura-haridra;
Usage ▶ spice
Parts Used ▶ rhizom
Distribution ▶ India, Pakistan, Bangladesh, also cultivated

Arora/Pandey 1996; Hanelt 2001; Rao et al. 1989; Schultze-Motel 1986; Seidemann 1993c; Shiva et al. 2002; Täufel et al. 1993; Teuscher 2003; Uphof 1968; Zeven/de Wet 1982

 Curcuma aromatica Salisb.

Common Names ▶ Bombay or Indian Arrowroot; wild turmeric, yellow zedoary; *German:* Aromatischer Blockzitwer, Wilde Gelbwurz, Wilder Kurkuma, Würzige Safranwurz(el); *Hindi:* jangli-haldi; *India:* kasturi majal, kasturi pasupu, kasturiarishina ran halada; *Javanese:* temo potre, temu puteri; *Korean:* kangwang; *Russian:* kurkuma aromatihaja
Usage ▶ spice
Parts Used ▶ rhizome
Distribution ▶ India, also cultivated
Note ▶ The tender shoots are eaten as a vegetable.

Arora/Pandey 1996; Bois 1934; Duke et al. 2003; Hager 4, 1992; Hanelt 2001; Hiller/Melzig 1999; Kumar 2001; Pochljobkin 1974, 1977; Schultze-Motel 1986; Shiva et al. 2002; Tao Guoda 1998; Täufelet al. 1993; Teuscher 2003; Zeven/de Wet 1982

 Curcuma caesia Roxb.

Common Names ▶ black zedoary; *German:* Blaugraue Kurkuma; *Hindi:* nar-kachura; *India:* kalahaldi, kalihalidi, manupasupa
Usage ▶ spice (rarely)
Parts Used ▶ fresh rhizome
Distribution ▶ India: Bengalia

Chopra et al. 1956; Erhardt et al. 2002; Hoppe 1949; Kumar 2001; Uphof 1968; Wealth of India 2, 1950

 Curcuma domestica Valeton

▶ *Curcuma longa* L.

 Curcuma longa L.

Synonyms ▶ *Amomum curcuma* Jacq., *Curcuma domestica* Valeton, *Curcuma purpurescens* Bl.,
Common Names ▶ turmeric, Indian saffron, curcuma rhizome, yellow ginger; *Arabic:* kur kum, zangabeel asfer; *Chinese:* jiang huang, yu-chin; *French:* curcuma, curcuma longue, safran des Indes, souchet des Indes; terre-mérite; *German:* Kurkuma(wurzel), Gelbwurzel, Gilbwurzel, Safranwurz(el), Gelber Ingwer; *Hindi:* haldi, halada; *Indonesian:* kunyit; *Italian:* curcuma longa, zafferano delle Indie, radice gialle; *Japanese:* ukon; *Javanese:* kunir; *Korean:* ulgum; *Malaysian:* kunyit; *Pilipino:* dilaw, duwaw, kalawag; *Portuguese:* açfrão da Índia, gengibre dourada; *Russian:* kurkuma, kurkuma dlinnaja; sheltyj koren'; *Sanskrit:* haridra, nisa; *Slovenian:* kurkuma; *Spanish:* cúrcuma larga, azafrán

CURCUMA L. - Saffron root - Zingiberaceae

▫ Curcuma longa, flowering

Melzig 1999; Hohmann et al. 2001; Hoppe 1949; Ibrahim bin Jantan et al. 1999; Jacquat 1990; Jentzsch et al. 1968; Karig 1975; Karl 2000; Krishnamurthy et al. 1976; Kumar 1991, 2001, 2003; Larsen et al. 1999; Leung 1991; Lewington 1990; Lösing et al. 1999; Lutomski et al. 1974; Magdan 1994; Maistre 1964; Malingré 1975; Manzan et al. 2003; McCarron et al. 1995b; Melchior/Kastner 1974; Morris/Mackley 1999; Nair et al. 1982; Nakayama et al. 1994; Norman 1991; Panja/De 2000; Peter 2001; Pfeifer et al. 2003; Pochljobkin 1974, 1977; Pruthi 1976; Pschyrembel 1998; Rakhunde et al. 1998; Ramprasad/Sirsi 1956; Rosengarten 1969; Ross 1999; Roth 1998; Sanagi/Ahmad 1993; Schröder 1991; Schultze-Motel 1986; Seidemann 1993c, 1998/2000; Seidemann/Siebert 1987; Sen et al. 1974; Shahi et al. 1994; Shankaracharya/Natarajan 1973; Sharma 2003; Siewek 1990; Small 1997; Sopher 1964; Staesche 1972; Tao Guoda 1998; Täufel et al. 1993; Taylor/Mc-Dowell 1992; Teufel 2003; Turova et al. 1987; Uhl 2000; Uphof 1968; Ye layyudhan et al. 2000; Verghese 1993; WHO 1990; Winkler/Lunau 1959; Wong 1999; Wüstenfeld/Haensel 1964; Wyk et al. 2004; Yusuf et al. 2002; Zeven/de Wet 1982

 Curcuma mangga Valeton et Zjip

▸ *Curcuma zedoaria (Christm.) Rosc.*

 Curcuma pallida Lour.

▸ *Curcuma zedoaria (Christm.) Rosc.*

 Curcuma pierreana Gagnep.

Usage ▸ spice; **product:** essential oil.
Parts Used ▸ rhizome
Distribution ▸ Asia: India (Assam), Malaysia, cultivated in Vietnam

Nguyen Xuan Dung et al. 1995; Uphof 1968; Zeven/de Wet 1982

 Curcuma purpurescens Bl.

▸ *Curcuma longa L.*

 Curcuma rotundae

▸ *Curcuma longa L.*

de la India, turmérico, zafrán molido; *Thai:* khamin, khamin-kaeng, khamin-chan, kheemin, min, taa-ya sa-ya; *Vietnamese:* ngheden
Usage ▸ spice, condiment ("curry powder"); **product:** essential oil
Parts Used ▸ rhizome
Distribution ▸ SE tropical Asia, naturalized (India?), also cultivated: India, Pakistan, Sri Lanka, Thailand, S China, Japan, Indonesia, Philippines, Madagascar, Réunion, Caribbean Islands: Jamaica, Haiti
Note ▸ The *Curcuma longa*-powder is the main constituent of curry powder. The powder is also wrongly sold as "Saffron" (powder). In the Middle Ages turmeric was known as *Crocus indicus* or *Indian saffron*. The cormel are *Radix Curcumae rotundae*.

Araujo et al. 2001; Berger 2001; Bois 1934; Burkill 5, 2000; Chane-Ming et al. 2002; Chatterjee et al. 2000; Craze 2002; Dastur 1954; Datta et al. 2001; Dalby 2000; Davidson 1999; Dudtschenko et al. 1989; Duke et al. 2003; Farnsworth/Bunyapraphatsara 1992; Duke et al. 2003; Farrell 1985; Ferrão 1992; Fintelmann/Wegener 2001; Garg et al. 2002; Govindarajan 1980; Gwosdz 1987; Hager 4, 1992; Hanelt 2001; He et al. 1998; Hepper 1992; Herrmann 1999; Hiller/

Curcuma rotunda L.

▷ *Boesenbergia rotundata (L.) Mansf.*

 Curcuma xanthorriza Roxb.

▷ *Curcuma zedoaria (Christm.) Rosc.*

 Curcuma zedoaria (Christm.) Rosc.

Synonyms ▶ Amomum zedoaria Christm., *Curcuma mangga* Valeton et Zjip, *Curcuma pallida* Lour., *Curcuma xanthorrhiza* Roxb., *Curcuma zerumbet* Roxb.

Common Names ▶ cochin-tumer, long zedoary, mango gingerround zedoary, Japanese turmeric, Javanese turmeric 'setwall', yellow turmeric; *Arabic:* zadwâr, zarunbâd; *Chinese:* peng e zhu; *French:* zédoaire, curuma zédoaire, temoé-lawag; *German:* Zittwer(wurzel), Javanische Gelbwurz(el); *Hindi:* banhaldi, kachura; *Italian:* zedoaria; *Japanese:* gajutsu; *Java:* temu lawak; *Korean:* achul; *Pilipino:* barak, bolon, lampoyang, langkauas; *Portuguese:* zedoária, zedoeira amarella, *Russian:* kurkuma zedoarija; *Spanish:* cedoaria, cedoaria redonda, cúrcuma de Java; *Thai:* phet buri, khamin khao; *Vietnamese:* nga truat, nghe den

Usage ▶ spice; **product:** essential oil

Parts Used ▶ rhizome

Distribution ▶ SE Asia, naturalized, also cultivated: India, China, Sri Lanka, Madagascar

Note ▶ The large fleshly rhizomes are rich in starch and used as a substitute for arrowroot.

Arora/Pandey 1996; Bois 1934; Dalby 2000; Dastur 1954; Duke et al. 2003; Farrell 1985; Hager 4, 1992; Hanelt 2001; Herklots 1972; Hiller/Melzig 1999; Hohmann et al. 2001; Hoppe 1949; Ibrahim bin Jantan et al. 1999; Jentzsch et al. 1968; Kumar 2001, 2003; Kyung Im Kim et al. 2002; Larsen et al. 1999; Maiwald/Schwante 1991; Maistre 1964; Malingré 1975; Mau et al. 2003; Melchior/Kastner 1974; Morris/Mackley 1999; Norman 1991; Pochljobkin 1974, 1977; Pschyrembel 1998; Rimpler et al. 1970; Roth/Kormann 1997; Schultze-Motel 1986; Schwante 1991; Seidemann 1993c; Sen et al. 1974; Staesche 1972; Täufel et al. 1993; Teuscher 2003; Uhl 2000; Uphof 1968; WHO 1990; Winkler/Lunau 1959; Wong 1999; Yusuf et al. 2002; Zeven/de Wet 1982

 Curcuma zerumbet Roxb.

▷ *Curcuma zedoaria (Christm.) Rosc.*

CUSPARIA Humb. ex R. Br. - Rutaceae

 Cusparia febrifuga Humb. ex DC.

Synonyms ▶ *Cusparia trifoliata* Engl., *Galipea officinalis* Hancock;

Common Names ▶ angostura; *Brazil (Portuguese):* amarelinho-da-serra, amarelo, angostura; *German:* Angostura(rinde)

Usage ▶ spice, special for hard liqueur (Angostura-Bitter, Aromatique etc.)

Common Names ▶ bark

Distribution ▶ West Indies, Brazil, Venezuela, Columbia

Charalambous 1994; Erhardt et al. 2002; Hiller/Melzig 1999; Hoppe 1949; Mors et al. 2000; Uphof 1968; Wüstenfeld/Haensel 1964

 Cusparia trifoliata Engl.

▷ *Cusparia febrifuga Humb ex DC.*

CYANOTIS G. Don - Commelinaceae

 Cyanotis tuberosa Schult. f.

Usage ▶ pot-herb
Parts Used ▶ leaf
Distribution ▶ India: W, E Peninsula region

Arora/Pandey 1996; Kumar 2003

CYMBOPETALUM (Dunal) Benth. - Annonaceae

 Cymbopetalum costaricense *(Don. Smith.) Fries*

Synonyms ▶ *Asimia costaricensis* Don. Smith.
Usage ▶ flavoring of drinking chocolate by natives
Parts Used ▶ floral leaves
Distribution ▶ Costa Rica

Uphof 1968

 Cymbopetalum penduliflorum *(Dunal) Baill.*

Common Names ▶ Mexican earflower; *German:* Mexikanische Schleifenblume; *Spanish:* flor de la oreja, xochinacathli (Mexico)
Usage ▶ flavoring
Parts Used ▶ floral leaves
Distribution ▶ mountains of Mexico, Guatamala and Belize
Note ▶ Flavoring of drinking chocolate and other beverages by Aztecs in pre-Columbian times.

Erhardt et al. 2002; Hanelt 2001; Rätsch 1998; Schultze-Motel 1986; Uphof 1968

CYMBOPOGON Spreng. - Lemongrass - Poaceae (Gramineae)

 Cymbopogon citratus *(DC.) Stapf*

Synonyms ▶ *Andropogon citratus* DC. *Andropogon citriodorum* Desf., *Andropogon schoenanthus* L.
Common Names ▶ citron grass, fever grass, fragrant thatch grass, geranium grass, West Indian lemon grass; *Cambodian:* slek krey sabou; *Chinese:* xiang mao; *French:* verweine de Indes, citronelle, herbe-citron; *German:* Indisches Zitronengras, Lemongras, Serehgrass; *Hindi:* gandhatrn; *India:* herva chaha, khawi, verveine; *Indonesian:* serai dapur, sereh; *Italian:* citronella, lemongrass; *Javanese:* sere; *Malaysian:* serai dapur, sereh, sereh makan; *Peru:* yerba Luisa; *Pilipino:* tanglad, salai, balioko; *Portuguese:* erva cidreira; *Slovenian:* vôňovec citrónový; *Spanish:* pasto cedrón, pasto limón, sontol, zacate elimón; *Thai:* cha khrai, khrai, soet-kroei; *Vietnamese:* cay xa, sa chanh, xa
Usage ▶ spice, pot-herb; **product:** essential oil (lemon grass oil; *French:* essence de verweine de Indes)
Parts Used ▶ leaf (grass)
Distribution ▶ unknown wild; cultivated in: W India, Sri Lanka, Vietnam, Java to Malaysia, Mauritius, Madagascar

Aedo et al. 2001; Bor 1953, 1954; Boruah et al. 1995; Bournot 1968; Brandares et al. 1987; Burkill 4, 1997; Carlini et al. 1986; Cheers 1998; Chopra 1956; Craze 2002; Davidson 1999; Duke et al. 2003; Erhardt et al. 2002; Faruq et al. 1994; Ferrua et al. 1994; Fuentes 1986; Hager 4, 1992; Hanelt 2001; Herklots 1972; Hiller/Melzig 1999; Hoppe 1949; Jagadishchandra 1975a, b; Juida de Carlini 1986; Kasali et al. 2001; Larkcom 1991; Lewington 1990; Matos et al. 1984; Melchior/Kastner 1974; Mikus/Schaser 1995; Mors/Rizzini 1961; Norman 1991; Oyedele et al. 2002; Oyen/Dung 1999; Pschyrembel 1998; Ross 1999; Roth/Kormann 1997; Schaneberg/Khan 2002; Seidemann 1993c, 1998/2000; Seidemann/Siebert 1987; Shiva et al. 2002; Sing-Sangwan et al. 1993; Small 1997; Soenarko 1977; Staesche 1972; Strauß 1969b; Täufel et al. 1993; Teuscher 2003; Torres 1993; Uhl 2000; Uphof 1968 Villamar et al. 1994; Virmani et al. 1979; Wealth of India 2, 1950; WHO 1990; Wiersema/León 1999; Wyk et al. 2004; Zeven/de Wet 1982

 Cymbopogon confertiflorus *(Steud.) Stapf*

▶ *Cymbopogon nardus (L.) Rendle var. confertiflorus (Steud.) Stapf ex Bor.*

 Cymbopogon exaltatus *Domin*

Common Names ▶ East India lemon grass; *German:* Ostindisches Lemongras, Hohes Lemongras
Usage ▶ spice
Parts Used ▶ fresh leaf (grass)
Distribution ▶ E India

Jagadishchandra 1975a, b

 Cymbopogon flexuosus *(Nees ex Steud.) Stapf*

Synonyms ▶ *Andropogon flexuosus* Nees ex Steud.
Common Names ▶ lemon grass, East Indian lemon grass, Malabar grass, Cochin grass; *Brazil (Portuguese):*

capim-limao, erva cidreira; *French:* herbe de Malabar, verveine des Indes; *German:* Malabargras, Ostindisches Lemongras, Cochingras; *India:* poolu; *Spanish:* pasto de Malabar

Usage ▶ spice, flavoring; **product:** essential oil (Malabar or Cochin oil)

Parts Used ▶ leaf (grass)

Distribution ▶ India, Indochina, SE Asia, Equatorial Africa and in the Caribbean also cultivated

Note ▶ Only the red-stemmed plant type is commercially cultivated for East Indian lemon grass oil (= Cochin lemon grass oil or Malabar grass oil). White-stemmed plants "Inchigrass" (sometimes separated at species level they belong to *Cymbopogon travancorensis* Bor from S India and are rarely cultivated. They are a source of essential oil (Inchigrass oil).

Boelens 1992, 1994; Bor 1993, 1994; Bournot 1968; Erhardt et al. 2002; Hanelt 2001; Hepper 1992; Jagadishchandra 1975a, b; Kulkarni/Ramesh 1992; Kumar 2001; Melchior/Kastner 1974; Morris/Mackley 1999; Moes/Rizzini 1961; Oyen/Dung 1999; Rajeswara et al. 1996; Roth/Kormann 1997; Schenck/Naundorf 1966; Seidemann 1993c; Sharma 2003; Shiva et al. 2002; Siewek 1990; Small 1997; Soenarko 1977; Srivastava et al. 2002; Täufel et al. 1993; Teuscher 2003; Uphof 1968; Zeven/de Wet 1982

Cymbopogon giganteus *(Hochst.) Chiov.*

Synonyms ▶ *Andropogon giganteus* Hochst.

Common Names ▶ tsauri grass; *French:* beignefata; *German:* Großes Lemongras;

Usage ▶ flavoring; **product:** essential oil (rich in phellandrine)

Parts Used ▶ leaf (grass)

Distribution ▶ W Africa: Senegal, the Gambia, Guinea, Ghana, Mozambique, Niger, Nigeria, Ivory Coast,

Burkill 1985; Hanelt 2001

Cymbopogon iwarancusa *(Jones) Schult. ex Roem. et Schult.*

Synonyms ▶ *Andropogon iwarancusa* W. Jones.

Common Names ▶ karnkusa grass, khavi grass; *Brazil (Portuguese):* capim cheiroso da India; *India:* jwarancusa;

Usage ▶ spice; **product:** essential oil

Parts Used ▶ leaf (grass)

Distribution ▶ India, Afghanistan, Himalayas

Note ▶ The essential oils contain more than 80% Piperitone.

Bournot 1968; Erhardt et al. 2002; Hanelt 2001; Jagadishchandra 1975a, b; Shiva et al. 2002; Uphof 1968

Cymbopogon martinii *(Roxb.) J.F. Wats. ex Atkinson*

Synonyms ▶ *Andropogon martinii* Roxb.

Common Names ▶ ginger grass, rosha grass, palmarosa gras; *German:* Ingwergras, Palmarosagras

Usage ▶ spice, flavoring; **product:** essential oil (Palmarosa oil, East Indian Geranium oil)

Parts Used ▶ leaf (grass)

Distribution ▶ N India, cultivated: India, Indonesia, Brazil, Madagascar, Zimabwe, S Africa

Boelens 1994; Charalambous 1994; Erhardt et al. 2002; Hager 4, 1992; Hanelt 2001; Hepper 1992; Jagadishchandra 975a, b; Kumar 2001; Oyen/Dung 1999; Roth/Kormann 1997; Sartoratto/Augusto 2003; Sharma 2003; Small 1997; Soenarko 1977; Srivastava/Satpute 2001; Srivastava et al. 2002; Small 1997; Teuscher 2003; Uphof 1968; Wealth of India 2, 1950; Zeven/de Wet 1982

Cymbopogon nardus *(L.) Rendle*

Common Names ▶ citronella, Ceylon citronella, nard grass; *French:* citronelle; *German:* (Ceylon) Zitronellgras, Lenabutagras; *Japanese:* kôsui gaya; *Portuguese:* citronela de Java; *Spanish:* zacate lemón

Usage ▶ spice, flavoring; **product:** essential oil (citronell oil)

Parts Used ▶ leaf (grass)

Distribution ▶ India, cultivated SE Asia, Sri Lanka; C, E Africa

Bois 1934; Bournot 1968; Charalambous 1994; Hager 4, 1992; Hanelt 2001; Hepper 1992; Jagadishchandra 1975a, b; Kumar 2001; Mahalwal/Ali 2003; Melchior/Kastner 1974; Moody et al. 1995; Mors/Rizzini 1961; Rajeswara et al. 1996; Roth/Kormann 1997; Seidemann 1993c; Shiva et al. 2002; Siewek 1990; Small 1997; Teuscher 2003; Uphof 1968; Zeven/de Wet 1982

Cymbopogon nardus (L.) Rendle var. confertiflorus (Steud.) Stapf ex Bor.

Synonyms ▶ *Andropogon confertiflorus* Steud.
Common Names ▶ citronella, citronelle grass, manna grass; *German:* Dichtblättriges Zitronellgras, Mannagras
Usage ▶ spice, flavoring; **product:** essential oil (citronelle oil)
Parts Used ▶ leaf (grass)
Distribution ▶ S India, Sri Lanka, also cultivated

Hanelt 2001; Jagadishchandra 1975a, b; Pursglove 1972; Small 1997

Cymbopogon schoenanthus (L.) Spreng.

Synonyms ▶ *Andropogon schoenanthus* L.
Common Names ▶ camel or geranium grass; *French:* herba à chameau, *German:* Kamelgras; *Spanish:* pasto de camellos
Usage ▶ spice, flavoring; **product:** essential oil
Parts Used ▶ leaf (grass)
Distribution ▶ N Africa: Morocco to Tunesia, Arabia, NW India, Afghanistan

Chopra et al. 1956; Erhardt et al. 2002; Hanelt 2001; Hepper 1992; Jagadishchandra 1975b; Lewington 1990; Mors et al. 2000; Uphof 1968

Cymbopogon travancorensis Bor

▶ *Cymbopogon flexuosus (Steud.) Stapf* (see Note)

Cymbopogon winterianus Jowitt ex Bor

Synonyms ▶ *Andropogon nardus* (L.) Hook.f. var. *mahapangiri* auct.
Common Names ▶ Java citronella, Java lemon grass; *French:* herba citron de Java; *German:* Echtes Zitronengras, Java-Zitronengras, Java-Zitronelle, Maha-Pengiri-Gras
Usage ▶ spice, flavoring; **product:** essential oil (Java citronell oil)
Parts Used ▶ leaf (grass)
Distribution ▶ Sri Lanka; only cultivated, especially in SE: China and S America: Guatamala
Note ▶ Java citronella oil contains 25–55% geraniol, about 35% citronellal. Strains with a citronellal content of 70% have been selected.

Boelens 1994; Bois 1934; Bor 1953, 1954; Bournot 1968; Charalambous 1994; Chopra et al. 1956; Erhardt et al. 2002; Hager 4, 1992; Hanelt 2001; Hiller/Melzig 1999; Jagadishchandra 1975a, b; Naqvi et al. 2002; Ng 1972; Oyen/Dung 1999; Pschyrembel 1998; Pursglove 1972; Roth/Kormann 1997; Sahoo/Debata 1955; Seidemann 1993c; Sharma 2003; Shiva et al. 2002; Sing et al. 1996; Small 1997; Soenarko 1977; Srivastava et al. 2002; Täufel et al. 1993; Teuscher 2003; Wealth of India 2, 1950; Zeven/de Wet 1982

Cyminosma odorata DC.

▶ *Acronchia odorata (Lour.) Baill.*

Cymonchia resinosa DC.

▶ *Acronchia puduculata (l.) Miq.*

Cymopterus littoralis A. Gray

▶ *Glehnia littoralis F. Schmidt ex Miq.*

CYNOMETRA L. - Fabaceae (Leguminosae)

Cynometra cauliflora L.

Common Names ▶ *Malaysian:* katak puru, nam nam (for the pods), puki, puru
Usage ▶ flavoring, spice; **product:** of chutney
Parts Used ▶ cooked fruits
Distribution ▶ India, Malaysia; cultivated in Java, Malaccas and India

Hanelt 2001; Schultze-Motel 1986; Uphof 1968

CYNOMORIUM L. - Balanophoraceae (Cynomoriaceae)

Cynomorium coccineum L.

Usage ▶ condiment by the Tuareg
Parts Used ▶ root (pulverized)
Distribution ▶ N Africa

Mabberley 1997; Uphof 1968

CYRTANDRA J.R. Forst et G. Forst - Gesneraceae

Cyrtandra pendula Bl.

Usage ▶ flavoring, spices
Parts Used ▶ sourish leaf
Distribution ▶ Java

Burkill 1966; Lemmens/Bunyapratsara 2003; Oyen/Dung 1999

CYTISUS Desf. - Fabaceae (Leguminosae)

Cytisus scoparius (L.) Link

Synonyms ▶ *Sarothamnus scoparius* (L.) Wimmer ex Koch, *Spartium scoparium* L.
Common Names ▶ broom, common broom, hogweed, scotch broom; *Chinese:* chin ch'iao; *French:* genêt, genêt à balais, genêt commun; *German:* Besenginster, Besenpfriem, Geißkapern; *Italian:* citiso dei carbonai, ginestra dei carbonai, genistra scopareccia; *Russian:* sarotamnus metel'tschatyj, sarotamnus metlistyj; *Spanish:* escobón, ginesta de escobas, retama negra;
Usage ▶ condiment; substitute for capers
Parts Used ▶ flower bud
Distribution ▶ C, E Europe

Aichele/Schwegler 2, 1994; Berger 1, 1949; 4, 1954; Bremness 2001; Cheers 1998; Coiciu/Racz (no year); Erhardt et al. 2002; Fleischhauer 2003; Gressner 1996, 1997; Hager 4, 1992; Hanelt 2001; Hiller/Melzig 1999; Hohmann et al. 2001; Hoppe 1949; Newall et al. 1996; Pschyrembel 1998; Rätsch 1998; Schönfelder 2001; Seidemann 1993c; Seidemann/Siebert 1987; Sharma 2003; Staesche 1972; Täufel et al. 1993; Teuscher 2003; Wyk et al. 2004; Zeven/de Wet 1982

DACRYDIUM Lamb. - Rimu - Podocarpaceae

Dacrydium franklinii *Hook.*

Common Names ▶ Huon pine, Macquarie pine
Usage ▶ flavoring; **product** of vanillin and huon pine oil
Parts Used ▶ wood
Distribution ▶ Australia, New Zealand, New Caledonia, Malaysian Archipelago, Borneo
Note ▶ The wood has a high methyl eugenol content.

Mabberley 1997; Uphof 1968

DAPHNE L. - Daphne - Thymelaeaceae

Daphne mezereum *L.*

Common Names ▶ (February) daphne, mezereon; *French:* bois gentil, bois joli, daphné morillon, garo sainbois; *German:* Bergpfeffer, Gemeiner Seidelbast, Kellerhals; *Italian:* mezereo camelea, laureola femmina; *Russian:* woltschje lyko; *Spanish:* dafne, loriguillo mezéreo, mezereón, torvidco
Usage ▶ spice
Parts Used ▶ fruit
Distribution ▶ Europe, W Asia, Canada; cultivated as an ornamental plant
Note ▶ Sporadically a substitute for pepper when not available! All parts of the plant are poisonous.

Cheers 1998; Erhardt et al. 2002; Hager 3, 1992; Heeger 1956; Hiller/Melzig 1999; Hoppe 1949; Schönfelder 2001; Seidemann 1993c

DAUCUS L. Carrot - Apiaceae (Cruciferae)

Daucus carota *L.* ssp. sativus *(Hoffm.) Schübl. et G. Martens*

Common Names ▶ garden carrot, carrot; *Arabic:* jazar; *Chinese:* he shi feng, hu luo bo, yeh lo po; *Dutch:* peen; *French:* carotte, racine jaune; *German:* Garten-Möhre, Möhre, Mohrrübe, Karotte, Gelbe Rübe; *Italian:* carota; *Japanese:* ninjin; *Korean:* hongdangmuu; *Malaysian:* lobak merah, karot; *Portuguese:* cenoura; *Russian:* morkov' kul'turnaja; *Spanish:* zanahoria
Usage ▶ spice: succade
Parts Used ▶ turnip-root
Distribution ▶ only cultivated world-wide
Note ▶ Substitute for orangade succade („Kandinat M") in the former German Democratic Republic (DDR, E Germany) after enzymatic treatment.

Aichele/Schwegler 3, 1995; Cheers 1998; Dudtschenko et al. 1989; Hanelt 2001; Kays/Dias 1995; Schönfelder 2001; Schultze-Motel 1986; Seidemann 1993c; Täufel et al 1993; Wyk et al. 2004; Zeven/de Wet 1982

Daucus visnaga *L.*

▶ *Ammi visnaga (L.) Lam.*

© Springer-Verlag Berlin Heidelberg 2005

DECALEPIS Wight et Arn. - Asclepiadaceae

 Decalepsis hamiltonii *Wight et Arn.*

Common Names ▶ mahali kizhangu
Usage ▶ spice (in India for meat) and condiment
Parts Used ▶ root
Distribution ▶ E India, Deccan Peninsula
Note ▶ The aromatic roots also pickled with lime.

Arora/Pandey 1996; Nagarajan/Rao 2003; Phadke et al. 1994; Seidemann 1993c; Thangadurai et al. 2002

DECASPERMUM Forst. et Forst.f. - Myrtaceae

 Decaspermum fruticosum *Forst.*

Synonyms ▶ *Nelitris paniculata* Lindl., *Psidium decaspermum* L.f.
Common Names ▶ shrubby decaspermum, tailor tree; *Malaysian:* na s'tuka, tuka benang, tukai benai
Usage ▶ condiment (Java)
Parts Used ▶ terminal shoots
Distribution ▶ tropical Asia to Australia and Polynesia

Uphof 1968

DEIANIRA Cham. et Schlecht. - Gentianaceae

 Deianira nervosa *Cham. et Schlecht.*

Synonyms ▶ *Callopsima amplexifolium* Mart.
Common Names ▶ *German:* Deianirakraut; *Brazil (Portuguese):* angelica bravo do matto
Usage ▶ spice; **product:** for bitter liqueurs
Parts Used ▶ herb
Distribution ▶ Brazil, also cultivated

Freise 1936; Hanelt 2001; Schultze-Motel 1986

DENDROBIUM Sw. - Orchidaceae

 Dendrobium salaccense *(Bl.) Lindl.*

Synonyms ▶ *Dendrobium gemellum* Ridley, *Grastidium salaccense* Bl.
Usage ▶ seasoning of rice
Parts Used ▶ leaf
Distribution ▶ tropical Asia, especially Malayan Peninsula

Oyen/Dung 1999; Uphof 1968

 Dendrolobium gemellum *Ridley*

▶ *Dendrolobium salaccense (Bl.) Lindl.*

DENNETTIA E.G. Baker - Annonaceae

 Dennettia tripetala *Baker f.*

Common Names ▶ pepper fruit
Usage ▶ spice
Parts Used ▶ young leaf, fruit
Distribution ▶ W Africa: Ivory Coast, Nigeria
Note ▶ The fruits have a peppery spice taste.

Burkill 1, 1985

DESPLATSIA Bocquillon - Tiliaceae

 Desplatsia dewevrei *(de Wild. et Durand) Burret*

Common Names ▶ *German:* Desplatsiafrucht
Usage ▶ spice for sauces
Parts Used ▶ fruit slime
Distribution ▶ tropical W Africa: Central African Republic, Nigeria

Burkhill 5, 2000; Neuwinger 1999

 Dialyanthera otoba *(H. et B.) Warb.*

▶ *Myristica otoba H. et B.*

DIANTHUS L. - Caryophyllaceae

 Dianthus caryophyllus *L.*

Synonyms ▶ *Dianthus coronarius* Lam.
Common Names ▶ carnation, clove pink, gilly flower; *French:* oeillet â bouquet, oeillet giroflée, oeillet de fleuristes *German:* Edelnelke, Gartennelke, Landnelke; *Italian:* garofano, garofano domestico; *Japanese:* kaneishon, rande nadeshiko; *Russian:* gvozdika gollandskaja, gvozdika sadovaja; *Spanish:* alheli, clavel, claveles
Usage ▶ flavoring: **product:** essential oil
Parts Used ▶ flower
Distribution ▶ cultivated in Europe: Italy, Sardinia, Sicily; cultivated in Spain, S France, the Netherlands, Caucasus and other countries

Bremness 2001; Cheers 1998; Davidson 1999; Erhardt et al. 2002; Hanelt 2001; Schultze-Motel 1986; Stör 1974; Tucker 1986; Uphof 1968; Villamar et al. 1994; Zeven/de Wet 1982

 Dianthus chinensis *L.*

Common Names ▶ carnation annual pink, Chinese pink, Indian pink; *Chinese:* qu mai; *French:* oeillet de Chine; *German:* Chinesische Nelke, Kaiser-Nelke; *Japanese:* kara nadeshike; *Korean:* phaeraengikkot; *Russian:* gvozdika
Usage ▶ flavoring
Parts Used ▶ flower
Distribution ▶ E China, Korea; in India and Korea cultivated

Cheers 1998; Chopra 1956; Fleischhauer 2003; Hanelt 2001; Hiller/Melzig 1999; NICPBP 1987; Roth/Kormann 1997; Schultze-Motel 1986

 Dianthus coronarius *Lam.*

▶ *Dianthus caryophyllus L.*

DICERANDRA Benth. - Florida mints, Dicerandra Balm - Lamiaceae (Labiatae)

 Dicerandra christmannii *Huck et Judd*

Usage ▶ flavoring; **product:** essential oil
Parts Used ▶ leaf
Distribution ▶ S USA

McCormick et al. 1993

 Dicerandra cornutissima *Huck*

Usage ▶ flavoring; **product:** essential oil
Parts Used ▶ leaf
Distribution ▶ S USA

McCormick et al. 1993

 Dicerandra frutescens *Benth.*

Usage ▶ flavoring; **product:** essential oil
Parts Used ▶ leaf
Distribution ▶ S USA
Note ▶ A high aromatic 'mint' plant.

Eisner et al. 1990; McCormick et al. 1993

 Dicerandra immaculate *Lakela*

Usage ▶ flavoring; **product:** essential oil
Parts Used ▶ leaf
Distribution ▶ S USA

McCormick et al. 1993

DICTAMNUS L. - Dittany - Rutaceae

Dictamnus albus L.

Synonyms ▸ *Dictamnus gymnostylis* Steven
Common Names ▸ burning bush, dittany, dittander; *Chinese:* bai xian pi; *French:* dictame blanc; herbe aux éclairs; *German:* Brennender Busch, Flammender Busch, Diptam, Hexenkraut, Pfefferkraut, Spechtwurzel, Weißer Diptam; *Italian:* dittamo bianco; *Russian:* jasenez
Usage ▸ flavoring and bitter (sporadically for alcoholic beverage)
Parts Used ▸ root
Distribution ▸ E, EC Europe, Iberian Peninsula, Germany, the Balkans

Aichele/Schwegler 3, 1995; Charalambous 1994; Cheers 1998; Dudtschenko et al. 1989; Erhardt et al. 2002; Fleischhauer 2003; Hager 4, 1992; Heeger 1956; Hepper 1992; Hiller/Melzig 1999; Kubeczka et al. 1990; Roth/Kormann 1997; Schönfelder 2001; Seidemann 1993c; Small 1997; Uphof 1968; Wüstenfeld/Haensel 1964

Dictamnus creticus J. Hill

 Origanum dictamnus L.

Dictamnus gymnostylis Steven

▸ *Dictamnus albus* L.

DICYPELLIUM Nees et Mart.- Lauraceae

Dicypellium caryophyllatum Nees

Common Names ▸ clove bark, pinkwood bark; *Brazil (Portuguese):* caneleira-cravo, craveiro-do maranhao, cravo-do-mato, cravinho, louro-cheirosa, louro-cravo, pau-cravo; *French:* cannelier giroflée; *German:* Nelkenzimt, Schwarzer Zimt; *Russian:* koriza gvosditschnaja
Usage ▸ spice, like cinnamon or glove bark; flavoring; **product:** essential oil
Parts Used ▸ bark
Distribution ▸ tropical America
Note ▸ The root was discovered in 1539 by Gozola Pizarro in the Upper Amazonas area.

Bois 1934; Erhardt et al. 2002; Hager 4, 1992; Hiller/Melzig 1999; Hoppe 1949; Kumar 2001; Mors et al. 2000; Seidemann 1993c; Seidemann/Siebert 1987; Staesche 1972; Teuscher 2003; Uphof 1968

DIGERA Forssk. - Amaranthaceae

Digera alternifolia Asch.

▸ *Digera muricata* (L.) Mart.

Digera arvensis Forssk.

▸ *Digera muricata* (L.) Mart.

Digera muricata (L.) Mart.

Synonyms ▸ *Digera alternifolia* Aschers, *Digera arvensis* Forssk.
Common Names ▸ wild rhubrab; *Arabic:* bud(d)jer, did(d)jar; *Hindi:* kanjero, latmhura, lasua tandla
Usage ▸ pot-herb (in India)
Parts Used ▸ leaf, tender shoot
Distribution ▸ S, SE Asia: India, Afghanistan, Indonesia, NE tropical Africa: Ethiopia, Arabia

Agarwal 1986; Arora/Pandey 1996; Hanelt 2001; Seshadri/Nambiar 2003; Uphof 1968

DILLENIA L. - Dilleniaceae

Dillenia indica L.

Common Names ▸ chalta tree, elephant apple; *German:* Chalthafrucht, Elephantenapfel; *Hindi:* chalta, cinar, hondapara, outenga, karambel; *Indone-*

◘ Dillenia indica, fruiting

sian: saimpol; *Japanese:* biwa modoki; *Javanese:* simpoh, chimpuh; *Malaysian:* chimpoh, (Indian) simpoh; *Pilipino:* katmon; *Portuguese:* frula estrela; *Thailand:* mataat; *Vietnamese:* sõ

Usage ▸ flavoring (acidic tasting for curries and jellies)
Parts Used ▸ fruit pulp
Distribution ▸ India, Sri Lanka, Indochina, Malaysia, W Africa: Cameroon, Ghana, Nigeria, Sierra Leone
Note ▸ The fruit should not be confused with the fruit of *Fernonia limonia* (L.) Swingle; the leaves have the aroma of anise seeds.

Bois 1934; Burkill 1966; 3, 1995; Cheers 1997; Dastur 1954; Engel/Phummai 2000; Erhardt et al. 2002; Hanelt 2001; Hoogland 1952; Kottegoda 1994; Kumar 2003; Lück 2000; Schenck/Naundorf 1966; Schultze-Motel 1986; Storrs 1997; Täufel et al. 1993; Uphof 1968; Wealth of India 3, 1952; Wiersema/León 1999

Diosma fragrans Sims

▸ *Adenandra fragrans (Sims) Roem. et Schult.*

Diosma odorata *(Wendl.) DC.*

▸ *Barosma crenulata (Torner) Hook.*

DIPLOTAXIS DC. - Wall rocket - Brassicaceae (Cruciferae)

Diplotaxis tenuifolia *(L.) DC.*

Common Names ▸ Lincoln's weed, sand rocket; *German:* Schmalblättiger Doppelsame, Doppelrauke, Ausdauernde Rucola
Usage ▸ pot-herb (locally)
Parts Used ▸ leaf
Distribution ▸ Europe

Aichele/Schwegler 3, 1995; Erhardt 2002; Täufel et al. 1993

DIPTERYX Schreb. - Tonka Bean - Fabaceae (Leguminosae)

Dipteryx odorota *(Aubl.) Willd.*

Synonyms ▸ *Baryosma longa* Gaertn., *Coumarouna odorate* Aubl.
Common Names ▸ tonka, tonka bean; *Brazil (Portuguese):* cumaru, cumaru-amarelo, cumaru-doamazonas; cumarurana, cumbaru, fèves tonka, muimapagé; *French:* coumarouna, fève de Tonka, tonquin; *German:* Tonkabohne; *Russian:* dipteriks
Usage ▸ spice, substitute for vanilla; flavoring of snuff and liqueurs; **product:** coumarin
Parts Used ▸ seed
Distribution ▸ N Brazil, Venezuela, Malaya, Guyana, Surinam
Note ▸ Seed is strongly bitter, the odor strong after coumarin; the coumarin is a carcinogenic substance and is forbidden in foods in many countries. The use of natural coumarin in food was banned in the USA and other countries (e.g. Germany) in 1954.

Boisvert/Hubert 2000; Davidson 1999; Duke et al. 2003; Ehlers et al. 1995; Erhardt et al. 2002; Guzman/Siemonsma 1999; Hanelt 2001; Hayashi/Thompson 1974; Hoppe 1949; Kumar 2001; Lorenzi 1992;

◘ Dipteryx odorata, seeds

Pound 1938; Mors/Rizzini 1961; Mors et al. 2000; Nakano/Suarez 1969; Pursglove 1968; Schröder 1991; Schultze-Motel 1986; Seidemann 1993c; 1994; Seidemann/Siebert 1987; Sullivan 1982; Täufel et al. 1993; Teuscher 2003; Uphof 1968; Wörner/Schreier 1991; Wüstenfeld/Haensel 1964; Zeven/de Wet 1982

 Dipteryx oppositifolia *(Aubl.) Willd.*

▶ *Taralea oppositifolia Aubl.*

 Dipteryx punctata *(Blake) Amshoff*

Common Names ▶ pointed tonka; *German:* Punktierte Tonkabohne
Usage ▶ spice, like tonka bean
Parts Used ▶ seed
Distribution ▶ Northern S America: Venezuela, Columbia, Caribbean

Hanelt 2001; Kumar 2001; Pursglove 1974; Schultze-Motel 1986; Seidemann 1993c

DISSOTIS Benth. - Melastomataceae

 Dissotis plumosa *(D. Don) Hook f.*

Synonyms ▶ *Dissotis rotundifolia* (Sm.) Triana
Common Names ▶ rock rose; *German:* Felsenrose
Usage ▶ spice for sauces, pot-herb
Parts Used ▶ leaf
Distribution ▶ C and W Africa: Sierra Leone to Zaire, Cameroon, Liberia

Ayensu 1978; Erhardt et al. 2002; Neuwinger 1998

 Dissotis rotundifolia *(Sm.) Hook f.*

▶ *Dissotis plumosa (D. Don) Hook*

 Dittrichia viscosa *(L.) Greut.*

▶ *Inula viscosa (L.) Ait.*

 Donacodes walang *Bl.*

▶ *Etlingera walang (Bl.) R.M. Smith*

DORSTENIA L. - Moraceae

 Dorstenia contrajerva *L.*

Synonyms ▶ *Dorstenia houstonii* L., *Dorstenia maculata* Lem., *Dorstenia quadrangularis* Stokes
Common Names ▶ Torn's herb, sanke wood; *French:* herbe aux serpents, racine de charchis; *German:* Dorstenie, Schlangenwurz; *Mexico:* barbadilla; *Spanish:* contra hierba, contra de jorba, contra-jerba
Usage ▶ flavoring
Parts Used ▶ leaf, rhizome
Distribution ▶ Brazil, S Mexico, West Indies, Africa; Java, Malaccas and in Africa and S America cultivated locally

Note ▶ The leaves and rhizome powder are used for improving the taste of cigarette tobacco.

Burkhill 1966; Hanelt 2001; Schultze-Motel 1986

 Dorstenia houstoni L.

❯ *Dorstenia contrajerva L.*

 Dorstenia maculata Lem.

❯ *Dorstenia contrajerva L.*

 Dorstenia opifera Mart.

Common Names ▶ *Brazil (Portuguese):* caapiá
Usage ▶ flavoring of tobacco
Parts Used ▶ rhizome
Distribution ▶ S America: Brazil

Kumar 2001

 Dorystenia quadrangularis Stokes

❯ *Dorystenia contrajerva L.*

DORYSTAECHAS Boiss. - Lamiaceae /Labiatae)

 Dorystaechus hastata Boiss. et Heldr. ex Benth.

Common Names ▶ Turkish lavender; *German:* Türkischer Lavendel
Usage ▶ flavoring; **product:** essential oil
Parts Used ▶ herb
Distribution ▶ Turkey: Antalya; W Asia

Başer/Öztürk 1992; Meriçili/Merçili 1986

DRACOCEPHALUM L. - D's Head - Lamiaceae (Labiatae)

 Dracocephalum ibericum Bieb.

Synonyms ▶ *Lallemantia iberica* (Bieb.) Fisch. et Mey.
Common Names ▶ *German:* Iberischer Drachenkopf
Usage ▶ pot-herb (in Iran)
Parts Used ▶ leaf
Distribution ▶ Asia Minor, Caucasus, Mesopotamia, Syria, Palestine

Uphof 1968

 Dracocephalum mexicanum H.B.K.

❯ *Agastache mexicana (Kunth) Lint et Epling*

 Dracocephalum moldavica L.

Synonyms ▶ *Moldavica punctata* Moench, *Moldavica suaveolens* Gilib.;
Common Names ▶ Moldavian balm, dragon's head; *Chinese:* xiang qing lan; *French:* dracocéphale moldavique, tête-de-dragon; *German:* Moldawischer Drachenkopf; Türkischer Drachenkopf, Moldauischer Drachenpfeffer, Türkische Melisse; *Russian:* melissa turezkaja, smeegolownik moldawskij, drakonogolownik, sinjawka
Usage ▶ spice, flavoring (liqueurs, worm wood wine, refreshment drinks)
Parts Used ▶ leaf, herb
Distribution ▶ E Europe, especially Moldavia, C Alps, S Siberia to Russian Far East, SW Asia, NW China

Böttcher/Günther 2003; Bois 1934; Dudtschenko et al. 1989; Erhardt et al. 2002; Halász-Zelenik et al. 1988; Hanelt 2001; Hiller/Melzig 1999; Holm et al. 1988a, b; Kakasy et al. 2002; Mikus/Schaser 1995; Pank et al. 1989; Pochljobkin 1974, 1977; Povilaityte/Venskutonis 2000; Povilaityte et al. 2001; Roth/Kormann 1997; Schultze-Motel 1986; Seidemann 1993c; Seidemann/Siebert 1987; Täufel et al. 1993; Teuscher 2003; Wealth of India 3, 1952

DRACONTOMELON Bl. - Anacardiacea

Dracontomelon dao (Blanco) Merr. et Rolfe

Synonyms ▶ *Dracontomelon mangiferum* (Blanco) Merr. et Rolfe; *Dracontomelon sylvestre* Bl.
Common Names ▶ argus pheasant; *Chinese:* yan min, yan mien chi; *Dutch:* drakeboom; *German:* Drachenapfel; *Malaysian:* asam kuang, sengkoewang; *Pilipino:* dao, batuan, kamarak, maliyau
Usage ▶ spice, flavoring curries
Parts Used ▶ fruit, leaf, flower
Distribution ▶ Asia: S China, Hong Kong; Myanmar, Thailand, Vietnam, Malaysia, Philippines
Note ▶ The lemon-like taste is very sour.

Burkill 1985; Hanelt 2001; Schultze-Motel 1986; Seidemann 1993c; Tigga/Sreekumar 1996; Uphof 1968; Wealth of India 3, 1952

Dracantomelon mangiferum (Blanco) Merr. et Rolfe

▶ *Dracantomelon dao* Bl.

Dracantomelon sylvestre Bl.

▶ *Dracantomelon dao* (Blanco) Merr. et Rolfe

DRIMYS J.R. Forst. et G. Forst. - Winter's Bark - Winteraceae (Magnoliaceae)

Drimys aromatica F. v. Muell.

▶ *Drimys lanceolata* (Poir.) Baill.

Drimys lanceolata (Poir.) Baill.

Synonyms ▶ *Drimys aromatica* F. v. Muell., *Tasmannia lanceolata* (Poir.) A.C. Sm.
Common Names ▶ pepper tree, Australian pepper tree, mountain pepper, Tasmania pepper; *German:* Pfefferbaum, Tasmaniapfeffer
Usage ▶ spice, like black pepper (pepper substitute); leaf: flavoring
Parts Used ▶ fruit
Distribution ▶ Australia: NS Wales, Victoria, Tasmania
Note ▶ The leaves of this plant also have a hot taste.

Erhardt et al. 2002; Mabberley 1997; Read/Menary 2001; Uphof 1968; Wiersema/León 1999

Drimys winteri J.R. Forst. et G. Forst.

Synonyms ▶ *Winterana aromatica* Desc.
Common Names ▶ drimys bark, Winter's bark, Magelhanisian caneel; *Brazil (Portuguese):* canela-amarga, casca-d'anta, cataia, melambo, paratudo, pau-para-tudo; *French:* Canelle de Magellan; *German:* Beiß-Canelo, Chilenischer Canelo, Karibischer Zimt, Magelhanischer Zimt, Winter's-Rinde; *Mexico:* yaga yiña
Usage ▶ spice
Parts Used ▶ bark
Distribution ▶ Chile to Cape Horn, S Argentina
Note ▶ The bark introduced by Captain John Winter 1578.

Berger 1, 1949; Bois 1934; Cortés/Oyarzún 1981; Erhardt et al. 2002; Hager 4, 1992; Hiller/melzig 1999; Hoffmann et al. 1992; Hoppe 1949; Lorenzi 1992; Mors et al. 2000; Rätsch 1998; Seidemann 1993c; Teuscher 2003; Uphof 1968; Wüstenfeld/Haensel 1964

DRYPETES Vahl - Euphorbiaceae

Drypetes aubrevillei Léandri

Common Names ▶ pepper bark, pepper stick, sting pepper; *German:* Pfefferrinde
Usage ▶ spice (rarely)
Parts Used ▶ bark
Distribution ▶ W Africa: Sierra Leone, Liberia, Ivory Coast
Note ▶ The bark has a peppery taste.

Burkill 2, 1994

DYSOXYLUM Bl. - Meliaceae

◘ Drimys winteri, flowering

Dysoxylum alliaceum *(Bl.) Bl.*

Synonyms ▶ *Dysoxylum costulatum* (Miq.) Miq., *Dysoxylum euphlebium* Merr., *Dysoxylum thyrsoideum* Hiern.
Usage ▶ flavoring
Parts Used ▶ young leaves (cooked)
Distribution ▶ Andaman Islands, Thailand, Indonesia, Malaysian Archipelago, N Australia, Solomon Islands, S Vietnam
Note ▶ All parts smell strongly of onions.

Guzman/Siemonsma 1999; Terra 1966

Dysoxylum costulatum *(Miq.) Miq.*

➤ *Dysoxylum alliaceum (Bl.) Bl.*

Dysoxylum euphlebium *Merr.*

➤ *Dysoxylum alliaceum (Bl.) Bl.*

Dysoxylum thyrsoideum *Hiern.*

➤ *Dysoxylum alliaceum (Bl.) Bl.*

Drypetes paxii *Hutch*

Usage ▶ spice (rarely)
Parts Used ▶ bark
Distribution ▶ W Africa: S Nigeria, W Cameroon and on to Gabun
Note ▶ The bark has a peppery taste.

Burkill 2, 1994

Drypetes pellegrinii *Léandri*

Usage ▶ spice (rarely)
Parts Used ▶ bark
Distribution ▶ W Africa: Ivory Coast and Ghana
Note ▶ The bark has a spicy taste.

Burkill 2, 1994

E

ECHINOCHLOA P. Beauv. - Cockspur - Poaceae (Gramineae)

 Echinochloa pyramidale *(Lam.) Hitchc. et Chase*

Common Names ▸ antelope grass, limpopo grass; *French:* antilope à herbe, antilope de graminées; *German:* Antilopengras, Pyramiden-Hühnergras
Usage ▸ condiment
Parts Used ▸ plant (grass)
Distribution ▸ tropical Africa: Nile, S Niger, Senegal, the Gambia, Nigeria
Note ▸ Produced as a vegetable salt from the plant ash; substitute for normal salt.

Burkill 2, 1994; Erhardt et al. 2002; Hanelt 2001; Uphof 1968

ECHINOPHORA L. - Apiaceae (Umbelliferae)

 Echinophora sibthorpiana *(Guss.) Tamasch.*

▸ *Echinophora tenuifolia* L.

Synonyms ▸ *Echiniphora sibthorpiana* (Guss.) Tamasch.
Common Names ▸ prickly parsnip; *French:* échinophore; *German:* Igelklette, Stacheldolde, *Russian:* koljutschesontitschnik; *Turkish:* çörtük

Usage ▸ pickling herb (fresh or dried); **product**: essential oil
Parts Used ▸ leaf, stalk
Distribution ▸ Balkan, Turkey (Anatolia), Iran, Azerbadijan, Armenia, Afghanistan

Baser et al. 1994b; Charalambous 1994; Özcan/Erkman 2001; Özcan et al. 2002

 Echites malabarica *Lam.*

▸ *Alstonia scholaris (L.) R.Br.*

 Echites scholaris *L.*

▸ *Alstonia scholaris (L.) R. Br.*

ELAEIS Jacq. - Oil Palm - Arecaceae

 Elaeis guineensis *Jacq.*

Common Names ▸ African oil palm, tall palm; *French:* palmier à huile; *German:* Afrikanische Ölpalme
Usage ▸ condiment
Parts Used ▸ leaf
Distribution ▸ W, E and C Africa, Angola
Note ▸ Produced as a vegetable salt from the plant ash, substitute for normal salt.

© Springer-Verlag Berlin Heidelberg 2005

Erhardt et al. 2000, 2002; Hiller/Welzig 1999; Neuwinger 1999; Täufel et al. 1993; Zeven/de Wet 1982

 Elaeodendron metabelica *Loes*

▶ *Cassine metabelica (Loes) Steedman*

 Elaeoselinum gummiferum *Desf.*

▶ *Margotia gummifera (Desf.) Lange*

ELETTARIA Maton - Cardamom - Zingiberaceae

 Elettaria cardamomum *(L.) Maton*

Synonyms ▶ *Alpinia cardamomum* Roxb., *Amomum cardamomum* L., *Amomum racemosum* Lam., *Cardamomum officinale* Salisb., *Elettaria repens* (Sonn.) Baill.

Common Names ▶ cardamom, small cardamom, Chester cardamom; *Arabic:* hab hab, heel; *Chinese:* dou kou hua, pai tou k'ou; *Dutch:* kardemomzaad; *French:* cardamome, cardamomier; *German:* Echter Kardamom, Malabar-Kardamom; *Hindi:* chotielachi; *Indonesian:* kapulaga; *Italian:* cardamomo (frutti); *Japanese:* karudoman; *Korean:* sodugu; *Malaysian:* buah pelaga; *Pilipino:* luk grawan; *Portuguese:* cardamomo; *Russian:* kardamom; *Sanskrit:* ela, upakunchika; *Slovenian:* kardamóm; *Spanish:* cardamomo

Usage ▶ spice; **product**: essential oil (cardamom oil)

Parts Used ▶ fruit (capsule), seed

Distribution ▶ S India, Sri Lanka, SE Asia

Note ▶ The seeds are sold (in London) as camphor seeds.

Berger 3, 1952; Berger 1958, 1964/65; Bois 1934; Boisvert/Hubert 2000; Bournot 1968; Burkill 5, 2000; Craze 2002; Dalby 2000; Davidson 1999; Domrös 1973; Dudtschenko et al. 1989; Duke et al. 2003; Farrell 1985; Ferrão 1992; Gopalakrishnan/Narayanan 1991; Govindarajan et al. 1982b; Guzman/Siemonsma 1999; Hager 4, 1992; Hanelt 2001; Hiller/Welzig 1999; Hoppe 1949; Kataoka et al. 1986; Maistre 1964; Matthies 1989; Melchior/Kastner 1974; Molegode 1938; Morris/Mackley 1999; Noleau et al. 1987; Norman 1991; Overdieck 1992; Peter 2001; Pochljobkin 1974, 1977; Pruthi 1976;

Pschyrembel 1998; Richard 1987; Rievals/Mansour 1974; Rosengarten 1969; Roth/Kormann 1997; Schröder 1991; Schultze-Motel 1986; Seidemann 1993c, 1998/2000; Seidemann/Siebert 1987; Shaban et al. 1987; Sharma 2003; Shiva et al. 2002; Small 1997; Staesche 1972; Täufel et al. 1993; Teuscher 2003; Tschirch 1892; Ophof 1968; Vogt 1998; Wyk et al. 2004; Zeven/de Wet 1982

 Elettaria cardamom *(L.) Maton var. major* *(Sm.) Thwaites*

▶ *Elattaria major Sm.*

 Elettaria cardamom *(L.) Maton* **var. minuscula** *Burkill*

Common Names ▶ Malabar cardamom; *German:* Malabar-Kardamom

Usage ▶ spice

Parts Used ▶ fruit

Distribution ▶ only cultivated

Guzman/Siemonsma 1999; ISO-Standard 1980

 Elettaria major *Sm.*

Synonyms ▶ *Elettaria cardamom* Maton var. *major* (Sm.) Thwaites

Common Names ▶ Ceylon cardamom, greater oblong cardamom, long cardamom, wild cardamom; *Arabic:* hab hal; *French:* cardamome; *German:* Ceylon-Kardamom, Langer Kardamom; *Italian:* cardamomo; *Russian:* cardamome; *Spanish:* cardamomo

Usage ▶ spice; **product**: essential oil

Parts Used ▶ fruit, seed

Distribution ▶ Sri Lanka

Note ▶ The seeds are seldom used as a spice. The species in *Elettarie cardamom* (L.) Maton are falsified.

Boir 1934; Domrös 1973; Hanelt 2001; Matthies 1989; Overdieck 1992b; Peter 2001; Schenck/Naundorf 1966; Schultze-Motel 1986; Seidemann 1993c; Uphof 1968; Wealth of India 3, 1952

 Elettaria repens *(Sonn.) Baill.*

▶ *Elettaria cardamomum (L.) Maton*

Elettaria cardamomum: a **flowering**, b **fruiting**

Elettaria solaris *Blume*

▶ *Etlingera solaris (Blume) R.M. Smith*

ELSHOLTZIA Willd. - Elsholtzia - Lamiaceae (Labiatae)

Elsholtzia ciliata *(Thunb.) Hyl.*

Synonyms ▶ *Elsholtzia cristata* Will., *Elsholtzia patrini* (Lepecin) Garcke
Common Names ▶ common elsholtzia; Vietnamese balm; *Chinese:* hsiang-ju, xiang ru; *German:* Echte Kamm-Minze, Elsholtzie, Vietnamesische Melisse, Würzminze; *Japanese:* naginata-koju; *Korean:* hyangyu; *Russian:* mjata prjanaja, ili el'sgol'zija, grebentschaja schandra, prjanyj issop; *Vietnamese:* gia to
Usage ▶ spice, specially for fish
Parts Used ▶ herb
Distribution ▶ C Asia: former Soviet Union, China, Taiwan, Japan, Korea; native in Europe

Dudtschenko et al. 1989; Górski et al. 1991; Kobold et al. 1987; Mikus/Schaser 1995; Pochljobkin 1974, 1977; Schultze-Motel 1986; Seidemann 1993c; Seidemann/Siebert 1987; Teubner 2001; Teuscher 2003; Zeven/de Wet 1982

Elsholtzia cristata *Will*

▶ *Elsholtzia ciliata (Thunb.) Hyl.*

Elsholtzia integrifolia *Benth.*

▶ *Nepeta tenuifolia Benth.*

Elsholtzia monostachys *H. Lévl. et Van.*

▶ *Agastache rugosa (Fisch. et C.A. Mey.) Kuntze*

Elsholtzia patrini *(Lepecin) Garcke*

▶ *Elsholtzia ciliata (Thunb.) Hyl.*

Elsholtzia stauntonii *Benth.*

Common Names ▶ mint bush, mint shrub, Staunton elsholtzia; *Chinese:* mu xiang ru; *German:* Chinesische Kamm-Minze, Chinesischer Gewürzstrauch
Usage ▶ spice
Parts Used ▶ leaf
Distribution ▶ C China: Ganru, Hebei, Henan, Shaanxi, Shanxi

Erhardt et al. 2000, 2002; Hanelt 2001; Small 1997; Teuscher 2003; Wiersema/León 1999

EMBELIA Burm. f. - Myrsinaceae

Embelia philippinensis *A. DC.*

Synonyms ▶ *Rhamnus lando* Llanos, *Ribesoides philippense* O. Kuntze, *Samara philippinensis* Vidal.
Common Names ▶ woody vine; *Pilipino:* dikai, lando, pongpong
Usage ▶ flavoring (for fish, meat, vegetables) and gives a sour taste to soup
Parts Used ▶ leaf
Distribution ▶ Philippines

Guzman/Siemonsma 1999; Oyen/Dung 1999; Uphof 1968

EMILIA Cass. - Asteraceae (Compositae)

 Emilia coccinea *(Sims) G. Don*

> *Emilia javanica (Burm.f.) C.B. Rob.*

 Emilia javanica *(Burm.f.) C.B. Rob.*

Synonyms ▸ *Emilia coccinea* (Sims) G. Don., *Hieracium javanicum* Burm.f.
Common Names ▸ tassel flower, Flora's paintbrush; *French:* goutte de sang; *German:* Javanische Troddelblume
Usage ▸ spice
Parts Used ▸ leaf, root
Distribution ▸ tropical Africa, India, S China, Java, E Africa and other tropical countries, in Poland cultivated and sometimes naturalized elsewhere
Note ▸ The taste is a delicate one, slightly acid with a touch of bitterness.

Erhardt et al. 2000, 2002; Hanelt 2001; Neuwinger 1999; Schultze-Motel 1986

ENDOSTEMON N.E. Br. - Lamiaceae (Labiatae)

 Endostemon tereticaulis *(Poir) M. Ashby*

Common Names ▸ *German:* Walzenförmiges Endostemonkraut
Usage ▸ condiment, flavoring
Parts Used ▸ herb
Distribution ▸ W Africa: Ghana, Senegal, Somalia, Uganda
Note ▸ The aromatic herb for meat in cooking.

Burkill 3, 1995

 Enteromorpha compressa *(L.) Grev.*

> *Ulva compressa* L.

ENYDRA Lour. - Asteraeae (Compositae)

 Enydra fluctuans *Lour.*

Synonyms ▸ *Cryphiospermum repens* Pal.
Usage ▸ spice
Parts Used ▸ leaf, stak
Distribution ▸ C, E Africa, India, SE Asia, cultivated in Java, Malaysia, Cambodia

Hanelt 2001; Schultze-Motel 1986; Uphof 1968

EREMOCHARIS Philippi - Apiaceae (Umbelliferae)

 Eremocharis radiata *(Wolff.) Johnston*

Common Names ▸ *German:* Strahliges Eremochariskraut
Usage ▸ pot-herb (for soups)
Parts Used ▸ herb
Distribution ▸ S America: Andes of N Chile and Peru

ERODIUM L'Herit ex Ait - Heron's Bill, Stork's Bill - Geraniaceae

 Erodium moschatum *(L.) L'Hérit. ex Ait.*

Synonyms ▸ *Geranium moschatum* L.
Common Names ▸ common Heron's bill, musky Stork's bill, musk clover, white stem filaree; *French:* érodion; *German:* Moschus-Reiherschnabel; *Russian:* aistnik, shuravel'nik
Usage ▸ spice (rarely)
Parts Used ▸ herb
Distribution ▸ N, C and S Europe: Iberia, France, from Russia to Siberia, N, S America, E, S Africa, Australia, New Zealand, elsewhere native

Aichele/Schwegler 3, 1995; Erhardt et al. 2002; Hanelt 2001; Hiller/Melzig 1999; Schultze-Motel 1986; Seidemann 1993c; Zeven/de Wet 1982

ERUCA Mill. - Rocket Salad - Brassicaceae (Cruciferae)

Eruca foetida Moench.

> *Eruca sativa* Mill.

Eruca sativa Mill.

Synonyms ▸ *Brassica eruca* L., *Eruca foetida* Moench., *Eruca vesicaria* (L.) Cav. ssp. *sativa* (Mill.) Thell.
Common Names ▸ garden or Roman rocket, rocket salad, rocket; *Dutch:* gekweckte eruca, tuimeruca; *French:* eruca, roquette, roquette des jardins; *German:* Jamba-Raps, Ölrauke, Persischer Senf, Ruke, Senfrauke; *Hindi:* taramira; *India:* taribed (Kashmir); *Italian:* eruca, rucola commune, ruchetta, ruca; *Japanese:* kibana-suzushiro, roketta; *Portuguese:* eruca, rúcula, pinchao; *Russian:* mindau, skuka posewnaja; *Sanskrit:* bhutaghna, daradharsha; *Spanish:* eruca, roqueta común
Usage ▸ condiment; **product** of mustard
Parts Used ▸ leaf, seed
Distribution ▸ Mediterranean regions: Spain, Greece, N and NE Africa, temperate Asia, India, Afghanistan, native elsewhere

Aichele/Schwegler 3, 1995; Bärtels 1997; Bremness 2001; Cheers 1997; Dudtschenko et al. 1989; Erhardt et al. 2002; Guzman/Siemonsma 1999; Hanelt 2001; Hepper 1992; Hiller/Melzig 1999; Kays/Dias 1995; Melchior/Kastner 1974; Miazawa et al. 2002; Schultze-Motel 1986; Seidemann 1993c; Seidemann/Siebert 1987; Täufel et al. 1993; Teuscher 2003; Uphof 1968; Villamar et al. 1994; Wiersema/León 1999; Zeven/de Wet 1982

Eruca vesicaria (L.) Cav. **sspec. sativa** (Mill.) Thell.

> *Eruca sativa* Mill.

ERYNGIUM L. - Sea Holly - Apiaceae (Umbelliferae)

Eryngium antihystericum Rottl.

> *Eryngium foetidum* L.

Eryngium foetidum L.

Synonyms ▸ *Eryngium antihystericum* Rottl.
Common Names ▸ culantro, eryngo, foreign coriander, sawtooth, sawtooth coriander, false coriander, fitweed, Mexican coriander; *Chinese:* jia yuan qian; *Cuban:* culantro, cimarrón; *French:* azier la fièvre, panicaut fétide, chardon étoilé, coulante; *German:* Langer Koriander, Stinkender Mannstreu, Stinkdistel; *Laos:* hom chin; *Malaysian:* daun ketumbar jawa; *Nicaragua:* culantrillo; *Portuguese:* coentro, coentro-bravo, coentro-da-colônia, coentro-da-caboclo; *Spanish:* cilantro, orégano de Cartagena; *Thai:* phak chee farang, pakchi farang, hom-pom-kula; *Vietnamese:* rau ngo gai, ngo tay, bat nga
Usage ▸ spice, condiment (flavoring)
Parts Used ▸ leaf, root
Distribution ▸ tropical and subtropical M, S America: Mexico, Cuba, Brazil, also cultivated; indroduced and cultivated in tropical Africa (Liberia) and E Asia: Cambodia and Thailand
Note ▸ Strictly speaking, the name "cilantro" refers to sawtooth coriander, but is often used for leaves of *Coriandrum sativum* L.

Aedo et al. 2001; Arora/Pandey 1996; Davidson 1999; Duke et al. 2003; Guzman/Siemonsma 1999; Koolhaas 1932; Kuebal/Tucker 1988; Leclercq et al. 1992; Mors et al. 2000; Lück 2000; Ogle et al. 2003; Roig 1974; Sankat/Maharaj 1994, 1996; Schultze-Motel 1986; Small 1997; Tucker 1986; Uphof 1968; Vidal 1964; Villamar et al. 1994; Wong et al. 1994

Erysimum barbarea L.

> *Barbarea vulgaris* R.Br.

Erysimum vernum Mill.

> *Baberea verna (Mill.) Aschers.*

Erythraea centaurum auct.

◉ *Centaurium erythraea Raf.*

ERYTHRINA L. – Coral Tree - Fabaceae (Leguminosae)

 Erythrina variegata L. **var.** *orientalis* (L.) Merr.

Synonyms ▶ *Erythrina indica* L., *Erythrina variegata* L.
Common Names ▶ Indian coral tree, moochy wood, tigeris claw; *Chinese:* hai tong pi; *German:* Indischer Korallenbaum; *Hindi:* chadap, mandara; *India:* mimbataru, mandalia, paribhadra; *Indonesian:* dadap belang, kalayana murikku; *Japanese:* deigo, diigu; *Malaysian:* chengkering, dedap batik; *Pilipino:* dap dap, karapdap, kabrab; *Sanskrit:* mandar, parijata; *Thai:* thong laang laai; *Vietnamese:* hai dong, vông nem
Usage ▶ pot-herb (Andaman Islands)
Parts Used ▶ leaf, young shoot
Distribution ▶ India: Andaman Islands; China, Malaysia, Philippines, Polynesia, Taiwan, Tanzania

Arora/Pandey 1996; Engle/Phummai 2000; Erhardt et al. 2000; Hager 2001; Kottegoda 1994; Quattrocchi 1999; Rätsch 1998; Schultze-Motel 1986; Sharma 2993

 Erythrochilus indicus Reinw. ex Bl.

◉ *Claoxylon indicum (Reinw. ex Bl.) Hassk.*

ERYTHROCOCCA Benth. - Euphorbiaceae

 Erythrococca atrovierens (Pax) Prain

Synonyms ▶ *Athroandra atrovirens* Pax et K. Hoffm., *Erythrococca flaccida* (Pax) Prain, *Erythrococca olearaceae* (Prain) Prain
Common Names ▶ *Congo, Zaire:* mascha, bindi dikil; *Ghana:* forowa, gyigyam; *Liberia:* bu yiddi pulu; *Senegal:* mbuhur mbéré; *Sierra Leone:* budelemi
Usage ▶ spice

Parts Used ▶ young leaf
Distribution ▶ tropical Africa: cultivated in Congo and Zaire

Burkill 2, 1994; Hanelt 2001; Schultze-Motel 1986; Radcliff-Smith 1987

 Erythrococca flaccida (Pax) Prain

◉ *Erythrococca atrovierens (Pax) Pain*

 Erythrococca olearaceae (Prain) Prain

◉ *Erythrococca atrovierens (Pax) Pain*

ESCOBEDIA Ruiz et Pav. - Scrophulariaceae

 Escobedia scabrifolia Ruiz et Pav.

Common Names ▶ saffron of Andes, saffron root; *Columbia:* azafrán, azafrán de raiz; *German:* Anden-Safran, Eskobedie; *Peru:* azafrán de los Andes, azafrán de montāna, palillo
Usage ▶ spice (saffron-like)
Parts Used ▶ root
Distribution ▶ Latin America: Upper Andes: Peru, Columbia, Venezuela

Davidson 1999; Hanelt 2001; Schultze-Motel 1986; Seidemann 1993c; Williams 1970

ETLINGERA Giseke - Torch Ginger - Zingiberaceae

 Etlingera elatior (Jack) R.M. Sm.

Synonyms ▶ *Alpinia eliator* Jack, *Alpinia speciosa* D. Dietr., *Alpinia speciosa* (Wendl.) K. Schum., *Nicolaia elatior* (Jack) Horan, *Phaeomeria magnifica* (Rosc.) K. Schum.
Common Names ▶ torch ginger, Philippine wax flower; *French:* sceptre de l'empereur; *German:* Malayi-

◘ Etlingera eliator, flowering

scher Fackelingwer, Heller Galgant, Kaiserzepter; *Indonesian:* combrang, honje (Sunda Island); *Javanese:* kĕchumbrang, kecombrang, tĕpus kampong, petikale; *Malaysian:* bunga kantan, bunga siantan; ubud udat; *Spanish:* antorcha, boca de dragón; *Thai:* kaalaa
Usage ▸ pot-herb and condiment (curries)
Parts Used ▸ floral bud, herb
Distribution ▸ Malaysia to Polynesia and Australia, Thailand, Indochina, China, Indonesia and Thailand also cultivated as an ornamental plant. Cultivated in Mexico

Cheers 1997; Duke et al. 2003; Erhardt et al. 2000, 2002; Guzman/Siemonsma 1999; Larsen et al. 1999; Moore 1962; Oyen/Dung 1999; Schultze-Motel 1986; Small 1997; Warren 1998; Wong 1999; Zeven/de Wet 1982

 Etlingera hemisphaerica *(Bl.) R.M. Sm.*

Synonyms ▸ *Nicolae hemisphaerica* (Bl.) Horan.
Usage ▸ condiment
Parts Used ▸ inflorescence
Distribution ▸ Sumatra, Java; in Java also cultivated

Dung 1999; Hanelt 2001; Larsen et al. 1999; Ochse et al. 1931; Schultze-Motel 1986; Wong 1999

 Etlingera rosea *Burtt et R.M. Sm.*

Synonyms ▸ *Amomum roseum* K. Schum.,
Common Names ▸ *Indonesian:* galoba papua, potmepini, gitipi tana (Moluccas)
Usage ▸ condiment (fish)
Parts Used ▸ aril
Distribution ▸ Indonesia

Guzman/Siemonsma 1999; Moore 1962

 Etlingera solaris *(Bl.) R.M. Smith*

Synonyms ▸ *Elettaria solaris* Blume, *Nicolaia solaris* (Bl.) Horan.
Common Names ▸ *Indonesian:* honje laka, honje warak
Usage ▸ a sour condiment
Parts Used ▸ fruit
Distribution ▸ Indonesia (Java)

Guzman/Sieonsma 1999; Moore 1962; Oyen/Dung 1999

 Etlingera walang *(Blume) R.M. Sm.*

Synonyms ▸ *Achasma walang* (Bl.) Val., *Amomum walang* (Bl.) Val., *Donacodes walang* Bl.
Common Names ▸ *Indonesian (Java):* walang, tepus walang
Usage ▸ spice
Parts Used ▸ leaf
Distribution ▸ Indonesia (W Java), also cultivated

Burkill 1966; Guzman/Siemonsma 1999; Hanelt 2001; Moore 1962

EUCALYPTUS L'Hér. – Gum - Myrtaceae

 Eucalyptus amygdalina *Labill.*

Synonyms ▸ *Eucalyptus salicifolia* Cav.
Common Names ▸ black peppermint, peppermint tree,

willowleaf eucalyptus; *German:* Pfefferminzbaum Schwarze Pfefferminze

Usage ▸ flavoring; **product**: essential oil
Parts Used ▸ leaf
Distribution ▸ Tasmania; cultivated in Chile, Zaire; former Soviet Union: W Georgia

Erhardt et al. 2002; Hanelt 2001; Uphof 1968; Zeven/de Wet 1982

Eucalyptus bridgesiana *Baker*

Common Names ▸ apple box, apple gum, apple-scented eucalyptus, but but; *German:* Apfel-Eukalyptus
Usage ▸ flavoring
Parts Used ▸ leaf
Distribution ▸ Australia

Eucalyptus citriodora *Hook.*

Common Names ▸ lemon eucalyptus, lemon-scented gum; *German:* Zitronen-Eukalyptus
Usage ▸ flavoring; **product**: essential oil
Parts Used ▸ leaf
Distribution ▸ Australia: Queensland; S America: Brazil

Cheers 1997; Kumar 2001; Roth/Kormann 1997; Schönfelder 2001

Eucalyptus dives *Schauer*

Common Names ▸ broad leaved peppermint-tree
Usage ▸ flavoring
Parts Used ▸ leaf
Distribution ▸ SE Australia, NS Wales, native in Zaire and S Africa
Note ▸ The essential oil has a high content of phellandrene and piperitone.

Erhardt et all. 2002; Hanelt 2001

Eucalyptus globulus *Labill.*

Common Names ▸ blue gum tree, fever tree, Tasmanian blue gum; *Arabic:* calibtus, hafor; *Chinese:* an ye; *French:* eucalyptus bleu, gommier bleu; *German:* Blaugummibaum, Fieberheilbaum, Kugeliger Eukalyptus; *Italian:* eucalipto; *Japanese:* yûkari-no-ki; *Russian:* eukaljpt; *Spanish:* eucalipto azul

Usage ▸ flavoring; **product**: essential oil (eucalyptus oil) and 1, 8 cineole (60–80%)
Parts Used ▸ leaf
Distribution ▸ SE Australia (Victoria), Tasmania, native in W America from California to Chile

Erhardt et al. 2000, 2002; Guzman/Siemmonsma 1999; Hanelt 2001; Hoppe 1949; Kumar 2001; Moors/Rizzini 1966; Rehm/Espig 1984; RothKormann 1997; Schönfelder 2001; Schultze-Motel 1986; Seidemann 1993c; Shiva et al. 2002; Uphof 1968; Wiersema/León 1966; Wyk et al. 2004

Eucalyptus macarthiri *Daene et Maiden*

Usage ▸ flavoring ; **product**: essential oil
Parts Used ▸ leaf
Distribution ▸ Australia: NS Wales
Note ▸ The essential oil has a high content of geranyl acetate (44–56%) and eudesmol (28–40%).

Erhardt et al. 2002; Hanelt 2001

Eucalyptus salicifolia *Cav.*

▸ *Eucalyptus amygdalina Labill.*

Eucalyptus staigeriana *F. Muell. ex F.M. Bailey*

Common Names ▸ lemon-scented ironbark
Usage ▸ flavoring; **product**: essential oil
Parts Used ▸ leaf
Distribution ▸ NE Australia (Queensland); cultivated in Brazil, Guatamala, Zaire, Indonesia (Java), Seychelles

Erhardt et al. 2000, 2002; Hanelt 2001; Kumar 2001; Mors/Rizzini 1966; Rehm/Espig 1984; Schultze-Motel 1986; Uphof 1968

EUCLEA L. - Ebenaceae

Euclea divinorum Hiern.

Common Names ▶ magic guarri
Usage ▶ condiment; **product**: as vegetable salt of the plant ash
Parts Used ▶ plant
Distribution ▶ S Africa
Note ▶ The plant ash is a substitute for edible salt.

Rosenkranz 2002

Eugenia aromatica (L.) Baill.

▶ *Syzygium aromaticum (L.) Merr. Et L.M. Perry*

Eugenia balsamea Ridley

▶ *Syzygium polyanthum (Wight) Walpers*

Eugenia caryophyllata Thunb.

▶ *Syzygium aromaticum (L.) Merr. et L.M. Perry*

Eugenia caryophyllus (Spreng.) Bullock ex S.G. Harrison

▶ *Syzygium aromaticum (L.) Merr. et L.M. Perry*

Eugenia nitida Duthie

▶ *Syzygium polyanthum (Wight) Walpers*

Eugenia pimenta DC.

▶ *Pimenta dioica (L.) Merr.*

Eugenia polyantha Wight

▶ *Syzygium polyanthum (Wight) Walpers*

Eugenia tabasco G. Don

▶ *Pimenta dioica (L.) Merr. var. tabasco*

EUPATORIUM L. - Asteraceae (Compositae)

Eupatorium cannabinum L.

Common Names ▶ bornaset, hemp agrimony, hemp weed, American thoroughwort (herb); *French:* eupatoire chanvrine, chanvre d'eau; *German:* Gewöhnlicher Wasserdost; Grundheil, Kunigundenkraut, Wasserhanf; *Italian:* canapa d'acqua; *Russian:* sedatsch, poskonnik konoplewyj; *Spanish:* eupatorio
Usage ▶ spice
Parts Used ▶ leaf, herb
Distribution ▶ Europe, Turkey, Lebanon, Israel, Caucasus, Iran(?), Himalayas, C Asia, N Africa: Morocco, Algeria

Aichele/Schwegler 4, 1995; Berger 4, 1954; Cheers 1997; Erhardt et al. 2000, 2002; Erichsen-Brown 1989; Hager 4, 1992; 5, 1993; Hiller/Welzig 1999; Hoppe 1949; Schönfelder 2001; Schultze-Motel 1986; Seidemann 1993c; Woerdenberg et al. 1991; Zygadlo et al. 1996

Eupatorium chinense L.

Synonyms ▶ *Buphthalmum oleraceum* Lour., *Eupatorium japonicum* Thunb. ex Murray
Common Names ▶ Chinesian hemp agrimony; *Chinese:* chenggan cao; *German:* Chinesischer Wasserdost; *Indonesian:* teklan gede; *Pilipino:* apanang-gubat
Usage ▶ seasoning, condiment especially for fish (in Annam)
Parts Used ▶ young leaf
Distribution ▶ Indochina, China, Korea, Taiwan, Japan, Philippines, also cultivated

Guzman/Siemonsma 1999; Uphof 1968

Eupatorium dalea L.

Usage ▶ substitute for vanilla
Parts Used ▶ herb
Distribution ▶ West Indies
Note ▶ The herb is strongly cumarin-scented.

Uphof 1968

 Eupatorium hemipteropodium Robinson

Common Names ▶ *Mexico:* chioplé
Usage ▶ flavoring of tobacco
Parts Used ▶ leaf
Distribution ▶ Mexico: Yucutan

Hanelt 2001; Schultze-Motel 1986

 Eupatorium japonicum Thunb. ex Murray

▶ *Eupatorium chinense* L.

 Eupatorium parviflorum Aubl.

▶ *Mikania perviflora* (Aubl.) Karst

 Eupatorium triplinerve Vahl

Common Names ▶ pool root, white snakeroot; *Bengalia:* ayapana
Usage ▶ flavoring
Parts Used ▶ leaf
Distribution ▶ tropical America, native India, Java, Mauritius; Brazil cultivated

Haelt 2001; Schultze-Motel 1986

EUPHORBIA L. - Spurge - Euphorbiaceae

 **Euphorbia hirta** L.

Synonyms ▶ *Euphorbia pilulifera* L.

Common Names ▶ bristly spurge, cat's hair, garden spurge; *German:* Borstige Wolfsmilch, Borstige Euphorbie, Pacharispflanze; *Russian:* ostschetinennij molotschaj
Usage ▶ seasoning
Parts Used ▶ leaf
Distribution ▶ tropical Africa: Benin, Cameroon, Ghana, Liberia, Madagascar, Mali, Nigeria, Senegal, Somalia, Tanzania, Zaire

Aedo et al. 2001; Ajao et al. 1985; Ayensu 1978; Blanc/de Saqui-Sannes 1972; Evans/Kinghorn 1975; Neuwinger 1998; Täufel et al. 1993; Villamar et al. 1994; Yoshida et al. 1990

 Euphorbia lathyris L.

Common Names ▶ caper spurge, gopher plant, mole plant, myrtle spurge; *Chinese:* hsü sui tzu, qian zi; *French:* euphorbe épurge; *German:* Spring-Wolfsmilch, Kreuzblättrige Wolfsmilch; *Italian:* escapuza, catapuzia; *Russian:* molotschaj-colnzerljad
Usage ▶ substitute for capers (*Capparis spinosa* L.) when not available
Parts Used ▶ unripened fruits
Distribution ▶ C, EC Europe, France, Iberia, Turkey, Caucasus, N America, CS America, China
Note ▶ The fruits are poisonous.

Erhardt 2000; Hanelt 2001; NICPBP 1987; Schultze-Motel 1986; Staesche 1972; Uphof 1968

 Euphorbia pulcherrima Willd. ex Klotzsch

Synonyms ▶ *Poinsettia pulcherrima* (Willd ex Klotzsch) R. Grah.
Common Names ▶ Christmas flower, dazzle, poinsettia; *Dutch:* poinsettia; *French:* euphorbe écarlate, poinsettia, étoile de Noel; *German:* Poinsettie, Weihnachtsstern; *Indonesian:* buga natal, paitan, pohon merah; *Javanese:* godong, ratjoon, ratjoonan; *Malaysian:* pohon merah, puring merah; *Pilipino:* pascuas; *Spanish:* flor de Pascua, flor de Santa Catarina; *Thai:* kismas; *Vietnamese:* trang nguyên
Usage ▶ seasoning
Parts Used ▶ leaf

Distribution ▶ in tropical countries of C America (Mexico), Africa and Asia; widely cultivated

Burkill 1965; Burkill 3, 1995; Engel/Phummai 2000; Erhardt et al. 2002; Hanelt 2001; Villamar et al. 1994; Wealth of India 3, 1952

EUPHRASIA L. - Eyebright - Scrophulariaceae

 Euphrasia officinalis L.

Synonyms ▶ *Euphrasia rostkoviana* Hayne
Common Names ▶ eufrasia, eyebright; *French:* casse-lunette, euphraise officinale, luminet, brise lunettes; *German:* Großblütiger Augentrost, Gewöhnlicher Augentrost, Wiesenaugentrost; *Italian:* eufrasia, erba degli occhi; *Russian:* otschnika aptetschnaja; *Spanish:* eufrasia
Parts Used ▶ pot-herb
Distribution ▶ C Europe, Turkey to W Siberia
Note ▶ In England the herb was used in the Middle Ages to spice soups.

Aichele/Schwegler 4, 1995; Berger 4, 1954; Bilgri/Adam 2000; Clair 1961; Erhardt et al. 2002; Hiller/Welzig 1999; Hohmann et al. 2001; Neidhardt 1947; Newall et al. 1996; Salama/Sticher 1974; Schönfelder 2001; Schultze-Motel 1986; Seidemann 1993c; Sharma 2003; Sticher/Salam 1981; Uphof 1968; Wyk et al. 2004

 Euphrasia rostkoviana *Hayne*

❯ *Euphrasia officinalis* L. ssp. *rostkoviana* (Hayne) Towns

 Eutrema wasabi *(Siebold) Maxim.*

❯ *Wasabi japonica* (Miq.) Matsum

EVODIA Lam. - Rutaceae

 Evodia amboinensis *Merr.*

Common Names ▶ *Indonesian:* gandaroora besar
Usage ▶ flavoring
Parts Used ▶ bark
Distribution ▶ Indonesia, especially Ambon
Note ▶ The bark is used as incense among the natives.

Uphof 1968

 Evodia lunu-ankenda *(Gaertn.) Merr.*

Synonyms ▶ *Evodia roxburghiana* Benth.
Common Names ▶ *Hindi:* kanalei, midaumabaphang, vanashempaga
Usage ▶ flavoring
Parts Used ▶ leaf
Distribution ▶ India: hills of E Himalayas, Deccan, Andaman Islands

Arora/Pandey 1996; Uphof 1968

 Evodia roxburghiana *Benth.*

❯ *Evodia lunu-ankenda* Merr.

F

FAGARA L. - Rutaceae

 Fagara arenaria *Engl.*

Usage ▶ flavoring (liqueur "Paratis")
Parts Used ▶ leaf
Distribution ▶ Brazil

Machado 1949

 Fagara avicennae *Lamk*

▶ *Zanthoxylum avicenna (Lamk) DC.*

 Fagara macrophylla *Engl.*

▶ *Fagara zanthoxyloides Lam.*

 Fagara rhetsa *Roxb.*

▶ *Zanthoxylum rhetsa* (Roxb.) DC.

 Fagara senegalensis *DC.*

▶ *Fagara zanthoxyloides* Lam.

 Fagara tessmannii *Engl.*

▶ *Zanthoxylum tessmannii (Engl.) Ayator*

 Fagara zanthoxyloides *Lam.*

Synonyms ▶ *Fagare macrophylla* Engl., *Zanthoxylum gilletii* (de Willd.) Waterm., *Zanthoxylum senegalensis* DC.
Common Names ▶ candle wood, fagara jaune, Senegal prickly ash; *German:* Senegalpfeffer, Senegal-Gelbholz; *Ghana:* faskori, kanfu, puom
Usage ▶ spice, applied like pepper for meat and sauces
Parts Used ▶ seed, leaf, bark
Distribution ▶ W Africa: Ghana, Guinea, Nigeria, Senegal
Note ▶ The fruit has a pepper scent and is inedible.

Ayensu 1978; Burkill 4, 1997; Chevalier 1957; Hanelt 2001; Irvoine 1961; Neuwinger 1994, 1999; Schultze-Motel 1986; Teuscher 2003

FEDIA Gaertn. African Valerian - Valerianaceae

 Fedia cornucopiae *(L.) Gaertn.*

Common Names ▶ African valerian, horn of plenty; *French:* corne d'abondance; *German:* Afrikanischer Baldrian
Usage ▶ pot-herb
Parts Used ▶ leaf

Distribution ▸ Europe: Iberia, France, Greece, Crete; N Africa

Erhardt et al. 2002; Uphof 1968

FERULA L. - Giant Fennel - Apiaceae (Umbelliferae)

Ferula assa-foetida L.

Synonyms ▸ *Ferula narthex* L., *Ferula rubicaulis* Boiss., *Ferula pseudalliacea* Rech.f., *Narthex polakii* Staph et Wettst.

Common Names ▸ asafetida, asant, devil's dung, food of gods; *Arabic:* andjudaan, haltit; *Chinese:* a wei; *Dutch:* duivelsdreck; *French:* ase fétide, asa-fœtida, férule persique; *German:* Stinkasant, Teufelsdreck; *India:* hîng, hingu; *Italian:* assafetida; *Portuguese:* assa fetida; *Russian:* ferula wonjutschaja, cmola wonjutschaja, durnoj dux, asmargok, ching; *Slovenian:* feruľa čertova; *Spanish:* asa fétida;

Usage ▸ spice

Parts Used ▸ the gum from the stem and the root

Distribution ▸ C Asia, Iran, Afghanistan

Note ▸ The gum has a very penetrating garlic-like odor and taste. At *Silphion* or *Laser pithium* or *silphium des anciens* concerns this plant. The German botanist Engelbert Kaempfer gave the first decription and illustration of the asafetida plant in 1712.

Appendino et al. 1994; Bois 1934; Boisvert/Huber 2000; Bordia/Arora 1975; Chamberlain 1977; Craze 2002; Dalby 2000; Davidson 1999; Dudtschenko et al. 1989; Duke et al. 2003; Erhardt et al. 2002; Farrell 1985; Guzman/Siemonsma 1999; Hanelt 2001; Hepper 1992; Hoppe 1949; Kajmoto et al. 1989; Lück 2000; Martinetz/Lohs 1987, 1988; Morris/Mackley 1999; Nassar 1994; Newall et al. 1996; Norman 1991; Pochljobkin 1974, 1977; Pruthi 1976; Rajanikanth et al. 1984; Rechinger 1987; Samimi/Unger 1979; Schultze-Motel 1986; Seidemann 1992b; 1993c; Seidemann/Siebert 1987; Siewek 1990; Strantz 1909; Täufel et al. 1993; Teuscher 2003; Tucker 1986; Uphof 1968; Wyk et al. 2004

Ferula foetida (Bunge) Regel

Common Names ▸ fetida; *French:* fétide; *German:* Stinkasant

Usage ▸ spice

Parts Used ▸ the gum from the stem and the root

Distribution ▸ C Asia

Note ▸ The gum has a very penetrating garlic-like odor and taste.

Chamberlain 1977; Erhardt et al. 2000, 2002; Hanelt 2001; Hepper 1992; Rechinger 1987; Schultze-Motel 1986; Seidemann/Siebert 1987; Uphof 1968; Wealth of India 4, 1956

Ferula galaniflua Boiss. et Buhse

▸ *Ferula gummosa* Boiss.

Ferula gummosa Boiss.

Synonyms ▸ *Ferula galbaniflua* Boiss. et Buhse

Common Names ▸ *German:* Galbanum, Galbanum-Gummi, Gummi-Asant; Mutterharz, Steckenkraut

Usage ▸ spice

Parts Used ▸ gum from stem

Distribution ▸ N Iran, C Asia: Afghanistan, Turkestan

Note ▸ The plant is the ancient 'slphion'(?)

▸ *Ferula asa-foetida* L.

Erhardt et al. 2002; Hepper 1992; Hoppe 1949; Mabberly 1997

Ferula narthex Boiss.

Common Names ▸ stinck fetida; *German:* Stab-Asant

Usage ▸ spice

Parts Used ▸ gum from the stem and the root

Distribution ▸ NW Himalayas

Erhardt et al. 2002; Hepper 1992; Sharma 2003; Siewek 1990 Uphof 1968

Ferula narthex L.

▸ *Ferula assa-foetida* L.

Ferula pseudalliacea Rech.f.

▸ *Ferula assa-foetida* L.

 **Ferula rubicaulis** *Boiss.*

▶ *Ferula assa-foetida L.*

FIBRAUREA Lour. - Menispermaceae

 Fibraurea recisa *Pierre*

Common Names ▶ tall woody vine; *Indochina:* dây vàng giang, nam hoàng liên
Usage ▶ flavoring; **product:** bitter liqueurs
Parts Used ▶ root
Distribution ▶ Indochina
Note ▶ The bitter root is a substitute for gention.

Mabberly 1997; Uphof 1968

 Ficaria verna *Huds*

▶ *Ranunculus ficaria L.*

FILIPENDULA Mill. - Dropwort, Meadow Sweet - Rosaceae

 Filipendula ulmaria *(L.) Maxim.*

Synonyms ▶ *Spirea ulmaria* L.
Common Names ▶ dropwort, meadow sweet, queen of the meadow; *French:* reine des prés, spirée ulmaire; *German:* Echtes Mädesüß, Sumpfspirea; *Italian:* filipendule, olmaria, regina dei prati; *Russian:* labasnik sestilepestnij, tawolga, sabasnik; *Spanish:* barba, reina de los prados, reina de cabra ulmaria
Usage ▶ flavoring (various beverages); **product:** essential oil
Parts Used ▶ flower
Distribution ▶ Europe, Caucasus, W and E Siberia, N Africa, native in N America

Berger 1, 1949; 4, 1954; Bilgri/Adam 2000; Bremness 2001; Cheers 1997; Coiciu/Racz (no year); Davidson 1999; Dudtschenko et al. 1989; Erhardt et al. 2002; Fleischhauer 2003; Hiller/Welzig 1999; Hoppe 1949; Lindeman et al. 1982; Newall et al. 1996; Pschyrembel 1998; Sanchez-Monge/Parellada 1981; Saifullina/Kozhina 1975; Scheer/Wichtl 1987; Schönfelder 2001; Schultze-Motel 1986; Seidemann 1993c; Täufel et al. 1993; Thieme 1966; Tucker 1986; Uphof 1968; Wyk et al. 2004

FLEMINGIA Roxb. ex Ait. - Fabaceae (Leguminosae)

 Flemingia congesta *Roxb. ex Ait. et Ait.f.*

▶ *Flemingia macrophylla (Willd.) Miq.*

 Flemingia macrophylla *(Willd.) Miq.*

Synonyms ▶ *Flemingia congesta* Roxb. et Ait. et Ait.f.
Common Names ▶ false saffron; *German:* falscher Safran; *Japanese:* enoki-mame
Usage ▶ seed, sporadically as a saffron substitute
Parts Used ▶ pulverized seed huils
Distribution ▶ SE Asia: tropical India, Africa

Erhardt et al. 2000, 2002; Hiller/Melzig 1999/2000

 Flueggea leucopyrus *Willd.*

▶ *Securinega leucopyrus (Willd.) Muell.-Arg.*

FOENICULUM Mill. - Fennel - Apiaceae (Umbelliferae)

 Foeniculum carri *Link*

▶ *carrum carri L.*

 Foeniculum dulce *DC. non Mill.*

▶ *Foeniculum vulgare var. dulce (DC.) Batt. et Trab.*

 Foeniculum officinale *All.*

▶ *Foeniculum vulgare Mill.*

Foeniculum piperitum *(Ucria) Sweet*

▶ *Foeniculum vulgare subsp. piperitum (Ucria) Cout.*

 Foeniculum vulgare *Mill.*

Synonyms ▶ *Anethum foeniculum* L., *Foeniculum capillaceum* Gilib., *Foeniculum officinale* All.

Common Names ▶ fennel; *Arabic:* acksoum, razianuj, sciamar, chemar; *Chinese:* shih lo, tzu mo lo, xiao hui xiang; *Dutch:* venkel; *French:* finnochio, fenouil commun; *German:* Fenchel, Bitterfenchel, Gartenfenchel, Gewürzfenchel; *Hindi:* saumph; *Indonesian:* adas; *Italian:* finocchio; *Javanese:* adas londo; *Japanese:* ui-kyô, uwiichô; *Korean:* hoehyang; *Malayan:* adas pedas, jintan manis; *Pilipino:* haras, anis; *Portuguese:* funcho; *Russian:* fenchel, apotetschnyj ukron, woloschckij ukron; *Sanskrit:* madhurika, misreya; *Slovenian:* fenikel; *Spanish:* finocchio, hinojo; *Thai:* mellet phong karee, thianklaep, phakchiduanha, yira

Usage ▶ spice; condiment: curries; **product:** essential oil

Parts Used ▶ fruit

Distribution ▶ Europe, frequent cultivated

Note ▶ Cultivated: *Foeniculum vulgare* var. *azoricum* (Mill.) Thell.: *German:* Gemüsefenchel, Knollenfenchel, Finocchino

Aichele/Schwegler 3, 1995; Ballarin/Ballarin 1972; Bernath et al. 1996a, b; Berger 3, 1952; Bilia et al. 2002; Bois 1934; Bournot 1968; Bremness 2001; Cheers 1997; Coiciu/Racz (no year); Coelho et al. 2003; Craze 2002; Czygan 1987; Davidson 1999; Dudtschenko et al. 1989; Ehlers et al. 2000; Farrell 1985; Guillen et al. 1994; Gupta et al. 1995; Guzman/Siemonsma 1999; Hager 5, 1993; Heeger 1956; Hiller/Melzig 1999; Hoffmann et al. 1992; Hohmann et al. 2001; Hondelmann 2002; Hoppe 1949; Kraus/Hammerschmidt 1980; Leung 1991; Melchior/Kastner 1974; Menard/Lehr 1987; Menghini et al. 1994; Morris/Mackley 1999; Norman 1991; Pank 1996; Pank et al. 2003a, b; Piccaglia/Marotti 2001; Plescher 1997; Pochljobkin 1974, 1977; Pschyrembel 1998; Rosengarten 1969; Schönfelder 2001; Schröder 1991; Schultze-Motel 1986; Seidemann 1993c; Sharma 2003; Shiva et al. 2002; Siewek 1990; Small 1997; Staesche 1972; Täufel et al. 1993; Teuscher 2003; Tindall 1983; Toth 1967a, b; Tucker 1986; Ubillos 1989; Uphof 1968; Venskutonis et al. 1996a; Villamar et al. 1994; Wiersema/León 1999; Wüstenfeld/Haensel 1964; Wyk et al. 2004; Zeven/de Wet 1982

 Foeniculum vulgare *Mill.* **ssp. vulgare var. azoricum** *(Mill.) Thell.*

▶ *Foeniculum vulgare* Mill.

 Foeniculum vulgare *Mill.* **var. dulce** *(DC.) Batt. et Trab.*

Synonyms ▶ *Foeniculum dulce* DC. non Mill.

Common Names ▶ sweet fennel, Florence fennel; *Arabic:* bisbas; *Chinese:* tian hui xiang; *Dutch:* knolvenkel; *French:* fenouil à bulbe, fenouil doux; *German:* Süßer Fenchel, Gewürzfenchel; *Italian:* finocchio domestica, finocchio dolce; *Japanese:* furohren-su-fennelu; *Portuguese:* funcho doce; funcho de Florença, fiolho; *Russian:* fenchel' ovoschtschnoj; *Spanish:* hinojo dulce, hinojo de Florencia; *Turkish:* tatli rezene, raiyane, irziyan

Usage ▶ spice; **product:** essential oil

Parts Used ▶ fruit

Distribution ▶ Europe, requently cultivated

Note ▶ In Germany used with anise, caraway and coriander fruits this is sold as "bread spice".

Bärtels 1997; Hager 5, 1993; Heeger 1956; Hohmann et al. 2001; Parzinger 1996; Schröder 1991; Schultze-Motel 1986; Seidemann 1993c; Seidemann/Siebert 1987; Täufel et al. 1993; Zeve/de Wet 1982

 Foeniculum vulgare *Mill.* **ssp. piperitum** *(Ucria) Cout.*

Synonyms ▶ *Anethum piperitum* Ucria, *Foeniculum piperitum* (Ucria) Sweet, *Meum piperitum* (Ucria) Spreng.

Common Names ▶ bitter fennel; *French:* fenouil amer; *German:* Bitterfenchel, Eselsfenchel, Pfefferfenchel; *Italian:* finocchio amaro, finocchione; *Spanish:* hinojo amargo

Usage ▶ spice

Parts Used ▶ fruit

Distribution ▶ Europe cultivated

Hager 5, 1993; Heeger 1956; Parzinger 1996; Pschyrembel 1998; Schultze-Motel 1986; Seidemann 1993c; Täufel et al. 1993; Zeghichi et al. 2003; Zeven/de Wet 1982

Foenum-graecum officinale Moench

▶ *Trigonella foenum-graecum* L.

FORTUNELLA Swingle - Kumquat - Rutaceae

Fortunella crassifolia Swingle

Common Names ▶ large round kumquat; *German:* Dickblättrige Zwergzitrone; *Chinese:* chin tan; *Japanese:* meiwa
Usage ▶ succade (flavoring)
Parts Used ▶ fruit
Distribution ▶ China, Japan; cultivated in Japan, India and USA

Hanelt 2001; Schultze-Motel 1986; Uphof 1968; Zeven/de Wet 1982

Fortunella hindsii (Champ. ex Benth.) Swingle

Synonyms ▶ *Atalantia hindsii* (Champ.) Oliv.
Common Names ▶ Hongkong wild kumquat, Formosan kumquat, Hongkong kumquat; *Chinese:* chin tou; *German:* Kinoto, Kumquat, Limequat, Minizitrone, Zwergzitrone
Usage ▶ succade (flavoring), and it serves as a spice
Parts Used ▶ fruit
Distribution ▶ Hongkong, China: Chekiang, Kwantung; cultivated in China and Japan

Bendel 2002; Erhardt et al. 2002; Hanelt 2001; Hume 1957; Ochse et al. 1961; Schultze-Motel 1986; Seidemann 1993c; Zeven/de Wet 1982

Fortunella japonica (Thunb.) Swingle

Synonyms ▶ *Citrus japonica* Thunb., *Citrus madurensis* Lour.
Common Names ▶ round kumquat, marumi-kumquat, calamondin, musk lime, China orange, golden mandarin; *Chinese:* chin kan, jin gan, szu kai kat; *French:* kumquat; *German:* Japanische Orange, Goldorange, Kalamondin, Marumi-Kumquat, Zwergorange; *India:* hazara; *Indonesian:* djerook kastoori; *Italian:* kumquat; *Japanese:* marumi, nagami, shikikikat, tonkinkan; *Pilipino:* aldonisis; *Russian:* kinkan, kumkwat; *Spanish:* kumquat redondo, quinoto; *Thai:* kam kat, ma-mao-wan; *Vietnamese:* quat, kim quat
Usage ▶ succade (flavoring), condiment
Parts Used ▶ fruit
Distribution ▶ S China; cultivated: E Asia, Japan, N America, origin in China

Bendel 2002; Cheers 1997; Davidson 1999; Engler/Phummai 2000; Erhardt et al. 2000, 2002; Hanelt 2001; Hume 1975; Purseglove 1968; Schultze-Motel 1986; Täufel et al. 1993; Uphof 1968; Verheij/Coronel 1991; Wealth of India 4, 1956; Zeven/de Wet 1982

Fortunella margarita (Lour.) Swingle

Synonyms ▶ *Citrus margarita* Lour.
Common Names ▶ oval kumquat; *Chinese:* chin chu, chin tsao, luo fou; *German:* Ovale Kumquat, Zwergorange; *Japanese:* nagami, nagu kinkan; *Peru:* naranjilla, tunguas; *Spanish:* kumquat; *Russian:* kin kan
Usage ▶ succade (flavoring)
Parts Used ▶ fruit
Distribution ▶ S China; cultivated: C and S America; West Indies, Sicily, Africa (Ethiopia, Zimbabwe), Hawaii, Malaysia

Davidson 1999; Hanelt 2001; Hume 1957; Purseglove 1968; Schultze-Motel 1986; Täufel et al. 1993; Uphof 1968; Wealth of India 4, 1956; Zeven/de Wet 1982

Fortunella polyandra (Ridl.) Tanaka

Synonyms ▶ *Atalantia polyandra* Ridl., *Fortunella swinglei* Tanaka
Common Names ▶ Malayan kumquat, hedge lime; *German:* Malaysische Kumquat, Zwergorange; *Malaysian:* limau pagar; *Spanish:* kumquat malayo; *Thai:* som chit
Usage ▶ succade (flavoring), spice
Parts Used ▶ fruit
Distribution ▶ Malaysia, also cultivated; S China (Hainan), Thailand

FORTUNELLA Swingle - Kumquat - Rutaceae

Hume 1957; Schultze-Motel 11986; Verheij/Coronel 1991; Wealth of India 4, 1956

Fortunella swinglei *Tanaka*

▸ *Fortunella polyandra (Ridl.) Tanaka*

G

Galanga major *Rumph.*

▶ *Alpinia galanga (L.) Willd.*

Galega purpurea *(L.) L.*

▶ *Tephrosia purpurea (L.) Pers.*

Galipea officinalis *Hancock*

▶ *Cusparia febrifuga Humb. ex DC.*

GALIUM L. - Bedstraw - Rubiaceae

Galium odora *Salisb.*

▶ *Galium odoratum (L.) Scop.*

Galium odoratum *(L.) Scop.*

Synonyms ▶ *Asperula odorata* L., *Galium odora* Salisb.
Common Names ▶ sweet woodruff, woodruff aspérule; *Dutch:* lieve vrouwe bedstro; *French:* aspérule odorante, muguet des bois, petit muguet, reine des bois, hépatique étoilee, belle étoile; *German:* Waldmeisterkraut, Duftlabkraut, Maikraut; *Italian:* asperula, caglio odorosa, stellina odorosa; *Russian:* jasmennik duschistij; *Spanish:* asperilla olorosa, aspérula

Usage ▶ flavoring
Parts Used ▶ fresh herb
Distribution ▶ N and C Europe, Turkey, Caucasus, N Iran, W, E Siberia, Japan, C Asia, N Africa, USA
Note ▶ In Germany and other Europe countries the use of this plant to produce essences for sale is prohibited (high cumarine content).

Aichele/Schwegler 3, 1995; Bilgri/Adam 2000; Bois 1934; Bremness 2001; Charalambous 1994; Cheers 1991; Clair 1961; Coiciu/Racz (no year); Davidson 1999; Dudtschenko et al. 1989; Erhardt et al. 2002; Fleischhauer 2003; Hager 5, 1993; Heeger 1956; Hiller/Melzig 1999; Hoppe 1949; Laub et al. 1982, 1985; Pschyrembel 1998; Roth/Kormann 1997; Schönfelder 2001; Šedo/Krejča 1983; Seidemann 1993c, 1994; Seidemann/Siebert 1987; Small 1997; Täufel et al. 1993; Teuscher 2003; Tucker 1986; Wörner/Schreier 1991; Wüstenfeld/Haensel 1964

GARCINIA L. - Mangosteen - Clusiaceae (Guttiferae)

Garcinia atroviridis *Griff. ex T. Anderson*

Common Names ▶ *Malaysian:* asam gelungor
Usage ▶ spice, substitute for tamarind pulp
Parts Used ▶ fruit
Distribution ▶ Malaysia

Burkill 1965; Hanelt 2001; Kumar 2003; Schultze-Motel 1986; Uphof 1968; Verheij/Coronel 1991; Zeven/de Wet 1982

Garcinia cambogia *(Gaertn.) Desr.*

Common Names ▶ Malabar tamarind, cambodge, goraga; *German:* Malabar-Tamarinde; *India:* gamboge; *Malaysian:* asam glugur, gelugur
Usage ▶ flavoring, locally as fruit spice
Parts Used ▶ fruit
Distribution ▶ India: Assam; Sri Lanka, Malaysia

Davidson 1999; Hanelt 2001; Krishnamurthy et al. 1981; Lewis/Neelakantan 1965; Peter 2001; Sánchez-Monge 1991; Uphof 1968; Verghese 1991

 Garcinia indica *(Thouars) Choisy*

Synonyms ▶ *Brindonia indica* Thouars, *Garcinia purpurea* Roxb.
Common Names ▶ brindonia tallow tree, kokam, Goa butter, fish tamarind; *French:* brindonnier; *German:* Kokum; *India:* kokam, kokum, ratambi
Usage ▶ spice, local as fruit spice; also for curries, chutneys and kebabs
Parts Used ▶ fruit peel
Distribution ▶ S West Indies, cultivated elsewhere an Mauritius, Réunion, Antilles
Note: ▶ The spice has a salty-sour taste

Arora/Pandey 1996; Dastur 1954; Davidson 1999; Hanelt 2001; Krishnamurthy et al. 1982; Lück 2000; Nawale et al. 1997; Peter 2001; Sampathu/Krishnamurthy 1982; Schultze-Motel 1986; Seidemann 1993c; Siewek 1990; Täufel et al. 1993; Uphof 1968; Wiersema/León 1999; Zeven/de Wet 1982

 Garcinia purpurea *Roxb.*

▶ *Garcinia indica (Thouars) Choisy*

 Garcinia tinctoria *(Choisy) W.F. Wight*

▶ *Garcinia xanthochymus Hook.f. ex T. Anderson*

 Garcinia xanthochymus *Hook.f. ex T. Anderson*

Synonyms ▶ *Garcinia tinctoria* (Choisy) W.F. Wight
Common Names ▶ *India:* gamboge, mundu
Usage ▶ spice, substitute for tamarind pulp
Parts Used ▶ fruit
Distribution ▶ India, Sri Lanka, China; cultivated in tropical regions of the New World

Cheers 1997; Wiersema/León 1999

GARDENIA Ellis - Gardenia - Rubiaceae

 Gardenia augusta *(L.) Merr.*

Synonyms ▶ *Gardenia grandiflora* Lour., *Gardenia jasminoides* Ellis, *Gardenia radicans* Thunb.
Common Names ▶ Cape jasmine, Cape jessamine, common gardenia, gardenia; *Chinese:* hi zu, zhi zi; *French:* gardénia; *German:* Kap-Gardenie, Kap-Jasmin, Jasminglanz; *Indonesian:* kaca piring; *Korean:* chijanamu; *Malaysian:* bunga cina; *Pilipino:* rosal; *Russian:* gardenia; *Thai:* phut chiin, phut son, khae thawaa, *Vietnamese:* dành dành
Usage ▶ flavoring (of tea); **product:** essential oil
Parts Used ▶ flower
Distribution ▶ origin E Asia, cultivated in India, China, Taiwan, Japan, cultivated elsewhere, Mexico
Note ▶ The dye in the fruits (wongsky) are used in the food industry and for coloring silk and other clothes.

Cheese 1997; Erhardt et al. 2000, 2002; Hiller/Melzig 1999; Jacquat 2000; Kottegoda 1994; Pfister et al. 1996; Roth/Kormann 1997; Schultze-Motel 1986; Seidemann 1993c; Storrs 1997; Turova et al. 1987; Villamar et al. 1994; Warren 1998; WHO 1990; Wiersema/León 1999; Zeven/de Wet 1982

 Gardenia grandiflora *Lour.*

▶ *Gardenia augusta (L.) Merr.*

 Gardenia jasminoides *Ellis*

▶ *Gardenia augusta (L.) Merr.*

Gardenia radicans *Thunb.*

▶ *Gardenia augusta (L.) Merr.*

 Gastrochilus pandurata *(Roxb.) Ridley*

▶ *Boesenbergia rotundata (L.) Mansf.*

GAULTHERIA Kalm ex L. - Shallon - Ericaceae

 Gaultheria procumbens *L.*

Synonyms ▶ *Gaultheria repens* Raf.
Common Names ▶ box berry, checker berry, alpine wintergreen, creeping wintergreen, mountain tea, tea berry; *French:* gaulthérie du Canada, thé de bois; *German:* Niedere Rebhuhnbeere, Niederliegende Scheinbeere; *Italian:* uva di monte
Usage ▶ flavoring; **product:** essential oil
Parts Used ▶ essential oil (Wintergreen oil)
Distribution ▶ E Canada, USA
Note ▶ The essential oil is used as a flavoring agent in beers, beverages, candies, chewing gums and soft drinks.

Duke 2002; Duke et al. 2003; Hanelt 2001; Hoppe 1949; Rätsch 1998; Roth/Kormann 1997; Uphof 1968; Wyk et al. 2004

 Gaultheria repens *Raf.*

▶ *Gaultheria procumbens L.*

GEIGERIA Griess. - Asteraceae (Compositae)

 Geigeria alata *(DC.) Oliv. et Hiern*

Common Names ▶ winged Geigeria; *German:* Geglügelte Geigerie
Usage ▶ spice (sauce by Arabs in Eritrea)
Parts Used ▶ herb
Distribution ▶ C, S tropical Africa and on into Arabia

Burkill 1, 1985

GENTIANA L. - Gentian - Gentianaceae

Ting-Non/Shang-wu 2001

 Gentiana asclepiadea *L.*

Common Names ▶ willow gentian; *French:* gentiane fausse asclépiade; *German:* Schwalbenwurz-Enzian;
Usage ▶ flavoring; as tonic and for bitters and bitter liqueurs (▶ *Gentiana lutea* L.)
Parts Used ▶ rhizome
Distribution ▶ Europe: Alps, Carpatian region

Aichele/Schwegler 3, 1995; Cheers 1997; Erhardt et al. 2002; Fleischhauer 2003; Hager 5, 1993; Hanelt 2001; Hiller/Menzig 1999; Schönfelder 2001; Seidemann 1993c; Wiersma/León 1997

 Gentiana chirayta *Roxb.*

▶ *Swertia chirata Buch.-Ham. ex Wall.*

 Gentiana kurroo *Royle*

Common Names ▶ Indian gentian; *German:* Indischer Enzian; *India:* karu, kutki;
Usage ▶ flavoring; substitute for *Gentiana lutea* L.
Parts Used ▶ rhizome
Distribution ▶ NW Himalayas

Chauhan 1999; Erhardt et al. 2002; Hanelt 2001; Schultze-Motel 1986; Sharma 2003; Uphof 1968; Wealth of India 4, 1956

 Gentiana lutea *L.* ◨

Common Names ▶ bitter root, yellow gentian; *French:* grande gentiane, gentiane jaune; *German:* Gelber Enzian, Bitterwurzel; *Italian:* genziana maggiore, genziana gialla; *Russian:* goretschavka scheltaja; *Spanish:* genciana amarilla
Usage ▶ flavoring, as tonic and for bitters and bitters

GENTIANA L. - Gentian - Gentianaceae

◘ Gentiana lutea, flowering

liqueurs (e.g.: Boonecamp, Stonsdorfer, Underberg)

Parts Used ▶ rhizome

Distribution ▶ mountains parts of S Europe, S Europe and E Europe to Asia minor.

Note ▶ The plant is protected in Germany and other countries. Predominantly wild growths are used. Mostly cultivated in trials or in small areas.

Aichele/Schwegler 3, 1995; Ariño et al. 1997; Brodt 1982; Cheers 1997; Chialva et al. 1986; Coiciu/Racz (no year); Dudtschenko et al. 1989; Fleischhauer 2003; Franz/Fritz 1975, 1978; Hager 5, 1993; Hanelt 2001; Heeger 1956; Hiller/Melzig 1999; Hohmann et al. 2001; Hoppe 1949; Newall et al. 1996; Pschyrembel 1998; Schier/Schultze1989; Schönfelder 2001; Schultze-Motel 1986; Seidemann 1993c; Seidemann/Siebert 1987; Sharma 2003; Täufel et. al. 1993; Uphof 1968; Verotta 1985; Viel 1979; Wagner/Münzing-Vasirian 1975; Wealth of India 4, 1956; Wüstenfeld/Haensel 1964; Wyk et al. 2004

Gentiana pannoniaca *Scop.*

Common Names ▶ Hungary gentian; *German:* Ungarischer Enzian;

Usage ▶ flavoring; as tonic and for bitters and bitter liqueurs (▶ *Gentiana lutea L.*)

Parts Used ▶ rhizome

Distribution ▶ high mountains of C Europe: Italy, Czech Republic, Slovakia

Aichele/Schwegler 3, 1995; Fleischhauer 2003; Hager 5, 1993; Hoppe 1949; Schönfelder 2001; Seidemann 1993c; Seidemann/Siebert 1987

Gentiana punctata *L.*

Common Names ▶ spotted gentian; *German:* Tüpfelenzian

Usage ▶ flavoring, as tonic and for bitters and bitters liqueurs (▶ *Gentiana lutea L.*)

Parts Used ▶ rhizome

Distribution ▶ high mountains S and EC Europe: France, Balkan Peninsula

Aichele/Schwegler 3, 1995; Fleischhauer 2003; Hager 5, 1993; Hoppe 1949; Schönfelder 2001; Seidemann 1993; Seidemann/Siebert 1987

Gentiana purpurea *L.*

Common Names ▶ purple gentian; *German:* Purpur-Enzian

Usage ▶ flavoring; as tonic and for bitters and bitter liqueurs (▶ *Gentiana lutea L.*)

Parts Used ▶ rhizome

Distribution ▶ high mountains of C Europe: France, Italy and Norway

Aichele/Schwegler 3, 1995; Fleischhauer 2003; Hager 5, 1993; Hoppe 1949; Schönfelder 2001; Seidemann 1993c; Seidemann/Siebert 1987; Sticher/Meier 1980

GENTIANELLA Moench - Felwort - Gentianaceae

 Gentianalla campestris (L.) Börner

Common Names ▸ Feld-Gentianelle
Usage ▸ flavoring for bitter liqueurs and substitute for hops (rarely)
Parts Used ▸ herb
Distribution ▸ Europe

Erhardt et al. 2003; Fleischhauer 2003; Schnelle 1999

GERANIUM L. - Crane's Bill - Geraniaceae

 Geranium crispum Berg

▸ *Pelargonium crispum (Berg.) L'Hérit ex Ait.*

 Geranium fragrans Poir.

▸ *Pelargonium fragrans (Poir.) Willd.*

 Geranium graveolens Thunb.

▸ *Pelargomium graveolens L'Hérit.*

 Geranium lugrubre Salisb.

▸ *Geranium macrorrhizum L.*

 Geranium macrorrhizum L.

Synonyms ▸ *Geranium lugubre* Salisb.
Common Names ▸ Bulgarian geranium; *French:* géranium des Balkans, géranium à grosses racines; *German:* Düsterer Storchschnabel, Felsen-Storchschnabel; *Russian:* geran prjamja
Usage ▸ flavoring; **product:** essential oil ("Zdrawetz oil")
Parts Used ▸ leaf, herb

Distribution ▸ Southern Alps and Apennines to the Balkan region, W Russia, Crimea

Arctander 1960; Erhardt et al. 2000, 2002; Hanelt 2001; Roth/Kormann 1997; Schultze-Motel 1986; Uphof 1968; Yeo 1988

 Geranium moschatum L.

▸ *Erodium moschatum (L.) L'Herit. ex Ait.*

 Geranium odoratissimum L.

▸ *Pelargonium odoratissimum (L.) L'Hérit. ex Ait*

 Geranium odoratum Burm.

▸ *Pelargonium odoratissimum (L.) L'Hérit. ex Ait.*

 Geranium revolutum Jacq.

▸ *Pelargonium radens Moor*

GEUM L. - Avens - Rosaceae

 Geum caryophyllea Gilib.

▸ *Geum urbanum L.*

 Geum laxmanni Gaertn.

▸ *Coluria geoides (Pall.) Ledeb.*

 Geum montanum L.

Common Names ▸ alpine avens; *German:* Berg-Nelkenwurzel, Tormentillwurzel
Usage ▸ spice
Parts Used ▸ root
Distribution ▸ Europe: Scandinavia and British Isles, mountains
Note ▸ The root has a clove-like taste and when cloves

are not available the root is used as a substitute for them.

Aichele/Schwegler 2, 1994; Erhardt et al. 2002; Hager 5, 1993; Seidemann 1993

Geum potentilloides *Ait.*

▶ *Coluria geoides* (Pall.) Ledeb.

Geum rivale *L.*

Common Names ▶ Indian chocolate root, purple avens, water avens; *German:* Bach-Nelkenwurzel

Usage ▶ spice

Parts Used ▶ root

Distribution ▶ Europe, Turkey, Caucasus, W, E Siberia, C Asia

Note ▶ The root has a like clove taste and when cloves are not available the root is used as a substitute for them.

Aichele/Schwegler 2, 1994; Cheers 1997; Dudtschenko et al. 1989; Erhardt et al. 2002; Erichsen-Brown 1989; Fleischhauer 2003; Hager 5, 1993; Hiller/Melzig 1999; Schönfelder 2001; Seidemann 1993c; Täufel et al. 1993; Uphof 1968

Geum urbanum *L.*

Synonyms ▶ *Caryophyllata urbana* Scop., *Caryophyllata officinalis* Moench, *Geum caryophyllea* Gilib.

Common Names ▶ avens root, clove root, herb bennet, wood avens; *French:* benoîte commune; *German:* Echte Nelkenwurzel, Urban-Nelkenwurzel; *Italian:* cariofillata, erba benedetta, garofanaia, ambretta; *Russian:* grawilat, grawilat antetschnij, grawilat gorodskoj, gosditschnik, grebennik, tschistez, benediktowa trawa, podlesnik; *Spanish:* cariofilata, radic de San benito

Usage ▶ spice

Parts Used ▶ root

Distribution ▶ Europe, Turkey, Syria, N Iraq, Caucasus, N Iran, W Siberia, W Himalayas, temperate Asia; NW Africa; native elsewhere

Note ▶ The root has a like clove taste and when cloves are not available the root is used as a substitute for them.

Aichele/Schwegler 2, 1994; Coiciu/Racz (no year); Dudtschenko et al. 1989; Erhardt et al. 2002; Erichsen-Brown 1989; Fleischhauer 2003; Hager 5, 1993; Heeger 1956; Hiller/Melzig 1999; Hoppe 1949; Newall et al. 1996; Pschyrembel 1998; Schönfelder 2001; Schultze-Motel 1986; Seidemann 1993c; Sharma 2003; Uphof 1968; Vollmann/Schultze 1995a, b; Wiersma/León 1997; Wüstenfeld/Haensel 1964; Wyk et al. 2004

GINGKO L. - Gingko - Gingkoaceae

Gingko biloba *L.*

Synonyms ▶ *Gingkyo biloba* Mayr, *Salisburia biloba* Hoffmannsegg

Common Names ▶ gingko, maidenhair tree; *Chinese:* pai-kuo, ying kuo; *French:* arbre aux quarante écus, ginkgo, noyer du Japon; *German:* Gingkobaum, Mädchenhaarbaum; *Japanese:* ginkyo, ginnan; *Korean:* khong; *Russian:* ginkgo; *Spanish:* arbol de los escudos

Usage ▶ spice for soups and vegetables dishes

Parts Used ▶ seeds, frequently roasted

Distribution ▶ SE China, Japan cultivated

Cheers 1997; Erhardt et al. 2002; Hanelt 2001; Lück 2000; Schultze-Motel 1986; Täufel et al. 1993; Uphof 1968; Wiersema/León 1997; Wyk et al. 2004

Gingkyo biloba *Mayr*

▶ *Gingko biloba* L.

GISEKIA L. - Aitzoaceae

Gisekia pharmnaceoides *L.*

Usage ▶ condiment

Parts Used ▶ herb

Distribution ▶ tropical W Africa: Mali, Niger, Nigeria, Senegal; E India

Burkill 1, 1985; Uphof 1968

GLECHOMA L. - Ground Ivy - Lamiaceae (Labiatae)

 Glechoma hederacea L.

Synonyms ▶ *Calamintha hederaceae* (L.) Scop., *Nepeta glechoma* Benth.
Common Names ▶ alehoof, gill over the ground, ground ivy; *Chinese:* lien ch'ien, ou huo xue dan, ts'ao; *French:* gléchome, courroie de Saint Jean, lierre terrestre, rondelette; *German:* Gundermann, Gundelrebe, Erd-Efeu; *Italian:* edera terrestre; *Japanese:* kakidoshi; *Russian:* budra, budea pljuschtschewidnaja; *Spanish:* hiedra terrestre, sumidad de hierba terreste
Usage ▶ spice (for soups)
Parts Used ▶ leaf, herb
Distribution ▶ Eurasia, temperate Asia, native N America
Note ▶ Use for flavoring beer like hops in the Middle Ages in England.

Aichele/Schwegler 4, 1995; Berger 4, 1954; Bilgri/Adam 2000; Cheers 1997; Clair 1961; Dahl 1994; Dudtschenko et al. 1989; Erhardt et al. 2002; Fleischhauer 2003; Hager 5, 1993; Hanelt 2001; Hiller/Melzig 1999; Hohmann et al. 2001; Newall et al. 1996; Schönfelder 2001; Schultze-Motel 1986; Seidemann 1993c; Stahl/Datta 1972; Tucker 1986; Uphof 1968; Wiersema/León 1999; Zieba 1973

GLEHNIA Schmidt ex Miq. - Apiaceae (Umbelliferae)

 Glehnia litoralis F. Schmidt ex Miq.

Synonyms ▶ *Cymopterus litoralis* A. Gray; *Phellopteris litoralis* Benth.
Common Names ▶ cork wing; *Chinese:* bei sha shen, shan hu cai; *Japanese:* hama-bôfu, yavya-böfu; *Korean:* kaetppangphung; *Russian:* glenija pribretschnaja
Usage ▶ condiment, resembling Angelica (*Angelica archangelica* L.) and Tarragon (*Artemisia dragunculus* L.); in Japan for salads, fish dishes, and sweet Japanese sake [toso])
Parts Used ▶ young leaf and leaf stalks
Distribution ▶ W, E coast of N Pacific: Japan, China, Sakhalin

Note ▶ The taste is like a cross between *Angelica* and *Tarragon*.

Davidson 1999; Facciola 1990; Hanelt 2001; Uphof 1968; Zeven/de Wet 1982

GLINUS L. - Molluginaceae

 Glinus lotoides L.

Synonyms ▶ *Mollugo hirta* Thunb.
Common Names ▶ hairy carpet weed, hairy glinus; *Arabic:* ghobbeira, mogheira; *Hindi:* gandibudi;
Usage ▶ pot-herb
Parts Used ▶ tender shoot
Distribution ▶ India, China

Arora/Padney 1996

 Glinus oppositifolius (L.) DC.

Synonyms ▶ *Mollugo oppositifolia* L.
Common Names ▶ *Pilipino:* malagosa, sarsalida; *Thai:* sa-doa-din, phakkhuang;
Usage ▶ pot-herb
Parts Used ▶ herb, leaf
Distribution ▶ Madagascar, also cultivated in tropical parts of Africa and Asia to N Australia

Hanelt 2001; Husain et al. 1992

GLOBBA L. - Zingiberaceae

 Globba marantina L.

Common Names ▶ *Indonesian:* kapulaga ambon, halia utan, bonelau; *Pilipino:* barak, bangliu
Usage ▶ seasoning
Parts Used ▶ bulbils (are spicy), herb
Distribution ▶ India: temperate Himalayas, Moluccas, Thailand, Philippines to New Guinea, Solomon Islands

Note ▶ The bulbils is also used as cardamom.

Arora/Padney 1996; Erhardt et al. 2002; Guzman/Siemonsma 1999; Kumar 2001; Oyen/Dung 1999

GLOCHIDION Forster et Forster f. - Euphorbiaceae

Glochidion llanosi *Müll.*

Synonyms ▶ *Phylanthus llanosi* Müll.
Usage ▶ spice (flavoring)
Parts Used ▶ young sprout, leaf (for fish dishes), bulbil
Distribution ▶ India: Khasi hills, region former Indochina, Philippines

Arora/Pandey 1996; Bois 1934; Hanelt 2001; Schultze-Motel 1986; Uphof 1968

GLYCINE Willd. - Soya Bean - Fabaceae (Leguminosae)

Glycine abrus *L.*

▶ *Abrus precatorius* L.

Glycine hispida *(Moench) Maxim*

▶ *Glycine max (L.) Merr.*

Glycine max *(L.) Merr.*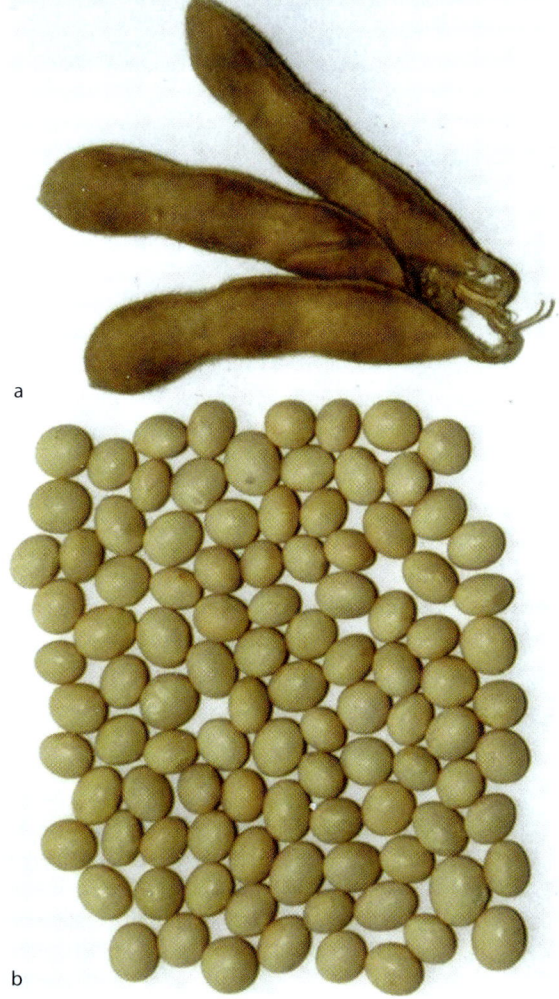

Synonyms ▶ *Glycine hispida* (Moench) Maxim., *Glycine soja* Siebold et Zucc., *Glycine ussuriensis* Regel et Maack, *Soja hispida* (Moench) Maxim.
Common Names ▶ soya, soya bean; *Arabic:* ful suyah; *Chinese:* mao dou; *Dutch:* sojaboon; *French:* soja, soya; *German:* Sojabohne; *Indonesian:* kacang kedelai; *Italian:* pianta di soia; *Japanese:* shoyu; *Korean:* khong; *Malaysian:* kacang, soya putih; *Portuguese:* soja; *Spanish:* soja, soya; *Russian:* soja, soewyj bob; *Spanish:* habas de soyas, frijol de soya, soya; *Thai:* thua rae.

Glycine max: a **fruits**, b **seeds**

Usage ▶ basis of condiment (soya sauces: miso, shoju, tsian je'u etc.)
Parts Used ▶ seed
Distribution ▶ origin E Asia; only cultivated
Note ▶ The seeds are the basis for the production (fermentation) of soya sauce and other soya products.

Bendel 2002; Beuchat 1984; Bois 1934; Brückner et al. 1989; Cheers 1997; Davidson 1999; Erhardt et al. 2000, 2002; Genovese/Lajolo 2002; Giami 1997, 2002; Hager 5, 1993; Hanelt 2001; Heeger 1956; Hoppe 1949; Kakade et al. 1972; Khader 1983; Kumar 2003; Lamboni et al. 1999; Lewington 1990; Liener 1994; Liu 1997; Markley 1950/51; Mazza/Ooham 2000; Pschyrembel 1998; Pursglove 1968; Schenck/Naundorf 1966; Seidemann 1993c; Song et al. 2003; Täufel et al. 1993; Teuscher 2003; Trueb 1999; Ueki et al. 1994; Uphof 1968;

Vogt 1987; Wiersma/León 1999; Wyk et al. 2004; Zeven/de Wet 1982

 Glycine soja *Siebold et Zucc.*

▶ *Glycine max (L.) Merr.*

 Glycine ussuriensis *Regel et Maack*

▶ *Glycine max (L.) Merr.*

GLYCYRRHIZA L. – Licorice, Liquorice, Sweetwood - Fabaceae (Leguminosae)

 Glycyrrhiza diochorides *Medik.*

▶ *Glycyrrhiza echinata L.*

 Glycyrrhiza echinata *L.*

Synonyms ▶ *Glycyrriza diochorides* Medik.
Common Names ▶ Roman liquorice; *German:* Römisches Süßholz, Russisches Süßholz;
Usage ▶ spice
Parts Used ▶ root
Distribution ▶ SE Europe, Turkey, Syria, Palestine, Caucasus, Iran, E Asia

Hager 5, 1993; Hanelt 2001; Heeger 1956; Morris/Mackley 1999; Schultze-Motel 1986; Seidemann 1993c; Steinlesberger 2004; Täufel et al. 1993; Willimot 1963; Zeven/de Wet 1982

 Glycyrrhiza echinate *Lepesch.*

▶ *Glycyrrhiza glabra L.*

 Glycyrrhiza glabra *L.*

Synonyms ▶ *Glycyrrhiza echinata* Lepesch., *Glycyrrhiza officinalis* Lepech., *Liquiritia officinarum* Medik., *Liquiritia officinalis* Moench
Common Names ▶ common licorice, liquorice, Spanish liquorice; *Chinese:* fen tsao; *French:* réglisse, glycyrrhize; *German:* Deutsches Süßholz, Griechisches Süßholz, Italienisches Süßholz, Spanisches Süßholz, Lakritze; *Hindi* mulhatti; *Italian:* liquirizia; *Korean:* minkamcho; *Portuguese:* alcaçuz; *Russian:* solodka golaja; *Sanskrit:* madhuka; *Slovenian:* sladké drievko; *Spanish:* alcazuz, licórize, orozuz, regalicia, regaliz
Usage ▶ spice, flavoring (tobacco, liqueurs); **product:** liquorice
Parts Used ▶ root
Distribution ▶ E, SE and SW Europe, Turkey, Iran, temperate Asia, India, China, Libya

Arystanova et al. 2001; Bielenberg 1998; Bullmann et al. 1990; Chandler 1985; Cheers 1997; Coiciu/Racz (no year); Craze 2002; Davidson 1999; Duke et al. 2003; Dudtschenko et al. 1989; Erhardt et al. 2002; Fleischhauer 2003; Gordon/Jing 1995; Hager 5, 1993; Hanelt 2001; Heeger 1956; Hiller/Melzig 1999; Hohmann et al. 2001; Hoppe 1949; Kwasniewski 1952; Leung 1991; Lewington 1990; Lutomski 1983; Mazza/Oomah 2000; Miething/Speicher-Brinker 1989; Morris/Mackley 1999; Newall et al. 1996; Pachaly 1990b; Pschyrembel 1998; Putscher 1968; Schönfelder 2001; Schultze-Motel 1986; Seidemann 1993c; Sharma 2003; Small 1997; Steinlesberger 2004; Täufel et al. 1993; Turova et al. 1987; Uphof 1968; Weinberg et al. 1993b; Wiersema/León 1999; Willimot 1963; Wüstenfeld/Haensel 1964; Wyk et al. 2004; Zeven/de Wet 1982

 Glycyrrhiza grandiflora *Tausch*

▶ *Glycyrrhiza uralensis* Fisch. ex DC.

 Glycyrrhiza lepidota *(Nutt.) Pursh*

Synonyms ▶ *Liquiritia lepidota* Nutt.
Common Names ▶ American licorice; *German:* Amerikanisches Süßholz
Usage ▶ spice
Parts Used ▶ root
Distribution ▶ America, SW Canada N Mexico
Note ▶ The fleshy roots are specially used by native Americans.

Hager 5, 1993; Hanelt 2001; Schultze-Motel 1986; Small 1997; Täufel et al. 1993; Uphof 1968

 Glycyrrhiza officinalis *Lepech.*

▶ *Glycyrrhiza glabra L.*

Glycyrrhiza uralensis *Fisch. ex DC.*

Synonyms ▶ *Glycyrrhiza grandiflora* Tausch
Common Names ▶ Chinese licorice, Manchurian liquorice; *Chinese:* gan zao, kann tsao; *German:* Chinesisches Süßholz, Russisches Süßholz, Anatolisches Süßholz; *Korean:* kamcho; *Russian:* solodka ural'skaja
Usage ▶ spice
Parts Used ▶ root
Distribution ▶ temperate Asia, India, Mongolia, E Europe, cultivated elsewhere

Dalby 2000; Hager 5, 1993; Hanelt 2001; Hoppe 1949; Leung 1991; Schultze-Motel 1986; Seidemann 1993c; Small 1997; Tan Tianwei et al. 2002; Täufel et al. 1993; Turova et al. 1987; Wiersema/Léon 1997

GNAPHALIUM L. - Cudweed - Asteraceae (Compositae)

 Gnaphalium indicum *L.*

Common Names ▶ Indian cudweed, tiny cudweed; *French:* gnaphale de l'Inde; *German:* Indisches Ruhrkraut; *Russian:* suscheniza
Usage ▶ pot-herb
Parts Used ▶ leaf
Distribution ▶ India: Bihar

Arora/Pandey 1996; Quattrocchi 1999

 Granatum punicum *St. Lag.*

▶ *Punica granatum* L.

 Grastidium salaccense *Bl.*

▶ *Dendrolobium salaccense (Bl.) Lindley*

GREVEA Baill. - Grossulariaceae (Montiniaceae)

 Grevea madagascariensis *Baill.*

Usage ▶ condiment
Parts Used ▶ fruit
Distribution ▶ E Africa, Madagascar

Mabberley 1997; Neuwinger 1999

GUAJACUM L. - Zygophyllaceae

 Guajacum bijugum *Stokes*

▶ *Guajacum officinale* L.

 Guajacum officinale *L.*

Synonyms ▶ *Guajacum bijugum* Stokes
Common Names ▶ lignum vitae; *French:* bois de vie, gaïac officinal, gaya; *German:* Guajakbaum, Franzosenholz, Pockholz; *Italian:* guaiaco, legno santo; *Russian:* gvajakovoe, bakaut; *Spanish:* guayacan negro, palosanto
Usage ▶ spice, flavoring (for herb liqueur)
Parts Used ▶ wood, resin
Distribution ▶ West Indies: Haiti, San Domingo; Mexico, Panama, Columbia, Venezuela
Note ▶ International trade with lignum vitae and the resin has been licenced since November 2002.

Berger 1, 1949; 3, 1952; Bourton 1968; Erhardt et al. 2000, 2002; Grow/Schaertzman 20001; Hager 5, 1993; Hiller/Melzig 1999; Hoppe 1949; Lewington 1990; Martinetz/Lohs 1991; McNay Wood 2003; Newall et al. 1996; Pschyrembel 1998; Rätsch 1992; Schultze-Motel 1986; Seidemann 1993c; Täufel et al. 1993; Uphof 1968; Vöttiner-Pletz 1990; Wiersema/León 1999; Wolters 1994; Wüstenfeld/Haensel 1964; Wyk et al. 2004

 Guajacum sanctum *L.*

Common Names ▶ holywood, lignum vitae; *German:* Heiliger Guajakbaum, Franzosenholz, Pockholz

GYMNEMA R.Br. - Asclepiadaceae

◘ Guajacum officinale, flowering

Usage ► spice, flavoring (for herb liqueur)
Parts Used ► wood, resin
Distribution ► S Mexico, West Indies, SE USA Florida, Caribbean, Meso-America
Note ► *Guajacum sanctum* was introduced to Europe in 1526 by San Domingo.

Berger 1, 1949; 3, 1952; Erhardt et al. 2000, 2002; Grow/Schwartzman 20001; Hager 5, 1993; Hiller/Melzig 1999; Hoppe 1949; Rätsch 1992; Seidemann 1993c; Uphof 1968; Villamar et al. 1994; Wiersema/León 1999; Wolters 1994

GUIZOTIA Cass. - Niger - Asteraceae (Compositae)

🍃 *Guizotia abyssinica* (L.f.) Cass.

Common Names ► niger seed, ramtil; *French:* guizotia, ramtil; *German:* Nigersaat, Ramtillkraut; *Russian:* nug masljtschnyj; *India:* jagni, kadellu, kalatil, ramtil, ramatila

Usage ► condiment for chutneys
Parts Used ► seed
Distribution ► E Africa: Ethiopia.; India, native in California

Burkill 1, 1985; Erhardt et al. 2002; Uphof 1968; Zeven/de Wet 1982

GYMNEMA R.Br. - Asclepiadaceae

🍃 *Gymnea affine* Dec.

▶ *Gymnea sylvestre* R. Br.

🍃 *Gymnema sylvestre* R. Br.

Synonyms ► *Gymnea affine* Dec.
Common Names ► miracle fruit; *Hindi:* gurmar, merasingi; *India:* adigam, podapatri, kavali, dhuleti
Usage ► pot-herb (in India)
Parts Used ► leaf
Distribution ► Peninsular India, China, Indochina, Malaysia, Australia, W tropical Africa Namibia, tropical and Southern Africa

Arora/Pandey 1996; Hanelt 2001; Irvine 1961; v. Koenen 1996; Uphof 1968; Wealth of India 4, 1956; Wiersema/León 1999

🍃 *Gymnosporia senegalensis* Loes.

▶ *Maytenus senegalensis* (Lam.) Exell

🍃 *Gynandropsis gynandra* (L.) Brig

▶ *Cleome gynandra* L.

🍃 *Gynandropsis pentaphylla* (L.) DC.

▶ *Cleome gynandra* L.

Gynopogon alyxia J.R. Forst

❯ *Alyxia lucida* Wall.

GYNURA Cass. – Velvet plant - Asteraceae (Compositae)

 Gynura procumbens *(Lour.) Merr.*

Synonyms ▸ *Gynura sarmentosa* (Blume) DC.
Common Names ▸ velvet, velvet plant; *German:* Niederliegende Samtpflanze
Usage ▸ flavoring, condiment
Parts Used ▸ leaf
Distribution ▸ tropical W Africa, tropical Asia: China, Thailand, Indonesia, Malaysia, Vietnam

Burkill 1, 1985; Davies 1978, 1979; Erhardt et al. 2002; Schultze-Motel 1986; Seidemann 1993c, 1998/2000; Uphof 1968

 Gynura pseudochina *(L.) DC.*

Synonyms ▸ *Gynura sinuata* DC.
Common Names ▸ velvet, velvet plant;
Usage ▸ spice
Parts Used ▸ leaf
Distribution ▸ tropical Africa, tropical E and SE Asia; cultivated in Malaysia, Vietnam and Java

Burkill 1, 1965; Davies 1979, 1981; Hanelt 2001; Schultze-Motel 1986; Seidemann 1993c, 1998/2000; Wealth of India 4, 1956

 Gynura sarmentosa *(Blume) DC.*

❯ *Gynura procumbens (Lour.) Merr.*

 Gynura sinuata *DC.*

❯ *Gynura pseudochina (L.) DC.*

GYROPHORA Miyoshi - Lichenaceae (Gyrophoraceae)

 Gyrophora esculenta *Miyoshi*

Synonyms ▸ *Umbilicaria esculenta* Hoffm.
Common Names ▸ *Japanese:* iwatake
Usage ▸ spice
Parts Used ▸ dry lichen
Distribution ▸ mountains in Japan, China

Lück 2004; Täufel et al 1993; Uphof 1968

H

HEDEOMA Pers. - Lamiaceae (Labiatae)

 Hedeoma pulegioides (L.) Pers.

Synonyms ▸ *Cunila pulegioides* L., *Melissa pulegioides* L., *Ziziphora pulegioides* (L.) Desf.
Common Names ▸ American (false) pennyroyal, mosquito plant, pennyroyal, squaw mint, tickweed; *German:* Flohkraut, Amerikanische Poley; *Mexico:* poleo;
Usage ▸ spice, flavoring liqueurs and food; **product:** essential oil
Parts Used ▸ leaf
Distribution ▸ EN America: Nebraska, Wisconsin to Quebec, New Brunswick; Arkansas and Georgia, E Canada

Bourton 1968; Charalambous 1994; Erhardt et al. 2002; Hager 5, 1993; Hanelt 2001; Hiller/Melzig 1999; Newall et al. 1974d, 1978; Roth/Kormann 1997; Schultze-Motel 1986; Seidemann 1993c; Small 1997; Tucker 1986; Uphof 1968; Wiersema/León 1999

HEDYCHIUM J. König - Ginger Lily - Zingiberaceae

 Hedychium aurantiacum *Roxb.*

▸ *Hedychium coccineum* Buch.-Ham. var. *aurantiacum* Roxb.

 Hedychium coccineum Buch.-Ham. **var. aurantiacum** *Roxb.*

Synonyms ▸ *Hedychium aurantiacum* Rosc.
Common Names ▸ red ginger lily, scarlet ginger lily; *German:* Roter Schmetterlingsingwer, Scharlachrote Kranzblume
Usage ▸ kitchen garden herb (rarely)
Parts Used ▸ herb
Distribution ▸ W Himalayas

Aurora/Pandey 1996; Erhardt et al. 2002; Gurib-Fakim et al. 2002; Kumar 2001; Yusuf et al. 2002

 Hedychium coronarium *König*

Common Names ▸ butterfly lily, garland flower, white ginger lily; *Brazil (Portuguese):* acucena, lírio-do-brejo; *Chinese:* tu giang huo; *German:* Weiße Schmetterlingslilie, Weiße Girlandenblume, Weiße Kranzblume; *Japanese:* hana shuku sha; *Thai:* mahaahong, krathaai, haanghong
Usage ▸ spice (rarely)
Parts Used ▸ flower
Distribution ▸ Himalyas, India, Sri Lanka, Malaysia, Thailand

Arora/Pandey 1996; Gurib-Fakim/Brendler 2004; Gurib-Fakim et al. 2002; Itokawa et al. 1988a, b; Jacquat 1990; Kumar 2001; Larsen et al. 1999; Mors et al. 2000; Rätsch/Müller-Ebeling 2003; Tao Guoda 1998; Uphof 1968; Wong 1999; Yusuf et al. 2002

© Springer-Verlag Berlin Heidelberg 2005

HEDYOSMUM O. Sw. - Chloranthaceae

Hedyosmum bonplandianum *Mart.*

▶ *Hedyosmum brasiliense Mart.*

Hedyosmum brasiliense *Mart.*

Synonyms ▶ Hedyosmum bonplandianum Mart.
Common Names ▶ *Brazil (Portuguese):* chá de bugre, chá-de-soldado, erva almíscar, erva-de-soldado, hortelã-do-brejo
Usage ▶ spice
Parts Used ▶ leaf
Distribution ▶ Brazil: Minas Gerais, São Paulo
Note ▶ The leaves have a peppermint-like taste.

Freise 1936; Hanelt 2001; Mors et al. 2000; Opdyke 1978; Schultze-Motel 1986; Uphof 1968

Helenium grandiflorum *Gilib.*

▶ *Inula helenium L.*

HELICHRYSUM Mill. - Everlasting Flower - Asteraceae (Compositae)

Helichrysum angustifolium *(Lam.) DC.*

▶ *Helichrysum italicum (Roth) D. Don*

Helichrysum italicum *(Roth) D. Don*

Synonyms ▶ Helichrysum angustifolium (Lam.) DC.; Helichrysum serotinum Boiss.
Common Names ▶ curry plant, Italian everlasting, immortelle; *French:* immortelle de Italiener; *German:* Currykraut, Italienische Sandstrohblume; *Russian:* zmin
Usage ▶ pot-herb; **product:** essential oil
Parts Used ▶ herb

Distribution ▶ SW and SE Europe, especially Italy, Iberia, S France, N Africa, W Asia
Note ▶ This plant and the ssp. *microphyllum* (Willd.) Nym., dwarf curry plant, and ssp. *serotinum* (Boiss.) P. Fourn., curry plant, have a slight curry taste.

Aichele/Schwegler 4, 1995; Angioni et al. 2003; Bärtels 1997; Bianchini et al. 2001; Charalambous 1994; Cheers 1997; Clebsch 1997; Erhardt et al. 2000, 2002; Norman 1998; Roussis et al. 1997; Satta et al. 1999; Teuscher 2003; Tucker 1986; Tucker et al. 1997; Wiersema/León 1999

Helichrysum serotinum *Boiss.*

▶ *Heliochrysum italicum (Roth) D. Don*

Helichrysum tianschanicum *Regel*

Common Names ▶ *French:* immortelle du Tian-shan; *German:* Maggikraut
Usage ▶ pot-herb
Parts Used ▶ herb
Distribution ▶ C Asia

Erhardt et al. 2002

HELIOTROPIUM L. - Heliotrope, Turnsole - Boraginaceae

Heliotropium arborescens *L.*

Synonyms ▶ Heliotropium corymbosum Ruiz. et Pav., Heliotropium peruvianum L.
Common Names ▶ cherry pie, helioptrope; *French:* héliotrope; *German:* Vanilleblume, Peruanische Sonnenwende, Strauchige Sonnenwende; *Peru:* docto vainilla; *Russian:* geliotron; *Spanish:* heliotropo
Usage ▶ flavoring
Parts Used ▶ leaf, flower
Distribution ▶ S America: Peru; cultivated elsewhere;

Cheers 1997; Erhardt et al. 2000, 2002; Hiller/Melzig 1999; Schultze-Motel 1986; Wiersema/León 1999

Heliotropium corymbosum *Ruiz. et Pav.*

▶ *Heliotropium arborescens L.*

 Heliotropium peruvianum *L.*

▶ *Heliotropium arborescens L.*

HEMEROCALLIS L. - Day lily - Hemerocallidaceae (Liliaceae)

 Hemerocallis aurantiaca *Baker*

Common Names ▶ orange (colored) day lily; *German:* Orangefarbene Taglilie
Usage ▶ flavoring
Parts Used ▶ flower bud, flower leaf
Distribution ▶ China, Korea

Davidson 1999; Erhardt 1988; Erhardt et al. 2002

 Hemerocallis citrina *Baroni*

Common Names ▶ lemon day lily; *French:* lis d'un jour; *German:* Zitronen- or Wilde Taglilie
Usage ▶ flavoring
Parts Used ▶ flower bud, flower leaf
Distribution ▶ China: Schensi

Erhardt et al. 2002

 Hemerocallis crocea *Lam.*

▶ *Hemerocallis fulva (L.) L.*

 Hemerocallis flava *(L.) L.*

▶ *Hemerocallis lilio-asphodelus L.*

 Hemerocallis fulva *(L.) L.*

Synonyms ▶ *Hemerocallis crocea Lam.*
Common Names ▶ fulvous day lily, golden needles, orange needles, tawn day lily; *Chinese:* hsuan ts'ao; *French:* hémérocalle brun-rouge; *German:* Bahnwärter-Taglilie, Rotgelbe Taglilie, Braunrote Taglilie; *Japanese:* yabu-kanzo, oni-kanzo; *Russian:* krasodnev ryschij; *Vietnamese:* hoa hien, kim chau, phac cham
Usage ▶ condiment, flavoring
Parts Used ▶ dry flower bud, flower leaf
Distribution ▶ origin Japan?, China, E Asia, native in Europe, N America, cultivated elsewhere

Aichele/Schwegler 5, 1996; Asen/Arisumi 1968; Bois 1934; Cheers 1997; Davidson 1999; Erhardt 1988; Erhardt et al. 2002; Fleischhauer 2003; Griesbach/Batdorf 1995; Inoue et al. 1990, 1994; Lück 2004; Robert et al. 2002; Schultze-Motel 1986; Seidemann 1993c; Täufel et al. 1993; Uphof 1968; WHO 1990; Wiersema/León 1999; Yuan 1993

 Hemerocallis graminea *Andr.*

▶ *Hemerocallis minor L.*

 Hemerocallis lilio-asphodelus *L.*

Synonyms ▶ *Hemerocallis flava* (L.) L.
Common Names ▶ lemon day lily, yellow day lily; *French:* hémérocalle jaune, lis asphodèle, lis jaune; *German:* Gelbe Taglilie, Wiesen-Taglilie; *Russian:* krasodnev scheltyi
Usage ▶ flavoring
Parts Used ▶ dry flower bud, flower leaf
Distribution ▶ Italy, Slovenia; SE Alps, native in Europe, W, E Siberia, Amur, China; cultivated elsewhere

Aichele/Schwergel 5, 1996; Asen/Arisumi 1968; Cheers 1997; Erhardt et al. 2002; Fleischhauer 2003; Robert et al 2002; Schultze-Motel 1986; Seidemann 1993c; Täufel et al. 1993; Wiersema/León 1999; Yuan 1993

 Hemerocallis minor *Mill.*

Synonyms ▶ *Hermerocallis graminea* Andr.
Common Names ▶ dwarf yellow day lily, grass leaf day lily, little day lily; *French:* hémérocalle naine; *German:* Kleine or Stern-Taglilie; *Russian:* krasodnev malyj

Usage ▶ spice
Parts Used ▶ flower bud, flower leaf
Distribution ▶ E Siberia, Mongolia, N China, Korea, also cultivated in China, Japan, Europe

Bois 1934; Erhardt 1988; Erhardt et al. 2002; Schultze-Motel 1986; Seidemann 1993c; Täufel et al. 1993; Yuan 1993

HEMIDESMUS R.Br. - Asclepiadeceae (Periplocaceae)

Hemidesmus indicus (Willd.) R.Br.

Common Names ▶ Indian sarsaparilla, nunnery root; *German:* Indische Sarsaparille; *Hindi:* anantumal, kapun, hindi salsa; *India:* durivel, karibandha, onontomulo, sogade, sugandhi pala, upalasari; *Sanskrit:* anantumala, naga jihva, sariva
Usage ▶ flavoring
Parts Used ▶ root extract and root syrup
Distribution ▶ India, Sri Lanka, cultivated in India

Hanelt 2001; Jirovetz et al. 2002; Karnick 1977; Nagarajan/Rao 2003; Nagarajan et al. 2001; Schultze-Motel 1986; Uphof 1968; Wealth of India 5, 1959

Heptapleurum aromaticum Seem.

▶ *Schefflaria aromatica* Harms

Heptapleurum venulosum Seem.

▶ *Schefflaria venulosa (Wright et Arn.)* Harms

HERACLEUM L. - Hogweed - Apiaceae (Umbelliferae)

Heracleum burmanicum Kurz

Common Names ▶ Burmain cow-parsley; *French:* berce de la Birmanie; *German:* Burmannischer Bärenklau, Burmannische Petersilie; *Russian:* borstschevik

Usage ▶ pot-herb, applied like parsley
Parts Used ▶ leaf
Distribution ▶ Himalayan region, Myanmar, N Thailand

Brondegaard 1990; Hanelt 2001; Schultze-Motel 1986; Seidemann 1993c

Heracleum persicum Michx.

Common Names ▶ Persian cow-parsley; *German:* Persischer Bärenklau, Persische Petersilie
Usage ▶ condiment (in pickles)
Parts Used ▶ seed
Distribution ▶ Iran

Uphof 1968

Herpestis rugosa Roth.

▶ *Limnophila rugosa (Roth.)* Merr.

HESPERETHUSA M. Roemer - Rutaceae

Hesperethusa crenulata (Roxb.) M. Roem.

Synonyms ▶ *Limonea crenulata* Roxb.
Common Names ▶ *Hindi:* beli, tondsha, tor elaga, nayvila, nayibullal; *India:* kattunarakan
Usage ▶ spice, for fish and meat dishes
Parts Used ▶ fruit
Distribution ▶ Indian, Myanmar, Thailand, SW China, Laos, Cambodia, Vietnam; in India and Java also cultivated

Hanelt 2001; Schultze-Motel 1968; Wealth of India 5, 1959

HETEROTHALAMUS Less. - Euphorbiaceae

Herterothalamus brunoides Less.

Synonyms ▶ *Marshallia brunoides* Less.

Usage ▶ flavoring
Parts Used ▶ plant
Distribution ▶ Argentina to Brazil

Uphof 1968

HETEROTHECA Cass. - Asteraceae (Compositae)

Heterotheca inuloides *Cass.*

Common Names ▶ false golden aster, Mexican arnica; *German:* Goldaster, Mexikanische Arnika
Usage ▶ spice (flavoring) sporadically for liqueurs and bitters
Parts Used ▶ flower
Distribution ▶ C America: Mexico
Note ▶ In Germany often used instead of

▶ *Arnica montana* L.

Hager 5, 1993; Hiller/Melzig 1999; Ochoa/Alonso 1966; Sagrero-Nieves/Bartley 1996; Saukel 1984; Villamar et al. 1994; Willuhn et al. 1985

HIBISCUS L. - Giant or Rose Mallow - Malvaceae

Hibiscus abelmoschus *L.*

▶ *Abelmoschus moschatus* ssp. *moschatus* Medik.

Hibiscus bifurcatus *Blanco*

▶ *Hibiscus surattensis* L.

Hibiscus cannabinus *Merr.*

▶ *Hibiscus radiatus* Cav.

Hibiscus lindley *Wall.*

▶ *Hibiscus radiatus* Cav.

Hibiscus moschatus *Salisb.*

▶ *Abelmoschus moschatus* ssp. *moschatus* Medik.

Hibiscus radiatus *Cav.*

Synonyms ▶ *Hibiscus cannabius* Merr., *Hibiscus lindley* Wall.
Common Names ▶ kenaf hibiscus; *Mexico:* clavelina
Usage ▶ pot-herb (in India)
Parts Used ▶ leaf
Distribution ▶ India: Khasi hills, cultivated in Brazil and in SE Asia; probably native to Asia tropics, but native widely in the Old and New world tropics.

Arora/Pandey 1996; Hanelt 2001; Heeger 1956; Schultze-Motel 1986; Small 1997; Zeven/Zhukovsky 1975; Uphof 1968

Hibiscus sabdariffa *L.*

Common Names ▶ Indian sorrel, Jamaica sorrel, red sorrel, roselle, sorrel; *Arabic:* karkadé; *Brazil (Portuguese):* azedinha, caruru-azedo, caruru-da-guiné, quibo-azedo, quiabo-róseo, quiabo-roxo, vinagreira; *Chinese:* lou shen kui; *Dutch:* roselle, Surinamese zuring; *French:* oseille de Guinée, karkadé; *German:* Afrikanische Malve, Rosella-Eibisch, Karkade, Rama, Roselle, Sudantee; *Hindi:* lalambari, patwa; *Italian:* carcadè, ibisco fiori, flor de Jamaica; *Japanese:* rohzelu; *Portuguese:* rosela, roseta, caruru azedo; *Russian:* gibiskuc sabdarifa, rosella; *Spanish:* acedera de Guinea, agrio de Guinea, hibisco, rosa de Jamaica serení; *Thai:* krachiap daeng; *Yemen:* kakakad
Usage ▶ spice, flavoring
Parts Used ▶ flower, leaf, calyx
Distribution ▶ probable origin in tropical Africa, widely cultivated
Note ▶ The fruits have a high vitamin C content. Dadawa bassoa, a local spice preparation in Northern Nigeria, is prepared from seeds of *Hibiscus sab-*

dariffa L. and *Hibiscus cannabinus* L. (Indian hemp).

Aedo et al. 2001; Bendel 2002; Berger 1, 1949; Bois 1934; Burkill 4, 1997; Chen et al 1998; Dashak et al. 2001; Davidson 1999; Dudtschenko et al. 1989; Duke et al. 2003; Engler/Phummai 2000; Erhardt et al. 2002; Griebel 1939; Hanelt 2001; Hiller/Melzig 1999; Hohmann et al. 2001; Menßen/Staesche 1974; Mors et al. 2000; Pi-Jen Tsai et al. 2002; Pouget et al. 1990; Pratt 1912; Pschyrembel 1998; Pursglove 1993c; Ross 1999; Schenck/Naundorf 1966; Schilcher 1976; Schönfelder 2001; Schultze-Motel 1986; Seidemann 1993c; Sharaf 1962; Siemonsma/Piluek 1993; Small 1997; Täufel et al. 1993; Udoyasekhara Rao 1996; Uphof 1968; Villamar et al. 1994; Wiersema/León 1999; Wilson/Menzel 1964; Wyk et al. 2004; Zeven/de Wet 1982

Hibiscus surattensis L.

Synonyms ▶ *Hibiscus bifurcatus* Blanco
Common Names ▶ *Pilipino:* ahimit, inabu, labong, sampinit
Usage ▶ pot-herb
Parts Used ▶ leaf
Distribution ▶ tropical Africa and Asia

Uphof 1968

Hieracium javanicum Burm.f.

▶ *Emilia javanica* (Burm.f.) C.B. Rob.

HIEROCHLOE R.Br. - Holy Grass - Poaceae (Gramineae)

Hierochloe odorata (L.) Beauv.

Common Names ▶ holy grass, manna grass, seneca grass, sweet grass, vanilla grass; *French:* houque odorante; *German:* Duftendes Mariengras, Duft-Mariengras, Vanillegras; *Russian:* ljadnik duschistyi, subrovka duschistaja, tschapolot'
Usage ▶ spice, flavoring of liqueurs
Parts Used ▶ leaf
Distribution ▶ Europe, temperate Asia: Caucasus, Alaska, Canada, USA

Aichele/Schwegler 5, 1996; Dudtschenko et al. 1989; Erhardt et al. 2002; Fleischhauer 2003; Hiller/Melzig 1999; Hoppe 1949; Rätsch 1998; Seidemann 1993c, 1994a; Täufel et al. 1993; Uphof 1968; Zainuddin et al. 2002

HILLIELLA (O. Schulz) Y.H. Zhang et H.W. Li - Brassicaceae (Cruciferae)

Hilliella shuangpaiensis H.W. Li

Common Names ▶ Chinese horseradish *German:* Chinesischer Meerrettich
Usage ▶ spice
Parts Used ▶ root
Distribution ▶ China, C Asia.
Note ▶ Various. Listed under *Cochlearia*.

Mabberley 1997

Hiroma undarioides Yendo

▶ *Undaria undarioides* Yendo

HONCKENYA Ehrh. - Caryophyllaceae

Honckenya peploides (L.) Ehrh.

Common Names ▶ sea purslane; *German:* Strandportulak
Usage ▶ pot-herb
Parts Used ▶ leaf and stalk
Distribution ▶ sandy coasts
Note ▶ The leaves and stalks have a borage-like taste.

Mabberley 1997

HOUTTUYNIA Thunb. - Fishwort - Saururaceae

 Houttuynia cordata *Thunb.*

Common Names ▶ fishwort, saururis, heart leaf; *Chinese:* ch'i, chou giao nai, ji cai, vap ca, yu xing cao, zhu bi kong; *French:* houttuynie, en coeur; *German:* Chamäleonblatt, Herzförmige Houttuynie; *Japanese:* gyoseiso, dokudami zoku, yoseiso, shih-yao, chung-yao; *Korean:* yakmemil, giáp ce, osaengch'o; *Russian:* chaulljunija serdcevidnaj; *Thai:* phak khaao thoj, pluu-kao; *Vietnamese:* giâp cá, la diep cá

Usage ▶ pot-herb, flavoring
Parts Used ▶ leaf, herb
Distribution ▶ India, China, Taiwan, S Himalayas, India, Cambodia, Vietnam, Japan, Korea, Java, cultivated USA, Argentina
Note ▶ The leaves have a very strong, soapy coriander or fishy taste and odor; the odor in the roots is particulary strong. Consider using it as a substitute for cilantro (coriander leaves). A tri-colored (red, green and white) variegated ornamental variety is availabe under several names.

Arora/Pandey 1996; Cheers 1997; Duke/Ayensu 1985; Erhardt et al. 2002; Kuebel/Tucker 1988; Kumara 2003; NICPBP 1987; Ogle et al. 2003; Pröbstle et al. 1992, 1994; Schultze-Motel 1986; Small 1997; Taylor 1976; Tucker 1986; Uphof 1968; WHO 1990; Wiersema/León 1999; Zeven/de Wet 1982

 Houttuynia cordata, flowering

HUA Pierre ex de Willd - Huaceae

 Hua gabonii *Pierre ex de Willd.*

Common Names ▶ Cameroon garlic; *Congo:* mufira; *German:* Kamerun-Knoblauch
Usage ▶ spice, condiment
Parts Used ▶ young sprouts and old leaves, seed
Distribution ▶ W Africa: Cameroon, Zaire
Note ▶ The plant parts are enveloped with meat and roasted. The seeds have a characteristic garlic aroma with fresh terpenic note.

Jirovetz et al. 2002; Mabberley 1997; Neuwinger 1999

HUMULUS L. - Hop - Cannabaceae

 Humulus lupulus *L.*

Common Names ▶ common hop, hop; *Chinese:* she-ma, pi jiu hua; *Dutch:* hop; *French:* houblon (commun), cônes de houblon, vigne du nord; *German:* (Gewöhnlicher) Hopfen; *Italian:* luppulo, luppulo stroboli; *Japanese:* karahana-sô; *Korean:* hophǔ; *Portuguese:* lúpulo; *Russian:* chmel', chmel' obyknovennyj; *Spanish:* lúpulo (común), estróbilos de lúpulo

Usage ▶ spice; **product:** pellets, essential oil (hop oil)
Parts Used ▶ female flower; **product:** pellets from crushed flowers, essential oil
Distribution ▶ temperate regions of Europe, N and C Asia, N America, cultivated in Germany, England, Belgium, France, former Czechoslovakia, Russia, China, N America, Australia
Note ▶ Origin of cultivation unknown, cultivated in W

Europe in the 8th to 9th Century in Germany an S France, 16th Century in England.

Albert/Muller 2000; Bart et al. 1994; Bendel 2002; Berger 1, 1949; Bois 1934; Borde et al. 1989; Bremness 2001; Cheers 1997; Clarke 1986; Coiciu/Racz (no year); Davidson 1999; Dudtschenko et al. 1989; Edwardson 1952; Farrell 1985; Field 1996; Fleischhauer 2003; Forster 1981; Guadagni et al. 1966; Hager 5, 1993; Hanelt 2001; Hänsel/Schulz 1986; Heeger 1956; Hiller/Melzig 1999; Hohmann et al. 2001; Hoppe 1949; Kač/Kovačevič 2000; Katsiodis et al. 1989; Kiewitt et al. 1983; Mayer 2003; Mizobuchi/Sato 1984, 1985; Neve 1991; Newall et al. 1996; Pschyrembel 1998; Roth/Kormann 1997; Schönfelder 2001; Schultze-Motel 1986; Seidemann 1993c; Seidemann/Siebert 1987; Sharma 2003; Sharpe/Laws 1981; Steinhaus/Schieberle 1999; Stiegler 1949; Täufel et al. 1993; Teuscher 2003; Tucker 1986; Trueb 1998; Ubillos 1989; Uphof 1968; Verzele 1986; Verschuere et al. 1992; Wiersma/León 1999; Wilding et al. 1983; Wyk et al. 2004; Zeven/de Wet 1982

HYDROCHARIS L. - Frogbit - Hydrocharitaceae

Hydrocharis chevelaiéri (de Wild.) Dandy

Common Names ▶ African frogbite; *French:* Africain morène; *German:* Afrikanischer Froschbiss
Usage ▶ condiment (vegetable salt)
Parts Used ▶ the plant ashes
Distribution ▶ C Africa
Note ▶ A source of edible salt. The salt is higher in potassium content than edible salt.

Hanelt 2001; Tisserant 1953

HYGROPHYLA R. Br. - Wisteria - Acanthaceae

Hygrophyla angustifolia R. Br.

Synonyms ▶ *Hygrophyla salicifolia* Nees
Common Names ▶ water wisteria; *German:* Schmalblättriger Wasserfreund, Schmalblättrige Wisterie
Usage ▶ pot-herb
Parts Used ▶ leaf
Distribution ▶ SE Asia

Arora/Pandey 1996; Erhardt et al. 2002

Hygrophila auriculata (Schum.) Heine

Synonyms ▶ *Baleria longifolia* L., *Baleria auriculata* Schum.
Common Names ▶ *German:* Kleinöhrige Wisterie
Usage ▶ condiment (vegetable ash)
Parts Used ▶ the plant ash
Distribution ▶ tropical W Africa: Cameroon, Congo, Zaire, also cultivated; India, Sri Lanka

Hanelt 2001; Schultze-Motel 1986

Hygrophyla salicifolia Nees

▶ *Hygrophyla augustifolia* R.Br.

HYPERICUM L. - St. John's Wort - Hypericaceae

Hypericum lanceolatum Lam.

▶ *Hypericum revolutum Vahl.*

Hypericum perforatum L.

Common Names ▶ (perforated) St. John's Wort, hardhay, khlawath weed, millepertius; *Dutch:* Sintjans-kruit; *French:* millepertuis perforé; herbe de la Saint-Jean, chasse diable; *German:* Echtes Johanniskraut, Johannisblut, Konradskraut, Tüpfelhartheu, Tüpfel-Johanniskraut; *Italian:* erba di San Giovanni comune, iperico, cacciadiavoli, pilatro; *Russian:* sweroboj pronsjonnolistnyj; *Spanish:* corazoncillo, hipérico, hierba de San Juan
Usage ▶ spice, sporadically of fish and bitter liqueurs; **product:** essential oil
Parts Used ▶ herb
Distribution ▶ Eurasia, NW Africa
Note ▶ In the 15th to 17th Century the herb was used to season beer.

Aichele/Schwegler 3, 1995; Albert/Muller 2000; Ang et al. 2002; Ayuga/Rebuelta 1986; Berger 4, 1954; Berghöfer/Hölzl 1986; Bourton 1968; Cheers 1997; Clair 1961; Coiciu/Racz (no year);

Cui/Ang 2002; Czygan 1993; Dorossiev 1985; Dudtschenko et al. 1989; Erdelmeier 1998; Erhardt et al. 2002; Erichsen-Brown 1989; Fleischhauer 2003; Fraschio 2003; Gray et al. 2000; Hager 5, 1993; Gruczczyk 2001; Hanelt 2000; Heeger 1956; Hiller/Melzig 1999; Hohmann et al. 2001; Hölzl/Ostrowski 1987; Hoppe 1949; Lürtz/Plescher 1998; Maisenbache/Kovar 1992; Mazza/Oomah 2000; Newall et al. 1996; Orth et al. 1999; Poginsky et al. 1988; Pschyrembel 1998; Roth 1990; Roth/Kormann 1997; Schilling 1969; Schönfelder 2001; Schultze-Motel 1986; Seidemann 1993c; Sharma 2003; Stochmal/Gruszczyk 1998; Täufel et al. 1993; Turova et al. 1987; Uphof 1968; Vanhaelen/Vanhaelen-Fastré 1983; Vickery 1981; Weyersdorf et al. 1995; Wyk et al. 2004

Hypericum revolutum *Vahl*

Synonyms ▸ *Hypericum lanceolatum* Lam.
Common Names ▸ curry bush, forest primrose; *German:* Currybusch
Usage ▸ condiment
Parts Used ▸ herb
Distribution ▸ S Africa

Joffe 1993

HYPTIS Jacq. - Lamiaceae (Labiatae)

Hyptis albida *H.B.K.*

Usage ▸ flavoring of food
Parts Used ▸ leaf
Distribution ▸ Mexico

Uphof 1968

Hyptis pectinata (L.) *Poit.*

Common Names ▸ comb hyptis, wild mint; *Spanish:* hierba de burro, xoltexnuk:
Usage ▸ flavoring
Parts Used ▸ leaf
Distribution ▸ W Africa: Ghana, Guinea to Cameroon, Ivory Coast, Nigeria; Mexico
Note ▸ In Ivory Coast and Upper Volta it is mixed with various spices (*Afromomum melegueta, Zingiber officinalis, Piper nigrum*, etc.).

Burkill 3, 1995; Dalziel 1957; Hanelt 2001; Irvine 1948; Malan et al. 1988; Pant et al. 1992; Pereda-Miranda et al. 1993; Uphof 1968

Hyptis spicigera *Lam.*

Synonyms ▸ *Nepeta americana* Aub.
Common Names ▸ black sesame, black beni seed, bush mint; *Chinese:* sui xu chang xiang; *German:* Schwarzer Sesam, Buschminze; *Portuguese:* mentrasto
Usage ▸ pot-herb, flavoring (in stews and sauces)
Parts Used ▸ seed (and it is also roasted)
Distribution ▸ Mexico to Brazil, Greater Antilles; widely native in tropical Africa, Madagascar and Asia
Note ▸ Cultivated there for its oily seeds. The plant is used by the Sakalava of Madagascar in the preparation of rum.

Burkill 3, 1995; Dalziel 1957; Erhardt et al. 2002; Irvine 1948; Kini et al. 1993; Kumar 2001; Mors/Rizzini 1961; Mors et al. 2000; Onayade et al. 1990; Pursglove 1968; Rehm/Espig 1996; Schultze-Motel 1986; Täufel et al. 1993; Uphof 1968; Zeven/de Wet 1982

Hyptis suaveolens (L.) *Poit.*

Synonyms ▸ *Ballota suaveolens* L., *Marrubium indicum* Blanco. non Burm.f.
Common Names ▸ tea-bush, Indian horehound, wild spikenard; *French:* gros baumes, hiptis à odeur, *German:* Indischer Andorn, Wohlriechender Andorn, Buschminze; *Hindi:* bilati tulsi, ganga tulsi; *Indonesian:* lampesan, jukut bau, mangkamang; *Japanese:* nioi-niga-kusa; *Malaysian:* malbar hutan, pokok kemangi; *Mexico:* chío gorda, chía grande; *Pilipino:* amotan, suob-kabayo, loko-loko; *Portuguese:* bamburral, mentrasto-grande; *Russian:* issop; *Spanish:* chao, hierba de las muelas, hortela do campo, orégano cimarron; *Thai:* kara, maeng lak kha
Usage ▸ spice, flavoring; **product:** essential oil
Parts Used ▸ shoot tip, leaf
Distribution ▸ native tropical America, naturalized pantropical, including SE Asia W Africa, also cultivated in Mexico and India
Note ▸ The plant has a mint flavor.

Ahmed et al. 1994; Arora/Padney 1996; Asekun/Ekundayo 2000; Azevedo et al. 2001; Bourton 1968; Fun/Svendsen 1990a; Guzman/Siemonsma 1999; Hanelt 2001; Iwu et al. 1990; Laily Din et al. 1988; Mallavaraou et al. 1993; Misra et al. 1981; Mors/Rizzini 1961; Oyen/Dung 1999; Pant et al. 1992; Schultze-Motel 1986; Täufel et al. 1993; Tucker 1986; Upadhyay et al. 1982; Uphof 1968; Villamar et al. 1994; Wiersema/Léon 1999; Wulff 1987; Zeven/de Wet 1982

HYSSOPUS L. – Hyssop - Lamiaceae (Labiatae)

Hyssopus angustifolius *M. Bieb.*

▶ *Hyssopus officinalis* L.

Hyssopus anisatus *Nutt.*

▶ *Agastache foeniculum (Pursh) Kuntze*

Hyssopus cinerascens *(Jord. et Fourr.) Brev.*

▶ *Hyssopus officinalis* L.

Hyssopus officinalis *L.*

Synonyms ▶ *Hyssopus angustifolius* M. Bieb., *Hyssopus cinerascens* (Jord. et Fourr.) Brev.

Common Names ▶ hyssop; *Chinese:* shen xiang cao; *French:* hysope, herbe sacrée; *German:* Ysop, Eisop, Isop, Josephskraut, Klosterysop; *Hindi:* jupha; *Italian:* issopo; *Portuguese:* hissopo; *Russian:* issop, issop obyknowennyj, gison, sison, jusefka, sinij sweroboj; *Sanskrit:* jupha; *Spanish:* hisopo, rabillo de gato;

Usage ▶ spice, pot-herb; **product:** essential oil and essence (for liqueurs: e.g. Benedictine, Chartreuse, Kartäuser etc.)

Parts Used ▶ leaf, flowering tops

Distribution ▶ S, C and E Europe, NW Africa, SW Asia: Caucasus, Anatolia, N Iran, to W Himalayas, native in parts of Europe and N America

Note ▶ Polymorphic species.

Aichele/Schwegler 4, 1995; Bärtels 1997; Berger 4, 1954; Bilgri/Adam 2000; Bremness 2001; Chalchat et al. 2001b; Cheers 1997; Clair 1961; Coiciu/Racz (no year); Davidson 1999; Erhardt et al. 2002; Fleischhauer 2003; Gerhardt 1981b; Hanelt 2001; Heeger 1956; Hilal et al. 1978; Hiller/Melzig 1999; Hoppe 1949; Joulain/Ragault 1976; Kerrola et al. 1994a; Piccaglia et al. 1999; Pochljobkin 1974, 1977; Pruthi 1976; Pschyrembel 1998; Renzini et al. 1999; Roth/Kormann 1997; Schönfelder 2001; Schulz/Stahl-Biskop 1991; Seidemann 1993c; Seidemann/Siebert 1987; Shah 1991; Sha et al. 1986; Sharma 2003; Sharma et al. 1963; Siewek 1990; Small 1997; Staesche 1972; Täufel et al. 1993; Teuscher 2003; Tsankova et al. 1993; Tucker 1986; Ubillos 1989; Wiersema/León 1999; Wüstenfeld/Haensel 1964; Wyk et al. 2004; Zeven/de Wet 1982

ILLICIUM L. – Star anise - Illiciaceae

 Illicium anisatum L.

Synonyms ▶ *Illicium religiosum* Sieb. et Zucc.
Common Names ▶ aniseed tree, bastard star anise, Japanese sacred anise poisonbay; *Chinese:* mang tsao, pa chio ksiang; *French:* anis de Chine, badiane de Chine; *German:* Shikimifrucht, Japanischer Sternanis, Heiliger Sternanis; *Japanese:* shikimi
Usage ▶ spice
Parts Used ▶ fruit, seed
Distribution ▶ Japan, Korea
Note ▶ The fruits are slightly poisonous.

Erhardt et al. 2000, 2002; Hanelt 2001; Rätsch 1998; Schultze-Motel 1986; Seidemann 1993c; Seidemann/Siebert 1987; Staesche 1972; Teuscher 2003; Uphof 1968; Villamar et al. 1994; Zänglein et al. 1989; Zeven/de Wet 1982

 Illicium anisatum Lour.

 *Illicium verum* J.D. Hook. f.

 Illicium cambodianum Hance

Common Names ▶ Cambodian star anise; *German:* Kambodscha Sternanis
Usage ▶ spice (locally); **product**: essential oil (Cambodian star anise oil, Cambodian anise oil)
Parts Used ▶ fruit, seed
Distribution ▶ Cambodia, Vietnam, Laos, Malaysia

Sánchez-Monge/Parellada 1981; Schultze-Motel 1986; Teuscher 2003; Uphof 1968

 Illicium floridanum Ellis

Common Names ▶ Florida star anise, purple anise; *German:* Florida Sternanis
Usage ▶ spice; **product**: essential oil (Florida star anise oil, Florida anise oil)
Parts Used ▶ fruit, seed
Distribution ▶ coastal plain of NW Florida to C Alabama, S Mississippi, and SE Lousiana to NE Mexico

Cheers 1997; Tucker/Maciarello 1999

 Illicium parviflorum Michx. ex Vent.

Common Names ▶ yellow star anise; *German:* Gelber Sternanis
Usage ▶ spice; **product**: essential oil from the bark and fruits (yellow star anise oil)
Parts Used ▶ fruit, seed
Distribution ▶ C Florida (local and rare)
Note ▶ The essential oil has a high safrole content (over 70 %)

Tucker/Maciarello 1999; Uphof 1968

 Illicium religiosum Sieb. et Zucc.

Illicium anisatum L.

Illicium stellatum L.

▸ *Illicium verum* Hook.f.

 Illicium verum Hook. f.

Synonyms ▸ *Illicium anisatum* Lour, *Illicium stellatum* L., *Badianifera officinarum* Kuntze

Common Names ▸ star anise, Chinese star anise, Chinese anis, badian, Indian anise; *Arabic:* albadyan; *Chinese:* ba jiao hui xian, ta hui hsiang; *Dutch:* sternanijs; *French:* badiane, anis étoilé, anis de la Chine; *German:* Badian, Sternanis, Chinesischer Anis, Indischer Anis, Sibrischer Anis; *Hindi:* chakriphool; *Indonesian:* bunga lawang, adas cina; *Italian:* anice stellato, badiana; *Malaysian:* bunga lawang, adas china; *Pilipino:* sanque, sanke; *Portuguese:* anis de China, anis estrelado; *Russian:* anisowoe derewo, kitajckij anis, swerdstschatyj anis, badian, badian tschabrez; *Slovenian:* badián; *Spanish:* anís de China, anís estrellado, badiána, illicium; *Thai:* chinpaetklip, dok chan, poy kak bua; *Turkish:* anason tchini; *Vietnamese:* cay hoy

Usage ▸ spice; **product:** essential oil (star anise oil, anise oil), and anethol

Parts Used ▸ fruit, seed

Distribution ▸ SE China, N Indochina: Laos, Vietnam; Korea, Japan, Taiwan, Hainan, Philippines (cultivated); the plant is not known wild.

Note ▸ Confusion with shikimi fruit (*Illicium anisatum* L.), Japanese star anise is slightly poisonous.

Berger 3, 1952; Bois 1934; Bourton 1968; Cheers 1997; Craze 2002; Cu et al. 1990b; Dalby 2000; Davidson 1999; Dudtschenko et al. 1989; Duke et al. 2003; Erhardt et al. 2000, 2002; Farrell 1985; Guzman/Siemonsma 1999; Hager 5, 1993; Hanelt 2001; Heeger 1956; Hiller/Melzig 1999; Hohmann et al. 2001; Kämpf/Steinegger 1974; Kataoka et al. 1986; Leung 1991; Melchior/Kastner 1974; Morris/Mackley 1999; Norman 1991; Porta et al. 1998; Pruthi 1976; Pschyrembel 1998; Roth/Kormann 1997; Schönfelder 2001; Schröder 1991; Schultze et al. 1990; Schultze-Motel 1986; Seidemann 1993c; Siewek 1990; Small 1997; Staesche 1972; Täufel et al. 1993; Teuscher 2003; Tuan/Ilangantileke 1997; Turova et al. 1987; Uhl 2000; Uphof 1968; Villamar et al. 1994; WHO 1990; Wiersema/León 1999; Wüstenfeld/Haensel 1964; Wyk et al. 2004; Zänglein/Schulze 1989; Zänglein et al. 1989; Zeven/de Wet 1982

▫ Illicium verum, fruits

 Imperatoria ostruthium L.

▸ *Peucedanum ostruthium* (L.) W.D.J. Koch

 Inga bigemia Willd.

▸ *Pithecellobium bigeminum* (L.) Mart.

 Inga dulcis Willd.

▸ *Pithecellobium dulce* (Roxb.) Dunal

INULA L. - Fleabane - Asteraceae (Compositae)

 Inula helenium L.

Synonyms ▸ *Aster helenium* Scop., *Aster officinalis* All., *Helenium grandiflorum* Gilib.

Common Names ▸ elecampane, scabwort, velvet dock, yellow starwort; *French:* aromate germanique, grande aunée, enule campane, héléne, inule; *German:* Echter Alant, Helenenwurzel, Oland, Ottwurzel; *Hindi:* pohkarmul; *Italian:* enula campana, elenio, inula; *Korean:* mokhyang; *Portuguese:* enula campana; *Russian:* dewjacil wysokij; *Sanskrit:* puskaramula; *Slovenian:* oman pravý; *Spanish:* énula campana, raíz del moro

Usage ▸ spice; **product:** essential oil

Parts Used ▸ rhizome

Distribution ▶ temperate Asia: Japan, E and SE Europe, native elsewhere

Aichele/Schwegler 4, 1995; Bärtels 1997; Bremness 2001; Cheers 1997; Clair 1961; Coiciu/Racz (no year); Craze 2002; Davidson 1999; Dudtschenko et al. 1989; Erhardt et al. 2002; Fleischhauer 2003; Gerhardt 1981a; Hager 5, 1993; Hanelt 2001; Heeger 1956; Hiller/Melzig 1999; Hoppe 1949; Newall et al. 1996; Pschyrembel 1998; Schönfelder 2001; Seidemann 1993c; Tucker 1986; Turova et al. 1987; Uphof 1968; Wiersema/León 1999; Wüstenfeld/Haeseler 1964; Wyk et al. 2004; Zeven/de Wet 1982

Inula viscosa (L.) Ait.

Synonyms ▶ *Dittrichia viscosa* (L.) Greut.
Common Names ▶ viscous elecampane, gluey elecampane; *German:* Klebriger Alant
Usage ▶ spice
Common Names ▶ rhizome
Distribution ▶ Mediterranean regions, N Africa, Canary Islands

Bärtels 1997; Hager 5, 1993; Usher 1968

IPHEION Raf. - Spring Starflower - Alliaceae (Liliaceae)

Ipheion uniflorum (Graham) Raf.

Common Names ▶ spring star flower; *French:* iphéion; *German:* Frühlingsstern, Vielblütige Sternblume
Usage ▶ spice, flavoring
Parts Used ▶ leaf
Distribution ▶ S Brazil, Argentina, Uruguay
Note ▶ Crushed leaves have a garlic aroma.

Cheers 1997; Erhardt et al. 2002

Ipomoea cymosa Roem. et Schult.

▶ *Merremia umbellata* (L.) Hallier f. ssp. *orientalis* Ooststr.

Ipomoea sinuata Ort.

▶ *Merremia dissecta* (Jacq.) Hall.

IRIS L. - Iris - Iridaceae

Iris florentina L.

▶ *Iris germanica* L. var. *florentina* Dykes

Iris germanica L.

Common Names ▶ flag iris, common iris, German iris, orris; *French:* iris commun, iris d'Allemagne; *German:* Deutsche Schwertlilie, Unechte Veilchenwurzel; *Italian:* giaggiolo, iris
Usage ▶ spice
Parts Used ▶ rhizome
Distribution ▶ SW Europe: Iberia, France, EC Europe, Turkey, Palestine, NW Africa, native in Europe, also cultivated
Note ▶ The plant is frequently cultivated in gardens.

Aichele/Schwergel 5, 1996; Bärtels 1997; Bourton 1968; Charalambous 1994; Cheers 1997; Cohen 1993b; Coiciu/Racz (no year); Dhar/Kalla 1973; Erhardt et al. 2002; Fleischhauer 2003; Heeger 1956; Hiller/Melzig 1999; Hoppe 1949; Pschyrembel 1993; Schönfelder 2001; Seidemann 1993c; Small 1997; Usher 1968; Weber 1997; Wiersema/León 1999; Wüstenfeld/Haensel 1964; Wyk et al. 2004; Zeven/de Wet 1982

Iris germanica L. var. florentina Dykes

Synonyms ▶ *Iris florentina* L.
Common Names ▶ Florentine iris, orris root; *Dutch:* duitse iris; *French:* iris de Florence; *German:* Deutsche Schwertlilie, Florentinische Schwertlilie, Florentinische Veilchenwurzel, Unechte Veilchenwurzel; *Italian:* giagiolo ianco; *Portuguese:* iris Florentino; *Spanish:* lirio blanco, lirio de florencia, iris Florentina
Usage ▶ spice
Parts Used ▶ rhizome
Distribution ▶ Europe: Iberia, Slovakia, Croatia, Crete, Cyprus, NW Africa, also cultivated

Aichele/Schwergel 5, 1996; Bärtels 1997; Bourton 1968; Cheers 1997; Dudtschenko et al. 1989; Erhardt et al. 2002; Hanelt 2001; Heeger 1956; Hiller/Melzig 1999; Hoppe 1949; Pschyrembel 1998; Roth/Kormann 1997; Schönfelder 2001; Schultze-Motel 1986; Seidemann 1993c; Uphof 1968; Wüstenfeld/Haensel 1964

Ixora L. - Rubiaceae

Ixora coccinea L.

Common Names ▶ flame-of-the-wood, ixora, Indian ixora, jungle flame; *German:* Dschungelbrand, Scharlachrote Ixorie; *India:* cetti, paranti, rajana, shetti; *Indonesian:* bunga soka; *Malaysian:* pecah periuk, bunga jarum, kaum kopi; *Pilipino:* dwarf santon, santon; *Thai:* khem (general name), khem farang; *Vietnamese:* Đòn

Usage ▶ flavoring

Parts Used ▶ flower

Distribution ▶ India, the Indian Ocean Islands (Madagascar, Seychelles, Comoros, Mascarenes)

Burkill 4, 1997; Engler/Phummai 2000; Erhardt et al. 2002; Gurib-Fakim/Brendler 2004; Plotkin 1994; Uphof 1968

J

 Jambolifera odorata *Lour.*

> *Acronchia odorata (Lour.) Baill.*

 Jambolifera resinosa *Lour.*

> *Acronchia peduculata (L.) Miq.*

JASMINUM L. - Jasmine, Jessamine - Oleaceae

 Jasminum floribundum *R. Br. ex Fresen*

> *Jasminum grandiflorum forma grandiflorum (L.) Kobuski*

 Jasminum grandiflorum *L*

> *Jasminum officinale L.f. grandiflorum*

 Jasminum humile *L.*

Common Names ▶ (Indian) yellow jasmine, Italian jasmine; *Chinese:* ai tan chun; *German:* Italienischer Jasmin, Indischer Jasmin, Gelber Jasmin
Usage ▶ flavoring
Parts Used ▶ flower
Distribution ▶ C Asia, Russian Middle Asia, India, SW China, Afghanistan, Myanmar

Chauhan 1999; Erhardt et al. 2002; Sharma 2003; Wiersema/León 1999

 Jasminum odoratissimum *L.*

Common Names ▶ yellow jasmine; *Japanese:* kin-sokei; *German:* Duft-Jasmin, Gelber Jasmin
Usage ▶ spice, flavoring of tea
Parts Used ▶ flower
Distribution ▶ Canary Islands, Madeira, cultivated in S France and Taiwan

Erhardt et al. 2002; Schultze-Motel 1986; Tamagami et al. 2001; Täufel et al. 1993; Trujillo et al. 1996; Uphof 1968

 Jasminum officinale *L.*

Common Names ▶ (common) white jasmine, jessamine,; *Chinese:* so-hsing, su fang hua, ye hsi ming; *Dutch:* jasmijnbloesen; *French:* jasmin commun; *German:* Echter Jasmin, Weißer Jasmin; *India:* chamba, chambeli; *Italian:* gelsomino; *Portuguese:* jasmin; *Russian:* shasmin; *Spanish:* jasmón blanco, jazmin blanco;
Usage ▶ spice, flavoring (for tea); **product:** essential oil (perfumery)
Parts Used ▶ flower
Distribution ▶ Himayas, India (Kashmir), SW China, Caucasus; Iran; native S Europe, widely cultivated and sometimes native

Bourton 1968; Cheers 1997; Erhardt et al. 2000, 2002; Grasse 1950; Muller 1965; Schultze-Motel 1986; Seidemann 1993c; Sharma

© Springer-Verlag Berlin Heidelberg 2005

2003; Täufel et al. 1993; Uphof 1968; Wiersema/León 1999; Zeven/de Wet 1982; Zhang et al. 2000;

Jasminum officinale L. f. grandiflorum (L.) Kobuski

Synonyms ▶ *Jasminum floribundum* R. Br. ex Fresen
Common Names ▶ Catalonian jasmine, Italian jasmine, jasmine; *French:* jasmine à grandes fleurs; *German:* Großblütiger Jasmin; *Italian:* gelsomino;
Usage ▶ spice (flavoring); **product:** essential oil
Parts Used ▶ flower
Distribution ▶ SW Arabia, also cultivated

Bourton 1968; Green 1986; Hiller/Melzig 1999; Oyen/Dung 1999; Roth/Kormann 1997; Shiva et al. 2002; Uphof 1968; Zeven/de Wet 1982

Jasminum paniculatum Roxb.

Common Names ▶ *German:* Rispiger Jasmin
Usage ▶ flavoring of tea
Parts Used ▶ flower
Distribution ▶ temperate China

Uphof 1968

Jasminum sambac (L.) Ait.

Synonyms ▶ *Nyctanthes sambac* L.
Common Names ▶ Arabian jasmine, biblical jasmine, samba, Tuscan jasmine; *Chinese:* mo li hua; *French:* jasmin d'Arabie; *German:* Arabischer Jasmin, Persischer Jasmin; *India:* bela, moghra, zambac; *Malaysian:* melati, melor; *Pilipino:* kampupot, sampa guita; *Portuguese:* bogarim, jasmin; *Spanish:* jazmín de Arabia; *Thai:* mali laa, mali son, maliwan
Usage ▶ spice, flavoring of tea; **product:** essential oil
Parts Used ▶ flower
Distribution ▶ India, Sri Lanka; widely cultivated in the Tropics
Note ▶ The 'Grand Duke of Tuscany' is a filled sort.

Arora/Pandey 1996; Burkill 4, 1997; Cheers 1997; Engel/Phummai 2000; Erhardt et al. 2002; Ito et al. 2002; Jacquat 1990; Kottegoda 1994; Lewington 1990; Schultze-Motel 1986; Seidemann 1993c; Täufel et al. 1993; Uphof 1968; Wiersema/León 1999; Zeven/de Wet 1982

Jatropha heudelotti (Baill.) Pierre ex Pax

▶ *Ricinodendron heudelotii* (Baill.) Pierre ex Pax

JUGLANS L. - Walnut - Juglandaceae

Juglans nigra L.

Common Names ▶ black walnut; *Chinese:* hei-che-tao; *French:* noyer noir, noyer commun; noyer noir d'Amerique; *German:* Schwarze Walnuss, Schwarznuss; *Portuguese:* nogueira preta; *Russian:* orech tschernij; *Spanish:* nogal americano
Usage ▶ flavoring
Parts Used ▶ shell
Distribution ▶ N America: Canada, USA, frequently cultivated

Aichele/Schwegler 2, 1994; Cheers 1997; Davidson 1999; Erhardt et al. 2000, 2002; Erichsen-Brown 1989; Garavel 1960; Hanelt 2001; Lewington 1990; Schultze-Motel 1986; Täufel et al. 1993; Uphof 1968; Wiersema/León 1999; Zeven/de Wet 1982

Juglans regia L.

Common Names ▶ English walnut, Madeira nut, Persian walnut; *Arabian:* joz, naksh souak, zouz; *Chinese:* che-tao, hu tao; *Dutch:* walnoot; *French:* noyer commun; *German:* Echte Walnuss, Welsche Walnuss; *Italian:* noce nostrana, noce comune, noce persiana; *Japanese:* chosen-gurumi; *Korean:* hodunamu; *Portuguese:* nogueira comun; *Russian:* orech grezkij; *Spanish:* escuerno, hoja de nogal, nogal común, nogal inglís
Usage ▶ flavoring in the liquor industry
Parts Used ▶ shell
Distribution ▶ Caucasus, Russia, C Asia, W Asia, India, SE Europe, widely native, elsewhere in temperate regions, Moldavia
Note ▶ Many varieties have been described.

Aichele/Schwegler 2, 1994; Bärtels 1997; Bendel 2002; Born 1991; Bois 1934; Bremness 2001; Çağlarımak 2003; Cheers 1997; Coiciu/Racz (no year); Dastur 1954; Davidson 1999; Debor 1974; Erhardt et al. 2002; Fleisch-hauer 2003; Hanelt 2001; Heeger 1956; Hepper 1992; Hiller/Melzig 1999; Hoffmann et al. 1992; Hohmann et al. 2001; Hoppe 1949; Jensen et al. 2003; Jirovetz et al. 1996; Lewington 1990; López et al. 1995; Pschyrembel 1993; Ravai 1992; Schaarschmidt 1988; Schönfelder 2001; Schultze-Motel 1986; Seidemann 1993c; Sharma 2003; Täufel et al. 1993; Uphof 1968; Villamar et al. 1994; Wiersema/León 1999; Wyk et al. 2004; Zeven/de Wet 1982

JUNIPERUS L. - Juniper - Cupressaceae

Juniperus communis L.

Common Names ▶ common juniper, juniper; *Dutch:* jeneverboom; *French:* baie de genièvre, genévrier (commun); *German:* (Gewöhnlicher) Wacholder, Machandel, Kranawitt, Kronawitt, Kaddig; *Italian:* bacca di ginepro, coccola di genipro; ginepro comune, ginepro nero; *Portuguese:* zimbreiro; *Russian:* moshshewel'nik, moshshewel'nik obyknowennyj, moshshucha, shenwr'e, bakkaut; *Slovenian:* borievka; *Spanish:* baya de enebro, enebrina, enebro (común), junipero
Usage ▶ spice, flavoring for liqueurs and hard spirits: Gin, Genever etc.; **product:** essential oil
Parts Used ▶ fruit
Distribution ▶ Europe, Turkey, Caucasus, Siberia, Korea, Japan, N Africa: Algeria, Morocco; Alaska, Canada, USA
Note ▶ Falsification with *Juniperus sabina* L: savin, Spanish savin; French: genévrieri sabine; German: Sadebaum, Stinkwacholder; Russian: moshshewel'nik.

Aichele/Schwegler 2, 1994; Alberts/Muller 2000; Angioni et al. 2003; Berger 3, 1952; Bois 1934; Bourton 1968; Bremness 2001; Caramiello et al. 1995; Chatzopoulou/Katsiotis 1993; Chatzopoulou et al. 2002; Cheers 1997; Coiciu/Racz (no year); Cosentino et a. 2003; Craze 2002; Da Cunha/Roque 1989; Davidson 1999; Duke et al. 2003; Dudtschenko et al. 1989; Erhardt et al. 2000, 2002; Farrell 1985; Fleischhauer 2003; Hager 5, 1993; Hanelt 2001; Heeger 1956; Hiller/Melzig 1999; Hohmann et al. 2001; Koukos/Papadopoulou 1997; Melchior/Kastner 1974; Morris/Mackley 1999; Newall et al. 1996; Norman 1991; Pochljobkin 1974, 1977; Pruthi 1976; Pschyrembel 1998; Rafique et al. 1993; Rätsch 1998; Roth/Kormann 1997; Sanchez de Medina et al. 1994; Schönfelder 2001; Schultze-Motel 1986; Seidemann 1993c; Seidemann/Siebert 1987; Sharma 2003; Shiva et al. 2002; Siewek 1990; Small 1997; Staesche 1972; Stassi et al. 1996; Täufel et al. 1993; Teuscher 2003; Tucker 1986; Wüstenfeld/Haensel 1964; Wyk et al. 2004

Juniperus phoenicea L.

Common Names ▶ Phoenicean juniper; *German:* Phönizischer Wacholder
Usage ▶ spice, flavoring
Parts Used ▶ fruit
Distribution ▶ Europe: Iberia, France; W Turkey, Cyprus, Palestine, NW Africa, Libya

Angioni et al. 2003; Erhardt et al. 2002; Hepper 1992; Schönfelder 2001

Juniperus virginiana L.

Synonyms ▶ *Sabina virginiana* (L.) Ant.
Common Names ▶ Virginian cedar, pencil cedar, eastern red cedar; *German:* Virginianischer Wacholder; *Italian:* gineprodella
Usage ▶ spice, flavoring
Distribution ▶ N America: New Brunswick to Georgia, N Dakota, E Texas

Rehm/Espig 1984; Rätsch 1998; Schultze-Motel 1986; Small 1997; Uphof 1968

Juniperus sabina L.

▶ *Juniperus communis* L.

JUSTICIA L. - Water Willow - Acanthaceae

Justicia gangetica L

▶ *Asystasia gangetica* (L.) T. Anders

Justicia quinquangularis Koenig ex Roxb.

Common Names ▶ quintuple justicia; *German:* Fünfkantige Justizie; *Portuguese:* justícia

Usage ▶ pot-herb
Parts Used ▶ leaf
Distribution ▶ India

Arora/Padney 1996; Kottegoda 1994

K

 Kadsura chinensis *Turcz.*

▸ *Schisandra chinensis Turcz.) Baill.*

KAEMPFERIA L. - Kaempferia - Zingiberaceae

 Kaempferia aethiopica *(Schweinf.) Benth.*

▸ *Siphonochilus aethiopicus (Schweinf.) B.L. Burtt*

 Kaempferia galanga *L.*

Synonyms ▸ *Kaempferia humilis* Salisb., *Kaempferia sessilis* Koenig

Common Names ▸ East Indian galan(gal), galanga, spice lily, resurrection lily; *German:* Chinesischer Galgant, Gewürzlilie, Kentjur, Thai-Ingwer; *Hindi:* chandra-mula; *India:* achoram, kacholum, kencur, tjekur; *Indonesian:* kencur; *Javanese:* chĕkur, chĕngkur, kĕnchur; *Malaysian:* cekur; *Pilipino:* disól, dotó, kisól; *Sanskrit:* chandramulika, sugandhavacha; *Thai:* pro hom, waan hom, waan teendin, khiey, krachai; *Vietnamese:* têu dâu

Usage ▸ spice (flavoring of rice, for samballans, etc.)

Parts Used ▸ rhizome, leaf

Distribution ▸ Indian, Sri Lanka; SE Asia: Java, Philippines, New Guinea, Africa: Sudan, also cultivated

Note ▸ The root powder also used as a perfume in cosmetics, e.g. shampoos.

Kaempferia galanga: a **flowering,** b **fresh rhizoms**

Alberts/Muller 2000; Arora/Pandey 1996; Bois 1934; Burkill 5, 2000; Davidson 1999; Duke/Ayeusu1985; Duke et al. 2003; Erhardt et al. 2002; Hiller/Melzig 1999; Jacquat 1990; Larsen et al. 1999; Lück 2004; Melchior/Kastner 1974; Norman 1991; Schultze-Motel 1986; Seidemann 1992, 1993c, 1998/2000; Seidemann/Siebert 1987; Sharma 2003; Siewek 1990; Tamaka/Nakao 1976; Tao Duoda 1998; Teuscher 2003; Turova et al. 1987; Uphof 1968; Wealth of India 5, 1959; WHO 1990; Wiersema/León 1999; Wong 1999; Wong et al. 1992; Wu et al. 2000; Yusuf et al. 2002; Zeven/de Wet 1982

 Kaempferia humilis Salisb.

▶ *Kaempferia galanga* L.

 Kaempferia longa Jacq.

▶ *Kaempferia rotunda* L.

 Kaempferia ovata Rosc.

▶ *Boesenbergia rotunda (L.) Mansf.*

 Kaempferia pandurata Roxb.

▶ *Boesenbergia rotunda (L.) Mansf.*

 Kaempferia rotunda L.

▶ *Boesenbergia rotunda (L.) Mansf.*

Kaempferia rotunda L.

Synonyms ▶ *Kaempferia longa* Jacq., *Kampferia versicolor* Salisb.
Common Names ▶ Chinese keys, resurrection lily, round-rooted galanggale (galangal), tropical lily; *German:* Runde Gewürzlilie, Gefleckte Gewürzlilie, Runder Kentjur; *Hindi:* bhuichampa; *India:* chandramalu; *Javanese:* kentjur, kuntji; *Sanskrit:* bhuchampaca, bhumichampa; *Thai:* kencur, waan hao non, krachai
Usage ▶ spice, flavoring
Parts Used ▶ young root, leaf
Distribution ▶ SE Asia (probably native), cultivated in numerous tropical Asia countries, e.g. India, Java, Malaysia, Thailand
Note ▶ The roots are also utilized for dying and as a sedative.

Cheers 1997; Duke et al. 2003; Erhardt et al. 2002; Hanelt 2001; Kumar 2001; Larsen et al. 1999; Melchior/Kastner 1974; Schultze-Motel 1986; Seidemann 1992, 1993c; Tanaka/Nakao 1976; Tao Duoda 1998; Wiemrsema/León 1999; Wong 1999; Yusuf et al. 2002; Zeven/de Wet 1982

 Kaempferia sessilis Koenig

▶ *Kaempferia galanga* L.

 Kaempferia versicolor Salisb.

▶ *Kaempferia rotunda* L.

KLAINEDOXA Pierre ex Engl. - Irvingiaceae (Simaroubaceae)

 Klainedoxa gabonensis Pierre ex Engl.

Common Names ▶ South African pepper tree, mountain seringa, (wild) pepper tree; *S Africa:* slaploot, wit sering
Usage ▶ spice
Parts Used ▶ seed
Distribution ▶ tropical W Africa: Ivory Coast, Cameroon, Congo, Zaire, Gabon

Bois 1934; Hanelt 1973; Mabberley 1997; Neuwinger 1999; Uphof 1968

 Kleinia tagetoides H.B.K.

▶ *Porophyllum tagetoides (Kunth) DC.*

 Kochia scoparia (L.) Schrader

▶ *Bassia scoparia (L.) A.J. Scott*

KNEMA Lour. - Myristicaceae

Knema bicolor Rat.

▶ *Knema corticosa Lour.*

Knema corticosa *Lour.*

Synonyms ▶ *Knema bicolor* Rat., *Myristica corticosa* Hock.f.
Common Names ▶ *French:* muscadier à suif; *German:* Dickrindiger Muskatbaum
Usage ▶ spice
Parts Used ▶ seed
Distribution ▶ Indochina, Myanmar, Thailand

Uphof 1968

KOSTELETZKAYA C. Presl. - Malvaceae

Kosteletzkaya adoensis *(Hochst.) Masters*

Common Names ▶ *German:* Kosteletzkablätter
Usage ▶ flavoring of palm wine
Parts Used ▶ leaf
Distribution ▶ Cameroon, Nigeria, Zaire

Burkill 4, 1997; Hanelt 2001; Neuwinger 1999

KRAMERIA L. ex Loefl. - Krameriaceae

Krameria lappacea *(Domb.) Burdet et B.B.*

Synonyms ▶ *Krameria triandra* Ruiz et Pav.
Common Names ▶ rhatany, Peruvian rhatany; *French:* kraméria, rhatania; *German:* (Rote) Ratanhia, Payta-Ratanhia; *Italian:* ratanhia; *Spanish:* ratania
Usage ▶ spice (for liqueurs)
Parts Used ▶ root
Distribution ▶ S America: Bolivia, Chile, Peru

Daems 1987; Erhardt et al. 2002; Hoppe 1949, 1, 1975; Rätsch 1998; Uphof 1968; Wüstenfeld/Haensel 1964; Wyk et al. 2004

Krameria triandra *Ruiz et Pav.*

▶ *Krameria lappacea (Domb.) Burdet et B.B.*

KYLLINGIA Rottb. - Cyperaceae

Kyllingia erecta *Schum.*

Common Names ▶ greater kyllingia, white kyllingia
Usage ▶ flavoring
Parts Used ▶ rhizome
Distribution ▶ W Africa: the Gambia, Mali, Niger, Senegal, Sierra Leone
Note ▶ The rhizom is aromtic with a rather bitter taste.

Burkill 3, 1995; Dalziel 1957

Kyllingia pumila *Michx.*

Usage ▶ flavoring
Parts Used ▶ root
Distribution ▶ temperate and tropical America, W Africa: Nigeria

Burkill 3, 1995; Dalziel 1957

Kyllingia squamulata *Rhonn.*

Usage ▶ flavoring
Parts Used ▶ culm-base
Distribution ▶ W Africa: Senegal, Tanzania, Uganda

Burkill 3, 1995

Kyllingia tenuifolia *Steud.*

Usage ▶ flavoring
Parts Used ▶ culm-base
Distribution ▶ W Africa: Nigeria

Burkill 3, 1995

 **Kyllingia umbellata** *Rott.*

 Mariscus alternifolia Vahl

L

LACTARIUS Fr. - Russulaceae

Lactarius helvus (Fr.) Fr.

Common Names ▶ *German:* Maggipilz, Bruchreizker
Usage ▶ flavoring of soups and salads
Parts Used ▶ fruitbodies (pulverized)
Distribution ▶ temperate zones in some countries of Europe

Gerhardt 1995; Klán 1981; Uphof 1968

LACTUCA L. - Lettuce - Asteraceae (Compositae)

Lactuca canadense L.

Common Names ▶ wild Canadian lettuce; *German:* Kanadischer Lattich
Usage ▶ pot-herb
Parts Used ▶ leaf
Distribution ▶ N America, W Indies

Uphof 1968

Lactuca runcinata DC.

Synonyms ▶ *Lactuca heyneana* DC.
Common Names ▶ *German:* Grobzähniger Lattich
Usage ▶ pot-herb
Parts Used ▶ leaf
Distribution ▶ India

Arora/Pandey 1993

Lagunea orientale (L.)

▶ *Polygonum orientale* L.

Lallemantia iberica (Bieb.) Fisch. Et Mey.

▶ *Dracocephalum ibericum* Bieb.

LAMINARIA J.V. Lamour. - Laminariaceae

Laminaria japonica J. Areschoug

Common Names ▶ kelp; *Chinese:* haidai, tai tai; *German:* Japanischer Blatttang, Japanische Braunalge, Kelp; *Japanese:* ma-konbu, yebisu me, shinori-kombu
Usage ▶ spice; **product:** used for seasoning
Parts Used ▶ thallus, seaweed
Distribution ▶ On the coasts from China to Japan and Europe: Norway and Isle of Man; also cultivated
Note ▶ *Laminaria spec.* used as food and as medicine (iodine) in China for more than 1500 years and has been cultivated in Japan since 1730. In Europa cultivation on the Isle of Man and at the Norwegian fjords.

© Springer-Verlag Berlin Heidelberg 2005

Hanelt 2001; Schultze-Motel 1986; Täufel et al. 1993, Turora et al. 1987

 Laminaria saccharina *Mart.*

▶ *Laminaria japonica J. Areschoug*

 Landolphia senegalensis *Kotsch. et Peyr.*

▶ *Saba senegalensis (DC.) Pichon*

 Languas conchigera *Burkill*

▶ *Alpinia conchigera Griff.*

 Languas galanga *(L.) Stuntz.*

▶ *Alpinia galanga (L.) Willd.*

 Languas officinarum *(Hance) Farw.*

▶ *Alpinia officinarum Hance*

 Languas vulgare *Koenig*

▶ *Alpinia galanga (L.) Willd.*

LANTANA L. - Lantana - Verbenaceae

 Lantana aculeata *L.*

▶ *Lantana camara L.*

 Lantana alba *Mill.*

▶ *Lippea alba (Mill.) N.E. Br.*

 Lantana camara *L.*

Synonyms ▶ *Camara vulgaris* Benth., *Lantana aculeata* L., *Lantana crocea* Jacq., *Lantana nivea* Vent., *Lantana sanguinea* Medik.

Common Names ▶ lantana, red sage, sherry pie, shrub verbena, wild sage, yellow sage; *Brazil (Portuguese):* camará, cambará, cambará-de-cheiro, cambará-de-chumu, chumbinho; *Chinese:* ma ying dan, wu se mei; *French:* lantanier; *German:* Wandelröschen; *India:* ghaneri, pulikampa, bara phulanoo, vaneri; *Indonesian:* ta ayam, waung; *Japanese:* rantana, shichi-henge; *Malaysian:* bunga tahi ayam bunga pagar; *Pilipino:* cinco negritos, kantutay, coronitas; *Thai:* phakaa krong; *Vietnamese:* bong oi, tram hoi

Usage ▶ spice, flavoring

Parts Used ▶ leaf

Distribution ▶ origin Neotropics; Mexico, tropical America, native Florida, Texas, Hawaii; cultivated in many warm countries, e.g. Sri Lanka

Ahmed et al. 1971; Barua et al. 1971; Burkill 5, 2000; Cheers 1997; Engel/Phummai 2000; Erhardt et al. 2002; Hiller/Melzig 1999; v. Koenen 1996; Kottegoda 1994; Louw 1943, 1948; Mors et al. 2000; Rajendran/Daniel 2002; Ross 1999; Schönfelder 2001; Schultze-Motel 1986; Storrs 1997; Täufel et al. 1993; Uphof 1968; Villamara et al. 1994; Warren 1998

 Lantana crocea *Jacq.*

▶ *Lantana camara L.*

 Lantana lilacina *Desf.*

Common Names ▶ *Brazil (Portuguese):* cambará rosa; *German:* Lilafarbenes Wandelröschen

Usage ▶ flavoring

Parts Used ▶ leaf, flower

Distribution ▶ Brazil: São Paulo to Rio Grande do Sol

Mors et al. 2000

Lantana microphylla *Mart.*

Common Names ▶ *Brazil (Portuguese):* alecrim-bravo, alecrim-do-campo; *German:* Kleinblättriges Wandelröschen
Usage ▶ flavoring
Parts Used ▶ leaf
Distribution ▶ S America: especially Brazil, cultivated in many countries

Uphof 1968

 Lantana nivea *Vent.*

▶ *Lantana camara L.*

 Lantana rhodesiensis *Mold.*

Common Names ▶ Rhodesian lantana; *German:* Rhodesisches Wandelröschen
Usage ▶ flavoring of foods and milk
Parts Used ▶ leaf, herb
Distribution ▶ W Africa: Ghana, Guinea, Nigeria, Senegal, Sierra Leone

Adegoke et al. 1968; Burkill 5, 2000

 Lantana sanguinea *Medik.*

▶ *Lantana camara L.*

 Lantana trifolia *L.*

Common Names ▶ three-leaved lantana; *German:* Dreiblättriges Wandelröschen; *Spanish:* orégano
Usage ▶ flavoring of milk und butter
Parts Used ▶ leaf, herb
Distribution ▶ E Africa: Ethiopa

Burkill 3, 1995; Husain et al. 1992; Neuwinger 1999; Rajendran/Daniel 2002; Tucker 1986

 Lapathum vesicarium *(L.) Moench*

▶ *Rumex vesicarius L.*

LAPORTEA Gaudich. - Bush nettle - Urticaceae

 Laportea crenulata *Gaudich.*

Common Names ▶ devil nettle, fever nettle; *German:* Gekerbte Laporte, Gekerbte Strauchnessel
Usage ▶ spice, like coriander fruit
Parts Used ▶ seed
Distribution ▶ India: tropical Himalayas

Arora/Padney 1996

LARREA Cav. - Zygophyllaceae

 Larrea glutinosa *Engelm.*

▶ *Larrea tridentata (Sesse et Moç.) Cav. ex DC.*

 Larrea mexicana *Moric.*

▶ *Larrea tridentata (Sesse et Moç.) Cav. ex DC.*

 Larrea tridentata *(Sesse et Moç.) Cav. ex DC.*

Synonyms ▶ *Larrea glutinosa* Engelm., *Larrea mexicana* Moric., *Zygophyllum tridentatum* Sesse et Moç.
Common Names ▶ creosote bush, greasewood; *German:* Kreosotstrauch; *Spanish:* gobernadora, hediondilla, hediondo, paloondo
Usage ▶ condiment
Parts Used ▶ floral-bud
Distribution ▶ SW USA, Mexico
Note ▶ Substitute for capers (*Capparis spinosa* L.).

Bernhard/Thiele 1981; Chirikdjian 1974; Hanelt 2001; Newall et al. 1996; Rätsch 1998; Schultze-Motel 1986; Uphof 1968; Villamar et al. 1994; Wiersema/León 1999

LASER P. Gaertn., B. Mey. et Scherb. - Apiaceae (Umbelliferae)

 Laser trilobum (L.) Borkh.

Synonyms ▶ *Laserpitium trilobum* L., *Siler trilobum* (L.) Crantz

Common Names ▶ laserwort, trivalve; *French:* cumin de chevaux; *German:* Echter Bergkümmel, Rosskümmel, Dreilappiger Rosskümmel; *Turkish:* kefe kimyonū

Usage ▶ spice, condiment

Parts Used ▶ seed

Distribution ▶ Europe: France, Spain, E Europe, Turkey, Lebanon, Caucasus, N Iran

Note ▶ In former times the plant was cultivated as a spice plant in gardens of C Europe.

Aichele/Schwegler 3, 1995; Akgül 1989b; Başer et al. 1993b; Brunke et al. 1991a; Erhardt et al. 2002; Fleischhauer 2003; Hanelt 2001; Heeger 1956; Kivanç/Akgül 1991; Mansfeld 1986; Schultze-Motel 1986; Seidemann 1993c; Siewek 1990; Täufel et al. 1993; Teuscher 2003

LASIA Lour. - Araceae

 Lasia aculeata Lour

▶ *Lasia spinosa* (L.) Thwaites

 Lasia spinosa (L.) Thwaites

Synonyms ▶ *Lasia aculeata* Lour.

Common Names ▶ Thwaites sampi; *Chinese:* ka guo, re yo; *German:* Dornige Zottelblume; *Indonesian:* ngamling, sambbeng; *Malaysian:* bekil, *Sri Lanka:* geli-geli; *Thai:* pa lang

Usage ▶ spice (for fish ponds)

Parts Used ▶ young leaf

Distribution ▶ Sri Lanka, India to China, Himalayas, Malaysia, Indonesia

Burkill 1965; Erhardt et al. 2002; Hanelt 2001; Kumar 2003; Ochse/van de Brinck 1931

LASERPITIUM L. - Laserwort - Apiaceae (Umbelliferae)

 Laserpitium siler L.

Synonyms ▶ *Siler montanum* Crantz

Common Names ▶ laserwort; *German:* Echter Bergkümmel, Rosskümmel; Berglaserkraut; *Russian:* gladysch

Usage ▶ spice

Parts Used ▶ seed, root (rarely)

Distribution ▶ S Europe, C Europe: mountains

Note ▶ In former times the plant was cultivated as a spice plant in gardens of C Europe mountains. Substitute for caraway (*Carum carvi* L.).

Erhardt et al. 2002; Fleischhauer 2003; Hanelt 2001; Heeger 1956; Hoppe 1949; Uphof 1968

 Laserpitium trilobium L.

▶ *Laser trilobium* (L.) Borkh.

LAURELIA Juss. - Monimiaceae

 Laurelia aromatica Juss.

Common Names ▶ Chilean laurel; *German:* Chilenischer Lorbeer

Usage ▶ spice (rarely in Peru)

Parts Used ▶ seed

Distribution ▶ Chile, Peru

Uphof 1968

 Laurelia semperivens (R. et P.) Tul.

Common Names ▶ Chilean nutmeg; *German:* Chilenischer Muskat, Immergrüner Muskat

Usage ▶ spice (rarely)

Parts Used ▶ fruit, seed

Distribution ▶ S America: Chile

Note ▶ Substitute for nutmeg (*Myristica fragrans* Houtt.).

Hager 5, 1993; Hiller/Melzig 1999; Hoffmann et al. 1992; Rätsch 1998; Warburg 1897

LAURENCIA Adans - Rhodomelaceae

Laurencia pinnatifida (Gmel.) Lamour

Common Names ▶ pepper dulce
Usage ▶ condiment
Parts Used ▶ pungent red algae
Distribution ▶ Atlantic Ocean, North Sea

Hanelt 2001; Uphof 1968

Laurocerasus officinalis M. Roem.

▶ *Prunus lauroccerasus* L.

LAURUS L. - Bay, Laurel - Lauraceae

Laurus albida Nutt.

▶ *Sassafras albidum* (Nutt.) Nees

Laurus eastivalis L.

▶ *Lindera benzoin* (L.) Bl.

Laurus benzoin L.

▶ *Lindera benzoin* (L.) Bl.

Laurus burmani Nees

▶ *Cinnamomum burmani* Nees

Laurus cassia L.

▶ *Cinnamomum aromaticum* Nees

Laurus cinnamomoides Mutis ex H.B.K.

▶ *Ocotea cymbarum* H.B.K.

Laurus cinnamomum Andr.

▶ *Cinnamomum aromaticum* Nees

Laurus cinnamon L.

▶ *Cinnamomon zeylanicum* Bl.

Laurus cubebea Lour.

▶ *Litsea cubeba* (Lour.) Pers.

Laurus culitlawan L.

▶ *Cinnamomum culitlawan* (L.) Kosterm.

Laurus dulcis Roxb.

▶ *Cinnamomum burmani* Bl.

Laurus nobilis L.

Synonyms ▶ *Laurus undulata* Mill.
Common Names ▶ laurel, bay laurel, sweet bay, noble laurel, Roman laurel, true laurel; *Arabic:* gar, ghâr (fruit); *Chinese:* yue gui zi, yueh kuei; *Dutch:* laurierblad; *French:* laurier common, laurel noble, feuille de laurier; laurier d'Apollon, laurier à jambon; *German:* Lorbeer, Lorbeerblatt, Gewürzlorbeer; *Italian:* alloro poetico, foglia di alloro, lauro poetico, lauro regale; *Japanese:* gekkeei ju; *Pilipino:* laurel, paminta dahon; *Portuguese:* loureiro; *Russian:* lawr, lawrowyj list, lawr blagorodnyj; *Spanish:* laurel, hoja de laurel; *Turkish:* defne ağ

Usage ▶ spice, condiments

Parts Used ▶ leaf, fruit

Distribution ▶ E Mediterranean region: Greece, C, E Mediterranean: Turkey, also cultivated

Agkül et al. 1989; Alberts/Muller 2000; Anac 1986; Bärtels 1997; Berger 3, 1952; Bois 1934; Borges et al. 1992; Bourton 1968; Braun/Meier 2002; Bremness 2001; Caredda et al. 2002; Cheers 1997; Clair 1961; Davidson 1999; Diaz-Maroto et al. 2002b; Dudtschenko et al. 1989; Duke et al. 2003; Erhardt et al. 2002; Farrell 1985; Guzman/Siemonsma 1999; Hager 4, 1992; Hanelt 2001; Heeger 1956, 1992; Hiller/Melzig 1999; Hoffmann et al. 1992; Hogg et al. 1974; Hokwerda et al. 1982; Hoppe 1949; Melchior/Kastner 1974; Morris/Mackley 1999; Nigam et al. 1960; Peter 2001; Pino et al. 1993b; Pochljobkin 1974, 1997; Pruthi 1976; Putievsky et al. 1984; Riaz et al 1989; Rosen-garten 1969; Sakar/Engelskowe 1985; Schönfelder 2001; Schultze-Motel 1986; Seidemann 1993c; Seidemann/Siebert 1987; Sfikas 1994; Siewek 1990; Small 1997; Staesche 1972; Tainter/Grenis 1993; Täufel et al. 1993; Teuscher 2003; Uphof 1968; Wyk et al. 2004; Zeven/de Wet 1982; Zola et al. 1977

Laurus quixos *Lam.*

▶ *Ocotea quixos (Lam.) Kosterm.*

Laurus undulata *Mill.*

▶ *Laurus nobilis L.*

LAVANDULA L. - Lavender - Lamiaceae (Labiatae)

Lavandula angustifolia *Mill.*

Synonyms ▶ *Lavandula officinalis* Chaix. *Lavandula spica* L., *Lavandula vera* DC.

Common Names ▶ English lavender, lavender; *Arabic:* khuzama; *Chinese:* xun yi cao; *French:* lavande véritable, lavande vrai; *German:* Lavande, Kleiner Lavendel, Echter Lavendel, Narde; *Italian:* lavanda; *Portuguese:* alfazema; *Russian:* lavanda, lavanda aptetschnaja, zvetnaja trava; *Spanish:* espliego común, flor de lavanda; *Turkish:* lavanta çiçeği

Usage ▶ spice, especially in Italian and France cuisine, flavoring; **product:** essential oil (lavender oil)

Parts Used ▶ flower

Distribution ▶ S Europe: France, Spain, native Crimea; also cultivated in warm countries of Europe, especially France

Note ▶ All the cultivars of common lavander are selections of ssp. *angustifolia*.

Agnes/Teisseire 1984; Aichele/Schwegler 4, 1995; Bärtels 1997; Benzinger 1986; Berger 1, 1949; Bilgri/Adam 2000; Boelens 1995; Bournot 1968; Bremness 2001; Charalambous 1994; Chaytor 1937; Cheers 1997; Clair 1961; Clebsch 1997; Coiciu/Racz (no year); Davidson 1999; Dudtschenko et al. 1989; Erhardt et al. 2002; Fleischhauer 2003; Guenther 1954; Hager 5, 1993; Heeger 1956; Hiller/Melzig 1999; Hohmann et al. 2001; Hoppe 1949; Lalande 1984; Lewington 1990; Meunier 1989, 1992; Ognyanov 1983/94; Oyen/Dung 1999; Pochljobkin 1974, 1977; Pschyrembel 1998; Roth/Kormann 1997; Schönfelder 2001; Schultze-Motel 1986; Seidemann 1993c; Seidemann/Siebert 1987; Sharma 2003; Shiva et al. 2002; Siewek 1990; Small 1997; Täufel et al. 1993; Teuscher 2003; Tucker 1986; Tucker/Hensen 1985; Ubillos 1989; Uphof 1968; Wiersema/León 1999; Wyk et al. 2004; Zeven/de Wet 1982

Lavandula dentata *L.*

Common Names ▶ French lavender; *German:* Französischer Lavendel, Zahnlavendel

Usage ▶ flavouring; **product:** essential oil

Parts Used ▶ flower

Distribution ▶ SW Europe: Spain, Balearic Islands, N Africa; native in Portugal, Italy, Sicily

Erhardt et al. 2002; Hanelt 2001; Small 1997

Lavandula hybrida *Reverchon*

Common Names ▶ lavandin, spike lavender; *German:* Lavender, Lavandine

Usage ▶ flavoring; **product:** essential oil (lavandin oil)

Parts Used ▶ fresh flower

Distribution ▶ Mediterranean regions; cultivated: especially S France, Spain, England

Note ▶ Hybrid from *Lavandula angustifolia* Mill. x *Lavandula latifolia* Medik.

Cheers 1997; Hanelt 2001; Hiller/Melzig 1999; Meunier 1989, 1992; Small 1997

Lavandula incana *Salisb.*

▶ *Lavandula stoechas L.*

Lavandula latifolia *Medik.*

Synonyms ▶ *Lavandula spica* auct., non L.
Common Names ▶ broadleaf lavender, spike lavender, spikenard, broad-leaf lavender; *Dutch:* spijk; *French:* spic, aspic; *German:* Großer Lavendel, Großer Speik, Speik-Lavendel, Spikenarde
Usage ▶ spice, flavoring; **product:** essential oil (spike lavender oil)
Parts Used ▶ flower
Distribution ▶ Europe: Iberia, France, Croatia

Benzinger 1986; Boelens 1986; Ckebsch 1997; Erhardt et al. 2002; Hanelt 2001; Heeger 1956; Hiller/Melzig 1999; Meunier 1989, 1992; Pascual et al. 1983, 1989; Roth/Kormann 1997; Schönfelder 2001; Schultze-Motel 1986; Small 1997; Tucker/Hensen 1985; Usher 1968; Zeven/de Wet 1982

 Lavandula officinalis *Chaix.*

▶ *Lavandula angustifolia Mill.*

 Lavandula spica *auct., non L.*

▶ *Lavandula latifolia Medik.*

 Lavandula spica *L.*

▶ *Lavandula angustifolia Mill.*

 Lavandula stoechadensis *St.-Lag.*

▶ *Lavandula stoechas L.*

 Lavandula stoechas *L.*

Synonyms ▶ *Lavandula incana* Salisb., *Lavandula stoechadensis* St.-Lag., *Stoechas arabica* Garsault
Common Names ▶ Arabian lavender, French lavender, Italian lavender, Spanish lavender; *Arabic:* hal-hal, meharga; *French:* stoechas arabique; *German:* Schopf-Lavendel, Welscher Lavendel, Arabischer Lavendel; *Portuguese:* alfazema; *Spanish:* romero santo
Usage ▶ flavoring; **product:** essential oil

Parts Used ▶ dry flower and herb
Distribution ▶ Mediterranean regions, C Italy, Dalmatia, N Africa; wild and cultivated in Spain; native S Australia

Hanelt 2001; Heeger 1956; Hepper 1992; Roth/Kormann 1997; Schultze-Motel 1986; Small 1997; Uphof 1968

 Lavandula vera *DC.*

▶ *Lavandula angustifolia Mill.*

LECANIODICUS Planch. ex Benth. - Sapindaceae

 Lecaniodicus cupanioides *Planch.*

Common Names ▶ *Ivory coast:* kringa, sataga
Usage ▶ flavoring (of water)
Parts Used ▶ flower
Distribution ▶ W Africa: Ghana, Ivory Coast, Nigeria, Sierra Leone, Togo

Ayensu 1978; Burkill 5, 2000; Uphof 1968

LEDUM L. - Labrador Tea - Ericaceae

 Ledum palustre *L.*

Synonyms ▶ *Rhododendrum palustre* (L.) Kron et Judd
Common Names ▶ wild rosemary, Labrador tea, marsh tea; *French:* lède, lédier, bois de savane, lédum des marais; *German:* Sumpfporst; Wilder Rosmarin, Brauerkraut, Labradorkraut, Porschkraut, Porst; *Russian:* barul'nik
Usage ▶ spice
Parts Used ▶ leaf
Distribution ▶ E Asia, Siberia, Russia Far East, E and W Canada, subarctic America, E, C and N Europe

Alberts/Muller 2000; Benoni 2000; Berger 1, 1950; 4, 1954; Erhardt et al. 2002; Erichsen-Brown 1989; Greve 1938; Hager 3, 1992; Heeger 1956; Hiller/Melzig 1999; Hoppe 1949; Lewington 1990;

Rätsch 1998; Sandermann 1980; Schönfelder 2001; Seidemann 1993b, c; Tattje/Bos 1981; Turova et al. 1987; Uphof 1968; Wiersema/León 1999; Wyk et al. 2004

LENTINUS Fr. - Polyporaceae

Lentinus edodes (Berk.) Sing

Synonyms ▶ *Agaricus edodes* Berk. *Cellus edodes* (Berk.) Ito et Imai; *Cortinellus shiitake* P. Henn., *Lepoita shiitake* Tanaka,
Common Names ▶ *Japanese:* shii-take; *German:* Chinapilz, Japanpilz, Pasaniapilz, Shii-take (Pilz)
Usage ▶ spice
Parts Used ▶ mushroom
Distribution ▶ E Asia: Japan, China to Indochina, also cultivated in many countries

Chen/Ho 1986; Hanelt 2001; Laatsch 1992; Liese 1948; Mansfeld 1962; Michael et al 1985; Morita/Kobashi 1966; Schultze-Motel 1986; Seidemann 1993d;

Leontodon taraxacum L.

▶ *Taraxacum officinale* agg. F.H. Wigg.

Leontodon vulgare Lam.

▶ *Taraxacum officinale* agg. F.H. Wigg.

LEOPOLDINIA Mart. - Araceae (Palmae)

Leopoldinia major Wall.

Common Names ▶ *Brazil (Portuguese):* iará, iará-uaçu, jará-açu, palmeira-iará
Usage ▶ spice
Parts Used ▶ ash from fruits (rarely)
Distribution ▶ S America: Brazil
Note ▶ Certain Indian tribes as a substitute for salt.

Uphof 1968

Lepianthes umbellatum (L.) Raf.

▶ *Piper umbellatum* (L.) Miq.

LEPIDIUM L. - Cress - Brassicaceae (Cruciferae)

Lepidium africanum (Burm.f.) DC. ssp. divaricatum (Ait.) Jonsell

Common Names ▶ pepper grass, pepper wort; *German:* Afrikanische Kresse; *S Africa:* kanariesaadgra, peper bossie, sterkkos, sterkgras
Usage ▶ spice
Parts Used ▶ green seed
Distribution ▶ tropical W Africa: Namibia

v. Koenen 1996

Lepidium campstre (L.) R. Br.

Common Names ▶ field cress; *German:* Feldkresse
Usage ▶ pot-herb, spice (rarely)
Parts Used ▶ fresh herb, seed
Distribution ▶ Europe, Turkey, Caucasus, native in N America

Erhardt et al. 2002; Fleischhauer 2003; Loch 1993

Lepidum densiflorum Schrad.

Synonyms ▶ *Lepidium meyenii* Walp
Common Names ▶ common pepper grass, prairie pepper grass, Peruvian ginger; *German:* Dichtblütige Kresse; *Russian:* kress, klopovnik; *Spanish:* maca
Usage ▶ pot-herb
Parts Used ▶ fresh herb
Distribution ▶ N America, native C, E Europe

Aichele/Schwegler 3, 1995; Ehrhardt et al. 2002; Erichsen-Brown 1989; Fleischhauer 2003; Hager 5, 1993; Loch 1993; Small 1997

Lepidium draba L.

▶ *Cardaria draba* (L.) Drev.

Lepidium fremontii S. Wats

Common Names ▶ desert pepperweed
Usage ▶ flavoring
Parts Used ▶ seed
Distribution ▶ SW USA
Note ▶ Used by the Indians of Arizona for flavoring.

Uphof 1968

Lepidium latifolium L.

Synonyms ▶ *Crucifera latifolia* E.H.L. Krause; *Nasturtium latifolium* Gillet et Magne; *Nasturtium latifolium* (L.) O. Kuntze
Common Names ▶ dittander, perennial peppergrass, poor man's pepper; *French:* grande passerage; *German:* Breitblättrige Kresse, Pfefferkresse, Breitblättriges Pfefferkraut
Usage ▶ pot-herb, spice
Parts Used ▶ fresh herb
Distribution ▶ Europe, Turkey, Caucasus, Iran, W Siberia, C Asia, Himalayas, Tibet, Morocco, Egypt

Aichele/Schwegler 3, 1995; Berger 2, 1950; Dudtschenko et al. 1989; Erhardt et al. 2002; Fleischhauer 2003; Hager 5, 1993; Hanelt 2001; Hiller/Melzig 1999; Pursglove 1968; Schultze-Motel 1986; Small 1997; Täufel et al. 1993; Zeven/de Wet 1982

Lepidium oleraceum Forsk.f.

Common Names ▶ cock's curry grass
Usage ▶ pot-herb
Parts Used ▶ fresh herb
Distribution ▶ S New Zealand

Uphof 1968

Lepidium sativum L.

Synonyms ▶ *Crucifera nasturtium* E.H.L. Krause

Common Names ▶ garden cress, land cress, pepper grass, pepperwort; *Arabic:* habb el-yashad, hab erche; carabu; *Chinese:* jia du xing cai; *Dutch:* sterkers, tuinkers; *French:* cresson alénois, cresson de jardin, cressonette, passerage cultivée; *German:* Gartenkresse, Pfefferkraut, Tellerkresse; *Hindi:* cansur, chausaur, halim; *India:* asadio; *Italian:* lepidio, crescione inglese, crescione di giardino, agretto, nastuerzu ortense; *Japanese:* gaaden kuresu; *Russian:* sabowij kres, kresssalat, kress posevnoj, peretschnik, chreniza, podchrennik; *Sanskrit:* candrasura, candrika; *Spanish:* berro alenois, berro de huerta, berro de tierra, lepidios, mastuerzo de huerta
Usage ▶ pot-herb
Parts Used ▶ fresh leaf
Distribution ▶ Egypt, Sudan, Ethiopia, Arabia, Turkey, Iran, Iraq, Pakistan, Syria, W Himalayas, cultivated worldwide in gardens

Bendel 2002; Berger 4, 1954; Bilgri/Adam 2000; Bremness 2001; Dhar/Dhar 2000; Dudtschenko et al. 1989; Erhardt et al. 2002; Gil/Mac Leod 1980; Hager 5, 1993; Hanelt 2001; Heeger 1956; Herrmann 1997a; Hiller/Melzig 1999; Hondelmann 2002; Jansen 1981; Leroy/Gillet 1964; Pochljobkin 1974, 1977; Schönfelder 2001; Schultze-Motel 1986; Small 1997; Täufel et al. 1993; Teuscher 2003; Tucker 1986; Winter/Willeke 1953a; Zeven/de Wet 1982

Lepidium virginicum L.

Common Names ▶ Virginia pepperweed; *German:* Virginische Kresse
Usage ▶ pot-herb (rarely)
Parts Used ▶ fresh herb
Distribution ▶ N America, native in Europe

Erhardt et al. 2002; Fleischhauer 2003; Small 1997

Lepiota shiitake Tanaka

▶ *Lentinus edodes* (Berk.) Sing.

LEPTOSPERMUM J.R. Forst et G. Forst. - Tea Tree - Myrtaceae

 Leptospermum citratum *Challinor, Cheel et Penfold*

Synonyms ▶ *Leptospermum petersonii* F.M. Bailey
Common Names ▶ lemon scented tea, tea tree; *French:* leptosperme; *German:* Zitronenmyrte; *Russian:* leptospermum
Usage ▶ spice, flavoring
Parts Used ▶ leaf
Distribution ▶ Guatamala, Kenya, Zaire, Georgia, Australia: NS Wales, Queensland, frequently cultivated

Cheers 1997; Erhardt et al. 2002; Hanelt 2001; Morrison 1958; Uphof 1968; Zeven/Zhukovsky 1975

 Leptospermum petersonii *F.M. Bailey*

 *Leptospermum citratum* Challinor

LEUCANTHEMUM Mill. - Ox-eye Daisy - Ateraceae (Compositae)

 Leucanthemum vulgare *Lam.*

Synonyms ▶ *Chrysanthemum leuvanthemum* L.
Common Names ▶ dog daisy, moon daisy, ox eye daisy; *French:* grande marguerite, leucanthème; *German:* Wiesenmargerite, Wiesen-Wucherblume; *Russian:* popewnik, niwjanik abyknowennyj, romanška lugiwaja
Usage ▶ pot-herb
Parts Used ▶ young leaf
Distribution ▶ Europe, Turkey, Caucasus, W, E Siberia, Amur, Sakhalin, Kamchatka, native N America

Clebsch 1997; Erhardt et al. 2002; Fleischhauer 2003; Hanelt 2001; Schnelle 1999; Uphof 1968; Wiersema/León 1997

LEUCAS R.Br. - Lamiaceae (Labiatae)

 Leucas aspera *(Willd.) Link*

Synonyms ▶ *Phlomis aspera* Willd.
Common Names ▶ rough leucas; *Chinese:* feng chao cao; *German:* Rauhes Brandkraut; *India:* kuba
Usage ▶ pot-herb
Parts Used ▶ plant
Distribution ▶ China, India: Gujarat and Rajasthan

Arora/Pandey 1996; Misra et al. 1992, 1993; Pradhan et al. 1990; Seshadr/Nambiar 2003; Singh 2001

 Leucas cephalotes *(Koen. ex Roth) Spr.*

Common Names ▶ *German:* Großköpfiges Brandkraut; *Hindi:* dhurpi-sag, goma, motapati
Usage ▶ pot-herb
Parts Used ▶ leaf, young shoots
Distribution ▶ India

Arora/Pandey 1996; Chauhan 1999; Sharma 2003; Singh 2001

 Leucas clarkei *Hook.*

Usage ▶ pot-herb
Parts Used ▶ leaf
Distribution ▶ India: in Chota Nagpur in Bihar

Arora/Pandey 1996; Singh 2001

 Leucas lanata *Wall. ex Benth.*

Common Names ▶ downy or wooly leucas; *German:* Wolliges Brandkraut
Usage ▶ pot-herb
Parts Used ▶ leaf, young shoots
Distribution ▶ India: in the plains and hills

Arora/Pandey 1996; Sharma 2003; Singh 2001

Leucas lavandulaefolia *Ress.*

Synonyms ▸ *Leucas linifolia* (Roth) Spreng., *Phlomis linifolia* Roth
Common Names ▸ line leaf leucas; *Chinese:* xian ye bai rong cao; *German:* Lavendelblättriges Brandkraut; *Hindi:* guma, halkusa, kumbha
Usage ▸ pot-herb
Parts Used ▸ leaf
Distribution ▸ India

Arora/Pandey 1996; Oyen/dung 1999; Singh 2001; Terra 1966

 Leucas linifolia *Spreng.*

 *Leucas lavandulaefolia Ress.*

 Leucas maritinicensis *(Jacq.) Ait.f.*

Common Names ▸ bobbin weed, tumble weed, ovate-leaf leucas; *German:* Martinisches Brandkraut; *Portuguese:* catinga-de-mulata, cordao-de-frade, cordão-de-sao-francisco, paude-praga
Usage ▸ pot-herb
Parts Used ▸ leaf
Distribution ▸ India, trop Africa: Ivory Coast, Nigeria, Senegal, Upper Volta

Arora/Pandey 1996; Burkill 3, 1995; Dalziel 1957; v. Koenen 1996; Mors et al. 2000; Singh 2001

 Leucas mollissima *Wall. ex Benth.*

Common Names ▸ white felt leucas; *Chinese:* bai rang cao; *German:* Weiches Brandkraut
Usage ▸ pot-herb (by the Santhals)
Parts Used ▸ leaf
Distribution ▸ India: in the plains and hills

Arora/Pandey 1996; Singh 2001

 Leucas zeylanica *(L.) R. Br.*

Synonyms ▸ *Phlomis zeylanica* L.
Common Names ▸ Ceylon leucas, admiration herb; *German:* Ceylanisches Brandkraut; *Chinese:* zhou mian cao, feng wo cao; *Malaysian:* ketumbak, ketumbit
Usage ▸ condiment (in Bali), flavoring
Parts Used ▸ leaf
Distribution ▸ India in the plains and hilly areas of Assam and Peninsular region

Arora/Pandey 1996; Oyen/Dung 1999; Singh 2001; Terra 1966; Uphof 1968

 Leucocasia gigantea *(Blume ex Hassk.) Schott*

▸ *Colocasia gigantea (Blume ex Hassk.) Hook.*

LEVISTICUM Hill. - Lovage - Apiaceae (Umbelliferae)

 Levisticum officinale *W.D.J. Koch*

Synonyms ▸ *Levisticum persicum* Freyn, *Ligusticum levisticum* L.
Common Names ▸ lovage, garden lovage, bladder seed; *Chinese:* dang gui; *Dutch:* lavas; *French:* livèche, ache de Montagne, céleri vivace; *German:* Liebstöckel, Maggikraut, Großer Eppich, Sauerstockkraut, Stecklaub, Suppenlob; *Italian:* levistico, ligustico, sedano di montagna; *Korean:* me-na-ri; *Portuguese:* levístico; *Russian:* srja, dudotschnik, dudtschataja, traba, ljubistok; ljubm, saborinam; *Spanish:* apio de montaña, levistico, ligústico;
Usage ▸ spice; **product:** essential oil (lovage oil)
Parts Used ▸ herb, root, seed
Distribution ▸ Iran, native in Europe, N America

Aichele/Schwegler 3, 1995; Bärtels 1997; Berger 3, 1952; 4, 1954; Bremness 2001; Clair 1961; Coiciu/Racz (no year); Cu et al. 1990a; Davidson 1999; Dudtschenko et al. 1989; Erhardt et al. 2002; Fleischhauer 2003; Hager 5, 1993; Hanelt 2001; Heeger 1956; Hiller/Melzig 1999; Hohmann et al. 2001; Hoppe 1949; Melchior/Kastner 1974; Pochljobkin 1974, 1977; Pruthi 1976; Roth/Kormann 1997; Schönfelder 2001; Schultze-Motel 1986; Seidemann 1993c; Seidemann/Siebert 1987; Small 1997; Staesche 1972; Täufel et al. 1993; Teuscher 2003; Toulemonde/Noleau 1988; Tucker 1986; Uhl 2000; Uphof 1968; Wüstenfeld/Haensel 1964; Wyk et al. 2004; Zeven/de Wet 1982

Levisticum persicum *Freyn*

▶ *Levisticum officinale W.D.J. Koch*

 Lexarza funebres *La Llave*

▶ *Quararibea funebris (La Llave) Vischer*

LIATRIS Gaertn. ex Schreb. - Button Snakerroot - Asteraceae (Compositae)

 Liatris odoratissima *Willd.*

Common Names ▶ blazing stare, button snakeroot, vanillaroot; *French:* liatride; *German:* Duft-Prachtscharte, Duft-Hirschzunge, Vanillewurzel; *Russian:* liatris
Usage ▶ spice, used in the tobacco industry
Parts Used ▶ leaf
Distribution ▶ USA: Virginia to Florida and Louisana

Berger 2, 1950; Hiller/Melzig 1999; Hoppe 1949; Seidemann 1993c

 Licaria puchury-major *(Mart.) Kosterm.*

▶ *Ocotea puchury-major Mart.*

 Licaria quixos *(Lam.) Kosterm.*

▶ *Ocotea quixos (Lam.) Kosterm.*

LIGUSTICUM L. - Lovage - Apiaceae (Umbelliferae)

 Ligusticum acutilobum *Sieb. et Zucc.*

▶ *Angelica acutiloba (Sieb. et Zucc.) Kitag*

 Ligusticum cuminum *(L.) Crantz*

▶ *Cuminum cyminum L.*

 Ligusticum levisticum *L.*

▶ *Levisticum officinale W.D.J. Koch*

 Ligusticum monnieri *Calest.*

▶ *Cnidium monnieri (L.) Cuss. ex Juss.*

 Ligusticum mutellina *(L.) Crantz*

Synonyms ▶ *Meum mutellina* Gaertn.
Common Names ▶ alpine lovage; *German:* Alpen-Mutterwurz
Usage ▶ flavoring
Parts Used ▶ root
Distribution ▶ C, S Europe: especially Alpine regions

Aichele/Schwegler 3, 1995; Erhardt et al. 2002; Fleischhauer 2003; Hanelt 2001; Schönfelder 2001; Uphof 1968

 Ligusticum scoticum *L.*

Common Names ▶ Scotch lovage, northern lovage; *German:* Schottischer Liebstock, Schottische Mutterwurz; *Japanese:* maruba toki; *Russian:* ligusticum scholandskij
Usage ▶ pot-herb (tastes like celery)
Parts Used ▶ herb
Distribution ▶ Far Eastern Russia, Japan, Korea

Erhardt et al. 2002; Facciola 1990; Hanelt 2001; Mabberly 1997; Schultze-Motel 1986; Small 1997; Tucker 1986; Uphof 1968

 Ligusticum tenuissimum *(Nakai) kilag*

▶ *Angelica tenuissima Nakai*

LILIUM L. - Lily - Liliaceae

 Lilium auratum *Lindl.*

Synonyms ▶ *Lilium wittei* Suring

Common Names ▶ gold-banded lily, golded-rayed lily of Japan, mountain lily; *French:* lis doré du Japon; *German:* Goldband-Lilie; *Japanese:* yama-yuri; *Russian:* lilija zolotistaja
Usage ▶ spice, flavoring
Parts Used ▶ cooked bulbs
Distribution ▶ Japan, Korea, also cultivated in Japan

Hanelt 2001; Schultze-Motel 1986; Uphof 1968; Zeven /de Wet 1982

Lilium concolor *Salisb.*

Synonyms ▶ *Lilium sinicum* Lindl.
Common Names ▶ Japanese red star lily, (morning) star lily; *Chinese:* hung hua tsai, hung pai ho, shan tan; *German:* Gleichfarbige Lilie, Morgenstern-Lilie; *Japanese:* ko-hime-yuri; *Russian:* lilija odnocvetnaja
Usage ▶ flavoring
Parts Used ▶ flower, bulb
Distribution ▶ China: Hunan, Hupeh, Yunnan; cultivated worldwide as an ornamental plant

Hanelt 2001; Schultze-Motel 1986; Seidemann 1993c

Lilium lancifolium *Thunb.*

Synonyms ▶ *Lilium tigrinum* Ker-Gawl.
Common Names ▶ devil lily, tiger lily; *Chinese:* chuan tan; *German:* Tigerlilie, Große Türkenbundlilie; *Japanese:* oni-yuri; *Korean:* chamnari
Usage ▶ flavoring
Parts Used ▶ flower, bulb
Distribution ▶ Japan, Korea, E China; cultivated worldwide as an ornamental plant

Erhardt et al. 2002; Hammer 1997; Heeger 1956; Hiller/Melzig 1999; Seidemann 1993c; Small 1997; Uphof 1968; Zeven/de Wet 1982

Lilium sinense *Lindl.*

▶ *Lilium concolor Salisb.*

Lilium tigrinum *Ker-Gawl.*

▶ *Lilium lancifolium Thunb.*

Lilium wittei *Suring*

▶ *Lilium auratum Lindl.*

LIMNOPHILA R.Br. - Scrophulariaceae

Limnophila aromatica *(Lam.) Merrill*

Synonyms ▶ *Ambulia aromatica* Lam., *Limnophyla punctata* Bl.
Common Names ▶ *German:* Reisfeldpflanze; *Vietnamese:* phak khayaeng, rau ngó
Usage ▶ pot-herb, flavoring
Parts Used ▶ leaf, herb
Distribution ▶ SE Asia: Malaysia, cultivated: Vietnam, S USA, Hawaii

Arora/Pandey 1996; Hanelt 2000; Kuebal/Tucker 1988; Small 1997; Wiersema/León 1999

Limnophila conferta *Benth.*

Common Names ▶ *Hindi:* munganari, muchriara
Usage ▶ pot-herb
Parts Used ▶ leaf
Distribution ▶ India

Arora/Pandey 1996

Limnophila gratioloides *R.Br*

▶ *Limnophila indica (L.) Druce*

Limnophila indica *(L.) Druce*

Synonyms ▶ *Limnophila gratioloides* R.Br., *Limnophila racemosa* Benth.

Common Names ▶ *German:* Indische Limnophilie; *Hindi:* ambuja, kuttra
Usage ▶ pot-herb (in Himalayas)
Parts Used ▶ leaf
Distribution ▶ India, tropical and subtropical Asia, Australia, Africa

Arora/Padney 1996; Erhardt et al. 2002

Limnophila punctata *Bl.*

▶ *Limnophily aromatica*

Limnophila racemosa *Benth.*

▶ *Limnophila indica (L.) Druce*

Limnophila roxborghii *auct., non G. Don.*

▶ *Limnophila rugosa (Roth) Merr.*

Limnophila rugosa *(Roth) Merr.*

Synonyms ▶ *Herpestis rugosa* Roth, *Limnophila roxborghii* auct., non G. Don.
Common Names ▶ *Indonesian:* hades, selaseh ayer, selaseh banyu; *Pilipino:* kalaoo, tala, tara-tara; *Thai:* kachom, om kop; *Vietnamese:* rau om
Usage ▶ condiment; herb is used to perfume
Parts Used ▶ leaf, herb
Distribution ▶ India, S China, Ryukyu Island, SE Asia to Fiji and Samoa
Note ▶ The leaves and stem smell of anise.

Guzman/Siemonsma 1999; Oyen/Dung 1999

Limnophila roxborghii *auct., non G. Don*

▶ *Limnophila regusa (Roth) Merr.*

Limonia acidissima *Houtt.*

▶ *Citrus aurantiifolia (Christm et Panz.) Swingle*

Limonia aurantiifolia *Christm. et Panz.*

▶ *Citrus aurantiifolia (Christm. et Panz.) Swingle*

Limonia crenulata *Roxb*

▶ *Hesperthusa crenulata (Roxb.) Roem.*

Limonia trichocarpa *Hance*

▶ *Poncirus trifoliata (L.) Raf.*

Limonia trifolia *Burm*

▶ *Triphasia trifolia (Burm.f.) P. Wilson*

Limonia trifoliata *L.*

▶ *Triphasia trifolia (Burm.f.) P. Wilson*

LINDERA Thunb. - Lauraceae

Lindera benzoin *(L.) Bl.*

Synonyms ▶ *Benzoin aestivale* (L.) Nees, *Laurus aestivalis* L., *Laurus benzoin* L.
Common Names ▶ Benjamin bush, spice bush, wild allspice; *French:* benjoin; *German:* Wohlriechender Fieberstrauch, Gewürzstrauch; *Russian:* bensojnoe
Usage ▶ spice (rarely used of the Cherokee Indians)
Parts Used ▶ leaf
Distribution ▶ E Canada: Ontario; E USA and Kansas, Texas, Florida

Bois 1934; Erhardt et al. 2002; Duke et al. 2003; Tucker 1986; Tucker et al. 1994; Tull 1999; Wofford 1983

LINUM L. - Flax - Linaceae

Linum arvense Neck

> *Linum usitatissimum* L.

Linum humile Mill.

> *Linum usitatissimum* L.

Linum sativum Hasselqu.

> *Linum usitatissimum* L.

Linum usitatissimum L.

Synonyms ▶ *Linum arvense* Neck, *Linum humile* Mill., *Linum sativum* Hasselqu., *Linum utile* Salisb.
Common Names ▶ common flax, crown flax, cultivated flax, flax, linseed; *Arabic:* kettan, ketten; *Chinese:* ya ma zi, chih ma; *Dutch:* vlas; *French:* lin, lin oléifère; grain de lin; *German:* Dreschlein, Flachs, Leinsaat, Öllein, Saat-Lein, Steppenflachs; *Italian:* lino, seme de lino; *Japanese:* ama; *Korean:* ama; *Russian:* ljon kudrjasch; *Spanish:* lino, linaza
Usage ▶ spice, roasted (rarely)
Parts Used ▶ roasted seed
Distribution ▶ cultivated in many countries, especially Canada; native in Iberia, France

Erhardt et al. 2002; Hanelt 2001; Heeger 1956; Hepper 1992; Hiller/Melzig 1999; Hoppe 1949; 1, 1975; Linke 1983; Schultze-Motel 1986; Turora 1987; Usher 1968; Wyk et al. 2004; Zeven/de Wet 1982

Linum utile Salisb.

> *Linum usitatissimum* L.

LIPPIA L. - Verbenaceae

Kintzios 2002

Lippea adoënsis Hochst.

Synonyms ▶ *Lippea grandiflora* Mart. et Schauer
Common Names ▶ Gambia tea bush; *French:* verveine d'Afrique; *German:* Gambia-Teestrauch
Usage ▶ pot-herb, flavoring
Parts Used ▶ leaf
Distribution ▶ tropical W Africa
Note ▶ Used as a tea substitute.

Hanelt 2001; Rabaté 1938; Schultze-Motel 1986; Uphof 1968; Walker 1953; Zeven/de Wet 1982

Lippia alba (Mill.) N.E. Br. ex Britt. et Wils.

Synonyms ▶ *Lantana alba* Mill., *Lippea asperifolia* A. Rich.
Common Names ▶ white oregano; *Brazil (Portuguese):* alecrim do campo, chá do tabulaeiro, cidrilha, salsa-limão, salva branca; *German:* Anisverbene, Weißer Oregano, Weißer Zitronenstrauch; *Spanish:* aguadiente de Espana, anise verbena; menta americana, salvia americana; anis de Espana, quita dolor
Usage ▶ spice; **product:** essential oil
Parts Used ▶ leaf, herb
Distribution ▶ in temperate and tropical areas of S America; native, and cultivated

Bahl et al. 2002; Berlin et al. 1974; Craveiro et al. 1981; Dellacassa et al. 1990; Frighetto et al. 1998; Fun/Svendsen 1990b; Gomes et al. 1993; Gurgol do Vale et al. 2002; Hanelt 2001; Matos et al. 1996; Mors et al. 2000; Rajendran/Daniel 2002; Retamar 1994; Schultze-Motel 1986; Siani et al. 2002; Small 1997; Villamar et al. 1994; Zoghbi et al. 1998

Lippea alba Mill.

> *Lippea alba (Mill.) N.E. Br. ex Britt. Et Wils.*

Lippea amentacea M.E. Jones

▶ *Lippea graveolens* H.B.K.

 Lippea asperifolia A. Rich.

▶ *Lippea alba* (Mill.) N.E. Brown

 Lippea asperifolia L.C. Rich.

▶ *Lippea javanica* (Burm.f.) Spreng.

 Lippia berlandieri Schauer

Common Names ▶ epazote; *German:* Berlandier-Oregano
Usage ▶ spice, seasoning
Parts Used ▶ leaf, herb
Distribution ▶ S America, Mexico

Hager 5, 1993; Seidemann/Siebert 1987; Uphof 1968; Yousif et al. 2000

 Lippia citriodora (Lam). Humb.

▶ *Lippea triphylla* (L'Hérit.) Kuntze

 Lippia dulcis Trev.

Synonyms ▶ *Phyla scaberrima* (Juss. ex Pers.) Mold.
Common Names ▶ Mexican lippea, yerba dulce; *German:* Aztekisches Süßkraut, Süßer Oregano, Süßer Zitronenstrauch, Mexikanisches Lippiakraut; *Mexico:* yerba dulce; *Spanish:* hierba dulce oregano, orozuz
Usage ▶ spice, flavoring
Parts Used ▶ leaf, herb
Distribution ▶ C America: Mexico to Panama

Berger 4, 1954; Craveiro et al. 1981; Erhardt et al. 2002; Hager 5, 1993; Hiller/Melzig 1999; Hoppe 1949; Lewington 1990; Rosengarten 1969; Schultze-Motel 1986; Small 1997; Uphof 1968; Villamar et al. 1994

 Lippea grandiflora Mart. et Schauer

▶ *Lippea adoënsis* Hochst.

 Lippia graveolens H.B.K.

Synonyms ▶ *Lippea amentacea* M.E. Jones
Common Names ▶ American oregano, Mexican oregano, Mexican sage, mintweed; *French:* origan marjolaine; *German:* Amerikanischer Oregano, Mexikanischer Oregano; *Spanish:* hierba dulce orogano, orégano cimarron
Usage ▶ spice
Parts Used ▶ leaf, herb
Distribution ▶ Mexico, El Salvador, S, C USA, MesoAmerica

Craveiro et al. 1981; Guzman/Siemonsma 1999; Hager 5, 1993; Melchior/Kastner 1974; Pino et al. 1989b, 1990, 1994; Rosengarten 1969; Seidemann 1993c; Seidemann/Siebert 1987; Small 1997; Tainter/Grenis 1993; Teuscher 2003; Tucker 1986; Tull 1999; Uphof 1968; Vernin et al. 2001; Villamar et al. 1994

 Lippia javanica (Burm.f.) Spreng.

Synonyms ▶ *Verbena javanica* Burm.f.
Synonyms ▶ *Lippea asperifolia* L.C. Rich., *Lippea scabra* Hochst.
Common Names ▶ Java thyme, fever tea, wild sage; *German:* Javanischer Thymian
Usage ▶ flavoring; **product:** essential oil
Parts Used ▶ leaf, herb
Distribution ▶ Africa, Java

Rajendran/Daniel 2002; Wiersema/León 1999

 Lippia lupulina Cham.

Common Names ▶ hop thyme; *German:* Hopfenthymian
Usage ▶ spice; **product:** essential oil
Parts Used ▶ leaf, herb
Distribution ▶ S America: Brazil

Gracas et al. 2002

◻ Lippea graveolens, flowering

 Lippia micromera *Schauer*

Common Names ▸ Spanish thyme, Puerto Rico oregano; *German:* Spanischer Thymian; *Spanish:* orégano del pais
Usage ▸ spice
Parts Used ▸ leaf, herb
Distribution ▸ Caribbean, Mexico, S America: Venzuela

Erhardt et al. 2002; Hager 5, 1993; Melchior/Kastner 1974; Seidemann 1993c; Schultze-Motel 1986; Small 1997; Täufel et al. 1993; Teuscher 2003; Tucker 1986; Tucker et al. 1993; Wiersema/León 1999

 Lippia multiflora *Mold.*

Common Names ▸ Gambian tea bush, *French:* thé de Gambi; *German:* Vielblütiger Organo, Gambia-Teebusch
Usage ▸ flavoring (partly with fufu and with sesam) for soups
Parts Used ▸ leaf (fresh and dried)
Distribution ▸ W Africa: the Gambia, Ghana, Guinea, Ivory Coast, Mali, Nigeria, Senegal, Sierra Leone, Togo, Upper Volta

Ayensu 1978; Burkill 5, 2000; Menut et al. 1995; Pellisier et al. 1994

 Lippea scabra *Hochst*

▸ *Lippia javanica (Burm.f.) Spreng.*

 Lippia triphylla *(L'Hérit.) Kuntze*

Synonyms ▸ *Aloysia citriodora* Ortega ex Pers., *Aloysia triphylla* (L'Hérit.) Britton, *Lippia citriodora* (Lam.) Humb., *Verbena triphylla* L'Hérit.
Common Names ▸ lemon verbena, Mexican majoram, Mexican oregano, cidron, herb Louisa; *Brazil (Portuguese):* cidrão, cidrilha, cidró, idreira, kerva, salva-limão; *French:* citron(n)elle verveine, verveine citronée, verveine des Indes, verveine odorante, verveine de Pérou, à trois feuilles, lippie, thé arabe; *German:* Mexikanischer Oregano, Echte Verbene, Punschkraut Verbenenkraut, Zitronenverbene, Zitronenstrauch; *Italian:* cedrina, erba cedrella, erba ingia, erba luigia, erba Luisa, limoncina, verbena odorosa; *Russian:* ladannik werbena limonnija; *Spanish:* hierba cidrera, hierba Luisa, kerva cidreira, verbene odorosa
Usage ▸ spice (meat, fish, pizza), condiment; **product:** essential oil
Parts Used ▸ leaf, herb
Distribution ▸ S America: Argentina, Chile, Uruguay; also cultivated in the Tropics, Spain and France

Bärtels 1997; Bellakhadar et al. 1994; Bremness 2001; Calpouzos 1954; Craveiro et al. 1981; Davidson 1999; El-Hamidi et al. 1983; Fleisher/Sneer 1982; Garland 1979; Gurib-Fakim/Brendler 2004; Hager 5, 1993; Hiller/Melzig 1999; Hoffmann et al 1992; Hoppe 1949; Kouhila et al 2001; Melchior/Kastner 1974; Mikus/Schaser 1995; Mikus et al. 1997; Mors et al. 2000; Newall et al. 1996; Rätsch 1998; Rajendran/Daniel 2002; Rosengarten 1969; Roth/Kormann 1997; Schönfelder 2001; Schultze-Motel 1986; Seidemann 1993c; Seidemann/Siebert 1987; Skaltsa/Shammas 988; Small 1977; Täufel et al. 1993; Teuscher 2003; Tucker 1986; Uhl 2000; Uphof 1968; Villamar et al. 1994; Wilkins/Madsen 1991; Wyk et al. 2004; Zeven/de Wet 1982

Liquiritia lepidota Nutt.

▶ *Glycyrrhiza lepidota (Nutt.) Pursh*

 Liquiritia officinalis Lepesch.

▶ *Glycyrrhiza glabra L.*

 Liquiritia officinarum Medik.

▶ *Glycyrrhiza glabra L.*

 Liriodendron figo Lour.

▶ *Michelia figo L.*

LITSEA Lam. - Lauraceae

 L

 Litsea citrata Bl.

▶ *Litsea cubeba (Lour.) Pers.*

 Litsea cubeba (Lour.) Pers.

Synonyms ▶ Laurus cubeba Lour., Litsea citrata Bl.
Common Names ▶ mountain pepper, pheasant-pepper; *Chinese:* bi cheng que, may chang picheng ch'ieb; *German:* Bergpfeffer; *India:* dieng si ing, mejankeri, sernam, tanghaercherking, terhilsok, zeng jil; *Javanese:* krangejan; *Vietnamese:* khao khinh, man tang, may chang, ta cham diang
Usage ▶ flavoring; **product:** essential oil and citral
Parts Used ▶ fruit
Distribution ▶ China, E Asia, India, Indochina

Bao Yipei 1995; Cheng/Cheng 1983; Gogoi 1997; Hager 5, 1993; Hanelt 2001; Kumar 2003; Lee et al. 1993; Nath et al. 1996b; Oyen/Dung 1999; Schultze-Motel 1986; Seidemann 1993c; Uphof 1968

 Litsea japonica (Thunb.) Juss.

Common Names ▶ *German:* Japanischer Bergpfeffer
Usage ▶ flavoring; **product:** essential oil and citral
Parts Used ▶ fruit
Distribution ▶ Japan, Riukiu Islands, S Korea

Cheng/Cheng 1983; Erhardt et al. 2002; Seidemann 1993c

 Litsea pipericarpa (Miq.) Kosterm.

Synonyms ▶ Lindera pipericarpa (Miq.) Boerl., *Polyadenia pepericarpa* Miq.
Common Names ▶ *Indonesian:* kulit antarsa, kulit pulaga; *Japanese:* hawa-biwa; *Malaysian:* medang serai
Usage ▶ flavoring
Parts Used ▶ fruit
Distribution ▶ Indonesia (Sumatra), Malaysian Peninsula
Note ▶ The fruits are used medicinally like cubebs (*Piper cubeba* L.f.) as a tonic.

Oyen/Dung 1999; Kostermans 1970

LOBELIA L. - Lobelia - Campanulaceae (Lobelioideae)

 Lobelia alsinoides Lam.

Synonyms ▶ Lobelia trigona Roxb.
Common Names ▶ *German:* Hain-Lobelie
Usage ▶ pot-herb
Parts Used ▶ leaf
Distribution ▶ India: in Chota Nagpur

Arora/Pandey 1996

 Lobelia trigona Roxb.

▶ *Lobelia alsinoides Lam.*

LONCHOCARPUS Kunth - Papilionaceae (Fabaceae)

 Lonchocarpus caynescens *(Schum. et Thonn.) Benth.*

Synonyms ▶ *Lonchocarpus philenoptera* Bentham, *Philenoptera laxiflora* (Guill. et Perrot) G. Roberts
Common Names ▶ African indigo, Yoruba indigo, *German:* Loncho; *Nigerian:* elu, njassi
Usage ▶ couscous spice
Parts Used ▶ leaf
Distribution ▶ C tropical Africa: Senegal to Sudan, Ethiopa, Kenya, Angola
Note ▶ Roots are an intense fish poison.

Ayensu 1978; Burkill 3, 1995; Neuwinger 1998, 1999; Uphof 1968; Wiersema/León 1999

 Lonchocarpus philenoptera *Bentham*

▶ *Lonchocarpus caynescens (Schum. et Thonn.) Benth.*

 Lophanthus rugosus *Fischer et C.A. Meyer*

▶ *Agastache rugosa (Fischer et C.A. Meyer) Kuntze*

LUCUMA Molina - Sapotaceae

Lucuma rivicoa *Gaertn.*

Synonyms ▶ *Lucuma nervosa* DC.
Common Names ▶ canistel
Usage ▶ spice (in Brazil)
Parts Used ▶ herb
Distribution ▶ C America, West Indies

Uphof 1968

LYCIUM L. - Teaplant - Solanaceae

 Lycium barbarum *L.*

Synonyms ▶ *Lycium vulgare* Dun.
Common Names ▶ barbary wolf berry, box thorn, Duke of Argyll's tea tree, matrimony vine; *Chinese:* Malaysia gou-gi-zi, ning xia gouqi; *French:* lyciet commun; *German:* Gewöhnlicher Bocksdorn, Chinesische Wolfsbeere; *Korean:* tankugijanamu
Usage ▶ pot-herb (in China);
Parts Used ▶ leaf
Distribution ▶ C China, native in Europe, Turkey, Caucasus, N Africa

D'Arcy 1986; Cheers 1997; Erhardt et al. 2002; Hager 5, 1993; Hanelt 2001; Schultze-Motel 1986

LYCOPERSICON Mill. - Tomato - Solanaceae

 Lycopersicon aethiopicum *(L.) Mill.*

▶ *Solanum aethiopicum L.*

 Lycopersicon esculentum Mill.

Synonyms ▶ *Lycopersicon lycopersicun* L., *Solanum lycopersicum* L.
Common Names ▶ tomato; *French:* tomate; *German:* Tomate; *Russian:* tomat
Usage ▶ spice: succade; **product:** catshup, chili sauces of ripe fruits
Parts Used ▶ unriped fruit
Distribution ▶ native in S America (Ecuador, Peru), also cultivated worldwide
Note ▶ Substitute for *citrus* succade ("Kandinat T") in the former German Democratic Republic (DDR, E Germany) after enzymatic treatment.

D'Arcy 1986; Bärtels 1997; Hager 5, 1993; Hiller/Melzig 1999; Kumar 2003; Schenck/Naundorf 1966; Seidemann 1993c; Täufel et al. 1993; Uphof 1968; Zeven/de Wet 1982

Lycopersicon lycopersicum L.

▶ *Lycopersicon esculentum* Mill.

 Lyperia atropurpurea Benth.

▶ *Sutera atropurpurea (Banks) Hiern.*

 Lyperia crocea Ecklon

▶ *Sutera atropurpurea (Banks) Hiern.*

LYSIMACHIA L. - Loosestrife - Primulaceae

 Lysimachia candida Lindl.

▶ *Lysimachia obovata* Buch-Ham.

 Lysimachia clethroides Duby

Common Names ▶ gooseneck loosestrife, clethra, loosestrife; *Chinese:* ai tao; *French:* lysimaque à feuilles de cléthra; *German:* Entenschnabel-Felberich
Usage ▶ condiment (in Tonkin)
Parts Used ▶ leaf
Distribution ▶ China, Japan, Indochina

Erhardt et a. 2002; Uphof 1968

 Lysimachia foenum-graecum Hance

Usage ▶ condiment
Parts Used ▶ leaf
Distribution ▶ China

Uphof 1968

 Lysimachia fortunei Maxim.

Common Names ▶ *German:* Fortuna-Gilbweiderich
Usage ▶ condiment
Parts Used ▶ leaf
Distribution ▶ China, Korea, Japan, Taiwan

Erhardt et al. 2002; Uphof 1968

 Lysimachia obovata Buch-Ham.

Synonyms ▶ *Lysimachia candida* Lindl.
Usage ▶ pot-herb, especially in Manipur
Parts Used ▶ herb
Distribution ▶ India

Uphof 1968

M

MACROPIPER Miq. - Piperaceae

Macropiper exelsium *(Forster f.) Miq.*

Common Names ▶ peppertree, kawa-kawa; *German:* Tahitipfeffer
Usage ▶ spice (locally)
Parts Used ▶ fruit
Distribution ▶ New Zealand

Cheers 1997; Mabberly 1997

MAERUA Forssk. - Capparaceae

Maerua angolensis *DC.*

Common Names ▶ bead bean, bead maerua
Usage ▶ condiment in sauces
Parts Used ▶ leaf
Distribution ▶ W Africa: Upper Volta

Burkill 1, 1985

MAESA Forssk. - Myrsinaceae

Maesa indica *Wall.*

Synonyms ▶ *Baobotry indica* Roxb., *Maesa morsha* Hamilt.
Common Names ▶ Indian maesa; *Chinese:* bao chuang ye; *India:* cu den; *Malaysian:* kasi hutan
Usage ▶ condiment by the Vietnamese
Parts Used ▶ leaf
Distribution ▶ China, India, Vietnam

Hanelt 2001; Uphof 1968

MAGNOLIA L. - Magnolia - Magnoliaceae

Magnolia figo *DC.*

▶ *Michelia figo L.*

Magnolia glauca *L.*

▶ *Magnolia virginiana L.*

Magnolia virginiana *L.*

Synonyms ▶ *Magnolia glauca* L.
Common Names ▶ laurel Magnolia, swamp laurel, Viri-

gian laurel, Virigian sweet bay; *German:* Sumpf-Magnolie, Blaugrüne Magnolie, Virgianische Magnolie

Usage ▶ spice (sporadically): meat and sauces
Parts Used ▶ leaf
Distribution ▶ NE, SE USA: Florida, Texas

Cheers 1997; Erhardt et al. 2002; Rätsch 1998; Tull 1999

Majorana crassa *Moench.*

▶ *Origanum syriacum* L.

Majorana cretica *Mill.*

▶ *Origanum onites* L.

Majorana dictamnus *(L.) Kostel.*

▶ *Origanum dictamnus* L.

Majorana hortensis *Moench.*

▶ *Origanum majorana* L.

Majorana majorica *(Cambess.) Briq.*

▶ *Origanum x majoricum* Cambess.

Majorana onites *(L.) Benth.*

▶ *Origanum onites* L.

Majorana paniculata *(Koch) Spenn.*

▶ *Origanum x majoricum* Cambess.

Majorana tomentosa *(Moench) Stokes*

▶ *Origanum dictamnus* L.

MALVA L. - Mallow - Malvaceae

Malva alchemillaefolia *Wall.*

▶ *Malva verticillata* L.

Malva erecta *Gilib.*

▶ *Malva sylvestris* L.

Malva mauritiana *L.*

▶ *Malva sylvestris* L.

Malva meluca *Graeb ex Medw.*

▶ *Malva verticillata* L.

Malva parviflora *L.*

Common Names ▶ cheeseweed, Egyptian mallow, little mallow, small mallow, small-flowered mallow; *Arabic:* kliobbeiza; *French:* grande mauve, mauve à petites, mauve d'Egypte, *German:* Ägyptische Malve, Kleinblütige Malve; *Hindi:* panirak; *Portuguese:* malva
Usage ▶ pot-herb (in India)
Parts Used ▶ plant
Distribution ▶ India, Pakistan, Caucasus, former Soviet middle Asia, Arabia, SE, SW Europe

Arora/Pandey 1996; Hager 5, 1993; Hanelt 2001; v. Koenen 1996; Schultze-Motel 1986; Täufel et. al.; 1993; Villamar et al. 1994; Wiersema/León 1999

Malva sylvestris *L.*

Synonyms ▶ *Malva erecta* Gilib., *Malva mauritiana* L.
Common Names ▶ blue mallow, cheeses, common mallow, high mallow, tall mallow; *French:* mauve des bois, mauve sauvage, fausse guimauve; *German:* Käsepappel, Rosspappel, Gemeine Malve, Wilde

Malve; *Hindi:* gulkhair, kunzi; *Italian:* malva comune; *Japanese:* zeni-aoi; *Korean:* khiauk, mogiyongauk; *Russian:* mal'va lewsnaja, prosvirnik lesnoj; *Spanish:* malva común

Usage ▶ pot-herb (Himalayas from Kashmir to Kumaon)

Parts Used ▶ plant

Distribution ▶ Europe, Turkey, Levante, Caucasus, C Asia, Himalayas, N Africa, native in N America, Australia, New Zealand, S America

Arora/Pandey 1996; Clair 1961; Erhardt et al. 2002; Fleischhauer 2003; Hager 5, 1993; Hanelt 2001; Heeger 1956; Hiller/Melzig 1999; Hoppe 1949; Pschyrembel 1998; Schönfelder 2001; Schultze-Motel 1986; Täufel et al. 1993; Uphof 1968; Wiersema/León 1999; Zeven/de Wet 1982

Malva verticillata L.

Synonyms ▶ *Malva alchemillaefolia* Wall., *Malva meluca* Graeb. ex Medw.

Common Names ▶ castillan marrow, curled mallow, cheeseweed; *Chinese:* dong kui zi; tung-k'uei; *French:* mauve crépue, mauve verticillée; *German:* Quirl-Malve; *Portuguese:* malva crespa, verdadeira; *Spanish:* malva crespa;

Usage ▶ condiment (Orient), pot-herb (India, Nilgiris)

Parts Used ▶ seed (Orient), plant

Distribution ▶ India: higher hills of NE and Nilgiris, Pakistan, Himalayas, China; cultivated and native elsewhere in temperate regions, e.g. Europe

Arora/Pandey 1996; Davidson 1999; Erhardt et al. 2002; Hager 5, 1993; Hanelt 2001; Heeger 1956; Kumar 2003; Larkcom 1991; Schultze-Motel 1986; Täufel et al. 1993; Uphof 1968; Wiersema/León 1999; Zeven/de Wet 1982

MAMMAEA L. - Guttiferae

Mammae americana L.

Synonyms ▶ *Potamocharis mamei* Rottb.

Common Names ▶ mammee, mammee-apple, mammey, Santo Domingo apricot, South American apricot; *Brazil (Portuguese):* abricó, abricó do pará; *French:* abricotier d'Amérique, abricotier de Saint Domingue; *German:* Mammey or Mammi-Apfel, Aprikose von St. Domingo; *Spanish:* mamey, Zapote mamey

Usage ▶ flavoring, e.g. for liqueurs (Eau de Créole)

Parts Used ▶ flower

Distribution ▶ West Indies, southern C America, N coast of S America

Hanelt 2001; Nowak/Schulz 1998; Schultze-Motel 1986; Uphof 1968; Usher 1974

Mammea longifolia Wall.

▶ *Ochrocarpos longifolius (Wall.) Benth. et Hook.f.*

MANGIFERA L. - Mango - Anacardiaceae

Mangifera ambas Forsk.

▶ *Mangifera indica L.*

Mangifera domestica Gaertn.

▶ *Mangifera indicaa L.*

Mangifera foetida Lour. x Mangifera indica L.

▶ *Mangifera x odorata Griff.*

Mangifera indica L.

Synonyms ▶ *Mangifera ambas* Forsk., *Mangifera domestica* Gaertn.

Common Names ▶ mango, Indian mano; *Arabic:* amba; *Brazil (Portuguese):* manga; *Dutch:* manga; *French:* manguier, mangue; *German:* Mango; *Hindi:* am; *India:* ambo, mamadi, amchoor, mangga, mannga arum manis, mangga golek; *Italian:* mango; *Japanese:* mangô; *Malaysian:* ampalam, males, mangas mempalam, amchur, pau; *Pilipino:* manga, mangga, mangang-kalabaw, pako; *Portuguese:* manguiera; *Russian:* mango, mangowoe; *Sanskrit:* amra, cuta, rasala; *Spanish:* mango; *Thai:* mamuang; *Vietnamese:* xoài;

Usage ▸ saucing agent for curries and chutneys

Parts Used ▸ unripe fruit

Distribution ▸ NE India, N Myanmar; cultivated in tropical Asia, Africa, Brazil

Note ▸ There are about 100 sorts. Amchur, aamchur or amchor is the powdery dry extract from the unripened fruits and this is used in Indian and Thai cuisine as seasoning for meat.

Arora/Pandey 1996; Askar et al. 1981; Bendel 2002; Bose 1985; Cheers 1997; Davidson 1999; Engel/Phummai 2000; Erhardt et al. 2002; Gangolly et al. 1957; Hanelt 2001; Hiller/Melzig 1999; Lewington 1990; Klostermans/Bombard 1993; Kottegoda 1994; Kumar 2003; Mitra et al. 2000; Morris/Mackley 1999; Mukherji 1949; Norman 1991; Pruthi 1976; Schenck/Naundorf 1966; Schultze-Motel 1986; Seidemann 1993c; Sharma 2003; Siewek 1990; Sing 1968; Soule 1951; Storrs 1997; Täufel et al. 1993; Uphof 1968; Usher 1974; Verheij/Coronel 1991; Villamar et al. 1994; Wiersema/León 1999; Zeven/de Wet 1982

 Mangifera odorata *Griff.*

Synonyms ▸ *Mangifera foetida* Lour. x *Mangifera indica* L.

Common Names ▸ kurwini mango, saipan mango; *French:* kuwein, manguier mango; *German:* Duft-Mango; *Indonesian:* bembem; *Malaysian:* kuini, kuwini, kweni, kwini

Usage ▸ spice for curries, chutneys, sauces

Parts Used ▸ green fruit

Distribution ▸ origin Malaysia(?), only cultivated in tropical countries

Note ▸ The fruit has a strongly sour taste.

Erhardt et al. 2000, 2002; Lück 2000; Schultze-Motel 1986; Täufel et al. 1993; Uphof 1968; Wiersema/León 1999; Zeven/de Wet 1982

 Mangifera pinnata *J. Koenig ex L.f.*

▶ *Spondias pinnata (J. Koenig ex L.f.) Kurz*

 Maranta galanga *L.*

▶ *Alpinia galanga (L.) Willd.*

 Maranta malaccensis *Burm.*

▶ *Alpinia malaccensis (Burm.) Rosc.*

MARASMIUS Fr. - Tricholomataceae (Agariaceae)

 Marasmius alliatus *Schaeff.*

▶ *Marasmius scorodonius (Schaeff) Fr.*

 Marasmius caryophylleus *(Schaeff.) Schroet.)*

▶ *Marasmius oreades (Bolt.) Fr.*

 Marasmius oreades *(Bolt.) Fr.*

Synonyms ▸ *Marasmius caryophylleus* (Schaeffer) Schroet.

Common Names ▸ fairy ring mushroom; *Dutch:* weidekringzwam; *French:* pied dur, faux mousseron, marasme montagnard, mousseron d'automne, nymphe des montagnes; bouton de guerte; *German:* Echter Krösling, Herbstmousseron, Nagelschwamm, Nelkenschwindling; *Italian:* gambe-secche, marasmio oreade; *Russian:* openok lugovoj, gwozdischnyj grib; *Spanish:* ninfa, camasec, crespilla, woxerno

Usage ▸ condiment

Parts Used ▸ mushroom

Distribution ▸ Europe

Singer 1986; Stowart 1978; Uphof 1968; Gmelin et al. 1976; Vidal et al. 1986; Wiersema/León 1999

 Marasmius scorodonius *(Schaeff.) Fr.*

Synonyms ▸ *Marasmius alliatus* Schaeff.

Common Names ▸ garlic marasmuius, garlic mushroom, true mousseron, *German:* Echter Mousseron, Knoblauchpilz, Echter Küchenschwindling

Usage ▸ condiment in soups and sauces

Parts Used ▸ mushroom

Distribution ▸ Europe

Farrell et al. 1984; Gmelin et al. 1976; Rapior et al. 1997; Singer 1986; Usher 1968

MARGOTIA Boiss - Apiaceae (Umbelliferae)

Margotia gummifera *(Desf.) Lange*

Synonyms ▶ *Elaeoselinum gummiferum* Desf.
Common Names ▶ *Spanish:* hinojo bastardo
Usage ▶ spice (rarely)
Parts Used ▶ fruit
Distribution ▶ W Mediterranean regions

MARISCUS Vahl - Cyperaceae

Mariscus alternifolia *Vahl*

Synonyms ▶ *Kyllingia umbellata* Rottb.
Usage ▶ flavoring
Parts Used ▶ rhizome
Distribution ▶ tropical Africa: the Gambia, Nigeria, Senegal, Sierra Leone to India

Burkill 3, 1995; Hanelt 2001; Irvine 1948; Zeven/Zhukovsky 1975

MARRUBIUM L. - Horehound - Lamiaceae (Labiatae)

Marrubium indicum *Thunb. Non Bum.f*

▶ *Hyptis suaeveolens (L.) Poiteau*

Marrubium vulgare *L.*

Common Names ▶ common horehound, hoarhound, marvel, white horehound; *Arabic:* marriout, morroubia, umm re-roubia, roubia; *Chinese:* cu xia zhi cao; *French:* marrube blanc, marrube vulgaire; *German:* Gewöhnlicher Andorn, Berghopfen, Weißer Dorant; *Italian:* marrobio, marrubio; *Portuguese:* marroio; *Russian:* chandra, chandra obyknowennaja; *Spanish:* marrubio blanco, marrubio común, malvarrubia
Usage ▶ spice, flavoring (liqueurs)
Parts Used ▶ leaf
Distribution ▶ Europe, Turkey, Levante, Caucasus, Iran, C Asia, India, N Africa, native N America

Aichele/Schwegler 4, 1995; Bärtels 1997; Berger 4, 1954; Bremness 2001; Cheers 1997; Clair 1961; Coiciu/Racz (no year); Davidson 1999; Dudtschenko et al. 1989; Erhardt et al. 2002; Fleischhauer 2003; Hager 5, 1993; Hanelt 2001; Heeger 1956; Henderson/MacCrindle 1969; Hiller/Melzig 1999; Hoffmann et al. 1992; Hohmann et al. 2001; Hoppe 1949; Kowalewski/Matlawska 1978; Nawwar et al. 1989; Newall et al. 1996; Ochoa/Alonso 1996; Pschyrembel 1998; Schnelle 1999; Schönfelder 2001; Schultze-Motel 1986; Seidemann 1993c; Sharma 2003; Small 1997; Uphof 1968; Villamar et al. 1994; Wiersema/León 1999; Wyk et al. 2004

■ *Marrubium vulgare*, flowering

MARSDENIA R. Br. - Condorvine - Asclepiadaceae

 Marsdenia cundurango Rchb.f.

Synonyms ► *Marsdenia reichenbachii* Triana;
Common Names ► condor plant, condurango, common condorvine, eagle-vine bark; *French:* condurango; *German:* Condurango, Cundurango, Gewöhnlicher Andenwein; *Italian:* condurango, vite aquilina; *Peru:* tucacsillu; *Russian:* kondurango; *Spanish:* condurango blanco
Usage ► spice, flavoring (liqueurs)
Parts Used ► bark of trunk and branch
Distribution ► S America Andes: Columbia, Ecuador, Peru
Note ► Earlier frequent listed under *Cundurango*.

Berger 1, 1949; Erhardt et al. 2000, 2002; Hager 5, 1993; Hiller/Melzig 1999; Hohmann et al 2001; Hoppe 1949; Koch 1981; Pschyrembel 1998; Rätsch 1992; Schultze-Motel 1986; Seidemann 1993c; Tschesche/Kohl 1968; Uphof 1968; Wiersema/León 1999; Wolters 1994; Wüstenfeld/Haensel 1964; Wyk et al. 2004

 Marsdenia reichenbachii *Triana*

▶ *Marsdenia cundurango* Rchb.

 Marshallia brunoides *Less.*

▶ *Heterothalamus brunoides* Less.

 Massoia aromatica *Becc.*

▶ *Cryptocaria aromatica (Becc.) Kosterm.*

 Maximowiczia chinensis *(Turcz.) Rupr.*

▶ *Schisandra chinensis (Turcz.) Baill.*

 Mayna serica *Spreng.*

▶ *Xylopia sericea St. Hil.*

MAYTENUS Molina - Celastraceae

 Maytenus senegalensis *(Lam.) Exell*

Synonyms ► *Celastrus senegalensis* Lam., *Gymnosporia senegalensis* Loes.
Common Names ► confetti tree, red spike thorn
Usage ► flavoring (Masai add to soup and broth)
Parts Used ► leaf
Distribution ► tropical and subtropical Africa: Angola, Cameroon, Ethiopia, the Gambia, Guinea, Kenya, Mali, Niger, Nigeria, Senegal, Somalia, Uganda, Zimbabwe; tropical and subtropical regions of India, Pakistan and Spain

Abraham ert al. 1971; Adegoke et al. 1968; Ayensu 1978; Brünning/Wagner 1978; Burkill 1, 1985; Gomez-Serraillos/Zaragoza 1980; Hager 5, 1993; Hiller/Melzig 1999; v. Koenen 1996; Neuwinger 1998; Usher 1968

MEDIASIA Pimenov - Apiaceae (Umbelliferae)

 Mediasia macrophylla *(Regel et Schmalh.) Pimenov*

Synonyms ► *Seseli macrophyllum* Regel et Schmalh.
Common Names ► *German:* Großblättrige Mediasie, Pamir; *Russian:* (Usbek.) alkor
Usage ► spice, flavoring; **product:** essential oil; for soft drinks
Parts Used ► fruit
Distribution ► Middle Asia: Altai mountains, Pamir, NE Afghanistan

Hanelt 2001; Rechinger 1987; Schultze-Motel 1986; Seidemann 1993c; Täufel et al. 1993

MEDICAGO L. - Medick - Fabaceae (Leguminosae)

 Medicago alba E.H.L. Krause

▶ *Melilotus albus* Medik.

 Medicago altissima E.H.L. Krause

▶ *Melilotus altissimus* Thuill.

 Medicago corniculata (L.) Trautv.

▶ *Trigonella corniculata* (L.) L.

 Medicago denticulata Willd.

▶ *Medicago hispida* Gaertn.

 Medicago hispida Gaertn.

Synonyms ▶ *Medicago denticulata* Willd., *Medicago polymorpha* L.
Common Names ▶ toothead bur clover, burr medick; *Chinese:* (nan) mu su; *German:* Rauher or Spanischer Schneckenklee; *Hindi:* maina; *Italian:* trifoglina; *Russian:* ljuzepna
Usage ▶ pot-herb in India
Parts Used ▶ plant
Distribution ▶ India: Himalayas, W Bengal, in Nilgiris and other hills of W Ghats, N India, W Europe, mediterranean countries; native worldwide, ruderal plant in Australia, S Africa

Arora/Pandey 1996; Hanelt 2001; Uphof 1968; Zeven/de Wet 1982

 Medicago officinalis E.H.L. Krause

▶ *Melilotus officinalis* Lam.

 Medicago polymorpha L.

▶ *Medicago hispida* Gaertn.

MELILOTUS Mill. - Melilot - Fabaceae (Leguminosae)

 Melilotus albus Medik.

Synonyms ▶ *Medicago alba* E.H.L. Krause, *Melilotus rugulosa* Willd., *Melilotus vulgaris* Willd., *Sertula alba* O. Kuntze, *Trifolium album* Lois., *Trifolium vulgare* Hayne
Common Names ▶ white melilot, white sweet clover, Bokhara clover, honey clover; *French:* mélilot blanc; *German:* Weißer Steinklee, Bokharaklee; *Portuguese/Spanish:* meliloto blanco
Usage ▶ spice, for herb cheese, condiments; **product:** tobacco sauces, liqueurs
Parts Used ▶ herb
Distribution ▶ N Africa, India, Indochina, Europe, native elsewhere
Note ▶ Still only rare.

Aichele/Schwegler 2, 1994; Dhar/Dhar 2000; Erhardt et al. 2002; Fleischhauer 2003; Hiller/Melzig 1999; Loch 1993; Schnelle 1999; Uphof 1968; Wiersema/León 1999; Zeven/de Wet 1982

 Melilotus altissimus Thuill.

Synonyms ▶ *Medicago altissima* E.H.L. Krause, *Melilotus giganteus* Wenderoth, *Melilotus macrorrhizus* (Waldst. et Kit.) Pers., *Melilotus officinalis* Lam., *Trifolium altissimum* Lois
Common Names ▶ tall meliot, tall yellow sweet, clover; *French:* grand méliot, trèfle musqué; *German:* Hoher Steinklee, Bokharaklee, Sumpf-Steinklee; *Italian:* meliloto gigantesco; *Portuguese:* trevo de cheiro; *Spanish:* meliloto gigante, trébol oloroso
Usage ▶ spice (flavoring)
Parts Used ▶ herb
Distribution ▶ Europe, W Siberia (Altai)
Note ▶ Still only rare.

Aichele/Schwegler 2, 1994; Erhardt et al. 2002; Fleischhauer 2003; Hanelt 2001; Heeger 1956; Hiller/Melzig 1999; Hoppe 1949; Loch

1993; Pschyrembel 1998; Schnelle 1999; Schönfelder 2001; Stevenson 1969; Uphof 1968; Wiersema/León 1999; Zeven/de Wet 1982

 Melilotus arvensis *Wallr.*

▶ *Melilotus officinalis Lam.*

 Melilotus caeruleus *Desv.*

▶ *Trigonella caerulea (L.) L.*

 Melilotus giganteus *Wenderoth*

▶ *Melilotus altissimus Thuill.*

 Melilotus macrorrhizus *(Waldst. et Kit.) Pers.*

▶ *Melilotus altissimus Thuill.*

 Melilotus officinalis *Lam.*

▶ *Melilotus altissimus Thuill.*

◘ Melilotus officinalis, flowering

 Melilotus officinalis *Lam.* ◘

Synonyms ▶ *Medicago officinalis* E.H.L. Krause, *Melilotus arvensis* Wallr., *Melilotus petitpierreanus* Willd., *Sertula arvensis* O. Kuntze, *Trifolium officinalis* L., *Trifolium petitpierreanum* Hayne

Common Names ▶ yellow melilot, yellow sweet clover, field melilot, ripped melilot; *Chinese:* hsun tsao, ling, ling hsiana; *French:* fleurie de mélilot, mélilot officinal, mélilot jaune; *German:* Echter Steinklee, Gelber Steinklee, Honigklee; *Italian:* meliloto giallo, trifoglio cavallino; *Korean:* norangjontongssari, yakjontongssari; *Portuguese:* trevocheiroso; *Russian:* donnik aptetschnyj; *Spanish:* meliloto amarillo, meliloto oficinal, cornilla real, trébol dorado, trébol dulce

Usage ▶ spice for herb cheese, condiments; **product:** liqueurs, tobacco, sauces

Parts Used ▶ herb

Distribution ▶ Europe wild widely spread, Caucasus, Iran, W Siberia, C Asia, Tibet, Himalayas, native N America

Note ▶ Still only rare.

Aichele/Schwegler 2, 1994; Ashton/Davies 1962; Berger 4, 1954; Bilgri/Adam 2000; Bremness 2001; Coiciu/Racz (no year); Dombrowicz et al. 1991; Dudtschenko et al. 1989; Ehlers et al 1997; Erhardt et al. 2002; Fleischhauer 2003; Hanelt 2001; Heeger 1956; Hohmann et al. 2001; Hoppe 1949; Koschtschejew 1990; Loch 1993; Pschyrembel 1998; Schnelle 1999; Schönfelder 2001; Seidemann 1993c; Seidemann/Siebert 1987; Stevenson 1969; Täufel et al. 1993; Uphof 1968; Wiersema/León 1999; Wörner/Schreier 1990; Wyk et al. 2004; Zeven/de Wet 1982

 Melilotus petitpierreanus *Willd.*

▶ *Melilotus officinalis Lam.*

 Melilotus rugulosa *Willd.*

▶ *Melilotus albus Medik.*

Melilotus vulgaris *Willd.*

◆ *Melilotus albus Medik.*

MELISSA L. - Balm - Lamiaceae (Labiatae)

 Melissa altissima *J.E. Smith*

◆ *Melissa officinalis L.*

 Melissa calamintha *L.*

◆ *Calamintha nepeta (L.) Savi*

 Melissa grandiflora *L.*

◆ *Calamintha grandiflora (L.) Moench*

 Melissa inodora *Bornm.*

◆ *Melissa officinalis L.*

 Melissa maxima *Ard.*

◆ *Perilla frutescens (L.) Britton*

 Melissa nepeta *L.*

◆ *Calamintha nepeta (L.) Savi*

 Melissa officinalis *L.*

Synonyms ▶ *Melissa altissima* J.E. Smith, *Melissa inodora* Bornm., *Melissa romana* Mill.

Common Names ▶ (bee) balm, lemon balm, lemon mint, melissa; *Arabic:* louiza, merzizou; *Chinese:* xiang feng hua; *Dutch:* melisse, melissa citroen; *French:* mélisse, mélisse officinale, citronelle, herbe au citron; *German:* Citronelle, Melisse, Zitronenmelisse, Bienenkraut, Herzkraut, Honigblatt; *Italian:* cedronella, erba limona; *Korean:* kyulhyangphul; *Portuguese:* erba cidreira; *Russian:* melissa, melissa lekarstvennaja, limonajamjata, matotschnik, roewnik, ptschel'nik, papotschnaja trawa; *Spanish:* balsamita major, citraria, melisa, toronjil

Usage ▶ spice, flavoring; **product:** essential oil (melissa oil, oil of balm) for liqueur: Benedictine, Chartreuse, melissa spirit)

Parts Used ▶ leaf, herb

Distribution ▶ N Africa, C and W Asia, Europe, Caucasus, widely cultivated and wild widespread in temperate regions

Aichele/Schwegler 4, 1995; Bärtels 1997; Berger 2, 1950; Bomme et al. 2002; Bremness 2001; Brieskorn/Krause 1974; Burgett 1980; Cheers 1997; Clair 1961; Coiciu/Racz (no year); Dorner 1985; Dudtschenko et al. 1989; Erhardt et al. 2002; Fleischhauer 2003; Guzman/Siemonsma 1999; Hager 5, 1993; Hanelt 2001; Heeger 1956; Hefendehl 1970; Hiller/Melzig 1999; Hoffmann et al. 1992; Hohmann et al. 2001; Hondelmann 2002; Hoppe 1949; Koch-Heitzmann/Schultze 1984, 1988; Mikus/Schaser 1995; Mulkens/Kapetanidis 1988; Pochljobkin 1974, 1977; Pschyrembel 1998; Richter 1993; Roberts 1997; Roth/Kormann 1997; Sarer/Kökdil 1991; Schönfelder 2001; Schultze-Motel 1986; Schulze 1989, 1995; Seidemann 1993c; Seidemann/Siebert 1987; Staesche 1972; Stobert 1978; Täufel et al. 1993; Teuscher 2003; Titel et al. 1982; Tucker 1986; Ubillos 1989; Uphof 1968; Wiersema/León 1999; Wüstenfeld/Haensel 1964; Wyk et al. 2004; Zänglein et al. 1995; Zeven/de Wet 1982

 Melissa officinalis *L.* **var. altissima** *(Sm.) Arcang*

Common Names ▶ Crete balm, Crete melissa; *German:* Kreta-Melisse;

Usage ▶ spice, flavoring

Parts Used ▶ leaf, herb

Distribution ▶ Mediterranean region, Caucasus

Dawson et al. 1988; Hanelt 2001; Teuscher 2003

 Melissa officinalis *L.* **var. hirsuta** *Pers.*

Common Names ▶ *German:* Rauhaarige Melisse

Usage ▶ spice, flavoring

Parts Used ▶ leaf

Distribution ▶ Mediterranean region, Caucasus

Hanelt 2001

Melissa officinalis *L.* ssp. parviflora *Benth.*

Common Names ▶ herb of tempera, small flowered melissa; *German:* Kleinblütige Melisse
Usage ▶ spice, flavoring
Parts Used ▶ leaf, herb
Distribution ▶ Himalayas

Hanelt 2001

Melissa parviflora *Benth.*

Common Names ▶ small flowered melissa; *German:* Kleinblütige Melisse
Usage ▶ flavoring of food products
Parts Used ▶ leaf, herb
Distribution ▶ temperate Himalayas

Pandey 1990; Small 1997

Melissa pulegioides *L.*

▶ *Hedeomea pulegioides (L.) Pers.*

Melissa romana *Mill.*

▶ *Melissa officinalis L.*

MELOTHRIA L. - Moccasin Grass - Cucurbitaceae

Melothria maderaspatana *(L.) Cogn.*

Synonyms ▶ *Mukia maderaspatana* (L.) M. Roem., *Mukia scabrella* Arn.
Common Names ▶ Madras apple; *German:* Madrasapfel, Madras-Melothrie
Usage ▶ pot-herb
Parts Used ▶ leaf
Distribution ▶ tropical Africa, Asia, Australia

Arora/Pandey 1996; Erhardt et al. 2002; Uphof 1968

MENTHA L. - Mint - Lamiaceae (Labiatae)

Lawton 2002; Marengo et al. 1991

Mentha alopecuroides *Hull.*

Common Names ▶ apple mint, bowl mint; *German:* Apfelminze, Breitblättrige Minze, Zottige Minze;
Usage ▶ flavoring
Parts Used ▶ leaf, herb
Distribution ▶ widely cultivated as a garden plant in Europe

Clement/Forster 1994; Hanelt 2001; Stace 1991; Täufel et al. 1993; Zeven/de Wet 1982

Mentha aquatica *L.*

Synonyms ▶ *Mentha hirsuta* Huds.
Common Names ▶ horse mint, marsh mint, water mint; *French:* menthe aquatique; *German:* Bachminze, Wasserminze; *Italian:* menta d'acqua; *Russian:* mjata wodjanoja; *Spanish:* hierba buena acuática
Usage ▶ spice (rarely), flavoring; **product:** essential oil (watermint or horsemint oil)
Parts Used ▶ leaf, herb
Distribution ▶ Europe, Turkey, Caucasus, W, C Asia, Africa, N America, native elsewhere

Aichele/Schwegler 4, 1995; Bärtels 1997; Berger 2, 1950; Bremness 2001; Clair 1961; Davidson 1999; Dudtschenko et al. 1989; Enríquez/Sand-Jensen 2003; Erhardt et al. 2002; Fleischhauer 2003; Hager 5, 1993; Hanelt 2001; Heeger 1956; Hiller/Melzig 1999; Hondelmann 2002; Hoppe 1949; Roberts 1997; Roberts/Plotto 2002; Rothe/Kormann 1997; Schönfelder 2001; Schultze-Motel 1986; Seidemann 1993c; Small 1997; Täufel et al. 1993; Teuscher 2003; Tucker 1986; Ubillos 1989; Uphof 1968; Wiersema/León 1999; Zeven/de Wet 1982

Mentha arvensis *L.*

Common Names ▶ corn mint, field mint, Japanese mint, tule mint, wild mint; *Chinese:* bo he, po ho, fan ho; *French:* menthe Japon, menthe Brésil, menthe des champs; *German:* Ackerminze, Feldminze, Japanische Minze, Österreichische Minze; *Itali-*

an: menta salvadeca, mentaster; *Malaysian:* daun pudina; *Russian:* mjata polevaja; *Spanish:* menta silvestre, menta japonesa; *Thai:* bai saranae; *Vietnamese:* bac he, buc ha nam

Usage ▶ flavoring; **product:** essential oil (Japanese mint oil, *German:* Japanisches Minzöl, Japanisches Pfefferminzöl, Po-Ho

Parts Used ▶ leaf, herb

Distribution ▶ Europe, Turkey, temperate Asia, N India, Thailand, Cambodia, Laos, Vietnam, Korea, Japan

Aichele/Schwegler 4, 1995; Berger 2, 1950; Bournot 1968; Cheers 1997; Dafferthofer 1981; Davidson 1999; Dudtschenko et al. 1989; Erhardt et al. 2002; Erichsen-Brown 1989; Fleischhauer 2003; Hager 5, 1993; Hanelt 2001; Heeger 1956; Hiller/Melzig 1999; Karasawa et al. 1995; Khyong et al. 1983; Kothari/Singh 1994; Peters 1927; Pruthi 1976; Pschyrembel 1998; Roberts 1997; Roth/Kormann 1997; Schönfelder 2001; Schultze-Motel 1986; Sharma 2003; Shiva et al. 2002; Small 1997; Täufel et al. 1993; Teuscher 2003; Turova et al. 1987; Uhl 2000; Uphof 1968; Villamar et al. 1994; WHO 1990; Wiersema/León 1999; Wyk et al. 2004; Zeven/de Wet 1982

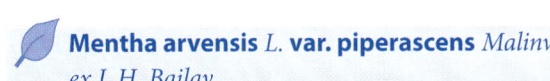
Mentha arvensis L. var. piperascens *Malinv. ex L.H. Bailay*

▶ *Mentha canadensis* L.

Mentha arvensis x M. spicata

▶ *Mentha x gracilis* Sole

Mentha asiatica *Borris*

▶ *Mentha longifolia* (L.) Nath. var. *asiatica* (Borris) Rech.f.

Mentha canadensis L.

Synonyms ▶ *Mentha arvensis* var. *piperascens* Malinv. ex L.H. Bailey

Common Names ▶ Chinese mint, Japanese (pepper)mint, corn mint; *German:* Chinesische Minze, Japanische Minze; *Japanese:* hakka

Usage ▶ spice; **product:** essential oil and pulegon

Parts Used ▶ herb

Distribution ▶ China, E Asia, Russian Far East, Indochina, Malaysia, Canada, USA, widely cultivated

Abad Farooqi et al. 1983; Berger 2, 1950; Bournet 1968; Davidson 1999; Dudtschenko et al. 1989; Duriyaprapan/Britten 1982; Erichsen-Brown 1989; Gašić et al. 1992; Hager 5, 1993; Hanelt 2001; Huet 1972; Lewington 1990; Pichitakul/Sthapitanonda 1977; Piper/Price 1975; Seidemann 1993c; Täufel et al. 1993; Tucker 1986; Uphof 1968; Villamar et al. 1994; Wiersema/León 1999; Zeven/de Wet 1982

Mentha x cardiaca *J. Gerard ex Baker*

▶ *Mentha x gracilis* Sole

Mentha citrata *Ehrh.*

▶ *Mentha x piperita* "citrata"

Mentha x cordifolia *Opiz ex Fresen*

Synonyms ▶ *Mentha spicata* L. var. *crispa*

Common Names ▶ Kentucky spearmint; *German:* Kentucky Krauseminze, Herzblättrige Minze; *Spanish:* yerba buena

Usage ▶ flavoring

Parts Used ▶ herb

Distribution ▶ Philippines, Java, cultivated in Thailand

Hanelt 2001; Hiller/Melzig 1999;

Mentha crispa L.

Common Names ▶ garden mint, spearmint; *German:* Gartenminze; Krause Minze; *Portuguese:* hortelã-de-folha-miúda, hortelã-de-panela, hortelã-de-rasteira

Usage ▶ spice, flavoring; **product:** essential oil (spearmint oil; *German:* Krauseminzöl)

Parts Used ▶ herb

Distribution ▶ S America: Brazil, cultivated in C Europe

Dudtschenko et al. 1989; Kil Jin Park et al. 2002; Hoppe 1949; Mors et al. 2000; Traxl 1975

Mentha x gentilis auct.

> *Mentha x gracilis* Sole

 Mentha x gracilis *Sole*

Synonyms ▸ a hybrid of *M. arvensis x M. spicata*; *Mentha x cardiaca* J. Gerard ex Baker; *Mentha x gentilis* auct.

Common Names ▸ Scotch mint, Scotch spearmint, ginger mint, red mint, American apple mint; *Brazil (Portuguese):* hortelã; *French:* menthe des jardins; *German:* Edelminze, Ingwerminze, Amerikanische Apfelminze; *Spanish:* almoradux;

Usage ▸ flavoring; **product**: essential oil

Parts Used ▸ leaf, herb

Distribution ▸ China, E Asia, Russia, Indochina, Malaysia, Canada, USA; widely cultivated

Bremness 2001; Davidson 1999; Erhardt et al. 2002; Heeger 1956; Kothari/Singh 1995; Lewington 1990; Mansfeld 1986; Nagasawa et al. 1975a, b; Small 1997; Surburg/Köpsel 1989; Täufel et al. 1993; Teuscher 2003; Tucker 1986; Uphof 1968; Wiersema/León 1999; Zeven/de Wet 1982

 Mentha gratissima *Weber*

> *Mentha x villosa* Huds.

 Mentha haplocalyx *(Briq.) Briq.*

Common Names ▸ Japanese peppermint; *Chinese:* bo he; *German:* Chinesische or Japanische Minze; *Hindi:* pudina; *Japanese:* hakkar; *Korean:* pakha; *Spanish:* menta japonesa;

Usage ▸ spice, flavoring for cigarettes; **product**: essential oil

Parts Used ▸ herb

Distribution ▸ E, S and SE Asia, New Zealand; after 1945 also grown in C and S America (Cuba, Mexiko, Brazil, Argentina), USA, Seychelles, India, Europe, especially Spain and Hungary; in Korea also cultivated

Note ▸ Tamba Yasuyori gives the first mention of the Japanese mint plant under the name "megusa" in 984 AD.

Ding/Sun 1983; Hager 5, 1993; Hanelt 2001; Leung 1991

 Mentha hirsuta *Huds.*

> *Mentha aquatica* L.

 Mentha insularis *Req.*

> *Mentha suaveolens* Ehrh.

 Mentha incana *Willd.*

> *Mentha longifolia* (L.) Huds.

 Mentha longifolia *(L.) Huds.*

Synonyms ▸ *Mentha incana* Willd., *Mentha spicata* L. var. *longifolia* L., *Mentha sylvestris* L.

Common Names ▸ biblical mint, silver mint, horse mint; *Chinese:* ou bo he; *German:* Rossminze, Silberminze, Langblättrige Minze

Usage ▸ spice, flavoring; **product**: essential oil (Scotch spearmint oil; *German:* Grünminzöl)

Parts Used ▸ herb

Distribution ▸ N W Africa, Caucasus, Siberia, W Asia, India, W, S Europe

Aichele/Schwegler 4, 1995; Berger 2, 1950; Bourwieg/Pohl 1973; Burkill 3, 1995; Chauhan 1999; Davidson 1999; Erhardt et al. 2002; Hager 5, 1993; Hanelt 2001; Heeger 1956, 1992; Hiller/Melzig 1999; Hondelmann 2002; Monfared et al. 2002; Small 1997; Roberts 1997; Schönfelder 2001; Schultze-Motel 1986; Sharma 2003; Small 1997; Soulèlès/Argyriadou 1988; Täufel et al. 1993; Teuscher 2003; Uphof 1968; Wiersema/León 1999; Zeven/de Wet 1982

 Mentha longifolia *(L.) Huds.* **var. asiatica** *(Boriss.) Rech.f.*

Synonyms ▸ *Mentha asiatica* Borris

Common Names ▸ Asiatic mint; *Dutch:* kruisement; *German:* Asiatische or Langblättrige asiatische Rossminze; *Zulu:* lohmlauge

Usage ▸ pot-herb (for sweets, joghurt or chutney); **product**: essential oil.

Parts Used ▸ herb

Distribution ▶ Asia: N India, Mongolia, cultivated in Turkmenistan

Dudtschenko et al. 1989; Hanelt 2001; Jaimand/Rezaee 2002; Karasawa et al. 1995; Wealth of India 6, 1962; Wyk 1997

Mentha longifolia (L.) Huds. **var. wissii** (Launert) L.E. Codd

Common Names ▶ Cape velvet mint; *German:* Weiche Kap-Minze
Usage ▶ spice, flavoring
Parts Used ▶ leaf
Distribution ▶ S Africa: Cape region;

Hanelt 2001; Husain 1992; Roberts 1997

Mentha longifolia (L.) Huds. x **Mentha suaveolens** Ehrh.

▶ *Mentha rotundifolia (L.) Huds.*

Mentha mollis Benth.

▶ *Minostachys mollis (Benth.) Griseb.*

Mentha patrini Lepech

Common Names ▶ comb mint; *German:* Kamm-Minze
Usage ▶ spice, flavoring
Parts Used ▶ leaf, herb
Distribution ▶ temperate Asia, Europe, N America; cultivated in E Europe and Austria
Note ▶ The plant has a peppemint-like flavor.

Lück 20004 Täufel et al. 1993

Mentha perilloides Lam.

▶ *Perilla frutescens (L.) Britt*

Mentha x piperita L.

Synonyms ▶ *Mentha x piperita* L. var. *vulgare* Sole
Common Names ▶ peppermint, brandy mint, mint flakes; *Arabic:* na 'an; *Chinese:* la bo he; *French:* menthe anglais, menthe poivrée; *German:* Pfefferminze, Englische Minze, Mutterkraut; *Italian:* menta piperita; *Korean:* huchubakha; *Portuguese:* hortăpimenta; *Russian:* mjata pereznaja, anglijckaja mjata; cholodnaja mjata, cholodjanka; *Spanish:* la menta (inglesa) piperita, hierba Luisa; *Turkish:* nane
Usage ▶ spice, flavoring; **product:** essential oil (peppermint oil) and menthol
Parts Used ▶ leaf, herb
Distribution ▶ origin in England, cultivated worldwide
Note ▶ The English pharmacist R. Ray gave the first mention of the peppermint plant in England in 1696.

Aichele/Schwegler 4, 1995; Amelunxen/Intert 1993; Bachthaler et al. 1976; Bärtels 1997; Berger 2, 1950; Bois 1934; Bournot 1968; Bouverat-Bernier 1992; Bremness 2001; Chalchat et al. 1997; Cheers 1997; Chialva 1993; Clair 1961; Clark/Menary 1979; Coiciu/Racz (no year); Court et al. 1993; Daffertshofer 1981; Dudtschenko et al. 1989; Erhardt et al. 2000, 2002; Erichsen-Brown 1989; Faleyx/Howland 1980; Farrell 1985; Franz et al. 1984; Frerot et al. 2002; Galle-Hoffmann/König 1998; Gasic et al. 1987; Grahle/Höltzel 1963; Guedon/Pasquier 1994; Gulati et al. 1975; Hager 5, 1993; Hanelt 2001; Heeger 1956; Hefendehl 1962; Hiller/Melzig 1999; Hoffmann et al. 1992; Hohmann et al. 2001; Hölzl et al. 1974; Hoppe 1949; Karasawa et al 1995; Kolak et al. 2001; Lawrence et al. 1986; 1989; Leung 1991; Lewington 1990; Maffei 1999; Melzer 1994; Pank et al. 1994; Peters 1927; Petry 1993; Pochljobkin 1974, 1977; Pruthi 1976; Pschyrembel 1998; Reverchon et al. 1994; Ringuelet et al. 2003; Rissanen et al. 2002; Roberts 1997; Rohloff 1999; Rojahn et al. 1977; Roth/Kormann 1997; Saha et al. 1986; Schönfelder 2001; Schultze-Motel 1986; Schulz/Krüger 1999; Seidemann 1993c; Seidemann/Siebert 1987; Seitz 1982; Sharma 2003; Shiva et al. 2002; Shotipruk et al. 2001; Small 1997; Stahn/Bomme 1997; Täufel et al. 1993; Teuscher 2003; Tucker 1986; Turova et al. 1987; Ubillos 1989; Uhl 2000; Uphof 1968; Voirin et al. 1990; Walker/Berattie 1980; Wiersema/León 1999; Wüstenfeld/Haensel 1964; Wyk et al. 2004; Zeven/de Wet 1982

Mentha x piperita 'citrata'

Synonyms ▶ *Mentha x piperita* L. var. *citrata* (Ehrh.) Brig.
Common Names ▶ bergamot mint, eau de Cologne mint, lemon mint, orange mint; *French:* Eau d' Cologne

menthe; *German:* Bergamotminze, Kölnisch-Wasser-Minze, Limonenminze, Orangenminze; *Spanish:* yerba buena
Usage ▶ flavoring; **product:** essential oil (lemon mint oil; *German:* Zitronenminzöl) for liqueurs
Parts Used ▶ fresh leaf
Distribution ▶ widely cultivated and sometimes native, cultivated: Italy, USA, Brazil

Berger 2, 1950; Erhardt et al. 2002; Hiller/Melzig 1999; Hoffmann et al. 1992; Mikus/Schaser 1995; Ochoa/Alonso 1996; Roberts 1997; Sartoratto/Augusto 2003; Schultze-Motel 1986; Shiva et al. 2002; Srinivasan et al. 1981; Täufel et al. 1993; Teuscher 2003; Villamar et al. 1994; Wiersema/León 1999

Mentha x piperita L. var. citrata *(Ehrh.) Brig.*

▶ *Mentha x piperita L. 'citrata'*

Mentha pulegium L.

Common Names ▶ English mint, European pennyroyal, pennyroyal (mint), pudding grass; *Arabic:* habaq, flayou, fliou, fulayya, fulayha; *Chinese:* chun ce bo he; *French:* baume peteit, herbe aux puces, menthe pouliot, pouliot; *German:* Poleiminze, Hirschminze, Flohkraut; *Italian:* menta puleggia, menta poleggio, menta pulezzo, menta romana; *Portuguese:* poejo; *Spanish:* póleo, póleo omún, póleo negro;
Usage ▶ flavoring; **product:** essential oil (pennyroal [mint] oil) and pulegon
Parts Used ▶ leaf, herb
Distribution ▶ N Africa, Caucasus, C and W Asia, Europe; cultivated elsewhere

Aichele/Schwegler 4, 1995; Akhila/Banthorpe 1980; Bärtels 1997; Bournot 1968; Bremness 2001; Cheers 1997; Clair 1961; Davidson 1999; Dudtschenko et al. 1989; Erhardt et al. 2002; Erichsen-Brown 1989; Fleischhauer 2003; Hager 5, 1993; Hanelt 2001; Heeger 1956; Hiller/Melzig 1999; Hoffmann et al. 1992; Hoppe 1949; Kokkini et al. 2002; Peters 1927; Schönfelder 2001; Small 1997; Roberts 1997; Roth/Kormann 1997; Schultze-Motel 1986; Seidemann 1993c; Teuscher 2003; Tucker 1986; Ubillos 1989; Uphof 1968; Villamar et al. 1994; Wyk et al. 2004; Zeven/de Wet 1982

Mentha requienii *Benth.*

Common Names ▶ Corsican mint, Spanish mint, menthella; *French:* crème de menthe; *German:* Korsische Minze, Kriechende Minze, Spanische Minze; *Italian:* mentha di requien
Usage ▶ flavoring (locally)
Parts Used ▶ leaf, herb
Distribution ▶ SE and SW Europe: Corsica, Sardinia, Monte Cristo

Bremness 2001; Davidson 1999; Erhardt et al. 2002; Mucciarelli/Sacco 1999; Roberts 1997; Small 1997; Teuscher 2003; Wiersema/León 1999

Mentha rotundifolia auct.

▶ *Mentha suaveolens Ehrh.*

Mentha x rotundifolia *(L.) Huds.*

Synonyms ▶ *Mentha longifolia* (L.) Huds. x *Mentha suaveolens* Ehrh.
Common Names ▶ false apple mint, round-leaved mint; *French:* menthe rorondefeuillée; *German:* Apfelminze, Rundblättrige Minze; *Russian:* mjata jablotschnaja, mjata kruglolistaja, mjata egipetckaja, dikij bal'sam; *Spanish:* matapulgas
Usage ▶ spice, flavoring
Parts Used ▶ leaf, herb
Distribution ▶ Europe
Note ▶ The cultivar *Variegata* is an especially pretty plant, used to decorate salads and known as a pineapple mint.

Berger 2, 1950; Heeger 1956; Herklots 1972; Hoffmann et al. 1992; Nagell/Hefendehl 1972; Polochljobkin 1974, 1977; Schultze-Motel 1986; Small 1997; Täufel et al. 1993; Uphof 1968; Villamar et al. 1994; Zeven/de Wet 1982

Mentha sachalinensis *(Briq.) Kudo*

Common Names ▶ Sakhalin mint; *Chinese:* dong bei bo he; *German:* Sachalinminze
Usage ▶ spice
Parts Used ▶ herb

Distribution ▶ E Russia: Sakhalin Island, Norway

Rohloff 2002

Mentha x smithiana *R.A. Graham*

Common Names ▶ red mint, bergamot mint; *German:* Rote Minze, Bergamotminze
Usage ▶ spice, flavoring
Parts Used ▶ leaf, herb
Distribution ▶ native British Isle, in SW, C and EC Europe

Erhardt et al. 2002; Hanelt 2001; Schultze-Motel 1986; Small 1997; Zeven/de Wet 1982

Mentha spicata *L.*

Synonyms ▶ *Mentha viridis* (L.) L.
Common Names ▶ crisp mint, garden mint, spearmint, green mint, lamb mint; *Arabic:* hana, nemdar na'na'; *Chinese:* liu lan xiang, xiang hua cai; *French:* menthe verte, menthe douce, menthe en épi, menthe de Notre-Dame, menthe crépu; *German:* Ährige Minze, Grüne Minze, Krauseminze, Grüne Rossminze, Marokkanische Minze; *Hindi:* pahari pudina, pudina; *India:* fudina; *Italian:* menta verde, menta gentile, menta romana, menta crespa; *Japanese:* midori-hakkada; *Korean:* rokpakha; *Portuguese:* hortelã, hortelã commun, hortelã verde; *Russian:* mjata kudrjawaja, nemzkaja mjata, kurtschabaja mjata, lugowaja mjata; *Spanish:* hierba buena, hierba santa, hierba sana mentha, menta romana, mentha crespa
Usage ▶ spice, flavoring (locally); **product:** essential oil
Parts Used ▶ leaf, herb
Distribution ▶ N Africa, W Asia, SE Europe; native in N temperate regions; cultivated: China, India, Russia, England, USA

Aggarwal et al. 2002; Aichele/Schwegler 4, 1995; Bärtels 1997; Berger 2, 1950; Bournot 1968; Bremness 2001; Cheers 1997; Clair 1961; Davidson 1999; Dudtschenko et al. 1989; Erhardt et al. 2002; Erichsen-Brown 1989; Farrell 1985; Fleischhauer 2003; Hager 5, 1993; Hanelt 2001; Heeger 1956; Herklots 1972; Hiller/Melzig 1999; Hondelmann 2002; Jirovetz et al. 2003; Kokkini/Vokou 1989; Kouhila et al. 2001; Leung 1991; Lewington 1990; Maffei et al. 1986; Misra et al. 1989; Ochoa/Alonso 1996; Peters 1927; Pino et al. 2001; Platin et al. 1994; Pochljobkin 1974, 1977; Pruthi 1976; Pschyrembel 1998; Rosengarten 1969; Roth/Kormann 1997; Sartoratto/Augusto 2003; Satyanarayana et al. 2001; Schönfelder 2001; Schultze-Motel 1986; Schulz/Krüger 1999; Seidemann 1993c; Seidemann/Siebert 1987; Sharma 2003; Shiva et al. 2002; Small 1997; Täufel et al. 1993; Teuscher 2003; Tucker 1986; Uhl 2000; Uphof 1968; Villamar et al. 1994; Warren 1997; Wiersema/León 1999; Wüstenfeld/Haensel 1964; Wyk et al. 2004; Zeven/de Wet 1982

Mentha spicata *L.* x Mentha suaveolens *Ehrh.*

▶ *Mentha x villosa Huds.*

Mentha spicata *L.* var. crispa

▶ *Menthax cordifolia Opitz ex Fresen*

Mentha spicata *L.* var. longifolia *L.*

▶ *Mentha longifolia (L.) Huds.*

Mentha suaveolens *Ehrh.*

Synonyms ▶ *Mentha insularis* Req., *Mentha rotundifolia* auct.
Common Names ▶ apple mint, pineapple mint, round-leaved mint; *German:* Ananas- or Apfelminze, Rundblättrige Minze, Süßminze; *Spanish:* mastranzo
Usage ▶ pot-herb, flavoring (locally)
Parts Used ▶ leaf, herb
Distribution ▶ N Africa, W Asia, SW Europe native elesewhere

Aichele/Schwegler 4, 1995; Bärtels 1997; Bremness 2001; Cheers 1997; Davidson 1999; Erhardt et al. 2002; Fleischhauer 2003; Hager 5, 1993; Schultze-Motel 1986; Small 1997; Täufel et al. 1993; Teuscher 2003; Tucker 1986; Villamar et al. 1994; Warren 1997; Wiersema/León 1999; Wyk et al. 2004; Zeven/de Wet 1982

Mentha sylvestris *L.*

▶ *Mentha longifolia (L.) Huds.*

Mentha x villosa Huds.

Synonyms ▶ *Mentha gratissima* Weber, *Mentha spicata* L. x *Mentha suaveolens* Ehrh.

Common Names ▶ bowl mint, apple mint, wooly mint; *Arabian:* mersita, timersidi, timersat; *Chinese:* yuang ye bo he; *German:* Ananasminze, Bowlenminze, Breitblättrige Minze, Hainminze, Fuchsschwanz-Minze

Usage ▶ spice, flavoring, pot-herb

Parts Used ▶ leaf

Distribution ▶ S and W Europe, frequent cultivated: England

Note ▶ The species is a morphological variable.

Bremness 2001; Davidson 1999; Erhardt et al. 2002; Hanelt 2001; Mansfeld 1986; de Pooter/Schamp 1987; Schultze-Motel 1986; Small 1997; Täufel et al. 1993; Teuscher 2003; Zeven/de Wet 1982

Mentha viridis (L.) L.

▶ *Mentha spicata* L.

◘ Menyanthes trifoliata, flowering

MENYANTHES L. - Bogbean, Marsh Trefoil - Menyanthaceae

Menyanthes trifoliata L. ◘

Common Names ▶ bogbean, bogmyrtle, buckbean, marsh-clover, marsh-trefoil, water trefoil; *Chinese:* shi cai, shui cai, ming tsai, tsui tsao; *French:* trèfle d'eau, ményanthe trifolié; *German:* Biberklee; Dreiblatt, Dreiblättriger Bitterklee, Fieberklee, Sumpfklee, Zottelblume, Zottenblume; *Italian:* menianto, trifoglio d'acqua, trifoglio fibrino; *Russian:* vachta trilistnaja, trifol'; *Spanish:* trébol de aquático

Usage ▶ spice (flavoring); **product:** liqueurs

Parts Used ▶ leaf

Distribution ▶ C, N and E Europe, Siberia; E Canada, subartic America

Aichele/Schwegler 3, 1995; Berger 2, 1950; Cheers 1997; Coiciu/Racz (no year); Dudtschenko et al. 1989; Erhardt et al. 2002; Erichsen-Brown 1989; Fleischhauer 2003; Giaceri 1972; Heeger 1956; Hiller/Melzig 1999; Hohmann et al. 2001; Hoppe 1949; Newall et al. 1996; Pschyrembel 1998; Schnelle 1999; Schönfelder 2001; Seidemann 1993c; Sharma 2003; Turova et al. 1987; Uphof 1968; Wiersema/León 1999; Wyk et al. 2004

MERCURIALIS L. - Mercury - Euphorbiaceae

Mercurialis annua L.

Common Names ▶ annual mercury, herb mercury, French mercury; *Arabic:* mourkeba, halbub, bou zenzir; *French:* foirolle, ramberge; *German:* Einjähriges Bingelkraut, Garten-Bingelkraut; *Russian:* proljesnik

Usage ▶ pot-herb (rarely)

Parts Used ▶ leaf

Distribution ▶ SE Europe, Turkey, Levante, Caucasus, Canary Islands, N Africa, W Asia; in Europe also cultivated earlier

Aichele/Schwegler 3, 1995; Erhardt et al. 2002; Fleischhauer 2003; Hanelt 2001; Sánchez-Monga 1951; Schönfelder 2001; Schultze-Motel 1986; Wiersema/Léon 1999

MERIANDRA Benth. - Lamiaceae (Labiatae)

Meriandra bengalensis *(Konja ex Roxb.) Benth.*

Synonyms ▶ *Salvia bengaliensis* König ex Roxb.
Common Names ▶ Bengal sage; *German:* Bengalischer Salbei; *Hindi:* kafurkapat
Usage ▶ flavoring
Parts Used ▶ leaf, herb
Distribution ▶ Ethiopia, Eritrea, native in the Indian Peninsula

Chopra et al. 1956; Hanelt 2001; Kumar 2003; Schultze-Motel 1986; Teuscher 2003; Torre et al. 1992; Wealth of India 6, 1962

MERREMIA Dennst. ex Endl. - Convolvulaceae

Merremia dissecta *(Jacq.) Hall. f.*

Synonyms ▶ *Convolvulus dissectus* Jacq., *Ipomoea sinuata* Ort.
Common Names ▶ alamo vine, dissected merremia; *Chinese:* duo lie huang cao; *German:* Queensland-Holzrose, Zerschlitzte Merremie
Usage ▶ flavoring, liqueurs, cake
Parts Used ▶ leaf (dry)
Note ▶ The plant has a bitter almond oil flavor.
Distribution ▶ S America, USA: SE, Texas, Florida; French Antilles, Guadeloupe; cultivated in India

Dragendorff 1898; Hanelt 2001; Hoppe 3 1987; Schultze-Motel 1986; Seidemann 1993c

Merremia emarginata *(Burm.f.) Hallier*

Common Names ▶ emarginate merremia; *Chinese:* shen ye shan zhu cai; *German:* Emporstrebende Merremie
Usage ▶ pot-herb

Parts Used ▶ leaf
Distribution ▶ Peninsular India

Arora/Pandey 1996

Merremia umbellata *(L.) Hallier f.* ssp. orientalis *Ooststr.*

Synonyms ▶ *Ipomoea cymosa* Roem. et Schult.
Common Names ▶ *German:* Kleindoldige Merremie
Usage ▶ pot-herb (by Santhals)
Parts Used ▶ leaf
Distribution ▶ India

Arora/Pandey 1996

MERTENSIA Roth - Bluebell - Boraginaceae

Mertensia maritima *(L.) Gray*

Common Names ▶ oyster plant, sea smooth lungwort, gromwell; *German:* Austernpflanze, Meer-Mertensie; *Russian:* mertensija
Usage ▶ pot-herb for salads, raw vegetables, spread
Parts Used ▶ frish herb
Distribution ▶ Europe: British Isles, Scandinavia, N Russia, E Siberia, Amur, Sachalin, kamtschatka, Korea, Japan

Erhardt 2000, 2002; Teuscher 2003; Uphof 1968

MESSERSCHMIDIA Hebenstr. - Boraginaceae

Messerschmidia argentea *(L. f.) Johnston*

Synonyms ▶ *Tournefortia argentea* L. f.
Common Names ▶ velvet; *German:* Silber-Messerschmidie;
Usage ▶ spice, flavoring
Parts Used ▶ leaf
Distribution ▶ tropical Asia, India: Nicobar Island
Note ▶ The plant has a parsley-like aroma.

Arora/Pandey 1996; Hanelt 2001; Uphof 1968; Zeven/de Wet 1982; Zeven/Zhukovski 1975

MEUM Mill. - Spignel - Apiaceae (Umbelliferae)

 Meum athamanticum *Jacq.*

Common Names ▶ baldmoney, bearwort, meu, spignel; *French:* baudremoine, cistre, fenouil des Alpes, meum; *German:* Bärwurz, Bärenfenchel, Bärkümmel, Dillblattwurz, Köppernickel; *Italian:* finocchio alpino, finochiello, meo; *Russian:* meum atamanskij; *Spanish:* finocha alpino
Usage ▶ spice, for herb spirits and liqueurs
Parts Used ▶ root
Distribution ▶ E, C European mountains

Aichele/Schwegler 3, 1995; Brandt/Schultze 1995; Cheers 1997; Erhardt et al. 2002; Fleischhauer 2003; Hager 5, 1993; Hiller/Melzig 1999; Hoppe 1949; König et al. 1996; Melchior/Karsten 1974; Schnelle 1999; Schönfelder 2001; Schultze-Motel 1986; Seidemann 1993c, 1995b; Stahl/Bohrmann 1967; Täufel et al. 1993; Uphof 1968; Walther 1969; Zeven/de Wet 1982

 **Meum mutellina** *Gaertn.*

▶ *Ligusticum mutellina (L.) Crantz*

 Meum piperitum *(Ucria) Spreng.*

▶ *Foeniculum vulgare Mill. ssp. piperitum (Ucria) Cout.*

MICHELIA L. - Magnoliaceae

 Michelia champaca *L.*

Synonyms ▶ *Michelia evonymoides* Burm.
Common Names ▶ golden champa, yellow champa; *Chinese:* chen po, chen pò ka, kim cuong moc, huang lan; *French:* champac; *German:* Chamapaka; *Hindi:* champaca, champa; *India:* campahah, campa, chamba, chempaka merah, tschampaka; *Indonesian:* cempaka, lecari; *Javanese:* chěmpaka, pěchari lochari; *Malaysian:* orang chempaka, cempaka merah, cempaka putih, cempaka kuning; *Pilipino:* champaca, tsampaka; *Thai:* cham phaa; *Vietnamese:* ngoc ian
Usage ▶ flavoring of tea; **product:** essential oil (chamapaca oil)
Parts Used ▶ flower
Distribution ▶ India, Malaysia, widely cultivated in the Tropics

Bourton 1968; Engel/Phummai 2000; Erhardt et al. 2002; Hanelt 2001; Hiller/Melzig 1999; Schultze-Motel 1986; Sharma 2003; Shiva et al. 2002; Uphof 1968

 Michelia evonymoides *Burm.*

▶ *Michelia champaca L.*

 Michelia figo *L.*

Synonyms ▶ *Liriodendron figo* Lour., *Michelia fuscata* (Andr.) Wall., *Magnolia figo* DC.
Common Names ▶ banana shrub, dwarf champac; *Chinese:* han xiad, han hsiao, chempaka ambon; *German:* Bananenstrauch; *Malaysian:* dwarf chempaka, chempaka ambon, kaum chempaka, pisang pisang; *Thai:* champii khaek; *Vietnamese:* tui' tiên
Usage ▶ flavoring of tea (rarely)
Parts Used ▶ flower
Distribution ▶ SE China, Georgia, coasts of the Black Sea, widely cultivated

Burkill 1966; Engler/Phummai 2000; Erhardt 2000, 2002; Hanelt 2001; Schultze-Motel 1986; Uphof 1968; Wealth of India 6, 1962; Wiersema/León 1999; Zeven/de Wet 1982

 Michelia fuscata *(Andrews) Wall.*

▶ *Michelia figo L.*

MICROMERIA Benth. - Lamiaceae (Labiatae)

Micromeria alternipilosa *K. Koch*

▶ *Satureja spicigera (K. Koch) Boiss.*

Micromeria fruticosa *(L.) Druce*

Common Names ▶ talmud widder hyssop, tea hyssop; *German:* Arabisches Bergkraut, Strauch-Felsenklippe
Usage ▶ flavoring, especially for lamb meat in the Arabic-Jewish cuisine; **product:** essential oil
Parts Used ▶ herb
Distribution ▶ Mediterranean region
Note ▶ The major component of the essential oil is pulegone (about 70%).

Dudai et al. 1993; Small 1997

Micromeria juliana *(L.) (Rchb.)*

Synonyms ▶ *Satureja juliana* L.
Common Names ▶ Julian savory; *French:* sarriette de Saint-Julian; *German:* Julian-Salbei
Usage ▶ spice
Parts Used ▶ herb
Distribution ▶ Mediterrean region

Small 1997

Micromeria montana *(L.) Rchb.*

▶ *Satureja montana* L.

Micromeria piperella Benth.

Synonyms ▶ *Thymus piperella* L.
Common Names ▶ *German:* Pfefferthymian; *Spanish:* pebreila, pebrinella, piperesa
Usage ▶ spice
Parts Used ▶ leaf
Distribution ▶ SW Europe

Micromeria spicigera *K. Koch*

▶ *Satureja spicigera (K. Koch) Boiss.*

Micromeria thymifolia *(Scop.) Fritsch*

Synonyms ▶ *Micromeria rupestris* (Wulfen ex Jacq.) Benth.
Common Names ▶ mountain mint; *German:* Thymianblättrige Bergminze
Usage ▶ flavoring (rarely)
Parts Used ▶ herb
Distribution ▶ Balkans: Croatia, Bosnia, S Italy, Hungary (rarely)

Erhardt et al. 2002; Kalogjera et al. 1993

MIKANIA Willd. - Asteraceae (Compositae)

Mikania parviflora *(Aubl.) Karst*

Synonyms ▶ *Eupatorium parviflorum* Aubl.
Common Names ▶ *French:* plante de l'étoile; *German:* Kleinblütige Mikanie; *Portuguese:* cipó-catinga, guaco; *Spanish:* guaco marda
Usage ▶ flavoring (rarely)
Parts Used ▶ flower
Distribution ▶ native to Amazonia, tropical C, S America
Note ▶ The flowers have a vanilla-like flavor.

Burkill 1985; Freise 1936; Hanelt 2001; Mors et al. 2000; Schultze-Motel 1986

Mimosa dulcis L.

▶ *Pithecellobium dulce (Roxb. Benth.*

Mimosa jiringa *Jack*

▶ *Archidendron jiringa (Jack) I. Nielson*

Mimosa biglobosa *Jacq.*

▶ *Parkia biglobosa (Jacq.) R. Br. ex G. Don*

MINTHOSTACHYS (Benth.) Spach. - Lamiaceae (Labiatae)

 Minthostachys mollis *(Benth.) Griseb.*

Synonyms ▶ *Mentha mollis* Benth., *Minthostachys spicata* (Benth.) Epling
Common Names ▶ Ecuadorian mint, tipo leaf ; *German:* Ecuadorische Minze, Argentinische Minze, Weiche Minze, Gewürzminze, Peperina; *Argentina:* peperina, piperina; *Ecuador:* muña, poleo, tipo
Usage ▶ spice (locally), condiment; **product:** essential oil
Parts Used ▶ leaf, herb
Distribution ▶ Andes: Venezuela, Colombia to Bolivia, and NW Argentina

Alingga/Feldeim 1985; Alkire et al 1994; Hanelt 2001; Schultze-Motel 1986; Täufel et al. 1993

 Minthostachys spicata *(Benth.) Epling*

▶ *Minthostachys mollis (Benth.) Griseb.*

MIRABILIS L. - Four O'Clock Plant - Nyctaginaceae

 Mirabilis jalapa *L.*

Synonyms ▶ *Mirabilis uniflora* Schrank
Common Names ▶ beauty of the night, false jalap, four o'clock plant, marvel of Peru; *Chinese:* zi mo li gen, zu mo li, yen chih; *French:* belle du nuit; *German:* Jalapa-Wunderblume; *Italian:* bella di notte; *Japanese:* oshiroi-bana; *Korean:* punkkot; *Malaysian:* kempang pukul empat, bunga pechah empet; *Mexico:* maravilla, Don Diego de node, arreboleva; *Spanish:* buenas tardes; clavellina tardes, hoja de xalape, maravilla
Usage ▶ spice, used like pepper (*Piper nigrum* L.)
Parts Used ▶ seed
Distribution ▶ possible origon in Mexico, Texas, Peru, native in N America, W Africa: Gabun, also cultivated in the Tropics

Aichele/Schwegler 2, 1994; Bärtels 1997; Cheers 1997; Erhardt et al. 2002; Hager 5, 1993; Hanelt 2001; Neuwinger 1999; Rätsch 1998; Uphof 1968; Villamar et al. 1994; Wiersema/León 1999; Zeven/de Wet 1982

 Mirabilis uniflora *Schrank*

▶ *Mirabilis jalapa L.*

 Moldavica punctata *Moench*

▶ *Dracocephalum moldavica L.*

 Moldavica suaveolens *Gilib.*

▶ *Dracocephalum moldavica L.*

MOLLUGO L. - Mollugine - Molluginaceae

 Mollugo hirta *Thunb.*

▶ *Glinus lotoides L.*

 Mollugo oppositifolia *L.*

▶ *Glinus oppositifolius (L.) DC.*

 Mollugo pentaphylla *L.*

Synonyms ▶ *Mollugo stricta* L.
Common Names ▶ carpet weed; *Chinese:* di ma huang; *German:* Fünfblättriges Weichkraut; *Japanese:* zakuro-sô; *Malaysian:* rumput belangkas, tapak burong
Usage ▶ pot-herb

Parts Used ▶ herb
Distribution ▶ China, India, Madagascar

Arora/Pandey 1996; Burkhill 1966; Neuwinger 1999; Uphof 1968; Wealth of India 6, 1962

Mollugo stricta L.

▶ *Molluga pentaphylla* L.

MOMORDICA L. - Bitter Cucumber - Cucurbitaceae

Momordica balsamina L.

Synonyms ▶ *Momordica schinzii* Cogn.
Common Names ▶ African cucumber, balsam apple; *Arabic:* mokah; *French:* margose, pomme de mercveille; *German:* Balsamapfel, Wunderapfel, Springkürbis, Warzengurke; *Nigerian:* garafuni; *Russian:* momordika
Usage ▶ pot-herb
Parts Used ▶ young leaf
Distribution ▶ tropical Africa: The Gambia, Ghana, Niger, Nigeria, Senegal, Sierra Leone; S Africa, W Asia, China, Malaysia, Archipelago, Australia, native in tropical America and Australia, cultivated in West Indies

Burkill 3, 1995; Hanelt 2001; Dalziel 1957; Erhardt et al. 2002; Hiller/Melzig 1999; v. Koenen 1996; Neuwinger 1998; Schultze-Motel 1986; Täufel et al. 1993; Tommasi et al. 1991; Wealth of India 6, 1962; Zeven/de Wet 1982

Momordica ceylanicum Mill.

▶ *Momordica charantia* L.

Momordica charantia L.

Synonyms ▶ *Momordica indica* L., *Momordica muricata* Willd., *Momordica ceylanicum* Mill.
Common Names ▶ balsam pear, bitter cucumber, bitter gourd, bitter melon, carilla plant; *Brazil (Portuguese):* Melao de St. Caetano; *Chinese:* ku gua, k'u kua, chin li chih, lai pu-tao; *French:* concombre africain, margose; *German:* Balsambirne, Amerikanische Springgurke, Bittere Springgurke; *Hindi:* karela; *Indonesian:* pepare; *Iran:* ginahang; *Japanese:* gôyâ, ki uri, naga-reishi; *Korean:* yuja; *Spanish:* bálsamo, balsamito cundeamor; *Thai:* mara, phak hoei
Usage ▶ spice, for fish and meat, curries
Parts Used ▶ pulp
Distribution ▶ native in SE USA, tropical countries, cultivated especially in E Asia

Arora/Pandey 1996; Ayensu 1978; Burkill 1985; Davidson 1999; Erhardt et al. 2002; Fatope et al. 1990; Gurip-Fakim/Brendler 2004; Hanelt 2001; Lück 2004; Neuwinger 1998; Schultze-Motel 1986; Sharma 2003; Uphof 1968; Villamar et al. 1994; Wealth of India 6, 1962; Wiersema/León 1999

Momordica indica L.

▶ *Momordica charantia* L.

Momordica muricata L.

▶ *Momordica charantia* L.

Momordica schinzii Cogn.

▶ *Momordica balsamina* L.

MONARDA L. - Indian nettle - Lamiaceae (Labiatae)

Monarda aristata Nutt.

▶ *Monarda citriodora* Cerv. ex Lag.

Monarda austromontana Epling

Common Names ▶ Mexican bergamot; *German:* Mexikanische Monarde, Mexikanische Bergamotte
Usage ▶ flavoring, pot-herb (rarely)
Parts Used ▶ leaf
Distribution ▶ Arizona, New Mexico, Mexico

Note ▶ The flavor is like oregano (*Origanum vulgare* L.).

Small 1997

Monarda citriodora *Cerv. ex Lag.*

Synonyms ▶ *Monarda aristata* Nutt., *Monarda dispersa* Sm.

Common Names ▶ lemon bergamot, lemon mint; *German:* Präriemonarde, Indianernessel, Zitronen-Monarde; *Mexico:* mount pima oregano

Usage ▶ spice, flavoring (tea and by the Hopi Indian spiced game meat); **product:** essential oil

Parts Used ▶ leaf

Distribution ▶ USA: Missiouri to Nebraska, S Florida, Alabama, Texas, New Mexico, Arizona and Mexico; cultivated in USA

Note ▶ In the Crimea cultivated experimentally for the essential oil.

Cheers 1997; Collins et al. 1994; Hanelt 2001; Mansfeld 1986; Mikus/Schaser 1995; Mikus et al. 1997; Scora 1967; Small 1997; Stobart 1974; Teuscher 2003; Tucker 1986; Uphof 1968; Wiersema/León 1999

▪ **Monarda didyma, flowering**

Monarda didyma *L.*

Common Names ▶ fragrant balm, Oswega bee balm, Oswega tea; *Chinese:* mei guo bo he; *French:* monarde; *German:* Goldmelisse; Scharlach-Bergamotte, Scharlach-Monarde, Pferdeminze, Rote Indianernessel, Rote Melisse

Usage ▶ spice, flavoring

Parts Used ▶ leaf, herb

Distribution ▶ USA: NE, C, SE; S America

Note ▶ The plant was introduced to Europe in 1656 as an ornamental plant.

Bremness 2001; Carnat et al. 1991a; Davidson 1999; Erhardt et al. 2002; Fleischhauer 2003; Hiller/Melzig 1999; Lück 2000; Schönfelder 2001; Small 1997; Täufel et al. 1993; Teuscher 2003; Uphof 1968

Monarda dispersa *Sm.*

▶ *Monarda citriodora Cerv. ex Lag.*

Monarda fistolosa *L.*

Common Names ▶ bee balm, horse bergamot, wild bergamot; *Chinese:* ni mei guo bo he; *French:* monarde fistuleuse; *German:* Goldmelisse, Späte Indianernessel, Röhrenblütige Monarde, Rosenmonarde, Wilde Bergamotte, Wilde Monarde; *Spanish:* oregano de la sierra

Usage ▶ pot-herb (salad), spice (locally)

Parts Used ▶ leaf

Distribution ▶ Canada, SW USA: California, Mexico

Berger 4, 1954; Cheers 1997; Chubey 1982; Davidson 1999; Dudtschenko et al. 1989; Erhardt et al. 2002; Erichsen-Brown 1989; Heinrich 1973, 1975; Mansfeld 1986; Marshall/Scora 1972; Mazza et al. 1987; Pfab et al. 1980; Schultze-Motel 1986; Seidemann 1993c; Small 1997; Teuscher 2003; Tucker 1986; Uphof 1968

Monarda pectinata *Nutt.*

Synonyms ▶ *Monarda penicillata* Gray

◘ Monarda fistulosa, flowering

Common Names ▶ pony beebalm, spotted beebalm; *German:* Getüpfelte Kamm-Monarde
Usage ▶ pot-herb
Parts Used ▶ leaf
Distribution ▶ USA: Nebraska, Colorada, S Arizona, Texas, and New Mexico
Note ▶ Sporadically included *Monarda pectinata* Nutt. in *Monarda citriodora* Cerv. ex Lag.

Hanelt 2001; Tucker 1986; Uper 1968; Usher 1974; Wiersema/León 1999

 Monarda punctata *L.*

Common Names ▶ American horse mint, dotted monarda, spotted beebalm; *German:* Punktierte Monarde
Usage ▶ pot-herb (rarely); **product:** essential oil
Parts Used ▶ leaf
Distribution ▶ N America, from E and C U.S. to northern Mexico
Note ▶ The plant is a source of thymol.

Mansfield 1986; Small 1997; Usher 1968

 Monarda penicillata *Gray*

❯ *Monarda pectinata* Nutt.

MONARDELLA Benth. – Lamiaceae (Labiatae)

 Monardella odoratissima *Benth.*

Common Names ▶ cloverhead horse mint, coyote mint, mountain pennyroyal, Pacific monardella, stinking horse mint; *German:* Stink-Pferdeminze
Usage ▶ flavoring (rarely)
Parts Used ▶ leaf, herb
Distribution ▶ WN America

Erhardt et al. 2002

MONDIA Skeels - Asclepiadaceae (Periplococeae)

 Mondia whitei *(Hook.f.) Skeels*

Synonyms ▶ *Chlorocodon whitei* Hook.f.
Usage ▶ spice for sauces; **product:** isovanillin
Parts Used ▶ root, rhizome
Distribution ▶ tropical Africa: Senegal to E Africa, S Africa (Natal)

Ficalho 1947; Koorbanally et al. 2000; Mabberley 1999; McCartan/Crouch 1998; Msonthi 1991; Neuwinger 1994, 1998

MONODORA Dunal - Annonaceae

Monodora angolensis *Welw.*

Common Names ▶ Angola calabash; *German:* Angola Kalabassenmuskat
Usage ▶ condiment
Parts Used ▶ seed

MONODORA Dunal - Annonaceae

Distribution ▶ tropical Africa, especially Angola

Uphof 1968

Monodora borealis Elliot

▶ *Monodora myristica (Gaertn.) Dunal*

Monodora brevipes Benth.

Common Names ▶ short-stemmed African nutmeg, yellow flowery nutmeg; *German:* Kurzstielige Kalabassen-Muskatnuss, Kalabassen-Macisbohne
Usage ▶ condiment, spice blend: mixed with 'country' pepper (?*Piper guinense* Schum. et Thonn) used for spice
Parts Used ▶ seed
Distribution ▶ W Africa: Ghana, Liberia, Nigeria, Sierra Leone

Burkill 1, 1985

Monodora grandiflora Benth.

▶ *Monodora myristica (Gaertn) Dunal*

Monodora myristica *(Gaertn.) Dunal*

Synonyms ▶ *Annona myristica* Gaertn., *Monodora borealis* Elliot; *Monodora grandiflora* Benth., *Xylopia undulata* P. Beauv.
Common Names ▶ African nutmeg, false nutmeg, calabash nutmeg, Jamaica nutmeg; *French:* fausse noix de muscade, muscade de Calabash, muscadier d'Afrique, muscadier de Calabasse; *German:* Kalabassen-Muskatnuss, Macisbohne, Afrikanische Gewürzbeere
Usage ▶ spice, condiment
Parts Used ▶ seed
Distribution ▶ W, C Africa, Angola, Ghana, Liberia, Nigeria, Sierra Leone, Uganda, W India
Note ▶ Very occasionally as a substitute for nutmeg; the taste is not like nutmeg.

▪ **Monodora myristica:** a **flowering,** b **seeds**

Bois 1934; Burkill 1966; Burkill 1, 1985; Ekundayao/Hammerschmidt 1988c; Erhardt et al. 2002; Guzman/Siemonsma 1999; Hager 5, 1993; Hanelt 2001; Hiller/Melzig 1999; Irvine 1961; Lamaty et al. 1987; Lammoureua 1975; Oboh/Ekperigin 2004; Okeke 1998; Omobuwajo et al. 2003; Onyenekwe et al. 1993; Schenck/Naundorf 1966; Schultze-Motel 1986; Seidemann 1996; Täufel et al. 1993; Teuscher 2003; Uphof 1968

Monodora tenuifolia *Benth.*

Common Names ▶ thin-leaved African nutmeg; *German:* Dünnblättrige Afrikanische Muskatnuss
Usage ▶ condiment
Parts Used ▶ seed
Distribution ▶ W Africa

Ayensu 1978; Burkill 1, 1985

MONTIA L. – Blink, Winter Purslane - Portulaccaceae

 Montia perfoliata *(Donn) Howell;*

Synonyms ▶ *Claytonia perfoliata* Donn; *Limnia perfoliata* (Donn) Haw.
Common Names ▶ miner's lettuce, winter purslane; *Chinese:* chuan ye chun mei cao; *French:* pourpier d'hiver, claytone de Cuba, claytonia perfoliée; *German:* Tellerkraut, Winterportulak, Winterpostelein; *Italian:* portulaca d'inverno; *Portuguese:* beldroega de Inverno, claitónia, *Russian:* monzija, Kubinskij schpinat; *Spanish:* verdolaga de invierno, verdolago de Cuba
Usage ▶ pot-herb
Parts Used ▶ fresh herb
Distribution ▶ Canada, USA: NW, Rocky Mountains, California; Mexico; native in W Europe, C Europe, Cuba

Bremness 2001; Erhardt et al. 2001; Hanelt 2001; Kays/Dias 1995; Terra 1966

MORINDA L. – Indian Mulberry - Rubiaceae

 Morinda bracteata *Roxb.*

▶ *Morinda citrifolia* L.

 Morinda citrifolia *L.*

Synonyms ▶ *Morinda bracteata* Roxb., *Morinda litoralis* Blanco, *Morinda macrophylla* Desf.
Common Names ▶ Indian mulberry, Indian painkiller, awl tree, brimestone tree; *French:* morinde, mûrier des Indes; *German:* Indische Maulbeere, Noni, Zitronenblättrige Morinda; *India:* bartundi, surangji; *Indonesian:* menkudu nenghudu, *Japanese:* yae-aoki; *Malaysian:* kattapitalavam, mannanatti; *Pacific region:* noni; *Spanish:* mora de la India; *Tahaiti:* nono; *Thai:* yo baan, mataasuea; *Vietnamese:* nhau, nhau rung
Usage ▶ pot-herb
Parts Used ▶ young leaf
Distribution ▶ native in Queensland (Australia), perhaps ranged to the South Sea Islands from eastern Polynesia to India, both wild and cultivated; Vietnam; naturalized in the New World in C to S America (Mexico, Panama, Columbia, Venezuela), and throughout the Caribbean, the Florida keys, the Indian Ocean Islands, and the West Indies common in coastal areas

Burkill 1966; Erhardt et al. 2002; Gurib-Fakim/Brendler 2004; Jacquat 1990; Hiller/Melzig 1999; Lück 2004; Ogle et al. 2003; Rätsch 1998; Schultze-Motel 1986; Seidemann 2001c; Storrs 1997; Uphof 1968; Wealth of India 6, 1962; WHO 1990; Wiersma/León 1999; Zeven/de Wet 1982

 Morinda litoralis *Blanco*

▶ *Morinda citrifolia* L.

 Morinda macrophylla *Desf.*

▶ *Morinda citrifolia* L.

MORINGA Adans. - Horse Radish Tree - Moringaceae

 Moringa erecta *Salisb.*

▶ *Moringa oleifera* Lam.

MORINGA Adans. - Horse Radish Tree - Moringaceae

◘ Morinda citrifolia, fruiting

◘ Moringa oleifera: a flowering, b fruit

Moringa oleifera Lam.

Synonyms ▶ *Moringa erecta* Salisb., *Moringa pterygosperma* Gaertn., *Moringa zeylanica* Pers.

Common Names ▶ drumstick, horse radish tree, radish tree; *French:* ben aillé, moringa ailel, pois quénique; *German:* Meerrettichbaum, Pferderettichbaum, Flügelsamiger Behennussbaum, Ölmoringie; *India:* sahijan, sangina, soajna, sigruh, moringu, murinna, murinkai, nunga; *Indonesian:* kelor; *Japanese:* wasabi-nok; *Malaysian:* emmanggai, kelok, kachang, kelir, lemunggai, morunggai; *Pilipino:* arunggai, kamalungai, malunggay; *Portuguese:* muringueiro; *Russian:* moringa; *Sanskrit:* sigru, s(h)obhanjana; *Spanish:* maranga, pareíso, pareíso blanco; *Swahili:* sojana; *Thai:* ma rum, phak-nuea-kai, phak eehum

Usage ▶ pot-herb (India); couscous and sauce spice (fruits and leaves)

Parts Used ▶ flower, leaf, fruit

Distribution ▶ India, cultivated elsewhere, e.g. Senegal

Note ▶ The flowers are also used as 'pot-herb'. The oil of the fruits is used for cosmetics and lamps.

Arora/Pandey 1996; Bendel 2002; Bois 1934; Burkill 4, 1997; Cheers 1997; Dastur 1954; Davidson 1999; Duke et al. 2003; Erhardt et al. 2002; Gurib-Fakim/Brendler 2004; Hager 5, 1993; Hanelt 1985, 2001; Hiller/Melzig 1999; Hutton 1998; Kottegoda 1994; Kumar 2003; Lalas 1998; Lück 2004; Mc-Navy/Wood 2003; Mansfeld 2003; Morton 1991; Neuwinger 1999; Oliveira et al. 1999; Oomen/Grubben 1978; Ramachandran et al. 1980; Schenck/Naundorf 1966; Schultze-Motel 1986; Seidemann 1993c; Seshadri/Nambiar 2003; Sharma 2003; Siemonsma/Piluek 1993; Small 1997; Somali et al. 1984; Storrs 1997; Täufel et al. 1993; Tsaknis et al. 1998, 1999; Uphof 1968; Villamar et al. 1994; Wealth of India 6, 1962; Wiersema/León 1999; Zeven/de Wet 1982

Moringa pterygosperma Gaertn.

▶ *Moringa oleifera* Lam.

Moringa zeylanica Pers.

▶ *Moringa oleifera* Lam.

Morisonia flexuosa L.

▶ *Capparis flexuosa* L.

Moschosma polystachon *Benth.*

▶ *Basilicum polystachon (L.) Moench*

 Mukia maderaspatana *(L.) M. Roem.*

▶ *Melothria maderaspatana (L.) Cogn.*

 Mukia scabrella *Arn.*

▶ *Melothria maderaspatana (L.) Cogn.*

MURRAYA König ex L. - Orange Jessamine - Rutaceae

 Murraya amoena *Salisb.*

▶ *Murraya paniculata (L.) Jack*

 Murraya exotica *L.*

▶ *Murraya paniculata (L.) Jack*

◘ **Murraya koenigii, plant**

 Murraya koenigii *(L.) Spreng.* ◘

Synonyms ▶ *Bergora koenigi* L., *Chalcas koenigii* Kurz, *Nimbo melioides* Dennst.
Common Names ▶ curry leaf, Indian bay; *French:* feuille de Murraya; *German:* Curryblätter, Curry-Orangenraute; *India:* kathnim, barsanga, bowala, gandhela, gendela, karipatta, karrinim, kasapiucha, limdo (mitho), mitha neem, kurry patta, *Indonesian:* daun kari; *Malaysian:* daun kari, karwa pale, kerupulai, garupillai; *Sanskrit:* surabhi nimbu; *Slovenian:* listy karí; *Sri Lanka:* karapincha; *Thai:* bai karee
Usage ▶ spice, pot-herb
Parts Used ▶ leaf
Distribution ▶ India, Sri Lanka, Thailand, Laos, Cambodia, Pacific Islands (Fiji), Bangladesh

Arora/Pandey 1996; Chauhan 1999; Cheers 1997; Craze 2002; Davidson 1999; Duke et al. 2003; Erhardt et al. 2002; Farrell 1985; Farhath Kanum et al. 2000; Hanelt 2001; Hohmann 1971; Kumar 2003; Lück 2004; MacLeod/Pieris 1982; MacCloed/Pieris 1982; Mallavapura et al. 1999; Morris/Mackley 1999; Norman 1990, 1991; Peter 2001; Prakash/Natarajan 1974; Raina et al. 2002; Saha/Chowdhury 1998; Schultze-Motel 1986; Seidemann 1993c, 1998/2000; Seidemann/Siebert 1987; Sharma 2003; Siewek 1990; Tachibana et al. 2001; Täufel et al. 1993; Teuscher 2003; Uhl 2000; Uphof 1968; Waßmuth-Wagner et al. 1992, 1995; Wiersema/León 1999; Zeven/de Wet 1982

Murraya paniculata *(L.) Jack*

Synonyms ▶ *Chalcas paniculata* L., *Murraya amoena* Salisb., *Murraya exotica* L., *Murraya sumatrana* Roxb.
Common Names ▶ bark tree, jasmine orange, Chinese myrtle, Chinese box-wood, Burmese box-wood, satinwood, cosmetic park tree, Hawaiian orange, orange jasmin; *Chinese:* chiu li xiang, chiu li hsiang tsao, kau lei heung; *French:* buis de Chine; *German:* Falscher Jasmin, Orangenraute; *Hindi:* kamina marchula, konji, nagagolunga, pandara; *India:* angara gida, ban mallika, konji; *Indonesian:* kemuning;

Japanese: gek-kitsu, gikiji; *Javanese:* djenar, kemoening; *Malaysian:* kĕmung, kemuning; *Pilipino:* banaasi, banati, kamuning; *Thai:* dawk kaeo, kêo

Usage ▶ flavoring (tea, confectionery)
Parts Used ▶ flower, (leaf, very sporadically)
Distribution ▶ N India, Sri Lanka, S China, Sumatra, Java, Philippines, Malaysia, New Caledonia, Australia, Pacific Islands

Cheers 1997; Dastur 1954; Davidson 1999; Engler/Pummai 2000; Erhardt et al. 2000, 2002; Kimoshita/Firman 1997; Kottegoda 1994; NICPBP 1987; Schultze/Motel 1986; Sharma 2003; Storrs 1997; Tao Duoda 1998; Uphof 1968; Villamar et al. 1994; Wiersema/León 1999; Zeven/de Wet 1982

Murraya sumatrana *Roxb.*

 Murraya paniculata (L.) Jack

MUSSAENDA L. - Rubiaceae

Mussaenda frondosa *L.*

Common Names ▶ *Hindi:* bebina, bedina; *Nepal:* asari
Usage ▶ pot-herb
Parts Used ▶ leaf
Distribution ▶ tropical Himalayas, cultivated in Indian gardens

Hanelt 2001; Schultze-Motel 1986; Uphof 1968

Myrodia funebris *(La Llave) Benth.*

 Quararibea funebres (La Llave) Vischer

MYRIANTHUS Pall. - Cecropiaceae

Myrianthus libericus *Rendle*

Usage ▶ pot-herb
Parts Used ▶ young leaf
Distribution ▶ Liberia

Burkill 1, 1985

MYRICA L. - Gale, Bog Myrtle - Myricaceae

Myrica cerifera *L.*

Common Names ▶ bay berry, candle berry, wax berry, wax myrtle; *French:* arbre à suif, cirier; *German:* Wachs-Myrte; *Italian:* corteccia
Usage ▶ spice, substitute for bay leaves in seasoning meats, sauces, soups
Parts Used ▶ leaf
Distribution ▶ NE, SE USA: Florida, Texas

Erhardt et al. 2002; Rätsch 1998; Tull 1999; Wyk et al. 2004

Myrica gale *L.*

Common Names ▶ bog myrtle, gale, meadow fern, sweet gale, sweet myrtle, wax berry; *French:* galé odorant, galé piment aquatique, myrte bâtard, myrte des Marais; *German:* Gagelstrauch, Heidemyrte, Moor-Gagelstrauch, Sumpfmyrte; *Italian:* mirto bastarda; *Russian:* mirt bolotnyj; woskowniza obyknowennaja; *Spanish:* mirto de brabante, mito holandés
Usage ▶ spice
Parts Used ▶ herb
Distribution ▶ E Asia, N America, Europe: France, England, Germany

Berger 4, 1954; Cheers 1997; Davidson 1999; Erhardt et al. 2002; Erichsen-Brown 1989; Fleischhauer 2003; Hiller/Melzig 1999; Ingels 1987; Lawrence/Weaver 1974; Rätsch 1998; Schantz/Kapétanidis 1971; Schnelle 1999; Seidemann 1993c; Uphof 1968; Wiersema/León 1999

Myrica pennsylvanica *Loisel.*

Synonyms ▶ *Myrica carolinensis* hort.
Common Names ▶ bay berry, candle berry, northern bayberry, waxy berry; *German:* Amerikanischer Gagelstrauch, Wachsbeere
Usage ▶ flavoring
Parts Used ▶ herb

Myrica gale: a **flowering**, b **fruiting**

Distribution ▶ E Canada, NE USA, Ohio; native in New Zealand, S England

Erhardt et al. 2002; Small 1997

MYRISTICA Gronov. - Nutmeg - Myristicaceae

 Myristica argentea Warb.

Common Names ▶ silver nutmeg; *French:* noix muscade mâle, noix muscade sauvage, noix muscade macassar, noix papoue; *German:* Papua-Muskat, , Makassar-Muskat, Langer Muskat, Wilder Muskat, Silber-Muskat, Pferde-Muskat
Usage ▶ spice
Parts Used ▶ seed, aril
Distribution ▶ New Guinea, where it is also occasionally cultivated

Bois 1934; Griebel 1909; Hager 5, 1993; Hiller/Melzig 1999; Mansfeld 1986; Melchior/Kastner 1974; Pruthi 1976; Rätsch 1998; Schenck/Naundorf 1966; Schultze-Motel 1986; Seidemann 1993c;
Seidemann/Siebert 1987; Siewek 1990; Teuscher 2003; Warburg 1897; Zeven/de Wet 1982

 Myristica aromatica *Lamk*

▶ *Myristica fragrans Houtt.*

 Myristica castaneifolia *A. Gray*

Synonyms ▶ *Myristca macrantha* A.C. Sm., *Myristica macrophylla* A. Gray
Common Names ▶ chestnut-leafy nutmeg; *German:* Großblütiger Muskat, Kastanienblättrige Muskat
Usage ▶ flavoring
Parts Used ▶ seed
Distribution ▶ SE Asia

Oyen/Dung 1999

 Myristica cinnamomea *King*

Common Names ▶ cinnamom nutmeg; *German:* Zimt-Muskat; *Malaysian:* mendarah, pala bukit
Usage ▶ spice (locally)
Parts Used ▶ seed
Distribution ▶ Malaysian Peninsula, Singapore, Borneo, Philippines

Guzman/Siemonsma 1999; Sinclair 1968

 Myristica contorta *Warb.*

▶ *Myristica dactyloides Gaertn.*

 Myristica corticosa *Hock.f.*

▶ *Knema corticosa Lour.*

 Myristica dactyloides *Wall.*

▶ *Myristica malabarica Lam.*

MYRISTICA Gronov. - Nutmeg - Myristicaceae

 Myristica dactyloides *Gaertn.*

Synonyms ▶ *Myristica laurifolia* Hook.f. et Thoms; *Myristica contorta* Warb.

Common Names ▶ Malabar nutmeg; *German:* Gefingerter Muskat, Malabar-Muskat; *Sri Lanka:* malaboda; *Tamil:* katjathikai

Usage ▶ spice (rarely)

Parts Used ▶ seed

Distribution ▶ India: S Madras, Malabar coast; Sri Lanka

Hanelt 2001; Hiller/Melzig 1999; Schultze-Motel 1986; Wealth of India 6, 1962

 Myristica fatua *Houtt.*

Synonyms ▶ *Myristica tomentosa* Thunb.

Common Names ▶ mountain nutmeg; *Dutch:* mannetjes nooten; *French:* muscadier de forêt ou montagne; *German:* Bergmuskat, Filziger Muskat, Unechter Muskat; *Malaysian:* pala laki-laki; palafuker, pala utan

Usage ▶ spice (local rarely)

Parts Used ▶ seed

Distribution ▶ India, Indonesia, Malaysia: Ambon, Banda, Java; cultivated: Java and Malaysia

Bois 1923; Hager 5, 1993; Rätsch 1998; Schultze-Motel 1986; Sinclair 1968; Warburg 1897

Myristica fragrans, fruits, partly open, visible the seeds with arillus (macis)

 Myristica fragrans *Houtt.*

Synonyms ▶ *Myristica aromatica* Lam., *Myristica moschata* Thunb., *Myristica officinalis* L. f.,

Common Names ▶ **nut:** (true) nutmeg, Banda nutmeg; *Arabic:* basbasah, gouz bouwa; *Chinese:* jon tou k'ou, rou dou kou; *Dutch:* nootmuskaat; *French:* muscadier commun, noix de muscade, musque; *German:* Echte Muskatnuss, Muskatnuss, Duftende Muskat(nuss); *Hindi:* jayphal, mada shaunda, taiphal; *Indonesian:* pala; *Italian:* noce moscata; *Japanese:* nikuzuki; *Javanese:* wohpala; *Malaysian:* buah pala; *Portuguese:* moscadeira; *Russian:* muskatnij orech; *Sanskrit:* jatikosa, jatiphala, taiphal; *Slovenian:* muskátový orech; *Spanish:* nogal moscado, nuez moscada; *Thai:* chan thet ▶ **aril: mace;** *Arabic:* basbasah; *French:* macis, noix muscade (capsule della); *German:* Mazis, Muskatblüte (= wrongly); *India:* tavitri; *Italian:* mace, macis, fiore de macis, fiore di noce moscata; *Portuguese:* flor de noz moscada; *Russian:* mazis; *Sanskrit:* jajipatri; *Slovenian:* muskátový kvet; *Spanish:* macis, maciá

Usage ▶ spice; **product:** essential oil

Parts Used ▶ seed, aril

Distribution ▶ It is not found wild, Malaysia?; cultivated elsewhere

Note ▶ The fruit peel with fruit flesh is used for yams and candied for cakes etc. The seeds used to be dipped in milk of lime to prevent the infestation of insects.

Baldry et al. 1976; Bittencourt et al. 2001; Bois 1934; Boisvert/Huber 2000; Borges/Pino 1993; Bournot 1968; Burkill 4, 1997; Cheers 1997; Craze 2002; Dalby 2000; Davidson 1999; Davis/Cooks 1982; Dudtschenko et al. 1989; Duke et al. 2003; Erhardt et al. 2002; Farrell 1985; Ferrão 1992; Forrest 1972; Gerhardt/Sundermann 1981; Guzman/Siemonsma 1999; Hager 5, 1993; Hanelt 2001; Hiller/Melzig 1999; Hohmann et al. 2001; Hoppe 1949; Joseph 1980; Kalbhen 1971; Krützfeld 2002; Lee/Caruso 1958; Lück 2004; Maistre 1964; McKee/Harden 1991; Melchior/Kastner 1974; Michaelis 1916; Milton 1999; Morris/Mackley 1999; Mors/Rizzini 1961; Norman 1991; Oberdieck 1989b; Ochse et al. 1961; Peter 2001; Pruthi 1976; Pschyrembel 1993; Pursglove 1968; Rosengarten 1969; Roth/Kormann 1997; Schenk/Lamparsky 1981; Schröder 1991; Schultze-Motel 1986; Seidemann 1993c, 1998/2000; Seidemann/Siebert 1987; Sharma 2003; Sherry et al. 1982; Shiva et al. 2002; Shulgin 1963; Siewek 1990; Simpson/Jackson 2002; Sinclair 1968; Small 1997; Staesche 1972; Strauß 1967; Täufel et al. 1993; Teuscher 2003; Tschirch 1892; Uhl 2000; Vaupel 2002b; Villamar et al. 1994; Warburg 1897; Weil 1965; Wiersema/León 1999; Wilhelm 1966; Wüstenfeld/Haensel 1964; Wyk et al. 2004; Zeven/de Wet 1982; Zizka/Fleckenstein 1985

 Myristica laurifolia Hook.f. & Thoms

▶ *Myristica dactyloides* Gaertn.

 macrantha A.C. Smith

▶ *Myristica castaneifolia* A. Gray

 Myristica macrophylla A. Gray

▶ *Myristica castaneifolia* A. Gray

 Myristica malabarica Lam.

Synonyms ▶ *Myristica dactyloides* Wall.
Common Names ▶ Bombay nutmeg, Malabar nutmeg; *German:* Bombay-Muskatnuss, Malabar-Muskatnuss
Usage ▶ spice
Parts Used ▶ seed, aril
Distribution ▶ India: Malabar coast, also cultivated in the garden
Note ▶ Sometimes is like an adulteration of *Myristica fragrans* Houtt.

Erhardt et al. 2002; Hager 5, 1993; Hiller/Melzig 1999; Melchior/Kastner 1974; Pruthi 1976; Rätsch 1998; Seidemann 1993c; Teuscher 2003; Uphof 1968; Warburg 1897; Wealth of India 6, 1962

 Myristica moschata Thunb.

▶ *Myristica fragrans* Houtt.

 Myristica officinalis L. f.

▶ *Myristica fragrans* Houtt.

 Myristica otoba H. et B.

Synonyms ▶ *Dialyanthera otoba* (H. et B.) Warb.
Common Names ▶ *German:* Otoba-Muskat
Usage ▶ spice
Parts Used ▶ seed
Distribution ▶ S America: Columbia

Teuscher 2003; Warburg 1897

 Myristica schefferi Warb.

▶ *Myristca succedanea* Bl.

 Myristica speciosa Warb.

Common Names ▶ Moluccan nutmeg; *French:* muscadier des Moluques; *German:* Batjan-Muskat, Molukken-Muskat; *Malaysian:* pala maba
Usage ▶ spice
Parts Used ▶ seed
Distribution ▶ Philippines: Batjan; Moluccas

Bois 1934; Hager 5, 1993; Rätsch 1998; Teuscher 2003; Warburg 1897

 Myristica succedanea Bl.

Synonyms ▶ *Myristica schefferi* Warb.
Common Names ▶ *Dutch:* halmaheiramuskaat; *French:* noix muscade à laquelle; *German:* Halmaheira-Muskat(nuss)
Usage ▶ spice, condiment, like *Myristica fragrans* Houtt.
Parts Used ▶ seed
Distribution ▶ N Moluccas; cultivated in a small region on Ternate.

Bois 1934; Burkill 19966; Hager 5, 1993; Hanelt 2001; Hiller/Melzig 1999; Mansfeld 1986; Melchior/Kastner 1974; Rätsch 1998; Schultze-Motel 1986; Seidemann 1993c; Uphof 1968; Wartburg 1897

Myristica tomentosa *Thunb.*

▶ *Myristica fatua* Houtt.

 Myristica womersleyi *J. Sinclair*

Common Names ▶ Womersleyie nutmeg; *German:* Womersleyie Muskat
Usage ▶ spice (locally); **product:** essential oil
Parts Used ▶ seed
Distribution ▶ NE New Guinea

Guzman/Siemonsma 1999; Oyen/Dung 1999; Warburg 1897

 Myrospermum peruiferae *Royle*

▶ *Myroxylon balsamum (L.) Harms var. peruifera (Royle) Harms*

 Myrospermum sonsonatense *Oerst.*

▶ *Myroxylon balsamum (L.) Harms var. peruiferae (Royle)Harms*

 Myrospermum toluiferum *DC.*

▶ *Myroxylon balsamum (L.) Harms var. balsamum*

MYROXYLON L.f. - Balsam - Fabaceae (Leguminosae)

 Myroxylon balsamum *(L.) Harms* **var. balsamum**

Synonyms ▶ *Myrospermum toluiferum* DC., *Myroxykon hanburyanum* Klotzsch, *Myroxylon tulifera* H.B.K., *Myroxylon toluiferum* A. Rich., *Toluifera balsamum* L., *Toluifera balsamum* var. *genuinum* Baill.
Common Names ▶ balsam of Tolu; *Brazil (Portuguese):* bálsamo-de-tolu; *French:* baumier de Tolu; *German:* Tolubalsam; *Italian:* balsamo di Tolù; *Russian:* bal'samowoe derewo; *Spanish:* bálsamo de Tolú, arbol de Tolú
Usage ▶ spice (rarely), flavoring of drinks, cakes, chocolate and chewing gum, joss drug
Parts Used ▶ balsam
Distribution ▶ S America: Guatamala, Columbia, Venezuela; Brazil; also cultivated

Bergemann 1950; Cheers 1997; Dalby 2000; Dusemund et al. 1991; Ecker-Schlimpf 1991; Erhardt et al. 2002; Glasl/Wagner 1974; Hager 5, 1993; Hanelt 2001; Hiller/Melzig 1999; Hoppe 1949; Langenheim 2003; Mansfeld 1959; Mors et al. 2002; Opdyke 1974a; Roth/Kormann 1997; Schultze-Motel 1986; Seidemann 1993c; Sharma 2003; Uphof 1968; Villamar et al. 1994; Wiersema/León 1999; Wyk et al. 2004; Zeven/de Wet 1982

 Myroxylon balsamum *(L.) Harms* **var. peruiferae** *(Royle) Harms*

Synonyms ▶ *Myrospermum peruiferae* Royle, *Myrospermum sonsonatense* Oerst., *Myroxylon peruiferae* Klotzsch, *Myroxylon peruiferum* L.f., *Toulifera peruiferae* Baill.
Common Names ▶ Balsam of Peru, black balsam, Peruvian balsam; *Brazil (Portuguese):* bálsamo-do-peru; *French:* arbor de balsamo, baumier du Pérou; *German:* Perubalsam, Salvadorbalsam; *Italian:* balsamo di Perú; *Russian:* miroksilon; *Spanish:* arbol de Pérou, balsamo de Pérou
Usage ▶ spice (rarely), flavoring of drinks, cakes, chocolate and chewing gum, joss drug
Parts Used ▶ balsam
Distribution ▶ S America: San Salvador, Guatamala, Peru
Note ▶ Cultivated by the Aztecs in the imperial garden in Mexico.

Bergemann 1950; Cheers 1997; Dalby 2000; Dusemund et al. 1991; Ecker-Schlimpf 1991; Erhardt et al. 2002; Glasl/Wagner 1974; Hager 5, 1993; Hanelt 2001; Hiller/Melzig 1999; Hoppe 1949; Langenheim 2003; Mansfeld 1959; Mors et al. 2002; Oliveira et al. 1978; Opdyke 1974a; Rätsch 1998; Roth/Kormann 1997; Schultze-Motel 1986; Seidemann 1993c; Sharma 2003; Uphof 1968; Villamar et al. 1994; Wiersema/León 1999; Wyk et al. 2004; Zeven/de Wet 1982

Myroxylon hanburyanum *Klotzsch*

▶ *Myroxylon balsamum (L.) Harms var. balsamum*

 Myroxylon peruiferae *Klotzsch*

▶ *Myroxylon balsamum (L.) Harms var. peruifera (Royle) Harms*

 Myroxylon toluifera *H.B.K.*

▶ *Myroxylon balsamum (L.) Harms. var. balsamum*

 Myroxylon toluiferum *A. Rich.*

▶ *Myroxylon balsamum (L.) Harms var. balsamum*

MYRRIS Mill. - Sweet Cicely - Apiaceae (Umbelliferae)

 Myrris aromatica *L.*

▶ *Myrris odorota (L.) Scop.*

 Myrris occidentalis *Benth. et Hook.f.*

Synonyms ▶ *Osmorhiza occidentalis* (Nutt.) Torr.
Common Names ▶ anise sweet, Western sweet root, sweet cicely; *German:* Aniswurzel
Usage ▶ spice, flavoring
Parts Used ▶ root
Distribution ▶ USA: California, SW Colorado

Hoppe 3, 1987

Myrris odorota *(L.) Scop.*

Synonyms ▶ *Myrris aromatica* L., *Scandix odorata* L.
Common Names ▶ anise chervil, garden myrrh, Spanish chervil, sweet cicely, sweet scented myrrh, giant sweet chervil; *Arabic:* samag albostan; *Chinese:* ou zhou mo yao; *Dutch:* roomse kervel; *French:* cerfeuil d'Espagne, cerfeuil musqué, cerfeuil odorant, myrris, persil d'anis; *German:* Aniskerbel, Myrrhenkerbel, Süßdolde, Wilder Anis; *Italian:* cerfoglio di spagua, finocchielle, mirride (delle Alpi), mirride odorota; *Portuguese:* cerefolho almiscarado; cerefolho anisado; *Russian:* kerbel' ispanckij, mnogoletnij kerbel', dikaja petruschka, duschistyj buten', ladan, mirris; *Spanish:* perifollo almizclado, perifollo oloroso
Usage ▶ spice
Parts Used ▶ fresh herb
Distribution ▶ C Europe, SE and SW Europe, native elsewhere

Aichele/Schwegler 3, 1995; Bois 1934; Bremness 2001; Brieskorn/Noble 1982; Cheers 1997; Clair 1961; Davidson 1999; Erhardt et al. 2002; Fleischhauer 2003; Hiller/Melzig 1999; Loch 1993; Mansfeld 1959; Pochljobkin 1974, 1977; Schnelle 1999; Schönfelder 2001; Schulz 1963; Schultze-Motel 1986; Seidemann 1993c; Täufel et al. 1993; Tucker 1986; Teuscher 2003; Uphof 1968; Wiendl/Franz 1994; Wiersema/León 1999; Zeven/de Wet 1982

MYRTUS L. - Myrtle - Myrtaceae

 Myrtus communis *L.*

Common Names ▶ common myrtle, myrtle; *Arabic:* adhera, guemmam, marsin, rihan; *French:* myrte, myrte comun; *German:* Echte Myrte, Brautmyrte, Myrthenkörner; *Italian:* mirto; *Russian:* mirt; *Slovenian:* myrta; *Spanish:* arrayán, mirto
Usage ▶ spice; fruits for sauces and meat, especially in Greece; **product:** essential oil
Parts Used ▶ leaf, fruit (berry)
Distribution ▶ Canary Islands, N Africa, W Africa, SE and SW Europe to NW Himalayas

Bärtels 1997; Berger 2, 1950; Bourton 1968; Bremness 2001; Craze 2002; Dalby 2000; Davidson 1999; Duke et al. 2003; Erhardt et al. 2000, 2002; Frau et al 2001; Hager 5, 1993; Hepper 1992; Hiller/Melzig 1999; Hoppe 1949; Jerkovic et al. 2002; Lück 2000; Roth/Kormann 1997; Schönfelder 2001; Schultze-Motel 1986; Seidemann 1993c; Small 1997; Täufel et al. 1993; Teuscher 2003; Uphof 1968; Wiersema/León 1999; Wyk et al. 2004; Zeven/de Wet 1982

Myrtus pimenta L.

› *Pimenta dioica (L.) Merr.*

N

NANNORHOPS H. Wendl. - Araceae (Palmae)

Nannorhops ritchiana *(Griff.) Aitch.*

Common Names ▸ marari palm; *German:* Mazaripalme
Usage ▸ pot-herb (by the natives)
Parts Used ▸ young leaf
Distribution ▸ India, Iran, Afghanistan, Pakistan

Erhardt et al. 2002; Uphof 1968

Narthex polakii *Staph et Wettst.*

▸ *Ferula assa-foetida L.*

Nardosmia japonica *Siebold et Zucc.*

▸ *Petasites japonica (Siebold et Zucc.) Maxim*

NARDOSTACHYS DC. - Nard, Spikenard - Valerianaceae

Nardostachys grandiflora *DC.*

Synonyms ▸ *Narthostachys jatamansi* (G. Don) DC.
Common Names ▸ Indian nard, spikenard; *French:* nardostachyde de l'Inde; *German:* Indische Narde, Nardenähre, Speicherähre; *India:* jatamansi
Usage ▸ flavoring; **product:** essential oil

Parts Used ▸ leaf, rhizome
Distribution ▸ N India to SW China, Himalayas

Chauhan 1999; Dalby 2000; Erhardt et al. 2002; Hager 5, 1993; Hanelt 2001; Hepper 1992; Hiller/Melzig 1999; Rätsch 1998; Schultze-Motel 1986; Sharma 2003; Uphof 1968; Warrier et al. 1995

Nardostachys jatamansi *(G. Don) DC.*

▸ *Narthostachys grandiflora DC.*

NASTURTIUM R.Br. - Water Cress - Brassicaceae (Cruciferae)

Nasturtium aquaticum *Wahlenb.*

▸ *Nasturtium officinale R.Br.*

Nasturtium armoracia *(L.) Fries*

▸ *Armoracia rusticana Gaertn. et B. Mey. et Scherb.*

Nasturtium latifolium *Gillet et Magne*

▸ *Lepidium latifolium L.*

Nasturtium latifolium (L.) O. Kuntze

▶ *Lepidium latifolium* L.

 Nasturtium microphyllum Boenn. ex Reichb.

Synonyms ▶ *Rorippa microphylla* (Boenn.) Hyl.
Common Names ▶ small leafy watercress; *German:* Kleinblättrige Brunnenkresse, Winterkresse
Usage ▶ pot-herb
Parts Used ▶ fresh leaf
Distribution ▶ Europe, E Europe, N Africa: Morocco; India; native elsewhere

Aichele/Schwegler 3, 1995; Erhardt et al. 2002; Hager 5, 1993; Small 1997; Schultze-Motel 1986; Small 1997; Täufel et al. 1993; Wiersema/León 1999; Zeven/de Wet 1982

 Nasturtium nasturtium Cockerell

▶ *Nasturtium officinale* R.Br.

 Nasturtium officinale R. Br.

Synonyms ▶ *Cardaminum nasturtium* Moench., *Crucifera fontana* E.H.L. Krause, *Nasturtium aquaticum* Wahlenb., *Nasturtium nasturtium* Cockerell, *Rorippa nasturtium-aquaticum* (L.) Hayek, *Sisymbrium nasturtium* Thunb.
Common Names ▶ common watercress, watercress; *Arabic:* guernech, harriqa, karsun mehi; *Chinese:* dou ban cai, xi yang cai gun; *Dutch:* waterkres; *French:* cresson d'eau, cresson de fontaine (d'eau); *German:* Echte Brunnenkresse, Bachkresse, Wasserkresse, Wassersenf, Wiesenkren; *Hindi:* pani sag; *Indonesian:* selada air; *Italian:* crescione aquatico, nasturcio, agretto aquatico; *Japanese:* mizu garashi, oranda garashi; *Portuguese:* agrião, agrião de água; *Russian:* kress wodjanoj, brunkress, kljuzewoj kress, wodjanoj kren, sherucha wodnaja; *Spanish:* berro de agua, berro de fuente, mastuerzo acuático; *Vietnamese:* xa lach, xong
Usage ▶ pot-herb
Parts Used ▶ fresh leaf
Distribution ▶ Europe, Canary Islands, N Africa, tropical Africa, Turkey, Caucasus, temperate Asia cosmopolitical, native in N America

Aichele/Schwegler 3, 1995; Arora/Pandey 1996; Bärtels 1997; Berger 4, 1954; Bilgri/Adam 2000; Cheers 1997; Davidson 1999; Dudtschenko et al. 1989; Erhardt et al. 2002; Fleischhauer 2003; Hager 5, 1993; Hanelt 2001; Heeger 1956; Hiller/Melzig 1999; Hoppe 1949; Koschtschejew 1990; Larkcom 1991; Leung 1991; Mansfeld 1962; Ogle et al. 2003; Pochljobkin 1974, 1977; Pschyrembel 1998; Pursglove 1968; Schönfelder 2001; Schultze-Motel 1986; Siemonesma/Piluek 1993; Small 1997; Šedo/Krejča 1983; Täufel et al. 1993; Teuscher 2003; Tindall 1983; Tull 1999; Uphof 1968; Villamar et al. 1994; Wiersema/León 1999; Wyk et al. 2004; Zeven/de Wet 1982

 Nasturtium officinale R.Br. x **Nasturtium microphyllum** Boenn.

▶ *Nasturtium sterile (Airy Shaw) Oefelein*

 Nasturtium x sterile (Airy Shaw) Oefelein

(= *Nasturtium officinale* R.Br. x *Nasturtium microphyllum* Boenn.)

Synonyms ▶ *Rorippa x sterilis* Airy Shaw
Usage ▶ pot-herb
Parts Used ▶ fresh leaf
Distribution ▶ C and W Europe, cultivated in England (rarely)

Hanelt 2001; Schultze-Motel 1986

NECTANDRA Roland ex Rottb. - Lauraceae

 Nectandra globosa (Aubl.) Mez

Common Names ▶ *Brazil (Portuguese):* canela-amarela, canela-da-capoeira, canela-mirim, canela-seca
Usage ▶ flavoring
Parts Used ▶ bark
Distribution ▶ Amazonia to Bahia

Mors et al. 2000

 **Nectandra cinnamoides** Nees

▶ *Ocotea cymbarum* H.B.K.

 **Nectandra cymbarum** *(H.B.K.) Nees*

▶ *Ocotea cymbarum* H.B.K.

 **Nectandra puchury-major** Nees et Mart.

▶ *Ocotea puchury-major* Mart.

NEPETA L. - Catmint - Lamiaceae (Labiatae)

 **Nepeta americana** *Aubl.*

▶ *Hyptis spicigera* Lam.

 Nepeta cataria *L.*

Synonyms ▶ *Calamintha albiflora* Van., *Nepeta citriodora* Becker, *Nepeta minor* Mill., *Nepeta vulgaris* Lam.
Common Names ▶ catmint, catnip, lemon catnip (America); *Chinese:* jing jie, jia jing jie; *French:* cataire, chataire, menhe de chat; *German:* Echtes Katzenkraut, Gewöhnliche Katzenminze, Steinmelisse; *Italian:* cataria, erba de gatta, gattaia comune, menthe die gatti; *Portuguese:* erva gateira; *Russian:* kotovnik koschaschij, koschatschja mjata; *Spanish:* calamento, hierba gatera, menta de gato
Usage ▶ spice (specially in France for seasoning sauces, soups etc.)
Parts Used ▶ herb
Distribution ▶ S, E Europe, Turkey, W Siberia, Amur, C Asia, India, China, Himalayas, Cuba and other Latin America countries; naturalized in W, C Europe, E Asia, N America, S Africa
Note ▶ The var. *citriodora* (Becker) Balb. contains much citronellol (ca. 50%) in the essential oil, used in food and the pharmaceutical industries and perfumery.

Aichele/Schwegler 4, 1995; Alberts/Muller 2000; Berger 2, 1950; Bourton 1968; Bremness 2001; Cheers 1997; Dellacassa et al 1997; Dudtschenko et al. 1989; Erhardt et al. 2002; Fleischhauer 2003; Hager 5, 1993; Hanelt 2001; Heeger 1956; Herron 2003; Hiller/Melzig 1999; Hoppe 1949; Kaczmarek 1957; Klimek et al. 2000; Mansfeld 1962; Mikus/Schaser 1995; Pooter et al. 1988; Roth/Kormann 1997; Schnelle 2003; Schönfelder 2001; Schultze-Motel 1986; Seidemann 1993c; Sharma 2003; Small 1997; Täufel et al. 1993; Teuscher 2003; Tucker 1986; Uphof 1968; Wealth of India 7, 1966; Wiersema/León 1999; Wyk et al. 2004; Zeven/de Wet 1982

 Nepeta citriodora *Becker*

▶ *Nepeta cataria* L.

 **Nepeta x faassenii** *Bergmans ex Stearn*

Common Names ▶ blue catmint; *French:* chataire, herbe aux chats; *German:* Blaue Katzenminze; Blauminze; Kleine Katzenminze
Usage ▶ spice
Parts Used ▶ herb
Distribution ▶ cultivated Europe, C Asia, Persia
Note ▶ Sterile hybrid.

Cheers 1997; Erhardt et al. 2002; Hanelt 2001; Pooter et al. 1988; Small 1997

 Nepeta glechoma *Benth.*

▶ *Glechoma hederacea* L.

Nepeta minor *Mill.*

▶ *Nepta cataria* L.

Nepeta mussinii *Spreng.*

▶ *Nepeta racemosa* Spreng.

 Nepeta racemosa *Lam.*

Synonyms ▶ *Nepeta mussinii* Spreng., *Nepeta reichenbachiana* Fisch. et C.A. Mey.

Common Names ▶ Mussin Catnip; *German:* Mussin's Katzenminze, Traubige Katzenminze; *Korean:* hyanghynŏgge
Usage ▶ spice (sporadically); **product:** essential oil (with high content of geraniol and citronellol)
Parts Used ▶ herb
Distribution ▶ Europe, Turkey, former Soviet Union, Iran, Korea

Başer et al. 1993a; Erhardt et al. 2002; Hanelt 2001; Schultze-Motel 1986; Wolf 1955

Nepeta reichenbachiana Fisch. et C.A. Mey.

▷ *Nepeta racemosa Lam.*

Nepeta tenuifolia Benth.

Synonyms ▶ *Elsholtzia integrifolia* Benth.
Common Names ▶ Japanese catnip, *Chinese:* jing jie; *Korean:* hyŏnggye
Usage ▶ spice (sporadically); **product:** essential oil
Parts Used ▶ herb
Distribution ▶ China, Japan, Korea

Hanelt 2001

Nepeta transcaucasia Grossh.

Common Names ▶ Transcaucasian catnip; *German:* Transkaukasische Katzenminze, *Russian:* kotownik sakawasskii
Usage ▶ spice (sporadically); **product:** essential oil
Parts Used ▶ herb
Distribution ▶ Transcaucasia: Azerbaidzhan

Khilik et al. 1977, 1979; Mishurova/Shikhiev 1977

Nepeta vulgaris Lam.

▷ *Nepeta cataria L.*

NEPTUNIA Lour. - Mimosaceae

Neptunia oleraceae Lour.

Synonyms ▶ *Neptunia prostrata* (Lam.) Bail.
Common Names ▶ water mimosa; *Bengal:* panilajak; *German:* Wasser-Mimose; *Hindi:* lajalu; *Malaysian:* keman ajer; *Thai:* phak krachet
Usage ▶ pot-herb (in India)
Parts Used ▶ sprout
Distribution ▶ tropical regions: India, Indochina, Malaysia, Australia, Mexico, S America, Africa

Arora/Pandey 1996; Chopra et al. 1956; Erhardt et al. 2002; Hanelt 2001; Usher 1968; Wealth of India 7, 1966; Wiersema/León 1999; Zeven/de Wet 1982

Neptunia prostrata (Lam.) Bail.

▷ *Neptunia olearacea Lour.*

Nicolaia elatior (Jack) Horan.

▷ *Etlingera elatior (Jack) R.M. Sm.*

Nicolaia solaris (Blume) Horan.

▷ *Etlingera solaris (Blume) R.M. Smith*

NIGELLA L. - Fennel Flower, Love-in-a-Mist - Ranunculaceae

Nigella arvensis L.

Synonyms ▶ *Nigella latifolia* Mill., *Nigella tenuifolia* Gilib.
Common Names ▶ wild fennel; *French:* nielle bâtarde, poivrette commun; *German:* Acker-Schwarzkümmel; *Russian:* tschernuschka polevaja
Usage ▶ spice, pot-herb
Parts Used ▶ seed, plant

Distribution ▶ C and E Europe, Turkey, Iraq, Levante, Caucasus, C Asia, Near East, N Africa

Aichele/Schwegler 2, 1994; Dudtschenko et al. 1989; Erhardt et al. 2002; Fleischhauer 2003; Hanelt 2001; Heeger 1956; Hooper 1937; Schnelle 1999; Schultze-Motel 1986; Täufel et al. 1993, Teuscher 2003

 Nigella coerula *Lam*

▶ *Nigella damascena* L.

 Nigella damascena *L.*

Synonyms ▶ *Nigella coerula* Lam., *Nigella pygmaea* Pers.
Common Names ▶ love-in-a-mist (plant), Damas black cumin, Jack-in-the-green, wild fennel; *Arabic:* habba souda, sinouj; *Dutch:* koijn; *French:* barbide, bâtarde, cheveux de Vénus, nigelle de Damas; *German:* Damaszener Schwarzkümmel, Türkischer Schwarzkümmel, Braut in Haaren, Gretel im Busch, Jungfer im Grünen (plant); *Italian:* anigella, damigella, capigilate, scapigilate; *Russian:* tschnuschka damasskaja
Usage ▶ spice; **product:** essential oil
Parts Used ▶ seed
Distribution ▶ E and S Europe, Turkey, Cyprus, Caucasus, Iran, NW Africa, Libya; naturalized in C Europe
Note ▶ The seed oil is used in medicine and the perfume industry; the plant is a very popular worldwide as an ornament.

Aichele/Schwegler 2, 1994; Arctander 1960; Bärtels 1997; Bois 1934; Bremness 2001; Cheers 1997; Dalby 2000; Dudtschenko et al. 1989; Erhardt et al. 2002; Fico et al. 2003; Fleischhauer 2003; Groom 1997; Hager 4, 1992; Hanelt 2001; Heeger 1956; Hepper 1992; Hoppe 1949; Oyen/Dung 1999; Paris et al. 1979; Redgrove 1933; Schweig 1999; Seidemann 1993c; Small 1997; Täufel et al. 1993; Teuscher 2003; Wiersema/León 1999; Zohary 1983

 Nigella indica *Roxb.*

▶ *Nigella sativa* L.

▣ **Nigella damascena, flowering and fruiting**

 Nigella latifolia *Mill.*

▶ *Nigella arvensis* L.

 Nigella pygmaea *Pers.*

▶ *Nigella damascena* L.

 Nigella sativa *L.*

Synonyms ▶ *Nigella indica* Roxb., *Nigella truncata* Viv.
Common Names ▶ black caraway, black cumin, black seed, fennel flower, nutmeg flower, Roman coriander; small fennel; *Arabic:* habba souda, habbatul barakah; *French:* cheveux de Vénus, cumin noir, nigelle, nigelle de Crète, toute épice; *German:* Echter Schwarzkümmel, Römischer Schwarzkümmel, Schwarzer Koriander, Schabasamen; *Hindi:* kalaunji; kalonji, kalajira, mugrela; *Italian:* nigella, cominella, melantio; *Malaysian:* jintam hitam; *Portuguese:* cominho negro; *Russian:* tschernuschka

posevnaja, tschernij tmin, mazok, nigella, rimskij koriandr; *Sanskrit:* kalajaji, krishnajiraka; *Slovenian:* černuška; *Spanish:* ajenuz, aranuel, neguilla, nigelia cultivada, pasionara; *Turkish:* çörek otu

Usage ▶ spice

Parts Used ▶ seed

Distribution ▶ EC Europe, Mediterranean regions, Caucasus, Iran, SW and C Asia, N, E Africa, native in Europe; cultivated from Mediterranean regions to C Asia

Note ▶ The plant name at 16th Century in Germany "Schwartzer Coriander".

Abel Alla El-Sayed et al. 1997; Aitzetmüller 1997; Al-Jassir 1992; Atta 2003; Babayan et al. 1978; Bärtels 1997; Bois 1934; Boisvert/Huber 2000; Bŭrits/Bucar 1998; Cheers 1997; Coiciu/Racz (no year); Craze 2002; Datta/Rabg 2000; Daukšas et al. 2002; Davidson 1999; Dudtschenko et al. 1989; Duke et al. 2003; El-Dakhakhny 1965; El-Dhaw/Abdel-Munuem 1996; Erhardt et al. 2000, 2002; Gad et al. 1963; Guzman/Siemonsma 1999; Hager 4, 1992; Hanafy/Hatem 1991; Hanelt 2001; Hasan et al 1989; Heeger 1956; Hepper 1992; Hiller/Melzig 1999; Hoppe 1949; Ihrig 1997; Jansen 1981; Lautenbacher 1997; Leroy/Gillet 1964; Mahfouz/El-Dakhahny 1960; Menounos et al. 1986; Morris/Mackley 1999; Nergiz/Ötles 1993; Norman 1991; Pochljobkin 1974, 1977; Ramadan/Mörsel 2002a, b, 2003; Salem 2001; Salama 1973; Saleh Al-Jassier 1992; Schönfelder 2001; Schweig 1999; Seidemann 1993c; Siewek 1990; Small 1997; Takruri/Dameh 1998; Täufel et al. 1993; Teuscher 2003; Toppozada et al. 1965; Tucker 1986; Türkay et al 1996; Uphof 1968; Ustum et al. 1990; Wiersema/León 1999; Wyk et al. 2004; Zeven/de Wet 1982; Zohary 1983

Nigella tenuiflora Gilib.

▶ *Nigella arvensis* L.

Nigella truncata Viv.

▶ *Nigella sativa* L.

Nigrina spicifera Lam.

▶ *Chloranthus spicatus (Thunb.) Makino*

Nima quassioides Buch-Ham.

▶ *Picrasma quassioides Benn.*

Nimbo melioides Dennst.

▶ *Murraya koenigii (L.) Spreng.*

Nothopanax fruticosum Miq.

▶ *Polyscias fruticosa (L.) Harms*

Nyctanthes sambac L.

▶ *Jasminum sambac (L.) Ait.*

Obione canescens *(Pursh) Moq.*

▶ *Atriplex canescens (Pursh) Nutt.*

OCHROCARPOS Noranta ex Du Petit-Thouars - Guttiferae

 Ochrocarpus longifolius *(Wall.) Benth. et Hook.f.*

Synonyms ▶ *Mammea longifolia* Wall.
Common Names ▶ *Hindi:* nag kesar, surgi, suringi, surampunna, surabunai
Usage ▶ spice (sporadically in India)
Parts Used ▶ flower, flower bud
Distribution ▶ West Indies

Chopra et al. 1956; Hanelt 2001; Mohan Rao et al. 2002; Uphof 1968

OCIMUM L. - Basil - Lamiaceae (Labiatae)

Hiltunen/Holm 19

 Ocimum africanum *Lour.*

▶ *Ocimum americanum L.*

 Ocimum album *L.*

▶ *Ocimum basilicum L.*

 Ocimum americanum *L.*

Synonyms ▶ *Ocimum africanum* Lour., *Ocimum canum* Sims; *Ocimum pilosum* Willd., *Ocimum simile* N.E. Br.
Common Names ▶ American basil, hoary basil, sweet basil; *Chinese:* hui luo le; *German:* Amerikanischer Basilikum, Zitronen-Basilikum, Kampferbasilikum, Herero Buschtee; *Hindi:* ajak, kala tulsi; *Malaysian:* kĕmangi; *Portuguese:* manjericão-branco; *Sanskrit:* ajaka; *Spanish:* albahaca velluda; *Thai:* maeng lak
Usage ▶ pot-herb; **product:** camphor
Parts Used ▶ leaf
Distribution ▶ tropical Africa, China, India, Indochina, Malaysia; former Soviet Union, Kenya, Pakistan, native in Europe, Australia and the Neotropics
Note ▶ In the essential oil are predominantly citral, lilanool and chavicol methylether.

Anjaneyalu/Gowda 1978; Arora/Pandey 1996; Ayensu 1978; Bois 1934; Bournot 1968; Darrah 1980; Dudtschenko et al. 1989; Ekundayo et al. 1989; Erhardt et al. 2002; Gupta 1994; Hanelt 2001; Herklots 1972; Hiller/Melzig 1999; v. Koenen 1996; Mikus et al. 1997; Ntezurubanza et al. 1985; Paton 1992; Plescher et al. 1997; Schultze-Motel 2003; Seidemann 1993c; Sharma 2003; Siemonsma/Piluek 1993; Sinha/Gulati 1990; Small 1997; Täufel et al. 1993; Teuscher 2003; Thappa et al. 1979; Upadhyay et al. 1991; Uphof 1968; Xaasan et al. 1980, 1981; Zeven de Wet 1982

© Springer-Verlag Berlin Heidelberg 2005

Ocimum americanum *Jacq.*

> *Ocimum basilicum* L.

 Ocimum americanum sensu *Pushpangadan et Sobti non* L.

> *Ocimum x citriodorum* Vis.

 Ocimum basilicum *L.*

Synonyms ▶ Basilicum citratum Rumph., *Ocimum album* L., *Ocimum americanum* Jacq., *Ocimum menthaefolium* Hochst. ex Benth.

Common Names ▶ common basil, lemon basil, monk's basil, sweet basil, basil; *Arabic:* kamahim, habag, raihān, rehan; *Brazil (Portuguese):* manjericão, alfavaca; *Dutch:* basilicum; *Chinese:* lo-le, lou le, hstang tsa i tzu tsaoi; *French:* basilic, basilic commun, grand basile, herbe royale; *German:* Basilienkraut, Basilikum, Braunsilge, Deutsches Pfefferkraut, Königskraut, Herrenkraut, Suppenbasil; *Hindi:* bavri, bavai; *India:* babui tulsi, sada tulas; *Italian:* basilico; *Japanese:* meböki; *Malaysian:* kemangi, pokoh, ruku, selasseh uteh; *Portuguese:* manjericão; *Russian:* basilik, duschki, duschistje wasil'ki, *Sanskrit:* barbari, munjariki, surasa, tungi, varvara; *Spanish:* alebega, albahaca; *Thai:* bai horapa; *Vietnamese:* rau (hung) que

Usage ▶ spice; **product:** essential oil

Parts Used ▶ leaf,

Distribution ▶ tropical Africa, tropical America, former Soviet Union; widely cultivated, possible origin in Africa (Madagascar) or NW India

Note ▶ This species is morphologically and chemically very heterogeneous.

Aichele/Schwegler 4, 1995; Akgül 1989; Ayensu 1978; Bärtels 1997; Berger 2, 1940; 4, 1954; Bois 1934; Bournot 1968; Bremness 2001; Brophy/Jogia 1986; di Cesare et al. 2002; Cheers 1997; Clair 1961; Coiciu/Racz (no year); Czygan 1997; Darrah 1980; Dudtschenko et al. 1989; Erhardt et al. 2000, 2002; Farrell 1085; Fatope/Takeda 1988; Fleisher 1981; Grayer et al. 1996; Guzman/Siemonsma 1999; Hager 5, 1993; Hall 1981; Hay/Waterman 1993; Heeger 1956; Herklots 1972; Hiller/Melzig 1999; Hoffmann et al. 1992; Hondelmann 2002; Hoppe 1949; Jain/Jain 1973; Jansen 1981; Javanmardi et al. 2002; Jayasinghe et al. 2003; Kumar 2003; Lachowicz et al 1996; Lawrence et al. 1972; Loughrin/Kasperbauer 2001, 2003; Mahesh-

◘ **Ocimum basilicum, flowering**

wari/Singh 1989; Marzell 1970; Metzger 1984; Ozcan/Chalchat 2002; Özek et al. 1995; Pääkkönen 1990; Peter 1978, 2001; Pino et al. 1994a; Plescher et al. 1998; Pochljobkin 1974, 1977; Pruthi 1976; Pschyrembel 1998; Pudalosi 1996; Randkawa/Gill 1995; Randriamiharisoa 1986; Rosengarten 1969; Roth/Kormann 1997; Sánchez et al. 1985; Schönfelder 2001; Schröder 1991; Schultze-Motel 1986; Seidemann 1993; Seidemann/Siebert 1987; Sharma 2003; Shen et al. 1991; Shiva et al. 2002; Siewek 1990; Small 1997; Staesche 1972; Stuart et al. 1994; Suchorska-Tropilo/Osinska 2001; Tada et al. 1996; Täufel et al. 1993; Teufel 2003; Teuscher 2003; Tucker 1986; Ubillos 1989; Uphof 1968; Vasconcelos et al. 2003; Villamar et al. 1994; Wyk et al. 2004; Zeven/de Wet 1982; Zole/Garnero 1973

 Ocimum basilicum *L.* **var. anisatum** *Benth.*

> *Ocimum x citriodorum* Vis.

 Ocimum basilicum *L.* **x Ocimum americanum** *L.*

> *Ocimum x citriodorum* Vis.

Ocimum basilicum *L.* ssp. basilicum

Synonyms ► *Ocimum bullatum* Lam., *Ocimum caryophyllatum* Roxb., *Ocimum citriodorum* Blanco, *Ocimum integerrimum* Willd., *Ocimum lanceolatum* Schum. et Thonn., *Ocimum medium* Mill., *Ocimum thyrsiflorum* L.

Common Names ► basil; *Arabic:* rehan; *Chinese:* luo de; *German:* Griechischer Buschthymian; *Hindi:* babui tulsi; *Italian:* basilico; *Malaysian:* sĕlasi; *Portuguese:* alfaraca; *Russian:* bazilik; *Spanish:* albahaca

Usage ► spice; **product:** essential oil, are used in the spice and perfumery industry

Parts Used ► leaf

Distribution ► cultivated in subtropical regions

Darrah 1980; Hanelt 2001; Plescher et al. 1997;

Ocimum basilicum *L.* ssp. minimum *(L.)* Danert

Synonyms ► *Ocimum minimum* L.
Common Names ► bush basil, Greek basil, little basil; *German:* Kleiner or Busch-Basilikum
Usage ► spice
Parts Used ► leaf (sporadically)
Distribution ► Asia: India, commonly cultivated

Bournot 1968; Bremness 2001; Hanelt 2001; Ozcan/Chalchat 2002; Plescher et al. 1997; Shaiva et al. 2002; Vasconcelos et al. 2003

Ocimum basilicum *L.* var. purpurescens *Benth.*

Common Names ► *German:* Rot(blättriger) Basilikum
Usage ► spice
Parts Used ► leaf (sporadically)
Distribution ► also cultivated

Bournot 1968; Bremness 2001; Darrah 1980; Hanelt 2001; Suchorska/Osinska 1992; Vasconcelos et al. 2003;

Ocimum bullatum *Lam.*

► *Ocimum basicilum. L.* ssp. *basilicum*

Ocimum campechianum *Mill.*

► *Ocimum micranthemum* Willd.

Ocimum canum *Sims*

► *Ocimum americanum* L.

Ocimum caryophyllatum *Roxb.*

► *Ocimum basilicum L.* ssp. *basilicum*

Ocimum citroidorum *Blanco*

► *Ocimum basilicum L.* ssp. *basilicum*

Ocimum x citriodorum *Vis.*

Synonyms ► *Ocimum americanum* sensu Pushpangadan et Sobti non L., *Ocimum basilicum* L. x *Ocimum americanum* L., *Ocimum basilicum* var. *anisatum* Benth., *Ozymum citratum* Rumph., *Ocimum dichotomum* Hochst. ex Benth.
Common Names ► lemon basil; *German:* Zitronenbasilikum
Usage ► spice, pot-herb
Parts Used ► leaf
Distribution ► NE tropical Africa, Arabia, tropical Asia China, W Asia, India, widely cultivated elsewhere in Africa and Asia

Bremness 2001; Darrah 1980; Hanelt 2001; Mikus et al. 1997; Paton 1992; Verma et al. 1989

Ocimum crispum *Thunb.*

► *Perilla frutescens (L.) Britton*

Ocimum dichotomum *Hochst ex Benth.*

► *Ocimum x citriodotum Vis.*

Ocimum febrifugum Lindl

▶ *Ocimum gratissimum* L.

 Ocimum forskolei Benth.

Synonyms ▶ *Ocimum menthiifolium* Hochst ex Benth.
Common Names ▶ *German:* Minzblättriger Basilikum; *Russian:* basilik m'jatolistnis
Usage ▶ spice, flavoring
Parts Used ▶ leaf
Distribution ▶ Arabia, Egypt, Yemen, Sudan, Somalia, Ethiopia, N, E Kenya

Dudtschenko et al. 1989; Hiltunen/Holm 1999

 Ocimum frutescens L.

▶ *Perilla frutescens (L.) Britton*

 Ocimum gratissimum L.

Synonyms ▶ *Ocimum febrifugum* Lindl., *Ocimum guineense* Schum. et Thonn., *Ocimum viride* Willd.
Common Names ▶ fever plant, clove or Russian basil, shrubby basil, tea bush; *Brazil (Portuguese):* alfavacão; *Chinese:* wu mao ding x luo le; *French:* menthe gabonaise, basilic de Ceylan; *German:* Afrikanischer Basilikum, Duft- or Nelken-Basilikum, Ostindischer Basilikum; *Hindi:* ram tulsi, vriadhatulasi; *Malaysian:* ruku-ruku hitam, sĕlasi bĕsar; *Spanish:* albahaca de clavo; *Thai:* kaphrao-chang, horapha-chang, yira; *Vietnamese:* huong nhu trang
Usage ▶ spice; **product:** essential oil
Parts Used ▶ leaf
Distribution ▶ Africa, India, Indochina, Malaysia, native in the Neotropics
Note ▶ Eugenol, thymol or citral are predominant by type.

Abani 1988; Aedo et al. 2001; Ayensu 1978; Berger 2, 1950; Bois 1934; Bournot 1968; Charles/Simon 1992b; Darrah 1980; Droh/Hefendehl 1974; Dudtschenko et al. 1989; Gurib-Fakim/Brendler 2004; Herklots 1972; Jiroretz et al. 1998; Khasla 1995; Kumar 2001; Mikus/Schaser 1995; Mikus et al. 1997; Ntezurubanza 1987; Oyen/Dung 1999; Paton 1992; Plescher et al. 1997; Saroratto/Augusto 2003; Schultze-Motel 1986; Seidemann 1993c; Sharma 2003; Small 1997; Sofowora 1970; Täufel et al. 1993; Teuscher 2003; Tucker 1986; Turova et al. 1987; Uphof 1968; WHO 1990; Wiersema/León 1999; Zeven/de Wet 1982

 Ocimum graveolens A. Br.

Common Names ▶ *German:* Duft-Basilikum
Usage ▶ condiment
Parts Used ▶ herb
Distribution ▶ Ethiopia

Uphof 1968

 Ocimum guineense Schum. et Thonn.

▶ *Ocimum gratissimum* L.

 Ocimum integerrimum Willd.

▶ *Ocimum basilicum* L. ssp. *basilicum*

 Ocimum kilimandscharicum Guerke

Common Names ▶ Kilimandscharo basil, camphor basil; *German:* Kilimandscharo-Basilikum, Kampfer-Basilikum; *Hindi:* kapur tulsi; *Russian:* basiliki kilimandshars'ki
Usage ▶ spice (locally); **product:** camphor
Parts Used ▶ leaf
Distribution ▶ tropical E Africa: Kenya, Uganda, Tanzania; introduced elsewhere in the Tropics
Note ▶ Cultivated during World War II for the preparation of camphor. The plant has an austere aroma of camphor and is a substitute for *Cinnamomum camphora* (L.) J. Presl.

Bekele/Hassanali 2001; Charles/Simon 1992b; Darrah 1980; Mansfeld 1986; Paton 1992; Plescher et al. 1997; Schultze-Motel 1986; Sharma 2003; Shiva et al 2002; Small 1997; Teuscher 2003; Tucker 1986; Uphof 1968; Wealth of India 7, 1991; Zeven/de Wet 1982

OCIMUM L. - Basil - Lamiaceae (Labiatae)

Usage ▸ spice (locally)
Parts Used ▸ leaf
Distribution ▸ S Florida, the Bahamas, West-India, from Mexico to Peru, Chile and Brazil
Note ▸ The plant is strongly aromatic.

Charles et al. 1990; Darrah 1980; Ferreira et al. 1992; Mors et al. 2000; Morton 1981; Small 1997; Teuscher 2003; Uphof 1968; Villamar et al. 1994

 Ocimum minimum L.

▸ *Ocimum basilicum ssp. minimum (L.) Danert*

 Ocimum pilosum Willd.

▸ *Ocimum americanum L.*

 Ocimum sanctum L.

▸ *Ocimum tenuflorum L.*

 Ocimum simile N.E. Br.

▸ *Ocimum americanum L.*

 Ocimum stamineum Sims

▸ *Ocimum canum Sims*

 Ocimum suave Willd.

Common Names ▸ African basil, sweet scented basil; *German:* Afrikanischer Basilikum;
Usage ▸ spice
Parts Used ▸ leaf, herb
Distribution ▸ tropical Africa: Guinea, N Nigeria, S Sudan, Ethiopia, Kenya, Tanzania, Mozambique, Uganda; Sri Lanka

Darrah 1980; Mikus et al 1997; Mikus et al. 1997; Paton 1992; Plescher et al. 1997; Schultze-Motel 1986; Tétényi et al. 1986; Teuscher 2003

 **Ocimum lanceolatum** Schum. et Thonn.

▸ *Ocimum basilicum L. ssp. basilicum*

 Ocimum medium Mill.

▸ *Ocimum basilicum L. ssp.basilicum*

 Ocimum menthiifolium Hochst. ex Benth.

▸ *Ocimum forskolei Benth.*

 Ocimum micranthemum Willd.

Synonyms ▸ *Ocimum campechianum* Mill.
Common Names ▸ Peru basil; *German:* Peruanischer Basilikum; *Mexico:* albahaca, albahaca de monte, kakaltum, albaca silvestre, salvaca; *Portuguese:* alfavaca, alfavaca-do-campo, mangericao-grande

Ocimum tenuiflorum L.

Synonyms ▸ *Ocimum sanctum* L.
Common Names ▸ holy basil, sacred basil, Thai basil, mosquito plant of South Africa; *Cambodian:* mrèah prĕu; *Chinese:* sheng luo le; *French:* basilic des moines, basilic sacré; *German:* Duftender Basilikum, Indischer Basilikum, Kleiner Basilikum, Heiliger Basilikum, Grüner Tulsi, Tulasi-Basilikum, Tulsipflanze; *Hindi.:* tulsi; *India:* kala tulasi; *Indonesian:* kemangi utan, lampes, ruku-ruku; *Korean:* kanuniphyangkkulphul; *Malaysian:* oku, ruku ruku, selash hitans, ulasi; *Pilipino:* balonoi, cologoco, loko loko, solasi; *Russian:* basilik swajaschenujy; *Sanskrit:* parnasa, suvasa tulasi, tulashi; *Spanish:* albahaca cimarrona, albahaca morada criolla; *Thai:* bai ga-prow, im-khim-lam, kaphrao, kom kodong; *Vietnamese:* e do, e tia, kuong nhu tai
Usage ▸ spice; **product:** essential oil
Parts Used ▸ leaf, herb
Distribution ▸ tropical Asia: India, Malaysia, native elsewhere in Tropics and widely cultivated
Note ▸ Eugenol, chavibetol or chavicol methyl ether are predominant types. Frequently cultivated in courtyards and around temples in India for about 3000 years.

Berger 4, 1954; Bois 1934; Brophy/Jogia 1984; Brophy et al. 1993; Cheers 1997; Chogo/Crank 1981; Darrah 1980; Dudtschenko et al. 1989; Erhardt et al. 2002; Gurib-Fakim/Brendler 2004; Guzman/Siemonsma 1999; Herklots 1972; Hiltunen/Holm 1999; Kelm et al. 2000; Kottegoda 1994; Lal et al. 1978; Lawrence et al. 1972; Narenda Singh/Sharma 1980; Plescher et al. 1997; Seidemann 1993c; Sharma 2003; Skaltsa-Diamantidis et al. 1990; Small 1997; Täufel et al. 1993; Teuscher 2003; Tucker 1986; Turova et al. 1987; Uphof 1968; Verma et al. 1989; WHO 1990; Wiersema/León 1999; Zeven/de Wet 1982

 Ocimum thyrsiflorum L.

▸ *Ocimum basilicum* L. ssp. *basilicum*

 Ocimum viride Will.

▸ *Ocimum gratissimum* L.

OCOTEA Aubl. - Lauraceae

 Ocotea cujumary Mart.

Common Names ▸ *Brazil (Portuguese):* cujumaru, cuimaru, cuimari, louro-cujumari
Usage ▸ flavoring (locally)
Parts Used ▸ bark, fruit
Distribution ▸ Amazonia

Mors et al. 2000

 Ocotea cinnamomoides (Mutis ex Kunth) Kosterm.

▸ *Ocotea cymbarum* H.B.K.

 Ocotea cymbarum H.B.K.

Synonyms ▸ *Laurus cinnamomoides* Mutis ex H.B.K., *Nectandra cinnamoides* Nees, *Nectandra cymbarum* (H.B.K.) Nees, *Ocotea cinnamomoides* (Mutis ex Kunth) Kosterm.
Common Names ▸ canela; *Portuguese (Brazil):* louro inamuty, louro mamory, koto; *Spanish:* canela do mato
Usage ▸ spice
Parts Used ▸ bark of young shoots, flower cup
Distribution ▸ tropical S America: Brazil, Ecuador
Note ▸ In Ecuador the inflorescence is also used as a spice.

Bois 1934; Bourton 1968; Gottlieb/Gottlieb 1980; Mors et al. 2000; Naranjo et al. 1981; Rätsch 1998; Schultze-Motel 1986; Seidemann 1993c; Teuscher 2003; Uphof 1968; Zeven/Zhukovsky 1975

 Ocotea puchury-major Mart.

Synonyms ▸ *Acrodiclidium puchury-major* (Nees et Mart. ex Nees) Mez., *Licaria puchury-major* (Mart.) Kosterm., *Nectandra puchury-major* Nees et Mart.
Common Names ▸ *Brazil:* louro-puxuri, pichuri, pixurim, puchuiri; *German:* Große Macisbohne, Pichurinuss, Pichurimbohne, Sassafrasnuss

Usage ▶ spice
Parts Used ▶ seed
Distribution ▶ S America: Amazonia, Brazil also cultivated
Note ▶ Spike is like the mace from nutmeg.

Hanelt 2001; Hiller/Melzig 1999; Mors et al. 2000; Rätsch 1998; Schultze-Motel 1986; Seidemann 1993c; Täufel et al. 1993; Teuscher 2003; Usher 1968; Wiersema/León 1999

Ocotea quixos *(Lam.) Kosterm.*

Synonyms ▶ *Laurus quixos* Lam., *Licaria quixos* (Lam.) Kosterm.
Common Names ▶ American cinnamon, ocotea, *German:* Amerikanischer Zimt; *Peru:* ispungu
Usage ▶ spice (for sweets)
Parts Used ▶ flores, fruit-cup
Distribution ▶ S America: Brazil, also cultivated for the cinnamon-like fragrance
Note ▶ The flowers and leaves have a cinnamon taste.

Hanelt 2001; Naranjo 1981; Naranjo et al. 1981; Rätsch 1998; Schultze-Motel 1986; Seidemann 1993c; Teuscher 2003

Ocotea pretiosa *(Nees) Mez*

Common Names ▶ false sassafras; *Brazil (Portuguese):* canela-sassafrás; *German:* Brasilianischer Sassafras
Usage ▶ spice, flavoring; **product:** essential oil
Parts Used ▶ bark, leaf
Distribution ▶ S America: Brazil

Hickey 1948; Kumar 2001; Mors et al. 1959, 2000

OENANTHE L. - Water Dropwort - Apiaceae (Umbelliferae)

Oenanthe javanica *(Bl.) DC.*

Synonyms ▶ *Oenanthe japonica* Miq., *Oenanthera linearis* DC., *Oenanthera stolonifera* Wall., *Sium javanicum* L., *Sium laciniatum* Bl.
Common Names ▶ Javan water dropwort, water celery, water parsley, water dropwort, oriental celery; *Arabic:* oshb el-maa almodala; *Chinese:* chin tsai, chu kuen, ku chin, shui ching; *Dutch:* torkruid; *French:* persil séri; *German:* Javanischer or Vietnamesischer Wasserfenchel, Wassersellerie; *Hindi:* saya; *Italian:* finocchio-acquatico; *Japanese:* seri, shijriba; *Javanese:* pampung; *Korean:* minari; *Malaysian:* pamponng, selom, selom piopo; *Portuguese:* funcho aquático, funcho oriental, cicuta; *Russian:* omaznik javanskij; *Spanish:* cicuta, enante, felandrio, hinojo de agua; *Thai:* phakchilom; *Vietnamese:* can nuoc
Usage ▶ spice, condiment in India, flavoring soups, chicken or fish dishes
Parts Used ▶ herb, young shoots
Distribution ▶ S, SE and E Asia from Pakistan to Japan, from Far E Russia to Australia: Queensland also cultivated
Note ▶ Culinary herb also in N America (Canada: Ontario).

Arora/Pandey 1996; Cheers 1997; Davidson 1999; Erhardt et al. 2002; Kuebal/Tucker 1988; Kumar 2003; Larkcom 1991; Lück 2004; Ogle et al. 2003; Schultze-Motel 1986; Small 1997; Tucker 1986; Uphof 1968; Wiersema/León 1999; Zeven/de Wet 1982

Oenanthe stolonifera *Wall.*

▶ *Oenanthe javanica (Bl.) DC.*

Oenanthera linearis *DC*

▶ *Oenanthe javanica (Bl.) DC.*

OLAX L. - Olacaceae

Olax scandens *Roxb.*

Usage ▶ spice
Parts Used ▶ leaf
Distribution ▶ SE Asia

Terra 1966

Olax viridis *Oliv.*

Usage ▸ spice
Parts Used ▸ seed
Distribution ▸ W Africa: Gabun; S Africa

Burkill 4, 1997; Jansen 1981; Mabberley 1997; Neuwinger 1999

 Olax zeylanica *L.*

Usage ▸ condiment (locally)
Parts Used ▸ leaf, fruit
Distribution ▸ W Himalayas, India, Sri Lanka
Note ▸ Leaves and fruits have a smell and a slight taste of garlic.

Mabberley 1997

 Olea fragrans *Thunb.*

▸ *Osmanthus fragrans (Thunb.) Lour.*

 Orellana americana *Kuntze*

▸ *Bixa orellana L.*

ORIGANUM L. - Majoram, Oregano - Lamiaceae (Labiatae)

Calpouzos 1954; Fleisher/Sneer 1982; Kintzios 2002; Wilkins/Madsen 1991;

 Origanum album *Salisb.*

▸ *Origanum onites L.*

 Origanum anglicum *Hill*

▸ *Origanum vulgare L. ssp. vulgare*

 Origanum x appli *(Domin) Boros*

▸ *Origanum x majoricum Cambess.*

 Origanum bucharicum *Bornm.*

▸ *Origanum vulgare L. ssp. gracile (Loch) Ietswart*

 Origanum compactum *Benth.*

▸ *Origanum vulgare L. ssp. vulgare*

 Origanum creticum *L.*

▸ *Origanum vulgare L. ssp. vulgare*

 Origanum dictamnus *L.*

Synonyms ▸ *Dictamnus creticus* J. Hill, *Majorana dictamnus* (L.) Kostel., *Majorana tomen-tosa* (Moench) Stokes, *Origanum saxatile* Salisb.
Common Names ▸ dittander, Crete dittany, dittany of Crete; *French:* dictame de Crete; *German:* Diptamdost(en), Kretischer Dost, Pfefferkraut; *Italian:* dittamo cretico
Usage ▸ spice, pot-herb; **product** for distilleries
Parts Used ▸ leaf
Distribution ▸ Greece: Crete; England, Italy, cultivated in USA, Crete, England

Berger 4, 1954; Bois 1934; Bosabalidis 1987, 1990a; Bosabalidis/Tsekos 1982; Charalambous 1994; Cheers 1997; Clebsch 1997; Davidson 1999; Erhardt et al. 2002; Hager 5, 1993; Hanelt 2001; Harvala et al. 198a; Heeger 1956; Hiller/Melzig 1999; Hoppe 1949; Ietswaart 1980; Møller et al. 1999; Schönfelder 2001; Schultze-Motel 1986; Seidemann 1993c; Small 1997; Täufel et al. 1993; Teuscher 2003

 Origanum floridum *Salisb.*

▸ *Origanum vulgare L. ssp. viride (Boiss.) Hayek*

ORIGANUM L. - Majoram, Oregano - Lamiaceae (Labiatae)

 **Origanum glandulosum** *Desf.*

▶ *Origanum vulgare L. ssp. vulgare*

 **Origanum gracile** *Koch*

▶ *Origanum vulgare ssp. gracile (Koch) Letswaart*

 **Origanum heracleoticum** *L.*

▶ *Origanum vulgare L. ssp. hirtum (Link) Letsw. and*
▶ *Origanum vulgare L. ssp. viride (Boiss.) Hayek.*

 **Origanum humile** *Mill.*

▶ *Origanum vulgare L. ssp. viride (Boiss.) Hayek*

 **Origanum indicum** *Roth*

▶ *Pogostemon heyneanus Benth.*

 **Origanum macrostachyum** *Hoffmanns & Link*

▶ *Origanum vulgare L. ssp. virens (Hoffmanns. et Link) Letswaart*

◘ Origanum majorana, flowering

Origanum majorana L. ◘

Synonyms ▶ *Majorana hortensis* Moench.

Common Names ▶ marjoram; sweet majoram, knotted majoram; *Arabic:* bardagoush, barsagusha, mardqouche, mizunjuske; *Dutch:* marjolein; *French:* marjolaine, origane; *German:* Gartendost, Majoran, Meiran, Wurstkraut; *Hindi:* marua, murwa, sathra; *Italian:* maggiorana, persia; *Portuguese:* mangerona; *Russian:* majoran, majoran sadowyj, duschiza sadowaja, kolbasnaja trawa, borstirochi; *Sanskrit:* kharapatra, marubaka; *Spanish:* mejorana; *Turkish:* makiron

Usage ▶ spice, pot-herb; **product**: essential oil (flavoring of foods, liqueurs, perfume)

Parts Used ▶ herb, leaf

Distribution ▶ N Africa, Cyprus, Anatolia, Egypt; Arabia, SW Asia, India, native in Tunesia, Spain, Corsica, Italy, France, Germany, Balkan Peninsula; also cultivated

Abou-Zied 1973; Bärtels 1997; Başer et al. 1993d; Berger 4, 1954; Bertelli et al. 2003; Bois 1934; Bremness 2001; Cheers 1997; Circella et al. 1995; Clair 1961; Coiciu/Racz (no year); Davidson 1999; Deans/Svoboda 1990; Dubial 1988; Dudtschenko et al. 1989; Farrell 1985; Fischer et al. 1985, 1987; Graner 1968; Guzman/Siemonsma 1999; Hager 5, 1993; Hanelt 2001; Heeger 1956; Heine 1993; Hepper 1992; Hiller/Melzig 1999; Hoffmann et al. 1992; Hoppe 1949; letswaart 1980; Karawya/Hifnawy 1976; Kawabata et al. 2003; Komaitis et al. 1992; Lagouri et al. 1993; Lossner 1967, 1968; Melchior/Kastner 1974; Novak et al. 2002; Nykänen 1986; Oberdick 1983, 1989, 1990; Omer et al. 1994; Peter 2001; Pochljobkin 1974, 1977; Pschyrembel 1998; Raghavan et al. 1997; Refaat et al. 1992/93; Roth/Kormann 1997; Ruberto et al. 1993; Şarer et al. 1982; Schönfelder 2001; Schröder 1991; Schultze-Motel 1986; Seidemann 1993c; Seidemann/Siebert 1987; Siewek 1990; Singh et al. 1996; Small 1997; Staesche 1972; Täufel et al. 1993; Teuscher 2003; Tucker 1986; Ubillos 1989; Vagi et al. 2002; Villamar et al. 1994; Zeven/de Wet 1982

Origanum x majorana Cambess.

[Origanum majorana x Origanum vulgare ssp. vulgare or Origanum majorana x vulgare ssp. virens]

Synonyms ▸ *Majorana hortensis* Moench, *Majorana majorica* (Cambess.) Briq., *Origanum x appli* (Domin) Boros, *Origanum paniculatum* Koch
Common Names ▸ Italian oregano, hardy marjoram; *German:* Italienischer Majoran
Usage ▸ pot-herb
Parts Used ▸ leaf, herb
Distribution ▸ SW Europe: Spain, Portugal, Balearic Islands, also cultivated in gardens of W, S and C Europe, USA, S America: Argentina, Brazil, Uruguay

Bremness 2001; Erhardt et al. 2002; Hanelt 2001; Lawrence 1984; Ietswaart 1980; Sartoratto/Augusto 2003; Seidemann 1993c; Small 1997; Täufel et al. 1993; Tucker 1986; Wilkins/Madsen 1991; Ydava/Saini 1991

 Origanum maru L.

▸ *Origanum syriacum* L.

 **Origanum microphyllum** (Benth.) Boiss.

Common Names ▸ microphylla oregano, small-leaved oregano; *French:* origan à petites feuilles; *German:* Schmalblättriger Oregano
Usage ▸ spice
Parts Used ▸ leaf, herb
Distribution ▸ native to Crete, cultivated in gardens in N America

Ietswaart 1980; Small 1997

 Origanum normale D. Don

▸ *Origanum vulgare* L. ssp. *viride* (Bois.) Hayek

 Origanum onites L.

Synonyms ▸ *Majorana cretica* Mill., *Majorana onites* (L.) Benth., *Origanum album* Salisb.
Common Names ▸ pot marjoram, Spanish hop; Turkish oregano, Cretan oregano; *French:* origan de chypre; *German:* Griechischer or Kretischer Dost, Ragani, Spanischer Hopfen, Französischer or Wilder Majoran; *Italian:* origano siciliano
Usage ▸ spice; **product:** essential oil (with a high carvacrol content)
Parts Used ▸ leaf, herb
Distribution ▸ S Europe: E Sicily, S Greece; W, S Anatolia, local cultivated in many Mediterranean countries.

Akgül/Bayrak 1987; Berger 4, 1954; Bois 1934; Bournot 1968; Bremness 2001; Cheers 1997; Davidson 1999; Erhardt et al. 2002; Griebel 1938; Hager 5, 1993; Hanelt 2001; Hiller/Melzig 1999; Ietswaart 1980; Kivanç/Akgül 1989; Melchior/Kastner 1974; Pizzale et al. 2002; Schultze-Motel 1986; Seidemann 1993c; Seidemann/Siebert 1987; Small 1997; Staesche 1972; Täufel et al. 1993; Teuscher 2003; Tucker 1986; Uphof 1968; Vokou et al. 1988

 Origanum orientale Mill.

▸ *Origanum vulgare* L. ssp. *vulgare*

 Origanum paniculatum Koch

▸ *Origanum x majorana* Cambess.

 Origanum saxatile Salisb.

▸ *Origanum dictamnus* L.

 Origanum smirnaeum Sibth. et Sm.

▸ *Origanum vulgare* L. ssp. *glandulosum* (Desf.) Ietswaart

 Origanum syriacum L.

Synonyms ▸ *Amaracus syriacus* (L.) Stokes, *Majorana crassa* Moench, *Origanum maru* L.

ORIGANUM L. - Majoram, Oregano - Lamiaceae (Labiatae)

Common Names ▶ Syrian oregano, white oregano; *Arabic:* za'atar; *French:* origan d'Egypte; *German:* Biblischer Oregano, Brauner Dost, Syrischer Dost, Arabischer Majoran, Syrischer Majoran, Echter Staudenmajoran; *Vietnamese:* rau kingh gioi;
Usage ▶ spice
Parts Used ▶ leaf, herb; **product:** essential oil (two chemotypes: carvacrol and thymol type)
Distribution ▶ E Mediterranean Region: S Anatolia, Cyprus, Syria, Lebanon, Israel, Jordan, Egypt, Sinai Peninsula
Note ▶ The plant is the biblical oregano.

Akgül/Bayrak 1987; Bärtels 1997; Baser et al. 2003; Berger 4, 1954; Bois 1934; Bourton 1968; Dudai et al. 1988; Fleisher/Fleisher 1988; Hager 5, 1993; Hanelt 2001; Hepper 1992; Hiller/Melzig 1999; Ietswaart 1980; Kamel et al. 2001; Ophof 1968; Small 1997; Täufel et al. 1993; Teuscher 2003; Tucker 1986; Tümen/Başer 1993

Origanum tytthantum Gontsch

▶ *Origanum vulgare L. ssp. gracile (Loch) Ietswaart*

Origanum virens Hoffmanns et Link

▶ *Origanum vulgare L. ssp. virens (Hoffmanns. et Link) Ietswaart*

Origanum viridulum Martrin-Donos

▶ *Origanum vulgare L. ssp. viride (Boiss.) Hayek*

Origanum vulgare L.

Synonyms ▶ *Thymus origanum* E.H.L. Krause
Common Names ▶ oregano, wild marjoram, pot marjoram, winter marjoram, winter sweet, pizza herb; *Arabic:* mardakosh; *Chinese:* niu zhi, tu xiang ru; *Dutch:* wilde marjolein; *French:* marjolaine sauvage, origan commun, origan vulgaire; *German:* Gewöhnlicher Dost, Wilder Majoran, Falscher Staudenmajoran, Orangenkraut, Oregano; Wintermajoran; *Hindi:* sathra; *Italian:* origano comune, maggiovana selvatica, regamo; *Japanese:* oregano; *Portuguese:* orégão; *Russian:* lutschiza, lutshitza abyknowennaja, materinka, ladanka, mazerdyschka, bloschiza, dushitsa, senowka, kara gynych; *Spanish:* orégano, oregano comun; *Turkish:* kekig orégano común, mercanköşk
Usage ▶ spice, pot-herb; **product:** essential oil
Parts Used ▶ herb
Distribution ▶ Europe, Turkey, Cyprus, Iraq, Iran, W and E Siberia, C Asia, Afghanistan, Pakistan; Himalayas, Mongolia, NW Africa, native in N America, China

Aichele/Schwegler 4, 1995; Arora/Pandey 1996; Bärtels 1997; Bendini et al. 2002; Bois 1934; Bremness 2001; Cervato et al. 2000; Cheers 1997; Clair 1961; Coiciu/Racz (no year); Davidson 1999; Dudtschenko et al. 1989; Erhardt et al. 2002; Guzman/Siemonsma 1999; Hager 5, 1993; Hanelt 2001; Heeger 1956; Hiller/Melzig 1999; Hondelmann 2002; Hoppe 1949; Kikuzaki/Nakatani 1989; Melchior/Kastner 1974; Miloš et al. 2000; Ondarza/Sanchez 1990; Pochljobkin 1974, 1977; Pruthi 1976; Pschyrembel 1998; Puertas-Mejia et al. 2002; Rodrigues et al. 2003; Rosengarten 1969; Roth/Kormann 1986; Salmeron et al. 1990; Schönfelder 2001; Seidemann 1993c; Seidemann/Siebert 1987; Sharma 2003; Siewek 1990; Small 1997; Staesche 1972; Teuscher 2003; Tucker 1986; Turova et al. 987; Ubillos 1989; Uhl 2000; Vekiari et al. 1993a, b; Villamar et al. 1994; Werker et al. 1985; Wyk et al. 2004; Zeven/de Wet 1982

Origanum vulgare L. ssp. glandulosum (Desf.) Ietswaart

Synonyms ▶ *Origanum glandulosum* Desf.; *Origanum smirnaeum* Sibth. et Sm.
Common Names ▶ *German:* Drüsiger Dost
Usage ▶ spice
Parts Used ▶ herb
Distribution ▶ N Africa: N Algeria, Tunesia

Hanelt 2001; Ietswaart 1980; Tucker 1986

Origanum vulgare L. ssp. gracile (Koch) Letsw.

Synonyms ▶ *Origanum bucharicum* Bornm., *Origanum gracile* Koch, *Origanum tyttantum* Gontsch
Common Names ▶ Russian oregano, Turkestan oregano; *German:* Kirgisischer Oregano, Schlanker Dost
Usage ▶ spice
Parts Used ▶ herb
Distribution ▶ E Anatolia, N Iraq, Iran, Pakistan, Afghanistan, C Asia: Pamiroalai, W Tien Shan

Berger 4, 1954; Hanelt 2001; Ietswaart 1980; Schultze-Motel 1986; Tucker 1986

Origanum vulgare L. subsp. hirtum (Link) Letsw.

Synonyms ▶ *Origanum heracleoticum* L., *Origanum smyrnaeum* Sibth. et Sm.
Common Names ▶ Greek or Italian oregano; *German:* Falscher Staudenmajoran; Griechischer Dost, Italienischer Dost, Pizza Oregano
Usage ▶ spice, pot-herb; **product:** essential oil
Parts Used ▶ herb
Distribution ▶ Europe: Italy, Sicily, Greece, Turkey

Akgül/Bayrak 1987; Baser et al. 1994a; Berger 4, 1954; Bremness 2001; Davidson 1999; Erhardt et al. 2002; Hanelt 2001; Hiller/Melzig 1999; Ietswaart 1980; Melchior/Kastner 1974; Milos et al. 2000; Schönfelder 2001; Schultze-Motel 1986; Täufel et al. 1993; Teuscher 2003; Tucker 1986; Veres et al. 2003

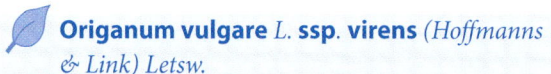

Origanum vulgare L. ssp. virens (Hoffmanns & Link) Letsw.

Synonyms ▶ *Origanum virens* Hoffmanns & Link, *Origanum macrostachyum* Hoffmanns et Link
Common Names ▶ wild majoram; *German:* Wilder Majoran
Usage ▶ spice
Parts Used ▶ herb
Distribution ▶ W Mediterranean: Azores, Canary Islands, Madeira, Iberian Peninsula, N Africa, Balearic Islands, also cultivated in Argentina (gardens)

Berger 4, 1954; Bournot 1968; Hanelt 2001; Hohmann 1968; Hiller/Melzig 1999; Ietswaart 1980; Tucker 1986

Origanum vulgare L. ssp. viride (Boiss.) Hayek

Synonyms ▶ *Origanum heracleoticum* L., *Origanum humile* Mill., *Origanum normale* G. Don, *Origanum viridulum* Martrin-Donos
Common Names ▶ wild marjoram; *German:* Falscher Staudenmajoran
Usage ▶ spice, condiment
Parts Used ▶ herb

Distribution ▶ S Europe (Italy) to E Asia: E China, also cultivated and in USA

Berger 4, 1954; Bois 1934; Erhardt et al. 2002; Hanelt 2001; Ietswaart 1980; Melchior/Kastner 1974; Schönfelder 2001; Tucker 1986

Origanum vulgare L. ssp. vulgare

Synonyms ▶ *Origanum angelicum* Hill., *Origanum compactum* Benth., *Origanum creticum* L., *Origanum floridum* Salisb., *Origanum latofolium* Mill., *Origanum orientale* Mill.
Common Names ▶ Oregano, wild marjoram; *French:* marjolaine, origan; *German:* Kretischer Dost, Oregano
Usage ▶ spice
Parts Used ▶ herb
Distribution ▶ N regions of the world: N Europe, N America

Berger 4, 1954; Bois 1934; Erhardt et al. 2002; Hanelt 2001; Hiller/Melzig 1999; Hoppe 1949; Ietswaart 1980; Maarse/van Ost 1973a, b, 1974; Melchior/Kastner 1974; Mockute et al. 20001; Schultze-Motel 1986; Uphof 1968; Veres et al. 2003

OSMANTHUS Lour. - Devil Wood, Sweet Olive - Oleaceae

Osmanthus asiaticus Nakai

▶ *Osmanthus fragrans (Thunb.) Lour.*

Osmanthus fragrans (Thunb.) Lour.

Synonyms ▶ *Olea fragrans* Thunb., *Osmanthus asiaticus* Nakai
Common Names ▶ fragant olive, sweet olive, tea olive; sweet osmanthus flower; *Chinese:* gui hua, yen kuei, mu hsi, *German:* Osmanthusblüten, Süße Duftblüte; *Japanese:* usugi mokusei, gin mokusei, kiu mokusei; *Vietnamese:* cay moc
Usage ▶ flavoring of tea and cake
Parts Used ▶ flower
Distribution ▶ Japan, SW China; E Himalayas

Bois 1934; Clebsch 1997; Duke/Ayensu 1985; Erhardt et al. 2002; Fleischmann et al. 2002; Hanelt 2001; Schultze-Motel 1986; Seidemann 1993c; Uphof 1968; Wealth of India 7, 1966; Wiersema/León 1999; Zeven/de Wet 1982

OSMORHIZA Michx. - Apiaceae (Umbelliferae)

 Osmorhiza claytonii *(Michx.) C.B. Clarke*

Common Names ▶ sweet jarvil
Usage ▶ condiment
Parts Used ▶ root, unripe seed
Distribution ▶ C, E Canada, USA
Note ▶ The plant has an anise-like aroma.

Small 1997

Osmorhiza longistylis *(Torr.) DC.*

Usage ▶ condiment
Parts Used ▶ leaf
Distribution ▶ W, N America
Note ▶ The plant has an anise-like aroma.

Mabberly 1997; Rätsch 1998

Osmorhiza occidentalis *(Nutt.) Torr.*

▶ *Myrrhis occidentalis* Benth. et Hook.f.

OSYRIS L. - Santalaceae

Osyris tenuifolia *Engl.*

Common Names ▶ African sandalwood; *German:* Afrikanisches Sandelholz
Usage ▶ flavoring
Parts Used ▶ wood
Distribution ▶ Africa, especially E Africa and Madagasca
Note ▶ Substitute for ▶ *Santalum album* L.

Hiller/Melzig 1999; Mabberly 1997; Uphof 1968

OTOSTEGIA Benth. - Lamiaceae (Labiatae)

Otostegia fruticosa *Forssk.*

Common Names ▶ *German:* Strauch-Otostegie
Usage ▶ spice, flavoring
Parts Used ▶ leaf
Distribution ▶ Egypt; neotropical Africa to C Asia
Note ▶ The essential oil is a substitute for thyme oil.

Aboutabl et al. 1995; Mabberly 1997

OTTELIA Pers. - Hydrocharitaceae

Ottelia alismoides *(L.) Pers.*

Synonyms ▶ *Ottelia condorensis* Gagn., *Ottelia japonica* Miq., *Ottelia javanaica* Miq.
Common Names ▶ water-plantain ottelia; *Dutch:* duikerbloom; *German:* Espada, Froschlöffelähnliche Ottelie; *Japanese:* mizu-ôba-ko; *Malaysian:* keladi ayer; *Pilipino:* espada; *Spanish:* tangila; *Thai:* santawah pak
Usage ▶ spice, for rice and vegetables
Parts Used ▶ leaf
Distribution ▶ Egypt, India, Sri Lanka, Thailand, China, Japan, N Australia

Cook et al. 1984; Erhardt et al. 2002; Hanelt 2001; Seidemann 1993c; Täufel et al. 1993; Wiersema/León 1999

Ottelia condorensis *Gagn.*

▶ *Ottelia alismoides* (L.) Pers.

Ottelia japonica *Miq.*

▶ *Ottelia alismoides* (L.) Pers.

Ottelia javanica *Miq.*

 Ottelia alismoides (L.) Pers.

Usage ▶ spice (of rice)
Parts Used ▶ herb
Distribution ▶ Madagascar, Zanzibar, Cape of Good Hope

Neuwinger 1999

OXALIS L. - Sorrel - Oxalidaceae

 Oxalis acetosella *L.*

Common Names ▶ common wood sorrel, wood sorrel, shamrock, alleluia; *French:* oseille de bûcheron, oxalide petite oseille, oxadille blanche, pain de coucou, surette; *German:* Waldsauerklee; *India:* khate meethi (Kashmir); *Italian:* acetosella, alleluja; *Russian:* kisliza obyknowennaja; *Spanish:* acederilla, oxadile blanca, trébol aceto
Usage ▶ spice
Parts Used ▶ leaf
Distribution ▶ Europe, Turkey, W and E Siberia, Himalayas, China, Japan, Alaska, Canada, NE, N, C, E and SE USA.

Aichele/Schwegler 3, 1995; Bendel 2002; Davidson 1999; Dhar/Dhar 2000; Dudtschenko et al. 1989; Erhardt et al. 2002; Erichsen-Brown 1989; Fleischhauer 2003; Hiller/Melzig 1999; Koschtschejew 1990; Schönfelder 2001; Schultze-Motel 1986; Schulz 1960; Seidemann 1993c; Sharma 2003; Stowart 1978; Täufel et al. 1993; Tucker 1986; Uphof 1968; Wiersema/León 1999

 Oxalis barrelieri *L.*

Synonyms ▶ *Oxalis sepium* A.St. Hil.
Common Names ▶ *Indonesian:* belimbing tanah
Usage ▶ pot-herb
Parts Used ▶ leaf;
Distribution ▶ native in tropical S America; S E Asia but it is only occasionally cultivated in Indonesia: Sumatra, Java, Malaysia, Papua and New Guinea
Note ▶ It was first observed in Java in 1888.

Guzman/Siemonsma 1999

 Oxalis caprina *L.*

Common Names ▶ goat's food, wood sorrel

 Oxalis corniculata *L.*

Synonyms ▶ *Oxalis javanica* Bl., *Oxalis lupulina* Kunth., *Oxalis repens* Thunb.
Common Names ▶ creeping wood sorrel, creeping oxalos, yellow sorrel; *Arabic:* hamd; *Brazil (Portuguese):* azedinha, pé-de-pombo, três-coraçoes, trevo; *Chinese:* cu jiang, cao, tsa chiang, hsiao suan tsai; *German:* Hornfrüchtiger Sauerklee, Hornsauerklee; *India:* ambuti, amrul, carngeri, puliyarai, puliyaral; *Italian:* carpigna; *Vietnamese:* me dat, me chua, aut
Usage ▶ spice
Parts Used ▶ leaf
Distribution ▶ Europe, Turkey, Levante, Caucasus region, Iran, C Asia, Himalayas, China, Japan, Amur, Sachalin, N Africa, cosmopolitical

Aichele/Schwegler 3, 1995; Dudtschenko et al. 1989; Erhardt et al. 2002; Erichsen-Brown 1989; Fleischhauer 2003; Hanelt 2001; Kumar 2003; Mors et al. 2000; Ogle et al. 2003; Schnelle 1999; Schultze-Motel 1986; Täufel et al. 1993; Tucker 1986; Villamar et al. 1994; Wiersema/León 1999

 Oxalis corymbosa *DC.*

Synonyms ▶ *Oxalis debilis* H.B.K., *Oxalis martiana* Zucc.
Common Names ▶ pink wood sorrel; *Brazil (Portuguese):* caruru-de-sapo; *German:* Doldentraubiger Sauerklee; *Indonesian:* kembang gelas; *Japanese:* muraseaki-katrabami, yafata; *Sumatra:* asam puja; *Vietnamese:* chua me (aas) hoa (or)
Usage ▶ pot-herb
Parts Used ▶ leaf
Distribution ▶ native to tropical S America; in SE Asia but it is only occasionally found in Indonesia: Java, W Sumatra; Malaysia and the Philippines
Note ▶ It was introduced into Java from Sydney before 1848.

Guzman/Siemonsma 1999; Hanelt 2001; Kumar 2003; Mors et al. 2000; Schultze-Motel 1986

 Oxalis debilis H.B.K.

▶ *Oxalis corymbosa* DC.

 Oxalis intermedia A. Rich.

▶ *Oxalis latifolia* Kunth.

 Oxalis javanica Bl.

▶ *Oxalis corniculata* L.

 Oxalis latifolia H.B.K.

Synonyms ▶ *Oxalis intermedia* A. Rich., *Oxalis mauritiana* Lodd.
Common Names ▶ purple garden oxalis, fish-tail oxalis; *German:* Breitblättriger Sauerklee; *Indonesian:* calingcing; *S Africa:* rooisuring, rooituisuring, suring, tuinsuring
Usage ▶ pot-herb
Parts Used ▶ leaf
Distribution ▶ C and trop S America, SE Asia it appears in Indonesia: Java

Aichele/Schwegler 3, 1995; Erhardt et al. 2002; Guzman/Siemonsma 1999; Hanelt 2001; Kottegoda 1994; Schultze-Motel 1986; Villamar et al. 1994; Wiersema/León 1999

 Oxalis livida Jacq.

Common Names ▶ bluish sorrel; *German:* Bläulicher Sauerklee;
Usage ▶ spice (of rice)
Parts Used ▶ herb
Distribution ▶ Madagascar, Zanzibar

Neuwinger 1999

 Oxalis lupulina Kunth

▶ *Oxalis corniculata* L.

 Oxalis martiana Zucc.

▶ *Oxalis corymbosa* DC.

 Oxalis mauritiana Lodd.

▶ *Oxalis corymbosa* DC.

 Oxalis repens Thunb.

▶ *Oxalis corniculata* L.

 Oxalis sepium A.St. Hil.

▶ *Oxalis barrelieri* L.

 Oxalis stricta L.

Common Names ▶ yellow sheep sorrel; *German:* Aufrechter Sauerklee
Usage ▶ pot-herb (rarely)
Parts Used ▶ fresh leaf
Distribution ▶ E Canada, USA, native in Europe and E Asia

Erhardt et al. 2002; Fleischhauer 2003

 Oxalis viridis Oliv.

Common Names ▶ green sorrel; *German:* Grüner Sauerklee
Usage ▶ spice
Parts Used ▶ seed
Distribution ▶ Gabun

Dudtschenko et al. 1989

Ozymum citratum *Rumph.*

> *Ocimum x citriodorum Vis.*

Padus mahaleb (l.) Borkh.

> Prunus mahaleb L.

Padus serotina (Ehrh.) Borkh.

> Prunus serotina Ehrh.

Palmaria palmata (L.) Kuntze

> Rhodymenia palmata (L.) Grev.

Panax fruticosum L.

> Polyscias fruticosa (L.) Harms

PANDANUS Parkinson - Screw Pine - Pandanaceae

Pandanus amaryllifolius Roxb.

Synonyms ▶ *Pandanus hasskarlii* Merr., *Pandanus latifolius* Hassk., *Pandanus odorus* Ridley

Common Names ▶ fragrant pandan, fragrant screwpine, pandan; *Cambodian:* taėy; *German:* Amarillisblättriger Schraubenbaum, Amarillisblättrige Pandanane; *Hindi:* flower: kewra, keora; *Indonesian:* pandan, pandan wangi, pandan rampe; *Malaysian:* daun pandan, pandan, pandan rampai, pandan wangi; *Pilipino:* pandan, pandan mabango; *Thai:* toey hom, panae-wo-nging; *Vietnamese:* la dua

Usage ▶ spice
Parts Used ▶ young leaves
Distribution ▶ Malaysia, cultivated Java, Molucca Islands

Arora/Pandey 1996; Davidson 1999; Erhardt et al. 2002; Guzman/Siemonsma 1999; Hanelt 2001; Hyndman 1984; Jezussek 2002; MacLeod/Pieris 1982; Oyen/Dung 1999; Salim et al. 203; Schultze-Motel 1986; Seidemann 1993c; Zeven/de Wet 1982

Pandanus hasskarlii Memill

> Pandanus amaryllifolius Roxb.

Pandanus latifolius Hassk.

> Pandanus amaryllifolius Roxb.

Pandanus odoratissimus L.f.

> Pandanus tectorius Parkins.

Pandanus odorifer (Forssk.) Kuntze

> Pandanus tectorius Parkins.

Pandanus odorus Ridley

> Pandanus amaryllifolius Roxb.

Pandanus tectorius Parkinson

Synonyms ▶ *Pandanus odoratissimus* L.f., *Pandanus odorifer* (Forssk.) Kuntze

Common Names ▶ fragrant screwpine, odour screwpine, pandanus leaf; *French:* baquois, vacouet; *German:* Duft-Schraubenbaum, Duft-Pandanane; *Hindi:* kewra; *Indonesian:* pandan pantai; *Malaysian:* mengkuang laut, mengkuang duri; *Pilipino:* pandan, pibis, pandan dagat; *Portuguese:* pântano; *Spanish:* bacua; *Thai:* bai toey hom, toei hum, toei daang

Usage ▶ spice

Parts Used ▶ leaf (rarely as curry leaf), ♂ inflorescence

Distribution ▶ coasts of India, Sri Lanka, Myanmar, Polynesia, Australia, Macarene Island

Note ▶ Its flowers are the source of keora or kewra.

Cheers 1997; Davidson 1999; Deshpande 1938; Dhingra et al. 1954; Dutta et al 1987; Engel/Phummai 2000; Erhardt et al. 2002; Hanelt 2001; Norman 1991; Schultze-Motel 1986; Seidemann 1993c, 1999, 2001d; Sharma 2003; Shiva et al. 2002; Siewek 1990; Stone 1976; Storrs 1997; Uhl 2000; Verheij/Coronel 1991; Wiersema/León 1999; Zeven/de Wet 1982

▪ Pandanus tectorius, flowering

PAPAVER L. - Poppy - Papaveraceae

Papaver somniferum L.

Common Names ▶ opium poppy, poppy, garden poppy, chessbolls; *Arabic:* chaschchâsch, abû ennûm, abû nû mân, khash khash, khesh kash; *Chinese:* ying sou, ying tzu shu; *Dutch:* blauwmaan, papaver; *French:* pavot, pavot à opium, pavot blanc, pavot des jardins, oeilette, oeilete bleue; *German:* Mohn, Gartenmohn, Schlafmohn, Klatschrose, Mohnblume; *Hindi:* āfim, afiyun, abini, aphin, kasa kasa, kasch kasch, khus-khus, postaka; *Italian:* papavero, papavero indiano, papavero domestico, rosolaccio; *Japanese:* chishi, keshi, kshi; *Korean:* aphyonkkot, khaskhasa; *Mexico:* adormidera, amapola dei opie; *Portuguese:* papoula; *Russian:* mak, zahrodny mak, makojcka; *Sanskrit:* ahiphena, āphūka, khaskhasa; *Slovenian:* mak siaty; *Spanish:* ababa, adormidera, semilla de amapola; *Turkish:* arfuin, koknar; *Vietnamese:* a phub dung, a phien, anh tuc

Usage ▶ spice (frequently roasted); **product:** fatty oil

Parts Used ▶ seed

Distribution ▶ native and cultivated in Asia, Europe

Aichele/Schwegler 2, 1994; Bärtels 1997; Berger 3, 1952; Bois 1934; Bremness 2001; Broszat 1992; Cheers 1997; Chizzola/Dobos 1997; Coiciu/Racz (no year); Craze 2002; Dalby 2000; Danert 1958; Davidson 1999; Erhardt et al. 2000, 2002; Farrell 1985; Grey-Wilson 2000; Grümmer 1976; Hager 3, 1992; Hammer 1981; Hammer/Fritsch 1977; Hanelt 2001; Heeger/Poethke 1947; Hepper 1992; Hiller/Melzig 1999; Hoppe 1949; Kapoor 1995; Köhnlein 2003; Lewington 1990; Meshedani et al. 1990; Morris/Mackley 1999; NICPBP 1987; Norman 1991; Peters 1927; Pruthi 1976; Rätsch 1998; Rosengarten 1969; Schönfelder 2001; Schultze-Motel 1986; Seidemann 1993c; Small 1997; Täufel et al. 1993; Teuscher 2003; Uhl 2000; Uphof 1968; Villamar et al. 1994; WHO 1990; Wiersema/León 1999; Wyk et al. 2004; Zeven/de Wet 1982

Papaya communis Noronha

 Carica papaya L.

Papaya vulgaris DC.

▶ *Carica papaya* L.

PARINARIA Aubl. - Chrysobalanceae

 Parinaria curatellifolium *Planch. ex Benth.*

Synonyms ▶ *Parinaria gardineri* Hemsl., *Parinaria mabolo* Oliv.
Common Names ▶ marbo cork tree, grys apple, mobola plum; *French:* parinaire, toutou blanc
Usage ▶ spice, applied piment
Parts Used ▶ fruit
Distribution ▶ Africa: Cameroon, from Senegal to Uganda, Sudan, southward to Zimbabwe; cultivated in tropical Africa and S America

Bois 1934; Hanelt 2001; v. Koenen 1996; Wiersema/León 1999

 Parinaria gardineri *Hemsl.*

▶ *Parinaria curatellifolium* Planch. ex Benth.

 Parinaria mabolo *Oliv.*

▶ *Parinaria curatellifolium* Planch. ex Benth.

PARKIA R.Br. - Mimosaceae (Leguminosae)

 Parkia africana *R. Br.*

▶ *Parkia biglobosa* (Jacq.) R. Br. ex G. Don

 Parkia biglobosa *(Jacq.) R. Br. ex G. Don*

Synonyms ▶ *Mimosa biglobosa* Jacq., *Parkia africana* R. Br., *Parkia clappertonia* Keay
Common Names ▶ (West) African locustbean, stink bean; *Dutch:* stinkboon; *French:* arbre à farine, mimosa pourpre; *German:* Afrikanische Locustbohne, Dawa, Dawabaum, Nittabaum, Sudan-Kaffee; *Indonesien:* peté; *Portuguese:* farroba; *W Africa:* kpalug, nere sun, netige, niri
Usage ▶ pot-herb, especially for soups
Parts Used ▶ seed
Distribution ▶ W Africa, Gabun, W Sudan, native in W India

Aké Assi/Guinko 1991; Davidson 1999; Erhardt et al. 2002; Hagos 1962; Hager 2001; Neuwinger 1999; Oboh/Ekperigin 2004Schultze-Motel 1986; Täufel et al. 1993; Wiersema/León 1999

 Parkia clappertonia *Keay*

▶ *Parkia biglobosa* (Jacq.) R.Br. ex G. Don.

 Parkia filicoidea *Welw.*

Common Names ▶ fernleafed nitta, West African locust bean; *German:* Farnblättriger Nitta, Westafrikanische Locustbohne
Usage ▶ flavoring for native dishes and soups
Parts Used ▶ seeds (pulverized)
Distribution ▶ tropical Africa

Uphof 1968

 Parkia macrocarpa *Miq.*

▶ *Parkia speciosa* Hassk.

Parkia speciosa *Hassk.*

Synonyms ▶ *Parkia macrocarpa* Miq.
Common Names ▶ peteh; *German:* Petehbohne; *Malaysian:* nyiring, petai; *Thai:* kato, sato, pa-tao
Usage ▶ spice
Parts Used ▶ seed
Distribution ▶ Malaysia, Indonesia (Java)
Note ▶ The seeds have a garlic taste.

Erhardt et al. 2002; Hagos 1962; Hager 2001; Schultze-Motel 1986; Seidemann 1993c; Seidemann/Siebert 1987; Siemonsma/Piluek

PASTINACA L. - Parsnip - Apiaceae (Umbelliferae)

◘ Parkia speciosa, fruits

1993; Staesche 1972; Strauß 1969; Täufel et al. 1993; Uphof 1968; Wiersema/León 1999; Zeven/de Wet 1982

carota rossa; *Japanese:* paasunitpu; *Korean:* hjangmuu, tanminari; *Portuguese:* pastinaca, chirivia; *Russian:* pasternak, pasternak posevnoj; *Spanish:* chirivía, pastinaca

Usage ▶ pot-herb

Parts Used ▶ leaf, root

Distribution ▶ Europe, Near East, Caucasus, Iran, E, W Siberia, native in N and S America, Australia, New Zealand

Note ▶ The wild type has a sour root.

Berger 4, 1954; Bilgri/Adam 2000; Bois 1934; Cheers 1997; Davidson 1999; Dudtschenko et al. 1989; Erhardt et al. 2002; Fleischhauer 2003; Hanelt 2001; Hiller/Melzig 1999; Körber-Grohne 1989; Kubeczka/Stahl 1975; Loch 1993; Schnelle 1999; Schönfelder 2001; Schultze-Motel 1986; Seidemann 1993c; Seidemann/Siebert 1987; Stahl/Kubeczka 1979; Täufel et al. 1993; Uphof 1968; Wiersema/León 1999; Zeven/de Wet 1982

 Pastinaca sylvestris *Mill.*

▶ *Pastinaca sativa L.*

 Pastinaca vulgaris *Bubani*

▶ *Pastinaca sativa L.*

 Paullinia asiatica *L.*

▶ *Toddalia asiatica (L.) Lam.*

PASTINACA L. - Parsnip - Apiaceae (Umbelliferae)

 Pastinaca esculenta *Salisb.*

▶ *Pastinaca sativa L.*

 Pastinaca sativa *L.*

Synonyms ▶ *Anethum pastinaca* Wibel, *Pastinaca esculenta* Salisb., *Pastinaca sylvestris* Mill., *Pastinaca vulgaris* Bulbani, *Peucedanum sativum* Benth. et Hook., *Selinum pastinaca* Crantz

Common Names ▶ parsnip; *Arabic:* gazar abiad; *Chinese:* mei guo fang feng; *Dutch:* pastenaak; *French:* panais, pastenaque; *German:* Hammelmöhre, Maggipflanze, Pastinak; *Italian:* pastinaca, pastriciani,

PECTIS L. - Asteraceae (Compositae)

 Pectis papposa *Harv. et A. Gray*

Common Names ▶ fetid marigold, chick weed; *German:* Stinkendes Mariegold

Usage ▶ flavoring by the Indians of New Mexico for meat

Parts Used ▶ flower

Distribution ▶ SW USA

Mabberly 1997; Uphof 1968

PEGANUM L. - Harmal, Harmel - Zygophyllaceae

 Peganum harmala L.

Common Names ▶ African rue, Syrian rue, mountain rue, wild rue; *Egyptian:* besasa; *French:* harmel, péganion, rue sauvage; *German:* Harmel, Harmalbohne, Steppenraute, Syrische Raute; *India:* gorakh amli, harmal, marmara; *Italian:* peganum; *Russian:* garmala, mogil'nik; *Sanskrit:* harmala; *Spanish:* alharma, gamarza; *Turkish:* uzarih

Usage ▶ spice

Parts Used ▶ seed

Distribution ▶ N Africa, W tropical Africa, temperate Asia: India, Pakistan, Mongola; E, SE Europe: Turkey, native elsewhere

Alberts/Muller 2000; Dastur 1954; Erhardt et al. 2002; Hager 5, 1993; Hanelt 2001; Hassan 1967; Hepper 1992; Hiller/Melzig 1999; Khubchandani/Srivastava 1989; Rätsch 1998; Seidemann 1993c; Sharma 2003; Uphof 1968; Wiegele 2000; Wiersema/León 1999; Wyk et al. 2004

PELARGONIUM L'Hérit. ex Ait. - Pelargonium, wrongly Geranium - Geraniaceae

About 100–120 fragrance *Pelargonium* sorts, varieties and hybrids used world wide.

Miller 1996; Rajeswara et al. 2001; Wiegele 2000

 Pelargonium abrotanum *(L.f.) Jacq.*

Common Names ▶ southern (wood) geranium; *German:* Wermutgeranie

Usage ▶ flavoring

Parts Used ▶ leaf

Distribution ▶ S Africa

Huxley et al. 1992

 Pelargonium alchemilloides *(L.) L'Hérit.*

Common Names ▶ peach geranium; *German:* Pfirsischpelargonie

Usage ▶ flavoring

Parts Used ▶ leaf, herb

Huxley et al. 1992

 Pelargonium asperum *Willd.*

▶ *Pelargonium x graveolens L'Herit. ex Ait.*

 Pelargonium capitatum *auct.*

▶ *Pelargonium capitatum (L.) Ait. x Pelargonium radens Moore*

 Pelargonium capitatum *(L.) Ait.* **x Pelargonium radens** *Moore*

Synonyms ▶ *Pelargonum capitatum* auct., *Pelargonium radens* auct., *Pelargonium roseum* auct.

Common Names ▶ rose scented geranium, rose geranium, rose pelargonium; *Chinese:* xiang ye; *French:* geranium rosat; *German:* Rosenduftgeranie; *Italian:* geranio rosa, giraniu; *Russian:* geran rosovaja; *Spanish:* malva rosa, geranio de olor

Usage ▶ flavoring; **product:** essential oil (rose-scented 'Geranium oil')

Parts Used ▶ leaf

Distribution ▶ S Africa

Note ▶ The species has a distinct fragrance of rose-like odor. The plant have been cultivated on the island of Réunion since 1886.

Bourton 1968; Bremness 2001; Davidson 1999; Dudtschenko et al. 1989; Erhardt et al. 2002; Hanelt 2001; Roth/Kormann 1997; Schultze-Motel 1986; Shiva et al. 2002; Small 1997; Uphof 1968; Wiegele 2000; Wiersema/León 1999; Zeven/de Wet 1982

Pelargonium x citrosum *Voigt ex Sprague*

Common Names ▶ citrosa geranium, mosquito plant, Robert's lemon rose; *German:* Citrosa-Geranie
Usage ▶ flavoring for tea, cakes
Parts Used ▶ leaf
Distribution ▶ only cultivated

Erhardt et al. 2002

 Pelargonium crispum *(Berg.) L'Hérit. ex Ait.*

Synonyms ▶ *Geranium crispum* Berg., *Geranium crispum* Kuntze
Common Names ▶ lemon geranium, curled leaved cranesbill; *German:* Orangenpelargonie, Zitronenpelargonie
Usage ▶ flavoring; **product:** essential oil
Parts Used ▶ leaf
Distribution ▶ S Africa, also cultivated
Note ▶ Cultivated for its lemon-scented oil from the leaves.

Bremness 2001; Cheers 1997; Erhardt et al. 2002; Hanelt 2001; Schultze-Motel 1986; Small 1997; Wiegele 2000; Wiersema/León 1999; Zeven/de Wet 1982

 Geranium crispum *Kuntze*

▶ *Geranium crispum* (Berg.) L'Herit. ex Ait.

 Pelargonium x fragrans *(Poir.) Willd.*

Synonyms ▶ *Geranium fragrans* Poir., *Pelargonium exstipulatum* (Cav.) L'Hér. ex Ait. x *Pelargonium odorarissimum* L'Hérit. ex Ait.
Common Names ▶ nutmeg scented geranium; *German:* Duftpelargonie, Muskatpelargonie; *Italian:* geranio odoroso
Usage ▶ flavoring
Parts Used ▶ leaf
Distribution ▶ S Africa: Capeland, also cultivated?

Bremness 2001; Erhardt et al. 2002; Hanelt 2001; Schultze-Motel 1986; Wiegele 2000

 Pelargonium x graveolens *L'Hérit. ex Ait.*

Synonyms ▶ *Geranium graveolens* Thunb., *Pelargonium asperum* Willd.
Common Names ▶ Bourbon geranium, rose geranium, rose pelargonium, sweet scented geranium; *French:* bec de grue, pelargonium rosat; *German:* Rosengeranie, Zitronenpelargonie; *Indonesian:* daun ambré; *Italian:* erba cancella, geranio odoroso; *Pilipino:* malvarosa;
Usage ▶ flavoring; **product:** essential oil
Parts Used ▶ leaf
Distribution ▶ S Africa, southern tropical Africa, also cultivated
Note ▶ The true species is mint scented.

Bourton 1968; Charalambous 1994; Cheers 1997; Demarne/van der Walt 1989; Demarne et al. 1993; Dudtschenko et al. 1989; Erhardt et al. 2002; Gupta et al. 2001; Hanelt 2001; Hiller/Melzig 1999; Manik/Sampath 1981; Oyen/Dung 1999; Prakasa Rao et al. 1995; Roth/Kormann 1997; Sangwan et al. 2002; Schultze-Motel 1986; Sharma 2003; Shiva et al. 2002; Small 1997; Webb 1984; Wiegele 2000; Wiersema/León 1999; Zeven/de Wet 1982

 Pelargonium x limoneum *Sweet*

Common Names ▶ English finger-bowl geranium
Usage ▶ flavoring; **product:** essential oil
Parts Used ▶ leaf
Distribution ▶ S Africa
Note ▶ The plant is a hybrid from lemon geranium (*Pelargonium crispum* [Berg.] L'Hérit. ex Ait.).

Small 1997

 Pelargonium odoratissimum *(L.) L'Hérit. ex Ait.*

Synonyms ▶ *Geranium odoratissimum* L., *Geranium odoratum* Burm.
Common Names ▶ apple scented geranium, nutmeg geranium; *French:* pélargonium citronne, geranium-rosat; *German:* Apfelduft-Pelargonie, Rosenstorchschnabel, Zitronenpelargonie; *Italian:* geranio odoroso, geranio incenso, malva d'Egitto; *Korean:* hyangkkotauk, yangauk; *Russian:* pelargonija duschistaja; *Spanish:* geranio malva, geranio de rosa, pelargonio

Usage ▶ flavoring; **product:** essential oil
Parts Used ▶ leaf
Distribution ▶ S Africa, cultivated: France, Spain, Brazil
Note ▶ The true species is apple-scented.

Bremness 2001; Cheers 1997; Davidson 1999; Erhardt et al. 2002; Hanelt 2001; Hiller/Melzig 1999; Kolodziej et al. 1995; Kumar 2001; Mors/Rizzini 1966; Schultze-Motel 1986; Small 1997; Teuscher 2003; Uphof 1968; Wiegele 2000; Wiersema/León 1999; Zeven/de Wet 1982

 Pelargonium papilionaceum Ait.

Common Names ▶ butterfly geranium; *Dutch:* vlinder pelargonium; *German:* Schmetterlings-Pelargonie
Usage ▶ flavoring
Parts Used ▶ leaf
Distribution ▶ S Africa
Note ▶ The plant has a fruity lemon-scented aroma

 Pelargonium radens auct.

▶ *Pelargonium capitatum (L.) Ait. x Pelargonium radens Moore*

 Pelargonium radens H.E. Moore

Synonyms ▶ *Geranium revolutum* Jacq.
Common Names ▶ mint geranium, balsam (scented) geranium; *German:* Minzpelargonie, Balsampelargonie
Usage ▶ flavoring; **product:** essential oil ('mawah oil' in Kenya)
Parts Used ▶ leaf
Distribution ▶ S Africa: Cape, cultivated in Kenya
Note ▶ The true species has a mint-scented odor.

Bremness 2001; Erhardt et al. 2002; Hanelt 2001; Kolodziej et al. 1995; Roth/Kormann 1997; Small 1997; Wiegele 2000; Zeven/de Wet 1982

 Pelargonium refeniforme Curtis

Common Names ▶ African umckaloebo
Usage ▶ flavoring; **product:** essential oil

Parts Used ▶ leaf
Distribution ▶ S Africa, Capeland
Note ▶ The essential oil has a high cumarin content (~0,5%). The root is used as a medicinal plant.

Hanelt 2001; Kolodziej et al. 1995; Miller 1996

 Pelargonium roseum auct.

▶ *Pelargonium capitatum (L.) Ait. x Pelargonium radens Moore*

 Pelargonium sidoides DC.

Common Names ▶ African umckaloeba
Usage ▶ flavoring; **product:** essential oil
Parts Used ▶ herb
Distribution ▶ S Africa, Capeland
Note ▶ The essential oil has a high content of cumarin (~0,5%). The root is used as a medicinal plant.

Hanelt 2001; Gensthaler 2003; Kolodziej et al. 1995; Wyk et al. 2004

 Pelargonium tomentosum Jacq.

Common Names ▶ peppermint scented geranium; *German:* Pfefferminzgeranie, Pfefferminzpelargonie
Usage ▶ flavoring; **product:** essential oil;
Parts Used ▶ leaf
Distribution ▶ S Africa, in W Europe, Morocco and N America cultivated
Note ▶ The peppermint-scented essential oil has a high mentone content.

Cheers 1997; Erhardt et al. 2002; Hanelt 2001; Kolodziej et al. 1995; Roth/Kormann 1997; Small 1997; Wiegele 2000; Wiersema/León 1999; Zeven/de Wet 1982

PENTADIPLANDRA Baill. - Pentadiplandraceae (Capparidaceae)

 Pentadiplandra brazzeana *Baill.*

Common Names ▶ *Congo:* kikamu, kikuolo, ngama, nguza
Usage ▶ sauce spice
Parts Used ▶ root (powder)
Distribution ▶ tropical W Africa, Cameroon, Congo, Gabon

Mabberly 1997; Neuwinger 1999

PEPEROMIA Ruiz et Pav. – Radiator Plant - Piperaceae (Peperomiaceae)

 Peperomia peepuloides *A. Dietr.*

 *Piper peepuloides (A. Dietr.) Roxb.*

 Peperomia pellucida *(L.) H.B.K.*

Synonyms ▶ *Piper concinnum* Haw., *Piper pellucidum* L.
Common Names ▶ pepper elder, rabbit ear; *Brazil (Portuguese):* comida-de-jaboti, maria-mole, ximbuí; *Cuban:* yerba de la plata; *German:* Durchscheinender Pfeffer; *Peru:* sacha-yuyu; *Pilipino:* olasiman ihalas, ikmong bata; *Thai:* phak krasang
Usage ▶ spice (in W Africa and Thailand)
Parts Used ▶ herb
Distribution ▶ tropical America, native in many tropical areas: W Africa, Thailand

Arora/Pandey 1996; Hanelt 2001; Mors et al. 2000; Schultze-Motel 1986; Villamar et al. 1994; Wierema/León 1999; Zeven/de Wet 1982

PERILLA L. - Perilla - Lamiaceae (Labiatae)

 Perilla arguta *Benth.*

Common Names ▶ *Japanese:* shiso
Usage ▶ condiment
Parts Used ▶ leaf
Distribution ▶ Japan, China

Small 1997; Uphof 1968

 Perilla citriodora *L.*

Common Names ▶ lemon egoma, lemon perilla; *German:* Zitronenperilla
Usage ▶ flavoring
Parts Used ▶ leaf
Distribution ▶ India
Note ▶ This species is listed as *Perilla frutescens* (L.) Britton var. *citriodora*.

Facciola 1990; Hanelt 2001; Honda et al. 1990, 1994, 1996; Ito/Honda 1996; Nitta et al. 2003; Seidemann 1993c; Uphof 1968; Wealth of India 7, 1966

 Perilla frutescens *(L.) Britton*

Synonyms ▶ *Melissa maxima* Ard., *Mentha perilloides* Lam., *Ocimum crispum* Thunb., *Ocimum frutescens* L., *Perilla nankins* Decne., *Perilla nankins* (Lour.) Spreng., *Perilla ocymoides* L.
Common Names ▶ beefsteak plant, perilla, perilla mint, purple mint, Chinese basil, Thai basil; *Chinese:* hsiang sui, su-tzu su, tzu su, zi su ye; *French:* pérille; *German:* Schwarznessel, Öl-Perilla, Thai-Basilikum; *Hindi:* bhanjira, hanshi; *Japanese:* aoshiso, shiso (green perilla), aka shiso egoma (red perilla); *Korean:* chajogi, tŭlkkae; *Russian:* perilla, sudsa; *Vietnamese:* rau tia to, tai to
Usage ▶ spice; **product:** essential oil, flavoring of sweet tobacco, sauces, chewing gum and candy
Parts Used ▶ leaf, herb
Distribution ▶ Himalayas, N India to N Myanmar, China, Japan, also cultivated in Ukraine, China, Ja-

pan, Korea, Vietnam, E USA, S Russia, Iran, occasionally SE and C Europe

Note ▶ A red *Akajiso* and green *Aojiso* variegated ornamental variety is available under several names. The roasted seed with or without chillies, also the flowering tops are ground into a paste and served as chutney.

Aritomi et al. 1985; Arora/Padney 1996; Byun et al. 1985; Cheers 1997; Clebsch 1997; Dönmez 2002; Erhardt et al. 2002; Fiegert-Seibt 2001; Fjita/Nakayama 1993; Guzman/Siemonsma 1999; Hanelt 2001; Hay/Waterman 1993; Hiller/Melzig 1999; Hondo et al. 1990, 1994, 1996; Ito/Honda 1996; Kameoka/Nishikawa 1976; Kang et al. 1992; Koezuka et al. 1986; Krauß et al. 2004; Kuebal/Tucker 1988; Larkom 1991; Longvah/Doesthale 1991; Lück 2000; Makino 1914; Misra/Husain 1987; Roth/Kormann 1997; Schultze-Motel 1986; Seidemann 1993c; Small 1997; Täufel et al. 1993; Tucker 1986; Turova et al. 1987; Uphof 1968; WHO 1990; Wiersema/León 1999; Yuba et al. 1995; Zeven/de Wet 1982

Perilla nankins *Decne.*

▶ *Perilaa frutescens (L.) Britton*

Perilla nankins *(Lour.) Spreng.*

▶ *Perilla frutescens (L.) Britton*

Perilla ocymoides *L.*

▶ *Perilla frutescens (L.) Britton*

PEROVSKIA Karel Perovskia - Lamiaceae (Labiatae)

Perovskia abrotanoides *Karel*

▣ Perovskia abrotanoides, flowering

Common Names ▶ Caspian sage, Russian sage; *Chinese:* fen yoa kua; *French:* pérovskia; *German:* Lederschnittige Perovskie, Russischer Salbei; *Russian:* perovskija polynnaja
Usage ▶ flavoring (sporadically); **product:** essential oil
Parts Used ▶ leaf
Distribution ▶ Himalayas, Afghanistan, Tibet, Turkmenistan; cultivated in the Crimea

Erhardt et al. 2002; Hiller/Melzig 1999; Nigam et al. 1969; Schultze-Motel 1986; Younos et al. 1972

Perovskaja atriplicifilia *Benth.*

Common Names ▶ Pamir sage, silver sage; *Chinese:* lin li ye, fen yoa kua; *French:* Perovskia à feuilles; *German:* Silberperovskie, Pamirsalbei
Usage ▶ flavoring (sporadically); **product:** essential oil.
Parts Used ▶ flower
Distribution ▶ Afghanistan, Pakistan, Himalayas, Tibet

Ali et al. 2001; Dabiri/Sefidkon 2001; Jassbi et al. 1999; Mucciarelli et al. 1993; Pourmortazavi et al. 2003; Rao 1926; Sefidkon et al. 1997; Younos et al. 1972

Perovskia scrophulariifolia *Bunge*

Common Names ▶ wrinkled perovskia; *French:* ridé pérovskia; *German:* Runzlige Perovskie

Usage ▶ flavoring; **product:** essential oil
Parts Used ▶ leaf
Distribution ▶ China, India, W Asia

Erhardt et al. 2002; Hanelt et al. 2001; Mucciarelli et al. 1993

PERSEA Mill. - Avocado - Lauraceae

Persea borbonia *(L.) Raf.*

Common Names ▶ red bay, sweet bay; *German:* Bourbon-Lorbeer
Usage ▶ condiment for soups
Parts Used ▶ leaf
Distribution ▶ S, SE USA

Uphof 1968

PERSICARIA (L.) Mill. - Smart weed - Polygonaceae

Persicaria hydropiper *(L.) Opiz*

Synonyms ▶ *Polygonum gracile* R.Br., *Polygonium hydropiper* L.
Common Names ▶ red cress, smartweed, water pepper; *French:* poivre d'eau; *German:* Wasserpfeffer-Knöterich, Pfefferknöterich, Pfefferkraut, Scharfer Knöterich, Vietnamesischer Koriander (very rarely); *Indonesian:* cacabean, si tuba sawah; *Japanese:* ma tade, yanagi tade; *Malaysian:* daun senahun, rumput tuboh, tube seluwang; *Pilipino:* agagat, buding, tuba; *Russian:* gorez peretschnyj, perez wodjanoj; *Thai:* pha chi mi, phak phai nam; *Vietnamese:* rau râm
Usage ▶ spice
Parts Used ▶ leaf, herb
Distribution ▶ Europe, N Africa: Morocco, Algeria, Turkey, Caucasus, Iran; Asia: Himalayas, China; Japan, Sri Lanka, Indochina, Malaysia; Australia; N America: Alaska, Canada
Note ▶ The leaves have a peppery taste.

Aichele/Schwegler 2, 1994; Berger 4, 1954; Bois 1934; Coiciu/Racz (no year); Davidson 1999; Dudtschenko et al. 1989; Erhardt et al. 2002; Guzman/Siemonsma 1999; hanelt 2001; Hiller/Melzig 1999; Hoppe 1949; Larkcom 1991; Schönfelder 2001; Schultze-Motel 1986; Seidemann 1993c; Small 1997; Tucker 1986; Turova et al. 1987; Wiersema/León 1999; Zeven/de Wet 1982

Persicaria maculata *(Rafin.) S.F. Gray*

▶ *Polygonum persicaria* L.

Persicaria odorata *(Lour.) Soják*

Synonyms ▶ *Polygonum odoratum* Lour.
Common Names ▶ laksa leaf, Cambodian mint, Vietnamese coriander, knotweed; *Cambodian:* chi krassang tomhom; *French:* renouée odorante; *German:* Vietnamesischer Koriander, Wohlriechender Knöterich; *Italian:* bistorrta; *Korean:* tunggulle; *Laos:* phak phè:w; *Malaysian:* daun laksa, daun kesum; *Thai:* chanchom, homchan, phak phai; *Vietnamese:* nghé, rau râm, chi krassang tomhom
Usage ▶ spice, pot-herb
Parts Used ▶ leaf
Distribution ▶ SE Asia, Indochina, cultivated in S Vietnam and Cambodia
Note ▶ The herb has a peppery taste.

Arora/Pandey 1996; Bois 1934; Cheers 1997; Davidson 1999; Erhardt et al. 2002; Guzman/Siemonsma 1999; hanelt 2001; Hutton 1998; Kuebal/Tucker 1998; Ogle et al. 2003; Schönfelder 2001; Schultze-Motel 1986; Tucker 1986; Uphof 1968; Vidal 1967; Wiersema/León 1999; Wyk et al. 2004; Zeven/de Wet 1982

Persicaria pubescens *(Blume) Hara*

Synonyms ▶ *Polygonum leptostachium* de Bruyn, *Polygonium pubescens* Blume, *Polygonium roettleri* Merr. non Roth.
Common Names ▶ *German:* Flaumiger Knöterich; *Indonesian:* siok-siok-rangan, tuboh lalap, tuboh perpancej; *Malaysian:* kelima paya, kesuma tebok selydang
Usage ▶ seasoning
Parts Used ▶ leaf
Distribution ▶ SE Asia (India), Taiwan, Japan, Java, Sumatra

Guzman/Siemonsma 1999

PETASITES Mill. - Butterbur - Asteraceae (Composites)

 Petasites albus (L.) Gaertn

Common Names ▶ white butterbur; *French:* pértasite blanc; *German:* Weiße Pestwurz
Usage ▶ spice (as salt when shortage)
Parts Used ▶ plant ash
Distribution ▶ Europe, Balkan countries, Turkey, Caucasus

Erhardt et al. 2002; Fleischhauer 2003; Loch 1993; Schnelle 1999

 Petasites hybridus G. Gaertn., B. Mey. et Scherb.

Synonyms ▶ *Petasites officinalis* Moench
Common Names ▶ butterbur, umbrella; *French:* pétasite vulgaire; *German:* Gemeine Pestwurz, Rote Pestwurz; *Italian:* farfaraccio; *Russian:* podbel, belokopytnik
Usage ▶ spice (as salt when shortage)
Parts Used ▶ plant ash
Distribution ▶ Europe, Turkey, Caucasus, N Iran, native in N America

Erhardt et al. 2002; Fleischmann 2003; Loch 1993; Wyk et al. 2004

 Petasites japonica (Sieb. et Zucc.) Maxim.

Synonyms ▶ *Nardosmia japonica* Sieb. et Zucc., *Petasites liukiuensis* Kitam
Common Names ▶ butterbur, bog rhubarb, Japanese butterbur, sweet coltsfoot; *Chinese:* feng dou cai; *Dutch:* Japans hoefblad; *French:* pétasitès japonais; *German:* Japanische Pestwurz; *Italian:* farfaraccio giapponese; *Japanese:* atikabuki, fuki; *Korean:* mowi; *Portuguese:* petaside japonès; *Spanish:* fárfara japonesa
Usage ▶ condiment
Parts Used ▶ flower bud
Distribution ▶ China, Japan, Korea, Sachalin, Ryukyu Island

Erhardt et al. 2002; Hanelt 2001; Kays/Dias 1995; Uphof 1968

 Petasites liukiuensis Kitam

▶ *Petasites japonica* (Sieb. et Zucc.) Maxim.

 Petasites officinalis Moench

▶ *Petasites hybridus* (L.) G. Gaertn., B. Mey. et Scherb.

 Petasites palmata Gray.

Common Names ▶ palmata butterbur, sweet coltsfoot; *German:* Süße Pestwurz
Usage ▶ spice (as salt)
Parts Used ▶ plant ash
Distribution ▶ N America

Uphof 1968

 Petasites paradoxus (Retz.) Baumg.

Common Names ▶ alpes butterbur; *German:* Alpen-Pestwurz
Usage ▶ spice (as salt if shortage)
Parts Used ▶ plant ash
Distribution ▶ Europe: Spain, France, Italy, Germany, Bosnia, Serbia, Romania, mountains

Erhardt et al. 2002; Fleischhauer 2003; Loch 1993

 Petasites speciosa (Nutt.) Piper

Usage ▶ spice (as salt)
Parts Used ▶ plant ash
Distribution ▶ British Columbia to California

Uphof 1968

 Petasites spurius (Retz.) Rehb.

Synonyms ▶ *Petasites tomentosus* DC.
Common Names ▶ like felt butterbur; *German:* Filzige Pestwurz
Usage ▶ spice (as salt if shortage)
Distribution ▶ Europe, W Siberia

Erhardt et al. 2002; Fleischhauer 2003; Loch 1993

 Petasites tomentosus DC.

 Petasites spurius (Retz.) Rehb.

PETIVERIA Bl. ex L. - Phytolaccaceae

 Petiveria alliaceae L.

Common Names ▶ erva-pipi, erva-de-guiné, erva-dàlho, guiné, pipi, tipi
Usage ▶ pot-herb
Parts Used ▶ herb
Distribution ▶ Amazonia to Rio de Janeiro and Mato Grosso
Note ▶ The plant has a garlic-like odor and taste.

Adesogan 1974; Mors et al. 2000; Szczpznski et al. 1972

PETROSELINUM Hill - Parsley - Apiaceae (Umbelliferae)

 Petroselinum crispum *(Mill.) Nym.*

Synonyms ▶ *Apium crispum* Mill., *Apium petroselinum* L., *Carum petroselinum* Benth. et Hook., *Petroselinum hortense* Hoffm., *Petroselinum vulgare* J. Hill,
Common Names ▶ parsley; *Arabic:* bagdunis, makdunis; *Chinese:* yang yan sui; *Dutch:* peterselie; *French:* persil; *German:* Gartenpetersilie, Krause Petersilie, Peterling, Petersilling; *Hindi:* ajmood, pitar saleri; *Italian:* prezzemolo; *Japanese:* paseri; *Korean:* hyangminari, kopsilminari; *Malaysian:* pasli; *Portuguese:* salsa, salsinha; *Russian:* petruschka; *S Africa:* pieterselie; *Spanish:* perejil; *Thai:* phak chee; *Turkish:* maïdanos; *Vietnamese:* rau mui tay
Usage ▶ spice, flavoring; **product:** essential oil. (is used in commercial food flavoring)
Parts Used ▶ leaf
Distribution ▶ the natural range as a wild plant is doubtful; originated probably in the Mediterranean region or W Asia, cultivated and native in Europe
Note ▶ Parsley is one of the most important culinary herbs. Beware: confusion with fool's parsley; *French:* petite ciguë; *German:* Hundpetersilie; *Russian:* kokorisch, sobatschja petruschka (*Aethusa cynapium* L.).

Aichele/Schwegler 3, 1995; Alberts/Muller 2000; Apeland 1971; Bärtels 1997; Berger 3, 1952; 4, 1954; Bois 1934; Broda et al. 2001; Bremness 2001; Cheers 1997; Clair 1961; Coiciu/Racz (no year); Danert 1959; Davidson 1999; Díaz-Maroto et al 2002a; Dudtschenko et al. 1989; Erhardt et al. 2000, 2002; Farrell 1985; Feldheim 1999; Grisebach/Billuber 1967; Guzman/Siemonsma 1999; Hager 3, 1992; Hanelt 2001; Heeger 1956; Heroklots 1972; Herrmann 1997; Hoppe 1949; Kasting et al. 1972; Lopez et al. 1999; MacLeod et al. 1985; Mansfeld 1962; Manderfeld et al. 1997; Masanetz/Grosch 1998; Melchior/Kastner 1974; Newall et al. 1966; Opdyke 1975b; Paillan-Legue 1987; Petropoulos et al. 2004; Pino et al. 1997b; Pochljobkin 1974, 1977; Porker 1989; Pruthi 1976; Pschyrembel 1998; Rätsch 1998; Rosengarten 1969; Roth/Kormann 1997; Schönfelder 2001; Schultze-Motel 1986; Seidemann 1993c; Seidemann/Siebert 1987; Sfikas 1999; Simon/Quinn 1988; Stahl/Jork 1964; Täufel et al. 1993; Teuscher 2003; Tucker 1986; Usher 1968; Vernon/Richard 1983; Villamar et al. 1994; Wagner/Hölzl 1968; Warncke 1994; Wiersema/León 1999; Wyk et al. 2004; Zeven/de Wet 1982

 Petroselinum crispum *(Mill.) Nym.* **var. crispum**

Synonyms ▶ *Petroselinum sativum* Hoffm.
Common Names ▶ common parsley, double curled parsley, *Arabic:* madanous; *Chinese:* yang yan sui; *Dutch:* peterselieblad; *French:* persil, persil commun; *German:* Blattpetersilie, Krause Petersilie; *Italian:* prezzemolo; *Japanese:* paserii; *Malaysian:* daun pasli; *Portuguese:* salsa, salsa commum, salsinha; *Russian:* petruschka listovaja; *Spanish:* perejil
Usage ▶ spice, seasoning, flavoring
Parts Used ▶ leaf fresh, dry, dehydrated or frozen
Distribution ▶ only cultivated
Note ▶ There is much further variability, especially in leaf size, shape and dissection patterns.

Bremness 2001; Erhardt et al. 2002; Farrell 1985; Francis/Isaksen 1989; Hanelt 2001; Ihrig 1993; Jung et al. 1992; Masanetz/Grosch 1998; Melchior/Kastner 1974; Pochljobkin 1974, 1977; Rätsch 1998; Schultze-Motel 1986; Seidemann 1993c; Small 1997; Täufel et al. 1993; Wiersema/León 1999

Petroselinum crispum *(Mill.) Hoffm.* **var. radicosum** *(Alef.) Danert*

Synonyms ▶ *Apium latifolium* Mill., *Apium tuberosum* Bernh. ex Rchb.
Common Names ▶ turnip-rooted parsley, Hamburg parsley; *Arabic:* madanous leefty; *Chinese:* gen xiang qin; *Dutch:* knolpeterselie; *French:* persil à grosse racine, persil tubéreux; *German:* Wurzelpetersilie, Knollenpeterilie; *Italian:* prezzèmolo tuberoso; *Japanese:* ne-paserii; *Portuguese:* salsa tuberosa; *Russian:* petruschka kornevaja; *Spanish:* perejil grande, perejil hamburgo, perejil tuberoso
Usage ▶ spice, flavoring.
Parts Used ▶ taproot, leaf
Distribution ▶ only cultivated

Erhardt et al. 2002; Hanelt 2001; Heeger 1956; Mansfeld 1962; Nitz et al. 1990; Paillan-Legue 1987; Rätsch 1998; Schultze-Motel 1986; Seidemann 1993c; Seidemann/Siebert 1987; Täufel et al. 1993; Tucker 1986

 Petroselinum hortense *Hoffm.*

▶ *Petroselinum crispum (Mill.) Nym.*

 Petroselinum sativum *Hoffm.*

▶ *Petroselinum crispum (Mill.) Nym. convar. crispum*

 Petroselinum vulgare *J. Hill*

▶ *Petroselinum crispum (Mill.) Nym.*

PEUCEDANUM L. - Hog's Fennel - Apiaceae (Umbelliferae)

 Peucedanum graveolens *(L.) Hiern.*

▶ *Anethum graveolens L.*

 Peucedanum nagpurensis *Prain*

Usage ▶ spice, applied like coriander
Parts Used ▶ leaf
Distribution ▶ India: Bilhar, Orissa, W Bengal

Arora/Pandey 1996; Uphof 1968

 Peucedanum ostruthium *(L.) W.D.J. Koch*

Synonyms ▶ *Imperatoria ostruthium* L.
Common Names ▶ hogfennel, master wort, pellitory of Spain; *French:* impératoire, benjoin des pays, peucédan ostrote; *German:* Meisterwurz(el), Kaiserwurz(el), Ostruz; *Italian:* imperatoria, erba rena, elafobosco erba rena; *Russian:* zarskij koren', kornewistsche zarskogo kostylja; *Spanish:* imperatoria;
Usage ▶ spice for green cheese and drinks
Parts Used ▶ rhizome
Distribution ▶ Europe: France, Iberia, C Europe, EC Europe: Crimea; N America

Aichele/Schwegler 3, 1995; Charalambous 1994; Erhardt et al. 2002; Fleischhauer 2003; Hanelt 2001; Hiller/Melzig 1999 Hoppe 1949; Mansfeld 1962; Schönfelder 2001; Schultze-Motel 1986; Seidemann 1993c; Usher 1968; Wüstenfeld/Haensel 1964; Zeven/de Wet 1982

 Peucadanum sativum *Benth. Et Hook.*

▶ *Pastinaca sativa L.*

 Peucadanum sowa *(Roxb. ex Fleming) Kurz*

▶ *Anethum graveolens L.*

 Peucedanum palustre *(L.) Moench*

Synonyms ▶ *Selinum palustre* L., *Selinum sylvestre* L.
Common Names ▶ milk parsley; *German:* Sumpf-Haarstrang
Usage ▶ spice
Parts Used ▶ root
Distribution ▶ Europe: Iberia; Asia, W Siberia

Note ► In some Slavic countries the roots are used as a substitute for ginger (*Zingiber officinalis* Rosc.).

Erhardt et al. 2002; Fleischhauer 2003; Schnelle 1999; Upof 1968

PEUMUS Molina - Boldo - Monimiaceae

Peumus boldus *Mol.*

Synonyms ► *Boldus boldus* (Molina) Lyons; *Peumus boldus* (Molina) Lyons, *Peumus chilensi* Schult. et Schult., *Peumus fragrans* Ruiz & Pav.
Common Names ► boldo (leaf); *French:* boldo, feuille de boldo; *German:* Boldoblätter; *Italian:* foglia di boldo; *Spanish:* hoja de boldo
Usage ► spice
Parts Used ► leaf
Distribution ► Chile, cultivated: Algeria

Charalambous 1994; Duke et al. 2003; Erhardt et al. 2000, 2002; Hanelt 2001; Hiller/Melzig 1999; Hoffmann et al. 1992; Hohmann et al. 2001; Hoppe 1949; Krug/Borkowski 1965; Mansfeld 1962; Newall et al. 1996; Ochoa/Alfonso 1996; Pschyrembel 1998; Rätsch 1992, 1998; Schultze-Motel 1986; Seidemann 1993c; Teuscher 2003; Uphof 1968; Urzúa/Acuña 1983; Wolters 1994; Wyk et al. 2004

Peumus boldus *(Molina) Lyons*

► *Peumus boldus* Molina

Peumus chilensis *Schult et Schult.*

► *Peumus boldus* Molina

Peumus fragrans *Ruiz & Pav.*

► *Peumos boldus* Molina

Phaeomeria magnifica *(Risc.) K. Schum.*

► *Etlingera eliator (Jack) R.M. Sm.*

Philenoptera laxiflora *(Guill. Et Perrot) G. Robert*

► *Lonchocarpus caynescens (Schum. et Thonn.) Benth*

PHLOMIS L. - Lamiacea (Labiatae)

Phlomis olivieri *Benth*

Common Names ► Jerusalem sage; *German:* Oliver-Brandkraut, Oliver-Strauchnessel; Jerusalem Salbei
Usage ► spice, flavoring; **product:** essential oil
Parts Used ► herb
Distribution ► Israel, Iran, Caucasus

Mirza/Nik 2003

Phyla scaberrima *(Juss. ex Pers.) Moldente*

► *Lippia dulcis Trevier.*

PHYLLANTHUS L. – Foliage Flower - Euphorbiaceae

Phyllanthus emblica *L.*

Synonyms ► *Cicca emblica* Kurz, *Emblica officinalis* Gaertn.,
Common Names ► emblic, emblic myrobalan, Indian gooseberry, Malacca tree; *Arabic:* amlag, as sanânir; *Chinese:* an mo le, yü kan tzu; *French:* myrobalan emblic; *German:* Ambla, Myrobalane; *Hindi:* amla, amlika, oanla; *Javanese:* kemloko, maloko; *Malaysian:* laka, melaka, asam melaka; *Sanskrit:* adiphala dhatri, amalaka; *Vietnamese:* kam lam, me rùng
Usage ► spice and for the production of vinegar
Parts Used ► fruit
Distribution ► tropical Asia, also cultivated in India, Sri Lanka, Thailand, Malaysia

Agarwal 1990; Encke et al. 2000, 2002; Hanelt 2001; Morton 1960;

Schenk/Naundorf 1966; Schultze-Motel 1986; Seidemann 1994c; Täufel et al. 1993; Uphof 1968

 Phyllanthus llanosi *Müll. Arg.*

▶ *Glochidion llanosi Müll.*

PHYSALIS L. Lantern plant - Solanaceae

 Physalis philadelphia *Lam.*

Synonyms ▶ *Physalis ixocarpa* auct.
Common Names ▶ tomatillo, ground cherry, jam cherry, husk tomatillo, purple gooseberry; *Chinese:* mao suan jiang; *French:* alkénge du Mexico, coquerct, tomate frais; *German:* Mexikanische Blasenkirsche, Tomatl; *Mexico:* tomatillo, miltomato, tomate de cáscara, tomate verde, tulumisi; *Portuguese:* nultomato: *Spanish:* tomate de cáscara, tomate verde
Usage ▶ spice (of sauces)
Parts Used ▶ ripe fruit
Distribution ▶ Mexico, Meso-America, widely cultivated and native

Anon. 2003; Erhardt et al. 2002; Hanelt 2001; Schultze-Motel 1986; Täufel et al. 1993; Villamar et al. 1994

 Physalis pubescens *L.*

Common Names ▶ downy ground cherry, dwarf cape gooseberry, ground cherry, strawberry tomato; *Brazil (Portuguese):* camapu, camaru, joá-poca; *Chinese:* ku zhi; *German:* Flaumige Blasenkirsche; *Mexico:* tomate de cáscara, guatomate, tomate silvestre, tomatillo de campo
Usage ▶ condiment, flavoring for sauces
Parts Used ▶ ripe fruit
Distribution ▶ N America: Canada, N USA; West Indies, tropical America, Mexico, Asia, native in S Russia, S Africa

Anon. 2003; Erhardt et al. 2002; Rätsch 1998; Uphof 1968

 Picraena excelsa *Lindl.*

▶ *Picrasma excelsa (Sw.) Planch.*

PICRASMA Bl. - Bitter wood - Simaroubaceae

 Picrasma ailanthoides *(Bunge) Planch.*

▶ *Picrasma quassioides (D. Don) Benn.*

 Picrasma excelsa *(Sw.) Planch.*

Synonyms ▶ *Picraena excelsa* Lindl.
Common Names ▶ bitterwood, Jamaica quassia, quassia wood, *French:* bois noyar, peste à poux; *German:* Jamaika-Bitterholz, Jamaika-Quassia; *Slovenian:* kvasia
Usage ▶ flavoring
Parts Used ▶ wood
Distribution ▶ Caribbean: Jamaica, Haiti, Puerto Rico, Meso-America, NS America, Venezuela

Charalambous 1994; Craze 2002; Erhardt et al. 2002; Hiller/Melzig 1999; Hoppe 1949; Uphof 1968; Wiersema/León 1999

 Picrasma quassioides *(D. Don) Benn*

Synonyms ▶ *Nima quassioides* Buch-Ham., *Picrasma ailanthoides* (Bunge) Planch.
Common Names ▶ bitterwood, quassia wood, *German:* Bitterholz; *India:* tithai, tithu, tutai; *Japanese:* nijaki,
Usage ▶ flavoring
Parts Used ▶ wood
Distribution ▶ India, N China, Korea, Japan, Taiwan

Erhardt et al. 2002; Hoppe 1949; NICPBP 1987; Uphof 1968

◘ Picrasma excelsa, flowering

PICRIS L. - Bitter herb, Oxtungue - Asteraceae (Compositae)

 Picris hieracioides L.

Common Names ▶ hawkweed, hawk picris; *German:* Gewöhnliches Bitterkraut
Usage ▶ pot-herb (India)
Parts Used ▶ aromatic young leaf, shoot
Distribution ▶ Europe, Turkey, E, W Siberia, Amur, Kamchatka, C Asia, Himalayas (Kashmir), China, India, Korea, Japan, native elsewhere

Arora/Padney 1996; Erhardt et al. 2002; Hager 4, 1992; Kumar 2003; Wiersema/León 1999

PILEA Lindl. – Artillery Plant - Urticaceae

 Pilea melastomoides *Bl.*

Synonyms ▶ *Pilea trinervia* Wight; *Urtica melastomoides* Poir.
Common Names ▶ black throated artillery plant; *German:* Schwarzschlundige Kanonierblume
Usage ▶ seasoning (in India)
Parts Used ▶ young leaf, and shoot
Distribution ▶ S India, Sri Lanka, China, E Asia, W Java
Note ▶ The young leaves have an aromatic taste.

Arora/Pandey 1996; Hanelt 2001; Schultze-Motel 1986; Wiersema/León 1999

 Pilea trinervia *Wight.*

▶ *Pilea melastomoides Bl.*

PIMENTA Lindl. - Allspice - Myrtaceae

 Pimenta acris *(Sw.) Kostel.*

▶ *Pimenta racemosa (Mill.) J.W. Moore*

 Pimenta dioica *(L.) Merr.*

Synonyms ▶ *Eugenia pimenta* DC., *Myrtus pimenta* L., *Pimenta officinalis* Lindl.; *Pimenta vulgaris* Lindl.
Common Names ▶ allspice, clove pepper, English spice, Jamaican pepper, pimento; *Arabic:* bahar; *Dutch:* pimenta, Jamaica peper; *French:* piment, piment des anglais, toute épice, piment de la Jamaique, piment des Anglais, quatre-épice, également appelé, poivre gireflé, poivre de la Jamaïque; *German:* Allerleigewürz, Englischgewürz, Jamaikapfeffer, Nelkenpfeffer, Neugewürz, Piment; *India:* kabab cheene; *Italian:* pimento, pimento inglese, pepe della Giamaica, pepe garofanato; *Portuguese:* pimenta da Jamaica; *Russian:* Jamajskij perez, wosditschnij perez; ormusch, piment; *Slovenian:* nové

korenie; *Spanish:* malequeta, pimiento de Jamaica, pimienta inglesa
Usage ▶ spice; **product:** essential oil (pimento oil)
Parts Used ▶ fruit
Distribution ▶ Mexico, C America, West Indies, Grenada, Guatamala, Honduras, Cuba, Brazil, also cultivated

Berger 3, 1952; Bois 1934; Boisvert/Huber 2000; Bourton 1968; Burkill 4, 1997; Charalambous 1994; Cheers 1997; Clair 1961; Craze 2002; Davidson 1999; Dudtschenko et al. 1989; Duke et al. 2003; Farrell 1985; Fuentes 1986; Guzman/Siemonsma 1999; Hanelt 2001; Hiller/Melzig 1999; Hoppe 1949; Lück 2000; Maistr 1964; Melchior/Kastner 1974; Morris/Mackley 1999; Norman 1991; Oberdieck 1989; Pino/Rosada 1996; Pino et al. 1989a; Pochljobkin 1974, 1977; Pruthi 1976; Pursglove 1968; Rosengarten 1969; Schröder 1991; Schultze-Motel 1986; Seidemann 1993c; Seidemann/Siebert 1987; Sharma 2003; Shiva et al. 2002; Siewek 1990; Smith/Beck 1984; Staesche 1972; Täufel et al. 1993; Teuscher 2003; Uhl 2000; Uphof 1968; Villamar et al. 1994; Wiersema/León 1999; Wüstenfeld/Haensel 1964; Zeven/de Wet 1982

Pimenta dioica (L.) Merr. var. tabasco

Synonyms ▶ *Eugenia tabasco* G. Don
Common Names ▶ Mexican pimento, Tabasco pimento; *French:* poivre de Chiappa; *German:* Mexico Piment, Tabasco-Piment;
Usage ▶ spice (locally)
Parts Used ▶ fruit
Distribution ▶ C America, SE Mexico

Melchior/Kastner 1974; Seidemann 1993; Teuscher 2003

Pimenta grandiflora St. Hil.

Common Names ▶ great piment; *Brazil (Portuguese):* cerrado; *German:* Großblütiger Nelkenpfeffer; *Portuguese:* pimenta de bugre
Usage ▶ spice (for meat)
Parts Used ▶ fruit
Distribution ▶ tropical America: Brazil

Mors/Rizzini 1966

Pimenta officinalis Lindl.

▶ *Pimenta dioica (L.) Merr.*

Pimenta racemosa (Mill.) J.W. Moore

Synonyms ▶ *Pimenta acris* (Sw.) Kostel.
Common Names ▶ bayrum (tree), West Indian bay; *German:* Bayrum(baum), Kronpiment
Usage ▶ falsification of all or English spice (*Pimenta dioica* [L.] Merr.)
Parts Used ▶ fruit; **product:** essential oil
Distribution ▶ Indonesia; W Indies: Jamaica, Puerto Rico; Venezuela, Guayana; Africa: Cameroon
Note ▶ Cultivated for essential oil distilled from the leaves.

Abaul et al. 1995; Duke et al. 2003; Erhardt et al. 2002; Hanelt 2001; Hiller/Melzig 1999; Schultze-Motel 1986; Seidemann 1993; Seidemann/Siebert 1987; Täufel et al. 1993; Teuscher 2003; Uphof 1968; Wiersema/León 1999; Zeven/de Wet 1982

Pimenta vulgaris Lindl.

▶ *Pimenta dioica (L.) Merr.*

PIMPINELLA L. - Burnet Saxifrage - Apiaceae (Umbelliferae)

Pimpinella anisetum Boiss. et Balansa

Common Names ▶ *German:* Anis-Bibernelle
Usage ▶ flavoring; **product:** essential oil
Parts Used ▶ fruit
Distribution ▶ Turkey

Davis 1972; Hanelt 2001; Schultze-Motel 1986;

Pimpinella anisoides V. Brig.

Synonyms ▶ *Pimpinella gussonii* Presl.
Common Names ▶ *Italian:* anice selvatico, cinninielli
Usage ▶ spice, flavoring
Parts Used ▶ fruit
Distribution ▶ E Italy, Sicily

Hammer et al. 2000; Hanelt 2001

PIMPINELLA L. - Burnet Saxifrage - Apiaceae (Umbelliferae)

 Pimpinella anisum L.

Synonyms ▶ *Anisum vulgare* Gaertn., *Anisum officinarum* Moench, *Apium anisum* (L.) Crantz, *Carum anisum* (L.) Baill.

Common Names ▶ anise, anise seed, aniseed, common anise, sweet cumin; *Arabic:* habba helwa, yanîsun; *Chinese:* huei hsiang; *French:* anis, anis vert, boucage, pimpinelle; *German:* Anis, Brotsame, Süßer Kümmel; *Indonesian:* jinten manis; *Italian:* anice verde; *Javanese:* mungfi, adismanis; *Malaysian:* jintan manis, jeramanis; *Russian:* anis, anizet, ganis; *Slovenian:* aníz; *Spanish:* anís verde; *Turkish:* anason

Usage ▶ spice; **product:** essential oil (anise oil), used in the spirit industry: "Aguardiente", "Anisette", "Ouzo", "Raki", "Kulüp", "Altinbis", "Yen", French "Pastis", etc.

Parts Used ▶ seed (fruit)

Distribution ▶ W Asia, Egypt, native in Europe

Note ▶ In Germany used with caraway, coriander and fennel and sold as "bread spice".

Aichele/Schwegler 3, 1995; Ballarin/Ballarin 1972; Berger 3, 1952; Bilgri/Adam 2000; Bohn et al. 1989; Bois 1934; Boisvert/Huber 2000; Bournot 1968; Bremness 2001; Chandler/Hawkes 1984; Charalambous 1994; Cheers 1997; Clair 1961; Coiciu/Racz (no year); Craze 2002; Czygan 1992; Dalby 2000; Davidson 1999; Dudtschenko et al. 1989; Embong et al. 1977; Erhardt et al. 2000, 2002; Fincke 1963; Fujimatu et al. 2003; Guzman/Siemonsma 1999; Hanelt 2001; Heeger 1956; Hepper 1992; Hiller/Melzig 1999; Hohmann et al. 2001; Hondelmann 2002; Kämpf/Steinegger 1974; Karaali/Başoğlu 1995; Kubeczka 1985; Kubeczka et al. 1976; Melchior/Kastner 1974; Morris/Mackley 1999; Newall et al. 1996; Norman 1991; Ondarza/Sachez 1990; Peter 2001; Pochljobkin 1974, 1977; Pruthi 1976; Pschyrembel 1998; Rodrigues et al. 2003; Rosengarten 1969; Roth/Kormann 1997; Santos et al. 1998; Schenck/Nauendorf 1966; Schönfelder 2001; Schröder 1991; Schultze-Motel 1986; Seidemann 1993c; Seidemann/Siebert 1987; Sharma 2003; Shiva et al. 2002; Siewek 1990; Small 1997; Staesche 1972; Täufel et al. 1993; Teuscher 2003; Tucker 1986; Turova et al. 1987; Ubillos 1989; Uphof 1968; Villamar et al. 1994; Wiersema/León 1999; Wüstenfeld/Haensel 1964; Wyk et al. 2004; Zeven/de Wet 1982

 Pimpinella bulbocastanum Jessen

▶ *Bunium bulbocastanum* L.

 Pimpinella heyneana Wall. ex Kurz

Common Names ▶ Heyne burnet; *German:* Heyne-Bibernelle

Usage ▶ condiment

Parts Used ▶ seed

Distribution ▶ in the hills of Deccan Island

Arora/Pandey 1996; Ochse/van den Brink 1931; Siemonsma/Piuek 1993

 Pimpinella major (L.) Huds.

Common Names ▶ greater burnet saxifrage; *French:* grand boucage; *German:* Große Bibernelle, Deutsche Theriakwurzel, Pfefferwurzel; *Italian:* pimpinella

Usage ▶ spice

Parts Used ▶ root

Distribution ▶ Europe, Caucasus

Erhardt et al. 2002; Fleischhauer 2003; Heeger 1956; Hiller/Melzig 1999; Loch 1993; Schönfelder 2001; Small 1997; Täufel et al. 1993; Teuscher 2003; Wyk et al. 2004

 Pimpinella saxifraga L.

Common Names ▶ common burnet saxifrage, garden burnet, burnet saxifrage; *French:* petit boucage, pimprenelle; persail de bouc; *German:* Kleine Bibernelle, Pfefferwurzel, Pimpernelle, Pimpinelle, Bockspetersilie, Deutsche or Weiße Theriakwurzel; *Italian:* pimpinella minore, tragoselino; *Russian:* bedrenekamnelomka, bodrez; *Spanish:* pimpinela blanca, saxifraga menor

Usage ▶ spice

Parts Used ▶ root

Distribution ▶ Europe, Turkey, Caucasus, W Iran, W and E Siberia, C Asia, native in N America, New Zealand

Aichele/Schwegler 3, 1995; Bohn 1991; Bois 1934; Cheers 1997; Coiciu/Racz (no year); Dudtschenko et al. 1989; Erhardt et al. 2002; Fleischhauer 2003; Hanelt 2001; Heeger 1956; Hiller/Melzig 1999; Koschtschejew 1990; Kubeszka/Bohn 1985; Loch 1993; Schönfelder 2001; Schultze-Motel 1986; Seidemann 1993c; Small 1997; Täufel et al. 1993; Tucker 1986; Uphof 1968; Wiersema/León 1999; Wüstenfeld/Haensel 1964

PINUS L. - Pine - Pinaceae

 Pinus maderiensis *Tenore*

▶ *Pinus pinea* L.

 Pinus pinea L.

Synonyms ▶ *Pinus maderiensis* Tenore

Common Names ▶ pine seed, pine nut, Indian nut, pignolies, pignolia-nut, stone pine, umbrella pine; *Arabic:* snober; *Dutch:* pijnkern; *French:* pignes, pin pignon, pin parasol; *German:* Pignoli, Pinienkern, Piniennuss, Piniensamen, Pinoli, Indianernuss, Schirmkiefer, Zirbelnuss; *Italian:* pino domestica, pino da pinola, pignolia; *Portuguese:* pinheiro manso; *Russian:* sosna pinja; *Spanish:* piñoero

Usage ▶ flavoring (meat, poultry, salad, cake, pudding) and food

Parts Used ▶ seed

Distribution ▶ W Asia, Europe: France, Iberian, Italian and Balkan Peninsulas, Turkey, cultivated: elsewhere in the Mediterranean region

Bärtels 1997; Coiciu/Racz (no year); Hanelt 2001; Harrison 1992; Hepper 1992; Lück 2000; Meremdi 1957; Schenck/Naundorf 1966; Seidemann 1993; Täufel et al. 1993; Teuscher 2003; Trueb 1998a; Uphof 1968; Wiersema/León 1999; Zeven/de Wet 1982

PIPER L. - Pepper - Piperaceae

 Piper aduncum L.

Synonyms ▶ *Artanthe adunca* (L.) Miq., *Piper angustifolium* Ruiz et Pav.; *Piper kuntzei* C. DC., *Piper multinervium* M.Martens & Galeotti

Common Names ▶ big pepper, Spanish elder, spiked pepper; *Brazil (Portuguese):* aperta-ruão, erva de jaboti, matico-falso, pimenta-longa, pimenta-de-macaco; *German:* Matico, Matico-Pfeffer, Gebogener Pfeffer, Gekrümmter Pfeffer; *Indonesian:* seuseureuhan; *Spanish:* cordoncillo; higuillo oloroso

Usage ▶ spice; **product:** essential oil (flavoring of beverages)

Parts Used ▶ fruit

Distribution ▶ C and S America: from Mexico to Brazil, and W Indies, in Malaysia native

Note ▶ In America a substitute for Indian long pepper. In Mexico as a spice for cocoa. It is the 'matico' or German Soldatenkraut of the European Pharmacy.

Berger 2, 1950; Erhardt et al. 2002; Guzman/Siemonsma 1999; Hanelt 2001; Hiller/Melzig 1999; Hoppe 1949; Mors et al. 2000; Orjaba et al. 1989; Rätsch 1992, 1998; Schultze-Motel 1986; Täufel et al. 1993; Teuscher 2003; Uphof 1968; Villamar et al. 1994; Wiersema/León 1999; Zeven/de Wet 1982

 Piper album

▶ *Piper nigrum* L.

 Piper angustifolium *Ruiz et Pav.*

▶ *Piper aduncum* L.

 Piper arborescens *Miq.*

▶ *Piper schmidtii* Hook

 Piper aromaticum *Lamk*

▶ *Piper nigrum* L.

 Piper auritum *Kunth*

Common Names ▶ alajan pepper; *Brazil (Portuguese):* hoja santa, makulan; *Belize:* bullhoof; *German:* Anispfeffer, Geröhrter Pfeffer, Mexikanischer Blattpfeffer, Zitronenmyrte; *Guatamala (Spanish):* acoyo, anisilla, anisillo, cordoncillo; *Mexico:* alajan

Usage ▶ spice, flavoring

Parts Used ▶ leaf

Piper auritum, flowering

Distribution ▶ S Mexico, Caribbean, Meso-America, S America; cultivated and native: Cuba

Duke et al. 2003; Hanelt 2001; Rätsch 1998; Schultze-Motel 1986; Täufel et al. 1993; Tucker 1986; Uphof 1968; Villamar et al. 1994; Wiersema/León 1999

 ### Piper baccatum *Bl.*

Common Names ▶ climbing pepper of Java; *Indonesian:* bodeh; *Javanese:* rinu, rinu manuk; *Pilipino:* sambanganai
Usage ▶ spice
Parts Used ▶ fruit, leaf
Distribution ▶ Indonesia, Malaysia
Note ▶ Adulterate for the cubebs.

Burkill 1966

Piper banksii *Miq.*

▶ *Piper canium Bl.*

 ### Piper borbonense *(Miq.) DC.*

Common Names ▶ *French:* poivre sauvage, cubebe du pays, betel marron
Usage ▶ spice
Parts Used ▶ fruit
Distribution ▶ Madagacsar, Réunion; Seyelles comores Mascarenes

Gurib-Fakim/Brendler 2004; Hoppe 1, 1975; Neuwinger 1999

 ### Piper caninum *Bl.*

Synonyms ▶ *Piper banksii* Miq., *Piper lauterbachii* DC., *Piper macrocarpum* DC.
Common Names ▶ common pepper vine; *German:* Gemeiner Pfeffer; *Indonesian:* mrican; *Malaysian:* sireh hutan, lada hantu, chambai; *Pilipino:* buyo-buyo
Usage ▶ flavoring
Parts Used ▶ fruit
Distribution ▶ Malaysia through New Guinea, to the Solomon Islands, Australia, also cultivated
Note ▶ Fruits are used as an adulterant for the cubebs and the leaves are chewed as a substitute for betel.

Guzman/Siemonsma 1999

 ### Piper capense *L.f.*

Common Names ▶ Cape pepper; *German:* Kap-Pfeffer; *S Africa:* matimati
Usage ▶ condiment
Parts Used ▶ fruit
Distribution ▶ Cameroon, Nigeria, S Africa

Burkill 4, 1997; Hager 4, 1992; Neuwinger 1999

Piper chaba Hunter

▶ *Piper retrofractum* Vahl

Piper clusii DC.

▶ *Piper guineense* Schum. et Thonn.

Piper colubrinum Link

Common Names ▶ *Brazil (Portuguese):* jaborandi-manso
Usage ▶ spice (locally) similar ginger rhizome (*Zingiber officinal* Rosc.)
Parts Used ▶ root
Distribution ▶ Amazonia to Rio de Janeiro

Mors et al. 2000

Piper concinnum Haw.

▶ *Peperomia pellucida* (L.) H.B.K.

Piper cubeba L.

Synonyms ▶ *Cubeba officinalis* Raff.
Common Names ▶ cubeb pepper, Java pepper, tailed pepper; *Arabic:* kabaha, kababa hindiya, kebbaba; *French:* cubebe, poivre cubèbe, poivre à queue; *German:* Javanischer Pfeffer, Kubeben, Kubebenpfeffer, Stielpfeffer, Schwanzpfeffer; *Hindi:* kababcini; *India:* cubab chinee; *Indonesian:* kemukus, lada berekur, chabai ekur, rinu; *Italian:* (pepe) cubebe, pepe a coda; *Korean:* philjŭnggadonggul; *Malaysian:* kemukus, lada berekur, chabai ekur; *Russian:* cubeba, perez cubeba; *Sanskrit:* gandhamarica, kankola; *Slovenian:* kubéba; *Spanish:* cubeba, pimienta cubeba
Usage ▶ flavoring, condiment (liqueur, ginger bread, honey bread); **product:** essential oil
Parts Used ▶ unripe fruit
Distribution ▶ Indonesia, Malaysia; cultivated elsewhere

Berger 3, 1952; Bournot 1968; Charalambous 1994; Cheers 1997; Clair 1961; Craze 2002; Dalby 2000; Davidson 1999; Erhardt et al. 2000, 2002; Guzman/Siemonsma 1999; Hiller/Melzig 1999; Hoppe 1949; Melchior/Kastner 1974; Norman 1991; Ochse et al. 1961; Pochljobkin 1974, 1977; Prabhu/Mulchandani 1985; Rätsch 1998; Roth/Kormann 1997; Schenck/Naundorf 1966; Schröder 1991; Schultze-Motel 1986; Seidemann 1993c; Seidemann/Siebert 1987; Shiva et al. 2002; Siewek 1990; Staesche 1972; Täufel et al. 1993; Teuscher 2003; Tschirch 1892; Wiersema/León 1999; Wüstenfeld/Haensel 1964

Piper elongatum Vahl

Common Names ▶ elangato pepper; *German:* Gestreckter Pfeffer, Matiko; *Spanish:* matico, cordoncillo, yerba soldado;
Usage ▶ spice, specially in Peru as cocoa spice
Parts Used ▶ leaf
Distribution ▶ S America: Peru, Argentina, Bolivia, Brazil

Duke 1970; Hanelt 2001; Mors et al. 2000; Rätsch 1998; Täufel et al. 1993; Teuscher 2003; Uphof 1968

Piper guineense Schum. et Thonn.

Synonyms ▶ *Piper clusii* DC.
Common Names ▶ Ashanti pepper, Benin pepper, Guinea cubeb, Guinea pepper, (black) West African pepper, African cubebs; *French:* poivre d'achantis, poivre d'Afrique, poivre de Guinée, poivre de kissi; *German:* Aschantipfeffer, Guineapfeffer, Falsche Kubeben, Afrikanische Kubeben, Kissipfeffer, Kongo-Kubeben; *Russian:* afrikanskij perez, gwinejckij perez, aschantijskij perez; *Portuguese:* pimento da "rabo"; *Spanish:* pimenta de Guinea, pimenta negra del país
Usage ▶ spice, additive (flavoring)
Parts Used ▶ fruit
Distribution ▶ tropical Africa: W Africa to E Africa, cultivated in Guinea and Zaire

Aedo et al. 2001; Ayensu 1978; Berger 3, 1952; Bois 1934; Boisvert/Huber 2000; Burkill 4, 1997; Dalby 2000; Davidson 1999; Ekundayo et al. 1988b; Erhardt et al. 2002; Ferrão 1992; Griebel 1948; Hiller/Melzig 1999; Jirovetz et al. 2002; Melchior/Kastner 1974; Pochljobkin 1974, 1977; Rätsch 1998; Schröder 1991; Schultze-Motel 1986; Seidemann 1993c; Seidemann/Siebert 1987; Siewek 1990; Staesche 1972; Täufel et al. 1993; Teuscher 2003; Usher 1968; Wiersema/León 1999; Zeven/de Wet 1982

Piper kuntzei C. DC.

▶ *Piper aduncum* L.

 Piper lanatum Roxb.

Common Names ▶ wooly piper; *German:* Wolliger Pfeffer; *Malaysian:* akar halong, chabai hutan
Usage ▶ condiment
Parts Used ▶ fruit
Distribution ▶ Malaysia

Burkhill 1966; Oyen/Dung 1999

 Piper lauterbacchii DC

▶ *Piper canium* Bl.

 Piper lolot C. DC.

Common Names ▶ lolot pepper; *French:* poivre de lolot; *German:* Lolotpfeffer; *Cambodian:* chaphlu; *Vietnamese:* cây lá lôt
Usage ▶ condiment, seasoning
Parts Used ▶ leaf
Distribution ▶ Indochina: Cambodia, Laos, Vietnam
Note ▶ Recently indtroduced to the USA (Hawaii and Texas) by Lao and Vietnamese refugees.

Guzman/Siemonsma 1999; Hanelt 2001; Kuebal/Tucker 1988; Uphof 1968: WH0 1990; Zeven/de Wet 1982

 Piper longifolium Ruiz et Pav.

Common Names ▶ long-leaved pepper, *German:* Langblättriger Pfeffer
Usage ▶ spice; **product:** essential oil
Parts Used ▶ leaf
Distribution ▶ SE Asia

Oyen/Dung 1999

▫ Piper longum, fruits

 Piper longum L. ▫

Common Names ▶ Indian long pepper; *Cambodian:* morech ansai; *French:* poivre long; *German:* Langer Pfeffer, Bengal-Pfeffer, Pipalpfeffer, Stangenpfeffer; *Hindi:* pipal; *Malaysian:* babek, chabai, kadok; *Russian:* dlinnyj perez; *Sanskrit:* pippali; *Thai:* phrik-hang
Usage ▶ spice
Parts Used ▶ fruit
Distribution ▶ E Himalayas, cultivated in India, Sri Lanka, Bangladesh

Arora/Pandey 1996; Atal/Ojha 1965; Berger 3, 1952; Bois 1934; Boisvert/Huber 2000; Dalby 2000; Davidson 1999; Erhardt et al. 2002; Guzman/Siemonsma 1999; Hiller/Melzig 1999; Hoppe 1949; Melchior/Kastner 1974; Norman 1991; Pochljobkin 1974, 1977; Pruthi 1976; Schmitt 1988; Schultze-Motel 1986; Seidemann 1963, 1993c; Shankaracharya et al. 1997; Sharma 2003; Shiva et al. 2002; Siewek 1990; Täufel et al. 1993; Teuscher 2003; Tewtrakul et al. 2000; Uphof 1968; Wiersema/León 1999; Zeven/de Wet 1982

Piper macrocarpum C. DC.

▶ *Piper canium* Bl.

 Piper mekongense C. DC.

Common Names ▶ wood wine; *German:* Mekong-Pfeffer
Usage ▶ condiment
Parts Used ▶ fruits
Distribution ▶ Indochina

Uphof 1986; Usher 1986

 Piper multinervium M. Martens & Galeotti

▶ *Piper aduncum* L.

 Piper nigrum L.

Synonyms ▶ *Piper aromaticum* Lamk
Common Names ▶ pepper, black and white pepper; *Arabic:* babary, filfil, fulpol aswad; *Chinese:* fou tsiao, hu chiao; *Dutch:* zwarte and witte peper; *French:* poivre, poivre noir, poivre blanc; *German:* Pfeffer, Schwarzer Pfeffer; *Hindi:* gol mirc, kali mirch; *Indonesian:* lada, merica and merica putih; *Italian:* pepe nero, pepe bianco; *Japanese:* kosho; *Korean:* huchunamu; *Malaysian:* lada hitam, lada putih; *Pilipino:* paminta, pamintaliso; *Portuguese:* pimenta, pimenteira; *Russian:* perez tschjornyj, tschernji perez, dlinnij perez, dolgij perez; *Sanskrit:* marica; *Slovenian:* čierne korenie; *Spanish:* pimentero, pimienta negra and blanca; *Thai:* phrik thai, phrik noi and prik kao; *Turkish:* kara biber
Usage ▶ spice
Parts Used ▶ fruit, unripe fruit
Distribution ▶ Native in India (Malabar Coast); cultivated in Sri Lanka, Malakka, Thailand, Vietnam, Korea, Malaysia, Indonesia, W Africa: Sierra Leone to Congo, Madagascar, S America, especially Brazil and Jamaica
Note ▶ Brazil is the greatest exporter in the world. *Green pepper* (French: poivre vert; German: Grüner Pfeffer) is unripened pepper and *white pepper* (French: poivre blanc; German: Penan-, Muntok- or Weißer Pfeffer) is peeled black pepper. *Pink pepper* is the fruits of ▶ *Schinus therebinthifolius* Raddi.

Ahlert/Kjer 2000; Ahlert et al. 1998; Bandyopadhyay et al. 1990; Berger 3, 1952; Bois 1934; Bournot 1968; Buckle et al. 1985; Burkill 4, 1997; Cheers 1997; Chevalier 1925; Craze 2002; Dalby 2000; Deng et al. 2003; Dobel 1886; Domrös 1973; Dudtschenko et al. 1989; Duke et al. 2003; Erhardt et al. 2002; Farrell 1985; Ferrão 1992; Ferreira et al. 1999; Geister 1989; Govindarajan 1977; Greene 1951; Gunther et al. 1966; Guzman/Siemonsma 1999; Hanelt 2001; Hiller/Melzig 1999; Hohmann et al. 2001; Hondelmann 2002; Hoppe 1949; Jagella/Grosch 1999; Jorovetz et al. 2002; Kollmannsberger et al. 1992; Krützfeld 2001; Lewington 1990; MacCarron et al. 1995; Maistre 1964; Mathew et al. 2001; Melchior/Kastner 1974; Menon et al. 2002; Morris/Mackley 1999; Murthy et al. 1999; NICPBP 1987; Norman 1991; Oberdieck 1992; Ochse et al. 1961; Orav er al. 2004; Paradkar et al. 2001; Parmar et al. 1997; Peter 2001; Pino et al. 1992; Pochljobkin 1974, 1977; Pradhan et al. 1999; Pruthi 1976; Pursglove 1968; Ravindran 2001; Rosen-garten 1969; Roth/Kormann 1997; Sanka 1989; Schmitt 1988; Schröder 1991; Schultze-Motel 1986; Seidemann 1963, 1993c, 1998/2000; Seidemann/Siebert 1987; Sharma 2003; Siewek 1989, 1990; Skapska et al. 2002; Small 1997; Staesche 1972; Täufel et al. 1993; Teuscher 2003; Tewtrakul et al. 2000; Tschirch 1892; Turova et al. 1987; Variyar/Bandyopadhyay 1994; Vaupel 2002b; WHO 1990; Wiersema/León 1999; Wüstenfeld/Haensel 1964; Wyk et al. 2004; Zeven/de Wet 1982

 Piper officinarum (Miq.) C. DC.

▶ *Piper retrofractum* Vahl

 Piper pachyphyllum Baker

Common Names ▶ mahalatsaka, voampirifery; *German:* Dickblättriger Pfeffer
Usage ▶ condiment
Parts Used ▶ seed
Distribution ▶ Madagascar

Neuwinger 1999; Uphof 1968

 Piper peepuloides (A.Dietr.) Roxb.

Synonyms ▶ *Chavica peepuloides* Miq., *Peperomia peepuloides* A. Dietr.
Common Names ▶ long pepper; *German:* Langer Pfeffer; *India:* savali peepul, pippul
Usage ▶ spice
Parts Used ▶ inflorescense

Distribution ▸ tropical Himalayas: Nepal to Bhutan, India (Assam), also cultivated

Berger 2, 1950; Hanelt 2001; Schultze-Motel 1986; Täufel et al. 1993; Teuscher 2003; Wealth of India 8, 1969

 **Piper pellucidum** L.

 Peperomia pellucida (L.) H.B.K.

 Piper pseudonigrum C. DC.

Common Names ▸ perennial vine; *German:* Falscher Schwarzer Pfeffer
Usage ▸ spice
Parts Used ▸ seed
Distribution ▸ Tonkin-China

Uphof 1968; Usher 1974

 Piper pyrifolium Vatke

Common Names ▸ *French:* linguie poivre
Usage ▸ spice
Parts Used ▸ fruit
Distribution ▸ Madagascar, Réunion, Seychelles, Comoros, Mascarenes

Gurib-Fakim/Brendler 2004; Neuwinger 1999

Piper retrofractum Vahl

Synonyms ▸ *Chavica retrofracta* (Vahl), *Piper chaba* Hunter, *Piper officinarum* (Miq.) Vahl
Common Names ▸ long pepper, Javanese long pepper; *Dutch:* javaanse lange peper; *French:* poivre long de Java; *German:* Chabarpfeffer, Java Pfeffer, Javanischer Pfeffer; *Indonesian:* cabe jawa; *Malaysian:* chabai jawa, bakek, kedawak; *Pilipino:* litlit; *Thai:* dipli, dipli-chuak
Usage ▸ spice
Parts Used ▸ immature spike, ripe fruit
Distribution ▸ Indonesia, Philippines, Vietnam to Malaysia, cultivated only on Java, Bali and some neighbouring islands

Note ▸ It resembles *Piper longum* L. The remedy also contains dried parts of rhizome or roots.

Burkill 1966; Erhardt et al 2002; Guzman/Siemonsma 1999; Hanelt 2001; Melchior/Kastner 1974; Pursglove 1968; Schröder 1991; Schultze-Motel 1986; Seidemann 1993c; Siewek 1990; Staesche 1972; Täufel et al. 1993; Teuscher 2003; Tewtrakul et al. 2000; Uphof 1968; Wiersema/León 1999; Zeven/de Wet 1982

 Piper ribesioides Wall.

Common Names ▸ *German:* Johannisbeer-ähnlicher Pfeffer; *Thai:* tha khaan lek
Usage ▸ spice
Parts Used ▸ fruit, fruit oil
Distribution ▸ Indochina, Malaysia

Berger 3, 1952; Hanelt 2001; Sánchez/Monge 1991; Seidemann 1993c; Täufel et al. 1993; Uphof 1968

 Piper saigonense DC.

Common Names ▸ Saigon pepper, *German:* Saigon-Pfeffer, Vietnamesischer Pfeffer; *Vietnamese:* lolo
Usage ▸ spice
Parts Used ▸ fruit
Distribution ▸ Vietnam, also cultivated

Bois 1934; Guzman/Siemonsma 1999; Erhardt et al. 2002; Hanelt 2001; Oyen/Dung 1999; Schultze-Motel 1986; Seidemann 1993c; Täufel et al. 1993; Teuscher 2003; Uphof 1968; Zeven/de Wet 1982

 Piper sanctum Schlecht ex Miq.

Common Names ▸ *Mexico:* acoyo, acuyo, cordonillo, xihuitl; *Spanish:* hierba sancts, hoja sancta
Usage ▸ flavoring (for soups and fish)
Parts Used ▸ herb
Distribution ▸ Mexico, Guatamala and C America, Mexico also cultivated
Note ▸ The leaves have an anise flavor.

Davidson 1999; Hanelt 2001; Rätsch 1998; Villamar et al. 1994

Piper sarmentosum *Roxb. ex Hunter*

Synonyms ▶ *Chavica sarmentosa* (Roxb. ex Hunter) Miq.

Common Names ▶ wild pepper; *Cambodian:* môrech ansai; *German:* Wildes Pfefferkraut, Thailändisches Pfefferkraut, Wurzelrankiger Pfeffer; *Indonesian:* kadok, karuk; *Javanese:* cabean; *Malaysian:* chabai, daun kaduk, kadok batu; *Pilipino:* patai-butu; *Thai:* chaa phluu, phluu ling, nom wa; *Vietnamese:* la lot

Usage ▶ spice, especially for soups

Parts Used ▶ infructescence, leaf

Distribution ▶ India to S China, Thailand, Vietnam and Philippines to the Moluccas

Arora/Pandey 1996; Guzman/Siemonsma 1999; Ogle et al. 2003; Rätsch 1998; Teuscher 2001; Schultze-Motel 1986

 ### Piper schmidtii *Hook.f.*

Synonyms ▶ *Piper arborescens* Miq.

Common Names ▶ Nilgiri pepper; *German:* Baumpfeffer, Nilgiri-Pfeffer, Schmidtscher Pfeffer

Usage ▶ spice

Parts Used ▶ fruit

Distribution ▶ India: Assam, W Ghats, Nilgiri hills; cultivated locally in the Nilgiri hills

Hanelt 2001; Schultze-Motel 1986; Seidemann 1993c

 ### Piper subpeltata *Willd.*

Synonyms ▶ *Pothomorphe subpeltata* (Willd.) Miq.

Common Names ▶ *German:* Schildpfeffer

Usage ▶ flavoring for fish

Parts Used ▶ leaf

Distribution ▶ Java, Philippines

Wealth of India 8, 1968

 ### Piper trichostachyon *(Miq.) C. DC.*

Common Names ▶ pouched pepper; *German:* Beutelpfeffer

Usage ▶ spice, used like *Piper nigrum* L.

Parts Used ▶ fruit

Distribution ▶ India: Myosore, Kerala, Madras, India cultivated

Hanelt 2001; Rätsch 1998; Schultze-Motel 1986; Täufel et al. 1993; Teuscher 2003; Wealth of India 8, 1969

 ### Piper umbellatum *L.*

Synonyms ▶ *Lepianthes umbullatum* (L.) Raf., *Pothomorphe umbellata* (L.) Miq.

Common Names ▶ shrubby pepper; *Brazil (Portuguese):* caapeba, capuba, lençol-de-santa Barbara, malvaísco, malvavisco, pariparoba; *German:* Dolden-Pfeffer; *Jamaica:* co foot; *Mexico:* mano de zopilote; *Spanish:* cordoncillo;

Usage ▶ flavoring

Parts Used ▶ fruit, basal part of the stem (Sierra Leone)

Distribution ▶ Mexico, S America, native in Paeleotropics, W Africa: Sierra Leone

Ayensu 1978; Barros et al. 1996; Burkill 4, 1997; Hanelt 2001; Kyjoa et al. 1980; Mors et al. 2000; Neuwinger 1999; Rätsch 1998; Schultze-Motel 1986; Seidemann 1993c; Wiersema/León 1999

 ### Piper unguiculatum *Ruiz et Pav.*

Common Names ▶ Perurian pepper; *German:* Peru-Pfeffer

Usage ▶ spice

Parts Used ▶ leaf

Distribution ▶ S America, especially Peru

Erhardt et al. 2002

 ### Piper verne

▶ *Piper nigrum* L.

 ### Pirola umbellata *L.*

▶ *Chimaphila umbellata* (L.) Bartona

PISTACIA L - Pistache, Pistachio - Anacardiaceae (Pistaciaceae)

Pistacia atlantica *Desf.* ssp. *cabulica*

▶ *Pistacia cabulica* (Stocks) Rchf.

Pistacia cabulica *(Stocks) Rchf.*

Synonyms ▶ *Pistacia atlantica* Desf. ssp. *cabulica*
Common Names ▶ Bombay mastic; *Afghanistan:* khinjuk, shurumma; *German:* Bombay-Mastix
Usage ▶ spice (rarely)
Parts Used ▶ resin
Distribution ▶ India, Afghanistan

Davidson 1999; Hanelt 2001; Uphof 1968; Wealth of India 8, 1969

Pistacia khinjuk *Stocks*

Common Names ▶ Bombay mastic; *German:* Bombay-Mastix
Usage ▶ spice (rarely); flavoring of milk jams, sheep's cheese *(fruit)*
Parts Used ▶ resin, fruit
Distribution ▶ India; S Arabia: Yemen

Qedan 1974; Schultze-Motel 1986; Uphof 1968; Wealth of India 8, 1969

Pistacia lentiscus *L.*

Synonyms ▶ *Pistacia massiliensis* Mill., *Terebinthus lentiscus* Moench
Common Names ▶ Chios mastic, lentisc, mastic; *Arabic:* dharou, derw, dirw, mustik; *French:* arbre au mastic, lentisquè; *German:* Mastix, Mastix-Pistazie; *Italian:* dentischio, lentisco, corno-capra, sondrio; *Portuguese:* almecegueira, avoeira; *Russian:* fistaschka-lentiskus; *Spanish:* lentisco
Usage ▶ spice, e.g. for liqueur: "Chio-Mastic Raki"
Parts Used ▶ resin
Distribution ▶ N Africa, Canary Islands, W Asia, SE and SW Europe, also cultivated

Pistacia lentiscus: a **flowering,** b **resin mastix**

Anon. 1993; Avanzato 2003; Bärtels 1997; Boisvert/Huber 2000; Dalby 2000; Davidson 1999; Duke et al. 2003; Erhardt et al. 2000, 2002; Hanelt 2001; Hepper 1992; Hiller/Melzig 1999; Hoppe 1949; Langenheim 2003; Lemaistre 1959; Magiatis et al. 1999; Schönfelder 2001; Schultze-Motel 1986; Seidemann 1993c; Uphof 1968; Whitehouse 1957; Wiersema/León 1999; Zeven/de Wet 1982

Pistacia massiliensis *Mill.*

▶ *Pistacia lentiscus* L.

Pistacia palaestina Boiss.

▶ *Pistacia terebinthus* L.

Pistacia terebinthus L.

Synonyms ▶ *Pistacia palaestina* Boiss., *Pistacia vera* Mill.
Common Names ▶ (Cyprus) turpentine, terebinth, Turk terbinth; *Arabic:* butm sâqis; *French:* térébinthe; *German:* Mastix, Therebinth-Pistazie, Türkischer Mastix; *Italian:* corno, terebinto; *Portuguese:* alfónsico, pistacheira; *Russian:* kevovoe, skipidaronoe derovo; *Spanish:* alfóncigo, pistachero;
Usage ▶ spice (for bread and sweets, liqueurs 'Mastica', 'Raki') and flavoring of wines
Parts Used ▶ resin
Distribution ▶ Mediterranean regions

Anon. 1993; Avanzato 2003; Bärtels 1997; Clare 1993; Couladis et al. 2003; Erhardt et al. 2000, 2002; Hager 5, 1993; Hanelt 2001; Hepper 1992; Langenheim 2003; Özcan 2004; Perikos 1993; Schultze-Motel 1986; Seidemann 1993c; Uphof 1968; Wiersema/León 1999; Zeven/de Wet 1982

Pistacia vera L.

Common Names ▶ green almond, pistache, pistachio; *Arabic:* fustik, fustuq, pista; *Chinese:* hu chen tzu, wu ming tzu, wu ming zi; *French:* pistachier cultivé; *German:* Echte Pimpernuss, Echte Alepponuss, Pistazie, Grüne Mandel, Pistazienmandel; *Italian:* pianta di (del) pistacchio; pistacchio, mandorla di pistacchio; *Japanese:* pisutachio; *Portuguese:* alfónsico, pistacheira; *Russian:* fistaschka; *Spanish:* alfóncigo, pistachero; *Turkish:* fistik
Usage ▶ spice
Parts Used ▶ fruit
Distribution ▶ E Mediterranean regions, W Asia, also cultivated in C Asia, N Iran, N Afghanistan, USA
Note ▶ About 4000 years ago first cultivated in Assyria; Trade sorts: Syrian (Aleppo), Italian pistache, Tunesian, Levantian or American pistache.

Anon. 1999; Ayfer 1967; Bärtels 1997; Bois 1934; Cheers 1997; Davidson 1999; Dudtschenko et al. 1989; Erhardt et al. 2000, 2002; Gebhardt 1977a; Hanelt 2001; Hepper 1992; Hiller/Melzig 1999; Kroon 1969; Küçüköner/Yurt 2003; Larue 1960; Lück 2000; Popov 1979; Reiner 1994; Schenck/Naundorf 1966; Schultze-Motel 1986; Seidemann 1993c; Spina 1962; Täufel et al. 1993; Uphof 1968; Whitehouse 1957; Wiersema/León 1999; Zeven/de Wet 1982; Zohary 1952

Pistacia vera Mill.

▶ *Pistacia terbinthus* L.

PISTIA L. - Water Lettuce - Araceae

Pistia africana Presl.

▶ *Pistia stratiotes* L.

Pistia spathulata Michx.

▶ *Pistia stratiotes* L.

Pistia stratiotes L.

Synonyms ▶ *Pistia africana* Presl., *Pistia spathulata* Michx.
Common Names ▶ Nile cabbage, water salad, water lettuce; *Chinese:* zu fu ping, lu ping, hang ping; *French:* laitue d'eau; *German:* Wassersalat; *Japanese:* botan-uki-kusa; *Malaysian:* daraido, darahuo, kambiang, loloan; *Pilipino:* kiapo; *Spanish:* lechuga de agua
Usage ▶ condiment
Parts Used ▶ in Africa formerly grown to produce salt from the ash
Distribution ▶ pantropic

Baker van de Brink 3, 1968; Burkill 1966; Duke et Ayensu 1985; Erhardt et al. 2002; Hanelt 2001; Pursglove 1972; Schultze-Motel 1986; Uphof 1968; Wiersema/León 1999; Zeven/de Wet 1982

PITHECELLOBIUM Mart. - Mimosaceae

 Pithecellobium affina *Baker ex Bent.*

❯ *Pithecellobium globosum Kosterm.*

 Pithecellobium angulatum *auct., non Benth.*

❯ *Archidendron fagifolium (Blume ex Miq.) I. Nielsen*

 Pithecellobium bigeminum *(L.) Mart.*

Synonyms ▸ *Inga begemia* Willd., *Pithecellobium monadelphum* Kosterm.
Common Names ▸ *Hindi:* kachlora; *Myanmar:* labrat; *Sanskrit:* avagvadha;
Usage ▸ spice (after cooking in the case of diabetes)
Parts Used ▸ seed
Distribution ▸ India, Sri Lanka, Myanmar, Malaysia, Philppines, Java
Note ▸ The seeds have a garlic odor.

Chopra 1956; Hanelt 2001; Schultze-Motel 1986; Uphof 1968; Wealth of India 8, 1969; Zeven/de Wet 1982; Zeven/Zhukovsky 1975

▫ **Pithecellobium dulce, fresh fruits**

 Pithecellobium dulce *(Roxb.) Benth.* ▫

Synonyms ▸ *Inga dulcis* Willd.; *Mimosa dulcis* L.
Common Names ▸ Manil(l)a tamarind, Madras thom, blackbead; *French:* pois sucré; *German:* Manila-Tamarinde, Mexikanischer Affenohrring; *India:* imli, hani-baul, *Indonesian:* asam belanda, asam kranji; asam londo; kamtsele; *Malaysian:* asam keranji; *Mexico:* costeno; *Pilipino:* camachile, demortis, kamatsele; *Spanish:* guanúchil, huamúchil, madre de flecha; *Thai:* ma khaam thet; *Vietnamese:* me keo
Usage ▸ flavoring (mixture added to curry powder)
Parts Used ▸ crushed seed
Distribution ▸ Mexico to tropical S America, India, Thailand, Philippines also cultivated
Note ▸ Seed coat as food (lemonade).

Arora/Pandey 1996; Engel/Pummai 2000; Erhardt et al. 2002; Hanelt 2001; Husain et al. 1992; Schultze-Motel 1986; Seidemann 1993c; Storrs 1999; Uphof 1968; Wiersema/León 1999; Zeven/de Wet 1982

 Pithecellobium fagifolium *Blume ex Miq.*

❯ *Archidendron fagifolium (Blume ex Miq.) I. Nielsen*

 Pithecellobium globosum *Kosterm.*

Synonyms ▸ *Pithecellobium affina* Baker ex Benth.
Usage ▸ condiment (curries and chutneys)
Parts Used ▸ fruit
Distribution ▸ India

Hanelt 2001

 Pithecellobium jiringa *(Jack) Prain ex King*

❯ *Archidendron jiringi (Jack) I.C. Nielsen*

Pithecellobium lobatum Benth.

> *Archidendron jiringa (Jack) I. Nielsen*

 Pithecellobium mindanaense Merr.

> *Archidendron fagifolium (Blume ex Miq.) I. Nielse*

PLATYCODON A. DC. - Ballon Flower - Campanulaceae

 Platycodon grandiflorus *(Jacq.) A. DC.*

Common Names ▶ balloon flower, Chinese bellflower, Japanese bellflower; *Chinese:* chieh keng, jie jeng, shi yong bin li; *French:* platycodon à grandes fleurs; *German:* Großblütige Ballonblume, Balsamstrauch; *Japanese:* kikyô, *Korean:* toraji, doraji; *Vietnamese:* cätcänh

Usage ▶ spice (rarely)
Parts Used ▶ leaf
Distribution ▶ Japan, N China, Mandschurei, Ussuri, N Korea

Erhardt et al. 2002; Hanelt 2001; Hiller/Melzig 1999; Kays/Dias 1995; Wiersema/Léon 1999; Wyk et al. 2004; Zeven/de Wet 1982

PLECTRANTHUS L'Herit. - Lamiaceae (Labiatae)

 Plectranthus amboinicus *(Lour.) Spreng.*

Synonyms ▶ *Coleus amboinicus* Lour., *Coleus aromaticus* Benth., *Plectranthus aromaticus* Benth.
Common Names ▶ Indian borage, Indian mint, cockspur flower, country borage, soup mint, French thyme, Spain thyme, Cuban or wild oregano; *French:* oeille, plecthranthe; *German:* Jamaika-Thymian, Cuba Thymian, Indischer Boretsch, Harfenstrauch, Mottenstrauch; *Hindi:* pathorchur; *Indonesian:* daun kambing, daunkucing; *Malaysian:* daun bangun-bangun, dacon ajenton; *Pilipino:* oregano, suganda, sildu; *Russian:* schporozvetnik;

◘ Plectranthus amboinicus, plant

Spanish: orégano brujo, orégano de España, orégano Frances, sugánda; *Vietnamese:* hung chanh, rau tan la day

Usage ▶ spice, pot-herb and condiment
Parts Used ▶ fresh leaf
Distribution ▶ India, Indonesia, Malaysia, Philippines, West Indies, E tropical Africa, S Africa, widely native in the Tropics
Note ▶ In Australia popularly known as "five-in-one"; in W Indies it is called "broad-leaf-thyme". The plant also a substitute for borage (*Borago officinalis* L.) and sage (*Salvia officinalis* L).

Berger 2, 1950; Brieskorn/Riedel 1977; Erhardt et al. 2002; Gurdip Singh et al. 2002; Gurib-Fakim/Brendler 2004; Guzman/Siemonsma 1999; Herklots 1972; Kuebal/Tucker 1988; Mallavarapu et al. 1999; Merrill 1937; Morton 1992; Pino et al. 1990; Prudent et al. 1995; Schultze-Motel 1986; Seidemann 1993c; Small 1997; Smith 1974; Täufel et al. 1993; Teubner 2001; Teuscher 2003; Tucker 1986; Verta et al. 1993; Wiersema/León 1999; Zeven/de Wet 1982

PLUCHEA Cass. - Pluche - Asteraceae (Compositae)

 Plectranthus aromaticus *Benth.*

▶ *Plectranthus amboinicus* (Lour.) Spreng.

 Plectranthus barbartus *Andr.*

Synonyms ▶ *Coleus barbatus* (Andr.) Benth.
Common Names ▶ *Brazil (Portuguese):* boldo, malva-santa; *German:* Bärtiger Boretsch
Usage ▶ flavoring, pot-herb (rarely)
Parts Used ▶ herb
Distribution ▶ E Africa, Madagascar, Southern Arab Peninsula, India, Sri Lanka, China

Castellón et al. 1987; Erhardt et al. 2002; Kelecom 1983; Kelecom et al. 1986; Mors et al. 2000; Uphof 1968; Wealth of India 2, 1950; Zelnik et al. 1977

 Plectranthus glandulosus *Hook.f.*

Synonyms ▶ *Plectranthus hylophilus* Guerke
Common Names ▶ glandy borage; *Cameroon:* avas; *German:* Drüsiger Boretsch; *Guinea:* fru-fru
Usage ▶ aromatic spice (locally)
Parts Used ▶ herb
Distribution ▶ Africa: Cameroon, Mali to Guinea, Nigeria; in E Cameroon cultivated

Burkill 3, 1995; Hanelt 2001; Schultze-Motel 1986; Westphal/Jansen 1989

 Plectranthus hylophilus *Guerke*

▶ *Plectranthus glandulosus* Hook.f.

PLUCHEA Cass. - Pluche - Asteraceae (Compositae)

 Pluchea foliosa *DC.*

▶ *Pluchea indica* Less.

 Pluchea indica *Less.*

Synonyms ▶ *Pluchea foliosa* DC.
Common Names ▶ Indian fleabane, Indian pluchea; *Chinese:* luan xi; *German:* Indische Pluche; *Hindi:* kukronda; *Japanese:* hiragi-giku; *Javanese:* lontas; *Malaysian:* beluntas; *Pilipino:* banig-banig, kalapini, lagunding late; *Vietnamese:* phat pha, cuc tan, tu bi
Usage ▶ pot-herb (in India); **product:** essential oil
Parts Used ▶ leaf
Distribution ▶ India: salt marches and swamps of Sunderbans to China, S Asia, Indo-Malaysia; cultivated in Thailand, Vietnam, Indonesia

Arora/Pandeys 1996; Bois 1934; Burkill 1966; Hanelt 2001; Ochse/van den Brink 1931; Uphof 1968; Wealth of India 8, 1969; Zeven/de Wet 1982

 Pluchea suaveolens *(Vell.) Kuntze*

Synonyms ▶ *Gnaphalium suaveolens* Vell.
Common Names ▶ *Argentina:* quitoc, lusera; *Portuguese:* caculucage, estoraque, mandecravo, quitoco, tabacarana
Usage ▶ flavoring for liqueurs
Distribution ▶ Argentina, Brazil, Paraguay, Uruguay

Hanelt 2001; Mors et al. 2000; Schultze-Motel 1986

POGA Pierre - Phytolaccaceae (Anisophyllaceae)

 Poga oleosa *Pierre*

Common Names ▶ African (Brazil) nut; *French:* erable d'àfrique; *German:* Afrikanische (Brasil-)Nuss
Usage ▶ condiment
Parts Used ▶ seed
Distribution ▶ W tropical Africa: Gabun
Note ▶ The taste resembles Brazil nut (*Bertholletia excelsa* Humb. et Bonpl.).

Jansen 1981; Neuwinger 1999; Uphof 1968

POGOSTEMON Desf. – Patchouly - Lamiaceae (Labiatae)

Pogostemon cablin (Blanco) Benth.

Synonyms ▶ *Pogostemon patchouli* Pellet;
Common Names ▶ American false pennyroyal, cablin patchouli, patchouly; *Chinese:* guang huo xiang, huo hsiang; *French:* patchouli de cablin; *German:* Indisches Patchouli; *Indonesian:* nilam wangi, singalon; *Korean:* hyangdulkkaephul, pachuri; *Malaysian:* dhalum wangi, tilam wangi; *Pilipino:* kabling, kadlum, pacholi, atluen; *Spanish:* cablan, pachuli; *Thai:* phimsen; *Vietnamese:* hoac huong
Usage ▶ spice for cakes and chewing-gum; product: essential oil (patchouli oil)
Parts Used ▶ leaf
Distribution ▶ China, Malaysia, Philippines, cultivated tropical Asia
Note ▶ In China where it was probably grown 2000 years ago

Akhila/Tawari 1984b; Berger 2, 1950; Bournot 1968; Charalambous 1994; Cheers 1997; Erhardt et al. 2002; Hasegawar et al. 1992; Hiller/Melzig 1999; Hoppe 1949; Kumar et al. 1986; Maeda/Miyake 1997; Nakahara et al. 1975; Oyen/Dung 1999; Raza Bhatti/Ingrouille 1997; Reglos/Guzman 1991; Robbins 1982; Roth/Kormann 1997; Schultze-Motel 1986; Sharma 2003; Shiva et al. 2002; Soepadyo/Tan 1968; Sugimura et al. 1990; Turora et al. 1987; Usher 1968; WHO 1990; Wiersema/León 1999; Wyk et al. 2004; Zeven/de Wet 1982

Pogostemon heyneanus Benth.

Synonyms ▶ *Origanum indicum* Roth
Common Names ▶ Indian patchouli, Javian patchouli; *French:* patchouli; *German:* Heyne-Patchouli, Javanisches Patchouli; *Hindi:* peholi; *Indonesian:* dilem (Sumatra), dilěm kěmbang, dhilep; *Malaysian:* boon khalif, nilam buket, pakochilam, rumput kuku; *Pilipino:* kadlum, lagumtum, malbaka; *Sri Lanka:* gan-kollan-kola
Usage ▶ spice for alcoholic beverages; product: essential oil (small quantities)
Parts Used ▶ leaf
Distribution ▶ Java; India, Indochina; Malaysia; cultivated elsewhere.

Berger 2, 1950; Cheers 1997; Maeda/Miyake 1997; Oyen/Dung 1999; Raza Bhatti/Ingrouille 1997; Schultze-Motel 1986; Soepadyo/Tan 1968; Sugimura et al. 1990; Wiersema/León 1999

Pogostemon patchouli Pellet

▶ *Pogostemon cablin (Blanco) Benth.*

Poinsettia pulcherrima (Willd. ex Klotzsch) R. Grah.

▶ *Euphorbia pulcherrima Willd. ex Klotzsch*

POLIOMINTHA Gray - Lamiaceae (Labiatae)

Poliomintha longiflora Gray

Common Names ▶ Mexican bush oregano; *German:* Mexikanischer Oregano
Usage ▶ spice (rarely); product: carvacrol
Parts Used ▶ herb
Distribution ▶ NE Mexico, native in the USA

Small 1997

POLYGALA L. - Milkwort - Polygalaceae

Polygala erioptera DC.

Common Names ▶ downy milkwort; *French:* laitier; *German:* Wollige Kreuzblume
Usage ▶ flavoring (tea, also in desserts in Rajasthan)
Parts Used ▶ leaf
Distribution ▶ W Indian Peninsula

Arora/Pandey 1996

POLYGONUM L. - Knotgrass - Polygonaceae

Polygonum aviculare L.

Common Names ▸ bird knotgras, knotweed, smartweed; *Arabic:* gordhab, qoddab; *Chinese:* bian xu, pien hsu, fen chieh tsao; *French:* renouée des oiseaux, trainasse; *German:* Acker-Knöterich, Vogeknöterrich, Weggras, Zerrgras; *India:* dreb (Kashmir); *Italian:* centinodia; *Japanese:* michi-yanagi; *Russian:* gusjatnica, sporysch, gorlez ptitschij; *Vietnamese:* rau dang
Usage ▸ pot-herb (rarely)
Parts Used ▸ leaf
Distribution ▸ Europe, Caucasus, W, E Siberia, C Asia

Chauhan 1999; Dhar/Dhar 2000; Erhardt et al. 2002; Fleischhauer 2003; Hanelt 2001; Hiller/Melzig 1999; Hoffmann et al. 1992; v. Koenen 1996; NICPBP 1987; Ogle et al. 2003; Schönfelder 2001; Schultze-Motel 1986; Tull 1999; Turora et al. 1987; Villamar et al. 1994; Wyk et al. 2004

Polygonum gracile R.Br

▸ *Persicaria hydropiper* L.

Polygonum hydropiper L.

▸ *Persicaria hydropiper (L.) Delarbre*

Polygonum leptostachyum de Bruyn

▸ *Persicaria pubescens (Blume) Hara*

Polygonum odoratum Lour.

▸ *Persicaria odorata (Lour.) Soják*

Polygonum orientale L.

Synonyms ▸ *Lagunea orientale* (L.) Nak.; *Persicaria orientalis* (L.) Spach
Common Names ▸ kiss me over the garden gate, prince's feather, smartweed, willow grass; *Chinese:* hong cao, hung tsao; *German:* Orientalischer Knöterich, Östlicher Knöterich; *Japanese:* o ke trade; *Korean:* noinjangdae, pulguntholjokkwi; *Thai:* phakuang
Usage ▸ pot-herb (by Garo and Khasia tribes in India)
Parts Used ▸ leaf
Distribution ▸ Iran, India, China, SE Asia, native in S Europe: Italy, Balkan Peninsula, Germany, USA, Brazil widely cultivated and native elsewhere
Note ▸ The shoots possess a sour taste.

Arora/Pandey 1996; Erhardt et al. 2002; Hanelt 2001; NICPBP 1987; Siemonsa/Piluek 1993; Wiersema/León 1999

Polygonum persicaria L.

Synonyms ▸ *Persicaria maculata* (Rafin.) S.F. Gray
Common Names ▸ lady's thumb, Persian knotgrass; *Chinese:* ma liao, ta liao; *French:* persicaire; *German:* Floh-Knöterich, Persischer Knöterich; *Russian:* gorez potschetschujnaja, trava potschetschujnaja
Usage ▸ pot-herb
Parts Used ▸ fresh leaf
Distribution ▸ temperate Asia, India, Malaysia, E Europe
Note ▸ Substitute for pepper when this is not otherwise available.

Berger 4, 1954; Bilgri/Adam 2000; Erhardt et al. 2002; Erichsen-Brown 1989

Polygonum pubescens Blume

▸ *Persicaria pubescens (Blume) Hara*

Polygonum roettleri Merr. non Roth.

▸ *Persicaria pubescens (Blume) Hara*

Polygonum salicifolium Brouss. ex Willd.

Common Names ▸ *German:* Weidenblättriger Knöterich
Usage ▸ spice, to manufacture salt in the Congo
Parts Used ▸ herb

Distribution ▶ tropical Africa e.g. Congo; tropical Asia; tropical America; Europe

Uphof 1968

Polygonum strigosum *R. Br.*

Common Names ▶ lean knotgrass; *German:* Schmächtiger Knöterich
Usage ▶ spice, to manufacture salt in the Congo
Parts Used ▶ herb
Distribution ▶ tropical Africa, e.g. Congo; tropical Asia; Australia

Uphof 1968

POLYPODIUM L. Polypody - Polypodiaceae

Polypodium vulgare *L.*

Common Names ▶ common polypody, golden maiden hair, rockbrake, wall fern; *Chinese:* shui long gu; *French:* polypode vulgaire; *German:* Engelsüß, Gewöhnlicher Tüpfelfarn; *Russian:* mnogonoshka
Usage ▶ spice, flavoring (rarely) of bitter liqueurs
Parts Used ▶ leaf (frond)
Distribution ▶ C, S Europe, Asia, N Africa, America

Fleischhauer 2003; Hiller/Melzig 1999; Rätsch 1998; Schnelle 1999; Uphof 1968

POLYSCIAS J.R. Forst et G. Forst - Araliaceae

Polyscias cumingiana *(Presl.) F.-Villar*

Synonyms ▶ *Anthrophyllum pinnatum* (Lam.) Clarke;
Common Names ▶ fern-leaf aralia; *German:* Cuming-Fiederaralie; *Pilipino:* bani
Usage ▶ potherb
Parts Used ▶ leaf
Distribution ▶ Malaysia, also cultivated

Polyscias fruticosa

Hanelt 2001; Schultze-Motel 1986; Wiersema/León 1999

Polyscias fruticosa *(L.) Harms*

Synonyms ▶ *Nothopanax fruticosum* Miq.; *Panax fruticosum* L.
Common Names ▶ ming aralie, tea tree; *German:* Strauchige Fiederaralie; *Japanese:* Taiwan momiji; *Javanese:* daoen grisik, tjakar kootjing; *Pilipino:* bani, makan, papua; *Sumatra:* ovang; *Vietnamese:* cay goi ca
Usage ▶ flavoring, spice (rarely);
Parts Used ▶ leaf, root
Distribution ▶ origin obscure; Malaysia, Polynesia; widely cultivated in tropical Asia
Note ▶ The leaves and roots have a prashnap taste.

Cheers 1997; Hanelt 2001; Schultze-Motel 1986; Seidemann 1993c; 1998/2000; Turova et al. 1987; Wiersema/León 1999; Zeven/de Wet 1982

Polyscias guilfoylei *(Cogn. et Marché) Bailey*

Synonyms ▶ *Aralia quilfoylei* Cogn. et Marché
Common Names ▶ geranium-leaf aralia, Guilfoyle polyscia, wild coffee; *German:* Guilfoyl Fiederaralie; *Japanese:* araiya; *Thai:* lep khrut bai yai
Usage ▶ flavoring (in Malaysia)
Parts Used ▶ leaf
Distribution ▶ SE Asia: Malaysia, Thailand; W Africa: Sierra Leone

Burkill 1, 1985; Engel/Phummai 2000; Schultze-Motel 1986

PONCIRUS Raf. – Bitter Orange - Rutaceae

 Poncirus trifoliata *(L.) Raf.*

Synonyms ▶ *Citrus trifoliata* L., *Citrus trifolia* Thunb., *Limonia trichocarpa* Hance
Common Names ▶ bitter orange, citrangequat, trifoliate orange, Japanese bitter orange; *Chinese:* chih shih, zhi shi, gou ju; *French:* orange amère, orange trifoliéle, oncir; *German:* Dreiblättrige Bitterorange; *Japanese:* karatachi; *Korean:* thaengjanamu; *Portuguese:* limoeiro trifoliado; *Spanish:* naranjo trébel
Usage ▶ spice
Parts Used ▶ fruit
Distribution ▶ Himalayas, C China; native in Japan, widely cultivated in temperate regions

Cheers 1997; Dugo/di Giacomo 2002; Erhardt et al. 2002; Roth/Kormann 1997; Schultze-Motel 1986; Uphof 1968; Wiersema/León 1999; Zeven/de Wet 1982

POPULUS L. - Poplar - Salicaceae

 Populus balsamifera *L.*

Common Names ▶ balsam poplar, hackmatack, tacamahaca poplar; *Chinese:* hai tung, tzu tung; *French:* peuplier baumier; *German:* Balsampappel; *Italian:* pioppo balsamico; *Russian:* topol' bal'samitscheskij; *Spanish:* álamo balsámico, chopo balsamifero
Usage ▶ spice, especially in the alcoholics industry (Boonekamp, Kartäuser, etc.)
Parts Used ▶ leaf bud
Distribution ▶ Alaska, Canada, USA, N America, Rocky mountains, cultivated in Europe

Aichele/Schwegler 3, 1995; Erhardt et al. 2000, 2002; Erichsen-Brown 1989; Hiller/Melig 1999; Hoppe 1949; Pschyrembel 1998; Schönfelder 2001; Seidemann 1993c; Sharma 2003; Uphof 1968; Wiersema/León 1999

 Populus nigra *L.*

Common Names ▶ black poplar, Lombardy poplar; *Arabic:* asafsaf, safsaf; *French:* peuplier noir, peuplier franc; *German:* Schwarz-Pappel, Pyramiden-Pappel; *Italian:* pioppo nero; *Russian:* topol' tschjornyj, osokor'; *Spanish:* alamo negro, chopo común;
Usage ▶ spice, especially in the alcoholics industry (Boonekamp, Kartäuser, etc.)
Parts Used ▶ leaf bud
Distribution ▶ Europe, Turkey, Caucasus, W, E Siberia, NW Africa, also widely cultivated

Aichele/Schwegler 3, 1995; Berger 1, 1949; Charalambous 1994; Cheers 1997; Coiciu/Racz (no year); Erhardt et al. 2002; Greenaway et al. 1990, 1992; Hepper 1992; Hiller/Melzig 1999; Hoppe 1949; Jerković/Mastelić 2003; Pschyrembel 1998; Roth/Kormann 1997; Schönfelder 2001; Seidemann 1993c; Uphof 1968; Wiersema/León 1999; Wüstenfeld/Haensel 1964

POROPHYLLUM Adans. - Asteraceae (Compositae)

 Porophyllum ellipticum *Cass.*

▶ *Porophyllum ruderale (Jacq.) Cass.*

Porophyllum macrophyllum *DC.*

▶ *Porophyllum ruderale (Jacq.) Cass.*

Porophyllum ruderale *(Jacq.) Cass.*

Synonyms ▶ *Cacalia porophyllum* L., *Porophyllum ellipticum* Cass., *Porophyllum macro-phyllum* DC.
Common Names ▶ *Bolivia (Spanish):* quillquina, killi
Usage ▶ pot-herb
Parts Used ▶ fresh leaves
Distribution ▶ tropical C and S America: Bolivian, W Indies
Note ▶ Tastes like coriander leaves.

Hanelt 2001; Mikus/Schaser 1995; Mikus et al. 1997

 Porophyllum ruderale *(Jacq.) Cass.* ssp. **macrocephalum**

Common Names ▶ *Spanish (Mexico):* papalo, papalo-quelite
Usage ▶ pot-herb
Parts Used ▶ fresh leaf
Distribution ▶ Mexico, USA: Texas

Villamar et al. 1994

 Porophyllum tagetoides *(Kunth) DC.*

Synonyms ▶ *Kleinia tagetoides* H.B.K.
Usage ▶ pot-herb
Parts Used ▶ fresh leaves or herb
Distribution ▶ Mexico, also cultivated

Burkill 1, 1985; Hanelt 2001

 Porrum ampeloprasum *(L.) Mill.*

▶ *Allium ampeloprasum* L. ssp. *ampeloprasum*

 Porrum commune *Rchb.*

▶ *Allium porrum* L.

 Porrum sativum *Mill.*

▶ *Allium porrum* L.

 Porrum sativum *(L.) Rchb.*

▶ *Allium sativum* L.

 Porrum scorodoprasum *(L.) Rchb.*

▶ *Allium scorodoprasum* L.

PORTULACA L. - Purslane - Portulacaceae

 Portulaca imbrica *Forssk.*

▶ *Portulacca quadrifida* L.

 Portulaca officinarum *Crantz*

▶ *Portilaca oleracea* L.

 Portulaca oleracea *L.*

Synonyms ▶ *Portulaca officinarum* Crantz, *Portulaca sylvestris* Montandon
Common Names ▶ kitchen garden purslane, pourslane, purslain, pursley; *Arabic:* bighel, blibcha, farfena, idsla, rigla; *Chinese:* ma chi xian, shih-yung-ma-ch'in-hsien; *Dutch:* postelein; *French:* pourpier, pourpier des potagers; *German:* Burgel, Bürzelkraut, Portulak; *Hindi:* kulfa, kulfa sag; ghol, *Indonesian:* krekot; *Italian:* portulaca, porcellana, portschellana, erba grassa; *Japanese:* suberi hiyu; *Korean:* soebirŭm, toejiphul; *Malaysian:* gelang pasier, segan jantan; *Pilipino:* golasiman, lungum, ngalug, sahihan, ulisiman; *Portuguese:* beldroega; *Russian:* donduri, portulak, portulak ovoschnoj; *Spanish:* verdolaga (común); *Vietnamese:* rau sam
Usage ▶ pot-herb
Parts Used ▶ leaf

Distribution ▶ origin unknown; distributed worldwide, also cultivated

Aichele/Schwegler 2, 1994; Arora/Pandey 1996; Bendel 2002; Berger 4, 1954; Bharuchta/Josh 1957; Bremness 2001; Byrne/McAndrews 1975; Cheers 1997; Clair 19961; Dhar/Dhar 2000; Davidson 1999; Dudtschenko et al. 1989; Erichsen-Brown 1989; Feng et al. 1961; Fleischhauer 2003; Hanelt 2001; Heeger 1956; Hepper 1992; Hiller/Melzig 1999; Jirovetz et al. 1993; Kiyoko/Cavers 1980; Koch 1988; v. Koenen 1996; Koschtschejew 1990; Körber-Grohne 1989; Kumar 2003; Lewington 1990; Liu et al. 2000; Lo ch 1993; McNavy Wood 2003; Mors et al. 2000; NICPBP 1987; Ogle et al. 2003; Ross 1999; Sala/Chemli 2004; Schenck/Naundorf 1966; Schnelle 1999; Schönfelder 2001; Schultze-Motel 1986; Seidemann 1993c; Sharma 2003; Siemonesma/Piluek 1993; Täufel et al. 1993; Turova et al. 1987; Uotila 1977; Uphof 1968; Villamar et a. 1994; WHO 1990; Wiersema/León 1999; Yammanura 1997; Zeven/de Wet 1982

Portulaca quadrifida *L.*

Synonyms ▶ *Portulacca imbrica* Forssk.
Common Names ▶ single-flowered purslane, wild purslane; *German:* Vierspaltiges Burzelkraut; *India:* luni, khati luni-ni-bhaji
Usage ▶ pot-herb (in India)
Parts Used ▶ plant
Distribution ▶ India, pantropical, but not in Australia and the Pacific E of Samoa; cultivated in western tropical Africa

Arora/Pandey 1996; Hanelt 2001; v. Koenen 1996; Ochse/van de Brink 1931; Schultze-Motel 1986; Uphof 1968

Portulaca sylvestris *Montandon*

▶ *Portilaca oleracea L.*

POTENTILLA L. - Cinquefoil, Five Finger - Rosaceae

Potentilla erecta *(L.) Raeusch.*

Synonyms ▶ *Potentilla sylvestris* Neck., *Potentilla tormentilla* Stokes, *Tormentilla erecta* L.
Common Names ▶ bloodroot, Shepherd's knot, tormentill(a); *French:* tormentille commune; *German:* Blutwurz, Aufrechtes Fingerkraut, Ruhrkraut,

◻ **Potentilla erecta, flowering**

Tormentilla, Wilder Gamander; *Italian:* tormentilla; *Russian:* kalgan laptschatka prjamaja; *Spanish:* tormentilla, siete en rama, consuelda roja
Usage ▶ flavoring for spirits and liqueurs
Parts Used ▶ rhizom
Distribution ▶ Europe, Turkey, Caucasus, W Siberia
Note ▶ Substitute for *Krameria lappacea* (Domb.) Burdet et B.B.

Ahn 1973; Aichele/Schwegler 2, 1994; Berger 4, 1954; Coiciu/Racz (no year); Dudtschenko et al. 1989; Erhardt et al. 2002; Erichsen-Brown 1989; Fleischhauer 2003; Hanelt 2001; Hiller/Melzig 1999; Hohmann et al. 2001; Hoppe 1949; Länger et al. 1993; Lund/Rimpler 1985; Pschyrembel 1998; Schönfelder 2001; Seidemann 1993c; Stachurski et al. 1995; Staesche 1968; Täufel et al. 1993; Turova et al. 1987; Uphof 1968; Wiersema/León 1999 Wüstenfeld/Haensel 1964; Wyk et al. 2004

Potentilla sylvestris *Neck.*

▶ *Potentilla erecta (L.) Raeusch.*

 **Potentilla tormentilla** Stokes

▶ *Potentilla erecta* (L.) Raeusch

 **Poterium officinale** (L.) A. Gray

▶ *Sanguisorba officinalis* L.

 **Poterium sanguisorba** L.

▶ *Sanguisorba minor* Scop.

 **Pothomorphe umbellata** (L.) Miq.

▶ *Piper umbellatum* L.

 **Poupartia dulcis** Bl.

▶ *Spondias acida* Bl.

 **Poupartia pinnata** Blanco

▶ *Spondias malayana* Kosterm.

PRIMULA L. – Primrose - Primulaceae

 **Primula officinalis** (L.) Hill

▶ *Primula veris* L.

 **Primula veris** L.

Synonyms ▶ *Primula officinalis* (L.) Hill
Common Names ▶ primrose, cowslip; *French:* primerolle, primevère officinale; *German:* Echte Schlüsselblume, Himmelschlüssel, Wiesen-Schlüsselblume; *Italian:* prima vera, primula; *Russian:* pervozvet, primula; *Spanish:* prímula
Usage ▶ spice for cakes, milk soups and salads
Parts Used ▶ flower

Distribution ▶ Europe: N Spain, S France, C Europe, native Europe to SE Russia, Crimea, Caucasus, W, E Siberia
Note ▶ The plant as a spice is so far only found locally.

Aichele/Schwegler 3, 1995; Berger 1, 1949; Bilgri/Asam 2000; Bremness 2001; Cheers 1997; Davidson 1999; Dudtschenko et al. 1989; Erhardt et al. 2002; Fleischhauer 2003; Grecu/Cucu 1975; Heeger 1956; Hiller/Melzig 1999; Hohmann et al. 2001; Kartnig/Ri 1973; Köhlein 1984; Mazza/Oohmah 2000; Mestenhauser 1961; Newall et al. 1996; Pschyrembel 1998; Schnelle 1999; Schönfelder 2001; Schultze-Motel 1986; Seidemann 1993c; Täufel et al. 1993; Uphof 1968; Wiersema/León 1999; Wyk et al. 2004

PROSTANTHERA Labill. - Australian Mint Bush - Lamiaceae (Labiatae)

Prostanthera rotundifolia R.Br.

Common Names ▶ Australian mint bush, round-leafed mint bush; *French:* menthe d'Australie, prostanthère; *German:* Australische Minze, Australischer Minzstrauch
Usage ▶ spice
Parts Used ▶ herb
Distribution ▶ Australia: NS Wales, Victoria, S Australia, Tasmania

Cheers 1997; Erhardt et al. 2002

PRUNUS L. - Cherry, Plum - Rosaceae

Prunus dasycarpa Ehrh.

▶ *Armeniaca dasycarpa* (Ehrh.) Borkh.

Prunus laurocerasus L.

Synonyms ▶ *Cerasus laurocerasus* (L.) Loisel, *Laurocerasus officinalis* M. Roem.
Common Names ▶ cherry laurel, laurel, *Arabic:* gurkarasi; *French:* laurier-amande, laurier-cerise; *German:* Kirschlorbeer, Lorbeerkirsche; *Italian:* lauroceraso, lauro regio; *Portuguese:* loureiro

cerejeira; *Russian:* lavrovischnja lekarstvennaja, lavrovischnja apteschnaja; *Spanish:* laurel cerezo, laurel real; *Turkish:* taflan

Usage ▸ spice, condiment

Parts Used ▸ fresh leaf

Distribution ▸ SE Europe: Jugoslavia, Greece, Bulgaria, Asia minor, Caucasus, N Iran; native in C Europe, native elsewhere

Note ▸ Fruits in India for perfumes and as a condiment for liqueurs (limited).

Ayaz 1997; Ayaz et al. 1995, 1997; Bärtels 1997; Berger 2, 1950; Bois 1934; Brondegaard 1991; Charalambous 1994; Cheers 1997; Dudtschenko et al- 1989; Erhardt et al. 2002; Fleischhauer 2003; Heeger 1956; Hoppe 1949; Hiller/Melzig 1999; Kadioglu/Yavru 1998; Kolayli et al. 2003; Schönfelder 2001; Seidemann 1993c; Small 1997; Tucker 1986; Uphof 1968; Wiersma/León 1999

Prunus mahaleb *L.*

Synonyms ▸ *Cereasus mahaleb* (L.) Mill., *Padus mahaleb* (L.) Borkh.

Common Names ▸ mahaleb cherry, perfumed cherry, St. Lucie cherry; *French:* cerisier de mahaleb, cerisier de Sainte Lucie, mahlebi, quénit; *German:* Felsenkirsche, Steinweichsel, Türkische Weichsel, Weichselrohr; *Italian:* ciliegio canino, malebo, ciliegiodi S. Lucia; *Russian:* tscheremycha-antipka, kutschina, magalepka; *Spanish:* cerezo de mahoma, cerezo de Santa Lucia; *Turkish:* mahlep, melhem, yabani kiraz

Usage ▸ spice

Parts Used ▸ leaf, seed

Distribution ▸ S Europe, Asia minor, Caucasus (Armanian), Syria; native in C Europe

Aichele/Schwegler 2, 1994; Bärtels 1997; Boisvert/Huber 2000; Brockman 1979; Charalambous 1994; Davidson 1999; Erhardt et al. 2002; Hanelt 2001; Hiller/Melzig 1999; Morris/Mackley 1999; Seidemann 1995c; Tucker 1986; Uphof 1968; Wiersma/León 1999; Zeven/de Wet 1982

Prunus nigra *Desf.*

▸ *Armeniaca dasycarpa* (Ehrh.) Borkh.

Prunus serotina *Ehrh.*

Synonyms ▸ *Padus serotina* (Ehrh.) Borkh., *Prunus virginiana* L

Common Names ▸ American bird cherry, black cherry, rum cherry; *French:* cerisier tardif, cerisier d'automne; *German:* Späte Traubenkirsche, Traubenkirsche, Wildkirsche; *Italian:* ciliegio tardivo; *Russian:* ceremucha pozdnjajo

Usage ▸ spice, flavoring

Parts Used ▸ leaf

Distribution ▸ Canada, N America: Ontario to N Dakota, Texas, Florida, Mexico, Guatamala, native in Europe

Note ▸ About 300 years ago introduction into Europe.

Aichele/Schwegler 2, 1994; Berger 1, 1949; Charalambous 1994; Cheers 1997; Erhardt et al. 2002; Erichsen-Brown 1989; Fleischhauer 2003; Hiller/Melzig 1999; 0choa/Alonso 1996; Rätsch 1998; Schnelle 1999; Schönfelder 2001; Seidemann 1993c; Uphof 1968; Villamar et al. 1994; Wiersma/León 1999; Zeven/de Wet 1982

Prunus virginiana *L.*

▸ *Prunus serotina* Ehrh.

Ptarmica atrata *(L.) DC.*

▸ *Achillea atrata* L.

Ptarmica clavenae *DC.*

▸ *Achillae clavenae* L.

PTERIDIUM Gled. ex Scop. - Bracke, Braken fern - Dennstaedtiaceae (Polypodiaceae, Pteridaceae)

Pteridium aquilinum *(L.) Kuhn*

Synonyms ▸ *Pteridium esculentum* (Forst.) Nakei, *Pteris aquilina* L.; *Pteris esculenta* Forst.

Common Names ▸ bracken fern, bracken, eagle fern, pasture brake; *French:* fougère aigle; *German:*

Adlerfarn; *Indonesian:* anam dangdeur, paku geulis; *Japanese:* warabi; *Javanese:* pakis gemblung; *Malaysian:* pakis gila; *Pilipino:* anamam sigpang; *Thai:* kut kia, kut kin

Usage ▶ condiment
Parts Used ▶ plant (herb)
Distribution ▶ Worldwide in all temperate and tropical regions. It is one of the most widely distributed vascular plants
Note ▶ This plant ash (salt) is higher in potassium content than common salt.

Bournot 1968; Erhardt et al. 2002; Fleischhauer 2003; Hanelt 2001; Hiller/Melzig 1999; Schnelle 1999; Schönfelder 2001; Schultze-Motel 1986; Uphof 1968

Pteridium esculentum *(Forst.) Nakei*

▶ *Pteridium aquilinum (L.) Kuhn*

Pteris aquilina *L.*

▶ *Pteridium aquilinum (L.) Kuhn*

Pteris esculenta *Forst.*

▶ *Pteridium aquilinum (L.) Kuhn*

PTEROCARPUS Jacq. - Rosewood - Fabaceae (Leguminosae)

Pterocarous indicus *Willd.*

▶ *Pterocarpus santalinus L.*

Pterocarpus santalinus *L.f.*

Synonyms ▶ *Pterocarpus indicus* Will.
Common Names ▶ red sandalwood, 'Red sanders', Anaman redwood, Burmese rosewood, Indian sandal wood; *Chinese:* tzu tan; *French:* bois de santal; *German:* Katliaturholz; Rote Flügelfrucht, Rotes Sandelholz; *Hindi:* lal chandan, rakhta chandan; *Indonesian:* angsana; *Javanese:* almug; *Malaysian:* angsana, sena; *Pilipino:* narra; *Russian:* krasnyi sandalovoe; *Sanskrit:* rakhta chandana; *Thai:* praduu laai

Usage ▶ spice, dye stuff
Parts Used ▶ wood
Distribution ▶ S India, Sri Lanka to the Phillipines
Note ▶ Called the national tree of the Philippines.

Berger 3, 1952; Dalby 2000; Dastur 1954; Engel/Phummai 2000; Erhardt et al. 2002; Hanelt 2001; Hepper 1992; Hiller/Melzig 1999; Hoppe 1949; Pschyrembel 1998; Rätsch 1998; Schultze-Motel 1986; Shankaranarayana/Kamala 1986: Seidemann 1993c; Seidemann/Siebert 1987; Storrs 1997; Täufel et al. 1993; Tschirch 1892; Uphof 1968; Wiersma/León 1999; Wüstenfeld/Haensel 1964

PUNICA L. - Pomegrate - Punicaceae

Punica granataum *L.*

Synonyms ▶ *Granatum punicum* St. Lag., *Punica spinosa* Lam.
Common Names ▶ pomegranate, anardana; *Arabic:* rommon, ruman, rommen; *Chinese:* shan shih lu, shi liu pi; *Dutch:* granaatappel; *French:* grenadier; *German:* Granatapfel, Punischer Apfel; *Hindi:* anar; *Indonesian:* delima; *Italian:* granado, mangrano, melograno; *Japanese:* zakuro; *Korean:* sokryunamu; *Malaysian:* delima; *Pilipino:* dalima granada; *Portuguese:* romã, romazeiro; *Russian:* granat; *Sanskrit:* dadima; *Spanish:* granada, mangrana; *Thai:* thap thim; *Vietnamese:* an thach, lu'u, mac liu

Usage ▶ spice (pulpa), condiment
Parts Used ▶ pulpa with seeds
Distribution ▶ SW, SE, C Asia, native Mediterranean region, Europe: S France, Switzerland
Note ▶ Used as a condiment in the acidification of chutneys and certain curries. The only related species is *Punica protopunica* Ralf., found wild on Socotra in the Indian Ocean.

Al-Maiman/Ahmad 2002; Arora/Pandey 1996; Avigad 1990; Bärtels 1007; Ben-Arie et al. 1984; Bendel 2002; Bose 1985; Cemeroglu et al. 1992; Charalambous 1994; Cheers 1997; Dastur 1954; Davidson 1999; Du et al. 1975; Dudtschenko et al. 1989; El-Nemr et al. 1992; Engel/Phummai 2000; Ewaida 1987; Farrell 1985; Hanelt 2001; Heeger 1956; Hepper 1992; Hernández et al. 1979; Hiller/Melzig

1999; Hodgson 1917; Hoppe 1949; Melgarejo et al. 2000; Morris/Mackley 1999; NICPBP 1987; Noda et al. 2002; Norman 1991; Ochse et al. 1961; Pruthi 1976; Saxena et al. 1987; Schenck/Naundorf 1966; Schönfelder 2001; Schultze-Motel 1986; Seidemann 1993c; Sfikas 1994; Sharma 2003; Silva 1983; Storrs 1997; Täufel et al. 1993; Uphof 1968; Verheij/Coronel 1991; Villamar et al. 1994; WHO 1990; Wiersma/León 1999; Wyk et al. 2004; Zeven/de Wet 1982

 Punica protopunica *Ralf.*

❯ *Punica granatum L.*

 Punica spinosa *Lam.*

❯ *Punica granatum L.*

PYCNANTHEMUM Michx. - Lamiaceae (Labiatae)

 Pycnanthemum pilosum *Nutt.*

Common Names ▶ hairy mountain mint; *German:* Amerikanische Bergminze, Weichhaarige Bergminze
Usage ▶ spice, flavoring
Parts Used ▶ leaf, herb
Distribution ▶ C, E America
Note ▶ The plant has an austere mint aroma.

Mabberley 1997

 Pycnanthemum virgianum *(L.) T. Durand et B.D. Jacks ex B.L. Rob. et Fernald*

Common Names ▶ Virginia mountain mint, Virginia thyme, wild basil; *French:* Virginia pycnanthème; *German:* Virginianische Bergminze; *Russian:* Virginia piknanthemum
Usage ▶ spice, flavoring
Parts Used ▶ leaf
Distribution ▶ N America

Small 1997; Tucker 1986; Uphof 1968; Wiersma/León 1999

PYCNANTHUS Warb. - Myristicaceae

 Pycnanthus angolensis *(Welw.) Warb.*

Synonyms ▶ *Pycnanthus kombo* (Baill.) Warb.
Common Names ▶ African nutmeg, Angolian nutmeg, false nutmeg; *Cameroon:* bakondo, ilimba, kiang, nasamba, tengé, tombe; *French:* arbre à suif, faux muscadier; *German:* Angolanische Muskat, Ilombanuss
Usage ▶ spice
Parts Used ▶ seed
Distribution ▶ S, W, and W, C tropical Africa, especially Angola

Ayensu 1978; Burkill 4, 1997; Neuwinger 1999; Warburg 1897; Wiersema/León 1999

 Pycnanthus kombo *(Baill.) Warb.*

❯ *Pycnanthus angolensis (Welw.) Warb.*

 Pyrethrum majus *(Desf.) Tzevel*

❯ *Tanacetum balsamite L.*

 Pyrus alnifolia *Lindl.*

❯ *Amelanchier alnifolia (Nutt.) Nutt.*

Q

QUARARIBEA Aubl. - Bombacaceae

Quararibea fieldii *Mills.*

Common Names ▸ *Mexico:* saha
Usage ▸ flavoring of chocolate
Parts Used ▸ flower
Distribution ▸ Yucután (Mexico)

Uphof 1968

Quararibea funebris *(La Llave) Vischer*

Synonyms ▸ *Lexarza funebris* La Llave, *Myrodia funebris* (La Llave) Benth.
Common Names ▸ *German:* Kakaoblütenbaum-Blüten, *Spanish:* flor de cacao, rosita de cacao, madre de cacao
Usage ▸ spice, flavoring of "pozonque" and "tejate"
Parts Used ▸ flower
Distribution ▸ C Mexico to NW Costa Rica

Alverson 1988; Guzman/Siemonsma 1999; Hanelt 2001; Raffauf/Zennie 1983; Rätsch 1998; Rosengarten 1969, 1977; Schultes 1957; Schultze-Motel 1986; Täufel et al. 1993

Quararibea turbinata *(Sw.) Poir.*

Common Names ▸ swizzlestick tree
Usage ▸ flavoring for chocolate beverages and tamale sauce (by the natives of the Dominican Republic)
Parts Used ▸ wood sticks
Distribution ▸ W Indies, tropical S America

Uphof 1968

QUASSIA L. - Bitterwood - Simaroubaceae

Quassia amara *L.*

Synonyms ▸ *Picrasma excelsa* (Sw.) Planch.
Common Names ▸ bitterwood, Jamaica wood, quassia wood, Surinam quassia; *Brazil (Portuguese):* amargo negro, chiriguaná, chuña-chuña, lucuma, marupá; *French:* bois amer, quassia de Surinam, quassier, quassier amer, quinine de Cayenne; *German:* Bitteresche, Bitterholz, Quassiaholz, Surinambitterholz; *Italian:* quassia; *Pilipino:* corales, kuasia; *Portuguese:* quassia; *Russian:* kvassija gor'kaja; *Slovenian:* kvasia; *Spanish:* crucete, cuasia amargo, guabito amargo, quassia; *Thai:* prathatchin
Usage ▸ spice (flavoring)
Parts Used ▸ wood
Distribution ▸ Mexico, West Indies, Guayana, N Brazil; cultivated elsewhere
Note ▸ Leaves in England earlier as a hop substitute.

Barbetti et al. 1987; Berger 1, 1949; Charalambous 1994; Cranze 2002; Engel/Phummai 2000; Erhardt et al. 2002; Hager 1, 1990; Hanelt 2001; Hiller/Melzig 1999; Hoppe 1949; Kottegoda 1994;

QUASSIA L. - Bitterwood - Simaroubaceae

◘ **Quassia amara, flowering**

 Quassia excelsa *Sw.*

Common Names ▶ Jamaica wood
Usage ▶ flavoring
Parts Used ▶ wood
Distribution ▶ W Indies, especially Jamaica
Note ▶ The bitter wood also used for the preparation of a certain aperatif.

Uphof 1968

Mors et al. 2000; Nestler 1979; Newall et al. 1996; Njar et al. 1993; Plotkin 1994; Pschyrembel 1998; Rätsch 1992; Schultze-Motel 1986; Seidemann 1993c; Täufel et al. 1993; Uphof 1068; Villamar et al. 1994; Wagner/Nestler 1978; Wiersema/León 1999; Wolters 1994; Wyk et al. 2004; Zeven/de Wet 1982

RANUNCULUS L. - Buttercup, Crowfoot - Ranunculaceae

 Ranunculus ficaria *L.*

Synonyms ► Ficaria verna Huds.
Common Names ► celandine, lesser celandine, pilewort; *French:* éclairette, ficaire; *German:* Scharbockskraut, Feigenwurz; *Russian:* tschistjak ljutitschnyj
Usage ► pot-herb, condiment
Parts Used ► leaf, flower bud
Distribution ► Europe, N Africa, Caucasus, Siberia, W Asia
Note ► The flower bud can be used as a substitute for capers.

Aichele/Schwegler 2, 1994; Bilgri/Adam 2000; Cheers 1997; Erhardt et al. 2002; Fleischhauer 2003; Hager 4, 1992; Heeger 1956; Hiller/Melzig 1999; Koschtschejew 1990; Newall et al. 1996; Schnelle 1999; Schönfelder 2001; Seidemann 1993c; Seidemann/Siebert 1987; Täufel et al. 1993; Teuscher 2003; Wiersema/León 1999

RAVENSARA Sonn. - Lauraceae

 Ravensara anisata *Danguy*

Common Names ► anise nutmeg; *German:* Anis-Muskat
Usage ► spice; **product:** essential oil (leaf)
Parts Used ► seed, bark, leaf
Distribution ► Comoros, Madagascar, Mascarenses, Seychelles

Gurib-Fakim/Brendler 2004; Neuwinger 1999

 Ravensara aromatica *Sonn.*

Synonyms ► *Agathophyllum aromaticum* (Sonn.) Willd.
Common Names ► clove nutmeg, Madagascar clove, Madagascar nutmeg; *French:* muscade de Madagascar, noix girofle, noix de Ravensara; *German:* Madagaskar-Muskat, Nägeleinnuss, Nelkennuss, Ravensara-Nuss
Usage ► spice, like nutmeg
Parts Used ► bark, leaf, seed
Distribution ► Madagascar, Réunion, Mauritius; introduced into Sri Lanka

Bois 1934; Hager 5, 1993; Hiller/Melzig 1999; Morton 1976; Neuwinger 1999; Rätsch 1998; Schenck/Naundorf 1966; Schultze-Motel 1986; Seidemann 1993c; Teuscher 2003; Uphof 1968; Usher 1968; Warburg 1897

RENANTHERA Lour. - Orchidaceae

 Renanthera moluccana *Bl.*

Common Names ► *Indonesian:* anggrek merah; bunga karang;
Usage ► spice
Parts Used ► young leaf
Distribution ► Indonesia, Papua, New Guinea
Note ► The leaves are sour in taste, with a slight suggestion of capers (*Capparis* spp.) and can be used as a flavoring, alone or with other food acids.

Burkill 1966; Guzman/Siemonsma 1999; Mahyar 1988; Oyen/Dung 1999; Uphof 1968

RENEALMIA L.f: - Zingiberaceae

 Renealmia alpinia *(Rottboell) Maas*

Synonyms ▶ *Amomum alpinia* Rottboell; *Renealmia exaltata* L.f.
Common Names ▶ mountain renenealmia; *German:* Berg-Renealmie; *Mexico:* ixquihit
Usage ▶ flavoring, used for wrapping a special dish
Parts Used ▶ leaf
Distribution ▶ tropical America: from Belize to Brazil, with the exception of the Greater Antilles

Hanelt 2001

 Renealmia aromatica *(Aubl.) Griseb.*

Synonyms ▶ *Alpinia aromatica* Aubl.
Usage ▶ flavoring
Parts Used ▶ leaf
Distribution ▶ S America: Guayana; C America: Cuba

 Renealmia exaltata *L.f.*

 Renealmia alpina (Rottboel) Maas

 **Renealmia occidentalis** *Sweet*

Synonyms ▶ *Renealmia domingensis* Horan
Usage ▶ flavoring
Distribution ▶ India

 Renealmia thyrsoidea *(Ruiz et Pav.) Poepp. et Endl.*

Synonyms ▶ *Amomum thyrsoideum* Ruiz et Pav.
Common Names ▶ *Colombia*: yei'; *Ecuador:* teentekage, teentemo, unkwisi
Usage ▶ spice

Parts Used ▶ fruit
Distribution ▶ S America: Costa Rica to Bolivia and Brazil; tropical Asia, specially Indonesia

Hanelt 2001; Sánchez-Monge/Parellada 1981; Schultze-Motel 1986; Vickers/Plowman 1984

 Rhabarbarum palmatum *(L.) Moench*

▶ *Rheum palmatum L.*

RESEDA L. - Mignonette - Resedaceae

 Reseda odorata *L.*

Common Names ▶ common mignonette, mignonette, sweet mignonette, sweet reseda; *French:* herbe d'amour, mignonette, réséda odorante; *Italian:* miglionet; *German:* Duft-Resede, Duft-Wau, Garten-Resede, Wohlriechender Wau; *Italian:* reseda, amorino, amoretti d'Egitto, miglionet; *Portuguese:* reseda de cheiro, erva de amor, minhonete; *Russian:* rezeda duschistaja; *Spanish:* réseda de odor, miñoneta duschistaja
Usage ▶ flavoring (of tea); **product:** essential oil
Parts Used ▶ flower
Distribution ▶ N Africa, cultivated world wide, especially in China and Europe: S France

Cheers 1997; Erhardt et al. 2002; Fleischhauer 2003; Hanelt 2001; Heeger 1956; Hiller/Melzig 1999; Kays/Dias 1995; Schultze-Motel 1986; Uphof 1968; Wealth of India 8, 1969; Zeven et Wet 1982

REYNOLDA Gray - Araliaceae

 Reynolda marchionensis *F. Br.*

Common Names ▶ *Polynesia:* pilohe pimata, pimata omoa
Usage ▶ scenting coconut oil; **product:** essential oil
Parts Used ▶ all plant parts
Distribution ▶ Polynesia

Uphof 1968

RHAMNUS L. - Buckthorn - Rhamnaceae

 Rhamnus celtifolia *Thunb.*

▶ *Rhamnus prinoides L'Hérit.*

 Rhamnus lando *Llanos*

▶ *Embelia philippinensis A. DC.*

 Rhamnus prinoides *L' Hérit.*

Synonyms ▶ *Rhamnus celtifolia* Thunb.
Common Names ▶ buckthorn, dogwood; *Ethiopia:* t'ado, tando; *French:* bois de nerorun; épine noire, noir prun; *German:* Gerbereichen-ähnlicher Kreuzdorn; *Italian:* ramno catatico, spina cervina; *Russian:* kuschina slabitel'naja; *S Africa:* blinkblaar, hondepis, kondepishout, mofifi, *Spanish:* espino cerval
Usage ▶ spice
Parts Used ▶ bark
Distribution ▶ E, S Africa: Ethiopia to S Africa and Nigeria, Angola, in Ethiopia also cultivated
Note ▶ Sporadically as a substitute for hops (*Humulus lupulus* L.), e.g. for beer: 'talla', and other alcoholic beverage: honey wine and mead ('teedj').

Hanelt 2001; Jansen 1981; Joffe 1993; Rätsch 1998; Schultze-Motel 1986; Seidemann 1993c; Uphof 1968

RHAPHIDOPHORA Hassk. - Araceae

 Rhaphidophora lobbii *Schott*

Common Names ▶ *Malaysian:* akar asam tebing paya
Usage ▶ spice, flavoring in curries
Parts Used ▶ leaf
Distribution ▶ Malaysia and Borneo

Rhamnus prinoides, fruiting

Burkill 1966; Guzman/Siemonsma 1999; Oyen/Dung 1999

RHEUM L. - Rhubarb - Polygonaceae

 Rheum officinale *Baill.*

Common Names ▶ Chinese rhubarb, medicinal rhubarb, Tibetan rhubarb; *Chinese:* chiang chun, chai ta-huang; *French:* rhubarbe officinal; *German:* Gartenrhabarber; Chinesischer Rhabarber, Ostindischer Rhabarber, Türkischer Rhabarber; *Italian:* rabarbaro; *Portuguese:* ruibarbo; *Russian:* reven' aptetschnyj, reven' russkij, reven' kitajskij; *Spanish:* ruibarbo de la China
Usage ▶ spice
Parts Used ▶ rhizome
Distribution ▶ W China, Indochina, E Tibet, commonly cultivated

Aichele/Schwegler 2, 1994; Blundstone/Dickinson 1964; Charalambous 1994; Cheers 1997; Dalby 2000; Erhardt et al. 2000, 2002; Hanelt 2001; Heeger 1956; Hiller/Melzig 1999; Hohmann et al. 2001; Hoppe 1949; Newall et al. 1996; Pschrempel 1998; Rätsch 1998; Schönfelder 2001; Schratz 1960; Schultze-Motel 1986; Seidemann 1993c; Small 1997; Treptow 1985; Uphof 1968; Wiersema/León 1999; Wüstenfeld/Haensel 1964; Zeven de Wet 1982

Rheum palmatum L.

Synonyms ▸ *Rhabarbarum palmatum* Moench
Common Names ▸ medicinal rhubarb, Turkey rhubarb; *French:* rhubarbe palmée; *German:* Chinesischer Rhabarber, Kanton-Rhabarber, Kron-Rhabarber, Medizinalrhabarber, Türkischer Rhabarber; *Italian:* rabarbaro; *Korean:* taehwang; *Portuguese:* rabárbo; *Spanish:* ruibarbo, ruibarbo de levante
Usage ▸ spice
Parts Used ▸ rhizome
Distribution ▸ mountains of the North area between China and Tibet, cultivated in China, Russia, Europe

Aichele/Schwegler 2, 1994; Charalambous 1994; Cheers 1997; Coiciu/Racz (no year); Dregus et al. 2004; Dudtschenko et al. 1989; Erhardt et al. 2000, 2002; Hanelt 2001; Heeger 1956; Hiller/Melzig 1999; Hohmann et al. 2001; Hoppe 1949; Lewington 1990; Miyazawa et al. 1996; Pschyrembel 1998; Rätsch 1998; Schönfelder 2001; Schratz 1960; Schultze-Motel 1986; Seidemann 1993c; Small 1997; Stahl et al. 1985; Täufel et al. 1993; Wiersema/León 1999; Wüstenfeld/Haensel 1964; Wyk et al. 2004; Zeven/de Wet 1982

Rheum rhabarbarum L.

Synonyms ▸ *Rheum rhaponticum* L., *Rheum undulatum* L.
Common Names ▸ garden rhubarb, Siberian rhubarb, rhapontic rhubarb, rhubarb; *Arabic:* rawend zahar; *Chinese:* shi yung ta huang; *French:* rhapontic, rhubarbe (anglaise); *German:* Gewöhnlicher Rhabarber, Krauser Rhabarber, Sibirischer Rhabarber, Rhapontik, Wellblatt-Rhabarber; *Italian:* rabarbaro, rapontico; *Japanese:* kara daio; *Korean:* taehwang, tungŭniptaehwang; *Portuguese:* ruibarbo; *Russian:* rapontik; *Spanish:* ruibarbo (francés), rapóntico
Usage ▸ spice
Parts Used ▸ rhizome

Distribution ▸ Bulgaria, Rhodope mountains, C Asia, S Siberia ?, only cultivated
Note ▸ Petioles are used in sauces and pies.

Aichele/Schwegler 2, 1994; Bendel 2002; Charalambous 1994; Dalby 2000; Davidson 1999; Engelshowe 1985; Erhardt et al. 2000, 2002; Hanelt 2001; Heeger 1956; Hiller/Melzig 1999; Hoppe 1949; Libert/Enghind 1989; Paneitz/Westendorf 1999; Rätsch 1998; Schratz 1960; Schultze-Motel 1986; Siemonsma/Piluek 1993; Täufel et al. 1998; Tindall 1983; Uphof 1968; Wüstenfeld/Haensel 1964; Zeven/de Wet 1982

Rheum rhaponticum L.

▸ *Rheum rhabarbarum L.*

Rheum undulatum L.

▸ *Rheum rhabarbarum L.*

Rhododendrum palustre (L.) Kron et Judd

▸ *Ledum palustre L.*

RHODYMENIA Grev. - Rhodomeniaceae

Rhodymenia palmata (L.) Grev.

Synonyms ▸ *Palmaria palmata* (L.) Kuntze
Usage ▸ condiment
Parts Used ▸ sea weed
Distribution ▸ Pacific and Atlantic Ocean

Uphof 1968

RHUS L. - Sumach - Anacardiaceae

Rhus aromatica Ait.

Synonyms ▸ *Rhus canadensis* Marshall non Mill.
Common Names ▸ fragrant sumach, lemon sumach,

skunkbush, polecat bush, Sicilian sumac *Arabic:* summak; *Dutch:* zuurkruid; *French:* sumac odorant, suma des corroyeurs; *German:* Duft-Sumach, Gewürz-Sumach, Süßer or Wohlriechender Sumach; *Hindi:* kankrasing; *Italian:* sommacco; *Turkish:* suma, somak;

Usage ▸ spice, condiment (for zahtar)
Parts Used ▸ fruit (beers)
Distribution ▸ Canada, USA, E, SE and SW Europe, N Africa, Caucasus, C and W Asia
Note ▸ The fruits have a sour taste.

Bozan et al. 2002; Cheers 1997; Erhardt et al. 2002; Guzman/Siemonsma 1999; Hanelt 2001; Heeger 1956; Hiller/Melzig 1999; Hoppe 1949; Roth/Kormann 1997; Rätsch 1998; Teuscher 2003; Wiersema/León 1999

Rhus canadensis Marshall non Mill.

▸ *Rhus aromatica* Ait.

Rhus commiphoroides Engl. et Gilg.

▸ *Rhus tenuinervis* Engl.

Rhus coriaria L.

Synonyms ▸ *Toxicondron coriaria* Kuntze
Common Names ▸ sumac, Sicilian sumac, lemonade tree; *Arabic:* sumaq; *French:* sumac; *German:* Tanner's Sumach, Gerbersumach; Sizilianischer Sumach, Sumak; *India:* samaka, sumak, tatrak, timtima; *Italian:* sommacco; *Portuguese:* sumagre; *Russian:* sumakh; *Slovenian:* sumach; *Spanish:* zumaque; *Turkish:* sumak, somak, tatari, tetri, tirimli
Usage ▸ spice, e.g. admixture for condiments (zahtar), flavoring for tobacco (leaf)
Parts Used ▸ seed, leaf
Distribution ▸ Mediterranean region; E, SE Europe, Caucasus, Iran, C, W Asia, also cultivated in Spain, Italy, W Asia

Bärtels 1997; Berger 2, 1950; Bois 1934; Brunke 1994a, b; Craze 2002; Davidson 1999; Dudtschenko et al. 1989; Duke et al. 2003; Erhardt et al. 2002; Erichsen-Brown 1989; Effenberger/Schilcher 1990; Guzman/Siemonsma 1999; Hanelt 2001; Hepper 1992; Hiller/Melzig 1999; Lück 2000; Morris/Mackley 1999; Norman 1991; Schultze-Motel 1986; Seidemann 1993c, 1995d; Seidemann/Siebert 1987; Siewek 1990; Staesche 1971, 1972; Täufel et al. 1993; Teuscher 2003; Uhl 2000; Wiersema/León 1999; Zeven/de Wet 1982

Rhus tenuinervis Engl.

Synonyms ▸ *Rhus commiphoroides* Engl. et Gilg.
Common Names ▸ *German:* Feinnerviger Sumach; *S Africa:* hyaena taaibos, morupapiri
Usage ▸ spice (of meat)
Parts Used ▸ leaf
Distribution ▸ southern Africa: Kalahari

v. Koenen 1996; Neuwinger 1999; Quattrocchi 2000

Ribesoides philippense O. Kuntze

▸ *Embelia philippensis* A. DC.

RICINODENDRON Mull. Arg. - Euphorbiaceae

Ricinodendron africanus Muell.

▸ *Ricinodendron heudelotii* (Baill.) Pierre ex Pax

Ricinodendron heudelotii (Baill.) Pierre ex Pax

Synonyms ▸ *Jatropha heudelotii* Baill., *Ricinodendron africanus* Muell.
Common Names ▸ African nut; *Angola:* minguella; *Cameroon:* andjejang; *French:* bois jasanga, essang; *Gabun:* engessam; *German:* Afrikanisches Mahagoni, Ojokbaum; *Ghana:* anwama; *Nigerian:* ekku; *Sierra Leone:* gbolei; *Spanish:* cuyo, ricino del país; *Uganda:* musodo;
Usage ▸ spice for sauce, condiment
Parts Used ▸ seed
Distribution ▸ tropical W and E Africa, cultivated in Cameroon

Aedo et al. 2001; Erhardt et al. 2002; Hanelt 2001; Neuwinger 1999;

Schultze-Motel 1986; Täufel et al. 1993; Uphof 1968; Wiersema/León 1999; Zeven/de Wet 1982

ROBINIA L. - Locust - Fabaceae (Leguminosae)

 **Robinia grandiflora** L.

❯ *Sesbania grandiflora (L.) Pers.*

 **Robinia pseudoacacia** L.

Common Names ▸ false acacia, black locust, robinia; *Chinese:* ci huai hua; *French:* faux acacia, robinier; *German:* Falsche Akazie, Schein-Akazie, Gemeine Robinie, Schotendorn, Silberregen; *Italian:* robinia, pseudo-acacia, false acacia; *Japanese:* hari-enju; *Korean:* akhasianamu; *Russian:* akazia belaja, robinia, loshno-akazija; *Spanish:* falsa acacia, robinia
Usage ▸ flavoring, e.g. fresh flowers added to omelettes
Parts Used ▸ flower
Distribution ▸ NC, NE and SE USA, widely cultivated and native in temperate regions, e.g. Europe

Aichele/Schwegler 2, 1994; Berger 1, 1949; Brockmann 1979; Cheers 1997; Dudtschenko et al. 1989; Erhardt et al. 2002; Fleischhauer 2003; Hager 3, 1992; 4, 1992; Hanelt 2001; Hiller/Melzig 1999; Hoppe 1949; Schnelle 1999; Schönefelder 2001; Schultze-Motel 1986; Seidemann 1993c; Täufel et al. 1993; Wiersema/León 1999; Zeven/de Wet 1982

RORIPPA Scop. - Water Cress - Brassicaceae (Crucifera)

 **Rorippa palustris** (L.) **Besser**

Common Names ▸ common water cress; *German:* Gewöhnliche Sumpfkresse
Usage ▸ pot-herb
Parts Used ▸ young leaf
Distribution ▸ Europe, Caucasus, Turkey, W, E Siberia, Amur, Sakhalin, Kamchatka, C, E Asia, Himalayas, Mongolia, N America, S America, Egypt, Australia

Erhardt et al. 2002; Fleischhauer 2003; Schnelle 1999

 Rorippa microphylla *(Boenn.) Hyl.*

❯ *Nasturtium microphyllum Boenn. ex Rchb.*

 Rorippa nasturtium-aquaticum *(L.) Hayek*

❯ *Nasturtium officinale R.Br.*

 Rorippa x sterilis *Airy Shaw*

❯ *Naturtium x sterile (Airy Shaw) Oefelein*

ROSA - Rose - Rosaceae

Cheers 1997; Dimov/Tsoutsoulova 1987; Markley 1997, 1999; Nissen 1992; Schirarend/Heimeyer 1996; Singer 1885; Tantau/Weinhausen 1956

 Rosa austriaca *Crantz*

❯ *Rosa gallica L.*

 Rosa x centifolia L.

Common Names ▸ cabbage rose, Holland rose, provence rose; *French:* rose à cent feuilles, rose de mai; *German:* Mairose, Zentifolie; *India:* devataruni, gulab, irosa, satapatri; *Japanese:* seiyô-ibara; *Mexico:* guie becohua, quije pecohua castilla; *Russian:* roza stolistaja; *Sanskrit:* devataruna
Usage ▸ flavoring; **product:** essential oil (rose oil)
Parts Used ▸ floral leaf
Distribution ▸ Caucasus, N Iran to Turkey, cultivated in France, Italy, Morocco, India, China

Cheers 1997; Hanelt 2001; Schultze-Motel 1986

Rosa chinensis *Jacq.*

Common Names ▶ China rose, Bengal rose; *German:* China-Rose; *Korean:* wŏlkyéhwa;
Usage ▶ flavoring; **product:** essential oil (rose oil)
Parts Used ▶ floral leaf
Distribution ▶ China, widely cultivated

NICPBP 1987; Oka et al. 1998; Tantau/Weinhausen 1956; Wiersema/León 1999

 Rosa x damascena *Mill.*

Synonyms ▶ *Rosa gallica* L. var. *damascena* Voss;
Common Names ▶ Damask rose, Portland rose, pink damask rose, Bulgarian rose; *Chinese:* du jue qiang wei, xia jin ying, chiang wei; *French:* rose de Damas, rose de tous les mois, rose de Puteaux; *German:* Damaszener Rose; Portland Rose; *Hindi:* fosli gulab, gulab ke phul; *Indonesian:* kembang eros, bunga ros; *Japanese:* kôshin-ibara; *Korean:* punhongkkothyangjangri; *Laos:* kuhlaab; *Malaysian:* bunga ayer mawar, ros, gul; *Portuguese:* rosa de Damasco, rose pálida; *Russian:* roza damasskaja; *Sanskrit:* shata patri; *Thai:* kulaap-on, kulaap mon, yee sun; *Vietnamese:* huong
Usage ▶ flavoring; **product:** essential oil (rose oil)
Parts Used ▶ floral leaf
Distribution ▶ Bulgaria, France, Italy, Morocco, Russia, Turkey
Note ▶ Cultivated only for essential oil production and as an ornamental shrub. The floral leaves are often candied and used in confectionery.

Bayrak/Akgül 1994; Berger 1, 1949; Bois 1934; Bourton 1968; Charalambous 1994; Davidson 1999; Dudteschenko et al. 1989; Erhardt et al. 2002; Eugster/Märki-Fischer 1991; Fishman 1984; Hager 2, 1991; Hiller/Melzig 1999; Hoppe 1949; Kalkman 1993; Krüssmann 1974; Lewington 1990; Oka et al. 1998; Oyen/Dung 1999; Rohner-Reinhard 1988; Roth/Kormann 1997; Schultze-Motel 1986; Seidemann 1993; Sharma 2003; Shiva et al. 2002; Singh/Deolia 1963; Small 1997; Tantau/Weinhausen 1956; Täufel et al. 1993; Touw 1982; Tucker 1986; Uphof 1968; Vogt 1991; Widrlechner 1981; Wiersema/León 1999; Zeven/de Wet 1982

 Rosa florida *Poir.*

▶ *Rosa multiflora* Thunb.

 Rosa gallica *L.*

Synonyms ▶ *Rosa austriaca* Crantz, *Rosa grandiflora* Salisb.
Common Names ▶ French rose; *Arabic:* ouard, *Chinese:* fa kuo, chiang wei; *French:* rose de Provence; *German:* Essigrose, Samtrose; *India:* gulab; *Portuguese:* roseira; *Russian:* roza francuzskaja
Usage ▶ flavoring, e.g. marzipan; **product:** essential oil (rose oil)
Parts Used ▶ floral leaf
Distribution ▶ S, C Europe, W Russia, Asia Minor; cultivated in Europe: S France, Spain, the Netherlands, England, former Soviet Union

Bremness 2001; Cheers 1997; Fleischhauer 2003; Hanelt 2001; Heeger 1956; Schultze-Motel 1986

 Rosa gallica *L.* **var. damascena** *Voss*

▶ *Rosa damascena* Mill.

 Rosa gallica *L.* **x Rosa moschata** *Herrm.*

▶ *Rosa moschata* Herrm.

 Rosa grandiflora *Salisb.*

▶ *Rosa gallica* L.

 Rosa moschata *Herrm.*

Synonyms ▶ *Rosa gallica* L. x *Rosa moschata* Herrm.
Common Names ▶ musk rose; *French:* neseri, rosier musqué; *German:* Moschusrose; *India:* kubjaha, kuji, kujai; *Javanese:* kembang rus; *Portuguese:* roseira selvagem; *Russian:* roza muskusnaja; *Spanish:* mosquiteta blanca
Usage ▶ flavoring; **product:** essential oil (rose oil)
Parts Used ▶ floral leaf

Distribution ▶ S and E Asia mountains, Ethiopa, only cultivated

Davidson 1999; Erhardt et al. 2002; Hager 2, 1991; Hanelt 2001; Hiller/Melzig 1999; Hoffmann et al. 1992; Lewinton 1990; Schultze-Motel 1986; Seidemann 1993c; Sharma 2003; Tantau/Weinhausen 1956; Täufel et al. 1993; Wiersema/León 1999; Zeven/de Wet 1982

Rosa multiflora *Thunb.*

Synonyms ▶ *Rosa florida* Poir., *Rosa polyantha* Sieb. et Zucc., *Rosa rubeoides* Andr.
Common Names ▶ baby rose, bramble rose, Japanese rose, pilar rose, seven sister rose; *Chinese:* chiáng wei chun; *German:* Japanrose, Noisetterose; *Japanese:* no-ibara; *Javanese:* sekar rus; *Malaysian:* kembang erus
Usage ▶ flavoring of tea and diets
Parts Used ▶ floral leaf
Distribution ▶ N China, Japan, Korea, India (Assam, Uttar Pradesh); cultivated in India, China, Java, Japan

Cheers 1997; Hanelt 2001; Schultze-Motel 1986;

Rosa polyantha *Sieb. et Zucc.*

▶ *Rosa multiflora Thunb.*

Rosa rubeoides *Andr.*

▶ *Rosa multiflora Thunb.*

ROSMARINUS L. - Rosemary - Lamiaceae (Labiatae)

Rosmarinus lavendulaceus *Noe*

▶ *Rosmarinus officinalis L.*

Rosmarinus officinalis *L.*

Synonyms ▶ *Rosmarinus lavandulaceus* Noe
Common Names ▶ rosemary, compass plant; *Arabic:* ekleel aljabal, iklil al-jabal, kelil; *Chinese:* mi tieh hsiang; *Dutch:* rozemarijn; *French:* romerino, incensier; rosmarin; *German:* Rosmarin, Kranzkraut; *India:* rusmari; *Italian:* ramerino, rosmarino; *Japanese:* mannenrû; *Pilipino:* dumero, romero, rosmir; *Portuguese:* alecrim, alerum; *Russian:* Rozmarin, morskaja rosa; *Spanish:* roméro; *Turkish:* biberye
Usage ▶ spice; **product**: essential oil (rosmary oil)
Parts Used ▶ leaf, herb
Distribution ▶ Mediterranean regions: Italy, Sardinia, Corsica, S France, cultivated worldwide

Aichele/Schwegler 4, 1995; Bano et al. 2003; Barbut et al. 1985; Bärtels 1997; Berger 2, 1950; Boelens 1985; Bourton 1968; Bremness 2001; Brieskorn et al. 1973; del Campo et al. 2000; di Cesare et al. 2001; Chalchat et al. 1993; Chang et al. 1977; Cheers 1997; Clair 1961; Coiciu/Racz (no year); Colins/Charles 1987; Davidson 1999; Dudtschenko et al. 1989; Elamrani et al. 2000; Fincke 1961; Flamini et al. 2002; Fournier et al. 1989; Guazzi et al. 2001; Guzman/Siemonsma 1999; Hager 1, 1990; Hanelt 2001; Heeger 1956; Henning et al. 2002; Hiller/Melzig 1999; Hof/Ammon 1989; Hoffmann et al. 1992; Hoppe 1949; Houlihan et al. 1985; Ibanez et al. 2000, 2003; Klimek et al. 2002; Lindberg et al. 1996; Melchior/Kastner 1974; Mule' et al 1994; Munné-Bosch et al. 2000; Nakatani/Inatani 1981, 1984; Newall et al. 1996; Ochoa/Alonso 1996; Opdyke 1974e; Ouahada/Benveniste 2000; Parnham/Kesselring 1985; Peter 2001; Pintore 2002; Pochljobkin 1974, 1977; Porte et al. 2000; Pruthi 1976; Pschyrembel 1998; Resche 1983; Reverchon/Senatore 1992: Rezzoug et al. 2000; Rosengarten 1969; Rosúa/García-Granados 1987; Roth/Kormann 1997; Salido et al. 2003; Schönfelder 2001; Schwarz/Ternes 1992a, b; Schwarz et al. 1992; Schultz-Motel 1986; Seidemann 1993c; Seidemann/Siebert 1987; Sfikas 1994; Sharma 2003; Shi et al. 2002; Shiva et al. 2002; Singh et al. 1996; Small 1997; Soliman et al. 1994; Staesche 1972; Tateo et al. 1988a, b; Täufel et al. 1993; Teuscher 2003; Tucker 1986; Tucker/Maciarello 1986; Ubillos 1989; Uphof 1968; Vareltzis et al. 1997; Verotta 1985; Villamar et al. 1994; Wiersema/León 1999; Wolski et al 2002; Wüstenfeld/Haensel 1964; Wyk et al. 2004; Zeven/de Wet 1982; Zimmermann 1980

RUMEX L. - Sorrel - Polygonaceae

Rumex acetosa *L.*

Synonyms ▶ *Acetosa pratensis* Mill.

Common Names ▶ common sorrel, garden sorrel, sorrel, sorrel dock, sour dock, sourgrass; *Arabic:* dierb, hummedha; *French:* grande oseille, oseille commune, oseille acide, oiseille de belleville, rhapontike des moines; *German:* Garten-Sauerampfer, Großer Sauerampfer, Wiesen-Sauerampfer; *Italian:* acetosa maggiore, atschetonsa, saleggiola; *Japanese:* sorelu, sukampo; *Portuguese:* azeda brava, azedinha da horta; *Russian:* stschewel' kislyj, stschavel' malyj, stschavelek; *Spanish:* acedera común, agrilla, vinagrera
Usage ▶ spice, for soups
Parts Used ▶ fresh leaf
Distribution ▶ N Africa, temperate Asia, India, Australia, Europe, native elsewhere

Aichele/Schwegler 2, 1994; Berger 4, 1954; Bilgri/Adam 2000; Bremness 2001; Cheers 1997; Erhardt et al. 2000, 2002; Erichsen-Brown 1989; Fleischhauer 2003; Heeger 1956; Hiller/Melzig 1999; Hoffmann et al. 1992; Hoppe 1949; Koschtschejew 1990; Schenck/Naundorf 1966; Schönfelder 2001; Small 1997; Täufel et al. 1993; Tucker 1986; Uhl 2000; Uphof 1968; Wiersema/León 1999; Zeven/de Wet 1982

Rumex acetosella L.

Common Names ▶ sheep sorrel; *German:* Kleiner Sauerampfer
Usage ▶ spice: for soups
Parts Used ▶ fresh leaf
Distribution ▶ Europe, Caucasus, W, E Siberia, Amur, Sakhalin, Kamchatka, native in N America

Erhardt et al. 2002; Fleischhauer 2003; Loch 1993; Schnelle 1999

Rumex Alpstris Jacq.

▶ *Rumex scutatus L.*

Rumex crispus L.

Common Names ▶ curled dock, dock, narrow dock; *French:* rumex crépu, patience crépue; *German:* Krauser Ampfer; *Italian:* romine crespo, lapazio; *Japanese:* nagaba-gishigishi; *Korean:* songguji, sorujaengi; *Peru:* moztaza; *Spanish:* acitosa, lengua de vaca
Usage ▶ condiment
Parts Used ▶ leaf
Distribution ▶ Europe; Turkey, C Asia, Mongolia, China, native N America, cosmopolitical

Erhardt et al. 2002; Erichson-Brown 1989; Fleischhauer 2003; Hanelt 2001; Heeger 1956; Schnelle 1999; Schönfelder 2001; Uphof 1968; Wyk et al. 2004

Rumex hastatus D.Don

Common Names ▶ speared sorrel; *German:* Spießiger Ampfer; *Hindi:* khatapalak;
Usage ▶ condiment (for chutney and pickles)
Parts Used ▶ leaf
Distribution ▶ India, W Himalayas: Kashmir

Arora/Pandey 1996; Uphof 1968

Rumex maritimus L.

Common Names ▶ golden dock; *German:* Ufer-Ampfer; *India:* jangli-palak, jal-palan, bunpaling, khattikan
Usage ▶ pot-herb (N India)
Parts Used ▶ leaf
Distribution ▶ Europe except Iberia, W, E Siberia, Amur, C Asia, N India, Mongolia, Japan, Algeria, America

Araora/Padney 1996; Erhardt et al. 2002; Kumar 2003

Rumex nemorosus Schrader

▶ *Rumex sanguisorbis L.*

Rumex obtusifolius L.

Common Names ▶ bitter dock, broadleaf dock; *French:* patience suvage: *German:* Stumpfblättriger Ampfer; *Italian:* romice dei prati; *Portuguese:* labaca; *Spanish:* vinagrillo, romaza de kojas grandes
Usage ▶ condiment
Parts Used ▶ young leaf
Distribution ▶ Europe, Turkey, Syria, Palestine, Cauca-

sus, N Iran, Canary Islands, Algeria, native in N America, S America, S Africa, Australia

Erhardt et al. 2002; Erichson-Brown 1989; Heeger 1956; Schönfelder 2001; Uphof 1968

Rumex sanguisorbis L.

Synonyms ▶ *Rumex nemorosus* Schrader
Common Names ▶ bloodred sorrel; *German:* Blut- or Hain-Ampfer;
Usage ▶ spice for soups (sporadically)
Parts Used ▶ leaf
Distribution ▶ temperate Europe, Caucasus, N Iran, NW Africa

Aichele/Schwegler 2, 1994; Erhardt et al. 2002

Rumex scutatus L.

Synonyms ▶ *Acetosa sculata* (L.) Mill., *Rumex alpstris* Jacq.
Common Names ▶ French sorrel, garden sorrel *French:* osielle aux feuilles ròndes, osielle bouclier, osielle ronde, patience écousson; *German:* Römischer Ampfer, Rundblättriger Ampfer, Schild-Sauerampfer; *India:* amrula, ambarati, changeri; *Italian:* acetosa romana, erba pan a vin; *Russian:* stschawel stschitkoviduij; *Spanish:* acedera con hojas redondas francesa, acedera romana, acedera redonda
Usage ▶ pot-herb, flavoring soups and dishes
Parts Used ▶ leaf
Distribution ▶ S Europe, Alps, C Europe, Asia minor, India: W Himalayas, Caucasus, sporadically cultivated
Note ▶ The leaves are lemon-sour.

Aichele/Schwegler 2, 1994; Arora/Pandey 1996; Bremness 2001; Erhardt et al. 2002; Fleischhauer 2003; Hanelt 2001; Heeger 1956; Schnelle 1999; Schultze-Motel 1986; Tucker 1986; Wiersema/León 1999; Zeven/de Wet 1982

Rumex vesicarius L.

Synonyms ▶ *Lapathum vesicarium* (L.) Moench
Common Names ▶ bladder dock, *French:* oseille d'Amérique; *German:* Blasiger Sauerampfer, Indischer Sauerampfer; *Hindi:* ambari, chukra, chuka, saluni, shakkankirai
Usage ▶ spice
Parts Used ▶ leaf
Distribution ▶ Europe: Greece, N Africa, Asia minor, Arabian Peninsula, W Asia, India, Malaysia

Arora/Pandey 1996; Chopra et al 1956; Erhardt et al. 2002; Hanelt 2001; Ochse/Bakhuizen van den Brink 1931; Schultze-Motel 1986; Wiersema/León 1999; Zeven/de Wet 1982

RUTA L. - Rue - Rutaceae

Ruta altera Mill.

▶ *Ruta graveolens* L.

Ruta angustifolia Pers.

▶ *Ruta chalepensis* L.

Ruta bracteosa DC.

▶ *Ruta chalepensis* L.

Ruta chalepensis L.

Synonyms ▶ *Ruta angustifolia* Pers., *Ruta bracteosa* DC., *Ruta fumariaefolia* Boiss. et Heldr.f.
Common Names ▶ Egyptian rue, fringent rue, Syrian rue; *Arabic:* fidjel, fidjla; *French:* rue d'Algérie; *German:* Aleppo-Raute, Gefranste Raute, Syrische Raute; *Hindi*: pismarum, sadab, satari
Usage ▶ spice
Parts Used ▶ herb
Distribution ▶ Canary Islands, Mediterranean region, Ethiopia, Arabia; cultivated C, S America, Mexico, W Indian Islands

Hanelt 2001; Hepper 1992; Hiller/Melzig 1999; Jansen 1981; Ochoa/Alonso 1996; Schultze-Motel 1986; Seidemann 1993c; Small 1997; Teuscher 2003; Uphof 1968; Vasudevan/Luckner 1968; Villamar et al. 1994; Zeven/de Wet 1982

Ruta fumariaefolia *Boiss. et Heldr.*

▶ *Ruta chalepensis L.*

 Ruta graveolens *L.*

Synonyms ▶ *Ruta altera* Mill., *Ruta officinalis* Pall.
Common Names ▶ common rue, fringed rue, herb of grace, herb of repentance, rue; *Arabic:* arudam fejan, *Chinese:* chou cao, yün xiang; *Dutch:* ruit; *French:* rue, rue de jardins, rue fétide, rue officinale, rue puante, herbe de grâce; *German:* Raute, Edelraute, Gartenraute, Weinraute, Gnadenkraut; *Hindi:* sitab; *Italian:* ruta, riccola, richetta; *Japanese:* henruda; *Javanese:* godong minggu; *Korean:* unhjang; *Portuguese:* arruda; *Russian:* ruta, ruta sadowaja; *Sanskrit:* sitaba, somalata; *Spanish:* ruda común; *Turkish:* sedefotou
Usage ▶ spice; **product:** essential oil
Parts Used ▶ herb
Distribution ▶ E, SE and SW Europe, widely native in Europe and N Africa, also cultivated

Aichele/Schwegler 3, 1995; Andorn et al. 1972; Bärtels 1997; Becela-Deller 1991; Berger 4, 1954; Bois 1934; Böttcher/Günther 2003; Bourton 1968; Bremness 2001; Charalambous 1994; Cheers 1997; Clair 1961; Classen/Knobloch 1985; Coiciu/Racz (no year); Davidson 1999; Erhart et al. 2002; Fleischhauer 2003; Hager 3, 1992; Hanelt 2001; Heeger 1956; Hepper 1992; Hiller/Melzig 1999; Hoffmann et al. 1992; Hohmann et al. 2001; Hondelmann 2002; Hoppe 1949; Jansen 1981; Nagel/Reinhard 1975; Novak et al. 1965; Ochoa/Alonso 1996; Pochljobkin 1974, 1977; Pschyrembel 1993; Rätsch 1998; Roth/Kormann 1997; Schönfelder 2001; Schultze-Motel 1986; Seidemann 1993c; Seidemann/Siebert 1987; Small 1997; Täufel et al. 1993; Teuscher 2003; Tucker 1986; Uphof 1968; Vasudevan/Luckner 1968; Villamar et al. 1994; Yaacob et al. 1989; Wiersema/León 1999; Wyk et al. 2004; Zeven/de Wet 1982

 Ruta graveolens *L.* **var. montana**

▶ *Ruta montana L.*

 Ruta montana *L.*

Synonyms ▶ *Ruta graveolens* L. var. *montana* L.
Common Names ▶ montain rue; *Arabic:* fidjla el-djebeli; *French:* rue de montagnes; *German:* Bergraute; *Spanish:* ruda de monte
Usage ▶ spice
Parts Used ▶ herb
Distribution ▶ SW Europe, N Africa, Mediterranean regions, also cultivated

Bourton 1968; Hanelt 2001; San Miguel 2003; Schultze-Motel 1986; Seidemann 1993; Teuscher 2003; Vasudevan/Luckner 1968

 Ruta officinalis *Pall.*

▶ *Ruta graveolens L.*

S

SABA Pichon - Apocynaceae

 Saba senegalensis *(DC.) Pichon*

Synonyms ▸ *Landolphia senegalensis* Kotsch. et Peyr.
Usage ▸ spice for sauces; condiment
Parts Used ▸ leaf
Distribution ▸ tropical W Africa: the Gambia, Guinea, Ivory Coast, Mali, Nigeria, Senegal
Note ▸ The plant is an appetite stimulant.

Burkill 1, 1985; Neuwinger 1998, 1999; Pelisier et al. 1996

 Salisburya biloba *Hoffmannsegg*

▸ *Gingko biloba* L.

SALVADORA L. - Salvadoraceae

 Salvadora indica *Wight*

▸ *Salvadora persica* L.

 Salavadora paniculata *Zucc. ex Steud.*

▸ *Salvadora persica* L.

 Salvadora persica *L.*

Synonyms ▸ *Cissus arborea* Forrsk., *Salvadora indica* Wight, *Salvadora paniculata* Zucc. ex Steud.
Common Names ▸ salt bush, toothbrush tea; *Arabic:* arak, miswak, siwak; *French:* arbre brosse à dents; *German:* Salzbusch, Senfbaum, Zahnbürstenstrauch, Löwenbusch; *India:* chota pilu, jhak, kharjal, kakkam, piludi, pilva, rhakhan
Usage ▸ condiment (raw leaves) for sauces and salads
Parts Used ▸ leaf, fruit
Distribution ▸ N and SW Africa: Mali, Niger, Nigeria, Senegal; NW India
Note ▸ The taste of the fruits "Senf der Bibel" is pungent or peppery. The peeled and dry miswak twigs are ued as tooth-brush in the arabian countries.

Arora/Pandey 1996; Burkill 5, 2000; Dalziel 1937; Hanelt 2001; Hiller/Melzig 1999; v. Koenen 1996; Roberty 1953; Uphof 1968; Wealth of India 9, 1972; Wyk et al. 2004

SALVIA L. - Sage - Lamiaceae (Labiatae)

Demirci et al. 2002; Yingrong Lu, Yeap Foo 2002

 Salvia aethiopis *L.*

Synonyms ▸ *Salvia lanata* Moench, *Sclarea aethiopis* (L.) Mill.
Common Names ▸ African sage, Mediterranean sage, wally sage; *German:* Afrikanischer Salbei,

Mohren-Salbei, Ungarischer Salbei; *Italian:* etiopide salvia
Usage ▶ spice, flavoring
Parts Used ▶ herb
Distribution ▶ S Europe, Iberia, S France, N Africa, Turkey, Asia minor, Transcaucasia, W Asia, native in N America

Boya/Valverde 1981; Chalchat et al. 2001a; Clebsch 1997; Erhardt et al. 2002; Hanelt 2001; Schultze-Motel 1986; Sutton 1999; Ulubelin/Uygur 1976; Wiersema/León 1999

Salvia aramiensis *Rech.*

Common Names ▶ Aramenian salve; *German:* Aramenischer Salbei
Usage ▶ flavoring
Parts Used ▶ herb
Distribution ▶ Armenia, Turkey

Demirci et al. 2002a; Sarer 1987

◘ Salvia argentea, plant

Salvia argentea *L.* ◘

Synonyms ▶ *Salvia candidissima* Guss.
Common Names ▶ silver sage; *French:* sauge argentée; *German:* Silberblatt-Salbei
Usage ▶ spice, flavoring
Parts Used ▶ leaf
Distribution ▶ Europe: S Spain, France, Balkans, Turkey, NW Africa

Clebsch 1997; Erhardt et al. 2002; Sutton 1999

Salvia bengaliensis *König ex Roxb.*

❯ *Meriandra bengaliensis (Konja ex Roxb.) Benth.*

Salvia camertoni *Regel*

❯ *Salvia elegans Vahl*

Salvia candidissima *Guss*

❯ *Salvia argentea L.*

Salvia carduacea *Benth.*

Common Names ▶ thistle sage; chia; *German:* Distelartiger Salbei, Distel-Salbei
Usage ▶ spice, condiment
Parts Used ▶ leaf
Distribution ▶ SE Asia, USA: California, Mexico

Clebsch 1997; Erhardt et al. 2002; Hanelt 2001; Kintzios 2000; Usher 1974

Salvia chudaei *Batt. et Trab.*

Usage ▶ condiment (in the Hogger region)
Parts Used ▶ plant
Distribution ▶ Sahel of Mali

Burkill 3, 1995; Maire 1933

Salvia clandestina L.

▶ *Salvia verbenaca L.*

Salvia clevelandii *(A. Gray) Greene*

Common Names ▶ blue sage, Jim sage; *German:* Cleveland Salbei, Blauer Salbei, Marzipan-Salbei
Usage ▶ spice
Parts Used ▶ leaf
Distribution ▶ native of California; Canada, northern USA

Clebsch 1997; Kintzios 2000; Small 1997

Salvia discolor *Kunth*

Common Names ▶ Peruvian sage; *German:* Peruanischer Salbei, Verschiedenfarbiger Salbei;
Usage ▶ pot-herb
Parts Used ▶ leaf
Distribution ▶ S America: Peru
Note ▶ the plant has a fruity, eucalyptus scent.

Cheers 1997; Clebsch 1997; Erhardt et al. 2002; Kintzios 2000; Small 1997; Sutton 1999

Salvia divinorum *Epling et Játiva*

Common Names ▶ herb of the Virgin; *German:* Azteken-Salbei, Wahrsager Salbei; *Mexico:* pipiltzintzinli; *Spanish:* hierba de la pastora, yerba de Maria
Usage ▶ flavoring
Parts Used ▶ herb, leaf
Distribution ▶ Mexico: Oaxaca

Erhardt et al. 2002; Hiller/Melzig 1999; Hanelt 2001; Rätsch 1998; Schultze-Motel 1986; Small 1997; Sutton 1999; Uphof 1968

Salvia dorisiana *Standley*

Common Names ▶ British Honduran sage, fruit sage; *German:* Fruchtsalbei, Goldsalbei, Pfirsisch-Salbei
Usage ▶ spice, flavoring: for fragrant concoctions and tea
Parts Used ▶ leaf
Distribution ▶ Honduras

Cheers 1997; Clebsch 1997; Halim/Collins 1975; Hanelt 2001; Kintzios 2000; Small 1997; Sutton 1999; Tucker 1986

Salvia elegans *Vahl*

Synonyms ▶ *Salvia camertoni* Regel, *Salvia incarnata* Cav., *Salvia rutilans* Carr.
Common Names ▶ tangerine sage, pineapple-scented sage (USA); *German:* Ananas-Salbei, Honigmelonen-Salbei
Usage ▶ pot-herb
Parts Used ▶ leaf, herb
Distribution ▶ C, S America: Mexico, Guatamala, only cultivated
Note ▶ The odor are few showed.

Bremness 2001; Cheers 1997; Clebsch 1997; Erhardt et al. 2002; Hanelt 2001; Kintzios 2000; Morton 1976; Small 1997; Sutton 1999; Teuscher 2003; Tucker 1986; Villamar et al. 1994

Salvia fruticosa *Mill.*

Synonyms ▶ *Salvia libanotica* Boiss. et Gaill., *Salvia triloba* L.f.
Common Names ▶ Greek sage, Turkish sage; *Arabic:* khayat el-djurhat; *Cretan:* faskomilia, sfakomilia; *French:* pomme de sauga, sauge trilobée, sauge à trois lobes; *German:* Fruchtsalbei, Griechischer Salbei, Kreuzsalbei; *Italian:* salvia trilobata; *Spanish:* salvia real, salvia triloba; *Turkish:* elma yagi; salvia real
Usage ▶ spice; **product:** essential oil
Parts Used ▶ leaf
Distribution ▶ Mediterranean regions: from S Balkans to Lebanon, Libya, S Italy, Albania, Dalmatia, Tur-

key; subspontaneously in NW Africa and Middle East; cultivated relict in S Spain and Algarve
Note ▶ The plant is also used as an adulterant of *Salvia officinalis* L.

Abdalla et al. 1983; Anon. 1993; Bärtels 1997; Berger 2, 1950; Brieskorn/Biechele 1971; Catsiotis/Iconomou 1984; Clebsch 1997; Erhardt et al. 2002; Hager 3, 1993; Hanelt 2001; Harvala et al. 1987b; Hepper 1992; Hiller/Melzig 1999; Hohmann et al. 2001; Karousou et al. 1998; Kintzios 2000; Pizzale et al. 2002; Pschyrembel 1998; Putievsky et al. 1986; Roth/Kormann 1997; Rhyu 1979; Schönfelder 2001; Schultze-Motel 1986; Small 1997; Sutton 1999; Teuscher 2003; Tucker 1986; Wiersema/León 1999

 Salvia glutinosa L.

Common Names ▶ Jupiter's distaff, sticky sage; *French:* sauge glutineuse; *German:* Klebriger Salbei, Gelber Salbei
Usage ▶ spice; **product:** essential oil
Parts Used ▶ herb
Distribution ▶ Europe, Turkey, Caucasus, N Iran

Clebsch 1997; Erhardt et al. 2002; Hoppe 1 1975Kintzios 2000; Sutton 1999

 Salvia grahamii Benth.

▶ *Salvia microphylla* Humb., Bonpl. et Kunth. var. *microphylla*

 Salvia hispanica L.

Synonyms ▶ *Salvia tetragona* Moench
Common Names ▶ Spanish sage; *French:* sauge de l'Espagne; *German:* Spanischer Salbei; *Guatamalan:* chián, chan; *Mexico:* chía, chía blanco, chía del campo
Usage ▶ spice, similar in sauces, salads, soups, beverages, diet food; **product:** essential oil (from leaves)
Parts Used ▶ seed, herb
Distribution ▶ SW USA: California to New Mexico; Mexico; spread to Cuba, Jamaica, and in the hills of Java
Note ▶ The species was cultivated widely by the Aztecs before the Spanish conquest. The seeds have a high fatty oil content.

Ahmed et al. 1994; Burkhill 1966; Bushway et al. 1984; Cahill 2003; Clebsch 1997; Estilai et al. 1990; Hanelt 2001; Kintzios 2000; Schultze-Motel 1986; Small 1997; Täufel et al. 1993; Uphof 1968; Villamar et al. 1994; Wiersema/León 1999

 Salvia hispanorum Lag.

▶ *Salvia lavandulifolia* Vahl

 Salvia horminum L.

▶ *Salvia viridis* L.

 Salvia incarnata Cav.

▶ *Salvia elegans* Vahl

 Salvia lanata Moench

▶ *Salvia aethiopis* L.

 Salvia lavandulifolia Vahl

Synonyms ▶ *Salvia hispanorum* Lag.
Common Names ▶ Spanish sage, lavander sage; *Dutch:* spaanse salie; *French:* sauge à feuilles de lavande; *German:* Lavendelblättriger Salbei, Spanischer Salbei; *Spanish:* salvia fina
Usage ▶ spice, flavoring for desserts, soft drinks, alcoholic beverage; **product:** essential oil
Parts Used ▶ leaf
Distribution ▶ Spain, France (Pyrenees), also cultivated
Note ▶ The plant is also used as adulterant of *Salvia officinalis* L.

Bärtels 1997; Bourton 1968; Bremness 2001; Charalambous 1994; Clebsch 1997; Erhardt et al. 2002; Hanelt 2001; Hiller/Melzig 1999; Kintzios 2000; Melchior/Kastner 1974; Perry et al. 2002; Roth/Kormann 1997; Schönfelder 2001; Schultze-Motel 1986; Seidemann 1993c; Small 1997; Sutton 1999; Tomás-Lorente et al. 1988; Tucker 1986; Villamar et al. 1994; Wiersema/León 1999

 Salvia libanotica Boiss. et Gaill.

▶ *Salvia fruticosa* Mill.

Salvia menthaefolia *Tenore*

Common Names ▸ Italian spring sage; *German:* Italienischer Frühlings-Salbei
Usage ▸ spice, flavoring
Parts Used ▸ leaf, herb
Distribution ▸ S Europe: Italy

Clebsch 1997; Kintzios 2000

Salvia microphylla Humb., *Bonpl. et Kunth.* var. microphylla

Synonyms ▸ *Salvia grahamii* Benth.
Common Names ▸ Graham salve, grapefruit sage; *German:* Graham's Salbei, Grapefruit-Salbei
Usage ▸ flavoring
Parts Used ▸ leaf
Distribution ▸ Mexico

Clebsch 1997; Erhardt et al. 2002; Hoppe 3, 1987; Sutton 1999

◘ Salvia officinalis, flowering

Salvia officinalis *L.* ◘

Synonyms ▸ *Salvia officinalis* L. ssp. *major* Gams, *Salvia tomentosa* Mill.
Common Names ▸ sage, common sage, garden sage; *Arabic:* maryamiya; mofassa, salima, salmya; *Chinese:* ching chieh; *French:* sauge, grande sauge, sauge officinale, thé de la Grèce, herbe sacrée; *German:* Echter Salbei, Garten-Salbei, Dalmatiner Salbei, Königssalbei; *Italian:* salvia; *Japanese:* sage; *Korean:* yakpulkkot; *Portuguese:* sálvia; *Russian:* schalfej aptetschny, schawlij, schal'wija; *Spanish:* salvia oficinal, salvia real; *Turkish:* adacayi
Usage ▸ spice, for various dishes, soups, sauces, salads, conserves, meat fish, cheese, wine etc.; **product:** essential oil
Parts Used ▸ leaf
Distribution ▸ S, SE Europe cultivated and native elsewhere: Mediterranean regions, Ukraine, Moldavia, C Europe, India, Indonesia (Java), Tanzania, southern Africa, Antilles, Canada, USA, Brazil
Note ▸ The species is very variable, especially in regard to leaf character. Many cultivars had been released.

Aichele/Schwegler 4, 1995; Andrews 1956; Asllani 2000; Bärtels 1997; Berger 2, 1950; Boelens/Boelens 1997; Bourton 1968; Bremness 2001; Brieskorn 1991; Brieskorn/Biechele 1971; BrieskornDömling 1969; Brieskorn/Fuchs 1962; di Cesare et al. 2001; Cheers 1997; Clair 1961; Clebsch 1997; Coiciu/Racz (no year); Cuvelier et al 1994; Davidson 1999; Dudtschenko et al. 1989; Erhardt et al. 2002; Farrell 1985; Grella/Picci 1988; Guzman/Siemonsma 1999; Hager 3, 1992; Hanelt 2001; Heeger 1956; Hiller/Melzig 1999; Hoffmann et al. 1992; Hohmann 1970; Hohmann et al. 2001; Hoppe 1949; Ivanic/Savin 1976; Jalsenjak et al. 1987; Kintzios 2000; Kouhila et al. 2001; Kreutzig 1982; Länger et al. 1993; Melchior/Kastner 1974; Miura et al. 2001; Newall et al. 1996; Ollanketo et al. 2002; Pachaly 1990a; Perry et al 1999; Peter 2001; Piccaglia et al. 1997; Pino et al. 2002; Pizzale et al. 2002; Pochljobkin 1974, 1977; Pruthi 1976; Pschyrembel 1998; Putjevsky et al. 1986; Reverchon et al. 1995; Rosengarten 1969; Roth/Kormann 1996; Santos-Gomes/Fernandes-Ferreira. 2003; Schönfelder 2001; Schultze-Motel 1986; Schwarz/Ternes 1992 a, b, 1993; Schwarz et al. 1992; Seidemann 1993c; Seidemann/Siebert 1987; Seyboldt 1998; Shi et al. 2002; Siewek 1990; Small 1997; Staesche 1972; Sutton 1999; Täufel et al 1993; Tucker 1986; Tucker et al 1980; Tucker/Maciarello 1990; Turova et al. 1987; Ubillos 1989; Ulubelen et al. 1981; Uphof 1968; Usher 1974; Vernin/Metzger 1986; Wealth of India 9, 1972; Wiersema/León 1999; Wüstenfeld/Haensel 1964; Yinrong Lu/Yeap Foo 2000, 2001; Wyk et al. 2004; Zeven/de Wet 1982

Salvia officinalis L. **ssp. major** Gams

> *Salvia officinalis* L.

 Salvia rutilans Carr.

> *Salvia elegans* Vahl

 Salvia sclarea L.

Synonyms ▶ *Sclarea vulgaris* Mill.
Common Names ▶ clary, clary sage, clary wort, English sage, muscat sage; *Arabic:* kaff ed-dubb; *French:* sauge sclarée, orvale, toute bonne; *German:* Clary-Salbei, Muskateller-Salbei, Scharlachkraut, Scharlei; *Italian:* erba moscatella sclarea; *Russian:* shafej muskatny; *Spanish:* hierba de los ojos
Usage ▶ spice, for liqueurs and wines (vermouth and muscatel), flavoring dishes and salads; **product:** essential oil (oil of clary); young leaves are eaten as a dessert (with orange sauce and sugar)
Parts Used ▶ leaf
Distribution ▶ SW Europe: Iberia, France, E Europe, Turkey, Syria, Palestine, Caucasus, Iran, C Asia, NW Africa, Kenya, native in Austria and Switzerland; frequently cultivated

Aichele/Schwegler 4, 1995; Bärtels 1997; Berger 2, 1950; Bremness 2001; Carrubba et al. 2002; Charalambous 1994; Cheers 1997; Clair 1961; Clebsch 1997; Coiciu/Racz (no year); Davidson 1999; Dudtschenko et al. 1989; Erhardt et al. 2002; Fleischhauer 2003; Hanelt 2001; Heeger 1956; Hiller/Melzig 1999; Ilieva 1979; Kintzios 2000; Maurer/Hauser 1983; Mikus/Schaser 1995; Pitarokili et al. 2002; Roth/Kormann 1997; Schultze-Motel 1986; Seidemann 1993c; Seidemann/Siebert 1987; Shiva et al. 2002; Small 1997; Sutton 1999; Täufel et al. 1993; Teuscher 2003; Tucker 1986; Ubillos 1989; Uphof 1968; Usher 1974; Wiersema/León 1999; Wl. 2004; Zámbori/Nyárádi-Szabady 1989; Zeven/de Wet 1982

 Salvia tetragona Moench

> *Salvia hispanica* L.

 Salvia tomentosa Mill.

> *Salvia officinalis* L.

 Salvia triloba L.f.

> *Salvia fruticosa* Mill.

 Salvia verbenaca L.

Synonyms ▶ *Salvia clandestina* L.
Common Names ▶ wild clary, verveine sage; *French:* sauge vervaine; *German:* Eisenkraut-Salbei
Usage ▶ spice, flavoring
Parts Used ▶ leaf
Distribution ▶ Asia minor, southern Mediterranean regions, S Europe, W Europe to Scotland, introduced to S Africa, S Australia and USA
Note ▶ Fresh leaves added to omelettes in Macronesia and SW Asia.

Aichele/Schwegler 4, 1995; Bärtels 1997; Cabo et al. 1986; Erhardt et al. 2002; Hanelt 2001; Kintzios 2000; Schultze-Motel 1986; Sutton 1999; Tucker 1986

 Salvia verticillata L.

Common Names ▶ *German:* Quirlblütiger Salbei, *Russian:* chalfej mutovtschatij
Usage ▶ spice, flavoring
Parts Used ▶ herb
Distribution ▶ S, SE Europe, France, Turkey, Asia minor, N Iraq, N Iran, Caucasus, W Siberia, native in C Europe, British Isles, Scandinavia, N America

Chalchat et al. 2001a; Clebsch 1997; Erhardt et al. 2000, 2002; Hanelt 2001; Schultze-Motel 1986

 Salvia viridis L.

Synonyms ▶ *Salvia horminum* L.
Common Names ▶ bluebeard sage, annual sage, Joseph sage, red topped sage; *Canada:* painted sage; *French:* sauge hormin or verte; *German:* Scharlachsalbei, Schopf-Salbei; *Italian:* gallitrico, chiarea, ormio
Usage ▶ spice for salads, soups and cooked vegetables; flavoring for beer and wine
Parts Used ▶ leaf
Distribution ▶ Morocco to Tunesia, S Europe, Turkey,

Iraq, Iran, Transcaucasia, Turkmenistan, also cultivated in C Europe

Cheers 1997; Clebsch 1997; Erhardt et al. 2002; Hanelt 2001; Hepper 1992; Kintzios 2000; Kokkalou et al. 1982; Schultze-Motel 1986; Small 1997; Sutton 1999; Täufel et al. 1993; Teuscher 2003; Tucker 1986; Ulubelen/Brieskorn 1975; Ulubelin et al. 1977; Uphof 1968; Usher 1974; Wiersema/León 1999; Zeven/de Wet 1982

Samara philippensis *Vidal.*

▶ *Embelia philippensis A.DC.*

SAMBUCUS L. - Elder - Caprifoliaceae

Sambucus nigra *L.*

Common Names ▶ black elder, common elder, European elder, elderberry; *Arabic:* okkez sidi moussa, bilasan, khelwan, kaman kabir; *French:* grand sureau, sureau commun, sureau noir; *German:* Fliederbeere, Holder, Hollerbeeren, Schwarzer Holunder; *Italian:* sambuco; *Portuguese:* sabugueiro negro; *Russian:* busina; *Spanish:* saúco (común), sambugo;
Usage ▶ spice, flavoring
Parts Used ▶ flower
Distribution ▶ Europe, Turkey, W Asia: Caucasus, N Iraq, W Iran, N Africa, also cultivated in Africa and Asia
Note ▶ Fresh flowers added to flavoring of omelettes.

Aichele/Schwegler 3, 1995; Becker 1981; Bendel 2002; Benk 1981; Berger 1, 1949; Bilgri/Adam 2000; Bremness 2001; Charalambous 1994; Cheers 1997; Coiciu/Racz (no year); Davidson 1999; Dudtschenko et al. 1989; Eichinger 1999; Erhardt et al. 2002; Erichsen-Brown 1989; Fleischhauer 2003; Heeger 1956; Hiller/Melzig 1999; Hoffmann et al. 1992; Hohmann et al. 2001; Hoppe 1949; Jensen/Nielsen 1973; Jorgensen et al. 2000; Newall et al. 1996; Pschyrembel 1998; Richter/Willuhn 1974; Schnelle 1999; Schönfelder 2001; Schultze-Motel 1986; Seidemann 1993c; Täufel et al. 1993; Toulemonde/Richard 1983; Turova et al. 1987; Wiersema/León 1999; Willuhn/Richter 1977; Wyk et al. 2004; Zeven/de Wet 1982

SANGUISORBA L. - Burnet - Rosaceae

Sanguisorba auriculata *Scop*

▶ *Sanguisorba officinalis L.*

Sanguisorba minor *Bertol.*

▶ *Sanguisorba minor Scop.*

Sanguisorba minor *Scop.*

Synonyms ▶ *Pimpinella minor* (Scop.) Lam., *Poteria sanguisorba* L., *Sanguisorba minor* Bertol.
Common Names ▶ burnet, salad burnet, small burnet; *French:* petite pimprenelle, pimprenelle, pimprenelle commune des prés, pimprinelle des jardins, prompenelle sanguisorbe; *German:* Kleiner Wiesenknopf, Kleine Bibernelle, Kleine Pimpinelle, Bockspetersilie, Steinpilzpetersilie, Husaren-Knopf; *Italian:* salvastrella, sanguisorba; *Russian:* tschernogolownik krowochllebkowyj; *Spanish:* pimpinela menor, pimpinela sanguisorba, salvastrella
Usage ▶ pot-herb
Parts Used ▶ fresh leaf
Distribution ▶ Europe, Turkey, Caucasus, N Iran, W Himalayas, Altai mountains, C Asia, Siberia, NW Africa, Libya; native elsewhere

Aichele/Schwegler 2, 1994; Bilgri/Adam 2000; Bremness 2001; Cheers 1997; Clair 1961; Davidson 1999; Dudtschenko et al. 1989; Erhardt et al. 2002; Fleischhauer 2003; Heeger 1956; Hiller/Melzig 1999; Hoppe 1949; Schnelle 1999; Schönfelder 2001; Schultze-Motel 1986; Seidemann 1993c; Small 1997; Teuscher 2003; Tucker 1986; Wiersema/León 1999

Sanguisorba officinalis *L.*

Synonyms ▶ *Poterium officinale* (L.) A. Gray; *Sanguisorba auriculata* Scop.,
Common Names ▶ burnet, garden burnet, great burnet; *Chinese:* ti-gü; *French:* grande pimprenelle, pimprenelle des près, sanguisorbe officinale; *German:* Blutkopf, Großer Wiesenknopf, Große Bibernelle,

Große Pimpinelle, Sperberkraut; *Italian:* salvastrella maggiore, sorbastrella, meloncella; *Japanese:* waremoko; *Russian:* krowochljobka; *Spanish:* pimpinela mayor
Usage ▶ pot-herb
Parts Used ▶ leaf
Distribution ▶ Europe, Caucasus, W, E Siberia, temperate Asia; N America, native elsewhere.

Aichele/Schwegler 2, 1994; Cheers 1997; Clair 1961; Erhardt et al. 2002; Fleischhauer 2003; Koschtschjew 1990; Heeger 1956; Hiller/Melzig 1999; Hoppe 1949; Newall et al. 1996; Nordborg 1967; Schnelle 1999; Schönfelder 2001; Schultze-Motel 1986; Seidemann 1993c; Small 1997; Täufel et al. 1993; Teuscher 2003; Tucker 1986; Usher 1968; Wyk et al. 2004; Zeven/de wet 1982

SANSEVIERIA Thunb. - Bowstring Hemp, Snake Plant - Dracaenaceae (Agavaceae)

 Sansevieria guineensis Gérome et Labroy

▶ *Sansevieria trifasciata* Prain

 Sansevieria trifasciata *Prain*

Synonyms ▶ *Aloe guineensis* Jacq., *Sansevieria guineesis* Gérome et Labroy
Common Names ▶ mother-in-law's tongue, African bowstring hemp, konje hemp, leopard lily, snake plant, tiger cat; *Chinese:* hu wei lan; *French:* chauvre d'Afrique, sansévière; *German:* Dreibändiger Bogenhanf, Schwiegermutterzunge; *Italian:* sanseveria; *Russian:* sanseverija trechputschkovaja, sansev'ra; *Spanish:* sansevieria; lengua de suegra
Usage ▶ flavoring, spice
Parts Used ▶ leaf
Distribution ▶ tropical W Africa, S Nigeria, native in E Asia
Note ▶ The littel flowers have an intense honey odor.

Agarwal 1990; Cheers 1997; Erhardt et al. 2002; Hanelt 2001; Schultze-Motel 1986; Wealth of India 9, 1972; Wiersema/León 1999; Zeven/de Wet 1982

SANTALUM L. – Sandalwood - Santalaceae

 Santalum album L.

Synonyms ▶ *Santalum myrtifolium* (L.) Roxb., *Santalum ovatum* R.Br., *Sirium myrtifolium* L.
Common Names ▶ East Indian sandalwood, white sandalwood, yellow sandalwood, white saunders; *Chinese:* chen tan, tan hsiang, tan xiang; *French:* bois santal, santal blanc, santal des Indes; *German:* Weißes Sandelholz; *Hindi:* safed candan, chandon; *Indonesian:* cendana, ai nitu, hau meni; *Italian:* sandalo blanco; *Malaysian:* chendana; *Portuguese:* sândalo blanco; *Russian:* sandalowoe derewo beloe; *Sanskrit:* candana, gandhasara, srikhanda; *Spanish:* leño de santalo citrino, sándalo blancao, sándalo Indias oríentales; *Thai:* chantana
Usage ▶ spice, flavoring e.g. ragouts; **product:** essential oil
Parts Used ▶ wood
Distribution ▶ Thailand, Indonesia, Malaysia, India, also cultivated in native range
Note ▶ Buddhists, Chinese and Muslims have used sandalwood as incense for its sweet fragrance in their ceremonies.

Barrett/Fox 1994a, b; Berger 3, 1952; Bourton 1968; Brunke/Schmaus 1995; Charalambous 1994; Cheers 1997; Dalby 2000; Dastur 1954; Davidson 1999; Erhardt et al. 2002; Harisetijono/Suriamihardja 1993; Hiller/Melzig 1999; Hoppe 1949; Kadambi 1954; Mathur 1961; Newall et al. 1996; Opdyke 1982; Oyen/Dung 1999; Piggott et al. 1997; Pschyrembel 1998; Rätsch 1998; Roth/Kormann 1997; Schultze-Motel 1986; Sharma 2003; Srinivasan et al. 1992; Wiersema/León 1999; Wüstenfeld/Haensel1964; Wyk et al. 2004; Zeven/de Wet 1982

 Santalum myrtifolium *(L.) Roxb.*

▶ *Santalum album* L.

 Santalum ovatum *R.Br.*

▶ *Santalum album* L.

SANTOLINA L. - Santolina - Asteraceae (Compositae)

Santolina rosemarinifolia *L.*

Synonyms ▶ *Santolina virens* Mill., *Santolina viridis* Willd.
Common Names ▶ green santolina; *German:* Grüne Heiligenblume, Grünes Heiligenkraut, Rosmarinblättriges Heiligenkraut
Usage ▶ flavoring
Parts Used ▶ (fresh) herb
Distribution ▶ SW Europe: Iberia, S France, Morocco
Note ▶ Sold in Algarve markets for flavoring.

Bremness 2001; Erhardt 2002; Mabberly 1997

Santolina virens *Mill.*

▶ *Santolina rosmarinifolia L.*

Santolina viridis *Willd.*

▶ *Santolina rosmarinifolia L.*

Saponaria vaccaria *L.*

▶ *Vaccaria hispanica (Mill.) Rauschert*

SAPOSHNIKOVA Schischk. - Apiaceae (Umbelliferae)

Saposhnikova divaricata *(Turcz.) Schischk.*

Synonyms ▶ *Siler divaricatum* (Turcz.) Benth. et Hook.f.
Common Names ▶ *Chinese:* fang-feng; *German:* Sperrige Saposhnikovie; *Japanese:* tosuke böfu; *Korean:* pangphung; *Russian:* sapožnikovija rasto-pyrennaja;
Usage ▶ spice, like parsley
Parts Used ▶ fresh leaves
Distribution ▶ NE China, E Siberia, Mongolia, Far Eastern Russia, Korea

Note ▶ The herb is a traditional condiment in Korea.

Baik et al. 1986; Hanelt 2001; Wiersema/León 1999

Sarothamnus scoparius *(L.) Wimmer ex Koch*

▶ *Cytisus scoparius (L.) Link*

SASSAFRAS Nees - Sassafras - Lauraceae

Sassafras albidum *(Nutt.) Nees*

Synonyms ▶ *Laurus albida* Nutt., *Sassafras officinalis* Nees et Eberm., Salvia varrifolium Salisb.
Common Names ▶ silky sassafras, cinnamo wood, fennel wood, saloop; *Arabic:* sasfras; *French:* sassafras blanc, bois de sassafras; *German:* Weißer Sassafras, Fenchelholzbaum, Nelkenzimtbaum; *Italian:* sassfrasso; *Spanish:* sasafrás
Usage ▶ spice, flavoring; **product:** essential oil (this oil has a high safrol content).
Parts Used ▶ bark, leaf
Distribution ▶ from E America and Canada to N Mexico
Note ▶ The powered leaves (*Filé-Powder*) by the Choctaw-Indians (southern USA) are used as spice for soups, sauces, stews etc. The root bark was also used in brewing root beer. The flowers were used as tea or brewed in beer.

Sassafras albidum, flowering

Alberts/Muller 2000; Berger 1, 1949; 3, 1952; Bourton 1968; Charalambous 1994; Cheers 1997; Davidson 1999; Duke et al. 2003; Erhardt et al. 2002; Erichsen-Brown 1989; Hager 7, 1995; Hiller/Melzig 1999; Hoppe 1949; Mors/Rizzini 1961; Pschyrembel 1998; Rätsch 1998; Roth/Kormann 1997; Sethi et al. 1976; Seidemann 1993c; Siewek 1990; Teuscher 2003; Tucker 1986; Uphof 1968; Wiersema/León 1999; Zwaving/Bos 1996

Sassafras officinalis Nees et M. Ebert

> *Sassafras albidum (Nutt.) Nees*

SATUREJA L. - Savory - Lamiaceae (Labiatae)

Satureja acinos (L.) Scheele

> *Acinos arvensis (Lam.) Dandy*

Satureja biflora (Buch.-Ham. ex D. Don) Briq.

Synonyms ▶ *Micromeria imbricata* Forssk.
Common Names ▶ lemon savory; *German:* Afrikanisches Zitronen-Bohnenkraut
Usage ▶ flavoring
Parts Used ▶ herb
Distribution ▶ Africa: Red Sea to Ethiopia
Note ▶ The plant has a strong citrus aroma.

Hoppe 3, 1987; Small 1997

Satureja calamintha (L.) Scheele

> *Calamintha menthifolia Host*

Satureja capitata L.

> *Thymus capitatus (L.) Hoffmanns & Link*

Satureja cuneifolia Ten.

Usage ▶ condiment, relish; **product:** essential oil
Parts Used ▶ herb
Distribution ▶ SE Europe: Croatia, Dalmatia

Milos et al. 2001; Mírjana et al. 2004

Satureja doughlasii (Benth.) Briq.

Synonyms ▶ *Micromeria doughlasii* Benth.
Common Names ▶ Indian mint, Oregon tea; *German:* Kaugummipflanze; *Spanish:* yerba buena
Usage ▶ flavoring
Parts Used ▶ leaf, herb
Distribution ▶ USA

Hoppe 2, 1977; Small 1997

Satureja grandiflora (L.) Scheele

> *Calaminthe grandiflora (L.) Moench*

Satureja hortensis L.

Synonyms ▶ *Satureja officinarum* Crantz, *Satureja viminea* Burm.
Common Names ▶ savory, garden savory, summer savory; *Arabic:* nadgh; *Chinese:* hsiang po-ho; *Dutch:* boonenkruid; *French:* sarriette annuelle, sarriette commune, savourée, sadrée, herbe de Saint Julien; *German:* (Sommer-) Bohnenkraut, Garten-Bohnenkraut, Garten-Quendel, Aalkraut, Kölle, Pfefferkraut, Jamaican savory; *Italian:* podverella, santoreggia domestica, satureia, savoreggia; *Japanese:* kidachi hakka; *Portuguese:* segurelha, alfavaca do campo, remédio do vaqueiro; *Russian:* tschaber, tschaber sadowuj, tscheber, tschobr, scheber; *Spanish:* ajedrea, ajedrea común, ajedrea de jardin, saboujas, abroso; *Turkish:* cibreska, zater
Usage ▶ spice, condiment in meat, cheese, and vegetable; **product:** essential oil
Parts Used ▶ leaf, herb (fresh, dried or frozen)
Distribution ▶ Europe, W Asia, also cultivated

Aichele/Schwegler 4, André 1998; 1995; Berger 4, 1954; Bremness

2001; Charalambous 1994; Cheers 1997; Clair 1961; Davidson 1999; Dudtschenko et al. 1989; Eger/Heine 1998: Erhardt et al. 2002; Farrell 1985; Ghannadi 2002; Guzman/Siemonsma 1999; Hanelt 2001; Hay/Waterman 1993; Heeger 1956; Herisset et al 1974; Hiller/Melzig 1999; Hondelmann 2002; Hoppe 1949; Lindberg et al. 1996; Melchior/Kastner 1974; Mikus et al. 1997; Pochljobkin 1974, 1977; Pruthi 1976; Rosengarten 1969; Roth/Kormann 1997; San Martín et al. 1973; Schönfelder 2001; Schultze-Motel 1986; Seidemann 1993c; Seidemann/Siebert 1987; Siewek 1990; Small 1997; Staesche 1972; Svoboda et al. 1990; Täufel et al. 1993; Teuscher 2003; Thieme et al. 1972; Tucker 1986; Ubillos 1989; Uphof 1968; Wiersema/León 1999; Zeven/de Wet 1982

Satureja hyssopifolia Bertol.

▶ *Satureja hortensis* L.

Satureja illyrica Host

▶ *Satureja montana* L.

Satureja juliana L.

▶ *Micromeria juliana (L.) Reichenb.*

Satureja montana L.

Synonyms ▶ *Micromeria montana* (L.) Rchb., *Satureja hyssopifolia* Bertol., *Satureja illyrica* Host, *Satureja subspicata* Bartl. ex Vis.
Common Names ▶ winter savory; *French:* sariette des montagnes, sariette vivace, sariette, savouree, poivre d'âne; *German:* Bergminze, Winter-Bohnenkraut, Karstbohnenkraut, Karstsaturei; *Italian:* santoreggia montana, erba peverella; *Russian:* tschaber tschimnij, al'pijskij tschaber, gornij tschaber; *Spanish:* hisopillo, saborija
Usage ▶ spice and culinary herb; **product:** essential oil
Parts Used ▶ leaf, herb
Distribution ▶ Europe; Crime; W Asia, cultivated in S and C Europe, also in India, S Africa and N America

Aichele/Schwegler 4, 1995; Bärtels 1997; Berger 4, 1954; Bois 1934; Bremness 2001; Charalambous 1994; Cheers 1997; Clair 1961; Davidson 1999; Dudtschenko et al. 1989; Erhardt et al. 2002; Hanelt 2001; Heeger 1956; Hiller/Melzig 1999; Konakchiev/Tsankova 2002; Lück 2004; Melchior/Kastner 1974; Mikus/Schaser 1995; Mikus et al. 1997; Milos et al. 2001; Pochljobkin 1974, 1977; Radonic/Milos 2003; Schönfelder 2001; Schultze-Motel 1986; Seidemann 1993c; Slavkovska et al. 1997, 2001; Small 1997; Täufel et al. 1993; Teuscher 2003; Thieme et al. 1972; Tucker 1986; Ubillos 1989; Uhl 2000; Uphof 1968; Wiersema/León 1999; Wyk et al. 2004; Zeven/de Wet 1982

Satureja montana L. ssp. citriodora

Common Names ▶ Slovenia citron herb, lemon-scented winter savory; *German:* Slovenisches Zitronenkraut
Usage ▶ spice; **product:** essential oil
Parts Used ▶ leaf, herb
Distribution ▶ Slovenia and other Balkan states

Satureja montana L. ssp. montana

Common Names ▶ Croatia savory; *German:* Kroatisches Bohnenkraut
Usage ▶ spice; **product:** essential oil
Parts Used ▶ leaf, herb
Distribution ▶ Croatia and other Balkan states

Radonic/Milos 2003

Satureja montana L. ssp. taurica

Common Names ▶ Crimea mountain savory; *German:* Krim-(Berg-)Bohnenkraut
Usage ▶ spice; **product:** essential oil
Parts Used ▶ leaf, herb
Distribution ▶ endemic to Crimea, cultivated in Transcaucasia

Burkill 2, 1994; Hanelt 2001; Schultze-Motel 1986

Satureja officinarum Crantz

▶ *Satureja hortensis* L.

Satureja origanoides L.

▶ *Cunila origanoides (L.) Britt.*

Satureja pilosa var. pilosa Velm

Common Names ▶ Balkan savory; *German:* Balkan-Bohnenkraut
Usage ▶ spice
Parts Used ▶ herb
Distribution ▶ Balkan mountains; Greece, Bulgaria

Konakchiev/Tsankova 2002

 Satureja repandra hort.

▶ *Satureja spicigera (K. Koch) Boiss.*

 Satureja subspicata Bartl. ex Vis

▶ *Satureja montana L.*

 Satureja spicigera (K. Koch) Boiss.

Synonyms ▶ *Micromeria alternipilosa* K. Koch, *Micromeria spicigera* K. Koch, *Satureja repanda* hort.
Common Names ▶ creeping savory; *German:* Ährentragendes Bohnenkraut, Ährentragender Salbei, Kriechendes Winterbohnenkraut; *Russian:* tschaber kolosonosnyj
Usage ▶ spice
Parts Used ▶ leaf, herb
Distribution ▶ NE Anatolia, W Caucasus to NW Iran

Bremness 2001; Erhardt et al. 2002; Hanelt 2001; Schultze-Motel 1986; Small 1997; Thieme et al. 1972

 Satureja subspicata Bartl. ex Vis.

▶ *Satureja montana L.*

 Satureja thymbra L.

Common Names ▶ Cretan savory, thryba; *Arabic:* za'atar franji, za'atar rumi; *German:* Persisches Bohnenkraut, Kretischer Thymian
Usage ▶ spice
Parts Used ▶ leaf

Distribution ▶ Mediterranean region: Greece, Sardinia, Spain, Turkey

Bärtels 1997; Bosabalidis 1990b; Capone et al. 1988; Davidson 1999; Hepper 1992; Kanias/Loukis 1992; Philianos et al. 1982, 1984; Small 1997; Teuscher 2003; Tucker 1986

 Satureja viminea Burm.

▶ *Satureja hortensis L.*

SAUSSUREA DC. - Alpine Saw Wort - Asterceae (Compositae)

 Saussurea costus (Falc.) Lipsch.

Synonyms ▶ *Saussurea lappa* (Decne.) C.B. Clarke
Common Names ▶ costus; *Chinese:* mu xiang; *German:* Gerippte Alpenscharte, Indische Costuswurzel
Usage ▶ spice
Parts Used ▶ root
Distribution ▶ N India, Kashmir, around Himalayas

Charalambous 1994; Davidson 1999; Erhardt et al. 2002; Hepper 1992; Hiller/Melzig 1999; Schultze-Motel 1986; Turora et al. 1987; Uphof 1968; Zeven/de Wet 1982

 Saussurea lappa (Decne.) C.B. Clarke

▶ *Saussurea costus (Falc.) Lipsch.*

 Scandix cerefolium L.

▶ *Anthriscus cerefolium (L.) G.F. Hoffm.*

 Scandix odorata L.

▶ *Myrrhis odorata (L.) Scop.*

SCHEFFLARIA J.R. Forst. et G. Forst.f. - Umbrella - Araliaceae

 Schefflaria aromatica *(Miq.) Harms*

Synonyms ▶ *Heptapleurum aromaticum* Seem.
Common Names ▶ lingkersap; *German:* Aromatische Schefflarie, Duft-Schefflarie, Aromatische Strahlenaralie; *Javanese:* djangjorang, klanting, sahang
Usage ▶ pot-herb
Parts Used ▶ leaf
Distribution ▶ Indonesia: Java

Hanelt 2001; Ochse/van den Brink 1931; Schultze-Motel 1986; Seidemann 1993; Uphof 1968

 Schefflaria venulosa *(Wight et Arn.) Harms*

Synonyms ▶ *Heptapleurum venulosum* Seem.
Common Names ▶ *German:* Geäderte Schefflarie; *Hindi:* karbot-sermul, kath-semul, kur-semul
Usage ▶ pot-herb
Parts Used ▶ herb
Distribution ▶ India

Burkill 1972; Erhardt et al. 2002; Wealth of India 9, 1972

SCHINUS L. - Peppertree - Anacardiaceae

 Schinus areira L.

▶ *Schinus molle* L. var. *areira* (L.) DC.

 Schinus molle L.

Synonyms ▶ *Schinus areira* L. var. *areira* (L.) DC.
Common Names ▶ „Californian" pepper, Peruvian pepper, Peruvian mastic tree, pink peppercorn; *Brazil (Portuguese):* árbol de la pimenta, aroeira, aroeira periquita, aroeira salso, corneita, pimenteiro; *French:* moleé de jardins, mollé, poivrier d'Amérique, poivrier du Pérou, faux-poivrier;

◘ Schinus molle, fruiting

German: Mollefrucht, Mollesaat, Peruanischer Pfeffer, Rosa Pfeffer, Schinuspfeffer; *Italian:* false peper; *Peruvian:* mulli, uchu; *Portuguese:* aroeira, aroeirinha, aroeira-mansa; *Russian:* schinus; *Spanish:* aguaribai, molle, pimientero falso, pirul
Usage ▶ spice
Parts Used ▶ fruit
Distribution ▶ tropical S and C America, especially Brazil, native in Mediterranean regions and tropical countries, e.g. S Africa, cultivated in W Africa
Note ▶ Colorful pepper (colored pepper, *French:* de coleurs poivre, *German:* Bunter Pfeffer) are a mixing of black, white and pink pepper.

Bärtels 1997; Bernard/Wrolstad 1963; Bernhard et al. 1983; Bois 1934; Charalambous 1994; Cheers 1997; Chialvi 1990; Davidson 1999; Erhardt et al. 2002; Hanelt 2001; Hiller/Melzig 1999; Kramer 1957; Lorenzi 1992; Melchior/Kastner 1974; Mors et al. 2000; Norman 1991; Pieribattesti et al. 1981; Pozzo-Balbi et al. 1978; Rätsch 1998; Schröder 1991; Schrutka-Rechtenstamm et al. 1988; Schultze-Motel 1986; Seidemann 1993c; Seidemann/Siebert 1987; Siewek 1990; Small 1997; Staesche 1972; Täufel et al. 1993; Teuscher 2003; Uphof 1968; Usher 1974; Villamar et al. 1994; Wiersema/León 1999; Zeven/de Wet 1982

Schinus terebinthifolius *Raddi*

Common Names ▶ Brazilian pepper, Christmas berry, Floriday holly, pink berry, pink pepper, red pepper; *French:* encens, baies roses de Bourbon, poivrier rose, sorbier, faux poivrier; *German:* Brasilianischer Pfeffer, Rosapfeffer, Turbitobaum; *Portuguese:* aroeira do brejo, aroeira do campo, aroreira do sertao, aroreira mansa, aroreira negra, aroeira precoce, terbinto; *Spanish:* copal, pimienta de Brazil

Usage ▶ sometimes equivalent to black pepper

Parts Used ▶ fruit

Distribution ▶ tropical C and S America, native in Mediterranean regions and tropical countries

Note ▶ The fruits have a skin stimulating effect (watch out!).

Bärtels 1997; Bauer/Silva 1973; Campello/Marsaioli 1974, 1975; Cheers 1997; Davidson 1999; Duke et al. 2003; Erhardt et al. 2002; Farrell 1985; Gurib-Fakim/Brendler 2004; Hanelt 2001; Hiller/Melzig 1999; Kaistha/Kier 1962; Lloyd et al. 1977; Lorenzi 1992; McNavy Wood 2003; Morris/Mackley 1999; Mors et al. 2000; Schrutka-Rechtenstamm et al. 1988; Schwenker/Skopp 1987; Seidemann 1993c; Skopp/Schwenker 1986; Skopp et al. 1987; Stahl 1982; Stahl et al 1983; Täufel et al. 1993; Teuscher 2003; Uphof 1968; Usher 1974; Wiersema/León 1999

SCHISANDRA Michx. - Schisandra - Schisandraceae

 Schisandra chinensis *(Turcz.) Baill.*

Synonyms ▶ *Kadsura chinensis* Turcz., *Maximowiczia chinensis* (Turcz.) Rupr.

Common Names ▶ Chinese magnolia vine, five-flavor fruit; *Chinese:* wu wei tzu; *German:* Chinesische Zitrone, Chinesische Limone, Chinesisches Spaltkörbchen; *Japanese:* chosen gmiski; *Korean:* omijanamu; *Russian:* limonnik kitajskij

Usage ▶ condiment; **product:** essential oil (bark)

Parts Used ▶ fruit, bark

Distribution ▶ China, E Asia, Japan, Russia (Amur region, Sakhalin), N Korea, Japan

Note ▶ The grated bark has a lemon odor with citral as a main constituent in the essential oil.

Berger 3, 1952; Dudtschenko et al. 1989; Erhardt et al. 2002; Hager 5, 1993; Hammer et al. 1997; Hiller/Melzig 1999; Kedzia 2002; Kump 2001; NICPBP 1987; Schultze-Motel 1986; Slanina et al. 1997; Turora et al. 1987; Wiersema/León 1999

 Schoenoprasum longifolium *Kunth*

▶ *Allium kunthii* G. Don

 Sclarea aethiopis *(L.) Mill.*

▶ *Salvia aethiopis* L.

 Sclarea vulgaris *Mill.*

▶ *Salvia sclarea* L.

SCORODOPHLOEUS Harms - Caesalpinniaceae (Leguminosae)

 Scorodophloeus zenkeri *Harms*

Common Names ▶ *Cameroon:* bobombi, essoun, olim, yomi; *Zaire*: bofidji, monajembe

Usage ▶ condiment

Parts Used ▶ bark

Distribution ▶ tropical Africa: Cameroon, Congo, Gabun, Zaire

Note ▶ The tree, especially the bark, has a garlic-like odor.

Burkill 3, 1995; Kouokam et al. 2002; Neuwinger 1999; Uphof 1968

SCYPHOCEPHALIUM Warb. - Myristicaceae

 Scyphocephalium mannii *(Benth.) Warb.*

Common Names ▶ *German:* Mann'sche Muskatnuss

Usage ▶ spice, like nutmeg (locally)

Parts Used ▶ fruit, seed

Distribution ▶ W Africa: Nigeria, Cameroon, Gabun

Burkill 4, 1997; Neuwinger 1999; Warburg 1897

Scyphocephalium ochocoa Warb.

Common Names ▶ ochoco nut(meg); *Cameroon:* issombo, osoko, sokwe, tsisombo; *German:* Ochoco (Muskat)nuss
Usage ▶ spice, like nutmeg (locally)
Parts Used ▶ seed
Distribution ▶ tropical Africa Cameroon, Gabun

Neuwinger 1999; Uphof 1968; Warburg 1897

SECURINEGA Comm. ex Juss. - Euphorbiaceae

Securinega leucopyrus *(Willd.) Muell. Arg.*

Synonyms ▶ *Flueggea leucopyrus* Willd.
Common Names ▶ *Hindi:* ainta, bilchuli, hartho, shinwi, tella-pulugudu, vellaippuanji, vorepuvan
Usage ▶ pot-herb (in India)
Parts Used ▶ leaf
Distribution ▶ NW India

Arora/Pandey 1996; Hanelt 2001; Schultze-Motel 1986; Wealth of India 9, 1972

SEDUM L. - Stonecrop - Crassulaceae

Sedum acre L.

Synonyms ▶ *Sedum ukrainae* hort.
Common Names ▶ goldmoss stonecrop, mossy stonecrop, wall pepper; *French:* orpin âcre, pain d'oiseau, poivre des murailles; *German:* Fetthenne, Scharfer Mauerpfeffer, Steinpfeffer, Vogelbrot; *Russian:* otschitok edkij
Usage ▶ pot-herb
Parts Used ▶ fresh herb
Distribution ▶ Europe to W Siberia, native in N America

Berger 4, 1954; Erhardt et al. 2002; Hiller/Melzig 1999; Hoppe 1949; Schönfelder 2001; Wiersema/León 1999

Sedum album L.

Common Names ▶ white stonecrop; *German:* Weißer Mauerpfeffer
Usage ▶ pot-herb
Parts Used ▶ fresh herb
Distribution ▶ Europe, Caucasus, Turkey, Lebanon, Iran, NW Africa, Libya

Erhardt et al. 2002; Fleischhauer 2003; Schnelle 1999

Sedum Alpstre *Vill.*

Common Names ▶ alpine stonecrop; *German:* Alpen-Fetthenne
Usage ▶ pot-herb
Parts Used ▶ fresh herb
Distribution ▶ Europe, Turkey

Erhardt et al. 2002; Fleischhauer 2003; Machatschek 1999; Schönfelder 2001;

Sedum annuum L.

Common Names ▶ annual stonecrop; *German:* Einjährige Fetthenne
Usage ▶ pot-herb
Parts Used ▶ fresh herb
Distribution ▶ Europe, Turkey, Caucasus, Iran, Greenland

Erhardt et al. 2002; Fleischhauer 2003; Machatschek 1999

Sedum atratum L.

Common Names ▶ blackish stonecrop, tawny stonecrop; *German:* Schwärzliche Fetthenne
Usage ▶ pot-herb
Parts Used ▶ fresh herb
Distribution ▶ Europe

Erhardt et al. 2002; Fleischhauer 2003; Machatschek 1999

Sedum cepaea L.

Common Names ▶ panicle stonecrop; *German:* Rispen-Fetthenne
Usage ▶ pot-herb
Parts Used ▶ fresh herb
Distribution ▶ Europe, Turkey, Syria, Algeria, Tunisia, Libya, native in the Netherlands, Germany

Erhardt et al. 2002; Fleischhauer 2003; Machatschek 1999

Sedum dasyphyllum L.

Common Names ▶ thick leaved stonecrop; *French:* orpin à feuilles glanduleuses; *German:* Dickblättrige Fetthenne
Usage ▶ pot-herb
Parts Used ▶ fresh herb
Distribution ▶ Europe, Romania, Turkey, NW Africa; native in Denmark,

Erhardt et al. 2002; Fleischhauer 2003; Machatschek 1999

Sedum hispanicum L.

Common Names ▶ Spain stonecrop; *German:* Spanische Fetthenne
Usage ▶ pot-herb
Parts Used ▶ fresh herb
Distribution ▶ Europe, Turkey, Lebanon, Palestine, Caucasus, N Iran; native in Sweden

Erhardt et al. 2002; Fleischhauer 2003; Machatschek 1999

Sedum reflexum L.

Synonyms ▶ *Sedum rupestre* L.
Common Names ▶ Jenny stonecrop, rock stonecrop, reflexed stonetrop, wall pepper, white stonecrop; *French:* orpin, orpin jaune, orpin réfléchie, trippe-madame, trique madame; *German:* Felsen-Fetthenne, Felsen-Mauerpfeffer, Tripmadam; *Italian:* erba grassa, erba pignola, sopravivolo dei muri; *Russian:* sajetsch'ja kanista, otschitok, otschitok otognutyi; *Spanish:* fabacrasa reflejada, siempreviva picante

Usage ▶ pot-herb to flavor soups and salads
Parts Used ▶ fresh herb
Distribution ▶ N, C Europe, Caucasus, native elsewhere

Aichele/Schwegler 2, 1994; Berger 4, 1954; Dudtschenko et al. 1989; Erhardt et al. 2002; Fleischhauer 2003; Hanelt 2001; Heeger 1956; Körber-Grohne 1989; Loch 1993; Schnelle 1999; Schönfelder 2001; Schultze-Motel 1986; Stephenson 1994; Stobart 1978; Wiersema/León 1999; Zeven/de Wet 1982

Sedum rubens L.

Common Names ▶ reddish stonecrop; *German:* Rötliche Fetthenne
Usage ▶ pot-herb
Parts Used ▶ fresh herb
Distribution ▶ Europe, Crimea, Canary Islands, Turkey, Levante, N Iran, NW Africa, Libya

Erhardt et al. 2002; Fleischhauer 2003; Machatschek 1999

Sedum rupestre L.

▶ *Sedum reflexum* L.

Sedum sexangulare L.

Common Names ▶ *French:* orpin de Boulogne; *German:* Goldmoos-Fetthenne; Milder Mauerpfeffer
Usage ▶ pot-herb
Parts Used ▶ fresh herb
Distribution ▶ Europe, except Iberia, Balkan Peninsula

Erhardt et al. 2002; Fleischhauer 2003; Loch 1993; Machatschek 1999; Schnelle 1999

Sedum spurium M. Bieb.

Common Names ▶ two row stonecrop; *French:* orpin bâtard; *German:* Kaukasus-Fetthenne
Usage ▶ pot-herb
Parts Used ▶ fresh herb
Distribution ▶ Caucasus, Turkey, N Iran, native in Europe

Erhardt et al. 2002; Fleischhauer 2003; Machatschek 1999; Schnelle 1999

Sedum telephium *L.*

Common Names ▶ *German:* Berg-Fetthenne, Purpur-Fetthenne
Usage ▶ pot-herb
Parts Used ▶ fresh herb
Distribution ▶ Europe, Turkey; native in N America

Erhardt et al. 2002; Fleischhauer 2003; Loch 1993; Machatschek 1999; Schnelle 1999; Schönfelder 2001

Sedum villosum *L.*

Common Names ▶ *German:* Behaarte Fetthenne, Sumpf-Fetthenne
Usage ▶ pot-herb
Parts Used ▶ fresh herb
Distribution ▶ Europe

Erhardt et al. 2002; Fleischhauer 2003; Machatschek 1999

Sedum ukrainae *hort.*

▶ *Sedum acre L.*

Selinum palustre *L.*

▶ *Peucedanum palustre (L.) Moench.*

Selinum sylvestre *L.*

▶ *Peucedanum palustre (L.) Moench.*

Selinum pastinaca *Crantz*

▶ *Pastinaca sativa L.*

Selinum monnieri *L.*

▶ *Cnidium monnieri (L.) Cuss. ex Juss.*

SENNA Mill. - Senna - Caesalpiniaceae (Leguminosae)

Senna obtusifolia *(L.) Irwin et Barmby*

Synonyms ▶ *Cassia obtusifolia* L., *Senna toroides* Roxb.
Common Names ▶ foetid cassia, sicklepod; *French:* casse puante, cassia fétide; *German:* Stumpfblättrige Senna; *India:* chakunota; *Javanese:* ketepeng
Usage ▶ pot-herb
Parts Used ▶ leaf
Distribution ▶ W Africa: Ghana, Liberia, Niger, Nigeria, Senegal; Mauritania; China cultivated
Note ▶ Fermented leaf is added to food as a condiment.

Burkill 3, 1995; Hager 4, 1992; Hanelt 2001; Irvine 1948

Serephidium balchanorum *(Krasch) Polj.*

▶ *Artemisia balchanorum Krasch*

Sertula alba *O. Kuntze*

▶ *Melilotus albus Medik.*

Sertula arvensis *O. Kuntze*

▶ *Melilotus officinalis Lam.*

SESAMUM L. - Sesame - Pedaliaceae

Sesamum alatum *Thonn.*

Synonyms ▶ *Sesamum capense* Burm.f.
Common Names ▶ *German:* Geflügelter Sesam; *Nigerian:* barewa, ri-din, *Sudan:* tacoutta
Usage ▶ pot-herb

SESAMUM L. - Sesame - Pedaliaceae

Parts Used ▸ dry plant
Distribution ▸ W Sudan, Ethiopia, E Africa, southern Africa

Burkill 4, 1997; Erhardt et al. 2002; Schultze-Motel 1986; Uphof 1968; Zeven/de Wet 1982

Sesamum brasiliense *Vell.*

▶ *Sesamum indicum L.*

Sesamum carpense *Burm.f.*

▶ *Sesamum alatum Thonn.*

Sesamum indicum *L.*

Synonyms ▸ *Anthadenia sesamoides* Lem., *Sesamum brasiliense* Vell., *Sesamum luteum* Retz., *Sesamum oleiferum* Moench, *Sesamum orientale* L., *Volkameria orientalis* O. Kuntze

Common Names ▸ sesame, beniseed, gingelly seed, oriental sesam, Indian sesam; *Arabic:* sem sum, simsim jelilan; *Chinese:* ching jang, hei, hu ma, zhi ma; *Dutch:* sesamzaad; *French:* sésame, till; *German:* Sesam, Indischer Sesam, Orientalischer Sesam; *India:* beni, gingelly, til; *Indonesian:* ijan; *Italian:* sesamo; *Japanese:* goma, uguma; *Korean:* hamkkae; *Portuguese:* gergelim; *Russian:* kunschut; *Slovenian:* sezam; *Spanish:* sésamo, ajonjolí; *Thai:* ngaa

Usage ▸ spice; foundation of many condiments
Parts Used ▸ seed
Distribution ▸ possible origin Ethiopia or India; widely cultivated: Middle East, China, Japan, Afghanistan, Asia minor and Iran, Turkey, Egypt, Mediterranean. area, E Africa

Arora/Pandey 1996; Bahkali et al. 1998; Bendel 2002; Bois 1934; Burkill 4, 1997; Charalambous 1994; Cheers 1997; Craze 2002; Davidson 1999; El-Tinay et al. 1976; Erhardt et al. 2002; Duke et al. 2003; Farrell 1985; Hanelt 2001; Hepper 1992; Hiller/Melzig 1999; Jomah et al. 2000; Johnson et al. 1979; Kottegoda 1994: Leung 1991; Linke 1983; Lyon 1972; Morris/Mackley 1999; Nakaumara et al. 1989; Nayar/Mehra 1970; Norman 1991; Rosengarten 1969; Schenck/Naundorf 1966; Schieberle 1996; Schultze-Motel 1986; Seidemann 1993c; Sharma 2003; Soliman et al. 1975; Täufel et al.

Sesamum indicum, flowering

1993; Teuscher 2003; Uhl 2000; Uphof 1968; Villamar et al. 1994; Weiss 1971; Westermann 1909; Wiersema/León 1999; Yen 1986; Yermanos et al. 1972; Yoshida 1994; Zeven/de Wet 1982

Sesamum luteum *Retz.*

▶ *Sesamum indicum L.*

Sesamum oleiferum *Moench*

▶ *Sesamum indicum L.*

Sesamum orientale *L.*

▶ *Sesamum indicum L.*

Sesban grandiflora *Poir.*

▶ *Sesbania grandiflora (L.) Pers.*

SESBANIA Scop. - Fabaceae (Leguminosae)

 Sesbania coccinera *Pers.*

▶ *Sesbania grandiflora (L.) Pers.*

 Sesbania grandiflora *(L.) Pers.*

Synonyms ▶ *Robinia grandiflora* L., *Sesban grandiflora* Poir., *Sesbania coccinea* Pers.
Common Names ▶ pea tree, scarlet wistera, vegetable humming bird, West Indian pea; *French:* fagotier; *German:* Großblütige Sesbanie; *Hindi:* agasti, basna; *India:* agati, males, turi; *Malaysian:* getih, kachana tur, kelur, turi; *Pilipino:* kambang-turi, katudai, kature *Thai:* khae baan tveri
Usage ▶ pot-herb
Parts Used ▶ leaf, green pod
Distribution ▶ India, Sri Lanka, Mauritius, SE Asia, N Australia, also cultivated in the Tropics of the Old and New Worlds

Agarwal 1986; Hanelt 2001; Irvine 1961; Mansfeld 1962; Schultze-Motel 1986; Uphof 1968; Wealth of India 9, 1972

 Sesbania tetraptera *Hochst. Baker*

Usage ▶ pot-herb
Parts Used ▶ leaf
Distribution ▶ tropical Africa

Uphof 1968

 Seseli amomum *Scop.*

▶ *Sison amomum L.*

 Seseli macrophyllum *Regel et Schmalh.*

▶ *Mediasia macrophylla (Regel et Schmalh.) Pimenov*

SILAUM Mill. - Pepper Saxifrage - Apiaceae (Umbelliferae)

 Silaum silaus *(L.)* **Schinz et Thell.**

Common Names ▶ pepper saxifrage; *French:* cumin dés pres, silaüs; *German:* Wiesensilge
Usage ▶ pot-herb
Parts Used ▶ fresh leaf
Distribution ▶ Europe, W Siberia

Erhardt et al. 2002; Fleischhauer 2003; Schnelle 1999

SILENE L. - Campion, Catchfly - Caryophyllaceae

 Silene inflata *Sm.*

▶ *Silene vulgaris (Moench) Garcke*

 Silene latifolia *(Mill.) Rendle et Britt.*

▶ *Silene vulgaris (Moench) Garcke*

 Silene vulgaris *(Moench) Garcke*

Synonyms ▶ *Silene inflata* Sm., *Silene latifolia* (Mill.) Rendle et Briitt.
Common Names ▶ bladder champion, blue root; *German:* Breitblättrige Lichtnelke, Gemeiner Traubenkropf; *Italian:* bubbolini, strigoli; *Russian:* smolewka chlopuschka, smolewka schirokolistnaja
Usage ▶ pot-herb (rarely)
Parts Used ▶ leaf
Distribution ▶ S Europe, Turkey, NW Africa

Hanelt 2001; Uphof 1968

 Siler montanum *Crantz*

▶ *Laserpitium silea L.*

Siler trilobium *(L.) Crantz*

▶ *Laserpitium trilobium (L.) Borkh.*

SILYBUM Vaill. ex Adans. - Milk Thistle - Asteraceae (Compositae)

 Silybum marianum *(L.) Gaertn.*

Synonyms ▶ *Carduus marianus* L.

Common Names ▶ milk thistle, hole thistle, our lady's thistle, source of silymarin, St. Mary's thistle; *Arabic:* shouk el-gamal, houk sinnari; *Dutch:* mariadistel; *French:* chardon Marie, silybe de Marie; *German:* Mariendistel; *Italian:* cardo mariano; *Korean:* ollukonggongkhwi; *Portuguese:* cardo leiteiro, cardo de María, cardosanto; *Russian.:* ostropëstro, rastoropscha; *Spanish:* cardo lechero, cardo mariano, cardo de la alameda

Usage ▶ spice, for herb spirits, pot-herb for salads (India)

Parts Used ▶ seed, leaf

Distribution ▶ native in SE and SW Europe, Near East; N Africa, widely native e.g. S Brazil

Aichele/Schwergel 4, 1995; Arora/Pandey 1996; Bärtels 1997; Braatz/Schneider 1976; Carrier et al. 2002; Cheers 1997; Czabajska et al. 1993; Davidson 1999; Diener 1994; Erhardt et al. 2002; Fleischhauer 2003; Hamid et al. 1983; Heeger 1956; Hepper 1992; Hetz et al. 1995; Hiller/Melzig 1999; Hohmann et al. 2001; Hoppe 1949; Leng-Peschlow/Strenge-1991; Merfort/Willuhn 1985; Mors et al. 2000; Pschyrembel 1998; Rätsch 1998; Schönfelder 2001; Schultze-Motel 1986; Seidemann 1993c; Sharma 2003; Täufel et al. 1993; Villamar et al. 1994; Wiersema/León 1999; Wiesel 1993; Wyk et al. 2004; Załeck/Kordana 1993; Zeven/de Wet 1982

SIMABA Aubl. - Simaroubiaceae

 Simaba paraensis *Ducke*

Usage ▶ spice, rarely (substitute for *Quassia amara L.*)
Parts Used ▶ bark
Distribution ▶ Brazil

Uphof 1968

SIMAROUBA Aubl. - Simaroubaceae

 Simarouba amara *Aubl.*

Synonyms ▶ *Simarouba glauca* Hemsley, *Simarouba officinalis* DC.

Common Names ▶ Jamaica bark, simaruba, *Brazil (Portuguese):* marupá, marupaís, marupaúba, paraíba, pe-deperdiz, simaruba; *German:* Bittere Ruhrrinde;

Usage ▶ spice (rarely), for bitter liqueurs etc.
Parts Used ▶ root bark
Distribution ▶ northern S America, Antilles; cultivated by the Kayopó Indians in Brazil. Caampo Cerrado

Hanelt 2001; Hiller/Melzig 1999; Lorenzi 1992; Mors etal. 2000; Polonsky et al. 1984; Uphof 1968; Usher 1974

 Simarouba glauca *Hmsley*

▶ *Simarouba amara Aubl.*

 Simarouba officinalis *DC.*

▶ *Simarouba amara Aubl.*

 Simarouba versicolor *St.-Hil.*

Common Names ▶ *Brazil (Portuguese):* marupaís, matabarata, paraíba, pau-paraíba
Usage ▶ spice (rarely), for bitter liqueurs
Parts Used ▶ root bark
Distribution ▶ Brazil

Mors et al. 2000

SINAPIS L. Mustard - Brassicaceae (Cruciferae)

Sinapis alba L.

Synonyms ▸ *Brassica hirta* Moench;
Common Names ▸ white mustard, yellow mustard, mustard; *Arabic:* khardal; *Chinese:* hu chieh, pai chieh; *Dutch:* mosterd; *French:* moutarde blanche, moutarde jaune; *German:* Englischer Senf, Gelb-Senf, Gewürz-Senf, Weißer Senf, Mostardkorn; *Hindi:* sufed rai; *Italian:* senape blanca; *Japanese:* shiro karashi; *Laos:* sômez sein; *Pilipino:* mustasa; *Portuguese:* mostarda blanca; *Russian:* gortschiza belaja; *Slovenian:* biela horcica; *Spanish:* mostaza blanca
Usage ▸ spice; **product** of mustard and essential oil
Parts Used ▸ seed
Distribution ▸ widely native, perhaps in Mediterranean regions and elsewhere in Eurasia
Note ▸ Trade sorts: Holland mustard, *German:* Holländischer Senf (*Sinapis alba* L. var. *bataviea* Jessen (Holland mustard); Rumänischer (Gelb-)Senf (*Sinapis alba* L. var. *melanosperma* Alef.).

 Sinapis alba, flowering

Aichele/Schwegler 3, 1995; Bärtels 1997; Bois 1934; Crasselt 1950; Craze 2002; Cui et al. 1992; Dalby 2000; Davidson 1999; Dudtschenko et al. 1989; Erhardt et al. 2002; Erichsen-Brown 1989; Gmelin 1969; Guzman/Siemonsma 1999; Hager 4, 1992; Hanelt 2001; Heeger 1956; Hermey/Ludi 1994; Hiller/Melzig 1999; Hohmann et al. 2002; Hondelmann 2002; Hoppe 1949; Leung 1991; NICPBP 1987; Pschyrembel 1998; Rosengarten 1969; Schönfelder 2001; Schröder 1991; Schultze-Motel 1986; Seidemann 1993c; Seidemann/Siebert 1987; Shanakaranarayana et al. 1971, 1972; Siebert 1990; Small 1997; Staesche 1972; Tainter/Grenis 1993; Täufel et al. 1993; Teuscher 2003; Tucker 1986; Uhl 2000; Uphof 1968; Vos/Blijleven 1988; Wiersema/León 1999; Zeven/de Wet 1982

Sinapis alba L. ssp. dissecta (Lag.) Bonn.

▸ *Sinapis dissecta* Lag.

Sinapis arvensis L.

Common Names ▸ charlock, field mustard, California rape, wild mustard; *French:* moutarde des champs, moutarde sauvage, sanve, sénevé; *German:* Ackersenf; *Italian:* senape, senape selvatica, serapino; *Portuguese:* moustarda dos campos; *Russian:* gortschiza polewaja; *Spanish:* mostaza de los campos, mostaza silvstre
Usage ▸ spice
Parts Used ▸ seed
Distribution ▸ Europe, N Africa, temperate Asia, India, widely native, propably native only in the Mediterranean region

Aichele/Schwegler 3, 1995; Bärtels 1997; Bois 1934; Dudtschenko et al. 1989; Erhardt et al. 2002; Erichsen-Brown 1989; Fleischhauer 2003; Hager 4, 1992; Hanelt 2001; Hepper 1992; Schnelle 1999; Schultze-Motel 1986; Täufel et al. 1993; Tucker 186; Wiersema/León 1999

Sinapis cernua Thunb.

▸ *Brassica cernua* (Thunb.) Forb. et Hemsl.

Sinapis dissecta Lag.

Synonyms ▸ *Sinapis alba* L. ssp. *dissecta* (Lag.) Bonn.

Common Names ▶ *German:* Gardalsenf, Zerschlitzter Senf
Usage ▶ spice (like)
Parts Used ▶ seed
Distribution ▶ from the Mediterranean areas up to the Ukraine
Note ▶ The plant can be found as a weed in flax fields.

Hanelt 2001; Heeger 1956; Melchior/Kastner 1974; Seidemann 1993c; Seidemann/Siebert 1987

Sinapis juncea L.

 Brassica juncea (L.) Czern.

Sinapis nigra L.

 Brassica nigra (L.) Koch

SIPHONOCHILUS J.W. Wood et Franks - Zingiberaceae

Siphonochilus aethiopicus *(Schweinf.) B.L. Burtt*

Synonyms ▶ *Cienskowskia aethiopica* Schweinf., *Kaempferia aethiopica* (Schweinf.) Benth., *Siphonochilus natalensis* (Schltt. et K. Schum.) Wood et Franks
Common Names ▶ African ginger, wild ginger; *French:* gingembre Africaine; *German:* Afrikanischer Ingwer, Wilder Ingwer; *Italian:* zenzero Africano; *Zulu:* isiphepketo, indungulo
Usage ▶ spice, flavoring
Parts Used ▶ rhizome, root
Distribution ▶ tropical Africa: Senegal, Niger to S Africa, E Africa
Note ▶ The rhizomes are slightly ginger in taste.

Burkill 5, 2000; Hanelt 2001; Tanaka/Nakao 1976; Uphof 1976

Silphonochiles natalensis *(Schltr. et K. Schum.) Wood et Franks*

 Silphonochilus aethiopicus (Schweinf.) B.L. Burtt.

Sirium myrtifolium L.

 Santalum album L.

SISON L. - Stone Parsley - Apiaceae (Umbelliferae)

Sison amomum L.

Synonyms ▶ *Cicuta amumum* Crantz; *Seseli amomum* Scop.; *Sium amomum* Roth; *Sium aromaticum* Lam.
Common Names ▶ hedge sison, honewort, stone parsley; *German:* Gewürzdolde, Herrnkümmel, Würzsilge; *Russian:* sison
Usage ▶ flavoring
Parts Used ▶ seed
Distribution ▶ S, SW Europe, Turkey, Caucasus

Erhardt 2002; Uphof 1968

Sisymbrium alliaria Scop.

 Alliaria petiolata (M. Bieb.) Cavara et Grande

Sisymbrium barbarea *(L.) Cr.*

 Barbarea vulgaris R.Br.

Sisymbrium nasturtium *Thunb.*

 Nasturtium officinale R.Br.

Sium amomum *Roth.*

 Sison amomum L.

Sium aromaticum *Lam.*

▶ *Sison amomum* L.

SIUM L. - Water Parsnip - Apiaceae (Umbelliferae)

Sium cicutaefolium *Schrank.*

Synonyms ▶ *Sium suave* Walter
Common Names ▶ water parsnip; *German:* Wasserpetersilie; Wasserschierlingblättriger Merk
Usage ▶ flavoring
Parts Used ▶ root, leaf (relish)
Distribution ▶ Canada, E, N America

Hanelt 2001; Uphof 1968; Wiersema/León 1999

Sium javanicum *L.* et Sium laciniatum *Bl.*

▶ *Oenanthera javanica (Bl.) DC.*

Sium suave *Walter*

▶ *Sium cicutaefolium Schrank.*

SKIMMIA Thunb. - Skimmia - Rutaceae

Skimmia arborescens *T. Anders*

Synonyms ▶ *Skimmia laureola* (DC.) Siebold et Zucc. ex Gambl.
Common Names ▶ *German:* Baumartige Skimmie, Lorbeer-Skimmie; *India:* barru, gurl pata, nehar
Usage ▶ flavoring (in curries); **product:** essential oil
Parts Used ▶ leaf
Distribution ▶ Nepal, Himalayas, Maynmar, China

Arora/Padney 1996

SMILACINA Desf. – False Salomon's Seal - Convallariaceae

Smilacina oleracea *Hook. f.*

Common Names ▶ false Salomon's seal; *French:* petit smilax; *German:* Gemüse-Schattenblume, Öl-Duftsiegel
Usage ▶ pot-herb
Parts Used ▶ herb
Distribution ▶ E Himalayas, from Nepal to Assam

Arora/Pandey 1996; Wealth of India 9, 1972

SMILAX L. - Sarsaparilla - Smilacaceae

Smilax aristolochiifolia *Mill.*

Synonyms ▶ *Smilax medica* Schlecht. et Cham.
Common Names ▶ Gray sarsaparilla, Mexican sarsaparilla, Veracruz sarsaparilla; *French:* salsepareille, seron épineux; *German:* Mexikanische Sarsaparille, Osterluzeiblättrige Stechwinde; *Italian:* salsapariglia smilace; *Russian:* sarsaparil'; *Spanish:* méjica zarzaparilla
Usage ▶ spice, flavoring (bark)
Parts Used ▶ rhizome, root, bark
Distribution ▶ C Mexico, Meso-America
Note ▶ Mexican sarsaparilla extracts have recently been used in baked goods, beverages (root beer), candies, and desserts.

Bown 1995; Hanelt 2001; Hiller/Melzig 1999; Hobbs 1988; Newall et al. 1996; Schultze-Motel 1986; Uphof 1968; Villamar et al. 1994; Wealth of India 9, 1972; Wiersema/León 1999; Wyk et al. 2004

Smilax medica *Schlecht. et Cham.*

▶ *Smilax aristolochiifolia Mill.*

SMYRNIUM L. - Alexanders - Apiaceae (Umbelliferae)

Smyrnium olusatrum L.

Common Names ▶ Alexanders, Alexander's plant, Alexandrian parsley, black lovage; *French:* maceron, ombrella jaune; *German:* Alisander, (Schwarze) Gelbdolde, Pferdeeppich, Schwarzer Liebstockel; *Italian:* erba smirnio; *Russian:* smirnija; *Spanish:* esmirnio

Usage ▶ spice

Parts Used ▶ seed

Distribution ▶ SW Europe, Algeria, Canary Islands native in S England (by Alexander the Great), Turkey

Note ▶ The seeds were used as a pepper substitute in the Middle Ages and in times of shortage. The plant was used in the 19. Century as celery (*Apium graveolens* L.).

Bärtels 1997; Bremness 2001; Davidson 1999; Erhardt et al. 2001; Small 1997; Stobart 1978; Täufel et al. 1993; Uphof 1968; Zeven/de Wet 1982

Smyrnium perfoliatum L.

Common Names ▶ biennial Alexanders; *German:* Stengelumfassende Gelbdolde

Usage ▶ spice (rarely)

Parts Used ▶ seed

Distribution ▶ Europe: Iberia, France, Italian Peninsula, Czech Republic; E, C Europe: Balkan Peninsula, Romania; Crimea, Turkey, Syria, Caucasus; native in British Isles, Denmark, C Europe

Erhardt et al. 2002; Small 1997

Soja hispida (Moench) Maxim

 *Glycine max* (L.) Merr.

SOLANUM L. - Nightshade - Solanaceae

Solanum aethiopicum L.

Synonyms ▶ *Lycopersicon aethiopicum* (L.) Mill., *Solanum gilo* Raddi

Common Names ▶ bitter berry, golden apple, scarlet egg plant, 'tomato of the Jews of Constantinople'; *Arabic:* el-tofah, el-dahabi; *French:* aubergine amère, aubergine gboma, tomate amère; *German:* Äthiopischer Nachtschatten, Äthiopische Tomate; *Nigerian:* osun

Usage ▶ pot-herb, bitter fruits as spice for sauces and soups

Parts Used ▶ leaf, fruit

Distribution ▶ tropical Africa: Ethiopia, Senegal, Sudan, Nigeria, tropical Asia

Burkill 5, 2000; Davidson 1999; Erhardt et al. 2002; Hanelt 2001; Hunziker 2001; Neuwinger 1999; Schultze-Motel 1986; Tindall 1983; Uphof 1968; Zeven/de Wet 1982

Solanum anguivi Lam.

Common Names ▶ children's tomato; *German:* Kinder-Tomate

Usage ▶ condiment for soups and sauces

Parts Used ▶ fruit

Distribution ▶ wild in tropical Africa: The Gambia, Guinea, Ivory Coast, Liberia, Nigeria, Togo; India, Sri Lanka, SE Asia

Burkill 5, 2000; Dalziel 1937; Hanelt 2001; Hunziker 2001; Irvine 1948; Schultze-Motel 1986

Solanum anomalum Thonn.

Synonyms ▶ *Solanum manni* Wright var. *compactum* Wright

Common Names ▶ children's tomato; *German:* Kinder-Tomate, Ungewöhnlicher Nachtschatten

Usage ▶ condiment (for sauces and soups), flavoring, added to stews

Parts Used ▶ herb, fruit

Distribution ▶ tropical Africa, also cultivated

Note ► The little red fruits have a bitter taste.

Hanelt 2001; Hunziker 2001; Schmelzer 1991; Schultze-Motel 1986; Uphof 1968; Zeven/de Wet 1982

Solanum anthrophagorum *Seem.*

Synonyms ► *Solanum uporo* Dunal
Common Names ► Cannibal's tomato; *Fiji:* boro dina; *German:* Kannibalen-Tomate, Menschenfresser-Tomate
Usage ► spice (sporadically)
Parts Used ► fruit
Distribution ► Pacific Islands: Fiji

Hanelt 2001; Hunziker 2001; Schenck/Naundorf 1966

Solanum dasyphyllum *Schum. et Thonn.*

Common Names ► raw nightshade; *German:* Rauher Nachtschatten
Usage ► flavoring, added to stews
Parts Used ► fruit
Distribution ► W Africa to Ethiopia and to S Africa

Erhardt et al. 2002; Hanelt 2001; Hunziker 2001

Solanum distichium *Schum. et Thonn.*

Common Names ► *German:* Gefüllter Nachtschatten
Usage ► flavoring (bitter taste), added to stews and for soups, sauces
Parts Used ► fruit
Distribution ► W Africa: Ghana, Nigeria and Uganda

Hanelt 2001; Schultze-Motel 1986

Solanum erianthum *D. Don*

Common Names ► big eggplant, China flower leaf, mullein nightshade, potato tree, tobacco tree; *Chinese:* jia yan ye shu; *German:* Große Eierpflanze, Wollblütiger Nachtschatten; *Japanese:* tabakugii, yanbaru-nasubi; *Sanskrit:* vidari; *Vietnamese:* ngoi, ca hoi, co sa lang

Usage ► flavoring, added to stews
Parts Used ► fruit
Distribution ► Mexico, S and C USA, Caribbean, Meso-America, native in India, China, Malaysia and Australia

Erhardt et al. 2002; Hanelt 2001; Hunziker 2001; Schultze-Motel 1986; Wealth of India 9, 1972; Wiersema/León 1999

Solanum gilo *Raddi*

▶ *Solanum aethiopicum L.*

Solanum lycopersicum L.

▶ *Lycopersicon esculentum Mill.*

Solanum macrocarpon L.

Common Names ► African eggplant, native eggplant, gboma egg plant; *Arabic:* bazengan africy; *Chinese:* fei zhou qie; *Dutch:* Afrikaanse aubergine; *French:* aubergine indigéne, fausse tomate, gboma, grosse angive; *German:* Afrikanische Aubergine, Großfrüchtige Aubergine; *Indonesian:* terong; *Italian:* melanzana africana; *Portuguese:* beringela africana; *Spanish:* berenjena africana
Usage ► pot-herb
Parts Used ► fresh (bitter) herb
Distribution ► E Africa, Madagascar, Mauritius, cultivated in Africa

Burkill 1965; Burkill 5 2000; Erhardt et al. 2002; Hanelt 2001; Hunziker 2001; Schultze-Motel 1986; Sn 1995; Seidemann 1995; Wiersema/León 1999; Zeven/de Wet 1982

Solanum manni *Wright* var. compactum *Wright*

▶ *Solanum anomalum Thonn.*

SOLANUM L. - Nightshade - Solanaceae

 Solanum macrocarpon, fruiting

xuan hua qie; *German:* Schraubenförmiger Nachtschatten, Spiraliger Nachtschatten
Usage ▶ condiment *(in Vietnam)*
Parts Used ▶ plant
Distribution ▶ India, N Vietnam, Laos

Hunziker 2001; Uphof 1968

 Solanum surattense Burm.f.

▶ *Solanum virgianum* L.

 Solanum uporo Dunal

▶ *Solanum anthropophagorum* Seem.

 Solanum xanthocarpum Schrad.

▶ *Solanum virginianum* L.

 Solanum virginianum L.

Synonyms ▶ *Solanum surattense* Burm.f., *Solanum xanthocarpum* Schrad.
Common Names ▶ yellow berried night-shade; *German:* Gelber Nachtschatten, Virgianischer Nachtschatten; *Hindi:* choti kateri, bhataktaiya; *Sanskrit:* kantakari, duhspar´ska, ksudra
Usage ▶ spice for Indian curries
Parts Used ▶ fruit
Distribution ▶ Asia: Indian, Indochina, Malaysia, native elsewhere
Note ▶ The fruit has a bitter taste.

Burkill 5, 2000; Chauhan 1999; Hanelt 2001; Hedayatulleh 1960; Hunziker 2001; Kottegoda 1994

Solanum nigrum L.

Common Names ▶ black nightshade, common nightshade, garden nightshade, night morella; *Chinese:* long kui, tien chieh tzu, tien pao tsao; *French:* morelle noire; *German:* Schwarzer Nachtschatten; *Hindi* makoy; *Indonesian:* leunca, ranti; *Italian:* solano nero; *Japanese:* inu-hôzuki, uwâguwâ kâtô; *Korean:* kamajung; *Portuguese:* erva-moura; *Sanskrit:* kākamāīci; *Spanish:* hierba mora, morella
Usage ▶ pot-herb
Parts Used ▶ leaf
Distribution ▶ Europe, cosmopolitical

Chauhan 1999; Erhardt 2002; Hanelt 2001; Heeger 1956; Hunziker 2001; Rätsch 1998; Schönfelder 2001; Uphof 1968;

Solanum spirale Roxb.

Common Names ▶ screw-shaped nightshade, coiled-flower nightshade, spiral nightshade; *Chinese:*

SOLENOSTEMON Thonn. - Painted Nettle - Lamiaceae (Labiatae)

Solenostemon monostachyus *(P. Beauv.) Briq.*

Common Names ▶ monkey's or hausa potato; *German:* Affenkartoffel, Einährige Buntnessel
Usage ▶ pot-herb (locally)
Parts Used ▶ leaf
Distribution ▶ tropical W and C Africa

Burkill 1985; Davidson 1999; Hanelt 2001; Schultze-Motel 1986; Täufel et al. 1993; Uphof 1968

SONNERATIA L. - Sonneratiaceae (Lythraceae)

Sonneratia caseolaris *(L.) Engl.*

Common Names ▶ red-flowered Pornupan mangrove; *German:* Käseartige Sonneratie
Usage ▶ flavoring of chutney and curries
Parts Used ▶ young fruit
Distribution ▶ Asia: Malaysia, Kalimantan, Borneo
Note ▶ The small trees common in mangrove swamps. The fruits have a cheese-like taste.

Erhardt et al. 2002; Mabberly 1997; Uphof 1968

SPARGANOPHORUS Boehm. - Asteraceae (Compositae)

Sparganophorus vaillantii *Crantz*

Usage ▶ condiment
Parts Used ▶ leaf
Distribution ▶ tropical Africa: Nigeria, Gold Coast, Cameroon, Congo, Toga, Niam-Niam, Fernando Po, and W Indies

Hanelt 2001; Uphof 1968

Spartium scoparium L.

▶ *Cytisus scoparius (L.) Link*

SPATHIPHYLLUM Schott. Peace Lily - Araceae

Spathiphyllum cannifolium *(Dryand.) Schott.*

Common Names ▶ spatheflower; *German:* Canna-Blattfahne
Usage ▶ flavoring tobacco
Parts Used ▶ leaf
Distribution ▶ Caribbean, tropical S America: Guinea, Venezuela, Colombia
Note ▶ The dried leaves have the scent of vanilla.

Erhardt et al. 2002; Uphof 1968

SPHAERANTHUS L. - Asteraceae (Compositae)

Sphaeranthus indicus L.

Common Names ▶ Indian globe thistle; *India:* atakkamaniyen, gorakh mundi, hapusa,
Usage ▶ pot-herb (in India)
Parts Used ▶ leaf
Distribution ▶ India, Indochina, Australia

Arora/Pandey 1996; Sharma 2003; Wiersema/León 1999

SPILANTHES Jacq. - Asteraceae (Compositae)

Spilanthes oleracea L.

Synonyms ▶ *Acmella oleracea* (L.) R.K. Jansen.
Common Names ▶ Brazilian cress, para cress, toothache plant; *Dutch:* Brazil vlakbloem; *French:* spilanthe des lieux humides, cresson du Brésil, cresson du Para, *German:* Parakresse, Husarenknopf (flow-

ers); *Japanese:* hokoso; *Malaysian:* pokok getang; *Portuguese:* agrião do Brasil, agrião do mato; *Russian:* kress brasil'skij; *Spanish:* berro de Pará, chisaca;

Usage ▶ spice (in Japanese dishes), pot-herb
Parts Used ▶ herb
Distribution ▶ West Indies, Brazil, India, cultivated in Japan (only the plant cultivated)

Bois 1934; Hiller/Melzig 1999; Schultze-Motel 1986; Seidemann 1993c; Täufel et al. 1993; Wiersema/León 1999; Zeven/de Wet 1982

Spirea ulmaria L.

 Filipendula ulmarie (L.) Maxim.

SPONDIAS L. - Mombin Plum - Anacardiaceae

Spondias acida Bl.

Synonyms ▶ *Poupartia dulcis* Bl.
Common Names ▶ acid mombin, *German:* Saure Balsampflaume, Saure Mombinpflaume
Usage ▶ acid flavoring
Parts Used ▶ young leaf, fruit
Distribution ▶ SE Asia

Oyen/Dung 1999

Spondias amara Lamk

 Spondias pinnata (J. Koenig ex L.f.) Kurz

Spondias mangifera Willd.

 Spondias pinnata (J. Koenig ex L.f.) Kurz

Spondias malayana Kosterm.

Synonyms ▶ *Poupartia pinnata* Blanco, *Spondias wirtgenii* Hassk.

Common Names ▶ Malaysian hog, Malaysian mombin plum; *Cambodian:* puen si, phlaè, mkak préi; *German:* Malayische Mombinpflaume; *Indonesian:* kadongdong, kloncing; *Malaysian:* amra; *Pilipino:* libás

Usage ▶ acid flavoring
Parts Used ▶ young leaf, inflorescensce, fruit
Distribution ▶ scattered in Malaysia, E Java

Guzman/Siemonsma 1999; Oyen/Dung 1999

Spondias novoguineensis Kosterm.

Common Names ▶ *Indonesian:* kanuris, ngaulo, uritchu
Usage ▶ acid flavoring
Parts Used ▶ fruit
Distribution ▶ Indonesia, New Guinea, Solomon Islands

Guzman/Siemonsma 1999; Oyen/Dung 1999

Spondias pinnata (J. Koenig ex L.f.) Kurz

Synonyms ▶ *Mangifera pinnata* J. Koenig ex L.f., *Spondias amara* Lambk., *Spondias mangifera* Willd.
Common Names ▶ wild mango, hog plum, yellow mombin, yellow plum; *Burmese:* gwe; *Cambodian*: mokak; *German:* Gelbe Balsampflaume, Mangopflaume, Wilder Mango; *Hindi:* amara, amna, amra; *Laos:* ko: k, ku: k; *Malaysian:* ĕmrah, kedondong, memberah; *Pilipino:* libás; *Sanskrit:* amrataka; *Thai:* ma-kok
Usage ▶ spice; flavoring (flower, leaf) fruit: made into chutney, pickles
Parts Used ▶ leaf, fruit
Distribution ▶ India, Himalayas, Andaman Island, Sri Lanka, Myanmar, introduced to and native in Thailand, Malaysia, Indochina
Note ▶ The ripe fruit has an odorless acid taste.

Arora/Pandey 1996; Blancke 2000; Erhardt et al. 2002; Guzman/Siemonsma 1999; Oyen/Dung 1999; Schultze-Motel 1986; Uphof 1968; Zeven/de Wet 1982

Spondias wirtgenii Hassk.

 Spondias malayana Kosterm.

STACHYS L. - Betony, Hedge Nettle, Woundwort - Lamiaceae (Labiatae)

Stachys byzantina K. Koch

Synonyms ▶ *Stachys lanata* Jacq
Common Names ▶ lamb's ears, lamb's lugs, lamb's tails, hedge nettle, wolly betony, wolly stachys; *French:* epiaire laineuse; *German:* Woll-Ziest, 'Helen von Stein'
Usage ▶ flavoring
Parts Used ▶ leaf
Distribution ▶ Asia minor, Crimea, Caucasus, N Iran

Erhardt et al. 2002; Mirza/Baher 2003

Stachys foeniculum Pursh.

▶ *Agastache foeniculum* (Pursh) Kuntze

Stachys lanata Jacq

▶ *Stachys byzantina* K. Koch

Stachys officinalis (L.) Trevis.

▶ *Betonica officinalis* L.

STACHYTARPHETA Vahl - False Vervain - Verbenaceae

Stachytarpheta cayennensis (L.C. Rich.) Schau.

Common Names ▶ Brazilian tea, rat tai vervain; *French:* petite vervaine queue de rat, petite queue de rat
Usage ▶ flavoring of sauces
Parts Used ▶ leaf
Distribution ▶ W Africa: E Cameroon, Nigeria

Burkill 5, 2000

Stachytarpheta indica (L.) Vahl

▶ *Stachytarpheta jamaicensis* Vahl

Stachytarpheta jamaicensis Vahl

Synonyms ▶ *Stachytarpheta indica* (L.) Vahl; *Valerianioides jamaicensis* (L.) Medic.; *Verbena jamaicense* L.
Common Names ▶ bastard vervain, Brazil tea, devil's coach whip, Jamaica vervain, light-blue snake, *Brazil (Portuguese):* gervao cheirosa; *Chinese:* jia ma bram; *German:* Bastard Vervaine; *Pilipino:* bolo-moros, kandi-kandi laan
Usage ▶ spice (rarely), in Java of flavoring of the stews
Parts Used ▶ leaf
Distribution ▶ Mexico, SE USA, S America, E Africa, Pempe; widely native in the Tropics e.g. Sri Lanka

Backer/van de Brinck 2, 1965; Burkill 1965; 5 2000; Hanelt 2001; Kottegoda 1994; Mors et al. 2000; Schultze-Motel 1986; Uphof 1968; Villamar et al. 1994; Weath of India 10, 1976; Wiersema/León 1999

STAPHYLEA L. - Bladder nut - Staphyleaceae

Staphylea pinnata L.

Common Names ▶ bladdernut, European bladdernut; *French:* staphylier; *German:* Klappernuss, Fiederspaltige Pimpernuss, Blasenstrauch, Paternosterstrauch; *Russian:* klakatschka, klokitschka;
Usage ▶ condiment
Parts Used ▶ flower bud
Distribution ▶ C, EC Europe: France, Italy; Turkey, Caucasus, frequently native
Note ▶ substitute for capers in Georgia and N America when in short supply.

Aichele/Schwegler 3, 1995; Dudtschenko et al. 1989; Fleischhauer 2003; Schnelle 1999; Seidemann 1993c; Teuscher 2003

STERCULIA L. - Sterculiaceae

Sterculia lanceaefolia Roxb.

Common Names ► longleafed sterculia; *German:* Langblättrige Sterkulie
Usage ► seasoning
Parts Used ► seed
Distribution ► India, SW China

Uphof 1968

Stilago bunius L.

❯ *Antidesma bunius (L.) Spreng.*

Stoechas arabica Garsault

❯ *Lavandula stoechas L.*

SUTERA Roth - Scrophulariaceae

Sutera atropurpurea (Banks) Hiern

Synonyms ► *Lyperia crocea* Ecklon, *Lyperia atropurpurea* Benth.
Common Names ► Cape saffron; *German:* Kapsafran;
Usage ► spice
Parts Used ► flower
Distribution ► S Africa: Transvaal, Namibia, Swaziland, Lesotho, Cape, Botswana
Note ► The entire flower is one of the sources of a commercial saffron (*Crocus sativus* L).

Erhardt et al. 2000; Hillard 1994; Schenck/Naundorf 1966; Seidemann 2003; Staesche 1972; Uphof 1968; Teuscher 2003

SWERTIA L. - Swertia - Gentianaceae

Swertia angustifolia Buch.-Ham. ex Wall.

Common Names ► beautiful swertia; *Chinese:* mei li zhang ya cai; *German:* Schmalblättriges Chirettakraut; Schmalblättiger Tarant
Usage ► spice, flavoring
Parts Used ► herb
Distribution ► China, India, Indochina

Berger 4, 1954; Husain et al. 1992; v. Koenen 1996; Wiersema/Léon 1999

Swertia chirata Buch.-Ham. ex Wall.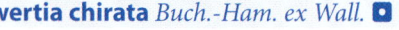

Synonyms ► *Gentiana chirayta* Roxb.
Common Names ► chirata, chireta; *German:* Chirata, Chirettakraut, Kirayakraut; *Hindi:* cirayta; *India:* chirata; *Sanskrit:* kiratatikta, bhunimba kirata
Usage ► spice, flavoring of bitter liqueurs and hard bitter spirits
Parts Used ► herb
Distribution ► Himalayan regions, cultivated in the mountains of N India (Himalayas) and Pakistan

Chauhan 1999; Erhardt et al. 2002; Hiller/Melzig 1999; Hoppe 1949; Schultze-Motel 1986; Seidemann 1993c; Sharma 2003; Uphof 1968; Wiersema/León 1999

Swertia manshurica (Komarow) Kitagawa

❯ *Swertia perennis L.*

Swertia perennis L.

Synonyms ► *Swertia manshurica* (Komarow) Kitagawa
Common Names ► felwort, marsh felwort; *Chinese:* mei wen dai, zhang ya cai; *German:* Blauer Sumpfstern, Blauer Tarant
Usage ► spice, flavoring of bitter liqueurs and bitter hard spirits
Parts Used ► herb
Distribution ► Himalayas, Europe cultivated

◘ Swertia chirata, flowering

Aichele/Schwegler 3, 1995; Erhardt et al. 2002; Täufel et al. 1993

SYZYGIUM Gaertn. - Jambos - Myrtaceae

🍃 **Syzygium aromatica** *Kuntze*

▶ *Syzygium aromaticum (L.) Merr. et L.M. Perry*

🍃 **Syzygium aromaticum** *(L.) Merr. et L.M. Perry*
◘

Synonyms ▶ *Caryophyllus aromaticus* L., *Eugenia aromatica* (L.) Baill., *Eugnia caryophyllata* Thunb. *Eugenia caryophyllus* (Spreng.) Bullock ex S.G. Harrison, *Syzygium romatica* Kuntze

Common Names ▶ clove, gilliflowers, nail of clove, Zanzibar redheads; *Arabic:* kermful, goronfel masamir, qaranful; *Cambodian:* khan phluu, khlam puu; *Chinese:* ding xiang, ting hsiang, ting tzu hsiang;

◘ Syzygium aromaticum, floral-buds

Dutch: kruidnagel; *French:* giroflé, clou de giroflé, giroflier; *German:* Gewürznelken, Nelken, Nägelein; *Hindi:* lavang, laung; *Indonesian:* cengkeh, cingkeh; *Italian:* chiòdo di garofano; *Java:* wohkaya lawqang; *Malaysian:* bunga chingkeh, chengkeh; *Laos:* do: k chan, ka: nz ph'u; *Pilipino:* klabong pako, clavo de comer; *Portuguese:* cravinto, cravo de India; *Russian:* gvosdika; *Sanskrit:* devakusuma, lavanga; *Slovenian:* klinček; *Spanish:* clavillo, clavo, clavo de especia, clavo de olor; *Sri Lanka:* karabu neti, karambu; *Thai:* garn ploo, kanphlu; *Turkish:* carenfil

Usage ▶ spice; **product:** essential oil

Parts Used ▶ floral-bud

Distribution ▶ originally Moluccas and Philippines, cultivated in tropical countries: Zanzibar, Pemba, Madagascar, Malaysia (Penang), Sri Lanka, Réunion, Mauritius, Martinique

Note ▶ The fruit, mother of clove, Anthophylli, (*German:* Mutternelken) has also been used as a spice in the original countries. More than 90%(?) of the cloves are used in tobacco to produce "kretek" cigarettes, which are smoked mainly in Indonesia. Some variation exists in and outside its centre of origin.

Berger 1, 1949; Bierther 1984; Bois 1934; Boisvert/Huber 2000; Bournot 1968; Burkill 4, 1997; Cheers 1998; Craze 2002; Dalby 2000; Davidson 1999; Deininger 1991; Dudtschenko et al. 1989; Duke et al. 2003; Erhardt et al. 2002; Farnsworth/Bunyapraphatsara 1992;

Farrell 1985; Ferrão 1992; Gopalakrishnan 1994; Gopalakrishnan/Hanti 1990; Guzman/Siemonsma 1999; Hanelt 2001; Hiller/Melzig 1999; Hohmann et al. 2001; Hoppe 1949; Koller 1979, 1981; Lawrence 1984; Lück 2000; Maistre 1964; Martin 1991; Morris/Mackley 1999; Mors/Rizini 1961; Newall et al. 1996; Norman 1991; Ochse et al. 1961; Peter 2001; Pruthi 1976; Pursglove 1968; Rosengarten 1969; Roth/Kormann 1997; Schröder 1991; Schultze-Motel 1986; Schweiheimer 1957; Seidemann 1993c, 1998/2000; Seidemann/Siebert 1987; Shriva et al. 2002; Siewek 1990; Staesche 1970; Täufel et al. 1993; Teuscher 2003; Tidbury 1949; Tschirch 1892; Vaupel 2002b; Villamar et al. 1994; Westphal/Jansen 1989; Wiersema/León 1999; Wüstenfeld/Haensel 1964; Wyk et al. 2004; Zeven/de Wet 1982

 Syzygium polyanthum *(Wight) Walp.*

Synonyms ▶ *Eugenia balsamea* Ridley, *Eugenia nitida* Duthie, *Eugenia polyantha* Wight,
Common Names ▶ salam, Indonesian bay-leaf; *Cambodian:* pring sratoab; *German:* Indischer Lorbeer, Indisches Lorbeerblatt, Friedensblatt, Salamblatt; *Indonesian:* daoen salam, daun salam; *Javanese:* manting; *Malaysian:* manting salam; *Thai:* daeng-kluai; dokmaeo, mak

Usage ▶ spice
Parts Used ▶ leaf
Distribution ▶ India: Nicobar Island, Myanmar, Indochina, Thailand, Malaysia, Indonesia: Java, Kalimantan, Sumatra
Note ▶ The use is comparable to that of *Laurus nobilis* L. leaves.

Davidson 1999; Guzman/Siemonsma 1999; Melchior/Kastner 1974; Schultze-Motel 1986; Seidemann 1993c; Seidemann/Siebert 1987; Staesche 1972; Strauß 1969d; Teuscher 2003

 Syzygium travancoricum *Gamble*

Usage ▶ flavoring; **product:** essential oil
Parts Used ▶ leaf
Distribution ▶ India
Note ▶ The essential oil has a typical aroma of raw mango.

Radha et al. 2002

T

TAGETES L. - Marigold - Asteraceae (Compositae)

 Tagestas anisata *Lillo*

▶ *Tagetes filifolia Lag.*

 Tagetes erecta *L.*

Common Names ▶ African marigold, Aztec marigold, big marigold, French marigold, saffron marygold; *Brazil:* cravo-de-defunto, rojao; *Chinese:* wan shou ju; *French:* tagète rose d'Inde; *German:* Aufrechte Studentenblume; Hohe Studentenblume; *Japanese:* senju-giku; *Mexico:* cempasóchil, cimpualxochite, cimpual, picosa, tiringuini, xkanlol; *Pilipino:* amarillo, ahito; *Portuguese:* maravilha; *Russian:* barchatcy prjamostojaschtschie; *Spanish:* flor de muerto
Usage ▶ spice, flavoring
Parts Used ▶ flower
Distribution ▶ Mexico, Meso-America, Brazil, widely cultivated, sometimes native

Aichele/Schwenker 4, 1995; Alan et al. 1968; Charalambous 1994; Dastur 1954; Erhardt et al. 2002; Espinar 1967; Heinrich 1996; Leung 1991; Mors et al. 2000; Neher 1968; Ochroa/Alosnso 1996; Schultze-Motel 1986; Seidemann 1993c; Shiva et al. 2002; Small 1997; Villamar et al. 1994; Wiersema/León 1999; Wolters 1998; Zeven/de Wet 1982

Tagetes erecta, flowering

 Tagetes filifolia *Lag.*

Synonyms ▶ *Tagetes anisata* Lillo, *Tagetes multifida* DC.
Common Names ▶ Irish lace marigold; *German:* Lakritzen-Studentenblume, Lakritzen-Tagetes; *Mexico:* flor de Santa María, hierba anís, manzanilla; *Spanish:* anicillo, tuna anís
Usage ▶ spice (flavoring)
Parts Used ▶ leaf

© Springer-Verlag Berlin Heidelberg 2005

Distribution ▶ Mexico, S America, Argentina

Espinar 1967; Ferraro 1955; Neher 1968; Small 1997; Villamar et al. 2000; Wiersema/León 1999; Wolters 1998

Tagetes glandulifera *Schrank*

▶ *Tagetes minuta* L.

Tagetes lucida *Cav.*

Common Names ▶ Mexican marigold mint, Mexican tarragon, sweet mace, sweet-scented marigold; *German:* Gewürztagetes, Glänzende Samtblume, Süße Studentenblume, "Winterestragon", Wolkenkraut; *Mexico:* pericón, quie laga zaa, yauhtli; *Russian:* barshatjstyj; *Spanish:* anisillo, pericón, periiquillo, hierba de Santa Maria
Usage ▶ condiment, flavoring
Parts Used ▶ herb
Distribution ▶ Mexico, Guatamala, Meso-America, India cultivated

Alberts/Muller 2000; Bichi et al. 1997; Bois 1934; Davidson 1999; Erhardt et al. 2002; Hiller/Melzig 1999; Neher 1968; Orth 2004; Rätsch 1998; Schultze-Motel 1986; Seidemann 1993c; Small 1997; Thappa et al. 1993; Tucker 1986; Villamar et al. 1994; Wiersema/León 1999; Wolters 1998

Tagetes lumulata *Ortega*

▶ *Tagetes patula* L.

Tagetes maxima *Kuntze*

Common Names ▶ great marigold; *German:* Große Studentenblume
Usage ▶ seasoning
Parts Used ▶ herb, flower
Distribution ▶ Mexico; S America: Bolivia, Guatamala, cultivated in Europe
Note ▶ The plant 'cempoalxóchtl' is contained in the head ornamentation of the Aztec goddess "Xoyolanhqui".

Espinar 1967; Lorenzo et al. 2002; Orth 2004

Tagetes minuta *L.*

Synonyms ▶ *Tagetes glandulifera* Schrank
Common Names ▶ Aztec marigold, dwarf marigold, stinking roger, wild marigold, Mexican marigold, Muster-John-Henry; *Brazil (Portuguese):* coari-bravo, cravo-de-defunto-miúdo; *German:* Kleine Studentenblume, Mexikanische Studentenblume, Stinkpeterle, Khakikraut; *Russian:* barchatcy melkie; *Spanish:* chinchilla enana
Usage ▶ condiment, flavoring
Parts Used ▶ flower, herb
Distribution ▶ C, S America, Brazil, cultivated and native in tropical S Europe, India

Atkinson et al. 1964; Bansal et al. 1999; Baser/Malyer 1996; Charalambous 1994; Chauhan 1999; Chisowa et al. 1998; Craveiro et al. 1988; Davidson 1999; Erhardt et al. 2002; Espinar 1967; Gardner et al. 1991; Garg/Mehta 1998; Graven et al. 1991; Handa et al. 1963; Kaul et al. 1998; v. Koenen 1996; Maria et al. 1997; Mors et al. 2000; Neher 1968; Ram et al. 1998; Rao et al. 1999, 2000; Zazdan et al. 1986; Schultze-Motel 1986; Seidemann 1993c; Sharma 2003; Singh et al. 1992, 1995, 2003; Small 1997; Thappa 1993; Wiersema/León 1999; Wolters 1998; Yannitsaros 1979; Zeven/de Wet 1982

Tagetes multifida *Lag.*

▶ *Tagetes filifolia* Lag.

Tagetes patula *L.*

Common Names ▶ French marigold, spreading marigold; *Chinese:* kong que cao, xi fan ju; *French:* oeillt d'Inde; *German:* Afrikanische Ringelblume, Abstehende Studentenblume; *Japanese:* kô-ô-sô; *Mexico:* clemole, clemolitos, iscoque, pastora, pastorcita; *Russian:* barchatcy otklonennye; *Spanish:* amapola, amarilla, copetillo
Usage ▶ spice, flavoring
Parts Used ▶ flower
Distribution ▶ Mexico, Guatamala, cultivated and native elsewhere; W Africa, Gabon; Philippines

Aichele/Schwenker 4, 1995; Bremness 2001; Burkill 1, 1985; Charalambous 1994; Dudtschenko et al. 1989; Erhardt et al. 2002; Espinar 1967; Hager 1, 1990; Hiller/Melzig 1999; Kottegoda 1994; Leung 1991; Neher 1968; Schultze-Motel 1986; Small 1997; Wiersema/León 1999; Wolters 1998; Zeven/de Wet 1982

Tagetes signata Bartl.

▶ *Tagetes tenuifolia* Cav.

 Tagetes tenuifolia Cav.

Synonyms ▶ *Tagetes signata* Bartl.
Common Names ▶ American saffron, signet marygold, slender leaf marigold, striped Mexican marigold, lemon gem, orange gem; *French:* tagète tachée; *German:* Gestreifte Mexikanische Studentenblume, Gewürz-Studentenblume, Schmalblättrige Studentenblume
Usage ▶ spice, flavoring
Parts Used ▶ flower, leaf, especially potato soups
Distribution ▶ Mexico, Ecuador, Meso-America, cultivated in Europe

Aichele/Schwenker 4, 1995; Erhardt et al. 2002; Espinar 1967; Neher 1968; Schultze-Marigold 1986; Small 1997; Wiersema/León 1999; Zeven/de Wet 1982

TAMARINDUS L. - Tamarind - Caesalpiniaceae (Leguminosae)

 Tamarindus indica L.

Synonyms ▶ *Tamarindus occidentalis* Gaertn., *Tamarindus officinalis* Hook, *Tamarindus umbrosa* Salisb.
Common Names ▶ tamarind, Indian tamarind; *Arabic:* homr, dakhar, tamre, tamre-nindi; *Chinese:* luo wang zi, suan jiao; *Dutch:* asam koening, tamarinde; *French:* tamarine, tamarinier; *German:* Tamarinde, Indische Dattel, Sauerdattel; *Hindi:* amli, anbli, imli; *Indonesian:* asem jawa; *Italian:* tamarindo; *Japanese:* tamarinde; *Javanese:* kemal, wit asem; *Malaysian:* asam jawa; *Mexico:* pachukuk, tamarindo; *Pilipino:* kamalagui, sampalik; *Portuguese:* tamarindeiro, tambarine; *Russian:* tamarind; *Slovenian:* tamarind; *Spanish:* tamarindo (indico); *Thai:* makhaam; *Vietnamese:* me, me chua;
Usage ▶ spice, condiment; **product:** essential oil (leaf)
Parts Used ▶ fruit, pulp; leaf
Distribution ▶ tropical Africa; cultivated in numerous tropical countries

◘ **Tamarindus indica:** a **flowering,** b **tamarind jam with seeds (from Egypt)**

Aké Assi/Guinko 1991; Arora/Pandey 1996; Benero et al. 1972; Berger 3, 1952; Bhattacharya et al. 1993, 1994a, b; Cheers 1997; Craze 2002; Dastur 1954; Davidson 1999; Duke et al. 2003; Erhardt et al. 2002; Farrell 1985; Giridharlai et al. 1958; Hager 1, 1990; Hanelt 2001; Hiller/Melzig 1999; Hoppe 1949; Jansen 1981; Kooinan 1961; Lee et al. 1975; Leung 1991; Lewis/Neelakantan 1964; Lück 2004; Marangoni et al. 1988; Morris/Mackley 1999; Nagaraja et al. 1975; Norman 1991; Peter 2001; Pruthi 1976; Pschyrembel 1998; Pursglove 1968; Rao 1995; Sagrero-Nieves et al. 1994; Schultze-Motel

Tamarindus indica, fruits

1986; Schwenck/Naundorf 1966; Seidemann 1993c, 1998/2000; Seidemann/Siebert 1987; Shankaracharya 1998; Sharma 2003; Siewek 1990; Staesche 1972; Stobart 1978; Storrs 1997; Strauß 1969; Täufel et al. 1993; Teuscher 2003; Tschirch 1892; Tsuda et al. 1994; Uhl 2000; Verheij/Coronel 1991; Villamar et al. 1994; Wiersema/León 1999; Zeven/de Wet 1982; Zhang/Ho 1990

 Tamarindus occidentalis *Gaertn.*

▶ *Tamarindus indica* L.

 Tamarindus officinalis *Hook.*

▶ *Tamarindus indica* L.

 Tamarindus umbrosa *Salisb.*

▶ *Tamarindus indica* L.

TANACETUM L. - Tansy - Asteraceae (Compositae)

 Tanacetum audibertii *(Req.) DC.*

▶ *Tanacetum vulgare* L.

 Tanacetum balsamite *L.*

Synonyms ▶ *Balsamita major* Desf., *Chrysanthemum balsamite* L., *Chrysanthemum majus* (Desf.) Aschers., *Pyrethrum majus* (Desf.) Tzvelev

Common Names ▶ alecost, costmary, mint geranium, sweet mary, bible leaf; *Chinese:* ju hua; *French:* balsamite, baume-coq, chrysanthème des fleuristes; *German:* Balsamkraut, Balsam-Margerite, Frauenminze, Kalufer, Marienblatt, Minzartiger Rainfarn, Römischer Balsam; *Italian:* balsamite, chrysantemo, osto; *Russian:* kalufer, kanuper, sarazinskaja mirtja, bal'samitscheskaja pjabinka; *Spanish:* chrysantema balsamita

Usage ▶ spice, flavoring for liqueurs

Parts Used ▶ herb

Distribution ▶ Turkey, Caucasus, N Iran, C Asia; naturalized in Spain, France, Italy, EC Europe, Russia; cultivated and naturalized elsewhere

Berger 4, 1954; Bois 1934; Bremness 2001; Charalambous 1994; Cheers 1997; Dalby 2000; Davidson 1999; Dudtschenko et al. 1989; Erhardt et al. 2002; Göckeritz 1968; Heeger 1956; Hoppe 1949; Pochljobkin 1974, 1977; Schönfelder 2001; Schultze-Motel 1986; Seidemann 1993c; Sharma 2003; Small 1997; Teuscher 2003; Tucker 1986; Uphof 1968; Wiersema/León 1999; Wyk et al. 2004

 Tanacetum vulgare *L.*

Synonyms ▶ *Chrysanthemum vulgare* (L.) Bernh.; *Tanacetum audibertii* (Req.) DC.

Common Names ▶ barbotine, button bitters, golden-button, common tansy, tansy; *Brazil (Portuguese):* catinga-de mulata, tasneira *French:* tanaisie vulgaire, athanase, barbotine indigène, ganelle, herbe aux vers; *German:* Rainfarn, Wurmkraut, Gänserich; *Italian:* ariceto, tanaceto, erba amara; *Portuguese:* tasneira; *Russian:* pishma; *Spanish:* atanasia, balsamite minor, hierba lombriguera, hierba de San Marcos, tanaceto

Usage ▶ spice, flavoring; **product:** essential oil; sporadically for liqueurs
Parts Used ▶ herb
Distribution ▶ temperate Asia, N America, Europe, native elsewhere
Note ▶ In the plant the thujon content is very high; therefore the plant is not used in the food industry.

Appendino 1982; Appendino et al. 1984; Berger 1, 1949; 4, 1954; Bois 1934; Bremness 2001; Charalambous 1994; Cheers 1997; Clair 1961; Collin et al. 1994; Davidson 1999; Dudtschenko et al. 1989; Erhardt et al. 2002; Heeger 1956; Hendriks et al. 1993; Hoppe 1949; Mors et al. 2000; Németh et al. 1994; Newall et al. 1996; Opdyke 1976; Pschyrembel 1998; Roth/Kormann 1997; Schiffer 1988; Schönfelder 2001; Schultze-Motel 1986; Seidemann 1993c; Sharma 2003; Small 1997; Teuscher 2003; Tucker 1986; Turova et al. 1987; Wiersema/León 1999; Wyk et al. 2004; Zeven/de Wet 1982

TARALEA Aubl. - Papilionaceae (Leguminosae)

Taralea oppositifolia *Aubl.*

Synonyms ▶ *Dipteryx oppositifolia* (Aubl.) Willd.
Common Names ▶ English tonka (bean); *German:* Englische Tonkabohne
Usage ▶ flavoring
Parts Used ▶ seed
Distribution ▶ Brazil, N, S and W America
Note ▶ Alternative substitute of the 'true' tonka bean (*Dipteryx odorata* [Aubl.] Willd.).

Schultze-Motel 1986; Seidemann/Siebert 1987; Wiersema/León 1999

TARAXACUM Web. ex Wigg. - Blowballs, Dandelion - Asteraceae (Compositae)

Taraxacum officinale *Wigg.*

Synonyms ▶ *Leontodon taraxacum* Kartsen, *Leontodon vulgare* Lam., *Taraxacum vulgare* Schrank
Common Names ▶ dandelion, lions-tooth, milk-gowan, pee in the bed, puffball; *Arabic:* tarak sahha; *French:* coq, dent le lion, laiteron, pissenlit; *German:* Butterblume, Gemeiner Löwenzahn, Kuhblume; *Hindi:* dudhal; *Italian:* capo di fratre, dente di leone, piscacane, radichiello, soffione; tarassaco; *Russian:* oduwantschik lekarstvennyj; *Sanskrit:* dugdhapheni, payasvini; *Spanish:* anagrón, diénte de león
Usage ▶ pot-herb
Parts Used ▶ fresh leaf
Distribution ▶ origin in Europe, Asia minor, widespread temperate weed; cultivated in W and C Europe, N America, India and Japan

Aichele/Schwenker 4, 1995; Bendel 2002; Berger 1, 1949; 4, 1954; Bilgri/Adam 2000; Charalambous 1994; Cheers 1997; Coiciu/Racz (no year); Davidson 1999; Dudtschenko et al. 1989; Erhardt et al. 2002; Fleischhauer 2003; Hager 4, 1992; Hanelt 2001; Hänsel et al. 1980; Hausen 1982; Heeger 1956; Hepper 1992; Hiller/Melzig 1999; Hohmann et al. 2001; Hoppe 1949; Leung 1991; Lewington 1990; Newall et al. 1996; Pschyrempel 1998; Schönfelder 2001; Schultze-Motel 1986; Seidemann 1993c; Sharma 2003; Small 1997; Täufel et al. 1993; Topf 1958; Tucker 1986; Turova et al 1987; Viillamar et al. 1994; Wiersema/León 1999; Wyk et al. 2004; Zeven/de Wet 1982

Taraxacum vulgare *Schrank*

▶ *Taraxacum officinale agg. F.H. Wigg.*

Tasmannia lanceolata *(Poir.) A.C. Sm.*

▶ *Drimys lanceolata (Poir.) Baill.*

TELFAIRIA Hook. - Cucurbitaceae

Telfairia occidentalis *Hook.f.*

Common Names ▶ fluted gourd, fluted pumkin, oyster nuts; *Chinese:* xi fei li; *German:* Faltenkürbis; Oroko, Riffelkürbis; *Portuguese:* sabina; *Spanish:* calabaza costillada
Usage ▶ pot-herb; food (oil/fat, vegetables)
Parts Used ▶ young branch and leaves
Distribution ▶ tropical E, S and W Africa

Bois 1934; Erhardt et al. 2002; Esuoso et al. 1998; Hanelt 2001; Kays/Dias 1995; Wiersema/León 1999; Zeven/de Wet 1982

TEPHROSIA Pers. - Hoary Pea - Fabaceae (Leguminosae)

 Tephrosia linearis *(Willd.) Pers.*

Common Names ▶ linear tephrosie, linear hoary pea; *French:* linéaire requiéne: *German:* Lineare Aschenwicke, Lineare Tephrosie; *Russian:* linejtschatyj tefrosia
Usage ▶ seasoning
Parts Used ▶ pulped leaves
Distribution ▶ from Senegal to Cameroon, E and S tropical Africa

Burkill 3, 1995; Hanelt 2001

 Tephrosia purpurea *(L.)* **Pers.**

Synonyms ▶ *Galega purpurea* (L.) L.
Common Names ▶ purple tephrosia, wild indigo, purple hoary pea, fish poison tree; *French:* pourpre requiénie; *German:* Wilder Indigo, Purpurne Aschenwicke, Pupurne Tephrosie, Surinam-Giftbaum; *Hindi:* bannilgach, carphonka, dhamasia; *Japanese:* nanban-fuji; *Sanskrit:* sharapunkha; *Russian:* tefrosija; *Tamil:* kolinchi
Usage ▶ flavoring (of milk in N Nigeria))
Parts Used ▶ root
Distribution ▶ tropical Africa and Asia: India, Australia

Burkill 3, 1995; Dalziel 19957; Erhardt et al. 2002; Hanelt 2001; v. Koenen 1996; Sharma 2003; Wealth of India 10, 1976; Zeven/de Wet 1982

 Terbentus lentiscus *Moench*

▶ *Pistacia lentiscus L.*

TETRAPLEURA Benth. - Fabaceae (Leguminosae)

 Tetrapleura tetraptera *(Schum. et Thonn.) Taub.*

Synonyms ▶ *Adenanthera tetraptera* Schum. et Thonn., *Tetrapleura thonningii* Benth.
Common Names ▶ *Cameroon:* akpa, dawo, essanga, sanga; *Congo:* badiok, eyaka, ezibil, kiaka; *German:* Tetrapleure; *Spanish:* frutos de enzisie
Usage ▶ spice, condiment
Parts Used ▶ fruit
Distribution ▶ W Africa, also cultivated

Adewunmi et al. 1982; Aedo et al. 2001; Essien et al. 1994; Hiller/Melzig 1999; Maillard et al. 1992; Neuwinger 1994, 1998; Ngassoum et al. 2001; Seidemann 2001b

TEUCRIUM L. - Germander - Lamiaceae (Labiatae)

 Teucrium canum *Fisch. et Mey.*

▶ *Teucrium chamaedrys L.*

 Teucrium chamaedrys *L.*

Synonyms ▶ *Chamaedrys officinalis* Moench, *Teucrium canum* Fisch. et Mey., *Teucrium leucophyllum* Benth., *Teucrium nuchense* C. Koch, *Teucrium officinale* Lam., *Teucrium pseudochamaedrys* Wendr., *Teucrium veronicae-folium* Salisb.
Common Names ▶ common germander, wall germander; *French:* calamandier petit chêne, chénette, germandrée chamaedrys, germandrée petit chêne; *German:* Edel-Gamander, Gamanderlein; *Italian:* abrotano maschino, calamandrea, querciola; *Russian:* dubrovnik purpurovyj, polyn'bosh'e derewo; *Spanish:* abrótano
Usage ▶ spice
Parts Used ▶ herb
Distribution ▶ Europe, N Africa, Caucasus, former Soviet Union, middle Asia, W Asia

TEUCRIUM L. - Germander - Lamiaceae (Labiatae)

◘ Tetrapleura tetraptera, fruits

Distribution ▶ S Europe: Croatia, Italy, Sardinia, Corsica; cultivated formerly in Germany

Beaupin et al. 1977; Bellesia et al. 1983; Berger 4, 1954; Cheers 1997; Erhardt et al. 2002; Heeger 1956; Hiller/Melzig 1999; Hoppe 1949; Schönfelder 2001; Schultze-Motel 1986; Seidemann 1993c; Zeven/de Wet 1982

Teucrium nuchense C. Koch

❯ *Teucrium chamaedrys* L.

Teucrium odorum Salisb.

❯ *Teucrium marum* L.

Teucrium officinale Lam.

❯ *Teucrium chamaedrys* L.

Aichele/Schwenker 4, 1995; Bellesia et al. 1983; Berger 4, 1954; Bremness 2001; Charalambous 1994; Cheers 1997; Dudtschenko et al. 1989; Erhardt et al. 2002; Gross et al. 1988; Heeger 1956; Hiller/Melzig 1999; Hoppe 1949; Marco et al. 1983; Schönfelder 2001; Seidemann 1993c; Wiersema/León 1999; Wyk et al. 2004; Zeven/de Wet 1982

Teucrium palustre Lam.

❯ *Teucrium scordium* L.

Teucrium pseudochamaedrys Wendr.

❯ *Teucrium chamaedrys* L.

Teucrium leucophyllum Benth.

❯ *Teucrium chamaedrys* L.

Teucrium scordioides Schreb.

❯ *Teucrium scordium* L.

Teucrium martinum Lam.

❯ *Teucrium marum* L.

Teucrium scorodium L. ◘

Teucrium marum L.

Synonyms ▶ *Chamaedry marum* Moench, *Teucrium maritinum* Lam., *Teucrium odorum* Salisb.
Common Names ▶ cat thyme; *French:* herbe mastiche, *German:* Amberkraut, Katzengamander, Katzenkraut, Moschuskraut; *Italian:* erba da gatte, maro
Usage ▶ spice
Parts Used ▶ herb

Synonyms ▶ *Teucrium palustre* Lam. *Teucrium scordioides* Schreb.
Common Names ▶ wall gamander, wall sage, wood germander; *Arabic:* bellout el-ard; *French:* germandrée scorodoine aquatique; *German:* Bergsalbei, Lauch-Salbei, Salbeigamander, Waldgamander, Waldsalbei, Wilder Gamander; *Russian:* dubrovnik skordievidny

◘ **Teucrium:** a **T. chamedrys, flowering,** b **T. scorodium, flowering**

Usage ▶ pot-herb (rarely)
Parts Used ▶ fresh herb
Distribution ▶ E, C Europe, W Siberia, C Asia, India (Kashmir)
Note ▶ The fresh herb has a garlic-like odor and taste.

Berger 4, 1954; Erhardt et al. 2002; Fleischhauer 2003; Heeger 1956; Hiller/Melzig 1999; Hoppe 1949; Jakupovic et al. 1985; Marco et al. 1983; Schönfelder 2001; Schultze-Motel 1986; Sevinate-Pinto/Antunes 1991

 Teucrium veronicae-folium Salisb.

❯ *Teucrium chamaedrys L.*

 Thlaspi bursa-pastoris L.

❯ *Capsella bursa-pastoris (L.) Medik.*

THONNINGIA Vahl - Balanophoraceae

 Thonningia dubia Hermsley

❯ *Thonningia sanguinea Vahl*

 Thonningia elegans Hermsley

❯ *Thonningia sanguinea Vahl*

 Thonningia sanguinea Vahl

Synonyms ▶ *Thonningia elegans* Hemsley; *Thonningia dubia* Hemsley
Common Names ▶ ground pineapple; *Congo:* litanda
Usage ▶ flavoring (soups)
Parts Used ▶ roots (no rhizomes!)
Distribution ▶ W Africa: Gabun, Ghana, Ivory Coast, Congo, Rwanda, Sierra Leone, Togo, Zaire

Adegoke et al. 1968; Ayensu 1978; Burkill 1, 1985; Dalby 2000; Hiller/Melzig 1999; Neuwinger 1998

 Thymbra spicata L.

❯ *Thymus capitatus (L.) Hoffmanns et Link*

THYMUS L. - Thyme - Lamiaceae (Labiatae)

 Thymus aestivus Willk.

❯ *Thymus vulgaris L.*

 Thymus angustifolius Pers.

❯ *Thymus serpyllum L.*

Thymus calamintha *Scop.*

▶ *Calamintha menthifolia Host*

 Thymus broussonetii *Boiss.*

Common Names ▶ Broussonet thyme; *German:* Broussenet-Thymian
Usage ▶ spice
Parts Used ▶ leaf, herb
Distribution ▶ N Africa: Moroccan rocks

Davidson 1999

 Thymus caespititius *Brot.*

Common Names ▶ Azores thyme, mountain thyme, tiny thyme; *German:* Azorenthymian
Usage ▶ spice, flavoring
Parts Used ▶ fresh leaves
Distribution ▶ NW Spain, Portugal, Canary Islands, Azores

Bärtels 1997; Berger 4, 1954; Bremness 2001; Davidson 1999; Guzman/Siemonsma 1999; Small 1997; Tucker 1986; Valverde 1986

 Thymus capitatus *(L.) Hoffmanns et Link*

Synonyms ▶ *Coridothymus capitatus* (L.) Rchb.f., *Satureja capitata* L., *Thymbra spicata* L., *Thymus cephalotos* L.
Common Names ▶ catir, conehead thyme, Cretean thyme, Senegal savory, Spanish oregano, zatir; *Arabic:* za'atar hommar, za'atar midbari; *French:* thym de Candie; *German:* Spanischer Oregano; *Italian:* timo arbustino; *Portuguese:* tomilho; *Spanish:* corido thyme, tomillo
Usage ▶ spice; especially essential oil (Spanish oregano oil), rich in carvacrol, used for food flavoring (baked goods, meats, ice-cream or candy), cosmetics
Parts Used ▶ flowering herb
Distribution ▶ Mediterranean region of Europe, Portugal to W and S Anatolia and to SW Asia, N Africa: Morocco to Tunesia, Israel

Akgül/Kivanç 1988; André 1998; Anon. 1993; Barberán et al. 1986; Bärtels 1997; Berger 4, 1954; Bourton 1968; Capone et al. 1988; Charalambous 1994; Cheers 1997; Fleisher/Fleisher 2002; Guzman/Siemonsma 1999; Hager 5, 1993; Hanelt 2001; Harcı et al. 2003; Hedhili et al. 2002; Herisset et al. 1974; Hiller/Melzig 1999; Hohmann 1968; Hoppe 3 1987; Kanias/Loukis 1992; Kustrak/Martinis 1990; Melchior/Kastner 1974; Miski et al. 1983; Philianos et al. 1982; Ravid/Putievsky 1985; Ruberto et al. 1992; Seidemann 1993c; Sendra/Cunat 1980a, b; Small 1997; Täufel et al. 1993; Tucker 1986; Tumen et al. 1994; Valverde 1986

 Thymus cephalotos *L.*

▶ *Thymus capitatus (L.) Hoffmanns & Link*

 Thymus chamaedrys *Fries*

▶ *Thymus pulegioides L.*

 Thymus x citriodorus *(Pers.) Schreb.*

Synonyms ▶ *Thymus pulegioides* x *Thymus vulgaris* L.
Common Names ▶ lemon thyme; *French:* thym citronné; *German:* Zitronenquendel, Zitronenthymian
Usage ▶ spice
Parts Used ▶ fresh leaves, fresh herb
Distribution ▶ the garden hybrid has been cultivated for a long time in Europe: England, Spain, Austria, Germany etc.
Note ▶ The name lemon thyme is also applied to lemon-scented chemotypes of other species e.g. bastard from *Thymus pulegioides* L. x Th. vulgaris L.) additionally orange-, coconut-, nutmeg-, lavender-, mint- and oregano-scented species and cultivars have been reported in the genus *Thymus*.

Berger 4, 1954; Bremness 2001; Cheers 1997; Erhardt et al. 2002; Guzman/Siemonsma 1999; Hanelt 2001; Hiller/Melzig 1999; Kustrak/Martinus 1990; Mikus/Schaser 1995; Schultze-Motel 1986; Seidemann 1993c; Small 1997; Stahl-Biskup/Holthuijzen 1995; Täufel et al. 1993; Teuscher 2003; Tucker 1986; Valverde 1986; Wiersema/León 1999; Wüstenfeld/Haensel 1964

 Thymus collinus *Salisb.*

▶ *Thymus vulgaris L.*

Thymus fedtschenkoi var. handelii
(Ronninger) Jalas

Usage ▸ condiment
Parts Used ▸ herb
Distribution ▸ Turkey, Asia minor

Başer et al. 2002

Thymus glandulosus Lag.

▸ *Thymus vulgaris* L.

Thymus grandiflorus (L.) Scop.

▸ *Calamintha grandiflora* (L.) Moench

Thymus herba-barona Lois.

Common Names ▸ caraway thyme; *German:* Kümmelthymian
Usage ▸ spice, especially in Great Britain for beef
Parts Used ▸ leaf
Distribution ▸ Corsica and Sardinia, cultivated in W Europe and N America

Cheers 1997; Davidson 1999; Guzman/Siemonsma 1999; Hanelt 2001; Juliano et al. 2000; Small 1997; Teuscher 2003; Tucker 1986; Usai et al. 2003; Valverde 1986; Wiersema/León 1999

Thymus hyemalis Lange

Common Names ▸ lemon thyme, winter thyme; *French:* verveine d'espagne; *German:* Spanische Verbene; Winterthymian
Usage ▸ spice; **product:** essential oil (*thyme lemon oil*)
Parts Used ▸ leaf
Distribution ▸ SE Iberian Peninsula

Adzet et al. 1976; Berger 4, 1967; Blanca et al. 1993; Cabo et al. 1986, 1987; Jordán et al. 2003; Sáez 1995; Valverde 1986

Thymus longicaulis Ronniger

Common Names ▸ *German:* Langstengeliger Thymian
Usage ▸ spice
Parts Used ▸ fresh leaf
Distribution ▸ SW Europe, SE Europe; native in Germany

Erhardt 2002; Fleischhauer 2003

Thymus mastichina L.

Synonyms ▸ *Thymus tomentosus* Willd.
Common Names ▸ mastic thyme, Spanish (wild) marjoram, Spanish thyme; *German:* Mastixthymian, Spanischer Thymian; *Portuguese:* bela luz; *Spanish:* mejorana silvestre, tomillo blanco;
Usage ▸ spice; **product:** essential oil (Spanish wild marjoram oil, with high cineol content), used in the food industry
Parts Used ▸ herb
Distribution ▸ SW Europe: Iberian Peninsula, introduced into Argentina

Bärtels 1997; Berger 4, 1954; Bourton 1968; Cheers 1997; Davidson 1999; Guzman/Siemonsma 1999; Hanelt 2001; Hiller/Melzig 1999; Mäckel 1944b; Melchior/Kastner 1974; Roth/Kormann 1987; Schultze-Motel 1997; Seidemann 1993c; Small 1997; Täufel et al. 1993; Tucker 1986; Ubillos 1989; Valverde 1986; Wiersema/León 1999

Thymus migricus Klokov et Des.-Shost

Usage ▸ condiment
Parts Used ▸ herb
Distribution ▸ Turkey, Asia minor

Başer et al. 2002

Thymus odoratissimus M. Bieb.

▸ *Thymus pallasianus* H. Braun

Thymus oenipontanus *H. Braun ex Borbás*

Common Names ▶ Austrian thyme; *German:* Österreichischer Thymian, Tiroler Thymian
Usage ▶ spice
Parts Used ▶ fresh leaf
Distribution ▶ Europe: France, Italy, Switzerland, Austria, SE Alps

Erhardt et al. 2002; Fleischhauer 2003

Thymus origanum *E.H.L. Krause*

▶ *Origanum vulgare* L.

Thymus pallasianus *H. Braun*

Synonyms ▶ *Thymus odoratissimus* M. Bieb.
Common Names ▶ *German:* Russischer Steppen-Thymian
Usage ▶ spice, specially for dessert sauces
Parts Used ▶ herb
Distribution ▶ former S Soviet Union

Bremness 2001; Erhardt et al. 2002; Teuscher 2003; Valverde 1986

Thymus pannonicus *Opiz*

Common Names ▶ *German:* Pannonischer Thymian, Steppen-Thymian
Usage ▶ spice
Parts Used ▶ fresh thyme
Distribution ▶ SE Europe: Austria, Hungary; native in Germany

Erhardt et al. 2002; Fleischhauer 2003

Thymus piperella *L.*

▶ *Micromeria piperella* Benth.

Thymus praecox *Opiz*

Common Names ▶ alba thyme, creeping thyme, hairy thyme; *French:* thym pecoce, thym rampant; *German:* Frühblühender Thymian
Usage ▶ spice, flavoring
Parts Used ▶ leaf, herb
Distribution ▶ Europe, Turkey, Caucasus, N Iran; available in Canada and the USA
Note ▶ The plant has a nutmeg scent.

Erhardt et al. 2002; Small 1997

Thymus przewalskii *(Kom.) Naka*

▶ *Thymus quinquecostatus* Čelak.

Thymus pubescens *Boiss. et Kotschy ex Celak*

Synonyms ▶ *Thymus xylorrhizus* Boiss. et Kotschy ex Boiss.
Common Names ▶ pilous thyme; *German:* Behaarter Thymian
Usage ▶ spice, flavoring; **product:** essential oil
Parts Used ▶ leaf, herb
Distribution ▶ Turkey, Iran

Sefidkon et al. 2002

Thymus pulegioides *L.*

Synonyms ▶ *Thymus chamaedrys* Fries
Common Names ▶ lemon thyme, caraway thyme, wild thyme, herba barona, Italian oregano; *German:* Arznei-Thymian, Feldthymian, Echter Quendel; Piemonteser Thymian
Usage ▶ spice; **product:** essential oil
Parts Used ▶ herb
Distribution ▶ Europe, also cultivated
Note ▶ The plant has a lemon odor.

Aichele/Schwenker 4, 1995; Berger 4, 1954; Bremness 2001; Davidson 1999; Erhardt et al. 2002; Fleischhauer 2003; Guzman/Siemonsma 1999; Hanelt 2001; Hiller/Melzig 1999; Mártonfi 1992; Mikus/Schaser 1995; Mockute/Bernotienne 1999; Schnelle 1999;

Schönfelder 2001; Senatore 1996; Small 1997; Uhl 2000; Täufel et al. 1993; Teubner 2001; Teuscher 2003; Tucker 1986; Ubillos 1989; Valverde 1986; Wiersema/León 1999

Thymus pulegioides L. x Thymus vulgare L.

▶ *Thymus x citriodorus (Pers.) Schreb.*

Thymus quinquecostatus Čelak.

Synonyms ▶ *Thymus przewalskii* (Kom.) Nakai, *Thymus quinquecostatus* var. *przewalskii* (Kom.) Ronniger, *Thymus serpyllum* L. var. *przewalskii* Kom.
Common Names ▶ Japanese thyme, five-ripped thyme; *Chinese:* di jiao; *German:* Japanischer Thymian; *Japanese:* ibuki-jakō-sō; *Korean:* paekriyhang
Usage ▶ spice, for soups, fish, eggs and meat dishes
Parts Used ▶ herb
Distribution ▶ NE and C China to Korea, Far E Russia, Japan, cultivated as a spice plant in Japan and as a medicinal plant in Korea

Fang et al. 1988; Guzman/Siemonsma 1999; Hanelt 2001; Schultze-Motel 1986; Seidemann 1993c; Small 1997; Teuscher 2003; Tucker 1986; Valverde 1986

Thymus quinquecostatus Čelak. var. przewalskii (Kom.) Ronninger

▶ *Thymus quinquecostatus Čelak.*

Thymus serpyllum L.

Synonyms ▶ *Thymus angustifolius* Pers.
Common Names ▶ creeping thyme, mother-of-thyme, wild thyme, Spanish origano; *Arabic:* zahtar, za'atar; *French:* serpolet, thym sauvage, thyme serpolet, *German:* Feldthymian, Feldpolei; Kunerle, Quendel, Sandthymian, Wilder Thymian; *Italian:* timo serpillo; *Russian:* tschabrez, tim'jan, tim'jan polsuutschik, obyknownnyj, trawa bogorodskaja; *Spanish:* sérpol
Usage ▶ spice; **product:** essential oil
Parts Used ▶ leaf, herb
Distribution ▶ C and NE Europe, Iceland, Siberia, NW Himalayas

Aichele/Schwenker 4, 1995; Arora/Pandey 1996; Arrebola et al. 1994; Berger 4, 1954; Bilgri/Adam 2000; Bremness 2001; Chalchat/Lamy 1997; Cheers 1997; Czygyan/Hänsel 1993; Davidson 1999; Dudtschenko et al. 1989; Erhardt et al. 2002; Fleischhauer 2003; Guzman/Siemonsma 1999; Hanelt 2001; Heeger 1956; Hiller/Melzig 1999; Hohmann et al. 2001; Hoppe 1949; Khan et al. 1988; Kustrak/Martinius 1990; Melchior/Kastner 1974; Oszagyan et al. 1996; Patáková/Chládek 1974; Pochljobkin 1974, 1977; Pruthi 1976; Pschyrembel 1998; Roth/Kormann 1997; Schnelle 1999; Schönfelder 2001; Seidemann 1993c; Sharma 2003; Small 1997; Täufel et al. 1993; Teubner 2001; Teuscher 2003; Tucker 1986; Turova et al. 1987; Uhl 2000; Valverde 1986; Wiersema/León 1999; Wüstenfeld/Haensel 1964

Thymus serpyllum L. var. przewalskii Kom.

▶ *Thymus quinquecostatus Čelak*

Thymus tenuifolius Mill.

▶ *Thymus vulgaris L.*

Thymus tomentosus Willd.

▶ *Thymus mastichina L.*

Thymus vulgaris L.

Synonyms ▶ *Thymus aestivus* Willk., *Thymus collinus* Salisb., *Thymus glandulosus* Lag., *Thymus tenuifolius* Mill.
Common Names ▶ common thyme, garden thyme; *Arabic:* sa'tar ramsi, za'ater; *Chinese:* ai hao, she xiang caotimjan; *Dutch:* tijm; *French:* farigoule, frigoule, thym commun, thym cultivé; *German:* Echter Thymian, Gartenthymian, Gewürzthymian, Kuttelkraut, Römischer Quendel, Thymian; *Italian:* timo, erbuccia, timo maggiore, timo volgare, pepollino; *Portuguese:* tomilho; *Russian:* tim'jan, tim'jan obyknowennyj, tim'jan duschistuj. tim'jan, timiamnik; *Spanish:* tomillo, tomillo común
Usage ▶ spice; **product:** essential oil (thyme oil) and thymol
Parts Used ▶ leaf, herb
Distribution ▶ W Mediterranean regions: Morocco, E Spain, S France, W Italy; widely cultivated in Eu-

rope and N America; also grown in other temperate and tropical areas

Note ▶ Sporadically falsification with *Thymus satureioides* Coss. et Balsana (German: Saturei-Thymian). This plant has a harsh odor and taste.

Aichele/Schwenker 4, 1995; Assouad/Valdeyron 1975; Bärtels 1997; Berger 4, 1954; Blázquez/ZafraPolo 1990; Bois 1934; Cheers 1997; Coiciu/Racz (no year); Czygan/Hänsel 1993; Davidson 1999; Delpit et al. 2000; Dommée et al. 1978; Dudtschenko et al. 1989; Echeverrigaray et al. 2001; Erhardt et al. 2002; Farrell 1985; Fleischhauer 2003; Gabel et al. 1962; Guzman/Siemonsma 1999; Hager 4, 1992; Hanelt 2001; Hay/Waterman 1993; Heeger 1956; Hiller/Melzig 1999; Hoffmann et al. 1992; Hohmann et al. 2001; Hondelmann 2002; Hudaib et al. 2002; Jackson/Hay 1994; Karawya/Hifnawy 1974; Kustrak/Martinius 1990; McGimpsey et al. 1994; Melchior/Kastner 1974; Newall et al. 1996; Nowak 1936; Opdyke 1974; Ouyon et al. 1986; Pachaly 1989; Pank/Krüger 2003c; Passet 1971; Patáková/Chládek 1974; Piccaglia/Marotti 1991; Pino et al. 1998; Pochljobkin 1974, 1977; Pruthi 1976; Rosen-garten 1969; Roth/Kormann 1997; Sameron et al. 1990; Schönfelder 2001; Schratz/Hörster 1971; Schultze-Motel 1986; Seidemann 1993c; Seidemann/Siebert 1987; Sharma 2003; Small 1997; Staesche 1972; Stahl-Biskup 1991, 2003; Stahl-Biskup/Sáez 2002; Täufel et al. 1993; Teuscher 2003; Tucker 1986; Turova et al. 1987; Ubillos 1989; Valverde 1986; Veskutonis et al. 1996c; Villamar et al. 1994; Wang et al. 1998; Weiss/Flück 1970; Wiersema/León 1999; Wyk et al. 2004; Zeven/de Wet 1982

 Thymus xylorrhizus *Boiss. et Kotschy ex Celak*

 Thymus pubescens Boiss. et Kotschy ex Celak

 Thymus zygis L.

Common Names ▶ Spanish thyme, sauce thyme; *German:* Spanischer Thymian, Südfranzösischer Thymian, Jochthymian

Usage ▶ spice; **product:** essential oil (thyme oil), source of thymol

Parts Used ▶ herb

Distribution ▶ SW Europe: native to Portugal, Spain and S France; N Africa: Morocco

Berger 4, 1954; Bourton 1968; da Cunha/Salgueiro 1991; Erhardt et al. 2002; Guzman/Siemonsma 1999; Hanelt 2001; Hiller/Melzig 1999; Mäckel 1994b; Melchior/Kastner 1974; Schönfelder 2001; Seidemann 1993c; Seidemann/Siebert 1987; Small 1997; Staesche 1972; Teuscher 2003; Tucker 1986; Ubillos 1989; Valverde 1986; Wiersema/León 1999; Youdim et al. 2002

TODDALIA Juss. - Rutaceae

 Toddalia aculeata *(Smith) Pers.*

 Toddalia asiatica (L.) Lam.

 Toddalia asiatica *(L.) Lam.*

Synonyms ▶ *Paullinia asiatica* L., *Toddalia aculeata* (Smith) Pers.; *Toddalia nitida* Lam.

Common Names ▶ Lopez fruit, wild orange tree; *Chinese:* fei long zhang zue; *German:* Lopez-Frucht, Stachelige Toddalie, Chinesische Toddalie; *Indonesian:* areuy beleketebek, duri kengkeng; *Japanese:* sara-kachû, saru-kake-mikan; *Malaysian:* akar kucing; *Pilipino:* dauag, subit, kaboat

Usage ▶ spice

Parts Used ▶ fruit

Distribution ▶ India: Nilgiri mountains, S China; Taiwan, SE Asia; Mauritius, Madagascar, Seychelles, Comoros, Mascarenes

Note ▶ Substitute for black pepper (*Piper nigrum* L.).

Gurib-Fakim/Brendler 2004; Guzman/Siemonsma 1999; Hiller/Melzig 1999; Seidemann 1993c

 Toddalia nitida LAM.

 Toddalia asiatica (L.) Lam.

 Toluifera balsamum L.

 Myroxylon balsamum (L.) Harms var. balsamum

 Toluifera balsamum var. Genuinum *Baill.*

 Myroxylon balsamum (L.) Harms var. balsamum

 Toluifera pereira *Baill.*

 Myroxylon balsamum (L.) Harms var. pereirae (Royle) Harms

Tormentilla erecta L.

> *Potentilla erecta* (L.) Räuschel

TORRESIA Allemão - Fabaceae (Leguminosae)

Torresia cearensis A.C. Smith

Synonyms ▶ *Amburana cearensis* (Fr. Alem.) A.C. Smith
Common Names ▶ *Brazil (Portuguese):* amburana, cumaru das caatingas, cumaru de cheiro
Usage ▶ flavoring; **product:** essential oil, c(o)umarin
Parts Used ▶ seed, bark
Distribution ▶ Brazil

Kumar 2001; Mors/Rizzini 1961

TORREYA Arn. - Nutmeg Yew - Taxaceae

Torreya california Torr.

Synonyms ▶ *Torreya myristica* Hook.
Common Names ▶ California nutmeg, California torreya, California yew; *German:* Kalifornische Muskatnuss, Kalifornische Nusseibe
Usage ▶ spice
Parts Used ▶ fruit
Distribution ▶ SW USA: Californian coastal region
Note ▶ The fruit has a terebinthic rather than a nutmeg-like aroma and is therefore not a nutmeg surrogate.

Brockmann 1979; Cherry 1997; Erhardt et al. 2002; Gruzman/Siemonsma 1999; Hager 5, 1993; Hiller/Melzig 1999; Rätsch 1998; Teuscher 1993; Warburg 1897; Wiersema/León 1999

Torreya myristica Hook.

> *Torreya california* Torr.

▫ Torreya california, fruits

Torreya taxifolia Arn.

Common Names ▶ Florida nutmeg, stinking cedar, stinking yew, leaved torreya; *German:* Florida Muskatnuss, Florida Nusseibe
Usage ▶ spice
Parts Used ▶ fruit
Distribution ▶ USA: Florida, Georgia
Note ▶ The fruit has a terebinthic rather than a nutmeg-like aroma and is therefore not a nutmeg surrogate.

Erhardt et al. 2002; Warburg 1897

Tournefortia argentea L.f.

> *Messerschmidia argentea* (L. f.) Johnston

Toxicondron coriaria Kuntze

> *Rhus coriaria* L.

TRACHYSPERMUM Link - Apiaceae (Umbelliferae)

 Trachyspermum ammi *(L.) Sprague*

Synonyms ▸ *Ammi copticum* L., *Bunium aromaticum* L., *Bunium copticum* Spreng., *Carum aromaticum* Spreng., *Carum copticum* (L.) Benth, *Trachyspermum copticum* (L.) Link

Common Names ▸ ajowain, bishop's weed, omum (plant), white cumin, Ethiopan caraway; *Arabic:* kamue muluki, choelle; *French:* ajowan, ammi de l'Inde; *German:* Ajowan, Adjowain, Ajowankümmel, ägyptischer Amm(e)i, Kretischer Kümmel, Schnabelsame; *Hindi:* ajovan, ajvyan; *India:* ajawa omum, ajouan, ajowan; *Malaysian:* mungsi; *Russian:* ashgon, aiowan, koptskij tmin, indijskij tmin, sira; *Sanskrit:* ajamoda, yavani

Usage ▸ spice, condiment for curries, pickles and bread

Parts Used ▸ fruit, seed

Distribution ▸ India, Pakistan, widely cultivated in S Asia, Indonesia, Near East: Jemen, Iran, SE Europe, and N Africa

Note ▸ Trade sorts in India: Desi Ajowain (large) and Nadiad Ajowain (small).

Ahmed I Ijaz et al. 1992; Ashok kumar et al. 1992; Bois 1934; Choudhury et al. 1998; Czupor 1970; Dalby 2000; Davidson 1999; Dudtschenko et al. 1989; Erhardt et al. 2002; Farrell 1985; Galakshmi et al. 2000; Hanelt 2001; Hiller/Menzel 1999; Ilyas 1980; Jansen 1981; Lockwood et al. 2002; Kambouche/El-Abed 2003; Lück 2000; Masoudi et al. 2002; Mehta/Zayas 1995; Morris/Mackley 1999; Norman 1991; Pochljobkin 1974, 1977; Pruthi 1976; Quadry/Atali 1967; Ranade 1997; Schultze-Motel 1986; Seidemann 1993c; Seidemann/Siebert 1987; Shiva et al. 2002; Siewek 1990; Small 1997; Teuscher 2003; Tucker 1986; Uphof 1968; Wagner/Hölzl 1968; Wiersema/León 1999; Zeven/de Wet 1982

 Trachyspermum copticum *(L.) Link*

▸ *Trachypermum ammi (L.) Sprague*

 Trachyspermum involucratum *(Roxb.) Wolff*

▸ *Trachyspermum roxburghianum (DC.) Craib*

 Trachyspermum roxburghianum *(DC.) Craib*

Synonyms ▸ *Apium involuncratum* Roxb. ex Flem., *Carum involucratum* (Roxb.) Baill., *Carum roxburghianum* Benth., *Trachyspermum involucratum* (Roxb.) Wolff

Common Names ▸ randhuni; *French:* làjmud des Indiens; *Hindi:* ajmud, ajmuda; *Indonesian:* surage, *Javanese:* pletikapu; *Pilipino:* kanuikui, malungkoi, *Thai:* phakchi-lom

Usage ▸ spice, seeds are used as a condiment for curries, chutneys, preserves, and pickles, and as a culinary herb

Parts Used ▸ seeds, herb, leaves

Distribution ▸ India cultivated, and widely cultivated in tropical Asia

Note ▸ The leaves are a culinary herb, a substitute for parsley.

Bois 1934; Davidson 1999; Erhardt et al. 2002; Guzman/Siemonsma 1999; Hanelt 2001; Schultze-Motel 1986; Seidemann 1993c; Täufel et al. 1993; Teuscher 2003; Tucker 1986; Uphof 1968; Zeven/de Wet 1982

TRECULIA Decne. ex Trecul - African Breadfruit - Moraceae

 Treculia africana *Decne.*

Common Names ▸ African bread fruit, African boxwood; *Cameroon:* boembe, bongo, bwembi, etoup, pusa, ziba; *Congo:* toum, bleblendou; *French:* arbre à pain d'Afrique; *German:* Afrikanische Brotfrucht, Okwa; *Portuguese:* saquente; *W Africa:* afon

Usage ▸ spice for sauces, soups, meat, and for aromatization of alcoholic drinks

Parts Used ▸ seed

Distribution ▸ tropical Africa, from Senegal to Mozambique, Madagascar

Note ▸ The fruits are cooked in the food and then removed.

Ayensu 1978; Bijttebier 1993; Burkill 4, 1997; Erhardt et al. 2002; Giami et al. 2000; Hanelt 2001; Lück 2000; Neuwinger 1999;

Schultze-Motel 1986; Seidemann 1995e; Täufel et al. 1993; Uphof 1968; Usher 1974; Wiersema/León 1999

TRIANTHEMA L. - Aizoaceae

Trianthema hydaspicea Edgew.

Usage ▶ pot-herb (in India)
Parts Used ▶ leaf
Distribution ▶ NW India

Arora/Pandey 1996

TRIBULUS L. - Zygophyllaceae

Tribulus alatus Delile

Common Names ▶ caltrap; *French:* herse; *German:* Geflügelter Burzeldorn; *Hindi:* gokhru-kalan; *Russian:* jakorzi
Usage ▶ pot-herb
Parts Used ▶ young plant
Distribution ▶ NW India

Arora/Padney 1996; Rätsch 1998

Trifolium album Lois.

▶ *Melilotus albus* Medik.

Trifolium altissimum Lois.

▶ *Melilotus altissimus* Thuill.

Trifolium coeruleum Moench

▶ *Trigonella caerulea* (L.) Sér.

Trifolium corniculata L.

▶ *Trigonella corniculata* (L.) L.

Trifolium officinalis L.

▶ *Melilotus officinalis* Lam.

Trifolium petitpierreanum Hayne

▶ *Melilotus officinalis* Lam.

Trifolium vulgare Hayne

▶ *Melilotus albus* Medik.

TRIGONELLA L. - Fenugreek - Fabaceae (Leguminosae)

Trigonella arabica Delile

Synonyms ▶ *Trigonella pecten* Schenk
Common Names ▶ Arabian fenugreek; *German:* Arabischer Bockshornklee
Usage ▶ spice
Parts Used ▶ seed
Distribution ▶ N Africa, especially in Arabia, Syria to NE Egypt

Agarwal 1986; Hanelt 2001; Wealth of India 10, 1976

Trigonella caerulea (L.) Sér.

Synonyms ▶ *Melilotus caeruleus* Desr., *Trifolium caeruleum* Moench., *Trigonella melilotus coerulea* (L.) Aschers. et Graebn.
Common Names ▶ sweet trifoil; *French:* baumier, mélilot bleu; trèfle blau, trèfle musque, trèfle odorant; *German:* Balsamklee, Bisamklee, Blauer Steinklee, Käseklee, Schabziegerklee, Zigeunerklee; *Italian:* melilotto azzurro; *Russian:* donnik cinij, goluboj

donnik, pashitnik goluboj, gun'ba, sinij kosij trilistnik

Usage ▶ spice (cheese and bread)

Parts Used ▶ seed, herb

Distribution ▶ E Mediterranean region, SE Europe; origin in the Mediterranean region, also cultivated and native

Note ▶ classified into a wild (ssp. *procumbens*) and a cultivated subspecies (ssp. *caerulea*).

Aichele/Schwegler 2, 1994; Berger 4, 1954; Bois 1934; Davidson 1999; Dudtschenko et al. 1989; Erhardt et al. 2002; Hanelt 2001; Heeger 1956; Hoppe 1949; Pochljobkin 1974, 1977; Seidemann 1993c; Täufel et al. 1993; Teuscher 2003; Wiersema/León 1999; Zeven/de Wet 1982

 Trigonella caerulea *(L.) Sér.* **ssp. caerulea**

Common Names ▶ sweet trefoil; *French:* baumier, trèfle bleu, trèfle musqué, trèfle odorant; *German:* Balsamklee, Blauer Steinklee, Schabziegerklee; *Italian:* balsamo, meliloto azzurro, *Russian:* pashitnik goluboj

Usage ▶ spice, condiment, flavoring

Parts Used ▶ herb

Distribution ▶ C, W and S Europe, N Africa, widely cultivated in gardens

Note ▶ The dried pulverized herb is used for the preparation or greenish herb cheese and for flavoring bread and other dishes. This species is classified into a wild (ssp. *procumbens*) and a cultivated subspecies (ssp. *caerulea*).

Erhardt et al. 2002; Hanelt 2001; Heeger 1956; Seidemann 1993c; Zeven/de Wewt 1982

 Trigonella corniculata *(L.) L.*

Synonyms ▶ *Medicago corniculata* (L.) Trautv., *Trifolium corniculata* L.

Common Names ▶ clustered trefoil; *German:* Traubiger Bockshornklee; *Hindi:* kasuri methi; *Sanskrit:* malya, piring sak

Usage ▶ spice, pot-herb

Parts Used ▶ seed

Distribution ▶ Mediterranean region, Near East countries

◘ **Trigonella foenum-graecum:** a **flowering,** b **seeds**

Agarwal 1986; Erhardt et al. 2002; Hanelt 2001; Heeger 1956; Mansfeld 1986; Täufel et al. 1993; Teuscher 2003; Wealth of India 10, 1976

Tainter/Grenis 1993; Täufel et al. 1993; Teuscher 2003; Tucker 1986; Wiersema/León 1999; Wyk et al. 2004; Zeven/de Wet 1982

 Trigonella esculenta *Willd.*

> *Trigonella corniculata (L.) L.*

 Trigonella foenum-graecum *L.*

Synonyms ▸ *Foenum-graecum officinale* Moench, *Trigonella graeca* St. Lag.
Common Names ▸ fenigrec, fenugreek, goat's horn, Greek clover, Greek hay-seed; *Arabic:* helbeh, hulbah; *Chinese:* hu lu ban; *Dutch:* fenegriek; *French:* fénugrec, foin grec, sénégrain, senegré, trigonelle; *German:* Griechischer Bockshornklee; Griechisches Heu, Kuhhorn, Kuhhornklee; Ziegenhornklee; *Hindi:* methi; *India:* methi ni bhaji; *Indonesian:* kelabet, klabet; *Italian:* fieno greco; *Korean:* horopha, khŭnnorangkkotjariphul; *Malaysian:* halba, kelabat; *Portuguese:* fenacho, fenogreco; *Russian:* fenugrek, ili pashitnik, fenum-grek, fenigrekowa trawa, gretscheskoe seno, kosij trilistnik; *Sanskrit:* methi, methika, pitabija; *Slovenian:* senovka grécka; *Spanish:* alholva, fenogreco, heno griego, *Turkish:* çemen otu; bouï tochouma
Usage ▸ spice
Parts Used ▸ seed
Distribution ▸ Caucasus, Ex-Soviet Union, C Asia, E Europe, cultivated and native elsewhere
Note ▸ Often classified into varieties *hausknechtii* Sierjaev and provar. *foenum-graecum*. Pre-historic remains from 4000–3000 BC in the Near East. Seeds preserved in Tutankhamun's tomb (1325 BC).

Aichele/Schwegler 2, 1994; Alagukannan et al. 1999; Bendel 2002; Bilgri/Adam 2000; Billaud/Adrian 2001; Bois 1934; Bremness 2001; Charalambous 1994; Clair 1961; Coiciu/Racz (no year); Craze 2002; Dastur 1954; Davidson 1999; Dudtschenko et al. 1989; Duke et al. 2003; Erhardt et al. 2002; Farrell 1985; Faril/Hardman 1968; Girardon et al 1986; Gupta et al. 1986; Guzman/Siemonsma 1999; Hanelt 2001; Heeger 1956; Hepper 1992; Hiller/Melzig 1999; Hohmann et al. 2001; Hoppe 1949; Kamal et al. 1987; Mazza/Oomah 2000; Morris/Mackley 1999; Newall et al. 1996; Norman 1991; Pochljobkin 1974, 1977; Pruthi 1976; Pschyrembel 1998; Rosengarten; Rouk/Mengesha 1963; Schenck/Naundorf 1966; Schönfelder 2001; Seidemann 1993c; Seidemann/Siebert 1987; Shalini Hooda/Sudesh Jood 2003; Sharma 2003; Siewek 1990; Small 1997; Staesche 1972;

 Trigonella graeca *St. Lag.*

> *Trigonella foenum-graecum L.*

 Trigonella melilotus caerulea *Aschers. et Graebn.*

> *Trigonella caerula (L.) Sér.*

 Trigonella pecten *Schenk*

> *Trigonella arabica Delile*

 Trigonella stellata *Forsk.*

Common Names ▸ *German:* Stern-Bockshornklee
Usage ▸ spice (rarely); **product:** essential oil
Parts Used ▸ seed, herb
Distribution ▸ N Africa: Arabia, Egypt, Tunisia, Algeria, Morocco, Canary Islands; W Asia: Iran, Iraq, Middle East, Israel, Lebanon, Kuwait

TRIPHASIA Lour. - Lime berry - Rutaceae

 Triphasia aurantiola *Lour.*

> *Triphasia trifolia (Burm.f.) P. Wilson*

 Triphasia trifolia *(Burm.f.) P. Wilson*

Synonyms ▸ *Limonia trifolia* Burm., *Limonia trifoliata* L., *Triphasia aurantiola* Lour.
Common Names ▸ lime berry, trifoliate lime, Chinese lime, myrtle lime; *Hindi:* chini naranghi; *Malaysian:* kelingket, limau kiah, limau kingkit; *Pilipino:* kalamansito, limoncito, sua-sua, tagimunau; *Spanish:* limoncito; *Thai:* manao tet
Usage ▸ succade (flavoring)

Parts Used ▶ fruit

Distribution ▶ tropical Asia, cultivated in many subtropical and tropical countries

Erhardt et al 2002; Hanelt 2001; Schultze-Motel 1986

TRITONIA Ker.-Gawl. Tritonia - Iridiaceae

 Tritonia aurea *(Hook.) Planch.*

Synonyms ▶ *Crocosmia aurea* (Pappe ex Hook.) Planch.
Common Names ▶ flame freesia, Cape saffron; *German:* Safranartige Tritonie, Kap-Safran
Usage ▶ spice
Parts Used ▶ flower
Distribution ▶ South Africa: Capeland
Note ▶ This plant frequent with *Crocosmia aurea (Pappe ex Hook.) Planch.*

Erhardt et al. 2002; Seidemann 2003

 Tritonia x crocosmiiflora *(Burbridge et Dean) N.E. Br.*

▶ *Crocosmia crocosmiiflora (Lemoine ex E. Morr.) N.E. Br.*

 Tritonia crocata *(L.) Ker-Gawl.*

▶ *Crocosmia aurea Planch.*

TROPAEOLUM L. - Nasturtium - Tropaeolaceae

 Tropaeolum dentatifolium *Stokes*

▶ *Tropaeolum minus L.*

 Tropaeolum elatum *Salisb.*

▶ *Tropaeolum majus L.*

◘ **Tropaeolum majus, flowering**

 Tropaeolum majus *L.* ◘

Synonyms ▶ *Tropaeolum elatum* Salisb., *Tropaeolum repandifolium* Stokes
Common Names ▶ Indian cress, (common) nasturtium, tall nasturtium; *Chinese:* chin lien hua, han lian hua; *French:* capucine grande, cresson d'Inde, cresson du Pérou, capucine; *German:* Große Kapuzinerkresse, Indianerkresse, Salatblume, Türkische Kresse; *Italian:* nasturzio comune, cappucina, fior, crescione del Perù; *Japanese:* nasutachûmu, kinrenka, nôzen haren; *Portuguese:* chagas, mastruço do Perú, flor de sangue, capuchinha grande; *Russian:* kapuzin, kapuzin-kress, indejckij kress; nasturzija; *Spanish:* capuchina, nasturcio, mastranzo de las Indias, Ilagas de Cristo
Usage ▶ substitute for capers (buds); flavoring leaves and ornamental flowers for salads
Parts Used ▶ bud, leaf, flower
Distribution ▶ S America: Columbia, Ecuador, Peru; cultivated in numerous countries

Aichele/Schwegler 3, 1995; Bendel 2002; Berger 1, 1949; Bois 1934; Bremness 2001; Chamisso 1987; Cheers 1998; Davidson 1999; Dudtschenko et al. 1989; Eichhorn/Winterhalter 2004; Erhardt et al. 2002; Fleischhauer 2003; Franz 1996; Halbeisen 1954; Hanelt 2001; Hiller/Melzig 1999; Lewington 1990; Lück 2000; Lykkesfeldt/Lindberg-Møller 1993; Pochljobkin 1974, 1977; Pschyrembel 1998; Schönfelder 2001; Schultze-Motel 1986; Seidemann 1993c; Seidemann/Siebert 1987; Small 1997; Staesche 1972; Täufel et al. 1993; Teuscher 2003; Tucker 1986; Villamar et al. 1994; Wiersema/León 1999; Wolters 1994; Wyk et al. 2004; Zeven/de Wet 1982

 Tropaeolum minus L.

Synonyms ▶ *Tropaeolum dentatifolium* Stokes, *Tropaeolum pulchellum* Salisb.
Common Names ▶ bush nasturtium, dwarf nasturtium, Indian cress; *French:* capucine; *German:* Kleine Kapuzinerkresse
Usage ▶ spice (locally); **product:** essential oil
Parts Used ▶ leaf
Distribution ▶ NW Latin America: Ecuador, Peru, also cultivated

Erhardt et al. 2002; Franz 1996; Hanelt 2001; Schultze-Motel 1986; Small 1997; Wiersema/León 1999

 Tropaeolum mucronatum Meyen

▶ *Tropaeolum tuberosum* Ruiz et Pav.

 Tropaeolum pulchellum Salisb.

▶ *Tropaeolum minus* L.

 Tropaeolum repandifolium Stokes

▶ *Tropaeolum majus* L.

 Tropaeolum tuberosum O. Kuntze

▶ *Tropaeolum tuberosum* Ruiz. et Pav.

 Tropaeolum tuberosum Ruiz et Pav.

Synonyms ▶ *Tropaeolum mucronatum* Meyen, *Tropaeolum tuberosum* O. Kuntze
Common Names ▶ bulbous nasturtium, tuberous nasturtium; *Chinese:* kuai jing lzan hua; *Dutch:* knof, Oostindische kers; *French:* capucine tubéreuse; *German:* Knollenkresse, Knollen-Kapuziner-kresse, Peruanische Kapuzinerkresse, Maca; *Italian:* nasturzio tuberoso; *Portuguese:* capuchinha tuberosa; *S America:* anu, maca, magua, mashua, maxua; *Spanish:* capuchina tuberosa, mashua, ysaño
Usage ▶ pot-herb
Parts Used ▶ young leaves
Distribution ▶ Bolivia, Peru; cultivated in the Andes of Chile to Columbia
Note ▶ In the Andes the edible rhizome tubers are cultivated

Cheers 1998; Erhardt et al. 2002; Hanelt 2001; Johns et al. 1982; Ramallo et al. 2004; Schultze-Motel 1986; Seidemann 2003; Small 1997; Täufel et al. 1993; Wiersema/León 1999; Zeven/de Wet 1982

TULBAGHIA L. - Society Garlic - Alliaceae (Liliaceae)

 Tulbaghia violacea Harv.

Common Names ▶ society garlic, wild garlic, wild knoflook; *Dutch:* wilde knoflok, wilde knoffel; *German:* Knoblauchs-Kaplilie, Veilchenblütiger Knoblauch, Zimmerknoblauch, Zimmerschnittlauch
Usage ▶ spice (locally)
Parts Used ▶ leaf
Distribution ▶ S Africa
Note ▶ The taste is like garlic. The Zulu of Natal often plant this species around their huts.

Cheers 1998; Davidson 1999; Erhardt et al. 2002; Small 1997; Teubner 2001; Wyk 1997; Zeven/de Wet 1982

TURNERIA L. - Turneraceae

 Turneria aphrodisiaca G.H. Ward

▶ *Turneria diffusa* Willd. ex Schult.

 Turneria diffusa Willd. ex Schult.

Synonyms ▶ *Turnera aphrodisiaca* G.H. Ward, *Turnera microphylla* Desv. ex Ham.

Common Names ► damiana, Mexican holly; *French:* damiana, thé bourrique; *German:* Schmalblättrige Damiana, Damianastrauch; *Italian:* damiana; *Mexico:* chac-mixib, chat; *Portuguese:* damiana; *Spanish:* damiana de Guerrero, misibcoc, hierba de la pastora

Usage ► flavoring (sweets, candy, liqueur among other drinks: 'Guadalajara' *(Mexico)*

Parts Used ► fruit

Distribution ► subtropical S America Mexico, Caribbean, Meso-America

Note ► Already in use by the Maya.

Alberts/Muller 2000; Auterhoff/Hänsel 1968; Berger 2, 50; Charalambous 1994; Erhardt et al. 2002; Hanelt 2001; Hiller/Melzig 1999; Mors et al. 2000; Newall et al. 1996; Ochoa/Alonso 1996; Pschyrembel 1998; Rätsch 1992; 1998; Spencer/Seigler 1980; Täufel et al. 1993; Villamar et al. 1994; Wiersema/León 1999; Wyk et al. 2004

Turneria microphylla Desv.

▶ *Turneria diffusa* Willd. ex Schult.

Tylostemon mannii Staph.

▶ *Beilschidia mannii (Meisn.) Benth. et Hook. f.*

TYPHA L. Bullrush, Reedmace - Typhaceae

Typha latifolia L.

Synonyms ► *Typha angustifolia* A. Rich., *Typha major* Curtis

Common Names ► bullrush, cat tail, marsh beetle, reedmace; *French:* massette à feuilles étroites, jonc de marais, quenoille; *German:* Breitblättriger Rohrkolben, Echter Rohrkolben; *Italian:* biodo, sala, schiancia, stianca; *Russian:* rogos; *Spanish:* enea junco

Usage ► spice of pancake (doughnut)

Parts Used ► pollen

Distribution ► Europe, Caucasus, Iran, W, E Sibiria, C Asia, Mongolia, N China, N America, Australia

Hanelt 2001; Hepper 1992; Mansfeld 1962; Schultze-Motel 1986; Stowart 1978

U

ULVA L.
Ulvaceae (Chlorophytae)

Ulva compressa L.

Synonyms ▶ *Enteromorpha compressa* (L.) Grev.
Common Names ▶ bowel seaweed; *Chinese:* kan-thai; *German:* Darmtang; *Japanese:* hira-ao-nori
Usage ▶ spice (for fish and meat)
Parts Used ▶ thalli
Distribution ▶ E Asia, cultivated in Japan
Note ▶ "Awonori" and "Okashi" are typical *Ulva* commercial products.

Hanelt 2001; Schultze-Motel 1986

UMBELLULARIA (Nees) Nutt. – California Bay or Laurel - Lauraceae

Umbellularia californica (Hook. et Am.) Nutt.

Common Names ▶ California bay, California laurel, pepper wood, headache tree, Oregon myrtle; *German:* Bergloorbeer, Kalifornischer (Berg-)Lorbeer, Oregon-Myrte
Usage ▶ condiment (use sparingly)
Parts Used ▶ leaf
Distribution ▶ NW and SW USA (California, Oregon)
Note ▶ Substitute for laurel, specially by the Hispanic inhabitants of America.

Bois 1932; Cheers 1998; Erhardt et al. 2002; Hiller/Melzig 1999; Small 1997; Tucker 1986; Uphof 1968

UNCARIA Schreb. - Gambier - Rubiaceae

Uncaria acida O. Kuntze

▶ *Uncaria gambir* (Hunter) Roxb.

Uncaria gambir (Hunter) Roxb.

Synonyms ▶ *Uncaria acida* O. Kuntze
Common Names ▶ gambir, gambier, pale catechu; *Chinese:* kou teng, tiao teng; *French:* gambier; *German:* Gambirpflanze, Betelbissen, Gelbes Katechu; *Malaysian:* gambir
Usage ▶ flavoring
Parts Used ▶ leaf and twig; **product**: juice (by boiling)
Distribution ▶ Malaysia, Sumatra, W Borneo, also cultivated in Indonesia
Note ▶ Thalli are extracted, thickened and dried (= gambir, gambir catechu).

Charalambous 1994; Erhardt et al. 2002; Hager 4, 1992; Hanelt 2001; Hiller/Melzig 1999; Mabberly 1997; Schenck/Nauendorfer 1966; Schultze-Motel 1986; Täufel et al. 1993; Uphof 1968; Wiersema/León 1999; Zeven/de et 1982

UNDARIA W.F.R.Suringar - Alariaceae

 Undaria distans *Miyabe et Okamura*

> *Undaria pinnatifida (W.H. Harvey) W.F.R. Suringar*

 Undaria pinnatifida *(W.H. Harvey) W.F.R. Suringar*

Synonyms ▶ *Alaria pinnatifida* Harvey, *Undaria distans* Miyabe et Okamura
Common Names ▶ wakame; *Chinese:* qun dai cai; *German:* Flügeltang; *Japanese:* kada-me, niki-me, wakame;
Usage ▶ spice, specially of rice
Parts Used ▶ thallus, fresh and dry
Distribution ▶ commercial cultivated in Japan, Korea, China
Note ▶ This is the most significantly cultivated *Undaria*-species.

Davidson 1999; Hanelt 2001; Jurkovic et al. 1995; Lück 2004; Schultze-Motel 1986; Seidemann 1993c; Uphof 1968; Urbano/Goñi 2002

 Undaria undarioides *(Yendo) Okamura*

Synonyms ▶ *Hirome undarioides* Yendo
Common Names ▶ hiroma; *German:* Japanischer Flügeltang; *Japanese:* hirome
Usage ▶ spice for rice
Parts Used ▶ thallus
Distribution ▶ Japan, Korea, China

Hanelt 2001; Schultze-Motel 1986

 Unona aethiopica *Dun.*

> *Xylopia aethiopica A. Rich.*

UROPHYLLUM Jack ex Wall. - Rubiaceae

 Urophyllum arboreum *(Reinw. ex Bl.) Korth.*

Common Names ▶ *Indonesian:* ki cengkeh
Usage ▶ spice
Parts Used ▶ bruised leaves
Distribution ▶ W Java
Note ▶ The odor smells strongly of cloves.

Guzman/Siemonsma 1999; Hiller/Melzig 1999

 Urtica melastomoides *Poir.*

> *Pilea melastomoides (Poir.) Wedd.*

 Uvaria aromatica *Lam.*

> *Xylopia aromatica (Lam.) Mart.*

 Uvaria odorata *Lam.*

> *Cananga odorata (DC.) Hook.f. et Thomps.*

VACCARIA Wolf - Cow herb - Caryophyllaceae

Vaccaria hispanica (Mill.) Rauschert

Synonyms ▶ *Saponaria vaccaria* L., *Vaccaria pyramidata* Medik., *Vaccaria vulgaris* Host.
Common Names ▶ cow cockle, cow herb, dairy pink; *Arabic:* foul el-'arab; *Chinese:* wang pu liu hsing; *French:* herbe aux vaches, saponaire des vaches; *German:* Kuhkraut; *Russian:* tysjatschegolov
Usage ▶ spice
Parts Used ▶ herb
Distribution ▶ Europe, Levante, Turkey, Iran, temperate Asia: Mongolia, China, N Africa; native in N America, Australia

Erhardt et al. 2002; Fleischhauer 2003; Schnelle 1999; Wiersema/León 1999

Vaccaria pyramidata Medik.

▶ *Vaccaria hispanica* (Mill.) Rauschert

Vaccaria vulgaris Host.

▶ *Vaccaria hispanica* (Mill.) Rauschert

VALERIANA L. Valerian - Valerianaceae

Valeriana celtica L.

Common Names ▶ Celtic nard, Celtic valerian; *German:* Echter Speik, Keltischer Baldrian, Keltische Narde, Gelber or Roter Speik
Usage ▶ flavoring
Parts Used ▶ root
Distribution ▶ S, W Europe: Austria, France, Italy and the Swiss Alps

Dalby 2000; Heeger 1956; Hiller/Melzig 1999; Uphof 1968

Valeriana officinalis L.

Synonyms ▶ *Valeriana pinnata* Gilib., *Valeriana sylvestris* Sadl., *Valeriana vulgaris* Rupr.
Common Names ▶ all-heal, common valerian, garden valerian, cat's valerian, setwall, valerian; *Chinese:* xie cao; *French:* herbe aux chats, guérittout, valériane (officinale); *German:* Arznei-Baldrian, Baldrian, Gewöhnlicher Baldrian, Katzenwurzel; *Italian:* amantilla, erva gatta, valeriana, valeriana silvestre; *Japanese:* kanakoso, kesso; *Korean:* yakpaguniamul; *Russian:* valeriana aptetschnaja; *Spanish:* valeriana (mayor)
Usage ▶ spice (rarely), flavoring; **product:** essential oil; (rarely in the liqueur industry)
Parts Used ▶ root
Distribution ▶ Europe, native and cultivated, Asia
Note ▶ The root was used as a condiment during medi-

eval times, and as a perfume during the XVI Century. At present it is used as a perfume in some oriental countries.

Aichele/Schwegler 3, 1995; Alberts/Muller 2000; Becker 1983, 1986; Bos et al. 1997; Bremness 2001; Cheers 1998; Clair 1961; Coiciu/Racz o. J.; Dudtschenko et al. 1989; Erhardt et al. 2000, 2002; Hanelt 2001; Hänsel/Schulz 1982; Heeger 1956; Hiller/Melzig 1999; Hoffmann et al. 1992; Hohmann et al. 2001; Honerlagen 1988; Hoppe 1949; Houghton 1988, 1994; Mayer 2003; Mazza/Ohham 2000; Newall et al. 1996; Pschyrembel 1998; Rätsch 1998; Roth/Kormann 1997; Schier/Schultze 1989; Schönfelder 2001; Schultze-Motel 1986; Seidemann 1993c; Sharma 2003; Titz/Titz 1982; Tucker 1986; Wagner et al. 1970; Wiersema/León 1999; Wüstenfeld/Haensel 1964; Wyk et al. 2004; Zeven/de Wet 1982

Valeriana pinnata *Gilib.*

> *Valeriana officinalis L.*

Valeriana sisymbrifolia *Schur.*

Synonyms ▶ *Valeriana tripteris* L.
Common Names ▶ Syrian nard; *German:* Syrische Narde, Narde
Usage ▶ flavoring; **product:** essential oil
Parts Used ▶ leaf
Distribution ▶ Turkey: Anatolia; Caucasus, Iran, Syria; C Asia

Valeriana sylvestris *Sadl.*

> *Valeriana officinalis L.*

Valeriana tripteris *L.*

> *Valeriana sisymbrifolia Schur.*

Valeriana vulgaris *Rupr.*

> *Valeriana officinalis L.*

Valerianoides jamaicensis *(L.) Medic.*

> *Stachytherpheta jamaicensis Vahl*

VANILLA Mill. - Vanilla - Orchidaceae

Vanilla abundiflora *J.J. Sm.*

Common Names ▶ Indonesian vanilla; *German:* Indonesische Vanille
Usage ▶ flavoring, like *Vanilla planifolia Andr.*
Parts Used ▶ fruit
Distribution ▶ Indonesia: Borneo; sporadically cultivated

Guzman/Siemonsma 1999; Hanelt 2001; Schultze-Motel 1986; Seidemann 1993c; Täufel et al. 1993; Uphof 1968

Vanilla albida *Bl.*

Synonyms ▶ *Vanilla griffithii* Rchb.f.
Common Names ▶ white vanilla; *German:* Weiße Vanille
Usage ▶ spice (locally)
Parts Used ▶ fruit
Distribution ▶ India, Thailand, Malaysia, Sumatra, Java; also cultivated

Hanelt 2001; Schultze-Motel 1986; Täufel et al. 1993

Vanilla aphylla *Bl.*

Common Names ▶ Java vanilla; *German:* Java-Vanille
Usage ▶ spice (locally)
Parts Used ▶ fruit
Distribution ▶ Myanmar, Malaysian Peninsula, Java

Erhardt et al. 2002

Vanilla aromatica *Willd.*

> *Vanilla planifolia Andr.*

Vanilla fragrans *(Salisb.) Ames*

▶ *Vanilla planifolia* Andr.

 Vanilla gardneri *Rolf.*

Common Names ▶ Brazil vanilla, Bahia vanilla, South American vanilla, Vanilla of Bahia; *German:* Brasilianische Vanille
Usage ▶ spice (locally)
Parts Used ▶ fruit
Distribution ▶ cultivated In Brazil

Guzman/Siemonsma 1999; Hoppe 1, 1977; Seidemann 1993c; Täufel et al. 1993; Uphof 1968

 Vanilla grandiflora *Lindl.*

▶ *Vanilla pompona* Schiede

 Vanilla griffithii *Rchb.f.*

▶ *Vanilla albida* Blume

 Vanilla guianensis *Splitg.*

▶ *Vanilla pompona* Schiede

 Vanilla lutescens *Moq.*

▶ *Vanilla pompona* Schiede

 Vanilla mexicana *P. Miller*

▶ *Vanilla planifolia* Andr.

 Vanilla palmarum *Lindl.*

Usage ▶ spice
Parts Used ▶ fruit
Distribution ▶ Guyana, Brazil

Bois 1934

 Vanilla phaeantha *Rchb.f.*

Synonyms ▶ *Vanilla planifolia* Griseb.
Usage ▶ spice
Parts Used ▶ fruit
Distribution ▶ origin: Cuba; St. Vincent, Florida, Bahamas, Antilles to Trinidad, Antilles also cultivated
Note ▶ Aldurant for *Vanilla planifolia* G. Jacks.

Bois 1934; Guzman/Siemonsma 1999; Schultze-Motel 1986; Uphof 1968; Wiersema/León 1999

 Vanilla planifolia *Andr.*

Synonyms ▶ *Vanilla aromatica* Willd.; *Vanilla fragrans* (Salisb.) Ames; *Vanilla mexicana* P. Miller, *Vanilla planifolia* G. Jacks., *Vanilla viridiflora* Bl.
Common Names ▶ Bourbon vanilla, Mexican vanilla, vanilla, vanilla fruit; *Arabic:* wanila; *Chinese:* hsiang ts'ao; *French:* vanille, vanille de Mexique, vanillier; *German:* (Echte) Vanille, Bourbon-Vanille, Vanille, Vanillefrucht, Vanilleschote (wrongly!); *Indonesian:* panila; *Italian:* vaniglia, vainiglia; *Japanese:* bauira; *Mexico:* siisbik, tlilxochitl; *Pilipino:* vanilla; *Portuguese:* baunilha; *Russian:* vanil'; *Slovenian:* vanilka; *Spanish:* vainilla, vainillero; lixóchitl, vainil (Mexico); *Thai:* wanila
Usage ▶ spice; **product:** (natural) vanillin
Parts Used ▶ fruit
Distribution ▶ Mexico to S America, West Indies, Mexico ; probably native elsewhere in C and S America, also widely cultivated in Sri Lanka
Note ▶ "Tlilxochitl" is the Aztec word for vanilla, meaning black pod. Trade sorts: Bourbon vanilla, Mexican vanilla, Tahiti vanilla, Java vanilla, Ceylon vanilla and the qualities: extra fine, fine, mifine and fendue extra.

Acat/Acat 2003; Adeji et al. 1993; Alberts/Muller 2000; Anklam/Müller 1993; Arana 1943; Archer 1989: Arnaud et al. 1981; Bergeron 1980; Bois 1934; Boisvert/Hubert 2000; Bouriguet 1954; Braatz/Lembke 1990; Bricout et al. 1974; Burkill 4, 1997; Busenberg et al. 1994; Busse 1899; Chadwick/Pope 1961; Cheers 1998; Correll 1953; Craze 2002; Dalang et al. 1982; Dalby 2000; Dauer 2002; Davidson 1999; Deltail 1897; Dignum et al. 2002, 2004; Dudtschenko et al. 1989; Duke et al. 2003; Ehlers 1999; Ehlers/Bartholomae 1993;

◘ Vanilla planifolia Andr., flowering

◘ Vanilla planifolia Andr., fruiting

Ehlers et al. 1994, 1995, 1999; Erhardt et al. 2002; Farrell 1985; Fayet et al. 1989; Fellous et al. 1992; Fraisse et al. 1984; Galetto/Hoffmann 1978; Guzman/Siemonsma 1999; Hanelt 2001; Hanum 1997; Hiller/Melzig 1999; Hoffmann W. (no year); Hoppe 1949; Kahan 1989; Kanisawa 1993; Kaunzinger et al. 1996; Kleinert 1963; Klimes/Lamparsky 1976; Klont 2000; Lambrecht et al. 1994; Leong et al. 1989a, b; Lewington 1990; Lhuguenol 1978; Maubert et al. 1988; Melchior/Kastner 1974; Morrell/Mackley 1999; Mosandl A. 1995; Norman 1991; Nguyen et al. 1991; Oberdieck 1998; Ochse et al. 1961; Odoux 2000; Pochljobkin 1974, 1977; Pruthi 1976; Ramachandra Rao/Ravishankar 2000; Ramarosan-Raonizafinimanana et al. 1997; Ranadive 1992; Ray et al. 1980; Rosengarten 1969; Roth/Kormann 1997; Rust 2003; Scharrer 2002; Scharrer/Mosandl 2001; Schröder 1991; Schultze-Motel 1986; Seidemann 1993c, 1998/2000; Seidemann/Siebert 1987; Shiva et al. 2002; Siewek 1990; Small 1997; Staesche 1972; Tabacchi et al. 1978; Täufel et al. 1993; Teuscher 2003; Uphof 1968; Vaupel 2002a, b, c; Villamar et al. 1994; Webster 1995; Westermann 1909; Wiersema/León 1999; Wildeisen 2001; Wildeman 1902; Wüstenfeld/Haensel 1964; Zeven/de Wet 1982

Vanilla planifolia Griseb.

▶ *Vanilla phaeantha Rchb.*

Vanilla planifolia G. Jacks. f. gigantea Hoehne

Common Names ▶ great vanilla; *Brazil:* vanilão; *German:* Große Vanille, Riesenvanille
Usage ▶ spice (locally)
Parts Used ▶ fruit
Distribution ▶ Mexico to S America (Brazil)

Hoppe 1, 1977; Mors/Rizzini 1961; Schultze-Motel 1986

Vanilla planifolia G. Jacks.

▶ *Vanilla planifolia Andr.*

Vanilla pompona *Schiede*

Synonyms ▶ *Vanilla grandiflora* Lindl., *Vanilla guianensis* Splitg., *Vanilla lutescens.* Miq., *Vanilla surinamensis* Rchb.

Common Names ▶ Pompon vanilla, West Indian vanilla; *French:* vanillon; *German:* Antillen-Vanille, Guadeloupe-Vanille, Pompon-Vanille, Vanillon, Westindische Vanille

Usage ▶ spice

Parts Used ▶ fruit

Distribution ▶ C America, SE Mexico, tropical North S America

Note ▶ The use of the fruits is predominently in the perfumery and tobacco industries.

Adedeji et al. 1993; Bois 1934; Davidson 1999; Ehlers/Pfister 1997; Erhardt et al. 2002; Hoffmann (no year); Hoppe 1, 1977; Melchior/Kastner 1974; Ochse et al. 1961; Pruthi 1976; Rey et al. 1980; Schröder 1991; Schultze-Motel 1986; Seidemann 1993c; Siewek 1990; Staesche 1972; Täufel et al. 1993; Teuscher 2003; Uphof 1968; Wildeisen 2001; Zeven/de Wet 1982

 Vanilla surinamsis *Reichb.*

▶ *Vanilla pompona* Schiede

 Vanilla sylvatica *McFarlane*

▶ *Vanilla tahitensis* J.W. Moore

 Vanilla tahitensis *J.W. Moore*

Synonyms ▶ *Vanilla sylvatica* McFarlane

Common Names ▶ Tahitan vanilla; *French:* vanille de Tahiti, vanille de Tiarei; *German:* Tahiti-Vanille; *Tahiti*: haapae, papanoe, teriiroa, tiarei

Usage ▶ spice

Parts Used ▶ fruit

Distribution ▶ Tahiti, native Hawaii; only cultivated Réunion, Mauritius

Note ▶ The fruits have a lower vanillin content, but the aroma is sweeter; best suited for cosmetics.

Adedeji et al. 1993; Davidson 1999; Ehlers et al. 1994; Erhardt et al. 2000, 2002; Guzman/Siemonsma 1999; Hiller/Melzig 1999; Hoppe 1977; Jack 1940; Lhuguenol 1978; Melchior/Kastner 1974; Ochse et al 1961; Pruthi 1976; Ramarosan-Raonizafinimanan et al. 1997; Rey et al 1980; Scharrer 2002; Schröder 1991; Schultze-Motel 1986; Seidemann 1993c; Siewek 1990; Staesche 1972; Tabacchi et al. 1978; Täufel et al. 1993; Teuscher 2003; Wiersoma/León 1999

 Vanilla viridiflora *Bl.*

▶ *Vanilla planifolia* Andr.

VERBASCUM L. Mullein - Scrophulariaceae

 Verbascum densiflorum *Bertol.*

Synonyms ▶ *Verbascum thapsiflorum* Schrad.

Common Names ▶ large flowered mullein, common mullein; *French:* bouillon blanc, fleur de St. Pierre, molène, *German:* Großblütige Königskerze, Großblütige Wollblume; *Italian:* bouillon blanc, verbasco maschio, tasso verbasso, candela della madonna, candela regia; *Korean:* pelpasŭcho; *Russian:* (korowjak) medweshe ucho; *Spanish:* gordolobo (común), verbasco

Usage ▶ flavoring (in the liqueur industry)

Parts Used ▶ flower

Distribution ▶ Caucasus, W Asia, Europe, native elsewhere

Aichele/Schwenker 4, 1995; Berger 1, 1949; 2, 1950; Bremness 2001; Erhardt et al. 2002; Fleischhauer 2003; Hanelt 2001; Heeger 1956; Hohmann et al. 2001; Hoppe 1949; Kraus/Franz 1987; Pschyrembel 1998; Schönfelder 2001; Schultze-Motel 1986; Seidemann 1993c; Sharma 2003; Swiatek et al. 1982; Wiersema/León 1999; Wüstenfeld/Haensel 1964; Zeven/de Wet 1982; Zwingenberger 1938

 Verbascum phlomoides *L.*

Common Names ▶ clasping mullein, orange mullein; *French:* bouillon blanc, molène faux-phlomis; *German:* Filz-Königskerze, Windblumen-Königskerze; *Italian:* barbarasco

Usage ▶ flavoring (especcially in the liqueur industry)

Parts Used ▶ flower

Distribution ▶ temperate Asia, Europe, native elsewhere

Aichele/Schwenker 4, 1995; Bärtels 1997; Berger 1, 1949; 2, 1950;

Erhardt et al. 2002; Fleischhauer 2003; Hanelt 2001; Heeger 1956; Hiller/Melzig 1999; Hoppe 1949; Kraus/Franz 1987; Pápay et al. 1980; Schönfelder 2001; Schultze-Motel 1986; Seidemann 1993c; Tschesche et al. 1980; Uphof 1968; Wiersema/León 1999; Wyk et al. 2004; Zwingenberger 1938

Dudtschenko et al. 1989; Ehrhardt et al. 2002; Erichsen-Brown 1989; Fleischhauer 2003; Heeger 1956; Hiller/Melzig 1999; Hoppe 1949; Neweall et al. 1996; NICPBP 1987; Rajendran/Daniel 2002; Rätsch 1998; Schönfelder 2001; Schultze-Motel 1986; Sharma 2003; Täufel et al. 1993; Uphof 1968; Weber 1995; Wyk et al. 2004

 Verbascum thapsiflomum *Schrad.*

▷ *Verbascum densiflorum Bertol.*

VERBENA L - Vervain, Verbena - Verbenaceae

 Verbena jamaicense *L.*

▷ *Stachytarpheta jamaicensis Vahl*

 Verbena javanica *Burm.f.*

▷ *Lippea javanica (Burm.f.) Spreng.*

 Verbena officinalis *L.*

Common Names ▶ common verbena, European vervain, lemon scented verbena, Simpler's joy, turkey grass, vervain; *Arabic:* ben nout, tronjia; *Chinese:* ma bian cao, ma pien tsao; *French:* citronelle, verveine, verveine odorante, verveine officinelle, herbe sacrée, herbe aux sorciers; *German:* Echtes Eisenkraut, Echte Verbene, Heiligkraut, Zitronen-Verbene; *Italian:* cedrina, verbena; *Japanese:* kumatsuzura; *Russian:* verbena; *Spanish:* verbena, verba de la princesa

Usage ▶ flavoring: **product:** essential oil

Parts Used ▶ leaf, herb

Distribution ▶ N Africa, temperate Asia, India, Europe, native worldwide

Note ▶ The essential oils are used as an alternative for lemon grass oil (*Cymbopogon citratus* [DC. ex Nees] Stapf), *Cymbopogon flexuosus* [Steud.] Stape).

Aichele/Schwenker 4, 1995; Berger 4, 1954; Bremness 2001; Charalambous 1994; Cheers 1998; Clair 1961; Davidson 1999;

VERNONIA Schreb. - Ironweed, Veronia - Asteraceae (Compositae)

 Vernonia albicans *Lees*

▷ *Vernonia cinerea Less.*

 Vernonia amygdalina *Delile*

Common Names ▶ almond veronia, bitter leaf; *Dutch:* bitterblad; *French:* vernonie; *German:*Bittere Veronie, Mandel-Veronie, Bittere Scheinaster; *Portuguese:* pau fede, fede, libó; *Russian:* vernonija mindal'naja; *Spanish:* hoja amarga, vernonia

Usage ▶ pot-herb, plant ash substitute for salt

Parts Used ▶ leaf

Distribution ▶ tropical Africa: Nigeria, Arabia, also cultivated in gardens

Note ▶ The refined plant ash, also of *Verbena conferta* Benth., are used as salt.

Aedo et al. 2001; Ayensu 1978; Burkill 1, 1985; Davidson 1999; Erhardt et al. 2002; Oomen/Grubben 1978; Schultze-Motel 1986; Seidemann 1993c; Uphof 1968; Zeven/de Wet 1982

 Vernonia anthelmintica *(L.) Willd.*

Synonyms ▶ *Baccharoides anthelmintica* (L.) Moench

Common Names ▶ kala jiri; *German:* Indischer Schwarzkümmel, Wurmscheinaster

Usage ▶ spice

Parts Used ▶ seed

Distribution ▶ Africa, Arab Peninsula, India, China, W Asia, Indochina

Chopra 1956; Erhardt et al. 2002; Hanelt 2001; Rätsch 1998; Schultze-Motel 1986; Uphof 1968; Wealth of India 10, 1976; Wiersema/León 1999; Zeven/de Wet 1982

Vernonia cinerea *Less.*

Synonyms ▶ *Vernonia albicans* DC.; *Vernonia leptophylla* DC.

Common Names ▶ little ironwood; *German:* Kleines Eisenholz, Kleine Scheinaster; *Indonesian:* maryuna; sasawi langit; *Malaysian:* rumput tahi babi, tambak-tambak; *Pilipino:* kolong-kugon, agas-moro, bulak-manok; *Thai:* kaan thuup, yaa dok khaao, yaa saam wan

Usage ▶ pot-herb (in India)

Parts Used ▶ leaf

Distribution ▶ Africa, China, E Asia, India, Indochina, Malaysia, Australia, native elsewhere

Arora/Pandey 1996; Wiersema/León 1999

 Vernonia conferta *Benth*

▶ *Vernonia amygdalina* Delile

 Vernonia leptophylla *DC.*

▶ *Vernonia cinerea* (L.) Less.

VETIVERIA Bory - Vetiver - Poaceae (Gramineae)

 Vetiveria odorota *Virey*

▶ *Vetivera zizanioides* (L.) Nash

 Vetiveria zizaniodes *(L.) Nash*

Synonyms ▶ *Andropogon festucoides* J.S. Presl., *Andropogon muricata* Retz., *Andropogon squarrosus* Hackel, *Andropogon zizanioides* (L.) Urban, *Vetiveria odorata* Virey

Common Names ▶ cus cus (grass), khus khus (grass), vetiver (grass); *Arabic:* izkhir; *Chinese:* xiang-geng-sao; *French:* vétivier, chiendent del Indes, chiendent odorant; *German:* Vetivergras, *Hindi:* khas-khas, vettiver, khus khus; *Malaysian:* nara wastu, akar wangi, kusu-kusu; nara setu; *Nigerian:* jema, so'dornde, so'mayo; *Sanskrit:* abhaya, bala, nalada, usiira; *Pilipino:* moras, amora, anis de moro, ilib, moras; *Spanish:* zacate violeta; *Sri Lanka:* sevendara, vettiver; *Thai:* faek, ya-faek-hom, ya-faeklum

Usage ▶ flavoring; **product:** essential oil (essence de Vetiver, vetiver oil)

Parts Used ▶ rhizome, root

Distribution ▶ S Asia: S India, Sri Lanka, China, Thailand, Malaysia, Philippines, Pakistan, Bangladesh and spread from SW Asia to tropical Africa: Zimbabwe, Kenya, Somalia, Nigeria; S America: Brazil, Colombia, Paraguay; widely cultivated in paleotropical lands

Bombarda et al. 1996; Bourton 1968; Chauhan 1999; Dastur 1954; Davidson 1999; Erhardt et al. 2002; Garnero 1971; Gottlieb/Iachan 1951; Grimshaw/Helfer 1995; Hager 5, 1993; Hanelt 2001; Hiller/Melzig 1999; Kumar 2001; Lavania 1991; Lück 2004; Maffei 2002; Mors/Rizzini 1961; Rao 1966; Robbins 1982; Roth/Kormann 1997; Sangat-Roemantyo 1990; Seidemann 1993c, 1998/2000; Sharma 2003; Shiva et al. 2002; Smadja 1990, 1991; Uphof 1968; Virmani/Datta 1975; Wiersema/León 1999; Zeven/de Wet 1982

VIOLA L. - Violet - Violaceae

 Viola cornuta *L.*

Common Names ▶ beddy pansy, horned pansy, horned violet, tufted pansy; *French:* violete cornue; *German:* Gehörntes Veilchen, Horn-Veilchen, Pyrenäen-Stiefmütterchen

Usage ▶ flavoring

Parts Used ▶ flower

Distribution ▶ SW Europe, native elsewhere, cultivated in gardens

Cheers 1997; Erhardt et al. 2000, 2002; Köhlein 1999; Wiersema/León 1999

 Viola odorata *L.*

Synonyms ▶ *Viola officinalis* Cr., *Viola sarmentosa* M. Bieb.

Common Names ▶ common violet, florist's violet, garden violet, sweet scented violet; *Arabic:* banaf sag; *French:* violet, violette de mars, violette odorante; *German:* März-Veilchen; Wohlriechendes Veilchen; *Hindi:* banaphsa; *India:* banef shah; *Italian:* viola mammola, viola zopa, mammolo roseviole, violetta; *Javanese:* antanan; *Russian:* fialka duschistaja; *Sanskrit:* nilapuspa; *Spanish:* violeta

Usage ▶ flavoring (milk pudding, ice cream) and as an edible decoration for dishes; **product:** frost, essence: sweets, liqueur "Parfait d'amour"; **product:** essential oil: perfumery industry

Parts Used ▶ flower, (and culinary) herb

Distribution ▶ N Africa, Caucasus, Europe, also cultivated

Note ▶ In early Greek and Roman times violets were used to flavor butter, oil, vinegar and wine. The floral leaves are often candied and used in confectionery.

Aichele/Schwegler 3, 1995; Berger 2, 1950; 4, 1954; Bourton 1968; Bremness 2001; Charalambous 1994; Cheers 1998; Cohnen 1993a; Coiciu/Racz (no year); Dastur 1954; Davidson 1999; Erhardt et al. 2002; Erichsen-Brown 1989; Fleischhauer 2003; Heeger 1956; Hiller/Melzig 1999; Hoffmann et al. 1992; Hoppe 1949; Köhlein 1999; Máñez/Viliar 1989; Rätsch 1998; Schönfelder 2001; Schultze-Motel 1986; Seidemann 1993c; Sharma 2003; Small 1997; Täufel et al. 1993; Tucker 1986; Villamar et al. 1994; Wiersema/León 1999; Zeven/de Wet 1982

Viola officinalis *Cr.*

▶ *Viola odorata L.*

Viola sarmentosa *M.Bieb.*

▶ *Viola odorata L.*

VITEX L. - Chaste Tree - Verbenaceae

Vitex agnus-castus *L.*

Common Names ▶ chaste pepper; *Arabic:* ghar, kaf mariyam, kherwa', fitex, shajarat; *Brazil (Portuguese):* alecrim de angola, alecrim do norte, pau de angola; *French:* gattilier agneau-chaste, poivre des moines; *German:* Mönchspfeffer, Keuschlamm, Abrahamstrauch; *India:* athlac, panjangusht, ranukabija, shambhaluka-bija; *Italian:* agnocasto, albero del pepe, pepe falso; *Russian:* awraamowo derewo

Usage ▶ spice

Parts Used ▶ fruit

Distribution ▶ Mediterranean region, Balkan, Asia minor, Iran, Caucasus to C Asia

Note ▶ Cultivated as a substitute and alternative for black pepper when not available.

Bärtels 1997; Bois 1934; Doğan/Mert 1998; Ekundayo et al. 1990; Erhardt et al. 2000, 2002; Feil et al. 2002; Griebel 1943b; Hepper 1992; Hiller/Melzig 1999; Kartnig 1986; Mors et al. 2000; Newall et al. 1996; Pschyrembel 1998; Rajendran/Daniel 2002; Schönfelder 2001; Schultze-Motel 1986; Seidemann 1993c; Senatore et al. 1996; Siewek 1990; Sørensen/Katsiotis 2000; Staesche 1972; Täufel et al. 1993; Teuscher 2003; Uphof 1968; Wiersema/León 1999; Wollenweber 1983; Wyk et al. 2004; Zeven/de Wet 1982

Vitex arborea *Desf.*

▶ *Vitex negundo L.*

Vitex bicolor *Willd.*

▶ *Vitex negundo L.*

Vitex chinensis *Mill*

▶ *Vitex negundo L.*

Vitex negundo *L.*

Synonyms ▶ *Vitex arborea* Desf., *Vitex bicolor* Willd., *Vitex chinensis* Mill.

Common Names ▶ Chinese chaste tree, horse shoe vitex; *Chinese:* kunang ching, kuang jing; *German:* Chinesischer Mönchspfeffer; *Malaysian:* lagundi, lemuning, peninchang

Usage ▶ spice (rarely)

Parts Used ▶ fruit

VITEX L. - Chaste Tree - Verbenaceae

◘ **Vitex agnus-castus:** a **flowering,** b **fruiting**

Distribution ► E Africa, Madagascar, India, Sri Lanka, Afghanistan, China, SE Asia, Philippines

Bois 1923; Chauhan 1999; Erhardt et al. 2002; Hanelt 2001; Rajendran/Daniel 2002; Schultze-Motel 1986; Sharma 2003

Vitis quadrangularis *Wall. et.*

❯ *Cissus quadrangularis L.*

Volkameria orientalis *O. Kuntze*

❯ *Sesamum indicum L.*

WARBURGIA Engl. - Canellaceae

 Warburgia salutaris *(Bertol.f.) Chiov.*

Common Names ▶ pepperbark tree; *German:* Heil-Warburgie; *Zulu:* isibhaha
Usage ▶ spice (locally)
Parts Used ▶ leaf, bark
Distribution ▶ E, S Africa
Note ▶ All plant parts have a peppery-hot taste.

Wyk et al. 2004

 Warburgia ugandensis *Spargue*

Common Names ▶ Uganda warburgia; *German:* Uganda-Warburgie
Usage ▶ flavoring
Parts Used ▶ bark
Distribution ▶ E Africa, especially Uganda
Note ▶ The bark has a cinnamon-like taste.

Wyk et al. 2004

WASABIA Matsum. - Wasabi - Brassicaceae (Cruciferae)

 Wasabia japonica *(Miq.) Matsum.*

Synonyms ▶ *Alliaria wasabi* Prantl., *Cochlearia wasabi* Sieb., *Eutrema wasabi* (Sieb.) Maxim, *Wasabi pungens* Matsum.
Common Names ▶ Japanese horseradish, mountain hollyhock, wasabi; *Chinese:* shan yu cai, shan kui; *French:* wasabi, raifort du japonais, raifort vert; *German:* Japanischer Meerrettich; *Japanese:* tsi, kiseseri, wasabi; *Korean:* maeunkochunaengi; *Portuguese:* wasabi, rábano japonés; *Russian:* vasabi; *Spanish:* wasábia, rábano japonés, rabanête japonés
Usage ▶ spice, like horse radish, a sushi condiment
Parts Used ▶ rhizome, herb, leaf stalk
Distribution ▶ Japan, E Siberia, E Asia
Note ▶ As a spice used raw, as a paste or as a dry powder.

Bois 1934; Chadwick et al. 1993; Davidson 1999; Depree et al 1999; Duke et al. 2003; Erhardt et al. 2002; Hanelt 2001; Hodge 1974; Hu 1991; Ina et al. 1989; Kling 1998; Kojima 1976; Kojima et al. 1982, 1985; Larkcom 1991; Mansfeld 1986; Ochi et al. 1995; Palmer 1990b; Schultze-Motel 1986; Seidemann 1993c; Small 1997; Sultana et al. 2003; Suzuki 1968; Täufel et al. 1993; Teuscher 2003; Tucker 1986; Uhl 2000; Uphof 1968; Watanabe et al. 2003; Wiersema/León 1999; Zeven/de Wet 1982

 **Wasabi pungens** *Matsum.*

▶ *Wasabi japonica (Miq.) Matsum.*

WEDELIA Jacq. - Goldcup - Asteraceae (Compositae)

 Wedelia biflora *(L.) DC.*

Synonyms ▸ *Wedelia scandens* C.B. Clarke; *Wollastonia biflora* (L.) DC.
Common Names ▸ *German:* Kletter Wedelie; *Indonesian:* saruni laut, cinga-cinga; *Pilipino:* hagonoi; *Thai:* phak khraat thale
Usage ▸ flavoring
Parts Used ▸ young leaf
Distribution ▸ tropical Africa to India, Indochina to Japan, Malaysia to tropical Australia

Arora/Padney 1996; Miles et al. 1993

 Wedelia scandens *C.B. Clarke*

▸ *Wedelia biflora DC.*

WEINMANNIA L. - Cunoniaceae

 Weinmannia fraxinea *Sm. ex D. Don*

▸ *Weinmannia sundaica Bl.*

 Weinmannia sundaica *Bl.*

Synonyms ▸ *Weinmannia fraxinea* Sm. ex D. Don
Common Names ▸ *German:* Eschenartige Weinmannie
Usage ▸ flavoring (foods)
Parts Used ▸ herb
Distribution ▸ Molucca Islands

Burkhill 1966; Oyen/Dung 1999

WESTRINGIA Sm. - Australian Rosemary - Lamiaceae (Labiatae)

 Westringia fruticosa *(Willd.) Druce*

Synonyms ▸ *Westringia rosmariniformis* Sm.
Common Names ▸ Australian rosemary; *German:* Australischer Rosmarin
Usage ▸ spice, like rosemary
Parts Used ▸ leaf, herb
Distribution ▸ Australia: Queensland, New South Wales

Cheers 1998; Erhardt et al. 2002; Small 1997

 Westringia rosmariniformis *Sm.*

▸ *Westringia fruticosa (Willd.) Druce*

 Winterana canella *L.*

▸ *Canella alba (L.) Murr.*

 Wintera aromatica *Desc.*

▸ *Drimys winteri J.R. et G. Forst.*

 Winterana canella *L.*

▸ *Canella winterana (L.) Gaertn.*

WRIGHTIA R.Br. - Apocynaceae

 Wrightia arborea *(Dennst.) Mabberley*

▸ *Wrightia tomentosa Roem. et Schult.*

 Wrightia hamiltoniana *Wall.*

▸ *Wrightia tomentosa Roem. et Schult.*

Wrightia tomentosa (Roxb.) *Roem. et Schult.*

Synonyms ▶ *Wrightia arborea* (Dennst.) Mabberley; *Wrightia hamiltoniana* Wall.
Usage ▶ pot-herb (by Santhals)
Parts Used ▶ leaf
Distribution ▶ India: Bihar, Kokna, Warli, Thailand

Arora/Padney 1996; Dastur 1954; Sharma 2003; Uphof 1968

X

 Xanthophthalmum coronarium (L.) Trehane

> *Chrysanthemum coronarium* L.

 **Xanthoxylum-Species**

> *Zanthoxylum-Species*

 Ximenia aegyptica L

> *Balanites aegyptica* (L.) Del.

XYLOPIA L. - African Pepper - Annonaceae

 Xylopia acutiflora (Dun.) A. Rich.

Common Names ▸ mountain spice (of Sierra Leone); *French:* elo à petites feuillec; *German:* Spitzblütiger Mohrenpfeffer
Usage ▸ spice, used like black pepper
Parts Used ▸ seed
Distribution ▸ W Africa: Ghana, Nigeria, Sierra Leone, W Cameroon
Note ▸ The ripe fruit with a coral-red pulp has a spice taste like several other *Xylopia* species.

Burkill 1, 1985

 Xylopia aethiopica (Dun.) A. Rich.

Synonyms ▸ *Unona aethiopica* Dun., *Xylopia eminii* Engl.
Common Names ▸ African pepper, Ethiopian pepper, Guinea pepper, kimba pepper, negro pepper, grains of Selim, spice tree; *Cameroon:* ebongo, ikola hindi kimba, mbonji, okolo; *Congo:* kani; *French:* piment noir de Guinée, poivre de Guinée, poivre d'Ethiopie, poivre negre; kani; *German:* Äthiopischer Pfeffer, Afrikanischer Pfeffer, Kumbapfeffer, Malaguettapfeffer, Mohrenpfeffer, Negerpfeffer, Kani-Körner von Selim; *Portuguese:* cabella kani (Ivory coast), malaguetta da Guine, pimenta da Guiné, malaguetta preta; *Russian:* psewdoloperzy (ksisopii), loschnye perzy, *Spanish:* maniguate, pimenta de Guinea, semillas del oyang
Usage ▸ spice, used like black pepper
Parts Used ▸ seed
Distribution ▸ tropical Africa from Senegal to Sudan and Uganda, southwards to Angola, Zaire, Zambia, and Mozambique, cultivated in the east of W Africa
Note ▸ Not to be confused with *Piper guineense* Schum. & Thonn; alternative for black pepper; the fruit contains a volatile aromatic oil.

Adegoke et al. 2003; Aedo et al. 2001; Ayedoun et al. 1996; Ayensu 1978; Barminas et al. 1999; Berger 3, 1952; Bois 1934; Burkill 1, 1985; Duke et al. 2003; Ekong/Ogan 1968; Erhardt et al. 2002; Ferrão 1992; Griebel 1943a, 1944; Harrigan et al. 1994; Hiller/Melzig 1999; Jirovetz et al. 1997; Karawya et al. 1979; Lamaty et al. 1987; Lück 2000; Melchior/Kastner 1974; Ngouda et al. 1998; Okeke 1998; Pochljobkin 1974, 1997; Schenck/Naundorf 1966; Schultze-Motel 1986; Seidemann 1993c; Seidemann/Siebert 1987; Siewek

1990; Small 1997; Staesche 1972; Tairu et al. 1999; Täufel et al. 1993; Teuscher 2003; Uphof 1968; Zeven/de Wet 1982

 Xylopia aromatica *(Lam.) Mart.*

Synonyms ▶ *Uvaria aromatica* Lam., *Xylopia granduflora* St. Hill., *Xylopia longifolia* DC.
Common Names ▶ *Brazil (Portuguese):* pachinhos, pimenta de macaco, pimenta de negro; *German:* Burropfeffer, Guinea-Pfeffer; *Peru:* mataro, omechuai caspí; *Portuguese:* embira, envireira, pimenta-de-árvone, pimenta-de-gentio, pimenta-de-negro, pimenta da-costa
Usage ▶ spice, condiment, used like pepper; **product:** essential oil
Parts Used ▶ seed
Distribution ▶ tropical America, S Brazil, Peru, Venzuela, Costa Rica
Note ▶ The seeds are reminiscent of black pepper (*Piper nigrum* L.)

Erhardt et al. 2002; Lorenzi 1992; Melchior/Kanert 1974; Moraes/Roque 1988; van Roosmalen 1985; Seidemann 1993c; Seidemann/Siebert 1987; Silva(Rocha 1981; Täufel et al. 1993; Teuscher 2003

 Xylopia brasiliensis *Spreng.*

Synonyms ▶ *Xylopia muricata* Velloso, *Xylopia parrifolia* Schlecht.
Common Names ▶ *German:* Matopfeffer; *Portuguese:* pimenta de mato, pinaíba, pindaubuna
Usage ▶ spice, rarely used like pepper
Parts Used ▶ seed
Distribution ▶ tropical S America: Brazil

de Costa 1959; Lorenzi 1992; Mors/Rizzini 1961

 Xylopia carminativa *(Aruda) Fries.*

Usage ▶ condiment
Parts Used ▶ seed
Distribution ▶ Brazil to British Guyana
Note ▶ The fruits and seeds have an odor and a taste of pepper.

van Roosmalen 1985; Uphof 1968

 Xylopia eminii *Engl.*

▶ *Xylpia aethiopica (Dun.) A. Rich.*

 Xylopia frutescens *Aubl.*

Common Names ▶ *German:* Halbstrauchiger Mohrenpfeffer; *Portuguese:* coagerucu, pindaíba, embira, envira, jererecou, pau de embira, pejerecum, pijerecu, pimento-de-gentio, pimenta-do-sertão, pindaúba
Usage ▶ condiment
Parts Used ▶ seed
Distribution ▶ tropical America especially Brazil

Mors et al. 2000; van Roosmalen 1985; Silva 1983; Uphof 1968

 Xylopia grandiflora *St. Hil.*

Common Names ▶ *German:* Großblütiger Mohrenpfeffer; *Portuguese:* malagueto
Usage ▶ **product:** essential oil
Parts Used ▶ flower
Distribution ▶ tropical S America, especially Brazil
Note ▶ The essential oils have a pepper-like flavor.

Uphof 1968

 Xylopia longifolia *DC.*

▶ *Xylopia aromatica (Lam.) Mart.*

 Xylopia parviflora *(A. Rich.) Benth.*

Common Names ▶ striped African pepper; *French:* fondé des rivières, poivres de Sédhiou; *German:* Sédhion Pfeffer, Kleinblütiger japanischer Pfeffer, Gestreifter Mohrenpfeffer; *Nigerian:* aghako, kimba, sesedo
Usage ▶ spice
Parts Used ▶ fruiting carpels
Distribution ▶ W Africa: Cameroon, Ghana, the Gambia, Nigeria, Senegal, Sierre Leone, Togo

Burkill 1, 1985; Jirovetz et al. 1997; Uphof 1968

XYLOPIA L. - African Pepper - Annonaceae   397

 Xylopia muricata Velloso

▶ *Xylopia brasiliensis* Spreng.

 Xylopia parvifolia Schlecht.

▶ *Xylopia brasiliensis* Spreng.

 Xylopia quintasii Pierre ex Engl. et Diels

▶ *Xylopia striata* Engl.

 Xylopia sericea St. Hil.

Synonyms ▶ *Mayna sericea* Spreng.
Common Names ▶ *German:* Seidiger Mohrenpfeffer; *Portuguese:* pindaíba-vermelha, pindaubuna-da-serra
Usage ▶ condiment
Parts Used ▶ seed
Distribution ▶ Brazil, Bolivia
Note ▶ The seeds used somewhat similar to black pepper (*Piper nigrum* L.)

Lorenzi 1992; Matos et al. 1984, 1985; Mors et al. 2000; van Roosmalen 1985; Uphof 1968

 Xylopia staudtii Engl. et Diels

Common Names ▶ bush pepper, Guinea pepper
Usage ▶ spice (rarely)
Parts Used ▶ fruit
Distribution ▶ W Africa: Ghana, Ivory Coast, Liberia, Sierra Leone

Burkill 1, 1985

 Xylopia striata Engl.

Synonyms ▶ *Xylopia quintasii* Pierre ex Engl. et Diels
Common Names ▶ negro pepper; *German:* Gestreifter Mohrenpfeffer, Kani
Usage ▶ spice
Parts Used ▶ fruit

 Xylopia: a **X. aethiopica, fruits**, b **X. striata, fruits**

Distribution ▶ W Africa

Erhardt et al. 2002; Griebel 1943; Melchior/Kastner 1974; Seidemann 1993c; Seidemann/Siebert 1987; Täufel et al. 1993; Teuscher 2003; Uphof 1968

 Xylopia undulata P. Beauv.

▶ *Monodora myristica* (Gaertn.) Dun.

Z

ZANTHOXYLUM L. - Yellow wood - Rutaceae

 Zanthoxylum acanthopodium *DC.*

Synonyms ▶ *Zanthoxylum acanthopodium* DC. var. timbor Hook.f.
Common Names ▶ Chinese pepper, tomar seed; *Chinese:* guo ma ka; *German:* Chinesischer Pfeffer; *Hindi:* darmar, tejphal; *Vietnamese:* sèn
Usage ▶ flavoring
Parts Used ▶ seed
Distribution ▶ Himalayas, N Bengal, Naga and Khasi hills, China, India, Vietnam
Note ▶ In S China cultivated and utilized for its pleasant flavor; the plant has a peculiar flavor of coriander; products of watara oil (from the fruits) are used in perfumery.

Hanelt 2001; Shiva et al. 2002; Teuscher 2003; Wealth of India 11, 1976

 Zanthoxylum acanthopodium *DC.* **var.** *timbor* Hook.f.

❯ *Zanthoxylum acanthopodium DC.*

 Zanthoxylum alatum *Roxb.*

❯ *Zanthoxylum armatum DC.*

 Zanthoxylum armatum *DC.*

Synonyms ▶ *Zanthoxylum alatum* Roxb.
Common Names ▶ Chinese pepper, Szechuan pepper; *Chinese:* hua jiao; *German:* Chinesischer Pfeffer, Szechuanpfeffer; *Hindi:* darmar, nepali dhaniya; *Japanese:* fuyu-sanshô; *Sanskrit:* tumburu, dhiva, gandhalu; *Vietnamese:* sengai;
Usage ▶ spice
Parts Used ▶ seed
Distribution ▶ India, Pakistan, Nepal, Himalayas, Malaysia, Philippines

Chauhan 1999; Dalby 2000; Davidson 1999; Erhardt et al. 2002; Facciola 1990; Hanelt 2001; Hiller/melzig 1999; Oyen/Dung 1999; Sharma 2003; Small 1997; Tucker 1986; Wealth of India 11, 1976

 Zanthoxylum avicennae *(Lam.) DC.*

Synonyms ▶ *Fagara avicennae* Lamk., *Zanthoxylum diversifolium* Warb., *Zanthoxylum tidorense* Miq.
Common Names ▶ *Indonesian:* karangeang, *Pilipino:* bagatambal
Usage ▶ spice: leaf resembling coriander leaves; seeds like anise; flavoring (fruit)
Parts Used ▶ leaf, seed, fruit
Distribution ▶ Thailand, China, Indonesia, Malaysia

Guzman/Siemonsma 1999; Oyen/Dung 1999; Uphof 1968

 Zanthoxylum bungeanum *Maxim*

❯ *Zanthoxylum simulans Hance*

© Springer-Verlag Berlin Heidelberg 2005

Zanthoxylum budrunga Wall. ex DC.

▸ *Zanthoxylum rhetsa* DC.

 Zamthoxylum bungei Planch.

▸ *Zingiber simulans* Hance

 Zanthoxylum clava-herculis L.

Common Names ▸ Hercules' club, pepper bark, Southern prickly ash, toothache tree, West-India yellow wood; *German:* Westindisches Gelbholz
Usage ▸ flavoring
Parts Used ▸ bark
Distribution ▸ C USA, except WC USA

Berger 1, 1949; Hiller/Melzig 1999; Newall et al. 1996; Rao/Davies 1986; Sharma 2003; Uphof 1968; Wiersema/León 1999

 Zanthoxylum coreanum Nakai

Common Names ▸ Korean yellow wood; *German:* Koreanisches Gelbholz; *Korean:* woangtschopinamu
Usage ▸ spice
Parts Used ▸ leaf, fruit, bark
Distribution ▸ Korea

Hanelt 2001

 Zanthoxylum diversifolium Warb.

▸ *Zanthoxylum avicennae (Lamk) DC.*

 Zanthoxylum fraxinoides Hemsl.

▸ *Zanthoxylum simulans* Hance

 Zanthoxylum gilletii (de Wild.) Waterm.

▸ *Fagara zanthoxyloides* Lam.

 Zanthoxylum limonella (Dennst.) Alston

▸ *Zanthoxylum rhetsa* (Roxb.) DC.

 Zanthoxylum nitidum Bunge

▸ *Zanthoxylum simulans* Hance

 Zanthoxylum oxyphyllum Edgew.

Common Names ▸ *Assam:* mezzenga; *Nepal:* timur, bhansi timur, szri
Usage ▸ condiment of curries
Parts Used ▸ fruit
Distribution ▸ Himalayan region, in India also cultivated

Hanelt 2001; Kumar 2003; Wealth of India 11, 1976

 Zanthoxylum parviflorum (A. Rich.) Benth.

Common Names ▸ small-flowered pepper; *German:* Kleinblütiger japanischer Pfeffer
Usage ▸ spice, used like pepper
Parts Used ▸ seed
Distribution ▸ Japan, China, Australia

 Zanthoxylum piperitum DC.

Synonyms ▸ *Fagara piperita* L.
Common Names ▸ Indian pepper, Chinese pepper, Sichuan pepper, Japanese prickly, Japanese pepper, Sichuan pepper, Szechuan pepper; *Chinese:* ch'uan chao, nan chiao, chiao mu; *French:* poivre Szetchuan, clavalier poivrier; *German:* Japanischer Pfeffer, (Echter) Szechuanpfeffer, Chinesischer Gelbholzbaum, Sichuanpfeffer; *Japanese:* sanshō, sansho; *Korean:* chophinamu, sanchonamu; *Russian:* japonckij perez, santoksiljum peretschnij, peretschnik; *Slovakian:* fagara; *Thai:* malar
Usage ▸ condiment (seed), pot herb (young leaf)
Parts Used ▸ seed, leaf
Distribution ▸ N China, Japan, Korea, Mongolia, also cultivated Hawaii

Adesina 1986; Bois 1934; Boisvert/Hubert 2000; Craze 2002; Da-

ZANTHOXYLUM L. - Yellow wood - Rutaceae

 Zanthoxylum piperitum, fruits

vidson 1999; Erhardt et al. 2002; Hanelt 2001; Hiller/Melzig 1999; Jiang et al. 2001; Jiang/Kubota 2004; Lihua Jiang et al. 2001; Kojima et al. 1997; Kusmoto et al. 1968; Lück 2000; Melchior/Kastner 1974; Norman 1991; Pfänder/Frohne 1987; Pochljobkin 1974, 1977; Sakai et al. 1968; Schulze-Motel 1986; Seidemann 1993c; Seidemann/Siebert 1987; Shimoda et al. 1997; Siewek 1990; Small 1997; Teuscher 2003; Tucker 1986; Uphof 1968; Yasuda et al. 1982; Yong-Doo Kim 2000; Zeven de Wet 1982

Zanthoxylum planispinum Sieb. et Zucc.

Common Names ▸ Chinese pepper; *Chinese:* hua chiao, zhu ye jiao; *French:* clavalier à épines planes; *German:* Breitdorniger Pfeffer, Chinesischer Pfeffer; *Hindi:* darmar, nepali dhaniya, tejphal; *India:* tezbal, timal, timbar, tundopoda, *Sanskrit:* dhiva, gandhalu tumburu

Usage ▸ spice

Parts Used ▸ fruit

Distribution ▸ China, Japan, Korea, Riukiu Islands, Taiwan

Bois 1934; Dalby 2000; Davidson 1999; Hanelt 2001; Schultze-Motel 1986

Zanthoxylum rhetsa *(Roxb.) DC.*

Synonyms ▸ *Fagara rhetsa* Roxb., *Zanthoxylum budrunga* Wall. ex DC., *Zanthoxylum limonella* (Dennst.) Alston

Common Names ▸ Indian prickly ash, Indian pepper, lemon pepper (tree), *French:* clavalier d'Inde; *German:* Indischer Pfefferbaum, Japanischer Pfefferbaum, Falsche Kubeben; *Hindi:* badrang; *India:* tambol, tirphal, rhetsamaramu, ilarangom, mullillam; *Javanese:* kadjeng siti, kaju lemal; *Malaysian:* hantar duri; *Sanskrit:* ashvaghra, atitejani, sutejasi, timur; *Sri Lanka:* katu-kina, rhetsu; *Thai:* kanchatton

Usage ▸ spice

Parts Used ▸ immature fruits, seed; **product:** essential oil (seeds)

Distribution ▸ India, Sri Lanka; also cultivated, Indonesia

Note ▸ Seeds as a substitute for pepper by the indigenous population.

Arora/Pandey 1996; Bois 1934; Davidson 1999; Hanelt 2001; Jirovetz et al. 1998; Lück 2004; Schultze-Motel 1986; Seidemann 1993c; Shiva et al. 2002; Small 1997; Teuscher 2003; Uphof 1968; Wealth of India 11, 1976; Wiersema/León 1999; Yasuda et al. 1982

Zanthoxylum senegalensis DC.

▸ *Fagare zanthoxyloides Lam.*

Zanthoxylum simulans Hance

Synonyms ▸ *Zanthoxylum bungeanum* Maxim, *Zanthoxylum bugei* Planch., *Zanthoxylum fraxinoides* Hemsl., *Zanthoxylum nitidum* Bunge,

Common Names ▸ Chinese pepper, Chinese prickly ash pepper; Szechuan pepper, Sichuan pepper; *Chinese:* ch'uan jiao; *French:* poivre chinois, poivre de la Chine, clavalier de Bunge; *German:* Chinesischer Pfeffer, Sech(z)uan-Pfeffer, Sichuan Pfeffer, Täuschende Stachelesche; *Vietnamese:* rau sung, trong

Usage ▸ spice, used like black pepper; leaf as seasoning

Parts Used ▸ seed, leaf

Distribution ▸ N and C China, E Asia; plants in the former Soviet Union

Note ▸ Often an adulterant for black pepper (*Piper nigrum* L.).

Bois 1934; Chen et al. 1994 b, 1995; Chyau et al. 1996; Dalby 2000; Davidson 1999; Ma Chuan-guo 2002; Ogle et al. 2003; Schultze-Motel 1986; Seidemann 1993c; Small 1997; Tirillini/Stoppini 1994; Tirillini et al. 1991; Tscheucher 2003; Uphof 1968; WHO 1990; Wiersema/Léon 1999; Wu/Chen 1993; Xiong et al. 1995; Zeven/de Wet 1982

 Zanthoxylum tessmannii *(Engl.) J.-F. Ayafor.*

Synonyms ▸ *Fagara tessmannii* Engl.
Common Names ▸ African pepper; *French:* poivre africain; *German:* Afrikanischer Pfeffer; *Cameroon:* nashou
Usage ▸ spice (for sauces)
Parts Used ▸ fruit
Distribution ▸ Equatorial Guinea, Congo, Cameroon; also cultivated

Hanelt 2001; Schultze-Motel 1986; Teuscher 2003; Westphal et al. 1980

 Zanthoxylum tidorense *Miq.*

▸ *Zanthoxylum avicennae (Lamk) DC.*

 Zerumbet speciosum *J.W. Wendl.*

▸ *Alpinia zerumbet (Pers.) B.L. Burtt et R.M. Sm.*

ZINGIBER Boehm. - Ginger - Zingiberaceae

 Zingiber amaricans *Bl.*

▸ *Zingiber zerumbet (L.) Rosc. ex Sm.*

 Zingiber aromaticum *Noronha*

▸ *Zingiber officinale Rosc.*

 Zingiber blancai *Hassk.*

▸ *Zingiber zerumbet (L.) Rosc. ex Sm.*

 Zingiber cassumunar *Roxb.*

Synonyms ▸ *Amomum montanum* Koenig, *Cassumunar roxburghii* Colla, *Zingiber luridum* Salisb., *Zingiber montanum* (Koenig) Dietrich, *Zingiber purpureum* Rosc.
Common Names ▸ Bengal ginger, cassumunar ginger; *French:* gingembre marron; *German:* Blockzitwer, Gelber Zitwer; *India:* jangliadrak; *Indonesian:* banglai; *Javanese:* bengle; *Malaysian:* bangle, bolai, bunglai, lampoyang; *Thai:* phlai, puloei, wanfai
Usage ▸ spice, condiment
Parts Used ▸ rhizome
Distribution ▸ Java, also cultivated, cultivated elsewhere tropical Asia: Cochin, China, Malaysia

Arora/Pandey 1996; Bois 1934; Guzman/Siemonsma 1999; Kumar 2001; Masuda/Jitoe 1994; Melchior/Kastner 1974; Schultze-Motel 1986; Seidemann 1993c; Seidemann/Siebert 1987; Teuscher 2003; Uphof 1968; Wu et al. 2000; Zevende Wet 1982

 Zingiber chrysostachys *Ridley*

Common Names ▸ *German:* Goldähriger Ingwer; *Malaysian:* lempui
Usage ▸ spice (locally)
Parts Used ▸ rhizome
Distribution ▸ Malaysia
Note ▸ Alternative for *Zingiber zerumbet* (L.) J.E. Sm.

Burkill 1966; Guzman/Siemonsma 1999; Holttum 1950; Kumar 2001; Theilade 1996

 Zingiber littorale *Val.*

▸ *Zingiber zerumbet (L.) Rosc. ex Sm.*

 Zingiber luridum *Salisb.*

▸ *Zingiber cassumunar Roxb.*

Zingiber mioga (Thunb.) Rosc.

Synonyms ▶ *Amomum mioga* Thunb.

Common Names ▶ mioga ginger, Japanese (wild) ginger; *Arabic:* zangabeel, *Chinese:* xiang he, jang ho; *French:* zédoaire; *German:* Japanischer Ingwer, Japaningwer; *Italian:* zenzero; *Japanese:* myouga, mioga; *Korean:* yangha; *Malaysian:* temu kuning; *Portuguese:* gengibre; *Russian:* imbir; *Spanish:* jengibre;

Usage ▶ spice

Parts Used ▶ rhizome

Distribution ▶ Japan, China, India, also cultivated in E Asia: China, Japan, Korea, Hawaii

Note ▶ The rhizomes have a bergamit-like flavor.

Abe et al. 2002; Bois 1934; Davidson 1999; Duke/Ayensu 1985; Erhardt et al. 2002; Guzman/Siemonsma 1999; Han et al 1983; Hanelt 2001; Jung-Hye Shin et al. 2002; Melchior/Kastner 1874; Schultze-Motel 1986; Seidemann 1993c; Seidemann/Siebert 1987; Small 1997; Staesche 1972; Täufel et al. 1993; Teuscher 2003; Uphof 1968; Wiersema/León 1999; Zeven/de Wet 1982

Zingiber montanum (Koenig) Dietrich

▶ *Zingiber cassumunar* Roxb.

Zingiber nigrum Gaertn.

▶ *Alpinia nigra (Gaertn.) Burtt*

Zingiber officinale Rosc.

Synonyms ▶ *Amomum angustifolium* Salisb., *Amomum zingiber* L., *Zingiber aromaticum* Noronha, *Zingiber zingiber* Karst.

Common Names ▶ ginger, common ginger; *Arabic:* al-zangabeel, zenjabil; *Chinese:* chiang, jiang; *Dutch:* gember; *French:* gingembre; *German:* Echter Ingwer, Ingwer; *Hindi.:* ada, ale, adrak, sonth; *Indonesian:* jahe, jae, lia; *Italian:* zenzero officinale; *Japanese:* shoga, shouga; *Korean:* saenggang, saeyang; *Malaysian:* haliya, jahi, atuja; *Pilipino:* luya, baseng, laya; *Portuguese:* gengibre, gengibre amarelo, gengibre das boticas; *Russian:* imbir', eljy koren', zingiber; *Sanskrit:* ardraka (fresh ginger), singabera, sunthi (dry ginger); *Slovakian:* dumbier; *Spanish:* jengibre, gengibre oficinal; *Thai:* khing, khing-daeng; *Turkish:* zentzephil; *Vietnamese:* gung

Usage ▶ spice, (fresh, dry, candied, shoots, and preserves); **product:** essential oil

Parts Used ▶ rhizome

Distribution ▶ frequent cultivated in the Tropics, specially in tropical Asia: China, India, Indonesia and Australia, W Africa; probable origin in tropical Asia

Note ▶ Fresh ginger-rhizome are often candied and used in confectionery.

▫ Zingiber officinale, fresh rhizom

Akhila/Tawari 1984a; Aris 2001; Atal et al. 1981; Bartley 1995; Bartley/Foley 1994; Bartley/Jacobs 2000; Beek et al 1987; Bois 1934; Boisvert/Huber 2000; Burkill 5, 2000; Chen/Ho 1988; Chen et al. 1986; Connell 1970; Connell/Jordan 1971; Craze 2002; Dake 1995; Dalby 2000; Das/Sarma 2001; Davidson 1999; Dudtschenko et al. 1989; Duke et al. 2003; Erler et al. 1988; Falch 1997; Farrell 1985; Ferrão 1992; Germer/Franz 1997; Gopalam/Ratnambal 1989; Govindarajan 1982a; Goyal/Korla 1997; Gurib-Fakim et al. 2002; Guzman/Siemonsma 1999; Hanelt 2001; Hartmann/Köstner 1994; Herklots 1972; Herrmann 1999; Hiller/Melzig 1999; Hohmann et al. 2001; Hoppe 1949; Hutton 1998; lee et al. 1986; Jansen 1981; Jia-Jiu/Jui-Sen 1994; Jung-Hye Shin et al. 2002; Kang-jin Cho et al. 2001; Kikuzaki et al. 1992; Kim et al. 1996; Krishnamurthy et al. 1977; Kumar 2001; Larkcom 1991; Larsen et al. 1999; Lawrence/Reynolds 1984; Leverington 1975; Macleod/Pieris 1984; Magda 1993; Mascolo et al. 1989; Maistre 1964; Mazza/Oham 2000; Melchior/Kastner 1974; Miyazawa/Kameoka 1988; Morris/Mackley 1999; Nair et al. 1982; Narasinga Rao (no year); Natarajan et al. 1972; Newall et al. 1996; Nishimura 1995; Norman 1991; Onyenekwe/Hashimoto 1999; Opdyke 1974c; Pérez-Gálvez/Mínguez-Mosquera 2001; Peter 2001; Pochljobkin 1974, 1977; Pruthi 1976; Pschyrembel 1998; Rosengarten 1969; Roth/Kormann 1997; Sakawura 1987; Schenck/Naundorf 1966; Schröder 1991; Schuhbaum/Franz 2000; Schultze-Motel 1986; Seidemann 1993c; Seidemann/Siebert 1987; Selbitschka 1991; Sharma 2003; Sharpnel 1967; Siewek 1990; Small 1997; Smith 1981; Staesche 1972; Steinegger/Stück 1982; Täufel et al. 1993; Teuscher 2003; Thode-Sonntag/Thode-Sonntag 1988; Tomlinson 1956; Uphof 1968; Vaupel 2002b; Villamar et al. 1974; Wagner 1992; Westermann 1909; WHO 1990; Wiersema/León 1999; Winterton/Richardson 1965; Wong 1999; Wu 1985; Wüstenfeld/Haensel 1964; Wyk et al. 2004; Yajing Shao et al. 2003; Yusuf et al. 2002; Zarate/Yeoman 1994; Zeven/de Wet 1982; Zhang et al. 1994, 2001

Zingiber purpureum Rosc.

▶ *Zingiber cassumunar Roxb.*

Zingiber spectabile (Griffith) Not.

Common Names ▶ black gingerwort, nodding ginger; *German:* Ansehlicher Ingwer, Nickender Ingwer; *Malaysian:* tepai, tepus halia, tepus tanah; *Thai:* changoe, dakngoe;
Usage ▶ flavoring (locally in Malaysia)
Parts Used ▶ rhizome
Distribution ▶ Malaysian Peninsula, Thai Peninsula

Erhardt et al. 2002; Guzman/Siemonsma 1999; Holttum 1950; Larsen et al. 1999; Wong 1999

Zingiber xanthorrhizum Moon

▶ *Boesenbergia rotunda (L.) Mansf.*

Zingiber zerumbet (L.) Rosc. ex Sm.

Synonyms ▶ *Amomum sylvestre* Lam., *Amomum zerumbet* L., *Zingiber amaricans* Bl., *Zingiber blancai* Hassk., *Zingiber littorale* Val.
Common Names ▶ wild ginger, zerumbet ginger; *American:* shampoo ginger; *French:* gingembre fou, gingerbre blanc; *German:* Bitterer Ingwer; *Indonesian:* lampuyang; *Japanese:* hana-shôga; *Malaysian:* lampoyang, mpojang, noronha; *Pilipino:* barik, langkawas, lampuyang; *Portuguese:* gengibre-amargo; *Spanish* jengibre amargo; *Thai:* kathue, kathue-pa, kawaen
Usage ▶ spice
Parts Used ▶ rhizome
Distribution ▶ SE Asia; probably indigenous to India, cultivated in India, Sri Lanka and China
Note ▶ In Java var. *amaricans* (Bl.) Theilade occurs wild and cultivated, var. *aromaticu*m (Val.) Theilade is found cultivated and sometimes wild or native, whereas var. *zerumbet* is only known cultivated.

◘ **Zingiber zerumbet, flowering**

Arona/Pandey 1996; Bois 1934; Dalby 2000; Davisdon 1999; Erhardt et al. 2002; Guzman/Siemonsma 1999; Holttum 1950; Kumar 2001, 2003; Larsen et al. 1999; Melchior/Kastner 1974; Schultze-Motel 1986; Seidemann 1993a, c; Seidemann/Siebert 1987; Shiobara et al. 1986; Staesche 1972; Täufel et al. 1993; Teuscher 2003; Uphof 1968; Wiersema/León 1999; Wong 1999; Wüstnfeld/Haensel 1964; Yusuf et al. 2002; Zeven/de Wet 1982

Zingiber zingiber Karst.

▶ *Zingiber officinale Rosc.*

ZIZIPHORA L. - Lamiaceae (Labiatae)

 Ziziphora pulegioides *(L.) Desf.*

▸ *Hedeoma pulegioides (L.) Pers.*

 Ziziphora tenuior *L.*

Common Names ▸ *German:* Judendorn; *Russian:* zizifora
Usage ▸ spice (flavoring) in Turkey for yoghurt
Parts Used ▸ leaf
Distribution ▸ C Asia: Turkey (Anatolia)

Dudtschenko et al. 1989; Hanelt 2001; Kulakovskaya 1976; Mabberly 1997; Sezik et al. 1991; Uphof 1968

ZOSTERA L. - Eelgrass - Zosteraceae (Potamogetonaceae)

 Zostera marina *L.*

Synonyms ▸ *Alga marina* Lam. *Zostera maritima* Gaertn.
Common Names ▸ eel grass, grass wrack; *Chinese:* hai dai; *French:* zostère marine; *German:* Gewöhnliches Seegras, Schmalblättriges Seegras; *Italian:* alga marina, aliga, allego; *Japanese:* ama-mo; *Russian:* trawa morskaja, sostera; *Spanish:* hierba marina
Usage ▸ spice of meat of the Seri-Indian
Parts Used ▸ root
Distribution ▸ N Atlantic and N Pacific coasts

Duke/Ayensu 1985; Erhardt et al. 20002; Fleischhauer 2003; Hanelt 2001; Schnelle 1999; Seidemann 1993c; Täufel et al. 1993; Uphof 1968

 **Zostera maritima** *Gaertn.*

▸ *Zostera marina L.*

ZYGOPHYLLUM L. - Zygophyllaceae

 Zygophyllum coccineum *L.*

Common Names ▸ Arabian pepper; *French:* poivre les Arabes; *German:* Arabischer Pfeffer
Usage ▸ spice (rarely)
Parts Used ▸ seed
Distribution ▸ N Africa, Arabia

Bois 1934

 Zygophyllum tridentatum *Sesse et Moç.*

▸ *Larrea tridentata (Sesse et Moç.) ex DC. Cav.*

 Zygophyllum fabago *L.*

Common Names ▸ bean caper, Syrian bean caper; *French:* fabagelle; *German:* Bohnenkaper; *Russian:* parnolistnik
Usage ▸ condiment
Parts Used ▸ flower bud
Note ▸ Substitute for capers (*Capparis spinosa* L.)
Distribution ▸ Caucasus, C and W Asia, India, E and SE Europe, native elsewhere

Dudtschenko et al. 1989; Hanelt 2001; Hiller/Melzig 1999; Rätsch 1998; Schultze-Motel 1986; Wiersema/León 1999

References

A

Abad Farooqi, A.H., A. Misra, A.A. Naqvi, 1983: Effect of plant age on quality and quantity of oil in Japanese mint. Indian Perfumer 27, 80–82

Abbiw, D.K., 1990: Useful plants of Ghana. West African uses of wild and cultivated plants. Intermediate Technol. Public. and the Royal Botanical Garden, Kew London, UK, 337 pp

Abdalla, M.F., N.A.M. Saleh, S. Gabr, A.M. Abu-Eyta, H. El-Said, 1983: Flavone glycosides of Salvia triloba. Phytochemistry 22, 2057–2060

Abe, M., Y. Ozawa, Y. Uda, Y. Yamada, Y. Morimitsu, Y. Nakamura, T. Osawa, 2002: Labdane-type diterpene dialdehyde, pungent principle of myoga, Zingiber mioga Roscoe. Biosci., Biotechnol., Biochem. 66, 2698–2700

Abel Alla El-Sayed, A.M., A. Husseney, A.I. Yassa, 1997: Bestandteile der Nigella sativa-Extrakte und Auswertung des Hemmeffektes auf Wachstum und Aflatoxin-Produktion von Aspergillus parasiticus. Dtsch. Lebensmittel-Rdsch. 93, 149–152

Aboutabl, E.A., N.M. Sokkar, R.M.A. Megid, H.L. de Pooter, H. Masoud, 1995: Composition and microbial activity of Otostegia fruticosa Forrsk oil. J. Ess. Oil Res. 7, 299–303

Abou-Zied, E.N., 1973: The seasonal variations of growth and volatile oil in two introduced types of Majorana hortensis Moench, grown in Egypt. Pharmazie 28, 55–56

Abraham, D.J., J. Trojanak, H.P. Muenzing, H.H.S. Fong, 1971: Structure elicidation of maytenonic acid, a new triterpene from Maytenus senegalensis. J. Pharm. Sci. 60, 1085–1087

Abraham, H., H. Seeber, 1995: Kulturversuche mit Achillea moschata Wulfen. Drogenreport 8, No. 13, 12–19

Abraham, K.O., M.L. Shakaranarayana, B. Raghavan, C.P. Natarajan, 1976: Allium-varieties, chemistry and analysis. Lebensmittel-Wiss. u. Technol. 9, 193–200

Acat, D., M. Acat, 2003: Vanilla, queen of spices. Manufacturing Confectioner 83, No. 5, 81–88

Adam, E., W. Mühlenbauer, A. Esper, W. Wolf, W. Spiess, 2000: Quality changes of onion (Allium cepa L.) as affected by drying process. Nahrung 44, 32–37

Adegoke, E.A., A. Akisanya, S.H.Z. Naqvi, 1968: Studies of Nigerian medicinal plants. I. A preliminary survey of plant alkaloids. J. West Afric. Assoc. 13, 13–33

Adegoke, G.O., B.J. Shura, 1994: Nutritional profile and antimicrobial spectrum of the spice Aframomumdanielli K. Schum. Plants Food for Human Nutril. 45, 175–182

Adegoke, G.O., O. Makinde. K.O. Falade, P.I. Uzo-peters, 2003: Extraction and characterization of antioxidants from Aframomum melegueta and Xylopia aethiopica. Er. Food. Res. Technol. 216 A, 526–528

Adesina, S.K., 1986: Further novel constituents of Zanthoxylum zanthoxyloides root and pericarp. J. Natur. Product 49, 715–716

Adesogan, E.K., 1974: Trithiolaniacin, a novel trthiolane from Petiveria alliacea. J. chem. Soc. Chem., Commun. (London) 906–907

Adewunmi, C.O., S.K. Adesina, V.O. Marquis, 1982: On

the laboratory and field trials of Tetrapleura tetraptera. Bull. Animal Health & Production in Africa. 30, 89–94

Aedo, C., RR.M. Valverde, M.T. Tellería, M. Velayas, 2001: Botánica y botánicos en Guinea ecuatorial. Real Jardín Botánica (CSIC), Agrencia Española de Coop. Internacial (AECI) Madrid, 257 pp

Adzet, T., R. Granger, J. Passet, R. San Martín, M. Simón, 1976: Chimiotypes de Thymus hyemalis Lange. Plantes Med. Phytol. 1, 6–15

Agarwal, V.S., 1990: Economic plants of India. Kailask Prakastan Calcutta (India), 419 pp

Aggarwal, K.K., S.P.S. Khanuja, Ateeque Ahmad, T.R.S. Kumar, V.K. Gupta, S. Kumar, 2002: Antimicrobial activity profiles of the two enantiomers of limonene and carvone isolated from the oils of Mentha spicata and Anethum sowa. Flavour & Fragrance J. 17, 59–63

Agnes, H., P. Teisseire, 1984: Essential oil of French lavender . Perfumer & Flavorist 9, 53–56

Ahlert, B., I. Kjer, 2000: Lebensmittelproduktion in der Entwicklungszusammenarbeit am Beispiel des Pfeffers. Z. Ernährungsökol. 1, 189–194

Ahlert, B., I. Kjer, B. Liebe, T. de Vries, R. Wedekind, 1998: Qualitätsverbesserung von schwarzen Pfeffer auf Erzeugerebene. Z. Arznei- & Gewürzpflanzen 3, No. 2, 15–20

Ahamed, G., F. El-Din, M. Jacques, 1975: Rôle de la temperature dans la mise à fleurs du Calamintha officinalis ssp. nepetoides Jordan. Compt. Rend. hebd. Séances l'Acad. Sci., D 280, 617–620

Ahmed Ijaz, Rai Yaqub, M.W. Akhtar, 1992: Distribution of fatty acids in the triglycerides of Carum copticum. Proc. Pakistan Acad. Sci. 29, 203–211

Ahmed, J., U.S. Shivhare, G. Singh, 2001: Drying characteristics and product quality of coriander leaves. Food & Bioproducts Processing 79 (C2), 103–106

Ahmed, K.M., A.M. Khattab, E.A.M. El-Khrisy, A.Z. Abdel-Amid, 1999: Constituents and molluscicidal activity of the aerial parts of Acacia saligna, Chrysanthemum coronarium and C. parthenium. Bull. Natl. Res. Centre Cairo 24, 13–25

Ahmed, M., R.W. Scora, I.P. Ting, 1994: Composition of leaf oil of Hyptis suaveolens (L.) Poit. J. Ess. Oil Res. 6, 571–575

Ahmed, M., I. Ting, R.W. Scora, 1994: Leaf oil composition of Salvia hispanica L. from three geographic areas. J. Ess. Oil Res. 6, 223–228

Ahamed, Z.F., A.M. El-Moghazy Shoaib, G.M. Wassel, S.M. El-Sayyad, 1971: Phytochemical study of Lantana camara. I. Planta med. 21, 282–288

Ahn, B., C.B. Yang, 191: Volatile flavor components of bangah (Agastache rugosa O. Kuntze) herb. Korean J. Food Sci. Technol. 23, 582–586

Ahn, B.-Z., 1973: Catechintrimeren aus der Eichenrinde und der Tormentillwurzel. Dtsch. Apotheker-Ztg. 113, 1466

Ahro, M. , M. Hakala, J. Sihvonen, J. Kauppinen, H. Kallio, 2001: Low-resolution gas-phase FT-IR method forthe determination of the limonene/carvone ratio in supercritical CO_2-extracted caraway fruit oils. J. Agric. Food Chem. 49, 3140–3144

Aichele, D., H.-W. Schwegler, 1994/1996: Die Blütenpflanzen Mitteleuropas. Franckh-Kosmos Verlags-GmbH & Co. Stuttgart; Bd. 1: 1994, 536 pp, Bd. 2:1994, 544 pp, Bd. 3: 1995, 576 pp, Bd. 4: 1995, 528 pp, Bd. 5: 1996, 527 pp

Aitzemüller, K., 1997: Schwarzkümmelöle. J.-Ber. Bundesanstalt f. Getreide-, Fett- u. Kartoffelforsch., 79 pp

Ajao, A.O., F. Emele, B. Femi-Onadeku, 1985: Antibacterial activity of Euphorbia hirta. Fitoterapia 55, 165–167

Akačić, B., D. Kuštrak, 1960: Versuchskulturen von Ammi visnaga (L.) Lam. und Ammi majus L. Planta med. 8, 203

Aké Assi, L., Sita Guinko, 1991: Plantes used in traditional medicine in West Africa. Edn. Roche, Basel, Switzerland, 151 pp

Akgül, A., 1989a: Volatile oil composition of sweet basil cultivating in Turkey. Nahrung 33, 87–88

Akgül, A., 1989b: A new spice from Turkey: Laser trilobum (L.) Borkh. Acta Alimentaria (Budapest) 18, 65–69

Akgül, A., 1996: A revived flavour: Capers (Capparis ssp.). Food (Gida) 21, 119–128 (in Turkish)

Akgül, A., A. Bayrak, 1987: Constituents of essential oils from Origanum species growing wild in Turkey. Planta med. 53, 114

Akgül, A., M. Kivanç, 1988: Inhibitory effects of six Turkish thyme-like spices on some common foodborne bacteria. Nahrung 32, 201–203

Akgül, A., M. Kivanc, A. Bayrak, 1989; Chemical composition and antimicrobial effect of Turkish laurel leaf oil. J. Ess. Oil Res. 1, 277–280

Akhila, A., D.V. Banthorpe, 1980: Biosynthesis of the

skeleton of pulegone in Mentha pulegium. Z. Pflanzen-physiol. 277–282

Akhila, A., R. Tewari, 1984a: Chemistry of ginger: a review. Current Res. Med. Aromatic Plants 6, 143–156

Akhira, A., R. Tewari, 1984b: Chemistry of patchouli oil: a review. Current Res. Med. Aromatic Plants 6, 38–54

Aksoy, H.A., 1983: Untersuchung der ätherischen Öle der Zwiebel türkischer Herkunft. Z. Lebensmittel-Unters. u. -Forsch. 177, 34–36

Alagukannan, G., M. Vijayakumar, 1999: Effect of plant grown substances on yield attributing parameters, yield and quality in fenugreek (Trigonella foenum graecum Linn.). South Indian Horticult. 47, 130–133

Alan, A.U., J.R. Couch, C.R. Creger, 1968: The carotenoids of the marigold, Tagetes erecta. Canad. J. Bot. 46, 1539–1541

Alberts, A., P. Mullen, 2000: Psychoaktive Pflanzen, Pilze und Tiere. Von Fliegenpilz und Teufelsbeere. Bestimmung, Wirkung, Verwendung. Franckh-Kosmos Verlags-GmbH & Co. Stuttgart, 270 pp

Alcaraztriza, F.J., 1988: Catalogo de las plantas aromaticas, condimentarias y medicinales de la region de Murica. Inst. Nacional de Investigac. Agrarias, Madrid, 156 pp; Monografias INIA No. 67

Ali, M.S., M. Saleem, A.W. Arian, 2001: A new acylated steroid glucoside from Perovskia atriplicifolia. Fitoterapie 72, 712–714

Aligiannis, N., E. Kalpoutzakis, S. Mitaku, I.B. Chinou, 2001: Composition and antimicrobial activity of the essential oils of two Origanum species. J. Agric. Food Chem. 49, 4168–4170

Alinga, T.J., W. Feldmann, 1985: Hemmung der Keimbildung bei gelagerten Kartoffeln durch das ätherische Öl der sudamerikanischen Muñapflanze (Minthostachys spp.). Ernährung/Nutrition 9, 254–256

Al-Jassir, M.S., 1992: Chemical composition and microflora of black cumin (Nigella sativa L.) seeds growing in Saudi Arabia. Food Chem. 45, 239–242

Alkämper, J., 1972: Capsicum-Anbau in Äthiopien für Gewürz- und Färbezwecke. Bodenkultur 23, 97–107

Alkire, B.H., A.O. Tucker, M.J. Maciarello, 1994: Tipo, Minthostachys mollis (Lamiaceae): an Ecuadorian mint. Econ. Bot. 48, 60–64

Alvarruiz, A., M. Rodrigo, J. Miquel, V. Giner, A. Feria, R. Vila, 1990: Influence of brining and packing conditions on product quality of capers. J. Food Sci. 55, 196–198, 227

Al-Maiman, S.A., D. Ahmad, 2002: Changes in physical and chemical properties during pomegranate (Punicagranatum L.) fruit maturation. Food Chem. 76, 437–441

Almela, L., J.M. López-Roca, M.E. Candela, M.D. Alcázar, 1991: Carotenoid composition of new cultivars ofred pepper for paprika. J. Agric. Food Chem. 39, 1606–1609

Alonso, G.L., M.R. Salinas, J. Garijo, 1998: Method to determine and authenticity of aroma on saffron (Crocus sativus L.). J. Food Prodect. 61, 1525–1528

Alonso, G.L., M.R. Salinas, F.J. Esteban-Infantes, M.A. Sánchez-Fernández, 1996: Determination of safranal from saffron (Crocus sativus L.) by thermal desorption-gas chromatography. J. Agric. Food Chem. 44, 185–188

Alonso, G.L., M.R. Salinas, J. Garijo, M.A. Sanchez-Fernandez, 2001: Composition of crocins and picrococinfrom Spanish saffron (Crocus sativus L.). J. Food Quality 24, 219–233

Al-Said, M.S., E.A. Abdelsattar, S.I. Khalifa, F.S. El-Feraly, 1988: Isolation and identification of an anti-inflammatory principle from Capparis spinosa. Pharmazie 43, 640–641

Alonso, G.L., M.A. Sánchez-Fernández, J.R. Sáez, A. Zalacain, M.R. Salinas, 2003: Evaluation of the color of Spanish saffron using tristimulus colorimetry. Ital. J. Food Sci. 15, 249–258

Alvarez-Castellanos, P.P., C.D. Bishop, M.J. Pascual-Villalobos, 2001: Antifungal activity of the essential oil of flowerheads of garland chrysanthemum (Chrysanthemum coronarium) against agricultural pathogens. Phytochemistry 57, 99–102

Alverson, W.S., 1988: A new subspecies of Quararibea funibres (Bombaceae) from Nicaragua. Ann. Missouri bot. Garden 74, 919–922

Ambasta, S.P. (ed), 1986: The useful plants of India. Council of Sci. & Ind. Res., New Delhi

Amelunxen, F., F. Intert, 1993: The filament bundle of Mentha piperita. Planta med. 59, 86–89

Anac, O., 1986: Essential oil contents and chemical composition of Turkish leaves. Perfumer & Flavorist 11, 73–75

Ananda Nayaki, D., S. Natarajan, 2000: Studies on heterosis for growth, flowering, fruit character and yield inchilli (Capsicum annuum L.) South Indian Horticult. 48, 53–55

Anderson Simon, G. G. Muraleedharan Nair, C. Amitabh, Morrison Errol, 1997: Supercritical fluid carbon dioxide extraction of annatto seeds and quantification of trans-bixin by high performance liquid chromatography. Phytochem. Anal. 8, 247–249

Andorn, T.M., N.V. Belova, G.A. Denisova, 1972: Essential oil and coumarin components of Ruta graveolens grown in Moldavia. Herba hung. 11, No. 2, 21–26

Andrews, A.C., 1956: Sage as a condiment in the Graeco-Roman aera. Econ. Bot. 10, 263–266

Andrews, J., 1985: Peppers. The domesticated capsicum. Univ. Texas Press Austin, Tex., 170 pp

Andrews, J., 1998: The pepper lady's pocket pepper primer. Univ. Texas Press Austin Tex., 184 pp

Ang, C.Y.W., Yanyan Cui, H.C. Chang, Wenhong Luo, T.M. Heinze, L.J. Lin, A. Mattia, 2002: Determination of St. John's wort components in dietary supplements and functional foods by liquid chromatography. J. AOAC internat. 85, 1360–1369

Angioni, A., A. Barra, M. Arlorio, J.D. Coisson, M.T. Russo, F.M. Pirisi, M. Satta, P. Cabras, 2003: Chemical composition, plant genetic differences, an antifugal activity of the essential oil of Helichrysum italicum G. Don ssp. microphyllum (Willd.) Nym. J. Agric. Food Chem. 51, 1030–1034

Angioni, A., A. Barra, M.T. Russo, V. Coroneo, S. Dessi, P. Cabras, 2003: Chemical composition of the essential oils of Juniperus from ripe and unripe berries and leaves and their antimicrobial activity. J. Agric. Food Chem. 51, 3073–3078

Anjaneyalu, Y.V., D.C. Gowda, 1978: Studies on the major acidic polysaccharide from the seed mucilage of Ocimum canum (Ocimum americanum). Current Sci. 47, 582–583

Anklam, E., A. Müller, 1993: Extraktion von Vanillin und Ethylvanillin aus Vanillezuckern verschiedener europäischer Provenienzen mit Hilfe von überkritischem Kohlendioxid. Dtsch. Lebensmittel-Rdsch. 89, 344–346

Annamalai, J.K., R.J. Patil, T.D. John, 1988: Improved curing methods for large cardamom. Spice India 4, 5–11

Anonym, (no year): Safran – Crocus sativus L. Museé de l'Alimentation Vevey (Switzerland). Druckschrift, 13 pp

Anonym, 1993: Spices and condiments – Botanical nomenclature. Internat. Organiz. Standardization Rom, ISO/DIS 676, 30 pp

Anonym, 1993: Wanderwege der Natur in Akamas. Fremdenverkehrszentrale Zypern, Fortswirtschaftl. Abt. Nikosia, Zypern 27 pp

Anonym 1998: Achillea – Botanik, Inhaltsstoffe, Analytik und Wirkung. (Workshop). Drogenreport 11, No. 19, pp 28–34

Anonym, 1999: Red Greek saffron, the most precious spice in the world. Food from Greece, No. 10, 82–85

Anonym, 1999: Pistazien aus Kalifornien. Naturwiss. Rdsch. 52, 65

Anonym, 2003: La familia Solanaceae en Jalisco. El género Physalis. Univ. De Guadalajara, Jalisco, Mexico, 127 pp

Apeland, J., 1971: Factors affecting respiration and colour during storage of parsley. Acta Horticult. 20, 43–63

Appendino, G., 1982: Crispoloide, an unusual hydroperoxysesquiterpene lactone from Tanacetum vulgare. Phytochemistry 21, 1099–1102

Appendino, G., P. Garibaldi, G.M. Nano, P. Tétényi, 1984: Tetrahydrofuran-type terpenoids from Tanacetum vulgare. Phytochemistry 23, 2545–2551

Appendino, G., S. Tagliapietra, G.M. Nano, J. Jakupovic, 1994: Sesquiterpene coumarin ethers from asafoetida. Phytochemistry 35, 183–186

Aquino, R., I. Behar, P. Garzarella, A. Dine, C. Pizza, 1985: Composizione chimicae proprietá biologichedella Erythraea centaurium Rafn. Boll. Soc. Ital. Biol. Sper. 61, 165–169

Arana, F.E., 1943: Action of a β-glucosidase in the curing of vanilla. Food Res. 8, 343–351

Araujo, C.C., L.L. Leon, 2001: Biological activities of Curcuma longa L. Mem. Inst. Ostwaldo Cruz. 96, 723–728

Archer, A.W., 1989: Analysis of vanille essences by high-performance liquid chromatography. J. Chromatography 462, 461–466

Arctander, S., 1960: Perfume and flavor materials of natural origin. Self pub., Elizabeth, NJ (USA), 736 pp

Argañosa, G.C., F.W. Sosulki, A.E. Slikard, 1998: Seed yields and essential oil of northern-grown coriander (Coriandrum sativum L.). J. Herbs, Spices, Med. Plants 6, 23–32

Ariña, A., I. Arberas, M.J. Leiton, M. de Renobales, J.B.

Dominguez, 1997: The extraction of yellow gentian root (Gentiana lutea L.). Z. Lebensmittel-Unters. u. -Forsch. 205 A, 295–299

Ariño, A., I. Arberas, G. Renobales, J.B. Domínguez, 1999: Influence of extraction method and storage conditions on the volatile oil of wormwood (Artemisia absinthium L.). Eur. Food Res. Technol. 209, 126–129

Aris, P., 2001: Rezepte und Ideen mit Ingwer. Time-Life Books B.V. Munich, 64 pp

Aritomi, M., T. Kumori, T. Kawasaki, 1985: Cyanogenic glycosides in leaves of Perilla frutescens. var. acuta. Phytochemistry 24, 2438–2439

Arnaud, N., J.-C. Bayle, M. Derbesy, 1983: Étude des gousses de vanille Madagascar – crières analitiques de la récolte 1981. Parfum, Cosmetics, Aromes 53, 99–101

Arnold, W.N., 1988: Vincent von Gogh and the thujone connection. J. Amer. Med. Assoc. 246, 42

Arora, R.K., A. Panday, 1996: Wild edible plants of India. Diversity, conservation and use. Indian Council agric. Res./National Bureau Plant Genetic Resources, New Delhi, 294 pp

Arrebola, M.L., M.C. Navarro, J. Jimenez, F.A. Ocama, 1994: Yield and composition of the essential oil of Thymus serphilloides subspec. serphylloides. Phytochemistry 36, 67–72

Arystanova, T.P., M.P. Irismetor, A.O. Sopbekova, 2001: Chromatographic determination of glycyrrhizinic acidin Glycyrrhiza glabra. Chem. Nat. Compound (Engl. Translation) 37, 89

Asekun, O.T., O. Ekundayo, 2000: Essential oil constituents of Hyptis suaveolens (L.) Poit. (bush tea) leaves from Nigeria. J. Ess. Oil Res. 12, 227–230

Asen, S., T. Arisumi, 1968: Anthocyanins from Hemerocallis. Proc. Amer. Soc. Hortic. Sci. 92, 641–645

Ashok Kumar, S.N. Naik, R.C. Maheswart, A.K. Gupta, 1992: Optimization of process conditions for isolation ofthymol enriched ajowan oil from ajowan seeds using carbondioxide. Indian Perfumer 36, 206–212

Ashton, W.M., E.G. Davies, 1962: Coumarin and related compounds in Anthoxanthum and Melilotus species, and their formation of dicoumarol. Biochem. J. 85, 22–23

Askar, A., M.G. Abd El-Fadeal, S.K. El-Samahy, 1981: Mango und Mangoprodukte. Flüssiges Obst 48, 186–189

Asllani, U., 2000: Chemical composition of Albanian sage oil (Salvia officinalis L.). J. Ess. Oil Res. 12, 79–84

Assouad, W., G. Valdeyron, 1975: Remarques sur la biologie du thym (Thymus vulgaris L.). Bull. Soc. Bot. France 122, 21–34

Atal, C.K., J.N. Ojha, 1965: Studies on the genus Piper. Part IV. Long peppers of Indian commerce. Econ. Bot. 29, 357–360

Atal, C.K., U. Zutshi, P.G. Rao, 1981: Scientific evidence on the role of ayurvedic herbals an bioavailability of drugs. Ethnopharmacology 4, 229–232

Atta, M.B., 2003: Some characteristics of nigella (Nigella sativa L.) seed cultivated in Egypt and its lipid profile. Food Chem. 83, 63–68

Atta, M.B., K. Imaizumi, 1998: Antioxidant activity of nigella (Nigella sativa L.) seeds extracts. J. Japan. Oil Chem. Soc. 47, 475–480

Atkinson, R.E., R.F. Curtis, G.T. Phillips, 1964: Bi-thienyl derivatives from Tagetes minuta L. Tetrahedron Letters 43, 3159–3162

Auterhoff, H., H.-P. Häufel, 1968: Inhaltsstoffe der Damiana-Droge. Arch. Pharmaz. 301, 537–544

Avallone, R., M. Plessi, M. Baraldi, A.J. Manzani, 1997: Determination of chemical composition of carob (Ceratonia siliqua): protein, fat, carbohydrates, and tannin. J. Food Composition Anal. 10, 166–172

Avanzato, D., 2003: Una forma momoica di Pistacia terebinthus L. scoperta in Bulgaria. Riv. Frutticolt. di ortofloricocolt. 65, No. 10, 55–53

Avigad, N., 1990: The inscribed pomegranate from the "House of the Lord". Biblical Archaeologist 53, 157–166

Ayaz, F.A., 1997: Studies on water soluble sugar and sugar alcohol in cultivar and wild forms of Lauroceraus officinalis Roem. Pakistan J. Bot. 29, 331–336

Ayaz, F.A., A. Kadioglu, M. Reunanen, M. Var, 1997: Phenolic acid and fatty acid composition in the fruit of Laurocerasus officinalis Roem. and its cultivars. J. Food Composition and Analysis 10, 350–357

Ayaz, F.A., M. Reunanen, M. Küçükislamoglu, M. Var, 1995: Seed fatty acid composition in wild form and cultivars of Laurocerasus officinalis Roem. Pakistan J. Bot. 276, 305–308

Ayedoun, A.M., B.S. Adeoti, P.V. Sossou, P.A. Leclercq, 1996: Influence of fruit conservation methods on the essential oil composition of Xylopia aethiopica (Dunal) A. Richard. from Benin. Flavour & Fragrance J. 11, 245–250

Ayensu, E.S., 1978: Medicinal plants of the West African. Reference Publ. Inc., Algonac, Mich. 330 pp

Ayfer, M., 1967: La culture du pistachier en Turquie. Fruits 22, 351–366

Ayuga, C., M.Rebuelta, 1986: A comparative study of phenolic acids of Hypericum caprifolium Boiss and Hypericum perforatum L. An. Real Acad. Farm. 52, 723–728

Azevedo, N.R., I.F.P. Campos, H.D. Ferreira, T.A. Portes, S.C. Santos, J.C. Seraphin, J.R. Paula, P.H Ferri, 2001: Chemical variabilty in the essential oil of Hyptis suaveolens. Phytochemistry 57, 733–736

B

Babayan, V.K., D. Kootungal, G.A. Halaby, 1978: Proximate analysis, fatty acid and amino acid composition of Nigella sativa L. seeds. J. Food Sci. 43, 1314–1319

Bachthaler, G., Ch. Franz, D. Fritz, J. Hölzl, A. Vömel, 1976: Vergleichende Untersuchungen verschiedener Herkünfte und Sorten von Mentha piperita. 2. Teil. Bayer. Landwirtsch. Jb. 53, 35–47

Badoc, A., A. Lamarti, 1991: A chemotaxonomic evaluation of Anethum graveolens L. (dill) of various origins. J. Ess. Oil Res. 3, 269–278

Baer, D.F., 1977: Systematics of the genus Bixa and geography of the cultivated annotto tree. Ph.D. thesis, Univ. of California, Los Angeles 239 pp; Dissertations Abstr. Internat. B37 (10), 4846 B

Bahkali, A.H., M.A. Hussain, AA.Y. Basahy, 1998: Protein and oil composition of sesame seeds (Sesamum indicum L.) grown in the Bizan area of Saudi Arabia. Internat. J. Food Sci. Nutrit. 49, 409–414

Bahl, C.P., T.R. Seshadri, T.N.C. Vedantham, 1971: Preparation of bixin and methyl bixin from Indian seeds of Bixa orellana. Curr. Sci 2, 27–28

Bahl, J.R., Shweta Sinha, A.A. Naqvi, R.P. Bansal, A.K. Gupta, Sushil Kumar, 2002: Linalool-rich essential oil quality variants obtained from irradiated stem nodes in Lippea alba. Flavour & Fragrance J. 17, 127–132

Baik, Mun-Chan, Ho-Dzan Hoang, K. Hammer, 1986: A check-list of Korean cultivated plants. Kulturpflanze 34, 69–144

Baldry, J., J. Dougan, W.S. Matthews, J. Nabney, G.R. Richering, 1976: Composition and flavour of nutmegoils. Internat. Flavours & Food Additives 7, 28–30

Ballarin, C., J. Ballarin, 1972: Zur dünnschichtchromatographischen Unterscheidung von Fenchel- und Anisöl. Pharmazie 27, 544

Bandyopadhyay, C., V.S. Narayan, P.S. Variyar, 1990: Phenolics of green pepper berries (Piper nigrum L.). J. Agric. Food Chem. 38, 1696–1699

Bano, M.J. del, J. Lorente, J. Castillo, O. Benavente-Garcia, J.A. del Rio, A. Ortuno, K.W. Quirin, D. Gerard, 2003: Phenolic diterpenes, flavones, and rosmarinic acid distribution during the development of leaves, flowers, stems, and roots of Rosmarinus officinalis. Antioxidant activity. J. Agric. Food Chem. 51, 4247–4253

Bansal, R.P., J.R. Bahl, S.N. Garg, A.A. Naqvi, S. Sharma, Muni Ram, Sushil Kumar, 1999: Variation in quality of essential oil distilled from vegetative and reproductive stages of Tagetes minuta crop grown in north India plains. J. Ess. Oil Res. 11, 747–752

Barberán, F.A.T., L. Hernández, F. Tomás, 1986: A chemotaxonomic study of flavonoids in Thyme capitata. Phytochemistry 25, 561–562

Barbetti, P., G. Grandolini, G. Fardella, I. Chiappini, 1987: Indole alkaloids from Quassia amara. Planta med. 53, 289–290

Barbut, S., D.B. Josephson, A. Maurer, 1985: Antioxidant properties of rosmary oleoresin in Turkey sausage. J. Food Sci. 50, 1356–1363

Baren, C.M., L. van Muschietti, P. di Leo Lira, J.D. Coussio, A.L. Bandoni, 2001: Volatile constituents from areal parts of Cunila spicata. J. Ess. Oil Res. 13, 351–353

Barminas, J.T., M.K. James, U.M. Abubakar, 1999: Chemical composition of seeds and oil of Xylopia aethiopica grown in Nigeria. Plant Foods Human Nutrit. 53, 193–198

Barnicoat, C., 1945: The reactions and properties of annatto as a cheese colour. J. Dairy Res. 8, 61–73

Barrau, J., 1961: Subsistance agriculture in Polynesia and Micronesia. Bernice P. Bishop Museum Bull., No. 23, 94 pp

Barrett, D.R., J.E.D. Fox, 1994a: Early growth of Santalum album in relation to shade. Austral. J. Bot. 42, 83–93

Barrett, D.R., J.E.D. Fox, 1994b: Santalum album: kernel composition, morphological and nutrient characteristics of pre-parasitic seedlings under various nutrient regimes. Ann. Bot. 79, 59–66

Bärtels, A., 1997: Farbatlas Mediterrane Pflanzen. Verlag E. Ulmer Stuttgart 400 pp

Barth, H.J., Chr. Klinke, C. Schmidt, 1994: The hop atlas. The history and geography of the of the cultivated plant. Fachverlag H. Carl Nürnberg, 384 pp

Bartley, J.P., 1995: A new method for the determination of pungent compounds in ginger (Zingiber officinale). J. Sci. Food Agric. 68, 215–222

Bartley, J.P., P. Foley, 1994: Supercritical fluid extraction of Australian-grown ginger (Zingiber officinale). J. Sci. Food Agric. 66, 365–371

Bartley, J.P., A.L. Jacobs, 2000: Effects of drying on flavour compounds in Australian-grown ginger (Zingiber officinalis). J. Sci. Food Agric. 80, 209–215

Bartschat, D., A. Mosandl, 1997: 3-Butyl(hexahydro)phthalide. Charakteristische Inhaltsstoffe des Seleries. GITLabor-Fachzeitschr., 874–876

Bartua, A.K., P. Chakrabarti, M.K. Chowdhury, A. Basak, K. Basu, 1976: Triterpenoids. The structure and stereochemistry of lantanolic acid – a new triterpenoid from Lantana camara. Tetrahedron 27, 1141–1147

Baruah, A., S.C. Nath, 2000: Morphology and essential oils of two Cinnamomum parthenoxylon variants growing in North East India. J. Med. & Aromatic Plants Sci. 22 (1 B) 370–376

Baruah, A., S.C. Nath, 2001: Cinnamomum assamicum S.C. & A. Barua – A new species of Lauraceae from North eastern India. J. Econ. & Taxon. Bot. 25, 27–32

Başer, K.H.C., T. Özek, 1993c: Composition of essential oil of Calamintha grandiflora. Planta med. 59, 390

Başer, K.H.C., T. Özek, N. Kirimer, 1993b: The essential oil of Laser trilobium fruit of Turkish origin. J. Ess. Oil Res. 5, 365–369

Başer, K.H.C., H. Malyer, 1996: Essential oil of Tagetes minuta L. from Turkey. J. Ess. Oil. Res. 8, 337–338

Başer, K.H.C., N. Öztürk, 1992: Composition of the essential oil of Dorystaechas hastata, a monotypic endemic from Turkey. J. Ess. Oil Res. 4, 369–374

Başer, K.H.C., T. Özek, A. Akgül, G. Tümen, 1993a: Composition of the essential oil of Nepeta racemosa Lam. J. Ess. Oil Res. 5, 215–217

Başer, K.H.C., T. Özek, G. Tümen, E. Seszik, 1993d: Composition of the essential oils of Turkish Origanum species with commercial importance. J. Ess. Oil Res. 5, 619–623

Başer, K.H.C., B. Demirci, A.A. Donmez, 2003: Composition of the essential oil of Perilla frutescens (L.) Brittonfrom Turkey. Flavour & Fragrance J. 18, 122–123

Başer, K.H.C., T. Özek, M. Kürkcüoglu, G. Tümen, 1994a: The essential oil of Origanum vulgare ssp. hirtum of Turkish origin. J. Ess. Oil Res. 6, 31–36

Başer, K.H.C., F.Z. Erdemgil, T. Özek, 1994b: Essential oil of Echinophora tenuifolia L. subsp. sibthor piana (Guss.) Tutin. J. Ess. Oil Res. 6, 399–400

Başer, K.H.C., M. Kürkcüoglu, T. Özek, 1992: Composition of the Turkish cumin seed oil. J. Ess. Oil Res. 4, 133–138

Başer, K.H.C., T. Özek, M. Tutas, 1995: Composition of cold-pressed bergamit oil from Turkey. J. Ess. Oil Res. 7, 341–342

Başer, K.H.C., M. Kurkcuolgu, B. Demirci, T. Ozek, 2003: The essential oil of Origanum syriacum L. var. sinaicum (Boiss.) Letswaart. Flavour & Fragrance J. 18, 98–99

Başer, K.H.C., B. Demirci, N. Kirimer, F. Satil, G. Tumen, 2002: The essential oil of Thymus migricus and T. fedtscheenkoi var. handelii from Turkey. Flavour & Fragrance J. 17, 41–45

Başer, K.H.C., T. Özek, T. Demirci, Y. Saritas, 2000: Essential oil of Crithmum maritimum L. from Turkey. J. Ess. Oil Res. 12, 424–426

Basker, D., M. Negbi, 1983; Uses of saffron. Econ. Bot. 37, 228–236

Basker, D., M. Negbi, 1985: Crocetin equivalent of saffron extracts: comparison of three extraction methods. J. Assoc. Publ. Analysis 23, 65–69

Baslas, B.K., R.K. Baslas, 1971: Chemical studies of the essential oil from the plants of Anethum graveolens and Anethum sowa (dill oils). Indian Perfumer 15, 27–29

Batten, A., 1986. Flowers of Southern Africa. Frandsen Publ. (PTY) Ltd. Johannesburg, S. Africa, pp 133–135

Bauer, L., G.A. de Assis Brasil e Silva, 1973: Os óleos essenciales de Chenopodium ambrosioides e Schinus terebinthifolius no Rio Grande do Sul. Rev. Brasil. Farm. 54, 240–242

Bayrak, A., A. Akgül, 1994: Volatile oil composition of Turkish rose (Rosa damascena). J. Sci. Food Agric. 64, 441–448

Baysal, T., D.A.J. Starmans, 1999: Supercritical carbon dioxide extraction of carvone and limonene from caraway seeds. J. Supercrit. Fluids 14, 225–234

Beaubaire, N.A., J.E. Simon, 1987: Ertragspotential von Borago officinalis L. 6. Internat. Symp. Med. & Aromatic Plants. The Hague, Acta Horticult., pp 101–113

Becela-Deller, Chr., 1991: Die Weinraute. Heilpflanze zwischen Magie und Wissenschaft. Dtsch. Apotheker-Ztg. 131, 2705–2709

Beck, T.A. van, M.A. Posthumus, G.P. Lelyveld, H.V. Phiet, B.T. Yen, 1987: Investigation of the essential oil of Vietnamese ginger. Phytochemistry 26, 3005–3010

Becker, E., 1981: Über die Verwendung der Blüten des Holunders als Lebensmittel. Riechstoffe, Aromen, Kosmetika 10, 330

Becker, H., 1983: Baldrian – eine viel bearbeitete Droge mit viel ungelösten Problemen. Dtsch. Apotheker-Ztg. 123, 2470–2473

Becker, H., 1986: Valeriana officinalis L. – Baldrian. Z. Phytother. 7, 149–152

Beier, R.C., E.H. Oertli, 1983: Psoralen and other linear furocoumarins as phytoalexins in celery. Phytochemistry 22, 2295–2597

Beier, R.C., G.W. Ivie, E.H. Oertli, D.L. Holt, 1983: HPLC analysis of linear furanocoumarins (psoralens) in healthy celery (Apium graveolens). Food Chem. Toxicol. 21, 163–165

Bekele, J., A. Hassanali, 2001: Blend effects in the toxicity of the essential oil constituents of Ocimum kilimandscharicum and Ocimum kenyense (Labiatae) on two post-harvest insect pests. Phytochemistry 57, 385–391

Bellakhdar, J., A.I. Idrissi, S. Canigueral, J. Iglesias, R. Vila, 1994: Composition of lemon verbena (Aloysia triphylla [L'Herit.] Britton) oil of Moroccan origin. J. Ess. Oil. Res. 6, 523–526

Bellarita, V., F. Schiafella, T. Mezzeti, 1974: Triterpenoids of Centaurum erythreae. Phytochemistry 13, 289

Bellesia, F., U.M. Pagnoni, A. Pinetti, R. Trave, 1983: The biosynthesis of dolichodial in Teucrium marum. Phytochemistry 22, 2197–2201

Ben-Arie, R., N. Segel, S. Guel fat-Reich, 1984: The maturation and ripening of the wonderful pomegranate. J. Amer. Soc. Horticult. Sci. 109, 898–902

Bendel, L., 202: Das große Früchte- und Gemüse-Lexikon. Patmos Verlag GmbH Düsseldorf, 411 pp

Bendini, A., T.G. Toschi, G. Lercker, 2002: Antioxidant activity of oregano (Origanum vulgare L.) leaves. Ital. J. Food Sci. 14, 17–24

Benero, J.R., A.J. Rodriguez, A. Collazo de Rivera, 1972: Tamarind. J. agric. Univ. Puerto Rico 56, No. 2, 185–186

Benk., E., 1981: Über die Verwendung der Blüten des Schwarzen Holunders als Lebensmittel. Riechstoffe, Aromen, Kosmetika 10, 330

Benoni, H., 2000: Enthält der Sumpfporst Alkaloide? Naturwiss. Rdsch. 53, 139–141

Benzinger, E., 1986: Naše lavandule (Yugoslav lavender). Farmac. Glasnik 42, No. 7, 193–201

Bergemann, L., 1950: Mytroxylon balsamum (L.) Harms var. pereirae (Royle) Baillon, der Persubalsam. Pharmazie 5, 341–347, 397–402, 450–455, 494–498, 548–553

Berger, F., 1958: Zur Samenanatomie der Zingiberaceen-Gattungen: Elettaria, Ammomum und Aframomom. Scii. Pharmac. (Vienna) 26, 224–258

Berger, F., 1954–1967: Handbuch der Drogenkunde. Erkennung, Wertbestimmung und Anwendung. Verlag für med. Wissenschaften, W. Maudrich Vienna. Bd. 1: Untersuchungsmethoden. Cortices – Flores, 401 pp; Bd. 2: Folia, 457 pp; Bd. 3: Fructus-Lignea, 1952, 558 pp; Bd. 4 Herbae, 1954, 609 pp; Bd. 5: Radices 1960, 626 pp; Bd. 6: Balsame, Harze, Gummiharze, Milchsäfte, Gummi, Extrakte, Samen, 1964, 489 pp; Bd. 7: Gesamtregister 1967, 144 pp

Berger, F., 1964/65: Neue Erkenntnisse auf dem Gebiet der Kardamomforschung. Gordian (Hamburg) 64, 836–839, 885–888, 922–924, 956–958; 960–961; 65, 24–27

Berger, R., 2001: Gelbwurzel hilft Magen und Leber. Curcuma longa – pflanzliches Arzneimittel mit Zukunft. PTA heute 15, No. 12, 42, 44, 46–47

Berger, W., 1984: Die Probenahme – eine wichtige Voraussetzung für die analytische Bewertung von Hopfen und Hopfenprodukten. Lebensmittelindustrie 31, 123–126

Bergeron, R., 1980: La vanilla: origin, culture, technologie, utilisation. Labo-Pharma-Problèms 229, 475–482

Berghöfer, R., J. Hölzl, 1986: Johanniskraut (Hypericum perforatum L.), Prüfung und Verfälschung. Dtsch. Apotheker-Ztg. 126, 2569–2573

Berlin, B., D.E. Breedlove, P.H. Raven, 1974: Principles of Tzeltal plant classification. An introduction to the botanied ethnography of botanied Mayan-speaking people of Highland Chiapas. Academic Press New York, London, 660 pp

Bernáth, J., E. Nemeth, A. Kaffaa, E. Hethely, 1996: Morphological and chemical evaluation of fennel (Foeniculum vulgare Mill.) populations of different origin. J. Ess. Oil Res. 8, 247–253

Bernhard, H.O., K. Thiele, 1981: Additional flavonoids from the leaves of Larrea tridentata. Planta med. 41, 100–103

Bernhard, R.A., 1970: Chemotaxonomy: Distribution studies of sulphur compounds in Allium. Phytochemistry 9, 2019–2927

Bernard, R.A., R. Wrolstad, 1963: The essential oil of Schinus molle, the terpene hydrocarbon fraction. J. Food Sci. 28, 59–63

Bernhard, R.A., T. Shibamoto, K. Yamaguchi, E. White, 1983: The volatile constituents of Schinus molle L. J. Agric. Food Chem. 31, 463–466

Berta, E., M.A. Bernal, J. Díaz, F. Pomar, F. Merino, 2000: Fruit development in Capsicum annuum: changes in capsaicin, lignin, free phenolics, and peroxidase pattern. J. Agric. Food Chem. 48, 6234–6239

Bertelli, D., M. Plessi, F. Mighetta, 2003: Effect of microwaves on volatile compounds in origanum. Lebensmittel-Wiss. u. -Technol. 36, 555–560

Beuchat, L.R., 1984: Fermented soybean foods. Food Technol. 38, 67–70

Bhalkar, S.V., P.J. Dubash, 1983: Methods of extraction of annatto from the seeds of Bixa orellana. Indian J. Dairy Sci. 36, 157–161

Bharuchta, F.R., G.V. Josh, 1957: Organic acid metabolism in Portulaca oleracea L.). Naturwissenschaften 44, 263

Bhat, J.V., R. Broker, 1953: Riboflavine and thiamine content of saffron, Crocus sativus L. Nature (London) 172, 544

Battacharjee, P., R.S. Singhal, A.S. Gholap, 2003: Supercritical carbon dioxide extraction for identification of aldultcration of black pepper with papaya seeds. J. Sci. Food Agric. 83, 783–786

Bhattacharya, S., S. Bal, R.K. Mukherjee, S. Bhattacharya, 1993: Some physical and engineering properties of tamarind (Tamarindus indica) seed. J. Food Engng. 18, 77–89

Bhattacharya, S., S. Bal, R.K. Mukherjee, S. Bhattacharya, 1994a: Studies on the characteristics of some products from tamarind (Tamrindus indicus) kernel. J. Food Sci. Technol. 31, 372–376

Bhattacharya, S., S. Bal, R.K. Mukherjee, S. Bhattacharya, 1994b: Functional and nutritional properties of tamarind (Tamarindus indica) kernel protein. Food Chem. 49, 1–9

Bianchina, A., P. Tomi, J. Costa, A.F. Bernardini, 2001: Composition of Helichrysum italicum (Roth) G. Donfil. subsp. italicum essential from Corsica (France) Flavour & Fragrace J. 16, 30–34

Bicchi, C., C. Frattini, 1987: Considerants and remarks on the analysis of Anthemis nobilis L. essential oil by capillary gas chromatography and "hyphenated" techniques. J. Chromatogr. 411, 237–249

Bicchi, C., M. Fresia, P. Rubiolo, D. Monti, C. Franz, I. Goehler, 1997: Constituents of Tagetes lucida Cav. ssp. lucida essential oil. Flavour & Fragrance J. 12, 47–52

Bicking, B., 1986: Die Zimtwirtschaft auf Sri Lanka (Ceylon). Anbau und Vermarktung, historische Bindung und aktuelle Perspektiven eines traditionsgebundenen Produktes. Mainzer geograph. Studien, No. 28, 148 pp

Bielenberg, J., 1998: Die Süßholzwurzel. Z. Phytother. 19, 197–208

Bielenberg, J., 2002: Die grüne Fee. Zentralnervöse Effekte durch Thujon. Österr. Apotheker-Ztg. 56, 566–569

Bierther, E., 1984: Die wohlriechende Eugenia. Dragoco ber. 29, 90–93

Biftu, T., 1981: Essential oil composition of Aframomum korarima. J. Chromatogr. 211, 280–283

Bilgri, A., B. Adam, 2000: Das Kloster Andechs Kräuterbuch. Wie Mönche Würzen und Heilen. St. Ulrich-Verlag Augsburg, 176 pp

Bilia, A.R., G. Flamini, V. Taglioli, I. Morelli, F.F. Vincieri, 2002: GC-MS analysis of essential oil of some commercial fennel teas. Food Chem. 76, 307–310

Billaud, C., J. Adrian, 2001: The importance of fenugreek in nutrition. Med. et Nutrit. (Paris) 37, 59–64

Bjeldanes, L.F., I. Kim, 1978: Sedative activity of celery oil constituents. J. Food Sci. 43, 143–144

Bijttebier, Fr.J., 1986: The Treculia africana, a tree with an alimentary purpose, with a very promising future in the Third World. 12. Congr. Internat. Cereal Chem. Hamburg, 23.–24.4., 220–222

Bittencourt Sprricigo, C., A. Bolzan, R.A.F. Machado, L.H. Castalan Carlson, J.C.P. Petrus, 2001: Separation of nutmeg essential oil and dense CO_2 with a cellulose acetate reverse osmosis membrane. J. Membrane Sci. 188, 173–179

Blanc, P., G. de Saqui-Sannes, 1972: Flavonoids of Euphorbia hirta. Plantes méd. Phytothér. 6, 106–109

Blanca, G., M. Cueto, L. Gutiérrez, M.J. Martínez, 1993: Two hybrids of Thymus hyemalis Lange (Labiatae) from Southeastern Spain. Folia geobot. Phytotax. (Praha) 28, 135–140

Blancke, R., 2000: Farbatlas exotische Früchte. Obst und Gemüse der Tropen und Subtropen. Verlag E. Ulmer Stuttgart, 286 pp

Blank, I., W. Grosch, 1991: Evaluation of potent odorants in dill seed and dill herb. Food Sci. 56, 63–67

Blank, R.J., E. Ramuntschack, Chr. Wilhelms, 1997: Kemiri (candle nut) und Kemiri-Produkte. Dtsch. Ges. techn. Zusammenarbeit (GTZ) GmbH Eschborn, Owner Edit. 12 pp

Blázquez, M.A., M.C. Zafa-Polo, 1990: A new chenotype of Thymus vulgaris ssp. aestivum (Renter ex Willd.) A. Bolós & O. Bolós. Pharmazie 45, 802–803

Block, E., 1985: The chemistry of garlic and onions. Sci. Amer. 252, 114–119

Block, E., 1992: The organosulfur chemistry of the genus Allium: implications for the organic chemistry of sulfur. Angew. Chem., Int. edn. Engl. 31, 1135–1178

Block, E., S. Naganathan, D. Putman, S.H. Zhao, 11992a: Allium chemistry: HPLC analysis of thiosulfinites from onion, garlic, wild garlic, leek, scallion, shallot, elephant (great-headed) garlic, chive, and Chinese chive; uniquely high allyl-to-methyl ratios in some garlic samples. J. agric. Food Chem. 40, 2418–2430

Block, E., D. Putman, SH. Zhao, 1992b: Allium chemistry: GC-MS analysis of thiosulfinates and related compounds from onion, leek, scallion, shallot, chive, and Chinese chive. J. Agric. Food Chem. 40 2431–2438

Blundstone, H.A.W., D. Dickinson, 1964: The chemistry of edible rhubarb. J. Sci. Food Agric. 15, 94–101

Bocchini, P., C. Andolò, R. Pozzi, G.C. Galletti, A. Antonelli, 2001: Determination of diallyl thiosulfinate (allicin) in garlic (Allium sativum L.) by high-performance liquid chromatography with a post-column photochemical reactor. Anal. Chim. Acta 441, 37–43

Boelens, M.H., 1985: The essential oil from Rosmarinus officinalis. Perfumer & Flavorist 10, 21–37

Boelens, M.H., 1986: The essential oil of spike lavender Lavandula latifolia Vill. (L. spica DC.). Perfumer & Flavorist 11, No. 5, 43–63

Boelens, M.H., 1991: A critical review on the chemical composition of the Citrus oils. Perfumer & Flavorist 16, 17–34

Boelens, M.H., 1992: Sensory and chemical evaluation of tropical grass oils. Perfumer & Flavorist 19, No. 2, 29– 36, 38–40. 42–44

Boelens, M.H., 1995: Chemical and sensory evaluation of Lavandula oils. Perfumer & Flavorist 20, No. 3, 23–51

Boelens, M.H., 1997: Chemical and sensory evaluation of three sage oils. Perfumer & Flavorist 22, No. 2, 19–40

Boelens, M.H., R. Jimenez, 1989: The chemical composition of the peel oils from unripe and ripe fruits of bitter orange, Citrus aurantium L. ssp. amara Engl. Flavour & Fragrance J. 4, 139–142

Boelens, M., P. de Valois, H. Wobben, A. van der Gen, 1971: Volatile flavour compounds from onion. J. Agric. Food Chem. 19, 984–991

Böttcher, H, I. Günther, 2003: Physiologisches Nachernteverhalten von Drachenkopf, Dracocephalum moldavica L . Drogenreport 16, No. 29, 9–13

Bohn, I.U., 1991: Pimpinella saxifraga und Pimpinella major – die kleine und große Pimpinelle. Z. Phytother. 12, 98–104

Bohn, I.U., K.-H. Kubeczka, W. Schultze, 1989: The essential oil of Pimpinella major. Planta med. 5, 489–490

Bois D., 1934: Les Plantes alimentaires chez tous les pouples et travers les ages. Histoire, utilisation, culture. Vol. III. Plantes a épices, a aromates, a condiments. Édit. P. Lechevalier Paris, 289 pp

Boisvert, C., P. Aucante, 1993: Saveurs du Safran. Édit. A. Michel S.A. Paris, 124 pp

Boisvert, C., A. Hubert, 1998: l'ABCdaire des Épices. d'Edn. Flammarion Paris, 119 pp

Bombarda, I.; E.M. Gaydou, J. Smadja, R. Faure, 1996: Sesquiterpenic expoxides and alcohols derived from hydrocarbons of vetiver essential oil. J. Agric. Food Chem. 44, 217–222

Bomme, U., 1989: Wirkstoffgehalte und Erträge von Ringelblumen-Sorten. Dtsch. Gartenbau 43, 2488–2489

Bomme, U., E. Feicht, R. Rinder, 2002: Ergebnisse aus mehrjährigen Leistungsprüfungen mit ausgewählten Herkünften von Zitronenmelisse (Melissa officinalis L.). Z. Arznei- u. Gewürzpflanzen 7, 368–376

Bonnet, B., 1976: Le poireau (Allium porrum L.): Aspects botaniques et agronimiques. Revue bibliographique. Saussurea 7, 121–155

Bonnländer, B., P. Winterhalter, 2000: 9-Hydroxypiperitone-β-D-glucopyranoside and other polar constituents from dill (Anethum graveolens L.) herb. J. Agric. Food Chem. 48, 4821–4825

Boo-Yong Lee, Hyeon-Son Choi, Se-Ryang Oh, Hyeong-Kyu Lee, 2001: Rheological characteristics of hot-water extract concentrates from Agastache rugosa O. Kuntze. Food Sci. Biotechnol. 10, 294–298

Bor, N.L., 1953, 1954: The genus Cymbopogon Spreng. in India, Burma and Ceylon. J. Bombay nat. History Soc. 51, 890–916, 52, 149–183

Borde, K., Ch. Chwoika, J. Petermann, U. Schmidt, 1989: Hopfen. VEB Deutscher Landwirtschaftsverlag Berlin, 188 pp

Borges, P., J. Pino, 1993a: The isolation of volatile oil from cumin seeds by steam destillation. Nahrung 37, 123–126

Borges, P., J. Pino, 1993b: Preparation of nutmeg oleoresin by alcohol extraction. Nahrung 37, 280–282

Borges, P., J. Pino, E. Sanchez, 1992: Isolation and chemical characterization of laurel leaf oil. Nahrung/Food 36, 494–496

Born, Chr., 1991: Die Walnuß. PTA heute 5, 509–510

Boruah, P., B.P. Misra, M.G. Pathak, A.C. Ghosh, 1995: Dynamics of essential oil of Cymbopogon citratus (DC.) Staph. under rust disease indices. J. Ess. Oil Res. 7, 337–338

Bos, R., H.J. Woerdenbag, H. Hendriks, J.J.C. Scheffer, 1997: Composition of the essential oils from underground parts of Valeriana officinalis L. s. l. and several closely related taxa. Flavour & Fragrance J. 12, 359–370

Bosabalidis, A.M., 1987: Morphometric evaluation of inclusion body containing leucoplasts in leaf epidermal cells of Origanum dictamnus L. Bot. Helvet. 97, 315–321

Bosabalidis, A.M., 1990a: Quantitative aspects of Origanum dictmanus L. glandulare scales. Bot. Helvet. 100, 199–206

Bosabalidis, A.M., 1990b: Glandular trichomes in Satureja thymbra leaves. Ann. Bot. 65, 71–78

Bosabalidis, A.M., 1996: Ontogenesis, ultrastructure, and morphometry of the petiole oil ducts of celery (Apiumgraveolens L.). Flavour & Fragrance J. 11, 269–274

Bosabalidis, A., I. Tsekos, 1982: Glandular scale development and essential oil secretion in Origanum dictamnus L. Planta 156, 496–504

Bose, T.K., 1985: Fruits of India. Tropical and subtropical. Naya Prokash Calcutta, India, 616 pp

Boss-Teichmann, C., Th. Richter, 2002: Bärlauch & Knoblauch. Sammeln und Anbau – Fitness und Gesundheit – Feine Rezepte. Verlag E. Ulmer Stuttgart, 79 pp

Böttcher, H., I. Günther, 2002: Physiologisches Nachernteverhalten von Wermut-Pflanzen (Artemisia absinthium L.). Drogenreport 15, No. 27, 3–8

Böttcher, H., I. Günther, 2003: Physiologisches Nachernteverhalten von Weinraute-Kraut (Ruta graveolens L.). Drogenreport 16, No. 30, 3–8

Bouriquet, G., 1954: Le vanillier et la vanille dans le monde. d'Edn. P. Lechevalier Paris, 748 pp

Bournot, K., 1968: Ätherische Öle. Lieferung 7: J.V. Wiesner: Die Rohstoffe des Pflanzenreichs. (Hrsg.: C. v. Regel.), Verlag J. Cramer Lehre, 5. Ed., 173 pp

Bourwieg, D., R. Pohl, 1973: Die Flavonoide von Mentha longifolia. Planta med. 24, 304–314

Bouverat-Bernier, J.P., 1989: Comparaison variétale de quatre Menthes poivrées pour la production d'huile essentielle. Herba Gallica (Chemillé) 1, 1–16

Bouverat-Bernier, J.P., 1992: Incidence des dates et des stades de récolte sur les rendements et qualités d'huile essentielle de la menthe poivre Mitcham. Herba Gallica 2, 61–77

Bouwmeester, H.A., A.R. Davies, H.G. Smid, R.S.A., 1995: Physiological limitations to carvone yield in caraway (Carum carvi L.) Ind. Crops Prod. 4, 39–51

Bovill, H., D. Reeve, 2003: Auf den Spuren der Citrusfrüchte. Eine Limette ist keine Zitrone. Flüssiges Obst 70, 270–271

Bowles, E.A., 1952: A handbook of Crocus and Colchicum for gardeners. The Bodley Head London, 2. edn., 222 pp

Bown, D., 1995: Encyclopedia of herbs and their uses. Dorling Kindersley, London, 424 pp

Bown, D., 1996: Du Mont's grosse Kräuter-Enzyklopädie. Dumont Buchverlag Cologne, 424 pp

Boya, M.T., S. Valverde, 1981: An orthoquinone isolated from Salvia aethiopis. Phytochemistry 20, 1367–1368

Bozan, B., M. Kosar, Z. Tunalier, N. Ozturk, K.H.C. Başer, 2003: Antioxidant and free radical seavenging activities of Rhus coriaria and Cinnamomum cassia extract. Acta Alimentarie 32, 53–61

Braatz, D., P. Lembke, 1990: Die Trüffeln der Tropen. Feinschmecker f. Ärzte, No. 2, 39–41

Braatz, R., C. Chr. Schneider (eds), 1976: Symposium on the pharmacodynamics of silymarin (from 29.–30.11.1974 in Cologne). Verlag Urban & Schwarzenberg Munich, 203 pp

Bramwell, D., 1997: Flora der Kanarischen Inseln; (Deutsche Ausgabe). Edn. Rueda Madrid, 216 pp

Brandares, M.F.T., A.M. Vuelban, B.B. Dar Juan, M.R. Ricalde, F.E. Anzaldo, 1987: Stability studies of essential oils from some Philippine plants. 2. Cymbopogon citratus (D.C.) Stapf. Philippine J. Sci. 116, 391–402

Braun, N.A., M. Meier, 2002: Lorbeer – vom Siegeskranz zum δ-Terpinylacetat. Dragoco-Rep.49, 64–76

Braunsdorf, R., U. Hener, S. Stein, A. Mosandl, 1993: Comprehensive CgC-IRMS analysis in the authenticity control of flavors and essential oils. I. Lemon oil. Z. Lebensmittel-Unters. u. -Forsch. 197, 137–141

Bremness, L., 2001: Das grosse Buch der Kräuter. Ein praktischer Führer für den Anbau, die Pflege und Verwendung von Kräutern. AT Verlag Arau Stuttgart, 286 pp

Brewster, J.L., 1994: Onions and other vegetable Alliums. CAB Internat., Wallingford, Oxon (UK), 256 pp

Bricout, J., J.-C. Fontes, L. Merlivat, 1974: Detection of synthetic vanilin in vanilla extracts by isotopic analysis. J. Assoc. Off. Anal. Chem. 57, 713–715

Brieskorn, C.H., 1991: Salbei – Seine Inhaltsstoffe und sein therapeutischer Wert. Z. Phytother. 12, 61–69

Brieskorn, C.H., W. Biechele, 1971: Zur Unterscheidung von Salvia officinalis und Salvia triloba. Dtsch. Apotheker-Ztg. 111, 141–142

Brieskorn, C.H., H.J. Dömling, 1969: Carnosolsäure, der wichtigste antioxidativ wirksame Inhaltsstoff des Rosmarin- und Salbeiblattes. Z. Lebensmittel-Unters. u. -Forsch. 141, 10–16

Brieskorn, C.H., A. Fuchs, 1962: Zur chemischen Identifizierung des offizinellen Salbeiblattes. Dtsch. Apotheker-Ztg. 102, 1268–1269

Brieskorn, C.H., W. Krause, 1974: Weitere Triterpene aus Melissa officinalis L. Arch. Pharmazie 307, 603–612

Brieskorn, C.H., P. Noble 1982: Inhaltsstoffe des ätherischen Öls der Myrrhe. Sesquiterpene und Furanose-squiterpene. Planta med. 44, 87–90

Brieskorn, C.H., W. Riedel, 1977: Flavonoide aus Coleus amboinicus. Planta med. 31, 308–310

Brieskorn, C.H., H.J. Michek, W. Biechele, 1973: Die Flavone des Rosmarinblattes. Dtsch. Lebensmittel-Rdsch. 69, 245–246

Broda, S., R. Habegger, A. Hanke, W.H. Schnitzler, 2001: Characterization of parsley by chemosensory and other analytical methods. J. Appl. Bot./Angew. Bot. 75, 201–206

Brodnitz, M.H., C. Pollock, P. Vallon, 1969: Flavour components of onion oil. J. Agric. Food Chem. 17, 760-763

Brodnitz, M.H., J.V. Pascale, L. van Derslice, 1971: Flavour components of garlic extract. J. Agric. Food Chem. 19, 273–275

Brodt, Ch., 1982. Botanik, Verwendung und Anbau von Gentiana lutea L. unter besonderer Berücksichtigung der Ökotyp-Umwelt-Interaction. Dipl.-Arbeit TU Weihenstephan, Lehrstuhl f. Gemüsebau

Brondegaard, V.J., 1990: Massenausbreitung von Bärenklau. Naturwiss. Rdsch. 43, 438–439

Brondegaard, V.J., 1991: Der Kirschlorbeer – ein blausäurehaltiger Zierstrauch. Naturwiss. Rdsch. 44, 482

Brophy, J., M.K. Jogia, 1984: Essential oils from two varieties of Fijian Ocimum sanctum (Tulsi). Fiji agric. J. 46, No. 1, 21–26

Brophy, J.J., M.K. Jogia, 1986: Essential oils from Fjian Ocimum basilicum L. Flavour & Fragrance J. 1, 52–55

Brophy, J.J., E.V. Lassak, 1987: The volatile leaf oils of Hyptis pectinata (L.) Poit., and Vitex trifolia L. var. bicolor (Willd.) Moldenke from Fiji. Flavour & Fragrance J. 2, 41–43

Brophy, J.J., R.J. Goldsack, J.R. Clarkson, 1993: The essential oil Ocimum tenuiflorum L. (Lamiaceae) growing in Northern Australia. J. Ess. Oil Res. 5, 459–461

Broszat, W., 1992: Der Mohn (Papaver somniferum L.). – Anbau und Markt einer wiederentdeckten Kulturpflanze. Tropenlandwirt. Beiheft 47, 170 pp

Brown, E.D., 1955: Cinnamom and cassia: sources,

Brown, E.D., 1956: Cinnamom and cassia: sources, production and trade. Part I: Cinnamom. Colonial Plant & Animal Products 5, 257–280

Brown, E.D., 1956: Cinnamom and cassia: sources, production and trade. Part II: Cassia. Colonia Plant & Animal Products 6, 96–116

Brückner, J., K. Hoppe, G. Mieth, 1989: Fermentierte und nicht fermentierte Lebensmittel auf Sojabohnenbasis. Lebensmittelindustrie 36, 57–61

Brückner, K., 1953: Zur Sorten- und Anbaufrage sowie Verwendung des Kultur-Löwenzahns (Taraxacum officinale Web.). Pharmazie 8, 1043–1051

Brummitt, R.K., C.E. Powell, 1992: Authors of plant names: a list of authors of scientific names of plants, with recommended standard forms of their names, including abbreviations. Royal Botanic Garden, Kew (UK), 732 pp

Brüning, R., H. Wagner, 1978: Übersicht über die Celasttraceen-Inhaltsstoffe: Chemie, Chemotaxonomie, Biosynthese, Pharmakologie. Phytochem. 17, 1821–1858

Brunke, E.-J., N. Fischer, F.-J. Hammerschmidt, G. Schmaus, 1993a: Sumach – ein orientalisches Gewürz. Über das ätherische Öl von Rhus coriaria L-Früchte. dragoco-ber. 38, 80–95

Brunke, E.J., F.J. Hammerschmidt, F.H. Köster, 1991a: Das ätherische Öl von Laser trilobium (Roßkümmel). dragoco-rep. 38, T

Brunke, E.J., F.J. Hammerschmidt, F.-H. Koester, P. Mair, 1991b: Constituents of dill (Anethum graveo lens L.) with sensory importance. J. Ess. Oil Res. 3, 257–267

Brunke, E.-J., F.-J. Hammerschmidt, G. Schmaus, A. Akgül, 1993b: The essential oil of Rhus coriaria L. fruits. Flavour & Fragrance J. 8, 209–214

Brunke, E.-J., G. Schmaus, 1995: Neue geruchsaktive Inhaltsstoffe von Sandelholzöl. dragoco-rep. 42, 195 217

Bruyn, J.W.de, G. Elzenga, M. Keuls, 1954: Selection of living angelica roots for volatile oil content. Euphytica 3, 147–153

Buckle, K.A., M. Rathnawathie, 1985: Compositional differences of black, green and white pepper oil from three cultivars. J. Food Technol. 20, 599–613

Bukasov, S.M. (БУКАСОВ, С.М.), 1930: ВОЗДЕЛЫВАЕМЫЕ РАСТНИЯ МЕКСИКИ, ГВАТЕМАЛЫ и КОЛУМБИИ (The cultivated plants of Mexico, Guatamala and Colombia). Институт Растениеводства Васхил Ленирад (Institut of plant cultivation of the Akad. of agric. Sci., Leningrad), 553 pp + 60 pp register)

Bullmann, J., P. Steinert, G. Galling, 1990: Eine einfache Methode zur Glycyrrhizingewinnung aus der Süßholzwurzel (Glycyrrhiza glabra L.). Chem. Mikrobiol. Technol. Lebensmittel, No. 12, 179–184

Burba, J.L., C.R. Galmarini, 1997: Proc. of the 1. Internat. Symposium on edible Alliaceae. Mendoza, Argentinia 1994, 14.–18.3. Internat. Soc. Horticult. Sci., 652 pp

Burgett, M. 1980: The use of lemon balm (Melissa officinalis) for attracting honey bee swarms. Bee World 61, No. 2, 44–46

Bûrits, M., F. Bucar, 1998: Studies of the antioxidative Wirkung von ätherischem Schwarzkümmelöl. Sci. Pharm. 66, 25–28

Burkill, I.H., 1966: A dictionary of the economic products of the Malay Peninsula. Goverments of Malaysia and Singapore by the Ministry of Agricult. and Co-Operatives, Kuala Lumpur, Malaysia, Vol I (A–H) and Vol. II (I–Z), 2444 pp

Burkill, I.H., 1985: The useful plants of West tropical Africa. Royal Botanic Garden Kew (UK), Vol. I, familes A–D, 1960 pp; 1994: Vol. II, families E–I, 636 pp; 1995, Vol. III, families J–L, 857 pp; 1997, Vol. IV, families M–R, 969 pp; 2000, Vol. V, families S–Z, 686 pp

Busenberg, B., 1944: Vanilla, chocolate and strawberry. The story of your favorite flavors. Lerner Publ. Co., Minneapolis, 122 pp

Bushway, A.A., A.M. Wilson, L. Houston, R.J. Bushway, 1984: Selected properties of the lipid and protein fractions from chia seed. J. Food Sci. 49, 555–557

Busse, W., 1899 : Ueber Gewürze. IV. Vanille. Arbeiten. Kaiserl. Gesundheits-Amte. 15, 1–72

Bussell, B.M., J.A. Considine, Z.E. Spadek, 1995: Flower and volatile oil ontogeny in Boronia megastigma. Ann. Bot. 76, 457–463

Buttery, R.S., R.M. Seifert, D.G. Guadasgni, L.C. Ling, 1969: Characterization of some volatile constituents of bell pepper. J. Agric. Food Chem. 17, 1322–1327

Byrne, R., J.H. MacAndrews, 1975: Pre-Columbian purslane (Portulacca oleracea L.) in the New World. Nature (London) 253, 726–727

Byun, G.L., S.M. Oh, J.S. Lee, S.J. Hahn, 1985: Agronomic characteristics of perilla. (Perilla ocymoides L.) and the selection of cultivars for leaf use. J. Korean Soc. horicult. Sci. 26, 113–121 (in Korean)

C

Cabo, J., M.E. Crespo, J. Jiménez, C. Navarro, 1986: A study of the essences from Thymus hyemalis collected in three different localities. Fitoterapia 57, 117–119

Cabo, J., J. Jiménez, M. Miró, C. Navarro, J. Ruiz, 1986: Salvia verbenaca. I. Macro- and micrimorphology. Fitoterapia 57, 409–413

Cabo, J., M.E. Crespo, J. Jiménez, C. Narro, S. Risco, 1987: Seasonal variation of essential oil yield and composition of Thymus hyemalis. Planta med. 53, 380–382

Çağlarırmak, N., 2003: Biochemical and physical properties of some walnut genotypes (Juglans regia L.). Nahrung/Food 47, 28–32

Cahill, J.P., 2003: Ethnobotany of Chia, Salvia hispanica L. (Lamiaceae). Econ. Bot. 57, 604–618

Calpouzos, L., 1954: Botanical aspects of oregano. Econ. Bot. 8, 222–233

Campanili, E., 1995: Helichrysum angustifolium DC. Esperienze cliniche nella psoriasi. Acta Phytotherp. 1, 8–10

Campello, J.P., A.J. Masaioli, 1974: Triterpenenes of Schinus terebinthifolius. Phytochemistry 13, 659–660

Campello, J.P., A.J. Marsaioli, 1975: Terbinthifolic acid and bauerenone: new triterpenoid ketones from Schinus terebinthifolius. Phytochemistry 14, 2300–2302

Camperi, L., D. Fraternale, D. Ricci, 2002: Essential oil composition and antioxidant activity of peels of Citrus sinensis blood and blood Citrus aurantium. Riv. Ital. EPPOS Ist. Tetrahedron 12, 21–28

Campo. J. del, M.J. Amiot, C. Nguyen-The, 2000: Antimicrobial effect of rosemary extract. J. Food Protection 63, 1359–1368

Capone, W., C. Mascia, M. Melis, L. Spanedda, 1988: Determination of terpenic compounds in the essential oil from Satureja thymbra L., growing in Sardinia. J. Chromatograph. 457, 427–430

Caramiello, R., A. Bocco, G. Buffa, M. Maffei. 1995: Chemotaxonomy of Juniperus communis, J. sibirica and J. intermedia. J. Ess. Oil Res. 7, 133–145

Caredda, A., B. Marongiu, S. Porcedda, C. Soro, 2002: Supercritical carbon dioxide extraction and characterization of Laurus nobilis essential oil. J. Agric. Food Chem. 50, 1492–1496

Carnat, A.P., J.L. Lamaison, A. Rémery, 1991a: Composition of leaf and flower essential oil from Monardadidyma L. cultivated in France. Flavour & Fragrance J. 6, 79–80

Carnat, A.P., A. Chossegros, J.L. Lamaison, 1991b: The essential oil of Satureja grandiflora (L.) Scheele-from France. J. Ess. Oil Res. 3, 361–362

Carnat, A.P., J.C. Chalchat, D. Fraisse, J.L. Lamaison, 2001: Isolation of a volatile concentrate of caraway seed. J. Ess. Oil Res. 13, 371–375

Carotenuto, A., V de Feo, E. Fattorusso, V. Lanzotti, S. Magno, C. Cicala, 1996: The flavonoids of Allium ursinum. Phytochemistry 41, 531–536

Carreiras, M.C., B. Rodríguez, R.E. López-García, R.M. Rabanal, 1987: A dimer of d-pinocarvone from Cedronella canariensis. Phytochemistry 26, 3351–3353

Carrier, D.J., T. Crowe, Sh. Sokhansanj, J. Wahab, B. Barl, 2202: Milk thistle, Silybium marianum (L.) Gaertn., flower head development and associated marker compound profile. J. Herbs, Spices & Medicin. Plants 10, 65–74

Carrubba, A., R. la Torre, R. Piccaglia, M. Marotti, 2002: Characterization of an Italian biotype of clary sage (Salvia sclarea L.) grown in a semi-arid Mediterrean environment. Flavour & Fragrance J. 17, 191–194

Carrubba, A., R. la Torre, A. Di Prima, F. Saiano, G. Alonzo, 2002: Statistical analyses on the essential oil of Italian coriander (Coriandrum sativum L.) fruits of different ages and origins. J. Ess. Oil Res. 14, 389–396

Carson, J.F., 1987: Chemistry and biological properties of onions and garlic. Food Rev. Internat. 3, 71–103

Castellón, A.F., M.R. Vale, M.I.L. Machado, F.J.A., Matos 1987: Alguns constiuintes químicos de Coleus barbatus Benth. cultivado em Forskolin (Coleonol). Cinc. E. Cult. (Sao Paulo) 39, Suppl., p 535

Castro, R.R., V.M. Nosti, 19987: El alcaparro (Capparis spinosa L.). Grasas y Aceites (Seville) 38, 173–175

Catsiotis, S., N.G. Iconomou, 1984: Qualitative and quantitative comparative gas-liquid-chromatographic analysis of the essential oil of Salvia triloba grown in Greece. Pharm. Acta Helv. 59, 29–32

Cemeroglu, B., N. Artik, S. Erbas, 1992: Extraction and composition of pomegranate juice. Flüssiges Obst 59, 335–340

Cervato, G., M. Carabelli, S. Gervasio, A. Cittera, R. Cazzola, B. Cestaro, 2000: Antioxidant properties of oregano (Origanum vulgare) leaf extracts. J. Food Biochem. 24, 453–465

Cesare, L.F. di, D. Viscardi, E.L. Fusari, R.C. Nani, 2001a: Study of the volatile fraction in basil and sage stored at –20°C. Ind. Aliment. (Ital.) 40, 1221–1225

Cesare, L.F. di, D. Viscardi, E.L. Fusari, R.C. Nani, 2001b: Study of the volatile composition of the raw and dried rosemary (Rosmarinus officinalis L.). Ind. Aliment. (Ital.) 40, 1343–1345

Cesare, L.F. di, R.C. Nani, E.L. Fusari, D. Viscardi, R. Vitale, 2001c: The influence of blanching, drying, and storage on the volatile fraction of the basil leaves. Ind. Aliment. (Ital.) 40, 1007–1013

Cesare, L.F. di, D. Viscardi, R. Nani, 2002: Influence of blanching with microwave and drying with dried air on the volatile composition of basil. Ind. Aliment. (Ital.) 41, 25–28

Chadwick, C.I., T.A. Lumpkin, L.R. Elberson, 1993: The botany, uses and production of Wasabi japonica (Miq.) (Cruciferae) Matsum. Econ. Bot. 47, 113–135

Chadwick, M.G.A., R. Pope, 1961: The market for Vanilla beans. Trop. Sci. 3, 174–183

Chalchat, J.C., R.P. Garry, A. Michet, B. Benjilali, J.L. Chabart, 1993: Essential oils of rosemary (Rosmarinus officinalis L.). The chemical composition of oils from various origins (Morocco, Spain, France). J. Ess. Oil Res. 5, 613–618

Chalchat, J.-C., J. Lamy, 1997a: Essential oils from Thymus serphilloides subspec. serphylloides and Thymus serphylloides subspec. gadorensis micropagated plants. J. Ess. Oil Res. 9, 527–533

Chalchat, J.-C., A. Michet, B. Pasquier, 1997b: Influence of harvesting time on chemical composition of Menthax piperita L. essential oil. Perfumer & Flavorist 22, No. 1, 17–21

Chalchat, J.C., M.S. Gorunovic, S.D.Petrovic, Z.A. Maksimovic, 2001a: Chemical composition of two wildspecies of the genus Salvia L. from Yugoslavia: Salvia aethiopis and Salvia verticillata. J. Ess. Oil Res. 13, 416–418

Chalchat, J.C., D. Adamovic, M.S. Gorunovic, 2001b: Composition of oils of three cultivated forms of Hyssopus officinalis endemic in Yugoslavia: f. albus Alef., f. cyaneus Alef. and f. ruber Mill. J. Ess. Oil Res. 13, 419–421

Chamberlain, D.F., 1977: The identity of Ferula asafoetida L. Notes royal bot. Garden, Edinburgh 35, 229–233

Chamblee, T.S., B.J. Clark, G.B. Brewster, T. Radford, G.A. Iacobucci, 1991: Quantitative analysis of the volatile constituents of lemon peel oil. Effects of silica gel chromatography on the composition of its hydrocarbon and oxygenated fractions. J. Agric. Food Chem. 39, 162–164

Chamisso, A.v., 1987: Illustriertes Heil-, Gift- und Naturpflanzenbuch. Verlag Reimer, Berlin

Chane-Ming, R. Vera, J.C. Chalchat, P. Cabassu, 2002: Chemical composition of essential oils from rhizomes, leaves and flowers of Curcuma longa L. from Reunion Island. J. Ess. Oil Res. 14, 249–152

Chandler, R.F., 1985: Licorice, more than just a flavour. Canad. pharmac. J. 118, 421–424

Chandler, R.F., D. Hawkes, 1984: Aniseed – a spice, a flavor, a drug. Canad. Pharm. J. 117, 28–29

Chang, St.S., B. Ostric-Matijaseric, O.A.L. Hsieh, 1977: Natural antioxidants from rosemary and sage. J. Food Sci. 42, 1102–1106

Charalambous, G. (ed), 1994: Spices, Herbs and edible Fungi. Development in Food Sci., No. 34. Elsevier Sci. B.V. Amsterdam, London, New York, Tokyo 764 pp

Charles, D.J., J.E. Simon, 1992a: Essential oil constituents of Ocimum kilimandscharicum Guerke. J. Ess. Oil Res. 4, 125–128

Charles, D.J., J.E. Simon, 1992b: A new geraniol chemotype of Ocimum gratissimum L. J. Ess. Oil Res. 4, 231–234

Charles, D.J., J.E. Simon, M.P. Widrlechner ,1991: Characterization of essential oil of Agastache species. J. Agric. Food Chem. 39, 1946–1949

Charles, D.J., J.E. Simon, K.V. Wood, 1990: Essential oil constituents of Ocimum micranthum Willd. J. Agric. Food Chem. 38, 120–122

Chatterjee, S., P.S. Variyar, A.S. Gholap, S.R. Padwal-Desai, D.R. Bongirwar, 2000: Effect of γ-irradiation on the volatile oil constituents of turmeric (Curcuma longa). Food Res. int. 33, 103–106

Chatzopoulou, P.S., S.T. Katsiotis, 1993: Study of the essential oil from Juniperus communis berries (cones) wild growing in Greece. Planta med. 59, 554–556

Chatzopoulou, P.S., A. Haan, S.T. de Katsiotis, 2002: Investigation on the supercritical CO_2 extraction of the volatile constituents from Juniperus communis obtained under different treatment of the 'berries' (cones). Planta med. 68, 827–831

Chahan, N.S., 1999: Medicinal and aromatic plants of Himachal Pradesh. Indus Publishing Comp., New Delhi, India, 632 pp

Chaytor, D.A., 1937: A taxonomic study of the genus Lavandula. J. Linnean Soc., Bot. 51, 153–204

Cheah, P.B., N.H. Abu Hasim, 2000: Natural antioxidant extract from galangal (Alpinia galanga) for minced beef. J. Sci. Food & Agric. 80, 1565–1571

Cheer, G. (ed), 1998: BOTANICA. Das Abc der Pflanzen. 10.000 Arten in Text und Bild. Könemann-Verlagsges. mbH. Köln 1997, 1007 pp

Chen, C.C., C.-T. Ho, 1986: Identification of sulphur compounds of shiitake (Lentinus edodes Sing.). J. Agric. Food Chem. 34, 830–833

Chen, C.C., C.-T. Ho, 1988: Gas chromatographic analysis of volatile components of ginger oil (Zingiber officinale Roscoe) extracted with liquid carbon dioxide. J. Agric. Food Chem. 36, 322–328

Chen, C.C., R.T. Rosen, C.T. Ho, 1986: Chromatographic analysis of gingerol compounds in ginger (Zingiber officinale Roscoe) extracted by liquid carbon dioxide. J. Chromatogr. 360, 163–173

Chen, C.C., M.C. Kuo, C.M. W., C.T. Ho, 1986; Pungent principles of ginger (Zingiber officinalis Roscoe) extracted by liquid carbon dioxide. J. Agric. Food Chem. 34, 477–480

Chen, I.-S., Y.-C. Lin, I.-L. Tsai, C.-M. Teng, F.-N. Ko, T. Ishakawa, H. Ishi, 1995: Coumarin and antiplatelet aggregation constituents from Xanthoxylum schinifolium. Phytochemistry 39, 1091–1097

Chen, S.-H., T.-C. Huang, C.-T. Ho, P.-J. Tsai, 1998: Extraction, analysis and study on the volatile in Rosella tea. J. Agric. Food Chem. 46, 1101–1105

Chen, I.S., S.J. Wu, I.L..Tsai, J.M. Pezzuto, M.C. Lu, H. Chai, N. Suh, C.M. Teng, 1994: Chemical and bioactive constituents from Zanthoxylum simulans. J. Nat. Prod. 57, 1206–1211

Cheng, M. Ch., Yu.Sh. Cheng, 1983: Oil of Litsea-sorts. J. Chinese Chem. Soc. (Taipei) 30, 59–62

Chevalier, A., 1906: Les Baobas (Adansonia) de l'Afrique continentale. Bull. Soc. Bot. France 53, 480–496

Chevalier, A., 1925: Le poivrier et sa culture en indochine. Goinermeenet Général de l'Indo-chin. Publ. Del'Agence Economique, Paris, No. 19, 31 pp

Chevalier, A., 1957: Une enquête sur les plants médicinales de l'afrique occidentale. I. Observations générales. Rev. internat. Bot. Appl. Agric. Trop. 17, 165–175

Chialva, F., 1993: Chemometric investigation on Italian peppermint oil. J. Agric. Food Chem. 41, 2028–2033

Chialvi, M., 1990: Volatile oil of Schinus molle. Flavour & Fragrance J. 5, 49–52

Chirikdjian, J.J., 1973: Isolation of kumatakenin and 4',5-dihydroxy-3-,3',7-trimethoxyflavone from Larrea tridentata. Z. Naturforsch. 28, 32–35

Chisholm, M.G., M.A. Wilson, G.M. Gaskey, 2003: Characterization of aroma volatiles in key lime essential oil (Citrus aurantifolia Swingle). Flavour & Fragrance J. 18, 106–115

Chisholm, M.G., J.A. Jell, D.M. Cass, jr., 2003: Characterization of the major odorants found in the peel oil of Citrus reticulata Blanco cv. Clementine using gas chromatographic-olfactometry. Flavour & Fragrance J. 18, 275–281

Chisowa, E.H., D.R. Hall, D.I. Farman, 1998: Chemical composition of the essential oil of Tagetes minuta L. from Zambia. J. Ess. Oil Res. 10, 183–184

Chizzola, R., G. Dobos, 1997: Mineralstoff- und Spurenelementgehalt von Mohnsamen. Ernährung/Nutrition 21, 55–57

Chopra, R.N., S.L. Nayar, I.C. Chopra, 1956: Glossary of Indian medical plants. Council Sci. & Industr. Res., New Delhi , 330 pp

Choudhury, S., R. Ahmed, P.B. Karjlal, 1998: Composition of the seed Trachyspermum ammi (L.) from North-East India. J. Ess. Oil Res. 10, 588–590

Choudhury, S.H., D.N. Bordoloi, 1986: Effect of sowing on the growth, yield and oil quality of Ocimum gratissimum Linn. Indian Perfumer 30, No. 1, 254–260

Chubey, B.B., 1982: Geraniol-rich essential oil from Monarda fistulosa L. Perfumer & Flavorist 7, No. 3, 32–34

Chuda, Y., H. Ono, M. Ohnishi-Kameyama, T. Nagata, T. Tsushida, 1996: Structural identification of two antioxidant quinic acid derivatives from garland (Chrysanthemum coronarium L.) J. Agric. Food Chem. 44, 2037–2039

Chuda, Y., M. Suzuki, T. Nagata, T. Tsushida, 1998: Content and cooking loss of three quinic acid derivatives from garland (Chrysanthemum coronarium L.). J. Agric. Food Chem. 46, 1437–1439

Chung HeeDon, Youn SunJoo, 1996: Comparison of chemical composition and taste of Korean native

Chinese chive leaves. J. Korean Soc. Horticult. Sci. 37, 611–616

Chunhui Deng, Guoxin Song, Yaoming Hu, Xiangmin Zhang, 2003: Determination of the volatile constituents of Chinese Coriandrum sativum L. by gas chromatography-mass spectrometry with solid-phase microextraction. Chromatographia 57, 357–361

Chyau, C.-C., J.-L. Mau, Ch.-M. Wu, 1996: Characteristics of the steam distilled oil and carbon dioxide extract of Zanthoxylum simulans fruits. J. Agric. Food. Chem. 44, 1096–1099

Cichewicz, R.H., M.G.Nair, 2002: Isolation and characterization of stelladerol, a new antioxidant naphthaleneglycoside, and other antioxidant glycosides from edible daylily (Hemerocallis) flowers. J. Agric. Food Chem. 50, 87–91

Circella, G., Ch. Franz, J. Novak, H. Resch, 1995: Influence of day length and leaf insertion on the composition of marjoram oil. Flavour & Fragrance J. 10, 371–374

Claire, C., 1961: Of Herbs & Spices. Abelard-Schumann London New York Toronto, 275 pp

Clare, C., 1993: Über die Weite des Meeres in den duftenden Orient. Der sechste Sinn des Christoph Columbus. dragoco-rep., 40, 263–260

Clark, R.J., R.C. Menary, 1979: Effects of photoperiod on the yield and composition of peppermint oil. J. Amer. Soc. horticult. Sci. 104, 699–702

Clarke, B.J., 1986: Hop products. J. Inst. Brewing 92, 123–130

Classen, B., K. Knobloch, 1985: Über die ätherischen Öle der Weinraute (Ruta graveolens L.) Z. Lebensmittel-Unters. u. -Forsch. 181, 28–31

Clebsch, B., 1997: A book of salvias. Sages for every garden. Timber Press, Inc. Portland, Oregon, 221 pp

Clement, E.J., M.C. Forster, 1994: Alien plants of the British Isles. Bot. Soc. British Isles, London, 590 pp

Codd, L.E., 1975: Plectranthus (Labiatae) and allied genera in Southern Africa. Bothalia 11, 371–442

Coelho, J.A.P., A.P. Pereira, R.L. Mendes, A.M.F. Palavra, 2003: Supercritical carbon dioxide extraction of Foeniculum vulgare volatile oil. Flavour & Fragrance Oil 18, 316–319

Cohnen, G., 1993a: Das Veilchen. dragoco-rep. 40, 73–77

Cohnen, G., 1993b: Iris. dragoco-rep. 40, 158–162

Coiciu, E., G. Rácz (no year) (1963?): Plante medicinale si aromatice. Ed. Acad. Republicii Populare Romine, Bucaresti 668 pp

Coit, J.E., 1951: Carob or St. John's bread. Econ. Bot. 5, 82–96

Coleman, W.M., B.M. Lawrence, 1992: Comparative automated static and dynamic quantitative head space analysis of coriander oil. J. Chomatogr. Sci. 30, 396–398

Colins, M.A., H.P.Charles, 1987: Antimicrobial activity of carnosol and ursolic acid: two antioxidant constituents of Rosmarinus officinalis L. Food Microbiol. 4, 311–315

Collin, G.J., H. Deslauriers, N. Pageau, M. Gagnon, 1993: Essential oil of tansy (Tanacetum vulgare L.) of Canadian origin. J. Ess. Oil Res. 5, 629–638

Collins, J.E., C.D. Bishop, S.G. Deans, K.P. Svoboda, 1994: Composition of the essential oil from the leaves and flowers of Monarda citriodora var. citriodora grown in the United Kingdom. J. Ess. Oil Res. 6, 27–29

Collins, R.P., A.F. Halim, 1971: Essential leaf oil in Calycanthus floridus. Planta med. 20, 241–243

Connell, D.W., 1970: The chemistry of the essential oil and oleoresin of ginger (Zingiber officinale Roscoe). Flavour Ind. 10, 677–693

Connell, D.W., R.A. Jordan, 1971: Composition and distinctive volatile flavour characteristics of the essential oil from Australian-grown ginger (Zingiber officinale) J. Sci. Food Agric. 22, 93–95

Cook, C.D.K., J.-J. Symonens, K. Urmi-König, 1984: A revision of the genus Ottelia (Hydrochirataceae). I. Generic consideration. Aquatic. Bot. 18, 263–274

Cook, F.E.M., 1995: Economy botany data collection standard. Royal Botanic Garden Kew, UK, 146 pp

Coomes, T.J., H.T. Islip, W.S.A. Matthews, 1955: Aframomum angustifolium seed from Zanzibar. Colonial Plant & Animal Products 5, 68–77

Corcilius, F., 1956: Zur Kenntnis der Inhaltsstoffe von Arnica montana L. Arch. Pharmazie 289, 75–81

Corner, E.J.H., 1988: Wayside trees of Malaya. The Malayan Nature Soc., Kuala Lumpur, Malaysia, 3. edn., 2 Vol., 774 pp

Correll, D.S., 1953: Vanilla. Its botany, history, cultivation and economic import. Econ. Bot. 7, 291–358

Cortés, M., M.L. Oyarzún, 1981: Tadeonal and isotadeonal from Drymis winterii. Fitoterapia 52, 33–35

Cortes-Eslava, J., S. Gomez-Arroyo, R. Villalobos-Pietrini, J.J. Espinosa-Aguirre, 2001: Metabolic actiation of three arylamines and two organophosphorus insecticides by coriander (Coriander sativum) a common edible vegetable. Toxicol. Letters 125, 39–49

Corti, P., E. Mazzei, S. Ferri, G.G. Franchi, E. Dreassi, 1996: High performance thin layer chromatographic quantitative analysis of picrocrocin and crocetin, active principles of saffron (Crocus sativus L., Iridaceae): A new method. Phytochem. Analysis 7, 201–203

Cosentina, S. A. Barra, B. Pisano, M. Cabizza, F.M. Piris, F. Palmas, 2003: Composition and antimicrobial properties of Sardinian Juniperus essential oils against foodborne pathogens and spoilage microorganisms. J. Food Protect. 66, 1288–1291

Costa, O.A. de, 1959: Pimenta do mato. Xilopia brasiliensis Spreng., Anonaceae. Arqu. Bromatol. (Rio de Janeiro) 7, 97–107

Cotton, C.M., L.V. Evans, J.W. Gramshaw, 1991: The accumulation of volatile oils in whole plants and cell cultures of tarragon (Artemisia dracunculus). J. Exp. Bot. 42, 365–375

Couladis, M., M. Özcan, O. Tzakou, A. Akgul, 2003: Comparitive essential oil composition of various parts of the turpentine tree (Pistacia terebeinthus L.) growing wild in Turkey. J. Sci. Food & Agric. 83, 136–138

Court, W.H., R.C. Roy, R. Pocs, 1993: Effect of harvest date on the yield and quality of the essential oil of peppermint. Canad. J. Plant Sci. 73, 815–824

Courter, J.W., A.M. Rhodes, 1969: Historical notes on horseradish. Econ. Bot. 23, 156–164

Craker, L.E., J.E. Simon, 1986–1989: Herbs, spices, and medicinal plants: Recent advances in botany horticulture, and pharmacology. Oryx Press Phoeniix Arizona. Vol. 1, 359 pp, Vol. 2, 225 pp, Vol. 3, 220 pp, Vol. 4, 267 pp

Crasselt, E., 1950: Über die physiologische Wirkung von Extrakten aus Sinapis alba. Arch. Phamazie 283, 275–280

Craveiro, A.A., W. Alencar, F.J. Matos, H.S. Andrade, M.I.L. Machado, 1981: Essential oil from Brazilian Verbenaceae Genus Lippea. J. Nat. Prod. 44, 598–601

Craveiro, A.A., F.J.A. Matos, M.I.L. Machado, J.W. Alencar, 1988: Essential oil of Tagetes minuta from Brazil. Perfumer & Flavor 13, No. 5, 35–36

Cravo, L., F. Périnau, A. Gaset, J.M. Bessiere, 1992: Study of the chemical composition of the essential oil, oleoresin and ist volatile product obtained from ambrette (Abelmoschus moschatus Moench) seed. Flavour & Fragrance J. 7, 65–67

Craze, R., 2002: The spice companion. The essential guide to using spices for your health and well being. Quintet Publishing Ltd. London (UK), 192 pp; Slovak Edn.: Koreniny. Základná príručka o využívaní korenínpre chuť, zdravie a pohodu. Fortuna Print Bratislava 2002, 192 pp

Crecchio, R. di, 1960: Lo zafferano. Italia Agricola 97, 629–649

Croft, J.R., D.N. Leach, 1985: New Guinea fern (Asplenium acrobyrum complex): identity, distribution and chemical composition of its salt. Econ. Bot. 39, 139–149

Crowe, A., 1997: A field guide to the native edible plants of New Zealand. Birkenhead, Auckland, New Zealand, 192 pp

Cserháti, T., E. Forgács, M.H. Morais, T. Mota, A. Ramos, 2000: Separation and quantitation of colour pigments of chili powder (Capsicum frutescens) by high-performance liquid chromatography-diode array detection) J Chromatogr. 896 A, 69–73

Cu, J.Q., Fan Pu, Yan Shi, F. Perineau, M. Delmas, A. Gaset, 1990a: The chemical composition of lovage headspace and essential oils produced by solvent extraction with various solvents. J. Ess. Oil Res. 2, 53–59

Cu, J.Q., F. Perineau, O. Goepfert, 1990b: GC-MC-analysis of star anise essential oil. J. Ess. Oil Res. 2, 91–92

Cufodontis, G., 1957: Bemerkenswerte Nutz- und Kulturpflanzen Aethiopiens. Botanische Ergebnisse der Expedition des Frobenius-Instituts der Univ. Frankfurt/M. nach Süd-Aethiopien 1954–1956. I. Senckenberg biol. 38, 405–415

Cui, W., N.A.M. Eskin, C, G. Biliaderis, 1992: Chemical and physical properties of yellow mustard (Sinapis alba L.) mucilage. Food Chem. 46, 169–176

Cui, Y., C.Y.W. Ang, 2002: Supercritical fluid extraction and high-performance liquid chromatographic determi-nation of phloroglucinols in St. John's wort (Hypericum perforatum L.). J. Agric. Food Chem. 50, 2755–2759

Cunha, A.P. da, L.R. Salgueiro, 1991: The chemical polymorphism of Thymus zygis ssp. sylvestris from central Portugal. J. Ess. Oil Res. 3, 409–412

Cuvelier, M.E., C. Berset, H. Richard, 1994: Antioxidant constituents in sage (Salvia officinalis). J. Agric. Food Chem. 42, 665–669

Czabajska, W., K. Kazmierczak, E. Ludwicz, 1993: Biologische Züchtungsuntersuchungen bei der Mariendistel, Silybium marianum (L.) Gaertn. Drogenreport 6, No. 10, 3–7

Czupor, L., 1970: Quantitative Bestimmung der γ-Pyrone in Fructus Ammi visnage. Dtsch. Apotheker-Ztg. 108, 1620–1624

Czygan, F.-C., 1987: Foeniculum vulgare – Der Fenchel. Z. Phytother. 8, 82–85

Czygan, F.-C., 1992: Anis (Anisi fructus DAB 10) – Pimpinella anisum L. Z. Phytother. 13, 101–106

Czygan, F.-C., 1993: Kulturgeschichte und Mystik des Johanniskrautes. Vom 2500 Jahre alten Apotropaikumzum aktuellen Antidepressivum. Z. Phytother. 14, 288–278

Czygan, F.-C., 1994: Das ätherische Öl der Schafgarbe. Dtsch. Apotheker-Ztg. 134, 228–229

Czygan, F.-C., 1997: Basilikum – Ocimum basilicum L. Z. Phytother. 18, 58–66

Czygan, F.-C., 1999: Die Sumpfdotterblume. Dtsch. Apotheker-Ztg. 139, 4032–4034

Czygan, F.-C., R. Hänsel, 1993: Thymian und Quendel – Arznei- und Gewürzpflanzen. Z. Phytother. 14, 104–110

D

Da Cunha, A.P., O.L.R. Roque, 1989: The chemical composition of the essential oil and alcoholic extract of Juniperus communis L. J. Ess. Oil Res. 1, 15–17

Dabiri, M., F. Sefidkon, 2001: Analysis of the essential oil from aerial parts of Perovskia atriplicifolia Benth. at different stages of plant growth. Flavour & Fragrance J. 16, 435–438

Daffertshofer, G., 1981: Mentha piperita und Mentha arvensis, Pfefferminzöle. Gordian (Hamburg) 81, No. 8, 10–11

Dahab, A.M.A., G.E. Fahmy, A.Z. Sarhan, I.M. Harridy, 1983a: Effect of chemical fertilizers on growth and essential oil of Calamintha officinalis var. "nepetoides". Ann. agricult. Sci., Ain Shams Univ. (Egypt) 28, 851–865

Dahab, A.M.A., G.E. Fahmy, A.Z. Sarhan, I.M. Harridy, 1983b: Effect of fertilizers on chemical constituents of Calamintha officinalis Jord var. nepetoides. Ann. agricult. Sci., Ain Shams Univ. (Egypt.) 28, 867–884

Dahl, J., 1994: Gundermann hat zwei Gesichter. Flora, Magazin für Haus u. Garten, No. 4, 65

Dake, G.N., 1995: Diseases of ginger (Zingiber officinale Roxb.) and their management. J. Spices & Aromatic Crops 4, 40–48

Dalang, F., E. Martin, J. Vogel, 1982: Examen des arómes de vanille et des produits alimentaires vanillés parchromatographie liquide à haute performance. Mitt. Gebiete Lebensmittel-Hyg. 73, 371–378

Dalby, A., 2000: Dangerous tastes. The story of spices. Univ. California Press Berkeley/Los Angeles, 184 pp

Dalziel, F., 1957: The useful plants of West tropical Africa. Grown agents for the colonies. Publishing S. Hutchinson London 256 pp

Danert, S., 1958: Zur Systematik von Papaver somniferum L. Kulturpflanze 6, 61–88

Danert, S., 1959: Zur Gliederung von Petroselinum crispum (Mill.) Nym. Kulturpflanze 7, 7–18

Daniel, P., A. Rajendran, 1966: Lippea alba (Mill.) N.E. Br. Distinct from Lippea javanica (Burm.f.) Sprengel (Verbenaceae). Bull. Inst. Bot. Survey of India 38, 19–24

D'Arcy, W.G. (ed), 1986: Solanaceae. Biology and Sytematics. Columbia Univ. Press New York NY., 603 pp

D'Arcy, W.G., W.H. Eshbaugh, 1974: New world pepper (Capsicum-Solanaceae) north of Columbia: a résumé. Baileya 19, 93–105

Darrah, H.H., 1980: The cultivated basils. Buckeye Printing Co., Independence, Missouri, 40 pp

Das, P., S.K. Sarma, 2001: Drying of ginger using solar cabinet dryer. J. Food Sci. Technol. (India) 38, 619–621

Dashak, D.A., M.L. Dawang, N.B. Lucas, 2001: An assessment of the proximate chemical composition of locally produced spices known as dadawa basso and dadawa kalwa from three markets in Pleteau State of Nigeria. Food Chem. 75, 231–235

Dastur, J.F., 1954: Useful plants of India and Pakistan. A popular handbook of trees and plants of industrial, economic and commercial utility. D.B. Taraporevala Sons & Co. Ltd. Bombay (India), 6. edn., 260 pp

Datta, A.K., S.K. Rang, 2000: Induced variable morphological mutation in Nigella sativa L. (black cumin). J. Hill Res. 13, 67–71

Datta, S, R. Chatterjee, P.K. Chattopadhyay, 2001:

Turmeric propagated from mother rhizome – its Morphological and yield characters. Environment & Ecology 19, 114–117

Dauer, A., 2002: Vanilla, Vanille – die Königin der Gewürze. Drogenreport 15, No. 27, 14–18

Daukšas, E., P.R. Venskutonis, B. Sivik, 1998: Extraction of lovage (Levisticum officinale Koch) root by carbondioxide. Effect of CO_2 parameters on the yield of the extract. J. Agric. Food Chem. 46, 4347–4351

Davidov, N.N., F.Ch Bachteeva, (Давидов Н.Н., Ф.Х Вахтеева): 1962: Botanical Dictionary: Russian-English-German-French-Latin. (Ботанический Словарь: Русско-Английско-Немецко-Французско-Латинский)., Redaktion Moskva. (Главная Редакция Иностраных Научно-Техничес- ких Словарей Физматгиза, Москва) Main redaction foreign sci.-technical dictionary Moskow, 335 pp (in Russian)

Davidson, A., 1999: The Oxford Companion to Food. Oxford Univ. Press, Oxford 1999, 892 pp

Davies, F.G., 1978: The genus Gynura (Compositae) in Africa. Kew Bull. 33, 335–342

Davies, F.G., 1979: The genus Gynura (Compositae) in Eastern Asia and the Himalayas. Kew Bull. 33, 629–640

Davies, F.G., 1981: The genus Gynura (Compositae) in Malaysia and Australia. Kew Bull. 35, 711–734

Davies, N.W., R.C. Menary, 1984: Volatile constituents of Boronia megastigma flowers. Perfumer & Flavorist 8, 5–8

Davis, D., 1992: Alliums. The ornamental onions. B.T. Batsford Ltd. London, 168 pp

Davis, D.V., R.G. Cooks, 1982: Direct characteristization of nutmeg constituents by mass spectrometry. J. Agric. Food Chem. 30, 495–504

Davis, F.S., L.G. Albrigo,1994: Citrus. CAB Internat. Wallingford, UK, 272 pp (Crop. Production Sci. in Horticult., Ser. No. 2)

Davis, P.H. (ed), 1972: Flora of Turkey and the Eastern Aegaen Islands. Univ. Press Edinburgh, UK, Vol. 4, 657 pp

Davis, W.N.L., 1970: The carob tree and its importance in the agricultural economy of Cyprus. Econ. Bot. 24, 460–470

Dawson, B.S., R. A. Franich, R. Meder, 1988: Essential oil of Melissa medicinalis L. subspec. altissima Arcang. Flavour & Fragrance J. 3, 167–170

Deans, S.G., K.P.Svoboda, 1990: The antimicrobial properties of majoram volatile oil. Flavour & Fragrance J. 5, 187–190

Debor, H.W., 1974: Bibliographie des internationalen Walnuß-Schrifttums. Bibliograph. Reihe Techn. Univ. Berlin, 148 pp

Degnan, A.J., J.H. von Elbe, R.W. Hartel, 1991: Extraction of annatto seed pigment by supercritical carbondioxide. J. Food Sci., 56, 1655–1659

Deininger, R., 1991: Gewürznelken (Syzygium aromaticum) und Nelkenöl – aktuelle Phytopharmaka. Z. Phytother. 12, 205–212

Delange, Y., 2003: Ces géants parmi les flores succulentes: lers Adansonia au baobabs. Succulentes (France) 26, No. 2, 3–9

Dellacassa, E., D. Lorenzo, L. Mondello, I. Stagno d'Alcontres, 1997: Chemical composition of the essential oil isolated from wild catnip Nepeta cataria L. cv. citriodora from the Drome region of France. J. Ess. Oil Res. 9, 523–525

Dellacassa, E., E. Soler, P. Menéndez, P. Moyna, 1990: Essential oils from Lippea alba (Mill.) N.E. Brown and Aloysia chamedrifolia Cham. (Verbenaceae) from Uruguay. Flavour & Fragrance J. 5, 105–108

Deltail, A., 1897: La vanilla, sa culture et sa préparation. Chaltansal, Bibliothèque d'Agriculture coloniale, Paris, 4. edn., 61 pp

Demarne, F., J.J.A. van der Walt, 1989: Origin of the rose-scented Pelargonium cultivar grown on Réunion Island. South Afric. J. Bot. 55, 184–191

Demarne, F., A.M. Viljoen, J.J.A. van der Walt, 1993: A study of the variation in the essential oil and morphology of Pelargonium capitatum (L.) L'Hérit. (Geraniaceae). Part 1. The composition of the oil. J. Ess. Oil Res. 5, 493–499

Demirci, B., K.H.C. Baser, G. Tumen, 2002a: Composition of the essential oil of Salvia aramiensis Rech. fil. growing in Turkey. Flavour & Fragrance J. 17, 23–25

Demirci, B., N. Tabanca, K.H.C. Baser, 2002b: Enantiometric distributation of some monoterpenes in the essential oils of some Salvia species. Flavour & Fragrance J. 17, 54–58

Deng, C., G. Song, Y. Hu, X. Zhang, 2003: Determination of the volatile constituents of Chinese Coriandrum sativum L. by gas chromatography-mass spectrometry with solid-phase microextraction. Chromatographia 57, 357–361

Depree, J.A., T.M. Howard G.P. Savage, 1999: Flavour and pharmaceutical properties of the volatile sulphur compounds of Wasabi (Wasabi japonica). Food Res. Internat. 31, 329–337

Deshpande, S.S., 1938: Essential oil from flowers of kewda – Pandanus odoratissimus Linn.f. J. Indian Chem. Soc. 15, 509–512

Dhalla, N.S., K.C. Gupta, M.S. Sastry, C.L. Malhotra, 1961: Chemical composition of the fruit of Momordicacharantia. Ind. J. Pharmac. 23, 128–129

Dhar, A.K., R. S. Dhar, 2000: Culinary and potherbs of Jammu and Kashmir. J. Herbs, Spices & Medicin. Plants 7, 7–18

Dhar, A.K., R. Sapru, K. Rekha, 1988: Studies on saffron in Kashmir. I. Variation in natural population and its cytological behaviour. Crop Improvement 15, 48–52

Dhar, A.K., A. Kalla, 1973: A new isoflavone from Iris germanica. Phytochemistry 12, 734–735

Dhingra, S.N., U.N. Shukla, G.N. Gupta, D.R. Dhingra, 1954: Essential oil of kewda – Pandanus odora tissimus Linn.f in the perfumery. Essent. Oil Res. 45, 219–222

Diarra, N.G., 1977: Quelques plantes vendues sur les marchés de bambako. J. Agric. Trop. Bot. Appl. 24, 41–42

Diáz-Maroto, M.C., M.S. Pérez-Coello, M.D. Cabezudo, 2002a: Effect of different drying methods on the volatile components of parsley (Petroselinum crispum L.). Eur. Food Res. Technol. A 215, 227–230

Diáz-Maroto, M.C., M.S. Pérez-Coello, M.D. Cabezudo, 2002b: Effect of drying method on the volatile in bay leaf (Laurus nobilis L.). J. Agric. Food Chem. 50, 4520-4524

Diederichsen, A. (ed), 1996: Coriander (Coriandrum sativum L.). Internat. Plant Genetic Resources Inst. Gatersleben, 83 pp

Diederichsen, A., K. Hammer, 1994: Vielfalt von Koriander im Weltsortiment der Genbank Gatersleben. Drogenreport 7, No. 11, 13–17

Diener, H., 1994: Die Mariendistel. PTA heute 8, 290–292

Dignum, M.J.W., J. Kerler, R. Verpoorte, 2002: Vanilla curing under laboratory conditions. Food Chem. 79, 165–171

Dignum, M.J.W., R. van der Heijden, J. Kerler, C. Winkel, R.Verpoorte, 2004: Identification of glucosides in green bean of Vanilla planifolia Andrews and kinetics of vanilla β-glucosidase. Food Chem. 85, 199–205

Dimov, N., A. Tsoutsoulova, 1987: Pattern recognition methods for discrimination of essential oils (Rose oils) by their gas chromatograms. Perfumer & Flavor. 12, No. 6, 45–50

Ding, D.S., H.D. Sun, 1983: Strukturbestimmung eines Prinzip des essentisches Öls von Mentha haplocalyx Briq. Acta Bot. Sinica 25, No. 1, 62–66 (in Chinese)

Dogan, Y., H.H. Mert, 1998: An autecological study on the Vitex agnus-castus. (Verbenaceae) distributed in West Anatolia. Turkish J. Bot. 25, 137–148

Dombrowicz, E., L. Swiatek, R. Gurye, R. Zadernowski, 1991: Phenolic acid in herb Melilotus officinalis. Pharmazie 46, 156–157

Dommée, B., M.W., Assouad, G. Valdeyron, 1978: Natural selection and gynodioecy in Thymus vulgaris L. Bot. J. Linn. Soc. 77, 17–28

Domrös, M., 1973: Die Gewürzpflanzen auf Ceylon – ihre kulturlandschaftliche und wirtschaftsgeographische Relevanz. Aachener geograph. Arbeiten 6, 135–157

Doneanu, C., G. Anitescu, 1998: Supercritical carbon dioxide extraction of Angelica archangelica L. root oil. J. Supercrit. Fluids 12, 59–67

Dönmez, A.A., 2002: Perilla: a new genus for Turkey. Turkish J. Bot. 26, 281–283

Dorner, W.G., 1985: Die Melisse – immer noch zu Überraschungen fähig. Pharmazie in unserer Zeit 14, 112–114

Dorossiev, I., 1985: Determination of flavonoids in Hypericum perforatum. Pharmazie 40, 585–586

Dragendorff, F. 1898: Die Heilpflanzen der verschiedenen Völker und Zeiten. F. Enke Verlag Stuttgart, 885 pp

Dregus, M., J. Barta, K.-H. Engel, 2004: Flüchtige Verbindungen von Rhabarber-Konzentrat. Lebensmittelchemie 58, 25

Dro, A.S., F.W. Hefendehl, 1974: Untersuchung des ätherischen Öls von Ocimum gratissimum L. Arch. Pharmaz. 307, 168–176

Du, C.T., P.L. Wang, F.J. Francis, 1975: Anthocyanins of pomegranate, Punica granatum. J. Food Sci. 40, 417–418

Dubiel, U., 1988: Der Majoran (Majorana hortensis Moench.) und sein Anbau im Einzugsbereich des VEB Majaoranwerk Aschersleben. Drogenreport 1, 65–72

Dudai, N., E. Werker, E. Putievsky, U. Ravid, 1988: Grandular hairs and essential oils in the leaves and flowers of Majorana syriaca. Israel J. Bot. 37, 11–18

Dudai, N., A. Poljakoff-Mayber, H.R. Lerner, E. Putievsky, U. Ravid, I. Katzier, 1993: Inhibition of germination and growth by volatile of Micromeria fruticosa. Acta Hort. (The Hague) 344, 123–130

Dudtschenko, A., G., A.S. Kos'jakov, V.V. Krivenko (Дудченко, А.Г., А.С. Коэьяков, В.В. Кривенко), 1989: Spicy-aromatic and spicy tasting plants. A encyclopedia. (Пряноароматические и пряновкусовые растения. Справочник). Scientific Publishers, Kiew (Киев Наукова Думка), 303 pp (in Russian)

Dugo, G.A., 1994: The composition of the volatile fraction of the Italian citrus essential oil. Perfumer & Flavorist 19, No. 6, 29–42

Dugo, G.A., A. di Giacomo, 2002: Citrus. The genus Citrus. Taylor & Francis London New York, 642 pp

Dugo, P., L. Mondello, K.D. Bartle, A.A. Clifford, D.G.P.A. Breen, G.A. Dugo, 1995: Deterpenation of sweet orange and lemon essential oils with supercritical carbon dioxide using silica gel as an adsorbent. Flavoring & Fragrance J. 10, 51–56

Duke, J.A., 1987: CRC Handbook of medical herbs. CRC Press Boca Raton, Florida, 677 pp

Duke, J.A., E.S. Ayensu, 1985: Medical plants of China. Reference Publ. Algonac, 2 Vol., 705 pp

Duke, J.A., M.Jo Bogenschutz-Godwin, J. duCellier, P.-A.K. Duke, 2003: CRC Handbook of medicinal Spices. CRC Press Boca Raton, Florida, 339 pp

Dung, N.X., T.D. Ching, D.D. Rang, P.A. Leclerq, 1994a: Constituents of the flower oil of Alpinia speciosa K. Schum. from Vietnam. J. Ess. Oil Res. 6, 433–434

Dung, N.X., T.D. Chinh, D.D. Rang, P.A. Leclercq, 1994b: Volatile constituents of the seed and fruit skin oils of Catimbium latilabre (Ridl.) Holtt. from Vietnam. J. Ess. Oil Res. 6, 541–543

Dung, N.X., L.D. Cu, N.H. Thai, L.D. Moi, L. van Hac, P.A. Leclercq, 1996: Constituents of the leaf and flower oils of Agastache rugosa (Fisch. et Mey.) O. Kuntze from Vietnam. J. Ess. Oil Res. 8, 135–138

Dürbeck, K., 1996: Die tropischen Verwandten der vier mitteleuropäischen Artemisia-Arten. Drogenreport 9, No. 14, 47–51

Duriyaprapan, S., E.J. Britten, 1982: The effect of age and location of leaf on quantity and quality of Japanese mint oil production. J. Exp. Bot. 33, 810–814

Dusemund, B., M. Goll, W. Grunow, T. Huwer, 191: Gesundheitliche Beurteilung von Perubalsam. Bundesgesundheitsblatt 34, 568–573

Dutta, P.K., H.O. Saxena, M. Braham, 1987: Kewda perfume industry in India. Econ. Bot. 41, 403–410

E

Echeverrigaray, S., G. Agostini, L. Atti-Serfini, N. Paroul, G.F. Pauletti, A.C. Atti dos Santos, 2001: Correlation between the chemical and genetic relationships among commercial thyme cultivars. J. Agric. Food Chem. 49, 4220–4223

Echim, T., 1993: Neue ausgefallene Basilikumsorten. Gemüse 29, 239–241

Ecker-Schlimpf, B., 1991: Perubalsam, wie ist er gesundheitlich zu beurteilen. Dtsch. Apotheker-Ztg. 132, 522–523

Edris, A.E., H.M. Fadel, 2002: Investigation of the volatile aroma components of garlic leaves essential oil. Possibility of utilization to enrich garlic bulb oil. Eur. Food Res. Technol. 214 A, 105–107

Edwardson, J.R., 1952: Hops – their botany, history, production and utilization. Econ. Bot. 6, 160–175

Effenberger, S., H. Schilcher, 1990: Gewürzsumchrinde. Z. Phytother. 11, 113–118

Eger, H., H. Heine, 1998: Inhaltsstoffe von Bohnenkrautsorten. Gemüse 36, 627–628

Ehlers, D., 1999: HPLC-Untersuchung von Handelsprodukten mit Vanille oder Vanillearoma. Dtsch. Lebensmittel-Rdsch. 95, 464–468

Ehlers, D., S. Bartholomae, 1993: Hochdruckflüssigkeitschromatographische Untersuchungvon Vanille-CO_2-Hochdruckextrakten. Z. Lebensmittel-Unters. u. -Forsch.197, 550–557

Ehlers, D., M. Pfister, 1997: Compounds of vanillons (Vanilla pompona Schiede). J. Ess. Oil Res. 9, 427–431

Ehlers, D., M. Pfister, S. Bartholomae, 1994: Analysis of Tahaiti vanilla by high-performance liquid chromatography. Z. Lebensmittel-Unters. u. -Forsch. 199, 38–42

Ehlers, D., M. Pfister, S. Bartholomae, 1995: HPLC-Untersuchung von natürlichen und künstlichen Vanille-Aromen. GIT, Fachzeitschr. Labor. 39, 765–768

Ehlers, D., J. Färber, A. Martin, K.-W. Qurin, D. Gerard, 2000: Untersuchung von Fenchelölen – Vergleich von CO_2-Extrakten und Wasserdampfdestillation. Dtsch. Lebensmittel-Rdsch. 96, 330–335

Ehlers, D., M. Pfister, W.-R. Bork, P. Toffel-Nadolny,

1995: HPLC analysis of tonka bean extracts. Z. Lebensmittel-Unters. u. -Forsch. 201, 278–282

Ehlers, D., S. Platte, W.-R. Bork, D. Gerard, K.-W Quirin, 1997: HPCL-Untersuchung von roten Steinklee-Extrakten. Dtsch. Lebensmittel-Rdsch. 93, 77–79

Ehlers, D., P. Schäfer, H. Doliva, J. Kirchhoff, 1999: Vanillingehalt und Inhaltsstoff-Verhältniszahlen von Vanilleschoten – Einfluss der Extraktionsbedingungen auf die ermittelten Werte – Diskussion von Richtwerten. Dtsch. Lebensmittel-Rdsch. 95, 123–129

Eichhorn, S., P. Winterhalter, 2004: Glucotropäinbestimmung in Kapuzinerkresse (Tropaeoleum majus L.). Entwicklung einer HPLC/Biosensorkopplung. Lebensmittelchemie 58, 4.

Eichinger, R., 1999: Holunder. G. Thieme Verlag Stuttgart, 128 pp

Eisner, T., K.D. McCormick, M. Sakaino, M. Eisner, S.R. Smedley, D.J. Aneshansley, M. Deyrup, R.L. Meinwald, 1990: Chemical defense of a rare mint plant. Chemoecology 1, 30–37

Ekong, D.E.U., A. U. Ogan, 1968: Constituents of Xylopia aethiopica. Structure of xylopix acid, a diterpene acid. J. Chem. Soc. (London) , 311–312

Ekundayo, O., F.J. Hammmerschmidt, 1988c: Constituents of essential oil of Monodora myristica seeds. Fitotherapia 59, 52–54

Ekundayo, O., F.J. Hammerschmidt, 1988a: Constituents of essential oil of Monodora myristica seeds. Fitoterapia 59, 52–54

Ekundayo, O., I. Laakso, R.M. Adegbola, B. Oguntimein, A. Sofowora, R. Hiltunen, 1988b: Essential oil constituents of ashanti pepper (Piper guineense) fruits (berries). J. Agric. Food Chem. 36, 880–882

Ekundayo, O., I. Laakso, R. Hiltunen, 1989: Constituents of the volatile oil from leaves of Ocimum canum Sims. Flavour & Fragrance J. 4, 17–18

Ekundayo, O., I. Laakso, M. Holopainen, R. Hiltunen, B. Oguntimain, V. Kauppinen, 1990: The chemical composition and antimicrobial activity of the leaf oil of Vitex agnus-castus L. J. Ess. Oil Res. 2, 115–119

Elamrani, A., S. Zrira, B. Benjilali, M. Berrada, 2000: A study of Moroccoan rosemary oils. J. Ess. Oil Res. 12, 487–495

El-Dakhakhny, M., 1965: Studies on Egyptian Nigella sativa L. Arzneimittel-Forsch. 15, 1227–1229

El-Dhaw, Z.Y., N.M. Abdel-Munaem, 1996: Chemical and biological values of black cumin seeds. J. Agric. Sci. Mansoura Univ. 21, 4149–4159

El-Hamidi, A., G. Richter, 1965: Preliminary investigation of Egyptian cumin oil by thinlayer chromatography. Lloydia 28, 252–256

El-Hamidi, A., S.S. Ahmed, F. Shaaraway, 1983: Lippea citriodora grown in Egypt. A new crop under development. Acta Horticult. 132, 31–33

El-Masry, S., A.H.A. Abou-Dania, F.A. Darwish, M.A. Abou-Karum, M. Grenz, F. Bohlmann, 1984: Sesquiterpene lactones from Chrysanthemum coronarium. Phytochemistry 29, 1467–1469

El-Nemr, S.E., I.A. Ismail, M. Ragab, 1992: The chemical composition of juice and seeds of pomegranate fruit. Fruit processing 2, 162–164

El-Tinay, A.H., A.H. Khattab, M.O. Khidir, 1976: Protein and oil composition of sesame seed. J. Amer. Oil Chem. Soc. 53, 648–653

Embong, M.B., D. Hadziyev, S. Molnar, 1977: Essential oils from spices grown in Alberta. Anise oil. J. Canad. Plant Sci. 57, 681–688

Engel, D., S. Phummai, 2000: A field guide to tropical plants in Asia. Times Edn. Singapore. 280 pp

Engel, R., A. Nahrstedt, F.J. Hammerchmidt , 1991: The essential oil of Cedronella canariensis and C. canariensis var. anisata. Planta med. 57, A80–A81

Engelbeen, M., 1946: Les Aleurites. Bull. agric. Congo Belge 37, 255–342

Engelshowe, R., 1985: Rhabarber, eine alte Droge – immer noch aktuell. Pharmazie in unserer Zeit 14, 40–49

Enríquez, S., K. Sand-Jensen, 2003: Variation of light absorption properties of Mentha aquatica L. as a function of leaf form: implications for plant growth. Internat. J. Plant Sci. 164, 125–136

Erdelmeier, C.A.J., 1998: Hyperforin, possibly the major nonnitrogenous secondary metabolite of Hypericumperforatum L. Pharmacopsychiatry 31 (S), 2–6

Erhardt, W., 1988: Hemerocallis-Taglilien. Verlag E. Ulmer Stuttgart, 169 pp

Erhardt, W., E. Götz, N. Bödeker, S. Seybold: 2000, 2002: ZANDER. Handwörterbuch der Pflanzennamen. Verlag E. Ulmer Stuttgart, 16. edn., 990 pp; 17. edn. 990 pp

Erichsen-Brown, Ch., 1989: Medicinal and other uses of North American plants. A historical survey with special reference to the Eastern Indian tribes. Dover Publications, Inc. New York, 511 pp

Erler, J., O. Vostrowsky, H. Strobel, K. Knobloch, 1988: Über das ätherische Öl des Ingwers, Zingiber officinalis Roscoe. Z. Lebensmittel-Unters. u. -Forsch. 231–234

Ernö, O., M. Gyorgy, B. Laszzló, 1985: A paprika. Akadémia Kiadó Budapest, 125 pp (in Hungarian)

Eshbaugh, W.H., 1970: A biosystematic and evolutionary study of Capsicum baccatum (Solanaceae). Brittonia22, 31–43

Espinar, L.A., 1967: Las especias de Tagetes (Compositae) de la region central Argentina. Kurtziana 4, 51–71

Essien, E.U., B.C. Izunwane, C.Y. Aremu, O.U. Eka, 1994: Significance for humans of the nutrient content of the dry fruit of Tetrapleura tetraptera. Plant Foods Human Nutrit. 45, 47–51

Estilai, A., A. Hashemi, K. Truman, 1990: Chromosome number and meiotic behavoir of cultivated chia, Salvia hispanica (Lamiaceae). Hort. Sci. 25, 1646–1647

Estrada, B., M.A. Bernal, J. Dáz, F. Pomar, F. Merino, 2000: Fruit development in Capsicum annuum: changes in capsaicin, lignin, free phenolics and peroxidase patterns. J. Agric. Food Chem. 48, 6234–6239

Esuoso, K., H. Lutz, M. Kutubuddin, E. Bayer, 1998: Chemical composition and potential of some under utilizedtropical biomass. I. Fluted pumpkin (Telfairia occidentalis). Food Chem. 61, 487–492

Eugster, C.H., E. Märki-Fischer, 1991: Chemie der Rosenstoffe. Angew. Chem. 103, 671–689

Evans, F.J., A.D. Kinghorn, 1975: The succulent Euphorbias of Nigeria. Lloydia 38, 363–365

Ewaida, E.H., 1987: Nutrient composition of "Taifi" pomegranate (Punica granatum L.) fragments and their suitability for the production of jam. Arab. Gulf J. Sci. Res., Agric. Biol. Sci. B 5, 367–378

F

Facciola, S., 1990: Cornucopia. A source book of edible plants. Kompong Publ., Vista, 677 pp

Fageria, M.S., P. Khandelwal, R.S. Dhaka, 2003: Effect of harvest stage and processing treatment on the quality of sun dried ker (Capparis decidua [Forsk] Edgew) fruit. J. Horticult. Sci. & Biotechnol. 78, 168–172

Faheid, S.M.M., 1998: Application of onion and garlic flavours in spaghetti manufacture. Dtsch. Lebensmittel-Rdsch. 94, 187–192

Fahmy, I.R., H. Abushady, A. Schönberg, A. Sina, 1947: A cristalline principle from Ammi majus L. Nature (London) 160, 468–469

Falch, B., 1997: Ingwer – nicht nur ein Gewürz. Dtsch. Apotheker-Ztg. 137, 4267–4278

Famini, G., E. Mastrorilli, P.L. Cioni, I. Morelli, F. Panizzi, 1999: Essential oil from Crithmium maritimum grown in Liguria (Italy): seasonal variation and antimicrobial activity. J. Ess. Oil Res. 11, 788–792

Fan, X., B.A. Niemira, K.J.B. Sokorai, 2003: Sensorial, nutritional and microbiological quality of fresh cilantro leaves as influenced by ionizing radiation and storage. Food Res. Internat. 36, 713–719

Faray, R.S., M.K. Shabana, H.A. Sallam, 1981: Biochemical studies one some chemical characteristics of slided Egyptian onions. J. Food Sci. 46, 1394–1399

Farhath Khanum, K.R.A., K.R. Sudarshana Krishna, K.R. Viswanathan, K. Santhanam, 2000: Anticarcinogenic effects of curry leaves in dimethylhydrazine-treated rats. Plant Foods Human Nutrit. 55, 347–355

Farkas, P., P. Hradský, M. Kovác, 1992: Novel flavour components identified in the staem destillate of onion (Allium cepa L.). Z. Lebensmittel-Unters. u. -Forsch. 459–462

Farnsworth, N.R., N. Bunyapraphatsara, 1992: Thai medical plants. Medicinical Plant Information Center, Bangkok, Thailand, 402 pp

Farrell, K. T., 1885: Spices, condiments, and seasonings. The AVI Publishing Comp., Inc. Westport, Conn., 396 pp

Faruq, M.O., J. Chowdhury, M. Yusuf, S.M. Farooque, M.M. Chowdhury, M.M. Khuda, 1994: TCL technique in the component characterization and quality determination of the Bangladesh lemongrass oil (Cymbopogon citratus Staph.). Bangladesh J. Sci. & Ind. Res. 29, No. 2, 27–38

Fatope, O.M., Y. Takeda, 1988: The constituents of the leaves of Ocimum basilicum. Planta med. 54, 190

Fatope, O.M., Y. Takeda, H. Yamashita, H. Okape, T. Yamauchi, 1990: New curcurbitane tripernoids from Mormodica charantia. J. Nat. Prod. 53, 1491–1497

Fattorusso, E., V. Lanzotti, O. Taglialatela-Scafati, M. di Rosa, A. Ianaro, 2000: Cytotoxic saponins from bulbs of Allium porrum L.J. Agric. Food Chem. 48, 3455–3462

Fattorusso, E. V. Lanzotti, O. Taglialatela-Scafati, C. Cicala, 2001: The flavonoids of leek, Allium porrum. Phytochemistry 57, 565–569

Fattorusso, E., M Iorizzi, V. Lanzotti, O. Taglialatela-Scafati, 2002: Chemical composition of shallot (Allium ascalonicum Hort.). J. Agric. Food Chem. 50, 5686–5690

Fayet, B., C. Tisse, M. Guerere, J. Estienne, 1987: Nouveaux critères analytiques dans l'étude des gousses devanille. Analysis 15, 217–226

Fazil, F.R.Y., R. Hardman, 1968: The spice fenugreek (Trigonella foenum-graecum L.): its commercial varieties of seed as source of diosgenin. Trop. Sci. 10, No. 2, 66–78

Fehr, D., 1979: Untersuchung über Aromastoffe von Sellerie (Apium graveolens L.). Pharmazie 34, 658–662

Feil, B., R. Seyfarth, C. Orlacchio, M. Messmer, P. Kade, B. Meier, B. Büter, 2002: Entwicklung und Ertrag verschiedener Vitex agnus-castus-Herkünfte im Feldanbau in Italien. Z. Arznei- u. Gewürzpflanzen 7, 363-367

Fellous, R., G. George, C. Schippa, 1992: Vanilline: anomalies dans le protocole d'identification par [13]C. Parfum, Cosmetics, Aromes 106, 95–97

Feng, P.C.M., L.J. Hynes, K.E. Magnus, 1961: High concentration of (–)-noradrenaline in Portulaca oleracea. Nature (London) 191, 1108

Fenwick, D., 2003: Crocosmia – a brief history. Bulbs 5, No. 1, 20–23

Fenwick, G.R., A.B. Hanley, 1985/986: The genus Allium. Part I–III. Crit. Rev. Food Sci. Nutrit. 22, 199–271, 273–377; 23, 1–73

Ferrão, J.E.M., 1992: A aventura das plantas e os descobrimentos Portugueses. Edição Inst. Investigação Cientificia Trop. Comissão para as Comemorações dos Desocobrimentos Portugueses & Fundação Berardo Lisboa, 241 pp

Ferrari, J.P., G. Allaud, 1971: Bibliographie du genre Capsicum. J. Agric. trop. Bot. Appl. 18, 385–479

Ferraro, P.M., 1955: Las species Argentinas del genero Tagetes. Bol. Soc. Argentina Bot. 6, 30–39

Ferreira, J.L.I., N.F. Roque, V.O. de Ferro, 1992: Ocimummicranthum Willd.-Manjericao of Brazil. Histologicoe quimico caracteristica. Rev. Inst. Adolfo Lutz, São Paulo 52, 47–50

Ferrua, F.Q., M.O.M. Marques, 1994: Lemongrass essential oil obtained by extraction with liquid CO_2. Cienc. e Tecnol. de Alimentos 14, (Suppl.), 83–92

Ficalho, C. de, 1947: Plantas Uteis das Africa Portuguesa. RP. MDC, Lisbon, 301 pp (In Portuguese)

Fico, G., A. Bader, G. Flamini, P.L. Cioni, I. Morelli, 2003: Essential oil of Nigella damscena L. (Ranunculaceae) seeds. J. Ess. Oil Res. 15, 57–58

Field, J.A., G. Nickerson, D.D. James, C. Heider, 1996: Determination of essential oils in hops by headspace solid-phase microextraction. J. Agric. Food Chem. 44, 1768–1772

Fincke, H., 1961: Über Rosmarin als verzierende Gebäck- und Marzipanzutat und seine sonstige Verwendung als würzende Pflanze und im Brauchtum. Gordian (Hamburg) 61, 22–25

Fincke, H., 1963: Vom Backwarengewürz Anis und seiner Geschichte. Gordian (Hamburg) 63, 582–590

Fintelmann, V., T. Wegener, 2001: Gelbwurzel – eine unterschätzte Heilpflanze. Dtsch. Apotheker-Ztg. 141, 3735–3743

Fischer, N., N. Nitz, F. Drawert, 1985: Über gebundene Aromastoffe in Pflanzen. 1. Mitt. Flüchtige Verbindungen aus gebundenen Vorstufen in Majorana hortensis Moench. Chem. Mikrobiol. Technol. Lebensmittel 9, 87–94

Fischer, N., N. Nitz, F. Drawert, 1987: Original flavour compounds and the essential oil composition of majoram. Flavour & Fragrance J. 2, 55–61

Fishman, M. 1984: The rose called "Kazanlik" . Rose letters 9, No. 6, 6–9

Flamini, G., E. Mastrorilli, P.L. Cioni, I. Morelli, L. Panizzi, 1999: Essential oil from Crithmum maritimum grown in Liguria (Italy): seasonal variation and microbial activity. J. Ess. Oil Res. 11, 788–792

Flamini, G., P.L. Cioni, I. Morelli, M. Macchia, L. Ceccarini, 2002: Main agronomic-productive characteristics of two ecotypes of Rosmarinus officinalis L. and chemical composition of their essential oils. J. Agric. Food Chem. 50, 3512–3517

Flamini, G., Cioni, P.L., I. Morelli, 2003: Differences in the fragrances of pollen, leaves, and floral parts of garland (Chrysanthemum coronarium) and composition of the essential oil from flowerheads and leaves. J. Agric. Food Chem. 51, 2267–2271

Fleischhauer, St. G., 2003: Enzyklopädie der essbaren Wildpflanzen. AT Verlag Aarau, Munich, 411 pp

Fleischmann, P., N. Watanabe, P. Winterhalter, 2002: Nachweis eines carotinoidspaltenden Enzyms in Blüten von Osmanthus fragrans L. Lebensmittelchemie 56, 36

Fleisher, A., 1981: Essential oils from two varieties of

Ocimum basilicum L. grown in Israel. J. Sci. Food Agric. 32, 1119–1122

Fleisher, A., Z. Fleisher, 1988: Identification of biblical hyssop and origin of the traditional use of oregano group-herbs in the Mediterranean region. Econ. Bot. 42, 232–241

Fleisher, Z., A. Fleisher, 2002: Volatiles of Coridothymus capitatus chemotypes growing in Israel: aromatic plants of the Holy Land and the Sinai. J. Ess. Oil Res. 14, 105–106

Fleisher, A., N. Sneer, 1982: Oregano spices and oreganum chemotypes. J. Sci. Food Agric. 33, 441–446

Flückiger, F.A., 1883: Die Chinarinde. R. Gaertner's Verlagsbuchhandl. Berlin, 79 pp

Fock-Heng, P.A., 1965: Cinnamom of the Seychelles. Econ. Bot. 19, 257–261

Forgacs, E., V. Kiss, T. Cserhati, J. Hollo, 1996: Determination of the moisture content of paprika (Capsicum annuum) powders: a comparative study. Food Sci. Technol. Internat. 2, 23–27

Foroumadi, A., A. Asadipour, F., Arabpour, Y. Amanzadeh, 2002: Composition of the essential oil of Bunium persicum (Boiss.) Fedtsch. from Iran. J. Ess. Oil Res. 14, 161–162

Forrest, J.E., 1972: Nutmeg and mace, the psychotropic spice from Myristica fragrans. Lloydia 35, 440–449

Forster, A., 1981: Über das Lagerverhalten verschiedener Hopfensorten und Hopfenprodukte. Brauwissenschaft 34, 429–439

Fosberg, F.R., 1965: Nomenclature of the horseradish (Cruciferae). Baileya 13, 1–4

Fossen, T., A.T. Petersen, Ø.M. Anderson, 1998: Flavonoids from red onion (Allium cepa). Phytochemistry 47, 281–285

Foster, L.J., 1962: Recent technical advances in the cultivation of the tung oil tree, Aleurites montana, in Nyasaland. Trop. Sci., Trinidad 39, No. 3, 169–187

Foster, N., L.S. Cordell, 1992: Chilies to chocolate. Food the Americans gave the world. Univ. Arizona Press Tucson & London, 191 pp

Fournier, G., J. Habib, A. Reguigui, F. Safta, S. Guetari, R. Chemli, 1989: Étude de divers échantillons d'huile essentielle de Rosmarin de Tunisie. Plantes méd. et Phytothér. 23, 180–185

Fournier, N., G. Vangheesdaele, 1980: Composition de la fraction glucidique de la Brassica juncea et de lamoutarde de Dijon. Rev. Franc. Corps Gras 27, 513–519

Foust, C.M., 1992: Rhubarb. The wondrous drug. Princeton Univ. Press Princeton, N.J,. 371 pp

Fraisse, D., F. Maquin, D. Stahl, K. Suon, J.C. Tabet, 1984: Analyse d'extraits de vanille par chromatographie en phase liquide, coulage chromatographie en phase gazeuse-spectrométrie de masse et spectrométrie demasse-spectrométrie de masse. Analysis 12, 63–71

Francis, G.W., M. Isaksen, 1989: Droplet counter current chromatigraphy of the carotenoids of parsley, Petroselinum crispum. Chromatographia 27, 549–551

Frank, C., A. Dietrich, U. Kremer, A. Mosandl, 1995: GC-IRMS in the authenticity control of the essential oil of Coriandrum sativum L. J. Agric. Food Chem. 43, 1634–1637

Franke, W., 1982: Vitamin C in sea fennel (Crithmum mariitmum), an edible wild plant. Econ. Bot. 36, 163–165

Franz, Ch., D. Fritz, 1975: Anbauversuche mit Gentiana lutea und Inhaltsstoffe einiger Ökotypen. Planta med. 28, 289–300

Franz, Ch., A. Ceylan, J. Hölzl, A. Vömel, 1984: Influence of the growing site on the quality of Mentha piperita L. -oil. Acta Horticult. 144, 145–150

Franz, Ch., D. Fritz, 1978: Cultivation aspects of Gentiana lutea L. Acta Horticult. 37, 289–314

Franz, G.: Kapuzinerkresse (Tropaeolum majus L.). Z. Phytother. 17, 255–262

Fraternale, D., L. Giamperi, D. Ricci, A. Manunta, 2001: Characterization of the essential oil of Calaminthanepeta (L.) Salvi ssp. nepeta cultured in vitro. Riv. Ital. EPPOS 11, 9–13

Frau, A., M. Cadinu, A. Repetto, A. Zeddo, 2001: Micropropagazione di cinque cloni di mirto sardo. Informatore Agrario (Italy) 57, 65–67

Freeburg, E.J., B.S. Mistry, G.A. Reineccius, 1994: Stability of citral-containing and citralless lemon oil in flavour emulsions and beverages. Perfumer & Flavorist 19, No. 4, 23–32

Freiburglaus, F., P. Meyer, H. Pfander, 1998: Geheimnisse des Safrans. Naturwiss. Rdsch. 51, 91–95

Freise, F.W., 1936: Brasilianische Medizinalpflanzen. Tropenpflanzer 39, 241–253, 330–339, 513–524

Frerot, E., A. Bagnoud, C. Vuilleumier, 2002: Menthofurolactone: a new p-menthane lactone in Mentha piperita L.: analysis, synthesis and olfactory properties. Flavour & Fragrance J. 17, 218–226

Frighetto, N., J.G. Oliveira, A.C. Siani, K.C. Chagas, 1998: Lippea alba (Mill.) N.E.Br. (Verbenaceae) as a source of linal oil. J. Ess. Oil Res. 10, 578–580

Fuchs, S., S. Sewenig, A. Mosandl, 2001: Monoterpene biosynthesis in Agathosma crenulata (Buchu). Flavour & Fragrance J. 16, 123–135

Fuentes, S., 1986: Contribution é l'étude de deux plantes aromatiques mexicaines Pimenta dioica Merill., Cymbopogon citratus Stapf. Thèse doct., 3e cycle Univ. Paris

Fujimatu, E., T. Ishikawa, J. Kitajima, 2003: Aromatic compound glucosides, alkyl glucoside and glucide from the fruit of anise. Phytochemistry 63, 609–616

Fujita, T., M. Nakayama, 1993: Monoterpene glucosides and other constituents from Perilla frutescens. Phytochemistry 34, 1545–1548

Fujita, T., H. Nishimura, K. Kaburagi, J. Mizutani, 1994: Plant growth inhibiting alpha pyrones from Alpinia speciosa. Phytochemistry 36, 23–27

Fun, C.E., A.B. Svendsen, 1990a: The essential oil of Hyptis suaveolens Poit. grown on Aruba. Flavour & Fragrance J. 5, 161–163

Fun, C.E., A.B. Svendsen, 1990b: The essential oil of Lippea alba (Mill.) N.E.Br. J. Ess. Oil Res. 2, 215–217

G

Gabel, E., K.H. Müller, I. Schoknecht, 1962: Die Wertbestimmung von Herba, Oleum und Tinctura Thymi. Dtsch. Apotheker-Ztg. 102, 293–295

Gad, A.M., M. El-Dakhakhny, M. Hassan, 1963: Studies on the chemical constitution of Egyptian Nigella sativa L. oil. Planta med. 11, 134–138

Galacci, R.R., 1989: Microscopic detection and identification of various substitute vegetable tissues in ground horse radish. J. Assoc. Off. Anal. Chem. 72, 619–621

Galakshmi, J.S., N.B. Shankaracharya, J. Pura Naik, L. Jagan Mohan Rao, 2000: Studies on chemical and technological aspects of ajowan (Trachyspermum ammi [L.] syn. Carum copticum Hiern.) seeds. J. Food Sci. Technol., Mysore 37, 277–281

Galambosi, B., Z. Sz-Galambosi, K.P. Svoboda, S.G. Deans, 1998: Blütenertrag und antioxidative Eigenschaften von Arnica montana L., kultiviert in Finnland. Drogenreport 11, No. 19, 10–13

Galetto, W.G., P.G. Hoffmann, 1978: Some benzyl esters present in the extract of Vanilla (Vanilla planifolia). J. Agric. Food Chem. 26, 195–197

Galindo-Cuspinera, V., M.B. Lubran, S.A. Rankin, 2002: Comparison of volatile compounds in water- and oil-soluble annatto (Bixa orellana L.) extracts. J. Agric. Food Chem. 50, 2010–2015

Galle-Hoffmann, U., W.-A. König, 1998: Ätherische Öle. 1. Pfefferminzöl. Dtsch. Apotheker-Ztg. 138, 3793–3799

Gandhi, M., V.K. Vinayak, 1990: Preliminary evaluation of extracts of Alstonia scholaris bark for in vitro antimalarial activity in mice. J. Ethnopharmacol. 29, 51–57

Gangolly, S.R., R. Singh, S.L. Katyal, D. Singh, 1957: The Mango. Indian Council of Agric. Res., New Delhi, 530 pp

Garnero, J., 1971: Review on the chemical composition of vetiver oil. Parfumery, Cosmetic., Sav. France 1, 569–589

Garavel, L., 1960: Le noyer noir d'Amerique (Juglans nigra). Rev. Forest. France 12, 362–373

Gardener, J.B., E.H. Graven, L. Webber, G. Benlans, M. Venter, 1991: Effect of soil type and nutrient status on the yield and composition of Tagetes oil (Tagetes minuta L.) J. Ess. Oil Res. 3, 303–307

Garg, S.N., V.K. Mehta, 1998: Acyclic monoterpenes from the essential oil of Tagetes minuta Flowers. Phytochemistry 48, 395–396

Garg, S.N., N. Mengi, N.K. Patra, Reena Charles, Sushil Lamar, 2002: Chemical examination of the leaf essential oil of Curcuma longa L. from the North Indian plains. Flavour & Fragrance J. 17, 103–104

Garland, S., 1979: The complete book of herb and spices. Viking Press, New York, NY, 288 pp

Garrabrants, N.L., L.E. Craker, 1987: Optimizing field production of dill. Acta Horticult. 28, 69–81

Gašić, O., N. Mimica-Dukić, D. Adamović, K. Borojevic, 1987: Variability of content and composition of essential oil in different genotypes of peppermint. Biochem. Sytematics a. Ecol. 15, 335–340

Gašić, O., N. Mimica-Dukić, D. Adamović, 1992: Variability of content and composition of essential oil of different Mentha arvensis L. var. piperascens cultivars. J. Ess. Oil Res. 4, 49–56

Gassmann, B., 1992: Knoblauch – Lebensmittel oder

Modedroge? Z. Ernährungs-Umsch. 39, 415–418, 444–449

Gebczynski, P., A. Korus, J. Slupski, 2001: Effect of storage conditions and the storage period on the content of nitrates and oxalates in dill (Anethum graveolens). Bormatol. I Chem. Toksykol. 34, 345-349 (in Polish)

Gebhardt, U., 1971: Estragon, ein Aristokratenkraut unter den Küchenkräutern. Fleischwirtschaft 57, 828–829

Gebhardt, U., 1977a: Grüne Pistazien zum Verzieren und Würzen. Fleischwirtschaft 57, 18–19

Gebhardt, U., 1977b: Wermut, ein sehr bitteres Gewürz. Fleischwirtschaft 57, 1230–1231

Gebhardt, U., 1977c: Beifuß ein beliebtes Gewürz für Braten. Fleischwirtschaft 57, 1926–1927

Gebhardt, U., 1979: Kapern. Gordian (Hamburg) 79, 72–74; Industrielle Obst- u. Gemüseverwert. 64, 298–300

Gebhardt, U., 1981a: Ysop (Hyssopus officinalis L.), das hübsche Kraut unter den Gewürzpflanzen. Fleischwirtschaft 61, 382–388

Gebhardt, U., 1981b: Alant (Inula helenium L.), eine häufig im Garten vorkommende Gewürzpflanze. Fleischwirtschaft 61, 1286–1287

Geister, H., 1989: Piperin – Schärfestoffe im Pfeffer und Pfefferoleoresin. Fleischwirtschaft 69, 1664, 1666

Genius, O.B., 1981: Radix Angelica archangelicae. Identitätsnachweis und Gehaltsbestimmung auf der Basis von Osthenol. Dtsch. Apotheker-Ztg. 121, 386–387

Genovese, M.I., F.M. Lajolo, 2002: Isoflavones in soya-based foods consumed in Brazil: levels, distribution, and estimated intake. J. Agric. Food Chem. 50, 5987–5993

Gerhardt, E., 1995: Pilze. BLV Verlagsges. mbH Munich Vienna Zürich, p 98

Gerhardt, U., 1979: Kapern. Gordian (Hamburg), 79, 72–74

Gerhardt, U., M. Sundermann, 1981: Mikroskopische Untersuchung von Gewürzen. Muskatnuß. Fleischwirtschaft. 61, 864

Germer, St., G. Franz, 1997: Ingwer – eine vielseitige Arzneidroge. Dtsch. Apotheker-Ztg. 137, 4260-4266

Ghannadi, A., 2002: Composition of the essential oil of Satureja hortensis L. seeds from Iran. J. Ess. Oil Res. 14, 35–36

Giaceri, G., 1972: Chromatographic identification of coumarin derivatives in Menyanthes trifoliata L. Fitoterapia 43, 134–138

Giami, S.Y., 1997: Evaluation of selected food characteristics of three advanced lines of Nigerian soya bean (Glycine max [L.] Merr.) Plant Foods Human Nutr. 50, 17–25

Giami, S.Y., 2002: Chemical composition and nutritional attributes of selected newly developed lines of soya bean (Glycine max [L.] Merr.). J. Sci. Food a. Agric. 82, 1735–1739

Giami, S.Y., M.N. Adindu, M.O. Akusu, J.N.T. Emelike, 2000; Compositional, functional and storage properties of flours from raw and heat processed African breadfruit (Treculia africana Decne) seeds. Plant Food Human Nutr. 55, 357–368

Giannetto, P., G. Romeo, M.C. Aversa, 1979: Calaminthadiol, a new 3,4-secotriterpenoid from Satureja calamintha (Calamintha sylvatica) and Satureja graeca (Micromeria graeca). Phytochemistry 18, 1203–1205

Gil, A., E.B. de la Fuente, A.E. Lenardis, M.L. Pereira, S.A. Suárez, A. Bandoni, C. van Baren, P.D.L. Lira, C.M Ghersa, 2002: Coriander essential oil composition from two genotypes grown in different environmental conditions. J. Agric. Food Chem. 50, 2870–2877

Gil, V., J. MacLeod, 1980: Glucosinoletes of Lepidium sativum and "Garden cress". J. Sci. Food Agric. 31, 739–741

Gilbert, J., H.E. Nursken, 1972: Volatile constituents of horseradish roots. J. Sci. Food Agric. 23, 527–539

Girardon, P., Y. Sauvaire, J.C. Baccou, J.M. Bessiere, 1986: Identification of 3-hydroxy-4,5-dimethyl-2(5H)-fura-nose in aroma volatiles of fenugreek seeds. Lebensmittel-Wiss. u. -Technol. 19, 44–46

Giridharlal, Das D.P., N.L. Jain, 1958: Tamarind beverage and sauce. Ind. Food Packer 12, 13–16

Giuffrida, D., F. Salvo, M. Ziino, G. Toscano, G. Dugo, 2002: Initial investigation on some chemical constituents of cappers (Capparis spinosa L.) from the Island of Salina. Ital. J. Food Sci. 14, 25–31

Glasl, H. Wagner, 1974: Perubalsam. Gaschromatographische Untersuchung und quantitative Bestimmung der Inhaltsstoffe von Balsamum peruvianum. Dtsch. Apotheker-Ztg. 114, 45–49

Göckeritz, D., 1968: Das ätherische Öl von Chrysanthemum balsamite. Pharmazie 23, 515–518

Gogoi, P., P. Baruah, S.C. Nath, 1997: Antifungal activity of the essential oil of Litsea cubeba Pers. J. Ess. Oil Res. 9, 213–215

Gomes, E.C., L.C. Ming, E.A. Moreira, O.G. Miguel, M.D. Miguel, A.V. Kerber, A. Conti, A. Weiss-Filho, 1993: Constituintes do óleo essencial de Lippea alba (Mill.) N.E.Br. (Verbenaceae). Rev. Brasil. Farmac. 74, 29–32

Gomez-Serranillos, M., F. Zaragoza, 1980: Phytochemical study of Maytenus senegalensis. An. Bromatol. 31, 180–188

Gonso, G.L., M.A. Sachez-Fernandez, J.R. Saez, A. Zalacain, M.R. Salinas, 2003: Evaluation of the color of Spanish saffron using tristimulus colorimetry. Ital. J. Food Sci. 15, 249–258

Gopalakrishnan, N., 1994: Studies on the storage quality of CO_2-extracted cardamom and clove bud oils. J. Agric. Food Chem. 42, 796–798

Gopalakrishnan, N., C.S. Narayanan, 1991: Supercritical carbon dioxide extraction of cardamom. J. Agric. Food Chem. 39, 1976–1978

Gopalakrishnan, N., P.P.V. Shanti, 1990: Composition of clove (Syzygium aromaticum) bud oil extracted using carbon dioxide. J. Sci. Food Agric. 50, 111–117

Gopalam, A., M. Ratnambal, 1989: Essential oils of ginger. Indian Perfumer 33, 63–69

Gordon, M.H., An Jing, 1995: Antioxidant activity of flavonoids from licorice. J. Agric. Food Chem. 43, 1784–1788

Gordon-Smith, Cl., 1996: Chillies, der besondere Geschmack. Droemersche Verlagsanstalt Th. Knauer Nachf. Munich, 64 pp

Gorkom, K.W. van, 1869: Die Chinacultur auf Java. Verlag W. Engelmann Leipzig, 61 pp

Gottlieb, O.R., A. Iachan, 1951: O vetiver do Brasil. Anais Assoc. Brasil. Quím. 10, 403–415

Gottlieb, O.R., M.T. Magalhães, 1960: Essential oil of the bark and wood of Aniba canellina. Perfumer & Ess. Oil Rec. 51, 69–70

Govindarajan, N.S., 1977: Pepper – chemistry, technology, and qualities. Crit. Rev. Food Sci. Nutrit. 9, 115–225

Govindarajan, V.S., 1980: Tumeric – chemistry, technology and quality. Crit. Rev. Food Sci. Nutrit. 12, 199–201

Govindarajan, V.S., 1982a: Ginger – chemistry, technology and quality evaluations. Crit. Rev. Food Sci. Nutrit. 17, 1–96, 189–275

Govindarajan, V.S., Santhhi Narasimhan, K.G. Raghaveer, Y.S. Lewis, 1982b: Cardamom – production, technology, chemistry, and quality. Crit. Rev. Food Sci. Nutrit. 17, 227–326

Govindarajan, V.S., 1985: Capsicum – Production, technology, chemistry and quality. Part I: History, botany, cultivation and primary processing. Crit. Rev. Food Sci. Nutrit. 22, 109–176

Govindarajan, V.S., 1985/86: Capsicum – Production, technology, chemistry and quality. Part II.: Processed products, standards, world production and trade. Crit. Rev. Food Sci. Nutrit. 23, 207–288

Govindarajan, V.S., 1986: Capsicum – Production, technology, chemistry and quality. Part III: Chemistry of the color, aroma and pungency stimili. Crit. Rev. Food Sci. Nutrit. 24, 245–355

Govindarajan, V.S., D. Rajalaksmi, N. Chand, 1988: Capsicum – Production, technology, chemistry and quality. Part IV: Evaluation of quality. Crit. Rev. Food Sci. Nutrit. 25, 185–282

Govindarajan, V.S., M.N. Sathyanarayana, 1990: Capsicum – Production, technology, chemistry and quality. Part V: Impact of physiology, pharmacology, nutrition, and metabolism, structure, pungency, pain, and desensitization sequences. Crit. Rev. Food Sci. Nutrit. 29, 435–474

Govindarajan, V.S., S. Narasimhan, K.G. Raghuveer, Y.S. Lewis, 1982: Cardamom – production, technology, chemistry and quality. Crit. Rev. Food Sci. Nutrit. 16, 229–326

Goyal, R.K., B.N. Korla, 1997: Changes in the fresh yield. Dry matter and quality of ginger (Zingiber officinale Rosc.) rhizomes during development. J. Food Sci. Technol. (Mysore) 34, 472–476

Gracza, L., 1987: Oxygen-containing terpene derivatives from Calendula officinalis. Planta med. 53, 227

Grahle, A., C. Höltzel, 1963: Photoperiodische Abhängigkeit der Bildung des ätherischen Öls bei Mentha piperita L. Naturwissenschaften 59, 552

Grainger, A., N. Winer, 1980: A bibliography of Ceratonia siliqua, the carob tree. Internat. Tree Corp. J. 1, 37–47

Graner, G., 1968: Herba Majoranae (Majoran). Präp. Pharmaz. 4, 68–91

Grasse, M.-Chr., 1950: Le Jasmin. Fleur de Grasse. Èdit. Parkstone, Bournemouth, Musée internat. de la Parfumerie, Grasse, France 143 pp

Graven, E.H., L. Webber, G. Benians, M. Venter, J.B. Gardner, 1991: Effect of soil type and nutrient status on the yield and composition of Tagetes oil (Tagetes minuta L.). J. Ess. Oil Res. 3, 3003–307

Gray, D.E., G.E. Rottinghaus, H.E.G. Garrett, S.G. Pallardy, 2000: Simultaneous determination of the predominant hyperforins and hypericins in St. John's wort (Hypericum perforatum L.) by liquid chromatography. J. Assoc. Offic. Analyt. Chem. Intern. 83, 944–949

Grayer, R.J., C.G. Kite, F.J. Goldstone, S.E. Bryan, A. Paton, E. Putievsky, 1999: Intraspecific taxonomy and essential oil chemotypes in sweet basil (Ocimum basilicum). Phytochemistry 43, 1033–1039

Grecu L., V. Cucu, 1975: Saponine aus Primula officinalis und Primula elatior. Planta med. 27, 247–253

Green, P.S., 1965: Studies in the genus Jasminum. 3. The species in cultivation in North America. Baileya 13, 137–172

Green, P.S., 1986: Jasminum in Arabia. Studies in the genus Jasminum L. (Oleaceae). 15. Kew Bull. 52, 933–947

Greene, L., 1951: Cultivation of black pepper. Technol. Collaboration Office of foreign Agric. relations. U.S. Dept. of Agricult. Washington, 47 pp

Greenaway, W., S. English, F.R. Whatley, 1990: Variations of in bud exudates composition of Populus nigra assessed by gas chromatography-mass spectrometry. Z. Naturforsch. 45c, 931–936

Greenaway, W., J. May, T. Scaysbrook, F.R. Whatley, 1992: Composition of bud and leaf exudates of some Populus species compared. Z. Naturforsch. 47c, 329–334

Greenberg, Sh., E. Lampert Ortiz, 1994: Good Gekruid. (The Spice of Life). Alles over specerijen. Uitgeverij het Spectrum Utrecht/Antwerpen, 192 pp

Grella, G.E., V. Picca, 1988: Variazioni stagnionali dell'olio essenziale di Salvia officinalis. Fitoterapia 54, 97–102

Gressner, G., 1997: Besenginster. Arznei- und Giftpflanze, Plantagenet und Teufelsbeeren. Dtsch. Apotheker-Ztg. 138, 807–813

Greve, P., 1938: Der Sumpfporst (Ledum palustre L.). Hannsischer Gildenverlag Hamburg, 121 pp

Greven, D., 1992: Safranernte in Andalusien: „Rotes Gold" von blauen Blüten. Rationelle Hauswirtsch. 24, No. 5, 10–13

Grey-Wilson, Chr., 2000: Poppies. A guide to the poppy family in the wild and in cultivation. Timber Press, Portland, Oregon, 2. edn., 256 pp

Griebel, C., 1909: Über den Nachweis der Papuasmacis. Z. Unters. Nahrungs- u. Genussmittel 18, 202–206

Griebel, C., 1938: „Spanischer Hopfen" als Würzkraut. Z. Unters. Lebensmittel 75, 568–572

Griebel, C., 1939: Hibiscus-„Blüten", eine zur Bereitung von Speisen und Getränken dienende Droge, ihr Hauptinhaltsstoff eine neue Säure von Fruchtsäurecharakter (Hibiscussäure). Z. Unters. Lebensmittel 77, 561–571

Griebel, C., 1943a: Über zwei afrikanische Ersatzpfefferarten: Paradieskörner und Mohrenpfeffer (Kani). Z. Unters. Lebensmittel 85, 426–436

Griebel, C., 1943b: Mönchspfeffer als Pfefferersatz. Z. Unters. Lebensmittel 85, 511–515

Griebel, C., 1944: Über Kani-Gewürz. Z. Unters. Lebensmittel-Unters. u. -Forsch. 87, 69–77

Griebel, C., 1948: Über Aschantipfeffer (Piper clusii DC.). Z. Lebensmittel-Unters. u. -Forsch. 88, 483–486

Griebel, C., A. Freymuth, 1916: Über Massoirinde. Z. Unters. Nahrungs- u. Genußmittel 31, 314–318

Griesbach, R.J., L. Batdorf, 1995: Flower pigments within Hemerocallis fulva L. fm. fulva, fm. rosea, and fm. disticha. Hort. Sci. 30, 353–354

Grimshaw, R.G., L. Helfer (ed), 1995: Vetiver grass for soil and water conservation, land rehabiltation and embankment stabilization. World Bank technic. Paper No. 273. World Bank, Washington, D.C. (USA), 281 pp

Grisebach, H., W. Billuber, 1967: Zur Biosynthese des Apigenins and Chrysoerioles in der Petersilie. Z. Naturforsch. 22 B, 746–751

Grob, K., P. Matile, 1980: Capilarry GC of glucosinolate-derived horseradish constituents. Phytochemistry 19, 1789–1793

Groom, N., 1997: The new perfume handbook. Blackie Acad. & Professional, London, 2. edn., 435 pp

Grow, S., E. Schwartzman, 2001: The status of Guaiacum species in trade. Med. Plant Conservations (Bonn) 7, No. 1, 19-21

Gruczczyk, M., 2001: Effect of rows distance and soil conditions on the growth and yield of St. John's wort, Hypericum perforatum L. herba polon. 47, 125–129

Grümmer, 1955: Der Mohn. A. Ziemsen-Verlag Lutherstadt Wittenberg, 40 pp

Guadagni, D.G., R.G. Buttery, J. Harris, 1966: Odour intensities of hope oil components. J. Sci. Food Agric. 17, 142–144

Guazzi, E., S. Maccioni, G. Monti, G. Flamini, P.L. Cioni, I. Morelli, 2001: Rosmarinus officinalis L. in the gravine of Palagianello (Taranto, South Italy). J. Ess. Oil Res. 13, 231–248

Guedon, D.J, B.P. Pasquier, 1994: Analysis and distribution of flavonoid glycosides and rosmarinic acid in 40 Mentha x piperita clones. J. Agric. Food Chem. 42, 679–684

Guenther, E., 1954: The French lavander and lavandin industry. Econ. Bot. 8, 166–173

Guillen, M.D., M.J. Manzanos, 1994: A contribution to study Spanish wild growing fennel (Foeniculum vulgare Mill.) as a source of flavour compounds. Chem., Mikrobiol., Technol. Lebensmittel 16, No. 5/6, 141–145

Gulati, B.C., S.P.S. Duhan, S.N. Garg, S.K. Roy, 1975: Performance of peppermint (Mentha piperita Linn.) varieties in Tarai of Uttar Pradesh. Indian Perfumer 19, No. 24–27

Günther, F., O. Burkhart, I. Oostinga, 1966: Ein Beitrag zur Untersuchung und Zusammensetzung von Pfeffersorten. Dtsch. Lebensmittel-Rdsch. 62, 212–215

Güntzel-Lingner, H., 1941: Der Knoblauch. Eigenverlag Zinser & Co. Leipzig, 76 pp

Gundidza, M., F. Chinyanganya, L. Chagonda, H.L. de Pooter, St. Mavi, 1994: Phytoconstituents and antimicrobial activity of the leaf essential oil of Clausenia anisata (Willd.) J.D. Hook ex Benth. Flavour & Fragrance J. 9, 299–303

Gupta, K., K.K. Thakral, S.K. Arora, D.S. Wagle, 1991: Studies on growth, structural carbohydrate and phytatein coriander. J. Sci. Food Agric. 54, 43–46

Gupta, K., K.K. Thakral, V.K. Gupta, S.K. Arora, 1995: Metabolic changes of biochemical constituents in developing fennel seeds (Foeniculum vulgare). J. Sci. Food Agric. 68, 73–76

Gupta, P.N., A. Magni, L.N. Misra, M.C. Nigam, 1984: Gas chromatographic evaluation of the essential oil of different strains of Amomum subulatum growing wild in Sikkim. Parfümerie u. Kosmetik 65, 528–529

Gupta, R.K., D.C. Jain, R.S. Thakur, 1986: Two furastanol saponins from Trigonella foenum-grecum. Phytochemistry 25, 2205–2207

Gupta, S.C., 1994: Ocimum canum Sim. A potential ource for linalool. Eurocosmetics 5, 36–38

Gurgel do Vale, T., E. Couto Furtado, J.G. Santos jr., G.S.B. Viana, 2002: Central effects of citral, myrcene and limonene, constituents of essential oil chemotypes from Lippea alba (Mill.) N.E. Brown. Phytomedicine 9, 709–714

Gurib-Fakim, A., N. Maudarbaccus, D. Leach, L. Doimo, H. Wohlgemuth, 2002: Essential oil composition of Zingiberaceae species from Mauritius. J. Ess. Oil Res. 14, 271–273

Gurudutt, K.N., J.P. Naik, P. Srinivas, B. Ravindranath, 1996: Volatile constituents of large cardamom (Amomum subulatum Roxb.). Flavour & Fragrance J. 11, 7–9

Guzman, C.C. de, J.S. Siemonsa (ed) 1999: Plant Resources of South-East Asia. No. 13: Spices. Backhuys Publishers Leiden (The Netherlands), 400 pp

Gwosdz, G.G., 1987: Currys und was dazu gehört. BLV Verlagsges. Munich, 117 pp

GyungHdon, H., J. HyunSae, K. YoungBae, 1999a: Composition and change of flavour compounds in garlic cloves (Allium sativum) by processing treatment. J. Korean Soc. Horticult. Sci. 40, 19–22

GyungHdon, H., J. HyunSae, K. YoungBae, 1999b: Effect of processing treatment on change in quantity of the functional components in garlic (Allium sativum L.). J. Korean Soc. Horticult. Sci. 40, 23–25

H

Habegger, R., W.H. Schnitzler, 2000: Aroma von Knollensellerie – ein Sortenvergleich. Obst-, Gemüse u. Kartoffelverarbeit. 85, 162–164

Haensch, G., G. Haberkamp de Antón, 1987: Wörterbuch der Landwirtschaft: Deutsch-Englisch-Französisch-Spanisch-Italienisch-Russisch. BLV-Verlagsges. Munich Vienna Zürich, 5. edn., 1264 pp

Hager's Handbuch der pharmazeutischen Praxis (Herausg.: Hänsel, R., K. Keller, H. Rimpler, G. Schneider): Vol. 4: Drogen A–D 1992, 1000 pp, Vol. 5: Drogen E–0 1993, 970 pp, Vol. 6: P–Z 1994; Folgeband 2 Drogen: A–K, 1998, 900 pp, Folgeband 3 Drogen L–Z, 1998; Springer-Verlag Heidelberg Berlin New York, 5. edn.

Haggag, M.Y., A.S. Shalaby, G. Verzahr-Petzri, 1975: Thinlayer and gaschromatographic studies on the essential from Achillea millefolium. Planta med. 27, 361–366

Hagmann, K., R. Knauss, 1998: Blutwurz- und Estragon-Ansatzschnäpse. Kleinbrennerei 50, No. 5, 6–8

Hagos, T.H., 1962: A revision of the genus Parkia R.Br. (Mim.) in Africa. Acta bot. Neerl. 11, 174–188

Hahn, G. und A. Mayer, 1985: Arnika. Palmengarten 49, No. 2, 102–104

Halász-Zelnik, K., L. Hornok, J. Domokos, 1988: Data on the cultivation of Dracocephalum moldavica L. in Hungary. Herba Hung. 27, 49–58

Halim, A.F., R.P. Collins, 1975: Essential oil of Salvia dorisiana Standley. J. Agric. Food Chem. 23, 506–510

Hall, R.L., 1981: The history, use and pharmacology of spices. Perfumer & Flavorist 6, No. 4, 1–11

Hälvä, S., L.E. Craker, J.E. Simon, D.J. Charles, 1992: Light levels, growth and essential oil in dill (Anethum graveolens L.). J. Herbs, Spices & Med. Plants 1, 47–58

Hälvä, S., R. Huopalahti, C. Franz, S. Mäkinen, 1988: Herb yield and essential oil of dill at different locations. J. Agric. Sci. Finland 60, 93–100

Hälvä, S., L.E. Craker, J.E. Simon, D.J. Charles, 1993: Growth and essential oil in dill. Anethum graveolens L., in response to temperature and photoperiod. J. Herbs, Spices & Med. Plant 1, 47–56

Halbeisen, Th., 1954: Untersuchungen über die antibiotischen Eigenschaften von Tropaeolum majus (Kapuzinerkresse). Naturwissenschaften 41, 378–379

Hamid, S., A. Saliv, S. Khan, P. Aziz, 1983: Experimental cultivation of Silybum marianum and chemical composition of its oil. Pakistan J. Sci. & Ind. Res. 26, 244–246

Hammer, K., 1981: Problems of Papaver somniferum-classification and some remarks on recently collected European poppy land-races. Kulturpflanze. 29, 287–296

Hammer, K., R. Fritsch, 1977: Zur Frage nach der Ursprungsart des Kulturmohns (Papaver somniferum L.). Kulturpflanze 25, 113–124

Hammer, K., H. Knüpffer, H.-D. Hoang, 1997: Koreanische Heilpflanzen – eine Liste der kultivierten Arten. Drogenreport 10, No. 16, 30 pp

Hammer, K., H. Krüger, 1995: Evaluierung der Dill (Anethum graveolens) – Kollektion der Genbank Gatersleben. Drogenreport 8, No. 13, 12–23

Hammer, K., G. Laghetti, S. Cifarelli, M. Spahillari, P. Perrino, 2000: Pimpinella anisoides Briganti. Genet. Resources Crop. Evolut. 47, 223–225

Hamon, N.W., 1987: Garlic and the genus Allium. Canad. Pharm. J. 120, 493–498

Han, H.R., J.I. Chang, Y.B. Park, 1983: Studies on ecology and cultur of Zingiber mioga. J. Korean Soc. Horticult. Sci 24,, No. 3, 200–206 (in Korean)

Hanafy, M.S.M., M.E. Hatem, 1991: Studies on the antimicrobial activity of Nigella sativa L. (black cumin). J. Ethnopharmacol. 34, 275–278

Hancı, S., S. Sahin, L. Yılmaz, 2003: Isolation of volatile oil from thyme (Thymbra spicata) by steam distillation. Nahrung/Food 47, 252–255

Hanelt, P., 1961: Zur Kenntnis von Carthamnus tinctorius L. Kulturpflanze 9, 114–145

Hanelt, P., 1963: Monographische Übersicht der Gattung Carthamnus. Feddes Rep. 67, 41–180

Hanelt, P., 1985: Zur Kenntnis kultivierter Arten der Moringa Adans (Moringaceae). Gleditschia. 13, 101–106

Hanelt, P., 1994: Die taxonomische Gliederung der Gattung Allium und ihre Kultur- und Nutzpflanzen. Drogenreport 7, No. 11, 17–25

Hanelt, P., 2001: Mansfeld's encyclopedia of agricultural and horticultural crops (excepted ornamentals). Springer Heidelberg Berlin New York, 6 parts, 3645 pp

Hanelt, P., H. Ohle 1978: Die Perlzwiebeln des Gaterslebener Sortiments und Bemerkungen zur Systematik und Karyologie dieser Sippe. Kulturpflanze 26, 339–348

Hanlidou, E., S. Kokkini, A.M. Bosabalidis, J. Bessière, 1991: Glandular trichomes and essential oil constituents of Calamintha menthifolia (Lamiaceae). Plant Systematics & Evolution 177, 17–26

Hänsel, R., J. Schultz, 1982: Valerensäuren und Valerenal als Leitstoffe des offiziellen Baldrians. Bestimmung mittels HPCL-Technik. 2. Mitt. Zur Qualitätssicherung von Baldrianpräparaten. Dtsch. Apotheker-Ztg. 122, 215–219

Hänsel, R., J. Schulz, 1986: Hopfenzapfen (Lupulus strobulus). DC-Prüfung auf Identität. Dtsch. Apotheker-Ztg. 126, 2347–2348

Hänsel, R., M. Kartarahardja, J.T. Huang, F. Bohlmann, 1980: Sesquiterpenlacton-β-D-Glucopyranoside sowie ein Eudesmanolid aus Taraxacum officinale. Phytochemistry 19, 857–861

Hansen, H., 1974: Content of glucosinolates in horseradish (Armoracia rusticana). Tidskr. Planteavl 73, 408–410

Handa, K.L., M.M. Chopra, M.C. Nigam, 1963: The essential oil of Tagetes minuta L. Perfum. Ess. Oil Res. 54, 372–374

Hanum, T., 1997: Changes in vanillin and activity of β-glucosidase and oxidases during post harvest processing of vanilla bean (Vanilla planifolia). Bull. Teknol. Ind. Pangan 8, 46–52

Hari, L., J. Bukuru, H.L. de Pooter, 1994: The volatile fraction of Aframomum sanguinerum. J. Ess. Oil Res. 6, 395–398

Harisetijono, Suriamihardja, 1993: Sandalwood in Nusa Tenggara Timur. Austral. Centre internat. Agric. Res., ACIAR Proc. 49, 39–43

Harrison, S.G., 1952: Edible pine kernels. Kew Bull. 371–375

Harten, A.M.van, 1970: Melegueta pepper. Econ. Bot. 24, 208–216

Hartmann, G., S. Köstner, 1994: Qualität von Ingwer und Ingwerextrakten. Einfluß verschiedener Verarbeitungsverfahren auf die Qualität unter besonderer Berücksichtigung des Gingerolgehaltes. Fleischwirtschaft 74, 313–315

Harvala, C., P. Menounos, N. Arggeiadou, 1987a: Essential oil from Origanum dictamnus. Planta med. 53, 107–110

Harvale, C., P. Menounos, N. Argyriadou, 1987b: Essential oil from Salvia triloba. Fitoterapia 58, 353–356

Hasan, C.M., M. Ahsan, N. Islam, 1989: In vitro antibacterial screening of the oils of Nigella sativa seeds. Bangladesh J. Bot. 18, 171–174

Hasegawa, Y., K. Tajma, N. Toi, Y. Sugimura, 1992: An additional constituent occurring in the oil from a patchouli cultivar. Flavour & Fragrance J. 7, 333–335

Hashimoto, S., M. Miyazawa, H. Kameoka, 1983: Volatile flavour components of chive (Allium schoenoprasum L.). J. Food Sci. 48, 1858–1859, 1897

Hassan, I., 1967: Some folk use of Peganum harmala in India and Pakistan. Econ. Bot. 21, 284

Haumann, L., 1925: Notes sur les genre Boussingaultia H.B.K. Anal. Musée nac. Hist. Nat. Buenos Aires 33, 347–359

Hausen, B.M., 1982: Taraxinsäure-1'-0-β-D-Glucopyranosid, das Kontaktallergen des Löwenzahns (Taraxacumofficinale). Dermatol., Beruf, Umwelt 30, 51–53

Hay, R.K.M., P.G. Waterman (ed), 1993: Volatile oil crops: their biology, biochemistry and production. Longman Sci. Technic., Avon, UK, 185 pp

Hayashi, T., R.H. Thompson, 1974: Isoflavones from Dipteryx odorota. Phytochemistry 13, 1943–1946

Hayes, A.J., B. Markovic, 2002: Toxicity of Australian essential oil of Backhousia citriodora (lemon myrtle). I. Antimicrobial activity and in vitro cytotoxicity. Food & Chem. Toxicol. 40, 535–543

He, X.-G., L.-Z. Lin, L.-Z. Lian, 1998: Liquid chromatography-electrospray mass spectrometric analysis of curcuminoids an sesquiterpenoids in turmeric (Curcuma longa). J. Chromatogr. A 818, 127–132

Hedayetullah, S., 1960: The need and possibility of cultivating rare medical plants. Pakistan J. Sci. & Ind. Res. 3, 171–178

Hedhili, L., M. Romdhane, A. Abderrabba, H. Planche, I. Cherif, 2002: Variability in essential oil composition of Tunisian Thymus capitatus (L.) Hoffmanns. et Link. Flavour & Fragrance J. 17, 26–28

Hee-Jun Park, Sang-Hyuk Kwon, Myung-Sun Lee, Gap-Tae Kim, Moo-Young Choi, Won-Tae Jung, 2000: Antimicrobial activity of the essential oil of the herbs of Agastache rugosa and ist composition. J. Korean Soc. Food Sci. Nutrit. 29, 1123–1126 (in Korean)

Heeger, E.F., 1956; Handbuch des Arznei- und Gewürzpflanzenanbaues. Drogengewinnung. Deutscher Bauernverlag 775 pp

Heeger, E.F., W.Poethke, 1947: Papaver somniferum L. Der Mohn. Anbau, Chemie, Verwendung. Pharmazie, Beiheft 4., pp 233–240

Heeger, E.F., W. Poethke, 1954: Gemeine Quecke, Ruchgras und Heublumen – Gewinnung, Inhaltsstoffe, Verwendung. Pharmazie 9, 131–138

Hefendehl, F.W., 1962: Zusammensetzung des ätherischen Öls von Mentha piperita im Verlauf der Ontogenese und Versuche zur Beeinflussung der Ölkomposition. Planta med. 10, 241–266

Hefendehl, F.W., 1970: Zusammensetzung der ätherischen Öle von Melissa officinalis L. und sekundäre Veränderungen der Ölkomposition. Arch. Pharmaz. 303, 345–349

Heikes, D.L., B. Scott, N.A. Gorzovalitis, 2001: Quantitation of volatile oils in ground cumin by supercritical fluid extraction and gas chromatography with flame ionization detection. J. Assoc. Offic. Analyt. Chem. 84, 1130–1134

Heine, H., 1993: Ergebnisse von Sortenprüfungen mit Arznei- und Gewürzpflanzen – Majoran. Drogenreport 6, No. 10, 33–34

Heinrich, M., 1996: Arzneipflanzen Mexikos. Ethnobotank, Phytochemie, Pharmakologie. Dtsch. Apotheker-Ztg. 136, 1739–1752

Heinrich, S., 1973: Entwicklung, Feinbau und Ölgehalt der Drüsenschuppen von Monarda fistulosa. Planta med. 23, 154–166

Heinrich, G., 1977: Die Feinstruktur und das ätherische Öl eines Drüsenhaares von Monarda fistulosa. Biochem. u. Physiol d. Pflanzen 171, 17–24

Heiser, C.B., B. Pickersgill, 1969: Names for the cultivated Capsicum species (Solanaceae). Taxon 18, 277 -283

Heiser, C.B., P.G. Smith, 1953: The cultivated Capsicum. Econ. Bot. 7, 214–227

Heisig, W., M. Wichtl, 1990: Bestimmung pflanzlicher Glykoside. Methodenkombination von zweidimensionaler DC und HPLC. 1. Mitt. Inhaltsstoffe aus Calendula officinalis flos. Dtsch. Apotheker-Ztg. 130, 2058–2062

Helm, J., 1956: Die zu Würz- und Speisezwecken kultivierten Arten der Gattung Allium. Kulturpflanze 4, 130–180

Helm, J., 1972: Apium graveolens L. Geschichte der Kultur und Taxonomie. Kulturpflanze 19, 73–100

Hendriks, H., R. Bos, H.J. Woerdenbag, 1993: Der Rainfarn – eine potentielle Arzneipflanze? Z. Phytother. 14, 333–336

Hendry, B.S., 1982: Composition and characteristics of dill. Perfumer & Flavorist 7, 39–44

Hennig, F., R. Kadner, W. Junghanns, B. Weinreich, 2002: Produktion von Rosmarinjungpflanzen (Ros marinus officinalis L.). Erste Untersuchungen zur In-vitro-Verklonung. Z. Arznei- u. Gewürzpflanzen 7, 377–381

Hepper, F.N., 1992: Pflanzenwelt der Bibel. Eine illustrierte Enzyklopädie. Deutsche Bibelgesellschaft Stuttgart, 192 pp

Herisset, A., J. Jolivet, A. Zoll, J.P. Chaumont, 1974: Nouvelles observations concernant les falsifications de lasarriette des jardins (Satureja hortensis L.). Plantes méd. et Phytotherp. 8, 287–294

Herklots, G.A.C., 1972: Vegetables in South-East Asia. G. Allen & Unwin Ltd. London, 525 pp

Hermey, B., R. Ludi, 1992: Senf – Senfsaat – Senfmehl in der Lebensmittelherstellung. Ernährungsindustrie, No. 6, 10–12

Hernándenz, F., P. Melgarejo, F.A. Thomás-Barberán, F. Artés, 1990: Evolution of juice on anthocyanins during ripening of new selected pomegranate (Punica granata) cloves. Eur. Food Res. & Technol. 210, 39–42

Heron, S., E. Lesellier, A. Tchapla, 1995: Analysis of triglycerols of borage oil by RPLC identification by coinjection. J. Liquid-Chomatogr. 18, 599–611

Hermann, J., 2001a: Cinchona-Arten. Z. Phytother. 22, 205–210

Hermann, J., 2001b: Chinarinde. Eine historische Reise um die Erde. Pharmaz. Ztg. 146, 1486 -1491

Herrmann, K., 1995: Inhaltsstoffe von Zwiebeln und Porree. 1. Inhaltsstoffe mit Ausnahme der schwefelhaltigen Aromastoffe. Ind. Obst- u. Gemüseverwert. 80, 342–348

Herrmann, K., 1995: Inhaltsstoffe von Zwiebeln und Porree. 2. Schwefelhaltige Aromastoffe. Ind. Obst- u. Gemüseverwert. 80, 348–353

Herrmann, K., 1997a: Die Inhaltsstoffe der Garten- und Brunnenkresse. Ind. Obst- u. Gemüseverwert. 82, 38–41

Herrmann, K., 1997b: Inhaltsstoffe des Meerrettichs. Ind. Obst- u. Gemüseverwert. 82, 274–275

Herrmann, K., 1997c: Inhaltsstoffe von Gemüsefenchel, Petersilie und Feldsalat. Ind. Obst- u. Gemüseverwert. 82, 278–279

Herrmann, K., 1999: Inhaltsstoffe von Ingwer und Kurkuma. Ind. Obst. u. Gemüseverwert. 84, 111–114

Herron, S., 2003: Catnip, Nepeta cataria, a morphological comparison of mutant and wild type specimens to gain an ethnobotanical perspective. Econ. Bot. 57, 135–142

Hese, S., 2002: Der Wermut – Artemisia absinthum L. Phytother. 23, 187–194

Hetz, E., R. Liersch, H. Jung-Heiliger, O. Schieder, 1995: Untersuchung von di- und polyploiden Linien der Arzneipflanze Silybium marianum (L.) Gaertn. unter dem Aspekt der Leistungs- und Ertragssteigerung. Drogenreport 8, No. 12, 17–23

Heyne, K., 1926: De nuttige planten van Nederlandsch Indië. Uitgave het Departm. Van Landbouw, Nijverheid & Handel in Nederlandsch-Indi, Buitenzorg, 1660 pp

Hidalgo, P.J., J.L. Ubera, J.A. Santos, F. LaFont, C. Castellanos, A. Palimino, M. Roman, 2002: Essential oil in Calamintha sylvatica Bromf. ssp. ascendens (Jordan) P.W. Ball: wild and cultivated productions and antifungal activity. J. Ess. Oil Res. 14, 68–71

Hilal, S.H., T.S. El-Alfy, M.M. El-Sherei, 1978: Investigations of the volatile oil of Hyssopus officinalis L. Egypt. J. Pharmac. Sci. 19, 177–184

Hiller, K., F. Melzig, 1999/2000: Lexikon der Arzneipflanzen und Drogen. Spektrum Akademischer Verlag Heidelberg/Berlin; Bd. 1 (1999): A–K 455 pp; Bd. 2 (2000): L–Z. 450 pp

Hilliard, O.M., 1994: The Manuleae, a tribe of Scrophulariaceae. Univ. Press. Edinburgh UK, 579 pp

Hills, L.D., 1980: The cultivation of the carob tree (Ceratonia siliqua). Internat. Tree Corp. J. 1, 15–36

Hiltunen, R., Y. Holm, 1999: Basil. The genus Ocimum. Harward Academic Publishing Amsterdam, 167 pp

Himeno, H., K. Sano, 1987: Synthesis of crocin, picrocin and safranal by saffron stigma-like structure proliferated in vitro. Agric. Biol. Chem. (Tokyo) 51, 2395–2400

Hirvi, T., E. Honkanan, 1985: Analysis of the volatile constituents of black chokeberry (Aronia melanocarpa Ell.). J. Sci. Food Agric. 36, 808–810

Ho Jung Youl, Kang Sang Ja, K. Changkil, K. Han Deuk, Hahn Sang Jung, 2001: Growth characteristics of Allium gray in Korea. J. Korean Soc. Horticult. Sci. 42, 177–183 (in Korean)

Hobbs, C., 1988: Sarsaparille – a literature review. Herbalgram 17, 10–15

Hobhouse, H., 2000: Fünf Pflanzen verändern die Welt. Chinarinde, Zucker, Tee, Baumwolle, Kartoffel. dtv/-Klett-Verlag Munich, 6. edn., 349 pp

Hodge, W.H., 1974: Wasabi – native condiment plant of Japan. Econ. Bot. 28, 118–129

Hodgson, R.W., 1917: The pomegranate. Calif. Agric. Experim. Stat. Bull. 276, 163–192

Hof, S., H.P.T. Ammon, 1989: Negative inotropic action of rosemary oil, 1,8-cineole, and bornyl acetate. Planta med. 55, 106–107

Hoffmann, A., C. Farga, J. Lastra, E. Veghazi, 1992: Plantas medicinales de uso comun en Chile. Edit. Fundacion Cl. Gay, Santagio, Chile, 273 pp

Hogg, J.W., S.J. Terhune, B.M. Lawrence,1974: Dehydro-1,8-cineole: a new monoterpene oxide in Laurus nobilis oil. Phytochem. Rep. 13, 868–869

Hohmann, B., 1968: Zwei weniger bekannte Gewürzkräuter: Origanum virens und Coridothymus capitatus. Z. Lebensmittel-Unters. u. -Forsch. 138, 212–216

Hohmann, B., 1970: Salbeiblätter – eine häufig verfälschte Droge. Dtsch. Apotheker-Ztg. 110, 1095–1097

Hohmann, B., 1971: Zur Mikroskopie von "Curry leaves", den Blättern von Murraya koenigii. Z. Lebensmittel. Unters. u. -Forsch. 145, 100–104

Hohmann, B., G. Reher, E. Stahl-Biskop, 2001: Mikroskopische Drogenmonographien der deutschsprachigen Arzneibücher. Wissenschaftl. Verlagsges. mbH Stuttgart, 636 pp

Hokwerda, H., R. Bos, D.H.E. Tattje, T.M. Malingre, 1982: Composition of essential oil of Laurus nobilis L. nobilis var. angustifolia and Laurus azorica. Planta med. 44, 116–119

Holm, Y., R. Hiltunen, I. Nykänen, 1988a: Capilary gas chromatographic mass spectrometric determination of the flavour composition of dragonhead (Dracocephalum moldavica L.). Flavour & Fragrance J. 3, 109–112

Holm, Y., B. Galambosi, R. Hiltunen, 1988b: Variation of the main terpenes in dragonhead (Dracocephalum moldavica L.). Flavour & Fragrance J. 3, 113–115

Holm, Y., P. Vuorela, R. Hiltunen, 1997: Enantiomeric composition of monoterpene hydrocarbons in n-hexane extracts of Angelica archangelica L. roots and seeds. Flavour & Fragrance J. 12, 397–400

Holttum, R.E., 1951: The Zingiberaceae of the Malay Peninsula. Garden's Bull. Singapore 13, 1–248

Hölzl, J., E. Ostrowski, 1987: Johanniskraut (Hypericum perforatum L.). HPCL-Analyse der wichtigsten Inhaltsstoffe und deren Variabilität in einer Population. Dtsch. Apotheker-Ztg. 127, 1227–1230

Hölzl, J., D. Fritz, C. Franz, L. Garte, 1974: Vergleichende Untersuchungen verschiedener Herkünfte und Sorten von Mentha piperita. 1. Teil. Dtsch. Apotheker-Ztg. 114, 513–514

Honda, G., M. Ito, M. Tabata, 1996: A new species of Perilla (Labiatae) from Japan. Japan. J. Bot. 71, 39–43

Honda, G., Y. Koezuka, M. Tabata, 1990: Genetic studies of color and hardness in Perilla frutescens. Japan. J. Breeding 40, 469–474

Honda, G., A. Yuba, T. Kojima, M. Tabata, 1994: Chemotaxonomic and cytogenetic studies on Perilla frutescens var. citriodora ("Lemon Egoma"). Natural Med. 48, 185–190

Hondelmann, W., 2002: Die Kulturpflanzen der griechisch-römischen Welt. Pflanzliche Ressourcen der Antike. Verlag Gebr. Bornträger Berlin Stuttgart, 134 pp

Honerlagen, L.H., 1988: Radix Valeriana officinale. Schweiz. Apotheker-Ztg. 126, 502–503

Hoogland, R.D., 1952: A revision of the genus Dillenia. Blumea 7, 1–145

Hoppe, H.A., 1949: Drogenkunde. Cram, de Gruyter & Co., Hamburg, 6. edn., 335 pp

Hoppe H.A., 1975: Drogenkunde. Verlag W. De Gruyter Berlin. 8. Ed., Vol. 1 Angiospermen, 1311 pp; Vol. 2: Gymnospermen, Kryptogamen, Tierische Drogen 1977, 366 pp; Vol. 3: Supplement 1987: 727 pp

Hornok, L., 1980: Effect of nutrition supply on yield of dill (Anethum graveolens L.) and the essential oil content. Acta Horticult. 96, 337–343

Hooper, D., 1937: Useful plants and drugs of Iran and Iraq. Field Museum Nat. Histor., Bot, Ser. 9, No. 3, 75–241

Hose, S., 2002: Der Wermut, Artemisia absinthium L. Z. Phytother. 23, 187–194

Houlihan, C.M., C.T. Ho, S.S. Chang, 1985: The structure of rosmarinquinone – a new antioxidant isolated from Rosmarinus officinalis L. J. Amer. Oil chem. Soc. 62, 96–99

Hu, M.F., 1991: The cultivation of wasabi. Taiwan agric. Experiment. Stat., Techn. Services 6, 21–26

Huang, Y.Z., S.Q. Chen, C.Y. He, Q.Y. Chen, Y.L. Wu, 1990: A study of chemical components of essential oil from Citrus bergamia and its close relatives and its taxonomy. Acta Phytotaxonomica Sinica 30, 239–244

Huet, R., 1972: L'huile essentielle de menthe au Bresil. Fruits 27, 469–472

Huit, R., C. Dupuis, 1968: L' huile essentielle de bergamote en Afrique et en Corse. Fruits 23, 301–311

Hume, H.H., 1957: Citrus fruits. Edn. MacMillan New York, NY, 444 pp

Hunziker, A.T., 2001: Genera Solanacearum. The genera of Solanaceae illustrated arranged according to a new system. A.R.G. Gantner Verlag KG Ruggell, 500 pp

Huopalahti, R., 1985: The content and composition of aroma compounds in three different cultivars of dill (Anethum graveolens L.). Z. Lebensmittel-Unters. u. -Forsch. 181, 92–96

Huopalathti, R., R. Lathinen, 19988: Studies on the essential oils of dills herbs. Flavour & Fragrance J. 3,121–125

Husain, A., O.P. Virmani, S.P. Popli, L.N. Misra, M.M. Gupta, G.N. Srivastava, Z. Absaham, A.K. Singh, 1992: Dictionary of Indian medical plants. Central Inst. Med. & Aromatic Plants, Lucknow, India, 2. edn., 546 pp

Hutton, I., 2002: Myth, reality and absinthe. Current Drug Discov. 9, 62–64

Hutton, W., 1998: Tropical Herbs & Spices of Thailand. Asia Books Co., Ltd. Bangkok (Thailand), 63 pp

Howard, L.R., P. Burma, A.B. Wagner, 1994: Firness and cell wall characteristics of pasteurized jalapeño pepper rings affected calcium chloride and acetic acid. J. Food Sci. 59, 1184–1186

Hudaib, M., E. Speroni, A.M. di Pietra, V. Cavrini, 2002: GC/MS evaluation of thyme (Thymus vulgaris L.) oilcomposition and variations during the vegetative cycle. J. Pharm. Biomed. Anal. 29, 691–700

Huxley, A., M. Griffiths, M. Levy, 1992: The New Royal Hortical Society. Dictionary of Gardening. The MacMillan Press Ltd., London & Stockton Press New York; Vol. A–C, 815 pp; Vol. 2 D–K, 747 pp; Vol. 3 L–Q, 790 pp; Vol. 4 R–Z, 888 pp

Hyang-Sook Choi, Mie-Soon Lee Kim, M. Sawamura, 2001: Constituent of the essential oil of Angelica tenuissima, an aromatic medicinal plant. Food Sci. Biotechnol. 10, 557–561

Hyndman, D.C., 1984: Ethnobotany of wopkaimin Pandanus: Significant Papua New Guinea plant resources. Econ. Bot. 38, 287–303

Hyun-Jung Kim, Bang-Sil Jun, Sung-Kyu Kim, Jae-Yonung Cha, Young-Su Cho, 2000: Polyphenolic compound content and antioxidative activities by extracts from seed, sprout und flower of safflower (Carthamus tinctorius L.). J. Korean Soc. Food Sci. Nutrit. 29, 1127–1132 (in Korean)

I

Ibanez, E., A. Cifuentes, A.L. Crego, F.J. Senorans, S. Cavero, G. Reglero, 2000: Combined use of supercritical fluid extraction, micellar electrokinetic chromatography, and reverse phase high performance liquid chromatography for the analysis of antioxidants from rosemary (Rosmarinus officinalis L.). J. Agric. Food Chem. 48, 4060–4065

Ibanez, E., A. Kubatova, F.J. Senorans, S. Cavero, G. Reglero, S.B. Hawthorne, 2003: Subcritical water extraction of antioxidant compounds from rosemary plants. J. Agric. Food Chem. 51, 375–382

Ibrahim bin Jantan, Abu Said Ahmad, Nor Azah Mohd, Abdul Rashih Ahmad, Halijah Ibrahim,

1999: Chemical composition of the rhizome oils of four Curcuma species from Malaysia. J. Ess. Oil Res. 11, 719–723

Ietswaart, J.H., 1980: A taxonomic revision of the genus Origanum (Labiatae). Leiden Bot., Ser. 4, Leiden Univ. Press, 156 pp

Ihrig, M., 1993: Identitätsprüfung von Petersilienkraut. Pharmaz. Ztg. 138, 3226–3227

Ihrig, M., 1997: Prüfung von Schwarzkümmelöl. Pharmaz. Ztg. 142, 1822–1824

Ijomah, J.U., E.C. Igwe, A. Audu, 2000: Nutrient composition of five "draw" leafy vegetables of Adamawa state, Nigeria. Global J. Pure & Appl. Sci. 6, 547–551

Ikeda, N., Shimizu, S. Udo, 1963: Studien on Mentha gentilis L. Japan. J. Breed 13, 31–41

Ikeda, R.M., L.A. Rolle, S.H. Vannin, W.L. Stanley, 1996: Isolation and identification of aldehyde in cold-pressed lemon oil. J. Agric. Food Chem. 10, 98–102

Illes, V., H.G. Daood, S. Perneczki, L. Szokonya, M. Then, 2000: Extraction of coriander seed oil by CO_2 and propane at super- and subcritical conditions. J. Supercrit. Fluids 17, 177–186

Ilieva, S., 1979: New Salvia sclarea L. cultivars developed by hybridization. Herba Hungar. 18, 197–203

Ilyas, M., 1976: Spices of India. I. Econ. Bot. 30, 273–280; 1978: II. 32, 238–263; III. 34, 236–259

Ina, K., H. Ino, M. Ueda, A. Yagi, I. Kishima, 1989: Omega-Methylthioalkyl-isothiocyanates in wasabi. Agric. Biol. Chem. 53, 537–538

Ingles, A., 1987: Gruit en Gruitbier. Voedingmiddelen-Technol. 20, No. 3, 12–16

Ingram, J.S., 1969: Saffron (Crocus sativus L.). Trop. Sci. 11, 177–184

Ingram, J.S., B.J. Francis, 1969: The anatto tree (Bixa orellana L.) – A guide to its occurrence, cultivation, preparation and uses. Trop. Sci. 9, 97–102

Inocencio, C., D. Rivera, F. Alcaraz, F.A. Tomás-Barberán, 2000: Flavonoid content of commercial capers (Capparis spinosa, C. sicula and C. orientalis) produced in mediterranean countries. Eur. Food Res. Technol. 212 A, 70–74

Inocencio, C., F. Alcaraz, F. Calderón, C. Obón, D. Rivera, 2002: The use of floral characters in Cap paris sect. Capparis to determine the botanical and geographical origin of capers. Eur. Food Res. Technol. 214 A, 335–339

Inoue, T., K. Iwagoe, T. Konishi, S. Kiyosawa, Y. Fujiwara, 1990: Novel 2,5-dihydrofuryl-γ-lactam derivatives from Hemerocallis fulva L. var. kwanzo Regel. I. Chem. Pharm. Bull. 38, 3187–3189

Inoue, T., T. Konishi, S. Kiyosawa, Y. Fujiwara, 1994: 2,5-dihydrofuryl-γ-lactam derivatives from Hemerocallis fulva L. var. kwanzo Regel. II. Chem. Pharm Bull. 42, 154–155

Irvine, A., 1961: Woody plants of Ghana with special reference to their uses. Univ. Press. Oxford, London, (UK), 868 pp

Irvine, F.R., 1948: The indigenous food-plants of West African peoples. New York Bot. Garden 49, 225–256

Isaac, O., 1992: Die Ringelblume. Botanik, Chemie, Pharmakologie, Toxikologie, Pharmazie und therapeutische Verwendung. Wiss.-schaftl. Verlagsges. mbH Stuttgart, 121 pp

Isaac, O., 1993: Chamaemelum nobile (L.) Allioni – Römische Kamille. Z. Phytother. 14, 212–222

Isaac, O., Calendula officinalis L. – Die Ringelblume. Z. Phytother. 15, 357–370

ISO-Standard, 1980: Saffron-Specification, No. 3632; 20 pp

Ito, M., G. Honda, 1996: A taxonomic study of Japanese wild Perilla (Labiatae). J. Physiogeography & Taxonom. 44, 43–52

Itokawa, H., M. Morita, S. Mitashi, 1981: Phenolic compounds from the rhizomes of Alpinia speciosa. Phytochemistry 20, 2303–2306

Itokawa, H., H. Morita, K. Takeya, M. Motidome, 1988: Driterpenes from rhizomes of Hedychium coronarium. Chem. Pharm. Bull. (Japan) 36, 2682–2684

Itokawa, H., H. Morita, I. Katon, K. Takeya, A.J. Cavalheiro, R.C.B. de Oliveira, M. Ishige, M. Motidome, 1988: Cytotoxic diterpenes from the rhizomes of Hedychium coronarium. Planta med. 54, 311–315

Ivanic, R., K. Savin, 1976: A comparative analysis of essential oils from several wild species of Salvia. Acta Physiol. Pharmacol. Bulgaria 10, 13–20

Iwu, M.M., C.O. Ezeugwu, C.O. Okunji, D.R. Sanson, M.S. Tempesta, 1990: Antimicrobial activity and terpenoids of the essential oil of Hyptis suaveolens. Internat. J. Crude Drug Res. 28, 73–76

J

Jack, H.W., 1940: Notes on curing Tahiti vanilla beans. Fiji agric. J. 11, No. 1, 22–23

Jackson, S.A.L., R.K.M. Hay, 1994: Characteristics of varieties of thyme (Thymus vulgaris L.) for use in the UK: oil content, composition and related characters. J. Horticult. Sci. 69, 275–281

Jacquat, Chr., 1990: Plants from the markets of Thailand. Editores Duang Kamaol, Bangkok, Thailand, 117 pp

Järvenpää, E., Z. Zhang, R. Huopalahti, J.W.King,1998: Determination of fresh onion (Allium cepa L.) volatiles by solid phase microextraction combined with gaschromatography-mass spectrometry. Z. Lebensmittel-Unters. u. -Forsch 207 A, 39–43

Jagadishchandra, K.S., 1975a: Recent studies on Cymbopogon Spreng. (aromatic grasses) with special reference to Indian taxa; cultivation and ecology: a review. J. Plantation Crops 3, 1–5

Jagadishchandra, K.S., 1975b. Recent studies on Cymbopogon Spreng (aromatic grasses) with special reference to Indian taxa: taxanomy, cytogenetics, chemistry, and sope. J. Plantation Crops 3, 43–57

Jagella, T., W. Grosch, 1999: Flavour and off-flavour compounds of black and white pepper (Piper nigrum L.). Evaluation of potent odorants of black pepper by dilution and concentrating techniques. Z. Lebensmittel-Unters. u. -Forsch. 209, 16–21

Jaimand, K., M.B. Rezaee, 2002: Chemical constituents of essential oils from Mentha longifolia (L.) Hudson var. asiatica (Boriss.) Rech.f. from Iran. J. Ess. Oil Res. 14, 107–108

Jain, S.R., M.L. Jain, 1973: Investigations on the essential oil of Ocimum basilicum. Planta med. 24, 286–289

Jakupovic, J., R.N. Baruah, F. Bohlmann, W. Quack, 1985: New clerodane derivatives from Teucrium scordium. Planta med. 33, 341–342

Jalsenjak, V., S. Peljujak, D. Kustrak, 1987: Microcapsules of sage oil: essential oil content and antimicrobialactivity. Pharmazie 42, 419–420

Jansen, P.C.M, 1981: Spices, condiments and medical plants in Ethiopa, their taxonomy and agricultural significance. Belmontia (N.S.) 12, 1–322 pp

Janssens, R.J., W.P. Vernooij, 2001: Calendula officinalis: a natural source for pharmaceutical, oleochemical and functional compounds. INFORM 12, 468–477

Jantan, I. Bin, I. Basni, Abu-Said Ahmad, Nor Azah Mohd Ali, Abdul-Rashih Ahmad, H. Ibrahim, 2001: Constituents of the rhizome oils of Boesenbergia pandurata (Roxb.) Schlecht from Malaysia, Indonesia and Thailand. Flavour & Fragrance J. 16, 110–112

Jassbi, A.R., V.U. Ahmad, R.B. Tareen, 1999: Constituents of the essential oil of Perovskia atriplicifolia Benth. Flavour & Fragrance J. 14, 38–40

Javanmardi, J., A. Khalighi, A. Kashi, H.P. Bais, J.M. Vivanco, 2002: Chemical characterization of Basil (Ocimum basilicum L.) found in local accessions and uses in traditional medicines in Iran. J. Agric. Food Chem. 50, 5878–5883

Jayaprakasha, G.K., L.J. Mohan Rao, K.K. Sakariah, 2003: Volatile constituents from Cinnamomum zeylanicum fruit stalks and their antioxidant activities. J. Agric. Food Chem. 51, 4344–4348

Jayasinghe, C., N. Gotoh, T. Aoki, S. Wada, 2003: Phenolic composition and antioxidant activity of sweet basil (Ocimum basilicum L.). J. Agric. Food Chem. 51, 4442–4449

Jeffrey, C., 1965: Artemisia maritima (Compositae) in the Indian sub-continent. Kew Bull. 19, 399–400

Jeker, M., O. Sticher, I. Calis, P. Rüedi, 1989: Allobetonicoside and 6-0-acetylmioporoside: two new iridoid glycosides from Betonica officinalis L. (Stachys officinalis). Helv. Chim. Acta 72, 1787–1791

Jensen, P.N., G. Sørensen, S.B. Engelsen, G. Bertelsen, 2001: Evaluation of quality changes in walnut kernels (Juglans regia L.) by VIS/NIR spectroscopy. J. Agric. Food Chem. 49, 5790–5796

Jensen, S.R., B.J. Nielsen, 27: Cyanogenic glucosides in Sambucus nigra L. Acta Chem. Scand. 27, 2661–2685

Jentzsch, K., P. Spiegel, R. Kanitz, 1968: Qualitative und quantitative Untersuchungen über Curcumafarbstoffe in verschiedenen Zingiberaceendrogen. Sci. Pharm. 36, 251–258

Jerković, I., J. Mastelić, 2003: Volatile compounds from leaf-buds of Populus nigra L. Phytochemistry 63, 109–113

Jerković, I., A. Radonic, I. Borcic, 2002: Comparative study of leaf, fruit and flower essential oils of Croatian Myrtus communis (L.) during a one-year vegetative cycle. J. Ess. Oil Res. 14, 266–270

Jerković, I., J. Mastelic, M. Milos, F. Juteau, V. Masotti, J. Viano, 2003: Chemical variability of Artemisia vulgaris L. essential oils originated from the Mediterranean area of France and Croatia. Flavour & Fragrance J. 18, 436–440

Jezussek, M., 2002: Zur Aromabildung beim Kochen

von Naturreis (Oryza sativa L.) sowie Blättern von Pandanus amaryllifolius Roxb. Diss. Univ. Munich

Jia-Jiu, Wu., Yang Jui-Sen, 1994: Effects of γ -irradiation on the volatile compounds of ginger rhizome (Zingiberofficinale Roscoe). J. Agric. Food Chem. 42, 2574–2577

Jiang, L., H. Kojima, K. Yamada, A. Kobayashi, K. Kubota, 2001: Isolation of some glycosides as aroma precursors in young leaves of Japanese pepper (Xanthoxylum piperitum DC.). J. Agric. Food Chem. 49, 5888–5894

Jiroretz, L., G. Buchbauer, R. Eberhardt, 2000: Analyse und Qualitätskontrolle von ätherischen Zimtölen (Rinden- und Blattöle) verschiedenen Ursprungs mittels GC, GC-MS und Olfaktometrie. Bestimmung des Cumarin- und Safrolgehaltes. Ernährung/Nutrition 24, 366–369

Jirovetz, L., G. Buchhauer, M. Ngassouen, 1997: Investigation of the essential oils from the dried fruits of Xylopia aethiopica (West African peppertree) and Xylopia parviflora from Cameroon. Ernährung/Nutrition 21, 324–325

Jirovetz, L., G. Buchbauer, A. Nikiforov, 1994: Vergleichende Inhaltsstoffanalyse verschiedener Dillkraut- und Dillsamenöle mittels GC/FL und GC/MS. Ernährung/Nutrition 18, 534–536

Jirovetz, L., G. Buchbauer, M. Hoeferl, 2003: Essential oil analysis of Hemidesmus indicus R.Br. roots from southern India. J. Ess. Oil Res. 14, 437–438

Jirovetz, L., G. Buchbauer, M.P Shafi, A. Saidutty, 1998: Analysis the aroma compounds of the essential oil of seeds of the spice plant Zanthoxylum rhetsa from Southern India. Z. Lebensmittel-Unters. u. - Forsch. 206 A, 228–229

Jirovetz, L., G. Buchbauer, M.B. Ngassoum, M. Geissler, 2002: Analysis of the headspace aroma compounds of the seeds of the Cameroonian "garlic plant" Hua gabbonii using SPME/GE/FID, SPME/GC/MS and olfactometry. Eur. Food Res. Technol. 214, 212–215

Jirovetz, L., G. Buchbauer, M.B. Ngassoum, M. Geissler, 2002: Aroma compound analysis of Piper nigrum and Piper guineense essential oils from Cameroon using solid-phase microextraction-gas chromatography, solid-phase microextraction-gas chromatography-mass spectrometry and olfactometry. J. Chromatography A 976, 265–275

Jirovetz, L., G. Buchbauer, M. Shahabi, M.B. Ngassoum, 2002: Comparative investigations of the essential oil and volatiles of spearmint. Perfumer & Flavorist 27, No. 6, 16, 18–22

Jirovetz, L., G. Buchbauer, R. Remberg, N. Winkler, A. Nikiforow, 1996: Analysis of the characteristic odor of walnut (Juglans regia L.) peels. Ernährung/Nutrition 20, 286–287

Jirovetz, L., H.P. Koch, W. Jaeger, G. Fritz, G. Resl, 1993: Purslane (Portulaca oleraccea L.): Investigation of the plant constituents by means of chromatic/spectroscopic system for the control of the described cholesterol reduction. Ernährung/Nutrition 17, 226–227

Jirovetz, L., G. Buchbauer, A.S. Stoyanova, E.V. Georgiev, S.T. Damianova, 2003: Composition, quality control, and antimicrobioal activity of the essential oil of long-time stored dill (Anethum graveolens L.) seeds from Bulgaria. J. Agric. Food Chem. 51, 3854–3857

Jirovetz, L., G. Buchbauer, A.S. Stoyanova, E.V. Georgiev, S.T. Damianova, 2003: Composition and antimicrobial activity of an essential oil of long-time stored fruits of "Ajowan" (Trachyspermum ammi) from Bulgaria. Ernährung/Nutrition 27, 463–466

Joffe, P., 1993: The gardener's guide to South Africa plants. Tafelberg Publishers Ltd. Cape Town, 367 pp

Johns, T., W.D. Kitts, F. Newsome, G.H. Towers, 1982: Anti-reproductive and other medicinal effects of Tropaeolum tuberosum. J. Ethnopharmacol. 5, 149–161

Johnson, L.A., T.M. Suleiman, E.W. Lucas, 1979: Sesame protein: a review and prospectus. J. Amer. Oil Chem. Soc. 56, 463–468

Jondiko, I.J.O., G. Pattenden, 1989: Terpenoids and an apocarotenoid from seeds of Bixa orellana. Phytochemistry 28, 3159–3162

Jones, H.A., L.K. Mann, 1963: Onions and their allies. Intersci. Publ., Inc. New York , N.Y., 271 pp

Jordan, M.J., R.M. Martínez, M.A. Cases, J.A. Sotomayor, 2003: Watering level effect on Thymus hyemalis Lange essential oil yield and composition. J. Agric. Food Chem. 51, 5420–5427

Jorgensen, U., M. Hansen, L. Christensen, K. Jensen, K. Kaack, 2000: Olfactory and quantitative analysis of aroma compounds in elder flower (Sambucus nigra L.) drink processed from five cultivars. J. Agric. Food Chem. 48, 2376–2383

Joseph, J., 1980: The nutmeg – its botany, agronomy,

production, composition and use. J. Plantation Crops 8, No. 2, 61–72

Joulain, D., M. Ragault, 1976: Sur quelques nouveaux constituants de l'huile essentielle d'Hyssopus officinalis L. Riv. Ital. Essenze, Profumi, Piante Offic., Aromi, Saponi, Cosmet., Aerosol 58, No. 3, 129–131

Juchelka, D., T. Beck, B. Maas, A. Dietrich, P. Kreis, A. Steil, K. Witt, D. Lüttkopf, S. Schmidt, A. Mosandl, 1997: Der Bitterorangebaum – eine Quelle für unterschiedliche ätherische Öle. Lebensmittelchemie 51, 140

Juida de Carlini, D.P., 1986: Pharmacology of lemongrass (Cymbopogon citratus). J. Ethnopharmacol. 17, 37–83

Juliano, C., A. Mattana, M. Usai, 2000: Composition and in vitro antimicrobial activity of the essential oil of Thymus herba-barona Loisel growing wild in Sardinia. J. Ess. Oil Res. 12, 516–522

Jung, H.P., A. Senn, W. Grosch, 1992: Evaluation of potent odorant in parsley leaves (Petroselinum crispum [Mill.] Nym. ssp. crispum) by aroma extract dilution analysis. Lebensmittel-Wiss. u. -Technol. 25, 55–60

Junior, P., 1989: Weitere Untersuchungen zur Verteilung und Struktur der Bitterstoffe von Meyanthes trifoliata. Planta med. 55, 83–86

Jurenitsch, J., W. Kubelka, K. Jentzsch, 1979: Identifizierung kultivierter Capsicum-Sippen; Taxonomie, Anatomie und Scharfstoffzusammensetzung. Planta med. 35, 174–183

Jurkovic, N., N. Kolb, I. Colic, 1995: Nutritive value of marine algae Liminaria japonica and Undaria pinnatifida. Nahrung/Food 45, 63–66

K

Kaak, K., L.P. Christensen, S.L. Hansen, K. Grevsen, 2004: Non-structural carbohydrates in processed soft fried onion (Allium cepa L.). Eur. Food. Res. Technol. 218 A, 372–379

Kač, M., M. Kovačevič, 2000: Presentation and determination of hop (Humulus lupulus L.) cultivation by amin-max model on composition of hop essential oil. Mschr. Brauerei 53, 180–184

Kaczmarek, F., 1957: Nepeta cataria var. citriodora als Droge zur Gewinnung von ätherischem Öl. Planta med. 5, 51–56

Kadambi, K., 1954: Inducing formation in the Indian sandal wood tree, Santalum album Linn. Indian Forest 80, 659–662

Kahan, S., 1989: Liquid chromatographic method for determination of vanillin and related flavor compounds in vanilla extract: collaborative study. J. Assoc. Off. Anal. Chem. 72, 614–618

Kaistha, K.K., L.B.Kier, 1962: Structural studies on terebinthone from Schinus terebinthifolius. J. Pharm. Sci. 51, 245–248, 1136–1139

Kajmoto, T., K. Yakiro, T. Nohara, 1989: Sesquiterpenoides and disulphides from Asafeotida. Phytochemistry 28, 1761–1763

Kakadem M.L., N.R. Simon, I.E. Liener, J.W. Lambert, 1972: Biochemical and nutritional assessment of different varieties of soybeans. J. Agric. Food Chem. 20, 87–90

Kakasy, A.Z., È. Lemberkovics, L. Kursinszki, G. Janicsak, È. Szöke, 2002: Data to the phytochemical evaluation of Moldavian dragonhead (Dragocephalum moldavica L., Lamiaceae). Herba polon. 48, 112–119

Kalbhen, D.A., 1971: Die Muskatnuß als Rauschdroge. Ein Beitrag zur Chemie und Pharmakologie der Muskatnuß (Myristica fragrans). Angew. Chem. 83, 392–396

Kalkman, C., 1973: The genus Rosa in Malesia. Blumea 21, 281–291

Kalogjera, Z., S. Pepeljnjak, S. Vladimir, 1993: Antibakterielle und antifungucide Aktivität von Micromeria thymifolia (Scop.) Fritsch. Pharmazie 48, 311–313

Kallio, H., K. Kerrola, 1992: Application of liquid carbon dioxide to the extraction of essential oil of coriander (Coriandrum sativum L.) fruits. Z. Lebensmittel-Unters. u. -Forsch. 195, 545–549

Kamal, R., R. Yadav, G.L. Sharma, 1987: Diosgenin content in fenugreek collected from different geographical regions of South India. J. Indian Agric. Sci. 45, 674–676

Kambouche, N., D. El-Abed, 2003: Composition of the volatile oil from the aerial parts of Trachypermum ammi (L.) Sprague from Oran (Algeria). J. Ess. Oil Res. 15, 39–40

Kamel, M.S., M.H. Assaf, H.A. Hasanean, K. Ohtani, R. Kasai, K. Yamasaki, 2001: Monoterpene glycosides from Origanum syriacum. Phytochemistry 58, 1149–1152

Kameoka, H., S. Hashimoto, 1983: Two sulphur constituents from Allium schoenoprasum L. Phytochemistry 22, 294–295

Kameoka, H. Iida, S. Hashimoto, M. Miyazawa, 1984: Sulphides and furanones from steam volatile oils of Allium fistolosum and Allium chinense. Phytochemistry 23, 155–158

Kämpf, R., E. Steinegger, 1974: Dünnschicht- und gaschromatographische Untersuchungen an Oleum anisi und Oleum anisi stellati. Pharmac. Acta Helv. 49, 87–93

Kang, R., R. Helms, M.J. Stout, H. Jaber, Z.Q. Chen, T. Nakatsu, 1992: Antimicrobial activity of the volatile constituents of Perilla frutescens and its synergistic effects with polygodial. J. Agric. Food Chem. 49, 2328–2330

Kang-Jin Cho, Jin-Weon Kim, In-Iok Choi, Jung-Bong Kim, Young-Soo Hwang, 2001: Isolation, identification and determination of antioxidant in ginger (Zingiber officinale) rhizome. Agric. Chem. & Biotechnol. 44, 12–15

Kanisawa, T., 1993: Flavor development in vanilla beans. Kouryou 180, 113–123

Kanthami, S. C.R. Narayanan, K. Venkataraman, 1960: Isolation of l-stachydrine and rutin from the fruit of Capparis. J. Sci. Technol. India 22, 430–440

Kaouadji, M., A.-M. Mariotte, 1986: Centaurium erythrea: analyse du contenu xanthonique des racines et considerations chimio-taxinomiques. Bull. Liaison-Groupe Polyphenols 13, 546–552

Kapoor, L.D., 1995: Opium poppy. Botany, Chemistry, and Pharmacology. Food Products Press (The Haworth Press Inc.), Binghamton, NY, 326 pp

Karaali, A., N. Başoğlu, 1995: Essential oil of Turkish anise and their use in the aromatization of raki. Z. Lebensmittel-Unters. u. -Forsch. 200, 440–442

Karasawa, D., S. Shatar, A. Erdenechimeg, Y. Okamoto, H. Tateba, S. Shimizu, 1995: A study on Mongolian mints. A new chemotype from Mentha asiatica Borris and constituents of M. arvensis L. and M. piperita L. J. Ess. Oil Res. 7, 255–260

Karawya, M.S., M.S. Hifnawy, 1974: Analytical study of the volatile oil of Thymus vulgaris L. growing in Egypt. J. Amer. offic. Analyt. Chemists 57, 997–1001

Karawya, M.S., M.S. Hifnawy, 1976: Egyptian marjoram oil. Egypt. J. Pharmac. Sci. 17, 329–334

Karawya, M.S., S.M. Abdel Wahab, M.S. Hifnawy, 1979: Essential oil of Xylopia aethiopica fruit. Planta med. 37, 57–59

Karig, F., 1975: Schnelle Kennzeichnung von Curcuma-Rhizomen mit dem TAS-Verfahren. Dtsch. Apotheker-Ztg. 115, 325–328

Karl, J., 2000: Die Arzneipflanze Curcuma – Gelbwurzel. Naturheilpraxis 1732–1736

Karnig, C.R., 1977: Ethnobotanical, pharmacognostical and cultivation studies of the Hemidesmus indicus R.Br. (Indian sarsaparilla). Herba Hung. 16, No. 2, 7–16

Karousou, R., V. Vokou, S. Kokkini, 1998: Variation of Salvia fruticosa on the Island of Crete (Greece). Bot. Acta 111, 250–254

Kartnig, Th., C.Y. Ri, 1973: Dünnschichtchromatographische Untersuchungen an den Zuckerkomponenten der Saponine aus den Wurzeln von Primula veris und P. elatior. Planta med. 23, 379–380

Kartnig, Th., 1986: Vitex agnus-castus – Mönchspfeffer oder Keuschlamm. Eine Arzneipflanze mit indirekt luteotroper Wirkung. Z. Phytother. 7, 199–212

Kasali, A.A., A.O. Oyedeji, A.O. Ashilokun, 2001: Volatile leaf oil constituents of Cymbopogon citratus (DC.) Stapf. Flavour & Fragrance J. 16, 377–378

Kasting, R., J. Andersson, E. v. Sydow, 1972: Volatile constituents in leaves of parsley. Phytochemistry 11, 2277–2282

Kasturi, T.R., B.H. Iyer, 1955: Fixed oil from Elettaria cardamomum seeds. J. Indian Inst. Sci. 37 A, 106–112

Kataoka, E., C. Tokue, W. Tanimura, 1986: Sterol, tocopherol and phospholipide in Cardamom and star anise. J. Agricult. Sci. (Tokyo), 31 189–196

Katsiodis, S.T., C.R. Langezaal, J.J.C. Scheffer, R. Verpoorte, 1989: Comparative study of the essential oils from hops of various Humulus lupulus L. cultivars. Flavour & Fragrance J. 4, 187–191

Katsouri, E., C. Demetzos, D. Perdetzoglou, A. Loukis, 2001: An interpopulation study of the essential oils of various parts of Crithmum maritimum L. growing in Amorgos Island, Greece. J. Ess. Oil Res. 13, 303–308

Kaul, P.N., A.K. Bhattacharya, B.R. Rajeswara Rao, K.V. Syamasundar, Srinivasaiyer Ramesh, 2003: Volatile constituents of essential oil isolated from different parts of cinnamon (Cinnamomum zeylanicum Bume). J. Sci. Food Agric. 83, 53–55

Kaul, P.N., K. Singh, B.R. Rajeswara Rao, G.R. Mallavarapu, S. Ramesh, 1998: Analyse des ätherischen Öls Tagetes minuta L. Parfümerie u. Kosmetik 79, 36–38

Kaunzinger, A., D. Juchelka, A. Mosandk, 1996: Echte Vanille, eine Herausforderung für die Lebensmittelindustrie. GIT Fachzeitschr., Labor. 40, 830–831, 834–836

Kawabata, J., K. Mizuhata, E. Sato, T. Nishioka, Y. Aoyama, T. Kasai, 2003: 6-Hydroxyflavonoids as α-glucosidase inhibitors from marjoram (Origanum majorana) leaves. Biosci., Biotechnol., Biochem. 67, 445–447

Kays, St. J., J.C.S. Dias, 1995: Common names of commercial cultivated vegetables of the world in 15 languages. Econ. Bot. 49, 115–152

Kazuma, K., T. Takahashi, K. Sato, H. Takeuchi, T. Matsumoto, T. Okuno, 2000: Quinochalcones and flavonoids from fresh florets in different cultivars of Carthamus tinctorius L. Biosci.. Biotechnol., Biochem. 64, 1588–1599

Kedzia, B., 2002: Biological and therapeutical properties of Schisandra chinensis fructus. Herba polon. 48,146–155

Kelecom, A., 1983: Isolation, structure determination, and absolute configuration of barbatusol, a new bioactive diterpene with a rearranged abietane skeleton from the labiate Coleus barbatus. Phytochemistry 23, 1677–1679

Kelecom, A. T.C. dos Santos, W.L.B. Medeiros, 1986: On the structure and absolute configuration of (–)-20-deoxocarnosol. An. Acad. Brasil. Cien. 58, 53–59

Keller, K., E. Stahl, 1982: Kalmus: Inhaltsstoffe und β-Asarongehalt bei verschiedenen Herkünften. Dtsch. Apotheker-Ztg. 122, 2463–2466

Keller, K., E. Stahl, 1983: Zusammensetzung des ätherischen Öles von β-asaronfreiem Kalmus. Planta med. 47, 71–74

Kelm, M.A., M.G. Nair, G.M. Strasburg, D.L. de Witt, 2000: Antioxidant and cyclooxygenase inhibitory phenolic compounds from Ocimum sanctum Linn. Phytomedicin 7, 7–13

Kennedy, A.I., S.G Deans, K.P. Svoboda, A.I. Gray, P.G. Waterman, 1993: Volatile oils from normal and transformed root of Artemisia absinthium. Phytochemistry 32, 1449–1451

Kerrola, K., H. Kallio, 1993: Volatile compounds and other characteristics of carbon dioxide extracts of coriander (Coriandrum sativum L.) fruit. J. Agric. Food Chem. 41, 785–790

Kerrola, K., B. Galambosi, H. Kallio, 1994a: Volatile components and odor intensivity of four phenotypes of hyssop (Hyssopus officinalis L.) J. Agric. Food Chem. 42, 776–781

Kerrola, K., B. Galambosi, H. Kallio, 1994b: Characterization of volatile composition and odor of Angelica (Angelica archangelica subspec. archangelica L.) root extracts. J. Agric. Food Chem. 42, 1979–1988

Keusgen, M., 2001: Die schwefelhaltigen Inhaltsstoffe von Knoblauch (Allium sativum L.) und deren Bedeutung für die Züchtungsforschung. Drogenreport 14, No. 25, 24–28

Khader, V., 1983: Nutritional studies on fermented, germinated and baked soyabean (Glycine max) preparations. J. Plant Foods 5, 31–37

Khan, M.B., Pazir Gul, Muhammad Farooq, F. Khan, 1988: Effect of locality on the oil yield and physicochemical characteristics of oil from Thymus serphyllum L. Pakistan J. Foresty 38, 223–229

Khilik, L.A., T.I. Bondarenko, V.A. Shlyapnikov, A.N. Fedorovich, I.M. Drevyatnikov, D.V. Raihman, 1977: Introduction of a new essential oil-bearing plant (Nepeta caucasia Grossh.). Perfumer & Flavorist 2, Nr. 5, 57–58

Khosla, M.K., 1995: Study of inter-relationship, phylogeny and evolutionary tendencies in genus Ocimum. Indian J. Genetics 55, 71–83

Khubchandani, M.L., S.K. Srivastava, 1989: A new anthrachinone glycoside from the seeds of Peganum harmale Linn. Current Sci. 58, 137–139

Kiewitt, K., K. Kolbe, R. Beining, K. Wilding, W. Flemming, 1983: Zur Herstellung von Hopfenpellets.1. Mitt. Grundlagen der Hopfenpelletierung. Lebensmittelindustrie 30, 73–76

Kikuzaki, H., N. Nakatani, 1989: Structure of a new antioxidative phenolic acid from oregano (Origanum vulgare L.). Agric. Biol. Chem. 53, 519–522

Kikuzaki, H., S.M. Tsai, 1982: Gingerdiol related compounds from the rhizomes of Zingiber officinale. Phytochemistry 31, 1783–1786

Kil Jin Park, Z. Vohnikova, F.P. Reis Brod, 2002: Evaluation of drying parameters and desorption isotherms of garden mint (Mentha crispa L.). J. Food Engng. 51, No. 3, 193–199

Kim, Myung-Kon, Byung-Eun Lee, Se-Eck Yun, Jai-Sik Hong, Young-Hoi Kim, Young-Kyu Kim, 1994: Changes in volatile constituents of Zingiber officinale Roscoe rhizomes during storage. Agric. Chem. Biotechnol. 37, 1–8. (in Korean)

Kim, T.H., J.H. Shin, H.H. Baek, H.J. Lee, 2001: Volatile flavour compounds in suspension culture of Agastacherugosa Kuntze (Korean mint). J. Sci. Food Agric. 81, 569–575

Kini, F., B. Kam, J.P. Aycard, E.M. Gaydou, I. Bombarda, 1993: Chemical composition of the essential oil of Hyptis spicigera Lam. from Burkina Faso. J. Ess. Oil Res. 5, 219–221

Kintzios, S.E., 2000: Sage. The Genus Salvia. harwood academic publishers Amsterdam (Netherlands), 297 pp

Kintzios, S.E., 2002: Oregano: the genera Origanum and Lippea. Taylor & Francis London, UK, 277 pp

Kirbaslar, F.G., S.I.Kirbaslar, 2003: Composition of cold-pressed bitter orange peel oil from Turkey. J. Ess. Oil Res. 15, 6–9

Kirbaslar, F.G., S.I. Kirbaslar, U. Dramur, 2001: The compositions of Turkish bergamot oils produced by cold-pressing and steam distillation. J. Ess. Oil Res. 13, 411–415

Kirimer, N., K.H.C. Başer, T. Özek, M. Kürkçüoglu, 1992: Composition of the essential oil of Calamintha nepeta subsp. glandulosa. J. Ess. Oil Res. 4, 189–190

Kirschbaum-Titze, P., C. Hiepler, E. Mueller-Seitz, M. Petz, 2002 a: Pungency in paprika (Capsicum annuum). 1. Decrease of capsicinoid content following cellular disruption. J. Agric. Food Chem. 50 a, 1260–1263

Kirschbaum-Titze, P., E. Mueller-Seitz, M. Petz, 2002 a: Pungency in paprika (Capsicum annuum). 2. Heterogenicity of capsaicinoid content in individual fruits from one plant. J. Agric. Food Chem. 50, 1264–1266

Kivanç, M., A. Akgül, 1989: Inhibitory effects of spice essential oils on yeast. Doğa, Türk Tarum ve Ormancilik Dergisi 13, 68–72

Kivanç, M., A. Akgül, 1991: Effect of Laser trilobium spice on natural microflora of köfte, a Turkish ground meat product. Nahrung 35, 149–154

Klán, J., 1981: Pilze. Artia-Verlag Prague, 174 pp

Kleinert, J., 1963: Vanille und Vanillin. Int. Fachschr. Schokoladen-Ind. 28, 246–261; Gordian (Hamburg) 63, 809–814, 840–854, 892–898

Klimek, B., T. Madja, J. Góra, J. Patora, 2000: Investigation of the essential oil from lemon catnip, Nepeta cataria L. var. citriodora, in comparison to the oil from lemon balm, Melissa officinalis L. Herba polon. 46, 226–234

Klimes, I., D. Lamparsky, 1976: Vanilla volatiles – a comprehensive analysis. Int. Flavors Food Addit. 7, 272–291

Kling, M., 1999: Mitsuba – ein japanisches Gewürz. Gemüse 35, 493–494

Klock, M., Th. Klock, 2002: Zitruspflanzen. Die schönsten Arten und Sorten. blv-Verlag Munich, 96 pp

Klock, P., 1998: Zitruspflanzen. Lagerstroemia-Verlag Hamburg, 120 pp

Klock, P., 2001: Zitruspflanzen für Wintergarten und Terrasse. Verlag E. Ulmer Vienna, 96 pp

Klont, R., 2000: Vanilla, composer and maestro. World Food Ingredient No. 9, 34, 36, 38, 40

Klostermans, A.J.G.H., J.M. Bombard, 1993: The Mangos. Their botany, nomenclature, horticulture and utilization. Academic Press, Harcourt Brace & Co., Publ. San Diago, New York, 233 pp

Kmiecik, W., P. Gebczynski, A. Korus, 2001: Effect of variety, type of usable part and the growing period on the content of nitrates, nitrites and oxalates in dill (Anethum graveolens). Bromatol. Chim. Toksykologiczna (Poland) 34, No. 3, 213–220

Knowles, P.F., 1955; Safflower – production, processing and utilization. Econ. Bot. 9, 273–299

Knowles, P.F., 1958: Safflower. Adv. Agronomy 10, 289–323

Kobata, K., T. Todo, S. Yazawa, K. Iwai, T. Watanab, 1998: Novel capsaicinoid-like substances, capsiate and dihydrocapsiate, from the fruits of a nonpungent cultivar, CH-19 sweet, of pepper (Capsicum annuum L.). J. Agric. Food Chem. 46, 1695–1697

Kobold, U., O. Vostrowsky, H.J. Bestmann, J.C. Bisht, A.K. Pant, A.B. Melkani, C.S. Mathela, 1987: Terpenoids from Elsholtzia species. II. Constituents of essential oil from a new chemotype of Elsholtzia cristata. Planta med. 53, 268–271

Koch, H., E. Steinegger, 1981/82: Zur Kenntnis der Inhaltsstoffe von Condurango Cortex. Pharm. Acta Helv. 56, 244–248; 57, 211–214

Koch, H.P., 1988: Portulak. Omega-3-Fettsäuren in einer alten Arzneipflanze. Dtsch. Apotheker-Ztg. 128, 2493

Koch, H.P., 1989: Kann man Knoblauch „geruchlos" machen? Dtsch. Apotheker-Ztg. 129, 1991–1996

Koch, H.P., 1994: Die Küchenzwiebel – eine zu Unrecht vernachlässigte Arzneipflanze. Pharmazie in unserer Zeit. 23, 333–339

Koch, H.P., 1995: „Moly" – der Zauberlauch der griechischen Mythologie. Geschichte der Pharmazie 47, 38-48; Beilage: Dtsch Apotheker-Ztg.

Koch, H.P., 1996: Der lange Weg zum „geruchlosen" Knoblauch. Pharmazie in unserer Zeit 25, 186–191

Koch, H.P., G. Hahn, 1988: Knoblauch. Grundlagen der therapeutischen Anwendung von Allium sativum L. Verlag Urban & Schwarzenberg Munich Vienna, 222 pp

Koch-Heitzmann, I., W. Schultze, 1984: Melissa officinalis. Dtsch. Apotheker-Ztg. 124, 2137–2145

Kodera, Y., A. Suzuki, O. Imada, Sh. Kasuga, I. Sumioka, A. Kanezawa, N. Taru, M. Fujikawa, Sh. Nagae, K. Masamoto, K. Maeshige, K. Ono, 2001: Physical, chemical and biological properties of S-allylcysteine, an amino acid derived from garlic. J. Agric. Food Chem. 50, 622–632

Koebnick, C., H.-J- Zunft, 2004: Potenzial von Carobballaststoffen in Prävention und Therapie der Hypercholesterinämie und des metabolischen Syndroms. Ernährungs-Umsch. 51, 46–50

Koenen, E. v., 1996: Heil-, Gift- und essbare Pflanzen in Namibia. Klaus Hess Verlag Göttingen, 336 pp

Koezuka, Y., G. Honda, M. Tabata, 1986: Genetic control of the chemical composition of volatile oils in Perilla frutescens. Phytochemistry 25, 859–863

Köhlein, F., 1984: Primeln. Verlag E. Ulmer Stuttgart, 406 pp

Köhlein, F., 1999: Viola. Veilchen, Stiefmütterchen, Hornveilchen. Verlag E. Ulmer Stuttgart, 124 pp

Köhlein, F., 2003: Mohn und Scheinmohn. Edit. Staudengärtnerei Gräfin von Zeppelin Sulzburg-Laufen, 192 pp

Kojima, M., 1976: Studies on the volatile components of wasabi (Eutrema) japonica using gas chromatography and head space technique. J. Japan. Soc. Food Sci. Technol. 23, 324–326

Kojima, M., H. Hamada, M. Yamashita, 1982: Studies on the stability of dried wasabi (Wasabia japonica) flour. J. Soc. Food Sci. Technol. 29, 232–237

Kojima, M., H. Hamada, N. Toshimitsu, 1985: Changes in isothiocynates of Wasabi japonica roots by drying. J. Japan. Soc. Food Sci. Technol. 32, 886–891

Kojoma, M., K. Kurihara, K. Yamada, S. Sekita, M. Satake, O. Iida, 2002: Genetic identification of cinnamom (Cinnamomum spp.) based on the trnL-trnF chloroplast DNA. Planta med. 68, 94–96

Kokkalou, E., A. Koedam, G. Phokas, 1982: Composition de l'huile essentielle de Salvia hormonium L. (Labiatae). Pharm. Acta. Helv. 57, 317–320

Kokkini, S., D. Vokou, 1989: Mentha spicata (Lamiaceae) chemotypes growing wild in Greece. Econ. Bot. 43,192–202

Kokkini, S., E. Hanlidou, R. Karousou, T. Lanaras, 2002: Variation of pulegone content in pennyroyal (Mentha pulegium L.) plants growing wild in Greece. J. Ess. Oil Res. 14, 224–227

Kolak, I., Z. Satovic, F. Rozic, 2001: Peppermint (Mentha piperita L.) Sjemenarstvo 18, 215–227 (in Croatia).

Kolayli, S., M. Kücük, C. Duran, F. Candan, B. Dincer, 2003: Chemical and antioxidant properties of Laurocerasus officinalis Roem. (Cherry laurel) fruit grown in the Black Sea region. J. Agric. Food Sci. 51, 7489–7494

Koller, W.D., 1979: Einige lagerungsbedingte Veränderungen bei gemahlenen Gewürznelken. Z. Lebensmittel-Unters. u. -Forsch. 169, 457–463

Koller, W.D., 1981: Identifizierung einiger flüchtiger Inhaltsstoffe von gemahlenen Gewürznelken. Z. Lebensmittel-Unters. u. -Forsch. 173, 99–100

Kollmannsberger, H., S. Nitz, F. Drawert, 1992: Über die Aromastoff-Zusammensetzung von Hochdruckextrakten. I. Pfeffer (Piper nigrum var. muntok). Z. Lebensmittel-Unters. u. -Forsch. 194, 545–551

Kolodziej, H., O. Kayer, M. Gutmann, 1995: Als Arznei verwendete Pelargonien aus Südafrika. Dtsch. Apotheker-Ztg. 135, 234–24

Komaitis, M.E., N. Infanti-Papatragianni, E. Melissari-Panagiotou, 1992: Composition of the essential oil of majoram (Origanum majorana L.). Food Chem. 45, 117–118

Kondo, A., H. Ohigashi, A. Murakami, J. Suratwadee, K. Koshimizu, 1993: 1'Acetoxychavicol acetate as apotent inhibitor of tumor promotor-induced Epstein-Barr virus activation from Languas galanga, a traditional Thai condiment. Biosci., Biotechnol., Biochem. 57, 1344–1345

König, W.A., A. Rieck, Y. Saritas, I.H. Hardt, K.-H. Kubeczka, 1996: Sesquiterpene hydrocarbone in the essential oil of Meum athamanticum. Phytochemistry 42, 461–464

Konvička, O., 1983: Knoblauch – eine Gewürz- und Heilpflanze. Naturwiss. Rdsch. 36, 209–215

Konvička, O., P. Würfl, 2001: Knoblauch (Allium sativum L.). Grundlagen der Biologie, und des Anbaues, Inhaltsstoffe und Heilwirkungen. Verlag Agrimedia GmbH Bergen/Dumme, 202 pp

Kooinan, P., 1961: The constitution of tamarindusamyloid. Rec. Trav. Chin. Pay-Bas 80, 849–865

Koolhaas, D.R., 1932: Das ätherische Öl aus Erygium foetidum L. Über das Vorkommen von Dodecen-(2)-al-1. Rev. Trav. Chin. Pays-Bas 51, 460–468

Körber-Grohne, U., 1989: Nutzpflanzen in Deutschland. Kulturgeschichte und Biologie. K. Theiss Verlag GmbH Stuttgart, 490 pp

Kostermans, A.J.G.H., 1970: Materials for a revision of Lauraceae. 3. Reinwardtia 8, No. 1, 21–196

Kothari, S.K., K.Singh, 1994: Chemical weed control in Japanese mint (Mentha arvensis L.). J. Ess. Oil Res. 6, 47–55

Kothari, S.K., K. Singh, 1995: The effect of row spacing and nitrogen fertilization on Scotch spearmint (Mentha gracilis Sole). J. Ess. Oil Res. 7, 287–297

Kotow, W.P., 1978: Chren (horse radish). Izdal. "Kolos", Leningrad 47 pp (in Russian)

Koschtschejew, A.K., 1990: Wildwachsende Pflanzen in unserer Ernährung. VEB Fachbuchverlag Leipzig

Kottegoda, S.R., 1994: Flowers of Sri Lanka. The Royal Asiatic Soc. of Sri Lanka, Colombo, 247 pp

Kouhila, M., A. Belghit, M. Daguenet, B.C. Boutaleb, 2001: Experimental determination of the sorption isotherms of mint (Mentha viridis), sage (Salvia officinalis) and verbena (Lippea citriodora). J. Food Engng. 47, 281–287

Koukos, P.K., K.I. Papadopoulous, 1997: Essential oil of Juniperus communis L. grown on Northern Greece: variation of fruit oil yield and composition. J. Ess. Oil Res. 9, 35–39

Kouokam, J.C., T. Jahns, H. Becker, 2002: Antimicrobial activity of the essential oil and some isolated sulfur rich compounds from Scorodophloeus zenkeri. Planta med. 68, 1082–1087

Kowalewski, Z., I. Matlawska, 1978: Zwiazki flawonoidowe w zielu szanty zwyczajnej (Marrubium vulgare). (Flavonoids in Marrubium vulgare herbage). Herba Polon. 24, 183–186 (in Polish)

Kramer, F.L., 1977: The pepper tree, Schinus molle L. Econ. Bot. 11, 322–326

Kraus, A., J.F. Hammerschmidt, 1980: Untersuchungen an Fenchelölen. dragoco-rep. 13, 31–40

Kraus, J., G. Franz, 1987: Schleimpolysaccharide aus Wollblumen. Dtsch. Apotheker-Ztg. 127, 665–669

Kreutzig, L., 1982: Untersuchungen zur Lagerstabilität von Salbeiblättern in Abhängigkeit von der Verpackung. Pharmaz. Ztg. 127, 893–894

Kraxner, U., D. Fritz, A. Wuensch, 1987: Bitterkeit in Meerrettich. Ind. Obst- u. Gemüseverwertung. 72, 103–104

Kremer, P., P. Jaeggi, 1995: Der Baum mit dem Wasserbauch. Kosmos 91, 88–95

Krishnamurthy, N., A.G. Mathew, E.S. Nambudivi, C.P. Natarajan, 1976: Oil and oleoresin of turmeric. Trop. Sci. 18, 37–45

Krishnamurthy, N., Y.S. Levis, .P. Natarjan, 1977: Ginger from hilly regions of India. J. Food Sci. Technol. (India) 14, 128–129

Krishnamurthy, N., Y.S. Lewis, B. Ravindranath, 1982: Chemical constitution of kokam fruit rind. J. Food Sci. Technol. 19, 97–100

Krishnamurthy, N., Y.S. Lewis, B. Ravindranath, 1981: On the structure of garcinol, isogarcinol and camboginol. Tetrahedron Letters 22, 793–795

Kroon, A.H.J., 1969: The pistachio nut (Pistacia vera L.). Trop. Abstr. 24, 73–74

Krstic, B., L.J. Merkulov, D. Gvozdenovic, S. Pajevic, 2001: Anatomical and physiological characteristics of seed in pepper (Capsicum annuum L.) varieties. Acta Agronimica Hungar. 49, 221–229

Krüger, H., K. Hammer, 1996: A new chemotype of Anethum graveolens L. J. Ess. Oil Res. 8, 205–206

Krüger, H., B. Zieger, 1993: Die Bestimmung etherischer Öle in Kümmelextrakten. Drogenreport 6, No. 10, 31–32

Krug, H., B. Borkowski, 1965: Neue Flavanol-Glykoside aus den Blättern von Peumus boldus Molina. Pharmazie 20, 692–698

Krüssmann, G., 1974: Rosen, Rosen, Rosen, unser Wissen über die Rose. Verlag P. Parey Berlin Hamburg, 447 pp

Krützfeld, K., 2001: Pfeffer als Gewürz und Arzneimittel. Dtsch. Apotheker-Ztg. 141, 6038–6044

Krützfeld, K., 2002: Muskat, die psychoaktive Nuss. Dtsch. Apotheker-Ztg. 142, 5622–5630

Krützfeld, K., 2002: Zimt – der Duft des Paradieses. Dtsch. Apotheker-Ztg. 142, 6254–6261

Krzaczak, T., 1998: Phenolic acids in Angelica sylvestris L. Herba polon. 44, 292–296

Krzaczak, T., R. Nowak, 2000: Sterols in the roots and fruits from Angelica sylvestris L. (Garden angelica). Herba polon. 46, 274–277

Kasandopulo, S.Yu, S.K. Mustafaev, T.P. Bazhina, 1995: Evaluation of natural heterogeneity in freshly harvested coriander fruits. Ицвестия Высжин Ухебныкн Цаведеннь, Пижхевая Технологиа,

Кпаснодар (Reports of University Food Technol., Krasnodar, No,1/2, 118; (in Russian)

Kubeczka, K.-H., 1985: Radix Pimpinella und ihre aktuellen Verfälschungen. Dtsch. Apotheker-Ztg. 125, 399–402

Kubeczka, K.-H., I.U. Bohn, 1985: Radix Pimpinellae und ihre aktuellen Verfälschungen. Nachweis von Verfälschungen durch DC- und GC. Strukturrevision der Hauptkomponenten des ätherischen Öles. Dtsch. Apotheker-Ztg. 125, 399–402

Kubeczka, K.-H., E. Stahl, 1975: Über ätherische Öle der Apiaceae (Umbelliferae). I. Das Wurzelöl von Pastinaca sativa. Planta med. 27, 235–241

Kubeczka, K.-H., V. Koch, E.M. Ney, 1990 Das Geheimnis des brennenden Busches. Das ätherische Öl von Dictamnus albus und seine Akkumulationsstrukturen. Dtsch. Apotheker-Ztg. 130, 2181–2185

Kubeczka, F., F.v. Massow, V. Formáček, M.A.R. Smith, 1976: A new type of phenolpropane from the essential oil of Pimpinella anisum L. Z. Naturforsch. 31 B, 283–284

Kuebel, K.R., A.O. Tucker, 1988: Vietnamese culinary herbs in the Unites States. Econ. Bot. 42, 413–419

Kulakovskaja, L.A., 1976: Anatomical-morphological studies of Ziziphora tenuior L. Trudy Inst. Bot. Kazach. SSR 35, 133–144

Kulkarni, R.N., S. Ramesh, 1992: Development of lemongrass clones with high oil content through population improvement. J. Ess. Oil Res. 4, 181–186

Kumar, A., A.K. Gauniyal, O.P. Virmani, 1986: Cultivation of Pogostemon patchouli for its oil. Current Res. on Med. and Aromatic Plants 8, 79–86

Kumar, S., 1991: Turmeric (Curcuma longa L.) and related taxa in Sikkim Himalaya. J. Econ. Taxon. Bot. 721–724

Kumar, S., 2001: Zingiberaceae of Sikkim. Deep Publications, New Delhi (India), 83 pp

Kumar, S., 2003: Leafy and edible plants of India. Scientific Publishers, Jodhpur (India), 132 pp

Kumar, S., D.C.S. Raju, 1989: Large cardamom and its wild relatives in Sikkim Himalayas. J. Hill. Res. 2, 102–107

Kump, A. 2001: Schisandra chinensis (Turcz) Baill. Ein Tertärrelikt gewinnt an Bedeutung. Ber. Dtsch. Ges. Qualiätsforsch. (Pflanzl. Nahrungsmittel) XXXVI. Vortragstagung: Gewürz- u. Heilpflanzen, 19.-20.3. in Jena, pp 217–220

Küçüköner, E., B. Yurt, 2003: Some chemical characteristics of Pistacia vera varieties produced in Turkey. Eur. Food Res. Technol. 217 A, 308–310

Kuo, M.-C., M. Chien, C.-T Ho, 1990: Novel polysulfides identified in the volatile components from Welsh onions (Allium fistulosum L. var. maichuon) and Scallions (Allium fistulosum L. var. caespitosum). J. Agric. Food Chem. 38, 1378–1381

Kustrak, D., Z. Martinis, 1990: Composition of essential oils of some thymus and thymbra species. Flavour & Fragrance J. 5, 227–231

Kusumoto, S., A. Ohsuka, M. Kotake, T. Sakai, 1968: Constituents of leaf oil from Japanese pepper. Bull. Chem. Soc. Japan 41, 1950–1953

Kwasniewski, V., 1952: Das Süßholz – Glycyrrhiza glabra L. Pharmazie 7, 444–448

Kyung Im Kim, Kwang Soon Shin, Woo Jin Jun, Bum Shik Hong, Doon Hoon Shin, Hong Yon Cho, Hyo Il Chang, Seong Mo Yoo, Han Chul Yang, 2001: Effects of polysaccharides from rhizomes of Curcuma zedoaria on macrophage functions. Biosci., Biotechnol., Biochem. 65, 2369–2377

L

Laatsch, H.,1992: Polysaccharide mit Antitumor-Aktivität aus Pilzen. Pharmazie in unserer Zeit 21, 159–166

Lachenmeier, D.W., W. Frank, C. Athanasakis, St.A. Padosch, B. Madea, M.A. Rothschild, L.U. Kröner, 2002: Absinth – ein Getränk kommt wieder in Mode: toxikologisch-analytische und lebensmittelchemische Bewertungen. Dtsch. Lebensmittel-Rdsch. 100, 117–129

Lachowicz, K.J., G.P. Jones, D.R. Briggs, F.E. Bienvenu, M.V. Palmer, S.S.T. Ting, M. Hunter, 1996: Characteristics of essential oil from basil (Ocimum basilicum L.) grown in Australia. J. Agric. Food Chem. 44, 877–881

Lagouri, V., G. Blekas, M. Tsimidou, S. Kokkini, D. Boskou, 1993: Composition and antioxidant activity of essential oils from oregano plants grown wild in Greece. Z. Lebensmittel-Unters. u. -Forsch. 197, 20–23

Laily Din, Zuriati Zakaria, Modh Wahid Samdudin, J. Brophy, R.F. Toia, 1988: Composition of the steam volatile oil from Hyptis suaveolens Poit. Pertanika 11, 239–242

Lal, R.N., T.K. Sen, M.C. Nigam, 1978: Gaschromatographie des ätherischen Öls von Ocimum sanctum L. Parfümerie u. Kosmetik 59, 230–231

Lalande, B., 1984: Lavender, lavandin and other French oils. Perfumer & Flavorist 9, No. 2, 117–121

Lalas, S., 1998: Qualita and stability characterization of Moringa oleifera seed oil. Ph. Thesis, Lincolnshire and Humberside Univ., U.K., 123 pp

Lallamand, H., N. Pirot, M. Dornier, M. Reynes, 2000: La cannelle: historique, production et principales caracteristiques. Fruits 55, 421–432

Lamaty, G., C. Menut, J.M. Bessiere, P.H. Amvan Zollo, 1987: Aromatic plants of tropical Central Africa. Volatile compounds of two Annonaceae from Cameroon: Xylopia aethiopica (Dunal) A. Rich. and Monodora myristica (Gaertn.) Dunal. Flavour & Fragrance J. 2, 91–94

Lamboni, C., K. Monkpoh, S. Konlani, A. Doh, 1999: Caractéristiques alimentaires du «Tonou», condiment àbase de soja ou de grains de néré. Med. et Nutrit. 2, 60–65

Lamoureux, C.H., 1975: Phenology and floral biology of Monodora myristica (Anacardiaceae) in Bogar, Indonesia. Ann. Bogoriensis 6, 1–25

Lambrecht, H., F. Pichlmayer, E.R. Schmid, 1994: Determination of the authenticity of vanilla extracts by stable isotope ratio and analysis and component analysis by HPLC. J. Agric. Food Chem. 42, 1722–1727

Landry, R. 1985: On l'appelle cannelle. Cuisine 408, 76–78

Langenheim, J.H., 2003: Plant Resins. Chemistry, Evolution, Ecology, and Ethnobotany. Timber Press Inc. Portland Oregon USA/Cambridge UK, 586 pp

Länger, R., 1990: Tausendgüldenkraut. Pharmakobotanische Untersuchungen an verschiedenen Arten der Gattung Centaurum Hill. Dtsch. Apotheker-Ztg. 130, 2366–2371

Länger, R., K. Winter, W. Bubelke, 1993: Zur Identitätsprüfung von Radix Tormentillae. Pharmazie 48, 776-769

Länger, R., Ch. Mechtler, H.O. Tanzler, J. Jurenitsch, 1993: Differences of the composition of the essential oil within an individuum of Salvia officinalis. Planta med. 59, A 635–A 636

Larkcom, J., 1991: Oriental vegetables. The complete guide for garden and kitchen. J. Murray Publishers Ltd., London 1991, 232 pp

Larue, M., 1960: Le pistachier en Iran. Fruits 15, 139–142

Laub, E., W. Olszowski, 1982: Über den Cumaringehalt in Waldmeister und seine DC-Bestimmun. Z. Lebensmittel-Unters. u. -Forsch. 175–179

Laub, E., W. Olszowski, R. Woller, 1985: Waldmeister und Maibowle. Dtsch. Apotheker-Ztg. 125, 848–850

Laul, M.S., S.D. Bhalenao, G.V. Muhmoley, G.R. Shah, V.S. Palal, 1984: Curing and storage of Nasik red onions (Allium cepa L.). Indian Food Packer 38, No. 6, 30–39

Lautenbacher, L.-M., 1997: Schwarzkümmelöl. Dtsch. Apotheker-Ztg. 137, 4602–4603

Lavania, U.C., 1991: Evaluation of an essential oil rich autotetraploid cultivar of vetiver. J. Ess. Oil Res. 3, 455–457

Lavy, G., 1987: Nutmeg intoxication in pregnancy. J. Reprod. Med. 32, 63–64

Lawrence, B.M., 1982: Coumarins and psorales in citrus oils. Perfumer & Flavorist 7, 57–65

Lawrence, B.M., 1984a: The botanical and chemical aspects of Oregano. Perfumer & Flavorist 9, 41–51

Lawrence, B.M., 1984b: Clove oil. Perfumer & Flavorist 9, 35–45

Lawrence, B.M., C.K. Shu, W.R. Harris, 1989: Peppermint oil differentiation. Perfumer & Flavorist 14, No. 6, 21–30

Lawrence, B.M., R.J. Reynolds, 1984: Major tropical spices – ginger (Zingiber officinalis Rosc.). Perfumer & Flavorist 9, 1–40

Lawrence, B.M., R.J. Reynolds, 1999: Progress in essential oils. Cananga oil. Perfumer & Flavorist 24, 31-36

Lawrence, B.M., K.M. Weaver, 1974: A comparative chemical composition of the essential oils of Myrica gale and Comptonia peregrino. Planta med. 25, 385–388

Lawrence, B.M., J.W. Hogg, S.J. Terhune, 1972: Essential oils and their constituents. The oils of Ocimum sanctum and Ocimum basilicum from Thailand. Flavour Ind., No. 1, 47–49

Lawrence, B.M., C.K. Shu, W.R Harris, 1986: Peppermint oil differentation. Perfumer & Flavorist 14, No. 6, 21–24, 26, 28–30

Lawton, B.P., 2002: Mints – A family of herbs and ornamentals. Taylor & Francis London, UK, 272 pp

Leberkovics, E., G. Petri, G. Vitanyi, L. Lelik, 1994: Essential oil composition of chervil growing wild in Hungary. J. Ess. Oil Res. 6, 421–422

Leclercq, P.A., N.X. Dung, V.N. Lo, N.V. Toanh, 1992: Composition of the essential oil of Eryngium foetidum from Vietnam. J. Ess. Oil Res. 4, 423–424

Lee, L.A., V.J. Caruso, 1958: Methods for distinguishing sources of nutmeg or mace. J. Assoc. Off. Agric. Chem. 41, 446–449

Lee, P.L., G. Swordes, G.L.K. Hunter, 1975: Volatile constituents of tamarind (Tamarindus indica L.). J. Agric. Food Chem. 23, 1195–1199

Lee, S.S., Y.J. Lin, C.K. Chen, K.C. Liu, C.H. Chen, 1993: Quaternary alkaloids from Litsea cubeba and Cryptocarya konishii. J. Nat. Products 56, 1971–1976

Lee, Y.B., Y.S. Kim, C.R. Ashmore, 1986: Antioxidant property in ginger rhizome and its application to meat products. J. Food Sci, 51, 20–23

Lehmann, H., 1982: Zur Eignung der Apfelbeere (Aronia melanocarpa) für die industrielle Verarbeitung. Lebensmittelindustrie 29, 175–177

Lehmann, H., 1990: Die Aroniabeere und ihre Verarbeitung. Flüssiges Obst 57, 746, 748–752

Leino, M.E., 1992: Effect of freezing, freeze-drying, and air drying on odor of chive. Characterized by head-spice gas chromatography and sensory analysis. J. Agric. Food Chem. 40, 1379–1384

Lemaistre, J., 1959: Le pistachier (Étude bibliographique). Fruits 14, 57–77

Lemberkovics, É., G. Petri, G. Vitanyi, L. Lelik, 1994: Essential oil composition of chervil growing wild in Hungary. J. Ess. Oil Res. 6, 421–422

Lenardis, A., E. de la Fuente, A. Gil, A. Tubia, 2000: Coriander (Coriandrum sativum L.) crop response to nitrogen availability. J. Herbs, Spices & Med. Plants 7, 47–58

Leng-Peschlow, E., A. Strenge-Hesse, 191: Die Mariendistel (Silybium marianum) und Silymarin als Lebertherapeutikum. Z. Phytother. 12, 162–174

Leong, G., A. Archavlis, M. Derbesy, 1989: Research on the glucoside fraction of the vanilla bean. J. Ess. Oil Res. 1, 33–41

Leong, G., A. Uzio, M. Derbesy, 1989: Synthesis, identification and determination of glucosides present in green vanilla beans (Vanilla planifolia Andrews). Flavour & Fragrance J. 4, 163–167

Leroy, J.F., H. Gillet, 1964: Sur deux plantes de la pharmacopée tchadienne (Lepidium sativum L.) et Nigella sativa L.). J. Agricult. Trop. Bot. Appl. 11, No. 5–7, 202–204

Leung, A.L., 1991: Chinesische Heilkräuter. E. Diedrichs Verlag, Munich, 2. Ed., 288 pp

Leverington, R.E., 1975: Ginger technology. Food Australia Technol. 27, 309–312

Lewington, A. 1990: Plants for people. Oxford Univ. Press, New York, 232 pp

Lewis, Y.S., S. Neelakantan, 1964: The chemistry, biochemistry and technology of tamarind. J. Sci. Ind. Res. 23, 204–206

Lewis, Y.S., S. Neelakantan, 1965: (–)-Hydroxycitric acid, the principal acid in the fruits of Garcinia cabogia Desr. Phytochemistry 4, 619–625

Libert, B., R. Enghind, 1989: Present distribution and ecology of Rheum rhaponticum (Polygonaeae). Willdenowie (Berlin) 19, 91–98

Liener, I.E., 1994: Implication of antinutrional components in soyabean foods. Crit. Rev. Food Sci. Nutrit. 34, 31–64

Liese, J., 1948: Der Shiitake-Pilz. Naturwissensch. u. Nahrung. No. 11/12, pp 23–25

Lihua Jiang, K. Kubota, 2001: Formation by mechanical stimulus of the flavor compounds in young leaves of Japanese pepper (Xanthoxylum piperitum DC.). J. Agric. & Food Chem. 49, 1353–1357

Lihua Jiang, H. Kojima, K. Yamada, A. Kobayashi, K. Kobota, 2001: Isolation of some glycosides as aroma precursors in young leaves of Japanese pepper (Xanthoxylum piperitum DC.) J. Agric. Food Chem. 49, 5888–5894

Lima, P., 1986: The Harrowsmith illustrated book of herbs. Camden House Publishing Ltd., Camden East, ON, Canada, 175 pp

Lin, X.H., R. Li, Z.T. Jiang, 2001: Study on the chemical composition of Armoracia lapathifolia essential oil. Food Sci, China 22, No. 3, 73–75

Lin Zheng-kui, Kua Ying-fang, Gu Yu-hong, 1986: The chemical constituents of the essential oil from the flowers, leaves and peels of Citrus aurantium. Acta Bot. Sinica 28, 635–640

Lindberg, M.H., L. Andersen, L. Christansen, P. Brockhoff, G. Berbelsen, 1996: Antioxidative activity of summer savory (Satureja hortensis L.) and rosemary (Rosmarinus officinalis L.) in mineral, cooked pork meat. Z. Lebensmittel-Unters. u. -Forsch. 203, 337–338

Lindeman, A., P. Jounela-Eriksson, M. Lounasmaa, 1982: The aroma composition the flower of meadow-sweet (Filipendula ulmaria [L.] Maxim.) Lebensmittel-Wiss. u. Technol. 15, 286–289

Lindner, W., 1971: Die Papayafrucht. Dtsch. Apotheker-Ztg. 111, 246–253

Lindner, M.W., 1951: Beitrag zur Kenntnis der Padang-Cassia. Pharmazie. 6, 352–354

Linke, H., 1983: Zählen getoastete Leinsamen uns Sesamkörner zu den Gewürzen? Fleischwirtschaft 63, 843

Liu, K., 1997: Soybeans: Chemistry, technology, and utilization. Chapman & Hall, New York, 553 pp

Liu, L., P. Howe, Y.-F. Zhou, Z.-Q. Xu, C. Hocart, R. Zhang, 2000: Fatty acid and β-carotene in Australian purslane (Portulacca oleracea) varities. J. Chromatogr. 893 A, 207–213

Lloyd, H.A., T.M. Jaouni, S.L. Evans, J.F. Morton, 1977: Terpenes of Schinus terbinthifolius. Phytochemistry 16, 1301–1302

Loch, W., 1993: Essbare und giftige Wildpflanzen in Mitteleuropa. Packpapierverlag Osnabrück, 178 pp

Lock, J.M., J.B. Hall, D.K. Abbiw, 1977: The cultivation of melegueta pepper (Aframomum melegueta) in Ghana. Econ. Bot. 21, 321–330

Lockwood, G.B., 1979: The major constituents of the essential oils of Cinnamomum cassia Blume growing in Nigeria. Planta med. 36, 380–381

Lösing, G., G. Albroscheit. M. Degener, G. Matheis, 1999: Bestimmung der farbgebenden Komponenten in Curcuma-Produkten. dragoco rep. 46, 110–115

Long-Solis, J., 1986: Capsicum y Cultura: La Historia del Chili. Fondo de Cultural Económica, México, 181 pp

Longvah, T., Y.G. Doestale, 12991: Chemical and nutrional studies on hanshi (Perilla frutescens), a traditional oilseed from northeast India. J. Amer. Oil Chem. Soc. 68, 781–784

Lockwood, G.B., G. Asghari, B. Hakimi, 2002: Production of essential oil constituents by cultured cells of Carum copticum. Flavour & Fragrance J. 17, 456–458

López, A., M.T. Pique, A. Romero, N. Aleta, 1995: Influence of cold-storage conditions on the quality of unshelled walnuts. Intern. J. Refrig. 18, 544–549

López-García, R.E., R.M. Rabanal, V. Darias, D. Martín-Herrera, M.C. Carreiras, B. Rodríguez, 1991: A preliminary study of Cedronella canariensis (L.) var. canariensis extracts for antiinflammatory and analgesic activity in rats and mice. Phytother. Res. 5, 273–275

López-García, R.E., M. Harnández-Pérez, R. Rabanal, V. Darias, D. Martin-Herera, A. Arias, J. Sanz, 1992: Essential oils and antimicrobial activity of two varieties of Cedronella canariensis (L.) W. et B. J. Ethnopharcol. 36, 207–211

Lorenzi, H. (ed) 1992 (I), 1998 (II): Árvores Brasileiras. Manual de identificacao e cultivo de planteas Arbóreas nativas do Brasil. Editora Plantarum LTDA Nova Odessa, Brasil, Vol. I, 370 pp, Vol. II, 2. edn., 368 pp

Lorenzo, D., I. Loayza, E. Dellacassa, 2002: Composition of the essential of Tagetes maxima Kuntze from Bolivia. Flavour & Fragrance J. 17, 115–118

Losser, G., 1967: Die Inhaltsstoffe von Majorana hortensis Moench. 3. Mitt. Das ätherische Öl. Pharmazie 22, 324–326

Lossner, G., 1968: Der Majoran – phytochemisch betrachtet. Planta med. 16, 54–57

Loughrin, J.H., M.J. Kasperbauer, 2001: Light reflected from colored mulches affects aroma and phenol content of sweet basil (Ocimum basilicum L.) leaves. J. Agric. Food Chem. 49, 1331–1335

Loughrin, J.H., M.J. Kasperbauer, 2003: Aroma content of fresh basil (Ocimum basilicum L.) leaves is affected by light reflected from colored mulches. J. Agric. Food Chem. 51, 2272–2276

Louw, P.G.J., 1943: Lantanin, the active principle of Lantana camara. I. Isolation and preliminary results on the determination of its constitution. Onderstepoort J. Vet. Sci. and Animal Ind. (South Africa) 18, 197–202

Louw, P.G.J., 1948: Lantadene A, the active principle of Lantana amara. II. Isolation of lantadene B, and the oxygen functions of lantadene A and lantadene B. Ondestepoort J. Vet. Sci. and Animal Ind. (South Africa) 23, 233–238

Lück, E., 2000: Von Abalone bis Zuckerwurz. Exotisches für Gourmets, Hobbyköche und Weltenbummler. Springer Verlag Berlin Heidelberg New York 194 pp

Lui, Y.C. , C.H. Ou, 1969: Revision of the Taiwan species of Cinnamom (Lauraceae). Quart. J. Chinese Forester 2, 1–83

Lund, E.D., R.E. Berry, C.J. Wagner, M.K. Veldhuis, 1972: Quantitative composition studies of water soluble aromatics from orange peel. J. Agric. Food Chem. 20, 685–689

Lund, E.D., W.L. Bryan, 1976: Composition of lemon oil distilled from commercial mill waste. J. Food Sci. 41, 1194–1197

Lund, K., H. Rimpler, 1985: Tormentillwurzel. Isolierung eines Ellagitannins und pharmakologisches screening. Dtsch. Apotheker-Ztg. 125, 105–108

Lunig, P.A., T. Ebbenhorst-Seller, T. de Rijk, J.P. Roozen, 1995: Effect of hot-air-drying on flavour compounds of bell-peppers (Capsicum annuum). J. Sci. Food Agric. 51, 355–365

Lurtz, U., A. Plescher, 1998: Einfluß von organischer und mineralischer Düngung auf Ertrag und Qualität von Johanniskraut. Drogenreport 10, No. 18, 28–31

Lutomski, J., B. Kedzia, W. Debska, 1974: Effect of alcoholic extract and active ingredients from Curcumalonga (L.) on bacteria and fungi. Planta med. 2, 9

Lutomski, J., 1983: Chemie und therapeutische Verwendung von Süßholz (Glycyrrhiza glabra). Pharmazie in unserer Zeit 12, 49–54

Ly, T.N., R. Yamauchi, M. Shimoyamada, K. Kato: Isolation and structural elucidation of some glycosides from the rhizomes of smaller galanga (Alpinia officinarum Hance). J. Agric. Food Chem. 50, 4919–4924

Ly, T.N., R. Shimoyamada, K. Kato, R. Yamauchi, 2003: Isolation and characterization of some antioxidative compounds from the rhizomes of smaller galanga (Alpinia officinarum Hance). J. Agric. Food Chem. 51, 4924–4929

Lykkesfeldt, J., B. Lindberg-Moller, 1993: Synthesis of benzylglucosinolate in Tropaeolum majus l.: Isothiocynates as potent enzyme-inhibitors. Plant Physiol 102, 609–613

Lyon, C.K., 1972: Sesame, current knowledge of composition and use. J. Amer. Oil. Chem. Soc. 49, 245–249

M

Maarse, H., 1974: Volatile oil of Origanum vulgare L. ssp. vulgare. III. Changes in composition during maturation. Flavour Ind. 5, 278–281

Maarse, H., F.H.L. van Ost, 1973a: Volatile oil of Origanum vulgare L. ssp. vulgare. I. Qualitative composition of the oil. Flavour Ind. 4, 477–480

Maarse, H., F.H.L. van Ost, 1973b: Volatile oil of Origanum vulgare L. ssp. vulgare. II. Oil contents and quantitative composition of the oil. Flavour Ind. 4, 481–484

Mabberley, D.J.: The plant book. A portable dictionary of the vascular plants. Cambridge Univ. Press Cambridge, UK, 2. edn. 1997, 858 pp

MacCarthy, D., 2002: Garlic at bay 'til May. Fresh Produce J., 8. (Febr.), 16–17

Ma Chuan-guo, 2002: Effect of microwave on the stability of Zanthoxylum bungeanum seed. J. Zhengzhou Inst. Technol. 23, 52–54 (in Chinese)

Machado, 149: Reference Rodriguesia (Rio de Janeiro) 12, No. 24, 79–118

MacHale, D., J.B. Sheridan, 1988: Detection and adulteration of cold-pressed lemon oil. Flavour & Fragrance J. 3, 127–133

Machatschek, M., 1999: Nahrhafte Landschaft. Ampfer, Kümmel, Wildspargel, Rapunzelgemüse, Speiselaub und andere wiederentdeckte Nutz- und Heilpflanzen. Verlag Böhlau Vienna Cologne Weimar 284 S.

Mackay, W.A., S.L. Kitto, 1987: Rapid propagation on French tarragon in vitro techniques. Acta Horticult. 28, 251–261

Mäckel, E., 1944a: Über den Mastix-Thymian, Thymus mastichina. Z. Lebensmittel-Unters. u. -Forsch. 87, 77–83

Mäckel, H.G., 1944b: Über den spanischen Thymian von Thymus zygis. Z. Lebensmittel-Unter. u. -Forsch. 87, 83–86

MacLeod, A.J., R. Islam, 1976: Volatile flavour components of coriander leaf. J. Sci. Food Agric. 27, 712–725

MacLeod, A.J., J.M. Ames, 1989: Volatile components of celery and celeriac. Phytochemistry 28, 1817–1824

MacLeod, A.J., N.M. Pieris, 1982: Analysis of the volatile essential oils of Murraya koenigii and Pandanus latifolius. Phytochemistry 21, 1653–1657

MacLeod, A.J., N.M. Pieris, 1984: Volatile aroma constituents of Sri Lanka ginger. Phytochemistry 23, 353–359

MacLeod, A.J., G. MacLeod, G. Subramanian, 1988: Volatile aroma constituents of celery. Phytochemistry 27, 373–375

MacLeod, A.J., C.H. Synder, G. Subramanian, 1985: Volatile aroma constituents of parsley leaves. Phytochemistry 24, 2623–2627

Madamba, P.S., R.H. Driscoll, K.A. Buckle, 1995: Models

for the specific head and thermal conductivity of garlic. Drying Technol. 13, 295–317

Madan, C.L., B.M. Kapur, U.S. Gupta, 1966: Saffron. Econ. Bot. 20, 377–385

Madulid, D., 1995: A pictorial cyclopedia of Philippine ornamental plants. Bookmark Inc., Makati City (Philippines), 2. edn., 389 pp

Maeda, E., H. Miyake, 1997: Leaf anatomy of patchouli (Pogostmenon patchouli) with reference to the disposition of mesophyll glands. Japan. J. Crop Sci. 66, 307–317

Maffei, M., 2002: Vetivera. The Genus Vetiveria. Taylor & Francis London and New York 2002, 191 pp

Maffei, M., A. Codignola, M. Fieschi, 1986: Essential oil from Mentha spicata L. (spearmint) cultivated in Italy. Flavour & Fragrance J. 1, 105–109

Maffei, M., 1999: Sustainable methods for a sustainable production of peppermint (Mentha x piperita L.) essential oil. J. Ess. Oil Res. 11, 267–282

Maga, J.A., 1975: Capsicum. Crit. Rev. Food Sci. Nutrit. 6, 177–199

Magda, R.R., 1993: Ginger. A pungent und biting tropical spice. Food Marketing & Technol. 7, No. 6, 12–13

Magda, R.R., 1994: Turmeric: a seasoning, dye and medicine. Food Marketing & Technol. 8, No. 5, 9–10

Maggioni, L., J. Keller, D. Aastley, 2001: European collections of vegetatively propagated Allium. Report of Workshop, 21–22 May

Magiatis, P., E. Melliou, A. Skaltsounis, I. Chinou, S. Mitaku, 1999: Chemical composition and antimicrobialactivity of the essential oils of Pistacia lentiscus var. chia. Planta med. 65, 749–752

Mahalwal, V.S., M. Ali, 2003: Volatile constituents of Cymbopogon nardus (Linn.) Rendle. Flavour & Fragrance J. 18, 73–76

Maheshwari, M.L., B.M. Singh, 1989: Introduction of European basils (Ocimum basilicum) in India. Indian Perfumer 33, 211–214

Mahfouz, M., M. E-Dakakhny, 1960: The isolation of crystaline active principle from Nigella sativa L. seeds. J. Pharmac. Sci. 1, 9–19

Mahyar, U.W., 1988: Observations on some species of the orchid genus Renanathera Loureiro. Reinwardtia 10, 399–418

Maillard, M., C.O. Adewunmi, K. Hostetman, 1991: A new triterpenoid compound of isolated from the fruit of Tetrapleura tetraptera. Planta med. 57, 174–175

Maire, H., 1933: Études sur la flora et la végétation du Sahara Central. Mém. Soc. Hist. Nat. Afr. Nord (Algier) 65, 186–187

Maischenbacher, P., K.A. Kovar, 1992: Analysis and stability of Hyperici oleum. Planta med. 58, 351-357

Maistre, J., 1964: Les plantes a épices. G.-P. Maisonneuve & Larose Paris, 289 pp

Maiwald, L., P.A. Schwantes, 1991: Curcuma xanthorrhiza Roxb. Z. Phytother. 12, 35–45

Makino, T., 1914: Perilla ocimoides Linn. Bot. Mag. 28. 180

Malan, K., Y. Pelissier, C. Marion, A. Blaise, J.M. Bessière, 1988: The essential oil of Hyptis pectinata. Planta med. 54, 531–532

Malingré, Th., 1975: Curcuma xanthorriza Roxb., temoe lawak, als plant met galdrijvende werking. Pharmac. Weekbl. 110, 601–610

Mallavarapu, G.R., Srinivasaiyer Ramesh, P.N. Kaul, A.K. Bhattacharya, B.R.R. Rao, 1993: The essential oil of Hyptis suaveolens (L.) Poit. J. Ess. Oil Res. 5, 321–323

Mallavarapu, G.R., Laxmi Rao, Srinivasaiyer Ramesh, 1999: Essential oil of Coleus aromaticus Benth. from India. J. Ess. Oil Res. 11, 742–744

Mallavarapu, G.R., S. Ramesh, K.V. Syamasundar, R.S. Chandrasekhara, 1999: Composition of Indian curry leaf oil. J. Ess. Oil Res. 11, 176–178

Manderfeld, M.M., H.W. Schäfer, P.M. Davidson, E.A. Zottola, 1997: Isolation and identication of antimicrobial furocoumarins from parsley. J. Food Protect. 60, 72–77

Máñez, S., A. Viliar, 1984: Flavonoids of the violales. Review article. Pharmazie 44, 250–255

Mani, A.K., V. Sampath, 1981: Seasonal influence on the oil content and quality of Geranium (Pelargonium graveolens). Indian Perfumer 25, 10, 41–44

Mann, L.K., W.T Stearn, 1960: Rakkyo or Ch'iao t'ou (Allium chinense G. Don, syn. A. bakeri Regel) a little known vegetable crop. Econ. Bot. 14, 69–83

Mansfeld, R., 1962: Vorläufiges Verzeichnis landwirtschaftlich oder gärtnerisch kultivierter Pflanzenarten (mit Ausschluss von Zierpflanzen). Akademie-Verlag Berlin 1962, 659 pp

Manzan, A.C.C.M., Toniolo, E. Bredow, N.C. Povh, 2003: Extraction of essential oil and pigments from Curcuma longa (L.) by steam distillation and extraction with volatile solvents. J. Agric. Food Chem. 51, 6802–6807

Maoka, T., Y. Fujiwara, K. Hashimoto, N. Akimoto,

2001: Isolation of a series of apocarotenoids from the fruits of the red paprika (Capsicum annuum L.). J. Agric. Food Chem. 49a, 1601–1606

Maoka, T., Y. Fujiwara, K. Hashimoto, N. Akimoto, 2001: Capsanthone-3,6-epoxide, a new carotenoid from the fruits of the red paprika (Capsicum annuum L.). J. Agric. Food Chem. 49, 3965–3968

Marangoni, A., I. Alli, S. Kermasha, 1988: Composition and properties of seeds of the tree legume Tamarindus indica L. J. Food Sci. 53, 1452–1455

Marco, J.L., B. Rodríguez, C. Pascual, G. Savona, 1983: Teuscorodin, teuscorodonin and 2-hydroxy-teuscorolide, neo-clerodane diterpenoids from Teucrium scorodonia. Phytochemistry 22, 727–731

Marengo, E., C. Baiocchi, M.C. Gennaro, P.L. Bertolo, 1991: Classification of essential mint oils of different geographic origin by applying pattern recognition methods to gas chromatographic data. Chemometrics & Intelligent Laboratory Systems 11, 75–88

Maria, L.T., V.Q.R. Marta, R.C. Guillermo, L.R. Abdala, 1997: Antimicrobial activity of flavonids from leaves of Tagetes minuta L. Ethnopharmacol. 56, 227–232

Marini, D., 1989: Borage (Borago officinalis): seed and oil. Prod. Chim. Aerosol Sci. 29, 50–53

Markley, K.S. (ed), 1950/51: Soybeans and soybean products. Intersci. Publishers, Inc. New York, 2 Vol., 1145 pp

Markley, R., 1997: Die BLV Rosen-Enzyklopädie. Geschichte, Botanik, Eigenschaften, Gestaltungsbeispiele, Pflanzung und Pflege, die besten Arten und Sorten, BLV Verlagsges. mbH Munich, 240 pp

Markley, R., 1999: Rosen. Arten und Sorten, Verwendung, Gestaltung, Pflege. BLV-Verlagsges. mbH Munich. 159 pp

Marotti, M., R. Piccaglia, 2002: Characterization of flavonoids in different cultivars of onion (Allium cepa L.). J. Food Sci. 67, 1229–1232

Marshall, H.H., R.W. Scora, 1972: A new chemical race of Monarda fistulosa (Labiatae). Canad. J. Bot. 50, 1845–1849

Martin, M.A., 1971: Introduction a l'ethnobotanique du Cambodge. Centre National Recherche Scientifique, Paris 257 pp

Martin, P.J., 1991: The Zanzibar clove industry. Econ. Bot. 45, 450–459

Martinetz, D., K.H. Lohs, 1987: Der Asant, ein vergessenes Heilmittel und Gewürz. Naturwiss. Rdsch. 40, 85–91

Martinetz, D., K.H. Lohs, 1988: Asa foetida – Heilmittel der asiatischen Volksmedizin. Pharmazie 43, 720–722

Martinetz, D., K.H. Lohs, 1991: Duft- und Reizhölzer. Wiss. u. Fortschritt 41, 278–280

Martini, G., I. Pinerolo, 1980: Estratti di aglio e cipolla e loro effetti battericide. Ind. Alimentari 19, 711–772

Martin-Lagos, R.A., M.F.O. Serrano, M.D. Ruiz-López, 1992: Comparative study by gas chromatography-mass spectrometry of methods for the extraction of sulfur compounds in Allium cepa L. Food Chem. 44, 305–308

Mártonfi, P., 1992: Polymorphism of essential oil in Thymus pulegioides subspec. chamaedrys in Slovakia. J. Ess. Oil Res. 4, 173–179

Marzell, H., 1970: Zur Geschichte des Basilikumkrautes (Ocimum basilicum). Regnum Vegetabile 71, 135–143

Masanetz, C., W. Grosch, 1998: Key odorants of parsley leaves (Petroselinum crispum Mill. Nym. ssp. crispum) by odour-activity values. Flavour & Fragrance J. 13, 115–124

Mascolo, N., R. Jain, S.C. Jain, F. Capasso, 1989: Ethnopharmacology investigation of ginger (Zingiber officinale). Ethnopharmacology 27, 129–140

Masoudi, S., A. Rustaiyan, N. Ameri, A. Monfared, 2002: Volatile oils of Carum copticum (L.) C.B. Clarke in Benth. et Hook. and Semenovia tragioides (Boiss.) Manden. from Iran. J. Ess. Oil Res. 14, 288–289

Masterova, I., Z. Grancaiova, S. Uhrinova, V. Suchy, K., Ubik, M. Nagy, 1991: Flavonoids in flower of Calendula officinalis L. Chem. Papers 45, 105–108

Masuda, T., A. Jitoe, 1994: Antioxidative and antiinflammatory compounds from tropical gingers: isolation, structure determination, and activities of cassumins A, B, and C, new complex curcuminoids from Zingiber cassumunar. J. Agric. Food Chem. 42, 1850–1856

Matos, M.E., F.J.A. Matos, R. Braz Filho, 1984, 1985: Constituintes químicios de Xylopia sericea St.-Hil. vel Affinis. Ciência e Cultura 36, No. 7, Suppl., p 528; 37, No. 7, Suppl., p 521

Matheis, G., G. Lösung, 1999: Zwiebelöl. Bedeutung, Zusammensetzung und Authentizitätsprüfung. dragoco-rep. 46, 165–169

Mathew, B., 1977: Crocus sativus and its allies. Plant Systematics & Evolution 128, 89–103

Mathew, B., 1996: A review of Allium section Allium. Royal Botanical Gardens Kew UK, 176 pp

Mathew, P.J., P.M. Mathew, Vijayaraghava Kumar, 2001: Graph clustering of Piper nigrum L. (black pepper). Euphytica 118, 257–264

Matthies, M., 1989: Kardamom (Elettaria cardamomum [L.] Maton und Elettaria major Smith), ein indisches Gewürz aus dem Mittelalterl. Braunschweig. Archaebotanik Dissertationes Botanic. 133, 191–200

Mathur, C., 1961: Artifical regeneration of Santalum album in Rajasthan. India Forest 87, 37–39

Matthews, R.T., 1987: Recovery and application of essential oils from oranges. Food Technol. 41, 57–61

Matos, F.J.A., M.I.L. Machado, A.A. Craveiro, J.W. Alencar, 1996: Essential oil composition of two chemotypes of Lippea alba grown in Northeast Brazil. J. Ess. Oil Res. 8, 695–698

Matsumura, T., T. Ishikawa, J. Kitajima, 2002: Water-soluble constituents of caraway: aromatic compound, aromatic compound glucoside and glucosides. Phytochemistry 61, 455–459

Mau, J.-L, E.Y.C. Lai, N.-P. Wang, C.-C Chen, C.-H. Chang, C.-C. Chyau, 2003: Composition and antioxidant activity of the essential oil from Curcuma zedoaria. Food Chem. 82, 583–591

Maubert, C., C. Guêrin, F. Mabon, G.J. Martin, 1988: Détermination de l'origine de la vanilline par analyse multidimensionnelle du fractionnement isotopique naturel spécifique de l'hydrogène. Analusis 16, 434–439

Maurer, B., A. Hauser, 1983: New sesquiterpenoids from clary sage oil (Salvia sclarea L.). Helv. chim. Acta 66, 2223–2235

Mayer, J.-G., 2003: Zur Geschichte von Baldrian und Hopfen. Z. Phytother. 24, 70–81

Mazza, G., 1980: Relative volatilities of some onion flavour components. J. Food Technol. 15, 35–41

Mazza, G., 1985: Gas chromatographic and mass spectrometric studies of the constituents of the rhizome of calamus. I. The volatile constituents of the essential oil. J. Chromatogr. 328, 179–194; II. The volatile constituents of alcoholic extracts. J. Chromatogr. 328, 195–206

Mazza, G., 1986: Etude sur la composition aromatique de l'huile essentielle de bergamotte (Citrus au rantium subspec. bergamia Risso et Poiteau Engl. J. Chromatogr. 362, 87–99

Mazza, G., 1998: Functional foods: Biochemical and processing aspects. Vol. I. CRC Press UK London, 460 pp

Mazza, G., 2002: Minor volatile constituents of essential oil and extracts of coriander (Coriandrum sativum L.) fruits. Sci. des Aliments 22, 617–627

Mazza, G., F.A. Kiehn, 1989: Essential oil of Agastache foeniculum, a potential source of methyl chavicol. J. Ess. Oil Res. 4, 295–299

Mazza, G., B.D. Oomah, 2000: Herbs, botanicals and teas. CRC Press UK London, 416 pp

Mazza, G., B.B. Chubey, F. Kiehn, 1987: Essential oil of Monarda fistolosa L. var. menthaefolia, a potential source of geraniol. Flavour & Fragrance J. 2, 129–132

Mazza, G., S. Ciaravolo, G. Chiricosta, S. Celli, 1992: Volatile flavour components from ripening and mature garlic bulbs. Flavour & Fragrance J. 7, 111–116

McCarron, M., A. Mills, D. Whittaker, Th. Kurian, J. Verghese, 1995a: Comparison between green and black pepper oils from Piper nirgrum L. berries of Indian an Sri Lankan origins. Flavour & Fragrance J. 10, 47–50

McCarron, M., A.J. Mills, D. Whittaker, T.P. Sunny, J. Verghese, 1995b: Comparison of the monoterpenes derived from green and fresh rhizomes of Curcuma longa L. from India. Flavour & Fragrance J. 10, 355–357

McCartan, S.A., N.R. Couch, 1998: In vitro culture of Mondia whitei (Periplocaceae), a threatened Zulu medicinal plant. South Afr. J. Bot. 64, 313–314

McCormick, K.D., M.A. Deyrup, E.S. Menges, S.R. Wallace, J. Meinwald, T. Eisner, 1993: Relevance of chemistry to conservation of isolated populations: the case of volatile leaf components of Dicerandra mints. Proc. Nation. Acad. Sci. USA 90, 7701–7705

McGimpsey, J.A., M.H. Douglas, J.W. van Klink, D.A. Beauregard, N.B. Perry, 1994: Seasonal variation in essential oil yield and composition from naturalized Thymus vulgaris L. in New Zealand. Flavour & Fragrance J. 9, 347–352

McKee, L.H., M.L. Harden, 1991: Nutmeg: a review. Lebensmittel-Wiss. u. Technol. 24, 198–203

McElnay, J.C., A. Li Wan Po, 1991: Garlic. Pharmac. J. 246, 324–326

Mehta, R., J.F. Zayas, Sh.-Sh. Yang, 1994: Ajowan as a source of natural lipid antioxidant. J. Agric. Food Chem. 42, 1420–1422

Meijer, T., 1940: The essential oil of massoi bark. Rec. Trav. Chim. Pays-Bas 59, 191–204

Meisenbacher, K., 1998: Die Papaya. Echte Heilpflanze oder „Modemelone"? PTA heute 12, 1164–1166, 1168–1170

Melchior, H., H. Kastner, 1974: Gewürze. Botanische und chemische Untersuchung. Bd. 2: Grundlagen und Fortschritte der Lebensmitteluntersuchung. Verlag P. Parey Berlin Hamburg, 290 pp

Melgarejo, P., D.M. Salazar, A.F. Amoros-A, 2000: Organic acid and sugar composition of harvested pomegranate fruit. Eur. Food Res. Technol. 211, 185–190

Melzer, A., Anwendung der Dünnschichtchromatographie, der Gaschromatographie und sensorische Untersuchungen zur Qualitätsbestimmung von Blättern, Aufgüssen und dem Öl der Pfefferminze, Mentha x piperita L. Drogenreport 7, No. 11, 31 (Aurorreferat zur Diplomarbeit)

Menard, U., C.-M. Lehr, 1987: Ätherischer Ölgehalt in Fenchelfrüchten. Dtsch. Apotheker-Ztg. 127, 1016–1017

Menghini, A., M. Capuccella, B. Romano, M. Foruiciari, 1994: Preliminary reports on the biorhythmus and seed production of fennel. Ann. Della Facol. Agraria, Univ. Degli Studi di Perugia (Italy) 45, 469–473

Menninger, E.A. (ed), 1977: Edible nuts of the world. Horticultural Books, Inc. Stuart, FL, 175 pp

Menon, N.A., K.P. Padmakumari, A.J. Jaylekshmy, 2002: Essential oil composition of four major cultivars of black pepper (Piper nigrum L.). J. Ess. Oil Res. 14, 84–86

Menounos, P., K. Staphylakis, D. Gegious, 1986: The sterols of Nigella sativa seed oil. Phytochemistry 25, 761–763

Menßen, H.G., K. Staesche, 1974: Hibiscusblüten, Hibiscus flos – Herkunft, Morphologie und Anatomie. Dtsch. Apotheker-Ztg. 114, 1211–1216

Menut, C., G. Lamaty, D.K. Sohounhloue, J. Dangou, J.M. Bessiere, 1995: Aromatic plants of tropical West Africa. III. Chemical composition of leaf essential oil of Lippea multiflora Moldenke from Benin. J. Ess. Oil Res. 7, 331–333

Merendi, A., 1957: Il pino domestico (Pinus pinea L.). Italia Agricola 94, 65–76

Merfort, I., D. Wendisch, 1988: Flavanolglucuronide aus den Blüten von Arnica montana. Planta med. 54, 247–250

Merfport, I., G. Willuhn, 1985: Mariendisteltee als Lebensmittel wenig sinnvoll. Dtsch. Apotheker-Ztg. 125, 695–696

Meriçili, F., A.H. Meriçili, 1986: The essential oil of Dorystaechas hastata. Planta med. 52, 506

Merrill, E.D., 1937: Coleus amboinicus. Addisonia 20, 11–12. p 646

Meshedani, T., J. Pokorny, J. Davidek, J. Panek, 1990: Effect of lipoxigenase in the lipidoxidation during the storage of poppy seed. Nahrung 34, 769–772

Mestanhauser, A., 1961: Die Inkulturnahme der Primula veris L. Wechselbeziehungen zwischen Saponin- und Stärkegehalt in der Wurzel. Pharmazie 16, 45–49

Metha, R.L., J.F. Zayas, 1995: Antioxydative effect of ajowan in a model system. J. Amer. Oil Chemists Soc. 72, 1215–1218

Meunier, Chr., 1989: Lavendel und Lavandinpflanze. dragoco rep. 34, 146–158

Meunier, Chr., 1992: Lavandes et lavandins. Édisud, Aix en Provence, France, 2. edn., 223 pp

Meyer, G., 2000: Die Ringelblume – Calendula officinalis. Kulturgeschichtliches Portrait einer Arzneipflanze. Z. Phytother. 21, 170–178

Meyer-Berge, A. 1991: Einfluß verschiedener Kulturmaßnahmen und Standortfaktoren auf die Blütenbildung und den Blütenertrag von Arnica montana L. Habilschr. Univ. Hohenheim, landwirtsch. Fakult., 188 pp

Meyer-Chlond, G., 1999: Arnika-Arzneipflanze mit Tradition und Zukunft. Dtsch. Apotheker-Ztg. 139, 3229–3232

Mheen, van der, H.J., H. Bosch, 1994: Teelt van dillekruid en dillezaad. Profestation voor de Akkerbouw en Groenteteelt in de Vollegrond, Informtieren Kenniscentrum, Lelystad, Netherlands, 37 pp

Michael, E., B. Henning, H. Leisel, 1985: Handbuch für Pilzfreunde. VEB G. Fischer-Verlag Jena, 6. Edn., Vol. 1, pp 109–110; Vol. 2, pp 376–377

Michaelis, A.A., 1916: Nux moschata. Die Muskatnuss als Gewürz und Heilmittel. Verlag A.O. Müller Heilbronn, 64 pp

Michaelis, K., O. Vostrowsky, H. Paulini, R. Zinte, K. Knobloch, 1982: Das ätherische Öl aus den Blüten von Artemisia vulgaris. Z. Naturforsch. 37 C, 3–4, 152–158

Miething, H., A. Speicher-Brinkler, 1989: Neolicurosid, ein neues Chalconglycosid aus den Wurzeln von Glycyrrhiza glabra. Arch. Pharmaz. 332, 141–143

Mikus, B., J. Schaser, 1995: Heil- und Gewürzpflanzen mit zitronenartigem Aroma. Drogenreport 8, No. 13, 32–38

Mikus, B., H. Krüger, B. Zeiger, M. Ebert, 1997: Chemische und aromatische Extreme in Gattungen etablierter und neuartiger Arznei-, Duft- und Gewürzpflanzen. Drogenreport 10, No. 18, p 68–73

Milbread, J., 1913: Über die Gattungen Afrostyrax Perk. et Gilg und Hura Pierre und die Knoblauchrinden Westafrikas. Bot. Jb. 49, 552–559

Miles, D.H., V. Chittawong, P.H. Hedin, U. Kokpol, 1993: Potential agrochemicals from leaves of Wedelia biflora. Phytochemistry 32, 1427–1429

Miller, F.M., L.M. Chow, 1975: Alkaloids of Achillea millefolium L. I. Isolation and characterization of achilleine. J. Amer. Chem. Soc. 76, 1353–1354

Miloš, M., J. Mastelic, J. Jerkovic, 2000: Chemical composition and antioxidant effect of glycosidically bound volatile compounds from oregano (Origanum vulgare L. ssp. hirtum). Food Chem. 71, 79–83

Milos, M., A. Radonic, N. Bezie, V. Dunkie, 2001: Localities and seasonal variations in the chemical composition of essential oils of Satureja montana L., and S. cuneifolia Ten. Flavour & Fragrance J. 16, 157–160

Milton, G., 1999: Muskatnuß und Musketen. P. Zsolnay Verlag Vienna, 448 pp

Minh Tu, N.T., L.X. Thanh, A. Une, H. Ukeda, M. Sawamura, 2002: Volatile constituents of Vietnamese pumelo, orange, tangerine and lime peel oils. Flavour & Fragrance J. 17, 169–174

Minija, J., E. Thoppil, 2001: Volatile oil constitution and microbicidal activities of essential oil of Coriandrum sativum L. J. Nat. Remedies 1, 147–150

Miraza, M.E., 1974: Indonesian cassia vera, production and quality improvement mandatory to revive marketing. Econ. Rev. (Indonesia) 60, 24–27

Mirazawa, M., T. Maehara, K. Kurose, 2002: Volatile components of leaves of rocket (Eruca sativa Mill.). Flavour & Fragrance J. 17, 187–190

Mirjana S., B. Nada, D. Valerija, 2004: Variability of Satureja cuneifolia Ten. Essential oils and their antimicrobial activity depending on the stage of development. Eur. Food Res. Technol. 218 A, 367–371

Mirza, M., Z.F. Baher, 2003: Essential oil of Stachys lanata Jacq from Iran. J. Ess. Oil Res. 15, 46–47

Mirza, M., Z.B. Nik, 2003: Volatile constituents of Phlomis olivieri Benth from Iran. Flavour & Fragrance J. 18, 131–132

Misharina, T.A., 2001: Influence of the duration and condition of storage on the composition of the essential oil from coriander seeds. Appl. Biochem. & Microbiol. 37, 622–628

Miski, M., A. Ulubelen, T.J. Mabry, 1983: 6-hydroxyflavones from Thymbra capitata. Phytochemistry 22, 2093–2094

Misra, L., A. Husain, 1987: The essential oil of Perilla ocimoides: a rich source of rosefuran. Planta med. 53, 379–380

Misra, L. N., B.R. Tyagi, R.S. Thakur, 1989: Chemotypic variation in Indian spearmint. Planta med. 55, 575–576

Misra, T.N., R.S. Singh, T.N. Ojha, J.Upadhyay, 1981: Chemical constituents of Hyptis suaveolens. I. Spectral and biological studies on a triterpene acid. J. Nat. Prod. 44, 735–738

Misra, T.N., R.S. Singh, H.S. Pandey, Satyendra Singh, 1992: Long-chain compounds from Laucas aspera. Phytochemistry 31, 1809–1810

Misra, T.N., R.S. Singh, Chandan Prasad, Satyendra Singh, 1993: Two aliphatic ketols from Leucas aspera. Phytochemistry 32, 199–201

Mitra, S., S. Kundu, S.K. Mitra, 2000: Studies on physical characteritics and chemical composition of some mano varieties grown in West Bengal. J. Interacademicia 4, 498–501

Mitsui, S., S. Kobayashi, S. Nagahori, A. Ogiso, 1976: Constituents from seeds of Alpinia galanga Willd. and their anti-ulcer activities. Chem. Pharm. Bull., Tokyo 24, 2377–2382

Miura, K., H. Kikuzaki, N. Nakatani, 2001: Apianane terpenoids from Salvia officinalis. Phytochemistry 58, 1171–1175

Miyake, Y., K. Yamamoto, T. Osawa, 1997: Isolation of eriocitrin (eriodictyol 7-rutinoside) from lemon fruit (Citrus limon Burm.f.) and its antioxidative activity. Food Sci. Technol. Int. Tokyo 3, 84–89

Miyake, Y., A. Murakami, Y. Sugiyama, M. Isobe, K. Koshimizu, H. Ohigashi, 1999: Identification of coumarins from lemon fruit (Citrus limon) as inhibitors of in vitro promotion and superoxide and nitric oxide generation. J. Agric. Food & Chem. 47, 3151–3157

Miyazawa, M., H. Kameoka, 1988: Volatile flavour components of Zingiberis rhizome (Zingiber officinale Roscoe). Agric. Biol. Chem. 52, 2961–2963

Miyazawa, M., Y. Minamino, H. Kameoka, 1996: Volatile compounds of rhizomes of Rheum palmatum L. Flavour & Fragrance J. 11, 57–60

Mizobuchi, S., Y. Sato, 1984: A new flavanone with antifugal activity isolated from hops. Agric. Biol. Chem. 49, 399–405

Mizobuchi, S., Y. Sato, 1984: Antifugal activities of hop bitter resins. Agric. Biol. Chem. 48, 2771–2775

Mockute, D., G. Bernotiene, A. Judzentiene, 2001: The essential oil of Origanum vulgare L. ssp. vulgare growing wild in Vilnius district (Lithuania). Phytochemistry 57, 65–69

Mohan Rao, L., J., H. Yada, H. Ono, M. Yoshida, 2002: Acylated and non acylated flavonol monoglycosides from the Indian minor spice nagkesar (Mammea longifolia). J. Agric. Food Chem. 50, 3143–3146

Molegode, W., 1938: Cardamoms. Trop. Agriculturist (Ceylon) 91, 325–328

Molino, J.-F., 1993: History and botany of Clausena anisum-olens (Blanco) Merr. cv. 'Clausanis' (Rutaceae), a promising essential oil crop plant. Acta Horticult. 331, 183–190

Molino, J.-F., 1995: Proposal to reject Illicium sanki Perr., a threat to Clausena anisum-olens (Blanco) Merr. (Rutaceae). Taxon 44, 427–428

Molino, R, 2000: The inheritance of leaf oil composition in Clausenia anisum-olens (Blanco) Merr. J. Ess. Oil Res. 12, 135–139

Molnár, P., J. Deli, Z. Matus, G. Tóth, A. Steck, H. Pfander, 2000: Isolation and characterization of mutatoxanthin-epimers from red paprika (Capsicum annuum). Eur. Food Res. Technol. 211 A, 396–399

Møller, J.K.S., H.L. Madsen, T. Aaltonen, L.H. Skibsted, 1999: Dittany (Origanum dictamnus) as a source of water-extractable antioxidants. Food Chem. 64, 215–218

Molter, T., 1975: Ungarischer Gewürzpaprika. Seine Geschichte von Mittelamerika bis Europa. Jb. der Wittheit zu Bremen 19, 199–205

Mondello, L., A. Casilli, P. Qu. Tranchida, L. Cicera, P. Dugo, G. Dugo, 2003: Comparison of fast and conventional GC analysis for citrus essential oils. J. Agric. Food Chem. 51, 5602–5606

Monfared, A., R. Nabid, A. Rustaiyan, 2002: Composition of a carvone chemotype of Mentha longifolia (L.) Huds. from Iran. J. Ess. Oil Res. 14, 51–52

Moody, J.O., S.A. Adeleye, M.G. Gundidza, G. Wyllie, O.O. Ajayi-Obe, 1995: Analysis of the essential oil of Cymbopogon nardus (L.) Rendle growing in Zimbabwe. Pharmazie 50, 74–75

Moon Jung Lee, Gung Pyo Lee, Kuen Woo Park, 2001: Effect of selenum on growth and quality in hydroponically-grown Korean mint (Agastache rugusa). J. Korean Soc. Horticult. Sci. 42, 483–486

Moore, H.E. jr., 1962: The correct name for the torch ginger. Bailleya 10, 121

Moraes, M.P.L., N.F. Roque, 1988: Diterpenes from Xylopia aromatica (Lam.) Mart. Phytochemistry 27, 3205–3208

Morais, H., A.C. Ramos, T. Cserháti, E. Forgágs, 2001: Effects of fluorescent light and vacuum packing on the rate of decomposition of pigments in paprika (Capsicum annuum) powder determined by reversed-phase high-performance liquid chromatography. J. Chromatogr. 936 A, 139–144

Mori, H., K. Kubota, A. Koblayashi, 1995: Potent aroma components of rhizomes from Alpinia galanga (Willd.) L. J. Japan. Soc. Food Sci. Technol. 42, 989–995

Morita, K., S. Kobashi, 1966: Isolation and synthesis of lenthionine, an odorous substance of shiitake, an edible mushroom. Tetrahederon Letters, 573–577

Morris, S., L. Mackley, 1999: The world encyclopedia of spices. The definitive cook's guide to spices and other aromatic ingredients. Lorenz Books London, 256 pp

Mors, W.B., C.T. Rizzini, 1961: Die industriell verwertbaren Pflanzen Brasiliens. Deutsch-Brasilian. Handelskammer, Rio de Janeiro, 143 pp

Mors, W.B., C.T. Rizzini, N.A. Pereira, 2000: Medicinal plants of Brazil. Reference Publications Ind., Algonac, Mich., 501 pp

Morteza-Semnani, K., M. Saeedi, 2003: Constituents of the essential oil of Commiphora myrrha (Nees) Engl. var. molmol. J. Ess. Oil Res.15, 50–51

Morton, J.F., 1960: The emblic (Phyllanthus emblica L.). Econ. Bot. 14, 119–128

Morton, J.F., 1976: Herbs and Spices. Golden Press New York (NY), 160 pp

Morton, J.F., 1991: The horseradish tree – Moringa pterygosperma (Moringaceae). A boom to arid lands? Econ Bot., 318–333

Morton, J.F., 1992: Country borage (Coleus amboinicus Lour.): a potent flavoring and medicinal plant. J. Herbs, Spices & Med. Plants 1, 77–90

Mosandl, A., 1995: Enantioselective capillary gas chromatography and stable isotope ratio mass spectrometry in the authenticity control of flavors an essential oils. Food Rev. Intern. 11, 597–664

Mosandl, A., D. Juchelka, 1997: Advances in the authenticity assessment of citrus oils. J. Ess. Oil Res. 9, 5–12

Mossberg, B., L. Sternberg, S. Ericson, 1992: Den nordiska Flora: Wahlström & Widstrand Publ. Stockholm, 696 pp

Moyler, D.A., M.A. Stephens, 1992: Counter current deterpenation of cold pressed sweet orange peel oil. Perfumer & Flavourist 17, No. 2, 37–38

Msonthi, J.D., 1991: A novel phenolic glycoside from Mondia whytei Skeels. Bull. Chem. Soc. Ethiopia 5, 107–110

Mucciarelli, M., M. Maffei, T. Sacco, 1993: Oli essenziali e tricomi ghiandolari di Perovskia atriplicifolia Benth. e Perovskia scrophulariaefolia Bunge. Riv. Ital. EPPOS/Ist. Tetrahedron 9, 3–7

Mucciarelli, M., T. Sacco, 1999: Glandular trichomes and essential oils of Mentha requieni Benth. J. Ess. Oil Res. 11, 759–764

Mukherji, D.K., 1973: Large cardamom. World Crops 25, 31–33

Mukherji, S., 1949: A monography on the genus Mangifera L. Lloydia 12, 73–136

Mule, A., G. Pirisino, M.D. Moretti, M. Satta, 1994: Caratteristiche dell' olio essenziale di Rosmarinus officinalis delle stazione di Cala Gonone (Sardegna). Presented at the Atti del Convegno internaz. "Coltivazione e miglioramento di piante officinali", Trento, Italy, June 2–3, pp 599–608

Mulkens, A., I. Kapetanidis, 1988: Eugenylglucoside, a new natural phenylpropanoid heteroside from Melissa officinalis. J. Natural Products 51, 496–498

Munné-Bosch, S., L. Alegre, K. Schwarz, 2000: The formation of phenolic diterpenes in Rosmarinus officinalis L. under Mediterranean climate. Eur. Food Res. Technol. 210, 263–267

Munshi, A.M., J.S. Sindhu, G.H. Baba, 1989: Improved cultivation practices for saffron. Indian Farming 39, No. 3, 27–30

Murakami, A., A. Nakamura, K. Koshimizu, H. Ohigashi, 1995: Glyceroglycolipides from Citrus hystrix, a traditional herb in Thailand, potently inhibit the tumor promoting activity of 12-0-tetradecanoylphorbol-13-acetate in mouse skin. J. Agric. Food Chem. 43, 2779–2783

Murakami, A., K. Toyota, S. Ohura, K. Koshimizu, H. Ohigashi, 2000: Structure-activity relationships of (1´S)-1´-acetoxychavicol acetate, a major constituent of a Southeast Asian condiment plant Languas galanga, on the inhibition of tumor-promotor-induced Epstein-Barr virus activation. J. Agric. Food Chem. 48, 1518–1523

N

Nagaraja, K.V., M.N. Manjunath, M.L. Nalini, 1975: Chemical composition of commercial tamarind juice concentrate. Ind. Food Packer 29, 17–20

Nagarajan, S., Jagan Mohan Rao, K.N. Gurudutt, 2001: Chemical composition of the volatiles Hemidesmusindicus R. Br. Flavour & Fragrance J. 16, 212–214

Nagasawa, T., K. Umemoto, T. Tsuneya, M. Shiga, 1975: (–)-1R-8-Hydroxy-4-p-methen-3-one isolated from essential oil of Mentha gentilis L. Agric. & Biol. Chem. 39, 553–554

Nagasawa, T., K. Umemoto, T. Tsuneya, M. Shiga, 1975: Studies on the wild mints of the Tokai district. VI. Absolute configuration of (+)-1-acetoxy-p-menth-3-one isolated from essential oil of Mentha gentilis L. Agric. & Biol. Chem. 39, 2083–2084

Nagel, M., E. Reinhard, 1975: Das ätherische Öl der Calluskulturen von Ruta graveolens. 2. Mitt. Physiologie zur Bildung des ätherischen Öles. Planta med. 27, 264–271

Nagell, A., F.W. Hefendehl, 1972: Strukturbestimmung von Diosphenolen. Phytochemistry 11, 3356–3361

Nair, M.K., T. Premkumar, P.N. Ravindran, Y.R. Sarma (ed), 1982: Proc. of the national seminar on ginger and turmeric, Calcutta 1980. Central Plantation Crops Res. Institute, Kasaragod (India), 253 pp

Nakahara, S., K. Kumatani, H. Kameoka, 1975: Acidic compounds in patchouli oil. Phytochemistry 14, 2712–2713

Nakamura, S., O. Nishimura, H. Masuda, 1989: Identification of volatile aroma components from the oil of roasted sesame seeds. Agric. Biol. Chem. (Tokyo) 53, 1891–1897

Nakanu, T., M. Suuarez, 1969: Neutral constituents

of the bark of Dipteryx odorata. Planta med. 18, 79–83

Nakatani, N., R. Inatani, 1981: Structure of rosmanol, a new antioxidant from rosemary. Agric. Biol. Chem. 45, 2385–2386

Nakatani, N., R. Inatani, 1984: Two antioxidative diterpenes from rosemary and revised structure from rosmanol. Agric. Biol. Chem. 2081–2085

Nakayama, R., Y. Tamura, H. Yamanaka, H. Kikuzaki, N. Nakatani, 1994: Two curcuminoid pigments from Curcuma domestica Val. Phytochemistry 33, 501–502

Nano, G.M., B. Bicchi, C. Frattini, M. Gallino, 1976: On the composition of some oils from Artemisia vulgaris. Planta med. 30, 211–215

Naqvi, A.A., S. Manal, A. Chattopadhyay, A. Prasad, 2002: Salt effect on the quality and recovery of essential oil of citronella (Cymbopogon winterianus Jowitt). Flavour & Fragrance J. 17, 109–110

Narendra Singh, R.K. Sharma,1980: Studies on the morphology and floral biology of tulsi (Ocimum sanctum L.). Plant Sci. 12, 26–27

Naranjo, P., A. Kjjoa, A.M. Giesbrecht, O.R. Gottlieb, 1981: Ocotea quixos, American cinnamom. J. Ethnopharmacol. 4, 233–236

Nassr, M.I., 1994: Spectral study of the farnesiferol B from Ferula asa foetida L. Pharmazie 49, 542–543

Natarajan, C.P., R.P. Bai, M.N. Krishnamurthy, B. Shankaracharya, S. Kuppuswamy, S. Govindarajan, Y.S. Lewis, 1972: Chemical composition of ginger varities and dehydratation studies on ginger. J. Food Sci. Technol. (Mysore) 9, 120–124

Nath, S.C., M.G. Pathak, A. Baruah, 1996a: Benzoylbenzoate, the major component of the leaf and stem bark oil of Cinnamomum ceylanicum Blume. J. Ess. Oil Res. 8, 327–328

Nath, S.C., A.K. Hazarika, A. Baruah, K.K. Sharma, 1996b: Essential oils of Litsea cubeba Pers. – An additional chemotype of potential industrial value from northern India. J. Ess. Oil Res. 8, 575–576

Nawale, R.N., Y.R. Parulekar, M.B. Magdum, 1997: Kokam (Garcinia indica Choisy). Cultivation in Kokam region of Maharashtra. Indian Cocoa, Arecanut & Spices J. 21, No. 2, 42–43

Nawwar, M.A.M., A.M.D. El-Mousallamy, H.H. Barakat, J. Buddrus, M. Linscheid, 1989: Flavonoid lactates from leaves of Marrubium vulgare. Phytochemistry 28, 3201–3206

Nayar, N.M., K.L. Mehra, 1970: Sesame: its use, botany, cytogenetics, and origin. Econ. Bot. 24, 20–31

Nee, T.Y., S. Cartt, M.R. Pallard, 1986: Seed coat components of Hibiscus abelmoschus. Phytochemistry 25, 2157–2161

Nefisa, A.H., H.M. Fadel. A.A.S. Ramadan, 1994: Effect of dehydration on some chemical constituents on onion. Egypt. J. Food Sci. 22, 369–372

Negbi, M., 1999: Saffron (Crocus sativus L.). Harward Academic Publishing Amsterdam. 154 pp

Negbi, M., B. Dagan, A. Doror, D. Basker, 1989: Growth, flowering, vegetative production and dormancy in the saffran crocus. Israel J. Bot. 38, No. 2–3, 95–113

Negei, K.S., K.C. Plant 1992: Less-known wild species of Allium L. (Amarillidaceae) from mountains region of India. Econ. Bot. 46, 112–114

Neher, R.T., 1968: The ethnobotany of Tagetes. Econ. Bot. 22, 317–325

Neidhardt, G., 1947: Euphrasia rostkoviana Hayne – der Augentrost. Pharmazie 2, Beiheft No. 1, 52 pp

Németh, É., 1998: Caraway. The genus carum. Harwood Academic Publ., Amsterdam, 200 pp

Németh, É., É. Héthelyi, J. Bernath, 1994: Comparison studies on Tanacetum vulgare L. chemotypes. J. Herbs, Spices & Med. Plants 2, No. 2, 85–92

Nergiz, C., S. Öthes, 1993: Chemical composition of Nigella sativa seeds. Food Chem. 48, 259–261

Nestler, Th., 1979: Neue Inhaltsstoffe von Lignum Quassiae und eine neue Gehaltsbestimmung der Quassia-Bitterstoffe. Diss. Univ. Munich, Fachbereich Chem. u. Pharmazie. 124 pp

Neumerkel, W., 2001: Wermut & Co. im Naturgarten Artern. Drogenreport 14, No. 25, 44–49

Newman, M., A. Lluillier, A.D. Poulsen, 2004: Checklist of the Zingiberaceae of Malesia. Blumea (Leiden), Suppl. 16, 165 pp

Neuwinger, H.D., 1994: Arzneipflanzen Schwarzafrikas. Dtsch. Apotheker-Ztg. 134, 453–464

Neuwinger, H.D., 1998: Afrikanische Arzneipflanzen und Jagdgifte. Chemie, Pharmakologie, Toxikologie. Wissenschaftl. Verlagsges. mbH Stuttgart, 2. edn., 690 pp

Neuwinger, H.D., 1999: private Mitteilungen

Neve, R.A., 1991: Hops. Chapman & Hall London New York, 266 pp

Newall, C.A., L.A. Anderson, J.D. Phillipson, 1996: Herbal Medicines. A guide for health-care professionals. The Pharmaceutical Press, London, 296 pp

Ng, T.T., 1972: Growth performance and production potential of some aromatic grasses in Sarawak – a preliminary assessment. Trop. Sci. 14, 47–58

Ngassoum, M.B., L. Jirovetz, G. Buchbauer, 2001: SMPE/GC/MS analysis of headspace aroma compounds of the Cameroonian fruit Tetraplaura tetraptera (Thonn.) Taub. Eur. Food Res. Technol. 213, 18–21

Ngouda, S., B. Nyasse, E. Tsamo, M.-Ch. Brochier, Chr. Morin, 1998: A trachylobane diterpenoid from Xylopia aethiopica. J. Natur. Prod. 61, 264–266

Nguyen, K., P. Barton, J.S. Spencer, 1991: Supercritical dioxide extraction of vanilla. J. Supercritic. Fluids 4, 40–46

Nguyen Xuan Dung, Nguyen Thi Bich Tuyet, P.A. Leclerq, 1995: Volatile constituents of the rhizome, stem and leaf oil of Curcuma pierreana Gagnep. from Vietnam. J. Ess. Oil Res. 7, 262–264

NICPBP (National Institute for the Control of Pharmaceutical and Biological Products) (ed), 1987: Colour atlas of Chinese traditional drugs. Acience Press Beijng (Peking), 300 pp

Nigam, J.C., G.N. Gupta, D.R. Dhingra, 1960: Essential oil from laurel berries. Perfum. Ess. Oil Rec. (London) 51, 351–354

Nigam, M.C., P.R. Rao, 1969: Gas chromatography of the essential oil of Perovskia abrotanoids. Parfümerie u. Kosmetik 50, 221–222

Nigg, H.N., J.O. Strandberg, R.C. Beier, H.D. Petersen, J.M. Harrison, 1997: Furanocoumarins in Florida celery varieties increased by fungucide treatment. J. Agric. Food Chem. 45, 1430–1436

Nin, S., P. Arfaiolim M. Bosetto, 1995: Quantitative determination of some essential oil components of selected Artemisia absinthum plants. J. Ess. Oil Res. 7, 271–275

Nishamura, O., 1995: Identification of the characteristic odorants in fresh rhizomes of ginger (Zingiber officinale Roscoe) using aroma extrakt dilution analysis and modified multidimensional gas chromatography-mass spectroscopy. J. Agric. Food Chem. 43, 2941–2945

Nissen, G., 1992: Alte Rosen. Verlag Boyens & Co. Heide (Germany), 7. edn., 124 pp

Nitta, Miyuki, Ju Kyong, Ohmi Ohnish, 2003: Asian Perilla Crops and their weedy forms: The cultivation, utilization and their genetic relationships. Econ. Bot. 57, 245–253

Nitz, S.M., M.H. Spraul, F. Dawert, 1990: Polyacetylenic alcohols as the major constituents in roots of Petroselinum crispum Mill. ssp. tuberosum. J. Agric. Food Chem. 38, 1445–1447

Njar, V.C.O., T.O. Alao, J.I. Okugon, H. Holland, 1993: Methoxycanthin-6-one: A new alkaloid from the stemwood of Quassia amara. Planta med. 59, 259–262

Noak, B., 1936: Zur Geschichte des Thymians. Großbetrieb für Dissertationsdruck R. Noske, Borna-Leipzig, 39 pp

Noda, Y., T. Kaneyuki, A. Mori, L. Packer, 2002: Antioxidant activities of pomegranate fruit extract and its anthocyanidins: delphinidin, cyanidin, and pelargonidin. J. Agric. Food. Chem. 50, 166–171

Noleau, I., B. Toulemonde, H. Richard, 1987: Volatile constituents of cardamom cultivated in Costa Rica. Flavour & Fragrance J. 2, 123–127

Nordborg, G., 1967: The genus Sanguisorba, section Poterium. Experimental studies and taxonomy. Opera Bot. 16, 166 pp

Norman, J., 1990: Spices. Dorling Kindersley Ltd. London, UK, 160 pp

Norman, J., 1991, 2004 (8. Edn.): Das grosse Buch der Gewürze. AT Verlag Aarau Stuttgart, 160 pp

Noro, T., T. Sekiya, M. Katoh, Y. Oda, T. Miyase, K. Kurayanagi, A. Ueno, S. Fukushima, 1988: Inhibitors of xanthine oxidase from Alpinia galanga. Chem. Pharm. Bull. Tokyo 36, 244–248

Nosti, V.M., R.R., Castro, 1987: Los constituyentes de las alcaparra y su variacion con el aderozo. Grasas Aceites (Seville) 38, 173–175

Novak, I., G. Buzas, E. Minker, M. Koltai, K. Szendrei, 1965: Alkaloide aus Ruta graveolens L. Pharmazie 20, 654–655

Novak, J., J. Langbehn, F. Pank, C.M. Franz, 2002: Essential oil compounds in a historical sample of majoram (Origanum majorana L., Lamiaceae). Flavour & Fragrance J. 17, 175–180

Nowak, B., B. Schulz, 1998: Tropische Früchte. Biologie, Verwendung, Anbau und Ernte. BLV Verlagsges. mbH Munich-Vienna-Zürich, 239 pp

Ntezurubanza, L., J.J. Scheffer, A. Looman, 1985: Composition of the essential oil of Ocimum canum grown in Rwanda. Pharmac. Weekbl. 7, 273–276

Ntezurubanza, L., J.J.C. Scheffer, A. Baerheim-Svendsen, 1987: Composition of the essential oil of Ocimum gratissimum grown in Rwanda. Planta Med. 53, 421–423

Nykänen, F., 1989: The effect of cultivation conditions

on the composition of basil oil. Flavour & Fragrance J. 4, 125–128

Nykänen, F., L. Nykänen, M. Alkio, 1991: Composition of angelica root oil obtained by supercritical CO_2 and steam distillation. J. Ess. Oil Res. 3, 229–236

Nykänen, I., 1986: High resolution gas chromatographic mass spectrometric determination of the flavour composition of majoram cultivated in Finland. Z. Lebensmittel-Unters. u. -Forsch. 183, 172–176

Nykänen, I., T. Holm, R. Hiltunen, 1989: Composition of the essential oil of Agastache foeniculum. Planta med. 55, 314–315

O

Oberdieck, R., 1981: Ein Beitrag zur Kenntnis und Analytik von Majoran (Majorana hortensis Moench.). Dtsch. Lebensmittel-Rdsch. 77, 63–74

Oberdieck, R., 1988: Paprika. Fleischwirtschaft. 68, 1086–1096

Oberdieck, R., 1989a: Piment. Fleischwirtschaft 69, 320–322, 327–330

Oberdieck, R., 1989b: Macis und Muskat. Fleischwirtschaft 69, 1648, 1653–1654, 1656–1658, 1660–1664

Oberdieck, R., 1990: Majoran. Fleischwirtschaft. 70, 391–396, 398

Oberdieck, R., 1991: Ein Beitrag zur Kenntnis und Analytik von Safran (Crocus sativus L.). Dtsch. Lebensmittel-Rdsch. 87, 246–252

Oberdieck, R., 1992a: Pfeffer. Fleischwirtschaft 72, 695–708

Oberdieck, R., 1992b: Cardamom. Fleischwirtschaft 72, 1657–1663

Oberdieck, R., 1998: Ein Beitrag zur Kenntnis und Analytik von Vanille. Dtsch. Lebensmittel-Rdsch. 94, 53–59

Ochi, H., N. Ramarathnam, M. Takeuchi, H. Sugiyama, 1995: Antioxidative activities of wasabi (Eutrema wasabi Maxim.) leaf, root and stem extracts. J. Japan. Soc. Nutrit. & Food Sci. 48, 236–238

Ochoa, F.L., C.M. Alonso, 1996: Plantas medicinales de México. Composición, usos y actividad biológica. Universidad Nacional Autónoma de México, 137 pp

Ochse, J.J., R.C. Bakhuizen van den Brink, 1931: Vegetables of the Dutch East Indias (edible tubers, bulbs, rhizomes and spices included). Dept. Landbouw, Nijverheid en Handel, Buitenzorg, 1005 pp

Ochse, J.J., M.J. Soule, M.J. Dijkman, C. Wehlburg, 1961: Tropical and subtropical Agriculture. MacMillan Corp. New York, Vol. I and II, 1446 pp

Odoux, E., 2000: Changes in vanillin and glucovanillin concentrations during the various stages of the process traditionally used for curing Vanilla fragrans beans in Réunion. Fruits 55, 119–125

Ogle, B.M., Ho Thi Tuyet, Hoang Nghia Duyet, Nguyen Nhut Xuan Dung, 2003: Food, feed or medicine: The multiple functions of edible wild plants in Vietnam. Econ. Bot. 57, 103–117

Ognyanov, I. 1983/84: Bulgarian lavender and Bulgarian lavender oil. Perfumer & Flavorist 8, No. 6, 29–41

Ohler, J.G.,1968: Bixa orellana L. Trop. Abstr. 24, 407–413

Oka, N., A. Ikegami, M. Ohki, K. Sakata, A. Yagi, N. Waranabe, 1998: Citronellyl disaccharide glycoside as anaroma precurser from rose flowers. Phytochemistry 47, 1527–1530

Okeke, E.C., 1998: The use and chemical content of some indigenous Nigerian spices. J. Herbs, Spices & Med. Plants 5, 51–63

Oliveira, A.B., M.I.L.M. Madruga, O.R. Gottlieb, 1978: Isoflavonids from Myroxylon balsamum. Phytochemistry 17, 593–595

Oleiveira, J.T.A., S.B. Silveira, I.M. Vasconcelos, B.S. Cavada, R.A. Moreira, 1999: Compositional and nutrional attributes of seeds from the multiple purpose tree Moringa oleifera Lamarck. J. Sci. Food Agric. 79, 815–820

Oliver, J.E., 2003: (S)(+)-Linalool from oil of coriander. J. Ess. Oil Res. 15, 31–33

Ollanketo, M., A. Peltoketo, K. Hartonen, R. Hiltunen, M.-L. Riekkola, 2002: Extraction of sage (Salvia officinalis L.) by pressurized hot water and conventional methods: antioxidant activity of the extracts. Eur. Food Res. Technol. 215 A, 158–163

Omer, E.A., H.E. Ouda, S.S. Ahmed, 1994: Cultivation of sweet majoram, Majorana hortensis, in newly reclaimed lands of Egypt. J. Herbs, Spices & Med. Plants 2, No. 2, 9-16

Omidbaigi, R., F. Sefidkon, 2003: Essential oil composition of Agastache foeniculum cultivated in Iran. J. Ess. Oil Res. 15, 52–53

Omidbaigi, R., G. Betti, B. Sadeghi, A. Ramezani, 2002: Einfluss des Zwiebelgewichts auf die Produktivität von Safran (Crocus sativus L.). Ergebnisse einer Anbaustudie in Khorasan (Iran). Z. Arznei- u. Gewürzpflanzen 7, 38–40

Omubuwajo, T.O., O.R. Omobuwajo, L.A. Sanni, 2003: Physical properties of calabash nutmeg (Monodoramyristica) seeds. J. Food Engng. 57, 375–381

Onayade, O.A., A. Looman, J.J.C. Scheffer, A.B. Svendsen, 1990: Composition of the herb essential oil of Hyptis spicigera Lam. Flavour & Fragrance J. 5, 101–105

Ondarza, M., A. Sachez, 1990: Steam distillation and supercritical fluid extraction of some Mexican spices. Chromatographia 30, 16–18

Onyenekwe, P.C., S. Hashimoto, 1999: The composition of the essential oil of dried Nigerian ginger (Zingiberofficinale Roscoe). Eur. Food Res. Technol. 209 A, 407–410

Onyenekwe, P.C., G.H. Ogbadu, H. Deslauriers, M. Gagnon, G.J. Collin, 1993: Volatile constituents of the essential oil of Monodora myristica (Gaertn.) Dunal. J. Sci. Food Agric. 61, 379–381

Ooi, L.S.M., Hong Yu, Chun-Mei Chen, S.S.M. Sun, V.E.C. Ooi, 2002: Isolation and characterization of abioactive mannose-binding protein from the Chinese chive Allium tuberosum. J. Agric. Food Chem. 50, 696–700

Oomen, H.A.P.C., G.J.H. Grubben, 1978: Tropical leaf vegetables in human nutrition. Koninklijk Instituut voor de Tropen Amsterdam, 140 pp

Opdyke, D.L.J., 1974a: Celery seed oil. Food, Cosmetics, Toxicol. 12, 849–859

Opdyke, D.L.J., 1974b: Chamomile oil roman. Food, Cosmetics, Toxicol. 12, 853

Opdyke, D.L.J., 1974c: Ginger oil in monographs on fragrance raw materials. Food, Cosmetics, Toxicol. 12, 901–902

Opdyke, D.L.J., 1974d: Pennyroyal oil European. Food, Cosmetics, Toxicol. 12, 949–950

Opdyke, D.L.J., 1974e: Rosemary oil. Food, Cosmetics, Toxicol. 12, 977–978

Opdyke, D.L.J., 1974f: Fragrance raw material monographs, Perubalsam, Perubalsam oil. Food, Cosmetics, Toxicol. 12, 951–954

Opdyke, D.L.J., 1974e: Sage of Dalmation. Food, Cosmetics, Toxicol. 12, 987–988

Opdyke, D.L.J., 1975a: Angelica root oil. Food, Cosmetics, Toxicol. 13 (Suppl.) 713

Opdyke, D.L.J., 1975b: Parsley seed oil. Food, Cosmetics, Toxicol. 13 (Suppl.) 897–898

Opdyke, D.L.J., 1977: Calamus oil. Food, Cosmetics, Toxicol. 15, 623–626

Opdyke, D.L.J., 1982: Sassafras oil. Food, Cosmetics, Toxicol. 20, 825–826

Orth, A., 2004: Die Dschungelapotheke der Gottkönige. In: G. Graichen (ed): Heilwissen versunkener Kulturen. Im Banne der grünen Götter. Econ Verlag Munich, pp 170–197

Orth, H.C.J., C. Rentel, P.C. Schmidt, 1999: Isolation, purity analysis and stability of hyperforin as a standard material from Hypericum perforatum L. J. Pharmac. Pharmacol. 51, 193–200

Orth, M., Th. van den Berg, F.-C. Czygan, 1994: Die Schafgarbe – Achillea millefolium. Z. Phytother. 15, 176–182

Orth, M., F.C. Czygan, V.P. Dedkov, 1999: Variation in essential oil composition and chiral monoterpenes of Achillea millefolium s.l. from Kaliningrad. J. Ess. Oil Res. 11, 681–687

Oruña-Concha, M.J., J. Simal-Lozano, M.J. González-Castro, M.E. Vázquez-Blanco, 1996: Determination of capsaicin and dihydrocapsaicin in Cayenne pepper and Padrón-pepper by HPLC. Dtsch. Lebensmittel-Rdsch. 92, 394–395

Osterburg, C., 1964: Die Vanille in der europäischen Süßwarenindustrie. Süßwaren 8, 600–602

Oszagyan, M., B. Simandi, J. Sawinsky, 1996: A comparison between the oil and supercritical carbondioxide extract of Hungarian wild thyme (Thymus serphyllum L.). J. Ess. Oil Res. 8, 333–335

Oszmianski, J., J.C. Sapis, 1988: Anthocyanes in the fruits of Aronia melanocarpa (chokeberry). J. Food Sci. 53, 1241–1242

Ouahada, S., B. Benveniste, 2000: Tunisian rosemary oil. Perfumer & Flavorist 25, No. 6, pp 24–25

Ouyon, P.H., Ph. M. Vernet, J.L. Guillerm, G. Valedyron, 1986: Polymorphismus and environment: the adaptive value of the oil polymorphism in Thymus vulgaris L. Heredity 57, 56–66

Oyedele, A.O., A.A. Gbolade, M.B. Sosan, F.B. Adewoyin, O.L. Soyelu, O.O. Orafidiya, 2002: Formulation of an effective mosquito-repellent topical product from lemongrass oil. Phytomedicine 9, 259–258

Oyen, L.P.A., N.X. Dung, 1999: Plant resources of South East Asia. No. 19: Essential oil plants. Backhuys Publishers Leiden, 279 pp

Özcan, M., 1999: Pickling and storage of caper berries. (Capparis ssp.). Z. Lebensmittel-Unters. u. -Forsch. 208, 379–382

Özcan, M., 2001: Pickling caper flower buds. Food Quality 24, 261–269

Özcan, M., 2002: A review on composition and processing of capers. (Capparis spp.) flowers bud and fruits. Obst-, Gemüse- u. Kartoffelverarbeit. 87, No. 4, pp 3–4

Özcan, M., A. Akgül, 1995: Capers (Capparis ssp.): Composition and pickling product. Workshop Med. & Aromatic Plants, 25–26 May, Ege Univ., Agric. Fac., Bornova, Izmir Turkey, pp 45–46

Özcan, M., A. Akgül, 1998: Influence of species, harvest data and size on composition of capers (Capparis spp.) flower buds. Nahrung/Food 42, 102–105

Özcan, M., A. Akgül, 2000: Physical and chemical properties of pickled capers (Capparis spp.) flower buds. Obst-, Gemüse u. Kartoffelverarbeit. 85, 165–167

Özcan, M., A. Akgül, K.H.C. Başcr, T. Özck, N. Tabanca, 2001: Essential oil of sea fennel (Crithmum martiimum) from Turkey. Nahrung/Food 45, 353–356

Özcan, M., J.C. Chalchat, 2002: Essential oil composition of Ocimum basilicum and Ocimum minimum L. in Turkey. Czech J. Food Sci. 20, 223–228

Özcan, M., O. Erkman, 2001: Antimicrobial activity of the essential oil of Turkish plant spices. Eur. Food Res. Technol. 212 A, 658–600

Özcan, M., A. Akgul, J.C. Chalchat, 2002: Volatile constituents of the essential oil of Acorus calamus L. grown in Konya Province (Turkey). J. Ess. Oil Res. 14, 366–368

Özek, T., S.H. Beis, D. Demircakmak, K.H.C. Baser, 1995: Composition of the essential oil of Ocimum basilicum L. in Turkey. J. Ess. Oil Res. 7, 203–205

P

Pachaly, P., 1989: Thymian und Thymianfluidextrakt. Dtsch. Apotheker-Ztg. 129, 2705–2706

Pachaly, P., 1990a: Salbeiblätter und dreilappiger Salbei. Dtsch. Apotheker-Ztg. 130, 169–170

Pachaly, P., 1990b: Süßholzwurzel. Dtsch. Apotheker-Ztg. 2264–2265

Padua, L.S. de, N. Bunyapraphatsara, R.H.M.J. Lemmens, 1999: Plant Resources of South-East Asia. No. 12. Medicinal and poisonous plants 1. Backhuys Publishers Leiden, 711 pp

Padua, L.S. de, N.Bunyapraphatsara, 2003: Plant Resources of South-Asia. No. 12; Medicinal and poisonous plants 3. Backhuys Publishers Leiden, 664 pp

Padulosi, S. (ed), 1996: Oregano. Proc. Internat. Plant Genetic Resources Inst./Internat. Workshop on Oregans. Valenzano (Bari), Italy, 8.–12.5., 176 pp

Pagni, A.M., S. Catalano, P.L. Cioni, C. Coppi, I. Morelli, 1990: Études morpho-anatomiques et phytochimiques de Calamintha nepeta (L.) Savi (Labiées). Plantas méd. et Phytothér. 24, 203–213

Paillan-Legure, J.H., 1987: Die Gehalte an etherischen Ölen in Blättern und Rüben von Petersilie (Petroselinum crispum [Miller] Nym. ex Hill.) und ihre möglichen Veränderungen in Abhängigkeit von Sorte, Düngungsvarianten und Ernteterminen. Diss. Univ. Hohenheim, 137 pp

Palevitch, D., L.E. Craker, 1995: Nutritional and medical importance of red peppers (Capsicum spec.). J. Herbs, Spices & Med. Plants 3, 55–83

Palmer, J., 1983: Saffron. Ann. J. Royal New Zealand Inst. Horticult. 11, 93–97

Palmer, J., 1990a: Mellow yellow: the saffron crocus. Growing today 3, No. 9, 14

Palmer, J., 1990b: Germination and growth of wasabi (Wasabi japonica [Miq.] Matsumara). New Zealand J. Crop & Horticult. Sci. 18, 161–164

Pande, V.K., A.V. Sonune, S.K. Philip, 2000: Solar drying of coriander and methi leaves. J. Food Sci. Technol., India 37, 592–595

Pandey, B.P., 1990: Spices and condiments. Shree Publishing House, New Delhi, India 120 pp

Paneitz, A., J. Westendorf, 1999: Anthranoid content of rhubarb (Rheum undulatum L.) and other Rheum species and their toxicological relevance. Eur. Food Res. Technol. 210 A, 97–101

Panja, B.N., D.K. De, 2000: Characterization of blue turmeric: a new hill collection. J. Interacademicia 4, 550–557

Pank, F., 1996: Fenchelanbau im Direktsaatverfahren. Z. Phythotherap. 17, Suppl. 1, pp 17–21

Pank, F., H. Krüger, 2003c: Sources of variability of thyme populations (Thymus vulgaris L.) and conclusions for breeding. Z. Arznei- u. Gewürzpflanzen 8, 117–124

Pank, F., D. Ennet, E. Buschbeck, 1989: Inkulturnahme von Drachenkopf (Dracocephalum moldavica L.) in der DDR. 1. Mitt. Pharmazeutische Verwendung, botanische Charakterisierung, Entwicklung des Produktionsverfahrens und Züchtungsziele. Drogenreport, 2, No. 2, 18–22

Pank, F., E. Schneider, H. Krüger, 2003a: Possibilities and limitations of estragole content reduction of Fennel (Foeniculum vulgare Mill.) and its preparations. Z. Arznei- u. Gewürzpflanzen 8, 165–172

Pank, F., W. Schmidt, O. Schrader, 1994: Qualität gegenwärtig genutzter Pfefferminzsorten (Mentha x piperita L.) und ihre Eignung für die Produktion von Teedrogen. Herba Germanica 2, 69–77

Pank, F., K. Taubenrauch, S. Pfeffer, H. Krüger, 2003b: Eigenschaften von Sorten und Herkünften des Fenchels (Foeniculum vulgare Mill. ssp. vulgare) im Vergleich. Z. Arznei- & Gewürzpflanzen 8, 68–73

Pant, A.K., A.K. Singh, C.S. Mathela, Rashmi Parihar, Vasu Dev, A.T. Nerio, A.T. Bottini, 1992: Essential oil of Hyptis suaveolens Poit. J. Ess. Oil Res. 4, 9–13

Pápay, V., L. Tóth, K. Osváth, Gy. Bujtás, 1980: Über die Flavanoide von Verbascum phlomoides L. Pharmazie 35, 334–335

Paradkar, M.M., R.S. Singhal, P.R. Kulkarni, 2001: A new TLC method to detect the presence of ground papaya seed in ground black pepper. J. Sci. Food Agric. 81, 1322–1325

Parain, Ch., 1986: Connaissances actuelles de la Canelle Cinnamomum zeylanicum Nees (Laura cées) et perspectives d'avenir. Thèse Pharm. Univ. Bordeaux

Pardo, J.E., A. Zalacain, M. Carmona, E. Lopez, A. Alvarruiz, G.L. Alonso, 2002: Influence of the type of dehydration process on the sensory properties of saffron spice. Ital. J. Food Sci. 14, 413–421

Paris, M., G. Clair, J. Nuger, 1979: Odour components of Nigella damascena seeds. Riv. Ital. EPPOS 61, 225–227

Parmar, V.S., H.N. Jha, S.K. Sanduja, R. Sanduja, 1982: Trigocumarin – a new cumarin of Trigonella foenum-graecum. Z. Naturforsch. 37 B, 521–523

Parmar, V.S., S.C. Jain, K.S. Bisht, R. Jain, P. Taneja, A. Jha, O.D. Tyagi, A.K. Prasad, J. Wengel, C.E. Olsen, P.M Boll, 1997: Phytochemistry of genus Piper. Phytochemistry 46, 597–673

Paroul, N., L. Rota, C. Frizzo, A.C. Atti dos Santos, P. Mayna, A. Escalona Gower, L. Atti Serafini, E. Cassel, 2002: Chemical composition of the volatile oil of angelica root obtained by hydrodistillation and supercritical CO_2 extraction. J. Ess. Oil Res. 14, 282–285

Partanen, R., M. Ahro, M. Hakala, H. Kallio, P. Forssell, 2002: Microencapsulation of caraway extract in β-cyclodextrin and modified starches. Eur. Food Res. Technol. 214 A, 242–247

Parzinger, R., 1996: Fenchel. Unterschiede der beiden Varietäten. Dtsch. Apotheker-Ztg. 136, 529–530

Pascual, T., J.de Ovejero, E. Caballero, C. Caballero, 1983: Constituents of the essential oil of Lavandula latifolia. Phytochemistry 22, 1033–1034

Pascual, T., J. de Ovejero, J. Anaya, E. Caballero, J.M. Hernández, M.C. Cabellero, 1989: Chemical composition of Spanish spike oil. Planta med. 55, 398–399

Passet, J., 1971: Thymus vulgaris L., chemotaxonomie et biogenese monoterpenique. C. R. Séances Acad. Agricult. France 57, 1197–1200

Patákova, D., M. Chládek, 1974: Über die antibakterielle Aktivität von Thymian- und Quendelölen. Pharmazie 29, 140–142

Patil, B.S., L.M. Pike, K.S. Yoo, 1995: Variation in the quercitin content in different colored onions (Allium cepa L.). Amer. Soc. Hortic. Sci. 120, 909–913

Paton, A., 1992: A synopsis of Origanum L. (Labiatae) in Africa. Kew Bull. 47, 403–435

Payens, J.D.P.W., 1967: A monograph of the genus Barringtonia (Lecythidaceae). Blumea 15, 157–263

Pech, B., Bruneton, 1982: Alkaloïdes du laurier noble, Laurus nobilis. J. Natur. Products 45, 560–563

Pelissier, Y., A. Malan, Y. Mahmout, J.-M. Bessiere, 1996: Fruit volatiles of Landolphia senegalensis (DC.) Kotschy et Peyr. and L. heudelitii DC. (Apocynaceae). J. Ess. Oil Res. 8, 299–301

Pelissier, Y., C. Marion, J. Casadebaig, M. Milhau, D. Kone, G. Loukou, Y. Nanga, J.M. Bessiere, 1994: A chemical, bacteriological, toxicological and clinical study of the essential oil of Lippea multiflora Mold. (Verbenaceae). J. Ess. Oil Res. 6, 623–630

Peng, J.P., X.S. Yao, H. Kobayashi, C. Ma, 1995: Novel furostanol glycosides from Allium macrostemon. Planta med. 61, 58–61

Peng, J.P., X.S. Yao, Y. Tezuka, T. Kikuchi, 1996: Furostanol glycosides from bulbs of Allium chinense. Phytochemistry 41, 283–285

Pérez-Gálvez, A., M.I. Mínguez-Mosquera, 2001: Structure-reactivity relationship in the oxidation of carotenoid pigments of the pepper (Capsicum annuum L.). J. Agric. Food Chem. 49, 4864–4869

Perikos, J., 1993: The chios gum mastic. Print All Ltd. Graphic Arts, 94 pp

Perineau, F., L. Ganou, J.M. Bessiere, 1991: Hydrodestillation du fruit de coriandre. Parfum, Cosmet. Aròmes 98, 79–84

Perry, N.B., R.E. Anderson, N.J. Breman, M.H. Douglas, A.J. Heany, J.A. MacGimpsey, B.M. Smallfield, 1999: Essential oils from Dalmatian sage (Salvia officinalis L.): Variations among individuals, plant parts, seasons, and sites. J. Agric. Food Chem. 47, 2048–2954

Perry, N.S.L., P.J. Houghton, P. Jenner, A. Keith, E.K. Perry, 2002: Salvia lavandulaefolia essential oil inhibits cholinesterase in vivo. Phytomedicine 9, 48–47

Perucka, I., W. Oleszek, 2000: Extraction and determination of capsaicinoids in fruits of hot pepper (Capsicum annuum L.) by spectrometry and high performance liquid chromatography. Food Chem. 71, 287–291

Peter, H.H., M. Remy, 1976: L'essence de basilic, étude comparative. Parfum., Cosmet. Aròmes 21, 61–68

Peter, K.V. (ed), 2001: Handbook of herbs and spices. Woodhead Publishing Ltd. Abingdon, Oxford (UK), 319 pp

Peters, H, 1927: Aus der Geschichte der Pflanzenwelt in Wort und Bild. Verlag A. Niemayer Mittenwald, 176 pp

Petri, G., É. Lemberkovics, L. Lelik, G. Vitanyi, 1994: Chervil – a wild growing aromatic plant in Hungary. Drogenreport 7, No. 11, p 58

Petry, R., 1993: Pfefferminze: Frische, Sauberkeit, Wohlgeschmack. Zucker- u. Süßwaren-Wirtsch. 46, 529–530

Pfänder, H.J., D. Frohne, 1987: Szechuanpfeffer. – Die Früchte von Zanthoxylum piperitum DC. (Rutaceae). Dtsch. Apotheker-Ztg. 127, 2381–2384

Pfänder, H.J., T. Wittmer, 1975: Untersuchungen zur Cartinoid-Zusammensetzung von Safran. Helv. Chim. Acta 58, 1608–1620, 2233–2236

Pfeiffer, E., S. Hoehle, A.M. Solyom, M. Metzler, 2003: Studies on the stability of turmeric constituents. J. Food Engng. 56, 257–259

Pfister, S., P. Meyer, A. Steck, H. Pfander, 1996: Isolation and structure elucidation of cartenoid-glycosyl esters in Gardenia fruits (Gardenia jasminoides Ellis) and saffron (Crocus sativus Linne). J. Agric. Food Chem. 44, 2612–2615

Phadke, N.Y., A.S. Gkolap, K. Ramakrishnan, G. Subbulakshmi, 1994: Essential oil of Decalepis hamiltonii as an antimicrobial agent. J. Food Sci. Technol. 31, 472–475

Philianos, S.M., T. Andriopoulou-Athanassoula, A. Loukis, 1982: Sur les constituents de l'essence du Thym. Capite (Thymus capitatus Hoffm. & Link, Coridothymus capitatus Reichb.f.) de diverses regions de la Grece. Biologica Gallo-Hellenica 9, No. 2, 285–290

Philianos, S.M., T. Andriopoulou-Athanassoula, A. Loukis, 1984: Constituents of the essential oil of Satureja thymbra L., from different regions of Greece. Internat. J. Crude Drug Res. 22, 145–149

Philip, M.P., N.P. Damodaran, 1985: Chemo-types of Ocimum sanctum Linn. Indian Perfumer 29, No. 1,2, 49–56

Philipp, O. und H.-D. Isengard, 1995: Eine neue Methode zur Identitätsprüfung von Zitronenölen mit HPLC. Z. Lebensmittel-Unters. u. -Forsch. 201, 551–554

Philippe, J., G. Suvarnalatha, R. Sankar, S. Suresh, 2002: Kessane in the Indian celery seed oils. J. Ess. Oil Res. 14, 276–277

Philosoph-Hadas, S., D. Jacob, S. Meir, N. Aharni, 1993: Model of action of CO_2 in delaying senescence of chervil leaves. Acta Horticult. 34, 117–122

Piccaglia, R., M. Marotti, 1991: Composition of the essential oil of an Italian Thymus vulgare L. ecotype. Flavour & Fragrance J. 6, 241–244

Piccaglia, R., M. Marotti, V. Dellacecca, 1997: Effect of planting density and harvest date on yield and chemical composition of sage oil. J. Ess. Oil Res. 9, 187–191

Piccaglia, R., L. Pace, F. Tammaro, 1999: Characterization of essential oils from three Italian ecotypes of hyssop (Hyssopus officinalis L. subsp. aristatus (Godron) Briq. J. Ess. Oil Res. 11, 693–699

Piccaglia, R. M. Marotti, 2001: Characterization of some Italian types of wild fennel (Foeniculum vulgare Mill.). J. Agric. Food Chem. 49, 239–244

Pichitakul, N., K. Sthapitanonda, 1977: The constituents of oil from different mint varieties. J. Nat. Res. Council Thail. 9, No. 2, 1–9 (in Thai)

Pieribattesti, J.-C., J.-Y. Cpnan, J. Gronden, E.-J. Vinvent, M. Gurere, 1981: Contribution a l'etude chimique desbaies roses de bourbon. Ann. Falsif. Exp. Chim. 74, 11–16

Piggott, M.J., E.L. Ghisalberti, R.D. Trengove, 1997: Western Australian sandalwood oil: extraction by different techniques and varieties of the major components in different sections of a single tree. Flavour & Fragrance J. 12, 43–46

Pi-Jen Tsai, J. McIntosh, P. Pearce, B. Camden, B.R. Jordan, 2002: Anthocyanin and antioxidant capacity in rosella (Hibiscus sabdariffa L.) extract. J. Food Sci. Nutrit. 7, 5–11

Pino, J.A., J. Aguero, V. Fuentes, 2002: Essential oil of Salvia officinalis L. ssp. altissima grown in Cuba. J. Ess. Oil Res. 14, 373–374

Pino, J.A., A. Rosado, 1996: Chemical composition of the leaf oil of Pimenta dioica L. from Cuba. J. Ess. Oil Res. 8, 321–332

Pino, J.A., P. Borges, E. Roncal, 1993a: Compositional differences of coriander fruit oils from various origins. Nahrung/Food 37, 119–122

Pino, J.A., P. Borges, E. Roncal, 1993b: The chemical composition of laurel leaf oil from various origins. Nahrung/Food 37, 592–595

Pino, J.A., P. Borges, R. Sanchez, 1992a: Alcohol deterpenation of black pepper oil. Internat. J. Food Sci. Technol. 27, 551–555

Pino, J.A., I.A. Hernández, E. Roncal, 1990: Comparison of isolation procedures for Mexican oregano oil. Nahrung/Food 34, 825–830

Pino, J.A., A. Rosado, V. Fuentes, 1997: Leaf of celery (Apium graveolens L.) from Cuba. J. Ess. Oil Res. 9, 719–720

Pino, J.A., A. Rosado, A. Gonzalez, 1989a: Analysis of the essential oil of pimento berry (Pimenta dioica). Nahrung/Food 33, 717–720

Pino, J.A., A., Rosado, A. Gonzalez, 1991: Volatile flavor components of garlic essential oil. Acta alimentary 20,163–171

Pino, J.A., V. Fuentes, M.T. Correa, 2001: Volatile constituents of Chinese chive (Allium tuberosum Rottl. Ex Sprengel) and rakkyo (Allium chinense G. Don). J. Agric. Food Chem. 49, 1328–1330

Pino, J.A., I.A. Hernández, M. Alvarenz, E. Roncal, 1994: Effect of grinding on quality of Mexican oregano. Alementaria No. 249, 53–54

Pino, J.A., E. Roncal, A. Rosado, I. Goire, 1994a: The essential oil of Ocimum basilicum L. from Cuba. J. Ess. Oil Res. 6, 89–90

Pino, J.A., E. Roncal, A. Rosado, I. Goire, 1995: Herb of dill (Anethum graveolens L.) grown on Cuba. J. Ess. Oil Res. 7, 219–220

Pino, J., A. Rosado, R. Baluja, P. Berges, 1989b: Analysis for the essential oil of Mexican oregano (Lippea graveolens H.B.K.). Nahrung/Food 33, 289–295

Pino, J.A., M. Sanchez, E. Roucal, 1993: Preparation and chemical composition of lemon oil concentrate. Nahrung/Food 37, 277–279

Pino, J.A., M. Sánchez, R. Sánchez, R. Roucal, 1992b: Chemical composition of orange oil concentrates. Nahrung/Food 36, 539–542

Pino, J.A., M. Estarrón, V. Fuentes, 1998: Essential oil of thyme (Thymus vulgaris L.) grown in Cuba. J. Ess. Oil Res. 9, 609–610

Pino, J.A., P. Borges, M. Martinez, M. Vargas, H. Flores, M. Estarron, V. Fuentes, 2001: Essential oil of Menthaspicata L. from Jalisco. J. Ess. Oil Res. 13, 409–410

Pintore, G., M. Usai, P. Bradesi, C. Juliano, G. Boatto, F. Tomi, M. Chessa, R. Cerri, J. Casenova, 2002: Chemical composition and antimicrobial activity of Rosmarinus officinalis L. oils from Sardinia and Corsica. Flavour & Fragrance J. 17, 15–19

Piper, T.J., M.J. Price, 1975: Typical oils from Mentha arvensis var. piperascens (Japanese mint) plants grown from seed. Internat. Flavours & Food Additives 6, 196–198

Pirce, H., 1994: Ahorne. Verlag E. Ulmer Stuttgart, pp 148–149

Pitarokili, D., M. Couladis, N. Petsikos-Panayotarou, O. Tzakou, 2002: Composition and antifugal activity on soil-borne pathogens of the essential oil of Salvia sclarea from Greece. J. Agric. Food Chem. 50, 6688–6691

Pizzale, L., R. Bortolomeazzi, St. Vichi, E. Überegger, L. S. Conte, 2002: Antioxidant activity of sage (Salvia officinalis and S. fruticosa) and oregano (Origanum onites and O. indercedens) extracts related to their phenolic compound content. J. Sci. Food Agric. 82, 1645–1651

Platin, S., Ö.E. Özer, U. Akman, Ö. Hortascu, 199: Equilibrium distributions of key components of spearmint oil in sub/supercritical carbon dioxide. J. Amer. Oil Chemists Soc. 71, 833–837

Plescher, A., 1997: Diagnose von Krankheiten und Beschädigungen an Fenchel (Foeniculum vulgare spp. vulgare Mill. Drogenreport 10, No. 18, 37–58

Plescher, A., I. Zobel, E. Siebeck, H. Kegler, 1998: Morphologie und Inhaltsstoffe gesunder und viruskranker Pflanzen verschiedener Ocimum-Arten. Drogenreport 10, No. 18, 59–73

Plocharski, K., K. Smolarz, 1987: La cultura d'Aronia et son utilité dans l'industrie alimentaire. Przemysl fermentacyjny i owocowo (Fermantation & Fruit-Industry), Warsaw 31, No. 4, 423–25

Plotkin, M.J., 1994: Der Schatz der Wayana. Abenteuer bei den Schamanen im Amazonas-Regenwald. Scherz Verlag Bern Munich Vienna, 288 pp

Pochljobkin, W.W., 1974: ВСЕ О ПРЯНОСТЯХ. Виды свойства применение. (All over spices. Sorts – properties – use). Пищевая Промышленность (Food Industry) 4. edn., Москва (Moskow), 207 pp

Pochljobkin, W.W., 1977: Alles über die Gewürze. Arten – Eigenschaften – Verwendung. Verlag MIR Moscow/VEB Fachbuchverlag Leipzig, 2. Edn., 179 pp

Poggendorf, A., D. Göckeritz, R. Pohloudek-Fabini, 1977: Der Gehalt an ätherischem Öl in Anethum graveolens L. Pharmazie 32, 607–613

Poginsky, B., J. Westendorf, N. Prosenc, M. Kuppe, H. Marquardt, 1988: Johanniskraut (Hypericum perforatum L.). Genotoxizität bedingt durch den Quercitingehalt. Dtsch. Apotheker-Ztg. 128, 1364–1366

Poiana, M., E. Reverchon, V. Sicari, B. Mincione, F. Crispo, 1994: Supercritical carbon dioxide extraction of bergamot oil: bergapten content in the extracts. Ital. J. Food Sci. 6, 459–466

Polonsky, J., J. Batnagar, C. Moretti, 1984: 15-Deacetylsergeolide, a potent antileukemic from Picrolemma pseudocoffea. J. Nat. Prod. 47, 994–996

Poole, S.K., C.F. Poole, 1994: Thin-layer chromatographic method for the determination of the principial polar aromatic flavour compounds of the cinnamons of commerce. Analyst 119, 113–120

Pooter, H.L. de, N.M. Schamp, 1987: The essential oil of Mentha x villosa nm. alopecuoides. Flavour & Fragrance J. 2, 163–165

Pooter, H.L. de, P. Goetghebeur, N. Schamp, 1987: Variability in composition of the essential oil of Calamintha nepeta. Phytochemistry 26, 3355–3356

Pooter, H.L. de, B. Nicolai, J. de Laet, L.F. de Buyck, N.M. Schamp, P. Goetghebeur, 1988: The essential oils of five Nepeta species. A preliminary evaluation of their use in chemotaxonomy by cluster analysis. Flavour & Fragrance J. 3, 155–159

Popov, K.P., 1979: Фисташка збеднеь асиа. (Pistachio in Middle Asia). Ісдаталъство "Улъм" Ашбад (Ascha-bad) 161 pp

Porsch, O., 1906: Die Duftentleerung der Boronia-Blüte. Verhandl. Zoolog.-Bot. Ges. Vienna 56, 605–607

Porta, D.G., R. Taddea, E. Reverchon, 1998: Isolation of clove bud and star anise essential oil by supercritical CO_2 extraction. Lebensmittel-Wiss. u.Technol. 31, 454–460

Porte, A., R.L. Godoy, D. Lopes, M. Koketsu, S.L. Goncavales, H.S. Torquilho: Essential oil of Rosmarinus officinalis L. (rosemary) from Rio de Janeiro, Brazil. J. Ess. Oil Res. 12, 577–580

Porter, N.G., 1989: Composition and yield of commercial essential oils from parsley. Flavour & Fragrance J. 4, 207–219

Potter, Th.L., 1996: Essential oil composition of cilantro. J. Agric. Food Chem. 44, 1824–1826

Potter, Th.L., I.S. Fagerson, 1990: Composition of coriander leaf volatiles. J. Agric. Food Chem. 38, 2054–2056

Pouget, M.P., B. Vennat, B. Lejeune, A. Porrat, 1990: Identification of anthocyanins of Hibiscus sabdariffa L. Lebensmittel-Wiss. u. Technol. 23, 101–102

Poulsen, A.D., J.M. Lock, 1999: A review of African forest Zingiberaceae. In: Tomberlake, J., S. Kativu (eds): African Plants: Biodiversity, Taxonomy and Uses. Royal Botanical Gardens, Kew, pp 51–64

Pound, F.-J., 1938: History, and cultivation of the tonka bean (Dipteryx odorota) with analysis of Trinidad, Venezuelan and Brazilian samples. Trop. Agricult. (Trinidad) 15, 4–9, 28–32

Pourmortazavi, S.M., F. Sefidkon, S.G. Hosseini, 2003: Supercritical carbon dioxide extraction of essential oils from Perovskaja atripleplicifolia Benth. J. Agric. Food Chem. 51, 5414–5419

Povilaityte, V., P.R. Venskutonis, 2000: Antioxidative of purple peril (Perilla frutescens L.), Moldavian dragon-head (Dragcocephalum moldavica L.), and Roman chamomile (Anthemis nobilis L.) extracts in rapeseed oil. J. Amer. Oil Chemists Soc. 77, 951–956

Povilaityte, V., M.E. Cuvelier, C. Berset, 2002: Antioxidant properties of Moldavian dragon head. J. Food Lipids 8, 45–64

Pozzo-Balbi, T., L.Nobile, G. Scapini, M. Cini, 1978: The triterpenoid acids of Schinus molle. Phytochemistry 13, 2107–2110

Prabhu, B.R., N.B. Mulchandani, 1985: Lignans from Piper cubeba. Phytochemistry 24, 329–331

Pradhan, B.P., D.K. Chakraborty, G.C. Subba, 1990: A

triterpenoid lactone from Leucas aspera. Phytochemistry 29, 1693–1695

Pradhan, K.J., P.S. Variyar, J.R. Bandekar, 1999: Antimicrobiological activity of novel phenolic compounds of green pepper (Piper nigrum L.). Food Sci. Technol./Lebensmittel-Wiss. 32, 121–124

Prakash, V., 1990: Leafy spices. CRC Press, Boca Raton, Florida, US, 114 pp

Prakash, V., C.P. Natarajan, 1974: Studies on curryleaf (Murraya koenigii L.). J. Food Sci. (Mysore) 11, 284–286

Pratt, D.S., 1912: Roselle. Phippine J. Sci. 7, 201–205

Preston, H., M. Rickard, 1980: Extraction and chemistry of annotto. Food Chem. 5, 47–56

Pröbstle, A., R. Bauer, 1992: Aristolactams and a 4,5-dioxoaporphine derivative from Houttuynia cordata. Planta med. 58, 568–569

Pröbstle, A., A. Neszmelyi, G. Jerkovich, H. Wagner, R. Bauer, 1994: Novel pyridine and 1,4-dihydropyridinealkaloids from Houttuynia cordata. Nat. Prod. Letter 4, 235–240

Prudent, D., F. Perineau, J.M. Bessière, G. Michel, 1993: Chemical analysis, bacteriostatic and fungistatic properties of the essential oil of the Atoumau from Martinique (Alpinia speciosa K. Schum.). J. Ess. Oil Res. 5, 255–264

Prudent, D., F. Perineau, J.M. Bessière, G. Michel, J.C. Baccou, 1995: Analysis of the essential oil of wild oregano from Martinique (Coleus aromaticus Benth.) – evaluation of its bacteriostatic and fungostatic properties. J. Ess. Oil Res. 7, 165–173

Pruthi, J.S., 1980: Spices and condiments. Chemistry, microbiology and technology. Academic Press New York, NY, 449 pp

Pschyrembel, 1998: Wörterbuch der Naturheilkunde und alternative Heilverfahren. Gondrom Verlag GmbH Bindlach, 324 pp

Pu, Fating, 1991: A revision of the genus Ligusticum (Umbelliferae) in China. Acta Phytotaxon. Sinica 29, 385–393, 525–548

Puertas-Mejia, M., S. Hillebrand, E. Stashenko, P. Winterhalter, 2002: In vitro radical scavenging activity of essential oils from Columbian plants and fractions from oregano (Origanum vulgare L.) essential oil. Flavour & Fragrance J. 17, 380–384

Pura naik, J., N. Balasubrahmanyam, S. Dhanaraj, K.N. Gurudutt, 2000: Packing and storage studies on fluecured large cardamom (Amomum subulatum Roxb.). J. Food Sci. Technol., India 37, 577–581

Pursglove, J.W., 1968: Tropical crops. Dicotyledons. 2 Vol., Longmans, Green and Co. Ltd. London, 719 pp

Pursglove, J.W., 1972: Tropical crops. Monocotyledons. 2 Vol., Longmans, Green and Co. Ltd. London, 607 pp

Putievsky, E., U. Ravid, N. Dudar, 1986: The influence of season and harvest frequency on essential oil and herbal yield from a pure clone of sage (Salvia officinalis) grown under cultivated conditions. J. Nat. Products 49, 326–329

Putievsky, E., U. Ravid, N. Snir, D. Sanderovich, 1984: Essential oil from cultivated bay laurel. Israel J. Bot. 33, 47–52

Putscher, M., 1968: Das Süßholz und seine Geschichte. Diss. Univ. Cologne, 418 pp

Q

Quadry, S.M.J.S., C.K. Atal, 1963: Studies on some umbelliferous fruits. II. Pharmacognosy of the fruits of Cuminum cyminum L. Planta med. 11, 427–431

Quadry, S.M.J.S., C.K. Atali, 1967: Studies on the umbelliferous fruits. III. Pharmacognosy of the fruits of Trachyspermum ammi Linn. (Carum copticum Benth. et Hook). Planta med. 15, 52–57

Quattrocchi, U., 1999/2000: Common names, scientific names, eponyms, synonyms, and etymology. CRC World Dictionary of Plant Names London, New York, Washington, Vol. I A–C1999, 714 pp; Vol. II D–L 1999, 857 pp; Vol. III M–Q 2000, 887 pp; Vol. IV R–Z 2000, 635 pp

R

Rabaté, J., 1938: Étude des essences de Lippea adoensis Hochst. Rev. Bot. Appl. Agric. Trop. 18, 350–354

Rabinowitch, H.D., J.L. Brewster, 1990: Onions and allied crops. Vol. I: Botany, Physiology, and genetics., 273 pp, Vol. II: Agronomy, biotic interactions, pathology and crop protection, 320 pp, Vol. III: Biochemistry, food science, and minor crops, 265 pp CRC Press, Inc. Boca Raton, Florida, USA

Rabinowitch, H.D., L. Currah, 2002: Allium Crop Science: Recent Advances. CABI Publishing Wallingford, UK, 515 pp

Radha, R., R. Latha, M.S. Swaminathan, 2002: Chemical composition and bioactivity of essential oil from Syzygium travancoricum Gamble. Flavour & Fragrance J. 17, 352–354

Radjabian, T., A. Saboora, H. Naderimanesh, H. Ebrahimzadeh, 2001: Comparative analysis of crocetin and its glycosyl esters from Crocus sativus L. and Crocus haussknechtii Bois. as an alternative source of saffron. J. Food Sci. Technol. (Myosore) 38, 324–328

Radonic, A., M. Milos, 2003: Chemical composition and antioxidant test of free and glycosidically bound volatile compounds of savory (Satureja montana L. subspec. montana) from Croatia. Nahrung/Food 47, 236–242

Raffauf, R.F., T.M. Zennie, 1983: The phytochemistry of Quararibea funebris. Bot. Museum leaflets Harvard Univ. 29, 151–158

Rafique, M., M. Hanif, F.M. Chaudhary, 1993: Evaluation and commercial exploitation of essential oil juniper berries of Pakistan. Pakistan J. Sci. & Ind. Res. 36, 107–109

Raghavan, B., K.O. Abraham, M.L. Shankaranarayana, W.D. Koller, 1994: Studies on the flavor changes during drying of dill (Anethum sowa Roxb.) leaves. J. Food Quality 17, 457–466

Raghavan, B., L.J. Rao, M. Singh, K.O. Abraham 1997: Effect of drying methods on the flavour quality of majoram (Oreganum majarana L.). Nahrung/Food 41, 159–161

Ramachandra Rao S., G.A. Ravishankar, 2000: Vanilla flavour: production by conventional and biotechnological routes. J. Sci. Food. Agric. 80, 289-304

Rath, S.P., C. Srinivasulu, S.N. Mahapatra, 1990: GC/MS analysis of essential oil from Bixa orellana Linn. seed. J. Indian Chem. Soc. 67, 86

Raina, V.K., S.K. Srivastava, K.K. Aggarwal, S. Ramesh, Sushil Kumar, 2001: Essential oil composition of Cinnamomum zeylanicum Blume leaves from little Andaman, India. Flavour & Fragrance J. 16, 374–376

Raina, V.K., R. Lal, Savita Tripathi, M. Khan, K.V. Syamasundar, S.K. Srivastava, 2002: Essential oil composition of genetically diverse stocks of Murraya koenigii from India. Flavour & Fragrance J. 17, 144–146

Raina, V.K., S.K. Srivastava, K.V. Syamasunder, 2002: The essential oil of greater galangal (Alpinia galanga [L.] Willd.) from the lower Himalayan region of India. Flavour & Fragrance J. 17, 358–360

Raina, V.K., S.K. Srivastava, K.V. Syamasunder, 2002: Essential oil composition of Acorus calamus L. from the lower region of the Himalayas. Flavour & Fragrance J. 18, 18–20

Raine, C., 1978: The onion. Memory. Oxford Univ. Press, Oxford, UK, 92 pp

Rajanikanth, B., B. Ravindranath, M.L. Shankaranarayana, 1984: Volatile polysulphides of Asa foetida. Phytochemistry 23, 899–900

Rajendran, A., P. Daniel, 2002: The Indian Verbenaceae. (A taxonomic revision). Bishen Singh Mahendra Pal Singh, Dehra Dun, India, 431 pp

Rajeswara Rao, B., P.N. Kaul, A.K. Bhattacharya, G.R. Mallavarapu, S. Ramesh, 1996: Yield and chemical composition of the essential oils of three Cymbopogon species suffering from Iron chlorosis. Flavour & Fragrance J. 11, 289–293

Rajeswara Rao, B., Arun Bhattacharya, Pran Kaul, S. Ramesh, 2001: Yield and chemical composition of rose-scented geranium (Pelargonium species) oil at different times of harvesting. J. Ess. Oil Res. 13, 456–459

Rakhunde, S.D., S.V. Munjal, S.R. Patile, 1998: Curcumin and essential oil constituents of some commonly grown turmeric (Curcuma longa L.) cultivars in Maharashtra. J. Food Sci. Technol. (Myosore) 35, 352–354

Ram, P., N.K. Patra, H.B. Singh, H.P. Singh, M. Ram, B. Kumar, S. Kumar, 1998: Effect of planting date and spacing on oil yield and major yield components of Tagetes minuta. J. Med. & Aromatic Plant Sci. 20, 742–745

Ramachandran, C., K.V. Peter, P.K. Gopalakishan, 1980: Drumstick (Moringa oleifera): A multipurpose Indian vegetable. Econ. Bot. 34, 276–283

Ramadan, M.F., J.-Th. Moersel, 2002a: Characterization of phospholipid composition of black cumin (Nigellasativa L.) seed oil. Nahrung/Food 46, 240–244

Ramadan, M.F., J.-Th. Moersel, 2002b: Neutral lipid classes of black cumin (Nigella sativa L.). Eur. Food Res. Technol. 214 A, 202–206

Ramadan, M.F., J.-Th. Moersel, 2002c: Oil composition of coriander (Coriandrum sativum L.) fruit seeds. Eur. Food Res. Technol. 215, 204-209

Ramadan, M.F., J.-Th. Moersel, 2003: Analysis of glycolipids from black cumin (Nigella sativa L.) coriander (Coriandrum sativum L. and niger (Guizotia abyssinica Cass.) oilseeds. Food Chem. 80, 197–204

Ramarosan-Raonizafinimanana, B., É.M. Gaydou, I. Bombarda, 1997: Hydrocarbons from three Vanilla beans: V. fragrans, V. madagascariensis and V. tahitensis. J. Agric. Food Chem. 45, 2542–2545

Ramón-Laca, L., 2003: The introduction of cultivated Citrus to Europe via northern Africa and the Iberian Peninsula. Econ. Bot. 57, 502–514

Ramprasad, C., M. Sirsi, 1956: Studies on Indian medicinal plants Curcuma longa Linn. In vitro antibacterial activity of curcumin and the essential oil. J. Sci. Res. Inst. 15, 239–242

Ranade, G.S., 1997: Essential oil profile of ajowan oil. Pafei J. 19, No. 4, 61–68

Ranadive, A.S., 1992: Vanillin and related flavor compounds in Vanilla extracts made from beans of various global origins. J. Agric. Food. Chem. 40, 1922–1924

Randkawa, G.S., B.S. Bill, 1995: Transplanting dates, harvesting stage, and yields of French basil (Ocimumbasilicum L.). J. Herbs, Spices & Med. Plants 3, 45–46

Randriamiharisoa, R., 1983: Contribution à l'etude analytique et structurale des differents grades d'huile essentielles d'Ylang-Ylang de Madagascar. (Canagium odoratum genuina). Thèse Docteur-Ing., Marseille

Randriamiharisoa, R., E.M. Gaydou, J.P.Bianchini, G. Raveljaona, G. Vernin, 1986: Etude de la variation de la composition chimique et classification des huiles essentielles de Basilic de Madagascar. Sci. Aliments (Paris) 6, 221–231

Rao, A.S., B. Rajanikanth, R. Seshandri, 1989: Volatile aroma components of Curcuma amada Roxb. J. Agric. Food Chem. 37, 740–743

Rao, E., V.S. Prakasa, K.V. Syamasundar, C.T. Gopinath, S. Ramesh, 1999: Agronomical and chemical studies on Tagetes minuta grown in a red soil of a semiarid tropical region in India. J. Ess. Oil Res. 11, 259–261

Rao, I.K.S., 1966: Vetiver (Vetiveria zizanioides) the aromatic plant of the East. Indian Farming 16, No. 3, 18–20

Rao, K.V., R. Davies, 1986: The ichthytoxic principles of Zanthoxylum clava-herculis. J. Nat. Prod. 49, 340–342

Rao, M.G, 1926: Essential oil from the flowerheads of Perovskia atriplicifolia Benth. Quart. J. Indian Chem. Soc. 3, 141–147

Rao, Prakasa E.V.S., R.S. Ganesha Rao, S. Ramesh, 1995: Seasonal variation in oil content and composition in two chemotypes of scented geranium (Pelargonium sp.). J. Ess. Oil Res. 7, 159–163

Rao, Prakasa E.V.S., 2000: Effect of nitrogen and harvest stage on the yield and oil quality of Tagetes minuta L. in tropical India. J. Herbs, Spices & Med. Plants 7, 19–24

Rao, Y.S., Anand Kumar, Sujatha Chatterjee, R. Naidu, C.K. George, 1993: Large cardamom (Amo mum subu-latum Roxb.) – A review. J. Spices & Aromatic Crops 2, No. 1/2, pp 1–15

Rao, Y.S., 1995: Tamarind economics. Spice India 8, 1–11

Rätsch, Chr., 1992: Indianische Heilkräuter. Tradition und Anwendung. E. Diedrichs Verlag Munich, 3. Edn., 319 pp

Rätsch, Chr., 1995: „Die grüne Fee". Absinth in der Schweiz. Jb. Ethnomed. u. Bewußtseins-Forsch. 4, 285–287

Rätsch, Chr., 1998: Enzyklopädie der psychoaktiven Pflanzen. Botanik, Ethnopharmakologie und Anwendung. AT Verlag Aarau/Wissenschaftl Verlagsges. mbH Stuttgart, 941 pp

Rätsch Chr., C. Müller-Ebeling, 2003: Lexikon der Liebesmittel. Pflanzliche, mineralische, tierische und synthetische Aphrodisiaka. AT Verlag Aarau, Switzerland 784 pp

Rätz, W., 1974: Dillkrautessenzen. Dtsch. Z. Lebensmitteltechnik 25, 264–265

Rau, G., 1994: Sri Lanka. Zimt, Gewürze, „grünes Gold". Dtsch. Apotheker-Ztg. 134, 2536–2539

Raunert, M., 1939: Der Paprika. Verpflegungstechnisch und diätisch. Verlag J.A. Barth Leipzig, 72 pp

Ravai, M., 1992: Quality characteristics of California walnuts. Cereal Foods World 37, 362, 365–366

Ravid, U., E. Putievsky, 1985: Composition of essential oils of Thymbra spicata and Satureja thymbra chemotypes. Planta med. 33, 337–338

Ravid, U., E. Putievsky, I. Katzir, 1994: Enantiomeric distribution of piperitone in essential oil of some Mentha spec., Calamintha incana (Sm.) Heldr. and Artemisia judaica L. Flavour & Fragrance J. 9, 85–87

Ravindran, P.N., 2000: Black Pepper, Piper nigrum.

Harwood Academic Publishers Amsterdam Singapore, 533 pp

Raza Bhatti, G., M. Ingrouille, 1997: Systematics of Pogostemon (Labiatae). Bull. Nat. History Museum London, Bot Ser. 27, 77–147

Razdan, T.K., R.K. Wanchoo, K.L. Dahar, 1986: Chemical composition and antimicrobial activity of the oil of Tagetes minuta Linn. Parfümerie u. Kosmetik 67, 52–55

Read, C., R. Menary, 2001: Leaf extract and polygodial yield in Tasmannia lanceolata (Poir.) A.C. Smith. J. Ess. Oil Res. 13, 348–350

Rechinger, K.H., (ed) 1987: Flora Iranica. No. 162. Umbelliferae. Akad. Druck- u. Verlagsanstalt Graz, 555 pp

Refaat, A.M., H.H. Baghdadi, H.E. Ouda, S.S. Ahmad, 1990: A comparative study between the Egyptian and Romanian sweet marjoram (Majorana hortensis). Planta med. 56, 527

Refaat, A.M., H.H. Baghdadi, E.H. Ouda, S.S. Ahmad, 1992/1993: A comparative study between the Egyptian and Romanian sweet Marjoram (Majorana hortensis Moench [Origanum majorana]). Egyptian J. Horticult. 19, 99–108

Reglos, R.A., C.C. de Guzman, 1991: Morpho-physiological modifications in patchouli, Pogostemon cablin (Blanco) Benth., under varying shade and nitrogen levels. Philippine Agriculturist 74, 429–435

Regrove, H.S., 1933: Spices and condiments. Isaac Pitman & Sons London, 361 pp

Rehm, S., G. Espig, 1996: Die Kulturpflanzen der Tropen und Subtropen. Verlag E. Ulmer Stuttgart-Hohenheim, 3. Edn., 528 pp

Reiner, H., 1994: Die Pistazie (Pistacia vera L.) – Eine Übersicht zur Warenkunde. Ernährung/Nutrition 18, 443–447

Renzini, G., F. Scazzocchino, M. Lu, G. Mazzanti, G. Salvatore, 1999: Antibacterial and cytotoxic activity of Hyssopus officinalis L. oils. J. Ess. Oil Res. 11, 649–654

Resche, A., 1983: Capillar gaschromatochraphie. Bestimmung der Rosmarinsäure in Blattgewürzen. Z. Lebensmittel-Unters. u. -Forsch. 176, 116–119

Retamar, J.A., 1994: Variaciones fitoquimicas de la especie Lippea alba (salvia morado) y sus aplicaciones en la quimica fina. Essenze deriv. Agrum. 12, 55–60

Reuter, D.H., 1986: Knoblauch (Allium sativum). Z. Phytother. 7, 99–106

Reverchon, E., F. Senatore, 1992: Isolation of rosemary oil: Comarism between hydrodistillation and supercritical CO_2-extraction. Flavour & Fragrance J. 7, 227–230

Reverchon, E., A. Ambruosi, F. Senatore, 1994: Isolation and peppermint oil using supercritical CO_2-extraction. Flavour & Fragrance J. 9, 19–23

Reverchon, E., R. Taddeo, G. Della Porta, 1995: Extraction of sage oil by supercritical CO_2: Influence of some process parameters. J. Supercrit. Fluids 8, 302–309

Rey, Ch., I. Slacanin, 1999: Domestication der echten Edelraute (Artemisia umbelliformis Lam. Drogenreport 11, No. 20, 39–41

Rey, S., F. Carbonel, M. van Doorn (F. Massy), 1980: Composition aromatique différentes espèces vanille. Ann. Falsific. exp. Chim. 73, 421–431

Rezzoug, S.A., N. Louka, K. Allaf, 2000: Effect of the main processing parameters of the instantaneous controlled pressure drop process and isolation from rosemary leaves. Kinetic aspects. J. Ess. Oil Res. 12, 336–344

Rhodes, A.M., S.G. Carmer, J.W. Courter, 1969: Measurement and classification of genetic variability in horse-radish. J. Amer. Soc. horticult. Sci. 94, 98–102

Rhyu, H.Y., 1979: Gas chromatographic characterization of sages of various geographic origins. J. Food Sci. 44, 758–762

Riaz, M., C.M. Ashraf, F.M. Chaudhary, 1989: Studies of the Pakistan Laurus nobilis L. in different season. Pakistan J. Sci. Res. 32, 33–35

Richard, H., 1987: Volatile constituents of cardamom cultivated in Costa Rica. Flavour & Fragrance J. 2, 123–127

Richter, Th., 1993; Melissa officinalis. Ein Leitmotiv für 1000 Jahre Pharmaziegeschichte. Dtsch. Apotheker-Ztg. 133, 3723–3726, 3729–3730

Richter, Th., 1999: Bärlauch in Medizin und Mythologie. Pharmaz. Ztg. 144, 2197–2198

Richter, W., G. Willuhn, 1974: Zur Kenntnis der Inhaltsstoffe von Sambucus nigra L. Ätherisches Öl, Alkane und Fettsäuren der Blüten. Dtsch. Apotheker-Ztg. 114, 947–951

Rimpler, H., R. Häusel, L. Kochendörfer, 1969: Xanthorrhizol, ein neues Sesquiterpen aus Curcuma xanthorrhiza. Z. Naturforsch. 25B, 9995–998

Ringuelet, J.A., E.L. Cerimele, C.P. Henning, M.S. Ré, M.I. Urrutia, 2003: Propagation methods and leaf yielding peppermint (Mentha x piperita L.). Herbs, Spices & Med. Plants 10, No. 3, 55–60

Risch, B., K. Herrmann, 1988: Hydroxyzimtsäure-Verbindungen in Citrus-Früchten. Z. Lebensmittel-Unters. u. -Forsch. 187, 530–534

Rissanen, K.S., A. Aflatuni, P. Tomperi, J.E. Jalonen, K.M. Laine, 2002: Herbage and essential oil yield and composition of Mentha piperita L. in different plant densities in northern Latitudes. J. Ess. Oil Res. 14, 243–246

Ritika Gupta, G.R. Mallavarapu, S. Banerjee, Sushil Kumar, 2001: Characteristics of an isomethone-rich somaclonal mutant isolated in a geraniol-rich rose-scented geranium accession of Pelargonium gra veolens. Flavour & Fragrance J. 16, 319–324

Rivals, P. A.H. Mansour, 1974: Sur les cardamomes de Malabar (Elettaria cardamomum Matón). J. Agric. Trop. Bot. Appl. 21, 37–43

Rivandran, P.N., 2000: Black pepper. Harvard Academic Publishing Kerala, India, 553 pp

Rivera, D., C. Inocencio, C. Obón, F. Alcaraz, 2003: Review of food and medicinal uses of Capparis L., sub-genus Capparis (Capparidaceae). Econ. Bot. 57, 515–534

Robards, K., Xia Li, M. Antolovich, St. Boyd, 1997: Characterisation of Citrus by chromatographic analysis of flavonoids. J. Sci. Food Agric. 75, 87–101

Robbins, S.R.J., 1982: Selected markets for the essential oils of patchouli and vetiver. Report Trop. Products Inst., No. G 167, 56 pp

Robbins, S.R.J., P. Greenhalgh 1979: The markets for selected herbaceous essential oils. Trop. Sci. 21, 63–70

Roberts, D., A. Plotto, 2002: A unique Mentha aquatica mint for flavor. Perfumer & Flavorist 27, No. 6, 24, 26–29

Roberts, M., 1997: Little book of mint. Southern Book publishers (Pty) Ltd. Halfway House, S. Africa, 68 pp

Roberty, G., 1953: Notes de botanique Quest-Africaine. VI. Plantes banales dans le Sahel die Nior. Bull. Inst. Franc. Afr. noire 15 A, 442–452

Rodrigo, M., M.J. Lazaro, A. Allvarruiz, V. Ginar, 1992: Composition of capers (Capparis spinosa): Influence of cultivar, size and harvest date. J. Food Sci. 57, 1152–1154

Rodrigues, M.R.A., E.B. Caramao, J.G. dos Santos, C. Dariva, J.V. Oliveira, 2003: The effects of temperature and pressure on the characteristics of the extracts from high-pressure CO_2 extraction of Majorana hortensis Moench. J. Agric. Food Chem. 51, 453–456

Rodrigues, V.M., P.T.V. Rosa, M.O.M. Marques, A.J. Petenate, M.A.A. Meireles, 2003: Supercritical of essential oil from aniseed (Pimpinella anisum L.) using CO_2: Solubility, kinetics, and composition data. J. Agric. Food Chem. 51, 1518–1523

Roersch, C., 1994: Plantas medicinales en el sur Andino del Peru. Koeltz Sci. Books Königstein, 2 Vol. 1188 pp

Roi y Mesa, J.T, 1974: Plantes medicinales, aromátices o venenosas de Cuba. La habana

Rohloff, J., 1999: Monoterpene composition of essential oil from peppermint (Mentha x piperita L.) with regard to leaf position using solid-phase microextraction and gas chromatography/mass spectrometry analysis. J. Agric. Food Chem. 47, 3782–3786

Rohloff, J., 2002: Essential oil composition of Sachalin mint from Norway detected by solid-phase microextraction and gas chromatography-mass spectrometry analysis. J. Agric. Food Chem. 50, 1543–1547

Rohner-Reinhard, H., 1988: Duftendes Morgenrot im Garten. Vom mythischen Rosenkeim zu modernen Gartenrosen. Dtsch. Apotheker-Ztg. 128, 2158–2164

Rojahn, W., F.J. Hammerschmidt, E. Ziegler, F.W. Hefendehl, 1977: Über das Vorkommen von Viridiflorol im Pfefferminzöl (Mentha piperita L.). drago-co-rep. 10, 230–232

Romeo, G., P. Giannetto, M.C. Aversa, 1980: A new 3,4-secopentacyclic triterpenoid from the genus Calamintha nepeta. Phytochemistry 19, 437–439

Rong Tsao, Qing Yu, J. Potter, M. Chiba, 2002: Direct and simultaneous analysis of sinigrin and allyl Isothiocyanate in mustard samples by high-performance liquid chromatography. J. Agric. Food Chem. 50, 4749–4753

Roosmalen, M.G.W. van, 1985: Fruits of the Guianan flora. Institute of Systematic Bot., Utrecht Univ. Netherlands, 483 pp

Rosa, J., G. Krugty, 1987: Essai d'utilization de fruits d'Aronia dans la production de vine rouge de fruit. Przemysl fermentacyjny i owocowo (Fermantation and Fruit Industry) Warschau 31, Nr., 4, 25–27

Rosengarten, F., 1969: The book of spices. Livingston Publishing Co. Wynnewood Penns., 479 pp

Rosengarten, F., 1977: An unusual spice from Oxaxaca: The flowers of Quararibea funebris. Bot. Museum Leaflets Harvard Univ. 25, 193–202

Rosenkranz, V., 2002: Pflanzenparadies Südafrika. Dtsch. Apotheker-Ztg. 142, 2126–2132

Rosenthal, C., 1954: Untersuchungen zur Sortendiagnostik von Estragon. Züchter 24, 40–47

Ross, I.A., 1999: Medicinal plants of the world. Chemical constituents, traditional and modern medicinal uses. Humanana Press Totowa, N.J. (USA), 415 pp

Rosúa, J.L., A. García-Granados, 1987: Analyse des huiles essentielles d'espèces du genre Rosmarinus L. et leur intérêt en tant que caractère taxonomique. Plantes méd. aromat. Phytothérapie 21, 138–143

Roth, I., 1990: Hypericum, Hypericin. Botanik, Inhaltsstoffe. Wirkung. Ecomed Verlagsges. mbH Landsberg, 158 pp

Roth, L., K. Kormann, 1997: Duftpflanzen – Pflanzendüfte. Ätherische Öle und Riechstoffe. Ecomed Verlagsges. mbH Landberg, 544 pp

Rothmaler, W. (ed), 1987: Exkursionsflora für die DDR und die BRD. Gefäßpflanzen, Bd. 2. 13. edn. Volk u. Wissen Volkseigener Verlag Berlin, 640 pp

Rouk, H.F., H. Mengesha, 1963: Fenugreek: its relationships, geography and economic importance. Imperial Ethiopian College of Agric. on mechan. Arts, Diredawa (Ethiopa), Experiment. Station Bull. No. 20

Roussis, V., M. Tsoukatou, I.B. Chinou, A. Ortiz, 1997: Composition and antibacterial activity of two Helichrysum species of Greek origin. Planta med. 63, 181–183

Ruberto, G., D. Biondi, M. Piatelli: Essential oil of Sicilian Thymus capitatus (L.) Hoffmanns et Link. J. Ess. Oil Res. 4, 417–418

Ruberto, G., D. Biondi, R. Meli, M. Piatelli, 1993: Volatile flavour components of Origanum onites L. Flavour & Fragrance J. 8, 197–200

Rubio, C., A. Hardisson, R.E. Martín, A. Báez, M.M. Martín, R. Álvarez, 2002: Mineral composition of the red and green pepper (Capsicum annuum) from Tenerife Island. Eur. Food Res. Technol. 214 A, 501–504

Rust, K., 2003: The beauty of buchu. Food Rev. 30, No. 6, 19, 21

Rust, K., 2003: Just plain vanilla? Food Rev. 30, No. 8, 33–34

S

Sagrero-Nieves, L., J.P. Bartley, A. Provis-Schwede, 1994: Supercritical fluid extraction of the volatile constituents from tamarind (Tamarindus indica L.). J. Ess. Oil Res. 6, 547–548

Sagrero-Nieves, L., J.P. Bartley, 1996: Volatile components from the leaves of Heterotheca inuloides Cass. Flavour & Fragrance J. 11, 49–51

Saha, B.N., A.K.S. Baruah, D.N. Bordoloi, R.K. Mathur, J.N. Baruah, 1986: Mentha piperita – a promising crop for Arunachal Pradesh. Indian Perfumer 30, 355–359

Saha, C., B.K. Chowdhury, 1998: Carbazoloquines from Murraya koenigii. Phytochemistry 48, 363–366

Sahasrabudhe, M.R., W.J. Mullin, 1980: Dehydratation of horse radish roots. J. Food Sci. 45, 1440–1443

Sahoo, S., B.K. Debata, 1995: Recent advances in breeding and biotechnology of aromatic plants: Cymbopogon species. Plant Breeding Abstr. 65, 1721–1731

Saifullina, N.A., S.I. Kozhina, 1975: Composition of essential oils from flowers of Filipendula ulmaria, F. denudata, and F. stepposa. Rastit. Resur 11, 542–544

Saini, S.S., G.N. Davis, 1969: Male sterility in Allium cepa and some hybrids. Econ. Bot. 23, 37–49

Sakai, T., K. Yoshihara, Y. Hirose, 1968: Constituents of fruit oil from Japanese pepper. Bull. Chem. Soc. Japan 41, 1945–1950

Sakamura, F.: Changes in volatile constituents of Zingiber officinale during storage and cultivation. Phytochemistry 26, 2207–2212

Sakar, M., R. Engelskowe, 1985: Monomere und dimere Gerbstoffvorstufen in Lorbeerblättern (Laurus nobilis L.). Z. Lebensmittel-Unters. u. -Forsch. 180, 494–495

Salama, O., O. Sticher, 1983: Iridoidglucoside von Euphrasia rostkoviana. Planta med. 45, 90–94

Salama, R.B., 1973: Sterols in the seed oil of Nigella sativa L. Planta med. 24, 375–377

Saleh Al-Jassir, 1992: Chemical composition and microflora of black cumin (Nigella sativa L.) seeds growing in Saudi Arabia. Food Chem. 45, 239–242

Salem, M.A., 2001: Effect of some heat treatment on

nigella seeds characteristics. I. Some physical and chemical properties of nigella seed oil. J. Agric. Res. Tanta Univ. 27, 471–486

Salido, S., J. Altarejos, M. Nogueras, A. Sanchez, P. Luque, 2000: Chemical composition and seasonal variations of rosemary oil from southern Spain. J. Ess. Oil Res. 15, 10–14

Salim, A.A., M.J. Garson, D.J. Craik, 2004: New alkaloids from Pandanus amaryllifolius. J. Nat. Products 67, 54–57

Salmeron, J., R. Jordano, R. Pozo, 1990: Antimycotic and antiflatoxigenic activity of oregano (Oreganum vulgare L.), and thyme (Thymus vulgaris L.). J. Food Protection 53, 697–700

Salveson, A., 1976: Gasliquid chromatographic separation and identification of the constituents of carawayseed oil. Planta med. 30, 93–96

Samarawira, J.S.E., 1964: Cinnamon. World Crops 16, 45–49

Sampathu, S.R., N. Krishnamurthy, 1982: Processing and utilisation of kokam (Garcinia indica). Indian Spices 19, No. 2, 15–16

Sampathu, S.R., S. Shivashankar, Y.S. Lewis, 1984: Saffron (Crocus sativus L.) – cultivation, processing, chemistry and standardisation. CRC Critical Rev. Food Sci Nutrit. 20, 123–157

Sanagi, M.M., E.K. Ahmad, 1993: Application of supercritical fluid extraction and chromatography to the analysis of turmeric. J. Chromatogr. Sci. 31, 20–25

Sánchez, A.H., A. de Castro, L. Rejano, 1992: Controlled fermentation of caper berries. J. Food Sci. 57, 675–678

Sánchez de Medina, F., M.J. Gamez, I. Jimenez, J. Jimenez, J.I. Osuna, A. Zarzuelo, 1994: Hypoglycemic activity of juniper berries. Planta med. 60, 197–200

Sánchez-Monge, 1991: Flora Agricola. Tom I. Min. de Agricult., Pesca y Alimentation, Madrid, 1294 pp

Sánchez-Monge y E. Parellada, 1981: Diccionario de plantes agricolas. Min. de Agricult. Madrid, 467 pp

Sandberg, F., A. Cronlund, 1982: A ethnopharmacological inventory of medical and toxic plants from Equatorial Africa. Ethopharmacology 5, 187–204

Sanders, W.J., J.L. Seidel, 1992: New synthesis of the pungent principles of ginger-zingerone and shoaol. J. Agric. Food Chem. 40, 264–265

Sandermann, W., 1980: Berserkerwut durch Sumpfporst-Bier. Brauwelt 120, 1870–1872

Sang, S., A. Lao, H. Wang, Z. Chen, 1999: Two new spirostanol saponins from Allium tuberosum. J. Nat. Prod. 62, 1028–1029

Sangat-Roemantyo, 1990: Ethnobotany of the Javanese incense. Econ. Bot. 44, 413–416

Sangwan, R.S., B. Arora, N.S. Sangwan, 2002: Spectral modulation of essential oil biogenesis in the scented Geranium, Pelargonium graveolens L. J. Herbs, Spices & Med. Plants 10, 85–91

Sankar, U.K., 1989: Studies on the physicochemical characteristics of volatile oil from pepper (Piper nigrum L.) extracted by supercritical carbon dioxide. J. Sci. Food Agric. 48, 483–493

Sankat, C.K., V. Maharaj, 1994: Drying the green herb shado beni (Eryngium foetidum L.) in a natural convection cabinet and solar driers. Food J. 9, 17–23

Sankat, C.K., V. Maharaj, 1996: Shelf life of the green herb shado beni (Eryngium foetidum L.) stored under refrigerated conditions. Postharvest Biol. & Technol. 7, 109–118

San Martín, R., R. Granger, T. Adzet, J. Passet, G. Teulade-Arbousset, 1973: Le polymorphisme chimique chez deux Labiées méditerranéennes: Satureia montana L. et Satureia obovata Lag. Plantes méd. et Phytothérapie 7, 95–103

San Miguel, E., 2003: Rue (Ruta L., Rutaceae) in traditional Spain: Frequency and distribution of its medicinal and symbolic applications. Econ. Bot. 57, 231–244

Santapau, H., 1951: A branched specimen of Costus speciosus Smith. J. Bombay Nat. Hist. Soc. 50, 527

Santos, P.M., A.C.Figueiredo, M.M. Oliveira, J.G. Barrosa, L.G. Pedro, S.G. Deans, A.K.M. Younus, J.J.C. Scheffer, 1998: Essential oil from hairy root cultures, and from fruits and roots of Pimpinella anisum. Phytochemistry 48, 455–460

Santos-Gomes, P.C., M. Fernandes-Ferreira, 2003: Essential oil produced by in vitro shoots of sage (Salvia officinalis L.). J. Agric. Food Chem. 51, 2260–2266

Sanz, J.F., E. Falco, J.A. Marco, 1990: New acetylenes from Chrysanthemum coronarium L. Liebigs Ann. Chem., NS. 3, 303–305

Şarer, E., 1987: Güney ve Iç Anadolu bölgelerinde yetişen bazı Salvia türlerinin uçuci yağlari üzerinda Arastirmalar. (Examination of essential oils of some wild grown Salvia sorts in south and inner Anatolia). Doğa Türk. Tip Ecz Derg 11, 97–103 (in Turkish)

Şarer, E., G. Kökdil, 1991: Constituents of the essential oil from Melissa officinalis. Planta med. 57, 89–90

Şarer, E., J.J.C. Scheffer, A.B. Svendsen, 1982: Monoterpenes in the essential oil of Origanum majorana. Planta med. 46, 236–239

Sarin, Y.K. S.G. Agarwal, R.K. Thappa, Kuldeep Singh, B.K. Kapahi, 1992: A high yielding citral-rich strain of Ocimum americanum L. from India. J. Ess. Oil Res. 4, 515–519

Sárkákany, S., J. Bernáth, P. Tétényi, 2001: A mák (Papaver somniferum L.). Akadémiai Kiadó, Budapest, 320 pp (in Hungarian)

Sartoratto, A., F. Augusto, 2003: Application of headspace solid phase microextraction and gas chromatography to the screening of volatile compounds from some Brazilian aromatic plants. Chromatographia 57, 351–356

Šatar, S. 1986: Chemische Charakterisierung ätherischer Öle mongolischer Arten der Artemisia L. Pharmazie 41, 819–820

Satta, M., C.I.G. Tuberoso, A. Angioni, F.M. Pirisi, P. Cabras, 1999: Analysis of the essential oil of Helichrysum italicum G. Don ssp. microphyllum (Willd.) Nym. J. Ess. Oil Res. 11, 711–715

Satyanarayana, A., N. Griridhar, K. Balaswamy, Shivaswamy, D.G. Rao, 2001: Studies on development of instant chutneys from pudina (mint, Mentha spicata) and gongura (Hibiscus sp). J. Food Sci. Technol. (Mysore) 38, 512–514

Satyanarayana, A., P.G. Prabhakara Rao, D.G. Rao, 2003: Chemistry, processing and toxicology of annotta (Bixa orellana L.). J. Food Sci. Technol. (India) 40, 131–141

Saukel, J., 184: Pharmakobotanische Untersuchungen von Arzneidrogen. III. Unterscheidungsmerkmale von Arnika montana L. und Heterotheca inuloide Cass. Sci. Pharm. (Vienna) 52, 35–46

Saunt, J., 2000: An illustrated guide: Citrus varieties of the world. Sinclair Internat. Ltd., Norwich, England, 160 pp

Sawadogo, M., A.M. Vidal-Tessier, P. Delaveau, 1985: L'oléo-resine du Canarium schweinfurthii Engler. Ann. Pharm. France 43, 89–96

Saxena, K., J.K. Manan, S.K. Berry, 1987: Pomegranate: post harvest technology, chemistry and processing. Indian Food Packer 4 (1), 43–60

Schaarschmidt, H., 1988: Die Walnussgewächse (Juglandaceae). Die neue Brehmbücherei. A. Ziemsen-Verlag Lutherstadt Wittenberg, 116 pp

Schaller, F., 1995: Untersuchung der Variabilität und Einfluß des Standortes auf das ätherische Öl verschiedener Populationen des Achillea millefolium Aggregats. Drogenreport 8, No. 13, 39–40

Schaneberg, B.T., I.A. Khan, 2002: Comparison of extraction methods for marker compounds in the essential oil of lemon grass By GC. J. Agric. Food Chem. 50, 1235–1349

Schantz, M.v., I. Kapétanidis, 1971: Qualitative und quantitative Untersuchung des ätherischen Öles von Myrica gale L. (Myricaceae). Pharm. Acta Helv. 46, 649–656

Scharrer, A., 2002: Vanille. Neues zur Authentizität. Diss. J.W. Goethe-Univ. Frankfurt/M., Fachbereich Chem. u. Pharmaz. Wiss., 115 pp

Scharrer, A., A. Mosandl, 2001: Reinvestigation of vanillin contents and component ratios of vanilla extracts using high-performance liquid chromatography and gas chromatography. Dtsch. Lebensmittel-Rdsch. 97, 449–456

Scheer, T., M. Wichtl, 1987: Zum Vorkommen von Kämpferol-4'-O-β-D-glucopyranoside in Filipendula ulmaria and Allium cepa. Planta med. 53, 573–574

Scheffer, J.J.C., A. Gani, A. Baerheim-Svendsen, 1981: Monoterpenes in the essential rhizome oil of Alpinia galanga (L.) Willd. Sci. Pharm. (Vienna) 49, 337–346

Schenck, E.-G., G. Naundorf, 1966: Lexikon der tropischen, subtropischen und mediterranen Nahrungs- und Genussmittel. Nicolaische Verlagsbuchhandl., Herford, 199 pp

Schenk, H.P., D. Lamparsky, 1981: Analysis of nutmeg oil using chromatographic methods. J. Chromatography 204, 391–395

Schieberle, P., 1996: Odour-active compounds in moderately roasted sesame. Food Chem. 55, 145–152

Schier, W., W. Schultze, 1989: Verfälschung von Arzneidrogen – Baldrianwurzel, Enzianwurzel und Nieswurzelstock. Dtsch. Apotheker-Ztg. 129, 1540–1542

Schiffer, E., 1988: Rainfarn – ein vergessenes Fleisch- und Wurstgewürz. Fleisch 42, 60–61

Schilcher, H., 1976: Vorschlag zur Wertbestimmung

von Hibiscusblüten (Calyx Hibisci sabdariffae). Dtsch. Apotheker-Ztg. 116, 1155–1159

Schilling, W., 1969: Die Gattung Hypericum unter besonderer Berücksichtigung von Hypericum perforatum L. Präparative Pharmaz. 5, 125–134

Schimmer, O., H. Mauthner, 1994: Centaurum erythraea Rath. – Tausendgüldenkraut. Z. Phytother. 15, 297–304

Schindler, H., 1953: Über den echten Ammei, Ammi visnaga (L.) Lam., eine Khellin-haltige, spasmolytisch wirksame mediterrane Droge. Pharmazie 8, 176–179

Schirarend, C., M. Heilmeyer (eds), 1996: Die goldenen Äpfel. Wissenswertes rund um die Zitrusfrüchte. Förderverein naturwiss. Museum Berlin e.V. Berlin, 95 pp

Schmelzer, G.H., 1991: Aubergines (Solanum spp.) les environs de Tai (Cote d'Ivoire). Bull. Mus. Nat. Hist. Nat. B. Adansonia 12, 281–292

Schmid, R., 1988: Rhabarber. Naturwiss. Rdsch. 41, 160–161

Schmidt, M., 1993: Die Ringelblume. Heilpflanze mit langer Tradition. PTA heute 7, 567–571

Schmitt, E., 1988: Pfeffer. Vom Reichtum an Gewürzen. Kultur & Technik 12, 214–220

Schneider, E., 1988: Cinnamomum verum – Der Zimt. Z. Phytother. 9, 193–196

Schneider, G., B. Mielke, 1979: Zur Analytik der Bitterstoffe Absinthin, Artabsin und Matrizin aus Artemisia absinthium L. I. Nachweis in der Droge und Dünnschichtchromatographie. Dtsch. Apotheker-Ztg. 118, 469–472

Schneider, G., B. Mielke, 1979: Zur Analytik der Bitterstoffe Absinthin, Artabsin und Matrizin aus Artemisia absinthium L. II. Isolierung und Gehaltsbestimmungen. Dtsch. Apotheker-Ztg. 119, 977–982

Schneider, K., J., 1992; Jurenitsch: Kalmus als Arzneidroge: Nutzen oder Risiko? Pharmazie 47, 79–85

Schnell, R., 1957: Plantes alimentaires et vie agricole de l'Afrique Noire. Essai de phytogéographie alimentaire. Edn. Larousse Paris, 233 pp

Schönfelder, I., P. Schönfelder, 2001: Der neue Kosmos Heilpflanzenführer. Über 600 Heil- und Giftpflanzen Europas. Franckh-Kosmos Verlags GmbH & Co. Stuttgart, 445 pp

Schratz, E., 1960: Zur Systematik des Genus Rheum. Planta med. 8, 299–321

Schratz, E., R. Rangoonwala, 1966: Die Variebilität des Capsicumgehaltes bei verschiedenen Capsicum-Arten und -Sorten. Sci. Pharmac. 34, 365–374

Schratz, H., H. Hörster, 1971: Zusammensetzung des ätherischen Öles von Thymus vulgaris und Thymus marschallianus in Abhängigkeit vom Blattalter und Jahreszeit. Planta med. 19, 160–176

Schratz, E., S.M.J.S. Quadry, 1966: The composition of the essential oil in Coriandrum sativum L. Oil composition during the course of ontogenesis. Planta med. 14, 436–442

Schröder, R., 1991: Kaffee, Tee und Kardamom. Verlag E. Ulmer Stuttgart, 255 pp

Schrutka-Rechtenstamm, R., E. Aurada, J. Jurenitsch, W. Kubelka, 1988: Identifizierung und Qualitätsprüfung von Schinusfrüchten. Ernährung/Nutrition 12, 541–547

Schuhbaum, H., G. Franz, 2000: Ingwer: Gewürz- und vielseitige Arzneipflanze. Z. Phytother. 21, 203–209

Schultes, R.E., 1957: The genus Quararibea in Mexico and the use of its flowers as a spice for chocolate. Bot. Museum Leaflets 17, 247–264

Schulz, B., 1960: Sauerklee. A. Ziemsen Verlag Lutherstadt Wittenberg, 84 pp

Schulz, B., 1963: Gold, Weihrauch und Myrrhe. Österr. Apotheker-Ztg. 17, 6–9

Schulz, G., E. Stahl-Biskop, 1991: Essential oils and glycosidic bound volatiles from leaves, stems, flowers and roots of Hyssopus officinalis L. (Lamiaceae). Flavour & Fragrance J. 6, 69–73

Schulz, H., H. Krüger, 1999: Zur Verbreitung, Züchtung und Verarbeitung von Pfefferminze und Krauseminze. dragoco rep. 46, 57–67

Schulz, H., H. Krüger, J. Liebman, H. Peterka, 1999: Vorkommen flüchtiger Schwefelverbindungen in verschiedenen Zwiebel- und Porreesorten sowie in Allium-Bastarden. Lebensmittelchemie 53, 2–3

Schultze, W., W.A. König, A. Hilkert, R. Richter, 1995: Melissenöle. Untersuchungen zur Echtheit mittels enantioselektiver Gaschromatographie und Isotopenverhältnis-Massenspektroskopie. Dtsch. Apotheker-Ztg. 135, 557–558, 563–564, 567–568, 571–575, 577

Schultze, W., A. Zänglein, R. Klose, K.-H. Kubeczka, 1989: Die Melisse. Dünnschichtchromatographische Untersuchung des ätherischen Öles. Dtsch. Apotheker-Ztg. 129, 155–163

Schultze, W., A. Zänglein, G. Lange, K.-H. Kubeczka, 1990: Sternanis und Shikimmi. Zur Unterscheidung der Früchte von Illicium verum Hook.f. und Illicium anisatum L. Teil 2: Phytochemische Unterscheidungsmerkmale. Dtsch. Apotheker-Ztg. 130, 1194–1201

Schultze-Motel, J. (ed), 1986: Rudolf Mansfeld's Verzeichnis landwirtschaftlicher und gärtnerischer Kulturpflanzen (ohne Zierpflanzen). Akademie Verlag Berlin, 2. edn., 1998 pp

Schwartz, H.F., S.K. Mohan, 1994: Compendium of onion and garlic diseases. The American Phytopathol. Soc., St. Paul, Minn., 90 pp

Schwarz, K., H. Ernst, 1996: Evaluation of antioxidative constituents from thyme. J. Sci. Food Agric. 20, 217–223

Schwarz, K., W. Ternes, 1992a,b: Antioxidative constituents of Rosmarinus officinalis and Salvia officinalis. Z. Lebensmittel-Unters. u. -Forsch. 195, 95–98; 195, 99–103

Schwarz, K., W. Ternes, 1993: Analytik und Charakterisierung von antioxidativ wirksamen Inhaltsstoffen aus Rosmarinus officinalis und Salvia officinalis. Lebensmittelchemie 47, 116–117

Schwarz, K., W. Ternes, E. Schmauderer, 1992: Antioxidative constituents of Rosmarinus officinalis and Salvia officinalis. Stability of phenolic diterpenes of rosemary extracts under thermal stress as required for technological processes. Z. Lebensmittel-Unters. u. -Forsch. 195, 104–107

Schweig, T., 1999: Schwarzkümmels kleine Körner im Kommen. Pharmaz. Ztg. 144, 2582–2587

Schweisheimer, W., 1957: More and more cloves are consumed in the world. Pungent spice from a equatorial countries. Perfumer Ess. Oil Res. 48, 599–600

Schwenker, G., G. Skopp, 1987: Phenolische Inhaltsstoffe in Rosa Pfeffer. Dtsch. Apotheker-Ztg. 127, 1345–1346

Scora, R.W., 1967: Study of the essential oil leaf oil of the genus Monarda (Labiatae). Amer. J. Bot. 54, 446–452

Scotter, M.J., L.A. Wilson, G.P. Appleton, L. Castle, 1998: Analysis of annatto (Bixa orellana). Food cowring formulations. I. Determination of coloring components and colored thermal degradation products by high-performance liquid chromatography with photodiode array detection. J. Agric. Food Chem. 46, 1031–1038

Seaforth, C.E., 1962: Cassia. Trop. Sci. 4, 159–162

Sedlakova, J., B. Kocourkova, V. Kuba, 2001: Determination of essential oil of content and composition in caraway (Carum carvi L.). Czech. J. Food Sci. 19, 31–36

Šedo, A., J. Krejča, 1983: Koreniny. (Gewürze). Príroda Bratislava, 253 pp (in Czech)

Sefidkon, F., L. Ahmadi, M. Mirza, 1997: Volatile components of Perovskia atriplicifolia Benth. J. Ess. Oil Res. 9, 101–103

Sefidkon, F., F. Askari, M. Ghorbanli, 2002: Essential oil composition of Thymus pubescens Boiss. et Kotschy ex Celak from Iran. J. Ess. Oil Res. 14, 116-117

Seidemann, J., 1961: Beitrag zur mikroskopischen Untersuchung der Rinde von Cassia lignea. Z. Lebensmittel-Unters. u. -Forsch. 116, 24–26

Seidemann, J., 1963: Eine einfache Methode zum Nachweis verschiedener Stärkearten und Piper longum in gemahlenem Pfeffer. Quart. J. Crude Drug Res. (Amsterdam) 2, 245–253

Seidemann, J., 1992a: Kentjur, eine wenig bekannte Gewürz- und Heilpflanze. Pharmazie 47, 636–639

Seidemann, J., 1992b: Durch Raubbau ausgerottete Pflanzen erst heute? Naturwiss. Rdsch. 45, 429–431

Seidemann, J., 1993a: Martinique-Ingwer, eine wenig bekannte Ingwerart. Dtsch. Lebensmittel Rdsch. 89, 117–120

Seidemann, J., 1993b: Sumpfporstblätter, früher ein bedenkenloser Hopfenersatz. Naturwiss. Rdsch. 46, 448–449

Seidemann, J., 1993c: Würzmittel-Lexikon. Ein alphabetisches Nachschlagwerk von Abelmoschussamen bis Zwiebeln. B. Behr's Verlag Hamburg, 620 pp

Seidemann, J., 1993 d: Zur Kenntnis von Shii-take. Dtsch. Lebensmittel-Rdsch. 89, 17–20

Seidemann, J., 1993 e: Die Aroniafrucht eine bisher wenig bekannte Obstart. Dtsch. Lebensmittel-Rdsch. 89, 149–151

Seidemann, J., 1994a: Sind Maibowle und Waldmeister gesundheitsschädlich? LMB-Lebensmittelbrief 5, 35–37

Seidemann, J., 1994b: Über eine bisher unbekannte Pfefferverfälschung. Dtsch. Lebensmittel-Rdsch. 90, 150–152

Seidemann, J.: 1994c: Zur Kenntnis von wenig bekannten exotischen Früchten. 3. Mitt. Ambla (Phyllanthus emblica L.). Dtsch. Lebensmittel-Rdsch. 90, 251–253

Seidemann, J., 1995a: Die Knoblauchsrauke, eine in

Vergessenheit geratene Gewürzpflanze. LMB-Lebensmittelbrief 6, 131

Seidemann, J., 1995b: Die Bärwurz, ein wenig bekanntes Gewürz. LMB-Lebensmittelbrief 6, 149–150

Seidemann, J., 1995c: Mahalebblätter. LMB-Lebensmittelbrief 6, 171–172

Seidemann, J., 1995d: Sumachfrüchte. LMB-Lebensmittelbrief 6, 187–188

Seidemann, J. ,1995e: Zur Kenntnis von wenig bekannten exotischen Früchten. 7. Mitt. Afrikanische Aubergine (Solanum macrocarpon L.). Dtsch. Lebensmittel-Rdsch. 91, 246–250

Seidemann, J., 1995f: Zur Kenntnis von wenig bekannten exotischen Früchten. 10. Mitt. Okwa – die afrikanische Brotfrucht. (Treculia africana Decne). Dtsch. Lebensmittel-Rdsch. 91, 345–349

Seidemann, J., 1996: Die Kalabassenmuskatnuß (Monodora myristica [Gaertn] Dun. – wird sie ihren Namen gerecht? Drogenreport 9, No. 15, 48–51

Seidemann, J., 1997a: Chili, Paprika oder Peperoni? Industr. Obst- u. Gemüseverwertung 82, 266–275

Seidemann, J., 1997b: Zeittafeln zur Geschichte von Gewürzen und anderen Würzmitteln. Self pub. 144 pp

Seidemann, J., 1998/2000: Gewürz- und Arzneipflanzen Sri Lankas. Drogenreport 11, No. 20, 68–75; 13, No. 22, 65–71

Seidemann, J., 1999: Der Schraubenbaum. Naturwiss. Rdsch. 52, 225–228

Seidemann, J., 2000: Saflor (Carthamus tinctorius L.) – eine wieder entdeckte Färbepflanze. Drogenreport 13, No. 24, 7–12

Seidemann, J., 2001a: Verfälschung von Gewürzen. Dtsch. Lebensmittel-Rdsch. 97, 28–30

Seidemann, J., 2001b: Tetrapleure (Tetrapleura tetraptera), eine unbekannte afrikanische Heil- und Gewürzpflanze. Z. Arznei- & Gewürzpflanzen 5, 59–63

Seidemann, J., 2001c: Noni – fragwürdiger Zauber aus der Südsee. Pharm. Ztg. 146, 3492–3496

Seidemann, J., 2001d: Mikroskopie der Blätter und Früchte des Schraubenbaumes (Pandanus tectorius [Parkins]). Dtsch. Lebensmittel-Rdsch. 97, 457–461

Seidemann, J., 2003: Kap-Safran – ein wenig bekanntes Gewürz. Drogenreport 16, No. 30, 14–16

Seidemann, J., 2003: Maca, ein pflanzliches Potenzmittel aus den Anden? Drogenreport 16, No. 30, 93–95

Seidemann, J., G. Siebert, 1987: Würzmittel. VEB Fachbuchverlag Leipzig 208 pp

Seitz, P., 1982: Die Pfefferminze und ihre Verwandten. Hessischer Obst- u. Gartenbau 37, 56–57

Sekiwa-Iijima, Y., Y. Aizawa, K. Kubota, 2001: Geraniol dehydrogenase activity related to aroma formation in ginger (Zingiber officinale Roscoe). J. Agric. Food Chem. 49, 5902–5906

Selbitschka, M., 1991: Inhaltsstoffe aus Ingwer (Zingiber officinale Roscoe). Diss. Univ. Bonn, math.-nat. F. 149 pp

Sen, A.R., P. Sengupta, N.G. Dastidar, 1974: Detection of Curcuma zedoaria and Curcuma aromatica in Curcuma longa (turmeric) by thin layer chromatography. Analyst (London) 99, 153–157

Senatore, F., V. de Feo, 1994: Essential oil of a possible new chemotype of Crithmum maritimum L. growing in Campania (Southern Italy). Flavour & Fragrance J. 9, 305–307

Senatore, F., 1996: Influence of harvesting time on yield and composition of the essential oil of a thyme (Thymus pulegioides L.) growing wild in Campania (southern Italy). J. Agric. Food Chem. 44, 1327–1332

Senatore, F., M. D'Agostino, I. Dini, 2000: Flavonoid glycosides of Barbarea vulgaris L. (Brassicacea). J. Agric. Food Chem. 48, 2659–2662

Seshadri, S, V.S. Nambiar, 2003: Kanjero (Digera arvensisi) and drumstick leaves (Moringa oleifera): nutrient profile and potential for human consumption. World Rev. Nutrit. Diet. 91, 41–59

Sethi, M.L., G.S. Rao, B.K. Chodhury, J.F. Morton, G.J. Kapadia, 1976: Identification of volatile constituents of Sassafras albidum root oil. Phytochemistry 15, 1773–1775

Sevinate-Pinto, I., T. Antunes, 1991: Glandular trichomes of Teucrium scorondonia L. Ultrastructure and secretion. Flora (Jena) 185, 207–213

Seyboldt, A., 1998: Die Gattung Salvia – Geschichte, Biologie, Anzucht und Verwendung. Drogenreport 11, No. 19, 35–36

Sezik, E., G. Tümen, K.H.C. Başer, 1991: Ziziphora tenuior L., a new source of pulegone. Flavour & Fragrance J. 6, 101–103

Sfikas, G., 1994: Medizinalpflanzen in Griechenland. Efstathiadis Group S.A., Athen, 140 pp

Shah, N.C., 1991: Chemical constituents of Hyssopus officinalis L.: "Zufe Yabis" a unani drug from U. P. Himalaya, India. Indian Perfumer 35, 49–52

Shah, N.C., A. P. Kahol, T. Sen, G.C. Uniyal, 1986: Gaschromatographic examination of oil of Hysso pus officinalis. Parfümerie u. Kosmetik 67, 116–118

Shaban, M.A.E., K.M. Kandeel, G.A. Yaout, S.E. Mehasab, 1987: The chemical composition of volatile oil of Elettaria cardamomum seeds. Pharmazie 42, 207–208

Shahi, R.P., B.G. Shahi, H.S. Yadava, 1994: Stability analysis for quality characters in turmeric (Curcuma longa L.). Crop. Res. 8, 112–116

Shalini Hooda, Sudesh Jood, 2003: Effect of soaking and germination on nutrient and antinutrient contents of fenugreek (Trigonella foenum graecum L. J. Food Biochem. 27, 165–176

Shankaracharya, N.B., 1998: Tamarind – chemistry, technological and uses. A critical appraisal. J. Food Sci. Technol. India 35, 193–208

Shankaracharya, N.B., C.P. Natarajan, 1971: Coriander: chemistry, technology and uses. Indian Spices 8, Nr 3, 4–13

Shankaracharya, N.B., C.P. Natarajan, 1973: Turmeric, chemistry, technology and uses. Indian Spices 10, No. 3, p 7; No. 8, p 8

Shankaracharya, N.B., L.J. Ro, J.P. Naik, S. Nagalakshmi, 1997: Characterisation of chemical constituents of Indian long pepper (Piper longum L.). J. Food Sci. Technol. (Myosore) 34, 73–75

Shankaranarayana, K.H., B.S. Kamala, 1997: Fragrant products from less odorous sandalwood oil. Perfumer & Flavorist 4, 19–20

Shankaranarayana, M.L., B. Raghavan, C.P. Natarajan, 1972: Mustard-varieties, chemistry and analysis. Lebensmittel-Wiss. u. -Technol. 5, 191–197

Sharaf, A., 1962: The pharmacological characteristics of Hibiscus sabdariffa L. Planta med. 10, 48–52

Sharma, H.M., B.M. Sharma, A.R. Devi, 1997: Contributions of the flora of Manipur. J. Econ. Taxon. Bot. 21, 233–238

Sharma, M.L., M.C. Nigan, V.L. Handa, 1963: The essential oil of hyssop. Riechstoffe, Aromen u. Körperpflegemittel 13, 33–44

Sharma, R., 2003: Medicinal plants of India – An encyclopaedia. Daya Publishing House Delhi, 302 pp

Sharpe, F.R., D.R.J. Laws, 1981: The essential hop oil – a review. J. Inst. Brewing 87, 96–107

Sharpnel, S., 1967: The technological development of the green ginger industry in Australia. Food Technol. Australia 19, 604–608

Shawl, A.S., S.K. Srivastava, K.V. Syamasundar, S. Tripathi, V.K. Raina, 2002: Essential oil composition of Achillae millefolium L. growing wild in Kashmir, India. Flavour & Fragrance J. 17, 165–168

Sheen, L.Y., Y.H.T. Ou, S.J. Tsai, 1991: Flavor characteristic compounds found in the essential oil of Ocimum basilicum L. with sensory evaluation and statistical analysis. J. Agric. Food Chem. 39, 939–943

Shen, X.Y., A. Isogai, K. Furihata, H.D. Sun, A. Suzuki, 1993: Two neo-clerodane diterpenoids from Ajugamacrosperma. Phytochemistry 33, 887–889

Sheppard, E.P., E.M. Boyd, 1970: Lemon oil as an expectorant inhalant. Pharmacol. Res. Commun. 2, 1–16

Shetty, R.S., R.S. Singhal, P.R. Kulkarni, 1994: Antimicrobial properties of cumin. World J. Microbiol. & Biotechnol. 10, 232–233

Shi, J., G. Mazza, M. le Maguer, 2002: Functional foods: biochemical and processing aspects. Vol. II. CRC Press UK London, 432 pp

Shimoda, M., Yin Wu, S. Nonaka, Y. Osajima, 1997: Cluster analysis of GC data on oxygenated terpenes of young leaf and green fruit samples of Japanese pepper (Xanthoxyum piperitum DC.). J. Agric. Food Chem. 45, 1325–1328

Shin, J.H., H.L. Chung, J.K. Seo, J.H. Sim, C.S. Huh, S.K. Kim, Y.J. Baek, 1951: Degradation kinetics of capsanthin in paprika (Capsicum annuum L.) as affected by heating. J. Food Sci. 66, 15–19

Shin, J.-H, Soo-Jung Lee, Nak-Ju Sung, 2002: Effects of Zingiber mioga, Zingiber mioga root and Zingiber officinalis on the lipid concentration in hyperlipidemic rats. J. Korean Soc. Food Sci. Nutrit. 31, 679– 684

Shiobara, Y., Y. Asakawa, M. Kodama, T. Takemoto, 1986: Zedoarol, 13-Hydroxygermancrone and curcuzeone, three sesquiterpenoides from Curcuma zedoaria. Phytochemistry 25, 1351–1353

Shiva, M.P., A. Lehri, A. Shiva, 2002: Aromatic & medicinal plants. Internat. Book Distributors Book Seller & Publishers Dehradun (India), 340 pp

Short, P.S., 1979: Apium L. sect. Apium (Umbelliferae) in Australasia. Adelaide Bot. Garden 1, 205–235

Shotipruk, A., P.B. Kaufman, H.Y. Wang, 2001: Feasibility study of repeated harvesting of menthol from biologically viable Mentha x piperitha using ultrasonic extraction. Biotechnol. Process 17, 924–928

Shuhama, I.K., M.L. Aguiar, W.P. Oliveira, L. A.P. Freitas, 2002: Experimental production of annatto powders in spouted bed dryer. J. Food Engng. 59, 93–97

Shulgin, A.T., 1963: Composition of the myristicin fraction from oil of nutmeg. Nature (London) 197, 379

Sianai, A.C., M.R.R. Tappin, M.F.S. Ramos, J.L. Mazzei, M. Conceição, K.V. Ramos, F.R. de Aquino Neto, N. Frighetto, 2002: Linalool from Lippea alba: Study of the reproducibility of the essential oil profile and the enantiomeric purity. J. Agric. Food Chem. 50, 3518–3521

Siemonsma, J.S., K. Piluek (ed), 1993: Plant resources of South-East Asia. No. 8. Vegetables. Pudoc Sci. Publishers Wageningen, 412 pp

Siewek, F., 1989: C02-Kaltvermahlung von Pfeffer-Erfahrungen aus der Praxis. Internat. Z. Lebensmittel-Technol. u. Verfahrenstechn. 40, 418

Siewek, F., 1990: Exotische Gewürze. Herkunft, Verwendung, Inhaltsstoffe. Birkhäuser Verlag Basel, 190 pp

Silva, J.B., 1983: Isolamento e identificação de alguns constituintes químicos e estudos farmacológicos preliminares Xylopia frutescens Aubl. (Annonaceae). Thesis Federal Univ. Paraiba (Brazil)

Silva, J.B., A.B. Rocha, 1981: Oleoresina do fruto de Xylopia aromatica (Lam.) Mart. Rev. Ciénc. Farm. (Sao Paulo) 3, 33–40

Silva, J.G. da, 1983: Punica granatum L. (Punicaceae). Diss. Univ. Rio de Janeiro, natur. Facult., 88 pp

Simandi, B., M. Oszagyan, E. Lemberlovics, G. Petri, A. Kery, S. Fejes, 1996: Comparison of the volatile composition of chervil oil obtained by hydrodistillation and supercritical fluid extraction. J. Ess. Oil Res. 8, 305–306

Simian, H., F. Robert, I. Blank, 2004: Identification and synthesis of 2-heptanethiol, a new flavor compound in bell pepper. J. Agric. Food Chem. 52, 306–310

Simon, J.E., J. Quinn, 1988: Characterization of essential oil of parsley. J. Agric. Food Chem. 36, 467–472

Simpson, G.I.C., Y.A. Jackson, 2002: Comparison of the chemical composition of East Indian, Jamaican and other West Indian essential oils of Myristica fragrans Houtt. J. Ess. Oil Res. 14, 6–9

Sinclair, J., 1968: The genus Myristica in Malaysia and outside Malaysia. Gardens Bull., Singapore 23, 1-540

Singer, M., 1961: Mushrooms and truffles. Botany, cultivation, and utilization. Hill Publ. London, 272 pp

Singer, M., 1885: Dictionnaire des roses ou guide général du rosiéres. d'Édit. Lebègue & Cie. Bruxelles, 2 tom. 36 Plaqua gravure sur bois

Singh, A, G.S. Randhawa, R.K. Mahey, 1987: Oil content and oil yield of dill (Anethum graveolens L.) herb under some agronomic practices. Acta Horticult. 208, 51–60

Singh, B., R.P. Sood, V. Singh, 1992: Chemical composition of Tagetes minuta L. oil from Himachal Pradesh (India). J. Ess. Oil Res. 4, 525–526

Singh, C.B., A.N. Asthana, K.L. Mehra, 1974: Evolution of Brassica juncea cultivars under domestication and natural selection in India. Genetica Agraria 28, 111–135

Singh, C.B., S.K. Deolia, 1963: Cultivation, extraction and economics of essential oil of roses (Rosa damascens Mill. and Rosa borboniana Desf.) Indian Perfumer 7, 76–80

Singh, D.B., 1978: Large cardamom. Cardamomum 10, No. 5, pp 3–15

Singh, Gurdip, Prakash Singh, Y.R. Prasid, 2002: Chemical and insecticidal investigations on leaf oil of Coleusamboinicus Lour. Flavour & Fragrance J. 17, 440–442

Singh, L.B., 1968: The Mango. Botany, cultivation, and utilization. Edn. J. Hill London (World Crop Books), 438 pp

Singh, M., B. Raghavan, K.O. Abraham, 1996: Processing of majoram (Majorana hortensis Moench.) and rosemary (Rosmarinus officinalis L.) Effect of blanching methods on quality. Nahrung/Food 40, 264–266

Singh, M., S. Ganesha Rao, E.V.S. Prakasa Rao, 1996: Effect of depth and method of irrigation on herb and oil yields of Java citronella (Cymbopogon winterianus Jowitt) under semiarid tropical conditions. J. Agronomy & Crop Sci. 177, 225–375

Singh, M.C., R.S. Tiwari, 1995: Yields and quality attributes of garlic (Allium sativum L.) genotypes. Haryans J. Horticult. Sci. 24, 46–49

Singh, S.S., S.K. Agarwal, S. Verma, M.S. Siddiqui, K. Sushil, 1998: Chemistry of garlic (Allium sati vum) with special reference to alliin and allicin – a review. J. Med. & Aromatic Plant Sci. 20, 93–100

Singh, V., 2001: Monograph on Indian Leucas R. Br. (Dronapushpi) Lamiaceae. Scientific Publishers Jodhpur (India), 208 pp

Singh, V., B. Singh, R.P. Sood, 1995: Herb, oil yield, oil content and constituent variation at different stages of Tagetes minuta. Indian Perfumer 39, No. 2, 102–106

Singh, V., B. Singh, V.K. Kaul, 2003: Domestication of wild marigold (Tagetes minuta L.) as a potential economic crop in western Himalaya and North Indian plains. Econ. Bot. 57, 535–544

Sing-Sangwan, N., R.S. Singwan, R. Luthra, R.S. Thakur, 1993: Geranial dehydrogenese: A determinant of essential oil of quality in lemon grass. Planta med. 59, 168–170

Sinha, G.K., B.C. Gulati, 1990: Study of essential oil from Ocimum americanum Linn. Philipp J. Sci. 119, 347-349

Sirirugsa, P., 1992: A revision of the genus Boesenbergia Kuntze (Zingiberaceae) in Thailand. Nat. History Bull. Siam Soc. 40, 67–90

Skaltsa, H., G. Shammas, 1988: Flavonoids from Lippea citriodora. Planta med. 54, 465

Skaltsa-Diamantidis, H., O. Tzakou, A. Loukis, N. Argyriadou, 1990: Analyse de l'huile essentielle d'Ocimum sanctum L. Nouveaux résultats. Plantes méd. et Phytothér. 24, No. 2, 79–81

Skapska, S., E. Kostrzewa, J. Jendrzejczak, K, Bal, K. Karłowski, M. Fon-Berg-Broczek, S. Porowski, A. Morawski, 2002: Effect of ultra high pressure (UHP) and temperature on the volatiles and piperine content of black pepper (Piper nigrum L.). Herba polon. 48, 120–129

Skopp, G., G. Schwenker, 1986: Biflavonoide aus Schinus terebinthifolius Raddi. Z. Naturforsch. 42 B., 1479–1482

Skopp, G., H.-J. Opferkuch, G. Schwenker, 1987: n-Alkylphenole aus Schinus terebinthifolius Raddi. Z. Naturforsch. 42 C, 7–16

Slanina, E., E. Taborska, L. Lojkova, 1997: Lignans in the seeds and fruits of Schisandra chinensis cultured in Europe. Planta med. 63, 277–280

Slavkovska, V., R. Jancic, S. Milosalejevic, D. Djokovic, 1997: Variability of the essential oil composition of the species Satureja montana L. (Lamiaceae). J. Ess. Oil Res. 9, 629–634

Slavkovska, V., R. Jancic, S. Bojovic, S. Milosavljevic, D. Djokoviv, 2001: Variability of essential oil of Satureja montana L. and Satureja kitaibelii Wierzb. ex Heuff. from the central part of the Balkan peninsula. Phytochemistry 57, 71–76

Smadja, J., 1991: Review on chemical composition of vetiver oil. J. Nature 3, No. 1, 3–17

Smadja, J, E.M. Gaydou, G. Lamaty, J.Y. Conan, 1990: Etude de facteurs de variation de la composition del'huile essentielle de vétyver Bourbon par analyse factorielle discriminante. Analusis 18, 343–351

Small, E. (ed), 1997: Culinary herbs. NRC Research Press Ottawa, 710 pp

Smith, J.R., 1996: Safflower. Amer. Organ. Chem. Soc. Press Champaign, Ill., 606 pp

Smith, P.G., C.B. Heiser, 1957: Taxonomy of Capsicum sinense Jacq. and the geographic distribution of the cultivated Capsicum species. Bull. Torrey Bot. Club. 84, 413–420

Smith, R.M., 1974: The essential oil of the medicinal plant Coleus amboinicus from Fiji. Fiji Agricult. J. 36, 21

Smith, R.: 1981: Zingiberaceae. Sinoptic keys to the tribes Zingibereae, Globbeae, Hedychieae, Alpineae. Royal Botanical Garden, Edinburgh (UK), 28 pp

Soenarko, S., 1977: The genus Cymbopogon Sprengel (Gramineae). Reinwardtia 9, 225–375

Soepadyo, R., H.T. Tan. 1968: Patchouli, a profitable catch crop. World Crops 3, 48–54

Sofowora, E.A., 1970: A study of variations in essential oils of cultivated Ocimum gratissimum. Planta med. 18, 173–176

Sohounhloue, K.D., J. Dangou, L. Djossou, A. Akoegninou, R. Adjobo, D.X. Garneau, H. Gagnon, F.I. Jean, 2002: Chemical composition of the leaf and flower oils of Aeollanthus pubescens Benth. from Benin. J. Ess. Oil Res. 14, 80–81

Soliman, M.M., S. Kinoshita, T. Yamanishi, 1975: Aroma of roasted sesame seeds. Agric. Bot. Chem. (Tokyo) 39, 973–977

Soliman, F.M., E.A. El-Kashoury, M.M. Fathy, M.H. Gonaid, 1994: Analysis and biological activity of essential oil of Rosmarinus officinalis L. from Egypt. Flavour & Fragrance J. 9, 29–33

Somali, M.A., M.A. Bajneid, S.S. At-Fhaimani, 1984: Chemical composition and characterization of Moringa peregrina seeds and seeds oil. J. Amer. Oil Chem. Soc. 61, 85–86

Somos, A., 1981: A Paprika. Akadémiai Kiadó Budapest, 396 pp

Somos, A., A. Kundt (eds),1984: The paprika. Kultura Hungarian Foreign Trade Co., and Akadémiai Kiadó Budapest 1984, 302 pp

Song, J.-Y., G.-H. An, C.-J. Kim, 2003: Color, texture, nutrient contents, and sensory value of vegetable soy-beans (Glycine max [L.] Merrill) as affected blanching. Food Chem. 83, 69–74

Sopher, D.E., 1964: Indigenous uses of turmeric (Curcuma domestica) in Asia and Oceania. Anthropos 59, 93–127

Sørensen, J.M., S.T. Katsiotis, 2000: Parameters influencing the yield and composition of the essential oil from Cretan Vitex agnus-castus fruits. Planta med. 66, 245–250

Soule, M.J., 1951: A bibliography of the mango (Mangifera indica L.). Univ. Miami Press, Coral Gables Florida, 87 pp

Soulélès, C., N. Argyriadou, 1990: The volatile constituents of Calamintha grandiflora. Planta med. 56, 234- 235

Soulélès, C., G. Shammas: Sur les constituants des feuilles et des fleurs de Calamintha nepeta (L.) Savi. Plantes méd. et Phytothérap. 19, 286–290

Soulélès, C., N. Argyriadou, S. Philianos, 1987: Constituents of essential oil of Calamintha nepeta. J. Nat. Products 50, 510–511

Soulélès, C., C. Harvala, J. Chinou, 1991: Flavonoids from Calamintha grandiflora. Internat. J. Pharmacogn. 29, 317–319

Souret, F.F., P.J. Weathers, 2000: The growth of saffron (Crocus sativus L.) in aeroponics and hydroponics. J. Herbs, Spices & Med. Plants 7, 25–35

Spapiro, R., J. Frances, 2001: Cascarilla bark essential oil of El Salvador: new source and standards. Perfumer & Flavorist 26, 22, 24–26

Spencer, K.C., D.S. Seigler, 1980: Deidaclin from Turner ulmifolia. Phytochemistry 19, 1863–1864

Spina, P., 1962: Il pistacchio. Italia Agricola 99, 477–492

Spinelli, P., 1951: Il „Bergamotto di Reggio Calabria" (Citrus bergamia Risso). Riv. Agric. Subtrop. & Trop. 45, 134–150

Srinivasan, P.S., S. Mohandass, S. Sambandamurthy, V. Sampath, 1981: A study on the growth and development of mint (Mentha piperita var. citrata Ehrh. Under Kodaikanal conditions. Indian Perfumer 25, No. 3-4, 60–63

Srinivasan, V.V., V.R. Sivaramakrishnan, C.R. Rangaswamy, H.S. Ananthapadmanabha, K.H. Shankaranarayana, 1992: Sandal (Santalum album L.). Indian Council Forestry Res. & Education. Dehra Dun, India 233 pp

Srinivasulu, C., S.N. Mahapatra, 1982: Chemical constituents of the essential oil from annatto (Bixa orellana) seeds. Indian Perfumer 26, 132–133

Srinivasulu, C., S.N. Mahapatra, 1989: A process for the isolation of bixin. Res. Ind. 34, No. 2,137–178

Srivastava, A., Y.N. Shukla, S.P. Jain, Sushil Kumar, 1999: Chemistry, pharmacology and uses of Bixa orellana – a review. J. Med. & Aromatic Plant Sci. 21, 1145–1154

Srivastava, N.K., G.K. Satpute, 2001: Monoterpene essential oil quantitative and qualitative component trait association in palmarosa (Cymbopogon martinii Wats.). J. Genetics & Breeding 55, 21–30

Srivastava, N.K., R. Kumer, D. Dixit, 2002: Interspecific variation in root grown in Cymbopogon in root boxes. J. Herbs, Spices & Med. Plants 10, 45–53

Stace, C., 1991: New flora of British Isles. Cambridge Univ. Press Cambridge, 1226 pp

Stachurski, L., E. Bednarek, J. Cz. Dobrowolski, H. Strzelecka, A.P. Mazurek, 1995: Tormentoside and two of its isomers obtained from the rhizomes Potentilla erecta. Planta med. 61, 94–95

Staesche, K., 1968: Vergleichende anatomische Untersuchungen an Drogen. Rhizoma Tormentillae und Rhizome verwandter Rosaceae. Dtsch. Apotheker-Ztg. 108, 329-332

Staesche, K., 1972: Gewürze. In: L. Acker, K.G. Berger, W. Diemair, W. Heimann, F. Kiermeier, J. Schormüller, S.W. Souci (eds) Handbuch der Lebensmittelchemie. Springer-Verlag Berlin Heidelberg New York, Bd. VI.: Alkaloidhaltige Genussmittel, Gewürze, Kochsalz, pp 426–610

Staesche, K., 1970: Sumach, ein türkisches Gewürz. Dtsch. Lebensmittel-Rdsch. 67, 202–204

Stahl, E.,1982: Rosa Pfeffer, ein gefährliches, exotisches Gewürz? Dtsch. Apotheker-Ztg. 122, 337–3409

Stahl, E., B. Bohrmann, 1967: Phthalide als Hauptbestandteile des ätherischen Öls der Früchte von Meumathamanticum Jacq. Naturwissenschaften 54, 118

Stahl, E., S.N. Datta, 1982: New sesquiterpenoids of the ground ivy (Glechoma hederaceae). J. Liebigs Ann. Chem. 757, 23–32

Stahl, E., D. Gerard, 1982: Entgiftung von Absinthkraut durch CO_2-Hochdruckextraktion. Planta med. 45, 147

Stahl, E., H. Jork, 1964: Chemische Rassen bei Arzneipflanzen. 1. Mitt. Untersuchung der Kulturvarietäten europäischer Petersilienherkünfte. Arch. Pharmaz. 297, 273–281

Stahl, E., K. Keller, 1981: Zur Klassifizierung handelsüblicher Kalmusdrogen. Planta med. 43, 128–140

Stahl, E., K. Keller, C. Blinn, 1983: Cardanol, a skin irritant in pink pepper. Planta med. 48, 5–9

Stahl, E., K.-H. Kubeczka, 1979: Das ätherische Öl der Apiaceae (Umbelliferae). Planta med. 37, 49–56

Stahl, E., H.G. Meußen, H. Jahn, 1985: Über den Gehalt der verschiedenen Hydroxyanthracenderivate in Rhabarberwurzeln und -zubereitungen. Dtsch. Apotheker-Ztg. 125, 1478–1480

Stahl-Biskup, E., 1991: The chemical composition of Thymus oils: a review of the literature 1960–1989. J. Ess. Oil Res. 3, 61–82

Stahl-Biskup, E., 2003: Biodiversität ätherischer Öle am Beispiel der Gattung Thymus. Drogenreport 16, No. 29, 3–7

Stahl-Biskup, E., J. Holthuijzen, 1995: Essential oil and glycosidically bound volatiles of lemon-scented Thymus x citriodorus (Pers.) Schreb. Flavour & Fragrance J. 10, 225–229

Stahl-Biskup, E., F. Sáez, 2002: Thyme – the Genus Thymus. Taylor and Francis Ltd. London 143 pp; Vol. 24: Medicinal and aromatic plants – Industrial profiles

Stahn, Th., U. Bomme, 1997: Beurteilung eines großen Sortiments von Mentha piperita-Herkünften. Untersuchungen zum Gehalt an ätherischem Öl, Geschmacks- und Wuchscharakter. Z. Arznei- u. Gewürzenpfl. 2, 164–172

Štajner, D., M. Milić-DeMarino, J. Canadanović-Brunet, 2003: Screening for antioxidant properties of leeks, Allium sphaerocephalon L. J. Herbs, Spices & Med. Plants 10, No. 3, 75–82

Stanley, W.L., S.H. Vannier, 1957: Chemical composition of lemon oil. I. Isolation of series of substituted coumarins. J. Amer. Chem. Soc. 79, 3488–3491

Starke, H., K. Herrmann, 1976: Flavanole und Flavone der Gemüsearten. VII. Flavanole des Porrees, Schnittlauchs und Knoblauchs. Z. Lebensmittel-Unters. u. -Forsch. 161, 25–30

Steinlesberger, H., 2004: Lakritze – Süßholzraspeln. Garten + Haus 5, No. 3, 35

Steinecker, E., K. Stück, 1982: Trennung und quantitative Bestimmung der Hauptscharfstoffe von Zingibergiberis rhizoma mittels kombinierter DC/HPLC. Pharm. Acta Helv. 57, 66–71

Steiner, M., I. Hochhausen, 1952: Abtrennung und Bestimmung ätherischer Öle aus Citrus-Fruchtschalen durch Destillation. Arzneimittel-Forsch. 2, 535–543

Steinhaus, M., P. Schieberle, 1999: Aromatische Verbindungen in getrockneten Hopfendolden der Varietät Spalter select. Lebensmittelchemie 53, 109–110

Stephanie, A., W. Baltes, 1991: Porreeöl-Aromen. Lebensmittelchemie 45, 14–15

Stephanie, A., W. Baltes, 1993: New aroma components of leek oil. Z. Lebensmittel-Unters. u. -Forsch. 94, 21–25

Stephenson, R., 1994: Sedum. Cultivated stone crops. Timber Press Inc. Portland, Oregon, 335 pp

Stevenson, G.A., 1969: An agranomic and taxonomic review of the genus Melilotus Mill. Canad. J. Plant Sci. 49, 1–20

Sticher, O., B. Meier, 1980: Quantitative Bestimmung der Bitterstoffe in Wurzeln von Gentiana lutea und Gentiana purpurea mit HPLC. Planta med. 40, 55–67

Sticher, O., O. Salama, 1981: Iridoid glucosides from Euphrasia rostkoviana. Planta med. 42, 122–123

Stiegler, K., 1949: Die deutsche Hopfenwirtschaft. Darstellung ihrer Entwicklung und Probleme. Ehrenwirth-Verlag Munich, 232 pp

Stochmal, A., M. Gruszczyk, 1998: Solid phase extraction and HPLC determination of hypericin in Hypericum perforatum L. Herba polon. 44, 315–323

Stone, B.C., 1976: The Pandanaceae of the New Hybrids, with an essay on interspecific variation in Pandanus tectorius. Kew Bull. 31, 47–70

Stör, D., 1974: Die Edelnelke. VEB Deutscher Landwirtschaftverlag Berlin, 2. Edn., 163 pp

Storrs, A., J. Storrs, 1997: Discovering trees and shrubs in Thailand & S.E. Asia. Craftsman Press Ltd. Bangkok (Thailand), 246 pp

Stowartt, T., 1978: Lexikon der Gewürze, Kräuter und Würzmittel. Hörnemann Verlag Bonn, 3 edn., 264 pp

Stoyanova, A., A. Konakchiev, O. Berov, 2002: Investigation on the essential oil of coriander from Bulgaria. Herba polon. 48, 67–70

Strantz, E., 1909: Zur Silphionfrage. Diss. Univ. Zürich, phil. F., 183 pp

Straubinger, M., B. Bau, S. Eckstein, M. Fink, P. Winterhalter, 1998: Identification of novel glycosidic aroma precursors in saffron (Crocus sativus L.). J. Agric. Food Chem. 46, 32–38

Strauß, D., 1967: Über Untersuchungen an Muskatnüssen und Muskatnusspulver. Dtsch. Lebensmittel-Rdsch. 63, 239–242

Strauß, D., 1969a: Über die Mikroskopie ostasiatischer Gewürze. I. Mitt. Tamarinde (Tamarindus indica L.). Dtsch. Lebensmittel-Rdsch. 65, 120–121

Strauß, D., 1969b: Über die Mikroskopie ostasiatischer Gewürze. II. Mitt. Sereh-Gras oder Zitronengras(Cymbopogon citratus Stapf, Andropogon citratus DC.). Dtsch. Lebensmittel-Rdsch. 63, 176-177

Strauß, D., 1969c: Über die Mikroskopie ostasiatischer Gewürze. III. Mitt. a) Kemiri-Nüsse (Aleurites moluccana Willd.); b) Peteh-Bohnen (Parkia speciosa Hassk.). Dtsch. Lebensmittel-Rdsch. 65, 210–211

Strauß, D., 1969d: Über die Mikroskopie ostasiatischer Gewürze. IV. Mitt. b) Indisches Lorbeerblatt (Eugenia polyantha Wight) „Daoen Salam". Dtsch. Lebensmittel-Rdsch. 65, 289–290

Strunz, M., G. Puschmann, V. Stephanie, D. Fritz, 1992: Chemische Varibialität von Dill (Anethumgraveolens). Gartenbauwissenschaft 57, 190–192

Stuart, G.R., J.V. de Oliveira, S.G. D'Avila, 1994: Extraction of essential oil from basil using carbon dioxide at high pressures. Cienc. e Tecnol. de Alimentos 14 (Suppl.) 11–16

Su, H.C.F., R. Hovart, 1988: Investigation of the main components in insect-active dill seed. J. Agric. Food Chem. 36, 752–753

Subban Nagarajan, L.J. Mohan Rao, 2003: Determination of 2-hydroxy-4-methoxybenzaldehyde in roots of Decalepis hamiltonii (Wight et Arn.) and Hemidesmus indicus R.Br. J. AOAC internat. 86, 564–567

Suichorska-Tropilo, K., E. Osinska, 2001: Morphological developmental and chemical analyses of 5 forms of sweet basil (Ocimum basilicum L.). Ann. Warsaw agric. Univ., Horticult. 22, 17–21

Sugimara, Y., Y. Ichikawa, K., Otsuji, M. Fujita, N. Toi, N. Kamata, R.M. del Rosario, G.R. Luingas, G.L. Tagan, 1990: Cultivarietal comparison of patchouli plants in relation to essential oil production and quality. Flavour & Fragrance J. 5, 109–114

Sulas, L., S. Caredda, 1997: Introduzone in coltura di nouve specie foraggere. Produttività e composizione bormatologica di Chrysanthemum coronarium L. (crisantemo) sottoposto a pascolamento simulato. Riv. Agronom. (Italy) 31, 1019–1026

Suleiman, A., S. Asgary, G.B. Lookwood, 1996: Volatile constituent of Achillea millefolium L. ssp. millefolium from Iran. Flavour & Fragrance J. 11, 265–267

Sullivan, G., 1982: Occurrence of umbelliferone in the seeds of Dipteryx odorota (Aubl.) Willd. J. Agric. Food Chem. 30, 609–610

Sultana, T., N.G. Porter, G.P. Savage, D.L. MacNeil, 2003: Comparison of isothiocynate yield wasabi rhizome tissues grown in soil or water. J. Agric. Food Chem. 51, 3586–3591

Surburg, H. M. Köpsel, 1989: New naturally occuring sesquiterpenes from Scotch spearmint oil (Mentha cardiaca Ger.). Flavour & Fragrance J. 4, 143–147

Surk, Y.-J., S.S. Lee, 1996: Capsaicin in hot chili pepper: Carcinogen, co-carcinogen or anticarcinogen? Food Chem. Toxicol. 34, 313–316

Sushila, M., 1987: Oils and fats in acrid plants with articular reference to Capparis decidua L. Transact. Indian Soc. Desert Technol. 12, 99–105

Sutton, J. 1999: The Gardener's guide to growing Salvias. David & Charles Newton Abbot/ Timber Press Portland, Oregon 160 pp

Suzuki, M, 1968: Wasabi (Manuscript). Kyoto Univ., Japan, 5 pp

Svoboda, A., R.K.M Hay, P.G. Waterman, 1990: Growing summer savory (Satureja hortensis) in Scotland: quantitative and qualitative analysis of volatile oil and factors influencing oil production. J. Sci. Food Agric. 53,193–222

Svoboda, K.P., J.Gough, J. Hampson, B. Galambosi, 1995: Analysis of the essential oils of some Agastache species grown in Scotland from various seed souces. Flavour & Fragrance J. 10, 139–145

Swiatek, L., O. Salama, O. Sticher, 1982: 6-0-β-D-Xylopyranosylcatalpol, a new iridoid glycosides from Verbascum thapsiforme. Planta med. 45, 153

Szczpznski, C. v., P. Zgorzelak, G.A. Hoyer, 1972: Isolierung, Strukturaufklärung und Synthese einer antimikrobiell wirksamen Substanz aus Petiveria alliacea L. Arzneimittelforschung 22, 1975–1976

T

Tachibana, Y., H. Kikuzaki, N.H. Lajis, N. Nakatani, 2001: Antioxidative activity of carbazons from Murraya koenogi leaves. J. Agric. Food Chem. 49, 5589–5594

Tada, H., Y. Murakami, T. Omoto, K. Shimomura, K.

Ishimura, 1996: Rosmarinic acid and related phenolics in hairy root culture of Ocimum basilicum. Phytochemistry 42, 431–434

Tae Hwan Kim, Joong Han Shin, Hyung Hee Baek, Hyong Joo Lee, 2001: Volatile flavour compounds in suspension culture of Agastache rugosa Kuntze (Korean mint). J. Sci. Food & Agric. 81, 569–575

Tainter, R.D. A.T. Grenis, 1993: Spices and seasoning. A food technology handbook. VCH Publishers, Inc. NewYork (NY), VCH Verlagsgesellsch. mbH Weinheim, 226 pp

Tairu, A.O., T. Hofmann, P. Schieberle, 1999: Characteristization of the key aroma compounds in dried fruits of the West African peppertree Xylopia aethiopica (Dunal) A. Rich. (Annonaceae) using aroma extract dilution analysis. J. Agric. Food Chem. 47, 3285–3287

Takenaka, M. T. Nagata, M. Yoshida, 2000: Stability and bioavalability of antioxydants in garland (Chrysanthemum coronarium L.). Biosci., Biotechnol., Biochem. 64, 2689–2691

Taniguchi, M., M. Yanai, Y.Q. Xiao, T. Kido, K. Babo, 1996: Three isocoumarins from Coriandrum sativum. Phytochemistry 42, 843–846

Takruri, H.R.H., M.A.F. Dameh, 1998: Study of nutritional value of black cumin seeds (Nigella sativa L.). J. Sci. Food Agric. 76, 404–410

Tantau, M., K. Weinhausen, 1956: Die Rose. Ihre Kultur und Verwendung. E. Ulmer Verlag Stuttgart, 2. edn., 146 pp

Tan Tianwei, Huo Qing, Ling Qianz, 2002: Purification of glycyrrhizin from Glycyrrhiza uralensis Fisch. with ethanol/phosphate aqueous two phase systems. Biotechnol Letters 24, 1417–1420

Tao Guoda (ed), 1998: Zhongguo-redai-yesheng-huahui. (Wild tropical plants in China). China Esperanto Press Peking, 172 pp

Tarantalis, P.A., M.G. Polissiou, 1997: Isolation and identification of the aroma components from saffron (Crocus sativus L.). J. Agric. Food Chem. 45, 459–462

Taskinen, J., L. Nykänen, 1975: Chemical composition of angelica root oil. Acta Chem. Scand. 29B, 57–764

Tassan, C.G., G.F. Russel, 1975: Chemical and sensory studies on cumin. J. Food Sci. 40, 1185–1187

Tateo, F., M. Fellin, E. Verderio, 1988: Production of rosemary oleoresin using supercritical carbon dioxide. Perfumer & Flavorist 13, 27–34

Tateo, F., M. Fellin, L. Santamaria, A. Bianchia, 1988b: Rosmarinus officinalis L. extract production antioxidant and antimutagenic activity. Perfumer & Flavorist 13, 48–54

Tattje, D.H.E., R. Bos, 1981: Composition of essential oil of Ledum palustre. Planta med. 41, 303–307

Täufel, A., L. Tunger, W. Ternes, M. Zobel, 1993: Lebensmittel-Lexikon. Verlag B. Behr's Hamburg, 2. edn., 2 Vol., 922 pp

Taylor, R.L., 1976: Houttuynia cordata, dokudami. Davidsonia 7, 63

Taylor, S.J., I.J. Natarajan, 1974: Curcuminoid pigments in turmeric (Curcuma domestica Val.) by reversed phase HPLC. Chromatographia 34, 73–77

Theisseire, P., A. Galfre, 1974: Contribution à la connaissance de la'huile essentielle d'Ylang-Ylang. 3ème adultératins des essences de basilic exotiques. Recherches 19, 269–274

Teleky-Vámossy, Gy., M. Petró-Turza, 1986: Evaluation of odour intensity versus concentration of natural garlic oil and some of its individual aroma compounds. Nahrung/Food 30, 775–782

Terpó, A., 1966: Kritische Revision der wildwachsenden Arten und der kultivierten Sorten der Gattung Capsicum L. Feddes Rep. 72, 155–191

Terra, G.J.A., 1966: Tropical vegetables. Commun. 54. Vegetable growing in the tropics and subtropics especially of indigenous vegetable. Royal Trop. Inst. Amsterdam, 107 pp

Tét´nyi, P., L. Ntezurubanza, F. Ayobangira, É. Héthelyi, L.V. Puyvelde, 1986: Essential oil variations of Ocimum suave in Rwanda. Herba Hungar. 25, 27–42

Teubner, Chr., S. Schönfeldt, U. Lundberg, 1985: A Paprika. Akadémia Budapest, 125 pp

Teuber, H., K. Herrmann, 1978: Flavanonolglykoside der Blätter und Früchte des Dills (Anethum graveolens L.). II. Gewürzphenole. Z. Lebensmittel-Unters. u. -Forsch. 167, 101–104

Teuscher, R., 2003: Gewürzdrogen. Ein Handbuch der Gewürze, Gewürzkräuter, Gewürzmischungen und ihrer ätherischer Öle. Wiss. Verlagsges. mbH Stuttgart, 468 pp

Thangadurai, D., S. Anitha, T. Pullaiah, N.P. Reddy, O. S. Ramachandraiah, 2002: Essential oil constituents invitro antimicrobial activity of Decalepis hamiltonii roots against foodborne pathogens. J. Agric. Food Chem. 50, 3147–3149

Thappa, R.K., M.S. Bhatia, S.G. Agarwal, K.L. Dhar, C.K.

Atal, 1979: Ocimin, a novel neolignan from Ocimum americanum. Phytochemistry 18, 1242

Thappa, R.K., S.G. Agarwal, N.K. Kalia, R. Kapoor, 1993: Changes in chemical composition of Tagetes minutaoil at various stage flowering and fruiting. J. Ess. Oil Res. 5, 375–379

Tharanathan, R.N., Y.V. Anjaneyalu, 1975: Structure of the acid-stable core-polysaccharide derived from the seed mucilage of Ocimum basilicum. Austral. J. Chem. 28, 1345–1350

Theilade, I., 1996: Revision of the genus Zingiberaceae in Peninsular Malaysia. Garden's Bull. Singapore 48, 207–236

Thieme, H., 1966: Isolierung eines neuen phenolischen Glykosids aus den Blüten von Filipendula ulmaria (L.) Maxim. Pharmazie 21, 123

Thieme, H., Thieme, H., Nguyen Thi Tam, 1968: Über eine Methode zur spektrophotometrischen Bestimmungvon Methylchavicol in Estragonölen. Pharmazie 23, 339–340

Thieme, H., Nguyen Thi Tam, 1972: Untersuchungen über die Akkumulation und die Zusammensetzung der ätherischen Öle von Satureja hortensis L., Satureja montana L. und Artemisia dragunculus L. im Verlauf der Ontogenese. 1. Mitt. Literaturübersicht, dünnschicht- und gaschromatograische Untersuchungen. Pharmazie 27, 255-265; 2. Mitt. Veränderungen des Ölgehaltes und der Ölzusammensetzung. Pharmazie 27, 324–331

Thode-Sonntag, I., J. Thode-Sonntag, 1988: Entdecken Sie Ingwer. Self pub. Hamburg, 47 pp

Thomas, C.A., 1960: Lakoocha (Artocarpus lakoocha L.) cultivation can give rich dividends. Indians Hortic. 5, 29–30

Thompson, P.H., 1976: The carissa in California. California Rare Fruit Grown Yearbook 8, 73–81

Tidbury, G,, 1949: The clove tree. Crosby & Lockwood London, 212 pp

Tigga, M., P.V. Sreekumar, 1996: Wild edible fruits of Andaman and Nicobar Islands. Bull. Inst. Bot. Survey of India 38, 25–37

Tindall, H.D., 1983: Vegetables in the Tropics. MacMillan Press London, 533 pp

Ting-nong, Ho, Liu Shang-wu, 2001: A worldwide monography of Gentiana. Science Press Beijing (China) New York, 685 pp

Tisserant, R.P.Ch., 1953: L'Agriculture dans les savanes de l'Oubangui. Bull. Inst. d'Etudes Central-Africaines, N.S. 6, 209–273

Tittel, G., H. Wagner, R. Bos, 1982: Über die chemische Zusammensetzung von Melissenölen. Planta med. 46, 91–98

Titz, W., E. Titz, 1982: Analyse der Formenmannigfaltigkeit der Valeriana officinalis-Gruppe im zentralen und südlichen Europa. Ber. dtsch. Bot. Ges. 95, 155–164

Tokitomo, Y., A. Kobashi, 1992: Isolation of the volatile components of fresh onion by thermal desorption cold trap capillary gas chromatography. Biosci., Biotechnol., Biochem. 56, 1865–1866

Tomás-Lorente, F., M. García-Grau, 1988: The wastes of the industrial treatment of Salvia lavandulaefolia as a source of biologically active flavonoids. Fitoterapia 59, 62–64

Tomlinson, P.B., 1956: Studies in the systematic anatomy of the Zingiberaceae. J. Linn. Soc., Bot. (London) 55, 547–592

Tommasi, N. de, F. de Simone, S. Piacenta, C. Pizza, N. Mahmood, 1995: Dipertenes from Mormordica balsamica. Nat. Prod. Letter 6, 261–268

Topf, H., 1958: Der Löwenzahn (Taraxacum officinale). A. Ziemsen Verlag Lutherstadt Wittenberg, 48 pp

Toppozada, H., H.A. Mazloum, M. El-Dakhakhny, 1965: The antibacterial properties of the Nigella sativa L. seeds. Active principle with clinical applications. J. Egypt. Med. Assoc. 48, 187–198

Torre, M.C. de la, M. Bruno, B. Rodríguez, G. Savona, 1992: Abietane and 20-nor-abietane diterpenoids from the root of Meriandra benghaliensis. Phytochemistry 31, 3953–3955

Torres, R.C., 1993: Citral from Cymbopogon citratus (DC.) Stapf (lemon grass) oil. Philippine J. Sci. 122, 269–287

Torres, R.C., R.R. Estralla, B.Q. Guevarra, 1994: Extraction and characterization of the essential oil of Philippine Cymbopogon citratus (DC.) Stapf. Philippine J. Sci. 123, 51–63

Tóth, L., 1967a: Untersuchungen über das ätherische Öl von Foeniculum vulgare. 1. Mitt. Die Zusammensetzung des Frucht- und Wurzelöles. Planta med. 15, 157–172

Tóth, L., 1967b: Untersuchungen über das ätherische Öl von Foeniculum vulgare. 2. Mitt. Veränderungen der verschiedenen Fenchelöle vor und nach der Ernte. Planta med. 15, 371–389

Toulemonde, B., I. Noleau, 1988: Volatile constituents of lovage. Flavours & Fragrance J. 4, 28

Toulemonde, B., H.M.J. Richmond, 1983: Volatile constituents of dry elder (Sambucus nigra L.) flowers. J. Agric. Food Chem. 31, 355–370

Toury, J., P. Lunven, R Giorgu, M. Jacquesson, 1957: Le baobab, arbre providental de l'African. Ann. Nutr. Aliment. 11, 99–102

Touw, M., 1982: Roses in Middle Ages. Econ. Bot. 36, 71–83

Tram Ngoc Ly, R. Yamauchi, K. Kato, 2001: Volatile components of the essential oil in galanga (Alpinia officinarum Hance) from Vietnam. Food Sci. Technol. Res. 7, 303–306

Traxl, V., 1975: Mata kaderava . (Mentha crispa). Nase Liecive Rastliny 12, No. 4, 97–98 (in Czech)

Trenkle, K., 1972: Recent studies on fennel (Foeniculum vulgare M.) 2. The volatile oil of the fruit, herbs and roots of fruit-bearing plants. Pharmazie 27, 319

Treptow, H., 1985: Rhabarber (Rheumarten) und seine Verwendung. Ernährung/Nutrition 9, 179–183

Trillini, B., A.M. Stoppini, 1994: Volatile constituents of the fruit secretory glands of Zanthoxylum bungeanum Maxim. J. Ess. Oil Res. 6, 249–252

Trillini, B., A. Manunta, A.M. Stoppini, 1991: Constituents of the essential oil of the fruits of Zanthoxylum buneanum. Planta med. 57, 90–91

Trueb, L.F., 1998a: Pinienkerne. Naturwiss. Rdsch. 51, 260–262

Trueb, L.F., 1998b: Der Hopfen. Naturwiss. Rdsch. 51, 469–472

Trujillo, J.M., J.M. Hernández, J.A. Pérez, H. Lépez, I. Frias, 1996: A secoiridoid glucoside from Jasminum odoratissimum. Phytochemistry 42, 553–554

Tsaknis, J., S. Lalas, V. Gergis, V. Dourtoglou, V. Spiliotis, 1998: Modifiecazioni nella qualità dell'olio di semi di Moringa oleifera, varietà Blyntyre, durante il processo di frittura. Riv. Ital. Sostanze Grasse 75, 181–190

Tsaknis, J., S. Lalas, V. Gergis, V. Dourtoglou, V. Spiliotis, 1999: Characterization of Moringa oleifera variety Mbololo seed oil of Kenya. J. Agric. Food Chem. 47, 4495–4499

Tsankova, E.T., A.N. Konaktchiev, E.M. Genova, 1993: Chemical composition of the essential oils of two Hyssopus officinalis taxa. J. Ess. Oil Res. 5, 609–611

Tscheche, R., H. Kohl, 1968: Kondurangoglycoside A, A1 und C, C1. Tetrahedron 24, 4359–4371

Tschesche, R., S. Sepulveda, Th. M. Braun, 1980: Über das Saponin der Blüten von Verbascum phlomoides L. Chem. Ber. 113, 1754–1760

Tschirch, A., 1892: Indische Heil- und Nutzpflanzen und deren Cultur. R. Gaertners Verlagsbuchhandl. Berlin, 223 pp

Tsuda, T., M. Watanabe, K. Oshima, A. Yamamoto, Sh. Kawakishi, T. Osawa, 1994: Antioxidative components isolated from the seeds of tamarind (Tamarindus indicus L.). J. Agric. Food Chem. 42, 2671–2674

Tuan, D.Q., S.G. Ilangantileke, 1997: Liquid CO_2 extraktion of essential oil from star anise fruits (Illicium verum L.). J. Food Engng. 31, 45–57

Tucker, A.O., 1986: Botanical nomenclature orf culinary herbs and potherbs. In: Craker, L.E., J.E. Simon (eds) Herbs, spices, and medical plants: recent advances in botany, horticulture, and pharmacology. Oryx Press Phoenix, Arizona, Vol. 2, pp 33–80

Tucker, A.O., K.J.W. Hensen, 1985: The cultivars of lavender and lavandin (Labiatae). Baileya 22, 168–177

Tucker, A.O., M.J. Maciarello, J.T. Howell, 1980: Botanical aspects of commercial sages. Econ. Bot. 34, 16–19

Tucker, A.O., M.J. Maciarello, 1986: The essential oils from Rosemary cultivars. Flavour & Fragrance J. 1, 137–142

Tucker, A.O., M.J. Maciarello, 1990: Essential oils of cultivars of Dalmatian sage (Salvia officinalis L.) J. Ess. Oil Res. 2, 139–144

Tucker, A.O., M.J. Maciarello; 1999: Volatile oils of Illicium floridanum and I. parviflorum (Illiciaceae) of the south-eastern United States and their potential economic utilization. Econ. Bot. 54, 435–438

Tucker, A.O., M.J. Maciarello, P.W. Burbage, G. Sturz, 1994: Spicebush (Lindera benzoin [L.] var. benzoin, Lauraceae): a tea, spice and medicine. Econ. Bot. 48, 333–336

Tucker, A.O., M.J. Maciarello, D.J. Charles, J.E. Simon, 1997: Volatile leaf oil of the curry plant (Helchrysum italicum [Roth] G. Don subsp. italicum) and dwarf curry plant (subspec. microphyllum (Willd.) Nym. in the North American herb trade. J. Ess. Oil Res. 11, 711–715

Tucker, A.O., M.J. Maciarello, J.R. Espaillat, E.-C. French, 1993: The essential oil of Lippea micromera Schauer in DC. (Verbenaceae). J. Ess. Oil Res. 5, 683–685

Tugrul, N., I. Doymaz, M. Pala, 2001: A study on drying characteristics of dill. Gida (Istanbul) 26, 403–407

Tull, D., 1999: Edible and useful plants of Texas and the Southwest. A practical guide. Univ. Texas Press, Austin TX, 518 pp

Tümen, G., K.H.C. Başer, 1993: The essential oil of Origanum syriacum L. var. bevanii (Holmes) letswaart. J. Ess. Oil Res. 5, 315–316

Tümen, G., N. Ermin, T. Ozek, M. Kurkcuoglu, K.H.C. Başer, 1994: Composition of essential oils from two varieties of Thymbra spicata L. J. Ess. Oil Res. 6, 463–468

Tunman, P., E. Mann, 1968: Über Inhaltsstoffe von Gewürzdrogen. 1. Mitt. Zum Vorkommen von Cumarinen, einem Wachs und Sterinen in Kraut von Artemisia dracunculus L. Z. Lebensmittel-Unters. u. -Forsch. 138, 146–150

Türkay, S., M.D. Burford, M.K. Sangün, E. Ekinci, K.D. Bartle, A.A. Clifford, 1996: Deacidification of black cumin seed oil by selective supercritical carbon dioxide extraction. J. Amer. Oil Chem. Soc. 73, 1265–1270

Turova, A.D., E.N. Saposhnikova, Vjen Dieok Li (Турова, А. Д., Э..Н. Саапожнникова и Вье Дыок Ли), 1987: Official plants of the USSR and Vietnam. (Лекарственные растения СССР и Вьетнама.) Publishers "Medicine", Moscow (Москва "Медицина"), 463 pp (in Russian)

U

Ubillos, A.M., 1989: Plantas aromáticas de la Espana peninsular. Edic. Mundi-Prensa, Madrid 112 pp

Udayasekhara Rao, P., 1996: Nutrient composition and biological evaluation of mesta (Hibiscus sabda riffa) seeds. Plant Food Human Nutrit. 49, 27–34

Ueki, T., Y. Nosa, Y. Teramoto, R. Ohba, S. Ueda, 1994: Practical soy sauce production using a mixed koji-making system. J. Fermentation & Bioengng. 78, 262–264

Uhl, S.R., 2000: Handbook of spices, seasonings & flavorings. Technomic Publishing Comp. Lancaster, PA, USA, 329 pp

Uhlig, J.W., A. Chang, J.J. Jen, 1987: Effect of phthalides on celery flavor. J. Food Sci. 52, 658–660

Ulubelen, A. T. Berkan, 1977: Triterpenic and steroidal compounds of Cnicus benedictus. Planta med. 31, 375–377

Ululeben, A., C.H. Brieskorn, 1975: Micromeric acid from Salvia horminum. Phytochemistry 14, 1450

Ulubelen, A., I. Uygur, 1976: Flavonoidal and other compounds of Salvia aethiopis. Planta med. 29, 318–320

Ulubelen, A., C.H. Brieskorn, N. Ozdemir, 1977: Triterpenoids of Salvia horminum, constitution of a new diol. Phytochemistry 16, 790–791

Ulubelen, A., M.Miski, T.J. Manry, 1981: A new acid from Salvia tomentosa. J. Nat. Prod. 44, 119–124

Ulubelen, A., A. Miski, P. Neuman, T.J. Mabry, 1979: Flavonoids of Salvia tomentosa (Labiatae). J. Nat. Prod. 42, 261–263

Umano, K., Y. Hagi, K. Nakahara, T. Shibamoto, 2000: Volatile chemicals identified in extracts from leaves of Japanese mugwort (Artemisia princeps Pamp.). J. Agric. Food Chem. 48, 3463–3469

Uotila, P., 1977: Notes on Portulacca olearacea (Portulaccaceae) in Finland. Mem. Soc. Fauna & Flora Fennica 53, 21–23

Upadhyay, R.K., L.N. Misra, Gurdip Singh, 1991: Sesquiterpene alcohols of the copane series from essential oil of Ocimum americanum. Phytochemistry 30, 691–693

Upadhyay, J., R.S. Singh, T.N. Misra, 1982: Chemical constituents of Hyptis suaveolens Poit. Indian J. Pharmac. Sci. 44, No. 2, 19–20

Uphof, J.C.Th., 1968: Dictionary of economic plants. Verlag J. Cramer Lehre, 2. edn., 591 pp

Urbano, M.G., I. Goni, 2002: Bioavailability of nutrients in rats fed on edible seaweeds, nori (Porphyra tenera) and wakame (Undaria pinnatifida), as a source of dietary fibre. Food Chem. 76, 281–286

Urzúa, A., R. Torres, 1983: Alkaloids from the bark of Peumus boldus. Fitoterapia 4, 175–177

Usai, M., A. Atzei, G. Pintore, I. Casanova, 2003: Composition and variability of the essential oil of Sardinian Thymus herba-barona Loisel. Flavour & Fragrance J. 18, 21–25

Usher, G., 1974: A dictionary of plants used by man. Constable & Co. Ltd. London, 619 pp

Üstun, G., L. Kent, N. Chekin, H. Civelekogiu, 1990: Investication of the technological properties of Nigella sativa (Black cumin) seed oil. J. Amer. Offic. Chem. Soc. 67, 958–960

V

Vagi, E., B. Simandi, H.G. Daood, A. Deak, J. Sawinsky, 2002: Recovery of pigments from Origanum majorana L. by extraction with supercritical carbon dioxide. J. Agric. Food Chem. 50, 2297–2301

Valentão, P., E. Fernandes, F. Carvalho, P.B. Andrade, R.M. Seabra; M.L. Bastos, 2001: Antioxidant activity of Centaurium erythrea infusion evidenced by its superoxide radical scaveging and xanthine oxidase inhibitory activity. J. Agric. Food Chem. 49, 3476–3479

Valentão, P., P.B. Andrade, E. Silva, A. Vicente, H. Santos, M.L. Bastos, R.M. Seabra, 2002: Methoxylated xanthones in the quality control of small centaury (Centaurium erythrea) flowering tops. J. Agric. Food Chem. 50, 460–463

Valkenburg, J.L.C-H., van, N. Bunyapraphatsara, 2001: Plant Resources of South-East Asia. No. 12, Medicinal and poisonous plants 3. Backhuys Publishers Leiden 782 pp

Valverde, M.R., 1986: Taxonomía de los géneros Thymus (excluda la sección Serphyllum) y Thymbra en la Península Iberica. Real Jardin Botánico (CSIC) Madrid, 324 pp

Vanhaelen-Fastré, R., 1973: Constitution et properties antibacterinnes de l'huile essentielle de Cnicus benedictus. Planta med. 24, 165–175

Vanhaelen-Fastré, R., 1974: Constituents polyacetyleniques de Cnicus benedictus. Planta med. 25, 47–59

Vanhaelen, M., R. Vanhaelen-Fastré, 1983: Quantitative determination of biologically active constituents in medical plant crude extracts by thin-layer chromatography-densitometry. J. Chromatogr. 281, 263–271

Vareltzis, K., D. Koufidis, E. Gavriilidou, E. Papavergou, S. Vasiliadou, 1997: Effectiviness of a natural rosemary (Rosemarinus officinalis) extract on the stability of fillet and minced fish during frozen storage. Z. Lebensmittel-Unter. u. -Forsch. 205 A, 93–209

Variyar, P.S., C. Bandyopadhyay, 1994: Estimation of phenolic compounds in green pepper berries (Piper nigrum L.) by high-performance liquid chromatography. Chromatographia 39, 743–746

Vasconcelos, S.M.G. de, F.J. de Abreu Matos, M.I. Lacerda Machado, A. Aragao Craveiro, 2003: Essential oils of Ocimum basilicum L., O. basilicum var. miniumum L. and O. basilicum var. purpurescens Benth. grown in north-eastern Brazil. Flavour & Fragrance J. 18, 13–14

Vasudevan, T.N., M. Luckner, 1968: Alkaloide aus Ruta angustifolia Pers., Ruta chalepensis L., Ruta graveolens L. und Ruta montana Mill. Pharmazie 23, 520–521

Vaugham, J.G., E.I. Gordon, 1973: A taxonomy study of Brassica juncea using the technique of electrophoresis, gas liquid chromatography and serology. Ann. Bot. 167–184

Vaugham, J.G., J.S. Hemingway, 1959: The utilization of mustards. Econ. Bot. 13, 196–204

Vaupel, E., 2002a: Betört von Vanille. Seit 500 Jahren begehrt und immer noch Forschungsthema. Kultur u. Technik 26, No. 1, 47–51

Vaupel, E., 2002b: Gewürze. Acht kulturhistorische Porträts. Deutsches Museum Munich, 144 pp

Vaupel, E., 2002c: Seit 500 Jahren als Gewürz begehrt. Pharmaz. Ztg. 147, 3628–3632, 3635

Vega, M.N., C. Ramos, 1987: Constituents of capers and changes during pickling. Grasas y Aceites 38, 173–175

Vekiari, S.A., V. Oreopoulou, C. Tzia, C.D. Thomopoulos, 1993a: Oregano flavonoids as lipid antioxidants. J. Amer. Oil Chem. 70, 483–486

Verghese, J., 1991: Garcinia cambogia (Desr.) Kodampuli. Indian Spices 28, No. 1, 19–21

Vernin, G., J. Metzger, D. Fraisse, C. Scharff, 1986: GC-MS(EI, PCI, NCI) Computer analysis of volatile sulfur compounds in garlic essential oils. Application of the mass fragmentometry SIM Technique. Planta med. 96–101

Vikiari, S.A., C. Tzia, V. Oreopoulou, C.D. Thomopoulos, 1993b: Isolation of natural antioxidants from oregano. Riv. Ital. Sostanze Grasse 70, 25–34

Venskutonis, P.R., A. Dapkevicius, T.A. van Beek, 1996a: Essential oils of fennel (Foeniculum vulgare Mill.) from Lithuania. J. Ess. Oil Res. 8, 211–213

Venskutonis, R., L. Poll, M. Larsen, 1996b: Effect of irradiation and storage on the composition of volatile compounds of basil (Ocimum basilicum L.). Flavour & Fragrance J. 11, 117–121

Venskutanis, R. L. Poll, M. Larsen 1996c: Influence of drying and irradiation on the composition of volatile compounds of thyme (Thymus vulgaris L.). Flavour & Fragrance J. 11, 123–128

Vera, R., J.M. Mondon, J.C. Pieribattesti, 1993: Chemical composition of the essential oil and aqueous extract of Plectranthus amboinicus. Planta med. 59, 182–183

Veres, K., E. Varga, A. Dobos, Z. Hajdu, I. Mathe, E. Nemeth, K. Szabo, 2003: Investigation of the composition and stability of the essential oil of Origanum vulgare ssp. vulgare L. and O. vulgare ssp. hirtum (L.) Ietswaart. Chromatographia 57, 95–98

Verghese, J., 1991: Cumin. Perfumer & Flavorist 16, 61–64

Verghese, J., 1993: Isolation of Curcumin from Curcuma longa L. rhizome. Flavour & Fragrance J. 8, 315–319

Verheij, E.W.M., R.E. Coronel (eds), 1991: Plant resources of South-East Asia. No. 2. Edible fruits. Pudoc Sci. Publishers Wageningen, 467 pp

Verma, P.K., M.S. Punia, G.D. Sharma, G. Talwar, 1989: Evaluation of different species of Ocimum for their herb and oil yield under Haryana conditions. Indian Perfumer 33, No. 2, 79–83

Vernin, G., J. Metzger, 1984: Analysis of basil oil by GC-MS data bank. Perfumer & Flavorist 9, 71–86

Vernin, G., J. Metzger, 1986: Analysis of sage oil by GC-MS-data bank. Perfumer & Flavorist 11, 79–84

Vernin, G., J. Metzger, D. Fraisse, C. Scharff, 1989: GC-MS (EI, PCI, NCI) Computer analysis of volatile sulfur compounds in garlic essential oils. Application of the mass fragmentometry SIM Technique. Planta med. 34, 96–101

Vernin, G., C. Lageot, E.M. Gaydou, C. Parkanyi, 2001: Analysis of the essential oil of Lippea graveolens HBK from El Salvador. Flavour & Fragrance J. 16, 219–226

Vernon, F., H.H. Richard, 1983: Untersuchung der flüchtigen Bestandteile des ätherischen Öles des Blattes von Kräuselpetersilie, Petroselinum hortense Hoffm. Lebensmittel-Wiss. u. -Technol. 16, 32–35

Verottea, L., 1985: Isolation and HPLC determination of the active principles of Rosmarinus officinalis and Gentiana lutea. Fitoterapia 56, 25–29

Verschuere, M., P. Sandra, F. David, 1992: Fractionation by SFE and microcolum analysis of the essential oil and the bitter principles of hops. J. chromatogr. Sci. 30, 388–391

Verzele, M., 1986: Centenary review. 100 years of hop chemistry and its relevance to brewing. J. Inst. Brewing 92, 32–48

Verzera, A., G. la Rosa, M. Zappala, A. Cotroner, 2000: Essential oil composition of different cultivars of bergamot in Sicily. Ital. J. Food Sci. 12, 493–501

Verzera, A., C. Russo, G. la Rosa, I. Bonaccorsi, A. Cotronea, 2001: Influence of cultivar lemon oil composition. J. Ess. Oil Res. 13, 343–347

Verzera, A., A. Tozzi, F. Gazea, G. Cicciarello, A. Cotroneo, 2003: Effects of rootstock on the composition of bergamot (Citrus bergamia Risso et Poiteau) essential oil. J. Agric. Food Chem. 51, 206–210

Vickers, W.T., T. Plowman, 1984: Useful plants of the Siona and Secoya Indians of Eastern Ecuador. Fieldiana, Not. (Chicago), N.S. No. 15, Publ. 1356, 63 pp

Vickery, A.R., 1981: Traditional uses and folklore of Hypericum in the British Isles. Econ. Bot. 35, 288–295

Vidal, J., 1967: Nothes ethnobotaniques alvégées sur quelques plantes du Cambodge. J. Agric. Trop. Bot. Appl. 14, No. 1–3, 21–66

Vidari, G., P. Vita-Finzi, 1971: Flavanols and quinones in stems of Aframomum gigantium. Phytochemistry 10, 3335–3359

Viel, P., 1979: Wissenswertes über Enzian. Ernährung/Nutrition 3, 210–211

Villalón, F.J. Dainello, D.A. Bender, 1994: 'Jaloro' hot yellow jalapeno pepper. Horticult. Sci. 29, 1092-1093

Villamar, A.A., L.M.C. Asseleih, M.E. Rodarte (eds), 1994: Atlas de las plantas de la medicina tradicional Mexicana. Inst. Nac. Indigenista, Col. Tlacopac, Mexiko, Vol. I–III, 1786 pp

Vines, R.A., 1960: Trees, shrubs, and woody vines of the Southwest. Univ. of Texas Press Austin, TX

Virmani, O.P., S.C. Datta, 1975: Vetivera zizanoides (L.) Nash. Indian Perfumer 19, 73–95

Virmani, O.P., R. Srivastava, S.G. Datta, 1979: Oil of lemongrass. Part 2: West Indian. World Crops 31, 120–121

Vöttiner-Pletz, P., 1990: Lignum sanctum. Zur therapeutischen Verwendung des Guajak vom 16.–20. Jahrhundert. Govi-Verlag Frankfurt/M., 312 pp

Vogel, G., 1995a: Winterzwiebel. TASPO-Gartenmagazin, No. 8, p 39

Vogel, G., 1995b: Schalotte. TASPO-Gartenmagazin 4, No. 9, p 50

Vogt, H.-H., 1991: Rosenwasser aus Taif. Naturwiss. Rdsch. 44, 65

Voirin, B., N. Brun, C. Bayet, 1990: Effect of day length on the monoterpene composition of leaves of Mentha x piperita L. Phytochemistry 29, 749–755

Vokou, D., S. Kokkini, M. Bessiere, 1989: Origanum onites (Lamiaceae) in Greece: distribution, volatile oil, yield, and composition. Econ. Bot. 43, 407–412

Vollmann, C., 1988: Levisticum officinale – Liebstöckel. Z. Phytother. 9, 128–132

Vollmann, C., W. Schultze, 1995a: Composition of the root essential oils of several Geum species and related members of the subtribus Geinae (Rosaceae). Flavour & Fragrance J. 10, 173–178

Vollmann, C., W. Schultze, 1995b: Nelkenwurz. Untersuchungen an Geum urbanum und der Handelsdroge Gei urbanei radix. Dtsch. Apotheker-Ztg. 135, 1238, 1241–1244, 127–148

Vogt, H.-H., 1987: Safran. Naturwiss. Rdsch. 40, 230

Vogt, H.-H., 1988: Kardamom. Naturwiss. Rdsch. 102–103

Vokou, D., S. Kokkini, J.M. Bessière, 1988: Origanum onites (Lamiaceae) in Greece: distribution, volatile oil yield, and composition. Econ. Bot. 42, 407–412

Vos, M.P. de, 1984: The African genus Crocosmia Planchon. J. S. Afric. Bot. 50, 463–502

Vos, R.H. de, W.G.H. Blijleven, 1988: The effect of processing on glucosinolates in cruciferous vegetables. Z. Lebensmittel-Unters. u. -Forsch. 187, 525–529

Vostrowsky, O., K. Michaelis, H. Ihm, K. Knobloch, 1984: Das ätherische Öl von Artemisia abrotanum. Z. Lebensmittel-Unters. u. -Forsch. 170, 125–128

Vostrowsky, O., K. Michaelis, H. Ihm, R. Zintl, K. Knobloch, 19881a: Über die Komponenten des ätherischen Öles aus Estragon (Artemisia dragunculus L.). Z. Lebensmittel-Unters. u. -Forsch. 173, 365–367

Vostrowsky, O., K. Michaelis, H. Ihm, R. Zintl, K. Knobloch, 1981b: Über die Zusammensetzung des ätherischen Öles aus Artemisia dracunculus L. während der Vegetationsperiode. Z. Naturforsch., 36 C, 724–727

Vostrowsky, O., Th. Brosche, H. Ihm, R. Zintl, K. Knobloch, 1981c: Über die Komponenten des ätherischen Öles aus Artemisia absinthium L. Z. Naturforsch. 36 C, 369–372

Vostrowsky, O., W. Garbe, H.J. Bestmann, J.G.S. Maia, 1990: Essential oil of alfavaca, Ocimum gratissimum, from Brazilian Amazon. Z. Naturforsch., Sect. C, Biosci. 45, 1073–1976

Vvedensky, A.I., 1944: The genus Allium in the USSR. Herbertia 11, 15–218

W

Wagner, H., J. Hölzl, 1968: Die mikroskopische und chromatographische Unterscheidung von Fructus Petroselini und Fructus Ajowani. Dtsch. Apotheker-Ztg. 108, 1620–1624

Wagner, H., K. Münzing-Vasirian, 1975: Eine chemische Wertbestimmung der Enziandroge. Dtsch. Apotheker-Ztg. 115, 1235–1239

Wagner, H., T. Nestler, 1978: N-Methoxy-1-vinyl-β-carbolin, ein neues Alkaloid Picrasma exelsa (Swartz) Planch. Tetrahydron Letters 31, 2777–2778

Wagner, H., A. Sendl, 1990: Bärlauch und Knoblauch. Dtsch. Apotheker-Ztg. 130, 1809–1815

Wagner, H., S. Bladt, K. Münzing-Vasirian, 1975: Dünnschichtchromatographie von Bitterstoffen. Pharmaz. Ztg. 120, 1262–1265

Wagner, K., 1992: Ein Scharfmacher. Pharmaz. Ztg. 137, 3588–3590

Wahlroos, Ö., 1965: Volatiles from chives (Allium schoenoprasum L.). Acta Scand. Chem. 19, 1327–1332

Walker, A.R., 1952: Usages pharmaceutique des plantas spontaneis du Gabon. Bull. Inst. Étrichs Central Africa, N.S. 4, 181–186

Walker, A.R., 1953: Usages pharmaceutique des plantes spontaneis du Gabon. Bull. Inst. Étrichs Central Africa, N.S. 5, 275–329

Walker, M.R., B.M. Beattie, 1980: Peppermint oil production in Tasmania. J. Agric. Tasmania 51, 5–8

Walther, H., 1969: Inhaltsstoffe von Meum athamanticum Jacq. Pharmazie 24, 781–782

Wang, M., J. Li, G.S. Ho, X. Peng, C.T.Ho, 1998: Isolation and identification of antioxidative flavonoid glycosides from thyme (Thymus vulgaris L.) J. Food Lipids 5, 313–321

Warburg, O., 1897: Die Muskatnuß. Ihre Geschichte, Botanik, Kultur, Handel und Verwertung sowie ihre Verfälschung und Surrogate. Verlag W. Engelmann Leipzig, 628 pp

Warncke, D., 1994: Petroselinum crispum – Die Gartenpeterilie. Z. Phytother. 15, 50–58

Warren, W., 1998: Tropical flowers of Thailand. Asia Books Co., Ltd. Bangkok (Thailand), 63 pp

Warrier, P.K., V.P.K. Nambiar, C. Ramankulty (eds), 1995: Indian medical plants. Orient Longman Madras (India) 444 pp

Wassenhove, F. van, P. Dirinck, G. Vulsteke, N. Schamp, 1990: Aromatic volatile composition of celery and celeriac cultivars. Hort. Sci. 25, 556–559

Waßmuth-Wagner, I., H. Jork, 1992: Strategy for the chromatographic separation of the essential oil of curry leaves. J. Planar Chromatogr.-Modern TCL 5, 383–387

Waßmuth-Wagner, I., H.-O. Kalinowski, H. Jork, 1995: Isolation and identification of 11-salinen-4α, 7β-ol and 10-aromadendranol in the essential oil of Murraya koenigii. Planta med. 61, 196–197

Watanabe, M., M. Ohata, S. Hayakawa, M. Isemura, S. Kumazawa, T. Nakayama, M. Furugori, N. Kinae, 2003: Identification of 6-methylsulfinylhexyl isothiocyanate as an apoptosis-inducing component in wasabi. Phytochemistry 62, 733–739

Way, R.M., 1985: Volatile oil analysis of cassia bark. J. Assoc. Off. Analyt. Chem. 68, 622–625

Wealth of India. A dictionary of raw materials. Vol. I A–B 1948, 258 pp; Vol. II C 1950, 426 pp; Vol. III Ca–Ci 1992, 683 pp; Vol. III D–E 1952, 236 pp; Vol. III, refid. edn.* I–III 1992, 683 pp + Index 119 pp; Vol. IV F–G* 1956, 287 pp; Vol. V H–K 1959, 332 pp; Vol. VI L–M* 1962 483 pp; Vol. VII N–Pe 1966, 330 pp; Vol. VIII Ph–Re 1969, 394 pp; Vol. iX Rh–So 1972, 472 pp; Publications & Information Directorate, CISR Delhi, India; * Council of Sci. & Ind. Res., New Delhi, India

Webb, W.J., 1984: The Pelarganium family: the species of Pelargonium, Monsonia and Sarcocaulon. Croom Helm London (UK), 104 pp

Weber, G. 2002: Qualitätsanforderungen an Meerrettichdauerwaren. Dtsch. Lebensmittel-Rdsch. 98, 220

Weber, R., 1995: Untersuchungen zum Inhaltsstoffspektrum und zur biologischen Aktivität von Verbena officinalis L. Verlag J. Cramer Berlin Stuttgart, 240 pp (Dissertations Bot. Bd. 252)

Weber, S., 1997: Iris. Die besten Arten und Sorten im Garten. E. Ulmer Verlag Stuttgart, 143 pp

Weber, U., 1951: Cuminum cyminum L., der Kreuzkümmel. Pharmazie 6, 22–26

Webster, T.M., 1995: New perspectives on vanilla. Cereal Foods World 40, 198–200: Weil, A.T., 1965: Nutmeg as a narcotic. Econ. Bot. 19, 194–217

Weinberg, D.S., M.L. Manier, M.D. Richardson, F.G. Haibach, 1993a: Identification and quantification of organosulfur compliance markers in a garlic extract. J. Agric. Food Chem. 41, 37–41

Weinberg, D.S., M.L. Manier, M.D. Richardson, F.G. Haibach, 1993b: Identification and quantification of isoflavonoid and triterpenoid compliance markers in a licorice-root extract powder. J. Agric. Food Chem. 41, 42–47

Weiss, B., H. Flück, 1970: Untersuchungen über die Variabilität von Gehalt und Zusammensetzung des ätherischen Öles in Blatt- und Krautdrogen von Thymus vulgaris L. Pharmac. Acta Helv. 45, 169–173

Weiss, E.A., 1971: Castor, sesame and safflower. Leonhard Hill London, 901 pp

Wen-lan Hu, Ji-ping Zhu, M. Hesse, 1989: Indole alkaloids from Alstonia angustifolia. Planta med. 55, 463–466

Werker, E., E. Putievsky, U. Ravid, 1985: The essential oils and glandular hairs in different chemotypes of Origanum vulgare L. Ann. Bot. 55, 793–801

Werker, E., E. Putievsky, U. Ravid, N. Dulai, I, Katzir, 1994: Glandular hairs, secretory cavities and the essential oil in leaves of tarragon (Artemisia dracunculus L.). J. Herbs, Spices & Med. Plants 2, 19–32

Westermann, D., 1909: Die Nutzpflanzen unserer Kolonien und ihre wirtschaftliche Bedeutung für das Mutterland. Verlag D. Reimer Berlin, 94 pp

Westphal, E., P.C.M. Jansen (eds), 1989: Plant resources in South-East Asia. A selection. Pudoc Wageningen, 322 pp

Weyersdorf, P., U. Splittgerber, H. Marschnall, 1995: Constituents of the leaf essential oil of Hypericum perforatum from India. Flavour & Fragrance J. 10, 365–370

Weyerstahl, P., H. Marschall, E. Manteuffel, S. Huneck, 1992: Volatile constituents of Agastache nigosa. J. Ess. Oil Res. 4, 585–587

Whitaker, J.R., 1976: Development of flavor, odor and purgency on onion and garlic. Adv. Food Res. 22, 73–133

Whitehouse, W.E., 1957: The pistachio nut. A new crop for the western United States. Econ. Bot. 11, 281–321

WHO Manila/Inst. Materia Medica Hanoi, 1990: Medicinal plants in Viet Nam. WHO Regional Publ., Western Pacific Ser., No. 3, 410 pp

Wickens, G.E., 1979: The uses of the baoba (Adansonia digitata L.) in Africa. In: Kunkel, G. (ed) Taxonomic aspects of African economic botany. Las Palmas de Gran Canaria, pp 27–34

Wickens, G.E., 1982: The baoba – Africa's upside-down tree. Kew Bull. 37, 173–209

Widder, S., Chr. Sabater, 2002: Aromabildung in der

Zwiebel: vom tränenden Auge zu einem der stärksten Aromastoffe. dragoco-rep. 49, No. 1, 4–19

Widerlechner, M.P., 1981: History and utilization of Rosa damascens. Econ Bot. 35, 42–58

Wiegele, M., 2000: Duftpelargonien. Anbau, Pflege, Sorte. Verlag E. Ulmer Stuttgart, 93 pp

Wiendl, R.M., G. Franz, 1994: Myrrhe. Neue Chemie einer alten Droge. Dtsch. Apotheker-Ztg. 134, 25–27, 31–32

Wiersema, J.H., B. León (eds), 1999: World economic plants. CRC Press New York, 749 pp

Wiesel, B., 1993: Untersuchungen zum Keimverhalten der Mariendistel (Silybium marianum [L.]) Gaertn. Drogenreport 6, No. 9, 8–11

Wijaya, C.H., H. Nishimura, T. Tanaka, J. Mizutani, 1991: Influence of drying method in volatile sulfur constituents of Caucasus (Allium victorialis L.). J. Food Sci. 56, 72–75

Wijesekera, R.O.B., 1978: The chemistry and technology of Cinnamon. Crit. Rev. Food Sci. Nutrit. 10, 1–30

Wildeisen, A., 2003: Vanille. Gewürz der Göttin. AT Verlag Aarau Munich, 2. Aufl, 128 pp

Wildeman, É. De, 1902: Les plantes tropicales de Grande culture. Maison d'Edn. A. Castaigne, Bruxelles, 303 pp

Wilding, K., W. Flemming, K. Kiewitt, K. Kolbe, R. Beining, N. Wetzel, F. Rosenhahn, 1983: Zur Herstellung von Hopfenpellets. 2. Mitt. Technologie der Hopfenpelletierung. Lebensmittelindustrie 30, 125–129

Wilhelm, C., 1966: Muskatnuß verschimmelt. Feinkostwirtschaft 9, 297–301

Wilkins, C.K, J. Madson, 1991: Oregano headspace constituents. Z. Lebensmittel-Unters. u. -Forsch. 192, 214–219

Wilkins, C.K., 1992: The influence of storage conditions on spice paprika quality. Lebensmittel-Wiss. u. Technol. 25, 219–223

Wilkinson, J.M., M. Hipwell, T. Ryan, H.M.A Cavanagh, 2003: Bioactivity of Backhousia citriodora: antibacterial and antifungal activity. J. Agric. Food Chem. 51, 76–81

Williams, L.O., 1970: A yellow food dye-Escobedia. Econ. Bot. 24, 459

Willimot, S.G., 1963: The culture and applications of liquorice. World crops. 15, 473–479

Willuhn, G., 1981: Neue Ergebnisse der Arnikaforschung. Pharmazie in unserer Zeit 10, 1–7

Willuhn, G., 1985a: Die Blüten von Arnica chamoissonis als Ersatzdroge für Arnica montana. Dtsch. Apotheker-Ztg. 125, 734–735

Willuhn, G., 1985b: Arnica montana L.: – Der Bergwohlverleih. Z. Phytother. 6, 94–102

Willuhn, G., J. Kresken, 1983: Zur Frage des Vorkommens von Pyrrolizidinalkaloiden in Arnika. Pharmaz. Ztg. 128, 515

Willuhn, G., G. Leven, 1991: Arnikablüten – Variabilität des Wirkstoffgehaltes in der von Arnica cha missonics Less. ssp. foliosa stammenden Drogen. Drogenreport. Sonderausgabe, Tagung „Arzneipflanze 1991", vom 9.–11.11.1991 in Erfurt, pp 17–22

Willuhn, G., W. Richter, 19977: Zur Kenntnis der Inhaltsstoffe von Sambucus nigra L. Planta med. 31, 328–343

Willuhn, G., R.-G. Westhaus, 1987: Loliolide (Calendin) from Calendula officinalis. Planta med. 53, 304

Willuhn, G., R Schneider, U. Matthiesen, 1985: Mexikanische Arnikablüten. Zusammensetzung des ätherischen Öles der Blütenkörbchen von Heterotheca inuloides. Dtsch. Apotheker-Ztg. 125, 1941–1944

Wilska-Jeszka, J., G. Anders, J. Los, M. Urabaniak-Olszewska, 1988: Influence des techniques de prétraitment des fruits d'Aronia melanocarpa sur la composition polyphénolique de vines. Proc. J. Internat. d'Etude du Groupe Polyphénole, Montpellier, France 13, p 43

Wilson, F.D., M.Y. Menzel, 1964: Kenaf (Hibiscus cannabinus) und Roselle (Hibiscus sabdariffa). Econ. Bot. 18, 80–91

Wilson, L.A., N.P. Senechal, M.P. Widrlechner, 1992: Headspace analysis of the volatile oils of Agas tache. J. Agric. Food Chem. 40, 1362–1366

Winer, N., 1980: The potential of the carob (Ceratonia siliqua). Internat. Tree Corp. J. 1, 15–26

Winkler, W., E. Lunau, 1959: Zur Unterscheidung der ätherischen Öle von Curcuma xanthorrhiza Roxb. und Curcuma longa L. durch Dünnschichtchromatographie. Pharmaz. Ztg. 104, 1407–1408

Winter, A.G., E. Willeke, 1953a: Untersuchungen über Antibiotika aus höheren Pflanzen. Gasförmige Hemmstoffe aus Lepidium sativum und ihr Verhalten im menschlichen Körper bei Aufnahme von Lepidium-Salat per os. Naturwissenschaften 40, 167–168

Winter, A.G., M. Hornborstel, 1953b: Untersuchungen über Antibiotika aus höheren Pflanzen. Gasförmige Hemmstoffe aus Cochlearia armoracia (Meerrettich) und ihr Verhalten im menschlichen Körper bei Aufnahme per os. Naturwissenschaften 40, 489–490

Winter, W.P. de, V.B. Amoroso (ed): 2003: Plant resources of South-East Asia. Vol. 15:2: Cryptogams: Ferns and fern allies. Backhuys Publishers Leiden, 268 pp

Winterhalter, P., M. Straubinger, 2000: Saffron – renewed interest in ancient spices. Food Rev. Inst. 16, 39–59

Winterston, D., K. Richardson, 199965: An investigation of the chemical constituents of Queensland grow ginger. Queensland J. Agric. Sci. 22, 205–214

Woerdenberg, H.J., H. Hendriks, R. Bos, 1991: Eupatorium cannabium L. – Der Wasserdost oder Wasserhanf. Neue Forschungen an einer alten Arzneipflanze. Z. Phytother. 12, 28–34

Wofford, G.E., 1983: A new Lindera (Lauraceae) from North America. J. Arnold Arboretum 64, 325–331

Wolf, G.P. de, 1955: Nepetes, notes on cultivated Labiatae. Baileya 3, 99–107

Wollenweber, E., K. Mann, 1983: Flavanols from fruits from Vitex agnus castus. Planta Med. 48, 126–127

Wolski, T., A. Ludwiczuk, A. Zwolan, M. Mardarowicz, 2000: GC/MS analysis of the content and composition of essential oil in leaves and gallenic preparations from rosemary (Rosmarinus officinalis L.). Herba polon. 46, 243–248

Wolters, B., 1994: Drogen, Pfeilgift und Indianermedizin. Arzneipflanzen aus Südamerika. Urs Freud Verlag Greifenberg, 286 pp

Wolters, B., 1996: Agave bis Zaubernuß. Heilpflanzen der Indianer Nord- und Mittelamerikas. Urs Freud Verlag Greifenberg 299 pp

Wolters, B., 1998: Zierpflanzen aus Mexiko und Südamerika. Dtsch. Apotheker-Ztg. 138, 4340–4343

Wong, K.C., K. Ong, C. Lim, 1992: Constitution of the volatile oil of Kaempferia galanga: Flavour & Fragrance J. 7, 263–266

Wong, K.M., 1999: Gingers of Peninsular Malaysia and Singapore. Natural History Publications (Borneo) Kota Kinabalu, 135 pp

Woodward, P., 1996: Garlic and friends. Hyland House Publ. Ltd. South Melbourne, Victoria, Australia, 248 pp

Wörner, M., P. Schreier, 1990: Flüchtige Inhaltsstoffe aus Steinklee (Melilotus officinalis [L.] Lam.) Z. Lebensmittel-Unters. u. -Forsch. 190, 425–428

Wörner, M., P. Schreier, 1991a: Flüchtige Inhaltsstoffe aus Tonkabohnen (Dipteryx odorota Willd.). Z. Lebensmittel-Unters. u. -Forsch. 193, 21–25

Wörner, M., P. Schreier, 1991b: Über die Aromastoff-Zusammensetzung von Waldmeister (Galium odoratum [L.] Scop.). Z. Lebensmittel-Unters. u. -Forsch. 190, 425–428

Wörner, M., M. Pflaum, P. Schreier, 1991c: Additional volatile constituents of Artemisia vulgaris L. herb. Flavour & Fragrance J., 257–260

Wright, C.W., 2002: Artemisia. Taylor & Francis London New York, 344 pp

Wu, S.J., I.S. Chen, 1993: Alkaloids from Zanthoxylum simulans. Phytochemistry 34, 1659–1661

Wu, Te-Lin, Qi-Gen Wu, Zhong-Yi Chen, 2000: Proceedings of the second symposium on the Family Zingiberaceae. Zhongshan Univ. Press Guangzhou, China 306 pp

Wulff, R.D., 1987: Effects of irradiance, temperature, and water status on grown and photosynthetic capacity in Hyptis suaveolens. Canad. J. Bot. 65, 2501–2506

Wüstenfeld, H., G. Haeseler, 1964: Trinkbranntweine und Liköre. Herstellung, Untersuchung und Beschaffenheit. Verlag P. Parey Berlin Hamburg, 4. edn., 623 pp

Wyk, B-E. van, C. Wink, M. Wink, 2004: Handbuch der Arzneipflanzen. Ein illustrierter Leitfaden. Wissenschaftl. Verlagsges. mbH Stuttgart, 480 pp

X

Xaasan, C.C., C.X. Ciilmi, M. Faarax, S. Passannanti, F. Piozzi, M. Paternostro, 1980: Unusual flavones from Ocimum canum. Phytochemistry 19, 2229–2230

Xaasan, C.C., A.D. Cabdulraxmaan, S. Passannanti, F. Piozzi, J.P. Schmid, 1981: Constituents of the essential oil of Ocimum canum. J. Natural Products 44, 752–753

Xiao-Jia Cai, P.C. Uden, E. Block, Xing Zhang, B.D. Quimby, J.J. Sullivan 1994: Allium chemistry: Identification of natural abundance organoselenum volatiles from garlic, elephant garlic, onion, and Chinese chive using headspace gaschromatography with atomic emission detection. J. Agric. Food Chem. 42, 2081–2084

Xiong, Qu., D. Shi, M. Mizuna, 1995: Flavanol glycosides in pericarps of Zanthoxylum bungeanum. Phytochemistry 39, 723–725

Y

Yaacob, K.C., Che M. Abdullah, D. Joulein, 1989: Essential oil of Ruta graveolens L. J. Ess. Oil Res. 1, 203–207

Yadava, R.N., V.K. Saini, 1991: Gas chromatographic examination of leaf oil of Majorana hortensis Moench. Indian Perfumer 35, 102–103

Ya jing Shaao, P. Marriott, R. Shellie, H. Huegel, 2003: Solid-phase microextraction – comprehensive two dimensional gas chromatography of ginger (Zingiber officinale) volatiles. Flavour & Fragrance J. 18, 5–12

Yamini, Y., F. Sefidkon, S.M. Pourmortazavi, 2002: Comparison of essential oil composition of Iranian fennel (Foeniculum vulgare) obtained by supercritical carbon dioxide extraction and hydrodistillation method. Flavour & Fragrance J. 17, 345–348

Yannitsaros, A., 1979: Tagetes minuta L. in Greece. Candollea 34, 99–107

Yasuda, I., K. Takeya, H. Itokawa, 1982: Distribution of unsaturated aliphatic amides in Japanese Zan thoxylum species. Phytochemistry 21, 1295–1298

Yayeh Zewdie, P.W. Bosland, 2000: Pungancy of Chile (Capsicum annuum L.) fruit is affected by node position. Hortic. Sci. 35, 1174

Yen, G.S., S.L. Shyu, J.S. Lin, 1986: Studies on protein and oil composition of sesame seeds. J. Agric. Forest. 35, 177–181

Yeo, P., 1988: Geranium. E. Ulmer Verlag Stuttgart, 235 pp

Yermanos, D.M., S. Hermstreet, W. Salab, C.K. Huszar, 1972: Oil content and composition of the seed in the world collection of sesame introduction. J. Amer. Oil Chem. Soc. 49, 20–23

Yinrong Lu, L. Yeap Foo, 2000: Flavonoid and phenolic glycoside from Salvia officinalis. Phytochemistry 55, 263–267

Yinrong Lu, L. Yeap Foo, 2001: Antioxidant activities of polyphenols from sage (Salvia officinalis). Food Chem. 75, 197–202

Yinrong Lu, L. Yeap Foo, 2002: Polyphenolics of Salvia – a review. Phytochemistry 59, 117–140

Yipei, B., 1995: Progress and status of research on Chinese Litsea cubeba oil. Chemistry & Industry of Forest Products 15, No. 2, 71–77

Yong-Doo Kim, Seong-Koo Kang, Ok-Ja Choi, Hong-Cheol Lee, Mi-Jeong Jang, Soo-Cheol Shin, 2000: Screening of antimicrobial activity of chopi (Zanthoxylum piperitum DC.) extract. J. Korean Soc. Food Sci. Nutrit. 29, 1116–1122 (in Korean)

Yoshida, H., 1994: Composition and quality characteristics of sesame (Sesamum indicum L.) oil roasted at different temperatures in an electric oven. J. Sci. Food Agric. 65, 331–336

Yoshida, T., O. Namba, L. Chen, T. Okuda, 1990: Euphorbin E, a hydrolyzable tannin dimer of highly oxidized structure from Euphrobia hirta. Chem. Pharm. Bull (Tokyo) 38, 1113–1115

Youdim, K.A., S.G. Deans, H.J. Finlayson, 2002: The antioxidant properties of thyme (Thymus zygis L.) essential oil: an inhibitor of liquid peroxidation and a free radical scavenger. J. Ess. Oil Res. 14, 210–215

Younos, Ch., M. Lorrain, J.M. Pelt, 1972: Chemistry and pharmacology of essential oil from Afghanistan Labiatae. II. Essential oils from two Perovskia species., P. abrotanoides and P. atriplicifolia. Plant Med. Phytother. 6, 178–182

Yousif, A.N., T.D. Durance, C.H. Scaman, B. Girard, 2000: Headspace volatiles and physical characteristics of vacuum-microwave, air, and freeze-dried oregano (Lippea berlandieri Schauer. J. Food Sci. 65, 926–930

Yu, P., Z. Xu, Y. Huang, 1985: The study on traditionally cultivated plants in Tai villages of Xishuangtana. Acta Bit. Yunnan 7, 169–186 (in Chinese)

Yu, T.H., Ch. Wu, Y.C. Liou, 1989: Volatile compounds from garlic. J. Agric. Food Chem. 37, 725–730

Yuan, Y., 1993: Blütenblatt-Kampagnen. Freies China 6, No. 1, 32–37

Yueh-Hsiung Kuo, Ping-Hung Lee, Yung-Shun Wein, 2002: Four new compounds from the seeds of Cassia fistula. J. Nat. Products 65, 1165–1167

Yusuf, M., M.A. Rahmann, J.U. Chowdhury and J. Begum, 2002: Zingibers in Bangladesh. J. Econ. Taxon. Bot. (India) 26, 566–570

Z

Zainuddin, A., J. Pokorný, R. Venskutonis, 2002: Antioxidant activity of sweetgrass (Hierochloe odorata [L.] P. Beauv.) extract in lard and rapeseed oil emulsions. Nahrung/Food 46, 15–17

Zalecki, R., St. Kordana, 1993: Anbau der Mariendistel (Silybium marianum [L.]) Gaertn. in Polen. Drogenreport 6, No. 10, 7–10

Zamboni, C.Q., M.B. Atui, H.I. Alvas, 1995: Adulteration in saffron and paprika. Cienc. e Tecnol. de Aliment. (Sao Paulo) 15, 124–127

Zámbori, É., J. Nyárádi-Szabady, 1989: Study of the relations between some morphological and bioproduction features of muscat sage. Herba Hungar. 28, 7–10

Zänglein, A., W. Schultze: Illicium verum – Sternanis. Z. Phytother. 10, 191–202

Zänglein, A., W. Schultze, K.-H. Kubeczka, 1989: Sternanis und Shikimi. Zur Unterscheidung der Früchte von Illicium verum Hook.f. und Illicium anisatum L. Teil 1: Morphologisch-anatomische Unterscheidungsmerkmale. Dtsch. Apotheker-Ztg. 129, 2819–2829

Zänglein, A., W. Schultze, E. Wolf, 1995: Melissenblätter – Ein Beitrag zur neuen Monographie im Europäischen Arzneibuch. Dtsch. Apotheker-Ztg. 135, 4623–4639

Zarate, R., M.M. Yeoman, 1994: Studies on the cellular localisation of the phenolic pungent principle of ginger, Zingiber officinale Roscoe. New Phytologist 126, 295–300

Zareena, A.V., P.S. Variyar, A.S. Gholap, D.R. Bongirwar, 2001: Chemical investigations of gamma-irrated saffron (Crocus sativus L.). J. Agric. Food Chem. 49, 687–691

Zawirska-Wojtasiak, R., E. Wąsowicz, 2002: Estimation of the main dill seeds odorant carvone by solid phasephase microextraction and gas chromatography. Nahrung/Food 46, 357–357

Zeghichi, S., S. Kallithraka, A.P. Simopoulos, Z. Kypriotakis, 2003: Nutritional composition of selected wild plants in the diet of Crete. World Rev. Nutrit. Diet. 91, 22–40

Zelnik, R., D. Lavie, E.C. Levy, A.H.-J. Wang, I.C. Paul, 1977: Barbatusin and cyclobarbatusin, two novel diterpenoids from Coleus barbatus Benth. Tetrahedron 33, 1457–1467

Zennie, Th.M., D. Ogzewella, 1977: Ascorbic acid and Vitamin A content of edible wild plants of Ohio and Kentucky. Econ. Bot. 31, 76–79

Zeven, A.C., J.M.J. de Wet, 1982: Dictionary of cultivated plants and their centres of diversity. Excluding ornamentals, forest trees and lower plants. Centre Agricult. Publishing and Documentation Wageningen, 219 pp

Zhang, D., Q., F.J. Lu, J.X. Tai, 2001: Study on the extraction of ginger oleoresin by supercritical CO_2 technique. Sci. & Technol. Food Ind. 22, 21–23

Zhang, X., W.T. Iwaoka, A.S. Huang, S.T. Namakota, R. Wong, 1994: Gingerol decreases after processing and storage of ginger. J. Food Sci. 59, 1338–1340, 1343

Zhang, X.F., X.H. Xiao, D. Ding, 2000: Study on extraction and application of essential oil of jasmine flowers. J. Wuhan polytechn. Univ., (China) No. 1, 18–20 (in Chinese)

Zhang, Y., C.T. Ho, 1990: Volatile components of Tamarind. J. Ess. Oil Res. 21, 197–198

Zhao, S.P., P.Z. Cong, L.H. Quan, 1991: Chemical examination of the essential oil of Foeniculum vulgare. Acta Bot. Sinica 33, 82–84 (in Chinese)

Zieba, J., 1973: Isolation and identication of flavonoids from Glechoma hederacea. Polish J. Pharmacol. Pharmac. 25, 593–597

Zimmermann, M., P. Schieberle, 2000: Vergleich wichtiger Aromastoffe in Edelsüß-Paprika aus Ungarn und Marokko. Lebensmittelchemie 54, 107

Zimmermann, V., 1980: Der Rosmarin als Heilpflanze und Wunderdroge. Ein Beitrag zu den mittelalterlichen Drogenmonographien. Sudhoffs Arch. Vjschr. Geschichte, d. Menschen, der Naturwiss., Pharmaz. u. Math. 64, 351–370

Zirvi, K.A., M. Ikram, 1975: Alkaloids of some of the plants of the Compositae. Pakistan J. Sci. & Ind. Res. 18, 93 -101

Zizka, G., S. Fleckenstein, 1985: Die Muskatnuß (Myristica fragrans Houtt.). Palmengarten 49, 89–91

Zobel, A.M., S.A. Brown, 1991: Furanocumarin concentrations in fruits and seeds of angelica. Environmenta Exp. Bot. 31, 447–452

Zoghbi, M.D.G.B., E.H.A. Andrade, A.S. Santos, M.H.L. Silva, J.G.S. Maia, 1998: Essential oils of Lippea alba (Mill.) N.E.Br. growing in the Brazilian Amazon. Flavour & Fragrance J. 13, 47–48

Zohary, M., 1952: A monographical study of the genus Pistacia. Palest. J. Bot. (Jerusalem) 5, 18–228

Zohary, M., 1960: The species of Capparis in the Mediterranean and the Near Eastern countries. Bull. Res. Counc. Israel 8D, 49–65

Zohary, M., 1983: The genus Nigella (Ranunculaceae): a taxonomic revision. Plant Systematics & Evolution 142, 71–107

Zola, A., J. Garnero, 1973: Contribution à létude de quelques essence de basilic de type européen. Parfum. Cosmet. Sarvons (France) 3, 15–19

Zola, A., J.P. le Vanda, F. Guthbrod, 1977: L'huile essentielle de laurier noble. Plant. Med. Phythother. 11, 241–246

Zoller, A., H. Nordwig, 1997: Heilpflanzen der ayurvedischen Medizin. K.F. Haug Verlag Heidelberg, 575 pp

Zollo, A.P.H., R. Abondo, L. Biyiti, C. Menut, J.M. Bessiere, 2002: Aromatic plants of tropical Central Africa XXXVIII. Chemical composition of the essential oils from four Aframomum species collected in Cameroon I. J. Ess. Oil Res. 14, 95–98

Zoschke, H., 1997: Das Chilli Pepper Buch. Anbau, Rezepte, Wissenswertes. Zoschke Data GmbH Schönberg, 224 pp

Zwaving, J.H., R. Boss, 1996: Composition of the essential oil from the root of Sassafras albidum (Nutt.) Nees. J. Ess. Oil Res. 8, 193–195

Zwingenberger, H., 1938: Zur Kenntnis des Königskerzenanbaues (Gewinnung der "Flores Verbasci"). Angew. Bot. 20, 1–61

Index of Illustrations

Abelmoschus moschatus   1
Acorus calamus   5
Aframomum melegueta   10
Alliaria petiolata   15
Allium cepa var. ascalonicum, aerial onions   17
Allium fistolosum   19
Allium victoralis   28
Alpinia calcarata   29
Alpinia zerumbet   32
Ammi visnaga, fruiting   34
Angelica archangelica   41
Anthoxanthum odoratum   44
Artemisia vulgaris   56
Bixa orellana   66
Boesenbergia rotunda   67
Borago officinalis   68
Calycanthus floridus   76
Canella winterana   78
Capparis spinosa   81
Carissa edulis   87
Centaurum erythrea   91
Ceratonia siliqua   92
Chenepodium bonus-henricus   96
Cinnamomum zeylanicum   104
Citrus medica   108
Coriandrum sativum   117
Costus speciosus   118
Crataeva tapia   121
Crocosmia x crocosmiflora   122
Curcuma longa   127
Dillenia indica   137
Dipteryx odorata   138
Drimys winteri   141
Elettaria cardamomum   145
Etlingera eliator   149

Gentiana lutea   164
Glycine max   168
Guajacum officinale   171
Houttuynia cordata   179
Illicium verum   184
Kaempferia galanga   191
Lippea graveolens   211
Marrubium vulgare   219
Melilotus officinalis   222
Menyanthes trifoliata   230
Monarda didyma   236
Monarda fistulosa   237
Monodora myristica   238
Morinda citrifolia   240
Moringa oleifera   240
Murraya koenigii   241
Myrica gale   243
Myristica fragrans   244
Nigella damascena   253
Ocimum basilicum   256
Ocimum kilimandscharicum   259
Origanum majorana   263
Pandanus tectorius   272
Parkia speciosa, fruits   274
Perovskia abrotanoides   279
Picrasma excelsa   286
Piper auritum   290
Piper longum   292
Pistacia lentiscus   296
Pithecellobium dulce   298
Plectranthus amboinicus   299
Polyscias fruticosa   303
Potentilla erecta   306
Quassia amara   312
Rhamnus prinoides   315

Salvia argentea   326
Salvia officinalis   329
Sassafras albidum   333
Schinus molle   337
Sesamum indicum   342
Sinapis alba   345
Solanum macrocarpon   350
Swertia chirata   355
Syzygium aromaticum   355
Tagetes erecta   357
Tamarindus indica   359, 360

Tetrapleura tetraptera   363
Teucrium   364
Torreya california   370
Trigonella foenum-graecum   373
Tropaeolum majus   375
Vanilla planifolia Andr.   384
Vitex agnus-castus   389
Zanthoxylum piperitum   401
Zingiber officinale   403
Zingiber zerumbet   404

Index of Common Names

a phien   272
a phub dung   272
à trois feuilles   85, 211
a wei   156
Aalkraut   334
aam haldi   126
ababa   272
abaiba   89
abé   77
abel   77
Abelmoschussamen   1
abelmosco   1
abelmosk seed   1
Abessinian mustard   69
Abessinischer Kardamom   9
Abessinischer Senf   69
abhaya   387
abini   272
Abrahamstrauch   388
abricó do pará   217
abricotier d'Amérique   217
abricotier de Saint Domingue   217
abrikos tschernij   49
abrikos volosistoplodnyj   49
abroso   334
abrotano   52, 362
abrotano camforata   52
abrótano macho   52
abrotano maschino   362
abrotone   52
absint-alsem   52
Absinth   52, 55
absinthe   52, 53, 54, 55, 56
absinthe arborscente   53
absinthe de mer   54
absinthe losna   52
absinthe romaine   55
absinthe, armoise   52
absinthe, grande   52
absintio   52
absinto   52
Abstehende Studentenblume   358
abû ennûm   272
abû nû mân   272

abu qaru   111
abu-el-misk   1
abu-el-mosk   1
acacie, false   318
açafrão   87, 122
açafrão bastardo   87
açafroa   65
açafroeira-da-terra   65
acaparrón   80
acedera común   321
acedera con hojas redondas francesa   322
acedera de Guinea   177
acedera redonda   322
acedera romana   322
acederilla   268
acetosa maggiore   320
acetosa romana   322
acetosella   268
açfra da Índia   126
ache de Montagne   205
ache-douce   47
achera   78
achillea millefoglio   3
achillée   3
achillée amère   3
achillée millefeuille   3
achillée noire   3
Achilleskraut   3
achiote   65, 66
achira   78
achoram   191
Achote   65
achul   128
acid mombin   352
acitosa   321
Ackerholler   7
Acker-Knoblauch   15
Acker-Knöterich   302
Ackerminze   224
Acker-Schwarzkümmel   252
Ackersenf   345
acksoum   158
acore à feuilles de graminée   5

acore calame   5
acore odorant   5
acore vrai   5
acoro   5
acoro verdadero   5
acoyo   289, 294
acucena   173
acuyo   294
ada   403
adacayi   329
Adam's apple   105
adas   158
adas china   40, 184
adas cina   184
adas londo   158
adas manis   40
adas pedas   158
adas pudus   40
adas sowa ender   40
adhera   247
adiant(um) tschjornaja   7
adigam   171
adiphala dhatri   284
adismanis   288
Adjowain   371
Adlerfarn   309
Adlerholz   48
Adlerholz, Malacca   48
Adlerholz, Khasi   48
admiration herb   205
adormidera   272
adrak   403
adraka   403
Affenbrotbaum   6
Affenkartoffel   351
Affenohrring, Malayischer   49
Affenohrring, Mexikanischer   298
āfim   272
afiyun   272
afon   371
Africain morène   180
African (brazil) nut   300
African basil   259
African bowstring hemp   332

African boxwood 371
African bread fruit 371
African calabash 6
African cubeb 291
African cucumber 235
African curry powder 63
African eggplant 349
African elemi 77
African false elm 91
African frogbite 180
African ginger 346
African indigo 213
African locustbean 273
African marigold 99, 357
African morène 180
African mustard 111
African myrrh 115
African nettle tree 91
African nut 300, 317
African nutmeg 238, 310
African oil palm 143
African onion 16
African pepper 192, 291, 395, 402
African rue 275
African sage 325
African sandalwood 267
African shallot 16
African spider flower 111
African turmeric 78
African umckaloebo 277
African valerian 155
African wormwood 52
Afrikaanse aubergine 349
Afrikanische (Brasil-)Nuss 300
Afrikanische Aubergine 349
Afrikanische Brotfrucht 371
Afrikanische Gewürzbeere 238
Afrikanische Kresse 202
Afrikanische Kubeben 291
Afrikanische Locustbohne 273
Afrikanische Malve 177
Afrikanische Ölpalme 143
Afrikanische Ringelblume 99, 358
Afrikanische Spinnenpflanze 111
Afrikanische Zwiebel 16
Afrikanischer Affenbrotbaum 6
Afrikanischer Baldrian 155
Afrikanischer Basilikum 258, 259
Afrikanischer Froschbiss 180
Afrikanischer Ingwer 346
Afrikanischer Pfeffer 395, 402
Afrikanischer Salbei 325
Afrikanischer Wermut 52
Afrikanischer Zürgelbaum 91
Afrikanisches Mahagoni 317
Afrikanisches Sandelholz 267
Afrikanisches Zitronen-
 Bohnenkraut 334

afrikanskij perez 291
afsantin 52
agagat 280
agar wood 48
agargarha 39
agas-moro 387
agastache 12
agasti 343
agati 343
agava 4
aggenko 79
aghako 396
agliaria 14
aglio 19, 24, 27
aglio d'India 24, 25
aglio domestico 24
aglio giapponense 19
aglio orsino 27
aglio romana 25
agnijal 110
agnocasto 388
agretto 203, 250
agretto aquatico 250
agrião 62, 84, 250, 352
agrião de água 250
agrião de horta 62
agrião do Brasil 352
agrião do mato 352
agrião dos prados 84
agrilla 321
agrimonia 13
agrio de Guinea 177
aguadiente de Espana 209
aguaribai 337
Ägyptische Malve 216
Ägyptische Zwiebel 23
Ägyptischer Amm(e)i 371
Ägyptischer Lauch 20
Ägyptischer Zahnbaum 61
ahimit 178
ahiphena 272
ahito 357
Ahorn, Großblättriger 2
Ahorn, Oregon 2
Ährentragender Salbei 336
Ährentragendes Bohnenkraut 336
Ährige Minze 229
ai hao 56, 368
ai nitu 332
ai tan chun 187
ai tao 214
aigremoine gariot 13
aïl 24
ail àcheval 15
ail à carène élegant 17
aïl blanc 24
ail chinoise 24
ail civette de Chine 24, 27

ail d'Orient 15
ail de ceuf 27
ail de Naples 21
ail des bois 26
ail des ours 27
ail doré 21
ail du Japon 19
ail du Turkestan 20
ail fistuleux 19
ail géant de l'Himalaya 19
ail ordinaire 24
ail penché 18
ail rocambole 24
ail sauvage 26
ail stérile 17
ainjarada 40
ainta 339
aiowan 371
aipo de cabeça 47
aipo de raíz 47
aipo hortense 46
air 5
air trostnikowyj 5
aistnik 146
ajaji 125
ajak 255
ajaka 255
ajamir 75
ajamoda 371
ajawa omum 371
ajedrea 334
ajedrea común 334
ajedrea de jardin 334
ajenja mayor 52
ajenjo común 52
ajenjo marino 54
ajenjo 52
ajenuz 254
ají 82, 83
ajmood 282
ajmud 46, 371
ajmuda 371
ajo 16, 20, 23
ajo común 24
ajo del Japon 20
ajo montaña 16
ajo pardo 25
ajo porro 16, 23
ajo puerro 23
ajonjolí 342
ajooras 31
ajouan 371
ajovan 371
ajowain 371
ajowan 371
Ajowan 371
Ajowankümmel 371
ajvyan 371

aka kina no ki 99
aka shiso egoma 278
akahara 39
akallaka 39
akar 2, 32, 292, 315, 369, 387
akar asam tebing paya 315
akar bagan 32
akar halong 292
akar kucing 369
akar wangi 387
akarakarabha 39
akarkara 39
Akawai nutmeg 6
akazia belaja 318
Akazie, Falsche 318
Akazie, Schein 318
akhasianamu 318
akpa 362
al'pijskij tschaber 335
al'pinia aptetschnaja 31
alajan 289
alajan pepper 289
álamo balsámico 304
alamo negro 304
alamo vine 231
Alant, Echter 184
Alant, Klebriger 185
alazor 87
alba thyme 367
albaca silvestre 259
albadyan 184
albahaca 255
albahaca 255, 256
albahaca cimarrona 260
albahaca de clavo 258
albahaca de monte 259
albahaca morada criolla 260
albahaca velluda 255
albero del pepe 388
albero di Giuda 93
alcaçuz 169
alcaparna 80
alcaparra 80, 81
alcaparreira 80, 81
alcaravea 87
alcarávia 87
alcazuz 169
alchemilla 14
alcoparras 81
alderleaf berry 33
aldonisis 159
ale 403
alebega 256
alecost 360
alecrim 197, 209, 320, 388
alecrim da parede 4
alecrim de angola 388
alecrim do campo 209

alecrim do norte 388
alecrim-bravo 197
alecrim-do-campo 197
alehoof 167
alekcandrijskaja polyn' 55
Alepponuss, Echte 297
Aleppo-Raute 322
alerum 320
alesta olorosa 45
Alexander's plant 348
Alexanders 348
Alexandrian parsley 348
alfaraca 257
alfarrobeira 93
alfavaca 256, 334
alfavaca do campo 259, 334
alfavacão 258
alfazema 200
alfóncigo 297
alfónsico 297
alga marina 405
algalia 1
algarrobo 93
alharma 275
alhel 135
alho 7, 15, 20, 23, 24, 27, 120
alho chinês 27
alho do Japão 20
alho francês 23
alho grosso de Espanha 25
alho porro 23
alho porro bravo 15
alho rocambole 25
alholva 374
alibangbang 64
alichimille 14
aliga 405
alim 48
Alisander 348
alkénge du Mexico 285
alkor 220
allego 405
alleluia 268
alleluja 268
Allerleigewürz 286
Allermannsharnisch 27, 28
Allermannsharnisch,
  Breitblättriger 28
allgood 96
all-heal 381
alliairé 14
alliaria 14
alligator pepper 9, 10, 11
Alligatorpfeffer 9, 10, 11
Alligatorpfeffer, Stengelloser 9
alloro poetico 199
allspice 75, 208, 286
allspice 208

almecegueira 296
almond, green 296
almond veronia 386
almoradux 226
almug 309
aloewood, Indian 48
Alpenbeifuß, Echter 55
Alpen-Fetthenne 339
Alpen-Mutterwurz 206
Alpen-Pestwurz 281
Alpenscharte, Gerippte 336
Alpen-Schnittlauch 25
Alpen-Steinquendel 4
alpes butterbur 281
alpine avens 165
alpine calamint 4
alpine chives 25
alpine leek 27
alpine lovage 206
alpine stonecrop 339
alpine wintergreen 163
alpine wormwood 55
alpine yarrow 3
alpinia merah 31
Altai onion 15
Altai-Zwiebel 15
altajskij luk 15
alyxia cinnamon 32
Alyxia-Zimtrinde 32
alzangabeel 403
am 30, 217
ama 209, 405
amalaka 284
amaltas 88
ama-mo 405
amantilla 381
amapola 272, 358
amapola dei opie 272
amara 105, 352, 360
amarelinho-da-serra 128
amarella 57, 128
amarelo 128, 137, 403
amargo negro 311
amarilla 99, 163, 358
Amarillisblättrige Pandanane 271
Amarillisblättriger
  Schraubenbaum 271
amarillo 222, 357
amati 45
amba 217
ambaitinga 89
ambarati 322
ambarcillo 1
ambari 322
ambarina 1
ambaúba 89
amber seed 1
Amberkraut 363

Ambla   284
ambo   217
ambora   71
amborasaha   71
Ambramalve   1
ambretta   1, 166
ambrette   1
Ambrette   1
ambrette seed   1
ambródia dos boticas   96
ambroisie du Mexique   95
ambroisine   95
Ambrose   95
Ambrosia   33, 95
ambrósia do México   95
ambuja   208
amburana   370
ambuti   268
amchoor   217
amchur   217
Ameisenbaum, Karibischer   89
amendoim mandobi   48
amendoine   48
ameo bastardo   33
ameo mayor   33
American (false) pennyroyal   173
American apple mint   226
American basil   255
American bird cherry   308
American bird pepper   82
American cinnamon   261
American cress   62
American dittany   125
American epazote   95
American false pennyroyal   301
American horsemint   237
American licorice   169
American nutmeg   124
American oregano   210
American saffron   359
American sea rocket   73
American spikenard   201
American stone mint   126
American thoroughwort   151
American wild onion   16
Amerikanische Apfelminze   226
Amerikanische Arnika   50
Amerikanische Bergminze   310
Amerikanische Muskatnuss   124
Amerikanische Poley   173
Amerikanische Springgurke   235
Amerikanische Steinminze   126
Amerikanischer Basilikum   255
Amerikanischer Diptam   125
Amerikanischer Gagelstrauch   242
Amerikanischer Meersenf   72
Amerikanischer Oregano   210
Amerikanischer Zimt   261

Amerikanisches Süßholz   169
Amla   284
amlag   284
amli   6, 64, 275, 359
amlika   284
Amm(e)i, Ägyptischer   371
Ammi   33, 34, 371
ammi   33, 34, 371
ammi bol'schaja   33
ammi de l'Inde   371
ammi élevé   33
ammi zubnaja   34
Ammi, Großer   33
amna   352
amome à grappe   35
amomum   10
amontau   107
amora   387
amoretti d'Egitto   314
amorino   314
amotan   181
ampalam   217
Ampfer, Blut   322
Ampfer, Hain   322
Ampfer, Kleiner   321
Ampfer, Krauser   321
Ampfer, Römischer   322
Ampfer, Rundblättriger   322
Ampfer, Spießiger   321
Ampfer, Sumpfblättriger   321
Ampfer, Ufer   321
amra   218, 352
amrataka   352
amrul   268
amrula   322
an mo le   284
an thach   309
an ye   150
an(n)atto   65
An(n)otta   65
anacycle   39
anagrón   361
anam dangdeu   309
anamam sigpang   309
Anaman redwood   309
Ananasminze   230
Ananas-Salbei   327
anantumal   176
anantumala   176
anar   309
anardana   309
anason   184, 288
anason tchini   184
Anatolisches Süßholz   170
anbli   359
Anden-Safran   148
Andenwein, Gewöhnlicher   220
andjejang   317

andjudaan   156
andong   116
Andorn, Gewöhnlicher   219
Andorn, Indischer   181
Andorn, Wohlriechender   181
aneldo   40
anepu   45
aneth   40
aneth odorant   40
aneto   40
angara gida   241
angélica   41
angelica bravo do matto   134
angelica, Chinese   43
angelica, wild   43
angelica, garden   41
angelica, seaside   42
Angelika   41
Angelika, Spitzlappige   41
Angelikawurz(el), Chinesische   43
Angelikawurzel, Zarte   43
angélique   41
angélique vraie   41
anggrek merah   313
anggrek merah   313
anglijckaja mjata   227
Angola calabash   237
Angola Kalabassenmuskat   237
Angolanische Muskat   310
Angolanische Zwiebel   16
Angolian nutmeg   310
angostura   128
Angostura(rinde)   128
angsana   309
anh tuc   272
anice selvatico   287
anice stellato   184
anice verde   288
anicillo   357
anigella   253
anis   87, 110, 125, 158, 183, 209, 247, 287, 313, 387
anis de China   184
anís de China   184
anis de Espana   209
anis de la Chine   184
anis de moro   387
anis des vosges   87
anis estrelado   184
anís estrellado   184
anis étoilé   184
anís verde   288
anis vert   288
Anis, Brotsame   288
Anis, Chinesischer   184
Anis, Indischer   184
Anis, Sibirischer   184
Anis, Wilder   247
Anis-Bibernelle   287

Anisblätter 110
Anis-Clausenie 110
anise 11, 183, 209, 247, 288, 313
anise chervil 247
anise common 288
anise hyssop 11
anise mint 11
anise nutmeg 313
anise seed 288
anise sweet 247
anise verbena 209
anise, Chinese 184
anise, purple 183
aniseed 183, 288
aniseed tree 183
anisilla 289
anisillo 289, 358
Aniskalmus 5
Aniskerbel 247
Anis-Muskat 313
anisowoe derewo 184
Anispfeffer 289
Anisverbene 209
Aniswurzel 247
Anisysop 11
aníz 288
anizet 288
Annamzimt 102
Annam-Zimt 102
annato 66
annual mercury 230
annual sage 330
annual stonecrop 339
anschelika 41
Ansehlicher Ingwer 404
anserina branca 95
ansérine 95
ansérine blanche 95
anserine vermifuge 95
antanan 388
antelope grass 143
Antidesmablätter 46
antidesme 45
Antillen-Vanille 385
antilope à herbe 143
antilope de graminées 143
Antilopengras 143
antorcha 149
anu 376
anwama 317
anysboegoe 7
anysbuchu 7
Anzurzwiebel 26
ao-shiso 278
apamarg 4
apamargamu 4
apang 4
apanang-gubat 151

apazote 95
apel'sina 109
apelcin kuslyj 106
aperta-ruão 289
Apfel, Punischer 309
Apfelbeere 51
Apfel-Chili 84
Apfelduft-Pelargonie 276
Apfel-Eukalyptus 150
Apfelminze 224, 226, 228
Apfelminze, Amerikanische 226
Apfelsine 109
aphin 272
āphūka 272
aphyonkkot 272
apio 46
apio blanco 46
apio de bulbo 47
apio de cortar 47
apio de montaña 205
apio de tallo 46
apio nabo 47
apio pequeño 47
apio rábano 47
aposote 95
apotetschnyj ukron 158
apple chili 84
apple gum 150
apple, golden 7
apple, Madras 224
apple mint 224, 226, 228
apple mint, American 226
apple scented geranium 276
apple-box 150
apple-scented eucalyptus 150
apricot, black 49
apricot, purple 49
apricot, Santo Domingo 217
apricot, South American 217
Aprikose von St. Domingo 217
apuy-apuyan 111
arabadha 88
Arabian fenugreek 372
Arabian jasmine 188
Arabian lavender 201
Arabian myrrh 115
Arabian pepper 405
Arabische Myrrhe 115
Arabischer Bockshornklee 372
Arabischer Jasmin 188
Arabischer Lavendel 201
Arabischer Majoran 265
Arabischer Pfeffer 405
Arabischer Schnittlauch 20
Arabisches Bergkraut 233
arachide 48
arachis 48
araiya 304

arak 325
Aramenian salve 326
Aramenischer Salbei 326
arancia 106
arancia forte 106
arancio amaro 106
arancio dolce 109
aranuel 254
árbol de la pimenta 337
arbol de los escudos 166
arbol de Pérou 246
arbol de Tolú 246
árbol dela quina 99
arbor de balsamo 246
arbre à cire 86
arbre à encens 69
arbre à farine 273
arbre à pain d'Afrique 371
arbre à suif 242, 310
arbre au mastic 296
arbre aux anémones 75
arbre aux quarante écus 166
arbre brosse à dents 325
arbre de Judée 93
arbre de melon 86
archangel 41
archangel, Japanese 42
archangelica 41
archangélique 41
arcuan 93
ardraka (fresh ginger) 403
arenddj 105
areuy beleketebek 369
arfuin 272
Argentinische Minze 234
argûân 93
argus pheasant 140
ariceto 360
armaise camphrée 52
armoise 52, 54, 56
armoise à fleurs laiteuses 54
armoise absinthe 52
armoise commune 56
armoise d'Afrique 52
armoise du Valais 56
armoise maritime 54
armorácia 50
arnica 51
árnica de montaña 51
arnica gornaja 51
arnica, Mexican 177
Arnika, Amerikanische 50
Arnika 51
Arnika, Echte 51
Arnika, Mexikanische 177
aroeira 337
aroeira do brejo 338
aroeira do campo 338

aroeira periquita 337
aroeira precoce 338
aroeira salso 337
aroeira-mansa 337
aroeirinha 337
aromate germanique 184
Aromatische Schefflarie 337
Aromatische Strahlenaralie 337
Aromatischer Blockzitwer 126
Aromatischer Kälberkropf 94
aronica du noir 51
aroreira do sertao 338
aroreira mansa 338
aroreira negra 338
arrayán 247
arrayan del campo 28
arreboleva 234
Arrowroot, Bombay 126
Arrowroot, Indian 126
Arrowroot, Indisches 78
Arrowroot, Queensland 78
Arrowroot, Westindisches 78
arruda 323
artamisa Méjico 55
artemisia 57
artillery plant, black throated 286
arudam fejan 323
arunggai 240
Arznei-Baldrian 381
Arznei-Thymian 367
as sanânir 284
asa fétida 156
asadio 203
asafar 122
asafetida 156
asa-fœtida 156
asafsaf 304
asam belanda 298
asam gelungor 161
asam glugur 162
asam jawa 359
asam keranji 298
asam koening 359
asam kranji 298
asam kuang 140
asam londo 298
asam melaka 284
asam puja 268
asant 156
Asant, Gummi 156
Asant, Stab 156
asarabacca 57
asaret d'Europe 57
asari 242
asaro baccaro 57
ásaro de Europa 57
ásarocomún 57
asatsuki 25

ascaloña 17
ascalonia 17
aschantijskij perez 291
Aschantipfeffer 291
Aschenwicke, Lineare 362
Aschenwicke, Purpurne 362
Aschgraue Laugenblume 119
Aschlauch 17
ase fétide 156
asem jawa 359
asfar 86
ash buttons 119
Ashanti pepper 291
ashgon 371
ashitaba 42
ashvaghra 401
ashweed 7
ashy cotula 119
Asiatic mint 226
Asiatische Rossminze 226
asistania zambo angenita 58
Askalonzwiebel 17
asmargok 156
asperilla olorosa 161
asperula 161
aspérula 161
aspérule odorante 161
aspic 201
assa fetida 156
Assam cinnamom 100
Assam-Zimt 100
assenzio 52
assenzio maggiore 52
assenzio Romano 52
Ästiger Lauch 24
asuri 69
asystasia 58
atakkamaniyen 351
atanasia 360
atasugi 17
athanase 360
Äthiopische Tomate 348
Äthiopischer Nachtschatten 348
Äthiopischer Pfeffer 395
athlac 388
atikabuki 281
atitejani 401
atluen 301
atschetonsa 320
atsuete 65, 66
atuja 403
aubergine amère 348
aubergine gboma 348
aubergine indigéne 349
aubergine, Afrikaanse 349
Aubergine, Afrikanische 349
Aubergine, Großfrüchtige 349
Aufrechte Boerhavie 67

Aufrechte Studentenblume 357
Aufrechter Sauerklee 269
Aufrechtes Fingerkraut 306
Aufsteigende Bergminze 74
Augentrost, Gewöhnlicher 153
Augentrost, Großblütiger 153
aun bawang 19
aurone 52
aurone des jardinsaurone mâle 52
Ausdauernde Rucola 137
Ausdauernder Boretsch 68
Austernpflanze 231
Australian blue ginger 29
Australian cinnamom 103
Australian fever 32
Australian lemon myrtle 61
Australian mint bush 307
Australian pepper 140
Australian rosemary 392
Australische Minze 307
Australische Zitronenmyrte 61
Australischer Minzstrauch 307
Australischer Rosmarin 392
Australischer Zimt 103
Austrian thyme 367
aut 268
auti 116
avagvadha 298
avas 300
avellana americana 48
avens root 166
avens, Siberian 114
avoeira 296
awl tree 239
awraamowo derewo 388
ayapana 152
azafrán 87, 122, 126, 148
azafrán bastardo 87
azafrán canario 87
azafrán de la India 126
azafrán de los Andes 148
azafrán de montāna 148
azafrán de raiz 148
azafran del pais 87
azeda brava 320
azedinha 177, 268, 320
azedinha da horta 320
azier la fièvre 147
Azorenthymian 365
Azores thyme 365
Aztec marigold 357, 358
Azteken-Salbei 327
Aztekisches Süßkraut 210
ba jiao hui xian 184
babagion 49
babary 293
babek 292
babui tulsi 256

baby rose 320
bac he 225
bacc 5
bacca 189
bacca di ginepro 189
baccha 5
bacela espinaca blanca 63
bach dâu khâu 35
Bachkresse 250
Bachminze 224
Bach-Nelkenwurzel 166
bacicci 120
bacile 120
bacua 272
badian 184
Badian 184
badián 184
badian tschabrez 184
badiana 184
badiána 184
badiane 183
badiane de Chine 183
badiok 362
badrang 401
bael 7
Baelbaum 7
bagal tikus 110
bagatambal 399
bagdunis 282
bagrjannik 93
bahar 286
Bahia vanilla 383
Bahnwärter-Taglilie 175
bai 19
bai ga-prow 260
bai hao 56
bai horapa 256
bai hua cai 111
bai karee 241
bai kuichai 27
bai ma-krut 107
bai pak 116
bai rang cao 205
bai saranae 225
bai toey hom 272
bai xian pi 136
bai zhi 41
baie de genièvre 189
baies roses de Bourbon 338
Baja 65
bakaut 170
bakek 294
Baker' garlic 18
bakkaut 189
bakondo 310
bal'samitscheskaja pjabinka 360
bal'samowoe derewo 246
bala 387

balabalanoyan 111
balah harara 61
balambing 59
balanites 61
baldmoney 232
Baldrian 155, 381
Baldrian, Afrikanischer 155
Baldrian, Arznei 381
Baldrian, Gewöhnlicher 381
Baldrian, Keltischer 381
balibamban 64
balioko 129
balisier comestible 78
balisierou faux-sucrier 78
Balkan savory 336
Balkan-Bohnenkraut 336
ball leek 26
ballon vine 85
Ballonblume, Großblütige 299
Ballonpflanze 85
Ballonrebe 85
balloon flower 299
balm 223
balm of Giliad 89, 115
balm, fragrant 236
balm, lemon 223
balm, melissa 223
balm, Moldavian 139
balm, Vietnamese 145
balok 48
balonoi 260
balsam (scented) geranium 277
balsam apple 235
balsam Makkah 115
balsam of Peru 246
balsam of Tolu 246
balsam pear 235
balsam poplar 304
balsam tree 115
balsam, Egyptian 61
Balsam, Römischer 360
Balsamapfel 235
Balsambirne 235
balsamier de la Mecque 115
balsamita major 223
balsamite 360
balsamite major 360
balsamite minor 360
balsamito cundeamor 235
Balsamklee 372, 373
Balsamkraut 360
Balsam-Margerite 360
Balsammyrrhe 115
balsamo 115, 246, 373
bálsamo 235, 246
balsamo de Pérou 246
bálsamo de Tolú 246
balsamo della Mecca 115

balsamo di Perú 246
balsamo di Tolù 246
bálsamo-do-peru 246
bálsamo-do-tolu 246
Balsampappel 304
Balsampelargonie 277
Balsampflaume, Gelbe 352
Balsampflaume, Saure 352
Balsamstrauch 89, 299
balung ayam 90
bamburral 181
bamia 1
bamija 1
ban 241, 250, 374
ban mallika 241
banaasi 242
banaf sag 388
banana shrub 232
Bananenstrauch 232
banaphsa 388
banati 242
Banda nutmeg 244
banef shah 388
bangah 12
bangkolon 85
banhaldi 128
banglai 402
bangle 30, 402
bangliu 167
bani 303
banig-banig 300
baniyuyo 46
bankuln 14
bannilgach 362
bao chuang ye 215
baobab 6
Baobab 6
baquois 272
Bär(en)-lauch 27
bara alachi 37
bara ilachai 37
bara nimbu 108
bara phulanoo 196
barak 128, 167
barakbolon 30
barakulanjar 30
baranjeira bergamota 106
barba 157
barbadilla 138
Barbaforte 50
Barbarakraut 62
Barbarakraut, Gewöhnliches 62
barbarasco 385
barbarea 62
barbarée herbe 62
barbari 256
barbary wolf berry 213
Barbenkraut 62

barbide   253
barbotine indigène   360
barbotine   360
barchatcy melkie   358
barchatcy otklonennye   358
barchatcy prjamostojaschtschie   357
bardagoush   263
Bärenfenchel   232
Bärenklau, Burmannischer   176
Bärenklau, Persischer   176
barewa   341
barhal   57
barik   404
bark tree   241
Bärkümmel   232
barlen   97
barna   120
barna bidasi   120
barru   347
barsagusha   263
barsanga   241
barshatjstyj   358
Bärtiger Boretsch   300
bartundi   239
barul'nik   201
baruna   120
Bärwurz   232
bas(s)al   17
basari   80
basbasah   244
baselle   63
baseng   403
basil   4, 112, 255, 278, 310
basil thyme   4
basil weed   112
basil, African   259
basil, American   255
basil, Chinese   278
basil, commun   256
basil, Greek   257
basil, hoary   255
basil, holy   260
basil, Kilimandscharo   258
basil, lemon   256, 257
basil, monk's   256
basil, Peru   259
basil, Russian   258
basil, sacred   260
basil, sweet   255
basil, Thai   260, 278
basil, wild   112, 310
basilic   256, 258, 260
basilic commun   256
basilic de Ceylan   258
basilic des moines   260
basilic sacré   260
basilic sauvage   4
basilico   256

basilicum   63, 256
Basilienkraut   256
basilik   256, 258, 260
basiliki kilimandshars'ki   258
basilik m'jatolistnis   258
basilik swajaschenujy   260
Basilikum   63, 255, 278
Basilikum, Afrikanischer   258
Basilikum, Amerikanischer   255
Basilikum, Busch   257
Basilikum, Duft   258
Basilikum, Duftender   260
Basilikum, Heiliger   260
Basilikum, Kilimandscharo   258
Basilikum, Kleiner   257, 260
Basilikum, Minzblättriger   258
Basilikum, Nelken   258
Basilikum, Ostindischer   258
Basilikum, Peruanischer   259
Basilikum, Rot(blättrig)er   257
Basilikum, Thai   260, 278
Basilikum, Tulasi   260
Basilikum, Vielähriger   63
Basilikum, Wilder   112
Basilikum, Zitronen   255
basna   343
basong   32
bas(s)al   17
bassal el-ankudy   19
bassal el-mustatere   19
bassal el-shifee   25
bastard cardamom   38
Bastard Malagetta   9
bastard meleguata   9
bastard mustard   111
bastard Siamese cardamom   38
bastard star anise   183
bastard vervain   353
Bastard-Kardamom   36, 38
Bastard-Meleguetapfeffer   9
Bastardsafran   87
Bastardsenf, Katzenschnurbart   111
Bastard-Vervaine   353
bat nga   147
batang pelaga   35
bâtard, myrte   242
bâtarde   252
Batatenzwiebel   17
Batavia cassia   100
Batavia cinnamon   100
Batavia-Kassie   100
Batjan-Muskat   245
bâton casse   88
battavinarinja   109
batuan   140
batung-china   48
baudremoine   232
Bauernsenf   50, 81

Bauhinia   64
Bauhinia, Malabar   64
Bauhinie, Purpurne   64
Bauhinie, Schmetterlings   64
bauira   383
Baumartige Skimmie   347
Baum-Chili   84
baume de galaad   89
baume peteit   228
baume-coq   360
Baumförmiger Meerkohl   119
baumier   246, 304, 372
baumier de la Jamaiquue   246
baumier de Tolu   246
baumier du Pérou   246
Baumpfeffer   295
Baum-Wermut   53
baunilha   383
bauri   80
bavai   256
bavri   256
bawang   17, 19, 23, 24, 27
bawang erah   17
bawang kucai   27
bawang merah   17
bawang pere   23
bawang puteh   24
bawang putih   24
bay berry   242
bay, California   279
bay laurel   199
bay, Indian   241
baya de enebro   189
bayberry   242
bayberry, northern   242
bay-leaf, Indonesian   356
bay-leaved capper   79
bayrum (tree)   287
Bayrum(baum)   287
bazengan africy   349
bazilik   257
bead bean   215
bead maerua   215
bean caper   405
bean caper, Syrian   403
bear's garlic   27
bears lime   107
bearwort   232
beautiful swertia   354
beauty of the night   234
bebina   242
bec brush, common   28
bec de grue   276
beddy pansy   387
bedi wunue   77
bedina   242
bedrenekamnelomka   288
(bee) balm   223, 237

beebalm   237
beebalm, Oswega   236
beebalm, pony   237
beebalm, spottet   237
beefsteak plant   278
Behaarte Fetthenne   341
Behaarter Thymian   367
Behennussbaum, Flügelsamiger   240
bei ai   56
bei sha shen   167
Beifuß, Bitterer   52
Beifuß, Blasser   55
Beifuß, Gemeiner   56
Beifuß, Gewürz   57
Beifuß, Gletscher   54
Beifuß, Indischer   54
Beifuß, Japanischer   55
Beifuß, Kamtschatka   56
Beifuß, Mexikanischer   55
Beifuß, Palästinenser   54
Beifuß, Pontischer   55
Beifuß, Rispen   56
Beifuß, Römischer   55
Beifuß, Walliser   56
Beifuß, Weißer   54
Beifuß-Gewürz   55
beignefata   130
Beilschmiede, Madagaskar   64
Beißbeere   82
Beiß-Canelo   140
Beißender Kardamom   34
bekil   198
bel   7
bel Indien   7
bela   7, 188, 366
bela luz   366
belangkas   234
Belbaum   7
beldroega   239, 305
beldroega de Inverno   239
bele   77
belethi   7
beli   7, 176
belimbing   2, 59, 268
belimbing asam   59
belimbing tanah   268
belimbing wuluh   59
bell pepper   82
bella di notte   234
belle du nuit   234
belle étoile   161
bellflower, Chinese   299
bellflower, Japanese   299
Bello-Isle cress   62
bellout el-ard   363
belokopytnik   281
Beltsville bunching onion   23
beluntas   300

belvedere   64
bembem   218
ben alé   240
ben nout   386
benarasi rai   70
bendera Española   78
benedikt aptetschnyj   112
benediktowa trawa   166
Benediktenkraut   112
Beng-Genipopp   56
Bengal cardamom   35, 36, 37
Bengal ginger   402
Bengal quince   7
Bengal rose   318
Bengal sage   231
Bengali cinnamom   102
Bengalische Quitte   7
Bengalischer Kardamom   35
Bengalischer Salbei   231
Bengal-Pfeffer   292
bengle   402
beni   342
beni seed, black   181
benibana   87
Benin pepper   291
beniseed   342
benito   166
Benjamin bush   208
benjoin   208, 283
benjoin des pays   283
benoîte commune   166
bensojnoe   208
berce de la Birmanie   176
bere lele   125
Bereifter Kardamom   10
bereksa   88
berenjena africana   349
bergamet   106
Bergamot   106, 227, 229, 235, 236
bergamot   106, 227, 229, 235, 236
bergamot mint   227, 229
bergamot orange   106
bergamot(t)e   106
bergamot, horse   236
bergamot, lemon   236
bergamot, Mexican   235
bergamot, wild   236
bergamota   106
bergamotier   106
Bergamotminze   228, 229
Bergamot-Orange   106
Bergamotte   106, 235, 236
Bergamotte, Mexikanische   235
Bergamotte, Scharlach   236
Bergamotte, Wilde   236
bergamotto (albero)   106
Berg-Fetthenne   341
Berg-Genipp   56

Berghopfen   219
Bergkraut, Arabisches   233
Bergkümmel   40, 198
Bergkümmel, Echter   198
Bergkümmel, Russischer   198
Berglaserkraut   198
Berglauch   26
Berglorbeer   365, 379
Bergminze   73, 74, 233, 310, 335
Bergminze, Amerikanische   310
Bergminze, Aufsteigende   74
Bergminze, Drüsige   74
Bergminze, Echte   74
Bergminze, Großblütige   74
Bergminze, Kreta   73
Bergminze, Thymianblättrige   233
Bergminze, Virginianische   310
Bergminze, Wald   74
Bergminze, Weichhaarige   310
Bergmuskat   244
Berg-Nelkenwurzel   165
Bergpfeffer   133, 212
Bergpfeffer, Japanischer   212
Bergraute   323
Berg-Renealmie   314
Bergsalbei   363
bergsteentijm   73
Bergthymian, Katzen   74
Bergwohlverleih   51
beringela africana   349
Berlandier-Oregano   210
bernagie   68
berrillo   62
berro   62, 203, 250, 352
berro alenois   203
berro de agua   250
berro de fuente   250
berro de huerta   203
berro de Pará   352
berro de tierra   203
berros de prado   85
bertalha   44
Bertholdskraut   96
Bertram   53
Bertram, Deutscher   39
Bertram, Römischer   39
Bertramskraut   53
berunai   45
besasa   275
Besenginster   132
Besenpfriem   132
Besen-Winde, Kanarische   115
besla   17
betek   86
betel marron   290
Betelbissen   379
bethu sag   95
betik   86

bétoine officinale 65
betonica 65
betónica oficiale 65
Betonie, Gemeine 65
bétonne 65
betony 65, 353
betony, wolly 353
betu 61
Beutelpfeffer 295
bhanjira 278
bhansi timur 400
bhataktaiya 350
bhuchampaca 192
bhuichampa 192
bhumichampa 192
bhunimba kirata 354
bhutaghna 147
bi cheng que 212
bian xu 302
biao 14
Biberklee 230
Bibernelle, Anis 287
Bibernelle, Große 288, 331
Bibernelle, Heyne 288
Bibernelle, Kleine 288, 331
biberye 320
bible frankincense 269
bible leaf 360
biblical jasmine 188
biblical mint 226
Biblischer Oregano 265
biela horcica 345
Bienenkraut 223
biengranada 96
biennial Alexanders 348
bieslook 19, 25
big eggplant 349
big leaf maple 2
big marigold 357
big pepper 289
bigarade 105
Bigarde 105
bigardier 105
bighel 305
bignai 45
bihuii nayi xtilla 108
bija 66, 388
bijapura 107
bijaura 108
biksa 66
bilasan 331
bilati tulsi 181
bilchuli 339
bilim 59
bilimbi 59
Bilimbi 59
billeri 84
bilva 7

bimblee 59
bindi dikil 148
Bingelkraut, Einjähriges 230
Bingelkraut, Garten 230
Binsenlauch 25
biodo 377
biondella 91
biranjasif 3
bird knotgras 302
bird pepper 82, 83
Birkenblättriger Bukkostrauch 62
birú manso 78
Bisamklee 372
Bisamkörner 1
Bisamkraut 3
bisbas 158
Bischofskraut 33
bishop's weed 7, 33, 371
Bishop's weed, false 33
Bishop's word 65
bistorrta 280
bitter berry 348
bitter cress 62, 84
bitter cucumber 235
bitter dock 321
bitter fennel 158
bitter gourd 235
bitter leaf 386
bitter melon 235
bitter orange 105, 304
bitter orange, Japanese 304
bitter root 163
bitter yarrow 3
bitterblad 386
bittercress 113
Bitterdistel 112
Bittere Ruhrrinde 344
Bittere Schafgarbe 3
Bittere Scheinaster 386
Bittere Springgurke 235
Bittere Veronie 386
Bitterer Beifuß 52
Bitterer Ingwer 404
Bitterer Kardamom 36
Bitteresche 311
Bitterfenchel 158
Bitterholz 285, 311
Bitterholz, Jamaika 285
Bitterklee, Dreiblättriger 230
Bitterkraut, Gewöhnliches 286
Bitterkresse 113
Bitterorange, Dreiblättrige 304
bitterwood 285, 311
Bitterwurzel 163
biwa modoki 137
bixa 66
black amomum 11
black apricot 49

black balsam 246
black beni seed 181
black caraway 71, 253
black cherry 308
black chokeberry 51
black cosmos 117
black cumin 71, 253
black currant 46
black elder 331
black galangal 30
black garlic 21
black gingerwort 304
black jack 65
black locust 318
black lovage 348
black mustard 70
black nightshade 350
black pepper 293
black peppermint 149
black poplar 304
black sassafras 103
black seed 253
black sesame 181
black spleenwort 7
black throated artillery plant 286
black walnut 188
black West African pepper 402
black zedoary 126
black zira 71
blackbead 298
black-edged yarrow 3
blackish stonecrop 339
bladder champion 343
bladder dock 322
bladder seed 205
bladdernut 353
bladdernut, European 353
Blasenkirsche, Flaumige 285
Blasenkirsche, Mexikanische 285
Blasenstrauch 353
Blasiger Sauerampfer 322
Blasser Beifuß 55
Blattang, Japanischer 195
Blattfahne, Canna 351
Blattpetzersilie 282
Blattpfeffer, Mexikanischer 289
Blattsellerie 47
Blaue Katzenminze 251
Blauer Salbei 327
Blauer Steinklee 372, 373
Blauer Sumpfstern 354
Blauer Tarant 354
Blaugraue Kurkuma 126
Blaugrüne Engelwurz 42
Blaugrüne Magnolie 216
Blaugummibaum 150
Blauhimmelstern 68
Bläulicher Sauerklee 269

Blauminze 251
blauwmaan 272
Blauzungenlauch 20
blazing stare 206
bleblendou 371
bleketupuk 110
blessed thistle 112
blibcha 305
blinkblaar 315
Blockzitwer 126, 402
Blockzitwer, Aromatischer 126
bloemriet, eetbar 78
bloemriet, Indisch 78
blood orange 109
bloodred sorrel 322
bloodroot 306
bloschiza 265
blue catmint 251
blue giant hyssop 11
blue gum tree 150
blue gum, Tasmanian 150
blue mallow 216
blue root 343
blue sage 327
blue weed 95
bluebeard sage 330
bluish sorrel 269
Blumée, Lanzettliche 66
Blumenohr, Indisches 78
Blut-Ampfer 322
Blutkopf 331
Blutorange 109
Blutroter Kardamom 8
Blutsverwandter Lauch 18
Blutwurz 306
Blutwurzel 306
bo he 224, 226, 230, 236
bo li qu 68
bobbin weed 205
bobombi 338
boca de dragón 149
Bocksdorn, Gewöhnlicher 213
Bockshornklee, Arabischer 372
Bockshornklee, Griechischer 374
Bockshornklee, Stern 374
Bockshornklee, Traubiger 373
Bockspetersilie 288, 331
bodeh 290
bodha 85
bodrez 288
boembe 371
boeni 45
boerelusern 99
boerhavia, erect 67
Boerhavie, Aufrechte 67
bofidji 338
bog myrtle 242
bog rhubarb 281

bogarim 188
bogbean 230
Bogenhanf, Dreibändiger 332
bogmyrtle 230
Bohnenkaper 405
Bohnenkraut (Sommer-) 333
Bohnenkraut, Ährentragendes 336
Bohnenkraut, Balkan 336
Bohnenkraut, Garten 333
Bohnenkraut, Krim 335
Bohnenkraut, Kroatisches 335
Bohnenkraut, Persisches 336
Bohnenkraut, Winter 336
bois amer 6, 311
bois cannon 89
bois de nerorun 315
bois de rose 43
bois de rose de Cayenne 43
bois de rose fewmelle 43
bois de santal 309
bois de sassafras 333
bois de savane 201 bois de vie 170
bois gentil 133
bois jasanga 317
bois joli 133
bois noyar 285
bois santal 332
boja 52
Bokhara clover 221
Bokharaklee 221
bol'schoj koren' 30
bolai 402
boldo 284, 300
Boldoblätter 284
bolo-moros 353
bolon 128
bolsa de pastor 81
bolsade-pastor 81
boltschez blagocloblennyj 112
Bombay Arrowroot 126
Bombay laurel 113
Bombay mastic 296
Bombay nutmeg 245
Bombay-Mastix 296
Bombay-Muskatnuss 245
bon Henri 96
bonelau 167
bong nga truat 67
bong oi 196
bongo 371
bonnet pepper 83
boon khalif 301
boonenkruid 334
bor(r)agine 68
Bor(r)etsch 68
borage 68, 299, 300
borage, country 299
borage, glandly 300

borage, Indian 299
borago 68
borasahr 71
bord-guene 109
Boretsch, Ausdauernder 68
Boretsch, Bärtiger 300
Boretsch, Drüsiger 300
Boretsch, Indischer 299
Boretsch, Zwerg 68
Borgelkraut 68
borievka 189
bornaset 151
boro dina 349
boronia 68
Boronie, Großnarbige 68
borotyo 90
borrach 68
bor(r)agine 68
borraja 68
borrana 68
borsa del pastore 81
borsacchina 81
Borstige Euphorbie 152
Borstige Leuchterblume 93
Borstige Wolfsmilch 152
borstirochi 263
borstschevik 176
bosh'e 52, 362
botan-uki-kusa 297
botri 96
bou shenaf 68
bou zenzir 230
boucage 288
bouï tochouma 374
bouillon blanc 385
Bourbon geranium 276
Bourbon vanilla 383
Bourbon-Lorbeer 280
Bourbon-Vanille 383
bourrache (officinale) 68
bourroche 68
bourse à pasteur 81
bousassal 68
bouton de guerte 218
bowala 241
bowel seaweed 379
bowl mint 224, 230
Bowlenminze 230
bowstring hemp, African 332
box berry 163
box thorn 213
box-wood, African 371
box-wood, Burmese 241
box-wood, Chinese 241
bracken 308
braken fern 308
bramble rose 320
Brandkraut, Ceylanisches 205

Brandkraut, Großköpfiges   204
Brandkraut, Lavendelblättriges   205
Brandkraut, Martinisches   205
Brandkraut, Rauhes   204
Brandkraut, Weiches   205
Brandschopf   90
Brandschopf, Silber   90
brandy mint   227
Brasilian nutmeg   124
Brasilianische Muskatnuss   124
Brasilianische Vanille   383
Brasilianischer Pfeffer   338
Brasilianischer Sassafras   261
Brasilianisches Rosenholz   43
Brasilpiment   76
Brauerkraut   201
Braunalge, Japanische   195
Braune Haselwurz   57
Brauner Dost   265
Brauner Senf   70
Braunrote Taglilie   175
Braunsilge   256
Braut in Haaren   253
Brautmyrte   247
(Brazil) nut   300
Brazil pimento   76
Brazil tea   353
Brazil vanilla   383
Brazil vlakbloem   351
Brazilian cress   351
Brazilian nutmeg   124
Brazilian pepper   338
Brazilian tea   353
bread fruit, African   371
breath of heaven   7
Breitblättige Limette   107
Breitblättrige Kresse   203
Breitblättrige Lichtnelke   343
Breitblättrige Minze   224, 230
Breitblättriger Allermannsharnisch   28
Breitblättriger Rohrkolben   377
Breitblättriger Sauerklee   269
Breitblättriges Pfefferkraut   203
Breitdorniger Pfeffer   401
Breitlauch   23
Brenndolde, Monnier's   112
Brennender Busch   136
bridal wreath   44
bright ginger   31
brimestone tree   239
brindonia tallow tree   162
brindonnier   162
brise lunettes   153
bristly spurge   152
British Honduran sage   327
broad leaf garlic   21
broad leaved peppermint-tree   150

broadleaf dock   321
broad-leaf lavender   201
broad-leaved garlic   21, 27
broad-leaved maple   2
broadleaf lavender   201
broom   132
Brotfrucht, Afrikanische   371
Brotkümmel   87
Brotsame   288
Broussenet-Thymian   365
Broussonet thyme   365
brown boronia   68
brown cardamom   37
brown mustard   69
Brown's pepper   83
Bruchreizker   195
brunkress   250
Brunnenkresse, Echte   250
Brunnenkresse, Kleinblättrige   250
Brustwurz   41
Brustwurzel   43
bu yiddi pulu   148
buah keras   14
buah pala   244
buah pelaga   144
bubbolini   343
bubga tantan   38
bubga tatan   38
buc ha nam   225
Bucco   62
buch   5, 62
buchu   27, 62, 63
buckbean   230
buckrams   27
buckthorn   315
bud(d)jer   136
Buddah's Hand   108
budea pljuschtschewidnaja   167
budelemi   148
buding   280
budra   167
buenas tardes   234
buffalo ear, lanceolated   66
Büffelohr, Lanzettliches   66
buga natal   152
bugle   13
bugle rampante   13
buglossa vera   68
buis de Chine   241
Bukkostrauch, Birkenblättriger   62
Bukkostrauch, Gekerbtblättiger   62
Bukkostrauch, Gesägtblättiger   63
bukwiza lekarstwennaja   65
bulak-manok   387
bulb onion   17
bulbocastano   71
bulbous chervil   94
bulbous nasturtium   376

Bulgarian geranium   165
Bulgarian rose   319
bullhoof   289
bullrush   377
bullwort   33
bunching onion, Japanese   19
búndei   45
buneh   45
bunga ayer mawar   319
bunga chingkeh   355
bunga cina   162
bunga jarum   186
bunga kantan   149
bunga karan   313
bunga kupu-kupu   64
bunga lawang   184
bunga pagar   196
bunga pechah empet   234
bunga ros   319
bunga siantan   149
bunga soka   186
bunga tahi ayam   196
bunga tanjong   38
bunganana   27
bunglai   402
buni   45
bunpaling   321
Buntnessel, Einährige   351
buona Enrico   96
buratschnik   68
buratschnik lekarstwennyj   68
Burgel   305
Burmain cow-parsley   176
Burmannische Petersilie   176
Burmannischer Bärenklau   176
Burmese box-wood   241
Burmese rosewood   309
burnet   288, 331
burnet saxifrage   288
burnet, garden   288, 331
burnet, great   331
burnet, Heyne   288
burnet, salad   331
burnet, small   331
burning bush   64, 136
burr medick   221
burro   79, 89, 181
Burropfeffer   396
Burzeldorn, Geflügelter   372
Bürzelkraut   305
Burzelkraut, Vierspaltiges   306
Busch, Brennender   136
Busch, Flammender   136
Busch-Basilikum   257
Buschminze   181
Buschtee, Herero   255
Buschthymian, Griechischer   257
bush basil   257

bush mint   181
bush nasturtium   376
bush oregano, Mexican   301
bush pepper   397
busina   331
but but   150
butau   49
butm sâqis   297
Butterblume   75, 361
butterbur   281
butterbur, Japanese   281
butterfly geranium   277
butterfly lily   173
butterfly tree   64
button snakeroot   206
buyo-buyo   290
bwembi   371
Byzantinischer Kälberkropf   94
ca hoi   349
caapeba   295
caapiá   139
cabaceira   120
cabaret   57
cabbage rose   318
cabbage, Nile   297
cabe   82, 83, 294
cabe bendot   84
cabe besar   82
cabe dieng   84
cabe gondol   84
cabe jawa   294
cabean   295
cabella kani   395
cablan   301
cablin patchouli   301
cabudula   75
cabuyau   107
cacabean   280
cacahouète   48
cacahuate   48
caccia febbre   91
cacciadiavoli   180
caculucage   300
café extranjero   1
cafoun   34
caggage, kaffir   111
caglio odorosa   161
cajueiro   6
calabaceira   6
calabash nutmeg   238
calabaza costillada   361
calamandier petit chên   262
calamandrea   362
calament   4, 74
calament à grande fleur   74
calament des Alpes   4
calamenta   74
calamento   74, 251

calamint   74, 112
calamint commun   74, 112
calamint, cushion   112
calamint, large flowered   74
calamint, lesser   74
calamint, showny   74
calaminta   74
calamís   5
calamo aromatico   5
cálamo arómatico   5
calamondin   159
calamus   5
calamus oil   5
calanga   30
calebassier du Sénégal   6
calendola   75
calendula   75
calibtus   150
California bay   379
California laurel   379
California nutmeg   370
California rape   345
California torreya   370
California yew   370
Californian pepper   537
calingcing   269
calisaya   99
calisaya blanca   99
calisaya morala   99
Calisya Chinarinde   99
calta   75
calta palustre   75
caltha des marais   75
caltrap   372
calumbán   14
cam   105
cam thao day   2
camachile   298
camapu   285
camará   196
camarão   6
camaru   285
camasec   218
cambará   6, 196
cambará de cheiro   6
cambará, rosa   196
cambará-de-cheiro   196
cambará-de-chumu   196
cambodge   162
Cambodia cinnamon   101
Cambodian cardamon   36
Cambodian chister cardamom   36
Cambodian mint   280
Cambodian star anise   183
camel grass   131
camel's foot tree purple   64
Cameroon cardamom   8, 9, 10
Cameroon garlic   179

Cameroon garlic tree   11
camias kalanias   59
camirio   14
camomila de jardin   95
camomila nobre   95
camomila odoranta   95
camomila verdadeira   95
camomilla romana   95
camomille romain   95
campa   232
campahah   232
campana   184
camphor absinthe   52
camphor basil   258
can nuoc   261
caña de das Indias   78
caña domestible   78
caña pistula   88
Canada onion   16
cana-da-índia   78
Canadian lettuce, wild   195
canafistula   88
cañafístula   88
canafistule   88
canang odorant   76
cananga   76
canapa d'acqua   151
canapistola   88
canary balm   89
Canary convolvulus   115
candana   332
candela della madonna   385
candela regia   385
candle berry   14, 242
candle nut   14
candle wood   155
candleberry   14
candrasura   203
candrika   203
cane reed   118
caneel, Magelhanisian   140
caneel, white   77
caneel, wild   77
canéfice   100
canéficier   88
canela   102, 104, 124, 140, 250, 260, 261
canela de Ceilán   104
canela de porco   124
canela de Saïgón   102
canela do mato   260
canela fogo   124
canela pimenta   124
canela pururuca   124
canela-amarela   250
canela-amarga   140
canela-da-capoeira   250
canela-mirim   250

canela-sassafrás   261
canela-seca   250
caneleira em ceilão   104
caneleira-cravo   136
canelero chino   100
canella   77, 100, 104
canella blanche   77
canella cinese   100
canelle   100, 102, 103, 104, 140
canelle casse   100
canelle d'Inde   104
canelle d'Indonésie   100
canelle de Batavia   100
canelle de Ceylan   104
canelle de Chine   100
canelle de Cochinchine   102
canelle de leds Philippines   103
canelle de Magellan   140
canelle de Padang   100
canelle de Saïgon   102
canelle de Timor   100
canelle de Viêtnam   102
canelle du Bengale   102
canelle fausse   100
canelle l'Australie   103
canellier   100, 104
canellier de Chine   100
canellila   43
canelo de la China   100
Canelo, Beiß   140
Canelo, Chilenischer   140
canistel   213
canna comestível   78
canna commestible   78
canna d'Indie   78
canna dolce   78
canna, edible   78
canna, Indian   78
Canna-Blattfahne   351
cannalam paranta   105
cannelier giroflée   136
cannella di Ceylon   104
Cannibal's tomato   349
cansur   203
cao   3, 4, 36, 38, 65, 151, 179, 182, 200, 204, 205, 219, 231, 239, 268, 302, 323, 358, 381, 386
cao guo   38
cao kou   36
Cap pepper   290
Cap safron   121, 354, 375
càparo   80
Cape jasmin   162
Cape jessamine   162
Cape pepper   290
Cape saffron   121, 354, 375
Cape velvet mint   227
caper   79, 80, 81, 152, 405

caper, Eastern   80
caper spurge   81, 152
caper, common   80
caper, Mariana   81
caper, Sicili   80
caper, spineless   80, 81
caper, spurge   81, 152
capigilate   253
capillaire boire   7
capim cheiroso da India   130
capim-limao   130
cap-in chop   123
capo di fratre   361
cappari   81
capparo   81
capper tree   79
càppero   80
cappucina   375
caprier   81
câprier (commun)   80
câprier sans feuilles   79
capsell   81
capselle   81
capselle à pasteur   81
capsico   83
capsicum pepper   82
capsicum, green   82
capuba   295
capuchina   375, 376
capuchina tuberosa   376
capuchinha grande   375
capuchinha tuberosa   376
capucine   375, 376
capucine grande   375
capucine tubéreuse   376
caquillier   73
carabu   203
carandas plum   86
caraway, (common)   87
caraway oil   87
caraway thyme   366, 367
caraway, black   71, 253
caraway, Egyptian   125
caraway, Ethiopan   371
caraway, Roman   125
Carbe   87
carcadè   177
cardaire   85
cardamina   84, 85
cardamina dos prados   84
cardamine à trois feuilles   85
cardamine des prés   84
cardamino dei prati   84
cardamom   144
cardamom nigra   36
cardamom, bastard   38
cardamom, Bengal   35, 36, 37
cardamom, Cambodian   36

cardamom, Cameroon   8, 9, 10
cardamom, Ceylon   144
cardamom, chester   35, 144
cardamom, Chinese   36
cardamom, East African   10
cardamom, cluster   35
cardamom, false   9, 35, 38
cardamom, greater   37
cardamom, Indian   37
cardamom, Indonesian   35
cardamom, Java   35, 37
cardamom, large   35, 37
cardamom, long   144
cardamom, Madagascar   8
cardamom, Malabar   38, 144
cardamom, Nepal   35, 37
cardamom, nigra   36
cardamom, round   35
cardamom, Siam   35, 36
cardamom, slender   36
cardamom, small   144
cardamom, Tavoy   38
cardamom, wild   8, 144
cardamom, wringed   37
cardamome   144
cardamome d'Ethiopie   9
cardamome de Cameroun   9
cardamome de Guinea   9
cardamome de Madagaskar   8
cardamome du Bengale   35
cardamome du Népal   37
cardamome krervanh   36
cardamome poilu de la Chine   38
cardamome rond   35
cardamom, tsao-ko   38
cardamomier   144
cardamomo   144
cardo mariano   344
cardo bendito   112
cardo bento   112
cardo de la alameda   344
cardo de María   344
cardo lechero   344
cardo leiteiro   344
cardo mariano   344
cardo santo   112
carenfil   355
carica   86
carilla plant   235
cariofilata   166
cariofillata   166
carissa   86
carissa, Egyptian   86
carisse   86
carnation   135
carnation annual pink   135
carngeri   268
carob   92

caroba 93
carobier 93
Carolina allspice 75
carota 133
carota rossa 274
carotte 133
carouge 93
carpet weed 167, 234
carphonka 362
carpigna 268
carrot 133
carrot, garden 133
carrubio 93
cartamo 87
cártamo 87
carthame 87
carthame de teinturiers 87
saffron, Cartwright 121
Cartwright-Safran 121
caruru azedo 177
caruru-azedo 177
caruru-da-guiné 177
caruru-de-sapo 268
carvi 87
casca preciosa 43
casca-acucena 124
casca-d'anta 140
cascade creeper 44
cascarilla 99, 123
cascarilla amarga 99
cascarilla crespilla 99
cascarilla de Cuba 123
cascarilla de la lomas 99
cascarilla de Trinidad 123
cascarilla del rey 99
cascarilla gallingo 99
cascarilla verde 99
cascarille 123
case weed 81
Caspian sage 279
casse 88, 100, 153, 341
casse fistuleuse 88
casse ligneuse 100
casse puante 341
casse-lunette 153
cassia 88, 100, 341
Cassia 88, 100, 341
cássia 100
cassia fétide 341
cassia lignea 100, 104
cassia vera 100
cassia, Batavia 100
cassia, Chinese 100
cassia, Culilawan 101
cassia, Indian 104
cassia, Indonesian 100
cassia, Java 103
cassia, Korintje 100

cassia, Korinjii 100
cassia, Padang 100
cassia, purging 88
cassia, Saïgon 102
cassia, Vietnamese 102
cassine, saffron 89
Cassineholz 89
cassumunar ginger 402
castagne di terra 71
castalda 7
castanet plant 123
castillan mallow 217
cat loi 118
cat tail 377
cat thyme 363
cat's hair 152
cat's valerian 381
cat's whiskers 111
cataia 140
cataire 251
Catalonian jasmine 188
catapuzia 152
cataria 251
Catawissa onion 23
Catawissazwiebel 23
cätcänh 299
catinga de mulata 205, 360
catir 365
catmint 251
catmint, blue 251
catnip 251
catnip, lemon 251
catnip, Mussin 252
catnip, Transcaucasian 252
cauhara 54
cáy cam 109
cây cu chóc 118
cay goi ca 303
cay hoy 184
cây lá lôt 292
cây lai 14
cay moc 266
cay xa 129
caya mouzambi 111
Cayenne pepper 82
Cayennepfeffer 82, 83
cebola 17
cebolha roxa 17
cebolinha francesa 25
cebolla 17
cebolleta 19, 25
cebolleta francesa 19, 25
cebollina común 25
cebollino 15, 19, 25
cebollino francés 25
cebollino inglés 19
cedar 65, 189, 370
cedar, eastern red 189

cedar, pencil 189
cedar, stinking 370
cedar, Virginian 189
cedoaria 128
cedoaria redonda 128
cedraat 108
cedrat 108
cédrat 108
cédrat digité 108
cédratier 108
cedrato 108
cedrina 211, 386
cedro (frutto) 108
cedrón de monte 28
cedronella 223
ceeping thyme 367, 368
cekur 191
celandine 313
céleri 46
céleri couper 47
céleri rave 47
céleri tubéreux 47
céleri vivace 205
celeriac 47
celery 46
celery, Chinese 47
celery, water 261
célosie 90
Celtic nard 381
Celtic valerian 381
çemen otu 374
cempaka 232
cempaka kuning 232
cempaka putih 232
cempasóchil 357
cendana 332
cengkeh 355, 380
cenizo blanco 95
cenoura 133
centaurea menor 91
centaurea minore 91
centaury 91
centinodia 302
Centorelle 91
cerefolho almiscarado 247
cerefolho anisado 247
cerefólio 45
ceremucha pozdnjajo 308
cerezo de mahoma 308
cerezo de Santa Lucia 308
cerfeuil 45, 94, 247
cerfeuil cultivé 45
cerfeuil d'Espagne 247
cerfeuil des fous 94
cerfeuil musqué 247
cerfeuil odorant 247
cerfoglio 45, 247
cerfoglio di spagua 247

cerisier d'automne   308
cerisier de mahaleb   308
cerisier de Sainte Lucie   308
cerisier tardif   308
černuška   254
cerrado   287
cetti   186
Ceylanisches Brandkraut   205
Ceylon cardamom   144
Ceylon cinnamon   104
Ceylon citronella   130
Ceylon leucas   205
Ceylon spinach   63
Ceylon Zitronellgras   130
Ceylon-Kardamom   144
Ceylon-Zimt   104
ch'i   178
ch'uan chao   400
ch'uan jiao   401
chá de bugre   174
chá do tabulaeiro   209
cha khrai   129
cha ran   97
chaa phluu   295
chabai   291, 294, 295
chabai ekur   291
chabai hutan   292
chabai jawa   294
Chabarpfeffer   294
chac-mixib   377
chadap   148
chá-de-soldado   174
chaemomila, Spanish   39
chagas   375
chai tahuang   315
chaitan   32
chajogi   278
chakriphool   184
chakunota   341
chalfej mutovtschatij   330
cha-loot   32
chalota chinesa   18
chalote   17
chalote chinesa   18
chalta tree   136
chalta   136
Chalthafrucht   136
cham phaa   232
Chamäleonblatt   179
chamapaka   232
chamba   187, 232
chambai   290
chambeli   187
chamburoto   84
chamere   59
chamizo   59
chamnari   207
chamot-ton   1

chamomile, garden   95
chamomile, noble   95
chamomile, Roman   95
champa   31, 232
champa, golden   232
champa, yellow   232
champac   232
champaca   232
champii khaek   232
champion, bladder   343
chan   75, 127, 184, 244, 328, 355
chan thet   244
chanchom   280
chandon   332
chandra   191, 219
chandra-mula   191
chandra obyknowennaja   219
chandramallika   97, 191
chandramalu   192
chandramulika   191
chang slang kuo   48
changeri   322
changoe   404
changrarro   89
chantana   332
chanvre d'eau   151
chao   181, 204, 400
chaphlu   292
charan   97
chardon bénit   112
chardon étoilé   147
chardon Marie   344
charlock   345
charrûb   92
charviel   45
chaschchâsch   272
chasse diable   180
chaste pepper   388
chaste tree, Chinese   388
chat   251, 377
chataire   251
chateigne de terre   71
chatiwa   32
chatra   40, 116
chattiyan   32
chatwan   32
chaulljunija serdcevidnaj   179
chausaur   203
chauvre d'Afrique   332
Chayenne   83
checkerberry   163
cheeses   216
cheeseweed   216, 217
chegadinka   8
chekkur   67
chĕkur   191
chemar   158
chĕmpaka   232

chempaka ambon   232
chempaka merah   232
chempaka, orang   232
chen p'o ka   232
chen pi   109
chen po   232
chen tan   332
chenanga   77
chendana   332
chénette   362
cheng   109, 212
chenggan cao   151
chengkeh   355
chengkering   148
chĕngkur   191
chenopodio ambrosioide   95
cherafes   47
cherry, black   308
cherry, laurel   307
cherry, mahaleb   308
cherry, perfumed   308
cherry, St. Lucie   308
cherry laurel   307
cherry pepper   82
cherry pie   174
cherry, ground   285
cherry, winter   85
chervil   45, 94, 124, 247
chervil, anise   247
chervil, bulbous   94
chervil, cow   94
chervil, garden   45, 97
chervil, parsnip   94
chervil, Spanish   247
chervil, sweet scented   247
chervil, turnip rooted   94
chervil, wild   124
chervil, woodland   45
chessbolls   272
chester cardamom   35, 36, 144
chestnut-leafy nutmeg   243
che-tao   188
cheveux de Vénus   253
chhe: tô:ch   110
chi krassang tomhom   280
chi tsai   81
chia   109, 326
chía   181, 328
chía blanco   328
chía del campo   328
chía grande   181
chián   328
chiang   268, 315, 319, 403
chiang chun   315
chiang wei   319
chiáng wei chun   320
chiang wie   320
chiantanella   93

chiao mu   400
chiappara   81
chiarea   330
chibt   40
chickweed   92, 274
chieh keng   299
chiendent del Indes   387
chiendent odorant   387
chiendent odorante   44
chih ma   209
chih shih   304
chijanamu   162
children's tomato   348
chile   82
chile apple   82
chile japonés   84
Chile laurel   198
chile manzana   84
Chile pepper   82
Chilean laurel   198
Chilean nutmeg   198
Chilenischer Canelo   140
Chilenischer Lorbeer   198
Chilenischer Muskat   198
Chili(s)   83
Chili (pepper)   82
Chili, Apfel   84
Chili, Baum   84
Chili peper   82, 83
chilipiquin   82
chilli manzana   84
chiltepe   82
chiltepin   82
chimpoh   137
chimpuh   137
chin ch'iao   132
chin chan hua   75
chin chi lo   99
chin chu   159
chin kan   158
chin li chih   235
chin lien hua   375
chin tan   159
chin tou   159
chin tsai   261
chin tsao   159
china bark   99
China flower leaf   349
China orange   159
China rose   318
china rossa   99
Chinabeifuß, Weißer   54
Chinalauch   24
Chinapilz   202
Chinarinde, Calisya   99
Chinarinde, Gelbe   99
Chinarinde, Rote   99
China-Rose   318

chinchilla enana   358
Chinese angelica   43
Chinese anis   184
Chinese basil   278
Chinese bellflower   299
Chinese box-wood   241
Chinese cardamom   36
Chinese cassia   100
Chinese celery   47
Chinese chaste tree   388
Chinese chive   23
Chinese chives   18, 27
Chinese cinnamom   100
Chinese fire leaf   116
Chinese garlic   19
Chinese giant hyssop   12
Chinese ginger   31
Chinese horse radish   178
Chinese key   67
Chinese keys   192
Chinese laurel   45, 46
Chinese leek   23, 27
Chinese lime   374
Chinese liquorice   170
Chinese magnolia vine   338
Chinese mint   225
Chinese mustard   69
Chinese myrtle   241
Chinese onion   18
Chinese parsley   116
Chinese pepper   83, 399, 400, 401
Chinese pink   135
Chinese prickly ash pepper   401
Chinese rhubarb   315
Chinese scallion   18
Chinese star anise   184
Chinese sweet grass   5
Chinese violt   58
Chinese wampee   111
Chinesian hemp agrimony   151
Chinesische Kamm-Minze   145
Chinesische Limone   338
Chinesische Angelikawurz(el)   43
Chinesische Minze   225, 226
Chinesische Nelke   135
Chinesische Paprika   83
Chinesische Toddalie   369
Chinesische Wolfsbeere   213
Chinesische Zimtrinde   100
Chinesische Zitrone   338
Chinesische Zwiebel   18
Chinesischer Anis   184
Chinesischer Galgant   191
Chinesischer Gelbholzbaum   400
Chinesischer Gewürzstrauch   145
Chinesischer Kardamom   36
Chinesischer Knoblauch   19
Chinesischer Lauch   24, 27

Chinesischer Meerrettich   178
Chinesischer Mönchspfeffer   388
Chinesischer Pfeffer   399, 401
Chinesischer Rhabarber   315, 316
Chinesischer Schnittlauch   18
Chinesischer Senf   69
Chinesischer Wasserdost   151
Chinesischer Zimt   100
Chinesisches Spaltkörbchen   338
Chinesisches Süßholz   170
ching   88, 156, 261, 329, 342, 388
ching chieh   329
ching jang   342
chini naranghi   374
chinnaja kora   99
chinois   109, 401
chinotto   109
Chinotto   109
chinpaetklip   184
chío gorda   181
chiòdo di garofano   355
chioplé   152
Chios mastic   296
chiote tree   65
chipil(e)   123
chipiliń de comerc   123
chirata   354
chireta   354
Chirettakraut, Schmalblättriges   354
chiriguaná   311
chirivía   274
chiruub   92
chisaca   352
chishi   272
chissum   52
chitvan   32
chiu   24
chiu kieou   24
chiu li hsiang tsao   241
chiu li xiang   241
chive   23, 25
chive garlic   25
Chives, Chinese   18, 27
Chloranthusblüten   97
chmel'   179
chmel' obyknovennyj   179
chocolate cosmos   117
chocolate root, Indian   166
choduguphul   36
choelle   371
chohore   42
chói mói   45
chollang   19
cholodjanka   227
cholodnaja mjata   227
chong kho   64
chonipthopphul   3
chophinamu   400

chopo balsamifero 304
chopo común 304
chora 41, 42
chordal 70
chosen gmiski 338
chosen-gurumi 188
chota pilu 325
choti kateri 350
chotielachi 144
chou cao 323
chou giao na 178
chren 50
chren guljavnikovyi 50
chren lugovoi 50
chreniza 203
chrinnizja krupkowidna 85
Christmas berry 338
Christmas flower 152
chrizantema uventschannaja 97
chrysantema balsamita 360
chrysantemo 360
chrysanthème a couronne 97
chrysanthème des fleuristes 360
chrysanthemum, cooking 97
chrysanthemum, garland 97
chü 109
chu kuen 261
chua me (aas) hoa 268
chuan tan 207
chuan ts'ai 63
chuan ye chun mei cao 239
chuete 66
chuka 322
chukra 322
chuksa 38
chulan 12, 97
chum het tai 93
chumbinho 196
chun ce bo he 228
chuña-chuña 311
chung-yao 179
chunsha 38
ci huai hua 318
ci shan gan 80
cibol 19
ciboule 17, 19
ciboule oignon patate 17
ciboulette 25, 27
ciboulette Chinoise 27
cibreska 334
cicanaga latyaga 104
cicely, sweet 247
cicuta 261
cidra 108
cidrão 108
cidreira 108, 129, 211, 223
cidrilha 209, 211
cidro 108

cidró 211
cidron 211
cientoenrama 3
čierne korenie 293
cilandrio 116
cilantro 116, 147
cili 83
cili padi 83
cili rawit 83
ciliegio canino 308
ciliegio tardivo 308
ciliegiodi S. Lucia 308
cimarrón 147
cimpual 357
cimpualxochite 357
cinar 136
cinchona 99
cinchona, red 99
cinchona, yellow 99
cinco negritos 196
cinéraire d'Afrique 99
cinga-cinga 392
cingkeh 355
Cinnamom 43, 100, 104, 243
cinnamom nutmeg 243
cinnamom tree 100
cinnamom wood 333
cinnamom, alyxia 32
cinnamom, American 261
cinnamom, Assam 100
cinnamom, Australian 103
cinnamom, Batavia 100
cinnamom, Bengali 102
cinnamom, Cambodia 101
cinnamom, Ceylon 104
cinnamom, Chinese 100
cinnamom, false 100
cinnamom, Japanese 102
cinnamom, mindanao 102
cinnamom, Oriniko 43
cinnamom, Philippine 103
cinnamom, Saigon 102
cinnamom, Sri Lanka 104
cinnamom, Taiwan 102
cinnamom, true 104
cinnamom, white 77
cinnamom, wild (of Japan) 102
cinnamon canela 104
cinninielli 287
cipó d'alho 7
cipó-catinga 233
cipoletta 19, 25
cipolla 17, 19
cipolla d'inverno 19
cipollina 25
cirayta 354
cirier 242
ciruela de Natal 86

cissus de Galam 105
cistre 232
citiso dei carbonai 132
citrangequat 304
citraria 223
citreno 107
citroen 107, 223
citroenkruid 52
citron 61, 105, 107, 129, 131, 211, 223, 335
citron backhousia 61
citron combara 107
citron de Medie 108
citron grass 129
citron myrtle 61
citron peel 108
citron vert 105, 107
citron(n)elle verveine 211
citronela de Java 130
citronella 129
Citronella 129
citronella, Ceylon 130
citronella, Java 131
citronelle 52, 129, 223, 386
Citronelle 52, 129, 223, 386
citronelle aurone 52
citronelle grass 131
citronnier 107
citrosa geranium 276
Citrosa-Geranie 276
cive 25
cive chino 27
civet 25
civette 24, 25, 27
civette de Chine 27
claitónia 239
clary 330
clary sage 330
clary wort 330
Clary-Salbei 330
clasping mullein 385
clausena 110
clausena, Hollywood 110
clausena, Indian 111
Clausenie, Anis 110
Clausenie, Indische 111
Clausenieblätter 111
clavalier à épines planes 401
clavalier d'Inde 401
clavalier de Bunge 401
clavalier poivrier 400
clavel 135
claveles 135
clavelina 177
clavellina tardes 234
clavillo 355
clavo 258, 355
clavo de comer 355

clavo de especia   355
clavo de olor   355
claytone de Cuba   239
claytonia perfoliée   239
clementine   109
clemole   358
clemolitos   358
clethra   214
Cleveland Salbei   327
climbing pepper of Java   290
clinopode   112
Clinopode, Gewöhnliche   112
clou de giroflé   355
clove   13, 114, 166, 258, 286, 313, 355
clove bar   136
clove basil   258
clove, Madagascar   313
clove nutmeg   313
clove oil plant   114
clove pepper   286
clove pink   135
clove root   114, 166
clover, Bokhara   221
clover, white sweet   221
cloverhead horse mint   237
cluster cardamomum   35
clustered trefoil   373
cmola wonjutschaja   156
cnicus   112
cnide   112
co foot   295
co hoi   95
co nha ngu   4
co sa lang   349
co tam giac   81
coagerucu   396
coari-bravo   358
coaxicó   124
coccola di genipro   189
Cochin grass   129
Cochingras   130
cochin-tumer   128
cochléaire   113
cochleária   113
cock ley bur   13
cock's curry grass   203
cockscomb   90
cockspur flower   299
coclearia   113
coentro   116, 147
coentro-bravo   147
coentro-da-caboclo   147
coentro-da-colônia   147
coiled-flower nightshade   350
collender   116
cologoco   260
colubrina   96
comb hyptis   181

comb mint   227
combara citron   107
combavas   107
combrang   149
comida-de-jaboti   278
cominella   253
cominho negro   253
comino   87, 125
comino de prado   87
common anise   288
common baobab   6
common basil   256
common bec brush   28
common broom   132
common burnet saxifrage   288
common calamint   74
common caper   80
common caraway   87
common centaury   91
common condorvine   220
common elder   331
common elsholtzia   145
common flax   209
common gardenia   162
common garlic   24
common germander   362
common ginger   403
common ginger of Australia   29
common Heron's bill   146
common hop   179
common horehound   219
common iris   185
common juniper   189
common laurel   199
common leek   23
common licorice   169
common mallow   216
common mignonette   314
common mullein   385
common myrrh   115
common myrtle   247
common nasturtium   375
common nightshade   350
common parsley   282
common pepper gras   202
common pepper vine   290
common polypody   303
common rue   323
common sage   329
common scurvy grass   113
common sorrel   320
common tansy   360
common thyme   368
common valerian   381
common verbena   386
common violet   387
common water cress   318
common watercress   250

common white jasmine   187
common winter cress   62
common wood sorrel   268
common wormwood   52
common yarrow   3
common, iris   185
commun, poivrette   253
compass plant   320
concombre Africain   235
condor plant   220
condorvine, common   220
condurango   220
Condurango   220
condurango blanco   220
cone pepper   82
conehead thyme   365
cônes de houblon   179
confetti tree   220
cong   19
Congo pump   89
consuelda roja   306
contra de jorba   138
contra hierba   138
contrajerba   138
convolvulus, Canary   115
cooking chrysanthemum   97
copal   338
copetillo   358
coq   361
coquerct   285
coral tree, Indian   148
corales   311
corazoncillo   180
cordao-de-frade   205
cordão-de-sao-francisco   205
cordoncillo   289, 291, 295
cordonillo   294
çörek otu   254
coriander   116, 147, 253, 280
coriander, false   147
coriander, foreign   147
coriander, Mexican   147
coriander, Roman   253
coriander, Russian   117
coriander, sawtooth   147
coriander, Vietnamese   280
coriander, wild   117
coriandolo   116
coriandr   116
coriandre   116
coriandro   116
corido thyme   365
cork tree   6, 273
cork wing   167
corn mint   224, 225
corne d'abondance   155
corneita   337
cornilla real   222

corno   296, 297
corno-capra   296
coronitas   196
Corsican mint   228
corteccia   242
corteccia di China   99
Cortex Caryophylloides ruber   101
corteza de quina   99
çörtük   143
cosmetic park tree   241
cosmos, black   117
cosmos, chocolate   117
cosmos, orange   118
cosmos, yellow   118
costeno   298
costmary   360
Costus   336
Costuswurzel, Indische   336
cotule à gris cendre   119
coulante   147
coumarouna   137
country borage   299
couronne de Saint Jean   56
courroie de Saint Jean   167
cow chervil   176
cow cockle   381
cow herb   381
cow parsley   45
cow-parsley, Burmain   176
cow, spice   94
cow-parsley, Persian   176
cowslip   307
coyote mint   237
crab's eye vine   2
cran   50
cranesbill, curled leaved   276
cranson   50, 113
crape ginger   118
cravanh   35
craveiro-da-terra   76
craveiro-do campo   76
craveiro-do maranhao   136
cravinho   136
cravinto   355
cravo de India   355
cravo-de-defunto   357
cravo-de-defunto-miúdo   358
cravo-do-campo   76
cravo-do-mato   136
creeped bugle   13
creeping oxalos   268
creeping savory   336
creeping thyme   367
creeping wintergreen   163
creeping wood sorrel   268
crème de menthe   228
cren   50
creosote bush   197

crescione aquatico   250
crescione dei prati   84
crescione del Perù   375
crescione di giardino   203
crescione inglese   203
crespilla   99, 218
cress, American   62
cress, Bello-Isle   62
cress, bitter   62, 84
cress, Indian   375, 376
cress, Peter's   120
cress, rocket   62
cress, upland   62
cress, winter   62
cresson alénois   203
cresson d'eau   250
cresson d'Inde   375
cresson de fontaine (d'eau)   250
cresson de jardin   62, 203
cresson de terre   62
cresson des prés   84
cresson du Brésil   351
cresson du Para   351
cresson du Pérou   375
cresson élégant   84
cresson(n)ette   84, 203
cresta de gallo   90
cresta di gallo   90
Cretan oregano   264
Cretan savory   336
Crete balm   223
crête de coq   90
Crete dittany   262
Crete melissa   223
Cretean thyme   365
Crimea mountain savory   335
crisantemo   97
crisp mint   229
crista de galo   90
critmo   120
Croatia savory   335
crocus   67, 121, 122
crocus tritonia   121
crocus, Scotch   121
crocus, tropical   67
Crocus-Tritonie   121
croton   116
crow garlic   28
crown daisy   97
crown flax   209
crucete   311
cú chóc   118
cu den   215
cu jiang   268
cu ngai   67
cu xia zhi cao   219
cuasia amargo   311
cubab chinee   291

Cuban oregano   299
Cuba-Thymian   299
cubeb pepper   291
cubeb, African   211
cubeba   291
cubebe   290
cubebe du pays   290
cubebe, Guinea   291
cuc tan   300
cuchay   27
cuckoo flower   84
cucumber, African   235
cucumber, bitter   235
cucumer tree   59
cudweed   54
cudweed, Indian   170
cuimari   260
cuimaru   260
cujumaru   260
culantrillo   147
culantro   116, 147
Culilawan cassia   101
Culilawan-Zimt   101
cultivated flax   209
cumaru   137, 370
cumaru das caatingas   370
cumaru de cheiro   370
cumaru-amarelo   137
cumaru-doamazonas   137
cumarurana   137
cumbaru   137
cumin   71, 87, 125, 198, 253, 288, 343, 371
cumin, black   253
cumin de chevaux   198
cumin dés pres   87, 124, 343
cumin noir   253
cumin tubéreux   71
cumin, black   71, 253
cumin, white   371
Cuming-Fiederaralie   303
cumino dei prati   87
cumino di Malta   125
cumino proprio   125
cumino romano   125
cumino tedesco   87
cumino vero   125
Cundurango   220
curassao   106
curassao peel   106
Curassaoschalen   106
curcuma   126
cúrcuma larga   126
cúrcuma de Java   128
curcuma longa   126
curcuma longue   126
curcuma rhizome   126
curcuma zédoaire   128

Index of Common Names

curled dock   321
curled leaved cranesbill   276
curled mallow   217
currant tree   45
curry bush   181
curry leaf   241
curry plant   174
Curryblätter   241
Currybusch   181
Currykraut   174
Curry-Orangenraute   241
cus cus (grass)   387
cushion calamint   112
cuta   218
cuyo   317
Cyprus turpentine   297
d'orange amère   105
da suan   24
da tou Suan   15
dacon ajenton   299
dadap belang   148
dadima   309
daengkluai   356
Daffodil garlic   21
dafne   133
dagbani nanzidow   33
dai toan   24
dai-dai   106
dairy pink   381
dakhar   359
dakngoe   404
dal-chini   104
dalima granada   309
Dalmatiner Salbei   329
dalupang   1
damanaka   57
Damas black cumin   253
Damask rose   319
Damaszener Rose   319
Damaszener Schwarzkümmel   253
Damiana   377
damiana   377
damiana de Guerrero   377
Damiana, Schmalblättrige   377
Damianastrauch   377
damigella   253
damong Maria   57
damsisa   33
dandelion   361
dang guai   43
dang gui   41, 43, 205
danglais   110
dành dành   162
Dänisches Löffelkraut   113
Danish scurvy grass   113
dantaharshana   106
dao   140
dao kou cao   4

daoen djeroek poeroet   107
daoen grisik   303
daoen hoeni   45
daoen salam   356
daopang   1
dap dap   148
daphne   133
daphné morillon   133
daphne, (February)   133
daradharsha   147
darahuo   297
daraido   297
darcini   104
dardilon   71
dark Crimson cardamom   8
darmar   399, 401
Darmtang   379
dasheen, great   114
datil pepper   83
Dattel, Indische   359
dattier du désert   61
dau giun   95
dauag   369
daun ambré   276
daun bangun-bangun   299
daun bawang   19
daun jeruk purut   107
daun juang   116
daun kaduk   295
daun kambing   299
daun kari   241
daun kesum   280
daun ketumbar   116, 147
daun ketumbar jawa   147
daun laksa   280
daun pandan   271
daun pasli   282
daun pudina   225
daun salam   356
daun senahun   280
daun sop   47
dauna   57
daunkucing   299
davana   55
Dawa   273
Dawabaum   273
dawk kaeo   242
dawo   362
day chi chi   2
day lily, fulvous   175
day lily, grassleaf   175
day lily, lemon   175
day lily, little   175
day lily, orange (coloured)   175
day lily, tawn   175
day lily, yellow   175
dây vàng giang   157
dayap   105

dazzle   152
dedap batik   148
defne ağ   199
Deianirakraut   134
deigo   148
delha   79
delima   309
demortis   298
děmpul   46
dent le lion   361
dente di leone   361
dentischio   296
deringu   5
derw   296
desert date   61
desert pepperweed   203
Desplatsiafrucht   134
Deutsche Schwertlilie   185
Deutsche Theriakwurzel   288
Deutscher Bertram   39
Deutscher Ingwer   5
Deutscher Kerbel   45
Deutsches Pfefferkraut   256
Deutsches Süßholz   169
devakusuma   355
devataruna   318
devataruni   318
devil lily   207
devil nettle   197
devil's coach whip   353
devil's dung   156
devil's tree   32
dewjacil wysokij   184
dhalum wangi   301
dhamasia   362
dhanya   116
dhanyaka   116
dharou   296
dhilep   301
dhiva   399, 401
dhuleti   171
dhurpi-sag   204
di jiao   368
di ma huang   234
diavoletto   83
Dichtblättriges Zitronellgras   131
Dichtblütige Kresse   202
Dickblättrige Fetthenne   340
Dickblättrige Zwergzitrone   159
Dickblättriger Pfeffer   293
Dickrindiger Muskatbaum   193
dictame blanc   136
dictame de Crete   262
did(d)jar   136
dieng si ing   212
diénte de león   361
dierb   320
diigu   148

djerool bodong   108
dikai   145
dikaja petruschka   247
dikij bal'sam   228
dikij luk   27
dikij tschesnok   22
dilaw   126
dilem   301
dilĕm kĕmbang   301
dill   40
Dill   40
dill, Indian   40
Dillblattwurz   232
dille   40
Dille   40
ding xiang   355
dinh lich   81
dipli   294
dipli-chuak   294
Diptam   125, 136
Diptam, Amerikanischer   125
Diptam, Maryland   125
Diptam, Weißer   136
Diptamdost(en)   262
dipteriks   137
dirw   296
disól   191
dissected merremia   231
Distelartiger Salbei   326
Distel-Salbei   326
dita bark   32
dita tangitang   32
Ditabaum   32
dittamo bianco   136
dittamo cretico   262
dittander   136, 203, 262
dittany   125, 136, 262
dittany of Crete   262
dittany, American   125
dittany, Maryland   125
dittany, mountain   125
djagil'nik   41
djagilL   41
djangjorang   337
djenar   242
djering   49
djeroek garoet   109
djerook kastoori   159
djĕrook pĕtjĕl   105
djerool bodong   108
djintan   125
djonge areuj   66
djoúmaira   75
dlinnij perez   293
dlinnyj perez   292
do:k chan   184, 355
dock   320, 322
docto vainilla   174

dog daisy   204
dog mint   112
dogwood   315
dòho   38
dok chan   184
dok kuichai   27
dokdoko   45
dokmaeo   356
dokudami zoku   179
Dolden-Pfeffer   295
Doldentraubiger Sauerklee   268
Doldiges Winterlieb   97
dolgij perez   293
Dòn   186
Don Diego de node   234
Doña Elvira   1
donduri   305
dong bei bo he   228
dong dang gui   41
dong guai   43
dong gui   43
dong kui zi   217
donnik aptetschnyj   222
donnik cinij   372
Doppelrauke   137
Doppelsame, Schmalblättriger   137
doradille noire   7
doraji   299
Dorant, Weißer   219
Dornenloser Kapernstrauch   81
Dornige Zottelblume   198
Dorstenie   138
Dost, Brauner   265
Dost, Drüsiger   265
Dost, Gewöhnlicher   264, 265
Dost, Griechischer   266
Dost, Italienischer   266
Dost, Kretischer   262, 266
Dost, Schlanker   265
Dost, Syrischer   265
dotó   191
dotted monarda   237
Dotterblume   75
Dotterblume, Sumpf   75
dou   93, 144, 168, 244, 250, 281
dou ban cai   250
dou kou hua   144
double curled persley   282
downy ground cherry   285
downy leucas   204
downy milkwort   301
Drachenapfel   140
Drachenkopf, Iberischer   139
Drachenkopf, Moldawischer   139
Drachenkopf, Türkischer   139
Drachenpfeffer, Moldawischer   139
dracocéphale Moldavique   139
draconcillo   53

dragon   53, 139
dragon sagewort   53
dragon's head   139
dragoncello   53
dragone   53
draguntrawa   53
Dragun-Wermut   53
drakeboom   140
drakonogolownik   139
dreb   302
Dreibändiger Bogenhanf   332
Dreiblatt   230
Dreiblätterkraut   125
Dreiblättrige Bitterorange   304
Dreiblättrige Fingeraralie   2
Dreiblättrige Zahnwurz   85
Dreiblättriger Bitterklee   230
Dreiblättriges Wandelröschen   197
Dreigriffeliger Hahnenkamm   90
Dreilappiger Rosskümmel   198
Dreschlein   209
drimys bark   140
dropwort   157, 261
dropwort, water   261
drumstick   26, 240
drumstick allium   26
Drüsige Bergminze   74
Drüsiger Boretsch   300
Drüsiger Dost   265
Dschungelbrand   186
du du xan   86
du jue qiang wei   319
dubrovnik purpurovyj   362
dubrovnik skordievidny   363
dudhal   361
dudotschnik   205
dudtschataja   205
Duft-Basilikum   258
Duftblüte, Süße   266
Duftende Korallenraute   68
Duftende Muskatnuss   244
Duftender Basilikum   260
Duftendes Mariengras   178
Duft-Glanzbaum   12
Duft-Hirschzunge   206
Duft-Jasmin   187
Duftlabkraut   161
Duftlauch   24
Duft-Mango   218
Duft-Mariengras   178
Duftnessel   11
Duftnessel, Mexikanische   12
Duft-Odermennig   13
Duftorange   109
Duft-Pandanane   272
Duftpelargonie   276
Duft-Prachtscharte   206
Duft-Resede   314

Index of Common Names   527

Duft-Schefflarie   337
Duft-Schraubenbaum   272
Duftsiegel, Öl   347
Duft-Sumach   317
Duft-Wau   314
dugdhapheni   361
duhspar´ska   350
duikerbloom   267
duitse iris   185
duivelsdreck   156
Duke of Argyll's tea tree   213
dumbier   403
dumero   320
dunna   19
Dünnblättrige Afrikanische Muskatnuss   239
dunnoe   86
duo lie huang cao   231
durandha   17
durasita   104
Durchscheinender Pfeffer   278
duri kengkeng   369
durivel   176
durnoj dux   156
duschevik kotovnikovyj   74
duschevik krupnocvetkovyj   74
duschevik mjatolistnyj   74
duschewka   4
duschistaja petruschka   46
duschistje wasil'ki   256
duschistyj buten'   247
duschiza sadowaja   263
duschki   256
dushitsa   265
Düsterer Storchschnabel   165
duwaw   126
dwarf borage   68
dwarf calamint   73
dwarf cape gooseberry   285
dwarf champac   232
dwarf chempaka   232
dwarf elder   7
dwarf marigold   358
dwarf nasturtium   376
dwarf santon   186
dwarf yellow day lily   175
dwojtschatka   65
dyer's saffron   86
e do   260
e tia   260
eabuyao   107
eagle fern   308
eagle wood   48
eagle wood, Malacca   48
eagle-vine bark   220
eal grass   405
earflower, Mexican   129
early yellow rocket   60

earth chestnut   71
East African cardamom   10
East India lemongrass   129
East Indian galanga   31, 191
East Indian root   31
East Indian sandalwood   332
Eastern caper   80
eastern red ceda   189
eau de Cologne menthe   227
eau de Cologne mint   227
Eberraute   52
Eberreis   52
ebong   395
échalote   17
echalotte chinoise   18
echalotte d'Afrique tropicale   16
echalotte d'Angole   16
échinophore   143
echite   32
echt rozenhout   43
Echte Aleppounuss   297
Echte Arnika   51
Echte Bergminze   74
Echte Brunnenkresse   250
Echte Edelraute   55
Echte Engelwurz(el)   41
Echte Kamm-Minze   145
Echte Kaper   80
Echte Königschinarinde   99
Echte Muskatnuss   244
Echte Myrrhe   115
Echte Myrte   247
Echte Nelkenwurzel   166
Echte Pimpernuss   297
Echte Schlüsselblume   307
Echte Verbene   211, 386
Echte Walnuss   188
Echte) Vanille   383
Echter Alant   184
Echter Alpenbeifuß   55
Echter Bergkümmel   198
Echter Galgant   30, 31
Echter Gewürzstrauch   75
Echter Ingwer   403
Echter Jasmin   187
Echter Kardamom   144
Echter Krösling   218
Echter Küchenschwindling   218
Echter Kümmel   87
Echter Lavendel   200
Echter Mousseron   218
Echter Quendel   367
Echter Rohrkolben   377
Echter Salbei   329
Echter Schwarzkümmel   253
Echter Sellerie   46
Echter Speik   381
Echter Staudenmajoran   265

Echter Steinklee   222
Echter Steinquendel   74
Echter Thymian   368
Echter Ziest   65
Echter Zimt   104
Echtes Eisenkraut   386
Echtes Johanniskraut   180
Echtes Katzenkraut   251
Echtes Löffelkraut   113
Echtes Mädesüß   157
Echtes Tausendgüldenkraut   91
Echtes Zitronengras   131
éclairette   313
écorce de quinquina   99
Ecuadorian mint   234
Ecuadorische Minze   234
Edel-Gamander   362
Edelminze   226
Edelnelke   135
Edelraute   53, 323
Edelraute, Echte   55
Edelraute, Gletscher   54
Edelraute, Schwarze   53
Edel-Schafgarbe   4
edera terrestre   167
edged garlic   16
edible canna   78
edible stemmed vine   105
eel grass   405
eetbar bloemriet   78
Efeu, Erd   167
également appelé   286
eggplant, African   349
eggplant, big   349
eggplant, native   349
egoma, lemon   278
egopode   7
Egypt leek   20
Egyptian balsam   61
Egyptian caraway   125
Egyptian carissa   86
Egyptian mallow   216
Egyptian myrobalane   61
Egyptian onion   23
Egyptian rue   322
Egyptian wood tree   123
Eibisch, Muskateller   1
Eibisch, Rosella   177
Eierkraut   53
Eierpflanze, Große   349
Einährige Buntnessel   351
Einblättrige Spinnenpflanze   112
Einjährige Fetthenne   339
Einjährige Paprika   82
Einjähriges Bingelkraut   230
Eisenholz, Kleines   387
Eisenkraut, Echtes   386
Eisenkraut-Salbei   330

Eisop   182
ekku   317
ekleel aljabal   320
el adkham   30
el galangal   30
ela   36, 144
elach   38
ela-ela   36
elafobosco erba rena   283
elangato pepper   291
elcho   37
el-dahabi   348
elder, black   331
elder, common   331
elder, European   331
elder, Spanish   289
elderberry   331
elecampane   184
elecampane, gluey   185
elecampane, viscous   185
elemi, African   77
elemier d'Afrique   77
elenio   184
elephant apple   136
Elephantenapfel   136
eljy koren'   403
elma yagi   327
elo à petites feuillec   395
elsholtzia, common   145
Elsholtzie   145
el-tofah   348
elu   213
emarginate merremia   231
embaúba   89
embira   396
emblic   284
emblic myrobalan   284
emmanggai   240
Emporstrebende Merremie   231
ĕmrah   352
en coeur   179
enante   261
encens   69, 338
ender   40
Endostemonkraut, Walzenförmiges   146
endro   40
Endständige Keulenlilie   116
Endständige Kolbenlilie   116
enea junco   377
enebrina   189
enebro (común)   189
eneldo   40
Engelsüß   303
engelwortel   41
Engelwurz(el), Echte   41
Engelwurz(el), Japanische   42
Engelwurz, Blaugrüne   42

Engelwurzel, Wald   43
engessam   317
Englische Minze   227
Englische Tonkabohne   361
Englischer Senf   345
Englischgewürz   286
English finger-bowl geranium   276
English lavender   200
English mint   228
English sage   330
English spice   286
English tonka (bean)   361
English walnut   188
enoki-mame   157
enredadera del mosquito   44
Entenschnabel-Felberich   214
énula campana   184
enule campane   184
envira   396
envireira   396
Enzian, Gelber   163
Enzian, Indischer   163
Enzian, Purpur   164
Enzian, Schwalbenwurz   163
Enzian, Ungarischer   164
Epazote   95
epazote   95, 210
epiaire laineuse   353
épinard de Malabar   63
épinard sauvage   96
épine noire   315
Eppich   46, 205
Eppich, Großer   205
erable d'àfrique   300
erandakarkati   86
erba moscatella sclarea   330
erba amara   360
erba benedetta   166
erba cancella   276
erba cedrella   211
erba cidreira   223
erba cocchiara   113
erba da gatte   363
erba de gatta   251
erba degli occhi   153
erba di San Giovanni comune   180
erba grassa   305, 340
erba ingia   211
erba limona   223
erba luigia   211
erba Luisa   211
erba moscatella sclarea   330
erba pan a vin   322
erba peverella   335
erba pignola   340
erba rena   283
erba San Pietro   120
erba Santa Barbara   62

erba smirnio   348
erbaka   57
erbuccia   368
Erdbeerspinat, Kopfiger   96
Erd-Efeu   167
Erdeichel   71
Erdkastanie   71
Erdnuss   48
erect boerhavia   67
erect spiderling   67
Eremochariskraut, Strahliges   146
erísimo   14
Erlenblättrige Felsenbirne   33
érodion   146
erra do bompastor   81
ertrudes   33
eruca   147
erumichinarakam   105
erva almíscar   174
erva cidreira   129, 130
erva de amor   314
erva de jaboti   289
erva do Espírito Santo   41
erva gateira   251
erva gatta   381
erva-dàlho   282
erva-de-guiné   282
erva-de-soldado   174
erva-moura   350
erva-pipi   282
ervate Santa Maria   95
eryngo   147
erythrée   91
Erzengelwurzel   41
escabeche   83, 84
escapuza   152
eschalot   17
Eschenartige Weinmannie   392
Eschlauch   17
escobón   132
escuerno   188
escuradentis   34
Eselsfenchel   158
eshbet mariam   52
esjalot   17
Eskobedie   148
esmirnio   348
espada   267
Espada   267
espinaca de Ceilán   63
espinaca de Malabar   63
espino cerval   315
espliego común   200
essang   317
essanga   362
Essbare Wachsbaumwurzel   86
essenzio ombrellifera   3
Essigrose   319

essoun   338
estafiata   55
estafiate   54
estoraque   300
estragão   53
estragon   53
Estragon   53
estragón   53
estragone   53
estróbilos de lúpulo   179
esutoragon   53
Etagenzwiebel   23
eta-kula   32
Ethiopan caraway   371
Ethiopan cardamom   9
Ethiopian pepper   395
Ethiopischer Kardamom   9
etiopide salvia   326
étoile de Noel   152
etoup   371
etrog citron   108
eucalipto azul   150
eucalyptus bleu   150
eucalyptus, apple-scented   150
eucalyptus, willowleaf   150
eufrasia   153
eukaljpt   150
Eukalyptus, Apfel   151
Eukalyptus, Kugeliger   150
Eukalyptus, Zitronen   150
eupatoire chanvrine   151
eupatorio   151
euphorbe écarlate   152
euphorbe épurge   152
Euphorbie, Borstige   152
euphraise officinale   153
Europäischer Meersenf   73
European arnica   51
European bladdernut   353
European elder   331
European pennyroyal   228
European vervain   386
European wild ginger   57
ever-ready onion   18
everlasting, Italian   174
eyaka   362
eyebright   153
ezibil   362
ezonegi   25
fa kuo   319
fabacrasa reflejada   340
fabagelle   405
Fackelingwer, Malayischer   148
Fackellilie   121
Fadhi-Myrrhe   115
faek   387
fafy   86
fagara   155, 400

fagara jaune   155
fagotier   343
Fagot-Zimt   100
fairy ring mushroom   218
falsa acacia   318
Falsche Akazie   318
Falsche Kubeben   291, 401
Falsche Winterrinde   99
Falscher Jasmin   241
Falscher Safran   87, 157
Falscher Schlangen-(Knob-)lauch   25
Falscher Schwarzer Pfeffer   294
Falscher Staudenmajoran   265, 266
Falscher Ylang-Ylang   51
false Acacia   318
false acacia   318
false apple mint   228
false bishop's weed   33
false cardamom   9, 35, 38
false cinnamom   100
false coriander   147
false garlic   21, 28
false golden aster   177
false jalap   234
false nutmeg   238, 310
false peper   337
false saffron   86, 157
false Salomon's seal   347
false sassafras   261
false Winter's bark   99
Faltenkürbis   361
fan ho   224
fan hung hua   122
fan kua   86
fan tou   48
fang feng   87, 274
fang-feng   333
Färberdistel   87
fárfara japonesa   281
farfaraccio   281
farfaraccio Giapponese   281
farferugine   75
farigoule   368
farmer mustard   81
Farnblättriger Nitta   273
farroba   273
faskomilia   327
faskori   155
father of moschus   1
fausse échalote   25
fausse guimauve   216
fausse noix de muscade   238
fausse tomate   349
faux acacia   318
faux aneth   125
faux anis   125
faux mousseron   218
faux muscadier   310

faux poireau   15
faux poivrier   338
feather geranium   96
february daphne   133
fede   386
feegel   50
fei long zhang zue   369
fei zhou qie   349
Feigenwurz   313
feijão   77
feijão bravo   77
Feinnerviger Sumach   317
felandrio   261
Felbrich, Entenschnabel   214
Feld-Gentianelle   165
Feldgarbe   3
Feldkresse   202
Feldkümmel   87
Feldminze   224
Feldpolei   368
Feldquendel   4
Feldthymian   367
Feldzwiebel   22
Felsenbirne, Erlenblättrige   33
Felsen-Fetthenne   340
Felsenkirsche   308
Felsenklippe, Strauch   233
Felsen-Mauerpfeffer   340
Felsenrose   138
Felsen-Storchschnabel   165
felwort   354
fen chieh tsao   302
fen tsao   169
fen yoa kua   279
fenacho   374
Fenchel   158
fenchel   158
Fenchel, Bitter   158
Fenchel, Süßer   158
fenchel' ovoschtschnoj   158
Fenchelholzbaum   333
fenegriek   374
feng chao cao   204
feng dou cai   281
feng pen   24
feng wo cao   205
fenigrec   374
fenigrekowa trawa   374
fenikel   158
fennel   120, 158, 252, 333
fennel flower   253
fennel wood   333
fennel, bitter   158
fennel, Florence   158
fennel, sea   120
fennel, small   253
fennel, sweet   158
fennel, wild   252

fenogreco 374
fenouil à bulbe 158
fenouil amer 158
fenouil bâtard 40
fenouil commun 158
fenouil de mer 120
fenouil des Alpes 232
fenouil doux 158
fenouil marin 120
fenouil puant 40
fern 308
fénugrec 374
fenugreek 372, 374
fenugrek 374
fenum-grek 374
fern-leaf aralia 303
fernleafed nitta 273
feruľa čertova 156
ferula wonjutschaja 156
férule persique 156
fetid marigold 274
fetida 156
fétide 147, 156, 323, 341
Fetthenne 339, 340
Fetthenne, Alpen 339
Fetthenne, Behaarte 341
Fetthenne, Berg 341
Fetthenne, Dickblättrige 340
Fetthenne, Einjährige 339
Fetthenne, Felsen 340
Fetthenne, Goldmoos 340
Fetthenne, Kaukasus 340
Fetthenne, Purpur 341
Fetthenne, Rispen 340
Fetthenne, Rötliche 340
Fetthenne, Schwärzliche 339
Fetthenne, Spanische 340
Fetthenne, Sumpf 341
feuille de boldo 284
feuille de laurier 199
feuille de Murraya 241
feuilles caya 111
fève de Tonka 137
fever grass 129
fever nettle 197
fever plant 258
fever tea 210
fever tree 150
feverwort 91
fèves tonka 137
few-flowered leek 22
fia fan 110
fialka duschistaja 388
ficaire 313
fidgel 50
fidjel 322
fidjla 322
fidjla el-djebeli 323

Fieberheilbaum 150
Fieberklee 230
Fieberrinde 99
Fieberstrauch, Wohlriechender 208
Fiederaralie, Cuming 301
Fiederaralie, Guilfoyl 304
Fiederaralie, Strauchige 303
Fiederblättriger Zweizahn 65
Fiederspaltige Pimpernuss 353
field cress 202
field garlic 22, 28
field melilot 222
field mint 224
field mustard 345
fieno greco 374
filfil 82, 293
filfil ahmar 82
filipendule 157
Filzige Pestwurz 281
Filziger Muskat 244
Filziger Paprika 84
Filz-Königskerze 385
finger citron 108
Fingeraralie, Dreiblättrige 2
Fingerkraut, Aufrechtes 306
Fingerwurzel 67
Finger-Zitrone 109
finnochio 158
finocchielle 247
finocchio 120, 158, 232, 261
finocchio alpino 232
finocchio amaro 158
finocchio dolce 158
finocchio domestica 158
finocchio marino 120
finocchio-acquatico 261
finocchione 158
finocha alpino 232
finochiello 232
fiolho 158
fior 75, 97, 375
fior d'ogni 75
fior d'oro 97
fiore de arnica 51
fiore de macis 244
fiore di noce moscata 244
fiori 106, 177
fiorrancio dei gardine 75
fish poison tree 362
fish tamarind 162
fish-tail oxalis 269
fishwort 178
fistaschka 296
fistaschka-lentiskus 296
fistik 297
fistula, purging 88
fitex 388
fitweed 147

five-flavour fruit 338
five-ripped thyme 368
Flachs 209
Flachsbaum, Lorbeerblättriger 45
flag iris 185
flame freesia 375
flame-of-the-wood 186
flame, jungle 186
Flammender Busch 136
Flaumige Blasenkirsche 285
Flaumiger Knöterich 280
flax 209
flayou 228
fleabane, Indian 300
fleur d'arnica 51
fleur de souci (des jardins) 75
fleur de St. Pierre 385
fleur de tous les mois 75
fleurie de mélilot 222
Fliederbeere 331
fliou 228
Flohblume 65
Floh-Knöterich 302
Flohkraut 173, 228
flor de árnica 51
flor de azahar 106
flor de cacao 311
flor de Jamaica 177
flor de la oreja 129
flor de lavanda 200
flor de merte 75
flor de muerto 357
flor de noz moscada 244
flor de Pascua 152
flor de pluma 3
flor de sangu 375
flor de Santa Catarina 152
flor de Santa María 357
Flora's-paintbrush 146
Florence fennel 158
Florentine iris 185
Florentinische Schwertlilie 185
Florentinische Veilchenwurzel 185
Florida nutmeg 370
Florida star anise 183
Florida Sternanis 183
Florida-Muskatnuss 370
Florida-Nusseibe 370
Floriday holly 338
florist's violet 387
flouve odorante 44
flower, fennel 253
flowerhead 95
flowering onion 21
Flügelfrucht, Rote 309
Flügelsamiger Behennussbaum 240
Flügeltang 380
Flügeltang, Japanischer 380

Index of Common Names 531

fluted gourd   361
fluted pumkin   361
fo shou kan   108
foetid cassia   341
foglia di alloro   199
foglia di boldo   284
foi faa   97
foin grec   374
foirolle   230
fondé des rivières   396
food of gods   156
foreign coriander   147
forest primrose   181
Formosan kumquat   159
forowa   148
Fortuna-Gilbweiderich   214
fosli gulab   319
fou tsiao   293
fougère aigle   308
foul el-'arab   381
four o'clock plant   234
fourwing salt bush   59
fragant olive   266
fragant sumach   316
fragment flowered garlic   24
fragrant balm   236
fragrant garlic   23
fragrant giant hyssop   11
fragant olive   266
fragrant pandan   271
fragrant screwpine   271, 272
fragrant sumach   316
fragrant thatch grass   129
fragrant yarrow   3
frankincense   69
Franzosenholz   170
Französischer Lavendel   200
Französischer Majoran   264
Französischer Senf   70
Frauenhaarfarn, Schwarzer   7
Frauenmantel, Gemeiner   14
Frauenmantel, Schimmernder   14
Frauenmantel, Silber   13
Frauenmantel, Spaltblättriger   13
Frauenmantel, Verwachsener   13
Frauenmantel, Zerschlitzter   13
Frauenminze   360
frède de Angola   63
freesia, flame   375
French lavender   200, 201
French marigold   357, 358
French mercury   230
French rose   319
French sorrel   322
French tarragon   53
French thyme   299
frenk kimionou   87
Friedensblatt   356

frigoule   368
frijol de soya   168
fringed rue   323
fringent rue   322
frogbite, African   180
Froschbiss, Afrikanischer   180
Froschlöffelähnliche Ottelie   267
frost flower   125
frost mint   125
Fruchtsalbei   327
fru-fru   300
Frühblühender Thymian   367
Frühe Winterkresse   62
Frühlings-Krokus   121
Frühlingslauch   17
Frühlings-Salbei, Italienischer   329
Frühlingsstern   185
Frühlingszwiebel   17
fruit aux cure-dents   34
fruit de khella   34
fruit sage   327
frula estrela   137
fruta bomba   86
fruto de cápsico   83
frutos de enzisie   362
fû-chô-sô   111
Fuchsschwanz-Minze   230
fudina   229
fuki   281
ful suyah   168
fula-fulfulde   33
fulayha   228
fulayya   228
fulpol aswad   293
fulvous day lily   175
funcho   120, 158, 261
funcho aquático   261
funcho de Florença   158
funcho do mar   120
funcho doce   158
funcho marino   120
funcho oriental   261
Fünfblättriges Weichkraut   234
Fünfkantige Justizie   189
furohrensu-fennelu   158
fustik   297
fustuq   297
fuyu-sanshô   399
fynblaar-saffraan   89
gaaden kuresu   203
gaai choy   70
Gaboon nut   119
Gabunnuss   119
gach mirichi   82
Gagelstrauch   242
Gagelstrauch, Amerikanischer   242
Gagelstrauch, Moor   242
gaïac officinal   170

gaiallone   122
gainier   93
gair   5
gajutsu   128
galagana   30
galanga   29, 30, 31, 191
galanga maior   30
galanga de l'Inde   30
galanga majeur   30
galanga mayor   30
galanga minore   31
galanga, East Indian   31
galanga, mussel   29
galangal   30, 192
galangal de la China   31
galangal officinal   31
galangal, black   30
galangal, Java   30
galangal, lesser   30
galangal, light   31
galangal, Malacca   30
galangowyj koren'   30
Galbanum   156
Galbanum-Gummi   156
gale   242
galé odorant   242
galé piment aquatique   242
galgan jávsky   30
galgan liečivý   31
Galgant, Chinesischer   191
Galgant, Echter   30, 31
Galgant, Gespornter   29
Galgant, Großer   30
Galgant, Heller   149
Galgant, Indischer   29
Galgant, Kleiner   31
Galgant, Malakka   30
Galgant, Muschel   29
Galgant, Pyramiden   30
Galgant, Schwarzer   30
Galgant, Siam   30
galingale   67
galingale, lesser   31
gallitrico   330
galoba merah   31
galoba papua   149
gamander, common   362
Gamander, Edel   362
Gamander, Katzen   363
Gamander, Wilder   306, 363
Gamanderlein   362
gamarza   275
gambe-secche   218
Gambian tea bush   209, 211
Gambia-Teebusch   211
Gambia-Teestrauch   209
Gambier   379
gambier   379

Gambir   379
gambir   379
Gambirpflanze   379
gamboge   162
gan zao   170
gandama   3
gandaroora besar   153
gandha   54
gandhalu   399
gandhalu tumburu   401
gandhamarica   291
gandhasara   332
gandhatrn   129
gandhela   241
gandibudi   167
gandrayan   42
ganelle   360
ganga tulsi   181
ganges river asystasia   58
ganis   288
gan-kollan-kola   301
Gänsefuß, Klebriger   96
Gänsefuß, Kopfiger   96
Gänsefuß, Weißer   95
Gänsefuß, Wohlriechender   95
Gänsekraut   57
Gänserich   360
Ganzblättriger Zürgelbaum   91
gao liang jiang   30, 31
gar   199
garafuni   235
Gardalsenf   346
garde robes   52
garden angelica   41
garden burnet   288, 331
garden carrot   133
garden chamomile   95
garden chervil   94
garden cress   203
garden lovage   205
garden mace   2
garden mint   225, 229
garden montbretia   121
garden myrrh   247
garden nightshade   350
garden onion   17
garden poppy   272
garden rhubarb   316
garden rocambole   24
garden rocket   147
garden sage   329
garden savory   334
garden sorrel   320, 322
garden spurge   152
garden thyme   368
garden valerian   381
garden violet   388
gardenia   162

gardénia   162
Gardenie, Kap   162
garengal   30
gares   108
garland   97, 173
garland chrysanthemum   97
garland daisy   97
garland flower   173
garlic   11, 14, 24, 27, 120, 179, 218, 376
garlic bark (tree)   11
garlic chives   27
garlic marasmuius   218
garlic mushroom   218
garlic mustard   14
garlic tree, Cameroon   11
garlic, Baker's tree   18
garlic, black   21
garlic, broad leaved   27
garlic, Cameroon   11, 179
garlic, Chinese   19
garlic, Daffodil   21
garlic, false   21, 28
garlic, giant   24, 25
garlic, hedge   14
garlic, Hooker   20
garlic, Japanese   19
garlic, keeled   16
garlic, naples   21
garlic, Neapolitan   21
garlic, Peking   24
garlic, serpent   24
garlic, society   376
garlic, Spanish   17, 23, 25
garlic, water   19
garlic, wild   27, 376
garlic, wood   27
garmala   275
garn ploo   355
garo sainbois   133
garofanaia   166
garofano   135, 355
garofano domestico   135
garrofa   93
Garten-Bingelkraut   230
Garten-Bohnenkraut   334
Gartendost   263
Gartenfenchel   158
Gartenheil   52
Gartenkerbel   45
Gartenkresse   203
Gartenminze   225
Gartenmohn   272
Garten-Möhre   133
Garten-Montbretie   121
Gartennelke   135
Gartenpetersilie   282
Garten-Quendel   334
Gartenraute   323

Garten-Resede   314
Gartenrhabarber   315
Garten-Salbei   329
Garten-Sauerampfer   320
Garten-Schafgarbe   2
Garten-Thymian   368
garupillai   241
gattaia comune   251
gattilier agneau-chaste   388
gaulthérie du Canada   163
gaya   130, 170
gazar abiad   274
gbako nisa   65
gbolei   317
gboma   348, 349
gboma egg plant   349
ge con   27
Geäderte Schefflarie   337
Gebogener Pfeffer   289
geelblom   99
geelwortel   30
Gefingerte Zitrone   108
Gefingerter Muskat   244
Gefleckte Gewürzlilie   192
Geflügelter Burzeldorn   372
Geflügelte Geigerie   163
Geflügelter Sesam   341
Gefranste Raute   322
Gefüllter Nachtschatten   349
Gehörntes Veilchen   387
Geigeria, winged   163
Geigerie, Geflügelte   163
Geißfuß   7
Geißkapern   132
Gekerbtblättriger Bukkostrauch   62
Gekerbte Laporte   197
Gekerbte Strauchnessel   197
Gekielter Lauch   16
gekkeei ju   199
gek-kitsu   242
Gekrümmter Pfeffer   289
gekweckte eruca   147
gelang pasier   305
Gelbdolde, (Schwarze)   348
Gelbdolde, Stengelumfassende   348
Gelbe Balsampflaume   352
Gelbe Chinarinde   99
Gelbe Kosmee   118
Gelbe Rübe   133
Gelbe Taglilie   175
Gelber Enzian   163
Gelber Ingwer   126
Gelber Jasmin   187
Gelber Nachtschatten   350
Gelber Salbei   328
Gelber Speik   389
Gelber Steinklee   222
Gelber Sternanis   183

Gelber Zitwer   126, 402
Gelbes Katechu   379
Gelbholz, Koreanisches   400
Gelbholz, Senegal   155
Gelbholz, Westindisches   400
Gelbholzbaum, Chinesischer   400
Gelb-Senf   345
Gelbwurz, Wilde   126
Gelbwurzel   126
Gelbwurzel, Javanische   128
g☐lenggang   93
geli-geli   198
geliotron   174
gelsomino   187
gelugur   162
gember   403
Gemeine Betonie   65
Gemeine Malve   216
Gemeine Pestwurz   281
Gemeine Robinie   318
Gemeine Schafgarbe   3
Gemeiner Beifuß   56
Gemeiner Frauenmantel   14
Gemeiner Kerbel   45
Gemeiner Löwenzahn   361
Gemeiner Pfeffer   290
Gemeiner Seidelbast   133
Gemeiner Traubenkropf   343
Gemeines Hirtentäschel   81
Gemüse-Pfeffer   82
Gemüse-Schattenblume   347
gen qin cai   47
gen xiang qin   283
genciana amarilla   163
gendela   241
gendola   63
genepi   3, 53
genepi blanc   3
genépi noir   53, 55
genêt   132
genêt à balais   132
genêt commun   132
genévrier (commun)   189
gengibre   403
gengibre amarelo   403
gengibre das boticas   403
gengibre dourada   126
gengibre oficinal   403
gengibre-amargo   403
genia ou gerru   53
genipi   54, 55
Genipikraut   3
Genipopp, Beng   56
Genipp, Walliser   56
genistra scopareccia   132
genseli   27
gentian, Hungary   164
gentian, Indian   163

gentian, purple   164
gentian, spotted   164
gentian, willow   163
gentian, yellow   163
gentiane fausse asclépiade   163
gentiane jaune   163
gentiane, grande   163
Gentianelle, Feld   165
genziana gialla   163
genziana maggiore   163
geran prjamja   165
geran rosovaja   275
Geranie, Bourbon   276
Geranie, Rosen   276
Geranie, Rosenduft   276
Geranie, Wermut   275
geranio de olor   275
geranio de rosa   276
geranio incenso   276
geranio malva   276
geranio odoroso   276
geranio rosa   275
géranium à grosses racines   165
géranium des Balkans   165
geranium   275
geranium grass   129, 131
geranium, mint   277
geranium, peppermint scented   277
geranium rosat   275, 276
geranium, Bulgarian   165
geranium-leaf aralia   304
Gerbereichen-ähnlicher
    Kreuzdorn   315
Gerbersumach   317
gergelim   342
Gerippte Alpenscharte   336
German iris   185
German Zigeunerknoblauch   27
germander, common   362
germander, wall   362
germandrée   362, 363
germandrée chamaedrys   362
germandrée petit chêne   362
germandrée scorodoine
    aquatique   363
gernotte   71
Geröhrter Pfeffer   289
gervao cheirosa   353
Gesägtblättriger Bukkostrauch   63
Gespornter Galgant   29
Gestielte Zwiebel   26
Gestreckter Pfeffer   291
Gestreifte Mexikanische
    Studentenblume   359
Gestreifter Mohrenpfeffer   396
getih   343
Getüpfelte Kamm-Monarde   237
Gewöhnliche Clinopode   112

Gewöhnliche Haselwurz   57
Gewöhnliche Katzenminze   251
Gewöhnliche Schafgarbe   3
Gewöhnliche Sumpfkresse   318
Gewöhnlicher Andenwein   220
Gewöhnlicher Andorn   219
Gewöhnlicher Augentrost   153
Gewöhnlicher Baldrian   381
Gewöhnlicher Bocksdorn   213
Gewöhnlicher Dost   265
Gewöhnlicher Giersch   7
Gewöhnlicher Hopfen   179
Gewöhnlicher Judasbaum   93
Gewöhnlicher Rhabarber   316
Gewöhnlicher Tüpfelfarn   303
Gewöhnlicher Wacholder   189
Gewöhnlicher Wasserdost   151
Gewöhnlicher Wiesenkerbel   45
Gewöhnliches Barbarakraut   62
Gewöhnliches Bitterkraut   286
Gewöhnliches Ruchgras   44
Gewöhnliches Seegras   405
Gewürzbeere, Afrikanische   238
Gewürz-Beifuß   57
Gewürzdolde   346
Gewürzfenchel   158
Gewürz-Guragi   9
Gewürz-Kälberkropf   94
Gewürzlilie   67, 191
Gewürzlilie, Gefleckte   192
Gewürzlilie, Runde   67, 192
Gewürzlorbeer   199
Gewürzminze   234
Gewürznelken   355
Gewürz-Senf   345
Gewürzstrauch   75, 145, 208
Gewürzstrauch, Chinesischer   145
Gewürzstrauch, Echter   75
Gewürz-Studentenblume   359
Gewürz-Sumach   317
Gewürztagetes   358
Gewürzthymian   368
ghafith   13
ghaneri   196
ghar   388
ghâr   199
ghobbeira   33, 167
ghol   305
gia to   145
giaggiolo   185
giagiolo ianco   185
giant garlic   24, 25
giant onion   19
giant sweet chervil   247
giâp cá   179
giáp ce   179
gibiskuc sabdarifa   177
giêng sàng   112

Giersch   7
Giersch, Gewöhnlicher   7
Giftbaum, Surinam   362
gikiji   242
gilam   108
Gilbweiderich, Fortuna   214
Gilbwurzel   126
gill over the ground   167
gilliflowers   355
gilly flower   135
gin mokusei   266
ginahang   235
ginepro comune   189
ginepro nero   189
gineprodella   189
ginesta de escobas   132
ginestra dei carbona   132
gingelly   342
gingelly seed   342
gingembre   346, 402, 403
gingembre Africaine   346
gingerbre blanc   404
gingembre fou   404
gingembre marron   402
ginger   29, 57, 67, 118, 126, 130, 148, 173, 202, 226, 346, 402
ginger African   346
ginger grass   130
ginger lily   118, 173
ginger lily, red   173
ginger lily, scarlet   173
ginger mint   226
ginger, Bengal   402
ginger, bright   31
ginger, cassumunar   402
ginger, Chinese   31
ginger, crape   118
ginger, Indian   29
ginger, Japanese (wild)   403
ginger, lesser   67
ginger, Malaysian   118
ginger, Mango   126
ginger, mioga   403
ginger, nodding   403
ginger, Peruvian   202
ginger, red   31
ginger, shampoo   403
ginger, Siamese   30
ginger, snap   29
ginger, wild   29, 57, 346, 404
ginger, yellow   126
ginger, zerumbet   404
gingerbre blanc   404
gingerwort, black   404
Gingkobaum   166
ginkgo   166
Ginkyo   166
ginnan   166

ginseng   43
giôi   110
gipsy onion   27
giraniu   275
girardina silvestre   7
Girlandenblume, Weiße   173
giroflé   355
giroflier   355
gison   182
gitipi tana   149
glacier wormwood   54
gladdon   105
gladysch   198
gladysch schtschetinistorolosistyj   198
glandly borage   300
Glanzbaum, Duft   12
Glänzende Samtblume   358
gléchome   167
Gleichfarbige Lilie   207
glenija pribretschnaja   167
Gletscher-Beifuß   54
Gletscher-Edelraute   54
glittering lady's mantle   14
globe thistle, Indian   351
gluey elecampane   185
glumachi   2
glycyrrhize   169
Gnadenkraut   323
gnaphale de l'Inde   170
Goa butter   162
goat's food   268
goat's horn   374
gobernadora   197
godong   152, 323
godong minggu   323
gokhru-kalan   372
gol mirc   293
golasiman   305
Goldähriger Ingwer   402
Goldaster   177
gold-banded lily   207
Goldband-Lilie   207
Goldblume   75, 97
golded-rayed lily of Japan   207
golden apple   7, 348
golden aster, false   177
golden champa   232
golden dock   321
golden maiden hair   303
golden mandarin   159
golden monbretia   121
golden needles   175
golden shower   88
golden-button   360
Gold-Kälberkropf   94
Goldlauch   21
Goldmelisse   236
Goldmontbretie   121

Goldmoos-Fetthenne   340
goldmoss stonecrop   339
Goldorange   159
Goldregen, Indischer   88
Goldsalbei   327
goloba koi   31
goluboj donnik   372
goma   204, 342
gommier bleu   150
goober   48
good king Henry   96
good luck plant   116
goose foot   95
gooseberry, Indian   284
gooseberry, purple   285
goosefoot   95
gooseneck loosestrife   214
gopher plant   152
gor'kij kress   113
goraga   162
gorakh amli   6, 275
gorakh chinch   6
gorakh mundi   351
gorbach   5
gordhab   302
gordolobo (común)   385
goretschavka scheltaja   163
gorez potschetschujnaja   302
gorez peretschnyj   280
gorlez ptitschij   302
gornij tschaber   335
gornyj luk   23
goronfel masamir   355
gorowoj luk   15
gorrinets   81
gortschiza belaja   345
gortschiza morskaja   73
gortschiza polewaja   345
gortschiza sareptskaja   69
gortschiza tschernaja   70
gosditschnik   166
gou ju   304
gou-gi-zi   213
gourd, bitter   235
gourd, fluted   361
goutte de sang   146
goutweed   7
gouz bouwa   244
gow choy   27
gôyâ   235
Graham salve   329
Graham's Salbei   329
grain de lin   209
graine de maniguette   10
graine de paradis   10
graines à vers   54
grains of paradise   9, 10
grains of Selim   395

grains, Guinea 10
grama de olor 45
grama olorosa 45
grana paradisi 10
granaatappel 309
granada 309
granado 309
granat 309
Granatapfel 309
grand basile 256
grand boucage 288
grand galanga 30
grand mélior 221
grand sureau 331
grande absinthe 52
grande aunée 184
grande gentiane 163
grande marguerite 204
grande mauve 216
grande oseille 320
grande passerage 203
grande pimprenelle 331
grande sauge 329
grani di meleguetta 10
grani paradisi 10
granos de paraíso 10
grapefruit sage 329
Grapefruit-Salbei 329
Graslauch 25
grass leaf sweet flag 5
grass wrack 405
grassleaf day lily 175
grawilat 166
grawilat antetschnij 166
grawilat gorodskoj 166
Gray sarsaparilla 347
Graziler Kardamom 36
greasewood 197
great burnet 331
great dasheen 114
great earthnut 71
great marigold 358
great pignut 71
great piment 287
great round-headed garlic 15
great taro 114
great vanilla 384
greater ammi 33
greater burnet saxifrage 288
greater cardamom 37
greater galangal 30
greater kyllingia 193
greater oblong cardamom 144
grebennik 166
grebentschaja schandra 145
greek basil 257
Greek clover 374
Greek hay-seed 374

Greek oregano 266
Greek sage 327
green almond 297
green capsicum 82
green mint 229
green pepper 82
green santolina 333
green sorrel 269
grenadier 309
Gretel im Busch 253
gretscheskoe seno 374
Griechischer Bockshornklee 374
Griechischer Buschthymian 257
Griechischer Dost 266
Griechischer Salbei 327
Griechisches Heu 374
Griechisches Süßholz 169
Grobzähniger Lattich 195
grof bieslook 19
gromwell 231
gros baumes 181
grosella China 59
Großblättrige Mediasie 220
Großblättriger Ahorn 2
Großblütige Ballonblume 299
Großblütige Bergminze 74
Großblütige Königskerze 385
Großblütige Sesbanie 343
Großblütige Wollkerze 385
Großblütiger Augentrost 153
Großblütiger Jasmin 188
Großblütiger Mohrenpfeffer 396
Großblütiger Muskat 243
Großblütiger Nelkenpfeffer 287
Großblütiger Steinquendel 74
grosse angive 349
Große Bibernelle 288, 331
Große Eierpflanze 349
Große Kapuzinerkresse 375
Große Knorpelmöhre 33
Große Macisbohne 260
Große Pimpinelle 332
Große Studentenblume 358
Große Türkenbundlilie 207
Große Vanille 384
Großer Ammi 33
Großer Eppich 205
Großer Galgant 30
Großer Kardamom 37
Großer Lavendel 201
Großer Odermennig 13
Großer Sauerampfer 320
Großer Speik 201
Großer Taro 114
Großer Wiesenknopf 331
Großes Lemongras 130
Großfrüchtige Aubergine 349
Großköpfiges Brandkraut 204

Großnarbige Boronie 68
Großsamiger Günzel 13
Großsamiger Kardamom 9
ground cherry 285
ground elder 7
ground ivy 167
ground nut 48
ground pinapple 364
Grundheil 151
Grüne Heiligenblume 333
Grüne Mandel 297
Grüne Minze 229
Grüne Rossminze 229
Grüner Sauerklee 269
Grüner Tulsi 260
Grünes Heiligenkraut 333
grys apple 273
guabito amargo 311
guaco 233
guaco marda 233
Guadeloupe-Vanille 385
guaiaco 170
Guajakbaum 170
Guajakbaum, Heiliger 170
guang huo xiang 301
guanúchil 298
guarumo 89
guatomate 285
guayacan negro 170
Guayana Muskatnuss 6
Guayana nutmeg 6
guchek 46
guela castilla 105
guela lima 105
guela xtilla 105
guemmam 247
guérittout 381
guernech 250
gui hua 266
gui zhan cao 65
gui zhi 100
guie becohua 318
Guilfoyl Fiederaralie 304
Guilfoyle polyscia 304
guiña castilla 104
guiña xtilla ticanaca 104
guindilla 82, 83
guiné 177, 282, 395
Guinea cardamom 9
Guinea cubeb 291
Guinea grains 10, 11
Guinea pepper 10, 291, 395, 397
Guinea-Kardamom 9
Guinea-Körner 11
Guinea-Pfeffer 10, 395, 396
guinejské korenie 10
Guinese laurel 45
guizotia 171

gul   319
gulab   318, 319
gulab ke phul   319
gulkhair   217
guma   205
Gummi-Asant   156
gun'ba   373
gunchak   46
gunchi   2
gunchin   46
Gundelrebe   167
Gundermann   167
gundhun   17
gung   403
gunja   2
Günzel, Großsamiger   13
Günzel, Kriechender   13
guo gu   37
guo ma ka   399
guo mai long lang   88
Guragi spice   9
Guragi-Gewürz   9
gurdum   86
gur-karasi   307
Gurkenbaum   59
Gurkenkraut   40, 68
gurl pata   347
gurmar   171
gusjatnica   302
Guter Heinrich   96
gvajakovoe   170
gvosdika   135, 355
gvozdika gollandskaja   135
gvozdika sadovaja   135
gwe   352
gwinejskij perez   10, 291
gwozdischnyj grib   218
gyigyam   148
gyôja-ninniku   27
gyoseiso   179
haanghong   173
haapae   385
Haarstrang, Sumpf   283
hab erche   203
hab hab   144
habag   256
Habanero   83
Habañero (pepper)   83
habaq   228
habas de soyas   168
habb el-yashad   203
habba helwa   288
habba souda   253
habbar   80
habbatul barakah   253
habhal hobashi   9
hackmatack   304
hades   208

hadsandhi   105
hafor   150
hagonoi   392
Hahnenkamm   90
Hahnenkamm, Dreigriffeliger   90
Hahnenkamm, Silber   90
hai dai   405
hai dong   148
hai tong pi   148
hai tung   304
haidai   195
Hain-Ampfer   322
Hain-Lobelie   212
Hainminze   230
hairy basil   253
hairy carpet weed   167
hairy glinus   167
hairy mountain mint   310
hairy thyme   367
hakka   225, 334
hakkar   226
halada   126
halba   253, 374
Halbstrauchiger Mohrenpfeffer   396
halbub   230
haldi   126
halhal   201
halia utan   167
halim   203
haliya   403
halkusa   205
Halmaheira-Muskatnuss   245
halmaheiramuskaat   245
haltit   156
hama-bôfu   167
hama-udo   42
Hamburg parsley   283
hamd   268
hamkkae   342
Hammelmöhre   274
han hsiao   232
han lian hua   375
han xiad   232
hana   173, 229, 404
hana shuku sha   173
hana-shôga   404
hanekam   90
hang ping   297
hanggasa   37
hangod   4
hangor   4
hanh   119
hanh la   17
hani-baul   298
hanshi   278
hantar duri   401
hapusa   351
harachampala   51

haras   158
hardhay   180
hardy marjoram   264
Harfenstrauch   299
haridra   126
hari-enju   318
harjora   105
harmal   275
harmala   275
Harmalbohne   275
harmel   275
Harmel   275
harriqa   250
harsha   68
hartho   339
Haselwurz, Braune   57
Haselwurz, Gewöhnliche   57
Haselwurz, Kanadische   57
Hasenpfeffer   57
hashesha almalak   41
hashisha thawmiyah   14
hau kha   30
hau meni   332
hausa potato   351
Hausknecht saffron   122
Hausknecht-Safran   122
hawa-biwa   212
Hawaiian good luck plant   116
Hawaiian orange   241
hawk picris   286
hawkweed   286
hay-seed, Greek   374
hazara   159
he shi feng   133
headache tree   379
heart leaf   178
heart pea   85
heart seed   85
heart-leaved pea   85
Heckenzwiebel   19
hedge garlic   14
hedge lime   159
hedge sison   346
hedgenettle   352
hediondilla   197
hediondo   197
heel   144
heglig   61
hei   342
hei-che-tao   188
Heidemyrte   242
heil   9, 391
Heiligenblume, Grüne   333
Heiligenkraut, Grünes   333
Heiligenkraut,
   Rosmarinblättriges   333
Heiliger Basilikum   260
Heiliger Guajakbaum   170

Index of Common Names

Heiliger Sternanis 183
Heiligkraut 386
Heil-Warburgie 391
Heilziest 65
helbeh 374
'Helen von Stein' 353
héléne 184
Helenenwurzel 184
heliotrope 174
héliotrope 174
heliotropo 174
Heller Galgant 149
hémérocalle brun-rouge 175
hémérocalle jaune 175
hémérocalle naine 175
hemp agrimony 151
hemp weed 151
heno griego 374
henruda 323
hen-yô-boku 113
hépatique étoilee 161
herabol myrrh 115
Herabol-Myrrhe 115
herb bennet 166
herb Gerard 7
herb Louisa 211
herb mercury 230
herb of grace 323
herb of repentance 323
herb of tempera 224
herb of the Virigin 327
herba à chameau 131
herba barona 367
herba citron de Java 131
herbe aux éclairs 136
herbe aux vaches 372
herbe à la cuiller 113
herbe à mille florins 91
herbe à neuf chemises 27
herbe à printemps 96
herbe à vers 95
herbe au citron 223
herbe aux charpentiers 3, 62
herbe aux chats 251, 381
herbe aux cuillers 113
herbe aux puces 228
herbe aux scorbut 113
herbe aux serpents 138
herbe aux sorciers 386
herbe aux vaches 381
herbe aux vers 360
herbe d'amour 314
herbe de grâce 323
herbe de la Saint Jean 180
herbe de Malabar 130
herbe de Saint Julien 334
herbe de Sainte Barbe 62
herbe de Saint-Pierre 120

herbe dragonne 53
herbe mastiche 363
herbe royale 256
herbe sacrée 182, 329, 386
herbe-citron 129
Herbstmousseron 218
Hercules'-club 400
Herero Buschtee 255
Herrenkraut 256
Herrnkümmel 346
herse 372
herva chaha 129
Herzblättrige Minze 225
Herzblättriger Meerkohl 119
Herzerbse 85
Herzförmige Houttuynie 179
Herzkraut 81, 223
Herzsame 85
Heu, Griechisches 374
heuníng gras 45
Hexenkraut 136
Hexenzwiebel 27
Heyne burnet 288
Heyne-Bibernelle 288
Heyne-Patchouli 301
hi zu 162
hibisco 177
hibiskus 1
hiedra terrestre 167
hierba anís 357
hierba buena 229
hierba buena acuática 224
hierba centella 75
hierba cidrera 211
hierba de burro 181
hierba de la pastora 327, 377
hierba de la princesa 28
hierba de las cucharas 113
hierba de las muelas 181
hierba de los ojos 330
hierba de San Juan 57, 180
hierba de San Marcos 360
hierba de Santa Bárbara 62
hierba de Santa Maria 358
hierba dulce oregano 210
hierba gatera 251
hierba hormi uera 95
hierba lombriguera 360
hierba Luisa 211, 227
hierba marina 405
hierba mora 350
hierba sana mentha 229
hierba sancts 294
hierba santa 229
hierbabuena acuática 224
higad-higad 4
high mallow 216
higo de mastuero 86

higuillo oloroso 289
hijid 6
Himmelschlüssel 307
Himmelsduft 7
hindi salsa 176
hîng 156
hingan 61
hingu 156
hinnos 99
hinojo 87, 120, 158, 219, 261
hinojo amargo 158
hinojo bastard 219
hinojo de agua 261
hinojo de Florencia 158
hinojo de prade 87
hinojo dulce 158
hinojo marino 120
hipérico 180
hiptis à odeur 181
hira-ao-nori 379
hiragi-giku 300
hiroma 380
hirome 380
Hirschminze 228
Hirschzunge, Duftende 206
Hirtentäschel, Gemeines 81
hisopillo 335
hisopo 182
hissopo 182
hjangmuu 274
hoa hien 175
hoa soi 97
hoac huong 301
hoarhound 219
hoary basil 255
hoary cress 85
hoary salt bush 59
hodengal 31
hodunamu 188
hoehyang 158
hog plum 352
hog, Malaysian 352
hog's garlic 27
hogfennel 283
hogweed 67, 132
Hohe Studentenblume 357
Hoher Steinklee 221
Hohes Lemongras 129
Hohllauch 19
Hohlsame, Strahlen 65
hoja amarga 386
hoja de boldo 284
hoja de laurel 199
hoja de nogal 188
hoja de xalape 234
hoja sancta 294
hoja santa 289
hôki-gi 64

hokoso   352
Holder   331
hole thistle   344
Holland mustard   345
Holland rose   318
Holländischer Senf   345
Hollerbeeren   331
Hollywood clausena   110
Holunder, Schwarzer   331
holy basil   260
holy grass   178
holy thistle   112
holywood   170
Holzkassia   100
Holzlauch   26
Holzrose, Queensland   231
Holzzimt   100, 103
Holzzimt, Indischer   104
Holzzimt, Java   103
hom chin   147
hombua   19
homchan   280
hom-pomkula   147
homr   359
hondapara   136
hondepis   315
honewort   124, 346
honewort, (Japanese)   125
honey clover   221
hong bi   111
hong cao   302
hong gua   113
hong hua   86
hong hua cai   86
hong lan hua   86
hongdangmuu   133
Hongkong kumquat   159
Hongkong wild kumquat   159
Honigblatt   223
Honigklee   222
Honigmelonen-Salbei   327
honje   149
honje laka   149
honje warak   149
Hooker garlic   20
Hooker-Zwiebel   20
hop   179, 210
hop thyme   210
hop, Spanish   264
Hopfen, (Gewöhnlicher)   179
Hopfen, Spanischer   264
Hopfenthymian   210
hophŭ   179
horapha-chang   258
horčica   70
horehound, common   219
horehound, Indian   181
horehound, white   219

horm daeng   17
horm lek   17
horn of plenty   155
horned pansy   387
horned violet   387
Hornfrüchtiger Sauerklee   268
Hornkraut, Sand   92
Hornsauerklee   268
Horn-Veilchen   387
horopha   374
horse bergamot   236
horse mint   224, 226
horse mint, American   237
horse mint, cloverhead   237
horse mint, stinking   236
horse shoe vitex   388
horsemint   237
horseradish   50, 391
horseradish tree   240
horseradish, Chinese   178
horseradish, Japanese   391
horsewood   110
hortãpimenta   227
hortelã   89, 174, 225, 226, 229
hortelã commun   229
hortelã de burro   89
hortela do campo   181
hortelã verde   229
hortelã-de-folha-miúda   225
hortelã-de-panela   225
hortelã-de-rasteira   225
hortelã-do-brejo   174
hot mint   280
hot pepper   83
houblon (commun)   179
houk sinnari   344
houque odorante   178
houttuynie   179
Houttuynie, Herzförmige   179
hsi sha tou   10
hsiang   90, 108, 112, 145, 184, 241,
        278, 288, 301, 320, 332, 334, 355, 383
hsiang po-ho   334
hsiang ssu   2
hsiang sui   278
hsiang ts'ao   112, 383
hsiang yuan   108
hsiang-ju   145
hsiao suan tsai   268
hsieh pai   17
hsien   95, 305
hsing sui   116
hsiug ya-li chiao   82
hstang tsa i tzu tsaoi   256
hsü sui tzu   152
hsuan ts'ao   175
hsun tsao   222
hu chen tzu   297

hu chiao   293
hu chieh   345
hu cong   17
hu lu ban   374
hu luo bo   133
hu ma   342
hu suan   25
hu sui   116
hu tao   188
hu wei lan   332
hua chiao   401
hua ha   31
hua jiao   399
huai hua ching   88
huai shu   93
huampit   111
huamúchil   298
huang lan   232
huang pi gua   111
huchubakha   227
huchunamu   293
huei hsiang   288
Hühnergras, Pyramiden   143
hui luo le   255
hui t'iao   95
hulbah   374
hulul   111
hummedha   320
hung chanh   299
hung hua tsai   207
hung pai ho   207
hung tou   2
hung tsao   302
Hungary gentian   164
huo   167, 173, 301
huo xiang   12, 301
Huon pine   133
huong   301, 319
huong nhu trang   258
hurhur   111
Husarenknopf   351
Husarenknopf (flowers)   351
husk tomatillo   285
hwangphinamu   111
hyaena taaibos   317
hyangdulkkaephul   301
hyanghynŏgge   252
hyangkkotauk   276
hyangminari   282
hyangyu   145
hyŏnggye   252
hysope   182
hysope anisée   11
hyssop   182, 233
hyssop, talmud widder   233
hyssop, tea   233
iará   202
iará-uaçu   202

Iberischer Drachenkopf   139
ibisco fiori   177
ibuki-jakō-sō   368
idalimbu   107
idreira   211
idsla   305
Igelklette   143
ijan   342
iklil al-jabal   320
ikmong bata   278
ikola hindi kimba   395
ilagas de Cristo   375
ilanga   76
ilang-ilang   76
Ilang-Ilang   76
ilara borotyo   401
ilarangom   401
ili el'sgol'zija   145
ili malagetta   10
ili pashitnik   374
ilib   387
ilimba   310
illicium   184
Ilombanuss   310
imbaúba   89
imbir   403
im-khim-lam   260
imli   298, 359
Immergrüner Muskat   198
immortelle   174
immortelle de Italiener   174
immortelle du Tian-shan   174
impératoire   283
imperatoria   283
inabu   178
incense tree   77
incensier   320
incernsi   69
incienso   69
Indean bread shot   78
indejckij kress   375
Indian aloewood   48
Indian anise   184
Indian Arrowroot   126
Indian bael   7
Indian bark   104
Indian bay   241, 287
Indian borage   299
Indian canna   78
Indian cardamom   37
Indian cassia   104
Indian cassia lignea   104
Indian chocolate root   166
Indian clausena   111
Indian coral tree   148
Indian cress   375, 376
Indian cudwee   170
Indian dill   40

Indian fleabane   300
Indian gentian   163
Indian ginger   29
Indian globe thistle   351
Indian gooseberry   284
Indian horehound   181
Indian ixora   186
Indian jasmine, yellow   187
Indian laburnum   88
Indian lemon   107, 129
Indian liquorice   2
Indian long pepper   292
Indian maesa   215
Indian mano   217
Indian mint   299, 334
Indian mugwort   54
Indian mulberry   239
Indian mustard   69
Indian nard   249
Indian nut   289
Indian painkiller   239
Indian patchouli   301
Indian pepper   400
Indian pink   135
Indian pluchea   300
Indian prickly ash   401
Indian saffron   126
Indian sandal wood   309
Indian sarsaparilla   176
Indian sesam   342
Indian shot   78
Indian simpoh   137
Indian sorrel   177
Indian spinach   62
Indian tamarind   359
Indian walnut   14
Indian yellow jasmine   187
Indianerkresse   375
Indianernessel   236
Indianernessel, Rote   236
Indianernessel, Späte   236
Indianernuss   289
indigo, African   213
indigo, wild   362
Indigo, Wilder   362
indigo, Yoruba   213
indijskij tmin   371
Indisch bloemriet   78
Indische Clausenie   111
Indische Costuswurzel   336
Indische Dattel   359
Indische Ingwerlilie   29
Indische Limnophilie   208
Indische Maulbeere   239
Indische Narde   249
Indische Pluche   300
Indische Sarsaparille   176
Indische Scharlachranke   113

Indischer Andorn   181
Indischer Anis   184
Indischer Basilikum   260
Indischer Beifuß   54
Indischer Boretsch   299
Indischer Enzian   163
Indischer Galgant   29
Indischer Goldregen   88
Indischer Holzzimt   104
Indischer Jasmin   187
Indischer Korallenbaum   147
Indischer Lorbeer   113, 356
Indischer Pfefferbaum   401
Indischer Sauerampfer   322
Indischer Schwarzkümmel   386
Indischer Senf   69, 70
Indischer Sesam   342
Indischer Spinat   63
Indischer Wunderstrauch   113
Indisches Arrowroot   78
Indisches Blumenohr   78
Indisches Lorbeerblatt   356
Indisches Moxakraut   54
Indisches Patchouli   301
Indisches Ruhrkraut   170
Indisches Zitronenblatt   107
Indisches Zitronengras   129
Indochina-Kardamom   36
Indonesian bay-leaf   356
Indonesian cardamom   35
Indonesian cassia   100
Indonesian vanilla   382
Indonesische Vanille   382
Indonesischer Boretsch   299
Indonesischer Zimt   100
indravalli   85
indungulo   346
ingudi   61
Ingwer   30, 118, 126, 191, 346, 402
Ingwer, Afrikanischer   346
Ingwer, Ansehlicher   404
Ingwer, Bitterer   404
Ingwer, Deutscher   5
Ingwer, Echter   403
Ingwer, Gelber   126
Ingwer, Goldähriger   402
Ingwer, Japanischer   403
Ingwer, Malaysischer   118
Ingwer, Martineque   31
Ingwer, Mango   126
Ingwer, Nickender   404
Ingwer, Roter   31
Ingwer, Siam   30, 31
Ingwer, Thai   30, 31, 191
Ingwer, Wilder   346
Ingwergras   130
Ingwerlilie   29, 31, 118
Ingwerlilie, Indische   29

Ingwerlilie, Porzellan 31
Ingwerminze 226
inondo 40
inu-hôzuki 350
inula 184
inule 184
iperico 180
iphéion 185
irfa 104
irga ol'cholistnaja 33
iris 185
iris commun 185
iris d'Allemagne 185
iris de Florence 185
iris florentina 185
iris Florentino 185
iris, common 185
iris, duitse 185
iris, Florentine 185
iris, German 185
Irish lace marigold 357
irnyj koren' 5
ironbark, lemon-scented 150
ironwood, little 387
irosa 318
irziyan 158
iscoque 358
isibhaha 391
isihlaza 90
isip 45
isiphepketo 346
Isop 182
ispungu 261
issombo 339
issop 145, 181
issop obyknowennyj 182
issopo 182
Italian everlasting 174
Italian jasmine 187, 188
Italian lavender 201
Italian oregano 264, 266, 367
Italian springe-sage 329
Italienische Sandstrohblume 174
Italienischer Dost 266
Italienischer Frühlings-Salbei 329
Italienischer Jasmin 187
Italienischer Majoran 264
Italienisches Süßholz 169
itkkot 87
Ivakraut 3
ivêche de Chine 112
ivrogne 52
ivy gourd 113
iwatake 172
ixora 186
ixora, Indian 186
Ixorie, Scharlachrote 186
ixquihit 314

izkhir 387
jaborandi-manso 291
Jackbohne, Meeres 77
Jack-by-the-hedge 14
Jack-in-the-green 253
Jacmalschalen 106
jae 403
jagaruwa 88
jagni 171
jahe 403
jahi 403
jajipatri 244
Jakobslauch 19
jakorzi 372
jalap, false 234
Jalapa-Wunderblume 234
jal-palan 321
Jamaica bark 344
Jamaica nutmeg 238
Jamaica peper 286
Jamaica pepper 286
Jamaica quassia 285
Jamaica sorrel 177
Jamaica vervain 353
Jamaica wood 311, 312
Jamaican savory 334
Jamaika-Bitterholz 285
Jamaikakaper 79
Jamaikapfeffer 286
Jamaika-Quassia 285
Jamaika-Thymian 299
Jamajskij perez 286
Jamba-Raps 147
jambira 107
jambiri 107
jam cherry 285
jang ho 403
jangli akhrot 14
jangli 126
jangliadrak 402
jangli-haldi 126
jangli-palak 321
Jantar 117
januprulisaru 46
Japanes pepper 400
Japanese archangel 42
Japanese bellflower 299
Japanese bitter orange 304
Japanese bunching onion 19
Japanese butterbur 281
Japanese catnip 252
Japanese cinnamon 102
Japanese garlic 19
Japanese honewort 125
Japanese horseradish 391
Japanese leek 19
Japanese mint 224
Japanese mugwort 55

Japanese parsley 125
Japanese (pepper)mint 225
Japanese peppermint 226
Japanese prickly 400
Japanese red star lily 207
Japanese rose 320
Japanese sacred anise poisonbay 183
Japanese scallion 18
Japanese sweet flag 5
Japanese thyme 368
Japanese turmeric 128
Japanese wild chervil 125
Japanese wild ginger 403
Japaningwer 403
Japanische Braunalge 195
Japanische Engelwurz(el) 42
Japanische Minze 224, 226
Japanische Orange 159
Japanische Pestwurz 281
Japanische Petersilie 125
Japanischer Beifuß 55
Japanischer Bergpfeffer 212
Japanischer Blatttang 195
Japanischer Flügeltang 380
Japanischer Ingwer 403
Japanischer Kalmus 5
Japanischer Knoblauch 19
Japanischer Meerrettich 391
Japanischer Pfeffer 396, 400
Japanischer Pfefferbaum 401
Japanischer Senf 69
Japanischer Sternanis 183
Japanischer Thymian 368
Japanischer Zimt 102
Japanpilz 202
Japanrose 320
Japans hoefblad 281
japonckij perez 400
jará-açu 202
jarak belanda 66
jasenez 136
jasmennik duschistij 161
jasmijnbloesen 187
jasmin 162, 187, 188, 241
Jasmin 162, 187, 188, 241
jasmin commun 187
jasmin d'Arabie 188
Jasmin, Arabischer 188
jasmin, Cape 241
Jasmin, Duft 187
Jasmin, Echter 187
Jasmin, Falscher 241
Jasmin, Gelber 187
Jasmin, Großblütiger 188
Jasmin, Indischer 187
Jasmin, Italienischer 187
Jasmin, Kap 162
Jasmin, Persischer 188

Jasmin, Rispiger 188
Jasmin, Weißer 187
jasmine 162, 187, 241
jasmine à grandes fleurs 188
jasmine orange 241
jasmine, Arabian 188
jasmine, biblical 188
jasmine, Catalonian 188
jasmine, Italian 187
jasmine, Tuscan 188
jasmine, white 187
jasmine, yellow 187
Jasminglanz 162
jasmón blanco 187
jatamansi 249
jatikosa 244
jatiphala 244
Java cardamom 35, 37
Java cassia 103
Java citronella 131
Java galangal 30
Java lemongrass 131
Java pepper 291
Java Pfeffer 294
Java thyme 210
Java vanilla 382
Java-Vanille 382
javaanse lange peper 294
Java-Holzzimt 103
Java-Kardamom 35, 37
Javan water dropwort 261
Javanese long pepper 294
Javanese turmeric 128
Javanische Gelbwurz(el) 128
Javanische Troddelblume 146
Javanischer Pfeffer 291, 294
Javanischer Thymian 210
Javanischer Wasserfenchel 261
Javanisches Patchouli 301
Java-Zitronelle 131
Java-Zitronengras 131
Javian patchouli 301
jawang nieng 49
jawer 5
jayphal 244
jazar 133
jazmin blanco 187
jazmín d'Arabia 188
jeera 125
jelutong 32
jema 387
jeneverboom 189
jenggar ayam 90
jengibre 403
jengibre amargo 404
jengkol 49
jengkolan 48
jenjuang 116

Jenny stonecrop 340
jequirity 2
jeramanis 288
jererecou 396
jering 48
jering goleng 48
jering tamarind 49
jeringan 5
jeringau 5
jĕrnuang 29
jeruk limau 105
jeruk limu 105
jeruk niois 105
jeruk sambal 105
jĕruk tangan 109
Jerusalem oak 96
Jerusalem sage 284
Jerusalem Salbei 284
Jerusalem tea 95
jessamine 162, 187
jessamine, Cape 162
Jesuit bark 99
jhak 325
ji cai 81, 178
ji guan hua 90
jiao dou shu 93
jia du xing cai 203
jia hu gua 85, 86
jia jing jie 251
jia ma bram 353
jia yan ye shu 349
jia yuan qian 147
jiang 30, 126, 268, 285, 403
jiang huang 126
jiao dou shu 93
jiao tou 18
jie cai 81
jie jeng 299
Jim sage 327
jimami 48
jin gan 159
jin ji le 99
jin zhan ju 75
jing jie 251, 252
jintam hitam 253
jintan 125, 158, 288
jintan manis 158, 288
jintan puteh 125
jinten bodas 125
jinten manis 288
jinten poteh 125
jinten putih 125
jira 87, 125
jiraka 125
jirakasafed zeera 125
jiring 49
joá-poca 285
Jochthymian 369

Johannisbeer-ähnlicher Pfeffer 294
Johannisblut 180
Johannisbrot 93
Johannisbrot, Wildes 93
Johanniskraut, Echtes 180
Johanniskraut, Tüpfel 180
jon tou k'ou 244
jonc de marais 377
jonc odorant 5
Joseph sage 330
Josephskraut 182
jou-kwei 104
joz 188
ju hua 360
juang 116
Judas tree 93
Judas's bag 6
Judasbaum 93
Judasbaum, Gewöhnlicher 93
Judendorn 405
Judenzwiebel 27
judino 93
judino derevo 93
jui-sheng-nu 109
jukut bau 181
Julian savory 233
Julian-Salbei 233
Jungfer im Grünen 253
junggukpuchu 18
jungle flame 186
juniper 189
juniper, common 189
juniper, Phoenicean 189
junipero 189
jupha 182
Jupiter's distaff 328
ju-ran 12
jusefka 182
justicia 189
justicia, quintuple 189
Justicie, Fünfkantige 189
juzzii 27
jwarancusa 130
k'u kua 235
ka busu 106
ka guo 198
ka tepus 38
kaalaa 149
kaan thuup 387
kabab cheene 286
kababa hindiya 291
kababcini 291
kabaha 291
kabar 79
kabara 80
kabare kindi 113
kabbar 80
kabling 301

kaboat 369
kabra 80
kabrab 148
kaca piring 162
kacang 168
kacang kedelai 168
kachana tur 343
kachang 240
kachang china 48
kachang gori 48
kachlora 298
kachnar 64
kacholum 191
kachom 208
kachura 126, 128
kada-me 380
kadaranarathai 108
Kaddig 189
kadellu 171
kadjeng siti 401
kadlum 301
kadok 292, 295
kadok batu 295
kadongdong 352
kae-aen 67
kaetppangphung 167
kaf mariyam 388
kaff ed-dubb 330
kaffir caggage 111
kaffir lime 107
Kaffir-Limette 107
kafurkapat 231
kaghzi mimboo 105
kair 79
kaisei to 106
Kaiser-Nelke 135
Kaisersalat 53
Kaiserwurz(el) 283
Kaiserzepter 149
kaju lemal 401
kakadani 80
kakakad 177
kakaltum 259
kākamāici 350
Kakaoblütenbaum-Blüten 311
kakidoshi 167
kakkam 325
kakri 80
kala 125, 255, 260, 386
kala tulasi 260
kala jiri 386
kala tulsi 255
Kalabassen-Macisbohne 238
Kalabassenmuskat, Angola 237
Kalabassenmuskat, Kurzstielige 238
Kalabassen-Muskatnuss 238
kalahaldi 126
kalajaji 254

kalajira 253
kalamansito 374
Kalamondin 159
kalaoo 208
kalapini 300
kalatil 171
kalaunji 253
kalawag 126
kalayana murikku 148
Kälberkropf, Aromatischer 94
Kälberkropf, Byzantinischer 94
Kälberkropf, Gewürz 94
Kälberkropf, Gold 94
Kälberkropf, Knolliger 94
kalgan laptschatka prjamaja 306
kali mirch 293
Kaliforniche Nusseibe 370
Kalifornische Muskatnuss 370
Kalifornischer (Berg-) Lorbeer 379
kalihalidi 126
kalingag 100
kalmoeswortel 5
Kalmus 5
Kalmus, Japanischer 5
Kalmus, Lakritze 5
kalonji 253
kaloo 2
kalp 6
Kalufer 360
kalufer 360
kalushniza bolotnaja 75
kalvari 80
kam kat 159
kam lam 284
kam(m)oun 125
kamahim 256
kamajung 350
kamalagui 359
kamalungai 240
kaman kabir 331
kamarak 140
kamatsele 199
kambang-turi 343
kambiang 297
Kambodscha-Kardamom 36
Kambodscha-Sternanis 183
Kambodscha-Zimt 101
kamcho 170
Kamelgras 131
kamennij luk 15
Kamerun-Kardamom 8, 9, 10
Kamerun-Knoblauch 179
Kamerun-Knoblauchbaum 11
kami 14, 100
Kamille, Römische 95
kamina marchula 241
kamkyulnamu 109
Kamm-Minze 145, 227

Kamm-Minze, Chinesische 145
Kamm-Minze, Echte 145
Kamm-Monarde, Getüpfelte 237
kam(m)oun 125
Kampferbasilikum 255, 258
Kampfer-Wermut 52
kampupot 188
Kamshatka wormwood 56
Kamtschatka-Beifuß 56
kamtsele 298
kamue muluki 371
kamun 125
kamuning 242
ka: nz ph'u 355
kan chü 109
Kanadische Haselwurz 57
Kanadische Steinpetersilie 124
Kanadische Zwiebel 16
Kanadischer Lattich 195
kanakoso 381
kanalei 153
kananga 77
kanariesaadgra 202
Kanarische (Besen-)Winde 115
kanchatton 401
kanda lasum 17
Kandeln 14
kander 80
kandi-kandi laan 353
Kaneel 77, 100, 104
kaneel 77, 100, 104
Kaneel, Weißer 77
Kaneelrinde 77
Kanehl 104
kaneishon 135
kanfu 155
kangwang 126
kanh ieu 17
kanh kho 17
Kani 395, 397
Kani-Körner von Selim 395
kanjero 136
kankola 291
kankrasing 317
kann tsao 170
kanna dandoku 78
kanna indijkaja 78
Kannibalen-Tomate 349
Kanonenbaum 89
Kanonierblume,
 Schwarzschlundige 286
kanphlu 355
kantakari 350
kan-thai 379
Kantiger Lauch 16
Kanton-Rhabarber 316
Kanton-Zimt 100
kantutay 196

kanuikui   371
kanuniphyangkkulphul   260
kanuper   360
kanuris   352
kapari   80
kapary   80
Kaper   80
Kaper, Echte   80
Kaper, Himalaya   80
Kaper, Jamaika   79
Kaper, Mariannen   81
Kaper, Nepal   80
Kaper, Orientalische   80
Kaper, Sizilianische   80
Kapernstrauch, Dornenloser   81
kapersi   79
kaperzy   80
Kap-Gardenie   162
kaphrao   260
kaphrao-chang   258
Kap-Jasmin   162
Kaplilie, Knoblauch   376
Kap-Minze, Weiche   227
kapol   35
kapper boom   80
Kap-Pfeffer   290
Kapsafran   121, 354
Kap-Safran   354, 375
kapulaga   35, 144, 167
kapulaga ambon   167
kapun   176
kapur tulsi   258
kapuzin   375
Kapuzinerkresse, Große   375
Kapuzinerkresse, Kleine   376
Kapuzinerkresse, Knollen   376
Kapuzinerkresse, Peruanische   376
kapuzin-kress   375
kara   135, 181, 265, 293, 316
kara biber   293
kara daio   316
kara gynych   264
kara nadeshike   135
karabu neti   355
karahana-sô   179
karaila   111
karambel   136
karambu   355
Karandapflaume   86
karangeang   399
karapdap   148
karapincha   241
karat   79
karatachi   304
karauya   87
karawiya   87
Karbensamen   87
karbot-sermul   337

kardamoenggve   35
kardamóm   144
Kardamom, Abessinischer   9
Kardamom, Bastard   36, 38
Kardamom, Beißender   34
Kardamom, Bengalischer   35
Kardamom, Bereifter   10
Kardamom, Bitterer   36
Kardamom, Blutroter   8
Kardamom, Ceylon   144
Kardamom, Chinesischer   36
Kardamom, Echter   144
Kardamom, Ethiopischer   9
Kardamom, Graziler   36
Kardamom, Großer   37
Kardamom, Großsamiger   9
Kardamom, Guinea   10
Kardamom, Java   35, 37
Kardamom, Kamerun   8, 9, 10
Kardamom, Korakima   9
Kardamom, Langer   35, 37, 38, 144
Kardamom, Madagaskar   8
Kardamom, Malabar   38, 144
Kardamom, Nepal   35, 37, 38, 144
Kardamom, Ostafrikanischer   10
Kardamom, Runder   35
Kardamom, Schlanker   36
Kardamom, Schwarzer   11, 36
Kardamom, Siam   35, 38
Kardamom, Sierra Leone   8
Kardamom, Sikkim   37
Kardamom, Trauben   35
Kardamom, Tsao-ko   38
kardemomzaad   144
kardobenedikt   112
Kardobenediktenkraut   112
karela   235
karibandha   176
Karibischer Ameisenbaum   89
Karibischer Zimt   140
karil   80
karipatta   241
Karkade   177
karkadé   177
karnkusa grass   130
karobbe   93
Karobe   93
karot   133
Karotte   133
karpura-haridra   126
karrinim   241
Karstbohnenkraut   335
Karstsaturei   335
karsun mehi   250
Kartäusertee   95
Kartoffelzwiebel   17
karu   163
karube   93

karudoman   144
karuk   295
karwa pale   241
kasa kasa   272
kasai   12
kasapiucha   241
kasch kasch   272
Käseartige Sonneratie   351
Käseklee   372
Käsepappel   216
kashia-keihi   100
kasi hutan   215
kasia   100
kasia sena   88
Kaskarille   123
Kaskarille, Saftlose   123
Kassia   100
Kassia, Korintji   100
kassie   100
Kassie, Batavia   100
Kastanienblättrige Muskat   243
kastuli   1
kasturi majal   126
kasturi pasupu   126
kasturiarishina ran halada   126
kasuri methi   373
kat   69, 159
katak puru   131
Katechu, Gelbes   379
katerbik   110
kathnim   241
kath-semul   337
kathue   404
kathue-pa   404
katifa orf el-deek   90
katjathikai   244
Katliaturholz   309
katmon   137
kato   273
katran serdcelistnyj   119
kattapitalavam   239
kátte-yelak-káy   37
kattra   64
kattunarakan   176
katudai   343
katu-kina   401
kature   343
Katzen-Bergthymian   74
Katzengamander   363
Katzenkraut   251, 363
Katzenkraut, Echtes   251
Katzenminze, Blaue   251
Katzenminze, Gewöhnliche   251
Katzenminze, Kleinblütige   74
Katzenminze, Kleine   251
Katzenminze, Mussin's   252
Katzenminze, Transkaukasische   252
Katzenminze, Traubige   252

Katzenschnurbart 111
Katzenwurzel 381
kau lei heung 241
Kaugummipflanze 334
Kaukasus-Fetthenne 340
kaum chempaka 232
kaum kopi 186
kavali 171
kawaen 404
kawa-kawa 215
kawa-midori 12
kaya teja 101
kayu lawang 103
kayu manis 100, 101, 104
kayu manis cina 100
kayu manis hutan 102
kayu salangan 103
kayumanis 110
kebbaba 291
keber 79
kebéré 80
kebere çiçegi 80
kebir 80
kĕchil 93
kĕchumbrang 149
kecombrang 149
kedawak 294
kedondong 352
keeled garlic 16
kefe kimyonü 198
keitó 90
kekig orégano común 265
Kekunanuss 14
kelabat 374
kelabet 374
keladi ayer 267
kelemoyang 29
kelil 320
kelima paya 280
kelingket 374
kelir 240
kella 34
Kellerhals 133
kelok 240
kelor 240
kelp 195
Kelp 195
Keltische Narde 381
Keltischer Baldrian 381
kelur 343
kemal 359
keman ajer 252
kemangi 103, 181, 256, 260
kĕmangi 255
kemangi utan 260
kembang eros 319
kembang erus 320
kembang gelas 268

kembang rus 319
kembiri 14
kemika 118
Kemiri 14
kemiri 14
Kemirinuss 14
kemloko 284
kemoening 242
kempang pukul empat 234
kemukus 291
kĕmung 242
kemuning 242
kenaf hibiscus 177
kenaga 77
kĕnchur 191
kencur 191
kenikir 118
Kentjur 191
kentjur 191
Kentjur, Runder 192
Kentucky Krauseminze 225
Kentucky spearmint 225
kenyong 78
kêo 242
keolar 64
keora 271
kĕpulaga 35
ker 79
kerbel 45, 247
Kerbel 45
Kerbel, Deutscher 45
Kerbel, Gemeiner 45
kerbel' ispanckij 247
Kerbelrübe 94
kermful 355
kerupulai 241
kerva 211
kerva cidreira 211
kervel 45, 247
Kerzennuss 14
kesar 122, 255
keshi 272
kesuma tebok selydang 280
kesumba 65
kesumba keling 66
ketepeng 341
ketmie musquée 1
kettan 209
ketten 209
ketumbak 205
ketumbar 116, 147
ketumbit 205
keu 118
Keulenlilie, Endständige 116
Keuschlamm 388
kevovoe 297
kewra 271, 272
key lime 105

keyok 88
keys, Chinese 192
kha 30, 110, 181
kha ka ai 110
kha paa 30
kha:x ta:dè:ng 30
khabbar 79
khae baan tveri 343
khae thawaa 162
khairwal 64
Khakikraut 358
kham saet 66
khamin chan 127
khamin khao 128
khamin-kaeng 127
kham-ngo 66
khan ko 17
khan phluu 355
khang namphung 110
khao khinh 212
khapparkadu 93
kharapatra 263
khardal 70, 345
kharjal 325
kharum 87
khash khash 272
Khasi-Adlerholz 48
khasi-eagle wood 48
khas-khas 387
khaskhasa 272
khatapalak 321
khate meethi 268
khati luni-ni-bhaji 306
khatta 106
khattikan 321
khavi grass 130
khawi 129
khayat el-djurhat 327
khazrîq 93
kheemin 127
khelal 34
khella 34
khelwan 331
khem 186
khem farang 186
khen chaai 47
kherwa' 388
khesh kash 272
khiauk 217
khiey 191
khilla baladi 34
khilla shitani 33
khing daeng 31
khing-daeng 403
khinjuk 296
khlam puu 355
khlawath weed 180
khong 166, 168

khrai 129
khulingan 31
khŭnnorangkkotjariphul 374
khus khus 272, 387
khus khus (grass) 387
khuun 88, 114
khuzama 200
khwarg 80
ki 99, 116, 150, 235, 256, 258, 380
ki cengkeh 380
ki uri 235
kiaka 362
kiang 310
kiapo 297
kibana kosumosu 118
kibana-suzushiro 147
kidachi hakka 334
kieou 24
kikamu 278
kikpoi 119
kiku-na 97
kikunori 97
kikuolo 278
kikyô 299
Kilimandscharo basil 258
Kilimandscharo-Basilikum 258
kilingiba 59
killa 33
kille sheitani 33
killi 305
kim chau 175
kim cuong moc 232
kim quat 159
kimba pepper 395
kimion 125
kin kan 159
Kinder-Tomate 348
kindsi klopownik 116
kingcup 75
kinh gioi dat 95
kinkan 159
kinnamon 99
Kinoto 159
kin-renka 375
kinsa 116
kin-sokei 187
kintsay 47
kiratatikta 354
Kirayakraut 354
Kirgisischer Oregano 265
kirmala 54
kirmani 54
Kirschlorbeer 307
kischnez 116
kiseseri 391
kisliza obyknowennaja 268
kismas 152
kişniş 116

kisól 191
kiss me over the garden gate 302
Kissipfeffer 291
kitajckaja koriza 100
kitajckij anis 184
kitajskaja oritscha 100
kitajskij kasia 100
kitajskij luk 19
kitamari 54
kitchen garden purslane 305
kittile 109
kiu mokusei 266
kiu ts'ai 27
kjintan 116
klabet 374
klabong pako 355
klakatschka 353
klanting 337
Klappernuss 353
klarroub 92
Klatschrose 272
Klebriger Alant 185
Klebriger Gänsefuß 96
Klebriger Salbei 328
Kleeblättriges Schaumkraut 85
Kleinblättrige Brunnenkresse 250
Kleinblättriges Wandelröschen 197
Kleinblütige Katzenminze 74
Kleinblütige Malve 216
Kleinblütige Melisse 224
Kleinblütige Mikanie 233
Kleinblütiger japanischer Pfeffer 396, 400
Kleinblütiges Wiesenschaumkraut 84
Kleindoldige Merremie 231
Kleine Bibernelle 288, 331
Kleine Kapuzinerkresse 376
Kleine Katzenminze 251
Kleine Pimpinelle 331
Kleine Scheinaster 387
Kleine Studentenblume 358
Kleine Taglilie 175
Kleiner Basilikum 260
Kleiner Galgant 31
Kleiner Lavendel 200
Kleiner Sauerampfer 321
Kleiner Wiesenknopf 331
Kleines Eisenholz 387
Kleinköpfiger Koriander 117
Kleinöhrige Wisterie 180
Kletter-Wedelie 392
Klimme, Vierkantige 105
Klimmtraube, Lederblättrige 51
Klimmtraube, Sechskronenblättige 51
klinček 355
kliobbeiza 216
klipsissie 7
kljuzewoj kress 250

Klöben 17, 19
klohor 88
klokitschka 353
kloncing 352
klopovnik 202
Klosterysop 182
Knoblauch 15, 19, 24, 27, 179, 376
Knoblauch, Acker 15
Knoblauch, Chinesischer 19
Knoblauch, Japanischer 19
Knoblauch, Kamerun 179
Knoblauch, Peking 24
Knoblauch, Pferde 15
Knoblauch, Sommer 15
Knoblauch, Veilchenblütiger 376
Knoblauch, Wilder 27
Knoblauch, Zimmer 376
Knoblauchbaum, Kamerun 11
Knoblauchbaum, Schuppenblättriger 11
Knoblauchbirne 120
Knoblauchpilz 218
Knoblauchs-Kaplilie 376
Knoblauchsrauke 14
Knobloch 24
knof 376
Knofel 24, 27
Knofel, Wilder 27
knoffel, wilde 376
knoflook 24, 376
Knollenbegonie 64
Knollen-Kapuzinerkresse 376
Knollenkerbel 94
Knollenkresse 376
Knollenkümmel 71
Knollen-Leuchterblume 93
Knollenpeterilie 283
Knollensellerie 47
Knolliger Kälberkropf 94
knolpeterselie 283
knolselderij 47
knolvenkel 158
Knorpelmöhre, Große 33
Knöterich, Acker 302
Knöterich, Flaumiger 280
Knöterich, Floh 302
Knöterich, Orientalischer 302
Knöterich, Östlicher 302
Knöterich, Persischer 302
Knöterich, Scharfer 280
Knöterich, Schmächtiger 303
Knöterich, Vogel 302
Knöterich, Wasserpfeffer 280
Knöterich, Weidenblättriger 302
Knöterich, Wohlriechender 280
knotgras, bird 302
knotgrass, Persian 302
knotted majoram 263

knotweed   280, 302
ko:k   352
kobon   43
Koch's Lauch   28
kochu   82
koendroro   116
ko-hime-yuri   207
Kohllauch   22
kohson   113
koijn   253
kôk namz man   14
kokam   162
kokan   90
koknar   272
Kokum   162
kokum   162
kolar   64
kolba   27
kolbasnaja trawa   263
Kolbenlilie, Endständige   116
kolinchi   362
kolintaso   78
koljandra   116
koljurija gravilatovidnaja   114
koljutschesontitschnik   143
Kölle   334
Kölnisch-Wasser-Minze   228
kolong-kugon   387
kolosok duschistyj   44
kom kodong   260
kondepishout   315
kondurango   220
kong que cao   358
Kongo-Kubeben   291
Königschinarinde, Echte   99
Königskerze, Filz   385
Königskerze, Großblütige   385
Königskerze, Windblumen   385
Königskraut   256
Königssalbei   329
konje hemp   332
konji   241
Konradskraut   180
kô-ô-sô   358
koper   40
Kopfiger Erdbeerspinat   96
Kopfiger Gänsefuß   96
Kopfiges Tausendgüldenkraut   91
kôpor   40
Köppernickel   232
kopsilminari   282
koptskij tmin   371
kopyten'europeyckij   57
Korallenbaum, Indischer   147
Korallenraute, Duftende   68
korarima cardamom   9
Korarima-Kardamom   9
Körbel   45

Korean mint   12
Korean yellow wood   400
Koreanische Minze   12
Koreanisches Gelbholz   400
kôreigus   82
koriander   147, 253, 280
Koriander   116, 147, 253, 280
Koriander, Kleinköpfiger   117
Koriander, Langer   147
Koriander, Russischer   117
Koriander, Schwarzer   253
Koriander, Vietnamesischer   280
Korianderblätter   116
koriandr   116, 254
korinjii cassia   100
korintje cassia   100
Korintji Kassie   100
koritschnik kitajskij   100
koriza gvosditschnaja   136
Körner, Guinea   11
kornewistsche zarskogo kostylja   283
(korowjak) medweshʼe ucho   385
korownik sladkijctwol   41
Korsische Minze   228
koryokyo   30
koschatschja mjata   251
kôshin-ibara   319
kosho   293
kosij trilistnik   373
Kosmee, Gelbe   118
Kosmee, Schwarze   117
kosmos tscheltys   118
Kosteletzkablätter   193
Kostwurz, Prächtige   118
kosu   116
kôsui gaya   130
koto   260
kotovnik koschaschij   251
kou hua   144
kou teng   379
kourrath   23
kou-yüan   108
kpalug   273
kra chai   67
krachai   67, 191, 192
krachiap daeng   177
krafis   46
kraméria   193
Kranawitt   189
krangejan   212
Kranzblume, Scharlachrote   173
Kranzblume, Weiße   173
Kranzkraut   320
Kranzraute   322
krasnyi sandalovoe   309
krasodnev malyj   175
krasodnev ryschij   175
krasodnev scheltyi   175

krathaai   173
krathiem   24
Krause Minze   225, 229
Krause Petersilie   282
Krauseminze, Kentucky   225
Krauser Ampfer   321
Krauser Rhabarber   316
krawan   36, 38
kreko krervanh   36
kreko shmol   36
krekot   305
Kren   50, 250
Kreosotstrauch   197
krervanh   36
kress   84, 113, 202, 250, 352, 375
kress brasilʼskij   352
kress posevnoj   203
kress wodjanoj   250
Kresse   202, 375
Kresse, Afrikanische   202
Kresse, Breitblättrige   203
Kresse, Dichtblütige   202
Kresse, Türkische   375
Kresse, Virginische   203
kress-salat   203
Kreta-Bergminze   73
Kreta-Melisse   223
Kretischer Dost   262, 266
Kretischer Kümmel   371
Kretischer Thymian   336
Kreuzblättrige Wolfsmilch   152
Kreuzblume, Wollige   301
Kreuzdorn, Gerbereichen-
   ähnlicher   315
Kreuzkümmel   125
Kreuzkümmel, Persischer   71
Kreuzkümmel, Schwarzer   71
Kreuzsalbei   327
krevanh   36
Kriechende Minze   228
Kriechender Günzel   13
Kriechendes Winterbohnenkraut   336
Krim-(Berg-)Bohnenkraut   335
kringa   201
krishnajiraka   254
kritmum morskoj   120
Kroatisches Bohnenkraut   335
Krokus, Frühlings   121
Krokus, Hausknecht   122
Kronawitt   189
Kronen-Wucherblume   97
Kron-Rhabarber   316
Kronpiment   287
Krösling, Echter   218
krowochljobka   332
krsnajiraka   87
kruidnagel   355
kruisement   226

kruku 122
Krullfarn, Schwarzstieliger 7
kshi 272
ksudra 350
ku chin 261
ku gua 235
ku zhi 285
ku:k 352
kuai jing lzan hua 376
kuan ye jiu 20
kuang jing 388
kuasia 311
kuba 204
kubéba 291
Kubebeb, Kongo 291
Kubeben 291, 401
Kubeben, Afrikanische 291
Kubeben, Falsche 291, 401
Kubebenpfeffer 291
Kubinskij schpinat 239
kubjaha 319
kucai 27
kuchai 24
Küchenschwindling, Echter 218
Küchenzwiebel 17
kue-ra 14
Kufel 45
Kugeliger Eukalyptus 150
Kugelköpfiger Lauch 26
Kugellauch 26
Kuhblume 361
Kuhhorn 374
Kuhhornklee 374
Kuhkraut 381
kuhlaab 319
kui chaai 27
kuini 218
kujai 319
kuji 319
kukronda 300
kûkwa 87
kulaap mon 319
kulaap-on 319
kulanjan 30
kulaya 111
kulfa 305
kulfa sag 305
kulikul 48
kúlinján 31
kulit antarsa 212
kulit pulaga 212
kulitlawang 101
Kuliwan-Zimt 101
kum 120, 126
kum bok 120
kuma-tsuzura 386
Kumbapfeffer 395
kumbha 205

Kumin 125
kumin 125
kŭmjanhwa 75
kumkwat 159
Kümmel, Echter 87
Kümmel, Kretischer 371
Kümmel, Römischer 125
Kümmel, Süßer 288
Kümmel, Wanzen 125
Kümmel, Türkischer 125
Kümmelthymian 366
Kumquat 159
kumquat 159
kumquat malayo 159
kumquat redondo 159
kumquat, Formosan 159
kumquat, Hongkong 159
kumquat, large round 159
kumquat, malayan 159
Kumquat, Malayische 159
kumquat, marumi 159
Kumquat, Marumi 159
kumquat, oval 159
Kumquat, Ovale 159
kumquat, round 159
kumumu 119
kun 89
kun'r' 45
kunain 99
kunang ching 388
kunayana 99
kunchir 46
kundri 113
Kunerle 368
Kunigundenkraut 151
kunir 126
kunkuma 122
kunni 2
kunri 2
kunschut 342
kuntji 192
kunyit 122, 126
kunyit kering 122
kunz 120
kunzi 217
kuong nhu tai 260
kur kum 126
kurkuma 126, 128
kurkuma aromatihaja 126
kurkuma dlinnaja 126
kurkuma zedoarija 128
Kurkuma, Blaugraue 126
Kurkuma(wurzel) 126
Kurkuma, Wilder 126
kurrat 20, 27
kurrat baladi 20
kurrat nabati 20
kurrat seeny 27

Kurratlauch 20
kurrel 79
kurry patta 241
kur-semul 337
kurtschabaja mjata 229
kurwini mango 218
Kurzstielige Kalabassenmuskat 238
kuschina slabitel'naja 315
kûshibâ 116
kushta 118
kushumba 87
kusu-kusu 387
kusum 87
kusumba 87
kut kia 309
kut kin 309
kutavira 82
kutki 163
kutsay 27
kutschina 308
Kuttelkraut 368
kuttra 208
kuwein 218
kuwini 218
kuzbara 116
kvasia 285, 311
kvassija gor'kaja 311
kwanggyulnamu 106
kweni 218
kwini 218
kyllingia, greater 193
kyllingia, white 193
kyulhyangphul 223
la menta (inglesa) piperita 227
la bo he 227
la chanch 107
la chiao 82
la diep cá 179
la dua 271
la gen 50
la lot 295
labaca 321
labasnik sestilepestnij 157
labong 178
Labrador tea 201
Labradorkraut 201
labrat 298
laburnum, Indian 88
lacoto 84
lada 82, 290, 293
lada berekur 291
lada besar 82
lada hantu 290
lada hitam 293
lada putih 293
ladan 247
ladanka 265
ladannik werbena 211

lady's lace   33
lady's leek   18
lady's love   52
lady's mantle   13
lady's smock   84
lady's thumb   302
laftaf   85
lag root   5
laga   2
lagumtum   301
lagundi   388
lagunding late   300
lahsan   24
lahsun   24
lai pu-tao   235
laimun   107
laiteron   361
laitier   301
laitue d'eau   297
laja   30
laja goa   30
laja gowah   30
lajalu   252
lajm   107
làjmud des Indiens   371
lajouma   95
laka   149, 284
lakeichi   57
lakooch   57
lakoocha   57
Lakritze   169
Lakritze-Kalmus   5
Lakritzen-Studentenblume   357
Lakritzen-Tagetes   357
laksa leaf   280
lal chandan   309
lal murghk   90
lalambari   177
lalatan   48
lalbachlu   63
lalmica   82
lalmirichi   82
lamb mint   229
lamb's ears   353
lamb's lugs   353
lamb's quarters   95
lamb's tails   353
lampes   260
lampesan   181
lampin budak   110
lampoyang   128, 402
lampuyang   404
lanceolated buffalo ear   66
land cress   62, 203
land kress   62
Landnelke   135
lando   145
Langblättrige asiatische
    Rossminze   226
Langblättrige Minze   226
Langblättrige Sterkulie   354
Langblättriger Pfeffer   292
Langdorniger Orangenbaum   107
Lange Siegwurzel   27
Langer Kardamom   35, 37, 144
Langer Koriander   147
Langer Muskat   243
Langer Pfeffer   292, 293
Langfaden, Traubiger   115
langgyulnamu   109
langkauas   30, 128
langkawas   30, 404
langkuas   30, 31
langkuas malaka   30
langkuas na pula   31
langsat burung   63
Langstengliger Thymian   366
lantana   196, 197
lantana, Rhodesian   197
lantana, three-leaved   197
lantanier   196
Lanzettliche Blumée   66
Lanzettliches Büffelohr   66
laos   14, 30, 120, 147, 280, 319, 345,
    352, 355
lapazio   321
laping pudak   110
Laporte, Gekerbte   197
laranja da terra   105
laranja doce   109
laranjeira   106, 109
laranjeira azêda   106
large cardamom   35, 38
large chives   25
large flowered calamint   74
large flowered mullein   385
large leaf maple   2
large round kumquat   159
lasafa   80
lasan   24
laserwort   198
laserwort, Russian   198
lasona   17
lassaf   79
lasua tandla   136
lasuna   24
latakasturi   1
latkan   65
latmhura   136
Lattich, Grobzähniger   195
Lattich, Kanadischer   195
lattssaf   79
Lauch, Ägyptischer   20
Lauch, Ästiger   24
Lauch, Blutsverwandter   18
Lauch, Chinesischer   24, 27

Lauch, Gekielter   16
Lauch, Kantiger   16
Lauch, Kugelköpfiger   26
Lauch, Nickender   22
Lauch, Rundköpfiger   26
Lauch, Schöner   17
Lauch, Schwarzer   21
Lauch, Seltsamer   22
Lauch, Spanischer   21
Lauch, Spitzer   21
Lauch, Steifer   26
Lauch, Trügerischer   26
Lauch, Vavilov   27
Lauch, Wilder   26
Lauch, Zwiebelreicher   21
Lauchhederich   14
Lauchkraut   14
Lauch-Salbei   363
Laugenblume, Aschgraue   119
laung   355
laurel   45, 113, 198, 215, 307, 379
laurel cerezo   308
laurel Magnolia   215
laurel noble   199
laurel real   308
laurel, bay   199
laurel, Bombay   113
laurel, California   379
laurel, cherry   307
laurel, Chilean   198
laurel, Chinese   45, 46
laurel, Guinese   45
laurel, noble   199
laurel, Roman   199
laurel, swamp   215
laurel, true   199
laurel, Virigian   214
laureola femmina   133
laurier à jambon   199
laurier common   199
laurier d'Apollon   199
laurier, de feuille   199
laurier-amande   307
laurier-cerise   307
laurierblad   199
lauro poetico   199
lauro precios   43
lauro regale   199
lauro regio   307
lauroceraso   307
lavanda   200
lavanda aptetschnaja   200
Lavande   200
lavande véritable   200
lavande vrai   200
lavander sage   328
Lavandin   200
Lavandine   200

Lavanta çiçeği 200
lavang 101, 355
lavanga 355
Lavangrinde 100
Lavang-Zimt 101
lavas 205
Lavendel, Arabischer 201
Lavendel, Echter 200
Lavendel, Französischer 200
Lavendel, Großer 201
Lavendel, Kleiner 200
Lavendel, Schopf 201
Lavendel, Speik 201
Lavendel, Türkischer 139
Lavendel, Welscher 201
Lavendelblättriger Salbei 328
Lavendelblättriges Brandkraut 205
lavender 139, 200
Lavender 139, 200
lavender, Arabian 201
lavender, English 200
lavender, French 200, 201
lavender, Italian 201
lavender, Spanish 201
lavender, spike 200, 201
lavender, Turkish 139
lavrovischnja apteschnaja 308
lavrovischnja lekarstvennaja 308
lawr 199
lawr blagorodnyj 199
lawrowyj list 199
lay yam 5
laya 403
laza 59
leaf celery 47
leaf mustard 69
leafy leopardsbane 50
lean knotgras 303
leaved torreya 370
lecari 232
lechosa 86
lechuga de agua 297
lède 201
ledebour onion 20
Ledebour-Zwiebel 20
Lederblättrige Klimmtraube 51
Lederschnittige Perovskie 279
ledger bark 99
lédier 201
lédum des marais 201
leech lime 107
leek 15, 23
leek, Lady's 18
leek mountain 26
leek, Chinese 23, 27
leek, Japanese 19
leela dhana 116
legno santo 170

Leinsaat 209
lelang 102
Lemon 61, 107, 129, 131, 150, 175, 204, 211, 223, 227, 236, 251, 256, 276, 278, 316, 334, 359, 365, 386, 401
lemon balm 223
lemon basil 256, 257
lemon bergamot 236
lemon catnip 251
lemon day lily 175
lemon egoma 278
lemon eucalyptus 150
lemon gem 359
lemon geranium 276
lemon hyssop 12
lemon mint 223, 227, 236
Lemon myrtle, Australian 61
lemon pepper (tree) 401
lemon perilla 278
lemon savory 334
lemon scented myrtle 61
lemon scented tea 204
lemon scented verbena 386
lemon sumach 316
lemon thyme 365, 367
lemon verbena 211
Lemon Ysop 12
lemon, Indian 107, 129
lemonade tree 317
Lemongras 129, 130
lemongras 129
lemongras, East Indian 129
Lemongras, Großes 130
Lemongras, Hohes 128
Lemongras, Ostindisches 129, 130
lemongrass 129
lemon-scented gum 150
lemon-scented ironbark 150
lemon-scented tea 204
lemon-scented winter savory 335
lempui 402
lemunggai 240
lemuning 388
Lenabutagras 130
lençol-de-santa Barbara 295
lengkuas 29
lĕngkuas 29
lĕngkuas kĕchil 29
lĕngkuas padang 29
lengua de suegra 332
lengua de vaca 321
leño de santalo citrino 332
lenstisquè 296
lentisc 296
lentisco 296
leopard lily 332
leopards bane 51
leopardsbane, leafy 50

lep khrut bai yai 304
lepidio 203
lepidios 203
leptosperme 204
leptospermum 204
lesnoj tschesnok 14
lesser Bishop's weed 34
lesser calamint 74
lesser celandine 313
lesser galangal 30
lesser galingale 31
lesser ginger 67
lettok 32
lettuce, miner's 239
lettuce, water 297
leucas, Ceylon 205
leucas, ovate-leaf 205
Leuchterblume, Borstige 93
Leuchterblume, Knollen 93
Leuchterblume, Rauhhaarige 93
leunca 350
leucanthème 204
levant garlic 15
Levantelauch 17
levístico 205
li mung 107
lia 403
liane reglisse 2
liatride 206
liatris 206
libás 352
libó 386
Lichternuss 14
Lichtnelke, Breitblättrige 343
licorice common 169
licorice mint 11
licórize 169
Liebstock, Schottischer 206
Liebstöckel 205
Liebstöckel, Schwarzer 348
lien ch'ien 167
liên tu 45
lierre terrestre 167
lieve vrouwe bedstro 161
light galangal 31
light-blue snake 353
lignum vitae 170
ligustico 205
ligústico 205
ligusticum scholandskij 206
like felt butterbur 281
Lilafarbenes Wandelröschen 196
Lilie, Gleichfarbige 207
Lilie, Goldband 207
Lilie, Morgenstern 207
Lilie, Tiger 207
lilija odnocvetnaja 207
lilija zolotistaja 207

lily, butterfly 173
lily, devil 207
lily leek 21
lily, leopard 332
lily of Japan 207
lily, mountain 207
lily, resurrection 191
lily, spice 191
lily, tiger 207
lily, tropical 192
lima 95, 105
lima ácida 105
limão 59, 105, 107, 209, 211
limão de caiena 59
limão galego 107
Lima-Tee 95
limau amkian 105
limau jari 109
limau kapas 105
limau kiah 374
limau kingkit 374
limau nipis 105
limau pagar 159
limau purut 107
limau susu 108
limbe 105
limdo (mitho) 241
lime 12, 105, 107, 110, 159, 374
lime berry 374
lime, Chinese 374
lime, kaffir 107
lime, leech 107
lime, makrut 107
lime, Mexican 105
lime, myrtle 374
lime, Persian 105, 107
lime, Tahiti 107
lime, trifoliate 374
lime, West Indian 105
Limequat 159
limero 105
limetta 107
limette 105, 107
limette acide 105, 107
Limette, Breitblättrige 107
Limette, Kaffir 107
limette, Mexican 107
limette, Persian 107
Limette, Persische 107
Limette, saure 105
limettier 105, 107
limiestruik 116
limmetje 105
Limnophile, Indische 208
limoeiro azedo 107
limoeiro trifoliado 304
limon 61, 105, 107
limón 61, 105, 107, 129

limón ceuti 105
limón myrto 61
limona 107, 223
limonajamjata 223
limoncina 211
limoncito 374
Limone 61, 105, 107, 338
limone 61, 105, 107, 338
Limone, Chinesische 338
Limonelle 105
Limonenminze 228
limonero 107
limonier 107
limonnija 211
limonnik kitajskij 338
limpopo grass 143
limu 105
limum 107
lin 209, 279
lin li ye 279
lin oléifère 209
linaza 209
Lincoln's weed 137
line leaf leucas 205
linéaire requiéne 362
linear hoary pea 362
linear tephrosie 362
Lineare Aschenwicke 362
Lineare Tephrosie 362
linejtschatyj tefrosia 362
ling 222, 295
ling hsiana 222
lingkersap 337
linguie poivre 294
lino 209
linseed 209
lions-tooth 361
lippea, Mexican 210
Lippeakraut, Mexikanisches 210
lippie 211
lipstick tree 65
liquirizia 169
liquorice 2, 169, 170
liquorice, American 169
liquorice, Chinese 170
liquorice, Manchurian 170
liquorice, Roman 169
liquorice, Spanish 169
lirio blanco 185
lirio de florencia 185
lírio-do-brejo 173
lis asphodèle 175
lis d'un jour 175
lis de marais 5
lis doré du Japon 207
lis jaune 175
lisban 70
listy karí 241

litanda 364
litlit 294
little basil 257
little day lily 175
little ironwood 387
little mallow 216
liu lan xiang 229
livèche 205
liverwort 13
lixóchitl 383
ljadnik duschistyi 178
ljon kudrjasch 209
ljubistok 205
ljubm 205
ljuzepna 221
Llajhua 7
lobak merah 133
Lobelie, Hain 212
locoto 83, 84
locust bean 92, 273
locust bean, Westafrican 273
Locustbohne 273
Locustbohne, Afrikanische 273
Locustbohne, Westafrikanische 273
Löffelkraut, Dänisches 113
Löffelkraut, Echtes 113
Löffelkresse 113
lohmlauge 226
lokhat 57
loko-loko 181, 260
lo-kuei 63
lo-le 256
lolo 294
loloan 297
lolot pepper 292
Lolotpfeffer 292
Lombardy poplar 304
lombok 83
lombok besar 82
Loncho 213
long beaked rattlepod 123
long buchu 63
long cardamom 144
long kui 350
long pepper 292, 293, 294
long pepper, Javanese 294
long zedoary 128
longleafed sterculia 354
long-leaved pepper 292
longroot onion 27
long-rooted garlic 27
lontas 300
loom 107
loosestrife 214
loosestrife, gooseneck 214
Lopez fruit 369
Lopez-Frucht 369
Lorbeer 113, 198, 280, 347, 356, 379

Lorbeer, Chilenischer   198
Lorbeer, Gewürz   199
Lorbeer, Indischer   113, 356
Lorbeer, Kalifornischer(Berg-)   379
Lorbeerblatt   199, 356
Lorbeerblättriger Flachsbaum   45
Lorbeerkirsche   307
Lorbeer-Skimmie   347
loriguillo mezéreo   133
loschetschnik   113
loshno-akazija   318
loschnye perzy   395
losna   52
lou le   256
lou shen kui   177
louiza   223
loureiro   199, 307
loureiro cerejeira   307
louro inamuty   260
louro mamory   260
louro-cheirosa   136
louro-cravo   136
louro-cujumari   260
louro-puxuri   260
lovage   205, 206, 348
lovage, alpin   206
lovage, black   348
lovage, garden   205
lovage, northern   206
lovage, Scotch   206
love tree   93
love-in-a mist   253
Löwenbusch   325
Löwenzahn, Gemeiner   361
lu ping   297
lu'u   309
luan xi   300
lucuma   311
Luftzwiebel   23
lugowaja mjata   229
lugowoj kress   84
luk   15, 16, 17, 19, 20, 21, 22, 23, 24, 25, 26, 27, 144
luk altajskij   15
luk angol'skij   16
luk batun   19
luk dlinno-ostrokonetschji   21
luk dutschistyj   24
luk gorowoj   23
luk grawan   144
luk kitaskij   19
luk kosoj   22
luk krupnotytschinkovyj   20
luk kurrat   20
luk medweshij   27
luk mnogojarusnyj   23
luk mongol'skij   15
luk nemezkii   17

luk oschanina   22
luk pobednij   27
luk porej   23
luk reptschatyj   17
luk rokambol'   25
luk salatnji   20
luk schalot   17
luk shemtschushnuj   15
luk sibirskij   15, 27
luk starejuschij   26
luk stebel'tschatyj   26
luk tatarka   19
luk tscheremscha   27
luk vavilova   27
luk vinogradnji   15
luk, dikij   27
luk, medveshij   27
lumban   14
lumbang   14
luminet   153
lunao de caiena   59
lundoh   45
lundu   45
lungum   305
luni   306
luo de   257
luo fou   159
luo hua shoag   48
luo wang zi   359
luppulo   179
luppulo stroboli   179
lúpulo   179
lúpulo (común)   179
lusera   300
lutschiza   265
lutshitza abyknowennaja   265
luya   403
lyciet commun   213
lysimaque à feuilles de cléthra   214
ma bian cao   386
ma ch'ing   42
ma chi xian   305
ma khaam thet   298
ma liao   302
ma pien tsao   386
ma rum   240
ma tade   280
ma ti ye   75
ma tum   7
ma ying dan   196
maak mia   116
maak phu   116
mac hau   35
mac liu   309
Maca   376
maca   376, 202
macacaporanga   43
macassa   8

mace   2, 358, 244
mace, sweet   358
macela   4, 95
macela dourada   95
maceron   348
Machandel   189
maciá   244
macis   244
Macisbohne   238, 260
Macisbohne, Große   260
Macisbohne, Kalabassen   238
Macquarie pine   133
mada shaunda   244
Madaeira vine   44
Madagascar cardamom   8
Madagascar clove   313
Madagascar nutmeg   313
Madagaskar-Beilschmiedie   64
Madagaskar-Kardamom   8
Madagaskar-Muskat   313
madan mast   51
madanous   282, 283
madanous leefty   283
Mädchenhaarbaum   166
Madeira nut   188
Madeira vine   44
Madeira Wein   44
Mädesüß, Echtes   157
madhuka   169
madhurika   158
Madras apple   224
Madras thom   298
Madrasapfel   224
Madras-Melothrie   224
madre de cacao   311
madre de flecha   298
maeng lak   255
maeng lak kha   181
maesa, Indian   215
maeunkochunaengi   391
magalepka   308
Magelhanischer Zimt   140
Magelhanisian caneel   140
Magenkraut   91
Maggikraut   174, 205
maggiorana   263
maggiovana selvatica   265
Maggipflanze   205, 274
Maggipilz   195
magic guarri   151
magimavu   6
Magnolie, Blaugrüne   216
Magnolie, Sumpf   216
Magnolie, Virigianische   215
magrood   105
magua   376
mahaahong   173
mahád   57

Mahagoni, Afrikanisches   317
Mahagoni, Zitronen   32
mahakadaeng   1
mahalatsaka   293
mahaleb cherry   308
mahali kizhangu   134
mahalunga   108
Maha-Pengiri-Gras   131
mahaphala   108
mahlebi   308
mahlep   308
Mährrettich   50
mai zong   103
maïdanos   282
maidenhair tree   166
Maikraut   161
main de Bouddha   108
maina   221
Mairose   318
maja   7
majoram, Mexican   211
majoram, knotted   263
majoram, sweet   263
majoram, wild   266
Majoran   263, 264, 265, 266
majoran   263, 264, 265, 266
majoran sadowyj   263
Majoran, Arabischer   265
Majoran, Falscher Stauden   266
Majoran, Französischer   264
Majoran, Italienischer   264, 265
Majoran, Stauden   265, 266
Majoran, Syrischer   265
Majoran, Wilder   265, 266
mak   272, 356
mak siaty   272
makan   303, 129
Makassar-Muskat   243
makdunis   282
makhaam   359
makiron   263
makojcka   272
ma-kok   352
ma-konbu   195
makoy   350
makrut lime   107
makulan   289
mal'va lewsnaja   217
malaavaca   30
Malabar cardamom   38, 144
Malabar grass   129
Malabar nightshade   63
Malabar nutmeg   244, 245
Malabar orchid   64
Malabar Spinat   63
Malabar tamarind   162
Malabar-Bauhinia   64
Malabargras, Ostindisches   130

Malabar-Kardamom   38, 144
Malabar-Muskat   244
Malabar-Muskatnuss   245
Malabarskaja koriiza   104
malabarskij schpinat   63
Malabar-Tamarinde   162
Malabar-Zimt   104
Malabargras   130
Malabathri bark   104
malaboda   244
Malacca eagle wood   48
Malacca galangal   30
Malacca tree   284
Malacca-Adlerholz   48
malagosa   167
malagueta   10
malagueto   396
malaguetta da Guine   395
malaguetta preta   395
Malaguettapfeffer   395
malakaw   86
Malakka-Galgant   30
malar   400
Malayan kumquat   159
Malayische Mombinpflaume   352
Malayischer Affenohrring   49
Malayischer Fackelingwer   148
Malaysia gou-gi-zi   213
Malaysian ginger   118
Malaysian hog   352
Malaysian mombin plum   352
Malaysische Kumquat   159
Malaysischer Ingwer   118
malbaka   301
malbar hutan   181
malebo   308
malequeta   287
males   217, 343
malgoat   10
mali laa   188
mali son   188
maliwan   188
maliyau   140
mallawetskij   10
mallika   241
mallow, blue   216
mallow, castillan   217
mallow, common   216
mallow, curled   217
mallow, Egyptian   216
mallow, high   216
mallow, little   216
mallow, small   216
mallow, tall   216
malmai   85
maloko   284
malunggay   240
malungkoi   371

malva   216, 275, 300
malva común   217
malva comune   217
malva crespa   217
malva d'Egitto   276
malva rosa   275
malvaísco   295
malvarosa   276
malvarrubia   219
malva-santa   300
malvavisco   295
Malve, Afrikanische   177
Malve, Ägyptische   216
Malve, Gemeine   216
Malve, Kleinblütige   216
Malve, Quirl   217
Malve, Wilde   216
malya   373
malyj koren'   31
mama cadela   70
mamadi   217
mamão   86
ma-mao-wan   159
mamey   217
mamitsa   52
mammee   217
mammee-apple   217
mammey   217
Mammey   217
Mammey-Apfel   217
Mammi-Apfel   217
mammolo roseviole   388
mamoeiro   86
mamón   86
mamuang   218
man tang   212
manam   111
manao   105, 374
manao tet   374
Manchurian liquorice   170
mandalia   148
mandar   148
mandara   148
mandarin, golden   159
Mandarine   109
mandecravo   300
Mandel, Grüne   297
Mandel-Veronie   386
mandorla di pistacchio   297
mang tsao   183
manga   217
manga   217
mangang-kalabaw   217
mangas mempalam   217
mangericao-grande   259
mangerona   263
mangga   217
mangga golek   217

mangkamang   181
Mango   126, 128, 217, 352
mango   126, 128, 217, 352
mangô   217
mango ginger   126
mango gingerround zedoary   128
Mango, Duft   218
mango, kurwini   218
mango, saipan   218
mango, wild   352
Mango, Wilder   352
Mango-Ingwer   126
Mangopflaume   352
mangowoe   218
mangrana   309
mangrano   309
mangrove, red-flowered
   Pornupan   351
mangue   217
manguier   217
manguier mango   218
manguiera   218
manguiera mango   218
maní   48
mani guetta   10
maniguate   395
maniguette   10
Manila tamarind   298
Manila-Tamarinde   298
manjericão   256
manjericão-branco   255
manjui   85, 86
Mann'sche Muskatnuss   338
manna grass   131, 178
Mannagras   131
mannanatti   239
mannenrû   320
mannetjes nooten   244
mannga arum manis   217
Mannstreu, Stinkender   147
mano de zopilote   295
mano, Indian   217
manobi   48
manoranjitan   51
manshetka   14
mansur   26
manteau de Notre-Dame   14
manting   356
manting salam   356
manul   24
manupasupa   126
manzanilla   95, 357
manzanilla fina   95
manzanilla romana   95
mao dou   168
mao soi   46
mao suan jiang   285
maqdunis afranji   45

mar belaja   95
mar duschistaja   96
mar obyknowennaja   95
mar'ambrosiewidnaja   95
mara   235
maranga   240
marapoto   1
marari palm   249
marasme montagnard   218
marasmio oreade   218
marasmuius, garlic   218
maravilha   357
maravilla   75, 234
marbo cork tree   273
marcala do campo   4
marcha   82
mardakosh   265
mardqouche   263
Margerite, Wiesen   304
margose   235
mariadistel   344
maria-mole   278
Mariana caper   81
Mariannen-Kaper   81
marica   293
marichiphalam   82
Mariegold, Stinkendes   274
Marienblatt   360
Mariendistel   344
Mariengras, Duftendes   178
Mariengras, Duft   178
marigold   75, 99, 274, 357, 358, 359
marigold flower   75
marigold, African   99, 357
marigold, Aztec   357, 358
marigold, big   357
marigold, French   357, 358
marigold, Irish lace   357
marigold, Mexican   358
marigold, spreading   358
marigold, wild   358
Marille, Rauhfrüchtige   49
Marille, Schwarze   49
marjolaine   263, 265, 266, 210
marjolaine sauvage   265
marjolein   263, 265
marjolein, sauvage   265
marjolein, wilde   265
marjoram   366, 263, 264, 265, 266
marjoram, hardy   264
marjoram, knotted   263
marjoram, pot   264, 265
marjoram, Spanish   366
marjoram, sweet   263
marjoram, wild   265, 266, 366
marjoram, winter   265
marmara   275
marmel   7

marmelos   7
maro   363
Marokkanische Minze   229
marriout   219
Marrobio   219
marroio   219
marrow, castillan   217
marrube blanc   219
marrube vulgaire   219
marrubio   219
marrubio blanco   219
marrubio común   219
marsh beetle   377
marsh felwort   354
marsh marigold   75
marsh mint   224
marsh-clover   230
marsh tea   201
marsh-trefoil   230
marsin   247
Martineque-Ingwer   31
Martinisches Brandkraut   205
marua   263
maruba toki   206
marubaka   263
marumi   159
marumi-kumquat   159
Marumi-Kumquat   159
marupá   311, 344
marupaís   344
marupaúba   344
marvel   219, 234
marvel of Peru   234
maryamiya   329
marygold   65, 357, 359
marygold, saffron   357
marygold, signet   359
Maryland Diptam   125
Maryland dittany   125
maryuna   387
Marzipan-Salbei   327
März-Veilchen   388
mascha   148
mashua   376
masljtschnyj   171
massarubee   111
Massarubee   111
massette à feuilles étroites   377
Massoi   124
Massoy   124
massoy   124
mastic   296
mastic, Bombay   296
mastic, chios   296
master wort   283
mastic thyme   366
Mastix   296, 297
Mastix, Bombay   296

Mastix, Türkischer 297
Mastix-Pistazie 296
Mastixthymian 366
mastranzo 229, 375
mastranzo de las Indias 375
mastruçio do Perú 375
mastuerzo 62, 85, 203, 250
mastuerzo acuático 250
mastuerzo de huerta 203
mastuerzo de prado 85
mataasuea 239
mataat 137
mata-barata 344
matapulgas 228
mataro 396
materinka 265
Matico 289, 291
matico 289, 291
matico-falso 289
Matico-Pfeffer 289
matiko 291
matimati 290
Matopfeffer 396
matotschnik 223
matrimony vine 213
Mauerpfeffer, Felsen 340
Mauerpfeffer, Milder 340
Mauerpfeffer, Scharfer 339
Mauerpfeffer, Weißer 339
Maulbeere, Indische 239
Mauritius papeda 107
mauve à petites 216
mauve crépue 217
mauve d'Egypte 216
mauve des bois 216
mauve, grande 216
mauve sauvage 216
mauve verticillée 217
maxua 376
may cam chia 109
may chang 212
may chang picheng ch'ieb 212
may man 32
may que 100
mayao 14
mayblob 75
Mazaripalme 249
mazerdyschka 265
Mazis 244
mazis 244
mazok 254
mbili 77
mbonji 395
mbuhur mbéré 148
mbuyu 6
me 38, 195, 205, 268, 284, 298, 302, 359, 380
me chua 268, 359

me dat 268
me keo 298
me rùng 284
me te ba 38
mè tré 36
meadow bright 75
meadow cress 84
meadow fern 242
meadow sweet 157
mebōki 256
Mecca myrrh 115
mědang kěmangi 102, 103
medang serai 212
Mediasie, Großblättrige 220
medicinal rhubarb 315, 316
Mediterran sage 325
Medizinalrhabarber 316
medweshij luk 27
mee yoi 18
Meerbeifuß 54
Meeres-Jackbohne 77
Meerfenchel 120
Meerkohl, Baumförmiger 119
Meerkohl, Herzblättriger 119
Meer-Mertensie 231
Meerrettich 50, 178
Meerrettich, Chinesischer 178
Meerrettich, Japanischer 391
Meerrettich, Wilder russischer 50
Meerrettichbaum 240
Meersenf 73
Meersenf, Amerikanischer 73
Meersenf, Europäischer 73
Meerwermut 54
meharga 201
mei guo bo he 236
mei guo fang feng 274
mei li zhang ya cai 354
mei ren jiao ger 78
mei wen dai 354
Meiran 263
Meisterwurz(el) 283
meiwa 159
mejankeri 212
méjica zarzaparilla 347
mejorana 263, 366
mejorana silvestre 366
Mekka-Myrrhe 115
Mekong-Pfeffer 293
melaka 284
melambo 140
melantio 253
melanzana africana 349
melao de St. Caetano 235
melati 188
Melbourne boronia 68
Melde, Weiße 95
melegueta 9, 10

melegueta pepper 10
Melegueta-Pfeffer 10
Meleguetapfeffer, Bastard 9
melhem 308
mélilot blanc 221
mélilot bleu 372
mélilot jaune 222
mélilot officinal 222
mélilot, fleurie de 222
melilot, field 222
melilot, ripped 222
melilot, tall 221
melilot, white 221
melilot, yellow 222
meliloto amarillo 222
meliloto azzurro 373
meliloto blanco 221
meliloto giallo 222
meliloto gigante 221
meliloto gigantesco 221
meliloto oficinal 222
meliloto-branco 221
meliot, white 221
melisa 223
melisa toronjil 223
melissa 139, 223
melissa citroen 223
melissa, crete 223
melissa lekarstvennaja 223
melissa, small flowered 224
melissa turezkaja 139
Melisse 112, 139, 145, 223, 236,
mélisse 223
mélisse officinale 223
Melisse, Kleinblütige 224
Melisse, Kreta 223
Melisse, Rauhaarige 223
Melisse, Rote 236
Melisse, Türkische 139
Melisse, Vietnamesische 145
Melisse, Wilde 112
Melisse, Zitronen 223
melkij perez 82
mellet phong karee 158
melograno 309
melon de tropiques 86
melón zapote 86
melon, bitter 235
meloncella 332
melone tree 86
Melonenbaum 86
melor 188
Melothrie, Madras 224
memberah 352
měmpělasari 32
me-na-ri 205
mendarah 243
mengkuang duri 272

mengkuang laut   272
menhe de chat   251
menianto   230
menkudu nenghudu   239
Menschenfresser-Tomate   349
menta americana   209
menta crespa   229
menta de gato   251
menta gentile   229
menta japonesa   225, 226
menta piperita   227
menta poleggio   228
menta puleggia   228
menta pulezzo   228
menta romana   228, 229
menta salvadeca   225
menta silvestre   225
menta verde   229
mentaster   225
mentha crespa   229
mentha d'aqua   228
mentha de montagne   74
mentha di requien   228
mentha gentile   229
mentha romana   229
mentha verde   229
menthe anglais   227
menthe aquatique   224
menthe Brésil   224
menthe, crème de   228
menthe crépu   229
menthe d'Australie   307
menthe, Eau d' Cologne   227
menthe de Notre-Dame   229
menthe des champs   224
menthe des jardins   226
menthe die gatti   251
menthe douce   229
menthe en épi   229
menthe gabonaise   258
menthe Japon   224
menthe poivrée   227
menthe pouliot   228
menthe roronde feuillée   228
menthe verte   229
menthella   228
mentrasto   181
mentrasto-grande   181
mentuccio commune   74
mentuccio maggiore   74
ményanthe trifolié   230
meo   232
meolangolo   106
meranang   30
merasingi   171
mercanköşk   265
mercury, annual   230
mercury, French   230

mérédic   50
merica   293
merica putih   293
Merk, Wasserschierlingsblättriger   347
merlimau   12
merremia, dissected   231
merremia, emarginate   231
Merremie, Emporstrebende   231
Merremie, Kleindoldige   231
Merremie, Zerschlitzte   231
mersita   230
Mertensie, Meer   231
mertensija   231
merú   78
merzizou   223
Messerschmidie, Silber   231
methi   373, 374
methi ni bhaji   374
methika   374
meu   232
meum   232
meum atamanskij   232
meutastro   89
Mexican arnica   177
Mexican bergamot   235
Mexican bush oregano   301
Mexican coriander   147
Mexican earflower   129
Mexican giant hyssop   12
Mexican holly   377
Mexican lime   105
Mexican limette   117
Mexican lippia   210
Mexican marigold mint   358
Mexican marigold   358
Mexican mugwort   55
Mexican onion   20
Mexican oregano   210, 211
Mexican parsley   116
Mexican pimento   287
Mexican sage   210
Mexican sarsaparilla   347
Mexican tarragon   358
Mexican tea   95
Mexican Vanilla   383
Mexico Piment   287
Mexicon majoram   265
Mexicon marigold   358
Mexikanische Arnika   177
Mexikanische Bergamotte   235
Mexikanische Blasenkirsche   285
Mexikanische Duftnessel   12
Mexikanische Monarde   235
Mexikanische Sarsaparille   347
Mexikanische Schleifenblume   129
Mexikanische Studentenblume   358
Mexikanische Zwiebel   20
Mexikanischer Affenohrring   298

Mexikanischer Beifuß   55
Mexikanischer Blattpfeffer   289
Mexikanischer Oregano   211, 301
Mexikanischer Tee   95
Mexikanischer Wermut   54
Mexikanischer Ysop   12
Mexikanisches Lippiakraut   210
Mexikanisches Traubenkraut   95
mezereo camelea   133
mezereon   133
mezereón   133
mezzenga   400
mi sui lan   12
mi tieh hsiang   320
mi zan lau   12
mia do   118
michi-yanagi   302
micocoulier   91
micocoulier africain   91
micocoulier d'Afrique   91
microphylla oregano   264
midaumabaphang   153
midori-hakkada   229
mierikswortel   50
miglionet   314
mignonette   314
mignonette, common   314
mignonette, sweet   314
Mikanie, Kleinblütige   233
Milder Mauerpfeffer   340
milefólia   3
milenrama   3
milfoil   3
mil-folhas   3
milhojas   3
milk parsley   283
milk thistle   344
milk wood   32
milk-gowan   361
milkwort, downy   301
millefeuille   3
millepertius   180
millepertuis perforé   180
miltomato   285
milv   7
mimbataru   148
mimboo   105
mimbro   59
mimosa pourpre   273
mimosa, water   252
Mimose, Wasser   252
min   127, 140
minari   261
mindanao cinnamom   102
Mindanao-Zimt   102
mindau   147
miner's lettuce   239
ming aralie   303

ming tsai 230
minguella 317
minhonete 314
Minizitrone 159
minkamcho 169
miñoneta duschistaja 314
mint bush 145, 307
mint bush, Australian 307
mint bush, round-leafed 307
mint flakes 227
mint geranium 277, 360
mint shrub 145
mint, apple 224, 226, 228, 229, 230
mint, asiatic 226
mint, bergamot 227, 229
mint, biblical 226
mint, bowle 230
mint, brandy 227
mint, bush 181
mint, Cambodian 280
mint, Chinese 225
mint, comb 227
mint, corn 224, 225
mint, Corsican 228
mint, coyote 237
mint, crisp 229
mint, dog 112
mint, eau de Cologne 227
mint, Ecuadorian 234
mint, English 228
mint, false apple 228
mint, field 224
mint, garden 225, 229
mint, ginger 226
mint, green 229
mint, horse 224, 226
mint, Indian 299, 334
mint, Japanese 223
mint, lamb 229
mint, lemon 223, 227, 236
mint, marsh 224
mint, mountain 233
mint, orange 227
mint, pineapple 229
mint, red 226, 229
mint, round-leaved 228
mint, Sakhalin 228
mint, silver 226
mint, soup 299
mint, Scotch 226
mint, Spanish 228
mint, stone 125, 126
mint, tule 224
mint, water 224
mint, wild 181, 224
mint, wooly 230
mintweed 210
Minz-Agastache 12

Minzartiger Rainfarn 360
Minzblättriger Basilikum 258
Minze, Ährige 229
Minze, Apfel 224, 226, 228, 229
Minze, Ananas 229
Minze, Argentinische 234
Minze, Australische 307
Minze, Bowlen 230
Minze, Breitblättrige 224, 230
Minze, Chinesische 225, 226
Minze, Ecuadorische 234
Minze, Englische 227
Minze, Fuchsschwanz 230
Minze, Gewürz 234
Minze, Grüne 229
Minze, Herzblättrige 225
Minze, Japanische 224, 225, 226
Minze, Kamm 227
Minze, Kölnisch-Wasser 228
Minze, Koreanische 12
Minze, Korsische 228
Minze, Krause 225
Minze, Kriechende 228
Minze, Langblättrige 226
Minze, Marokkanische 229
Minze, Österreichische 224
Minze, Rote 229
Minze, Rundblättrige 228
Minze, Spanische 228
Minze, Süße 229
Minze, Wasser 228
Minze, Weiche 234
Minze, Zottige 224
Minzpelargonie 277
Minzstrauch, Australischer 307
mioga 403
mioga ginger 403
miracle fruit 171
miri 14
miroksilon 246
mirra 115
mirride (delle Alpi) 247
mirride odorota 247
mirris 247
mirt bolotnyj 242
mirte 247
mirto 61, 242, 247
mirto bastarda 242
mirto dal ptofumi di limone 61
mirto de brabante 242
misibcoc 377
misreya 158
miswak 325
mitha neem 241
mito holandés 242
mitsuba 47, 125
mitzuba 125
mizu garashi 250

mizunjuske 263
mizu-ôba-ko 267
mjata egipetckaja 228
mjata jablotschnaja 228
mjata kruglolistaja 228
mjata kudrjawaja 229
mjata pereznaja 227
mjata polevaja 225
mjata prjanaja 145
mjata wodjanoja 224
mkak préi 352
mnogoletnij kerbel' 247
mnogonoshka 303
mo kuei 100
mo li hua 188
mo yao 115, 247
mobola plum 273
mock lime 12
modang sanggar 103
modjo 7
mofassa 329
mofifi 315
mogheira 167
moghra 188
mogil'nik 275
mogiyongauk 217
mohari 69
Mohn 272
Mohnblume 272
Möhre 133
Mohrenpfeffer 395
Mohrenpfeffer, Gestreifter 396, 397
Mohrenpfeffer, Großblütiger 396
Mohrenpfeffer, Halbstrauchiger 396
Mohrenpfeffer, Seidiger 397
Mohrenpfeffer, Spitzblütiger 395
Mohren-Salbei 326
Mohrrübe 133
moinson 71
mokah 235
mokak 352
mokhyang 184
Moldavian balm 139
Moldavique, dracocéphale 139
Moldawischer Drachenkopf 139
Moldawischer Drachenpfeffer 139
mole plant 152
moleé de jardins 337
molène 385
molène faux-phlomis 385
molle 337
mollé 337
Mollefrucht 337
Mollesaat 337
molotschaj-colnzerljad 152
Molucces nutmeg 245
Molukken-Muska 245
moly 21

Molyzwiebel 21
Mombinpflaume, Malayische 352
mombin plum, Malaysian 352
mombin, yellow 352
Mombinpflaume, Saure 352
momordika 235
monajembe 338
monarda, dotted 237
monarde 235
monarde fistuleuse 236
Monarde, Getüpfelte Kamm 237
Monarde, Mexikanische 235
Monarde, Punktierte 237
Monarde, Röhrenblütige 236
Monarde, Rosen 236
Monarde, Scharlach 236
Monarde, Wilde 236
Monarde, Zitronen 236
monardella, Pacific 237
Mönchspfeffer 388
Mönchspfeffer, Chinesischer 388
móng bò hoa tíma 64
mongol'skij luk 15
mongphali 48
monk's basil 256
monkey 57, 351
monkey bread tree 6
monkey jack 57
monkey nut 48
monkey's potato 351
Monnier's Brennndolde 112
montain rue 323
montbretia, garden 121
Montbretie, Garten 121
monzija 239
moochy wood 148
moon daisy 204
Moor-Gagelstrauch 242
mora de la India 239
morang elaichi 35
moras 387
morech ansai 292
môrech ansai 295
morella 350
morelle noire 350
morène, Africain 180
Morgenstern-Lilie 207
Morinda, Zitronenblättrige 239
morinde 239
moringa 240
moringa ailel 240
moringu 240
morkov' kul'turnaja 133
morning star lily 207
morroubia 219
morskaja rosa 320
morskoj calat 113
morunggai 240

morupapiri 317
moscadeira 244
moscado do Brasil 124
Moschuskraut 363
Moschus-Reiherschnabel 146
Moschusrose 319
Moschus-Schafgarbe 3
moshshewel'nik 189
moshshewel'nik obyknowennyj 189
moshshucha 189
mosquiteta blanca 319
mosquito plant 173, 276, 260
mosquito plant of South Africa 260
mossy stonecrop 339
mostarda blanca 345
mostarda preta 70
Mostardkorn 345
mostaza blanca 345
mostaza de Indias 69
mostaza de los campos 345
mostaza negra 70
mostaza silvstre 345
mosterd 69, 345
motapati 204
motéveldodé 37
mother heart 81
mother of thyme 4
mother-in-law's tongue 332
mother-of-thyme 368
motherwort 56
Mottenstrauch 299
mount pima oregano 236
mountain arnica 51
mountain dittany 125
mountain hollyhock 391
mountain leek 26
mountain lily 207
mountain mint 233, 310
mountain mint, hairy 310
mountain mint, Virgianan 310
mountain nutmeg 244
mountain pennyroyal 237
mountain pepper 140, 212
mountain reneamia 314
mountain rue 275
mountain seringa 192
mountain spice (of Sierra Leone) 395
mountain tea 163
mountain thyme 365
mountain tobacco 51
mountain wormwood 56
mourkeba 230
mouse-ear chickweed 92
moussanga-voulou 118
mousseron d'automne 218
Mousseron, Echter 218
mousseron, faux 218
Mousseron, Herbst 218

mousseron, true 218
moustarda dos campos 345
moutarde à feuilles 69
moutarde blanche 345
moutarde de Chine 69
moutarde de Sarepte 69
moutarde des champs 345
moutarde jaune 345
moutarde noire 70
moutarde sauvage 345
mouzambe 111
mowi 281
moxa 57
Moxakraut, Indisches 54
mozambé 111
moztaza 321
mpapai 86
mpojang 404
mrèah prèu 260
mrican 290
mu cua 32
mu hsi 266
mu ts'ung 19
mu xiang 145, 336
mu xiang ru 145
mu yao 115
muchriara 207
mufira 179
mugrela 253
muguet des bois 161
mugwort 52, 54, 56
mugwort, Indian 54
mugwort, Japanese 55
mugwort, Mexican 55
mugwort, white 54
muimapagé 137
mulberry, Indian 230
mulhatti 169
mullein, clasping 385
mullein, common 385
mullein, large flowered 385
mullein nightsolade 349
mullein, orange 385
mulli 337
mullillam 401
muña 234
muncang 14
mundu 162
mung phali 48
munganari 207
mungfi 288
mungsi 371
munjariki 256
muran 12
murasaki soshin ka 64
muraseaki-katrabami 268
mûrier des Indes 239
murikku 148

muringueiro   240
murinkai   240
murinna   240
murwa   263
musambi   109
musan ô maru bushu-kan   108
muscade de Calabash   238
muscade de Madagascar   313
muscadier à suif   193
muscadier commun   244
muscadier d'Afrique   238
muscadier de calabasse   238
muscadier de forêt ou montagne   244
muscadier des Moluques   245
muscat sage   330
Muschelgalgant   29
mushk dana   1
mushroom, fairy ring   218
mushroom, garlic   218
musk clover   146
musk lime   159
musk mallow   1
musk milfoil   3
musk okra   1
musk rose   319
musk yarrow   3
Muskat, Anis   313
Muskat, Batjan   245
Muskat, Bombay   245
Muskat, Chilenischer   198
Muskat, Filziger   244
Muskat, Großblütiger   243
Muskat, Halmaheira   245
Muskat, Immergrüner   198
Muskat, Kastanienblättrige   243
Muskat, Langer   243
Muskat, Madagaskar   313
Muskat, Makassar   243
Muskat, Molukken   245
Muskat, Otoba   245
Muskat, Papua   243
Muskat, Pferde   243
Muskat, Silber   243
Muskat, Unechter   244
Muskat, Wilder   114, 243
Muskat, Womersleyie   246
Muskat, Zimt   243
Muskat, Gefingerter   244
Muskatbaum, Dickblättriger   193
Muskatblüte   244
Muskateller-Eibisch   1
Muskateller-Salbei   330
muskatnij orech   244
Muskatnuss   6, 124, 238, 244, 338, 370
Muskatnuss, Afrikanische   238, 239
Muskatnuss, Amerikanische   124
Muskatnuss, Angolanische   310
Muskatnuss, Bombay   245

Muskatnuss, Brasilianische   124
Muskatnuss, Duftende   244
Muskatnuss, Echte   244
Muskatnuss, Florida   370
Muskatnuss, Guayana   6
Muskatnuss, Halmaheira   245
Muskatnuss, Kalifornische   370
Muskatnuss, Kalabassen   237, 238
Muskatnuss, Malabar   244, 245
Muskatnuss, Mann'sche   338
Muskatnuss, Ochoco   339
muskátový kvet   244
muskátový orech   244
Muskat-Pelargonie   276
musky Stork's bill   146
musodo   317
musque   244, 372
mussambé cor de rosa   111
mussambé de cinco folhas   111
mussanga-vulu   118
mussel galanga   29
Mussin's Katzenminze   252
mustard   81, 111, 345
mustard green   69
mustard, Abessinian   69
mustard, African   111
mustard, bastard   111
mustard, black   70
mustard, brown   69
mustard, Chinese   69
mustard, farmer   81
mustard, field   345
mustard, Holland   345
mustard, Indian   69
mustard, Sarepta   69
mustard, true   70
mustard, white   345
mustard, wild   345
mustard, yellow   345
mustasa   70, 345
Muster-John-Henry   358
mustik   296
mutasia   110
Mutterharz   156
Mutterkraut   227
Mutterkümmel   125
Mutterwurz, Alpen   206
Mutterwurz, Schottische   206
Mutterwurz, Zarte   43
Mutterzimt   100
muyonza   86
myouga   403
myrobalan d'Egypt   61
myrobalan, emblic   284
Myrobalane   61, 284
myrobalane, Egyptian   61
myrrh   115, 247
myrrh, African   115

myrrh, Arabian   115
myrrh, common   115
myrrh, herabol   115
myrrh, Mecca   115
myrrh, Somali   115
myrrhe   115
Myrrhe, Arabische   115
Myrrhe, Echte   115
Myrrhe, Fadhi   115
Myrrhe, Herabol   115
Myrrhe, Mekka   115
Myrrhe, Somalia   115
Myrrhe, Süße   115
Myrrhenkerbel   247
myrris   247
myrta   247
myrte   61, 242, 247, 379
myrte bâtard   242
myrte citronée   61
myrte comun   247
myrte des Marais   242
Myrte, Echte   247
Myrte, Oregon   379
Myrtenblättrige Pomeranze   109
Myrtenkörner   247
myrtle   5, 61, 109, 152, 241, 247
374, 379
myrtle, bog   242
myrtle flag   5
myrtle grass   5
myrtle lime   374
myrtle, Chinese   241
myrtle, Oregon   379
myrtle, spurge   152
myrtle, sweet   5, 242
myrtle-leaf orange   109
na 'an   227
na s'tuka   134
nabati   20
Nabelkraut   97
Nabelorange   109
Nachtschatten, Äthiopischer   348
Nachtschatten, Gelber   350
Nachtschatten, Gefüllter   349
Nachtschatten, Rauher   349
Nachtschatten, Schraubenförmer   350
Nachtschatten, Schwarzer   350
Nachtschatten, Spiraliger   350
Nachtschatten, Ungewöhnlicher   348
Nachtschatten, Virginischer   350
Nachtschatten, Wollblütiger   349
nadgh   334
naeng-i   81
nag kesar   255
naga jihva   176
nagaba-gishigishi   321
nagadamani   57
nagagolunga   241

nagami   159
naga-reishi   235
nagdauna   57
Nägelein   355
Nägeleinnuss   313
Nagelschwamm   218
naginata-koju   145
nagu kinkan   159
nail of clove   355
naksh souak   188
nalada   387
nam hoàng liên   157
nam nam   131
nan chiao   400
nan mu su   221
nanban saikachi   88
nanban-fuji   362
nane   227
naples garlic   21
nara wastu   387
nara setu   387
narangi   109
narangita   109
naranja (dulce)   109
naranja mirtifolia   109
naranjilla   159
Naranjille   120
naranjo   106, 304
naranjo agrio   106
naranjo amargo   106
naranjo trébel   304
nard grass   130
nard, Celtic   381
nard, Indian   249
nard, Syrian   382
Narde   200, 249, 381
Narde, Indische   249
Narde, Keltische   381
Narde, Syrische   382
Nardenähre   249
nardostachyde de l'Inde   249
nar-kachura   126
narra   309
narenddj   105
narrow dock   321
nasamba   310
nashou   402
nastojastschaja gortschiza   70
nastuerzu ortense   203
nasturcio   250, 375
nasturtium   375, 376
nasturtium, bulbous   376
nasturtium, bush   376
nasturtium, (common)   375
nasturtium, dwarf   376
nasturtium, tall   375
nasturtium, tuberous   376
nasturzija   375

nasturzio comune   375
nasturzio tuberoso   376
nasutachûmu   375
native eggplant   349
native myrtle   61
natna   95
Natternknoblauch   23, 24
navel orange   109
nayibullal   176
nayvila   176
Neapel-Zwiebel   21
Neapolitan garlic   21
nebu   106
Negerpfeffer   395
negi   19, 23
negro pepper   395, 397
neguilla   254
nehar   347
Nelke, Chinesische   135
Nelke, Kaiser   135
Nelken   285, 355
Nelken–Basilikum   258
Nelkennuss   313
Nelkenpfeffer   75, 286
Nelkenpfeffer, Großblütiger   287
Nelkenschwindling   218
Nelkenwurzel, Bach   166
Nelkenwurzel, Berg   165
Nelkenwurzel, Echte   166
Nelkenwurzel, Sibirische   114
Nelkenwurzel, Urban   166
Nelkenzimt   136
Nelkenzimtbaum   333
nemdar na'na'   229
nemezkaja, romaschka   39
nemzkaja mjata   229
nemezkii, luk   17
Neopolitanische Zwiebel   21
Nepal cardamom   35, 37
nepali dhaniya   401
nepali dhaniyo   399
Nepal-Kardamom   35, 37
ne-paserii   283
nepitella   74
nere sun   273
neseri   319
netige   273
nettle, devil   197
nettle, fever   197
nettle, hedge   352
nettle, onion   14
Neugewürz   286
New Guinea salt fern   58
New Guinea Salzfarn   58
nga truat   67, 128
ngaa   342
ngai diep   57
ngai num kho   67

ngalug   305
ngama   278
ngamling   198
ngaulo   352
nghe   122, 127
nghé   280
nghe den   128
ngo   66, 116, 147
ngo tay   147
ngoc ian   232
ngoi   349
ngun phung khao   110
nguru tat nam   4
nguza   278
nha ngai   57
nhau   239
nhau rung   239
ni mei guo bo he   236
niam om   97
nibu   106
Nickender Ingwer   404
Nickender Lauch   22
Niedere Rebhuhnbeere   163
Niederliegende Samtpflanze   172
Niederliegende Scheinbeere   163
nielle bâtarde   252
nigelia cultivada   254
nigella   253, 254
nigelle   253
nigelle de Crète   253
nigelle de Damas   253
Nigersaat   171
niger seed   171
night morella   350
nightshade, black   350
nightshade, coiled-flower   350
nightshade, common   350
nightshade, garden   350
nightshade, Malabar   63
nightshade, mullein   349
nightshade, raw   349
nightshade, screw-shaped   350
nightshade, spiral   350
nijaki   285
niki-me   380
nikkai   103
nikuzuki   244
nilam buket   301
nilam wangi   301
nilapuspa   388
Nile cabbage   297
Nilgiri pepper   295
Nilgiri-Pfeffer   295
nilum   57
nimbu   107, 241
nimikitan   4
nimma   105
niñarupá   28

ninfa   218
ning mêng   105, 107
ning xia gouqi   213
ninjin   133
ninniku   24, 27
nioi-niga- kusa   181
nira   23, 24, 27
nira-negi   23
niri   273
nisa   65, 126
nishi-yomogo   57
Nitta, Farnblättriger   273
Nittabaum   273
niu zhi   265
niwjanik abyknowennyj   204
njassi   213
no-biru   20
noble chamomile   95
noble laurel   199
noble milfoil   4
noble yarrow   4
noce comune   188
noce moscata   244
noce nostrana   188
noce persiana   188
nocuana   59
nodding ginger   404
nodding onion   18
nofua   66
nogal americano   188
nogal común   188
nogal inglís   188
nogal moscado   244
no-geito   90
nogotki   75
nogueira comun   188
nogueira preta   188
no-ibara   320
noinjangdae   302
noir galanga   30
noir prun   315
Noisetterose   320
noix de Bancoul   14
noix de muscade   238, 244
noix de Ravensara   313
noix de terre   71
noix des Indes   14
noix des Moluques   14
noix girofle   313
noix muscade (capsule della)   244
noix muscade à laquelle   245
noix muscade macassar   243
noix muscade mâle   243
noix muscade sauvage   243
noix papoue   243
nokogiro-sô-zoku   3
nolaiali   45
nom wa   295

Noni   239
noni   239
nono   239
nootmuskaat   244
norangjontongssari   222
noronha   404
northern bayberry   242
northern lovage   206
nose bleed   3
noukha   34
nové korenie   287
noyer commun   188
noyer du Japon   166
noyer noir   188
noyer noir d'Amerique   188
noz da Índia   14
nôzen haren   375
noz-moscada-do-brasil   124
nubti   92
nuez moscada   244
nug   171
nug masljtschnyj   171
nultomato   285
nunga   240
nunnery root   176
Nusseibe, Florida   370
Nusseibe, Kalifornische   370
nut, Gaboon   119
nut, Madeira   188
nut, Persian   188
nutmeg   6, 9, 58, 114, 124, 198, 238,
    243, 253, 276, 310, 313, 370
nutmeg cardamom   9
nutmeg flower   253
nutmeg geranium   276
nutmeg scented geranium   276
nutmeg, African   238, 239, 310
nutmeg, Akawai   6
nutmeg, (South) American   124
nutmeg, Angolian   310
nutmeg, anise   313
nutmeg, Banda   244
nutmeg, Bombay   245
nutmeg, Brasilian   124
nutmeg, calabash   238
nutmeg, California   370
nutmeg, chestnut-leafy   243
nutmeg, Chilean   198
nutmeg, cinnamom   243
nutmeg, clove   313
nutmeg, false   238, 310
nutmeg, Florida   370
nutmeg, Jamaica   238
nutmeg, Madagascar   313
nutmeg, Malabar   244
nutmeg, Moluccan   245
nutmeg, mountain   244
nutmeg, plum   58

nutmeg, silver   243
nutmeg, true   244
nutmeg, wild   114
nutmeg, Wormersleyi   246
nutmeg, yellow flowery   238
nyaix   77
nyiring   273
nymphe des montagnes   218
o ke trade   302
oanla   284
ob cheuy   100
oblique onion   22
obyknownnyj   368
ochoco (nut)meg   339
Ochoco-(Muskat)nuss   338
ocotea   261
Odermennig, Duft   13
Odermennig, Großer   13
odiandro   71
odjom   8
odour agrimony   13
odour screwpine   272
oduwantschik lekarstvennyj   361
oeilete bleue   272
oeilette   272
oeille   299
oeillet â bouquet   135
oeillet de Chine   135
oeillet de fleuristes   135
oeillet giroflée   135
oeillt d'Inde   358
oignon   19, 23, 25
oignon catawissa   23
oignon d'Egypte   23
oignon d'Espagne   25
oignon d'hiver   19
oignon sous terre   17
oil of Boronia   68
oil palm, African   143
oiseille de belleville   320
Ojokbaum   317
oka-zeri   112
okkez sidi moussa   331
okolo   395
okpha   17
oku   260
Okwa   371
olaia   93
Oland   184
olasiman ihalas   278
old man storenwood   52
Öl-Duftsiegel   347
olibanum tree   69
olim   338
olive, fragant   266
olive, sweet   266
olive, tea   266
Oliver's bark   103

Oliver-Brandkraut   284
Oliver-Strauchnessel   284
Öllein   209
ollukonggongkhwi   344
olmaria   157
Ölmoringie   240
Ölpalme, Afrikanische   143
Öl-Perilla   278
Ölrauke   147
om kop   208
oman pravý   184
omaznik javanskij   261
ombrella jaune   348
omechuai caspí   396
omijanamu   338
omum (plant)   371
oncir   304
ondo   58
ondoko   58
oni-kanzo   175
onion Catawissa   23
onion nettle   14
onion, African   16
onion, Altai   15
onion, American wild   16
onion, Beltsville bunching   23
onion, bulb   17
onion, Canada   16
onion, Chinese   18
onion, Egyptian   23
onion, ever-ready   18
onion, flowering   21
onion, garden   17
onion, giant   19
onion, gipsy   27
onion, Japanese bunching   19
onion, ledebour   20
onion, longroot   27
onion, Mexican   20
onion, nodding   18
onion, oblique   22
onion, Oriental   18
onion, pearl   23
onion, potato   17
onion, silverskin   19
onion, Spanish   17
onion, spring   19
onion, Wakegi   23
onion, Welsh   19
onion, wild   18, 28
onion, yellow   20
oni-udo   42
oni-yuri   207
onkom   66
onontomulo   176
Oostindische kers   376
op choei   100, 102
openok lugovoj   218

opium poppy   272
Opoponax-Harz   115
oranda garashi   250
oranda mutsuba   46
orang chempaka   232
Orange   105, 109, 118, 159, 175, 227, 241, 304, 359, 369, 385
orange (coloured) day lily   175
orange (douce)   109
orange amère   105, 304
orange cosmos   118
orange gem   359
orange jasmin   241
orange mint   227
orange mullein   385
orange needles   175
orange, bitter   105, 304
orange, blood   109
orange, China   159
Orange, Hawaiian   241
Orange, Japanische   159
orange, navel   109
orange, Seville   105
orange, sour   105
Orange, süße   109
orange, sweet   109
orange, trifoliate   304
orange, trifoliôle   304
orange, Valencia   109
Orangefarbene Taglilie   175
Orangenbaum, Langdorniger   107
Orangenkraut   265
Orangenminze   228
Orangen-Pelargonie   276
Orangenraute   241
oranger doux   109
oranger du Malabar   7
orangio dolce   109
oranjeappel   105
orchid tree   12, 64
orchid, Malabar   64
Orchidee, Perlen   97
orech grezkij   188
orech tschernij   188
Oregano   209, 236, 264, 299, 301, 365, 367
oregano   209, 236, 264, 299, 301, 365, 367
orégano   147, 181, 197, 210, 265, 299
orégano brujo   299
orégano cimarron   181, 210
orégano de Cartagena   147
orégano de España   299
oregano de la sierra   236
orégano del pais   211
orégano Frances   299
oregano, American   210
Oregano, Amerikanischer   210

Oregano, Berlandier   210
Oregano, Biblischer   265
oregano, Cretan   264
oregano, Greek   266
oregano, Italian   264, 266, 367
oregano, Kirgisischer   265
oregano, Mexican   210, 211
Oregano, Mexikanischer   211, 301
oregano, microphylla   264
oregano, mount pima   236
oregano, Pizza   266
oregano, Puerto Rico   211
oregano, Russian   265
Oregano, Schmalblättriger   264
oregano, small-leaved   264
Oregano, Spanischer   365
Oregano, Süßer   210
oregano, Syrian   265
oregano, Turkestan   265
oregano, Turkish   264
Oregano, Vielblütiger   211
Oregano, Weißer   209
oregano, white   299
oregano, wild   209, 265
Oregon Ahorn   2
Oregon maple   2
Oregon myrtle   379
Oregon tea   334
oregon, Mexicon   20
Oregon-Myrte   379
oreille d'home   57
oriental caper   80
oriental celery   261
oriental garlic   27
Oriental onion   18
oriental sesam   342
Orientalische Kaper   80
Orientalischer Knöterich   302
Orientalischer Sesam   342
origan   210, 263, 266
origan à petites feuilles   264
origan commun   265
origan d'Egypte   265
origan de chypre   264
origan marjolaine   210
origan vulgaire   265
origano comune   265
origano siciliano   264
orignon sous terre   17
Oriniko cinnamom   43
Orinoko-Zimt   43
Orleansamen   65
ormio   330
ormusch   286
Oroko   361
orozuz   169, 210
orpin   339
orpin à feuilles glanduleuses   340

orpin âcre   339
orpin bâtard   340
orpin de Boulogne   340
orpin jaune   340
orpin réfléchie   340
orris   185
orris root   185
oruga del jardin   62
orvale   330
osaengch'o   179
Oschanin-Zwiebel   22
oseille acide   320
oseille commune   320
oseille d'Amérique   322
oseille de bûcheron   268
oseille de Guinée   177
oshb el-maa almodala   261
oshiroi-bana   234
osielle aux feuilles ròndes   322
osielle bouclier   322
osielle ronde   322
Osmanthusblüten   266
osoko   339
osokor'   304
Ostafrikanischer Kardamom   10
Osterluzeiblättrige Stechwinde   347
Österreichische Minze   224
Österreichischer Thymian   367
Ostindischer Basilikum   258
Ostindischer Rhabarber   315
Ostindisches Lemongras   130
Östlicher Knöterich   302
osto   360
ostropëstro   344
Ostruz   283
ostschetinennij molotschaj   152
osun   348
Oswega beebalm   236
Oswega tea   236
Otoba-Muskat   245
Otostegie, Strauch   267
otschitok   339
otschitok edkij   339
otschitok otognutyi   340
otschnika aptetschnaja   153
Ottelie, Froschlöffelähnliche   267
Ottwurzel   184
ou bo he   226
ou huo xue dan   167
ou zhou mo yao   247
ouard   319
oud kameira   48
our lady's thistle   344
outenga   136
oval buchu   62
oval kumquat   159
Ovale Kumquat   159
ovang   303

ovate-leaf leucas   205
ox eye daisy   204
oxadile blanca   268
oxadille blanche   268
oxalide petite oseille   268
oyster nuts   361
oyster plant   231
pa chio ksiang   183
pa dooc   38
pa lang   198
paasunitpu   274
pacar keling   66
Pacharispflanze   152
pachinhos   396
pacholi   301
pachukuk   359
pachuli   301
pachuri   301
Pacific berry   33
Pacific maple   2
Pacific monardella   237
pacing   118
padaganghwal   41
Padang cassia   100
Padang-Zimt   100
padudugnamul   125
paechohyang   12
paekriyhang   368
pahari buch   5
pahari pudina   229
pahinro kirap   80
pai chieh   345
pai tou k'ou   36, 144
paico   95
pai-kuo   166
pain d'oiseau   339
pain de coucou   268
pain de singe   6
painkiller, Indian   239
paintbrush, Flora's   146
painted sage   330
paitan   152
pakchi farang   147
pakha   226
pakis gemblung   309
pakis gila   309
pako   217, 355
pakochilam   301
paku geulis   309
pala   176, 243
pala bukit   243
pala laki-laki   244
pala maba   245
pala utan   244
palafuker   244
palandu   17
Palästinenser Beifuß   54
pale catechu   379

paleino odoroso   44
paléo odorosa   44
pal-ha   30
palillo   148
pali-mari   32
palla langkuas   30
palm lily   116
palm, marari   249
palm, tall   143
palmarosa gras   130
Palmarosagras   130
palmata butterbur   281
palmeira-iará   202
palmier à huile   143
palmira alstonia   32
palo amerillo   28
palo de burro   79
paloondo   197
palosanto   170
paminta   199, 293
paminta dahon   199
pamintaliso   293
pamir   220, 279
Pamir sage   279
Pamir-Salbei   279
pamponng   261
pampung   261
pan y agua   79
panacée des Montagnes   51
panae-wo-nging   271
panais   274
panajachel   123
panasa   34
panasan   34
pancal kidang   12
pandan   271
pandan dagat   272
pandan mabango   271
pandan pantai   272
pandan rampai   271
pandan rampe   271
pandan wangi   271
Pandanane, Amarillisblättrige   271
Pandanane, Duft   272
pandanus leaf   272
pandara   241
Pandrón-pepper   82
pane   34
pane di santo Giovanni   93
panetro khafkhader   80
pangaphul   12
pangphung   333
pani sag   250
panicaut fétide   147
panicle stonecrop   340
panila   383
panilajak   252
paniquesillo   81

panirak   216
panjangusht   388
Pannonischer Thymian   367
pansy, beddy   387
pansy, horned   387
pansy, tufted   387
pântano   272
páo rosa   43
papaia   85, 86
papalo   305
papaloquelite   305
papanoe   385
papaver   272
papavero   272
papavero domestico   272
papavero indiano   272
papaya   86
Papaya   86
papayer   86
papayero   86
papayo   86
papeda, Mauritius   107
papeeta   86
papita   86
papjja   86
Papo Canary tree   77
papotschnaja trawa   223
papoula mak   272
Pappel, Balsam   304
Pappel, Pyramiden   304
Pappel, Schwarz   304
paprika   82, 83
paprika čili   83
Paprika, Büschel   82
Paprika, Chinesische   83
Paprika, Einjährige   82
Paprika, Filziger   84
papua   149, 243, 303
Papua-Muskat   243
para cress   351
Paradieskörner   9, 10
parahuln   36
paraíba   344
Parakresse   351
paranti   186
paratudo   140
paravati padi   85
pareíso   240
pareíso blanco   240
paribhadra   148
parifollo   45
parijata   148
parinaire   273
pariparoba   295
parnasa   260
parnolistnik   405
parra de Madeira   44
parsley   45, 99, 116, 124, 176, 261, 282, 346, 348
parsley, Alexandrian   348
parsley, Chinese   116
parsley, cow   45
parsley, Hamburg   283
parsley, Japanese   125
parsley, Mexican   116
parsley, milk   283
parsley, water   261
parsley, wild   99
parsley, wild stone   124
parsnip   41, 94, 143, 274, 347
parsnip chervil   94
parsnip, prickly   143
parsnip, water   347
parsnip, wild   41
paru   23
Pasaniapilz   202
pascuas   152
paserii   282, 283
pashitnik goluboj   373
pasionara   254
pasli   282
passerage cultivée   203
passerage drave   85
pastenaak   274
pastenaque   274
pasternak   274
pasternak posevnoj   274
pastinaca   274
Pastinak   274
pasto cedrón   129
pasto de camello   131
pasto de Malabar   130
pasto limón   129
pastora   327, 358, 377
pastorcita   358
pastriciani   274
pasture brake   308
pastuschija sumka   81
pa-tao   273
patae d'Amériqua   44
patai-butu   295
patal tumbari   93
Patchouli   301
patchouli de cablin   301
Patchouli, Heyne   301
patchouli, Indian   301
Patchouli, Indisches   301
Patchouli, Javanisches   301
patchouli, Javian   301
Paternostererbse   2
Paternosterstrauch   353
pathorchur   299
patience crépue   321
patience écousson   322
patience suvage   321
patminari   46
patul parolan   85
patwa   177
pau   5, 120, 136, 140, 217, 344, 386, 388, 396
pau de angola   388
pau de embira   396
pau fede   386
pau-cravo   136
pau-d'alho   120
paude-praga   205
pau-paraíba   344
pau-para-tudo   140
pavot   272
pavot à opium   272
pavot blanc   272
pavot des jardins   272
paw-paw   86
payagua   120
Payagua   120
payasvini   361
Payta-Ratanhia   193
pazote   95
pea tree   343
pea, heart   85
pea, heart-leaved   85
peach geranium   275
peanut   48
pearl onion   23
pearl orchid flower   97
pebreila   233
pebrinella   233
pecah periuk   186
pěchari lochari   232
pe-deperdiz   344
pé-de-pombo   268
pee in the bed   361
peen   133
péganion   275
peganum   275
peholi   301
pehuij nayi castillo   108
pejerecum   396
Peking garlic   24
Peking-Knoblauch   24
pelaga   35, 144
Pelargonie, Apfelduft   276
Pelargonie, Balsam   277
Pelargonie, Duft   276
Pelargonie, Muskat   276
Pelargonie, Orangen   276
Pelargonie, Pfefferminz   277
Pelargonie, Pfirsisch   275
Pelargonie, Schmetterling   277
Pelargonie, Zitronen   276
pelargonija duschistaja   276
pelargonio   276
pélargonium citronne   276
pelargonium rosat   276

pelargonium, vlinder   277
pelitre   39
pellitory of Spain   283
pellitory, Roman   39
pellitory, Spanish   39
pelpasůcho   385
pencil cedar   189
peng e zhu   128
penghao   97
peninchang   388
penjilang   116
pennyroyal   173
pennyroyal (mint)   228
pennyroyal, American (false)   173
pepare   235
pepasan   113
pepaya   86
pepe a coda   291
pepe bianco   293
pepe cubebe   291
pepe della Giamaica   286
pepe falso   388
pepe garofanato   286
pepe nero   293
peper bossie   202
peper, chili   83
peper, spaanse   82, 83
Peperina   234
peperone   82
Peperoni   82
pepollino   368
pepper   9, 82, 134, 140, 192, 199, 202, 212, 225, 278, 280, 286, 289, 337, 343, 379, 388, 395, 399, 405
pepper bark   140, 400
pepper dulce   199
pepper elder   278
pepper fruit   134
pepper grass   202
pepper saxifrage   343
pepper stick   140
pepper tree   40
pepper wood   379
pepper wort   202
pepper, African   402
pepper, American bird   82
pepper, Arabian   405
pepper, Ashanti   291
pepper, Australian   140
pepper, bark   140
pepper, bell   82
pepper, Benin   291
pepper, bird   82, 83
pepper, black   293
pepper, bonnet   83
pepper, Brazilian   338
pepper, Brown's   83
pepper, bush   397

pepper, Californian   387
pepper, cape   290
pepper, capsicum   82
pepper, Cayenne   82, 83
pepper, cherry   82
pepper, Chile   82
pepper, Chinese   83, 399, 400
pepper, clove   286
pepper, cone   82
pepper, cubebe   291
pepper, datil   83
pepper, eleganto   291
pepper, Ethiopian   395, 397
pepper, green   82
pepper, Guinea   10, 291, 395, 397
pepper, habanero   83
pepper, hot   83
pepper, Indian   400
pepper, Japanese   400
pepper, Java   291
pepper, kimba   395
pepper, lolot   292
pepper, melegueta   10
pepper, mountain   140, 212
pepper, negro   395, 397
pepper, Nilgiri   295
pepper, Pandrón   82
pepper, Peruvian   83, 295, 337
pepper, pheasant   212
pepper, pink   338
pepper, pouched   295
pepper, red   82, 338
pepper, Saigon   294
pepper, Sichuan   400
pepper, small flowered   400
pepper, spiked   280
pepper, squash   83
pepper, stick   240
pepper, stripped African   396
pepper, sweet   82
pepper, Szechuan   399, 400
pepper, tabasco   83
pepper, Tasmannia   144
pepper, Tahiti   215
pepper, white   293
pepper, water   280
pepper, West African   291
pepper, wild   295
pepperbark   391, 400
pepperbark tree   391
peppercorn, pink   337
peppermint   149, 226, 277
peppermint scented geranium   277
peppermint tree   149, 150
peppermint, black   149
peppermint, Japanese   226
peppertree   215
peppertree, South African   192

pepperweed, desert   203
pepperweed, Virginia   203
pepperwort   203
perangimulik   82
pereirorá   43
perejil   282, 283
perejil tuberoso   283
perejil, grande   283
perejil, hamburgo   283
perennial peppergrass   203
perennial vine   294
peretschnik   203, 400
perez wodjanoj   280
perez cubeba   291
perez strukovyj   82, 83
perez tschjornyj   293
(perforated) St. John's Wort   180
perfumed cherry   308
peria bulan   85
pericón   358
perifollo almizclado   247
perifollo oloroso   247
periiquillo   358
perilla   278
perilla mint   278
Perilla, Öl   278
pérille   278
Perlen-Orchidee   97
Perlknoblauch   28
Perllauch   23
Perlzwiebel   23, 24
pérovskia   279
Perovskia à feuilles   279
Perovskie, Ledershnittige   279
Perovskie, Runzlige   279
Perovskie, Silber   279
perovskija polynnaja   279
perre pierres   120
persail de bouc   288
Perse limette   107
persia   263
Persian cow-parsley   176
Persian knotgras   302
Persian lemon   105, 107
Persian lime   105, 107
Persian walnut   188
persicaire   302
persil   116, 247, 261, 282
persil à grosse racine   283
persil arabe   116
persil commun   282
persil d'anis   247
persil séri   261
persil tubéreux   283
Persische Petersilie   176
Persische Limette   107
Persischer Bärenklau   176
Persischer Jasmin   188

Persischer Knöterich   302
Persischer Kreuzkümmel   71
Persischer Senf   147
Persisches Bohnenkraut   336
pértasite blanc   281
Peru basil   259
Peruanische Kapuzinerkresse   376
Peruanische Sonnenwende   174
Peruanischer Basilikum   259
Peruanischer Pfeffer   83, 337
Peruanischer Salbei   327
Perubalsam   246
Peru-Pfeffer   295
Peru-Portulak   44
Peruvian balsam   246
Peruvian bark   99
Peruvian ginger   202
Peruvian mastic tree   337
Peruvian pepper   83, 295, 337
Peruvian rhatany   193
Peruvian sage   327
pervozvet   307
peste à poux   285
Pestwurz, Alpen   281
Pestwurz, Filzige   281
Pestwurz, Gemeine   281
Pestwurz, Japanische   281
Pestwurz, Rote   281
Pestwurz, Süße   281
Pestwurz, Weiße   281
petai   273
petaside japonès   281
pétasite vulgaire   281
pétasitès japonais   281
peté   273
peteh   273
Petehbohne   273
Peter's cress   120
Peterling   282
peterselie   282
peterselieblad   282
Petersilie, Blatt   282
Petersilie, Burmannische   176
Petersilie, Japanische   125
Petersilie, Krause   282
Petersilie, Persische   176
Petersilie, Steinpilz   331
Petersilie, Wasser   347
Petersilie, Wilde   99
Petersilling   282
petikale   149
petit boucage   288
petit galanga   31
petit muguet   161
petit oignon   15
petit porreau   25
petit smilax   347
petite absinthe   55

petite centaurée   91
petite consoule   13
petite pimprenelle   331
petite queue de rat   353
petite vervaine queue de rat   353
petits doigts   67
petruschka   46, 247, 282
petruschka kornevaja   283
petruschka listovaja   282
petuschij predeschok   90
peucédan ostrote   283
peuplier baumier   304
peuplier franc   304
peuplier noir   304
Pfeffer   10, 82, 278, 289, 290, 295, 337, 395, 399, 405
Pfeffer, Afrikanischer   395, 402
Pfeffer, Arabischer   405
Pfeffer, Äthiopischer   395
Pfeffer, Bengal   292
Pfeffer, Brasilianischer   338
Pfeffer, Breitdorniger   401
Pfeffer, Californischer   387
Pfeffer, Cayenne   82
Pfeffer, Chabar   294
Pfeffer, Chinesischer   399, 401
Pfeffer, Dickblättriger   293
Pfeffer, Dolden   295
Pfeffer, Durchscheinender   278
Pfeffer, Gebogener   289
Pfeffer, Gekrümmter   289
Pfeffer, Gemeiner   290
Pfeffer, Gemüse   82
Pfeffer, Geröhrter   289
Pfeffer, Gestreckter   291
Pfeffer, Guinea   10, 396
Pfeffer, Japanischer   400, 401
Pfeffer, Java   294
Pfeffer, Javanischer   291, 294
Pfeffer, Johannisbeer-ähnlicher   294
Pfeffer, Kap   290
Pfeffer, Kleinblütiger japanischer   396, 400
Pfeffer, Langblättriger   292
Pfeffer, Langer   292, 293
Pfeffer, Lolot   294
Pfeffer, Matico   289
Pfeffer, Mekong   293
Pfeffer, Melegueta   395
Pfeffer, Peru   295
Pfeffer, Peruanischer   83, 295, 337
Pfeffer, Pipal   292
Pfeffer, Rosa   337, 338
Pfeffer, Saigon   294
Pfeffer, Schild   295
Pfeffer, Schmidtscher   295
Pfeffer, Schwarzer   293
Pfeffer, Sech(z)uan   400, 401

Pfeffer, Sédhion   396
Pfeffer, Sichuan   401
Pfeffer, Spanischer   82
Pfeffer, Süßer   82
Pfeffer, Tabasco   83
Pfeffer, Tasmania   140
Pfeffer, Türkischer   82
Pfeffer, Ungarischer   82
Pfeffer, Vietnamesischer   294
Pfeffer, Wolliger   292
Pfefferbaum   140, 401
Pfefferbaum Japanischer   401
Pfefferbaum, Indischer   401
Pfefferfenchel   158
Pfefferkadok, Wurzelrankiger   295
Pfefferknöterich   280
Pfefferkraut   136, 203, 256, 280, 295, 334
Pfefferkraut, Breitblättriges   203
Pfefferkraut, Deutsches   256
Pfefferkraut, Thailändisches   295
Pfefferkraut, Wildes   295
Pfefferkresse   203
Pfefferminzbaum   150
Pfefferminze   227
Pfefferminze, Schwarze   150
Pfefferminze, Stink   236
Pfefferminzpelargonie   277
Pfefferminzperanie   277
Pfefferrinde   140
Pfefferthymian   233
Pfefferwurzel   50, 288
Pfeilkresse   85
Pferdeeppich   348
Pferde-Knoblauch   15
Pferdeminze   236
Pferde-Muskat   243
Pferderettichbaum   240
Pfirsischpelargonie   275
Pfirsisch-Salbei   327
Pflaumenmuskat   58
pha   19, 280, 300
pha chi mi   280
phac cham   175
phaeraengikkot   135
phak phai nam   280
phak chee   47, 116, 147, 282
phak chee farang   147
phak chee lom   47
phak eehum   240
phak hoei   235
phak khaao thoj   197
phak khayaeng   207
phak khraat thale   392
phak krachet   252
phak krasang   278
phak phè:w   280
phak phai   280

phak phai nam 280
phak sian 111
phak-nueakai 240
phakaa krong 196
phakchi 40, 116, 261, 371
phakchi lao 40
phakchiduanha 158
phakchi-lom 261, 371
phakhom 116
phakhom-noi 116
phakkhuang 167
phakuang 302
pharan 19
phat pha 300
phât tu 109
phayaa sattaban 32
pheasant-pepper 212
phet buri 128
Philippine cinnamon 103
Philippine wax flower 148
Philippinischer Zimt 103
philjŭnggadonggul 291
phimsen 301
phlaè 352
phlai 402
phluu ling 295
Phoenicean juniper 189
Phönizischer Wacholder 189
phothisat 14
phrik kheefa 83
phrik noi 293
phrik thai 293
phrik-hang 292
phut chiin 162
phut son 162
pi jiu hua 179
pianta di (del) pistacchio 297
pianta di soia 168
pibis 272
pichuri 260
Pichurimbohne 260
Pichurinuss 260
pick tooth 34
picosa 357
pie de leon 14
pied dur 218
Piemonteser Quendel 367
pien hsu 302
pies-ansyr 23
pieterselie 282
pignes 289
Pignoli 289
pignolia 289
pignolia-nut 289
pignolies 289
pignut, great 71
pigweed 95
pijerecu 396

pijnkern 289
piknanthemum, Virgiana 310
pilar rose 320
pilatro 180
pile murghka 90
pilewort 313
pilohe pimata 314
pilous thyme 367
piludi 325
pilva 325
pimata omoa 314
Piment 82, 96, 242, 286, 395
piment 82, 96, 242, 286, 395
piment botris 96
piment chien 83
piment de Cayenne 83
piment de la Jamaique 286
piment des Anglais 286
piment doux 82
piment noir de Guinée 395
piment oiseau 83
Piment, Mexico 287
Piment, Tabasco 83, 287
pimenta 83, 124, 286, 289, 291, 293, 337, 395
pimenta da Guiné 395
pimenta da Jamaica 286
pimenta da-costa 396
pimenta de bugre 287
pimenta de Guinea 291, 395
pimenta de macaco 396
pimenta de mato 396
pimenta malagueta 83
pimenta negra del país 291
pimenta-de árvore 396
pimenta-de-gentio 396
pimenta-de-macaco 289
pimenta-denegro 396
pimenta-do-sertão 396
pimenta-longa 289
pimenteira 293
pimenteiro 337
pimentero 293
pimento 76, 82, 286, 291, 396
pimento da „rabo" 291
pimento inglese 286
pimento, Brazil 76
pimento, Mexican 287
pimento, Tabasco 287
pimento-de-gentio 396
pimienta blanca 286
pimienta cubeba 291
pimienta de Brazil 338
pimienta de Cayena 83
pimienta inglesa 287
pimienta negra 293
pimienta picante 83
pimientero falso 337

pimiento 82, 287
pimiento de Jamaica 287
Pimpernelle 288
Pimpernuss, Echte 297
Pimpernuss, Fiederspaltige 353
pimpinela blanca 288
pimpinela mayor 332
pimpinela menor 331
pimpinela sanguisorba 331
pimpinella 288
pimpinella minore 288
pimpinelle 288, 331
Pimpinelle 288, 331
Pimpinelle, Große 332
Pimpinelle, Kleine 331
pimprenelle 288, 331
pimprenelle commune des prés 331
pimprenelle des près 331
pimprinelle des jardins 331
pimprinelle, petite 331
pin parasol 289
pin pignon 289
pinchao 147
pindaíba 396, 397
pindaíba-vermelha 397
pindaúba 396
pindaubuna-da-serra 397
pine nut 289
pine seed 289
pineapple mint 229
pineapple, ground 364
pineapple-scented sage 327
pinheiro manso 289
Pinienkern 289
Piniennuss 289
Piniensamen 289
pink berry 338
pink centaury 91
pink damask rose 319
pink lime-berry 110
pink pepper 338
pink peppercorn 337
pink porcelain lily 31
pink wood sorrel 268
pink, Chinese 135
pink, clove 135
pink, Indian 135
pinksterbloem 84
pinkwood bark 136
pinkwood sorrel 268
pino da pinola 289
pino domestica 289
piñoero 289
Pinoli 289
pioppo balsamico 304
pioppo nero 304
pipal 292
Pipalpfeffer 292

Piper Melegueta 10
piper, wooly 292
piperesa 233
piperina 234
pipi 282
pipiltzintzinli 327
pippali 292
pippul 293
piretro da África 39
piretro romano 39
piring sak 373
piri-piri (pepper) 83
piris 83
pirul 337
pisang pisang 232
pisang sebiak 78
piscacane 361
pishma 360
pismarum 322
pissenlit 361
pista 297
pistacchio 297
pistache 297
pistacheira 297
pistacheira 297
pistachero 297
pistachier cultivé 297
pistachio 297
Pistazie 296
Pistazie, Mastix 296
Pistazie, Therebinth 297
Pistazienmandel 297
pisutachio 297
pitabija 374
pitar saleri 282
pityouda absinthe 56
pixurim 260
piyaz 17
pizza herb 265
Pizza Oregano 266
pla ko 38
plante de l'étoile 233
platycodon à grandes fleurs 299
plecthranthe 299
pletikapu 371
Pluche, Indische 300
pluchea, Indian 300
plum nutmeg 58
plum, carandas 86
plum, yellow 352
pluu-kao 179
po ho 224
Pockholz 170
podagra 7
podagraire 7
podapatri 171
podbel 281
podchrennik 203

podlesnik 166
podverella 334
poejo 126, 228
pohkarmul 184
pohon merah 152
poi 63, 111
poinsettia 152
Poinsettie 152
pointed beebalm 237
pointed tonka 138
poireau 23
poireau d'été 15
pois à Chapelet 2
pois bord de mer 77
pois de coeur 85
pois maritime 77
pois quénique 240
pois sucré 298
poivre 9, 82, 280, 286, 290, 293, 335, 339, 388, 395, 400, 405
poivre à queue 291
poivre africain 402
poivre blanc 293
poivre chinois 401
poivre cubèbe 291
poivre d'achantis 291
poivre d'Afrique 10, 291
poivre d'âne 335
poivre d'eau 280
poivre d'Espagne 82
poivre d'Ethiopie 9, 395
poivre de Cayenne 82, 83
poivre de Chiappa 287
poivre de Ethiopie 395
poivre de Guinée 10, 291, 395
poivre de kissi 291
poivre de la Chine 401
poivre de la Jamaïque 286
poivre de lolot 292
poivre des moines 388
poivre des murailles 339
poivre giréflé 286
poivre les Arabes 405
poivre long 292, 294
poivre long de Java 294
poivre negre 395
poivre noir 293
poivre rouge 83
poivre sauvage 290
poivre Szetchuan 400
poivres de Sédhiou 396
poivrette commun 252
poivrier d'Amérique 337
poivrier du Pérou 337
poivrier rose 338
pokoh 256
pokok cherek 110
pokok getang 352

pokok kemangi 181
pokok kĕmantu 110
polang 30
polecat bush 316
Poleiminze 228
poleo 173, 234
póleo 228
póleo negro 228
póleo omún 228
polewaja gortschiza 84
Poley, Amerikanische 173
polyn' al'nijskaja 55
polyn' balchanov 53
polyn' blednejusschtschaja 55
polyn' gor'kaja 52
polyn' limmonaja 53
polyn' metel'tschataja 56
polyn' pontijskaja 55
polyn' stragon 53
polyn'boshe derewo 362
polyn'obyknovennaja 57
polyn'rumskaja 55
polypode vulgaire 303
polypody, common 303
polystachyous basilicum 63
pomegranate 310
pomeranez 106
Pomeranze 109
Pomeranze, Myrtenblättrige 109
pomme d'orange 109
pomme de mercveille 235
pomme de sauga 327
Pompon vanilla 385
Pompon-Vanille 385
poncil 108
pongpong 145
ponkan 109
Pontischer Beifuß 55
Pontischer Wermut 55
pony beebalm 237
pool root 152
poole batoo 31
poolu 130
poor man's pepper 203
pop-a-gun 89
popewnik 204
poplar, black 304
poplar, Lombardy 304
poppy 272
populage 75
porcellana 305
porraccio 15
porrandello 15
porreau 23
Porree 23
Porree, Wilder 25
porro 23
porro selvatico 15

Porschkraut   201
Porst   201
Portland rose   319
Portland Rose   319
portschellana   305
portulaca   239, 305
portulaca d'inverno   239
portulak   44, 305
Portulak   44, 305
portulak ovoschnoj   305
Portulak, Winter   239
Porzellan-Ingwerlilie   31
poskonnik konoplewyj   151
postaka   272
postelein   305
pot marigold   75
pot marjoram   265
potato onion   17
potato, hausa   351
potato, monkey's   351
potato-tree   349
potmepini   149
pouched pepper   295
pouliot   228
pourpier   239, 305
pourpier d'hiver   239
pourpier des potagers   305
pourpre requiénie   362
pourslane   305
poy kak bua   184
požlt farbiarsky   87
Prächtige Kostwurz   118
Prachtscharte, Duftende   206
praduu laai   309
prairie pepper gras   202
Präriemonarde   236
prathatchin   311
prayong paa   12
prezzemolo   282
prezzèmolo tuberoso   283
prickly ash, Senegal   155
prickly parsnip   143
pricklychaff flower   4
prik kao   293
prima vera   307
primerolle   307
primevère officinale   307
primrose   307
primrose, forest   181
primula   307
prince's feather   302
prince's pine   97
pring sratoab   356
pritschecnotschnyj   25
prjanij tmin   125
prjanyj issop   145
pro hom   191
proljesnik   230

prompenelle sanguisorbe   331
prostaja polyn' tschernobyl'nik   57
prostanthère   307
prosvirnik lesnoj   217
provence rose   318
pseudo-acacia   318
psewdoloperzy (ksisopii)   395
Pskemenser Zwiebel   23
pskemskij luk   23
ptetschnyj   41
ptitschii perez   82
ptschel'nik   223
puar   30, 35
puchu   24
puchuiri   260
pudding grass   228
pudding pipe tree   88
pudina   225, 226, 229
pudina fudina   226
puding mas   113
puen si:phlaè   352
puerro   23
puerro agreste   15
puerro silvestre   15
Puerto Rico oregano   211
puffball   361
puki   131
pukyakssuk   52
pulai   32
pulang sili   82
pulasari   32
pulé   32
pulguntholjokkwi   302
pulikampa   196
puliyarai   268
puliyaral   268
pullamurasi gida   46
puloei   402
pumkin, fluted   361
punag champa   31
punaise mâle   116
punhongkkothyangjangri   319
Punischer Apfel   309
punkkot   234
Punktierte Monarde   237
Punktierte Tonkabohne   138
Punschkraut   211
puom   155
Pupurne Tephrosie   362
purat   14
purging cassia   88
purging fistula   88
puring merah   152
purole-flower garlic   25
purple anise   183
purple apricot   49
purple avens   166
purple garden oxalis   269

purple gentian   164
purple gooseberry   285
purple hoary pea   362
purple mint   278
purple tephrosia   362
Purpur-Enzian   164
Purpur-Fetthenne   341
Purpurne Aschenwicke   362
Purpurne Bauhinie   64
purret   23
purslane, garden   305
purslane, sea   178
purslane, winter   239, 305
pursley   305
puru   131
pusa   371
puskaramula   184
putat gajah   63
putat hutan   63
putat tuba   63
puwar pelaga   35
pyaj   17
pycnanthème, Virgiana   310
Pyramiden-Galgant   30
Pyramiden-Hühnergras   143
Pyramiden-Pappel   304
Pyrenäen-Stiefmütterchen   387
pyréthre d'Afrique   39
pyrethrum   39
qaranful   355
qian zi   152
qiao tou   18
qin cai   46
qing xiang zi   90
qoddab   302
qu mai   135
qua su   57
quabar   79
quassia   285, 311
quassia de Surinam   311
quassia wood   285, 311
quassia, Jamaica   285
Quassia, Jamaica   285
quassia, Surinam   311
Quassiaholz   311
quassier   311
quassier amer   311
quat   109, 159
quat thuc   109
quatre-épice   286
que   100, 212, 256, 358
qué lá trà   100
Queen Anne's lace   45
queen of the meadow   157
Queensland Arrowroot   78
Queensland-Holzrose   231
Quendel   334, 367
Quendel, Echter   367

Quendel, Garten   334
Quendel, Piemonteser   367
Quendel, Römischer   368
Quendel, Zitronen   365
quénit   308
quenoille   377
querciola   362
quiabo-róseo   177
quiabo-roxo   177
quibo-azedo   177
quie laga zaa   358
quije pecohua castilla   318
quillquina   305
quilquina   7
quimg hsiang   90
quina   99
quina amarela   99
quina amarilla   99
quina do Amazonas   99
quina negra   99
quinine   99, 311
quinine de Cayenne   311
quinine tree   99
quinoto   159
quinquina   99
quinquina (jaune)   99
quinquina rouge   99
quintuple justicia   189
Quirlblütiger Salbei   330
Quirl-Malve   217
quit   109
quita dolor   209
quitoc   300
quitoco   300
Quitte, Bengalische   7
qun dai cai   380
qurfa   104
raam   97
rabanête japonés   391
rábano japonés   391
rábano picanto   50
rábano rusticano   50
rábano rustico   50
rábano silvestre   50
rabarbaro   315
rabárbo   316
rabbit ear   278
rabillo de gato   182
racine de charchis   138
racine jaune   133
radice gialle   126
radichiello   361
radish tree   240
rafano rusticano   50
Ragani   364
rai   69, 345
raifort du japonais   391
raifort officinal   113

raifort sauvage   50
raifort vert   391
raihān   256
Rainfarn   360
Rainfarn, Minzartiger   360
raiyane   158
raíz del Epíritu Santo   41
raíz del moro   184
raiz forte   50
raja pruk   88
rajah kayu   88
rajana   186
rajika   69
rakhta chandana   309
rakkyō   18
rakkyu   63
raktamarica   82
ram tulsi   258
Rama   177, 306
ramatila   171
ramberge   230
ramerino   320
ramno catatico   315
ramp   27
Rams   17, 27
ramsons   27
Ramsons   27
ramtil   171
Ramtillkraut   171
rande nadeshiko   135
randhuni   371
rangamali   65
rantana   196
ranti   350
ranting   29
ranukabija   388
ranzusskaja gortschiza   70
rape, California   345
rapontico   316
rapóntico   316
rapontik   316
Raps, Jamba   147
Ras el Hanout   10
rasala   218
rasca   87
rascovec   125
rasona   24
rastoropscha   344
rat tai vervain   353
ratambi   162
ratanhia   193
Ratanhia, Payta   193
Ratanhia, Rote   193
ratania   193
rati   2
ratjoon   152
ratjoonan   152
rau (hung) que   256

rau cai   69
rau dang   302
rau he   24
rau kingh gioi   265
rau mui tay   282
rau ngó   207
rau ngo gai   147
rau nuoi dai   95
rau om   208
rau râm   280
rau sam   305
rau sung   401
rau tan la day   299
rau thia la   40
rau tia to   278
Rauher Nachtschatten   349
Rauher Schneckenklee   221
Rauhes Brandkraut   204
Rauhfrüchtige Marille   49
Rauhhaarige Leuchterblume   93
Rauhhaarige Melisse   223
Rauhklette   4
rauskoe serno   10
Raute   54, 275, 322
Raute, Aleppo   322
Raute, Garten   323
Raute, Gefranste   322
Raute, Syrische   275, 322
Raute, Weiße   54
Ravensara-Nuss   313
raw nightshade   349
rawend zahar   316
rayan   69
razianuj   158
re yo   198
Rebhuhnbeere, Niedere   163
red bark   99
red bay   280
red bud   93
red chili   83
red cinchona   99
red cress   280
red fox   90
red ginger   31
red ginger lily   173
red mint   226, 229
red pepper   82, 338
red perilla   278
red Peruvian   99
red sage   196
red sandalwood   309
Red sanders   309
red sorrel   177
red spike thorn   220
red spinach   90
red topped sage   330
reddish stonecrop   340
red-flowered Pornupan

mangrove   351
redwood, Anaman   309
redheads, Zanzibar   355
reedmace   377
reflexed stonetrop   340
regalicia   169
regaliz   169
regamo   265
regina dei prati   157
réglisse   169
reglisse sauvage   2
rehan   256, 257
Reiherschnabel, Moschus   146
reina de cabra ulmaria   157
reina de los prados   157
reine des bois   161
reine des prés   157
Reisfeldpflanze   207
rejang   32
remédio do vaqueiro   334
remon   107
renealmia, mountain   314
Renealmie, Berg   314
renouée des oiseaux   302
renouée odorante   280
reo   38
repollo chino   69
requiéne, linéaire   362
rĕsah   37
reseanez   25
reseda de campo   28
reseda de cheiro   314
réseda de odor   314
réséda odorante   314
reseda, amorino   314
Resedablättriges Schaumkraut   85
Resede, Duft   314
Resede, Garten   314
Resedenwein   44
resurrection lily   191, 192
retama negra   132
reven' aptetschnyj   315
reven' kitajskij   315
reven' russkij   315
rezeda duschistaja   314
Rhabarber, Chinesischer   315, 316
Rhabarber, Gewöhnlicher   316
Rhabarber, Kanton   316
Rhabarber, Krauser   316
Rhabarber, Kron   316
Rhabarber, Medizinal   316
Rhabarber, Ostindischer   315
Rhabarber, Sibirischer   316
Rhabarber, Türkischer   315, 316
Rhabarber, Wellblatt   316
rhakhan   325
rhapontic   316
rhapontic rhubarb   316

Rhapontik   316
rhapontike des moines   320
rhatania   193
rhatany   193
rhatany, Peruvian   193
rhetsamaramu   401
rhetsu   401
Rhodesian lantana   197
Rhodesisches Wandelröschen   197
rhubarb   281, 315
rhubarbe (anglaise)   316
rhubarbe officinal   315
rhubarbe palmée   316
rhubarb, bog   281
rhubarb, Chinese   315
rhubarb, garden   316
rhubarb, medicinal   315
rhubarb, Siberian   316
rhubarb, Tibetan   315
rhubarb, Turkey   316
rhubarb, wild   136
riccola   323
richetta   323
ricino del país   317
ridé pérovskia   279
ri-din   341
rieng   30
rieng am   30
riêng âm   36
rieng nep   30
rieng nuoc   29
rieng rung   29
rieng thuoc   31
rièng tiá   31
Riesenvanille   384
Riesen-Zwiebel   19
Riffelkürbis   361
rigla   305
rihan   247
rimskaja romaschka   95
rimskij koriandr   254
rimskij tmin   125
Ringelblume   75, 99, 358
Ringelblume, Afrikanische   99, 358
Ringelrose   75
rinu   290, 291
rinu manuk   290
ripped melilot   222
Rispen-Beifuß   56
Rispen-Fetthenne   340
Rispiger Jasmin   188
rizzomolo   33
rjabina cernoplodnaja   51
Robert's lemon rose   276
robinia   318
Robinie, Gemeine   318
robinier   318
rocambola   25

rocambole   24
rocambole, garden   24
rock rose   138
rock samphire   120
rock stonecrop   340
rockbrake   303
Rockenbole   23
Rockenbolle   24, 25
rocket   62, 73, 137, 147
rocket cress   62
rocket salad   147
rocket, early yellow   62
rocket, garden   147
rocket, Roman   147
rocket, sand   137
rocket, winter   147
ročná paprika   82
rocotillo   83
rocoto   84
rocouyer   65
roewnik   223
rogos   377
Röhrenblütige Monarde   236
Röhrenkassia   88
Rohrkolben, Breitblättriger   377
Rohrkolben, Echter   377
rohzelu   177
rojao   357
rojmari   3
Rokambole   23
roketta   147
rokpakha   229
romã   309
Roman caraway   125
Roman chamomile   95
Roman chamomile oil   95
Roman coriander   253
Roman laurel   199
Roman liquorice   169
Roman pellitory   39
Roman rocket   147
Roman wormwood   55
romanška lugiwaja   204
romaschka nemezkaja   39
romaza de kojas grandes   321
romazeiro   309
romerillo   28
romerino   320
romero   201, 320
roméro   320
romero santo   201
romice dei prati   321
romine crespo   321
Römische Kamille   95
Römischer Ampfer   322
Römischer Balsam   360
Römischer Beifuß   55
Römischer Bertram   39

Römischer Kümmel   125
Römischer Quendel   368
Römischer Schwarzkümmel   253
Römisches Süßholz   169
rommen   309
rommon   309
ronde kardemom   35
rondelette   167
rooamari   3
rooisuring   269
rooituisuring   269
roomse kervel   247
root celery   47
roqueta común   147
roquette   62, 73, 147
roquette de mer   73
roquette des jardins   62, 147
ros   319
rosa de Damasco   319
rosa de Jamaica sereni   177
Rosa Pfeffer   337
rosal   162
Rosapfeffer   337, 338
rosary pea   2
rose à cent feuilles   318
rose de Damas   319
rose de mai   318
rose de Provence   319
rose de Puteaux   319
rose de tous les mois   319
rose geranium   275
rose pálida   319
rose pelargonium   275
rose scented geranium   275
rose wood   43
rose, baby   320
rose, Bengal   318
rose, bramble   320
rose, Bulgarian   319
rose, cabbage   318
rose, China   318
Rose, China   318
Rose, Dalmaszener   319
rose, Damask   319
rose, French   319
rose, Holland   318
rose, Japanese   320
rose, musk   319
rose, pilar   320
rose, Portland   319
Rose, Portland   319
rose, provence   318
rose, seven sister   318
roseira   319
roseira selvagem   319
rosela   177
rosella   177
Rosella-Eibisch   177

roselle   177
Roselle   177
rosemary   123, 201, 320, 392
rosemary, Australian   392
rosemary, wild   123, 201
Rosenduftgeranie   275
Rosengeranie   276
Rosenholz, Brasilianisches   43
Rosenmonarde   236
Rosenstorchschnabel   276
roseta   177
rosewood, Burmese   309
rosha grass   130
roshkoboe   93
rosier musqué   319
rosita de cacao   311
Rosmarin   123, 201, 320, 392
rosmarin   320
Rosmarin, Australischer   392
Rosmarin, Wilder   123, 201
Rosmarinblättriges Heiligenkraut   333
rosmarino   320
rosmir   320
rosolaccio   272
Rosskümmel   198
Rosskümmel, Dreiblättriger   198
Rosslauch   22
Rossminze   226, 229
Rossminze, Asiatische   226
Rossminze, Grüne   229
Rosspappel   216
Rot(blättriger) Basilikum   257
Rote Chinarinde   99
Rote Flügelfrucht   309
Rote Indianernessel   236
Rote Melisse   236
Rote Minze   229
Rote Pestwurz   281
Rote Ratanhia   193
Roter Ingwer   31
Roter Pfeffer   337, 338
Roter Schmetterlingsingwer   173
Roter Senf   70
Roter Speik   381
Rotes Sandelholz   309
Rotgelbe Taglilie   175
Rötliche Fetthenne   340
rou dou kou   244
rou gui   100
roubia   219
Roucon-Samen   65
rough chaff tree   4
rough leucas   204
round cardamom   35
round Chinese cardamom   36
round kumquat   159
round cardamom   35
round-leaf buchu   62

round-leaved mint   228, 229
round-leaved mint bush   307
round-rooted galanggale   192
roza damasskaja   319
roza francuzskaja   319
roza muskusnaja   319
roza stolistaja   318
rozemarijn   320
rozmarin   320
Rübe, Gelbe   133
ruca   147
ruchetta   147
Ruchgras, Gewöhnliches   44
rucola commune   147
Rucola, Ausdauernde   137
rúcula   147
ruda común   323
ruda de monte   323
ruddles   75
rue   275, 322
rue d'Algérie   322
rue de jardins   323
rue de montagnes   323
rue fétide   323
rue officinale   323
rue puante   323
rue sauvage   275
rue, African   275
rue, common   323
rue, Egyptian   322
rue, fringent   322
rue, montain   275, 323
rue, Syrian   275, 322
rue, wild   275
Ruhrkraut   170, 306
Ruhrkraut, Indisches   170
Ruhrrinde, Bittere   344
ruibarbo   315
ruibarbo (francés)   316
ruibarbo de la China   315
ruibarbo de levante   316
ruit   323
Ruke   147
ruku   256, 258, 260
ruku ruku   260
ruku-ruku hitam   258
Rukusamen   65
rum cherry   308
ruman   309
Rumänischer Senf   345
rumex crépu   321
rumput   29, 234, 280, 301, 387
rumput belangkas   234
rumput kuku   301
rumput tahi babi   387
rumput tuboh   280
Rundblättrige Minze   228, 229
Rundblättriger Ampfer   322

Runde Gewürzlilie 67, 192
Runder Kardamom 35
Runder Kentjur 192
Rundköpfiger Lauch 25
Runzlige Perovskie 279
Runzliger Ysop 12
rusmari 320
Russian basil 258
Russian coriander 117
Russian laserwort 198
Russian oregano 265
Russian sage 279
Russian wild horseradish 50
Russischer Bergkümmel 198
Russischer Koriander 117
Russischer Salbei 279
Russischer Senf 69
Russischer Steppen-Thymian 367
Russisches Süßholz 169
Russisches Süßholz 169
russkaja gortschiza 69
ruta 323
ruta sadowaja 323
Rutenkohl 69
ryû kyû-tororo-aoi 1
s(h)obhanjana 240
sa chanh 129
sa nhan 38
sa nhon 36
sa'afaran 122
sa'tar ramsi 368
Saat-Lein 209
sabasnik 157
sabel'nik 5
sabina 361
saborija 335
saborinam 205
saboujas 334
sabowij kres 203
sabugueiro negro 331
Sachalinminze 228
sacha-yuyu 278
sacred barma 120
sacred basil 260
sacred garlic pear 120
sacrodactyle 108
sada tulas 256
sadab 322
sadoa-din 167
sadrée 334
saenggang 403
saeyang 403
saf(f)ran 122
sāfarān 122
safed candan 332
safflower 86
saffraan 89, 122
saffraan, fynblaar 89
saffron 86, 89, 121, 122, 126, 148, 157, 354, 357, 359, 375
saffron cassine 89
saffron marygold 357
saffron of Andes 148
saffron root 148
saffron thistle 86
saffron, cape 121, 374, 375
saffron, Cartwright 121
saffron, dyer's 86
saffron, false 86, 157
saffron, Hausknecht 122
saffron, Indian 126
saffron, small-leaved 89
saffron, Spanish 122
saffron, wild 121, 122
Saflor 87
saflor krasil'nyj 87
Safran 87, 89, 121, 122, 126, 148, 157, 375
šafran 122
safrān 122
safran bâtard 87
safran des Indes 126
Safran, Anden 148
Safran, Cartwright 121
Safran, Echter 122
Safran, Falscher 87, 157
Safran, Hausknecht 122
Safran, Kap 354, 375
Safran, Schmalblättriger 89
Safran, Touristen 87
Safranartige Tritonie 375
safranon 87
Safranwurz(el) 126
Safranwurz(el), Würzige 126
safre 87
safsaf 304
Saftlose Kaskarille 123
safuran 122
saga 2
saga betina 2
sagasein 77
sage 54, 169, 210, 231, 279, 284, 325, 363
sage, African 325
sage, annual 330
sage, Bengal 231
sage, blue 327
sage, bluebeard 330
sage, British Honduras 327
sage, Caspian 279
sage, clary 330
sage, common 329
sage, English 330
sage, fruit 327
sage, garden 329
sage, Graham 329
sage, grapefruit 329
sage, Greek 327
sage, Italian spring 329
sage, Jerusalem 284
sage, Jim 327
sage, Joseph 330
sage, lavander 328
sage, Mediterran 325
sage, Mexican 210
sage, muscat 330
sage, painted 330
sage, Pamir 279
sage, Peruvian 327
sage, pineapple-scented 327
sage, red 196
sage, red topped 330
sage, Russian 279
sage, silver 279, 326
sage, Spanish 328
sage, sticky 328
sage, tangerine 327
sage, thistle 326
sage, Turkish 327
sage, vervaine 330
sage, wally 325
sage, wild 196, 210
sage, yellow 196
sagebrush 56
saha 311
sahang 337
sahihan 305
sahijan 240
Saïgon cinnamon 102
Saigon pepper 294
Saïgon-cassia 102
Saïgon-Pfeffer 294
Saïgon-Zimt 102
Sailer's tobacco 56
saimpol 137
saipan mango 218
sajetsch'ja kanista 340
Sakhalin mint 228
sala 377
salad burnet 331
salad, water 297
salad lek 20
salai 129
salakat 101
salam 356
salam, manting 356
salamander tree 45
Salamanderbaum 45
Salamblatt 356
Salatblume 375
Salatlauch 20
Salbei, Afrikanischer 325
Salbei, Ährentragender 336
Salbei, Ananas 327

Salbei, Aramenischer 326
Salbei, Azteken 327
Salbei, Bengalischer 231
Salbei, Blauer 327
Salbei, Clary 330
Salbei, Cleveland 327
Salbei, Dalmatiner 329
Salbei, Distel 326
Salbei, Distelartiger 326
Salbei, Echter 329
Salbei, Eisenkraut 330
Salbei, Frucht 327
Salbei, Garten 329
Salbei, Gelber 328
Salbei, Gold 327
Salbei, Graham's 329
Salbei, Grapefruit 329
Salbei, Griechischer 327
Salbei, Honigmelonen 327
Salbei, Italienischer Frühling 329
Salbei, Jerusalem 284
Salbei, Julian 233
Salbei, Klebriger 328
Salbei, König 329
Salbei, Lavendelblättriger 328
Salbei, Marzipan 327
Salbei, Mohren 326
Salbei, Muskateller 330
Salbei, Pamir 279
Salbei, Peruanischer 327
Salbei, Pfirsich 327
Salbei, Quirlblütiger 330
Salbei, Russischer 279
Salbei, Scharlach 330
Salbei, Schopf 330
Salbei, Silberblatt 326
Salbei, Spanischer 328
Salbei, Ungarischer 326
Salbei, Verschiedenfarbiger 327
Salbei, Wahrsager 327
Salbeigamander 363
saleggiola 320
salima 329
salmya 329
saloop 333
salsa 176, 282
salsa commum 282
salsa tuberosa 283
salsa-limão 209
salsapariglia smilace 347
salsepareille 347
salsinha 282
salt bush 59, 325
salt fern, New Guinea 58
saluni 322
salva branca 209
salvaca 259
Salvadorbalsam 246

salva-limão 211
salvastrella 331
salvastrella maggiore 332
salve, Aramenian 326
sálvia 329
salvia americana 209
salvia fina 328
salvia oficinal 329
salvia real 327, 329
salvia trilobata 327
Salzbusch 325
Salzmelde 59
samag albostan 247
samaka 317
samba 188
sambanganai 290
sambbeng 198
sambuco 331
sambugo 331
sampa guita 188
sampalik 359
samphire, (sea) 120
samphire, rock 120
sampinit 178
Samtblume, Glänzende 358
Samtpflanze, Niederliegende 172
Samtrose 319
san lo-po 45
san nien 59
san yeh lan 12
sanchonamu 400
sand and leek 25
sand leek 23, 25
sand rocket 137
sandal wood, Indian 309
sándalo blancao 332
sandalo blanco 332
sândalo blanco 332
sandalo della India occidental 332
sandalo des las Antilles 332
sándalo Indias oríentales 332
sandalowoe derewo beloe 332
sandalwood 267, 309, 332
Sandalwood, African 267
sandalwood, East Indian 332
sandalwood, red 309
sandalwood, white 332
sandalwood, yellow 332
Sandelholz, Afrikanisches 267
Sandelholz, Rotes 309
Sandelholz, Weißes 332
Sandelholz, Westindisches 332
Sand-Hornkraut 92
Sandlauch 25
Sandstrohblume, Italienische 174
Sandthymian 368
sanga 362
sangina 240

sangkhriat 12
sanguisorba 331
sanguisorbe officinale 331
sanhân âm 38
sanke 184
sanke wood 138
Sanktoriakraut 91
sanque 184
sansev'ra 332
sanseveria 332
sanseverija trechputschkovaja 332
sansévière 332
sansevieria 332
sansho 400
sanshō 400
santal blanc 332
santal des Indes 332
santara 109
santara orange 109
santawah pak 267
santhoninssuk 54
Santo Domingo apricot 217
santoksiljum peretschnij 400
sanatolina, green 353
santon 186
santoreggia domestica 334
santoreggia montana 335
sanve 345
sao chou tsao 64
saphuran 122
saponaire des vaches 381
Saposhnikovie, Sperrige 333
sapožnikovija rasto-pyrennaja 333
saptaparna 32
saquente 371
sara-kachû 369
sarazinskaja mirtja 360
Sarepta mustard 69
Sareptasenf 69
sariette 74, 334, 335
sariette des montagnes 335
sariette népéta 74
sariette vivace 335
sariva 176
sarmisan songino 22
sarotamnus metel'tschatyj 132
sarotamnus metlistyj 132
sarpmoola 49
sarriette à grandes fleurs 74
sarriette annuelle 334
sarriette commune 334
sarriette de Saint-Julian 233
sarriette népéta 74
sarsalida 167
sarsa 7
sarsapareille 347
sarsaparil' 347
Sarsaparilla 176, 347

sarsaparilla, Gray 347
sarsaparilla, Indian 176
sarsaparilla, Mexican 347
sarsaparilla, Veracruz 347
Sarsaparille, Indische 176
Sarsaparille, Mexikanische 347
sarson 69
sarsva 69
saru-kake-mikan 369
saruni laut 392
sarwari 90
sasafrás 333
sasawi langit 387
sasfras 333
saskatoon 33
sassafras, black 103
Sassafras, Brasilianischer 261
sassafras, false 261
Sassafras, Weißer 333
Sassafrasnuss 260
sassfras blanc 333
sassfrasso 333
sataga 201
satapatri 318
satapuspi 40
satari 322
sathagudi 109
sathra 263, 265
satinwood 241
sato 273
Saturei-Thymian 369
satureia 334
sauce thyme 369
saúco (común) 331
Sauerampfer, Blasiger 322
Sauerampfer, Garten 320
Sauerampfer, Großer 320
Sauerampfer, Indischer 322
Sauerampfer, Kleiner 321
Sauerampfer, Krauser 321
Sauerampfer, Schild 322
Sauerampfer, Wiesen 320
Sauerdattel 359
Sauerklee, Aufrechter 269
Sauerklee, Bläulicher 269
Sauerklee, Breitblättriger 269
Sauerklee, Doldentraubiger 268
Sauerklee, Grüner 269
Sauerklee, Hornfrüchtiger 268
Sauerstockkraut 205
sauge 326
sauge à feuilles de lavande 328
sauge à trois lobes 327
sauge argentée 326
sauge de l'Espagne 328
sauge glutineuse 328
sauge hormin 330
sauge officinale 329

sauge sclarée 330
sauge trilobée 327
sauge verte 330
sauge vervaine 330
saumph 158
saunders, white 332
Saure Balsampflaume 352
Saure Limette 105
Saure Mombinpflaume 352
saururis 178
savali peepul 293
savoreggia 334
savory 334
savory creeping 336
savory, Croatia 335
savory, Jamaican 334
savory, Julian 233
savory, Senegal 365
savory, showy 74
savory, summer 334
savory, winter 335
savouree 335
savourée 334
sawtooth 147
sawtooth coriander 147
saxifraga menor 288
saya 261
scabwort 184
scallion, Chinese 18
scallion, Japanese 18
scalogno 17
scapigilate 253
scarlet egg plant 348
scarlet ginger lily 173
scarlet wistera 343
scented boronia 68
scented vernal grass 44
sceptre de l'empereur 148
Schabasamen 253
Schabziegerklee 372, 373
Schafgarbe, Bittere 3
Schafgarbe, Edel 4
Schafgarbe, Garten 2
Schafgarbe, Gemeine 3
Schafgarbe, Gewöhnliche 3
Schafgarbe, Moschus 3
Schafgarbe, Schwarze 3
Schafgarbe, Schwarzrandige 3
Schafgarbe, Süße 2
Schafgarbe, Wiesen 3
schal'wija 329
schalfej aptetschny 329
schalot luk 17
Schalotte 17
Scharbockskraut 113, 313
Scharfer Knöterich 280
Scharfer Mauerpfeffer 339
Scharlach-Bergamotte 236

Scharlachkraut 330
Scharlach-Monarde 236
Scharlachranke, Indische 113
Scharlachrote Ixorie 186
Scharlachrote Kranzblume 173
Scharlachsalbei 330
Scharlei 330
schatta 83
Schattenblume, Gemüse 347
Schaumkraut, Kleeblättriges 85
Schaumkraut, Resedablättriges 85
schawlij 329
scheber 334
Schefflarie, Aromatische 337
Schefflarie, Duft 337
Schefflarie, Geäderte 337
Schein-Akazie 318
Scheinaster, Bittere 386
Scheinaster, Kleine 387
Scheinbeere, Niederliegende 163
schernomorskaja polyn' 55
schiancia 377
Schiefe Zwiebel 22
Schildpfeffer 295
Schild-Sauerampfer 322
Schimmernder Frauenmantel 14
schinus 337
Schinuspfeffer 337
Schirmkiefer 289
schischetschnik 71
schischnik 71
Schittling 25
schiwit 40
Schlafmohn 272
Schlangen(knob-)lauch 24
Schlangen(knob-)lauch, Wilder 25
Schlangenlauch 23
Schlangenwurz 138
Schlangenwurzel, Virginische 49
Schlanker Dost 265
Schlanker Kardamom 36
Schleifenblume, Mexikanische 129
Schlotte 17
Schlüsselblume, Echte 307
Schlüsselblume, Wiesen 307
Schmächtiger Knöterich 303
Schmalblättriger Doppelsame 137
Schmalblättriger Safran 89
Schmalblättriger Tarant 354
Schmalblättrige Damiana 377
Schmalblättrige Studentenblume 359
Schmalblättrige Wisterie 180
Schmalblättriger Oregano 264
Schmalblättriger Wasserfreund 180
Schmalblättriges Chirettakraut 354
Schmalblättriges Seegras 405
Schmetterlings-Bauhinie 64
Schmetterlingsingwer, Roter 173

Schmetterlingslilie, Weiße 173
Schmetterlings-Pelargonie 277
Schmidtscher Pfeffer 295
Schmuckkörbchen, Schwarzes 117
Schmuckkörbchen,
    Schwefelgelbes 118
Schnabelsame 371
Schneckenklee, Rauher 221
Schneckenklee, Spanischer 221
Schnittknoblauch 18, 27
Schnittlauch 20, 25
Schnittlauch, Alpen 25
Schnittlauch, Arabischer 20
Schnittlauch, Chinesischer 18
Schnittlauch, Sibirischer 25
Schnittlauch, Zimmer 376
schnitt-luk 25
Schnittsellerie 47
Schnittzwiebel 19
Schöner Lauch 17
Schokoladenblume 117
Schopf-Lavendel 201
Schopf-Salbei 330
Schotendorn 318
Schotenpfeffer 82
Schottische Mutterwurz 206
Schottischer Liebstock 206
schporozvetnik 299
Schraubenbaum,
    Amarillisblättriger 271
Schraubenbaum, Duft 272
Schraubenförmiger
    Nachtschatten 350
Schuppenblättiger
    Knoblauchbaum 11
Schuppige Spitzenblume 49
Schwalbenwurz-Enzian 163
Schwanzpfeffer 291
Schwarze Edelraute 53
Schwarze Gelbdolde 348
Schwarze Kosmee 117
Schwarze Marille 49
Schwarze Pfefferminze 150
Schwarze Schafgarbe 3
Schwarze Walnuss 188
Schwarzer Frauenhaarfarn 7
Schwarzer Galgant 30
Schwarzer Holunder 331
Schwarzer Kardamom 11, 36
Schwarzer Koriander 253
Schwarzer Kreuzkümmel 71
Schwarzer Lauch 21
Schwarzer Liebstockel 348
Schwarzer Nachtschatten 350
Schwarzer Pfeffer 293
Schwarzer Pfeffer, Falscher 294
Schwarzer Senf 70
Schwarzer Sesam 181

Schwarzer Zimt 136
Schwarzes Schmuckkörbchen 117
Schwarzkümmel, Acker 252
Schwarzkümmel, Damaszener 253
Schwarzkümmel, Echter 253
Schwarzkümmel, Indischer 386
Schwarzkümmel, Römischer 253
Schwarzkümmel, Türkischer 253
Schwärzliche Fetthenne 339
Schwarznessel 278
Schwarznuss 188
Schwarz-Pappel 304
Schwarzrandige Schafgarbe 3
Schwarzschlundige
    Kanonierblume 286
Schwarzstieliger Krullfarn 7
Schwefelgelbes
    Schmuckkörbchen 118
Schwertbohne 77
Schwertlilie, Deutsche 185
Schwiegermutterzunge 332
Schwindelkorn 116
sceptre de l'empereur 148
sciamar 158
scorbute grass 113
scorodoine aquatique 363
scotch bonnet 83
scotch broom 132
Scotch crocus 121
Scotch lovage 206
Scotch mint 226
Scotch spearmint 226
screw-shaped nightshade 350
scrub myrtle 61
se lu zi 87
se vona 118
sea ambrosia 33
sea fennel 120
sea purslane 178
sea rocket, American 73
sea samphire 120
sea smooth lungwort 231
sea watch 42
sea wormwood 54
seaside angelica 42
seaside jack bean 77
seaside sword bean 77
sebeh 78
Sech(z)uan Pfeffer 401
Sechskronenblättrige
    Klimmtraube 51
sedanino 47
sedano 46, 205
sedano da coste 46
sedano da erbucce 47
sedano da taglio 47
sedano di montagna 205
sedano rapa 47

sedatsch 151
sedefotou 323
Sédhion-Pfeffer 396
sedri 65
Seefenchel 120
Seegras, Gewöhnliches 405
Seegras, Schmalblättriges 405
segan jantan 305
segurelha 334
sehara 64
Seidelbast, Gemeiner 133
Seidiger Mohrenpfeffer 397
seiyo hanazuo 93
seiyô-ibara 318
sekar rus 320
sekisho 5
sel'derej 46, 47
sel'derej korneplodnyj 47
sel'derej listovoj 47
selada air 250
selaseh ayer 208
selaseh banyu 208
selash hitans 260
sĕlasi 257
sĕlasi bĕsar 258
selasseh uteh 256
selderi 47
selderij 46
seledri 47
sellerej 46
Sellerie 46, 47
Sellerie, (Echter) 46
selom 261
selom piopo 261
Seltsamer Lauch 22
sem sum 342
semance de paradis 10
seme de lino 209
sĕmeru 110
semilla de amapola 272
semillas del oyang 395
sèn 399
sena 88, 309
senape 69, 345
senape blanca 345
senape indiana 69
senape selvatica 345
senape vera 70
seneca grass 178
Senegal prickly ash 155
Senegal savory 365
Senegal-Gelbholz 155
Senegalpfeffer 155
sénégrain 374
senegré 374
sénévé 70, 345
sénévé noire 70
senevra 70

Senf, Abessinischer 69
Senf, Brauner 70
Senf, Chinesischer 69
Senf, Englischer 345
Senf, Französischer 70
Senf, Gelb 345
Senf, Gewürz 345
Senf, Holländischer 345
Senf, Indischer 69, 70
Senf, Japanischer 69
Senf, Persischer 147
Senf, Roter 70
Senf, Rumänischer 345
Senf, Russischer 69
Senf, Schwarzer 70
Senf, Weißer 345
Senf, Zerschlitzter 346
Senfbaum 325
Senfkaper 111
Senfrauke 147
Senfkohl 70
sengai 399
sengkoewang 140
senjuang 116
senju-giku 357
Senna, Stumpfblättrige 341
senovka grécka 374
senowka 265
sepat 46
serai dapur 129
serapino 345
serdetschnik 85
sere 129
sereh 129
sereh makan 129
Serehgrass 129
seri 261
serkkom 36
serkom 36
sernam 212
seron épineux 347
serpent garlic 24
serpentary, Virginia 49
serpnik 120
sérpol 368
serpolet 368
seruiatku 47
Sesam 181, 341
Sesam, Geflügelter 341
sesam, Indian 342
Sesam, Indischer 342
sesam, oriental 342
Sesam, Orientalischer 342
Sesam, Schwarzer 181
sesame 181, 342
sésame 342
sesame, black 181
sesamo 342

sésamo 342
sesamzaad 342
sesawi 69
Sesbanie, Großblütige 343
sesedo 396
setawar 118
setwall 128, 381
seuseureuhan 289
seven sister rose 320
sevendara 387
Seville orange 105
Seychellen-Zimt 104
sezam 342
sfakomilia 327
sha jin ko 38
sha ren 38
shad scale 59
shadjrat mariam 52
shadloo 64
shafej muskatny 330
shafellah 79
shafran 122
shafran posewnoj 122
shagaret mariam 53
shaitan 32
shajarat 388
shakkankirai 322
shalari 46
shalott 17
shambhaluka-bija 388
shampoo ginger 404
shamrock 268
shan cong 27
shan hu cai 167
shan huang pi 110
shan kui 391
shan shih lu 309
shan tan 207
shan yu cai 391
shanar bahariya 120
shan-chu-tsai 87
sharapunkha 362
shasmin 187
shata patri 319
shawrina 86
she ch'uang tzu 112
she xiang caotimjan 368
shebet 40
sheep sorrel 269, 321
sheitani 33
shell ginger 31
sheltyj koren' 126
she-ma 179
shemtschushnuj, luk 15
shen xiang cao 182
shen ye shan zhu cai 231
sheng luo le 260
shenwr'e 189

sheperd's knot 306
shepherd's purse 81
sherry pie 196
sherucha wodnaja 250
shetti 186
shi 52, 67, 82, 105, 133, 230, 299, 304, 309, 316
shi cai 230
shi chang pu 5
shi li zi 14
shi liu pi 309
shi rumi 52
shi yong bin li 299
shi yung ta huang 316
shiatan wood 32
shiba 53
shibit 40
shibutschka dubhiza 13
shibutschka polsutschij 13
shichi-henge 196
shih lo 158
shih luo zi 40
shih-yao 179
shih-yung-ma-ch'in-hsien 305
shii-take 202
Shii-take (Pilz) 202
shijriba 261
shikikikat 159
shikimi 183
Shikimifrucht 183
shinori-kombu 195
shinwi 339
shirane-senkyū 43
shiro akoza 95
shiro karashi 345
shiso 278
shiso (green perilla) 278
s(h)obhanjana 240
shôbu 5
shoga 403
shokuyô kanna 78
short buchu 62
short-stemed African nutmeg 238
shouga 403
shouk el-gamal 344
showy calamint 74
showy savory 74
shoyu 168
shrub verbena 196
shrubby basil 258
shrubby decaspermum 134
shrubby pepper 295
shu kann 108
shu kua 86
shui cai 230
shui ching 261
shui long gu 303
shul 7

shungiku 97
shuravel'nik 146
shurnizo 45
shurumma 296
si tuba sawah 280
siah seerah 87
siah zeera 71
Siam cardamom 35, 36
Siamese ginger 30
Siam-Galgant 30
Siam-Ingwer 30, 31
Siam-Kardamom 35
sianama 95
siao suan 19
Siberian avens 114
Siberian rhubarb 316
Sibirische Nelkenwurzel 114
Sibirische Zwiebel 15
Sibirischer Rhabarber 316
Sibirischer Schnittlauch 25
sibirskaja tschermscha 27
sibirskij dinkij luk 15
Sibrischer Anis 184
sibuyas 17
sibuyas tagalog 17
sibuyasna mura 19
Sichuan pepper 400, 401
Sichuan Pfeffer 400, 401
Sicili capper 80
Sicilian sumac 316, 317
sicklepod 341
Siegwurzel, Lange 27
siempreviva picante 340
Sierra Leone-Kardamom 8
siete caldos 84
siete en rama 306
signet marygold 359
sigru 240
sigruh 240
siisbik 383
Sikkim-Kardamom 37
siläüs 343
Silberblatt-Salbei 326
Silber-Brandschopf 90
Silber-Frauenmantel 13
Silber-Hahnenkamm 90
Silber-Messerschmidie 231
Silberminze 226
Silber-Muskat 243
Silber-Perovskie 279
Silberraute 54
Silberregen 318
Silberzwiebel 19, 23
sildu 299
sili 82
silky sassafras 333
silver lady's mantle 13
silver mint 226

silver nutmeg 243
silver sage 279, 326
silver wormwood 53
silverskin onion 19
silybe de Marie 344
simaruba 344
simaruba 344
Simpler's joy 386
simpoh 137
simsim jelilan 342
sinara 88
Sinau 14
singabera 403
singalon 301
single-flowered purslane 306
single-leave cleome 112
sinij kosij trilistnik 373
sinij sweroboj 182
sinouj 253
Sintjans-kruit 180
sintok 103
siok-siok-rangan 280
sira 40, 371
sireh hutan 290
siriphal 7
sisaja gortschiza 69
sison 182, 346
sissie 7
sitab 323
sitaba 323
sitampu 110
siunya 110
siwak 179, 325
siyah 71, 87, 125
siyah zira 87, 125
Sizilianische Kaper 80
Sizilianischer Sumach 317
Skimmie, Baumartige 347
Skimmie, Lorbeer 347
skipidaronoe derovo 297
Skorbutkraut 113
škorica 104
skoroda 25
skryptnica japonskaja 125
skuka posewnaja 147
skunkbush 316
sladké drievko 169
slag garlic 28
slangenlook 25
slaploot 192
slek krey sabou 129
slender cardamom 36
slender leaf marigold 359
slijim appelboom 7
slimy anserine herb 96
slizun 22
Slovenia citron herb 335
Slovenisches Zitronenkraut 335

sluon 24
small absinth 55
small burnet 331
small cardamom 144
small fennel 253
small galangal 31
small leafy watercress 250
small mallow 216
smallage 47
small-flowered mallow 216
small-flowered melissa 224
small-flowered pepper 400
small-leaved oregano 264
small-leaved saffron 89
smartweed 280, 302
smeegolownik 139
smirnija 348
smojanka 85
smolewka chlopuschka 343
smolewka schirokolistnaja 343
snake plant 332
snake wood 89
snakeroot, button 206
snakeroot, Virginia 49
snakeroot, white 152
snap ginger 29
snedok 45
snober 289
snyt'obyknovennija 7
sõ 137
so'dornde 387
so'mayo 387
soajna 240
soap berry 61
society garlic 376
sodugu 144
soebirŭm 305
soet-kroei 129
soewyj bob 168
soffione 361
sogade 176
soh shi 67
sohoehyang 40
so-hsing 187
soi 40, 46, 97
soja 168
Sojabohne 168
sojaboon 168
sojana 240
sokchangpho 5
sokryunamu 309
sokwe 339
solano nero 350
solasi 260
solodka golaja 169
solodka ural'skaja 170
solototycjatschnik malyj 91
som chit 159

som ma fai   111
som mu'   108
somak   317
somalata   323
Somali myrrh   115
Somalia-Myrrhe   115
somchaba   1
sômez sein   345
sommacco   317
(Sommer-) Bohnenkraut   334
Sommer-Knoblauch   15
Sommerzwiebel   17, 18
Sommer-Zypresse   64
sondrio   296
songguji   321
Sonnenwende   75, 174
Sonnenwende, Peruanische   174
Sonnenwende, Strauchige   174
Sonneratie, Käseartige   351
sonth   403
sontol   129
sontschina   15
sopravivolo dei muri   340
sorbastrella   332
sorbier   338
sorelu   320
sorrel   59, 177, 268, 320
sorrel dock   320
sorrel, bloodred   322
sorrel, bluish   269
sorrel, common   320
sorrel, French   322
sorrel, garden   320, 322
sorrel, green   269
sorrel, Indian   177
sorrel, Jamaica   177
sorrel, red   177
sorrel, sheep   269, 321
sorrel, speared   321
sorrel, wood   268
sorrel, yellow   268, 269
sorujaengi   321
sosna pinja   289
sostera   405
souchet des Indes   126
souci d'eau   75
souci officinal   75
soukoum   79
soup celery   47
soup mint   299
sour dock   320
sour lime   105, 107
sour orange   105
source of silymarin   344
sourgrass   320
South African pepper tree   192
South American apricot   217
South American nutmeg   124

South American vanilla   383
southern (wood) geranium   275
southern (worm) wood   52
Southern prickly ash   400
sowa   40
soy   40
soya   168
soya bean   168
soya putih   168
spaanse peper   82, 83
spaanse salie   328
Spain hop   264
Spain stonecrop   340
Spain thyme   299
Spaltblättriger Frauenmantel   13
Spaltkörbchen, Chinesisches   338
Spanische Fetthenne   340
Spanische Minze   228
Spanische Verbene   366
Spanischer Hopfen   264
Spanischer Lauch   21
Spanischer Oregano   365
Spanischer Pfeffer   82
Spanischer Salbei   328
Spanischer Schneckenklee   221
Spanischer Thymian   211, 366, 369
Spanisches Süßholz   169
Spanish (wild) marjoram   366
Spanish chaemomila   39
Spanish chervil   247
Spanish elder   289
Spanish garlic   17, 23, 25
Spanish garlic rocambole   25
Spanish hop   264
Spanish lavender   201
Spanish liquorice   169
Spanish mint   228
Spanish needles   65
Spanish onion   17
Spanish oregano   365, 368
Spanish pellitory   39
Spanish saffron   87, 122
Spanish sage   328
Spanish thyme   211, 366, 369
Späte Indianernessel   236
Späte Traubenkirsche   308
spatheflower   351
speared sorrel   321
spearmint   225, 229
spearmint, Kentucky   225
spearmint, Scotch   226
Spechtwurzel   136
Speicherähre   249
Speik, Echter   381
Speik, Gelber   381
Speik, Keltischer   381
Speik, Roter   381
Speik, Weißer   3

Speik-Lavendel   201
Speisezwiebel   17
Sperberkraut   332
Sperrige Saposhnikovie   333
spic   201
spice bush   208
spice cow   94
spice lily   191
spice tree   395
spicy berry   49
spicy cedar   65
spider flower, African   111
spider herb   111
spider wisp   111
spiderling, erect   67
Spießiger Ampfer   321
spignel   232
spike lavder   200
spike lavender   200
spiked pepper   289
spikenard   181, 201, 249
Spikenard   181, 201, 249
spikenard, American   201
spikenard, wild   181
Spikenarde   201
spilanthe des lieux humides   351
spina cervina   315
spinach, Ceylon   63
spinach, Indian   62
Spinat, Indischer   63
Spinat, Malabar   63
spindlepod   112
spineless caper   80, 81
spinks   84
Spinnenbaum   120
Spinnenpflanze, Afrikanische   111
Spinnenpflanze, Einblättrige   112
spiral ginger   118
spiral nightshade   350
Spiraliger Nachtschatten   350
Spiralingwer   118
spirée ulmaire   157
Spitzblütiger Mohrenpfeffer   395
Spitzenblume, Schuppige   49
Spitzer Lauch   21
Spitzlappige Angelika   41
spoonwort   113
sporysch   302
spotted gentian   164
spottet beebalm   237
spreading marigold   358
spring onion   19
spring savory   4
spring star flower   185
Springgurke, Amerikanische   235
Springgurke, Bittere   235
Springkürbis   235
Spring-Wolfsmilch   152

spur pepper   83
spurge, caper   81, 152
spurge, garden   152
spurge, myrtle   152
squaw mint   173
squash pepper   83
Sri Lanka cinnamom   104
srikhanda   332
srja   205
St. Jansbrood   93
St. John's bread   92, 93
St. John's Wort   180
St. Lucie cherry   308
St. Mary's thistle   344
Stab-Asant   156
Stacheldolde   143
Stachelesche, Täuschende   401
Stachelige Toddalie   369
stachys, wolly   353
stammed vine   105
Stangenpfeffer   292
Stangensellerie   46
staphylier   353
star anise   183
star anise, bastard   183
star anise, Cambodian   183
star anise, Chinese   184
star anise, Florida   183
star anise, yellow   183
star lily, Japanese   207
star lily, morning   207
starejuschij luk   26
statjing   118
Staudenmajoran, Echter   265
Staudenmajoran, Falscher   265, 266
Staunton elsholtzia   145
Stechwinde, Osterluzeiblättrige   347
Steckenkraut   156
Stecklaub   205
Steifer Lauch   26
Steinklee, Blauer   372, 373
Steinklee, Echter   222
Steinklee, Gelber   222
Steinklee, Hoher   221
Steinklee, Sumpf   221
Steinklee, Weißer   221
Steinmelisse   251
Steinminze   125
Steinminze, (Amerikanische)   126
Steinpetersilie   124
Steinpetersilie, Kanadische   124
Steinpfeffer   339
Steinpilz-Petersilie   331
Steinquendel   4
Steinquendel, Alpen   4
Steinquendel, Echter   74
Steinquendel, Großblütiger   74
Steinquendel, Wald   74

Steinraute   3
Steinweichsel   308
stellina odorosa   161
Stengelloser Alligartorpfeffer   9
Stengelumfassende Gelbdolde   348
Steppenflachs   209
Steppenraute   275
Steppen-Thymian   367
sterculia, longleafed   354
sterkers   203
sterkgras   202
sterkkos   202
Sterkulie, Langblättrige   354
sternanijs   184
Sternanis   183
Sternanis, Florida   183
Sternanis, Gelber   183
Sternanis, Heiliger   183
Sternanis, Japanischer   183
Sternanis, Kambodscha   183
Sternblume, Vielblütige   185
Stern-Bockshornklee   374
Stern-Taglilie   175
stianca   377
sticky sage   328
Stiefmütterchen, Pyrenäen   387
Stiellauch   26
Stielpfeffer   291
stinck fetida   156
sting pepper   140
stink bean   273
Stinkasant   156
stinkboon   273
Stinkdill   116
Stinkdistel   147
Stinkender Mannstreu   147
Stinkendes Mariegold   274
stinking cedar   370
stinking horse mint   237
stinking roger   358
stinking yew   370
Stinkpeterle   358
Stink-Pferdeminze   237
stoechas arabique   201
stolowij perez   82
stone mint   125
stone mint, American   126
stone parsley   124, 346
stone pine   289
stonecrop, alpine   339
stonecrop, annual   339
stonecrop, blackish   339
stonecrop, goldmoss   339
stonecrop, Jenny   340
stonecrop, mossy   339
stonecrop, panicle   340
stonecrop, reflexed   340
stonecrop, reddish   340

stonecrop, rock   340
stonecrop, Spain   340
stonecrop, tawny   339
stonecrop, thick leaved   340
stonecrop, two row   340
stonecrop, white   339, 340
Storchschnabel, Düsterer   165
Storchschnabel, Felsen   165
stragon   53
Strahlen-Hohlsame   65
Strahliges Eremochariskraut   146
Strand-Ambrosie   33
Strandbeifuß   54
Strandportulak   178
Strauch-Felsenklippe   233
Strauchige Fiederaralie   303
Strauchige Sonnenwende   174
Strauchnessel, Gekerbte   197
Strauchnessel, Oliver   284
Strauch-Otostegie   267
strawberry blight   96
strawberry tomato   285
strigoli   343
striped African pepper   396
striped Mexican marigold   359
stschavel' malyj   321
stschavelek   321
stschawel stschitkoviduij   322
stschebruschka   4
stschewel' kislyj   321
Studentenblume, Abstehende   358
Studentenblume, Aufrechte   357
Studentenblume, Gestreifte
   Mexikanische   359
Studentenblume, Gewürz   359
Studentenblume, Große   358
Studentenblume, Hohe   357
Studentenblume, Kleine   358
Studentenblume, Lakritzen   357
Studentenblume, Mexikanische   358
Studentenblume, Schmalblättrige   359
Studentenblume, Süße   358
Stumpfblättrige Senna   341
Stumpfblättriger Ampfer   321
su   20, 57, 65, 187, 221, 278
su fang hua   187
suan jiao   359
suan ning mêng   105
suan-ch'eng   105
sua-sua   374
suber hiyu   305
subit   369
subrovka duschistaja   178
succade   108
Sudan-Kaffee   273
Sudantee   177
Südfranzösischer Thymian   369
sudsa   278

suédois   82
sufaid murga   90
sufed rai   345
suganda   299
sugánda   299
sugandha   30
sugandhavacha   191
sugandhi pala   176
sui xu chang xiang   181
sukade   108
sukampo   320
suma   317
suma des corroyeurs   317
sumac   316
sumac odorant   316
sumac, Sicilian   316
sumach   316
Sumach, Duft   317
Sumach, Feinnerviger   317
sumach, fragant   316
Sumach, Gewürz   317
sumach, lemon   316
Sumach, Sizilianischer   317
Sumach, Süßer   317
Sumach, Tanner's   317
Sumach, Wohlriechender   317
sumagre   317
Sumak   317
sumak   317
sumakh   317
sumaq   317
sumidad de hierba terreste   167
summak   316
summer cypres   64
summer savory   334
Sumpfdotterblume   75
Sumpf-Fetthenne   341
Sumpf-Haarstrang   283
Sumpfklee   230
Sumpfkresse, Gewöhnliche   318
Sumpf-Magnolie   216
Sumpfmyrte   242
Sumpfspirea   157
Sumpf-Steinklee   221
Sumpfstern, Blauer   354
sundali   88
sunthi (dry ginger)   403
suob-kabayo   181
Suppenbasil   256
Suppenlauch   23, 25
Suppenlob   205
surabhi nimbu   241
surabunai   255
surage   371
suramarit   100
surampunna   255
surangji   239
surasa   256

sureau noir   331
sureau commun   331
surepiza   62
surepka obyknowennaja   62
surette   268
surgi   255
Surinam quassia   311
Surinambitterholz   311
Surinamese zuring   177
Surinam-Giftbaum   362
suring   269
suringi   255
surva   40
suscheniza   170
Süßdolde   247
Süße Duftblüte   266
Süße Myrrhe   115
süße Orange   109
Süße Pestwurz   281
Süße Schafgarbe   2
Süße Studentenblume   358
Süßer Fenchel   158
Süßer Kümmel   288
Süßer Oregano   210
Süßer Pfeffer   82
Süßer Sumach   317
Süßer Zitronenstrauch   210
Süßholz, Amerikanisches   169
Süßholz, Anatolisches   170
Süßholz, Chinesisches   170
Süßholz, Deutsches   169
Süßholz, Griechisches   169
Süßholz, Italienisches   169
Süßholz, Römische   169
Süßholz, Russisches   169, 170
Süßholz, Spanisches   169
Süßkraut, Aztekisches   210
Süßminze   229
sutejasi   401
su-tzu su   278
suva ni bhaji   40
suvarnaka   88
suvasa tulasi   260
suzuka-zeri   43
swamp laurel   215
swarte mosterd   70
Swato orange   109
sweet basil   255, 256
sweet bay   199, 216, 280
sweet bay, Virigian   216
sweet cicely   247
sweet clover   221
sweet coltsfoot   281
sweet cumin   288
sweet fennel   158
sweet flag   5
sweet gale   242
sweet grass   5, 178

sweet hearts   65
sweet jarvil   267
sweet mace   358
sweet majoram   263
sweet mary   360
sweet mignonette   314
sweet myrtle   5, 242
sweet nancy   2
sweet olive   266
sweet orange   109
sweet osmanthus flower   266
sweet pepper   82
sweet reseda   314
sweet scented basil   259
sweet scented geranium   276
sweet scented marigold   358
sweet scented myrrh   247
sweet scented violet   388
sweet trefoil   373
sweet trifoil   372
sweet verbene myrtle   61
sweet vernal grass   44
sweet woodruff   161
sweet yarrow   2
swerdstschatyj anis   184
sweroboj pronsjonnolistnyj   180
swizzlestick tree   311
syahjira   87
Syrian bean caper   405
Syrian nard   382
Syrian oregano   265
Syrian rue   275, 322
Syrische Narde   382
Syrische Raute   275, 322
Syrische Zwiebel   17
Syrischer Dost   265
Syrischer Majoran   265
Szechuan pepper   399, 400
Szechuanpfeffer   399, 400
szri   400
szu kai kat   159
t'ado   315
t'ung hao   97
ta ayam   196
ta cham diang   212
ta hui hsiang   112, 184
ta hui hsiang ts'ao   112
ta liao   302
ta suan   25
taa-ya sa-ya   127
tabacarana   300
tabakugii   349
tabasco (pepper)   83
Tabasco (Pfeffer)   83
tabasco pimento   287
Tabasco-Pfeffer   83
Tabasco-Piment   287
tabel   116

tacamahaca popular   304
tacoutta   341
taehwang   316
taėy   271
taflan   308
tagète rose d'Inde   357
tagète tachée   359
Tagetes, Lakritzen   357
tagimunau   374
taglak babae   30
Taglilie, Bahnwärter   175
Taglilie, Braunrote   175
Taglilie, Gelbe   175
Taglilie, Kleine   175
Taglilie, Orangefarbene   175
Taglilie, Rotgelbe   175
Taglilie, Stern   175
Taglilie, Wiesen   175
Taglilie, Wilde   175
Taglilie, Zitronen   175
tagpo   49
Tahiti lime   107
Tahitian vanilla   385
Tahitipfeffer   215
Tahiti-Vanille   385
tai tai   195
tai to   278
tail grape   51
tailed pepper   291
tailor tree   134
taiphal   244
Taiwan cinnamon   102
Taiwan momiji   303
Taiwanischer Zimt   102
tala   208
talas padang   114
talewort   68
taling pling   59
talingkup   110
tall mallow   216
tall meliot   221
tall nasturtium   375
tall palm   143
tall woody vine   157
tall yellow   221
tallae   20
talmud widder hyssop   233
tam lung   113
tamanegi   17
tamarang   59
tamarind   49, 162, 298, 359
tamarind pulp   161
tamarind, fish   162
tamarind, Manil(l)a   298
Tamarinde   162, 298, 359
tamarinde   162, 298, 359
Tamarinde, Malabar   162, 298
Tamarinde, Manila   162, 298

tamarindeiro   359
tamarindo   359
tamarindo (indico)   359
tamarine   359
tamarinier   359
tambak-tambak   387
tambarine   359
tambol   401
tamoe koentji   67
tampang   57
tamre   359
tamre-nindi   359
tan hsiang   332
tan xiang   332
tanaceto   360
tanaisie vulgaire   360
tando   315
tang jiao shu   32
Tangarine   109
tangarine   109
tangerine sage   327
tanghaercherking   212
tangila   267
tangkhong   48
tang-kuei   41
tanglad   129
tanglu   12
tankugijanamu   213
tanminari   274
Tanner's Sumach   317
tannoun   33
tansy   360
tansy, common   360
tantandok   111
tapak burong   234
tapak kud   64
tápana   80
tàpara   81
tapiá   120
tapia fruit   120
Tapiafrucht   120
tapioca, wild   78
tapira coinana   88
tapodhana   57
tarak sahha   361
taramira   147
Tarant, Blauer   354
Tarant, Schmalblättriger   354
tarassaco   361
tara-tara   208
tarchun   53
tardes, buenas   234
tardes, clavenilla   234
targone   53
taribed   147
tarkhun   53
taro, great   114
Taro, großer   114

tarragon   53, 358
tarragon, Mexican   358
Täschel   81
Taschenkraut   81
Tasmanian blue gum   150
Tasmannia pepper   140
Tasmanniapfeffer   140
tasneira   360
tassel flower   146
tasso verbasso   385
tatari   317
tatarka, luk   19
tatarskoe sel'e   5
tatli rezene   158
tatrak   317
Täuschende Stachelesche   401
Tausendblatt   3
Tausendgüldenkraut, Echtes   91
Tausendgüldenkraut, Kopfiges   91
tavitri   244
Tavoy cardamom   38
tawar tawar   118
tawn day lily   175
tawny stonecrop   339
tawolga   157
té de Méjico   95
tè del Méssico   95
te thai   81
tea berry   163
tea bush   209, 211, 258
tea hyssop   233
tea olive   266
tea tree   204, 213, 303
tea-bush   181
teborak zachun   61
tebu   118
teentekage   314
teentemo   314
Teestrauch, Gambia   211
tefrosija   362
teja   101
teja   101
tejpat   104
tejphal   399, 401
teklan gede   151
telacucha   113
tella-pulugudu   339
Tellerkraut   239
Tellerkresse   203
telur belangkas   12
temo potre   126
temoé-lawag   128
tempalang   63
Tempelbaum   120
Tempelpflanze   120
temple plant   120
temple tree   120
temu kunchi   67

temu kuning   403
temu lawang   128
temu puteri   126
tengé   310
tengguli   88
těngguli   88
tep pirou   101
tepai   404
tephrosie, linear   362
Tephrosie, Lineare   362
tephrosie, purple   362
Tephrosie, Purpurne   362
tepus   38, 149, 404
tepus batu   37
tepus halia   404
kampong kampong   149
tepus tanah   404
tepus walang   149
terbinto   338
terebinth   297
Terebinth   297
terebinth, Turk   297
térébinthe   297
terebinto   297
terfaq   13
terhilsok   212
teriiroa   385
terong   349
terra noce   71
terre-mérite   126
terre-noix   71
tête-de-dragon   139
Tetrapleure   362
tetri   317
têu dâu   191
Teufelsbaum   32
Teufelschnoblech   27
Teufelsdreck   156
Texas white bush   28
tezbal   401
tha khaan lek   294
thaengjanamu   304
Thai basil   260, 278
Thai-Basilikum   278
Thai-Ingwer   30, 31, 191
Thailändisches Pfefferkraut   295
thale   392
thao qua   35
thap thim   309
thé arabe   211
thé bourrique   377
thé de bois   163
thé de Gambi   211
thé de la Grèce   329
thé des Canaries   89
thé du Mexique   95
Therebinth-Pistazie   297
Theriakwurzel   41, 288

Theriakwurzel, Deutsche   288
Theriakwurzel, Weiße   288
thian khaopluak   40
thian tatakkataen   40
thianklaep   158
thick leaved stonecrop   340
thidin   65
thin-leaved African nutmeg   239
thistle sage   326
thistle, blessed   112
thistle, holy   112, 344
thistle, milk   344
thom   24, 298
thom, Madras   298
thong laang laai   148
thong bach   19
three-leaved lantana   197
thryba   336
thua rae   168
thum   24
Thwaites sampi   198
thym citronné   365
thym commun   368
thym cultivé   368
thym de Candie   365
thym pecoce   367
thym rampant   367
thym sauvage   368
thyme lemon   366, 367
thyme serpolet   368
thyme, Azores   365
thyme, Broussonet   365
thyme, Austrian   367
thym, caraway   366, 367
thyme, cat   363
thyme, common   368
thyme, conehead   365
thyme, creeping   367, 368
thyme, Cretan   365
thyme, five-ripped   368
thyme, French   299
thyme, garden   368
thyme, hairy   367
thyme, Japanese   368
thyme, Java   210
thyme, mastic   366
thyme, pilous   367
thyme, sauvage   368
thyme, Spain   299
thyme, Spanish   211, 366, 369
thyme, tiny   365
thyme, wild   367, 368
thyme, winter   366
Thymian   210, 299, 336, 365
Thymian, Arznei   367
Thymian, Behaarter   367
Thymian, Broussenet   365
Thymian, Cuba   299

Thymian, Echter   368
Thymian, Feld   367, 368
Thymian, Frühblühender   367
Thymian, Garten   368
Thymian, Gewürz   368
Thymian, Jamaika   299
Thymian, Japanischer   368
Thymian, Javanischer   210
Thymian, Joch   369
Thymian, Kretischer   336
Thymian, Kümmel   366
Thymian, Langstengliger   366
Thymian, Österreichischer   367
Thymian, Pannonischer   367
Thymian, Pfeffer   233
Thymian, Piemonteser   367
Thymian, Spanischer   211, 366, 369
Thymian, Steppen   367
Thymian, Südfranzösischer   369
Thymian, Tiroler   367
Thymian, Wilder   368
Thymian, Zitronen   365
Thymianblättrige Bergminze   233
ti fu   64
tian hui xiang   158
tian jia hu ci   86
tian-cheng   109
tiao teng   379
tiarei   385
Tibetan rhubarb   315
tickweed   173
ticed bub   55
tien chieh tzu   350
tien pao tsao   350
tiger cat   332
tiger lily   207
tigeris claw   148
Tigerlilie   207
ti-gü   331
tijm   368
tikas tikas   78
tikasag   120
tikusan   110
til   342
tilam wangi   301
till   342
Tille   40
tim'jan   368
tim'jan duschistuj   368
tim'jan obyknowennyj   368
tim'jan polsuutschik   368
timal   401
timbar   401
timersat   230
timersidi   230
timiamnik   368
timo   365, 368
timo arbustino   365

timo maggiore   368
timo serpillo   368
timo volgare   368
timon   87, 125
timtima   317
timur   400, 401
tin ped   32
ting hsiang   355
ting tzu hsiang   355
tinisas   57
tiny cudweed   170
tiny thyme   365
tipi   282
tipo   234
tipo leaf   234
tirimli   317
tiringuini   357
Tiroler Thymian   367
tirphal   401
tithai   285
tithu   285
titipati   57
titka sak   120
tjakar kootjing   303
tjekur   191
tlilxochitl   383
tmin   87, 125, 254, 371
tmin obyknowennyj   87
tmin tminowyj   125
tô gara shi   82
tô kin sen ka   75
tobacco-tree   349
Toddalie, Chinesische   369
Toddalie, Stachlige   369
toei daang   272
toei hum   272
toejiphul   305
toemae kontji   67
toey hom   271
toí   24
toko   66
Tolubalsam   246
tôm   24
tomanag   48
tomar seed   399
tomat   213
tomate   213, 285, 348
Tomate   213, 285, 348
tomate amère   348
tomate de cáscara   285
tomate frais   285
tomate silvestre   285
tomate verde   285
Tomate, Äthiopische   348
Tomate, Kannibalen   349
Tomate, Kinder   348
Tomate, Menschenfresser   349
tomatillo   285

tomatillo de campo   285
Tomatl   285
tomato   213, 285, 348
tomato of the Jews of
    Constantinoble   348
tomato, Cannibal's   349
tomato, children's   348
tombak-tombak   66
tombe   310
tomhom   280
tomilho   365, 368
tomillo   365, 368
tomillo blanco   366
tomillo común   368
ton horm   19
tondsha   176
tongibuchu   27
Tonka   137, 361
tonka bean   137
Tonkabohne   137, 361
Tonkabohne, Englische   361
Tonkabohne, Punktierte   138
tonkinkan   159
Tonkin-Zimt   102
tonquin   137
toothache plant   351
toothache tree   400
toothbrush tea   325
toothead bur clover   221
top onion   23
topol' bal'samitscheskij   304
topol' tschjornyj   304
tor elaga   176
toraji   299
torch ginger   148
torkruid   261
tormentill(a)   306
Tormentilla   306
tormentilla   306
tormentille commune   306
Tormentillwurzel   165
Torn's herb   138
toroni   29
toronja   106
toronjil   12, 223
torvidco   133
torreya, California   370
torreya, leaved   370
tosuke böfu   333
totoji   12
toum   371
toumba   77
Touristen-Safran   87
toute bonne   96, 330
toute épice   253, 286
toutou blanc   273
toy wort   81
traba   205

tragoselino   288
trainasse   302
tram hoi   196
trang nguyên   152
trangguli   88
tranpetier   89
Transcaucasian catnip   252
Transkaukasische Katzenminze   252
trapiá   120
Trauben-Kardamom   35
Traubenkirsche   308
Traubenkirsche, Späte   308
Traubenkopf, Gemeiner   343
Traubenkraut   95
Traubenkraut, Mexikanisches   95
Traubige Katzenminze   252
Traubiger Bockshornklee   373
Traubiger Langfaden   115
trava potschetschujnaja   302
trawa bogorodskaja   368
trawa morskaja   405
trébel   304
trébol aceto   268
trébol de aquático   230
trébol dorado   222
trébol dulce   222
trébol oloroso   221
tree of kings   116
tree wormwood   53
tree, onion   23
tree, sorrel   59
trèfle blau   372
trèfle bleu   373
trèfle d'eau   230
trèfle musqué   221, 372
trèfle odorant   372
tregano   93
três-coraçoes   268
trevo   221, 268
trevo de cheiro   221
trevocheiroso   222
trifoglina   221
trifoglio cavallino   222
trifoglio d'acqua   230
trifoglio fibrino   230
trifol'   230
trifoliate bitter cress   85
trifoliate lime   374
trifoliate orange   304
trigonelle   374
Tripmadam   340
trippe-madame   340
trique madame   340
tritonia   121
Tritonie, Crocus   121, 376
Tritonie, Safranartige   375
trivalve   198
Troddelblume, Javanische   146

Trompetenbaum 89
trong 401
tronjia 386
tropical crocus 67
tropical lily 192
true cinnamom 104
true laurel 199
true mousseron 218
true mustard 70
true nutmeg 244
Trügerischer Lauch 26
trumpet tree 89
ts'ao 112, 167, 175, 383
ts'ung 17, 19
tsa chiang 268
tsampaka 232
tsao ko 38
tsao-ko cardamom 38
Tsao-ko Kardamom 38
tsauri grass 130
tschaber 334
tschaber kolosonosnyj 336
tschaber sadowuj 334
tschaber tschimnij 335
tschabrez 184, 368
tschapolot' 178
tschampaka 232
tscheber 334
tscheremscha luk 27
tscheremycha-antipka 308
tschermscha 27
tschernij tmin 254
tschernji perez 293
tschernobyl'nik 57
tschernogolownik krowochllebkowyj 331
tschernomorskaja polyn' 55
tschernuschka polevaja 252
tschernuschka posevnaja 253
tschesnoik 24
tschesnotschnaja trawa 14
tschesnotschnik 14
tschesnotschniza 14
tschistez 166
tschistjak ljutitschnyj 313
tschnuschka damasskaja 253
tschobr 334
tse suan 27
tsi 391
tsin tsin 55
tsisombo 339
tsui tsao 230
ts'ung 19
tu bi 300
tu giang huo 173
tu tou 48
tu xiang ru 265
tuba 63, 280

tube seluwang 280
tuberous nasturtium 376
tuboh lalap 280
tuboh perpancej 280
tucacsillu 220
tufted pansy 387
tui' tiên 232
tuimeruca 147
tuinkers 203
tuinsuring 269
tuka benang 134
tukai benai 134
tulashi 260
Tulasi-Basilikum 260
tule mint 224
tŭlkkae 278
Tulsi 20
Tulsi 181, 255, 260
Tulsi, Grüner 260
Tulsipflanze 260
tulumisi 285
tum el-emlak 25
tum rlkhabazeen 18
tumbar 116
tumble weed 205
tumburu 399
tum-zu el-raas 15
tuna anís 357
tundopoda 401
tunggulle 280
tunggulpha 17
tungi 256
tung-k'uei 217
tungkadi pari 116
tunguas 159
tungüniptaehwang 316
Tüpfelenzian 164
Tüpfelfarn, Gewöhnlicher 303
Tüpfelhartheu 180
Tüpfel-Johanniskraut 180
turanj 108
Turbitobaum 338
turi 343
Turk terbinth 297
Türkenbundlilie, Große 207
Türkenkresse 85
Turkestan oregano 265
Turkestanische Zwiebel 20
turkey grass 386
Turkey rhubarb 316
Türkische Kresse 85, 375
Türkische Melisse 139
Türkische Weichsel 308
Türkischer Drachenkopf 139
Türkischer Kümmel 125
Türkischer Lavendel 139
Türkischer Mastix 297
Türkischer Pfeffer 82

Türkischer Rhabarber 315, 316
Türkischer Schwarzkümmel 253
Turkish lavender 139
Turkish oregano 264
Turkish sage 327
turmeric 78, 126, 128
turmeric, African 78
turmeric, Japanese 128
turmeric, Javanese 128
turmeric, wild 126
turmeric, yellow 128
turmérico 127
turnip rooted celery 47
turnip rooted chervil 94
turnip-rooted parsley 94, 283
turpentine, (Cyprus) 297
Tuscan jasmine 188
tutai 285
tuttabuona 96
tvak 104
twisled leaf 22
two row stonecrop 340
tysjatschegolov 381
tysjatschelistnik obyknowennyj 3
tzu 2, 112, 152, 158, 256, 272, 278, 284, 297, 304, 309, 338, 350, 355
tzu mo lo 158
tzu su 278
tzu tan 309
tzu tung 304
uampi galumpi 111
uban kayu 85
ubi gereda 78
ubud udat 149
uchu 83, 337
ueang phet maa 118
Ufer-Ampfer 321
Uganda warburgia 391
Uganda-Warburgie 391
uguma 342
ui 17, 158
ui-kyô 158
ukon 30, 126
ukpe 66
ukron 40
ulasi 260
ulgum 126
ulisiman 305
umbrella 281, 289
umbrella pine 289
umckaloeba, African 277
umm re-roubia 219
umtao 46
Unechte Veilchenwurzel 185
Unechter Muskat 244
Ungarischer Enzian 164
Ungarischer Pfeffer 82
Ungarischer Salbei 326

Ungewöhnlicher Nachtschatten 348
unhjang 323
unkwisi 314
unsuy 116
upakunchika 144
upalasari 176
upland cress 62
Urban-Nelkenwurzel 166
uritchu 352
urucu 65
urucum 65
usiira 387
usonsumbi 115
usugi mokusei 266
uva di monte 163
uwâguwâ kâtô 350
uwiichô 158
uzarih 275
uzina 85
vaca ugragandha 5
vachta trilistnaja 230
vacouet 272
vainiglia 383
vainil 383
vainillero 383
vaj 5
vajra valli 105
Valais wormwood 56
Valencia orange 109
valerian 381, 155
valerian, African 155
valeriana 381
valeriana (mayor) 381
valeriana aptetschnaja 381
valeriana silvestre 381
valériane (officinale) 381
vampi 111
vanashempaga 153
vaneri 196
vaniglia 383
vanil' 383
vanilão 384
vanilka 383
vanilla 178, 382
vanilla fruit 383
vanilla grass 178
vanilla of Bahia 383
vanilla, Bahia 383
vanilla, Bourbon 383
vanilla, Brazil 383
vanilla, great 384
vanilla, Indonesian 382
vanilla, Java 382
vanilla, Pompon 385
vanilla, South American 383
vanilla, tahitan 385
vanilla, West Indian 385
vanilla, white 382

vanillaroot 206
vanille 382
Vanille 382, 383
vanille de Mexique 383
vanille de Tahiti 385
vanille de Tiarei 385
Vanille, Antillen 385
Vanille, Bourbon 383
Vanille, Brasilianische 383
Vanille, Echte 383
Vanille, Große 384
Vanille, Guadeloupe 385
Vanille, Indonesische 382
Vanille, Pompon 385
Vanille, Tahiti 385
Vanille, Weiße 382
Vanille, Westindische 385
Vanilleblume 174
Vanillefrucht 383
Vanillegras 178
Vanilleschote (wrongly !) 383
Vanillewurzel 206
vanillier 383
vanillon 385
Vanillon 385
vap ca 178
varuna 120
varvara 256
vasabi 391
vash 5
Vavilov leek 27
vavilova luk 27
Vavilov-Lauch 27
vegetable humming bird 343
Veilchen, Gehörntes 387
Veilchen, Horn 387
Veilchen, März 388
Veilchen, Wohlriechendes 388
Veilchenblütiger Knoblauch 376
Veilchenwurzel, Florentinische 185
Veilchenwurzel, Unechte 185
vellaippuanji 339
velvet 172, 184, 227, 231
velvet dock 184
velvet plant 172
venkel 158
Veracruz sarsaparilla 347
verba de la princesa 386
verbasco 385
verbasco maschio 385
verbena 196, 209, 211, 386
verbena odorosa 211
verbena, common 386
verbena, lemon scented 386
Verbene, Echte 211, 386
Verbene, Spanische 366
Verbene, Zitronen 211, 386
Verbenenkraut 211

verdadeira 95, 217
verdolaga (común) 305
verdolaga de invierno 239
verdolago de Cuba 239
vernonia 386
vernonija mindal'naja 386
veronia, almond 386
Veronie, Bittere 386
Veronie, Mandel 386
Verschiedenfarbiger Salbei 327
vervain 353, 386
vervain, bastard 353
vervain, European 386
vervain, Jamaica 353
verveine de Pérou 211
verveine 129, 209, 211, 330, 366, 386
verveine citronée 211
verveine d'Afrique 209
verveine d'espagne 366
verveine des Indes 130, 211
verveine odorante 211, 386
verveine officinelle 386
verveine sage 330
Verwachsener Frauenmantel 13
verweine de Indes 129
vetiver (grass) 387
Vetivergras 387
vétivier 387
vettiver 387
vidari 349
Vielähriger Basilikum 63
Vielblütige Sternblume 185
Vielblütiger Organo 211
Vierkantige Klimme 105
Vierspaltiges Burzelkraut 306
Vietnamese balm 145
Vietnamese cassia 102
Vietnamese coriander 280
Vietnamesische Melisse 145
Vietnamesischer Koriander 280
Vietnamesischer Pfeffer 294
Vietnamesischer Wasserfenchel 261
Vietnam-Zimt 102
vigne de Bakel 105
vigne du nord 179
vilayaiti lasson 23
vilayati afsantin 52
vinagreira 177
vinagrera 321
vinagrillo 59, 321
vine, Madeira 44
viola mammola 388
viola zopa 388
violet 387
violet, common 387
violet, garden 388
violet, sweet scented 388
violeta 387, 388

violete cornue   387
violetta   388
violette de mars   388
violette odorante   388
violo da pesci   84
Virgiana laurel   215
Virgiana mountain mint   310
Virgiana piknanthemum   310
Virgiana pycnanthème   310
Virgiana, serpentary   49
Virgiana, snakeroot   49
Virgiana sweet bay   216
Virgiana thyme   310
Virgianische Bergminze   310
Virgianische Magnolie   216
Virgianischer Nachtschatten   350
Virgianischer Wacholder   189
Virginia pepperweed   203
Virginia serpentary   49
Virginia snakeroot   49
Virginian cedar   189
Virginische Kresse   203
Virginische Schlangenwurzel   49
visalatvak   32
viscous elecampane   185
visnaga   33, 34
visnaga maggiore   33
vite aquilina   220
viznaga   34
vlas   209
vlinder pelargonium   277
voampirifery   293
Vogelbrot   339
Vogelknöterich   302
Vogelpfeffer   82, 83
volantincillo   111
volatín   111
vông nem   148
vôňovec citrónový   129
vorepuvan   339
vriadhatulasi   258
waan hao non   192
waan hom   191
waan teendin   191
Wacholder, Gewöhnlicher   189
Wacholder, Phönizischer   189
Wacholder, Virgianischer   189
Wachsbaumwurzel, Essbare   86
Wachsbeere   242
Wachs-Myrte   242
Wahrsager-Salbei   327
wakame   380
wakegi   23
Wakegi onion   23
walang   149
Warburgie, Heil   391
Warburgie, Uganda   391
Wald-Engelwurzel   43

Wald-(knob-)lauch   27
Waldbergminze   74
Walddolde   97
Waldgamander   363
Waldmeisterkraut   161
Waldsalbei   363
Waldsauerklee   268
Wald-Steinquendel   74
wall fern   303
wall gamander   363
wall germander   362
wall pepper   339
wall sage   363
Walliser Beifuß   56
Walliser Genipp   56
Walliser Wermut   56
wally sage   325
walnoot   188
Walnuss, Echte   188
Walnuss, Schwarze   188
Walnuss, Welsche   188
walnut, black   188
walnut, English   188
walnut, Persian   188
Walzenförmiges
   Endostemonkraut   146
wampee   111
wampee Chinese   111
wampi   111
wampoi   111
wan shou guo   86
Wandelröschen   196, 197
Wandelröschen, Dreiblättriges   197
Wandelröschen, Kleinblättriges   197
Wandelröschen, Lilafarbenes   196
Wandelröschen, Rhodesisches   197
wan-fai   402
wang pu liu hsing   381
wanila   383
wan shou ju   357
wan-phraathit   67
wansui   116
Wanzendill   116
Wanzenkümmel   116, 125
warabi   309
waremoko   332
warmnot   52
warucha   113
Warzengurke   235
wasabi   240, 391
wasabi-nok   240
wasábia   391
Washerman's plant   4
Wasserdost, Chinesischer   151
Wasserdost, Gewöhnlicher   151
Wasserfenchel   120, 261
Wasserfenchel, Javanischer   261
Wasserfenchel, Vietnamesischer   261

Wasserfreund, Schmalblättriger   180
Wasserhanf   151
Wasserknoblauch   19
Wasserkresse   250
Wasser-Mimose   252
Wasserminze   224
Wasserpetersilie   347
Wasserpfeffer-Knöterich   280
Wassersalat   297
Wasserschierlingblättriger Merk   347
Wassersellerie   261
Wassersenf   250
water avens   166
water celery   261
water cress, common   318
water dropwort   261
water garlic   19
water lettuce   297
water mimosa   252
water mint   224
water parsnip   347
water parsley   261
water pepper   280
water salad   297
water trefoil   230
water wisteria   180
watercress   250
waterkres   250
water-plantain ottelia   267
Wau, Duft   314
Wau, Wohlriechender   314
waung   196
wax flower, Philippine   148
wax myrtle   242
wax-berry   242
Wedelie, Kletter   392
Weggras   302
Weiche Kap-Minze   227
Weiche Minze   234
Weiches Brandkraut   205
Weichhaarige Bergminze   310
Weichkraut, Fünfblättriges   234
Weichsel, Türkische   308
Weichselrohr   308
weidekringzwam   218
Weidenblättriger Knöterich   302
Weihnachtsstern   152
Weihrauch   115
Weihrauch-Baum   69
Weinberg-Lauch   28
Weinlauch   15
Weinmannie, Eschenartige   392
Weinraute   323
Weiße Girlandenblume   173
Weiße Kranzblume   173
Weiße Melde   95
Weiße Pestwurz   281
Weiße Raute   54

Weiße Schmetterlingslilie   173
Weiße Theriakwurzel   288
Weiße Vanille   382
Weißer Beifuß   54
Weißer Chinabeifuß   54
Weißer Diptam   136
Weißer Dorant   219
Weißer Gänsefuß   95
Weißer Jasmin   187
Weißer Kaneel   77
Weißer Mauerpfeffer   339
Weißer Oregano   209
Weißer Sassafras   333
Weißer Senf   345
Weißer Speik   3
Weißer Steinklee   221
Weißer Wermut   54
Weißer Zimt   77
Weißer Zitronenstrauch   209
Weißes Sandelholz   332
Wellblatt-Rhabarber   316
Welsche Walnuss   188
Welscher Lavendel   201
Welschlauch   23
Welsh onion   19
Wermut   52
Wermut, Afrikanischer   52
Wermut, Dragun   53
Wermut, Kampfer   53
Wermut, Pontischer   55
Wermut, Mexikanischer   54
Wermut, Walliser   56
Wermut, Weißer   57
Wermut, Wilder   57
Wermut, Zitronen   53
Wermutgeranie   275
West African locustbean   273
West African pepper   291
West Indian bay   287
West Indian lemongrass   129
West Indian lime   105
West Indian pea   343
Westafrican locust bean   273
Westafrikanische Locustbohne   273
western mugwort   54
western serviceberry   33
Western sweet root   247
Western yarrow   3
West-India yellow wood   400
Westindian vanilla   385
Westindische Vanille   385
Westindisches Arrowroot   78
Westindisches Gelbholz   400
Westindisches Sandelholz   332
white butterbur   281
white caneel   77
white cheese wood   32
white cinnamon   77

white costas   118
white cumin   371
white felt leucas   205
white genipi   55
white ginger lily   173
white horehound   219
white jasmine   187
white kyllingia   193
white melilot   221
white mugwort   54
white mustard   345
white oregano   209, 265
white pepper   293
white sage   54
white sandalwood   332
white saunders   332
white snakeroot   152
white stem filaree   146
white stonecrop   339, 340
white sweet clover   221
white vanilla   382
whitetop   85
Wiesenaugentrost   153
Wiesenkerbel, Gewöhnlicher   45
Wiesenknopf, Großer   331
Wiesenknopf, Kleiner   331
Wiesenkren   250
Wiesenkresse   84
Wiesenkümmel   87
Wiesenmargerite   204
Wiesen-Sauerampfer   320
Wiesen-Schafgarbe   3
Wiesenschaumkraut   84
Wiesenschaumkraut, Kleinblütiges   84
Wiesen-Schlüsselblume   307
Wiesensilge   343
Wiesen-Taglilie   175
Wiesen-Wucherblume   204
wild allspice   208
wild angelica   43
wild basil   112, 310
wild bergamot   236
wild Canadian lettuce   195
wild caneel   77
wild cardamom   8, 144
wild chervil   124
wild chervil, Japanese   125
wild cinnamon   77
wild cinnamon (of Japan)   102
wild clary   330
wild coffee   304
wild coriander   117
wild fennel   252, 253
wild garlic   27, 376
wild ginger   57, 346, 404
wild indigo   362
wild knoflook   376
wild kumquat, Hongkong   159

wild leek   15, 26
wild look   15
wild mango   352
wild marigold   358
wild marjoram   265
wild mint   181, 224
wild mustard   345
wild nutmeg   114
wild onion   16, 18, 28
wild orange tree   369
wild oregano   299
wild parsley   99
wild parsnip   41
wild pepper   295
wild pepper tree   192
wild purslane   306
wild rhubrab   136
wild rosemary   123, 201
wild rue   275
wild saffron   122
wild sage   196, 210
wild Siamese cardamom   38
wild spikenard   181
wild stone parsley   124
wild tapioca   78
wild thyme   367, 368
wild turmeric   126
wild) pepper tree   192
Wilde Bergamotte   236
Wilde Gelbwurz   126
wilde knoffel   376
Wilde Malve   216
wilde marjolein   265
Wilde Melisse   112
Wilde Monarde   236
Wilde Petersilie   99
wilde prei   15
Wilde Taglilie   175
wildeals   52
Wilder (japanischer) Zimt   102
Wilder Anis   247
Wilder Basilikum   112
Wilder Gamander   306, 363
Wilder Indigo   362
Wilder Ingwer   346
Wilder Knoblauch   27
Wilder Knofel   27
wilder knoflok   27
Wilder Kurkuma   126
Wilder Lauch   26
Wilder Majoran   265, 266
Wilder Mango   352
Wilder Muskat   114, 243
Wilder Porree   25
Wilder Rosmarin   123, 201
Wilder russischer Meerrettich   50
Wilder Thymian   368
Wilder Wermut   57

Wildes Johannisbrot   93
Wildes Pfefferkraut   295
Wildkirsche   308
willow gentian   163
willow grass   302
willowleaf eucalyptus   150
Windblumen-Königskerze   385
Winde, Kanarische   115
wing scale   59
winged Geigeria   163
Winter- Bohnenkraut   335
winter cherry   85
winter cress   62
winter marjoram   265
winter purslane   239
winter savory   335
winter sweet   265
winter thyme   366
Winter's bark   140
Winter's bark, false   99
Winter's-Rinde   140
Winterbohnenkraut, Kriechendes   336
wintercress, common   62
Winterestragon   358
wintergreen, alpine   163
wintergreen, creeping   163
Winterkresse   62, 250
Winterkresse, Frühe   62
Winterlauch   23
Winterlieb, Doldiges   97
Wintermajoran   265
Winterportulak   239
Winterpostelein   239
Winterrinde, falsche   99
Winterthymian   366
Winterzwiebel   19
Wirbeldost   112
wisteria, scarlet   343
wisteria, water   180
Wisterie, Kleinöhrige   180
Wisterie, Schmalblättrige   180
wit asem   359
wit sering   192
witte peper   293
woangtschopinamu   400
wodjanoj kren   250
wohkaya lawqang   355
Wohlgemut   68
Wohlriechender Andorn   181
Wohlriechender Fieberstrauch   208
Wohlriechender Gänsefuß   95
Wohlriechender Knöterich   280
Wohlriechender Sumach   317
Wohlriechender Wau   314
Wohlriechendes Veilchen   388
wohpala   244
wolf berry, barbary   213
Wolfsbeere, Chinesische   213

Wolfsmilch, Borstige   152
Wolfsmilch, Kreuzblättrige   152
Wolfsmilch, Spring   152
Wolkenkraut   358
wŏlkyéhwa   318
Wollblume, Großblütige   385
Wollblütiger Nachtschatten   349
Wollige Kreuzblume   301
Wolliger Pfeffer   292
Wolliges Brandkraut   204
wolly betony   353
wolly stachys   353
Woll-Ziest   353
woloschckij ukron   158
woltschje lyko   133
Womersleyi nutmeg   246
Womersleyie Muskat   246
womg pa   111
wong pei   111
wong poi   111
wood avens   166
wood garlic   27
wood germander   363
wood leek   26
wood sorrel   268
wood wine   293
wood, rose   43
woodland chervil   45
woodruff asperule   161
woodruff, sweet   161
woody betony   65
woody vine   145, 157
wooly leucas   204
wooly mint   230
wooly piper   292
wormdrijvende ganzevoet   95
wormseed   95
wormwood   52, 56
wormwood, African   52
wormwood, alpine   55, 56
wormwood, glacier   54
wormwood, Kamshatka   56
wormwood, mountain   56
wormwood, Roman   55
wormwood, sea   54
wormwood, silver   53
wormwood, tree   53
wormwood, Valais   56
wosditschnij perez   286
woskowniza obyknowennaja   242
wounderwort   3
woxerno   218
wrĕsah   37
wringed cardamom   37
wrinkled giant hyssop   12
wrinkled perovskia   279
wu mao ding x luo le   258
wu ming tzu   297

wu ming zi   297
wu se mei   196
wu wei tzu   338
Wucherblume, Kronen   97
Wucherblume, Wiesen   204
Wunderapfel   235
Wunderblume, Jalapa   234
Wunderstrauch, Indischer   113
Wurmkraut   360
Wurmscheinaster   386
Wurstkraut   263
Wurzelpetersilie   283
Wurzelrankiger Pfeffer   295
Wurzelsellerie   47
Würzige Safranwurz(el)   126
Würzminze   145
Würzsilge   346
xa lach   250
xà sàng   112
xai ye qing hao   53
xang song   66
xi fan ju   358
xi fei li   361
xi yang cai gun   250
xi ye ai   56
xia jin ying   319
xian ye bai rong cao   205
xiang   2, 12, 32, 87, 90, 108, 129, 139, 145, 158, 181, 223, 229, 241, 265, 275, 283, 301, 323, 332, 336, 355, 368, 387, 403
xiang feng hua   223
xiang he   403
xiang hua cai   229
xiang mao   129
xiang pi mu   32
xiang qing lan   139
xiang ru   145, 265
xiang si zi   2
xiang ye   275
xiang-geng-sao   387
xiao   19, 25, 63, 158
xiao guan xun   63
xiao hui xiang   158
xiao suan   25
xie   18
xie ba   17
xie bai   19
xie cao   381
xihuitl   294
xiito ajo   24
ximbuí   278
xinh pau chu   5
xkanlol   357
xoài   218
xochinacathli   129
xochipelli   118
xoltexnuk   181

xonacat 21
xong 250
xpayumak 79
xu ong song 66
xuan hua qie 350
xuân xa 38
xun yi cao 200
ya êrh chin 125
ya faran 122
ya ma zi 209
ya zhu na 116
yaa dok khaao 387
yaa saam wan 387
yabani kiraz 308
yabu-kanzo 175
yae-aoki 239
ya-faekhom 387
ya-faeklum 387
yafata 268
yaga naraxo 109
yaga yiña 140
yagrumo 89
yai 17
yakjontongssari 222
yakmemil 179
yaknŭngjaengi 95
yakpaguniamul 381
yakpulkkot 329
yama-yuri 207
yan mien chi 140
yan min 140
yanagi tade 280
yanbaru-nasubi 349
yang ch'un sha 38
yang cong 17
yang di li 78
yang shi cao 3
yang ts'ung 17
yang yan sui 282
yangauk 276
yangha 403
yangpha 17
yanîsun 288
yao shui su 65
yarrow (herb) 3
yatcha-sei 48
yauhtli 358
yavani 371
yavya-böfu 167
ye chi kuan 90
ye hsi ming 187
yebisu me 195
yee sun 319
yeera 125
yeh lo po 133
yeh-suan 19
yeh-tsin-tsai 19
yei' 314

yellow bark 99
yellow berried night-shade 350
yellow champa 232
yellow cinchona 99
yellow cosmos 118
yellow day lily 175
yellow flowery nutmeg 238
yellow gentian 163
yellow ginger 126
yellow jasmine 187
yellow melilot 222
yellow mombin 352
yellow mustard 345
yellow onion 21
yellow plum 352
yellow rocket 62
yellow sage 196
yellow sandalwood 332
yellow sheep sorrel 269
yellow sorrel 268
yellow squash pepper 83
yellow star anise 183
yellow starwort 184
yellow sweet clover 222
yellow wood, Korean 400
yellow zedoary 126
yen chih 234
yen kuei 266
yerba buena 225, 228, 334
yerba de la plata 278
yerba de Maria 327
yerba dulce 210
yerba Luisa 129
yerba soldado 291
ying kuo 166
ying sou 272
ying tzu shu 272
yira 158, 258
yiu-tu-chih 7
ylang-ylang 76
Ylang-Ylang 76
Ylang-Ylang, Falscher 51
yo baan 239
yomi 338
yomogi 57
yompuchu 18
yoruba indigo 213
yoseiso 179
ysaño 376
Ysop 182
Ysop, Lemon 12
Ysop, Mexikanischer 12
Ysop, Runzliger 12
yü kan tzu 284
yu xing cao 179
yuan sui 87
yuang ye bo he 230
yu-chin 126

yue gui zi 199
yueh kuei 199
yuen sai 116
yuja 235
yûkari-no-ki 150
yün xiang 323
yuzu 107
za'atar 265, 336, 365, 368
za'atar franji 336
za'atar hommar 365
za'atar midbari 365
za'atar rumi 336
za'ater 368
za'ferān 122
zaafraran 122
zacate elimón 129
zacate lemón 130
zacate violeta 387
Zachunbaum 61
zadwâr 128
zafferano 122, 126
zafferano delle Indie 126
zafferano domestica 122
zaffrone 87, 122
zafrán molido 127
Zahnbaum, Ägyptischer 61
Zahnbürstenstrauch 325
Zahnlavendel 200
Zahnstocher-Kraut 34
Zahnwurz, Dreiblättrige 85
zahrodny mak 272
zahtar 368
zakuro 234, 309
zakuro-sô 234
zambac 188
zamzariq 93
zan 40
zan hunh hua 122
zanahoria 133
zang hui xiang 87
zangabeel 126, 403
zangabeel asfer 126
Zanzibar redheads 355
Zapote mamey 217
zarskij koren' 283
Zarte Angelikawurzel 43
Zarte Mutterwurz 43
zarunbâd 128
zarza 44
zater 334
zatir 365
Zedernfrucht 108
zédoaire 128, 403
zedoária 128
zedoary 126, 128
zedoary, black 126
zedoary, long 128
zedoary, yellow 126

zedoeira amarella   128
zédoire   54
Zedratzitrone   108
zeevenkel   120
zejsonskaja koriza   104
zeler   46
zemba   78
zeml anojorech   48
zeng jil   212
zeni-aoi   217
zenjabil   403
Zentifolie   318
zentzephil   403
zenzero   346, 403
zenzero Africano   346
zenzero officinale   403
zerleg sarmis   20
Zerrgras   302
Zerschlitzte Merremie   231
Zerschlitzter Frauenmantel   13
Zerschlitzter Senf   346
zerumbet ginger   404
zhang liu tou   118
zhang ya cai   354
zhi   41, 100, 105, 162, 219, 265, 285, 304, 342
zhi ma   342
zhi shi   105, 304
zhi zi   162
zhou mian cao   205
zhu bi kong   179
zhu lem   97
zhu ye jiao   401
zi mo li gen   234
zi su ye   278
ziba   371
zibeline   59
Ziegenhornklee   374
Ziegenlauch   22
Zier-Paprika   82
Ziest, Echter   65
Ziest, Woll   352
Zigeunerklee   372
Zigeunerknoblauch, German   27
Zigeunerlauch   27
zimbreiro   189
Zimmerknoblauch   376
Zimmerschnittlauch   376
Zimt, Amerikanischer   261
Zimt, Annam   102
Zimt, Assam   100
Zimt, Australischer   103
Zimt, Ceylon   104
Zimt, Chinesischer   100
Zimt, Culiwan   101
Zimt, Echter   104
Zimt, Fagot   100
Zimt, Indonesischer   100

Zimt, Japanischer   102
Zimt, Java   103
Zimt, Kambodscha   101
Zimt, Kanton   100
Zimt, Karibischer   140
Zimt, Kuliwan   101
Zimt, Lavang   100
Zimt, Magelhanischer   140
Zimt, Malabar   104
Zimt, Mindanao   102
Zimt, Orinoko   43
Zimt, Padang   100
Zimt, Philippinischer   103
Zimt, Schwarzer   136
Zimt, Saïgon   102
Zimt, Seychellen   104
Zimt, Taiwanischer   102
Zimt, Tonkin   102
Zimt, Vietnam   102
Zimt, Wilder   102
Zimt, Weißer   77
Zimtkassia   100
Zimt-Muskat   243
Zimtrinde, Alyxia   32
Zimtrinde, Chinesische   100
zingiber   403
Zip(p)eln   17
Zipolle   17
Zirbelnuss   289
zireh siyah   71
zitron   108
Zitronat   108
Zitronatzitrone   108
Zitrone   107, 109, 338
Zitrone, Chinesische   338
Zitrone, Finger   108
Zitrone, Gefingerte   108
Zitronellgras, Ceylon   130
Zitronellgras, Dichtblättriges   131
Zitronenbasilikum   257
Zitronenblatt, Indisches   107
Zitronenblättrige Morinda   239
Zitronen-Bohnenkraut, Afrikanisches   334
Zitronen-Eukalyptus   150
Zitronengras, Echtes   131
Zitronengras, Indisches   129
Zitronengras, Java   131
Zitronenkraut   52, 335
Zitronenkraut, Slovenisches   335
Zitronen-Mahagonie   32
Zitronenmelisse   223
Zitronen-Monarde   236
Zitronenmyrte   61, 204, 289
Zitronenmyrte, Australische   61
Zitronenpelargonie   276
Zitronenperilla   278
Zitronenquendel   365

Zitronenstrauch   209, 211
Zitronenstrauch, Süßer   210
Zitronenstrauch, Weißer   209
Zitronen-Taglilie   175
Zitronenthymian   365
Zitronen-Verbene   211, 386
Zitronen-Wermut   53
Zittwer(wurzel)   128
Zitwer, Gelber   126, 402
zizifora   405
zmejuyj koren   49
zmin   174
zoete djeroek   109
zong guan   103
zong hai   103
zostère marine   405
Zottelblume   198, 230
Zottelblume, Dornige   198
Zottenblume   230
Zottige Minze   224
zouz   188
zu fu ping   297
zu lan hua   97
zu mo li   234
zumaque   317
Zürgelbaum, Afrikanischer   91
Zürgelbaum, Ganzblättriger   91
zurrón de pastor   81
zuurkruid   316
zvetnaja trava   200
zwarte peper   293
Zweizahn, Fiederblättriger   65
Zwerg-Boretsch   68
Zwergorange   159
Zwergzitrone   159
Zwergzitrone, Dickblättrige   159
Zwiebel, Afrikanische   16
Zwiebel, Ägyptische   23
Zwiebel, Altai   15
Zwiebel, Angolanische   16
Zwiebel, Askalon   17
Zwiebel, Bataten   17
Zwiebel, Catawissa   23
Zwiebel, Chinesische   18
Zwiebel, Frühlings   17
Zwiebel, Gestielte   26
Zwiebel, Hecken   19
Zwiebel, Hooker   20
Zwiebel, Kanadische   16
Zwiebel, Kartoffel   17
Zwiebel, Küchen   17
Zwiebel, Ledebour   20
Zwiebel, Mexikanische   20
Zwiebel, Neapel   21
Zwiebel, Neopolitanische   21
Zwiebel, Oschanin   22
Zwiebel, Pskemenser   23
Zwiebel, Riesen   19

Zwiebel, Schiefe   22
Zwiebel, Sibirische   15
Zwiebel, Silber   19
Zwiebel, Sommer   17, 18
Zwiebel, Speise   17
Zwiebel, Syrische   17
Zwiebel, Turkestanische   20
Zwiebel, Winter   19
Zwiebelreicher Lauch   21
Zypresse, Sommer   64